MW01675273

INSTRUCTOR'S SOLUTIONS MANUAL

JEFFERY A. COLE
Anoka-Ramsey Community College

INTERMEDIATE ALGEBRA

EIGHTH EDITION

Margaret L. Lial
American River College

John Hornsby
University of New Orleans

Terry McGinnis

PEARSON
Addison Wesley

Boston San Francisco New York
London Toronto Sydney Tokyo Singapore Madrid
Mexico City Munich Paris Cape Town Hong Kong Montreal

Reproduced by Pearson Addison-Wesley from electronic files supplied by the author.

Publishing as Pearson Addison-Wesley, 75 Arlington Street, Boston, MA 02116.

ISBN 0-321-28568-9

1 2 3 4 5 6 OPM 08 07 06 05

PEARSON
Addison
Wesley

Preface

This *Instructor's Solutions Manual* contains solutions to all exercises in the text *Intermediate Algebra, Eighth Edition* by Margaret L. Lial, John Hornsby, and Terry McGinnis. A student version of this manual is also available—it contains solutions to all margin exercises, the odd-numbered exercises in each section, all Relating Concepts exercises, as well as solutions to all the exercises in the review sections, the chapter tests, and the cumulative review sections. It is my hope that by browsing through the solutions, professors will save time in determining appropriate assignments for their particular class.

I appreciate feedback concerning errors, solution correctness or style, and manual style. Any comments may be sent directly to me at the address below, at jeff.cole@anokaramsey.edu, or in care of the publisher, Pearson Addison-Wesley.

I would like to thank Ken Grace, of Anoka-Ramsey Community College, and Jeannine Grace, for typesetting the manuscript and providing assistance with many features of the manual; Mary Johnson, of Inver Hills Community College, and Marv Riedesel, for their careful accuracy checking and valuable suggestions; Jim McLaughlin, for his help with the entire art package; and the authors and Maureen O'Connor and Lauren Morse, of Pearson Addison-Wesley, for entrusting me with this project.

Jeffery A. Cole
Anoka-Ramsey Community College
11200 Mississippi Blvd. NW
Coon Rapids, MN 55433

Table of Contents

CHAPTER 1 REVIEW OF THE REAL NUMBER SYSTEM

1.1 Basic Concepts

1.1 Margin Exercises

1. $\{0, 10, \frac{3}{10}, 52, 98.6\}$

(a) The natural numbers are $\{1, 2, 3, 4, 5, 6, \ldots\}$. In the given set, the elements 10 and 52 are natural numbers.

(b) The whole numbers are $\{0, 1, 2, 3, 4, 5, \ldots\}$. In the given set, the elements 0, 10, and 52 are whole numbers.

2. **(a)** $\{x \mid x \text{ is a whole number less than } 5\}$

The whole numbers are $\{0, 1, 2, 3, 4, 5, 6, \ldots\}$. The whole numbers less than 5 would be the set $\{0, 1, 2, 3, 4\}$.

(b) $\{y \mid y \text{ is a whole number greater than } 12\}$

$\{13, 14, 15, \ldots\}$ is the set of whole numbers greater than 12.

3. **(a)** $\{0, 1, 2, 3, 4, 5\}$

In set-builder notation, one answer is $\{x \mid x \text{ is a whole number less than } 6\}$.

(b) $\{7, 14, 21, 28, \ldots\}$

In set-builder notation, one answer is $\{x \mid x \text{ is a multiple of 7 greater than } 0\}$.

4. **(a)** $\{-4, -2, 0, 2, 4, 6\}$

These are the even integers between and including -4 and 6. Place dots for $-4, -2, 0, 2, 4$, and 6 on a number line.

(b) $\{-1, 0, \frac{2}{3}, \frac{5}{2}\}$

Place dots at -1 and 0 on a number line. To graph $\frac{2}{3}$, divide the interval between 0 and 1 into three parts. $\frac{2}{3}$ is located two-thirds of the distance from 0 to 1. To graph $\frac{5}{2}$, first change the fraction to a mixed number.

$$\frac{5}{2} = 2\frac{1}{2}$$

Divide the interval between 2 and 3 into two equal parts. $\frac{5}{2}$ is located halfway between 2 and 3.

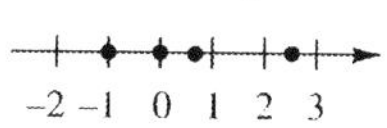

(c) $\{5, \frac{16}{3}, 6, \frac{13}{2}, 7, \frac{29}{4}\}$

Place dots at 5, 6, and 7 on a number line. To locate the fractions, first change each to a mixed number and place dots at $\frac{16}{3} = 5\frac{1}{3}$, $\frac{13}{2} = 6\frac{1}{2}$, and $\frac{29}{4} = 7\frac{1}{4}$.

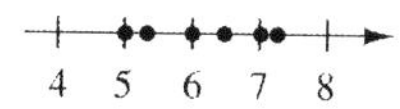

5. **(a)** -6 is a *rational* number because

$$-6 = -\frac{6}{1}.$$

-6 is also a *real* number.

(b) 12 is a *whole* number, a *rational* number $\left(12 = \frac{12}{1}\right)$, and a *real* number.

(c) $.\overline{3}$, a repeating decimal, is a *rational* number and a *real* number.

(d) $-\sqrt{15}$ is an *irrational* number and a *real* number.

(e) π, a nonrepeating, nonterminating decimal, is an *irrational* number and a *real* number.

(f) $\frac{22}{7}$ is a *rational* number and a *real* number.

(g) 3.14, a terminating decimal, is a *rational* number and a *real* number.

6. **(a)** The statement "All whole numbers are integers" is *true*. As shown in Figure 4 in the textbook, the set of integers includes the set of whole numbers.

(b) The statement "Some integers are whole numbers" is *true* since 0 and the positive integers are whole numbers. The negative integers are not whole numbers.

(c) The statement "Every real number is irrational" is *false*. Some real numbers such as $\frac{4}{9}$, $.\overline{3}$, and $\sqrt{16}$ are rational. Others such as $\sqrt{2}$ and π are irrational.

7.

	Number	Additive Inverse
(a)	9	-9
(b)	-12	12
(c)	$-\frac{6}{5}$	$\frac{6}{5}$
(d)	0	0
(e)	1.5	-1.5

8. **(a)** $|6| = 6$

(b) $|-3| = -(-3) = 3$

(c) $-|5|$

Evaluate the absolute value first. Then find the additive inverse.

$$-|5| = -(5) = -5$$

(d) $-|-2| = -(2) = -2$

(e) $-|-7| = -(7) = -7$

(f) $|-6| + |-3|$

Evaluate each absolute value first, and then add.

$$|-6| + |-3| = 6 + 3 = 9$$

(g) $|-9| - |-4| = 9 - 4 = 5$

(h) $-|9 - 4| = -|5| = -5$

9. Household audio/video equipment: $|-4.2| = 4.2$

Computer/data processing services: $|4.9| = 4.9$

Since $4.9 > 4.2$, the computer/data processing services will show the greater change.

10. **(a)** $3 < 7$ since 3 is to the left of 7 on a number line.

(b) $9 > 2$ since 9 is to the right of 2.

(c) $-4 > -8$ since -4 is to the right of -8.

(d) $-2 < -1$ since -2 is to the left of -1.

(e) $0 > -3$ since 0 is to the right of -3.

11. **(a)** $-2 \leq -3$ is *false*.

Since -2 is to the right of -3 on a number line, it must be larger.

(b) $8 \leq 8$ is *true*.

The statement is "8 is less than *or* equal to 8." Since 8 is equal to 8, the statement is true.

(c) $-9 \geq -1$ reads "-9 is greater than or equal to -1." This is *false*, since -9 is to the left of -1 on a number line.

(d) $5 \cdot 8 \leq 7 \cdot 7$

Since $5 \cdot 8 = 40$, $7 \cdot 7 = 49$, and $40 \leq 49$, we must have $5 \cdot 8 \leq 7 \cdot 7$. The statement is *true*.

(e) $3(4) > (2)6$

Since $3(4) = 12$ and $2(6) = 12$, the statement becomes $12 > 12$, which is *false*.

1.1 Section Exercises

1. $\{x \mid x \text{ is a natural number less than } 6\}$
The set of natural numbers is $\{1, 2, 3, \dots\}$, so the set of natural numbers less than 6 is $\{1, 2, 3, 4, 5\}$.

2. $\{m \mid m \text{ is a natural number less than } 9\}$ is the set $\{1, 2, 3, 4, 5, 6, 7, 8\}$.

3. $\{z \mid z \text{ is an integer greater than } 4\}$
The set of integers is $\{\dots, -3, -2, -1, 0, 1, 2, 3, \dots\}$, so the set of integers greater than 4 is $\{5, 6, 7, 8, \dots\}$.

4. $\{y \mid y \text{ is an integer greater than } 8\}$ is the set $\{9, 10, 11, 12, \dots\}$.

5. $\{a \mid a \text{ is an even integer greater than } 8\}$
The set of even integers is $\{\dots, -2, 0, 2, 4, \dots\}$, so the set of even integers greater than 8 is $\{10, 12, 14, 16, \dots\}$.

6. $\{k \mid k \text{ is an odd integer less than } 1\}$ is the set $\{\dots, -7, -5, -3, -1\}$.

7. $\{x \mid x \text{ is an irrational number that is also rational}\}$
Irrational numbers cannot also be rational numbers. The set of irrational numbers that are also rational is $\emptyset$.

8. $\{r \mid r \text{ is a number that is both positive and negative}\}$ is $\emptyset$.

9. $\{p \mid p \text{ is a number whose absolute value is } 4\}$
This is the set of numbers that lie a distance of 4 units from 0 on the number line. Thus, the set of numbers whose absolute value is 4 is $\{-4, 4\}$.

10. $\{w \mid w \text{ is a number whose absolute value is } 7\}$ is the set $\{-7, 7\}$.

11. $\{2, 4, 6, 8\}$ can be described by $\{x \mid x \text{ is an even natural number less than or equal to } 8\}$.

12. $\{11, 12, 13, 14\}$ can be described by $\{x \mid x \text{ is an integer between 10 and } 15\}$.

13. $\{4, 8, 12, 16, \dots\}$ can be described by $\{x \mid x \text{ is a multiple of 4 greater than } 0\}$.

14. $\{\dots, -6, -3, 0, 3, 6, \dots\}$ can be described by $\{x \mid x \text{ is an integer multiple of } 3\}$.

15. Graph $\{-3, -1, 0, 4, 6\}$.
Place dots for $-3, -1, 0, 4$, and 6 on a number line.

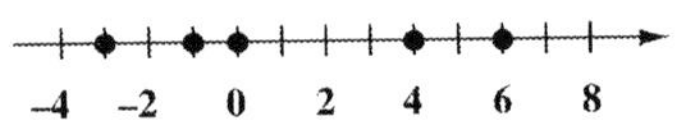

16. Graph $\{-4, -2, 0, 3, 5\}$.
Place dots for $-4, -2, 0, 3$, and 5 on a number line.

17. Graph $\{-\frac{2}{3}, 0, \frac{4}{5}, \frac{12}{5}, \frac{9}{2}, 4.8\}$.

Place dots for $-\frac{2}{3} = -.\overline{6}$, 0, $\frac{4}{5} = .8$, $\frac{12}{5} = 2.4$, $\frac{9}{2} = 4.5$, and 4.8 on a number line.

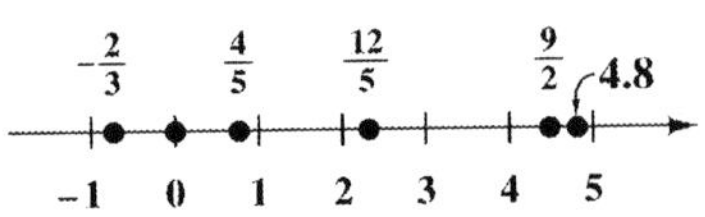

18. Graph $\left\{-\frac{6}{5}, -\frac{1}{4}, 0, \frac{5}{6}, \frac{13}{4}, 5.2, \frac{11}{2}\right\}$.

Place dots for $-\frac{6}{5} = -1.2$, $-\frac{1}{4} = -.25$, 0, $\frac{5}{6} = .8\overline{3}$, $\frac{13}{4} = 3.25$, 5.2, and $\frac{11}{2} = 5.5$ on a number line.

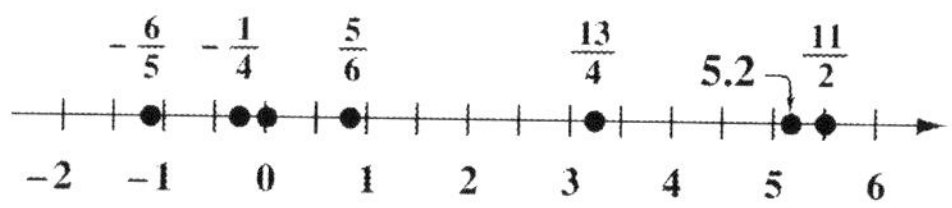

19. $\left\{-8, -\sqrt{5}, -.6, 0, \frac{3}{4}, \sqrt{3}, \pi, 5, \frac{13}{2}, 17, \frac{40}{2}\right\}$

(a) The elements $5, 17,$ and $\frac{40}{2}$ (or 20) are natural numbers.

(b) The elements $0, 5, 17,$ and $\frac{40}{2}$ are whole numbers.

(c) The elements $-8, 0, 5, 17,$ and $\frac{40}{2}$ are integers.

(d) The elements $-8, -.6, 0, \frac{3}{4}, 5, \frac{13}{2}, 17,$ and $\frac{40}{2}$ are rational numbers.

(e) The elements $-\sqrt{5}$, $\sqrt{3}$, and π are irrational numbers.

(f) All the elements are real numbers.

20. $\left\{-9, -\sqrt{6}, -.7, 0, \frac{6}{7}, \sqrt{7}, 4.\overline{6}, 8, \frac{21}{2}, 13, \frac{75}{5}\right\}$

(a) The elements $8, 13,$ and $\frac{75}{5}$ (or 15) are natural numbers.

(b) The elements $0, 8, 13,$ and $\frac{75}{5}$ are whole numbers.

(c) The elements $-9, 0, 8, 13,$ and $\frac{75}{5}$ are integers.

(d) The elements $-9, -.7, 0, \frac{6}{7}, 4.\overline{6}, 8, \frac{21}{2}, 13,$ and $\frac{75}{5}$ are rational numbers.

(e) The elements $-\sqrt{6}$ and $\sqrt{7}$ are irrational numbers.

(f) All the elements are real numbers.

21. The statement "Every rational number is an integer" is *false*. Some are integers, but others, like $\frac{3}{4}$, are not.

22. The statement "Every natural number is an integer" is *true*. The integers include the natural numbers, their negatives, and zero.

23. The statement "Every irrational number is an integer" is *false*. Irrational numbers have decimal representations that neither terminate nor repeat, so no irrational numbers are integers.

24. The statement "Every integer is a rational number" is *true*. Every integer q can be written as $\frac{q}{1}$.

25. The statement "Every natural number is a whole number" is *true*. The whole numbers consist of the natural numbers and zero.

26. The statement "Some rational numbers are irrational" is *false*. No rational numbers are irrational numbers.

27. The statement "Some rational numbers are whole numbers is *true*. Every whole number is rational.

28. The statement "Some real numbers are integers" is *true*. Every integer is a real number.

29. The statement "The absolute value of any number is the same as the absolute value of its additive inverse" is *true*. The distance on a number line from 0 to a number is the same as the distance from 0 to the additive inverse of that number.

30. The statement "The absolute value of any nonzero number is positive" is *true*. Every nonzero number is a positive distance from 0.

31. **(a)** The additive inverse of 6 is -6.

(b) $6 > 0$, so $|6| = 6$.

32. **(a)** The additive inverse of 8 is -8.

(b) $8 > 0$, so $|8| = 8$.

33. **(a)** The additive inverse of -12 is $-(-12) = 12$.

(b) $-12 < 0$, so $|-12| = -(-12) = 12$.

34. **(a)** The additive inverse of -15 is $-(-15) = 15$.

(b) $-15 < 0$, so $|-15| = -(-15) = 15$.

35. **(a)** The additive inverse of $\frac{6}{5}$ is $-\frac{6}{5}$.

(b) $\frac{6}{5} > 0$, so $\left|\frac{6}{5}\right| = \frac{6}{5}$.

36. **(a)** The additive inverse of $.13$ is $-.13$.

(b) $.13 > 0$, so $|.13| = .13$.

37. $|-8|$
Use the definition of absolute value.

$$-8 < 0, \quad \text{so} \quad |-8| = -(-8) = 8.$$

38. $|-11| = -(-11) = 11$

39. $\frac{3}{2} > 0$, so $\left|\frac{3}{2}\right| = \frac{3}{2}$.

40. $\frac{7}{4} > 0$, so $\left|\frac{7}{4}\right| = \frac{7}{4}$.

41. $-|5| = -(5) = -5$

42. $-|17| = -(17) = -17$

43. $-|-2| = -[-(-2)] = -(2) = -2$

44. $-|-8| = -[-(-8)] = -(8) = -8$

45. $-|4.5| = -(4.5) = -4.5$

46. $-|12.6| = -(12.6) = -12.6$

47. $|-2| + |3| = 2 + 3 = 5$

48. $|-16| + |12| = 16 + 12 = 28$

49. $|-9| - |-3| = 9 - 3 = 6$

50. $|-10| - |-5| = 10 - 5 = 5$

51. $$\begin{aligned}|-1| + |-2| - |-3| &= 1 + 2 - 3\\ &= 3 - 3\\ &= 0\end{aligned}$$

52. $|-6| + |-4| - |-10| = 6 + 4 - 10 = 0$

53. **(a)** The greatest absolute value is $|2.80| = 2.80$.

Therefore, Mumbai (Bombay) had the greatest change in population. The population increased 2.80%.

(b) The smallest absolute value is $|.02| = .02$.

Therefore, Beijing had the smallest change in population. The population increased .02%.

54. **(a)** The greatest absolute value is $|-7552| = 7552$. Therefore, China had the greatest discrepancy between exports and imports. Imports exceeded exports by \$7552 million.

(b) The smallest absolute value is $|96| = 96$. Therefore, Turkey had the smallest discrepancy between exports and imports. Exports exceeded imports by \$96 million.

55. Compare the depths of the bodies of water. The deepest, that is, the body of water whose depth has the greatest absolute value, is the Pacific Ocean ($|-12{,}925| = 12{,}925$) followed by the Indian Ocean, the Caribbean Sea, the South China Sea, and the Gulf of California.

56. Compare the heights of the mountains. The shortest, that is, the smallest value, is Point Success followed by Rainier, Matlalcueyetl, Steele, and McKinley.

57. True; the absolute value of the depth of the Pacific Ocean is

$$|-12{,}925| = 12{,}925$$

which is greater than the absolute value of the depth of the Indian Ocean,

$$|-12{,}598| = 12{,}598.$$

58. False; the absolute value of the depth of the Gulf of California is

$$|-2375| = 2375$$

which is *less than* the absolute value of the depth of the Caribbean Sea,

$$|-8448| = 8448.$$

59. True; since -6 is to the left of -2 on a number line, -6 is less than -2.

60. True; since -4 is to the left of -3 on a number line, -4 is less than -3.

61. False; since -4 is to the left of -3 on a number line, -4 is *less* than -3, not greater.

62. False; since -2 is to the left of -1 on a number line, -2 is *less* than -1, not greater.

63. True; since 3 is to the right of -2 on a number line, 3 is greater than -2.

64. True; since 5 is to the right of -3 on a number line, 5 is greater than -3.

65. True; since $-3 = -3$, -3 is greater than *or equal to* -3.

66. True; since $-4 = -4$, -4 is less than *or equal to* -4.

67. "7 is greater than y" can be written as $7 > y$.

68. "-4 is less than 12" can be written as $-4 < 12$.

69. "5 is greater than or equal to 5" can be written as $5 \geq 5$.

70. "-3 is less than or equal to -3" can be written as $-3 \leq -3$.

71. "$3t - 4$ is less than or equal to 10" can be written as $3t - 4 \leq 10$.

72. "$5x + 4$ is greater than or equal to 19" can be written as $5x + 4 \geq 19$.

73. "$5x + 3$ is not equal to 0" can be written as $5x + 3 \neq 0$.

74. "$6x + 7$ is not equal to -3" can be written as $6x + 7 \neq -3$.

75. $-6 < 7 + 3$?
$-6 < 10$ *True*

The last statement is true since -6 is to the left of 10 on a number line.

76. $-7 < 4 + 2$?
$-7 < 6$ *True*

77. $2 \cdot 5 \geq 4 + 6$?
$10 \geq 10$ *True*

78. $8 + 7 \leq 3 \cdot 5$?
$15 \leq 15$ *True*

79. $-|-3| \geq -3$?
$-3 \geq -3$ *True*

80. $-|-5| \leq -5$?
$-5 \leq -5$ *True*

81. $-8 > -|-6|$?
$-8 > -6$ *False*

82. $-9 > -|-4|$?
$-9 > -4$ *False*

83. The 2001 egg production in IA was 8691 million and in CA it was 5998 million. Thus, the egg production in IA was greater than in CA in 2001.

84. In 2002, the states that had egg production greater than 6500 million eggs were Iowa (9910), Ohio (7940), and Pennsylvania (6520).

85. For which states is the bar longer (and number larger) for 2002 than 2001?

IA: $9910 > 8691$; OH: $7940 > 7900$; CA: $6124 > 5998$; TX: $4774 > 4734$; MN: $3124 > 3112$

Thus, 2002 egg production was greater than 2001 production for Iowa, Ohio, California, Texas, and Minnesota.

86. 2002: TX $= x = 4774$, OH $= y = 7940$
Since $4774 < 7940$, $x < y$ is true.

87. 2002: IN $= x = 5973$, MN $= y = 3124$
Since $5973 > 3124$, $x > y$ is true.

1.2 Operations on Real Numbers

1.2 Margin Exercises

1. Add the absolute values of the numbers. The answer is negative since two negative numbers are being added.

(a) $$-2 + (-7) = -(|-2| + |-7|) = -(2+7) = -9$$

(b) $-15 + (-6) = -(15 + 6) = -21$

(c) $-1.1 + (-1.2) = -(1.1 + 1.2) = -2.3$

(d) $$-\frac{3}{4} + \left(-\frac{1}{2}\right) = -\left(\frac{3}{4} + \frac{1}{2}\right) = -\left(\frac{3}{4} + \frac{2}{4}\right) = -\frac{5}{4}$$

2. **(a)** $12 + (-1)$

Subtract the smaller absolute value from the larger absolute value $(12 - 1)$, and take the sign of the number with the larger absolute value.

$$12 + (-1) = 12 - 1 = 11$$

(b) $3 + (-7)$

Subtract the smaller absolute value from the larger absolute value $(7 - 3)$, and take the sign of the larger.

$$3 + (-7) = -(7 - 3) = -4$$

(c) $-17 + 5 = -(17 - 5) = -12$

(d) $-\frac{3}{4} + \frac{1}{2} = -\frac{3}{4} + \frac{2}{4} = -\frac{1}{4}$

(e) $-1.5 + 3.2 = 1.7$

3. **(a)** $9 - 12$

Change the sign of the second number and add.

$$9 - 12 = 9 + (-12) = -3$$

(b) $-7 - 2 = -7 + (-2) = -9$

(c) $-8 - (-2)$

Change the sign of the -2, then add.

$$-8 - (-2) = -8 + 2 = -6$$

(d) $-6.3 - (-11.5) = -6.3 + 11.5 = 5.2$

(e) $12 - (-5) = 12 + 5 = 17$

4. **(a)** $-6 + 9 - 2$

Add and subtract in order from left to right.

$$-6 + 9 - 2 = (-6 + 9) - 2 = 3 - 2 = 1$$

(b) $$12 - (-4) + 8 = (12 + 4) + 8 = 16 + 8 = 24$$

(c) $$-6 - (-2) - 8 - 1 = -6 + 2 - 8 - 1 = -4 + (-8) - 1 = -12 - 1 = -13$$

(d) $$-3 - [(-7) + 15] + 6 = -3 - 8 + 6 = -11 + 6 = -5$$

5. **(a)** $-7(-5) = 7 \cdot 5 = 35$

The numbers have the same sign, so the product is positive.

(b) $-.9(-15) = 13.5$

(c) $\left(-\frac{4}{7}\right)\left(-\frac{14}{3}\right) = \frac{4 \cdot 2 \cdot 7}{7 \cdot 3} = \frac{4 \cdot 2}{3} = \frac{8}{3}$

(d) $7(-2) = -14$

The numbers have different signs, so the product is negative.

(e) $-.8(.006) = -.0048$

(f) $\frac{5}{8}(-16) = -\frac{5 \cdot 2 \cdot 8}{8} = -10$

(g) $-\frac{2}{3}(12) = -\frac{2 \cdot 4 \cdot 3}{3} = -8$

6.

	Number	Reciprocal
(a)	15	$\frac{1}{15}$
(b)	-7	$\frac{1}{-7} = -\frac{1}{7}$
(c)	$\frac{8}{9}$	$\frac{9}{8}$
(d)	$-\frac{1}{3}$	$-\frac{3}{1} = -3$
(e)	$.125 = \frac{1}{8}$	$\frac{8}{1} = 8$

7. **(a)** $\frac{9}{0}$ is undefined.

(b) $\frac{0}{9} = 0$

(c) $\frac{-9}{0}$ is undefined.

(d) $\frac{0}{-9} = 0$

8. **(a)** $\frac{-16}{4} = -16\left(\frac{1}{4}\right) = -4$

The numbers have opposite signs, so the quotient is negative.

(b) $\frac{8}{-2} = 8\left(\frac{1}{-2}\right) = -4$

(c) $\frac{-15}{-3} = -15\left(-\frac{1}{3}\right) = 5$

(d) $\frac{\frac{3}{8}}{\frac{11}{16}} = \frac{3}{8} \div \frac{11}{16} = \frac{3}{8} \cdot \frac{16}{11} = \frac{3 \cdot 2 \cdot 8}{8 \cdot 11} = \frac{6}{11}$

9. **A.** $\frac{3}{5}$ **B.** $\frac{3}{-5} = -\frac{3}{5} = \frac{-3}{5}$

C. $-\frac{3}{5} = \frac{-3}{5}$ **D.** $\frac{-3}{-5} = \frac{3}{5}$

B and **C** are equal to $\frac{-3}{5}$.

1.2 Section Exercises

1. The sum of a positive number and a negative number is 0 if the numbers are additive inverses. For example, $4 + (-4) = 0$.

2. The sum of two positive numbers is a positive number. For example, $18 + 6 = 24$.

3. The sum of two negative numbers is a negative number. For example, $-7 + (-21) = -28$.

4. The sum of a positive number and a negative number is negative if the negative number has a larger absolute value. For example, $-14 + 9 = -5$.

5. The sum of a positive number and a negative number is positive if the positive number has a larger absolute value. For example, $15 + (-2) = 13$.

6. The difference between two positive numbers is negative if the number with the larger absolute value is subtracted from the one with the smaller absolute value. For example, $5 - 12 = -7$.

7. The difference between two negative numbers is negative if the number with smaller absolute value is subtracted from the one with larger absolute value. For example, $-15 - (-3) = -12$.

8. The product of two numbers with like signs is positive. For example, $(-2)(-8) = 16$ and $2 \cdot 8 = 16$.

9. The product of two numbers with unlike signs is negative. For example, $-5(15) = -75$.

10. The quotient formed by any nonzero number divided by 0 is undefined, and the quotient formed by 0 divided by any nonzero number is 0. For example, $\frac{-17}{0}$ is undefined and $\frac{0}{42}$ is equal to 0.

11. $13 + (-4) = 13 - 4 = 9$

12. $19 + (-13) = 19 - 13 = 6$

13. $-6 + (-13) = -(6 + 13) = -19$

14. $-8 + (-15) = -(8 + 15) = -23$

15. $-\frac{7}{3} + \frac{3}{4} = -\frac{28}{12} + \frac{9}{12} = -\frac{19}{12}$

16. $-\frac{5}{6} + \frac{3}{8} = -\frac{20}{24} + \frac{9}{24} = -\frac{11}{24}$

17. The difference between 2.3 and .45 is 1.85. The number with the larger absolute value, -2.3, is negative, so the answer is negative. Thus, $-2.3 + .45 = -1.85$.

18.
$$\begin{aligned} -.238 + 4.55 &= |4.55| - |-.238| \\ &= 4.55 - .238 \\ &= 4.312 \end{aligned}$$

19. $-6 - 5 = -6 + (-5) = -(6 + 5) = -11$

20. $-8 - 13 = -8 + (-13) = -(8 + 13) = -21$

21. $8 - (-13) = 8 + 13 = 21$

22. $13 - (-22) = 13 + 22 = 35$

23. $-16-(-3)=-16+3=-13$

24. $-21-(-8)=-21+8=-13$

25. $$\begin{aligned}-12.31-(-2.13)&=-12.31+2.13\\&=-(12.31-2.13)\\&=-10.18\end{aligned}$$

26. $$\begin{aligned}-15.88-(-9.22)&=-15.88+9.22\\&=-(15.88-9.22)=-6.66\end{aligned}$$

27. $\frac{9}{10}-\left(-\frac{4}{3}\right)=\frac{9}{10}+\frac{4}{3}=\frac{27}{30}+\frac{40}{30}=\frac{67}{30}$

28. $\frac{3}{14}-\left(-\frac{1}{4}\right)=\frac{3}{14}+\frac{1}{4}=\frac{6}{28}+\frac{7}{28}=\frac{13}{28}$

29. $-2-|-4|=-2-4=-2+(-4)=-6$

30. $9-|-13|=9-13=9+(-13)=-4$

31. The product of two numbers with *different* signs is *negative*, so
$$5(-7)=-35.$$

32. The product of two numbers with *different* signs is *negative*, so
$$6(-6)=-36.$$

33. The product of two numbers with the *same* sign is *positive*, so
$$-8(-5)=40.$$

34. The product of two numbers with the *same* sign is *positive*, so
$$-10(-4)=40.$$

35. The product of two numbers with the *same* sign is *positive*, so
$$-10\left(-\frac{1}{5}\right)=2.$$

36. $-\frac{1}{2}(-12)=6$

37. The product of two numbers with *different* signs is *negative*, so
$$\frac{3}{4}(-16)=-\frac{3}{4}\cdot 4\cdot 4=-12.$$

38. The product of two numbers with *different* signs is *negative*, so
$$\frac{4}{5}(-35)=-\frac{4}{5}\cdot 5\cdot 7=-28.$$

39. The product of two numbers with the *same* sign is *positive*, so
$$-\frac{5}{2}\left(-\frac{12}{25}\right)=\frac{5\cdot 2\cdot 6}{2\cdot 5\cdot 5}=\frac{6}{5}.$$

40. $-\frac{9}{7}\left(-\frac{35}{36}\right)=\frac{9\cdot 5\cdot 7}{7\cdot 4\cdot 9}=\frac{5}{4}$

41. The product of two numbers with the *same* sign is *positive*, so
$$-\frac{3}{8}\left(-\frac{24}{9}\right)=\frac{3\cdot 3\cdot 8}{8\cdot 9}=1.$$

42. $-\frac{2}{11}\left(-\frac{99}{4}\right)=\frac{2\cdot 9\cdot 11}{11\cdot 2\cdot 2}=\frac{9}{2}$

43. $-2.4(-2.45)=5.88$

44. $-3.45(-2.14)=7.383$

45. $3.4(-3.14)=-10.676$

46. $5.66(-2.1)=-11.886$

In Exercises 47–58, a number and its reciprocal always have the same sign.

47. The reciprocal of $6=\frac{6}{1}$ is $\frac{1}{6}$.

48. The reciprocal of $8=\frac{8}{1}$ is $\frac{1}{8}$.

49. The reciprocal of $-7=-\frac{7}{1}$ is $-\frac{1}{7}$.

50. The reciprocal of $-11=-\frac{11}{1}$ is $-\frac{1}{11}$.

51. The reciprocal of $-\frac{2}{3}$ is $-\frac{3}{2}$.

52. The reciprocal of $-\frac{7}{8}$ is $-\frac{8}{7}$.

53. The reciprocal of $\frac{1}{5}$ is $\frac{5}{1}=5$.

54. The reciprocal of $\frac{1}{4}$ is $\frac{4}{1}=4$.

55. The reciprocal of $.02=\frac{2}{100}=\frac{1}{50}$ is $\frac{50}{1}=50$.

56. The reciprocal of $.45=\frac{45}{100}=\frac{9}{20}$ is $\frac{20}{9}$.

57. The reciprocal of $-.001=-\frac{1}{1000}$ is -1000.

58. The reciprocal of $-.0003=-\frac{3}{10{,}000}$ is $-\frac{10{,}000}{3}$.

59. The quotient of two nonzero real numbers with *different* signs is *negative*, so
$$\frac{-14}{2}=-14\cdot\frac{1}{2}=-2\cdot 7\cdot\frac{1}{2}=-7.$$

60. The quotient of two nonzero real numbers with *different* signs is *negative*, so
$$\frac{-26}{13}=-26\cdot\frac{1}{13}=-2\cdot 13\cdot\frac{1}{13}=-2.$$

61. The quotient of two nonzero real numbers with the *same* sign is *positive*, so

$$\frac{-24}{-4} = 24 \cdot \frac{1}{4} = 4 \cdot 6 \cdot \frac{1}{4} = 6.$$

62. The quotient of two nonzero real numbers with the *same* sign is *positive*, so

$$\frac{-36}{-9} = 36 \cdot \frac{1}{9} = 4 \cdot 9 \cdot \frac{1}{9} = 4.$$

63. The quotient of two nonzero real numbers with *different* signs is *negative*, so

$$\frac{100}{-25} = -100 \cdot \frac{1}{25} = -4 \cdot 25 \cdot \frac{1}{25} = -4.$$

64. $\frac{300}{-60} = -300 \cdot \frac{1}{60} = -5 \cdot 60 \cdot \frac{1}{60} = -5$

65. $\frac{0}{-8} = 0 \cdot \left(-\frac{1}{8}\right) = 0$

66. $\frac{0}{-10} = 0 \cdot \left(-\frac{1}{10}\right) = 0$

67. Division by 0 is undefined, so $\frac{5}{0}$ is *undefined.*

68. Division by 0 is undefined, so $\frac{12}{0}$ is *undefined.*

69. The quotient of two nonzero real numbers with the *same* sign is *positive*, so

$$-\frac{10}{17} \div \left(-\frac{12}{5}\right) = \frac{10}{17} \cdot \frac{5}{12} = \frac{2 \cdot 5 \cdot 5}{17 \cdot 2 \cdot 6} = \frac{25}{102}.$$

70. $-\frac{22}{23} \div \left(-\frac{33}{4}\right) = \frac{22}{23} \cdot \frac{4}{33} = \frac{2 \cdot 11 \cdot 4}{23 \cdot 3 \cdot 11} = \frac{8}{69}$

71. $\frac{\frac{12}{13}}{-\frac{4}{3}} = \frac{12}{13} \div \left(-\frac{4}{3}\right) = \frac{12}{13}\left(-\frac{3}{4}\right)$

$$= -\frac{3 \cdot 4 \cdot 3}{13 \cdot 4} = -\frac{9}{13}$$

72. $\frac{\frac{5}{6}}{-\frac{1}{30}} = \frac{5}{6} \div \left(-\frac{1}{30}\right) = \frac{5}{6}\left(-\frac{30}{1}\right)$

$$= -\frac{5 \cdot 5 \cdot 6}{6} = -25$$

73. $-\frac{27.72}{13.2} = -2.1$

74. $\frac{-126.7}{36.2} = -3.5$

75. $\frac{-100}{-.01} = \frac{100}{\frac{1}{100}} = 100 \div \frac{1}{100} = 100 \cdot 100$

$$= 10{,}000$$

76. $\frac{-50}{-.05} = \frac{50}{\frac{5}{100}} = 50 \div \frac{5}{100} = 50 \cdot \frac{100}{5}$

$$= 10 \cdot 100 = 1000$$

77. $-7 + 5 - 9 = (-7 + 5) - 9 = -2 - 9 = -11$

78. $-12 + 13 - 19 = 1 - 19 = -18$

79. $6 - (-2) + 8 = 6 + 2 + 8 = 8 + 8 = 16$

80. $7 - (-3) + 12 = 7 + 3 + 12 = 10 + 12 = 22$

81. $-9 - 4 - (-3) + 6 = (-9 - 4) + 3 + 6$

$$= -13 + 3 + 6$$
$$= -10 + 6$$
$$= -4$$

82. $-10 - 5 - (-12) + 8 = -10 + (-5) + 12 + 8$

$$= -15 + 12 + 8$$
$$= -3 + 8$$
$$= 5$$

83. $-4 - [(-4 - 6) + 12] - 13$

$$= -4 - (-10 + 12) - 13$$
$$= -4 - 2 - 13$$
$$= -4 + (-2) + (-13)$$
$$= -6 + (-13)$$
$$= -19$$

84. $-10 - [(-2 + 3) - 4] - 17$

$$= -10 - (1 - 4) - 17$$
$$= -10 - (-3) - 17$$
$$= -10 + 3 - 17$$
$$= -7 - 17$$
$$= -24$$

85. To find the difference between these two temperatures, subtract the low temperature from the high temperature.

$$90° - (-22°) = 90° + 22° = 112°\text{F}$$

86. $120° - (-29°) = 120° + 29° = 149°$
The difference is 149°F.

87. Represent the loss of \$24.4 million as −24.4. Subtract the \$13.1 million profit from the previous year.

$$-24.4 - 13.1 = -(24.4 + 13.1) = -37.5$$

Expressed as a negative amount, the difference is −\$37.5 million.

88. $37.58 - (-9.10) = 37.58 + 9.10 = 46.68$
The difference is 46.68%.

89.

Year	Difference (in billions)
2000	\$538 − \$409 = \$129
2010	\$916 − \$710 = \$206
2020	\$1479 − \$1405 = \$74
2030	\$2041 − \$2542 = −\$501

90. The cost of Social Security will exceed revenue in 2030 by \$501 billion.

91. $-15{,}393 - 1350 = -(15{,}393 + 1350)$
$= -16{,}743$

The difference between the 1995 and 1994 trade balances was $-\$16{,}743$ million.

92. $1350 - 1664 = 1350 + (-1664)$
$= -314$

The difference between the 1994 and 1993 trade balances was $-\$314$ million.

93. For 1997, use $-14{,}549 \approx -14{,}500$.
For 1996, use $-17{,}506 \approx -17{,}500$.

An estimate of the difference is

$-14{,}500 - (-17{,}500) = -14{,}500 + 17{,}500$
$= 3000,$

which is choice **C.**

94. For 2002, use $-37{,}146 \approx -37{,}000$.
For 2001, use $-30{,}041 \approx -30{,}000$.

An estimate of the difference is

$-37{,}000 - (-30{,}000) = -37{,}000 + 30{,}000$
$= -7000,$

which is choice **D.**

1.3 Exponents, Roots, and Order of Operations

1.3 Margin Exercises

1. **(a)** $\underbrace{3 \cdot 3 \cdot 3 \cdot 3 \cdot 3}_{\text{5 factors of 3}} = 3^5$

(b) $\frac{2}{7} \cdot \frac{2}{7} \cdot \frac{2}{7} \cdot \frac{2}{7} = \left(\frac{2}{7}\right)^4$

(c) $(-10)(-10)(-10) = (-10)^3$

(d) $(.5)(.5) = (.5)^2$

(e) $y \cdot y \cdot y \cdot y \cdot y \cdot y \cdot y \cdot y = y^8$

2. **(a)** $5^3 = 5 \cdot 5 \cdot 5$
$= 25 \cdot 5 = 125$

(b) $3^4 = 3 \cdot 3 \cdot 3 \cdot 3$
$= 9 \cdot 3 \cdot 3$
$= 27 \cdot 3 = 81$

(c) $(-4)^5 = (-4)(-4)(-4)(-4)(-4)$
$= 16(-4)(-4)(-4)$
$= -64(-4)(-4)$
$= 256(-4)$
$= -1024$

(d) $(-3)^4 = (-3)(-3)(-3)(-3) = 81$

(e) $(.75)^3 = (.75)(.75)(.75)$
$= (.5625)(.75)$
$= .421875$

(f) $\left(\frac{2}{5}\right)^4 = \frac{2}{5} \cdot \frac{2}{5} \cdot \frac{2}{5} \cdot \frac{2}{5} = \frac{16}{625}$

3.

		Exponent	Base	Value
(a)	7^3	3	7	$7^3 = 343$
(b)	$(-5)^4$	4	-5	$(-5)^4 = 625$
(c)	-5^4	4	5	$-5^4 = -625$
(d)	$-(.9)^5$	5	.9	$-(.9)^5 = -.59049$

4. **(a)** $\sqrt{9} = 3$ since 3 is positive and $3^2 = 9$.

(b) $\sqrt{49} = 7$ since 7 is positive and $7^2 = 49$.

(c) $-\sqrt{81} = -9$ since the negative sign is outside the radical sign.

(d) $\sqrt{\frac{121}{81}} = \frac{11}{9}$ since $\left(\frac{11}{9}\right)^2 = \frac{121}{81}$.

(e) $\sqrt{.25} = .5$ since $(.5)^2 = .25$.

(f) $\sqrt{-9}$ is not a real number.

(g) $-\sqrt{-169}$ is not a real number.

5. **(a)** $5 \cdot 9 + 2 \cdot 4 = 45 + 8$ *Multiply.*
$= 53$ *Add.*

(b) $4 - 12 \div 4 \cdot 2 = 4 - 3 \cdot 2$ *Divide.*
$= 4 - 6$ *Multiply.*
$= -2$ *Subtract.*

6. **(a)** $(4 + 2) - 3^2 - (8 - 3)$
$= 6 - 3^2 - 5$ *Add and subtract inside parentheses.*
$= 6 - 9 - 5$ *Evaluate the power.*
$= -3 - 5$ *Subtract from left to right.*
$= -8$

(b) $6 + \frac{2}{3}(-9) - \frac{5}{8} \cdot 16$
$= 6 + (-6) - 10$ *Do the multiplications first.*
$= 0 - 10$ *Add.*
$= -10$ *Subtract.*

7. **(a)** $\dfrac{10-6+2\sqrt{9}}{11\cdot 2-3(2)^2}$

$= \dfrac{10-6+2(3)}{11\cdot 2-3(4)}$ *Evaluate powers and roots.*

$= \dfrac{10-6+6}{22-12}$ *Multiply.*

$= \dfrac{10}{10}$ *Add and subtract.*

$= 1$ *Reduce.*

(b) $\dfrac{-4(8)+6(3)}{3\sqrt{49}-\frac{1}{2}(42)} = \dfrac{-4(8)+6(3)}{3(7)-\frac{1}{2}(42)}$ *Evaluate the root.*

$= \dfrac{-32+18}{21-21}$ *Multiply.*

$= \dfrac{-14}{0}$, which is undefined.

8. Let $w=4$, $x=-12$, $y=64$, and $z=-3$.

(a) $5x-2w = 5(-12)-2(4)$
$= -60-8$
$= -68$

(b) $-6(x-\sqrt{y}) = -6\left(-12-\sqrt{64}\right)$
$= -6(-12-8)$
$= -6(-20)$
$= 120$

(c) $\dfrac{5x-3\cdot\sqrt{y}}{x-1} = \dfrac{5(-12)-3\sqrt{64}}{-12-1}$

$= \dfrac{5(-12)-3(8)}{-12-1}$

$= \dfrac{-60-24}{-12-1}$

$= \dfrac{-84}{-13}$

$= \dfrac{84}{13}$

(d) $w^2+2z^3 = 4^2+2(-3)^3$
$= 16+2(-27)$
$= 16+(-54)$
$= -38$

9. $2.2023x - 4356.6$

1990: $2.2023(1990)-4356.6 \approx 26.0$
2000: $2.2023(2000)-4356.6 = 48.0$

In 1990, 26.0% of U.S. households invested in mutual funds. In 2000, 48.0% of U.S. households invested in mutual funds.

1.3 Section Exercises

1. $-4^6 = (-4)^6$ is *false*.
$-4^6 = -(4\cdot4\cdot4\cdot4\cdot4\cdot4) = -4096$,
whereas
$(-4)^6 = (-4)(-4)(-4)(-4)(-4)(-4)$
$= 4096.$

2. $-4^7 = (-4)^7$ is *true*. $(-4)^7 = -4^7$ since $(-4)^7$ has an odd number of negative factors, which yields a negative product.

3. The statement "$\sqrt{16}$ is a positive number" is *true*. The symbol $\sqrt{}$ always gives a positive square root when the radicand is positive.

4. $3+5\cdot6 = 3+(5\cdot6)$ is *true*. The order of operations says that multiplication should be done before addition.

5. The statement "$(-2)^7$ is a negative number" is *true*. $(-2)^7$ has an odd number of negative factors, so the product is negative.

6. The statement "$(-2)^8$ is a positive number" is *true*. $(-2)^8$ has an even number of negative factors, so the product is positive.

7. The statement "The product of 8 positive factors and 8 negative factors is positive" is *true*. The product of 8 positive factors is positive and the product of 8 negative factors is positive, so the product of these two products is positive.

8. The statement "The product of 3 positive factors and 3 negative factors is positive" is *false*. The product of 3 positive factors is positive and the product of 3 negative factors is negative, so the product of these two products is negative.

9. The statement "In the exponential -3^5, -3 is the base" is *false*. The base is 3, not -3. If the problem were written $(-3)^5$, then -3 would be the base.

10. The statement "$\sqrt{a}$ is positive for all positive numbers a" is *true*.

11. **(a)** $8^2 = 64$

(b) $-8^2 = -(8\cdot8) = -64$

(c) $(-8)^2 = (-8)(-8) = 64$

(d) $-(-8)^2 = -(64) = -64$

12. **(a)** $4^3 = 64$

(b) $-4^3 = -(4\cdot4\cdot4) = -64$

(c) $(-4)^3 = (-4)(-4)(-4) = -64$

(d) $-(-4)^3 = -(-64) = 64$

13. $8 \cdot 8 \cdot 8 = 8^3$

14. $10 \cdot 10 \cdot 10 \cdot 10 = 10^4$

15. $\frac{1}{2} \cdot \frac{1}{2} = \left(\frac{1}{2}\right)^2$

16. $\frac{3}{4} \cdot \frac{3}{4} \cdot \frac{3}{4} \cdot \frac{3}{4} \cdot \frac{3}{4} = \left(\frac{3}{4}\right)^5$

17. $(-4)(-4)(-4)(-4) = (-4)^4$

18. $(-9)(-9)(-9) = (-9)^3$

19. $z \cdot z \cdot z \cdot z \cdot z \cdot z \cdot z = z^7$

20. $a \cdot a \cdot a \cdot a \cdot a \cdot a = a^6$

21. $4^2 = 4 \cdot 4 = 16$

22. $2^4 = 2 \cdot 2 \cdot 2 \cdot 2 = 16$

23. $.28^3 = (.28)(.28)(.28) = .021952$

24. $.91^3 = (.91)(.91)(.91) = .753571$

25. $\left(\frac{1}{5}\right)^3 = \frac{1}{5} \cdot \frac{1}{5} \cdot \frac{1}{5} = \frac{1}{125}$

26. $\left(\frac{1}{6}\right)^4 = \left(\frac{1}{6}\right)\left(\frac{1}{6}\right)\left(\frac{1}{6}\right)\left(\frac{1}{6}\right)$
$= \frac{1}{6 \cdot 6 \cdot 6 \cdot 6} = \frac{1}{1296}$

27. $\left(\frac{7}{10}\right)^3 = \left(\frac{7}{10}\right)\left(\frac{7}{10}\right)\left(\frac{7}{10}\right) = \frac{343}{1000}$

28. $\left(\frac{4}{5}\right)^4 = \left(\frac{4}{5}\right)\left(\frac{4}{5}\right)\left(\frac{4}{5}\right)\left(\frac{4}{5}\right)$
$= \frac{4 \cdot 4 \cdot 4 \cdot 4}{5 \cdot 5 \cdot 5 \cdot 5} = \frac{256}{625}$

29. $(-5)^3 = (-5)(-5)(-5) = -125$

30. $(-3)^5 = (-3)(-3)(-3)(-3)(-3) = -243$

31. $(-2)^8 = (-2)(-2)(-2)(-2)(-2)(-2)(-2)(-2)$
$= 256$

32. $(-3)^6 = (-3)(-3)(-3)(-3)(-3)(-3) = 729$

33. $-3^6 = -(3 \cdot 3 \cdot 3 \cdot 3 \cdot 3 \cdot 3) = -729$

34. $-4^6 = -(4 \cdot 4 \cdot 4 \cdot 4 \cdot 4 \cdot 4) = -4096$

35. $-8^4 = -(8 \cdot 8 \cdot 8 \cdot 8) = -4096$

36. $-10^3 = -(10 \cdot 10 \cdot 10) = -1000$

37. In $(-4.1)^7$, the exponent is 7; the base is -4.1.

38. In $(-3.4)^9$, the exponent is 9; the base is -3.4.

39. In -4.1^7, the exponent is 7; the base is 4.1.

40. In -3.4^9, the exponent is 9; the base is 3.4.

41. $\sqrt{81} = 9$ since 9 is positive and $9^2 = 81$.

42. $\sqrt{64} = 8$ since 8 is positive and $8^2 = 64$.

43. $\sqrt{169} = 13$ since 13 is positive and $13^2 = 169$.

44. $\sqrt{225} = 15$ since 15 is positive and $15^2 = 225$.

45. $-\sqrt{400} = -(\sqrt{400}) = -(20) = -20$

46. $-\sqrt{900} = -(\sqrt{900}) = -(30) = -30$

47. $\sqrt{\frac{100}{121}} = \frac{10}{11}$ since $\frac{10}{11}$ is positive and $\left(\frac{10}{11}\right)^2 = \frac{100}{121}$.

48. $\sqrt{\frac{225}{169}} = \frac{15}{13}$ since $\frac{15}{13}$ is positive and $\left(\frac{15}{13}\right)^2 = \frac{225}{169}$.

49. $-\sqrt{.49} = -\sqrt{(.7)^2} = -(.7) = -.7$

50. $-\sqrt{.64} = -\sqrt{(.8)^2} = -(.8) = -.8$

51. There is no real number whose square is negative, so $\sqrt{-36}$ is *not a real number*.

52. There is no real number whose square is negative, so $\sqrt{-121}$ is *not a real number*.

53. **(a)** $\sqrt{144} = 12$; choice **B**

(b) $\sqrt{-144}$ is not a real number; choice **C**

(c) $-\sqrt{144} = -12$; choice **A**

54. There is no real number whose square is negative, so $\sqrt{-900}$ is not real.

55. If a is a positive number, then $-a$ is a negative number. Therefore, $\sqrt{-a}$ is not a real number, so $-\sqrt{-a}$ is *not a real number*.

56. If a is a positive number, then $\sqrt{a}$ is also a positive number, and $-\sqrt{a}$ is a *negative* number.

57. $12 + 3 \cdot 4 = 12 + 12$ *Multiply.*
$= 24$ *Add.*

58. $15 + 5 \cdot 2 = 15 + 10$ *Multiply.*
$= 25$ *Add.*

59. $2[-5 - (-7)] = 2(-5 + 7)$
$= 2(2) = 4$

60. $3[-8 - (-2)] = 3(-8 + 2)$
$= 3(-6) = -18$

61. $-12\left(-\frac{3}{4}\right) - (-5) = 9 - (-5)$
$= 9 + 5 = 14$

62. $-7\left(-\frac{2}{14}\right) - (-8) = 1 - (-8)$
$= 1 + 8 = 9$

63. $6 \cdot 3 - 12 \div 4 = 18 - 12 \div 4$ *Multiply.*
$= 18 - 3$ *Divide.*
$= 15$ *Subtract.*

64. $9 \cdot 4 - 8 \div 2 = 36 - 8 \div 2$ *Multiply.*
$= 36 - 4$ *Divide.*
$= 32$ *Subtract.*

65. $10 + 30 \div 2 \cdot 3 = 10 + 15 \cdot 3$ *Divide.*
$= 10 + 45$ *Multiply.*
$= 55$ *Add.*

66. $12 + 24 \div 3 \cdot 2 = 12 + 8 \cdot 2$ *Divide.*
$= 12 + 16$ *Multiply.*
$= 28$ *Add.*

67. $-3(5)^2 - (-2)(-8)$
$= -3(25) - (-2)(-8)$ *Evaluate power.*
$= -75 - 16$ *Multiply.*
$= -91$ *Subtract.*

68. $-9(2)^2 - (-3)(-2)$
$= -9(4) - (-3)(-2)$ *Evaluate power.*
$= -36 - 6$ *Multiply.*
$= -42$ *Subtract.*

69. $5 - 7 \cdot 3 - (-2)^3$
$= 5 - 7 \cdot 3 - (-8)$ *Evaluate power.*
$= 5 - 21 + 8$ *Multiply; change subtraction to addition.*
$= -16 + 8$ *Subtract.*
$= -8$ *Add.*

70. $-4 - 3 \cdot 5 + 6^2$
$= -4 - 3 \cdot 5 + 36$ *Evaluate power.*
$= -4 - 15 + 36$ *Multiply.*
$= -19 + 36$ *Subtract.*
$= 17$ *Add.*

71. $-7\left(\sqrt{36}\right) - (-2)(-3)$
$= -7(6) - (-2)(-3)$ *Evaluate root.*
$= -42 - 6$ *Multiply.*
$= -48$ *Subtract.*

72. $-8\left(\sqrt{64}\right) - (-3)(-7)$
$= -8(8) - (-3)(-7)$ *Evaluate root.*
$= -64 - 21$ *Multiply.*
$= -85$ *Subtract.*

73. $\dfrac{-8 + (-16)}{-3} = \dfrac{-24}{-3} = 8$

74. $\dfrac{-9 + (-11)}{-2} = \dfrac{-20}{-2} = 10$

75. $\dfrac{\left(-5 + \sqrt{4}\right)(-2^2)}{-5 - 1}$
$= \dfrac{(-5 + 2)(-4)}{-6}$ *Evaluate root and power; subtract.*
$= \dfrac{(-3)(-4)}{-6}$ *Work inside parentheses.*
$= \dfrac{12}{-6}$ *Multiply.*
$= -2$ *Divide.*

76. $\dfrac{\left(-9 + \sqrt{16}\right)(-3^2)}{-4 - 1}$
$= \dfrac{(-9 + 4)(-9)}{-5}$ *Evaluate root and power; subtract.*
$= \dfrac{(-5)(-9)}{-5}$ *Work inside parentheses.*
$= -9$ *Reduce.*

77. $6|4 - 5| - 24 \div 3$
$= 6|-1| - 24 \div 3$ *Simplify within absolute value bars.*
$= 6(1) - 24 \div 3$ *Take absolute value.*
$= 6 - 8$ *Multiply and divide.*
$= -2$ *Subtract.*

78. $-4|2 - 4| + 8 \cdot 2$
$= -4|-2| + 8 \cdot 2$ *Simplify within absolute value bars.*
$= -4(2) + 8 \cdot 2$ *Take absolute value.*
$= -8 + 16$ *Multiply.*
$= 8$ *Add.*

79. $\dfrac{2(-5) + (-3)(-2)}{-8 + 3^2 - 1}$
$= \dfrac{-10 + 6}{-8 + 9 - 1}$ *Evaluate power; multiply.*
$= \dfrac{-4}{1 - 1}$ *Add.*
$= \dfrac{-4}{0}$ *Subtract.*

Since division by 0 is undefined, the given expression is *undefined.*

80. $\dfrac{3(-4) + (-5)(-8)}{2^3 - 2 - 6}$
$= \dfrac{-12 + 40}{8 - 2 - 6}$ *Evaluate power; multiply.*
$= \dfrac{28}{6 - 6}$ *Add; subtract.*
$= \dfrac{28}{0}$ *Subtract.*

Since division by 0 is undefined, the given expression is *undefined.*

81. $3a + \sqrt{b} = 3(-3) + \sqrt{64}$
$= 3(-3) + 8$
$= -9 + 8$
$= -1$

82. $-2a - \sqrt{b} = -2(-3) - \sqrt{64}$
$= 6 - 8$
$= -2$

83. $\sqrt{b} + c - a = \sqrt{64} + 6 - (-3)$
$= 8 + 6 + 3$
$= 14 + 3 = 17$

84. $\sqrt{b} - c + a = \sqrt{64} - 6 + (-3)$
$= 8 - 6 + (-3)$
$= 2 + (-3) = -1$

85. $4a^3 + 2c = 4(-3)^3 + 2(6)$
$= 4(-27) + 12$
$= -108 + 12 = -96$

86. $-3a^4 - 3c = -3(-3)^4 - 3(6)$
$= -3(81) - 18$
$= -243 - 18 = -261$

87. $\frac{2c + a^3}{4b + 6a} = \frac{2(6) + (-3)^3}{4(64) + 6(-3)}$
$= \frac{12 + (-27)}{256 + (-18)}$
$= \frac{-15}{238} = -\frac{15}{238}$

88. $\frac{3c + a^2}{2b - 6c} = \frac{3(6) + (-3)^2}{2(64) - 6(6)}$
$= \frac{18 + 9}{128 - 36}$
$= \frac{27}{92}$

89. $(v \times .5485 - 4850) \div 1000 \times 31.44$
$= (100{,}000 \times .5485 - 4850) \div 1000 \times 31.44$
$= (54{,}850 - 4850) \div 1000 \times 31.44$
$= 50{,}000 \div 1000 \times 31.44$
$= 50 \times 31.44$
$= 1572$

The owner would pay $1572 in property taxes.

90. $(v \times .5485 - 4850) \div 1000 \times 31.44$
$= (150{,}000 \times .5485 - 4850) \div 1000 \times 31.44$
$= (82{,}275 - 4850) \div 1000 \times 31.44$
$= 77{,}425 \div 1000 \times 31.44$
$= 77.425 \times 31.44$
$= 2434.242 \approx 2434$

The owner would pay $2434 in property taxes.

91. $(v \times .5485 - 4850) \div 1000 \times 31.44$
$= (200{,}000 \times .5485 - 4850) \div 1000 \times 31.44$
$= (109{,}700 - 4850) \div 1000 \times 31.44$
$= 104{,}850 \div 1000 \times 31.44$
$= 104.85 \times 31.44$
$= 3296.484 \approx 3296$

The owner would pay $3296 in property taxes.

92. **(a)** number of oz $\times$ % alcohol $\times .075 \div$ body weight in lb $-$ hr of drinking $\times .015$

$= 48 \times 3.2 \times .075 \div 190 - 2 \times .015$

(b) $48 \times 3.2 \times .075 \div 190 - 2 \times .015$
$= 153.6 \times .075 \div 190 - 2 \times .015$
$= 11.52 \div 190 - 2 \times .015$
$\approx .061 - 2 \times .015$
$= .061 - .03$
$= .031$

93. number of oz $\times$ % alcohol $\times .075 \div$ body weight in lb $-$ hr of drinking $\times .015$

$= 36 \times 4.0 \times .075 \div 135 - 3 \times .015$
$= 144 \times .075 \div 135 - 3 \times .015$
$= 10.8 \div 135 - 3 \times .015$
$= .08 - 3 \times .015$
$= .08 - .045$
$= .035$

94. BAC for the man
$= 48 \times 3.2 \times .075 \div (190 + 25) - 2 \times .015$
$\approx .024$

BAC for the woman
$= 36 \times 4.0 \times .075 \div (135 + 25) - 3 \times .015$
$\approx .023$

Increased weight results in lower BACs.

95. Decreased weight will result in higher BACs.

BAC for the man
$= 48 \times 3.2 \times .075 \div (190 - 25) - 2 \times .015$
$\approx .040$

BAC for the woman
$= 36 \times 4.0 \times .075 \div (135 - 25) - 3 \times .015$
$\approx .053$

96. **(a)** $3.31714x - 6597.86$

Year (x)	Education Spending (in billions of dollars)
1997	26.5
1998	29.8
1999	$3.31714(1999) - 6597.86 = 33.1$
2000	$3.31714(2000) - 6597.86 = 36.4$
2001	$3.31714(2001) - 6597.86 = 39.7$

(b) Federal spending on education has increased (about \$3.3 billion per year) during these years.

1.4 Properties of Real Numbers

1.4 Margin Exercises

1. **(a)** Let $a = 8$, $b = m$, and $c = n$ in the statement of the distributive property. Then,

$$8(m + n) = 8m + 8n.$$

(b)
$$\begin{aligned} -4(p-5) &= -4[p + (-5)] \\ &= -4(p) + (-4)(-5) \\ &= -4p + 20 \end{aligned}$$

(c) Use the second form of the distributive property.

$$3k + 6k = (3 + 6)k = 9k$$

(d) Use the second form of the distributive property.

$$\begin{aligned} -6m + 2m &= (-6 + 2)m \\ &= -4m \end{aligned}$$

(e) $2r + 3s$

Since there is no common number or variable here, the distributive property cannot be used.

(f)
$$\begin{aligned} 5(4p - 2q + r) &= 5(4p) + 5(-2q) + 5(r) \\ &= 20p - 10q + 5r \end{aligned}$$

2. **(a)**
$$\begin{aligned} 14 \cdot 5 + 14 \cdot 85 &= 14(5 + 85) \\ &= 14(90) \\ &= 1260 \end{aligned}$$

(b)
$$\begin{aligned} 78 \cdot 33 + 22 \cdot 33 &= (78 + 22)33 \\ &= 100(33) \\ &= 3300 \end{aligned}$$

3. **(a)** The number that must be added to 4 to get 0 is -4.

(b) The number that must be added to -7.1 to get 0 is 7.1.

(c) The sum of -9 and 9 is 0.

(d) The number that must be multiplied by 5 to get 1 is the reciprocal of 5, which is $\frac{1}{5}$.

(e) The number that must be multiplied by $-\frac{3}{4}$ to get 1 is the reciprocal of $-\frac{3}{4}$, which is $-\frac{4}{3}$.

(f) The product of 7 and $\frac{1}{7}$ is 1.

4. **(a)**
$$\begin{aligned} p - 3p &= 1p - 3p && \textit{Identity property} \\ &= (1 - 3)p && \textit{Distributive property} \\ &= -2p \end{aligned}$$

(b)
$$\begin{aligned} &r + r + r \\ &= 1r + 1r + 1r && \textit{Identity property} \\ &= (1 + 1 + 1)r && \textit{Distributive property} \\ &= 3r \end{aligned}$$

(c)
$$\begin{aligned} &-(3 + 4p) \\ &= -1(3 + 4p) && \textit{Identity property} \\ &= -1(3) + (-1)(4p) && \textit{Distributive property} \\ &= -3 + (-4p) \\ &= -3 - 4p \end{aligned}$$

(d)
$$\begin{aligned} &-(k - 2) \\ &= -1(k - 2) && \textit{Identity property} \\ &= -1(k) + (-1)(-2) && \textit{Distributive property} \\ &= -k + 2 \end{aligned}$$

5. **(a)**
$$\begin{aligned} &12b - 9b + 4b - 7b + b \\ &= (12 - 9 + 4 - 7 + 1)b && \textit{Distributive property} \\ &= b \end{aligned}$$

(b)
$$\begin{aligned} &-3w + 7 - 8w - 2 \\ &= -3w - 8w + 7 - 2 && \textit{Commutative and associative properties} \\ &= -11w + 5 && \textit{Combine like terms.} \end{aligned}$$

(c) $-3(6 + 2t) = -18 - 6t$ *Distributive property*

(d)
$$\begin{aligned} &9 - 2(a - 3) + 4 - a \\ &= 9 - 2a + 6 + 4 - a && \textit{Distributive property} \\ &= -2a - a + 9 + 6 + 4 && \textit{Commutative and associative properties} \\ &= -3a + 19 && \textit{Combine like terms.} \\ &\text{or } 19 - 3a \end{aligned}$$

(e)
$$\begin{aligned} &(4m)(2n) \\ &= [(4m)(2)]n && \textit{Order of operations} \\ &= [4(m \cdot 2)]n && \textit{Associative property} \\ &= [4(2m)]n && \textit{Commutative property} \\ &= [(4 \cdot 2)m]n && \textit{Associative property} \\ &= (8m)n && \textit{Multiply.} \\ &= 8(mn) && \textit{Associative property} \\ &= 8mn \end{aligned}$$

6. **(a)** Using the multiplication property of zero, $197 \cdot 0 = 0$.

(b) Using the multiplication property of zero,

$$(0)\left(-\frac{8}{9}\right)=0.$$

(c) $0\cdot \underline{\quad\quad} = 0$

Any real number will be a solution, since zero times any number will equal zero.

1.4 Section Exercises

1. The identity element for addition is 0 since, for any real number a, $a+0=0+a=a$. Choice **B** is correct.

2. The identity element for multiplication is 1 since, for any real number a, $a\cdot 1=1\cdot a=a$. Choice **C** is correct.

3. The additive inverse of a is $-a$ since, for any real number a, $a+(-a)=0$ and $-a+a=0$. Choice **A** is correct.

4. The multiplicative inverse of a is $\frac{1}{a}$ since, for any *nonzero* real number a, $a\cdot\frac{1}{a}=1$ and $\frac{1}{a}\cdot a=1$. Choice **D** is correct.

5. The multiplication property of 0 states that the <u>product</u> of 0 and any real number is <u>0</u>.

6. The commutative property is used to change the <u>order</u> of two terms or factors.

7. The associative property is used to change the <u>grouping</u> of three terms or factors.

8. Like terms are terms with the <u>same</u> variables raised to the <u>same</u> powers.

9. When simplifying an expression, only <u>like</u> terms can be combined.

10. The coefficient in the term $-8yz^2$ is <u>-8</u>.

11. Using the second form of the distributive property,

$$\begin{aligned}5k+3k&=(5+3)k\\&=8k.\end{aligned}$$

12. $$\begin{aligned}6a+5a&=(6+5)a\\&=11a\end{aligned}$$

13. $$\begin{aligned}-9r+7r&=(-9+7)r \quad \textit{Distributive property}\\&=-2r\end{aligned}$$

14. $-4n+6n=(-4+6)n=2n$

15. $-8z+4w$
Since there is no common variable factor here, we cannot use the distributive property to simplify the expression.

16. $-12k+3r$
Since there is no common variable factor here, we cannot use the distributive property to simplify the expression.

17. $$\begin{aligned}-a+7a&=-1a+7a && \textit{Identity property}\\&=(-1+7)a && \textit{Distributive property}\\&=6a\end{aligned}$$

18. $$\begin{aligned}-s+9s&=-1\cdot s+9s && \textit{Identity property}\\&=(-1+9)s && \textit{Distributive property}\\&=8s\end{aligned}$$

19. Using the distributive property,

$$2(m+p)=2m+2p.$$

20. $3(a+b)=3a+3b$

21. $$\begin{aligned}-5(2d-f)&=-5[2d+(-f)]\\&=-5(2d)+(-5)(-f)\\&=-10d+5f\end{aligned}$$

22. $$\begin{aligned}-2(3m-n)&=-2[3m+(-n)]\\&=-2(3m)+(-2)(-n)\\&=-6m+2n\end{aligned}$$

23. $$\begin{aligned}96\cdot 19+4\cdot 19&=(96+4)19\\&=(100)19\\&=1900\end{aligned}$$

24. $$\begin{aligned}27\cdot 60+27\cdot 40&=27(60+40)\\&=27(100)\\&=2700\end{aligned}$$

25. $$\begin{aligned}58\cdot\frac{3}{2}-8\cdot\frac{3}{2}&=(58-8)\frac{3}{2}\\&=(50)\frac{3}{2}=\frac{50}{1}\cdot\frac{3}{2}\\&=\tfrac{150}{2}=75\end{aligned}$$

26. $$\begin{aligned}\frac{8}{5}\cdot 17+\frac{8}{5}\cdot 13&=\frac{8}{5}(17+13)\\&=\tfrac{8}{5}(30)\\&=48\end{aligned}$$

27. $$\begin{aligned}4.31(69)+4.31(31)&=4.31(69+31)\\&=4.31(100)\\&=431\end{aligned}$$

28. $$\begin{aligned}\frac{4}{5}(17)+\frac{4}{5}(23)&=\frac{4}{5}(17+23)\\&=\tfrac{4}{5}(40)\\&=32\end{aligned}$$

29. $$\begin{aligned}&-12y+4y+3+2y\\&=-12y+4y+2y+3\\&=(-12+4+2)y+3\\&=-6y+3\end{aligned}$$

30. $-5r-9r+8r-5$
$=(-5-9+8)r-5$
$=-6r-5$

31. $-6p+11p-4p+6+5$
$=(-6+11-4)p+(6+5)$
$=p+11$

32. $-8x-5x+3x-12+9$
$=(-8-5+3)x-12+9$
$=-10x-3$

33. $3(k+2)-5k+6+3$
$=3k+6-5k+6+3$
$=3k-5k+6+6+3$
$=(3-5)k+6+6+3$
$=-2k+15$

34. $5(r-3)+6r-2r+4$
$=5r-15+6r-2r+4$
$=5r+6r-2r-15+4$
$=(5+6-2)r-11$
$=9r-11$

35. $-2(m+1)+3(m-4)$
$=-2(m)+(-2)(1)+3(m)+(3)(-4)$
$=-2m-2+3m-12$
$=-2m+3m-2-12$
$=(-2+3)m+[-2+(-12)]$
$=m-14$

36. $6(a-5)-4(a+6)$
$=6a+6(-5)+(-4)(a)+(-4)(6)$
$=6a-30-4a-24$
$=6a-4a-30-24$
$=2a-54$

37. $.25(8+4p)-.5(6+2p)$
$=.25(8)+.25(4p)+(-.5)(6)+(-.5)(2p)$
$=2+p-3-p$
$=p-p+2-3$
$=(1-1)p+2-3$
$=0p-1$
$=-1$

38. $.4(10-5x)-.8(5+10x)$
$=.4(10)+.4(-5x)+(-.8)(5)+(-.8)(10x)$
$=4-2x-4-8x$
$=-10x$

39. $-(2p+5)+3(2p+4)-2p$
$=-2p-5+6p+12-2p$
$=(-2+6-2)p+(-5+12)$
$=2p+7$

40. $-(7m-12)-2(4m+7)-8m$
$=-1(7m-12)-2(4m+7)-8m$
$=-7m+12-8m-14-8m$
$=-23m-2$

41. $2+3(2z-5)-3(4z+6)-8$
$=2+6z-15-12z-18-8$
$=6z-12z+2-15-18-8$
$=(6-12)z-39$
$=-6z-39$

42. $-4+4(4k-3)-6(2k+8)+7$
$=-4+16k-12-12k-48+7$
$=16k-12k-4-12-48+7$
$=4k-57$

43. $5x+8x=(5+8)x$ *Distributive property*
$=13x$

44. $9y-6y=(9-6)y$ *Distributive property*
$=3y$

45. $5(9r)=(5\cdot 9)r$ *Associative property*
$=45r$

46. $-4+(12+8)$
$=(-4+12)+8$ *Associative property*
$=8+8=16$

47. $5x+9y=9y+5x$ *Commutative property*

48. $-5(7)=7(-5)$ *Commutative property*
$=-35$

49. $1\cdot 7=7$ *Identity property*

50. $-12x+0=-12x$ *Identity property*

51. $8(-4+x)$
$=8(-4)+8x$ *Distributive property*
$=-32+8x$

52. $3(x-y+z)=3x-3y+3z$ *Distributive property*

53. Answers will vary. Commutative: one example is washing your face and brushing your teeth. The activities can be carried out in either order.

54. Answers will vary. One example is waking up and going to sleep.
Another example is tying and untying your shoes. These activities are inverses.

55. The terms have been grouped using the *associative property* of addition.

56. The terms have been regrouped using the *associative property* of addition.

57. The order of the terms inside the parentheses has been changed using the *commutative property* of addition.

58. The terms have been regrouped using the *associative property* of addition.

59. The common factor, x, has been distributed over the expression in parentheses using the *distributive property*.

60. The numbers in parentheses have been added to simplify the expression.

Chapter 1 Review Exercises

1. $\{-4, -1, 2, \frac{9}{4}, 4\}$

Place dots for $-4, -1, 2, \frac{9}{4} = 2.25$, and 4 on a number line.

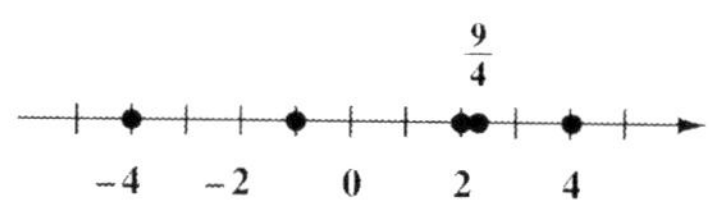

2. $\{-5, -\frac{11}{4}, -.5, 0, 3, \frac{13}{3}\}$

Place dots for $-5, -\frac{11}{4} = -2.75, -.5, 0, 3$, and $\frac{13}{3} = 4.\overline{3}$ on a number line.

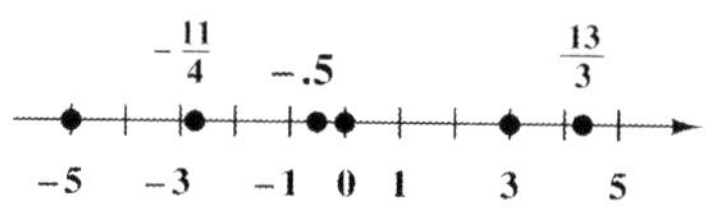

3. $|-16| = -(-16) = 16$

4. $|23| = 23$

5. $-|-4| = -[-(-4)] = -(4) = -4$

6. $|-8| - |-3| = -(-8) - [-(-3)]$
$= 8 - [3]$
$= 5$

In Exercises 7–10,

$$S = \left\{-9, -\frac{4}{3}, -\sqrt{4}, -.25, 0, .\overline{35}, \frac{5}{3}, \sqrt{7}, \sqrt{-9}, \frac{12}{3}\right\}.$$

7. The elements 0 and $\frac{12}{3}$ (or 4) are whole numbers.

8. The elements $-9, -\sqrt{4}$ (or -2), 0, and $\frac{12}{3}$ (or 4) are integers.

9. The elements $-9, -\frac{4}{3}, -\sqrt{4}$ (or -2), $-.25, 0, .\overline{35}, \frac{5}{3}$, and $\frac{12}{3}$ (or 4) are rational numbers. (Remember that terminating and repeating decimals are rational numbers.)

10. All the elements in the set are real numbers except $\sqrt{-9}$.

11. $\{x \mid x \text{ is a natural number between 3 and 9}\}$
The natural numbers between 3 and 9 are $4, 5, 6, 7$, and 8. Therefore, the set is $\{4, 5, 6, 7, 8\}$.

12. $\{y \mid y \text{ is a whole number less than } 4\}$
The whole numbers less than 4 are $0, 1, 2$, and 3. Therefore, the set is $\{0, 1, 2, 3\}$.

13. $4 \cdot 2 \le |12 - 4|$
$8 \le |8|$
$8 \le 8$ *True*

14. $2 + |-2| > 4$
$2 + 2 > 4$
$4 > 4$ *False*

15. $4(3 + 7) > -|40|$
$4(10) > -40$
$40 > -40$ *True*

16. The longest bar represents the greatest change. Hyundai had the greatest change of 50%.

17. The shortest bar represents the smallest change. General Motors had the smallest change of −5%.

18. Honda: $|11| = 11$

Toyota: $|-11| = 11$

These are equal, not greater than. The answer is *false*.

19. Hyundai > 4(Honda) ?
$50 > 4(11)$?
$50 > 44$ *True*

20. $$-\frac{5}{8} - \left(-\frac{7}{3}\right) = -\frac{5}{8} + \frac{7}{3}$$
$$= -\frac{15}{24} + \frac{56}{24} = \frac{41}{24}$$

21. $$-\frac{4}{5} - \left(-\frac{3}{10}\right) = -\frac{4}{5} + \frac{3}{10}$$
$$= -\frac{8}{10} + \frac{3}{10}$$
$$= -\frac{5}{10} = -\frac{1}{2}$$

22. $-5 + (-11) + 20 - 7$
$= -16 + 20 - 7$
$= 4 - 7$
$= -3$

23. $-9.42 + 1.83 - 7.6 - 1.9$
$= -7.59 - 7.6 - 1.9$
$= -15.19 - 1.9$
$= -17.09$

24. $-15 + (-13) + (-11)$
$= -28 + (-11)$
$= -39$

25. $-1 - 3 - (-10) + (-7)$
$= -4 + 10 + (-7)$
$= 6 + (-7)$
$= -1$

26. $\frac{3}{4}-\left(\frac{1}{2}-\frac{9}{10}\right)=\frac{3}{4}-\left(\frac{5}{10}-\frac{9}{10}\right)$
$=\frac{3}{4}-\left(-\frac{4}{10}\right)$
$=\frac{3}{4}+\frac{4}{10}$
$=\frac{15}{20}+\frac{8}{20}=\frac{23}{20}$

27. $-\frac{2}{3}-\left(\frac{1}{6}-\frac{5}{9}\right)=-\frac{2}{3}-\left(\frac{3}{18}-\frac{10}{18}\right)$
$=-\frac{2}{3}-\left(-\frac{7}{18}\right)$
$=-\frac{2}{3}+\frac{7}{18}$
$=-\frac{12}{18}+\frac{7}{18}$
$=-\frac{5}{18}$

28. $-|-12|-|-9|+(-4)-|10|$
$=-(12)-9+(-4)-10$
$=-21+(-4)-10$
$=-25-10$
$=-35$

29. $11{,}049-(-282)=11{,}049+282$
$=11{,}331$

The difference is 11,331 feet.

30. $2(-5)(-3)(-3)=(-10)(-3)(-3)$
$=(30)(-3)$
$=-90$

31. $-\frac{3}{7}\left(-\frac{14}{9}\right)=\frac{3}{7}\cdot\frac{2\cdot 7}{3\cdot 3}=\frac{2}{3}$

32. $-4.6(2.48)=-11.408$

33. $\frac{75}{-5}=15\cdot 5\left(\frac{1}{-5}\right)=-15$

34. $\frac{-2.3754}{-.74}=3.21$

35. $\frac{5}{7-7}=\frac{5}{0}$, which is *undefined.*
$\frac{7-7}{5}=\frac{0}{5}=0$

36. $10^4=10\cdot 10\cdot 10\cdot 10=10{,}000$

37. $\left(\frac{3}{7}\right)^3=\frac{3}{7}\cdot\frac{3}{7}\cdot\frac{3}{7}=\frac{27}{343}$

38. $(-5)^3=(-5)(-5)(-5)=-125$

39. $-5^3=-(5\cdot 5\cdot 5)=-125$

40. $(1.7)^2=(1.7)(1.7)=2.89$

41. $\sqrt{400}=20$, because 20 is positive and $20^2=400$.

42. $-\sqrt{196}=-\left(\sqrt{196}\right)=-(14)=-14$

43. $\sqrt{\frac{64}{121}}=\frac{8}{11}$ since $\frac{8}{11}$ is positive and $\left(\frac{8}{11}\right)^2=\frac{64}{121}$.

44. $-\sqrt{.81}=-\sqrt{(.9)^2}=-(.9)=-.9$

45. Since there is no real number whose square is -64, $\sqrt{-64}$ is *not a real number.*

46. $-14\left(\frac{3}{7}\right)+6\div 3=-2(3)+(6\div 3)$
$=-6+2=-4$

47. $-\frac{2}{3}\left[5(-2)+8-4^3\right]$
$=-\frac{2}{3}[5(-2)+8-64]$ *Evaluate the power.*
$=-\frac{2}{3}[-10+8-64]$
$=-\frac{2}{3}[-66]=\frac{2}{3}(22\cdot 3)=44$

48. $\frac{-5(3^2)+9\left(\sqrt{4}\right)-5}{6-5(-2)}$
$=\frac{-5(9)+9(2)-5}{6+10}$
$=\frac{-45+18-5}{16}$
$=\frac{-32}{16}=-2$

In Exercises 49–51, let $k=-4$, $m=2$, and $n=16$.

49. $4k-7m=4(-4)-7(2)$
$=-16-14$
$=-30$

50. $-3\sqrt{n}+m+5k$
$=-3\left(\sqrt{16}\right)+2+5(-4)$
$=-3(4)+2-20$
$=-12+2-20$
$=-30$

51. $\frac{4m^3-3n}{7k^2-10}=\frac{4(2)^3-3(16)}{7(-4)^2-10}$
$=\frac{4(8)-3(16)}{7(16)-10}$
$=\frac{32-48}{112-10}$
$=\frac{-16}{102}=-\frac{8}{51}$

52. **(a)** 6 ft 1 in. $= (6 \cdot 12 + 1)$ in. $= 73$ in.

704 × (weight in pounds) ÷ (height in inches)^2

$$\begin{aligned} &= 704 \times 190 \div 73^2 \\ &= 704 \times 190 \div 5329 \\ &= 133{,}760 \div 5329 \\ &\approx 25 \end{aligned}$$

Carlos Beltran's BMI is 25.

(b) Answers will vary.

53. $2q + 19q$
$= (2 + 19)q$ *Distributive property*
$= 21q$

54. $13z - 17z$
$= (13 - 17)z$ *Distributive property*
$= -4z$

55. $-m + 6m$
$= -1m + 6m$ *Identity property*
$= (-1 + 6)m$ *Distributive property*
$= 5m$

56. $5p - p$
$= 5p + (-1)p$ *Identity property*
$= [5 + (-1)]p$ *Distributive property*
$= 4p$

57. $-2(k + 3)$
$= -2(k) + (-2)(3)$ *Distributive property*
$= -2k - 6$

58. $6(r + 3)$
$= 6(r) + 6(3)$ *Distributive property*
$= 6r + 18$

59. $9(2m + 3n)$
$= 9(2m) + 9(3n)$ *Distributive property*
$= 18m + 27n$

60.
$$\begin{aligned} &-(-p + 6q) - (2p - 3q) \\ &= -1(-p + 6q) + (-1)(2p - 3q) \\ &= -1(-p) + (-1)(6q) \\ &\quad + (-1)(2p) + (-1)(-3q) \\ &= p - 6q - 2p + 3q \\ &= p - 2p - 6q + 3q \\ &= -p - 3q \end{aligned}$$

61.
$$\begin{aligned} &-3y + 6 - 5 + 4y \\ &= -3y + 4y + 6 - 5 \\ &= y + 1 \end{aligned}$$

62.
$$\begin{aligned} &2a + 3 - a - 1 - a - 2 \\ &= 2a - a - a + 3 - 1 - 2 \\ &= 0 \end{aligned}$$

63.
$$\begin{aligned} &-3(4m - 2) + 2(3m - 1) - 4(3m + 1) \\ &= -12m + 6 + 6m - 2 - 12m - 4 \\ &= -12m + 6m - 12m + 6 - 2 - 4 \\ &= -18m \end{aligned}$$

64. $2x + 3x = (2 + 3)x$ *Distributive property*
$= 5x$

65. $-4 \cdot 1 = -4$ *Identity property*

66. $2(4x) = (2 \cdot 4)x$ *Associative property*
$= 8x$

67. $-3 + 13 = 13 + (-3)$ *Commutative property*
$= 10$

68. $-3 + 3 = 0$ *Inverse property*

69. $5(x + z) = 5x + 5z$ *Distributive property*

70. $0 + 7 = 7$ *Identity property*

71. $8 \cdot \frac{1}{8} = 1$ *Inverse property*

72. $3a + 5a + 6a$
$= (3 + 5 + 6)a$ *Distributive property*
$= 14a$

73. $\frac{9}{28} \cdot 0 = 0$ *Multiplication property of 0*

74. **[1.2]** For 2000 (in millions of dollars):

$|178{,}941 - 230{,}838| = |-51{,}897| = 51{,}897$

The balance of trade (exports minus imports) is *negative* since the amount of money spent on imports is greater than the amount of money received for exports.

75. **[1.2]** For 2001 (in millions of dollars):

$|163{,}424 - 216{,}268| = |-52{,}844| = 52{,}844$

The balance of trade (exports minus imports) is *negative* since the amount of money spent on imports is greater than the amount of money received for exports.

76. **[1.2]** For 2002 (in millions of dollars):

$|160{,}923 - 209{,}088| = |-48{,}165| = 48{,}165$

The balance of trade (exports minus imports) is *negative* since the amount of money spent on imports is greater than the amount of money received for exports.

77. **[1.3]** $\left(-\frac{4}{5}\right)^4 = \left(-\frac{4}{5}\right)\left(-\frac{4}{5}\right)\left(-\frac{4}{5}\right)\left(-\frac{4}{5}\right)$
$= \frac{256}{625}$

78. **[1.2]** $-\frac{5}{8}(-40) = -\frac{5}{8} \cdot \frac{-40}{1}$
$= 5 \cdot 5 = 25$

79. **[1.3]** $-25\left(-\frac{4}{5}\right) + 3^3 - 32 \div \sqrt{4}$
$= -25\left(-\frac{4}{5}\right) + 27 - 32 \div 2$
$= 20 + 27 - 16$
$= 31$

80. **[1.2]** $-8 + |-14| + |-3| = -8 + 14 + 3$
$= 9$

81. **[1.3]** $\frac{6 \cdot \sqrt{4} - 3 \cdot \sqrt{16}}{-2 \cdot 5 + 7(-3) - 10}$
$= \frac{6 \cdot 2 - 3 \cdot 4}{-2 \cdot 5 + 7(-3) - 10}$
$= \frac{12 - 12}{-10 - 21 - 10}$
$= \frac{0}{-41} = 0$

82. **[1.3]** $-\sqrt{25} = -(5) = -5$

83. **[1.2]** $-\frac{10}{21} \div -\frac{5}{14} = -\frac{10}{21} \cdot -\frac{14}{5}$
$= \frac{2 \cdot 5}{3 \cdot 7} \cdot \frac{2 \cdot 7}{5}$
$= \frac{2 \cdot 2}{3} = \frac{4}{3}$

84. **[1.2]** $.8 - 4.9 - 3.2 + 1.14$
$= -4.1 - 3.2 + 1.14$
$= -7.3 + 1.14$
$= -6.16$

85. **[1.3]** $-3^2 = -(3 \cdot 3) = -9$

86. **[1.2]** $\frac{-38}{-19} = 38\left(\frac{1}{19}\right) = 2 \cdot 19\left(\frac{1}{19}\right) = 2$

87. **[1.4]** $-2(k-1) + 3k - k$
$= -2k + 2 + 3k - k$
$= -2k + 3k - k + 2$
$= (-2 + 3 - 1)k + 2$
$= 0k + 2 = 2$

88. **[1.3]** Since there is no real number whose square is -100, $-\sqrt{-100}$ is *not a real number.*

89. **[1.3]** $-m(3k^2 + 5m) = -2[3(-4)^2 + 5(2)]$
Let $k = -4$, $m = 2$.
$= -2[3(16) + 5(2)]$
$= -2[48 + 10]$
$= -2[58]$
$= -116$

90. **[1.3]** In order to evaluate $(3+2)^2$, you should work within the parentheses first.

Chapter 1 Test

1. $\{-3, .75, \frac{5}{3}, 5, 6.3\}$
Place dots at $-3, .75, \frac{5}{3} = 1.\overline{6}, 5$, and 6.3.

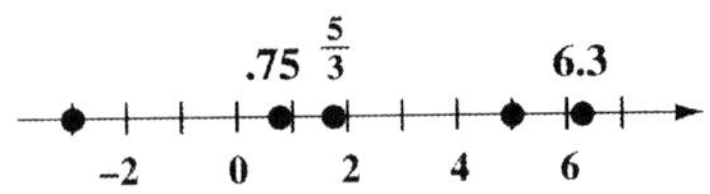

In Exercises 2–5,

$$A = \left\{-\sqrt{6}, -1, -.5, 0, 3, \sqrt{25}, 7.5, \tfrac{24}{2}, \sqrt{-4}\right\}.$$

2. The elements $0, 3, \sqrt{25}$ (or 5), and $\frac{24}{2}$ (or 12) are whole numbers.

3. The elements $-1, 0, 3, \sqrt{25}$ (or 5), and $\frac{24}{2}$ (or 12) are integers.

4. The elements $-1, -.5, 0, 3, \sqrt{25}$ (or 5), 7.5, and $\frac{24}{2}$ (or 12) are rational numbers.

5. All the elements in the set are real numbers except $\sqrt{-4}$.

6. $-6 + 14 + (-11) - (-3)$
$= 8 + (-11) + 3$
$= -3 + 3 = 0$

7. $10 - 4 \cdot 3 + 6(-4)$
$= 10 - 12 + (-24)$
$= -2 + (-24) = -26$

8. $7 - 4^2 + 2(6) + (-4)^2$
$= 7 - 16 + 12 + 16$
$= 19$

9. $\frac{10 - 24 + (-6)}{\sqrt{16}(-5)}$
$= \frac{-14 + (-6)}{4(-5)}$
$= \frac{-20}{-20} = 1$

10. $\frac{-2[3 - (-1 - 2) + 2]}{\sqrt{9}(-3) - (-2)}$
$= \frac{-2[3 - (-3) + 2]}{3(-3) - (-2)}$
$= \frac{-2[8]}{-9 - (-2)}$
$= \frac{-16}{-7} = \frac{16}{7}$

11. $\dfrac{8\cdot 4-3^2\cdot 5-2(-1)}{-3\cdot 2^3+1}$

$=\dfrac{8\cdot 4-9\cdot 5-2(-1)}{-3\cdot 8+1}$

$=\dfrac{32-45+2}{-24+1}$

$=\dfrac{-11}{-23}=\dfrac{11}{23}$

12. $17{,}400-(-32{,}995)=17{,}400+32{,}995$
$=50{,}395$

The difference between the height of Mt. Foraker and the depth of the Philippine Trench is 50,395 feet.

13. $14{,}110-(-23{,}376)=14{,}110+23{,}376$
$=37{,}486$

The difference between the height of Pikes Peak and the depth of the Java Trench is 37,486 feet.

14. $-24{,}721-(-23{,}376)=-24{,}721+23{,}376$
$=-1345$

The Cayman Trench is 1345 feet deeper than the Java Trench.

15. $\sqrt{196}=14$, because 14 is positive and $14^2=196$.

16. $-\sqrt{225}=-\left(\sqrt{225}\right)=-(15)=-15$

17. Since there is no real number whose square is -16, $\sqrt{-16}$ is *not a real number*.

18. **(a)** If a is positive, then $\sqrt{a}$ will represent a positive number.

(b) If a is negative, then $\sqrt{a}$ will not represent a real number.

(c) If a is 0, then $\sqrt{a}$ will be 0.

In Exercises 19–20, let $k=-3$, $m=-3$, and $r=25$.

19. $\sqrt{r}+2k-m=\sqrt{25}+2(-3)-(-3)$
$=5+2(-3)-(-3)$
$=5-6-(-3)$
$=5-6+3$
$=-1+3$
$=2$

20. $\dfrac{8k+2m^2}{r-2}=\dfrac{8(-3)+2(-3)^2}{25-2}$

$=\dfrac{8(-3)+2(9)}{23}$

$=\dfrac{-24+18}{23}$

$=\dfrac{-6}{23}$ or $-\dfrac{6}{23}$

21. $-3(2k-4)+4(3k-5)-2+4k$
$=-3(2k)+(-3)(-4)+4(3k)$
$+4(-5)-2+4k$
$=-6k+12+12k-20-2+4k$
$=-6k+12k+4k+12-20-2$
$=10k-10$

22. When simplifying

$$(3r+8)-(-4r+6),$$

the subtraction sign in front of $(-4r+6)$ changes the sign of the terms $-4r$ and 6.

$(3r+8)-(-4r+6)$
$=3r+8-(-4r)-6$
$=3r+8+4r-6$
$=3r+4r+8-6$
$=7r+2$

23. $6+(-6)=0$
The answer is **B**, *Inverse property*.
The sum of 6 and its inverse, -6, equals zero.

24. $4+5=5+4$
The answer is **E**, *Commutative property*.
The order of the terms is reversed.

25. $-2+(3+6)=(-2+3)+6$
The answer is **D**, *Associative property*.
The order of the terms is the same, but the grouping has changed.

26. $5x+15x=(5+15)x$
The answer is **A**, *Distributive property*.
This is the second form of the distributive property.

27. $13\cdot 0=0$
The answer is **F**, *Multiplication property of* 0.
Multiplication by 0 always equals 0.

28. $-9+0=-9$

The answer is **C**, *Identity property.*
The addition of 0 to any number does not change the number.

29. $4 \cdot 1 = 4$

The answer is **C**, *Identity property.*
Multiplication of any number by 1 does not change the number.

30. $(a+b)+c=(b+a)+c$

The answer is **E**, *Commutative property.*
The order of the terms a and b is reversed.

CHAPTER 2 LINEAR EQUATIONS AND APPLICATIONS

2.1 Linear Equations in One Variable

2.1 Margin Exercises

1. To decide if a given number is a solution, substitute that number for the variable in the equation to see if the resulting statement is true or false.

(a) $3k = 15;\ 5$

The number 5 is a solution since $3 \cdot 5 = 15$ and $15 = 15$ is true.

(b) $r + 5 = 4;\ 1$

The number 1 is not a solution since $1 + 5 = 6$ and $6 = 4$ is false.

(c) $-8m = 12;\ \frac{3}{2}$

The number $\frac{3}{2}$ is not a solution since $-8\left(\frac{3}{2}\right) = -12$ and $-12 = 12$ is false.

2. **(a)**

$3p + 2p + 1 = -24$	*Original equation*
$5p + 1 = -24$	*Combine terms.*
$5p + 1 - 1 = -24 - 1$	*Subtract 1.*
$5p = -25$	*Combine terms.*
$\frac{5p}{5} = \frac{-25}{5}$	*Divide by 5.*
$p = -5$	*Proposed solution*

Check by substituting -5 for p in the *original* equation.

$3p + 2p + 1 = -24$		*Original equation*
$3(-5) + 2(-5) + 1 = -24$	?	*Let $p = -5$.*
$-15 - 10 + 1 = -24$	?	
$-24 = -24$		*True*

The true statement indicates that $\{-5\}$ is the solution set.

(b)

$3p = 2p + 4p + 5$	*Original equation*
$3p = 6p + 5$	*Combine terms.*
$3p - 6p = 6p + 5 - 6p$	*Subtract 6p.*
$-3p = 5$	*Combine terms.*
$\frac{-3p}{-3} = \frac{5}{-3}$	*Divide by −3.*
$p = -\frac{5}{3}$	*Proposed solution*

Check by substituting $-\frac{5}{3}$ for p in the *original* equation.

$3p = 2p + 4p + 5$		*Original equation*
$3\left(-\frac{5}{3}\right) = 2\left(-\frac{5}{3}\right) + 4\left(-\frac{5}{3}\right) + 5$	?	*Let $p = -\frac{5}{3}$.*
$-5 = -\frac{10}{3} - \frac{20}{3} + 5$	?	
$-5 = -\frac{30}{3} + 5$	?	
$-5 = -10 + 5$	?	
$-5 = -5$		*True*

Solution set: $\left\{-\frac{5}{3}\right\}$

(c)

$4x + 8x = 17x - 9 - 1$	*Original equation*
$12x = 17x - 10$	*Combine terms.*
$12x - 17x = 17x - 10 - 17x$	*Subtract 17x.*
$-5x = -10$	*Combine terms.*
$\frac{-5x}{-5} = \frac{-10}{-5}$	*Divide by −5.*
$x = 2$	*Proposed solution*

Check by substituting 2 for x in the *original* equation.

$4x + 8x = 17x - 9 - 1$		*Original equation*
$4(2) + 8(2) = 17(2) - 9 - 1$	?	*Let $x = 2$.*
$8 + 16 = 34 - 10$	?	
$24 = 24$		*True*

Solution set: $\{2\}$

(d)

$$-7 + 3t - 9t = 12t - 5$$
$$-7 - 6t = 12t - 5$$

Combine terms.

$$-7 - 6t + 6t + 5 = 12t - 5 + 6t + 5$$

Add 6t; add 5.

$$-2 = 18t$$
$$\frac{-2}{18} = \frac{18t}{18}$$

Divide by 18.

$$-\frac{1}{9} = t$$

Proposed solution

We will use the following notation to indicate the value of each side of the original equation after we have substituted the proposed solution and simplified.

Check $t = -\frac{1}{9}$: $-\frac{19}{3} = -\frac{19}{3}$

Solution set: $\left\{-\frac{1}{9}\right\}$

3. **(a)**

$$\begin{aligned} 5p+4(3-2p) &= 2+p-10 \\ 5p+12-8p &= 2+p-10 && \textit{Distributive property} \\ 12-3p &= p-8 && \textit{Combine terms.} \\ 12-3p+3p+8 &= p-8+3p+8 && \textit{Add 3p; add 8.} \\ 20 &= 4p && \textit{Combine terms.} \\ \frac{20}{4} &= \frac{4p}{4} && \textit{Divide by 4.} \\ 5 &= p && \textit{Proposed solution} \end{aligned}$$

Check $p=5$: $-3=-3$

Solution set: $\{5\}$

(b) $3(z-2)+5z=2$

$$\begin{aligned} 3z-6+5z &= 2 && \textit{Distributive property} \\ 8z-6 &= 2 && \textit{Combine terms.} \\ 8z-6+6 &= 2+6 && \textit{Add 6.} \\ 8z &= 8 && \textit{Combine terms.} \\ \frac{8z}{8} &= \frac{8}{8} && \textit{Divide by 8.} \\ z &= 1 && \textit{Proposed solution} \end{aligned}$$

Check $z=1$: $2=2$

Solution set: $\{1\}$

(c) $-2+3(x+4)=8x$

$$\begin{aligned} -2+3x+12 &= 8x && \textit{Distributive property} \\ 3x+10 &= 8x && \textit{Combine terms.} \\ 3x+10-3x &= 8x-3x && \textit{Subtract 3x.} \\ 10 &= 5x && \textit{Combine terms.} \\ \frac{10}{5} &= \frac{5x}{5} && \textit{Divide by 5.} \\ 2 &= x && \textit{Proposed solution} \end{aligned}$$

Check $x=2$: $16=16$

Solution set: $\{2\}$

(d)

$$\begin{aligned} 6-(4+m) &= 8m-2(3m+5) \\ 6-4-m &= 8m-6m-10 && \textit{Distributive property} \\ 2-m &= 2m-10 && \textit{Combine terms.} \\ 2-m+m+10 &= 2m-10+m+10 && \textit{Add m; add 10.} \\ 12 &= 3m && \textit{Combine terms.} \\ \frac{12}{3} &= \frac{3m}{3} && \textit{Divide by 3.} \\ 4 &= m && \textit{Proposed solution} \end{aligned}$$

Check $m=4$: $-2=-2$

Solution set: $\{4\}$

4. **(a)** $\frac{2p}{7}-\frac{p}{2}=-3$

Multiply each side by the LCD, 14.

$$\begin{aligned} 14\left(\frac{2p}{7}-\frac{p}{2}\right) &= 14(-3) \\ 14\left(\frac{2p}{7}\right)-14\left(\frac{p}{2}\right) &= 14(-3) && \textit{Distributive property} \\ 4p-7p &= -42 && \textit{Multiply.} \\ -3p &= -42 && \textit{Combine terms.} \\ \frac{-3p}{-3} &= \frac{-42}{-3} && \textit{Divide by } -3. \\ p &= 14 && \textit{Proposed solution} \end{aligned}$$

Check $p=14$: $-3=-3$

Solution set: $\{14\}$

(b) $\frac{k+1}{2}+\frac{k+3}{4}=\frac{1}{2}$

Multiply each side by the LCD, 4, and use the distributive property.

$$\begin{aligned} 4\left(\frac{k+1}{2}\right)+4\left(\frac{k+3}{4}\right) &= 4\left(\frac{1}{2}\right) \\ 2(k+1)+1(k+3) &= 2 \\ 2k+2+k+3 &= 2 \\ 3k+5 &= 2 \\ 3k &= -3 && \textit{Subtract 5.} \\ k &= -1 && \textit{Divide by 3.} \end{aligned}$$

Check $k=-1$: $\frac{1}{2}=\frac{1}{2}$

Solution set: $\{-1\}$

5. **(a)** $.04x + .06(20 - x) = .05(50)$

Multiply each side by 100, and use the distributive property.

$$\begin{aligned} 4x + 6(20 - x) &= 5(50) \\ 4x + 120 - 6x &= 250 \\ -2x + 120 &= 250 \\ -2x &= 130 && \textit{Subtract 120.} \\ x &= -65 && \textit{Divide by } -2. \end{aligned}$$

Check $x = -65$: $2.5 = 2.5$

Solution set: $\{-65\}$

(b) $.10(x - 6) + .05x = .06(50)$

Multiply each side by 100, and use the distributive property.

$$\begin{aligned} 10(x - 6) + 5x &= 6(50) \\ 10x - 60 + 5x &= 300 \\ 15x - 60 &= 300 \\ 15x &= 360 && \textit{Add 60.} \\ x &= 24 && \textit{Divide by 15.} \end{aligned}$$

Check $x = 24$: $3 = 3$

Solution set: $\{24\}$

6. **(a)**
$$\begin{aligned} 5(x + 2) - 2(x + 1) &= 3x + 1 \\ 5x + 10 - 2x - 2 &= 3x + 1 \\ 3x + 8 &= 3x + 1 \\ 3x + 8 - 3x &= 3x + 1 - 3x \\ & \qquad \textit{Subtract } 3x. \\ 8 &= 1 \qquad \textit{False} \end{aligned}$$

Since the result, $8 = 1$, is *false*, the equation has no solution and is called a *contradiction*.

Solution set: $\emptyset$

(b) $\dfrac{x + 1}{3} + \dfrac{2x}{3} = x + \dfrac{1}{3}$

Multiply each side by the LCD, 3, and use the distributive property.

$$\begin{aligned} 3\left(\frac{x+1}{3}\right) + 3\left(\frac{2x}{3}\right) &= 3\left(x + \frac{1}{3}\right) \\ x + 1 + 2x &= 3x + 1 \\ 3x + 1 &= 3x + 1 \end{aligned}$$

This is an *identity*. Any real number will make the equation true.

Solution set: {all real numbers}

(c)
$$\begin{aligned} 5(3x + 1) &= x + 5 \\ 15x + 5 &= x + 5 \\ 14x + 5 &= 5 && \textit{Subtract } x. \\ 14x &= 0 && \textit{Subtract 5.} \\ x &= 0 && \textit{Divide by 14.} \end{aligned}$$

This is a *conditional equation*.

Check $x = 0$: $5 = 5$

Solution set: $\{0\}$

2.1 Section Exercises

1. **A.** $3x + x - 1 = 0$ can be written as $4x - 1 = 0$, so it is linear.

   C. $6x + 2 = 9$ is in linear form.

2. **B.** $8 = x^2$ is not a linear equation because the variable is squared.

   D. $\frac{1}{2}x - \frac{1}{x} = 0$ is not a linear equation because there is a variable in the denominator of the second term.

3.
$$\begin{aligned} 3(x + 4) &= 5x && \textit{Original equation} \\ 3(6 + 4) &= 5 \cdot 6 \ ? && \textit{Let } x = 6. \\ 3(10) &= 30 \ ? && \textit{Add.} \\ 30 &= 30 && \textit{True} \end{aligned}$$
Since a true statement is obtained, 6 is a solution.

4.
$$\begin{aligned} 5(x + 4) - 3(x + 6) &= 9(x + 1) && \textit{Original equation} \\ 5(-2 + 4) - 3(-2 + 6) &= 9(-2 + 1) \ ? && \textit{Let } x = -2. \\ 5(2) - 3(4) &= 9(-1) \ ? && \textit{Add.} \\ 10 - 12 &= -9 \ ? && \textit{Multiply.} \\ -2 &= -9 && \textit{False} \end{aligned}$$

   Since a false statement is obtained, -2 is not a solution.

5. Suppose your last name is Lincoln. Then $x = 7$ and both sides are evaluated as -48. The equation is an identity, so any number is a solution.

6. The final line of the check, $-11 = -11$, does not give the solution, only a confirmation that the solution found, in this case -3, is correct.

7. **(a)** $3x = 6$ is an *equation* because it contains an equals sign.

   (b) $3x + 6$ is an *expression* because it does not contain an equals sign.

   (c) $5x + 6(x - 3) = 12x + 6$ is an *equation* because it contains an equals sign.

   (d) $5x + 6(x - 3) - (12x + 6)$ is an *expression* because it does not contain an equals sign.

8. There is no way to add both 9 and 8 to the same expression $(6x)$ and get equal quantities.

9. The solution contains a sign error when the distributive property was applied. The left side of the second line of the solution should be $8x - 4x + 6$. This gives us $4x + 6 = 3x + 7$ and then $x = 1$. Thus, the correct solution is 1.

10. $-(2m - 4) = -1(2m - 4)$, so the $-$ sign represents -1.

$$\begin{aligned} -5m - (2m - 4) + 5 &= -5m - 2m + 4 + 5 \\ &= -7m + 9 \end{aligned}$$

In the following exercises, we do not show the checks of the solutions. To be sure that your solutions are correct, check them by substituting into the original equations.

11.
$$\begin{aligned} 7x + 8 &= 1 \\ 7x + 8 - 8 &= 1 - 8 && \textit{Subtract 8.} \\ 7x &= -7 \\ \frac{7x}{7} &= \frac{-7}{7} && \textit{Divide by 7.} \\ x &= -1 \end{aligned}$$

Solution set: $\{-1\}$

12.
$$\begin{aligned} 5x - 4 &= 21 \\ 5x - 4 + 4 &= 21 + 4 && \textit{Add 4.} \\ 5x &= 25 \\ \frac{5x}{5} &= \frac{25}{5} && \textit{Divide by 5.} \\ x &= 5 \end{aligned}$$

Solution set: $\{5\}$

13.
$$\begin{aligned} 5x + 2 &= 3x - 6 \\ 5x + 2 - 3x &= 3x - 6 - 3x && \textit{Subtract 3x.} \\ 2x + 2 &= -6 \\ 2x + 2 - 2 &= -6 - 2 && \textit{Subtract 2.} \\ 2x &= -8 \\ \frac{2x}{2} &= \frac{-8}{2} && \textit{Divide by 2.} \\ x &= -4 \end{aligned}$$

Solution set: $\{-4\}$

14.
$$\begin{aligned} 9p + 1 &= 7p - 9 \\ 9p + 1 - 7p &= 7p - 9 - 7p && \textit{Subtract 7p.} \\ 2p + 1 &= -9 \\ 2p + 1 - 1 &= -9 - 1 && \textit{Subtract 1.} \\ 2p &= -10 \\ \frac{2p}{2} &= \frac{-10}{2} && \textit{Divide by 2.} \\ p &= -5 \end{aligned}$$

Solution set: $\{-5\}$

15.
$$\begin{aligned} 7x - 5x + 15 &= x + 8 \\ 2x + 15 &= x + 8 && \textit{Combine terms.} \\ 2x &= x - 7 && \textit{Subtract 15.} \\ x &= -7 && \textit{Subtract x.} \end{aligned}$$

Solution set: $\{-7\}$

16.
$$\begin{aligned} 2x + 4 - x &= 4x - 5 \\ x + 4 &= 4x - 5 && \textit{Combine terms.} \\ -3x + 4 &= -5 && \textit{Subtract 4x.} \\ -3x &= -9 && \textit{Subtract 4.} \\ x &= 3 && \textit{Divide by } -3. \end{aligned}$$

Solution set: $\{3\}$

17.
$$\begin{aligned} 12w + 15w - 9 + 5 &= -3w + 5 - 9 \\ 27w - 4 &= -3w - 4 && \textit{Combine terms.} \\ 30w - 4 &= -4 && \textit{Add 3w.} \\ 30w &= 0 && \textit{Add 4.} \\ w &= 0 && \textit{Divide by 30.} \end{aligned}$$

Solution set: $\{0\}$

18.
$$\begin{aligned} -4t + 5t - 8 + 4 &= 6t - 4 \\ t - 4 &= 6t - 4 && \textit{Combine terms.} \\ -5t - 4 &= -4 && \textit{Subtract 6t.} \\ -5t &= 0 && \textit{Add 4.} \\ t &= 0 && \textit{Divide by } -5. \end{aligned}$$

Solution set: $\{0\}$

19.
$$\begin{aligned} 3(2t - 4) &= 20 - 2t \\ 6t - 12 &= 20 - 2t && \textit{Distributive property} \\ 8t - 12 &= 20 && \textit{Add 2t.} \\ 8t &= 32 && \textit{Add 12.} \\ t &= 4 && \textit{Divide by 8.} \end{aligned}$$

Solution set: $\{4\}$

20.
$$\begin{aligned} 2(3 - 2x) &= x - 4 \\ 6 - 4x &= x - 4 && \textit{Distributive property} \\ 6 - 5x &= -4 && \textit{Subtract x.} \\ -5x &= -10 && \textit{Subtract 6.} \\ x &= 2 && \textit{Divide by } -5. \end{aligned}$$

Solution set: $\{2\}$

21.
$$\begin{aligned} -5(x + 1) + 3x + 2 &= 6x + 4 \\ -5x - 5 + 3x + 2 &= 6x + 4 && \textit{Distributive property} \\ -2x - 3 &= 6x + 4 && \textit{Combine terms.} \\ -3 &= 8x + 4 && \textit{Add 2x.} \\ -7 &= 8x && \textit{Subtract 4.} \\ -\tfrac{7}{8} &= x && \textit{Divide by 8.} \end{aligned}$$

Solution set: $\{-\frac{7}{8}\}$

22. $5(x+3)+4x-5=4-2x$

$5x+15+4x-5=4-2x$ *Distributive property*

$9x+10=4-2x$ *Combine terms.*

$11x+10=4$ *Add 2x.*

$11x=-6$ *Subtract 10.*

$x=-\frac{6}{11}$ *Divide by 11.*

Solution set: $\{-\frac{6}{11}\}$

23. $2(x+3)=-4(x+1)$

$2x+6=-4x-4$ *Remove parentheses.*

$6x+6=-4$ *Add 4x.*

$6x=-10$ *Subtract 6.*

$x=\frac{-10}{6}=-\frac{5}{3}$ *Divide by 6.*

Solution set: $\{-\frac{5}{3}\}$

24. $4(t-9)=8(t+3)$

$4t-36=8t+24$ *Remove parentheses.*

$-4t-36=24$ *Subtract 8t.*

$-4t=60$ *Add 36.*

$t=-15$ *Divide by –4.*

Solution set: $\{-15\}$

25. $3(2w+1)-2(w-2)=5$

$6w+3-2w+4=5$ *Remove parentheses.*

$4w+7=5$ *Combine terms.*

$4w=-2$ *Subtract 7.*

$w=\frac{-2}{4}=-\frac{1}{2}$ *Divide by 4.*

Solution set: $\{-\frac{1}{2}\}$

26. $4(x-2)+2(x+3)=6$

$4x-8+2x+6=6$

$6x-2=6$

$6x=8$

$x=\frac{8}{6}=\frac{4}{3}$

Solution set: $\{\frac{4}{3}\}$

27. $2x+3(x-4)=2(x-3)$

$2x+3x-12=2x-6$

$5x-12=2x-6$

$3x=6$

$x=\frac{6}{3}=2$

Solution set: $\{2\}$

28. $6x-3(5x+2)=4(1-x)$

$6x-15x-6=4-4x$

$-9x-6=4-4x$

$-5x=10$

$x=\frac{10}{-5}=-2$

Solution set: $\{-2\}$

29. $6p-4(3-2p)=5(p-4)-10$

$6p-12+8p=5p-20-10$

$14p-12=5p-30$

$9p=-18$

$p=-2$

Solution set: $\{-2\}$

30. $-2k-3(4-2k)=2(k-3)+2$

$-2k-12+6k=2k-6+2$

$4k-12=2k-4$

$2k=8$

$k=4$

Solution set: $\{4\}$

31. $2[w-(2w+4)+3]=2(w+1)$

$2[w-2w-4+3]=2(w+1)$

$2[-w-1]=2(w+1)$

$-w-1=w+1$ *Divide by 2.*

$-1=2w+1$ *Add w.*

$-2=2w$ *Subtract 1.*

$-1=w$ *Divide by 2.*

Solution set: $\{-1\}$

32. $4[2t-(3-t)+5]=-(2+7t)$

$4[2t-3+t+5]=-(2+7t)$

$4[3t+2]=-(2+7t)$

$12t+8=-2-7t$

$19t+8=-2$ *Add 7t.*

$19t=-10$ *Subtract 8.*

$t=-\frac{10}{19}$ *Divide by 19.*

Solution set: $\{-\frac{10}{19}\}$

33. $-[2z-(5z+2)]=2+(2z+7)$

$-[2z-5z-2]=2+2z+7$

$-[-3z-2]=2+2z+7$

$3z+2=2z+9$

$z=7$

Solution set: $\{7\}$

34. $-[6x-(4x+8)]=9+(6x+3)$
$$-[6x-4x-8]=9+6x+3$$
$$-(2x-8)=6x+12$$
$$-2x+8=6x+12$$
$$-8x=4$$
$$x=\frac{4}{-8}=-\frac{1}{2}$$

Solution set: $\{-\frac{1}{2}\}$

35. $-3m+6-5(m-1)=-5m-(2m-4)+5$
$$-3m+6-5m+5=-5m-2m+4+5$$
$$-8m+11=-7m+9$$
$$-m+11=9$$
$$-m=-2$$
$$m=2$$

Solution set: $\{2\}$

36. $4(k+2)-8k-5=-3k+9-2(k+6)$
$$4k+8-8k-5=-3k+9-2k-12$$
$$-4k+3=-5k-3$$
$$k=-6$$

Solution set: $\{-6\}$

37. $-3(x+2)+4(3x-8)=2(4x+7)+2(3x-6)$
$$-3x-6+12x-32=8x+14+6x-12$$
$$9x-38=14x+2$$
$$-38=5x+2$$
$$-40=5x$$
$$-8=x$$

Solution set: $\{-8\}$

38. $-7(2x+1)+5(3x+2)=6(2x-4)-(12x+3)$
$$-14x-7+15x+10=12x-24-12x-3$$
$$x+3=-27$$
$$x=-30$$

Solution set: $\{-30\}$

39. The denominators of the fractions are 3, 4, and 1. The least common denominator is $(3)(4)(1)=12$, since it is the smallest number into which each denominator can divide without a remainder.

40. Yes, the coefficients will be larger, but you will get the correct solution. As long as you multiply each side of the equation by the *same* nonzero number, the resulting equation is equivalent and the solution does not change.

41. **(a)** We need to make the coefficient of the first term on the left an integer. Since $.05=\frac{5}{100}$, we multiply by 10^2 or 100. This will also take care of the second term.

(b) We need to make .006, .007, and .009 integers. These numbers can be written as $\frac{6}{1000}$, $\frac{7}{1000}$, and $\frac{9}{1000}$. Multiplying by 10^3 or 1000 will eliminate the decimal points (the denominators) so that all the coefficients are integers.

42. $.06(10-x)(100)$
$$=.06(100)(10-x)$$
$$=6(10-x)$$
$$=60-6x$$ Choice **B** is correct.

43. $$\frac{m}{2}+\frac{m}{3}=5$$
Multiply each side by the LCD, 6.
$$6\left(\frac{m}{2}+\frac{m}{3}\right)=6(5)$$
$$6\left(\frac{m}{2}\right)+6\left(\frac{m}{3}\right)=30 \quad \textit{Distributive property}$$
$$3m+2m=30$$
$$5m=30 \quad \textit{Add.}$$
$$m=6 \quad \textit{Divide by 5.}$$

Check $m=6$: $5=5$

Solution set: $\{6\}$

44. $$\frac{x}{5}-\frac{x}{4}=1$$
Multiply each side by the LCD, 20.
$$20\left(\frac{x}{5}-\frac{x}{4}\right)=20(1)$$
$$20\left(\frac{x}{5}\right)-20\left(\frac{x}{4}\right)=20 \quad \textit{Distributive property}$$
$$4x-5x=20$$
$$-x=20 \quad \textit{Subtract.}$$
$$x=-20 \quad \textit{Multiply by } -1.$$

Check $x=-20$: $1=1$

Solution set: $\{-20\}$

45. $$\frac{3x}{4}+\frac{5x}{2}=13$$
Multiply each side by the LCD, 4.
$$4\left(\frac{3x}{4}+\frac{5x}{2}\right)=4(13)$$
$$4\left(\frac{3x}{4}\right)+4\left(\frac{5x}{2}\right)=4(13) \quad \textit{Distributive property}$$
$$3x+10x=52$$
$$13x=52 \quad \textit{Combine terms.}$$
$$x=4 \quad \textit{Divide by 13.}$$

Check $x=4$: $13=13$

Solution set: $\{4\}$

46. $$\frac{8x}{3} - \frac{2x}{4} = -13$$
Multiply each side by the LCD, 12.
$$12\left(\frac{8x}{3} - \frac{2x}{4}\right) = 12(-13)$$
$$12\left(\frac{8x}{3}\right) - 12\left(\frac{2x}{4}\right) = 12(-13) \quad \textit{Distributive property}$$
$$32x - 6x = -156$$
$$26x = -156$$
$$x = -6 \quad \textit{Divide by 26.}$$

Check $x = -6$: $-13 = -13$

Solution set: $\{-6\}$

47. $$\frac{1}{5}x - 2 = \frac{2}{3}x - \frac{2}{5}x$$
Multiply each side by the LCD, 15, and use the distributive property.
$$15\left(\tfrac{1}{5}x\right) - 15(2) = 15\left(\tfrac{2}{3}x\right) - 15\left(\tfrac{2}{5}x\right)$$
$$3x - 30 = 10x - 6x$$
$$3x - 30 = 4x$$
$$-30 = x \quad \textit{Subtract 3x.}$$

Check $x = -30$: $-8 = -8$

Solution set: $\{-30\}$

48. $$\frac{3}{4}x - \frac{1}{3}x = \frac{5}{6}x - 5$$
Multiply each side by the LCD, 12, and use the distributive property.
$$12\left(\tfrac{3}{4}x\right) - 12\left(\tfrac{1}{3}x\right) = 12\left(\tfrac{5}{6}x\right) - 12(5)$$
$$9x - 4x = 10x - 60$$
$$5x = 10x - 60$$
$$-5x = -60 \quad \textit{Subtract 10x.}$$
$$x = 12 \quad \textit{Divide by } -5.$$

Check $x = 12$: $5 = 5$

Solution set: $\{12\}$

49. $$\frac{x-8}{5} + \frac{8}{5} = -\frac{x}{3}$$
Multiply each side by the LCD, 15, and use the distributive property.
$$15\left(\frac{x-8}{5}\right) + 15\left(\frac{8}{5}\right) = 15\left(-\frac{x}{3}\right)$$
$$3(x-8) + 3(8) = -5x$$
$$3x - 24 + 24 = -5x$$
$$3x = -5x$$
$$8x = 0 \quad \textit{Add 5x.}$$
$$x = 0 \quad \textit{Divide by 8.}$$

Check $x = 0$: $0 = 0$

Solution set: $\{0\}$

50. $$\frac{2r-3}{7} + \frac{3}{7} = -\frac{r}{3}$$
Multiply each side by the LCD, 21, and use the distributive property.
$$21\left(\frac{2r-3}{7}\right) + 21\left(\frac{3}{7}\right) = 21\left(-\frac{r}{3}\right)$$
$$3(2r-3) + 3(3) = 7(-r)$$
$$6r - 9 + 9 = -7r$$
$$6r = -7r$$
$$13r = 0 \quad \textit{Add 7r.}$$
$$r = 0 \quad \textit{Divide by 13.}$$

Check $r = 0$: $0 = 0$

Solution set: $\{0\}$

51. $$\frac{3x-1}{4} + \frac{x+3}{6} = 3$$
Multiply each side by the LCD, 12.
$$12\left(\frac{3x-1}{4} + \frac{x+3}{6}\right) = 12(3)$$
$$3(3x-1) + 2(x+3) = 36$$
$$9x - 3 + 2x + 6 = 36$$
$$11x + 3 = 36$$
$$11x = 33$$
$$x = 3$$

Check $x = 3$: $2 + 1 = 3$

Solution set: $\{3\}$

52. $$\frac{3x+2}{7} - \frac{x+4}{5} = 2$$
Multiply each side by the LCD, 35.
$$35\left(\frac{3x+2}{7} - \frac{x+4}{5}\right) = 35(2)$$
$$5(3x+2) - 7(x+4) = 70$$
$$15x + 10 - 7x - 28 = 70$$
$$8x - 18 = 70$$
$$8x = 88$$
$$x = 11$$

Check $x = 11$: $5 - 3 = 2$

Solution set: $\{11\}$

53. $$\frac{4t+1}{3}=\frac{t+5}{6}+\frac{t-3}{6}$$

Multiply each side by the LCD, 6.

$$\begin{aligned} 6\left(\frac{4t+1}{3}\right) &= 6\left(\frac{t+5}{6}+\frac{t-3}{6}\right) \\ 2(4t+1) &= (t+5)+(t-3) \\ 8t+2 &= 2t+2 \\ 6t &= 0 \\ t &= 0 \end{aligned}$$

Check $t=0$: $\frac{1}{3}=\frac{5}{6}-\frac{3}{6}$

Solution set: $\{0\}$

54. $$\frac{2x+5}{5}=\frac{3x+1}{2}+\frac{-x+7}{2}$$

Multiply each side by the LCD, 10.

$$\begin{aligned} 10\left(\frac{2x+5}{5}\right) &= 10\left(\frac{3x+1}{2}+\frac{-x+7}{2}\right) \\ 2(2x+5) &= 5(3x+1)+5(-x+7) \\ 4x+10 &= 15x+5-5x+35 \\ 4x+10 &= 10x+40 \\ -6x &= 30 \\ x &= \tfrac{30}{-6} = -5 \end{aligned}$$

Check $x=-5$: $-1=-7+6$

Solution set: $\{-5\}$

55. $.05x+.12(x+5000)=940$

Multiply each side by 100.

$$\begin{aligned} 5x+12(x+5000) &= 100(940) \\ 5x+12x+60{,}000 &= 94{,}000 \\ 17x &= 34{,}000 \\ x &= 2000 \end{aligned}$$

Check $x=2000$: $100+840=940$

Solution set: $\{2000\}$

56. $$.09k+.13(k+300)=61$$

Multiply each side by 100.

$$\begin{aligned} 100[.09k+.13(k+300)] &= 100(61) \\ 100(.09k)+100(.13)(k+300) &= 6100 \\ 9k+13(k+300) &= 6100 \\ 9k+13k+3900 &= 6100 \\ 22k &= 2200 \\ k &= \tfrac{2200}{22} = 100 \end{aligned}$$

Check $k=100$: $9+52=61$

Solution set: $\{100\}$

57. $.02(50)+.08r=.04(50+r)$

Multiply each side by 100.

$$\begin{aligned} 2(50)+8r &= 4(50+r) \\ 100+8r &= 200+4r \\ 4r &= 100 \\ r &= 25 \end{aligned}$$

Check $r=25$: $1+2=3$

Solution set: $\{25\}$

58. $$.20(14{,}000)+.14t=.18(14{,}000+t)$$

Multiply each side by 100.

$$\begin{aligned} 100[.20(14{,}000)+.14t] &= 100[.18(14{,}000+t)] \\ 20(14{,}000)+14t &= 18(14{,}000+t) \\ 280{,}000+14t &= 252{,}000+18t \\ 28{,}000 &= 4t \\ t &= 7000 \end{aligned}$$

Check $t=7000$: $2800+980=3780$

Solution set: $\{7000\}$

59. $.05x+.10(200-x)=.45x$

Multiply each side by 100.

$$\begin{aligned} 5x+10(200-x) &= 45x \\ 5x+2000-10x &= 45x \\ 2000-5x &= 45x \\ 2000 &= 50x \\ 40 &= x \end{aligned}$$

Check $x=40$: $2+16=18$

Solution set: $\{40\}$

60. $.08x+.12(260-x)=.48x$

Multiply each side by 100.

$$\begin{aligned} 8x+12(260-x) &= 48x \\ 8x+3120-12x &= 48x \\ -4x+3120 &= 48x \\ 3120 &= 52x \\ x &= \tfrac{3120}{52} = 60 \end{aligned}$$

Check $x=60$: $4.8+24=28.8$

Solution set: $\{60\}$

61. $.006(x+2)=.007x+.009$

Multiply each side by 1000.

$$\begin{aligned} 6(x+2) &= 7x+9 \\ 6x+12 &= 7x+9 \\ 3 &= x \end{aligned}$$

Check $x=3$: $.03=.021+.009$

Solution set: $\{3\}$

62. $.004x + .006(50 - x) = .004(68)$
Multiply each side by 1000.

$$\begin{aligned} 4x + 6(50 - x) &= 4(68) \\ 4x + 300 - 6x &= 272 \\ -2x + 300 &= 272 \\ -2x &= -28 \\ x &= 14 \end{aligned}$$

Check $x = 14$: $.056 + .216 = .272$

Solution set: $\{14\}$

63. A conditional equation has at least one solution, an identity has infinitely many solutions, and a contradiction has no solutions.

64. By dividing the equation $8x = 7x$ by x, he possibly divided the equation by 0. He should have subtracted $7x$ from each side of the equation to get $x = 0$. The solution set is $\{0\}$.

65. **(a)** $5 = 5$ is true and the original equation has solution set {all real numbers}, choice **B**.

(b) $x = 0$ indicates the original equation has solution set $\{0\}$, choice **A**.

(c) $5 = 0$ is false and the original equation has solution set $\emptyset$, choice **C**.

66. Each equation in choices **A**, **B**, and **D** is an identity and has {all real numbers} as its solution set. The equation in choice **C**, $4x = 3x$, has $\{0\}$ as its solution set.

67.
$$\begin{aligned} -2x + 5x - 9 &= 3(x - 4) - 5 \\ 3x - 9 &= 3x - 12 - 5 \\ 3x - 9 &= 3x - 17 \\ -9 &= -17 \quad \textit{False} \end{aligned}$$

The equation is a *contradiction.*

Solution set: $\emptyset$

68.
$$\begin{aligned} -6x + 2x - 11 &= -2(2x - 3) + 4 \\ -4x - 11 &= -4x + 6 + 4 \\ -4x - 11 &= -4x + 10 \\ -11 &= 10 \quad \textit{False} \end{aligned}$$

The equation is a *contradiction.*

Solution set: $\emptyset$

69.
$$\begin{aligned} -11x + 4(x - 3) + 6x &= 4x - 12 \\ -11x + 4x - 12 + 6x &= 4x - 12 \\ -x - 12 &= 4x - 12 \\ 0 &= 5x \\ 0 &= x \end{aligned}$$

This is a *conditional* equation.

Solution set: $\{0\}$

70.
$$\begin{aligned} 3x - 5(x + 4) + 9 &= -11 + 15x \\ 3x - 5x - 20 + 9 &= -11 + 15x \\ -2x - 11 &= -11 + 15x \\ -17x &= 0 \\ x &= 0 \end{aligned}$$

This is a *conditional* equation.

Solution set: $\{0\}$

71.
$$\begin{aligned} -2(t + 3) - t - 4 &= -3(t + 4) + 2 \\ -2t - 6 - t - 4 &= -3t - 12 + 2 \\ -3t - 10 &= -3t - 10 \end{aligned}$$

The equation is an *identity.*

Solution set: {all real numbers}

72.
$$\begin{aligned} 4(2d + 7) &= 2d + 25 + 3(2d + 1) \\ 8d + 28 &= 2d + 25 + 6d + 3 \\ 8d + 28 &= 8d + 28 \end{aligned}$$

The equation is an *identity.*

Solution set: {all real numbers}

73.
$$\begin{aligned} 7[2 - (3 + 4x)] - 2x &= -9 + 2(1 - 15x) \\ 7[2 - 3 - 4x] - 2x &= -9 + 2 - 30x \\ 7[-1 - 4x] - 2x &= -7 - 30x \\ -7 - 28x - 2x &= -7 - 30x \\ -7 - 30x &= -7 - 30x \end{aligned}$$

The equation is an *identity.*

Solution set: {all real numbers}

74.
$$\begin{aligned} 4[6 - (1 + 2x)] + 10x &= 2(10 - 3x) + 8x \\ 4[6 - 1 - 2x] + 10x &= 20 - 6x + 8x \\ 4(5 - 2x) + 10x &= 20 + 2x \\ 20 - 8x + 10x &= 20 + 2x \\ 20 + 2x &= 20 + 2x \end{aligned}$$

The equation is an *identity.*

Solution set: {all real numbers}

2.2 Formulas

2.2 Margin Exercises

1. **(a)** To solve $I = prt$ for p, treat p as the only variable.

$$\begin{aligned} I &= prt \\ I &= p(rt) && \textit{Associative property} \\ \frac{I}{rt} &= \frac{p(rt)}{rt} && \textit{Divide by rt.} \\ \frac{I}{rt} &= p \quad \text{or} \quad p = \frac{I}{rt} \end{aligned}$$

(b) To solve $I = prt$ for r, treat r as the only variable.

$$I = prt$$
$$\frac{I}{pt} = \frac{r(pt)}{pt} \qquad \textit{Divide by pt.}$$
$$\frac{I}{pt} = r \quad \text{or} \quad r = \frac{I}{pt}$$

2. **(a)** Solve $P = a + b + c$ for a.

$$P - (b + c) = a + (b + c) - (b + c) \qquad \textit{Subtract } (b + c).$$
$$P - b - c = a \quad \text{or} \quad a = P - b - c$$

(b) Solve $m = 2k + 3b$ for k.

$$m - 3b = 2k + 3b - 3b \qquad \textit{Subtract 3b.}$$
$$m - 3b = 2k$$
$$\frac{m - 3b}{2} = \frac{2k}{2} \qquad \textit{Divide by 2.}$$
$$\frac{m - 3b}{2} = k \quad \text{or} \quad k = \frac{m - 3b}{2}$$

3. Solve $y = \frac{1}{2}(x + 3)$ for x.

$$2y = x + 3 \qquad \textit{Multiply by 2.}$$
$$2y - 3 = x \qquad \textit{Subtract 3.}$$

4. Solve $A = 2HW + 2LW + 2LH$ for W.

$$A - 2LH = 2HW + 2LW \qquad \textit{Subtract 2LH.}$$
$$A - 2LH = W(2H + 2L) \qquad \textit{Distributive property}$$
$$\frac{A - 2LH}{2H + 2L} = \frac{W(2H + 2L)}{2H + 2L}$$
$$\frac{A - 2LH}{2H + 2L} = W \quad \text{or} \quad W = \frac{A - 2LH}{2H + 2L}$$

5. **(a)** Use the formula for the area of a triangle. Solve for h.

$$A = \frac{1}{2}bh$$
$$A\left(\frac{2}{b}\right) = \frac{1}{2}bh\left(\frac{2}{b}\right) \qquad \textit{Multiply by } \frac{2}{b}.$$
$$\frac{2A}{b} = h \quad \text{or} \quad h = \frac{2A}{b}$$

Now substitute $A = 36$ and $b = 12$.

$$h = \frac{2(36)}{12} = 6$$

The height is 6 in.

(b) Use $d = rt$. Solve for r.

$$\frac{d}{t} = \frac{rt}{t} \qquad \textit{Divide by t.}$$
$$\frac{d}{t} = r \quad \text{or} \quad r = \frac{d}{t}$$

Now substitute $d = 500$ and $t = 20$.

$$r = \frac{500}{20} = 25$$

The rate is 25 mph.

(c) Use $d = rt$. Solve for t.

$$\frac{d}{r} = \frac{rt}{r} \qquad \textit{Divide by r.}$$
$$\frac{d}{r} = t \quad \text{or} \quad t = \frac{d}{r}$$

Now substitute $d = 500$ and $r = 156.291$.

$$t = \frac{500}{156.291} = 3.199$$

His time was about 3.199 hr.

6. **(a)** The given amount of mixture is 20 oz. The part that is oil is 1 oz. Thus, the percent of oil is

$$\frac{\text{amount}}{\text{base}} = \frac{1}{20} = .05 = 5\%.$$

(b) Let x represent the amount of commission earned.

$$\frac{x}{12{,}000} = .08 \qquad \frac{a}{b} = \textit{percent}$$
$$x = .08(12{,}000) \qquad \textit{Multiply by 12,000.}$$
$$x = 960$$

The salesman earns \$960.

7. Let x represent the amount spent on pet supplies/medicine.

$$\frac{x}{29.7} = .256 \qquad \textit{25.6\%} = .256$$
$$x = .256(29.7) \qquad \textit{Multiply by 29.7.}$$
$$x = 7.6032$$

Therefore, about \$7.6 billion was spent on pet supplies/medicine.

2.2 Section Exercises

1. **(a)** $x = \frac{5x + 8}{3}$

$$3x = 3\left(\frac{5x + 8}{3}\right)$$
$$3x = 5x + 8$$

(b) $t = \frac{bt + k}{c} \ (c \neq 0)$

$$ct = c\left(\frac{bt + k}{c}\right)$$
$$ct = bt + k$$

2. (a) $3x - 5x = 5x + 8 - 5x$
$3x - 5x = 8$

(b) $ct - bt = bt + k - bt$
$ct - bt = k$

3. Use the distributive property in each case.

(a) $-2x = 8$ (b) $(c - b)t = k$

4. (a) $\frac{-2x}{-2} = \frac{8}{-2}$
$x = -4$

(b) $\frac{(c - b)t}{c - b} = \frac{k}{c - b}$
$t = \frac{k}{c - b}$

5. The restriction $b \neq c$ must be applied. If $b = c$, the denominator becomes 0 and division by 0 is undefined.

6. To solve an equation for a particular variable, such as solving the second equation for t, go through the same steps as you would in solving for x in the first equation. Treat all other variables as constants.

7. Solve $A = LW$ for W.

$$A = LW$$
$$\frac{A}{L} = \frac{LW}{L} \qquad \textit{Divide by } L.$$
$$\frac{A}{L} = W \quad \text{or} \quad W = \frac{A}{L}$$

8. Solve $d = rt$ for t.

$$\frac{d}{r} = \frac{rt}{r} \qquad \textit{Divide by } r.$$
$$\frac{d}{r} = t \quad \text{or} \quad t = \frac{d}{r}$$

9. Solve $P = 2L + 2W$ for L.

$$P = 2L + 2W$$
$$P - 2W = 2L$$
$$\frac{P - 2W}{2} = \frac{2L}{2}$$
$$\frac{P - 2W}{2} = L \quad \text{or} \quad L = \frac{P}{2} - W$$

10. Solve $A = bh$ for b.

$$\frac{A}{h} = b \qquad \textit{Divide by } h.$$

11. (a) Solve for $V = LWH$ for W.

$$V = LWH$$
$$\frac{V}{LH} = \frac{LWH}{LH}$$
$$\frac{V}{LH} = W \quad \text{or} \quad W = \frac{V}{LH}$$

(b) Solve for $V = LWH$ for H.

$$V = LWH$$
$$\frac{V}{LW} = \frac{LWH}{LW}$$
$$\frac{V}{LW} = H \quad \text{or} \quad H = \frac{V}{LW}$$

12. (a) Solve $P = a + b + c$ for b.

$$P - (a + c) = a + b + c - (a + c)$$
Subtract $(a + c)$*.*
$$P - a - c = b$$

(b) Solve $P = a + b + c$ for c.

$$P - (a + b) = a + b + c - (a + b)$$
Subtract $(a + b)$*.*
$$P - a - b = c$$

13. Solve $C = 2\pi r$ for r.

$$C = 2\pi r$$
$$\frac{C}{2\pi} = \frac{2\pi r}{2\pi} \qquad \textit{Divide by } 2\pi.$$
$$\frac{C}{2\pi} = r$$

14. Solve $A = \frac{1}{2}bh$ for h.

$$\frac{2A}{b} = \frac{2}{b}\left(\frac{1}{2}bh\right) \qquad \textit{Multiply by 2; divide by } b.$$
$$\frac{2A}{b} = h$$

15. (a) Solve $A = \frac{1}{2}h(b + B)$ for h.

$$2A = h(b + B) \qquad \textit{Multiply by 2.}$$
$$\frac{2A}{b + B} = h \qquad \textit{Divide by } b + B.$$

(b) Solve $A = \frac{1}{2}h(b + B)$ for B.

$$2A = h(b + B) \quad \text{or} \quad 2A = hb + hB$$
$$\frac{2A}{h} = b + B \qquad 2A - hb = hB$$
$$\frac{2A}{h} - b = B \quad \text{or} \quad \frac{2A - hb}{h} = B$$

16. Solve $S = 2\pi rh + 2\pi r^2$ for h.

$S - 2\pi r^2 = 2\pi rh$ *Subtract $2\pi r^2$.*

$\frac{S - 2\pi r^2}{2\pi r} = \frac{2\pi rh}{2\pi r}$ *Divide by $2\pi r$.*

$\frac{S - 2\pi r^2}{2\pi r} = h$ or $\frac{S}{2\pi r} - r = h$

17. Solve $F = \frac{9}{5}C + 32$ for C.

$$F - 32 = \tfrac{9}{5}C$$
$$\tfrac{5}{9}(F - 32) = \tfrac{5}{9}(\tfrac{9}{5}C)$$
$$\tfrac{5}{9}(F - 32) = C$$

18. Solve $C = \frac{5}{9}(F - 32)$ for F.

$\frac{9}{5}C = \frac{9}{5} \cdot \frac{5}{9}(F - 32)$ *Multiply by $\frac{9}{5}$.*

$\frac{9}{5}C = F - 32$

$\frac{9}{5}C + 32 = F$ *Add 32.*

19. $A = \frac{1}{2}bh$

$2A = 2(\frac{1}{2}bh)$ *Multiply by 2.*

$2A = bh$

$\frac{2A}{b} = \frac{bh}{b}$ *Divide by b.*

$\frac{2A}{b} = h$

$\frac{2A}{b} = \frac{2}{1} \cdot \frac{A}{b}$

$= 2\left(\frac{A}{b}\right)$ *This choice is* **A**.

$= 2A\left(\frac{1}{b}\right)$ *This is choice* **B**.

To get choice **C**, divide $A = \frac{1}{2}bh$ by $\frac{1}{2}b$.

$$\frac{A}{\frac{1}{2}b} = \frac{\frac{1}{2}bh}{\frac{1}{2}b} \text{ gives us } h = \frac{A}{\frac{1}{2}b}.$$

Choice **D**, $h = \frac{\frac{1}{2}A}{b}$, can be multiplied by $\frac{2}{2}$ on the right side to get $h = \frac{A}{2b}$, so it is *not* equivalent to $h = \frac{2A}{b}$. Therefore, the correct answer is **D**.

20. The formula is not solved for L because L should not appear on each side of the final equation. Solve for L as follows:

$$A = 2HW + 2LW + 2LH$$
$$A - 2HW = 2LW + 2LH$$
$$A - 2HW = L(2W + 2H)$$
$$\frac{A - 2HW}{2W + 2H} = L$$

21. Solve $2k + ar = r - 3y$ for r.
Get the "r-terms" on one side and the other terms on the other side.

$ar - r = -2k - 3y$

$(a - 1)r = -2k - 3y$ *Distributive property*

$r = \frac{-2k - 3y}{a - 1}$ *Divide by a–1.*

The answer can also be written as

$$r = \frac{2k + 3y}{1 - a},$$

which would occur if you took the r-terms to the right side in your first step.

22. Solve $4s + 7p = tp - 7$ for p.

$$7p - tp = -4s - 7$$
$$(7 - t)p = -4s - 7$$
$$p = \frac{-4s - 7}{7 - t} \text{ or } p = \frac{4s + 7}{t - 7}$$

23. Solve $w = \frac{3y - x}{y}$ for y.

$wy = 3y - x$ *Multiply by y.*

$x = 3y - wy$ *Get y-terms on one side.*

$x = (3 - w)y$ *Distributive property*

$\frac{x}{3 - w} = y$ *Divide by 3–w.*

Equivalently, we have

$$y = \frac{-x}{w - 3}.$$

24. Solve $c = \frac{-2t + 4}{t}$ for t.

$$ct = -2t + 4$$
$$ct + 2t = 4$$
$$(c + 2)t = 4$$
$$t = \frac{4}{c + 2} \text{ or } t = \frac{-4}{-c - 2}$$

25. Solve $d = rt$ for t.

$$t = \frac{d}{r}$$

To find t, substitute $d = 500$ and $r = 133.870$.

$$t = \frac{500}{133.870} \approx 3.735$$

His time was about 3.735 hours.

26. Solve $d = rt$ for t.

$$t = \frac{d}{r}$$

Replace d by 450 and r by 138.518.

$$t = \frac{450}{138.518} \approx 3.249$$

His time was about 3.249 hours.

27. Use the formula $F = \frac{9}{5}C + 32$.

$$\begin{aligned} F &= \tfrac{9}{5}(40) + 32 \quad \textit{Let } C = \textit{40.} \\ &= 72 + 32 \\ &= 104 \end{aligned}$$

The corresponding temperature is 104°F.

28. Use the formula $C = \frac{5}{9}(F - 32)$.

$$\begin{aligned} C &= \tfrac{5}{9}(-4 - 32) \quad \textit{Let } F = -4. \\ &= \tfrac{5}{9}(-36) \\ &= -20 \end{aligned}$$

The corresponding temperature is about -20°C.

29. Solve $P = 4s$ for s.

$$s = \frac{P}{4}$$

To find s, substitute 920 for P.

$$s = \frac{920}{4} = 230$$

The length of each side is 230 m.

30. Use $V = \pi r^2 h$.
Replace r by 46 and h by 220.

$$\begin{aligned} V &= \pi(46^2)(220) \\ &\approx 1{,}462{,}474.2 \end{aligned}$$

To the nearest tenth, the volume is 1,462,474.2 m^3.

31. Use the formula $C = 2\pi r$.

$$\begin{aligned} 480\pi &= 2\pi r \quad \textit{Let } C = 480\pi. \\ \frac{480\pi}{2\pi} &= r \quad \textit{Divide by } 2\pi. \end{aligned}$$

So the radius of the circle is 240 inches and the diameter is twice that length, that is, 480 inches.

32. $d = 2r = 2(2.5) = 5$
The diameter is 5 inches.
$C = 2\pi r = 2\pi(2.5) = 5\pi$
The circumference is 5π inches.

33. Solve $V = LWH$ for L.

$$L = \frac{V}{WH}$$

To find L, substitute $V = 128$, $W = 4$, and $H = 4$.

$$L = \frac{128}{(4)(4)} = 8$$

The stack is 8 ft long.

34. One cord contains 128 ft^3, so 1.5 cords contains

$$1.5(128) = 192 \text{ ft}^3.$$

Answers will vary. Since

$$4 \cdot 6 \cdot 8 = 192,$$

one set of possible dimensions is 4 ft by 6 ft by 8 ft. Another set is 2 ft by 6 ft by 16 ft (since $2 \cdot 6 \cdot 16 = 192$).

35. The mixture is 36 oz and that part which is alcohol is 9 oz. Thus, the percent of alcohol is

$$\tfrac{9}{36} = \tfrac{1}{4} = \tfrac{25}{100} = 25\%.$$

The percent of water is

$$100\% - 25\% = 75\%.$$

36. Let $x =$ the amount of pure acid in the mixture. Then x can be found by multiplying the total amount of the mixture by the percent of acid given as a decimal (.35).

$$x = 40(.35) = 14$$

There are 14 L of pure acid. Since there are 40 L altogether, there are $40 - 14$, or 26 L of pure water in the mixture.

37. Find what percent \$6300 is of \$210,000.

$$\tfrac{6300}{210{,}000} = .03 = 3\%$$

The agent received a 3% rate of commission.

38. Solve $I = prt$ for r.

$$\begin{aligned} r &= \frac{I}{pt} \\ r &= \frac{221}{3400(1)} \\ &= .065 = 6.5\% \end{aligned}$$

The interest rate on this deposit is 6.5%.

In Exercises 39–42, use the rule of 78.

$$u = f \cdot \frac{k(k+1)}{n(n+1)}$$

39. Substitute 700 for f, 4 for k, and 36 for n.

$$u = 700 \cdot \frac{4(4+1)}{36(36+1)}$$
$$= 700 \cdot \frac{4(5)}{36(37)} \approx 10.51$$

The unearned interest is \$10.51.

40. Substitute 600 for f, 12 for k, and 36 for n.

$$u = 600 \cdot \frac{12(12+1)}{36(36+1)}$$
$$= 600 \cdot \frac{12(13)}{36(37)} \approx 70.27$$

The unearned interest is \$70.27.

41. Substitute 380.50 for f, 8 for k, and 24 for n.

$$u = (380.50) \cdot \frac{8(8+1)}{24(24+1)}$$
$$= (380.50) \cdot \frac{8(9)}{24(25)} \approx 45.66$$

The unearned interest is \$45.66.

42. Substitute 450 for f, 9 for k, and 24 for n.

$$u = 450 \cdot \frac{9(9+1)}{24(24+1)}$$
$$= 450 \cdot \frac{9(10)}{24(25)} \approx 67.50$$

The unearned interest is \$67.50.

43. **(a)** Chicago:

$$\text{Pct.} = \frac{W}{W+L} = \frac{17}{17+13} = \frac{17}{30} \approx .567$$

(b) Minnesota had the same record as Chicago, so their winning percentage is also about .567.

(c) Detroit:

$$\text{Pct.} = \frac{W}{W+L} = \frac{15}{15+16} = \frac{15}{31} \approx .484$$

(d) Cleveland:

$$\text{Pct.} = \frac{W}{W+L} = \frac{13}{13+18} = \frac{13}{31} \approx .419$$

44. **(a)** Houston:

$$\text{Pct.} = \frac{W}{W+L} = \frac{20}{20+11} = \frac{20}{31} \approx .645$$

(b) Chicago:

$$\text{Pct.} = \frac{W}{W+L} = \frac{18}{18+13} = \frac{18}{31} \approx .581$$

(c) Cincinnati:

$$\text{Pct.} = \frac{W}{W+L} = \frac{16}{16+15} = \frac{16}{31} \approx .516$$

(d) St. Louis:

$$\text{Pct.} = \frac{W}{W+L} = \frac{16}{16+16} = \frac{16}{32} = .500$$

45. $\frac{43.7 \text{ million}}{106.7 \text{ million}} \approx .41$

In 2003, about 41% of the U.S. households that owned at least one TV set owned at least 3 TV sets.

46. $\frac{51.2 \text{ million}}{106.7 \text{ million}} \approx .48$

In 2003, about 48% of the U.S. households that owned at least one TV set received premium cable television.

47. In 1982, 3% of the television audience watched cable.

$$3\% \text{ of } 50{,}000 = .03(50{,}000) = 1500$$

48. $21\% \text{ of } 110{,}000 = .21(110{,}000) = 23{,}100$

49. In 1999, 41% of the television audience watched cable.

$$41\% \text{ of } 35{,}000 = .41(35{,}000) = 14{,}350$$

50. $12\% \text{ of } 65{,}000 = .12(65{,}000) = 7800$

51. $.33(161{,}430) = 53{,}271.90$

To the nearest dollar, \$53,272 will be spent to provide housing.

52. $.07(161{,}430) = 11{,}300.10$

To the nearest dollar, \$11,300 will be spent for health care.

53. $\frac{\$29{,}000}{\$161{,}430} \approx .1796$

So the food cost is about 18%, which agrees with the percent shown in the graph.

2.3 Applications of Linear Equations

2.3 Margin Exercises

1. **(a)** "9 added to a number" translates as

$$9 + x \quad \text{or} \quad x + 9.$$

(b) "The difference between 7 and a number" translates as

$$7 - x.$$

Note: $x - 7$ is the difference between a number and 7.

(c) "Four times a number" translates as

$$4 \cdot x \quad \text{or} \quad 4x.$$

(d) "The quotient of 7 and a nonzero number" translates as

$$\frac{7}{x}.$$

2. **(a)** The sum of a number and 6 is 28.

The sum of a number and 6	is	28.
↓	↓	↓
$x + 6$	$=$	28

An equation is $x + 6 = 28$.

(b)

If twice a number	decreased	by 3,	the result is	17.
↓	↓	↓	↓	↓
$2x$	$-$	3	$=$	17

An equation is $2x - 3 = 17$.

(c)

The product of a number and 7	is	twice the number	plus	12.
↓	↓	↓	↓	↓
$7x$	$=$	$2x$	$+$	12

An equation is $7x = 2x + 12$.

(d)

The quotient of a number and 6,	added to	twice the number,	is	7.
↓	↓	↓	↓	↓
$\frac{x}{6}$	$+$	$2x$	$=$	7

An equation is $\frac{x}{6} + 2x = 7$.

3. **(a)** $5x - 3(x + 2) = 7$ is an *equation* because it has an equals sign.

(b) $5x - 3(x + 2)$ is an *expression* because there is no equals sign.

4. *Step 2*
The length and perimeter are given in terms of the width W. The length L is 5 cm more than the width, so

$$L = W + 5.$$

The perimeter P is 5 times the width, so

$$P = 5W.$$

Step 3
Use the formula for perimeter of a rectangle.

$$P = 2L + 2W$$
$$5W = 2(W + 5) + 2W \quad \textit{P = 5W; L = W + 5}$$

Step 4
Solve the equation.

$$5W = 2W + 10 + 2W \quad \textit{Distributive property}$$
$$5W = 4W + 10 \quad \textit{Combine terms.}$$
$$W = 10 \quad \textit{Subtract 4W.}$$

Step 5
The width is 10 and the length is

$$L = W + 5 = 10 + 5 = 15.$$

The rectangle is 10 cm by 15 cm.

Step 6
15 is 5 more than 10 and $P = 2(10) + 2(15) = 50$ is five times 10.

5. *Step 2*
Let $x =$ the number of home runs for Sosa.
Then $x + 119 =$ the number of home runs for Bonds.

Step 3
The sum of their home runs is 1197, so an equation is

$$x + (x + 119) = 1197.$$

Step 4
Solve the equation.

$$2x + 119 = 1197$$
$$2x = 1078 \quad \textit{Subtract 119.}$$
$$x = 539 \quad \textit{Divide by 2.}$$

Step 5
Sosa had 539 home runs and Bonds had $539 + 119 = 658$ home runs.

Step 6
658 is 119 more than 539, and the sum of 539 and 658 is 1197.

6. **(a)** Let x be the number, which is increased by 15% of x, or $.15x$. Then an equation is

$x + .15x = 287.5.$	
$1x + .15x = 287.5$	*Identity property*
$1.15x = 287.5$	*Combine terms*
$x = 250$	*Divide by 1.15.*

The number is 250.

(b) Let x be the amount she earned before deductions. Then 10% of x, or $.10x$, is the amount of her deductions. An equation is

$x - .10x = 162.$	
$1x - .10x = 162$	*Identity property*
$.90x = 162$	*Combine terms*
$x = 180$	*Divide by .90.*

She earned $180 before deductions were made.

7. **(a)** Let $x =$ the amount invested at 5%.
Then $72{,}000 - x =$ the amount invested at 3%.

Use $I = prt$ with $t = 1$.
Make a table to organize the information.

Rate (as a Decimal)	Principal	Interest
.05	x	$.05x$
.03	$72{,}000 - x$	$.03(72{,}000 - x)$
Totals	72,000	3160

The last column gives the equation.

$.05x + .03(72{,}000 - x) = 3160$	
$.05x + 2160 - .03x = 3160$	*Distributive property*
$.02x + 2160 = 3160$	*Combine terms.*
$.02x = 1000$	*Subtract 2160.*
$x = 50{,}000$	*Divide by .02.*

The woman invested $50,000 at 5% and $72,000 – $50,000 = $22,000 at 3%.

(b) Let $x =$ the amount invested at 5%.
Then $34{,}000 - x =$ the amount invested at 4%.

Use $I = prt$ with $t = 1$.

Make a table to organize the information.

Rate (as a Decimal)	Principal	Interest
.05	x	$.05x$
.04	$34{,}000 - x$	$.04(34{,}000 - x)$
Totals	34,000	1545

The last column gives the equation.

$.05x + .04(34{,}000 - x) = 1545$	
$.05x + 1360 - .04x = 1545$	*Distributive property*
$.01x + 1360 = 1545$	*Combine terms.*
$.01x = 185$	*Subtract 1360.*
$x = 18{,}500$	*Divide by .01.*

The man invested $18,500 at 5% and $34,000 – $18,500 = $15,500 at 4%.

8. **(a)** Let $x =$ the number of liters of the 10% solution. Then $x + 60 =$ the number of liters of the 15% solution.

Make a table to organize the information.

Percent (as a Decimal)	Number of Liters	Liters of Pure Solution
$10\% = .10$	x	$.10x$
$25\% = .25$	60	$.25(60)$
$15\% = .15$	$x + 60$	$.15(x + 60)$

The last column gives the equation.

$.10x + .25(60) = .15(x + 60)$	
$.10x + 15 = .15x + 9$	*Distributive property*
$15 = .05x + 9$	*Subtract .10x.*
$6 = .05x$	*Subtract 9.*
$120 = x$	*Divide by .05.*

120 L of 10% solution should be used.

(b) Let $x =$ the amount of $8 per lb candy. Then $x + 100 =$ the amount of $7 per lb candy.

Make a table to organize the information.

Price per Pound	Number of Pounds	Value
$8	x	$8x$
$4	100	400
$7	$x + 100$	$7(x + 100)$

The last column gives the equation.

$8x + 400 = 7(x + 100)$	
$8x + 400 = 7x + 700$	*Distributive property*
$x + 400 = 700$	*Subtract 7x.*
$x = 300$	*Subtract 400.*

300 lb of candy worth $8 per lb should be used.

9. **(a)** Let $x =$ the number of liters of pure acid.

Percent (as a Decimal)	Number of Liters	Liters of Pure Acid
$100\% = 1$	x	x
$30\% = .30$	6	$.30(6)$
$50\% = .50$	$x + 6$	$.50(x + 6)$

The last column gives the equation.

$$\begin{aligned} x + .30(6) &= .50(x + 6) \\ 1x + 1.8 &= .5x + 3 \\ .5x + 1.8 &= 3 && \textit{Subtract .5x.} \\ .5x &= 1.2 && \textit{Subtract 1.8.} \\ x &= 2.4 && \textit{Divide by .5.} \end{aligned}$$

2.4 L of pure acid are needed.

(b) Let $x =$ the number of liters of water.

Percent (as a Decimal)	Number of Liters	Liters of Pure Antifreeze
$0\% = 0$	x	0
$50\% = .50$	20	$.50(20)$
$40\% = .40$	$x + 20$	$.40(x + 20)$

The last column gives the equation.

$$\begin{aligned} 0 + .50(20) &= .40(x + 20) \\ 10 &= .4x + 8 \\ 2 &= .4x && \textit{Subtract 8.} \\ 5 &= x && \textit{Divide by .4.} \end{aligned}$$

5 L of water are needed.

2.3 Section Exercises

1. **(a)** 12 more than a number $\underline{x + 12}$

(b) 12 is more than a number. $\underline{12 > x}$

2. **(a)** 3 less than a number $\underline{x - 3}$

(b) 3 is less than a number. $\underline{3 < x}$

3. **(a)** 4 smaller than a number $\underline{x - 4}$

(b) 4 is smaller than a number. $\underline{4 < x}$

4. **(a)** 6 greater than a number $\underline{x + 6}$

(b) 6 is greater than a number. $\underline{6 > x}$

5. 20% can be written as

$.20 = .2 = \dfrac{20}{100} = \dfrac{2}{10} = \dfrac{1}{5}$, so "20% of a number" can be written as $.20x$, $.2x$, or $\dfrac{x}{5}$. We see that "20% of a number" cannot be written as $20x$, choice **D**.

6. $13 - x$ is the translation of "x less than 13." The phrase "13 less than a number" is translated $x - 13$.

7. Twice a number, decreased by 13 $\underline{2x - 13}$

8. The product of 6 and a number, decreased by 12 $\underline{6x - 12}$

9. 12 increased by three times a number $\underline{12 + 3x}$

10. 12 more than one-half of a number $\underline{\frac{1}{2}x + 12}$

11. The product of 8 and 12 less than a number $\underline{8(x - 12)}$

12. The product of 9 more than a number and 6 less than the number $\underline{(x + 9)(x - 6)}$

13. The quotient of three times a number and 7 $\underline{\frac{3x}{7}}$

14. The quotient of 6 and five times a nonzero number $\underline{\frac{6}{5x}\ (x \neq 0)}$

15. The sentence "the sum of a number and 6 is -31" can be translated as

$$\begin{aligned} x + 6 &= -31. \\ x &= -37 && \textit{Subtract 6.} \end{aligned}$$

The number is -37.

16. The sentence "the sum of a number and -4 is 12" can be translated as

$$\begin{aligned} x + (-4) &= 12. \\ x &= 16 && \textit{Add 4.} \end{aligned}$$

The number is 16.

17. The sentence "if the product of a number and -4 is subtracted from the number, the result is 9 more than the number" can be translated as

$$\begin{aligned} x - (-4x) &= x + 9. \\ x + 4x &= x + 9 \\ 4x &= 9 \\ x &= \tfrac{9}{4} \end{aligned}$$

The number is $\frac{9}{4}$.

18. The sentence "if the quotient of a number and 6 is added to twice the number, the result is 8 less than the number" can be translated as

$$\begin{aligned} 2x + \frac{x}{6} &= x - 8. \\ 12x + x &= 6x - 48 && \textit{Multiply by 6.} \\ 13x &= 6x - 48 \\ 7x &= -48 \\ x &= -\tfrac{48}{7} \end{aligned}$$

The number is $-\frac{48}{7}$.

19. The sentence "when $\frac{2}{3}$ of a number is subtracted from 12, the result is 10" can be translated as

$$12 - \tfrac{2}{3}x = 10.$$
$$36 - 2x = 30 \quad \textit{Multiply by 3.}$$
$$-2x = -6 \quad \textit{Subtract 36.}$$
$$x = 3 \quad \textit{Divide by } -2.$$

The number is 3.

20. The sentence "when 75% of a number is added to 6, the result is 3 more than the number" can be translated as

$$6 + .75x = x + 3.$$
$$600 + 75x = 100x + 300 \quad \textit{Multiply by 100.}$$
$$600 - 25x = 300 \quad \textit{Subtract 100x.}$$
$$-25x = -300 \quad \textit{Subtract 600.}$$
$$x = 12 \quad \textit{Divide by } -25.$$

The number is 12.

21. $5(x+3) - 8(2x-6)$ is an *expression* because there is no equals sign.

22. $-7(y+4) + 13(y-6)$ has no equals sign, so it is an *expression.*

23. $5(x+3) - 8(2x-6) = 12$ has an equals sign, so this represents an *equation.*

24. $-7(y+4) + 13(y-6) = 18$ has an equals sign, so it is an *equation.*

25. $\frac{r}{2} - \frac{r+9}{6} - 8$ is an *expression* because there is no equals sign.

26. $\frac{r}{2} - \frac{r+9}{6} = 8$ has an equals sign, so it is an *equation.*

27. *Step 1*
We are asked to find the number of patents each university secured.

Step 2
Let $x =$ the number of patents MIT secured.
Then $x - 38 =$ the number of patents Stanford secured.

Step 3
A total of 230 patents were secured, so

$$\underline{x} + \underline{x - 38} = 230.$$

Step 4

$$2x - 38 = 230$$
$$2x = 268$$
$$x = \underline{134}$$

Step 5
MIT secured 134 patents and Stanford secured $134 - 38 = \underline{96}$ patents.

Step 6
The number of Stanford patents was 38 fewer than the number of MIT patents and the total number of patents was $134 + \underline{96} = \underline{230}$.

28. *Step 1*
We are asked to find the number of book buyers at each type of bookstore.

Step 2
Let $x =$ the number of book buyers at large chain bookstores. Then
$x - 70 =$ the number of book buyers at small chain/independent bookstores.

Step 3
A total of 442 book buyers shopped at these two types of stores, so

$$\underline{x} + \underline{(x - 70)} = 442.$$

Step 4

$$2x - 70 = 442$$
$$2x = 512$$
$$x = \underline{256}$$

Step 5
There were 256 large chain bookstore shoppers and $256 - 70 = \underline{186}$ small chain/independent shoppers.

Step 6
The number of large chain shoppers was 70 more than the number of small chain/independent shoppers, and the total number of these shoppers was $256 + \underline{186} = \underline{442}$.

29. *Step 2*
Let $W =$ the width of the base. Then $2W - 65$ is the length of the base.

Step 3
The perimeter of the base is 860 feet. Using $P = 2L + 2W$ gives us

$$2(2W - 65) + 2W = 860.$$

Step 4

$$4W - 130 + 2W = 860$$
$$6W - 130 = 860$$
$$6W = 990$$
$$W = \tfrac{990}{6} = 165$$

Step 5
The width of the base is 165 feet and the length of the base is $2(165) - 65 = 265$ feet.

Step 6
$2L + 2W = 2(265) + 2(165) = 530 + 330 = 860$, which is the perimeter of the base.

30. *Step 2*
Let $x =$ the length of one of the sides of equal length.

Step 3
The perimeter of the triangle is 931.5 feet. Using $P = a + b + c$ gives us

$$x + x + 438 = 931.5$$

Step 4

$$\begin{aligned} 2x + 438 &= 931.5 \\ 2x &= 493.5 \quad \textit{Subtract 438.} \\ x &= 246.75 \quad \textit{Divide by 2.} \end{aligned}$$

Step 5
The two walls are each 246.75 feet long.

Step 6
The answer checks since
$246.75 + 246.75 + 438 = 931.75$, which is the correct perimeter.

31. *Step 2*
Let $x =$ the length of the middle side. Then the shortest side is $x - 75$ and the longest side is $x + 375$.

Step 3
The perimeter of the Bermuda Triangle is 3075 miles. Using $P = a + b + c$ gives us

$$x + (x - 75) + (x + 375) = 3075.$$

Step 4

$$\begin{aligned} 3x + 300 &= 3075 \\ 3x &= 2775 \quad \textit{Subtract 300.} \\ x &= 925 \quad \textit{Divide by 3.} \end{aligned}$$

Step 5
The length of the middle side is 925 miles. The length of the shortest side is $x - 75 = 925 - 75 = 850$ miles. The length of the longest side is $x + 375 = 925 + 375 = 1300$ miles.

Step 6
The answer checks since $925 + 850 + 1300 = 3075$ miles, which is the correct perimeter.

32. *Step 2*
Let $L =$ the length of the top floor. Then $\frac{1}{2}L + 20$ is the width of the top floor.

Step 3
The perimeter of the top floor is 520 feet. Using $P = 2L + 2W$ gives us

$$2L + 2\left(\tfrac{1}{2}L + 20\right) = 520.$$

Step 4

$$\begin{aligned} 2L + L + 40 &= 520 \\ 3L + 40 &= 520 \\ 3L &= 480 \\ L &= 160 \end{aligned}$$

Step 5
The length of the top floor is 160 feet and the width of the top floor is $\frac{1}{2}(160) + 20 = 100$ feet.

Step 6
$2L + 2W = 2(160) + 2(100) = 320 + 200 = 520$, which is the perimeter of the top floor.

33. *Step 2*
Let $x =$ the height of the Eiffel Tower.
Then $x - 804 =$ the height of the Leaning Tower of Pisa.

Step 3
Together these heights are 1164 ft, so

$$x + (x - 804) = 1164.$$

Step 4

$$\begin{aligned} 2x - 804 &= 1164 \\ 2x &= 1968 \\ x &= 984 \end{aligned}$$

Step 5
The height of the Eiffel Tower is 984 feet and the height of the Leaning Tower of Pisa is $984 - 804 = 180$ feet.

Step 6
180 feet is 804 feet shorter than 984 feet and the sum of 180 feet and 984 feet is 1164 feet.

34. *Step 2*
Let $x =$ the number of visitors to Monticello.
Then $x + 327 =$ the number of visitors to Mount Vernon (both numbers in thousands).

Step 3
Together there were 1257 thousand visitors, so

$$x + (x + 327) = 1257.$$

Step 4

$$\begin{aligned} 2x + 327 &= 1257 \\ 2x &= 930 \\ x &= 465 \end{aligned}$$

Step 5
In 2003, there were 465 thousand visitors to Monticello and $465 + 327 = 792$ thousand visitors to Mount Vernon.

Step 6
792 thousand is 327 thousand more than 465 thousand and the sum of 792 thousand and 465 thousand is 1257 thousand.

35. *Step 2*
Let $x =$ the Yankees' payroll (in millions).
Then $x - 32.8 =$ the Mets' payroll (in millions).

Step 3
The two payrolls totaled \$266.6 million, so

$$x + (x - 32.8) = 266.6$$

Step 4

$$\begin{aligned} 2x - 32.8 &= 266.6 \\ 2x &= 299.4 \\ x &= 149.7 \end{aligned}$$

Step 5
In 2003, the Yankees' payroll was \$149.7 million and the Mets' payroll was $149.7 - 32.8 =$ \$116.9 million.

Step 6
\$116.9 million is \$32.8 million less than \$149.7 million and the sum of \$116.9 million and \$149.7 million is \$266.6 million.

36. *Step 2*
Let $x =$ the number of hits Williams got. Then $x + 276 =$ the number of hits Hornsby got.

Step 3
Their base hits totaled 5584, so

$$x + (x + 276) = 5584.$$

Step 4

$$\begin{aligned} 2x + 276 &= 5584 \\ 2x &= 5308 \\ x &= 2654 \end{aligned}$$

Step 5
Williams got 2654 base hits, and Hornsby got $2654 + 276 = 2930$ base hits.

Step 6
2930 is 276 more than 2654 and the total is $2654 + 2930 = 5584$.

37. Let $x =$ the percent increase.

$$\begin{aligned} x &= \frac{\text{amount of increase}}{\text{base score}} \\ &= \frac{21.0 - 20.6}{20.6} = \frac{.4}{20.6} \approx .019 = 1.9\% \end{aligned}$$

The percent increase was approximately 1.9%.

38. Let $x =$ the percent increase.

$$\begin{aligned} x &= \frac{\text{amount of increase}}{\text{base number}} \\ &= \frac{1{,}070{,}000 - 817{,}000}{817{,}000} = \frac{253{,}000}{817{,}000} \\ &\approx .310 = 31.0\% \end{aligned}$$

The percent increase was approximately 31.0%.

39. Let $x =$ the 2002 population.

The 2002 population was 104.5% of the 1990 population.

$$\begin{aligned} x &= (104.5\%)(358{,}648) \\ &= 1.045(358{,}648) \\ &= 374{,}787.16 \end{aligned}$$

The 2002 population was about 374,787.

40. Let $x =$ the CPI in 1995. Then $.181x$ represents the 18.1% increase from 1995 to 2002.

The 1995 CPI plus the 18.1% increase equals the 2002 CPI. Thus,

$$\begin{aligned} x + .181x &= 179.9. \\ 1.181x &= 179.9 \\ x &= \frac{179.9}{1.181} \\ x &\approx 152.3 \end{aligned}$$

The CPI was approximately 152.3 in 1995.

41. Let $x =$ the amount of the receipts excluding tax. Since the sales tax is 9% of x, the total amount is

$$\begin{aligned} x + .09x &= 2725 \\ 1x + .09x &= 2725 \\ 1.09x &= 2725 \\ x &= \frac{2725}{1.09} = 2500 \end{aligned}$$

Thus, the tax was $.09(2500) =$ \$225.

42. Let $x =$ the amount of commission. Since x is 6% of the selling price,

$$x = .06(159{,}000) = 9540.$$

So after the agent was paid, he had $159{,}000 - 9540 =$ \$149,460.

43. Let $x =$ the amount invested at 3%. Then $12{,}000 - x =$ the amount invested at 4%. Complete the table. Use $I = prt$ with $t = 1$.

Rate (as a Decimal)	Principal	Interest
.03	x	$.03x$
.04	$12{,}000 - x$	$.04(12{,}000 - x)$
Totals	12,000	440

The last column gives the equation.

$$\begin{array}{ccccc} \text{Interest at 3\%} & + & \text{interest at 4\%} & = & \text{total interest.} \\ .03x & + & .04(12{,}000 - x) & = & 440 \end{array}$$

$$\begin{aligned} 3x + 4(12{,}000 - x) &= 44{,}000 \quad \textit{Multiply by 100.} \\ 3x + 48{,}000 - 4x &= 44{,}000 \\ -x &= -4000 \\ x &= 4000 \end{aligned}$$

He should invest \$4000 at 3% and
$12{,}000 - 4000 = \$8000$ at 4%.

Check \$4000 @ 3% = \$120 and
\$8000 @ 4% = \$320; \$120 + \$320 = \$440.

44. Let $x =$ the amount invested at 2%. Then
$60{,}000 - x =$ the amount invested at 3%.
Complete the table. Use $I = prt$ with $t = 1$.

Rate (as a Decimal)	Principal	Interest
.02	x	$.02x$
.03	$60{,}000 - x$	$.03(60{,}000 - x)$
Totals	60,000	1600

The last column gives the equation.

$$\begin{array}{ccccc} \text{Interest at } 2\% & + & \text{interest at } 3\% & = & \text{total interest.} \\ .02x & + & .03(60{,}000 - x) & = & 1600 \end{array}$$

$$\begin{aligned} 2x + 3(60{,}000 - x) &= 160{,}000 \quad \textit{Multiply by 100.} \\ 2x + 180{,}000 - 3x &= 160{,}000 \\ -x &= -20{,}000 \\ x &= 20{,}000 \end{aligned}$$

She invested \$20,000 at 2% and
$60{,}000 - x = 60{,}000 - 20{,}000 = \$40{,}000$ at 3%.

Check \$20,000 @ 2% = \$400 and
\$40,000 @ 3% = \$1200; \$400 + \$1200 = \$1600.

45. Let $x =$ the amount invested at 4.5%. Then
$2x - 1000 =$ the amount invested at 3%.
Use $I = prt$ with $t = 1$. Make a table.

Rate (as a Decimal)	Principal	Interest
.045	x	$.045x$
.03	$2x - 1000$	$.03(2x - 1000)$
Total		1020

The last column gives the equation.

$$\begin{array}{ccccc} \text{Interest at } 4.5\% & + & \text{interest at } 3\% & = & \text{total interest.} \\ .045x & + & .03(2x - 1000) & = & 1020 \end{array}$$

$$\begin{aligned} 45x + 30(2x - 1000) &= 1{,}020{,}000 \quad \textit{Multiply by 1000.} \\ 45x + 60x - 30{,}000 &= 1{,}020{,}000 \\ 105x &= 1{,}050{,}000 \\ x &= \frac{1{,}050{,}000}{105} = 10{,}000 \end{aligned}$$

She invested \$10,000 at 4.5% and
$2x - 1000 = 2(10{,}000) - 1000 = \$19{,}000$ at 3%.

Check \$10,000 @ 4.5% = \$450 and
\$19,000 @ 3% = \$570; \$450 + \$570 = \$1020.

46. Let $x =$ the amount invested at 3.5%. Then
$3x + 5000 =$ the amount invested at 4%.
Use $I = prt$ with $t = 1$. Make a table.

Rate (as a Decimal)	Principal	Interest
.035	x	$.035x$
.04	$3x + 5000$	$.04(3x + 5000)$
Total		1440

The last column gives the equation.

$$\begin{array}{ccccc} \text{Interest at } 3.5\% & + & \text{interest at } 4\% & = & 1440. \\ .035x & + & .04(3x + 5000) & = & 1440 \end{array}$$

$$\begin{aligned} 35x + 40(3x + 5000) &= 1{,}440{,}000 \quad \textit{Multiply by 1000.} \\ 35x + 120x + 200{,}000 &= 1{,}440{,}000 \\ 155x &= 1{,}240{,}000 \\ x &= \frac{1{,}240{,}000}{155} = 8000 \end{aligned}$$

She invested \$8000 at 3.5% and
$3x + 5000 = 3(8000) + 5000 = \$29{,}000$ at 4%.

Check \$8000 @ 3.5% = \$280 and
\$29,000 @ 4% = \$1160; \$280 + \$1160 = \$1440.

47. Let $x =$ the amount of additional money to be invested at 2%.
Use $I = prt$ with $t = 1$. Make a table.
Use the fact that the total return on the two investments is 3%.

Rate (as a Decimal)	Principal	Interest
.05	29,000	.05(29,000)
.02	x	$.02x$
.03	$29{,}000 + x$	$.03(29{,}000 + x)$

The last column gives the equation.

$$\begin{array}{ccccc} \text{Interest at } 5\% & + & \text{interest at } 2\% & = & \text{interest at } 3\%. \\ .05(29{,}000) & + & .02x & = & .03(29{,}000 + x) \end{array}$$

$$\begin{aligned} 5(29{,}000) + 2x &= 3(29{,}000 + x) \quad \textit{Multiply by 100.} \\ 145{,}000 + 2x &= 87{,}000 + 3x \\ 58{,}000 &= x \end{aligned}$$

They should invest \$58,000 at 2%.

Check \$29,000 @ 5% = \$1450 and
\$58,000 @ 2% = \$1160;
\$1450 + \$1160 = \$2610, which is the same as
(\$29,000 + \$58,000) @ 3%.

48. Let $x =$ the amount of additional money to be invested at 4%.
Use $I = prt$ with $t = 1$. Make a table.
Use the fact that the total return on the two investments is 5.5%.

Rate (as a Decimal)	Principal	Interest
.06	15,000	.06(15,000)
.04	x	$.04x$
.055	$x + 15{,}000$	$.055(x + 15{,}000)$

The last column gives the equation.

$$\begin{array}{ccccc} \text{Interest} & + & \text{interest} & = & \text{interest} \\ \text{at 6\%} & & \text{at 4\%} & & \text{at 5.5\%.} \\ .06(15{,}000) & + & .04x & = & .055(x + 15{,}000) \end{array}$$

$$\begin{aligned} 60(15{,}000) + 40x &= 55(x + 15{,}000) && \textit{Multiply by 1000.} \\ 900{,}000 + 40x &= 55x + 825{,}000 \\ 75{,}000 &= 15x \\ 5000 &= x \end{aligned}$$

He should deposit \$5000 at 4%.

Check \$15,000 @ 6% = \$900 and \$5000 @ 4% = \$200; \$900 + \$200 = \$1100, which is the same as (\$15,000 + \$5000) @ 5.5%.

49. Let $x =$ the number of liters of 10% acid solution needed. Make a table.

Percent (as a Decimal)	Liters of Solution	Liters of Pure Acid
.04	10	$.04(10) = .4$
.10	x	$.10x$
.06	$x + 10$	$.06(x + 10)$

$$\begin{array}{ccccc} \text{Acid} & + & \text{acid} & = & \text{acid} \\ \text{in 4\%} & & \text{in 10\%} & & \text{in 6\%.} \\ .4 & + & .10x & = & .06(x + 10) \end{array}$$

$$\begin{aligned} .4 + .10x &= .06x + .6 && \textit{Distributive property} \\ .04x &= .2 && \textit{Subtract .06x and .4.} \\ x &= 5 && \textit{Divide by .04.} \end{aligned}$$

Five liters of the 10% solution are needed.

Check 4% of 10 is .4 and 10% of 5 is .5; $.4 + .5 = .9$, which is the same as 6% of $(10 + 5)$.

50. Let $x =$ the number of liters of 14% alcohol solution needed. Make a chart.

Percent (as a Decimal)	Liters of Solution	Liters of Pure Alcohol
.14	x	$.14x$
.50	20	$.50(20) = 10$
.30	$x + 20$	$.30(x + 20)$

$$\begin{array}{ccccc} \text{Alcohol} & + & \text{alcohol} & = & \text{alcohol} \\ \text{in 14\%} & & \text{in 50\%} & & \text{in 30\%.} \\ .14x & + & 10 & = & .30(x + 20) \end{array}$$

$$\begin{aligned} 14x + 1000 &= 30(x + 20) && \textit{Multiply by 100.} \\ 14x + 1000 &= 30x + 600 \\ -16x &= -400 && \textit{Subtract 30x and 1000.} \\ x &= 25 && \textit{Divide by } -16. \end{aligned}$$

25 L of 14% solution must be added.

Check 14% of 25 is 3.5 and 50% of 20 is 10; $3.5 + 10 = 13.5$, which is the same as 30% of $(25 + 20)$.

51. Let $x =$ the number of liters of the 20% alcohol solution. Make a table.

Percent (as a Decimal)	Liters of Solution	Liters of Pure Alcohol
.12	12	$.12(12) = 1.44$
.20	x	$.20x$
.14	$x + 12$	$.14(x + 12)$

$$\begin{array}{ccccc} \text{Alcohol} & + & \text{alcohol} & = & \text{alcohol} \\ \text{in 12\%} & & \text{in 20\%} & & \text{in 14\%.} \\ 1.44 & + & .20x & = & .14(x + 12) \end{array}$$

$$\begin{aligned} 144 + 20x &= 14(x + 12) && \textit{Multiply by 100.} \\ 144 + 20x &= 14x + 168 && \textit{Distributive property} \\ 6x &= 24 && \textit{Subtract 14x and 144.} \\ x &= 4 && \textit{Divide by 6.} \end{aligned}$$

4L of 20% alcohol solution are needed.

Check 12% of 12 is 1.44 and 20% of 4 is .8; $1.44 + .8 = 2.24$, which is the same as 14% of $(12 + 4)$.

52. Let $x =$ the number of liters of 10% alcohol solution. Make a chart.

Percent (as a Decimal)	Liters of Solution	Liters of Pure Alcohol
.10	x	$.10x$
.50	40	$.50(40) = 20$
.40	$x + 40$	$.40(x + 40)$

$$\begin{array}{ccccc} \text{Alcohol} & + & \text{alcohol} & = & \text{alcohol} \\ \text{in 10\%} & & \text{in 50\%} & & \text{in 40\%.} \\ .10x & + & 20 & = & .40(x + 40) \end{array}$$

$$\begin{aligned} 10x + 2000 &= 40(x + 40) && \textit{Multiply by 100.} \\ 10x + 2000 &= 40x + 1600 \\ -30x &= -400 && \textit{Subtract 40x and 2000.} \\ x &= \tfrac{40}{3} \text{ or } 13\tfrac{1}{3} && \textit{Divide by } -30. \end{aligned}$$

$13\frac{1}{3}$ L of 10% solution should be added.

Check 50% of 40 is 20 and 10% of $\frac{40}{3}$ is $\frac{4}{3}$; $20 + \frac{4}{3} = 21\frac{1}{3}$, which is the same as 40% of $\left(\frac{40}{3} + 40\right)$.

53. Let x = the amount of pure dye used (pure dye is 100% dye). Make a table.

Percent (as a Decimal)	Gallons of Solution	Gallons of Pure Dye
1	x	$1x = x$
.25	4	$.25(4) = 1$
.40	$x + 4$	$.40(x + 4)$

Write the equation from the last column in the table.

$x + 1 = .4(x + 4)$
$x + 1 = .4x + 1.6$ *Distributive property*
$.6x = .6$ *Subtract .4x and 1.*
$x = 1$ *Divide by .6.*

One gallon of pure (100%) dye is needed.

Check 100% of 1 is 1 and 25% of 4 is 1; $1 + 1 = 2$, which is the same as 40% of $(1 + 4)$.

54. Let x = the number of gallons of water. Make a chart.

Percent (as a Decimal)	Gallons of Solution	Gallons of Pure Insecticide
0	x	$0(x) = 0$
.04	6	$.04(6) = .24$
.03	$x + 6$	$.03(x + 6)$

Insecticide in water + insecticide in 4% = insecticide in 3%.
$0 + .24 = .03(x + 6)$

$0 + 24 = 3(x + 6)$ *Multiply by 100.*
$24 = 3x + 18$ *Distributive property*
$6 = 3x$ *Subtract 18.*
$2 = x$ *Divide by 3.*

2 gallons of water should be added.

Check 4% of 6 is .24, which is the same as 3% of $(2 + 6)$.

55. Let x = the amount of \$6 per lb nuts. Make a table.

Cost per lb	Pounds of Nuts	Total Cost
\$2	50	$2(50) = 100$
\$6	x	$6x$
\$5	$x + 50$	$5(x + 50)$

The total value of the \$2 per lb nuts and the \$6 per lb nuts must equal the value of the \$5 per lb nuts.

$$100 + 6x = 5(x + 50)$$
$$100 + 6x = 5x + 250$$
$$x = 150$$

He should use 150 lb of \$6 nuts.

Check 50 pounds of the \$2 per lb nuts are worth \$100 and 150 pounds of the \$6 per lb nuts are worth \$900; \$100 + \$900 = \$1000, which is the same as $(50 + 150)$ pounds worth \$5 per lb.

56. Let x = the number of ounces of 2¢ per oz tea. Make a table.

Cost per oz	Ounces of Tea	Total Cost
2¢ or .02	x	$.02x$
5¢ or .05	100	$.05(100) = 5$
3¢ or .03	$x + 100$	$.03(x + 100)$

Cost of 2¢ tea + cost of 5¢ tea = cost of 3¢ tea.
$.02x + 5 = .03(x + 100)$

$2x + 500 = 3(x + 100)$ *Multiply by 100.*
$2x + 500 = 3x + 300$ *Distributive property*
$200 = x$ *Subtract 2x and 300.*

200 oz of 2¢ per oz tea should be used.

Check 200 oz of 2¢ per oz tea is worth \$4 and 100 oz of 5¢ per oz tea is worth \$5; \$4 + \$5 = \$9, which is the same value as $(200 + 100)$ oz of 3¢ per oz tea.

57. We cannot expect the final mixture to be worth more than each of the ingredients. Answers will vary.

58. Let x = the number of liters of 30% acid solution. Make a chart.

Percent (as a Decimal)	Liters of Solution	Liters of Pure Acid
.30	x	$.30x$
.50	15	$.50(15) = 7.5$
.60	$x + 15$	$.60(x + 15)$

Acid in 30% + acid in 50% = acid in 60%.
$.30x + 7.5 = .60(x + 15)$

$3x + 75 = 6(x + 15)$ *Multiply by 10.*
$3x + 75 = 6x + 90$ *Distributive property*
$-3x = 15$ *Subtract 6x and 75.*
$x = -5$ *Divide by −3.*

The solution, -5, is impossible since the number of liters of 30% acid solution cannot be negative. Therefore, this problem has no solution.

59. **(a)** Let x = the amount invested at 5%.
$800 - x$ = the amount invested at 10%.

(b) Let y = the amount of 5% acid used.
$800 - y$ = the amount of 10% acid used.

60. Organize the information in a table.

(a)

Percent (as a Decimal)	Principal	Interest
.05	x	$.05x$
.10	$800 - x$	$.10(800 - x)$
.0875	800	$.0875(800)$

The amount of interest earned at 5% and 10% is found in the last column of the table, $.05x$ and $.10(800 - x)$.

(b)

Percent (as a Decimal)	Liters of Solution	Liters of Pure Acid
.05	y	$.05y$
.10	$800 - y$	$.10(800 - y)$
.0875	800	$.0875(800)$

The amount of pure acid in the 5% and 10% mixtures is found in the last column of the table, $.05y$ and $.10(800 - y)$.

61. Refer to the tables for Exercise 60. In each case, the last column gives the equation.

(a) $.05x + .10(800 - x) = .0875(800)$

(b) $.05y + .10(800 - y) = .0875(800)$

62. In both cases, multiply by 10,000 to clear the decimals.

(a)
$$\begin{aligned} .05x + .10(800 - x) &= .0875(800) \\ 500x + 1000(800 - x) &= 875(800) \\ 500x + 800{,}000 - 1000x &= 700{,}000 \\ -500x &= -100{,}000 \\ x &= 200 \end{aligned}$$

Jack invested \$200 at 5% and $800 - x = 800 - 200 = \$600$ at 10%.

(b)
$$\begin{aligned} .05y + .10(800 - y) &= .0875(800) \\ 500y + 1000(800 - y) &= 875(800) \\ 500y + 800{,}000 - 1000y &= 700{,}000 \\ -500y &= -100{,}000 \\ y &= 200 \end{aligned}$$

Jill used 200 L of 5% acid solution and $800 - y = 800 - 200 = 600$ L of 10% acid solution.

63. The processes used to solve Problems A and B were virtually the same. Aside from the variables chosen, the problem information was organized in similar tables and the equations solved were the same.

2.4 Further Applications of Linear Equations

2.4 Margin Exercises

1. Let $x =$ the number of dimes.
Then $26 - x =$ the number of half-dollars.

Denomination	Number of Coins	Total Value
\$.10	x	$.10x$
\$.50	$26 - x$	$.50(26 - x)$
Total		8.60

Multiply the number of coins by the denominations, and add the results to get 8.60.

$$\begin{aligned} .10x + .50(26 - x) &= 8.60 \\ 1x + 5(26 - x) &= 86 \quad \textit{Multiply by 10.} \\ 1x + 130 - 5x &= 86 \\ -4x &= -44 \\ x &= 11 \end{aligned}$$

The cashier has 11 dimes and $26 - 11 = 15$ half-dollars.

2. Let $x =$ the amount of time needed for the cars to be 420 mi apart.

Make a table. Use the formula $d = rt$, that is, find each distance by multiplying rate by time.

	Rate	Time	Distance
Northbound Car	60	x	$60x$
Southbound Car	45	x	$45x$
Total			420

The total distance traveled is the sum of the distances traveled by each car, since they are traveling in opposite directions. This total is 420 mi.

$$\begin{aligned} 60x + 45x &= 420 \\ 105x &= 420 \\ x &= \tfrac{420}{105} = 4 \end{aligned}$$

The cars will be 420 mi apart in 4 hr.

3. Let $x =$ the time it takes Clay to catch up to Elayn. Then $x + \frac{1}{2} =$ Elayn's time.

Make a table. Use the formula $d = rt$, that is, find each distance by multiplying rate by time.

	Rate	Time	Distance
Elayn	3	$x + \frac{1}{2}$	$3(x + \frac{1}{2})$
Clay	5	x	$5x$

The distance traveled by Elayn is equal to the distance traveled by Clay.

$$\begin{aligned} 3\left(x+\tfrac{1}{2}\right) &= 5x \\ 3x+\tfrac{3}{2} &= 5x \\ 6x+3 &= 10x \quad \textit{Multiply by 2.} \\ 3 &= 4x \\ \tfrac{3}{4} &= x \end{aligned}$$

It takes Clay $\frac{3}{4}$ hr or 45 min to catch up to Elayn.

4. Let $x =$ the measure of the second angle.
Then $x + 15 =$ the measure of the first angle, and $2x + 25 =$ the measure of the third angle.

The sum of the three measures must equal 180°.

$$\begin{aligned} x+(x+15)+(2x+25) &= 180 \\ 4x+40 &= 180 \\ 4x &= 140 \\ x &= 35 \end{aligned}$$

The angles measure 35°, $35 + 15 = 50°$, and $2(35) + 25 = 95°$.

2.4 Section Exercises

1. The total amount is

$$\begin{aligned} 38(.05)+26(.10) &= 1.90+2.60 \\ &= \$4.50. \end{aligned}$$

2. Use $d = rt$, or $t = \dfrac{d}{r}$.
Substitute 7700 for d and 480 for r.

$$t = \frac{7700}{480} \approx 16.04$$

Its travel time is approximately 16 hours.

3. Use $d = rt$, or $r = \dfrac{d}{t}$. Substitute 520 for d and 10 for t.

$$r = \frac{520}{10} = 52$$

His rate was 52 mph.

4. Use $P = 4s$ or $s = \dfrac{P}{4}$.
Substitute 40 for P.

$$s = \frac{40}{4} = 10$$

The length of each side of the square is 10 in. This is also the length of each side of the equilateral triangle. To find the perimeter of the equilateral triangle, use $P = 3s$.
Substitute 10 for s.

$$P = 3(10) = 30$$

The perimeter would be 30 inches.

5. The problem asks for the distance Jeff travels to the workplace, so we must multiply the rate, 10 mph, by the time, $\frac{3}{4}$ hr, to get the distance, 7.5 mi.

6. Begin by subtracting 36° from 180°. Then find half of this difference to get the measure of each of the equal angles.

7. Let $x =$ the number of pennies. Then x is also the number of dimes, and $44 - 2x$ is the number of quarters.

Denomination	Number of Coins	Value
.01	x	$.01x$
.10	x	$.10x$
.25	$44-2x$	$.25(44-2x)$
Totals	44	4.37

The sum of the values must equal the total value.

$$\begin{aligned} .01x+.10x+.25(44-2x) &= 4.37 \\ x+10x+25(44-2x) &= 437 \\ &\textit{Multiply by 100.} \\ x+10x+1100-50x &= 437 \\ -39x+1100 &= 437 \\ -39x &= -663 \\ x &= 17 \end{aligned}$$

There are 17 pennies, 17 dimes, and $44 - 2(17) = 10$ quarters.

Check The number of coins is $17 + 17 + 10 = 44$ and the value of the coins is $\$.01(17) + \$.10(17) + \$.25(10) = \$4.37$, as required.

8. Let $x =$ the number of nickels and the number of quarters. Then $2x$ is the number of half-dollars.

Denomination	Number of Coins	Value
.05	x	$.05x$
.25	x	$.25x$
.50	$2x$	$.50(2x)$
Total		2.60

The sum of the values must equal the total value.

$$\begin{aligned} .05x+.25x+.50(2x) &= 2.60 \\ 5x+25x+50(2x) &= 260 \\ &\textit{Multiply by 100.} \\ 5x+25x+100x &= 260 \\ 130x &= 260 \\ x &= 2 \end{aligned}$$

continued

She found 2 nickels, 2 quarters, and $2(2) = 4$ half-dollars.

Check The number of nickels, 2, is the same as the number of quarters. The number of half-dollars, 4, is twice the number of quarters. The value of the coins is
$\$.05(2) + \$.25(2) + \$.50(4) = \$2.60$, as required.

9. Let $x =$ the number of quarters. Then $47 - x$ is the number of half-dollars.

Denomination	Number of Coins	Value
.25	x	$.25x$
.50	$47 - x$	$.50(47 - x)$
Totals	47	17.00

The sum of the values must equal the total value.

$$.25x + .50(47 - x) = 17.00$$
$$25x + 50(47 - x) = 1700 \quad \textit{Multiply by 100.}$$
$$25x + 2350 - 50x = 1700$$
$$-25x = -650$$
$$x = 26$$

She has 26 quarters and $47 - 26 = 21$ half-dollars.

Check The number of coins is $26 + 21 = 47$ and the value of the coins is
$\$.25(26) + \$.50(21) = \$17.00$, as required.

10. Let $x =$ the number of pennies. Then $39 - x$ is the number of dimes.

Denomination	Number of Coins	Value
.01	x	$.01x$
.10	$39 - x$	$.10(39 - x)$
Totals	39	2.64

The sum of the values must equal the total value.

$$.01x + .10(39 - x) = 2.64$$
$$1x + 10(39 - x) = 264 \quad \textit{Multiply by 100.}$$
$$x + 390 - 10x = 264$$
$$-9x = -126$$
$$x = 14$$

He has 14 pennies and $39 - 14 = 25$ dimes.

Check The number of coins is $14 + 25 = 39$ and the value of the coins is
$\$.01(14) + \$.10(25) = \$2.64$, as required.

11. Let $x =$ the number of \$10 coins. Then $41 - x$ is the number of \$20 coins.

Denomination	Number of Coins	Value
10	x	$10x$
20	$41 - x$	$20(41 - x)$
Totals	41	540

The sum of the values must equal the total value.

$$10x + 20(41 - x) = 540$$
$$10x + 820 - 20x = 540$$
$$-10x = -280$$
$$x = 28$$

He has 28 \$10 coins and $41 - 28 = 13$ \$20 coins.

Check The number of coins is $28 + 13 = 41$ and the value of the coins is
$\$10(28) + \$20(13) = \$540$, as required.

12. Let $x =$ the number of two-cent pieces. Then $3x$ is the number of three-cent pieces.

Denomination	Number of Coins	Value
.02	x	$.02x$
.03	$3x$	$.03(3x)$
Total		2.42

The sum of the values must equal the total value.

$$.02x + .03(3x) = 2.42$$
$$2x + 3(3x) = 242 \quad \textit{Multiply by 100.}$$
$$2x + 9x = 242$$
$$11x = 242$$
$$x = 22$$

She has 22 two-cent pieces and $3(22) = 66$ three-cent pieces.

Check The number of three-cent pieces, 66, is three times the number of two-cent pieces, 22. The value of the coins is
$\$.02(22) + \$.03(66) = \$2.42$, as required.

13. Let $x =$ the number of floor tickets. Then $550 - x =$ the number of balcony tickets.

Cost of Ticket	Number Sold	Amount Collected
\$40	x	$40x$
\$28	$550 - x$	$28(550 - x)$
Totals	550	\$20,800

$$40x + 28(550 - x) = 20{,}800$$
$$40x + 15{,}400 - 28x = 20{,}800$$
$$12x = 5400$$
$$x = 450$$

450 floor tickets and $550 - 450 = 100$ balcony tickets were sold.

Check The number of tickets sold was $450 + 100 = 550$. The amount collected was $\$40(450) + \$28(100) = \$20{,}800$, as required.

14. Let $x =$ the number of student tickets sold. Then $410 - x =$ the number of nonstudent tickets sold.

Cost of Ticket	Number Sold	Amount Collected
\$3	x	$3x$
\$7	$410 - x$	$7(410 - x)$
Totals	410	\$1650

Write the equation from the last column of the table.

$$\begin{aligned} 3x + 7(410 - x) &= 1650 \\ 3x + 2870 - 7x &= 1650 \\ -4x &= -1220 \\ x &= 305 \end{aligned}$$

305 student tickets were sold; $410 - 305 = 105$ nonstudent tickets were sold.

Check The number of tickets sold was $305 + 105 = 410$. The amount collected was $\$3(305) + \$7(105) = \$1650$, as required.

15. $d = rt$, so

$$r = \frac{d}{t} = \frac{100}{12.65} \approx 7.91$$

Her rate was about 7.91 m per sec.

16. $d = rt$, so

$$r = \frac{d}{t} = \frac{400}{53.02} \approx 7.54$$

Her rate was about 7.54 m per sec.

17. $d = rt$, so

$$r = \frac{d}{t} = \frac{400}{47.50} \approx 8.42$$

His rate was about 8.42 m per sec.

18. $d = rt$, so

$$r = \frac{d}{t} = \frac{400}{43.84} \approx 9.12$$

His rate was about 9.12 m per sec.

19. Let $t =$ the time until they are 110 mi apart. Make a table. Use the formula $d = rt$, that is, find each distance by multiplying rate by time.

	Rate	Time	Distance
First Steamer	22	t	$22t$
Second Steamer	22	t	$22t$
Total			110

The total distance traveled is the sum of the distances traveled by each steamer, since they are traveling in opposite directions. This total is 110 mi.

$$\begin{aligned} 22t + 22t &= 110 \\ 44t &= 110 \\ t &= \tfrac{110}{44} = \tfrac{5}{2} \text{ or } 2\tfrac{1}{2} \end{aligned}$$

It will take them $2\frac{1}{2}$ hr.

Check Each steamer traveled $22(2.5) = 55$ miles for a total of $2(55) = 110$ miles, as required.

20. Let $t =$ the time it takes for the trains to be 315 km apart.
Make a table. Use the formula $d = rt$, that is, find each distance by multiplying rate by time.

	Rate	Time	Distance
First train	85	t	$85t$
Second train	95	t	$95t$
Total			315

The total distance traveled is the sum of the distances traveled by each train, since they are traveling in opposite directions. This total is 315 km.

$$\begin{aligned} 85t + 95t &= 315 \\ 180t &= 315 \\ t &= \tfrac{315}{180} = \tfrac{7}{4} \text{ or } 1\tfrac{3}{4} \end{aligned}$$

It will take the trains $1\frac{3}{4}$ hr before they are 315 km apart.

Check The first train traveled $85(1.75) = 148.75$ km and the second train traveled $95(1.75) = 166.25$ km. The sum is 315 km, as required.

21. Let $t =$ Mulder's time.
Then $t - \frac{1}{2} =$ Scully's time.

	Rate	Time	Distance
Mulder	65	t	$65t$
Scully	68	$t - \frac{1}{2}$	$68\left(t - \frac{1}{2}\right)$

The distances are equal.

$$\begin{aligned} 65t &= 68\left(t - \tfrac{1}{2}\right) \\ 65t &= 68t - 34 \\ -3t &= -34 \\ t &= \tfrac{34}{3} \quad \text{or} \quad 11\tfrac{1}{3} \end{aligned}$$

Mulder's time will be $11\frac{1}{3}$ hr. Since he left at 8:30 A.M., $11\frac{1}{3}$ hr or 11 hr 20 min later is 7:50 P.M.

Check Mulder's distance was $65\left(\frac{34}{3}\right) = 736\frac{2}{3}$ miles. Scully's distance was $68\left(\frac{34}{3} - \frac{1}{2}\right) = 68\left(\frac{65}{6}\right) = 736\frac{2}{3}$, as required.

22. Let $x =$ Lois' travel time.
Since Clark leaves 15 minutes after Lois, and $\frac{15}{60} = \frac{1}{4}$ hr, $x - \frac{1}{4} =$ time for Clark.
Complete the table using the formula $rt = d$.

	Rate	Time	Distance
Lois	35	x	$35x$
Clark	40	$x - \frac{1}{4}$	$40\left(x - \frac{1}{4}\right)$

Since Lois and Clark are going in opposite directions, we add their distances to get 140 mi.

$$\begin{aligned} 35x + 40\left(x - \tfrac{1}{4}\right) &= 140 \\ 35x + 40x - 10 &= 140 \\ 75x &= 150 \\ x &= 2 \end{aligned}$$

Lois' time will be 2 hours. They will be 140 mi apart at 8 A.M. + 2 hr = 10 A.M.

Check Lois' distance was $35(2) = 70$. Clark's distance was $40\left(2 - \frac{1}{4}\right) = 40\left(\frac{7}{4}\right) = 70$. The sum is 140 miles, as required.

23. Let $x =$ Latrella's speed when she rides. Then $x - 10 =$ her speed walking.

	Rate	Time	Distance
Riding	x	$\frac{15}{60} = \frac{1}{4}$ hr	$\frac{1}{4}x$
Walking	$x - 10$	$\frac{45}{60} = \frac{3}{4}$ hr	$\frac{3}{4}(x - 10)$

The distances are equal.

$$\begin{aligned} \tfrac{1}{4}x &= \tfrac{3}{4}(x - 10) \\ x &= 3(x - 10) && \textit{Multiply by 4.} \\ x &= 3x - 30 \\ -2x &= -30 \\ x &= 15 && \textit{Divide by } -2. \end{aligned}$$

Her speed when she rides is 15 mph.

Check The distance she travels to school when riding is $\frac{1}{4}(15) = 3.75$ miles. The distance when walking is $\frac{3}{4}(15 - 10) = 3.75$ miles, as required.

24. Let $x =$ Dewayne's speed when he drives his car. Use the formula $d = rt$ with 30 min $= \frac{1}{2}$ hr and 45 min $= \frac{3}{4}$ hr.

	Rate	Time	Distance
Car	x	$\frac{1}{2}$	$\frac{1}{2}x$
Bus	$x - 12$	$\frac{3}{4}$	$\frac{3}{4}(x - 12)$

The distance by car and the distance by bus are the same.

$$\begin{aligned} \tfrac{1}{2}x &= \tfrac{3}{4}(x - 12) \\ 2x &= 3(x - 12) && \textit{Multiply by 4.} \\ 2x &= 3x - 36 \\ 36 &= x \end{aligned}$$

Dewayne's speed by car is 36 mph. The distance he travels to work by car is $\frac{1}{2}(36) = 18$ mi.

Check The distance he travels to work by bus is $\frac{3}{4}(36 - 12) = \frac{3}{4}(24) = 18$ miles, as required.

25. Let $x =$ Anne's time. Then $x + \frac{1}{2} =$ Johnny's time.

	Rate	Time	Distance
Anne	60	x	$60x$
Johnny	50	$x + \frac{1}{2}$	$50\left(x + \frac{1}{2}\right)$

The total distance is 80.

$$\begin{aligned} 60x + 50\left(x + \tfrac{1}{2}\right) &= 80 \\ 60x + 50x + 25 &= 80 \\ 110x &= 55 \\ x &= \tfrac{55}{110} = \tfrac{1}{2} \end{aligned}$$

They will meet $\frac{1}{2}$ hr after Anne leaves.

Check Anne travels $60\left(\frac{1}{2}\right) = 30$ miles. Johnny travels $50\left(\frac{1}{2} + \frac{1}{2}\right) = 50$ miles. The sum of the distances is 80 miles, as required.

26. Let x = Aimee's rate (speed) during the first part of the trip. Then $x - 25$ = her rate during rush hour traffic. Make a table using the formula $rt = d$.

	Rate	Time	Distance
First Part	x	2	$2x$
Rush Hour	$x - 25$	$\frac{1}{2}$	$\frac{1}{2}(x-25)$

The total distance was 125 miles.

$$\begin{aligned} 2x + \tfrac{1}{2}(x-25) &= 125 \\ 4x + x - 25 &= 250 \quad \textit{Multiply by 2.} \\ 5x &= 275 \\ x &= 55 \end{aligned}$$

The speed during the first part of the trip was 55 mph.

Check The distance traveled during the first part of the trip was $55(2) = 110$ miles. The distance traveled during the second part of the trip was $(55 - 25)(.5) = 15$ miles. The sum of the distances is 125 miles, as required.

27. The sum of the measures of the three angles of a triangle is 180°.

$$\begin{aligned} (x-30) + (2x-120) + \left(\tfrac{1}{2}x + 15\right) &= 180 \\ \tfrac{7}{2}x - 135 &= 180 \\ 7x - 270 &= 360 \\ &\textit{Multiply by 2.} \\ 7x &= 630 \\ x &= 90 \end{aligned}$$

With $x = 90$, the three angle measures become

$$\begin{aligned} (90-30)^\circ &= 60^\circ, \\ (2\cdot 90 - 120)^\circ &= 60^\circ, \\ \text{and} \quad \left(\tfrac{1}{2}\cdot 90 + 15\right)^\circ &= 60^\circ. \end{aligned}$$

28. The sum of the measures of the three angles of a triangle is 180°.

$$\begin{aligned} (x+15) + (10x-20) + (x+5) &= 180 \\ 12x &= 180 \\ x &= 15 \end{aligned}$$

With $x = 15$, the three angle measures become

$$\begin{aligned} (15+15)^\circ &= 30^\circ, \\ (10\cdot 15 - 20)^\circ &= 130^\circ, \\ \text{and} \quad (15+5)^\circ &= 20^\circ. \end{aligned}$$

29. The sum of the measures of the three angles of a triangle is 180°.

$$\begin{aligned} (3x+7) + (9x-4) + (4x+1) &= 180 \\ 16x + 4 &= 180 \\ 16x &= 176 \\ x &= 11 \end{aligned}$$

With $x = 11$, the three angle measures become

$$\begin{aligned} (3\cdot 11 + 7)^\circ &= 40^\circ, \\ (9\cdot 11 - 4)^\circ &= 95^\circ, \\ \text{and} \quad (4\cdot 11 + 1)^\circ &= 45^\circ. \end{aligned}$$

30. The sum of the measures of the three angles of a triangle is 180°.

$$\begin{aligned} (2x+7) + (x+61) + x &= 180 \\ 4x + 68 &= 180 \\ 4x &= 112 \\ x &= 28 \end{aligned}$$

With $x = 28$, the three angle measures become

$$\begin{aligned} (2\cdot 28 + 7)^\circ &= 63^\circ, \\ (28+61)^\circ &= 89^\circ, \quad \text{and} \quad 28^\circ. \end{aligned}$$

31. The sum of the measures of the angles of a triangle is 180°.

$$\begin{aligned} x + 2x + 60 &= 180 \\ 3x + 60 &= 180 \\ 3x &= 120 \\ x &= 40 \end{aligned}$$

The measures of the unknown angles are 40° and $2x = 80^\circ$.

32. The sum of the measures of the marked angles, $60^\circ + y^\circ$, must equal 180°. Thus, the measure of the unknown angle is 120°.

33. The sum of the measures of the unknown angles in Exercise 31 is $40^\circ + 80^\circ = 120^\circ$. This is equal to the measure of the angle in Exercise 32.

34. The sum of the measures of angles ① and ② is equal to the measure of angle ③.

35. Vertical angles have equal measure.

$$\begin{aligned} 8x + 2 &= 7x + 17 \\ x &= 15 \end{aligned}$$

$$8\cdot 15 + 2 = 122 \quad \text{and} \quad 7\cdot 15 + 17 = 122.$$

The angles are both 122°.

36. Vertical angles have equal measure.

$$\begin{aligned} 9 - 5x &= 25 - 3x \\ 9 &= 25 + 2x \\ -16 &= 2x \\ -8 &= x \end{aligned}$$

$9 - 5(-8) = 49$ and $25 - 3(-8) = 49$.

The angles are both $49°$.

37. The sum of the two angles is $90°$.

$$\begin{aligned} (5x - 1) + 2x &= 90 \\ 7x - 1 &= 90 \\ 7x &= 91 \\ x &= 13 \end{aligned}$$

The angles are $(5 \cdot 13 - 1)° = 64°$ and $(2 \cdot 13)° = 26°$.

38. Supplementary angles have an angle measure sum of $180°$.

$$\begin{aligned} (3x + 5) + (5x + 15) &= 180 \\ 8x + 20 &= 180 \\ 8x &= 160 \\ x &= 20 \end{aligned}$$

With $x = 20$, the two angle measures become

$$(3 \cdot 20 + 5)° = 65°$$
$$\text{and} \quad (5 \cdot 20 + 15)° = 115°.$$

39. Let $x =$ the first consecutive integer. Then $x + 1$ will be the second consecutive integer, and $x + 2$ will be the third consecutive integer.

The sum of the first and twice the second is 17 more than twice the third.

$$\begin{aligned} x + 2(x + 1) &= 2(x + 2) + 17 \\ x + 2x + 2 &= 2x + 4 + 17 \\ 3x + 2 &= 2x + 21 \\ x &= 19 \end{aligned}$$

Since $x = 19$, $x + 1 = 20$, and $x + 2 = 21$. The three consecutive integers are 19, 20, and 21.

40. Let $x =$ the first integer. Then $x + 1$, $x + 2$, and $x + 3$ are the next three consecutive integers. The sum of the first three integers is 54 more than the fourth.

$$\begin{aligned} x + (x + 1) + (x + 2) &= (x + 3) + 54 \\ 3x + 3 &= x + 57 \\ 2x &= 54 \\ x &= 27 \end{aligned}$$

The four consecutive integers are 27, 28, 29, and 30.

41. Let $x =$ the current age. Then $x + 1$ will be the age next year.
The sum of these ages will be 103 years.

$$\begin{aligned} x + (x + 1) &= 103 \\ 2x + 1 &= 103 \\ 2x &= 102 \\ x &= 51 \end{aligned}$$

If my current age is 51, in 10 years I will be

$$51 + 10 = 61 \text{ years old.}$$

42. Let $x =$ the page number on one page. Then $x + 1$ is the page number on the next page.
The sum of the page numbers is 189.

$$\begin{aligned} x + (x + 1) &= 189 \\ 2x + 1 &= 189 \\ 2x &= 188 \\ x &= 94 \end{aligned}$$

The page numbers are 94 and 95.

Summary Exercises on Solving Applied Problems

1. Let $x =$ the width of the rectangle. Then $x + 3$ is the length of the rectangle.

If the length were decreased by 2 inches and the width were increased by 1 inch, the perimeter would be 24 inches. Use the formula $P = 2L + 2W$, and substitute 24 for P, $(x + 3) - 2$ or $x + 1$ for L, and $x + 1$ for W.

$$\begin{aligned} P &= 2L + 2W \\ 24 &= 2(x + 1) + 2(x + 1) \\ 24 &= 2x + 2 + 2x + 2 \\ 24 &= 4x + 4 \\ 20 &= 4x \\ 5 &= x \end{aligned}$$

The width of the rectangle is 5 inches, and the length is $5 + 3 = 8$ inches.

2. Let $x =$ the width. Then $2x$ is the length (twice the width).
Use $P = 2L + 2W$ and add one more width to cut the area into two parts. Thus, an equation is

$$\begin{aligned} 2L + 2W + W &= 210. \\ 2(2x) + 2x + x &= 210 \\ 7x &= 210 \\ x &= 30 \end{aligned}$$

The width is 30 meters, and the length is $2(30) = 60$ meters.

3. Let $x=$ the regular price of the item.
The sale price after a 32% (or .32) discount was \$21.05, so an equation is

$$\begin{aligned} x - .32x &= 21.05. \\ .68x &= 21.05 \\ x &\approx 30.96 \end{aligned}$$

To the nearest cent, the regular price was \$30.96.

4. Let $x=$ the regular price of the videodisc player.
The sale price after a discount of 40% (or .40) was \$255, so an equation is

$$\begin{aligned} x - .40x &= 255. \\ .60x &= 255 \\ x &= 425 \end{aligned}$$

The regular price of the videodisc player was \$425.

5. Let $x=$ the amount invested at 4%.
Then $2x$ is the amount invested at 5%.
Use $I=prt$ with $t=1$ yr. Make a table.

Rate (as a Decimal)	Principal	Interest
.04	x	$.04x$
.05	$2x$	$.05(2x)=.10x$
Total		112

The last column gives the equation.

$$\begin{array}{ccccc} \text{Interest at 4\%} & + & \text{interest at 5\%} & = & \text{total interest.} \\ .04x & + & .10x & = & 112 \end{array}$$

$$\begin{aligned} 4x + 10x &= 11{,}200 \quad \textit{Multiply by 100.} \\ 14x &= 11{,}200 \\ x &= 800 \end{aligned}$$

\$800 is invested at 4% and 2(\$800) = \$1600 at 5%.

Check \$800 @ 4% = \$32 and
\$1600 @ 5% = \$80; \$32 + \$80 = \$112

6. Let $x=$ the amount invested at 3%.
Then $x+2000$ is the amount invested at 4%.
Use $I=prt$ with $t=1$ yr. Make a table.

Rate (as a Decimal)	Principal	Interest
.03	x	$.03x$
.04	$x+2000$	$.04(x+2000)$
Total		920

The last column gives the equation.

$$\begin{array}{ccccc} \text{Interest at 3\%} & + & \text{interest at 4\%} & = & \text{total interest.} \\ .03x & + & .04(x+2000) & = & 920 \end{array}$$

$$\begin{aligned} 3x + 4(x+2000) &= 92{,}000 \quad \textit{Multiply by 100.} \\ 3x + 4x + 8000 &= 92{,}000 \\ 7x &= 84{,}000 \\ x &= 12{,}000 \end{aligned}$$

\$12,000 is invested at 3% and
\$12,000 + \$2000 = \$14,000 is invested at 4%.

Check \$12,000 @ 3% = \$360 and
\$14,000 @ 4% = \$560; \$360 + \$560 = \$920

7. Let $x=$ the number of Emmy awards won by *The Simpsons.* Then $2x-9=$ the number of Emmys won by *Frasier.* The total number of Emmys won by both was 51.

$$\begin{aligned} x + (2x-9) &= 51 \\ 3x - 9 &= 51 \\ 3x &= 60 \\ x &= 20 \end{aligned}$$

The Simpsons won 20 Emmys and *Frasier* won $2(20)-9=$ 31 Emmys.

8. Let $x=$ the amount grossed by *Star Wars: Episode IV—A New Hope.* Then $x+139.8=$ the amount grossed by *Titanic* (in millions). Together they grossed \$1061.8 million.

$$\begin{aligned} x + (x+139.8) &= 1061.8 \\ 2x + 139.8 &= 1061.8 \\ 2x &= 922 \\ x &= 461 \end{aligned}$$

Star Wars: Episode IV—A New Hope grossed \$461 million and *Titanic* grossed
$461+139.8=$ \$600.8 million.

9. Let $x=$ Cheruiyot's time.
Then $x+.25=$ Zakharova's time.

	Rate	Time	Distance
Cheruiyot	11.98	x	$11.98x$
Zakharova	10.74	$x+.25$	$10.74(x+.25)$

The distances are equal.

$$\begin{aligned} 11.98x &= 10.74(x+.25) \\ 1198x &= 1074(x+.25) \quad \textit{Multiply by 100.} \\ 1198x &= 1074x + 268.5 \\ 124x &= 268.5 \\ x &\approx 2.17 \end{aligned}$$

Cheruiyot's time was about 2.17 hours and Zakharova's time was about $2.17+.25=2.42$ hours.

10. Use the formula for the volume of a box.

$$\begin{aligned} V &= LWH \\ 75 &= 5(1.5)H \\ 75 &= 7.5H \\ 10 &= H \end{aligned}$$

The height is 10 ft.

11. Let $x =$ the side length of the square cut out of each corner. Then the width is $12 - 2x$ and the length is $16 - 2x$. We want the length to be 5 cm less than twice the width.

$$\begin{aligned} \text{length} &= 2(\text{width}) - 5 \\ 16 - 2x &= 2(12 - 2x) - 5 \\ 16 - 2x &= 24 - 4x - 5 \\ 16 - 2x &= 19 - 4x \\ 2x &= 3 \\ x &= \tfrac{3}{2} \text{ or } 1\tfrac{1}{2} \end{aligned}$$

The square should be $1\frac{1}{2}$ cm on each side.

Check The width is $12 - 2\left(\frac{3}{2}\right) = 9$ and the length is $16 - 2\left(\frac{3}{2}\right) = 13$. Two times the width is $2(9) = 18$, which is 5 more than the length, 13.

12. Let $x =$ the length of the shortest side. Then $2x$ is the length of the middle side and $3x - 2$ is the length of the longest side.

The perimeter is 34 inches. Using $P = a + b + c$ gives us

$$\begin{aligned} x + 2x + (3x - 2) &= 34. \\ 6x - 2 &= 34 \\ 6x &= 36 \\ x &= 6 \end{aligned}$$

The lengths of the three sides are 6 inches, $2(6) = 12$ inches, and $3(6) - 2 = 16$ inches.

Check The sum of the lengths of the three sides is $6 + 12 + 16 = 34$ inches, as required.

13. Let $x =$ the cost of one year at the community college. Then $x + 2367 =$ the cost of one year at the public university. The total cost for two years at the community college and two years at the public university is \$10,250.

$$\begin{aligned} 2x + 2(x + 2367) &= 10{,}250 \\ 2x + 2x + 4734 &= 10{,}250 \\ 4x &= 5516 \\ x &= 1379 \end{aligned}$$

The cost of one year at the community college is \$1379 and the cost of one year at the public university is $1379 + 2367 = \$3746$.

Check \$3746 is \$2367 more than \$1379. Two years at the community college would cost $2(\$1379) = \$2758$ and two years at the public university would cost $2(\$3746) = \$7492$, so the total is $\$2758 + \$7492 = \$10{,}250$.

14. Let $x =$ the smallest integer. Then $x + 1$ is the middle integer and $x + 2$ is the largest integer.

"The sum of the smallest and largest of three consecutive integers is 32 more than the middle integer" translates to

$$\begin{aligned} x + (x + 2) &= 32 + (x + 1). \\ 2x + 2 &= x + 33 \\ x &= 31 \end{aligned}$$

The three consecutive integers are 31, 32, and 33.

Check The sum of the smallest and largest integers is $31 + 33 = 64$, which is the same as 32 more than the middle integer.

15. The sum of the measures of the three angles of a triangle is 180°.

$$\begin{aligned} x + (6x - 50) + (x - 10) &= 180 \\ 8x - 60 &= 180 \\ 8x &= 240 \\ x &= 30 \end{aligned}$$

With $x = 30$, the three angle measures become

$$(6 \cdot 30 - 50)^\circ = 130^\circ, \quad (30 - 10)^\circ = 20^\circ, \text{ and } 30^\circ.$$

Chapter 2 Review Exercises

1.

$$\begin{aligned} -(8 + 3x) + 5 &= 2x + 6 \\ -8 - 3x + 5 &= 2x + 6 \\ -3x - 3 &= 2x + 6 \\ -5x &= 9 \\ x &= -\tfrac{9}{5} \end{aligned}$$

Solution set: $\left\{-\frac{9}{5}\right\}$

2.

$$\begin{aligned} -(r + 5) - (2 + 7r) + 8r &= 3r - 8 \\ -r - 5 - 2 - 7r + 8r &= 3r - 8 \\ -7 &= 3r - 8 \\ 1 &= 3r \\ \tfrac{1}{3} &= r \end{aligned}$$

Solution set: $\left\{\frac{1}{3}\right\}$

3. $$\frac{m-2}{4}+\frac{m+2}{2}=8$$

Multiply each side by the LCD, 4.

$$4\left(\frac{m-2}{4}+\frac{m+2}{2}\right)=4(8)$$
$$(m-2)+2(m+2)=32$$
$$m-2+2m+4=32$$
$$3m+2=32$$
$$3m=30$$
$$m=10$$

Solution set: $\{10\}$

4. $$\frac{2q+1}{3}-\frac{q-1}{4}=0$$
$$4(2q+1)-3(q-1)=0 \quad \textit{Multiply by 12.}$$
$$8q+4-3q+3=0$$
$$5q+7=0$$
$$5q=-7$$
$$q=-\tfrac{7}{5}$$

Solution set: $\{-\frac{7}{5}\}$

5. $$5(2x-3)=6(x-1)+4x$$
$$10x-15=6x-6+4x$$
$$10x-15=10x-6$$
$$-15=-6 \quad \textit{False}$$

This is a false statement, so the equation is a *contradiction.*

Solution set: $\emptyset$

6. $$-3x+2(4x+5)=10$$
$$-3x+8x+10=10$$
$$5x+10=10$$
$$5x=0$$
$$x=0 \quad \textit{Divide by 5.}$$

Solution set: $\{0\}$

7. $$-\frac{3}{4}x=-12$$
$$-3x=-48 \quad \textit{Multiply by 4.}$$
$$x=16$$

Solution set: $\{16\}$

8. $$.05x+.03(1200-x)=42$$

Multiply by 100 to clear all decimals.

$$5x+3(1200-x)=4200$$
$$5x+3600-3x=4200$$
$$2x+3600=4200$$
$$2x=600$$
$$x=300$$

Solution set: $\{300\}$

9. Solve each equation.

A. $x-5=5$
$$x=10 \quad \textit{Add 5.}$$

Solution set: $\{10\}$

B. $4x=5x$
$$0=x \quad \textit{Subtract 4x.}$$

Solution set: $\{0\}$

C. $x+3=-3$
$$x=-6 \quad \textit{Subtract 3.}$$

Solution set: $\{-6\}$

D. $6x-6=6$
$$6x=12 \quad \textit{Add 6.}$$
$$x=2 \quad \textit{Divide by 6.}$$

Solution set: $\{2\}$

Equation **B** has $\{0\}$ as its solution set.

10. Solve $-2x+5=7$.

Begin by subtracting 5 from each side. Then divide each side by -2.

11. $$7r-3(2r-5)+5+3r=4r+20$$
$$7r-6r+15+5+3r=4r+20$$
$$4r+20=4r+20$$
$$20=20 \quad \textit{True}$$

This equation is an *identity.*

Solution set: {all real numbers}

12. $$8p-4p-(p-7)+9p+13=12p$$
$$8p-4p-p+7+9p+13=12p$$
$$12p+20=12p$$
$$20=0 \quad \textit{False}$$

This equation is a *contradiction.*

Solution set: $\emptyset$

13. $$-2r+6(r-1)+3r-(4-r)=-(r+5)-5$$
$$-2r+6r-6+3r-4+r=-r-5-5$$
$$8r-10=-r-10$$
$$9r=0$$
$$r=0$$

This equation is a *conditional* equation.

Solution set: $\{0\}$

14. Solve $V=LWH$ for L.

$$\frac{V}{WH}=\frac{LWH}{WH} \quad \textit{Divide by WH.}$$
$$\frac{V}{WH}=L \quad \text{or} \quad L=\frac{V}{WH}$$

15. Solve $A = \frac{1}{2}h(b+B)$ for b.

$$2A = h(b+B) \quad \text{or} \quad 2A = hb + hB$$
$$\frac{2A}{h} = b + B \quad \text{or} \quad 2A - hB = hb$$
$$\frac{2A}{h} - B = b \quad \text{or} \quad \frac{2A - hB}{h} = b$$

16. Solve $C = \pi d$ for d.

$$\frac{C}{\pi} = \frac{\pi d}{\pi} \quad \textit{Divide by } \pi.$$
$$\frac{C}{\pi} = d \quad \text{or} \quad d = \frac{C}{\pi}$$

17. Use the formula $V = LWH$ and substitute 180 for V, 6 for L, and 5 for W.

$$180 = 6(5)H$$
$$180 = 30H$$
$$6 = H$$

The height is 6 feet.

18. Divide the amount of decrease by the original amount.

$$\frac{45{,}104 - 40{,}758}{45{,}104} = \frac{4346}{45{,}104} \approx .096$$

The percent decrease is about 9.6%.

19. Use the formula $I = prt$, and solve for r.

$$\frac{I}{pt} = \frac{prt}{pt}$$
$$\frac{I}{pt} = r$$

Substitute 30,000 for p, 7800 for I, and 4 for t.

$$r = \frac{7800}{30{,}000(4)} = \frac{7800}{120{,}000} = .065$$

The rate is 6.5%.

20. Use the formula $C = \frac{5}{9}(F - 32)$ and substitute 77 for F.

$$C = \tfrac{5}{9}(77 - 32)$$
$$= \tfrac{5}{9}(45) = 25$$

The Celsius temperature is 25°.

21. The projected number of people in the Hispanic category is approximately

$$.117(148{,}847{,}000) = 17{,}415{,}099.$$

22.
$$C = 2\pi r$$
$$200\pi = 2\pi r \quad \textit{Substitute } 200\pi \textit{ for } C.$$
$$\frac{200\pi}{2\pi} = r \quad \textit{Divide by } 2\pi.$$
$$100 = r$$

The radius is 100 mm.

23. "One-third of a number, subtracted from 9" is written

$$9 - \tfrac{1}{3}x.$$

24. "The product of 4 and a number, divided by 9 more than the number" is written

$$\frac{4x}{x+9}.$$

25. Let $x =$ the width of the rectangle. Then $2x - 3 =$ the length of the rectangle.

Use the formula $P = 2L + 2W$ with $P = 42$.

$$42 = 2(2x - 3) + 2x$$
$$42 = 4x - 6 + 2x$$
$$48 = 6x$$
$$8 = x$$

The width is 8 meters and the length is $2(8) - 3 = 13$ meters.

26. Let $x =$ the length of each equal side. Then $2x - 15 =$ the length of the third side.

Use the formula $P = a + b + c$ with $P = 53$.

$$53 = x + x + (2x - 15)$$
$$53 = 4x - 15$$
$$68 = 4x$$
$$17 = x$$

The lengths of the three sides are 17 inches, 17 inches, and $2(17) - 15 = 19$ inches.

27. Let $x =$ the number of kilograms of peanut clusters. Then $3x$ is the number of kilograms of chocolate creams.
The clerk has a total of 48 kg.

$$x + 3x = 48$$
$$4x = 48$$
$$x = 12$$

The clerk has 12 kilograms of peanut clusters.

28. Let $x =$ the number of liters of the 20% solution. Make a table.

Percent (as a Decimal)	Liters of Solution	Liters of Pure Chemical
.20	x	$.20x$
.50	15	$.50(15) = 7.5$
.30	$x + 15$	$.30(x + 15)$

The last column gives the equation.

$$.20x + 7.5 = .30(x + 15)$$
$$.20x + 7.5 = .30x + 4.5$$
$$3 = .10x$$
$$30 = x$$

30 L of the 20% solution should be used.

29. Let $x =$ the number of liters of water.

Percent (as a Decimal)	Liters of Solution	Liters of Pure Acid
.40	30	$.40(30) = 12$
.00	x	$.00(x) = 0$
.30	$30 + x$	$.30(30 + x)$

The last column gives the equation.

$$\begin{aligned} 12 + 0 &= .30(30 + x) \\ 12 &= 9 + .3x \\ 3 &= .3x \\ 10 &= x \end{aligned}$$

10 L of water should be added.

30. Let $x =$ the amount invested at 6%. Then $x - 4000 =$ the amount invested at 4%.

Rate (as a Decimal)	Principal	Interest
.06	x	$.06x$
.04	$x - 4000$	$.04(x - 4000)$
Total		\$840

The last column gives the equation.

$$\begin{aligned} .06x + .04(x - 4000) &= 840 \\ 6x + 4(x - 4000) &= 84{,}000 \quad \textit{Multiply by 100.} \\ 6x + 4x - 16{,}000 &= 84{,}000 \\ 10x &= 100{,}000 \\ x &= 10{,}000 \end{aligned}$$

Anna should invest \$10,000 at 6% and \$10,000 − \$4000 = \$6000 at 4%.

31. Use the formula $d = rt$ or $r = \frac{d}{t}$. Here, d is about 400 mi and t is about 8 hr. Since $\frac{400}{8} = 50$, the best estimate is choice **A**.

32. Use the formula $d = rt$.

(a) Here, $r = 53$ mph and $t = 10$ hr.

$$d = 53(10) = 530$$

The distance is 530 miles.

(b) Here, $r = 164$ mph and $t = 2$ hr.

$$d = 164(2) = 328$$

The distance is 328 miles.

33. Let $x =$ the time it takes for the trains to be 297 mi apart.

Make a table. Use the formula $d = rt$.

	Rate	Time	Distance
Passenger Train	60	x	$60x$
Freight Train	75	x	$75x$
Total			297

The total distance traveled is the sum of the distances traveled by each train.

$$\begin{aligned} 60x + 75x &= 297 \\ 135x &= 297 \\ x &= 2.2 \end{aligned}$$

It will take the trains 2.2 hours before they are 297 miles apart.

Check $2.2(60) + 2.2(75) = 297$

34. Let $x =$ the speed of the faster car and $x - 15 =$ the speed of the slower car. Make a table. Use the formula $d = rt$.

	Rate	Time	Distance
Faster Car	x	2	$2x$
Slower Car	$x - 15$	2	$2(x - 15)$
Total			230

The total distance traveled is the sum of the distances traveled by each car.

$$\begin{aligned} 2x + 2(x - 15) &= 230 \\ 2x + 2x - 30 &= 230 \\ 4x &= 260 \\ x &= 65 \end{aligned}$$

The faster car travels at 65 km per hr, while the slower car travels at $65 - 15 = 50$ km per hr.

Check $2(65) + 2(50) = 230$

35. Let $x =$ amount of time spent averaging 45 miles per hour. Then $4 - x =$ amount of time at 50 mph.

	Rate	Time	Distance
First Part	45	x	$45x$
Second Part	50	$4 - x$	$50(4 - x)$
Total			195

From the last column:

$$\begin{aligned} 45x + 50(4 - x) &= 195 \\ 45x + 200 - 50x &= 195 \\ -5x &= -5 \\ x &= 1 \end{aligned}$$

The automobile averaged 45 mph for 1 hour.

Check 45 mph for 1 hour = 45 miles and 50 mph for 3 hours = 150 miles; $45 + 150 = 195$.

36. Let $x =$ the average speed for the first hour. Then $x - 7 =$ the average speed for the second hour. Using $d = rt$, the distance traveled for the first hour is $x(1)$ miles, for the second hour is $(x - 7)(1)$ miles, and for the whole trip, 85 miles.

$$\begin{aligned} x + (x - 7) &= 85 \\ 2x - 7 &= 85 \\ 2x &= 92 \\ x &= 46 \end{aligned}$$

The average speed for the first hour was 46 mph.

Check 46 mph for 1 hour = 46 miles and $46 - 7 = 39$ mph for 1 hour = 39 miles; $46 + 39 = 85$.

37. **[2.1]**
$$\begin{aligned} (7 - 2k) + 3(5 - 3k) &= k + 8 \\ 7 - 2k + 15 - 9k &= k + 8 \\ -11k + 22 &= k + 8 \\ -12k + 22 &= 8 \\ -12k &= -14 \\ k &= \tfrac{-14}{-12} = \tfrac{7}{6} \end{aligned}$$

Solution set: $\left\{\frac{7}{6}\right\}$

38. **[2.1]**
$$\frac{4x + 2}{4} + \frac{3x - 1}{8} = \frac{x + 6}{16}$$

Clear fractions by multiplying by the LCD, 16.

$$\begin{aligned} 4(4x + 2) + 2(3x - 1) &= x + 6 \\ 16x + 8 + 6x - 2 &= x + 6 \\ 22x + 6 &= x + 6 \\ 21x &= 0 \\ x &= 0 \end{aligned}$$

Solution set: $\{0\}$

39. **[2.1]**
$$\begin{aligned} -5(6p + 4) - 2p &= -32p + 14 \\ -30p - 20 - 2p &= -32p + 14 \\ -32p - 20 &= -32p + 14 \\ -20 &= 14 \qquad \textit{False} \end{aligned}$$

The equation is a *contradiction.*

Solution set: $\emptyset$

40. **[2.2]** Let $x =$ the length of the shortest side. Then $2x =$ the length of the middle side, and $3x - 4 =$ the length of the longest side.

The perimeter is 68 inches. Add the lengths of the three sides.

$$\begin{aligned} x + 2x + (3x - 4) &= 68 \\ 6x - 4 &= 68 \\ 6x &= 72 \\ x &= 12 \end{aligned}$$

If $x = 12$, then $2x = 2(12) = 24$, and $3x - 4 = 3(12) - 4 = 32$.

The sides have lengths 12 in., 24 in., and 32 in.

41. **[2.2]** Solve $ak + bt = 6t - sk$ for k.

Get the "k-terms" on one side.

$$\begin{aligned} ak + sk &= 6t - bt \\ (a + s)k &= 6t - bt \\ k &= \frac{6t - bt}{a + s} \end{aligned}$$

If we take the k-terms to the right side in the first step, we get

$$k = \frac{bt - 6t}{-a - s}.$$

42. **[2.1]**
$$\begin{aligned} 5(2r - 3) + 7(2 - r) &= 3(r + 2) - 7 \\ 10r - 15 + 14 - 7r &= 3r + 6 - 7 \\ 3r - 1 &= 3r - 1 \\ 3r &= 3r \\ 0 &= 0 \qquad \textit{True} \end{aligned}$$

Solution set: {all real numbers}

43. **[2.2]** Let $x =$ the length of each side of the original square; $x + 4 =$ the length of each side of the enlarged square.

The original perimeter is $4x$. The perimeter of the enlarged square is $4(x + 4)$. The perimeter of the enlarged square is 8 in. less than twice the perimeter of the original square.

$$\begin{aligned} 4(x + 4) &= 2(4x) - 8 \\ 4x + 16 &= 8x - 8 \\ 16 &= 4x - 8 \\ 24 &= 4x \\ 6 &= x \end{aligned}$$

The length of a side of the original square is 6 in.

44. **[2.3]** Let $x =$ the number of electoral votes received by Bush. Then $x - 34 =$ the number of electoral votes received by Kerry. There were a total of 538 votes.

$$\begin{aligned} x + (x - 34) &= 538 \\ 2x - 34 &= 538 \\ 2x &= 572 \\ x &= 286 \end{aligned}$$

Bush received 286 votes and Kerry received $286 - 34 = 252$ votes.

45. **[2.4]** Let $x =$ the time traveled by eastbound car. Then $x - 1 =$ the time traveled by westbound car.

	Rate	Time	Distance
Eastbound Car	40	x	$40x$
Westbound Car	60	$x - 1$	$60(x - 1)$

Their total distance is 240 mi.

$$\begin{aligned} 40x + 60(x - 1) &= 240 \\ 40x + 60x - 60 &= 240 \\ 100x - 60 &= 240 \\ 100x &= 300 \\ x &= 3 \end{aligned}$$

The eastbound car traveled for 3 hr and the westbound car traveled for $3 - 1 = 2$ hr.

46. **[2.3]** Let $x =$ the amount invested at 4%. Then $x + 500 =$ the amount invested at 5%.

Rate (as a Decimal)	Principal	Interest
.04	x	$.04x$
.05	$x + 500$	$.05(x + 500)$

The total interest is $142.

$$\begin{aligned} .04x + .05(x + 500) &= 142 \\ .04x + .05x + 25 &= 142 \\ .09x + 25 &= 142 \\ .09x &= 117 \\ x &= 1300 \end{aligned}$$

$1300 was invested at 4% and $1300 + 500 = \$1800$ was invested at 5%.

47. **[2.1]** $.08x + .04(x + 200) = 188$

$$\begin{aligned} 8x + 4(x + 200) &= 18{,}800 \quad \textit{Multiply by 100.} \\ 8x + 4x + 800 &= 18{,}800 \\ 12x + 800 &= 18{,}800 \\ 12x &= 18{,}000 \\ x &= 1500 \end{aligned}$$

Solution set: $\{1500\}$

48. **[2.2]** $Ax + By = C$ for x

$$\begin{aligned} Ax &= C - By \quad \textit{Subtract By.} \\ x &= \frac{C - By}{A} \quad \textit{Divide by A.} \end{aligned}$$

Chapter 2 Test

1.

$$\begin{aligned} 3(2x - 2) - 4(x + 6) &= 4x + 8 \\ 6x - 6 - 4x - 24 &= 4x + 8 \\ 2x - 30 &= 4x + 8 \\ -2x - 30 &= 8 \\ -2x &= 38 \\ x &= -19 \end{aligned}$$

Solution set: $\{-19\}$

2.

$$\begin{aligned} .08x + .06(x + 9) &= 1.24 \\ 8x + 6(x + 9) &= 124 \quad \textit{Multiply by 100.} \\ 8x + 6x + 54 &= 124 \\ 14x + 54 &= 124 \\ 14x &= 70 \\ x &= 5 \end{aligned}$$

Solution set: $\{5\}$

3. $$\frac{x + 6}{10} + \frac{x - 4}{15} = 1$$

Multiply each side by the LCD, 30.

$$\begin{aligned} 3(x + 6) + 2(x - 4) &= 30 \\ 3x + 18 + 2x - 8 &= 30 \\ 5x + 10 &= 30 \\ 5x &= 20 \\ x &= 4 \end{aligned}$$

Solution set: $\{4\}$

4. **(a)**

$$\begin{aligned} 3x - (2 - x) + 4x + 2 &= 8x + 3 \\ 3x - 2 + x + 4x + 2 &= 8x + 3 \\ 8x &= 8x + 3 \\ 0 &= 3 \quad \textit{False} \end{aligned}$$

The false statement indicates that the equation is a *contradiction.*

Solution set: $\emptyset$

(b) $$\frac{x}{3} + 7 = \frac{5x}{6} - 2 - \frac{x}{2} + 9$$

Multiply each side by the LCD, 6.

$$\begin{aligned} 2x + 42 &= 5x - 12 - 3x + 54 \\ 2x + 42 &= 2x + 42 \\ 0 &= 0 \quad \textit{True} \end{aligned}$$

This equation is an *identity.*

Solution set: {all real numbers}

(c)
$$\begin{aligned} -4(2x-6) &= 5x+24-7x \\ -8x+24 &= -2x+24 \\ 24 &= 6x+24 \\ 0 &= 6x \\ 0 &= x \end{aligned}$$

This is a *conditional equation.*

Solution set: $\{0\}$

5. Solve $-16t^2+vt-S=0$ for v.

$$\begin{aligned} vt &= S+16t^2 \\ v &= \frac{S+16t^2}{t} \\ \text{or}\quad v &= \frac{S}{t}+16t \end{aligned}$$

6. Solve $ar+2=3r-6t$ for r.
Get the "r-terms" on one side.

$$\begin{aligned} ar-3r &= -2-6t \\ r(a-3) &= -2-6t \\ r &= \frac{-2-6t}{a-3} \end{aligned}$$

If we take the r-terms to the right side in the first step, we get

$$r=\frac{2+6t}{3-a}.$$

7. Solve $d=rt$ for t and substitute 500 for d and 142.971 for r.

$$t=\frac{d}{r}=\frac{500}{142.971}\approx 3.497$$

Burton's time was about 3.497 hr.

8. Use $I=Prt$ and substitute \$2281.25 for I, \$36,500 for P, and 1 for t.

$$\begin{aligned} 2281.25 &= 36{,}500r(1) \\ r &= \frac{2281.25}{36{,}500}=.0625 \end{aligned}$$

The rate of interest is 6.25%.

9. $\dfrac{27{,}876}{38{,}123}\approx .731$

73.1% were classified as post offices.

10. Let $x=$ the amount invested at 3%. Then $28{,}000-x=$ the amount invested at 5%.

Rate (as a Decimal)	Principal	Interest
.03	x	$.03x$
.05	$28{,}000-x$	$.05(28{,}000-x)$
Totals	\$28,000	\$1240

From the last column:

$$\begin{aligned} .03x+.05(28{,}000-x) &= 1240 \\ 3x+5(28{,}000-x) &= 124{,}000 \\ &\quad \textit{Multiply by 100.} \\ 3x+140{,}000-5x &= 124{,}000 \\ -2x &= -16{,}000 \\ x &= 8000 \end{aligned}$$

He invested \$8000 at 3% and \$28,000 − \$8000 = \$20,000 at 5%.

11. Let $x=$ the speed of the faster car. Then $x-15=$ the speed of the slower car.

Make a table. Use the formula $d=rt$.

	Rate	Time	Distance
Slower Car	$x-15$	6	$6(x-15)$
Faster Car	x	6	$6x$
Total			630

The total distance traveled is the sum of the distances traveled by each car.

$$\begin{aligned} 6(x-15)+6x &= 630 \\ 6x-90+6x &= 630 \\ 12x &= 720 \\ x &= 60 \end{aligned}$$

The faster car traveled at 60 mph, while the slower car traveled at $60-15=45$ mph.

12. The sum of the three angle measures is 180°.

$$\begin{aligned} (2x+20)+x+x &= 180 \\ 4x+20 &= 180 \\ 4x &= 160 \\ x &= 40 \end{aligned}$$

The three angle measures are 40°, 40°, and $(2\cdot 40+20)^\circ=100^\circ$.

13. $A = \dfrac{24f}{b(p+1)}$

$A = \dfrac{24(200)}{1920(24+1)}$ *Let f = 200, b = 1920, p = 24.*

$= \dfrac{4800}{48{,}000}$

$= .1$

The approximate annual interest rate is 10%.

14. $A = \dfrac{24f}{b(p+1)}$

$A = \dfrac{24(740)}{3600(36+1)}$ *Let f = 740, b = 3600, p = 36.*

$= \dfrac{17{,}760}{133{,}200}$

$\approx .1333$

The approximate annual interest rate is 13.33%.

15. 21% of 5000 $= .21(5000) = 1050$

We would expect 1050 white-collar workers in a group of 5000 stockholders.

Cumulative Review Exercises (Chapters 1–2)

Exercises 1–6 refer to set A.

Let $A = \left\{-8, -\frac{2}{3}, -\sqrt{6}, 0, \frac{4}{5}, 9, \sqrt{36}\right\}$.

Simplify $\sqrt{36} = 6$.

1. The elements 9 and 6 are natural numbers.

2. The elements 0, 9, and 6 are whole numbers.

3. The elements -8, 0, 9, and 6 are integers.

4. The elements $-8, -\frac{2}{3}, 0, \frac{4}{5}, 9,$ and 6 are rational numbers.

5. The element $-\sqrt{6}$ is an irrational number.

6. All the elements in set A are real numbers.

7. $-\dfrac{4}{3} - \left(-\dfrac{2}{7}\right) = -\dfrac{4}{3} + \dfrac{2}{7}$

$= -\dfrac{28}{21} + \dfrac{6}{21}$

$= -\dfrac{22}{21}$

8. $|-4.2| + |5.6| - |-1.9| = 4.2 + 5.6 - 1.9$

$= 9.8 - 1.9$

$= 7.9$

9. $(-2)^4 + (-2)^3 = 16 + (-8) = 8$

10. $\sqrt{25} - \dfrac{\sqrt{100}}{2} = 5 - \dfrac{10}{2}$

$= 5 - 5$

$= 0$

11. $(-3)^5 = (-3)(-3)(-3)(-3)(-3) = -243$

12. $\left(\dfrac{6}{7}\right)^3 = \dfrac{6}{7} \cdot \dfrac{6}{7} \cdot \dfrac{6}{7} = \dfrac{216}{343}$

13. $4^6 = 4 \cdot 4 \cdot 4 \cdot 4 \cdot 4 \cdot 4 = 4096$

14. $-4^6 = -(4 \cdot 4 \cdot 4 \cdot 4 \cdot 4 \cdot 4) = -4096$

15. $-\sqrt{36} = -(6) = -6$

$\sqrt{-36}$ is *not a real number*.

16. $\dfrac{4-4}{4+4} = \dfrac{0}{8} = 0$

$\dfrac{4+4}{4-4} = \dfrac{8}{0}$, which is *undefined*.

For Exercises 17–20, let $a = 2$, $b = -3$, and $c = 4$.

17. $-3a + 2b - c = -3(2) + 2(-3) - 4$

$= -6 - 6 - 4$

$= -16$

18. $-2b^2 - c^2 = -2(-3)^2 - 4^2$

$= -2(9) - 16$

$= -18 - 16$

$= -34$

19. $-8(a^2 + b^3) = -8[2^2 + (-3)^3]$

$= -8[4 + (-27)]$

$= -8(-23)$

$= 184$

20. $\dfrac{3a^3 - b}{4 + 3c} = \dfrac{3(2)^3 - (-3)}{4 + 3(4)}$

$= \dfrac{3(8) - (-3)}{4 + 3(4)}$

$= \dfrac{24 + 3}{4 + 12}$

$= \dfrac{27}{16}$

21. $-7r + 5 - 13r + 12$

$= -7r - 13r + 5 + 12$

$= (-7 - 13)r + (5 + 12)$

$= -20r + 17$

22. $-(3k + 8) - 2(4k - 7) + 3(8k + 12)$

$= -3k - 8 - 8k + 14 + 24k + 36$

$= -3k - 8k + 24k - 8 + 14 + 36$

$= 13k + 42$

23. $(a+b)+4=4+(a+b)$
The order of the terms $(a+b)$ and 4 have been reversed. This is an illustration of the *commutative property*.

24. $4x+12x=(4+12)x$
The common variable, x, has been removed from each term. This is an illustration of the *distributive property*.

25. $-9+9=0$

The sum of a number and its opposite is equal to 0. This is an illustration of the *inverse property*.

26.
$$\begin{aligned} -4x+7(2x+3) &= 7x+36 \\ -4x+14x+21 &= 7x+36 \\ 10x+21 &= 7x+36 \\ 3x &= 15 \\ x &= 5 \end{aligned}$$

Solution set: $\{5\}$

27.
$$\begin{aligned} -\frac{3}{5}x+\frac{2}{3}x &= 2 \\ 3(-3x)+5(2x) &= 15(2) \quad \textit{Multiply by 15.} \\ -9x+10x &= 30 \\ x &= 30 \end{aligned}$$

Solution set: $\{30\}$

28.
$$\begin{aligned} .06x+.03(100+x) &= 4.35 \\ 6x+3(100+x) &= 435 \quad \textit{Multiply by 100.} \\ 6x+300+3x &= 435 \\ 9x &= 135 \\ x &= 15 \end{aligned}$$

Solution set: $\{15\}$

29. Solve $P=a+b+c$ for c.
$$P-(a+b)=a+b+c-(a+b)$$
Subtract (a + b).
$$P-a-b=c$$

30.
$$\begin{aligned} 4(2x-6)+3(x-2) &= 11x+1 \\ 8x-24+3x-6 &= 11x+1 \\ 11x-30 &= 11x+1 \\ -30 &= 1 \quad \textit{False} \end{aligned}$$

Solution set: $\emptyset$

31.
$$\frac{2}{3}x+\frac{5}{8}x=\frac{31}{24}x$$
Multiply by the LCD, 24.
$$\begin{aligned} 8(2x)+3(5x) &= 31x \\ 16x+15x &= 31x \\ 31x &= 31x \quad \textit{True} \end{aligned}$$

Solution set: {all real numbers}

32. Let $x=$ the amount of pure alcohol that should be added.

Percent (as a Decimal)	Liters of Solution	Liters of Pure Alcohol
1.00	x	$1.00x$
.10	7	$.10(7)$
.30	$x+7$	$.30(x+7)$

The last column gives the equation.

$$\begin{aligned} 1.00x+.10(7) &= .30(x+7) \\ 10x+1(7) &= 3(x+7) \quad \textit{Multiply by 10.} \\ 10x+7 &= 3x+21 \\ 7x &= 14 \\ x &= 2 \end{aligned}$$

2 L of pure alcohol should be added to the solution.

33. Let $x=$ the number of nickels. Then $x-4=$ the number of quarters. The collection contains 29 coins, so the number of pennies is

$$29-x-(x-4)=33-2x.$$

	Number of Coins	Denomination	Value
Pennies	$33-2x$	.01	$.01(33-2x)$
Nickels	x	.05	$.05x$
Quarters	$x-4$	.25	$.25(x-4)$
Totals	29		\$2.69

From the last column:

$$\begin{aligned} .01(33-2x)+.05x+.25(x-4) &= 2.69 \\ 1(33-2x)+5x+25(x-4) &= 269 \\ & \textit{Multiply by 100.} \\ 33-2x+5x+25x-100 &= 269 \\ 28x-67 &= 269 \\ 28x &= 336 \\ x &= 12 \end{aligned}$$

There are $33-2(12)=9$ pennies, 12 nickels, and $12-4=8$ quarters.

34. Let $x =$ the amount invested at 5%. Then $x + 2000 =$ the amount invested at 6%.

Rate (as a Decimal)	Principal	Interest
.05	x	$.05x$
.06	$x + 2000$	$.06(x + 2000)$

The total interest is $670.

$$\begin{aligned} .05x + .06(x + 2000) &= 670 \\ .05x + .06x + 120 &= 670 \\ .11x + 120 &= 670 \\ .11x &= 550 \\ x &= 5000 \end{aligned}$$

$5000 was invested at 5% and $5000 + 2000 =$ $7000 was invested at 6%.

35. Let $x =$ the time for Jack to be $\frac{1}{4}$ mile ahead of Jill.

	Rate	Time	Distance
Jack	7	x	$7x$
Jill	5	x	$5x$

Jack's distance is $\frac{1}{4}$ mile more than Jill's distance.

$$\begin{aligned} 7x &= 5x + \tfrac{1}{4} \\ 2x &= \tfrac{1}{4} \\ x &= \tfrac{1}{8} \end{aligned}$$

Jack will be $\frac{1}{4}$ mile ahead of Jill in $\frac{1}{8}$ hr.

36. Clark's rule:

$$\frac{\text{Weight of child in pounds}}{150} \times \begin{matrix}\text{adult}\\\text{dose}\end{matrix} = \begin{matrix}\text{child's}\\\text{dose}\end{matrix}$$

If the child weighs 55 lb and the adult dosage is 120 mg, then

$$\frac{55}{150} \times 120 = 44.$$

The child's dosage is 44 mg.

37. 6 feet, 3 inches $= 6(12) + 3 = 75$ inches

$$\begin{aligned} \text{BMI} &= \frac{704 \times (\text{weight in pounds})}{(\text{height in inches})^2} \\ &= \frac{704 \times 205}{75^2} = \frac{144{,}320}{5625} \approx 25.7 \end{aligned}$$

His BMI is about 25.7.

38. **(a)** 1975: 1756
2000: 1480

$1756 - 1480 = 276$

The number decreased by 276 newspapers.

(b) $\frac{276}{1756} \approx .157$ or 15.7%.

The number decreased by approximately 15.7%.

39. In the figure, the two angles are supplementary, so their sum is 180°.

$$\begin{aligned} (10x + 7) + (7x + 3) &= 180 \\ 17x + 10 &= 180 \\ 17x &= 170 \\ x &= 10 \end{aligned}$$

The two angle measures are $10(10) + 7 = 107°$ and $7(10) + 3 = 73°$.

CHAPTER 3 LINEAR INEQUALITIES AND ABSOLUTE VALUE

3.1 Linear Inequalities in One Variable

3.1 Margin Exercises

1. **(a)** $x < -1$ is written in interval notation as $(-\infty, -1)$. The parenthesis next to -1 means that -1 is not in the interval and is not part of the graph.

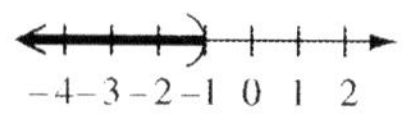

(b) $x \geq -3$ is written in interval notation as $[-3, \infty)$. The bracket next to -3 means that -3 is in the interval and is part of the graph.

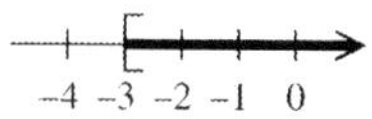

(c) $-4 \leq x < 2$ is written in interval notation as $[-4, 2)$. Here, -4 is in the interval and 2 is not.

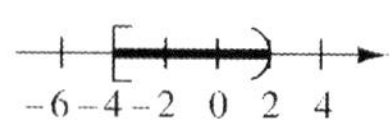

2. **(a)**
$$\begin{aligned} p + 6 &< 8 \\ p + 6 - 6 &< 8 - 6 \quad \textit{Subtract 6.} \\ p &< 2 \end{aligned}$$

Check: Substitute 2 for p in the equation $p + 6 = 8$.

$$\begin{aligned} p + 6 &= 8 \\ 2 + 6 &= 8 \quad ? \quad \textit{Let } p = 2. \\ 8 &= 8 \quad \textit{True} \end{aligned}$$

This shows that 2 is the boundary point. Now test a number on each side of 2 to verify that numbers *less than* 2 make the *inequality* true. We choose 0 and 3.

$$p + 6 < 8$$

$0 + 6 < 8$? *Let* $p = 0$.	$3 + 6 < 8$? *Let* $p = 3$.
$6 < 8$ *True*	$9 < 8$ *False*
0 is in the solution set.	3 is not in the solution set.

The check confirms that $(-\infty, 2)$ is the solution set.

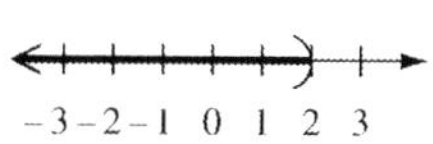

(b)
$$\begin{aligned} 8x &< 7x - 6 \\ 8x - 7x &< 7x - 6 - 7x \quad \textit{Subtract } 7x. \\ x &< -6 \end{aligned}$$

Check: Substitute -6 for x in $8x = 7x - 6$.

$$\begin{aligned} 8x &= 7x - 6 \\ 8(-6) &= 7(-6) - 6 \quad ? \quad \textit{Let } x = -6. \\ -48 &= -42 - 6 \quad ? \\ -48 &= -48 \quad \textit{True} \end{aligned}$$

This shows that -6 is the boundary point. Now test a number on each side of -6.

$$8x < 7x - 6$$

Let $x = -8$.	*Let* $x = 0$.
$8(-8) < 7(-8) - 6$?	$8(0) < 7(0) - 6$?
$-64 < -56 - 6$?	$0 < -6$ *False*
$-64 < -62$ *True*	
-8 is in the solution set.	0 is not in the solution set.

The check confirms that $(-\infty, -6)$ is the solution set.

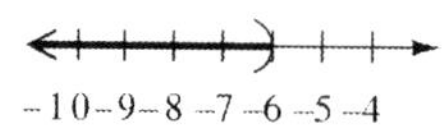

3.
$$\begin{aligned} 2k - 5 &\geq 1 + k \\ 2k - k &\geq 5 + 1 \quad \textit{Subtract } k\textit{; add 5.} \\ k &\geq 6 \end{aligned}$$

Check: Substitute 6 for k in $2k - 5 = 1 + k$.

$$\begin{aligned} 2k - 5 &= 1 + k \\ 2(6) - 5 &= 1 + 6 \quad ? \quad \textit{Let } k = 6. \\ 12 - 5 &= 7 \quad ? \\ 7 &= 7 \quad \textit{True} \end{aligned}$$

So 6 satisfies the equality part of $\geq$. Choose 0 and 7 as test points.

$$2k - 5 \geq 1 + k$$

Let $k = 0$.	*Let* $k = 7$.
$2(0) - 5 \geq 1 + 0$?	$2(7) - 5 \geq 1 + 7$?
$-5 \geq 1$ *False*	$14 - 5 \geq 1 + 7$?
	$9 \geq 8$ *True*
0 is not in the solution set.	7 is in the solution set.

The check confirms that $[6, \infty)$ is the solution set.

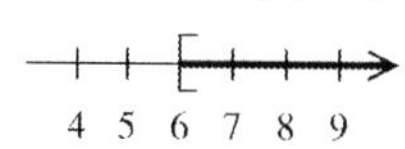

4. This exercise is designed to reinforce the idea that multiplying each side of an inequality by a negative number reverses the direction of the inequality symbol.

(a) $7 < 8$

$7(-5) > 8(-5)$ *Multiply by -5; reverse the inequality symbol.*

$-35 > -40$ *True*

(b) $-1 > -4$

$-1(-5) < -4(-5)$ *Multiply by -5; reverse the inequality symbol.*

$5 < 20$ *True*

5. **(a)** $2x < -10$

$\frac{2x}{2} < \frac{-10}{2}$ *Divide by $2 > 0$; do not reverse the inequality.*

$x < -5$

Check that the solution set is the interval $(-\infty, -5)$.

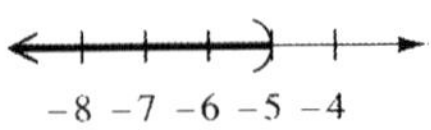

(b) $-7k \geq 8$

$\frac{-7k}{-7} \leq \frac{8}{-7}$ *Divide by $-7 < 0$; reverse the inequality.*

$k \leq -\frac{8}{7}$

Check that the solution set is the interval $(-\infty, -\frac{8}{7}]$.

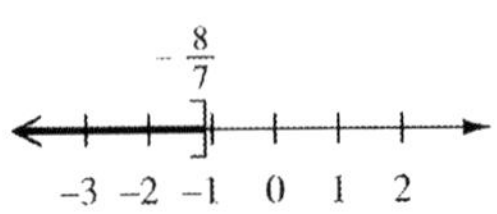

(c) $-9m < -81$

$\frac{-9m}{-9} > \frac{-81}{-9}$ *Divide by $-9 < 0$; reverse the inequality.*

$m > 9$

Check that the solution set is the interval $(9, \infty)$.

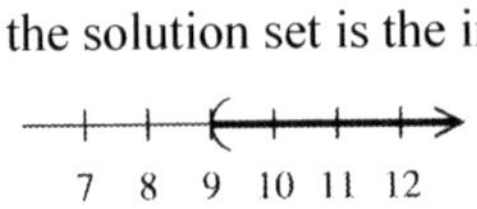

6. **(a)**

$5 - 3(m-1) \leq 2(m+3) + 1$

$5 - 3m + 3 \leq 2m + 6 + 1$ *Clear parentheses.*

$8 - 3m \leq 2m + 7$

$8 - 5m \leq 7$ *Subtract 2m.*

$-5m \leq -1$ *Subtract 8.*

$\frac{-5m}{-5} \geq \frac{-1}{-5}$ *Divide by -5; reverse the inequality.*

$m \geq \frac{1}{5}$

Check that the solution set is the interval $[\frac{1}{5}, \infty)$.

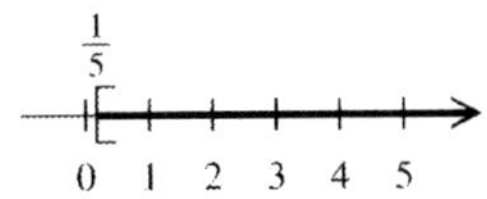

(b) $\frac{1}{4}(m+3) + 2 \leq \frac{3}{4}(m+8)$

$4\left[\frac{1}{4}(m+3) + 2\right] \leq 4\left[\frac{3}{4}(m+8)\right]$ *Multiply by 4.*

$(m+3) + 8 \leq 3(m+8)$

$m + 11 \leq 3m + 24$

$-2m + 11 \leq 24$ *Subtract 3m.*

$-2m \leq 13$ *Subtract 11.*

$m \geq -\frac{13}{2}$ *Divide by -2; reverse symbol.*

Check that the solution set is the interval $[-\frac{13}{2}, \infty)$.

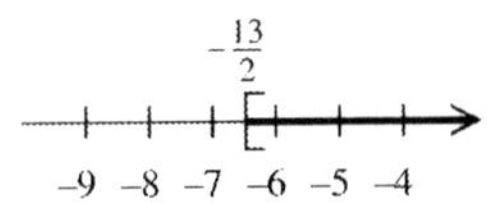

7. **(a)** $-3 \leq x - 1 \leq 7$

$-2 \leq x \leq 8$ *Add 1 to each part.*

Check that the solution set is the interval $[-2, 8]$.

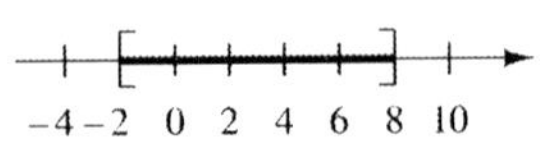

(b) $5 < 3x - 4 < 9$

$9 < 3x < 13$ *Add 4 to each part.*

$\frac{9}{3} < \frac{3x}{3} < \frac{13}{3}$ *Divide each part by 3.*

$3 < x < \frac{13}{3}$

Check that the solution set is the interval $(3, \frac{13}{3})$.

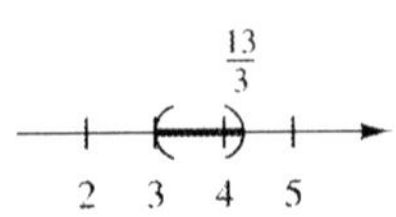

8. *Step 2*
Let $h =$ the number of hours she can rent the leaf blower.

Step 3
She must pay \$5, plus $\$1.75h$, to rent the leaf blower for h hours, and this amount must be *no more than* \$26.

Cost of renting	is no more than	26 dollars.
↓	↓	↓
$5 + 1.75h$	$\leq$	26

Step 4

$$\begin{aligned} 1.75h &\leq 21 && \textit{Subtract 5.} \\ h &\leq 12 && \textit{Divide by 1.75.} \end{aligned}$$

Step 5
She can use the leaf blower for a maximum of 12 hr. (She may use it for less time, as indicated by the inequality $h \leq 12$.)

Step 6
If Dona uses the leaf blower for 12 hr, she will spend $5 + 1.75(12) = 26$, the maximum amount.

9. Let $x =$ the grade the student must make on the fourth test.

To find the average of the four tests, add them and divide by 4. This average must be at least 90, that is, greater than or equal to 90.

$$\begin{aligned} \frac{92 + 90 + 84 + x}{4} &\geq 90 \\ \frac{266 + x}{4} &\geq 90 \\ 266 + x &\geq 360 && \textit{Multiply by 4.} \\ x &\geq 94 && \textit{Subtract 266.} \end{aligned}$$

The student must score at least 94 on the fourth test.

Check: $\dfrac{92 + 90 + 84 + 94}{4} = \dfrac{360}{4} = 90$

A score of 94 or more will give an average of at least 90, as required.

3.1 Section Exercises

1. $x \leq 3$

In interval notation, this inequality is written $(-\infty, 3]$. The bracket indicates that 3 is included. The answer is choice **D**.

2. $x > 3$

In interval notation, this inequality is written $(3, \infty)$. The parenthesis indicates that 3 is not included. The answer is choice **C**.

3. $x < 3$

In interval notation, this inequality is written $(-\infty, 3)$. The parenthesis indicates that 3 is not included. The graph of this inequality is shown in choice **B**.

4. $x \geq 3$

In interval notation, this inequality is written $[3, \infty)$. The bracket indicates that 3 is included. The graph of this inequality is shown in choice **A**.

5. $-3 \leq x \leq 3$

In interval notation, this inequality is written $[-3, 3]$. The brackets indicates that -3 and 3 are included. The answer is choice **F**.

6. $-3 < x < 3$

In interval notation, this inequality is written $(-3, 3)$. The parentheses indicate that neither -3 nor 3 is included. The answer is choice **E**.

7. Use a parenthesis when an endpoint is not included; use a bracket when it is included.

8. Simplify each side separately using the distributive property and combining terms. Get all variable terms on one side of the inequality and numbers on the other side. Get the variable with a coefficient of 1 on one side of the inequality sign and a number on the other side. The inequality sign must be reversed when multiplying or dividing each side by a negative number.

For Exercises 9–30 and 37–45, the check is shown for Exercises 9–11 and left for the student for the other exercises.

9.
$$\begin{aligned} 4x + 1 &\geq 21 \\ 4x &\geq 20 && \textit{Subtract 1.} \\ \frac{4x}{4} &\geq \frac{20}{4} && \textit{Divide by 4.} \\ x &\geq 5 \end{aligned}$$

Check: Substitute 5 for x in the *equation* $4x + 1 = 21$.

$$\begin{aligned} 4(5) + 1 &= 21 \quad ? \\ 21 &= 21 \quad \textit{True} \end{aligned}$$

So 5 satisfies the equality part of $\geq$. Choose 0 and 6 as test points.

$$4x + 1 \geq 21$$

Let x = 0.	*Let x = 6.*
$4(0) + 1 \geq 21$?	$4(6) + 1 \geq 21$?
$1 \geq 21$ *False*	$25 \geq 21$ *True*
0 is not in the solution set.	6 is in the solution set.

The check confirms that $[5, \infty)$ is the solution set.

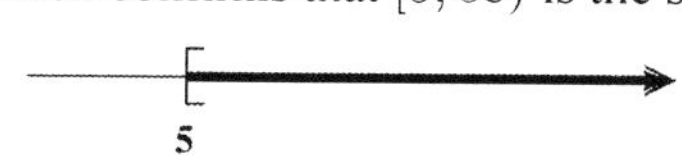

10. $5t + 2 \geq 52$

$$\begin{aligned} 5t &\geq 50 && \textit{Subtract 2.} \\ \frac{5t}{5} &\geq \frac{50}{5} && \textit{Divide by 5.} \\ t &\geq 10 \end{aligned}$$

Check: Substitute 10 for t in the *equation* $5t + 2 = 52$.

$$\begin{aligned} 5(10) + 2 &= 52 \quad ? \\ 52 &= 52 \qquad \textit{True} \end{aligned}$$

So 10 satisfies the equality part of $\geq$. Choose 0 and 20 as test points.

$$5t + 2 \geq 52$$

Let t = 0.	*Let t = 20.*
$5(0) + 2 \geq 52$?	$5(20) + 2 \geq 52$?
$2 \geq 52$ *False*	$102 \geq 52$ *True*
0 is not in the solution set.	20 is in the solution set.

The check confirms that $[10, \infty)$ is the solution set.

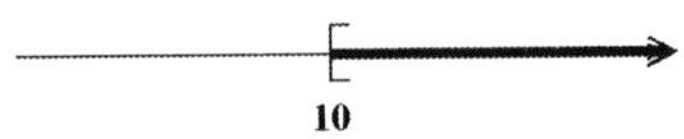

11. $\dfrac{3k-1}{4} > 5$

$$\begin{aligned} 4\left(\frac{3k-1}{4}\right) &> 4(5) && \textit{Multiply by 4.} \\ 3k - 1 &> 20 \\ 3k &> 21 && \textit{Add 1.} \\ k &> 7 && \textit{Divide by 3.} \end{aligned}$$

Check: Let $k = 7$ in the *equation* $\dfrac{3k-1}{4} = 5$.

$$\begin{aligned} \frac{3(7)-1}{4} &= 5 \quad ? \\ \frac{20}{4} &= 5 \quad ? \\ 5 &= 5 \qquad \textit{True} \end{aligned}$$

This shows that 7 is the boundary point. Now test a number on each side of 7. We choose 0 and 10.

$$\frac{3k-1}{4} > 5$$

Let k = 0.	*Let k = 10.*
$\dfrac{3(0)-1}{4} > 5$?	$\dfrac{3(10)-1}{4} > 5$?
$\dfrac{-1}{4} > 5$ *False*	$\dfrac{29}{4} > 5$ *True*
0 is not in the solution set.	10 is in the solution set.

The check confirms that $(7, \infty)$ is the solution set.

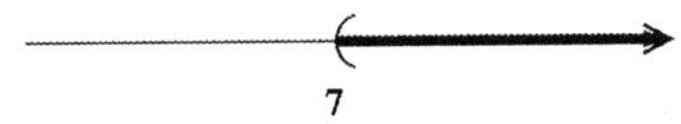

12. $\dfrac{5z-6}{8} < 8$

$$\begin{aligned} 8\left(\frac{5z-6}{8}\right) &< 8 \cdot 8 && \textit{Multiply by 8.} \\ 5z - 6 &< 64 \\ 5z &< 70 && \textit{Add 6.} \\ z &< 14 && \textit{Divide by 5.} \end{aligned}$$

Check that the solution set is the interval $(-\infty, 14)$.

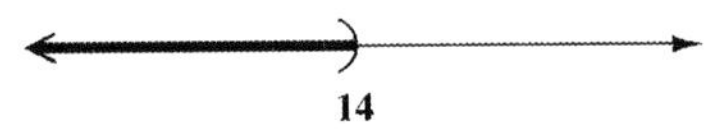

13. $-4x < 16$

Divide each side by -4 and reverse the inequality symbol.

$$\begin{aligned} \frac{-4x}{-4} &> \frac{16}{-4} \\ x &> -4 \end{aligned}$$

Check that the solution set is the interval $(-4, \infty)$.

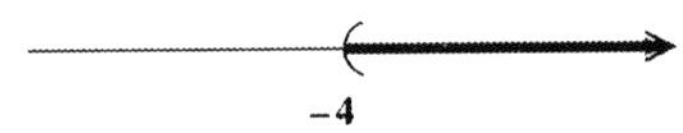

14. $-2m > 10$

Divide each side by -2 and reverse the inequality symbol.

$$\begin{aligned} \frac{-2m}{-2} &< \frac{10}{-2} \\ m &< -5 \end{aligned}$$

Check that the solution set is the interval $(-\infty, -5)$.

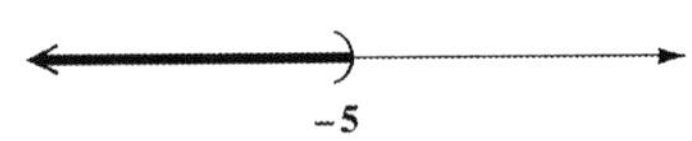

15. $-\dfrac{3}{4}r \geq 30$

Multiply each side by $-\frac{4}{3}$ and reverse the inequality symbol.

$$\begin{aligned} -\tfrac{4}{3}\left(-\tfrac{3}{4}r\right) &\leq -\tfrac{4}{3}(30) \\ r &\leq -40 \end{aligned}$$

Check that the solution set is the interval $(-\infty, -40]$.

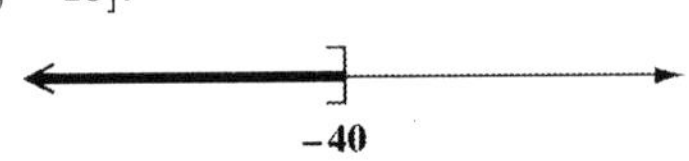

16. $-\frac{2}{3}x \le 12$

Multiply each side by $-\frac{3}{2}$ and reverse the inequality symbol.

$-\frac{3}{2}\left(-\frac{2}{3}x\right) \ge -\frac{3}{2}(12)$

$x \ge -18$

Check that the solution set is the interval $[-18, \infty)$.

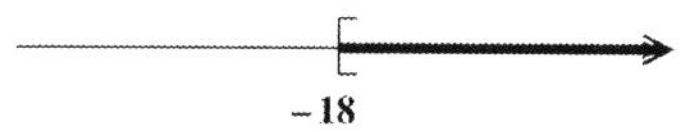

17. $-1.3m \ge -5.2$

Divide each side by -1.3 and reverse the inequality symbol.

$\frac{-1.3m}{-1.3} \le \frac{-5.2}{-1.3}$

$m \le 4$

Check that the solution set is the interval $(-\infty, 4]$.

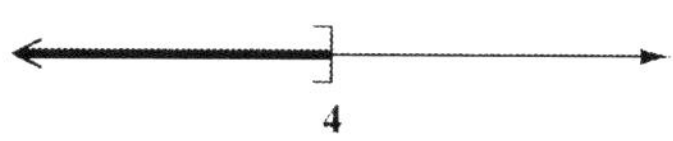

18. $-2.5x \le -1.25$

Divide each side by -2.5 and reverse the inequality symbol.

$\frac{-2.5x}{-2.5} \ge \frac{-1.25}{-2.5}$

$x \ge .5$

Check that the solution set is the interval $[.5, \infty)$.

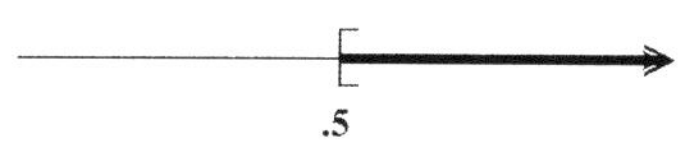

19. $\frac{2k-5}{-4} > 5$

Multiply each side by -4 and reverse the inequality symbol.

$-4\left(\frac{2k-5}{-4}\right) < -4(5)$

$2k - 5 < -20$

$2k < -15$ *Add 5.*

$k < -\frac{15}{2}$ *Divide by 2.*

Check that the solution set is the interval $\left(-\infty, -\frac{15}{2}\right)$.

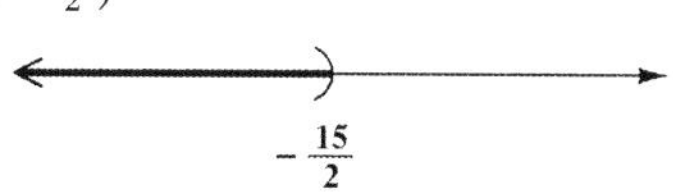

20. $\frac{3z-2}{-5} < 6$

Multiply each side by -5 and reverse the inequality symbol.

$3z - 2 > -30$

$3z > -28$ *Add 2.*

$z > -\frac{28}{3}$ *Divide by 3.*

Check that the solution set is the interval $\left(-\frac{28}{3}, \infty\right)$.

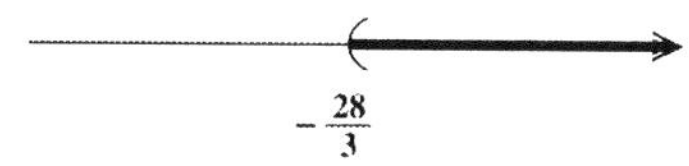

21. $x + 4(2x-1) \ge x$

$x + 8x - 4 \ge x$

$9x - 4 \ge x$

$9x \ge x + 4$ *Add 4.*

$8x \ge 4$ *Subtract x.*

$x \ge \frac{1}{2}$ *Divide by 8.*

Check that the solution set is the interval $\left[\frac{1}{2}, \infty\right)$.

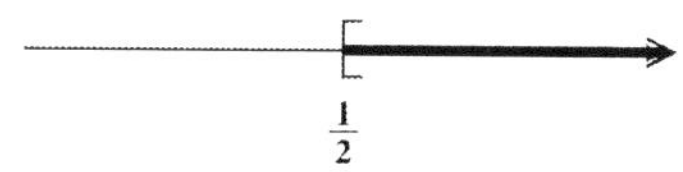

22. $m - 2(m-4) \le 3m$

$m - 2m + 8 \le 3m$

$-m + 8 \le 3m$

$8 \le 4m$ *Add m.*

$2 \le m$ or $m \ge 2$

Check that the solution set is the interval $[2, \infty)$.

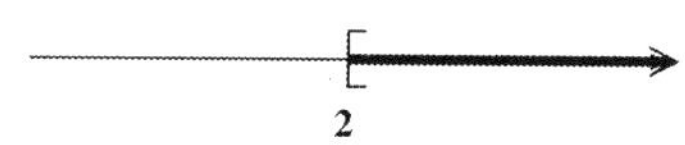

23. $-(4+r) + 2 - 3r < -14$

$-4 - r + 2 - 3r < -14$ *Distributive property*

$-4r - 2 < -14$ *Combine terms.*

$-4r < -12$ *Add 2.*

Divide each side by -4 and reverse the inequality symbol.

$r > 3$

Check that the solution set is the interval $(3, \infty)$.

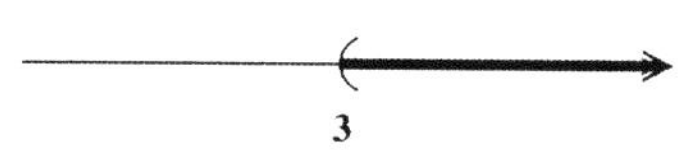

24.
$$\begin{aligned}-(9+k)-5+4k &\geq 4\\ -9-k-5+4k &\geq 4\\ -14+3k &\geq 4\\ 3k &\geq 18\\ k &\geq 6\end{aligned}$$

Check that the solution set is the interval $[6, \infty)$.

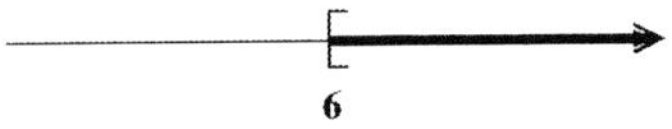

25. $-3(z-6) > 2z-2$

$-3z+18 > 2z-2$ *Distributive property*

$-5z > -20$ *Subtract 2z; subtract 18.*

Divide each side by -5 and reverse the inequality symbol.

$z < 4$

Check that the solution set is the interval $(-\infty, 4)$.

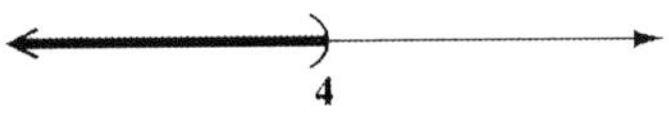

26.
$$\begin{aligned}-2(x+4) &\leq 6x+16\\ -2x-8 &\leq 6x+16\\ -8x &\leq 24\\ x &\geq -3\end{aligned}$$
Reverse symbol.

Check that the solution set is the interval $[-3, \infty)$.

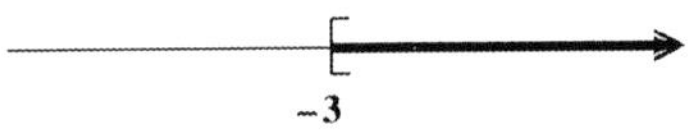

27. $\frac{2}{3}(3k-1) \geq \frac{3}{2}(2k-3)$

Multiply each side by 6 to clear the fractions.

$6 \cdot \frac{2}{3}(3k-1) \geq 6 \cdot \frac{3}{2}(2k-3)$

$4(3k-1) \geq 9(2k-3)$

$12k-4 \geq 18k-27$ *Distributive property*

$-6k \geq -23$ *Subtract 18k; add 4.*

Divide each side by -6 and reverse the inequality symbol.

$k \leq \frac{23}{6}$

Check that the solution set is the interval $(-\infty, \frac{23}{6}]$.

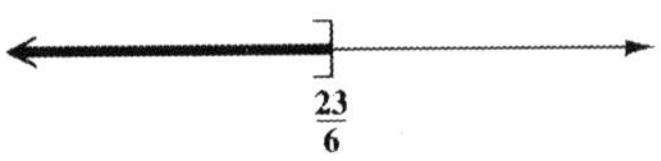

28.
$$\begin{aligned}\frac{7}{5}(10m-1) &< \frac{2}{3}(6m+5)\\ 15 \cdot \tfrac{7}{5}(10m-1) &< 15 \cdot \tfrac{2}{3}(6m+5)\\ 21(10m-1) &< 10(6m+5)\\ 210m-21 &< 60m+50\\ 150m &< 71\\ m &< \tfrac{71}{150}\end{aligned}$$

Check that the solution set is the interval $(-\infty, \frac{71}{150})$.

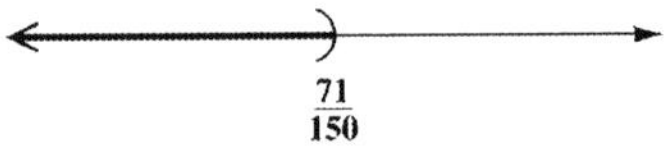

29. $-\frac{1}{4}(p+6)+\frac{3}{2}(2p-5) < 10$

Multiply each term by 4 to clear the fractions.

$$\begin{aligned}-1(p+6)+6(2p-5) &< 40\\ -p-6+12p-30 &< 40\\ 11p-36 &< 40\\ 11p &< 76\\ p &< \tfrac{76}{11}\end{aligned}$$

Check that the solution set is the interval $(-\infty, \frac{76}{11})$.

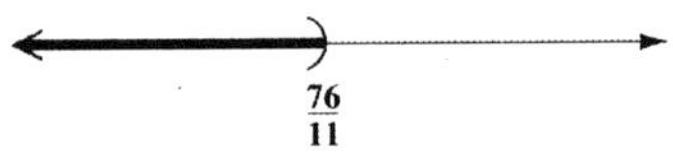

30.
$$\begin{aligned}\frac{3}{5}(k-2)-\frac{1}{4}(2k-7) &\leq 3\\ 20 \cdot \tfrac{3}{5}(k-2)-20 \cdot \tfrac{1}{4}(2k-7) &\leq 20 \cdot 3\\ 12(k-2)-5(2k-7) &\leq 60\\ 12k-24-10k+35 &\leq 60\\ 2k &\leq 49\\ k &\leq \tfrac{49}{2}\end{aligned}$$

Check that the solution set is the interval $(-\infty, \frac{49}{2}]$.

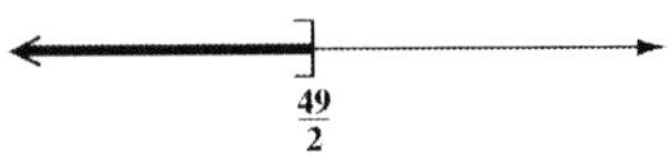

31.
$$\begin{aligned}5(x+3)-2(x-4) &= 2(x+7)\\ 5x+15-2x+8 &= 2x+14\\ 3x+23 &= 2x+14\\ x &= -9\end{aligned}$$

Solution set: $\{-9\}$

The graph is the point -9 on a number line.

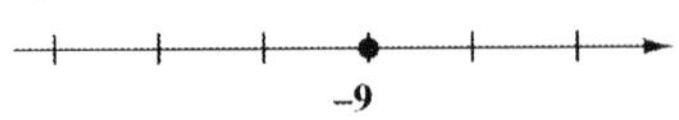

32. $5(x+3)-2(x-4) > 2(x+7)$
$5x+15-2x+8 > 2x+14$
$3x+23 > 2x+14$
$x > -9$

Check that the solution set is the interval $(-9, \infty)$. The graph extends from -9 to the right on a number line; -9 is not included in the graph.

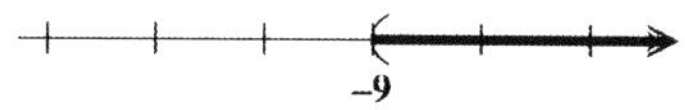

33. $5(x+3)-2(x-4) < 2(x+7)$
$5x+15-2x+8 < 2x+14$
$3x+23 < 2x+14$
$x < -9$

Check that the solution set is the interval $(-\infty, -9)$.
The graph extends from -9 to the left on a number line; -9 is not included in the graph.

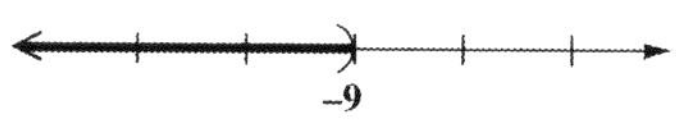

34. If we graph all the solution sets from Exercises 31–33; that is, $\{-9\}$, $(-9, \infty)$, and $(-\infty, -9)$, on the same number line, we will have graphed the set of all real numbers.

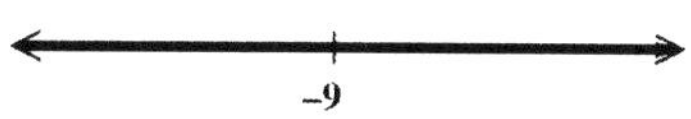

35. The solution set of the given equation is the point -3 on a number line. The solution set of the first inequality extends from -3 to the right (toward ∞) on the same number line. Based on Exercises 31–33, the solution set of the second inequality should then extend from -3 to the left (toward $-\infty$) on the number line. Complete the statement with $\underline{(-\infty, -3)}$.

36. $-2 < x$ is the same as $x > -2$. This inequality represents all real numbers greater than -2. Its graph is shown in choice **A**.

37. $-4 < x-5 < 6$
Add 5 to each part of the inequality to isolate the variable x.
$-4+5 < x-5+5 < 6+5$
$1 < x < 11$

Check that the solution set is the interval $(1, 11)$.

38. $-1 < x+1 < 8$
Subtract 1 from each part to isolate the variable x.
$-1-1 < x+1-1 < 8-1$
$-2 < x < 7$

Check that the solution set is the interval $(-2, 7)$.

39. $-9 \le k+5 \le 15$
Subtract 5 from each part.
$-9-5 \le k+5-5 \le 15-5$
$-14 \le k \le 10$

Check that the solution set is the interval $[-14, 10]$.

40. $-4 \le m+3 \le 10$
$-7 \le m \le 7$ *Subtract 3.*

Check that the solution set is the interval $[-7, 7]$.

41. $-6 \le 2(z+2) \le 16$
$-6 \le 2z+4 \le 16$ *Distributive property*
$-10 \le 2z \le 12$ *Subtract 4.*
$-5 \le z \le 6$ *Divide by 2.*

Check that the solution set is the interval $[-5, 6]$.

42. $-15 < 3(p+2) < -12$
$-15 < 3p+6 < -12$ *Distributive property*
$-21 < 3p < -18$ *Subtract 6.*
$-7 < p < -6$ *Divide by 3.*

Check that the solution set is the interval $(-7, -6)$.

43. $-16 < 3t+2 < -10$
$-18 < 3t < -12$ *Subtract 2.*
$-6 < t < -4$ *Divide by 3.*

Check that the solution set is the interval $(-6, -4)$.

44. $-1 \le \dfrac{2x-5}{6} \le 5$

$-6 \le 2x - 5 \le 30$ *Multiply by 6.*

$-1 \le 2x \le 35$ *Add 5.*

$-\frac{1}{2} \le x \le \frac{35}{2}$ *Divide by 2.*

Check that the solution set is the interval $\left[-\frac{1}{2}, \frac{35}{2}\right]$.

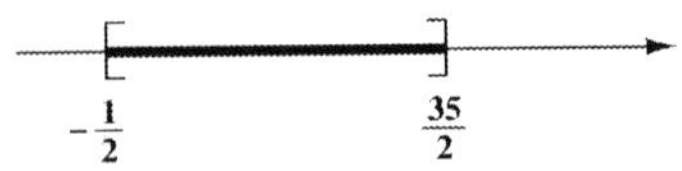

45. $-3 \le \dfrac{3m+1}{4} \le 3$

$-12 \le 3m + 1 \le 12$ *Multiply by 4.*

$-13 \le 3m \le 11$ *Subtract 1.*

$-\frac{13}{3} \le m \le \frac{11}{3}$ *Divide by 3.*

Check that the solution set is the interval $\left[-\frac{13}{3}, \frac{11}{3}\right]$.

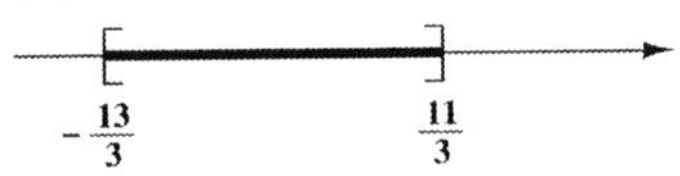

46. Draw a horizontal line at the 90°F mark. It intersects the *upper* boundary of the forecasted highs in two places. The temperature is expected to be at least 90°F from about 2:30 P.M. to 6:00 P.M.

47. Draw a horizontal line at the 70°F mark. It intersects the *lower* boundary of the forecasted highs in two places. The temperature is predicted to be no more than 70°F from about 8:00 A.M. to 10:15 A.M. and after about 9:00 P.M.

48. Locate the dot at 5:39 P.M. and draw a vertical line through it. The line will intersect the lower and upper boundaries of the forecasted highs at about 84°F and 91°F, so the range of predicted temperatures is 84°F–91°F.

49. Locate the dot at 8:45 A.M. and draw a vertical line through it. The line will intersect the lower and upper boundaries of the forecasted highs at about 65°F and 67°F, so the range of predicted temperatures is 65°F–67°F.

50. Let $x =$ her score on the third test.
Her average must be at least 84 (≥ 84). To find the average of three numbers, add them and divide by 3.

$$\frac{90+82+x}{3} \ge 84$$

$\dfrac{172+x}{3} \ge 84$ *Add.*

$172 + x \ge 252$ *Multiply by 3.*

$x \ge 80$ *Subtract 172.*

She must score at least 80 on her third test.

51. Let $x =$ his score on the third test.
His average must be at least 90.

$$\frac{92+96+x}{3} \ge 90$$

$92 + 96 + x \ge 270$ *Multiply by 3.*

$188 + x \ge 270$ *Add.*

$x \ge 82$ *Subtract 188.*

He must score at least 82 on his third test.

52. Let $x =$ the number of miles driven.
The cost of renting from Ford is \$15 plus the mileage cost of $.14x$, while the cost of renting from Chevrolet is \$14 plus the mileage cost of $.16x$.

Cost from Chevrolet $>$ Cost from Ford

$14 + .16x > 15 + .14x$

$1400 + 16x > 1500 + 14x$ *Multiply by 100.*

$2x > 100$

$x > 50$

After 50 mi, the price to rent from Chevrolet exceeds the price to rent from Ford.

53. Let $x =$ the number of miles driven. Then $108 + .14x$ is the Toyota price, and $7(28) = 196$ is the Avis price.

Avis price $<$ Toyota price

$196 < 108 + .14x$

$88 < .14x$

$x > \dfrac{88}{.14} \approx 628.6$

After approximately 628.6 mi, the Avis price is less than the Toyota price.

54. Cost $C = 20x + 100$; Revenue $R = 24x$
The business will show a profit only when $R > C$. Substitute the given expressions for R and C.

$R > C$

$24x > 20x + 100$

$4x > 100$

$x > 25$

The company will show a profit upon selling 26 tapes.

55. $C = 3x + 2300$, $R = 5.50x$
To make a profit, $R > C$.

$5.50x > 3x + 2300$

$2.5x > 2300$

$x > 920$

The company will show a profit after making 921 deliveries.

56. $\text{BMI} = \dfrac{704 \times (\text{weight in pounds})}{(\text{height in inches})^2}$

(a) Let the height equal 72.

$$\begin{aligned} 19 \le \ &\text{BMI} \le 25 \\ 19 \le \ &\frac{704w}{72^2} \le 25 \\ 19(72^2) \le \ &704w \le 25(72^2) \\ \frac{19(72^2)}{704} \le \ &w \le \frac{25(72^2)}{704} \\ (\approx 139.91) \le \ &w \le (\approx 184.09) \end{aligned}$$

According to the BMI formula, the healthy weight range (rounded to the nearest pound) for a person who is 72 inches tall is 140 to 184 pounds.

(b) Answers will vary.

57. **(a)** Let $A = 35$.

$$\begin{aligned} .7(220 - A) &\le THR \le .85(220 - A) \\ .7(220 - 35) &\le THR \le .85(220 - 35) \\ .7(185) &\le THR \le .85(185) \\ 129.5 &\le THR \le 157.25 \end{aligned}$$

The range is about 130 to 157 beats per minute.

(b) Answers will vary.

58. Six times a number is between -12 and 12.

$$\begin{aligned} -12 < 6x &< 12 \\ -2 < x &< 2 \quad \textit{Divide by 6.} \end{aligned}$$

This is the set of all numbers between -2 and 2, that is, $(-2, 2)$.

59. Half a number is between -3 and 2.

$$\begin{aligned} -3 < \tfrac{1}{2}x &< 2 \\ -6 < x &< 4 \quad \textit{Multiply by 2.} \end{aligned}$$

This is the set of all numbers between -6 and 4, that is, $(-6, 4)$.

60. When 1 is added to twice a number, the result is greater than or equal to 7.

$$\begin{aligned} 2x + 1 &\ge 7 \\ 2x &\ge 6 \quad \textit{Subtract 1.} \\ x &\ge 3 \quad \textit{Divide by 2.} \end{aligned}$$

This is the set of all numbers greater than or equal to 3, that is, $[3, \infty)$.

61. If 8 is subtracted from a number, the result is at least 5.

$$\begin{aligned} x - 8 &\ge 5 \\ x &\ge 13 \quad \textit{Add 8.} \end{aligned}$$

This is the set of all numbers greater than or equal to 13, that is, $[13, \infty)$.

62. One third of a number is added to 6, giving a result of at least 3.

$$\begin{aligned} 6 + \tfrac{1}{3}x &\ge 3 \\ \tfrac{1}{3}x &\ge -3 \quad \textit{Subtract 6.} \\ x &\ge -9 \quad \textit{Multiply by 3.} \end{aligned}$$

This is the set of all numbers greater than or equal to -9, that is, $[-9, \infty)$.

63. Three times a number, minus 5, is no more than 7.

$$\begin{aligned} 3x - 5 &\le 7 \\ 3x &\le 12 \quad \textit{Add 5.} \\ x &\le 4 \quad \textit{Divide by 3.} \end{aligned}$$

This is the set of all numbers less than or equal to 4, that is, $(-\infty, 4]$.

3.2 Set Operations and Compound Inequalities

3.2 Margin Exercises

1. **(a)** Let $A = \{3, 4, 5, 6\}$ and $B = \{5, 6, 7\}$.

The set $A \cap B$, the intersection of A and B, contains those elements that belong to both A *and* B; that is, the numbers 5 and 6. Therefore,

$$A \cap B = \{5, 6\}.$$

(b) Let $N = \{s, d, c, g\}$ and $O = \{i, m, h, g\}$.

The set $N \cap O$ contains those elements that belong to both N *and* O; that is, symptom g. Therefore,

$$N \cap O = \{g\}.$$

2. **(a)** $x < 10$ and $x > 2$
Graph each inequality.

The word *and* means to take the values that satisfy both inequalities; that is, the numbers between 2 and 10, not including 2 and 10.

Solution set: $(2, 10)$

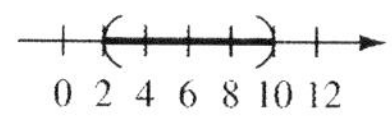

(b) $x + 3 \leq 1$ and $x - 4 \geq -12$
Solve each inequality.

$$\begin{aligned} x + 3 &\leq 1 \quad &\text{and}\quad x - 4 &\geq -12 \\ x + 3 - 3 &\leq 1 - 3 \quad &\text{and}\quad x - 4 + 4 &\geq -12 + 4 \\ x &\leq -2 \quad &\text{and}\quad x &\geq -8 \end{aligned}$$

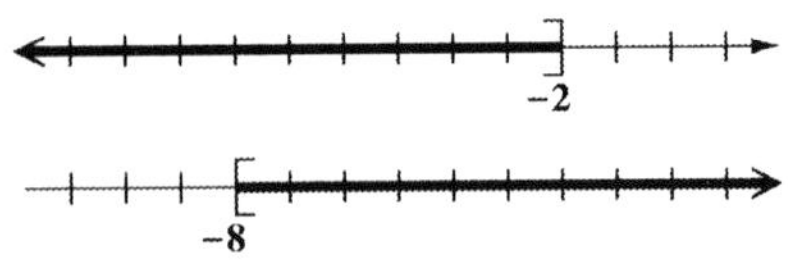

The values that satisfy both inequalities are the numbers between -8 and -2, including -8 and -2.

Solution set: $[-8, -2]$

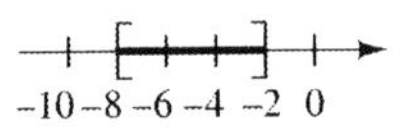

3.

$$\begin{aligned} 2x &\geq x - 1 \quad &\text{and}\quad 3x &\geq 3 + 2x \\ 2x - x &\geq x - 1 - x \quad &\text{and}\quad 3x - 2x &\geq 3 + 2x - 2x \\ x &\geq -1 \quad &\text{and}\quad x &\geq 3 \end{aligned}$$

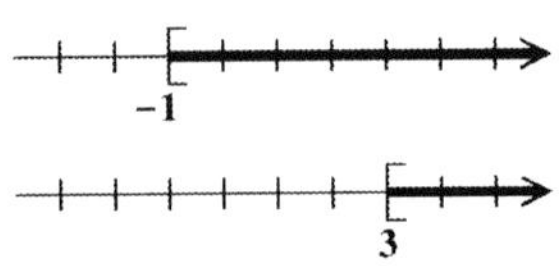

The overlap of the two graphs consists of the numbers that are greater than or equal to -1 and are also greater than or equal to 3; that is, the numbers greater than or equal to 3.

Solution set: $[3, \infty)$

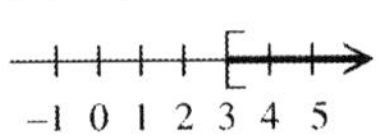

4. **(a)** $x < 5$ and $x > 5$

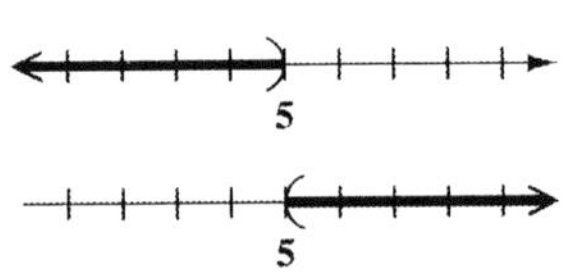

The numbers that satisfy both inequalities must be less than 5 *and* greater than 5. There are no such numbers.

Solution set: $\emptyset$

(b) $x + 2 > 3$ and $2x + 1 < -3$

$$\begin{aligned} & & 2x &< -4 \\ x &> 1 \quad \text{and} & x &< -2 \end{aligned}$$

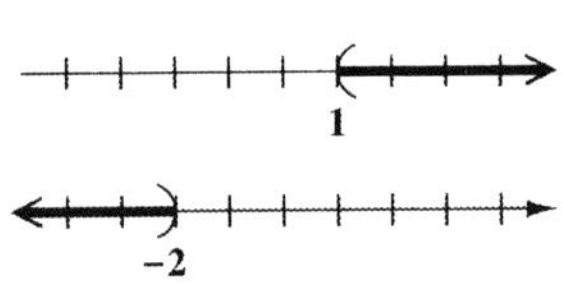

The numbers that satisfy both inequalities must be greater than 1 *and* less than -2. There are no such numbers.

Solution set: $\emptyset$

5. **(a)** Let $A = \{3, 4, 5, 6\}$ and $B = \{5, 6, 7\}$.

The set $A \cup B$, the union of A and B, consists of all elements in either A *or* B (or both). Start by listing the elements of set A: $3, 4, 5, 6$. Then list any additional elements from set B. In this case, the elements 5 and 6 are already listed, so the only additional element is 7. Therefore,

$$A \cup B = \{3, 4, 5, 6, 7\}.$$

(b) The set $N \cup O$ consists of all elements in either N or O (or both).

$$\begin{aligned} N \cup O &= \{s, d, c, g\} \cup \{i, m, h, g\} \\ &= \{s, d, c, g, i, m, h\} \end{aligned}$$

6. **(a)** $x + 2 > 3$ or $2x + 1 < -3$

$$\begin{aligned} & & 2x &< -4 \\ x &> 1 \quad \text{or} & x &< -2 \end{aligned}$$

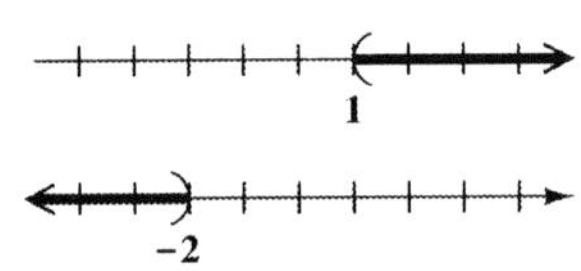

The graph of the solution set consists of all numbers greater than 1 *or* less than -2.

Solution set: $(-\infty, -2) \cup (1, \infty)$

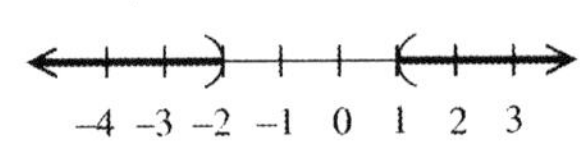

(b) $x - 1 > 2$ or $3x + 5 < 2x + 6$

$$x > 3 \quad \text{or} \quad x < 1$$

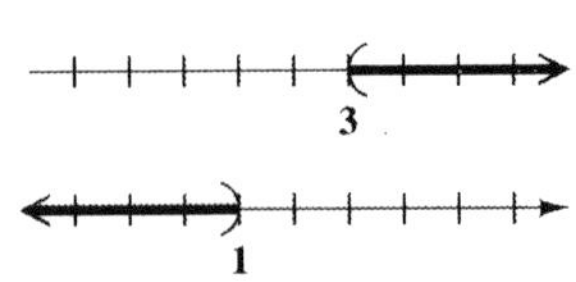

The graph of the solution set consists of all numbers greater than 3 *or* less than 1.

Solution set: $(-\infty, 1) \cup (3, \infty)$

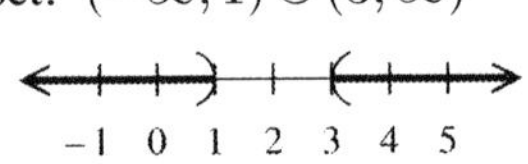

7. **(a)** $2x + 1 \le 9$ or $2x + 3 \le 5$
$2x \le 8$ or $2x \le 2$
$x \le 4$ or $x \le 1$

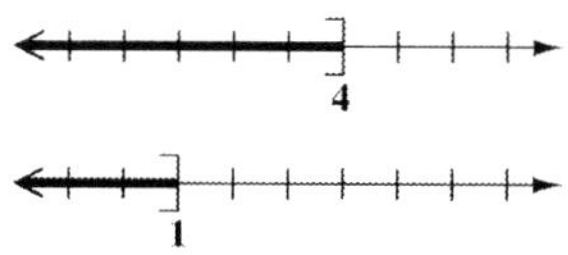

The solution set is all numbers less than or equal to 4 *or* less than or equal to 1. Note that this is simply the first inequality, $x \le 4$.

Solution set: $(-\infty, 4]$

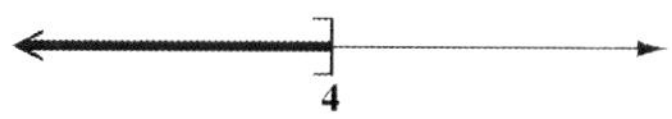

(b) $3x - 2 \le 13$ or $x + 5 \ge 7$
$3x \le 15$
$x \le 5$ or $x \ge 2$

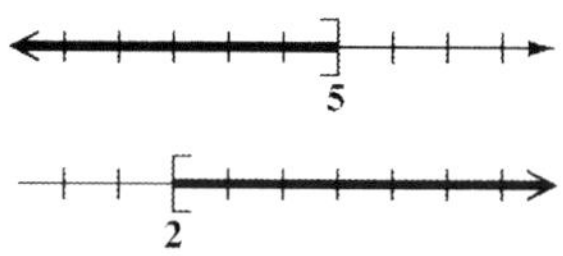

The solution set is all numbers that are either less than or equal to 5 *or* greater than or equal to 2. All real numbers are included.

Solution set: $(-\infty, \infty)$

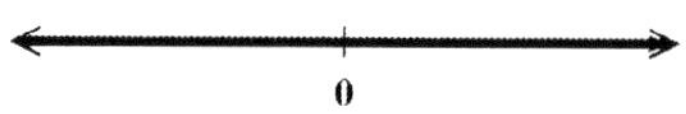

8. **(a)** All films had admissions greater than 130,000,000, but no films had a gross income of less than \$500,000,000. Thus, there are no elements (films) that satisfy each set, so the required set is the empty set, symbolized by $\emptyset$.

(b) Since all the films had admissions greater than 130,000,000 and we have an *or* statement, the second condition doesn't have an effect on the solution set. The required set is the set of all films; that is, {*Gone with the Wind, Star Wars, The Sound of Music, E.T., The Ten Commandments*}.

3.2 Section Exercises

1. **(i)** $2x + 1 = 3$
$2x = 2$
$x = 1$ **Solution set:** $\{1\}$

(ii) $2x + 1 > 3$
$2x > 2$
$x > 1$ **Solution set:** $(1, \infty)$

(iii) $2x + 1 < 3$
$2x < 2$
$x < 1$ **Solution set:** $(-\infty, 1)$

The union of the solution sets $\{1\}$, $(1, \infty)$, and $(-\infty, 1)$ is the set of all real numbers; that is, $(-\infty, \infty)$. The statement is *true*. (See Section 3.1, Exercises 31–35, for a discussion of this concept.)

2. Let $A = \{x \mid x \ge 5\}$ and $B = \{x \mid x \le 5\}$. The element 5 belongs to both A and B, so the intersection of A and B is *not* empty. The intersection is $\{5\}$, so this statement is *false*.

3. The union of $(-\infty, 6)$ and $(6, \infty)$ is $(-\infty, 6) \cup (6, \infty)$, which is all numbers except 6. This statement is *false*.

4. Because of the brackets, 6 is an element of the set $[6, \infty)$ and also of the set $(-\infty, 6]$. The intersection of the sets is $\{6\}$. This statement is *true*.

In Exercises 5–12, let $A = \{1, 2, 3, 4, 5, 6\}$, $B = \{1, 3, 5\}$, $C = \{1, 6\}$, and $D = \{4\}$.

5. The intersection of sets A and D is the set of all elements in both set A and D. Therefore,

$$A \cap D = \{4\} \quad \text{or} \quad \text{set } D.$$

6. 1 is the only element in both sets B and C, so $B \cap C = \{1\}$.

7. The intersection of set B and the set of no elements (empty set), $B \cap \emptyset$, is the set of no elements or $\emptyset$.

8. No element is common to both sets, so $A \cap \emptyset = \emptyset$, the empty set.

9. The union of sets A and B is the set of all elements that are in either set A or set B or both sets A and B. Since all numbers in set B are also in set A, the set $A \cup B$ will be the same as set A.

$$A \cup B = \{1, 2, 3, 4, 5, 6\} \quad \text{or} \quad \text{set } A$$

10. A union of sets contains all elements that belong to either set (or both).

$$B \cup D = \{1, 3, 4, 5\}$$

11. The set $B \cup C$ is made up of all numbers that are either in set B *or* in set C (or both).

$$B \cup C = \{1, 3, 5, 6\}$$

12. Note that $C \cup B = B \cup C$, so from Exercise 11,

$$C \cup B = \{1, 3, 5, 6\}.$$

13. The first graph represents the set $(-\infty, 5)$. The second graph represents the set $(0, \infty)$. The intersection includes the elements common to both sets, that is, $(0, 5)$.

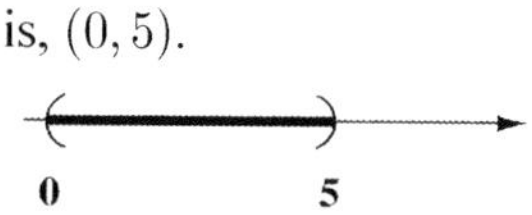

14. The first graph represents the set $(-\infty, 5]$. The second graph represents the set $(-\infty, 2]$. The intersection includes the elements common to both sets, that is, $(-\infty, 2]$.

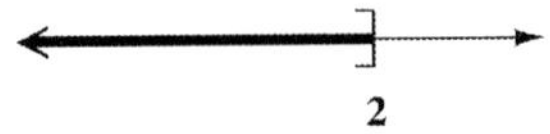

15. The first graph represents the set $[3, \infty)$. The second graph represents the set $[6, \infty)$. The intersection includes the elements common to both sets, that is $[6, \infty)$.

16. The first graph represents the set $(-\infty, -5]$. The second graph represents the set $[6, \infty)$. The union includes all elements in either set, or in both, that is, $(-\infty, -5] \cup [6, \infty)$.

17. The first graph represents the set $(-\infty, 1]$. The second graph represents the set $(-\infty, 8]$. The union includes all elements in either set, or in both, that is, $(-\infty, 8]$.

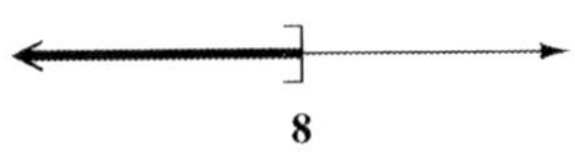

18. The first graph represents the set $[1, \infty)$. The second graph represents the set $[8, \infty)$. The union includes all elements in either set, or in both, that is, $[1, \infty)$.

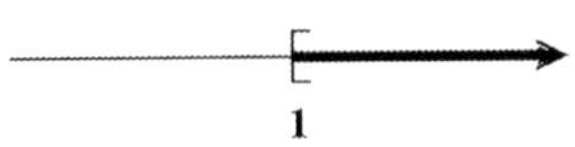

19. Answers will vary. One example is: The intersection of two streets is the region common to *both* streets.

20. When the word *and* is used, it means that the solution will include all numbers that satisfy both inequalities, so you will use the *intersection* of the solution sets to graph the solution set.

When the word *or* is used, it means that the solution will include all numbers that satisfy either one of the inequalities (or both), so you will use the *union* of the solution sets to graph the solution set.

21. $x < 2$ and $x > -3$

The graph of the solution set will be all numbers that are both less than 2 and greater than -3. The solution set is $(-3, 2)$.

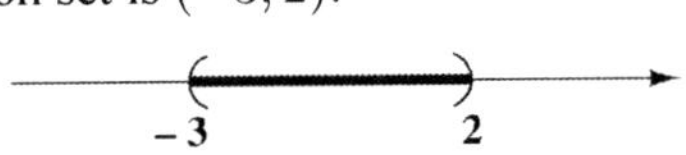

22. $x < 5$ and $x > 0$

And means intersection. The graph of the solution set will be all numbers that are both less than 5 and greater than 0. The overlap is the numbers between 0 and 5, not including 0 and 5. The solution set is $(0, 5)$.

23. $x \leq 2$ and $x \leq 5$

The graph of the solution set will be all numbers that are both less than or equal to 2 and less than or equal to 5. The overlap is the numbers less than or equal to 2. The solution set is $(-\infty, 2]$.

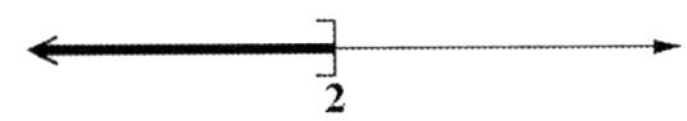

24. $x \geq 3$ and $x \geq 6$

The graph of the solution set will be all numbers that are both greater than or equal to 3 and greater than or equal to 6. This will be all numbers greater than or equal to 6. The solution set is $[6, \infty)$.

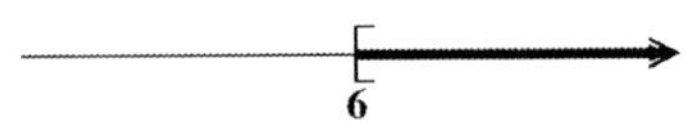

25. $x \leq 3$ and $x \geq 6$

The graph of the solution set will be all numbers that are both less than or equal to 3 and greater than or equal to 6. There are no such numbers. The solution set is $\emptyset$.

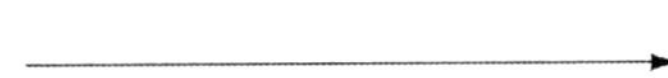

26. $x \leq -1$ and $x \geq 3$

The graph of the solution set will be all numbers that are both less than or equal to -1 and greater than or equal to 3. There are no such numbers. The solution set is $\emptyset$.

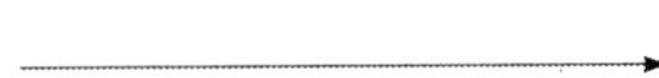

27. $x - 3 \leq 6$ and $x + 2 \geq 7$
$x \leq 9$ and $x \geq 5$

The graph of the solution set is all numbers that are both less than or equal to 9 and greater than or equal to 5. This is the intersection. The elements common to both sets are the numbers between 5 and 9, including the endpoints. The solution set is $[5, 9]$.

28. $x + 5 \leq 11$ and $x - 3 \geq -1$
$x \leq 6$ and $x \geq 2$

The graph of the solution set is all numbers that are both less than or equal to 6 and greater than or equal to 2. This is the intersection. The elements common to both sets are the numbers between 2 and 6, including the endpoints. The solution set is $[2, 6]$.

29. $3x - 4 \leq 8$ and $4x - 1 \leq 15$
$3x \leq 12$ and $4x \leq 16$
$x \leq 4$ and $x \leq 4$

Since both inequalities are identical, the graph of the solution set is the same as the graph of one of the inequalities. The solution set is $(-\infty, 4]$.

30. $7x + 6 \leq 48$ and $-4x \geq -24$
$7x \leq 42$
$x \leq 6$ and $x \leq 6$

Since both inequalities are identical, the graph of the solution set is the same as the graph of one of the inequalities. The solution set is $(-\infty, 6]$.

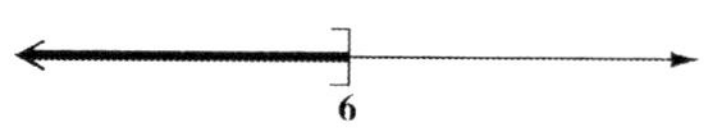

31. $x \leq 1$ or $x \leq 8$

The graph of the solution set will be all numbers that are either less than or equal to 1 or less than or equal to 8. The solution set is $(-\infty, 8]$.

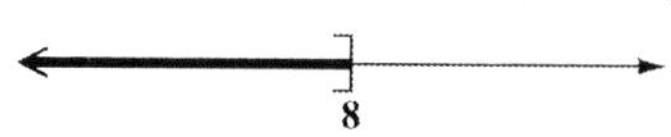

32. $x \geq 1$ or $x \geq 8$

The graph of the solution set will be all numbers that are either greater than or equal to 1 or greater than or equal to 8. The solution set is $[1, \infty)$.

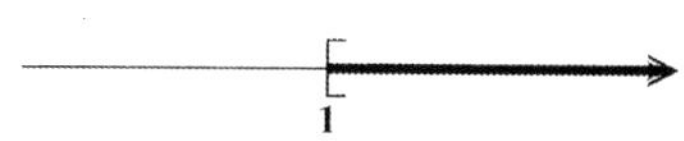

33. $x \geq -2$ or $x \geq 5$

The graph of the solution set will be all numbers that are either greater than or equal to -2 or greater than or equal to 5. The solution set is $[-2, \infty)$.

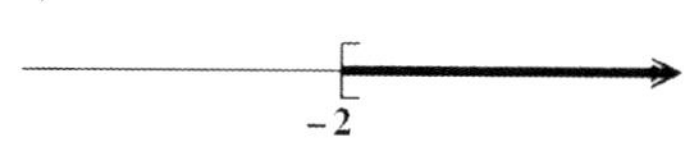

34. $x \leq -2$ or $x \leq 6$

The graph of the solution set will be all numbers that are either less than or equal to -2 or less than or equal to 6. The solution set is $(-\infty, 6]$.

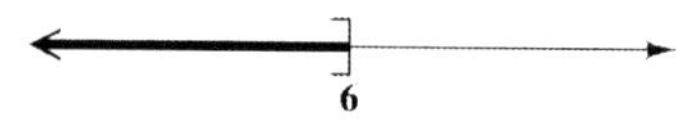

35. $x + 3 \geq 1$ or $x - 8 \leq -4$
$x \geq -2$ or $x \leq 4$

The graph of the solution set will be all numbers that are either greater than or equal to -2 or less than or equal to 4. This is the set of all real numbers. The solution set is $(-\infty, \infty)$.

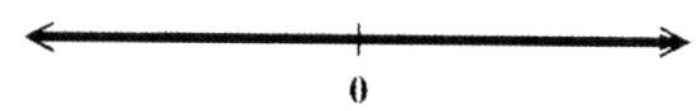

36. $x + 6 \geq 11$ or $x - 4 \leq 3$
$x \geq 5$ or $x \leq 7$

The graph of the solution set will be all numbers that are either greater than or equal to 5 or less than or equal to 7. The solution set is $(-\infty, \infty)$.

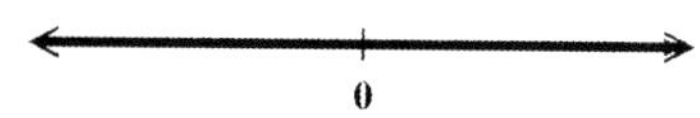

37. $x + 2 > 7$ or $x - 1 < -6$
$x > 5$ or $x < -5$

The graph of the solution set is all numbers either greater than 5 or less than -5. This is the union. The solution set is $(-\infty, -5) \cup (5, \infty)$.

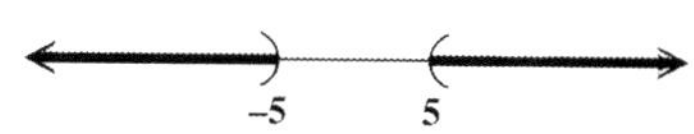

38. $x + 1 > 3$ or $x + 4 < 2$
$x > 2$ or $x < -2$

The graph of the solution set is all numbers either greater than 2 or less than -2. This is the union. The solution set is $(-\infty, -2) \cup (2, \infty)$.

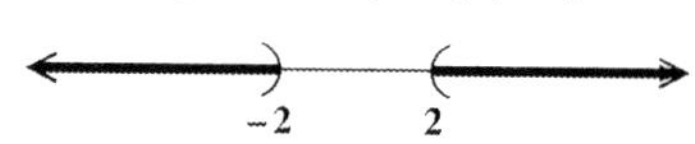

39. $4x - 8 > 0$ or $4x - 1 < 7$
$4x > 8$ or $4x < 8$
$x > 2$ or $x < 2$

The graph of the solution set is all numbers either greater than 2 or less than 2. This is all real numbers except 2. The solution set is $(-\infty, 2) \cup (2, \infty)$.

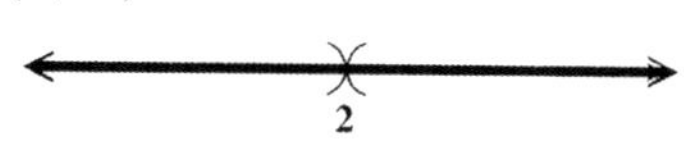

40.
$$\begin{aligned} 3x &< x + 12 &\text{or}&& 3x - 8 &> 10 \\ 2x &< 12 &\text{or}&& 3x &> 18 \\ x &< 6 &\text{or}&& x &> 6 \end{aligned}$$

The graph of the solution set is all numbers less than 6 or greater than 6 (but not equal to 6). The solution set is $(-\infty, 6) \cup (6, \infty)$.

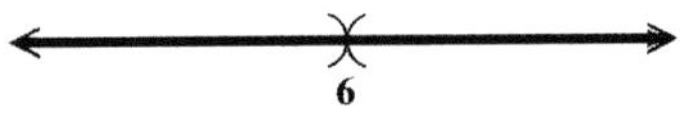

41. $(-\infty, -1] \cap [-4, \infty)$

The intersection is the set of numbers less than or equal to -1 and greater than or equal to -4. The numbers common to both original sets are between, and including, -4 and -1. The simplest interval form is $[-4, -1]$.

42. $[-1, \infty) \cap (-\infty, 9] = (-\infty, 9] \cap [-1, \infty)$

The intersection is the set of numbers less than or equal to 9 and greater than or equal to -1. The numbers common to both original sets are between, and including, -1 and 9. The simplest interval form is $[-1, 9]$.

43. $(-\infty, -6] \cap [-9, \infty)$

The intersection is the set of numbers less than or equal to -6 and greater than or equal to -9. The numbers common to both original sets are between, and including, -9 and -6. The simplest interval form is $[-9, -6]$.

44. $(5, 11] \cap [6, \infty)$

The intersection is the set of numbers between 5 and 11, including 11 but not 5, and greater than or equal to 6. The numbers common to both original sets are between, and including, 6 and 11. The simplest interval form is $[6, 11]$.

45. $(-\infty, 3) \cup (-\infty, -2)$

The union is the set of numbers that are either less than 3 or less than -2, or both. This is all numbers less than 3. The simplest interval form is $(-\infty, 3)$.

46. $[-9, 1] \cup (-\infty, -3)$

The union is the set of numbers between -9 and 1, including -9 and 1, or less than -3. This is all numbers less than, and including 1. The simplest interval form is $(-\infty, 1]$.

47. $[3, 6] \cup (4, 9)$

The union is the set of numbers between, and including, 3 and 6, or between, but not including, 4 and 9. This is the set of numbers greater than or equal to 3 and less than 9. The simplest interval form is $[3, 9)$.

48. $[-1, 2] \cup (0, 5)$

The union is the set of numbers between, and including, -1 and 2, or between, but not including, 0 and 5. This is the set of numbers greater than or equal to -1 and less than 5. The simplest interval form is $[-1, 5)$.

49. $x < -1$ and $x > -5$

The word "and" means to take the intersection of both sets. $x < -1$ and $x > -5$ is true only when

$$-5 < x < -1.$$

The graph of the solution set is all numbers greater than -5 *and* less than -1. This is all numbers between -5 and -1, not including -5 or -1. The solution set is $(-5, -1)$.

50. $x > -1$ and $x < 7$

This is an intersection. The solution set is $(-1, 7)$.

51. $x < 4$ or $x < -2$

The word "or" means to take the union of both sets. The graph of the solution set is all numbers that are either less than 4 *or* less than -2, or both. This is all numbers less than 4. The solution set is $(-\infty, 4)$.

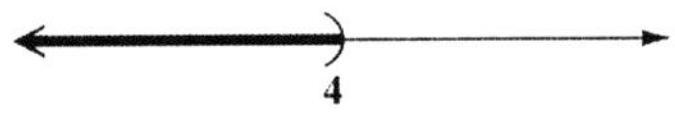

52. $x < 5$ or $x < -3$

This is a union. The solution set is $(-\infty, 5)$.

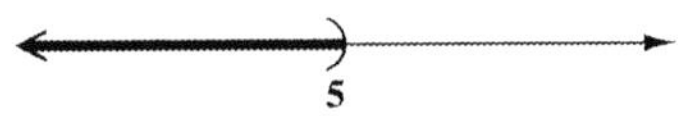

53.
$$\begin{aligned} x + 1 &\geq 5 &\text{and}&& x - 2 &\leq 10 \\ x &\geq 4 &\text{and}&& x &\leq 12 \end{aligned}$$

The word "and" means to take the intersection of both sets. The graph of the solution set is all numbers that are both greater than or equal to 4 *and* less than or equal to 12. This is all numbers between, and including, 4 and 12. The solution set is $[4, 12]$.

54. $2x - 6 \leq -18$ and $2x \geq -18$
$2x \leq -12$
$x \leq -6$ and $x \geq -9$

This is an intersection.
The solution set is $[-9, -6]$.

55. $-3x \leq -6$ or $-3x \geq 0$
$x \geq 2$ or $x \leq 0$

The word "or" means to take the union of both sets. The graph of the solution set is all numbers that are either greater than or equal to 2 *or* less than or equal to 0. The solution set is $(-\infty, 0] \cup [2, \infty)$.

56. $-8x \leq -24$ or $-5x \geq 15$
$x \geq 3$ or $x \leq -3$

This is a union.
The solution set is $(-\infty, -3] \cup [3, \infty)$.

For Exercises 57–60, find the area and perimeter of each of the given yards.

For Luigi's, Mario's, and Than's yards, use the formulas $A = LW$ and $P = 2L + 2W$.

Luigi's yard
$A = 50(30) = 1500 \text{ ft}^2$
$P = 2(50) + 2(30) = 160 \text{ ft}$

Mario's yard
$A = 40(35) = 1400 \text{ ft}^2$
$P = 2(40) + 2(35) = 150 \text{ ft}$

Than's yard
$A = 60(50) = 3000 \text{ ft}^2$
$P = 2(60) + 2(50) = 220 \text{ ft}$

For Joe's yard, use the formulas $A = \frac{1}{2}bh$ and $P = a + b + c$.

Joe's yard
$A = \frac{1}{2}(40)(30) = 600 \text{ ft}^2$
$P = 30 + 40 + 50 = 120 \text{ ft}$

To be fenced, a yard must have a perimeter $P \leq 150$ ft. To be sodded, a yard must have an area $A \leq 1400 \text{ ft}^2$.

57. Find "the yard can be fenced *and* the yard can be sodded."

A yard that can be fenced has $P \leq 150$. Mario and Joe qualify.

A yard that can be sodded has $A \leq 1400$. Again, Mario and Joe qualify.

Find the intersection. Mario's and Joe's yards are common to both sets, so Mario and Joe can have their yards both fenced and sodded.

58. Find "the yard can be fenced *and* the yard cannot be sodded."

A yard that can be fenced has $P \leq 150$. Mario and Joe qualify.

A yard that cannot be sodded has $A > 1400$. Luigi and Than qualify.

Find the intersection. There are no yards common to both sets, so none of them qualify.

59. Find "the yard cannot be fenced *and* the yard can be sodded."

A yard that cannot be fenced has $P > 150$. Luigi and Than qualify.

A yard that can be sodded has $A \leq 1400$. Mario and Joe qualify.

Find the intersection. There are no yards common to both sets, so none of the qualify.

60. Find "the yard cannot be fenced *and* the yard cannot be sodded."

A yard that cannot be fenced has $P > 150$. Luigi and Than qualify.

A yard that cannot be sodded has $A > 1400$. Again, Luigi and Than qualify.

Find the intersection. Luigi's and Than's yards are common to both sets, so Luigi and Than qualify.

61. The set of expenses that are less than \$2500 for public schools *and* are greater than \$3000 for private schools is {Tuition and fees}.

62. The set of expenses that are less than \$2500 for public schools *and* are less than \$3000 for private schools is {Board rates, Dormitory charges}.

63. The set of expenses that are less than \$2200 for public schools *or* are greater than \$3000 for private schools is {Tuition and fees, Board rates}.

64. The set of expenses that are greater than \$10,000 *or* are less than \$2000 is {Tuition and fees}.

3.3 Absolute Value Equations and Inequalities

3.3 Margin Exercises

1. **(a)** $|x| = 3$

The solution set consists of the numbers that are 3 units from 0; that is, the numbers 3 and -3.

Solution set: $\{-3, 3\}$

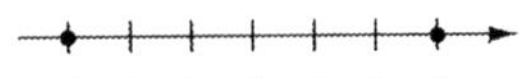

(b) $|x| > 3$

The solution set consists of the numbers that are more than 3 units from 0. This is all numbers greater than 3 *or* all numbers less than -3.

Solution set: $(-\infty, -3) \cup (3, \infty)$

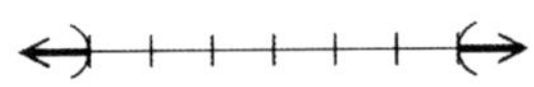

(c) $|x| < 3$

The solution set consists of the numbers that are less than 3 units from 0. This is all numbers between -3 and 3.

Solution set: $(-3, 3)$

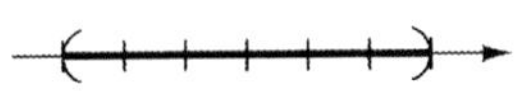

2. **(a)** $|x + 2| = 3$

$$x + 2 = 3 \quad \text{or} \quad x + 2 = -3$$
$$x = 1 \quad \text{or} \quad x = -5$$

Check $x = 1$: $|3| = 3$ *True*

Check $x = -5$: $|-3| = 3$ *True*

Solution set: $\{-5, 1\}$

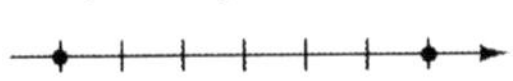

(b) $|3x - 4| = 11$

$$3x - 4 = 11 \quad \text{or} \quad 3x - 4 = -11$$
$$3x = 15 \quad \text{or} \quad 3x = -7$$
$$x = 5 \quad \text{or} \quad x = -\tfrac{7}{3}$$

Check $x = 5$: $|11| = 11$ *True*

Check $x = -\frac{7}{3}$: $|-11| = 11$ *True*

Solution set: $\{-\frac{7}{3}, 5\}$

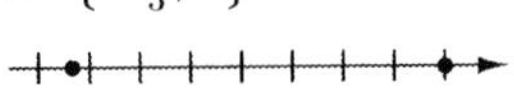

3. **(a)** $|x + 2| > 3$

$$x + 2 > 3 \quad \text{or} \quad x + 2 < -3$$
$$x > 1 \quad \text{or} \quad x < -5$$

We'll assume that we have found the correct boundary points, -5 and 1. These points partition the real number line into 3 intervals. We'll pick a point in each region and check it in the original absolute value inequality, $|x + 2| > 3$.

Check $x = -6$: $|-4| > 3$ *True*
Check $x = 0$: $|2| > 3$ *False*
Check $x = 2$: $|4| > 3$ *True*

Solution set: $(-\infty, -5) \cup (1, \infty)$

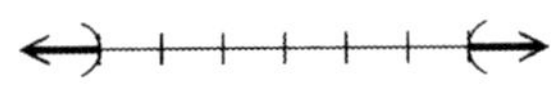

(b) $|3x - 4| \geq 11$

$$3x - 4 \geq 11 \quad \text{or} \quad 3x - 4 \leq -11$$
$$3x \geq 15 \quad \text{or} \quad 3x \leq -7$$
$$x \geq 5 \quad \text{or} \quad x \leq -\tfrac{7}{3}$$

Check $x = -3$, 0, and 6 in $|3x - 4| \geq 11$.

Check $x = -3$: $|-13| \geq 11$ *True*
Check $x = 0$: $|-4| \geq 11$ *False*
Check $x = 6$: $|14| \geq 11$ *True*

Solution set: $\left(-\infty, -\frac{7}{3}\right] \cup [5, \infty)$

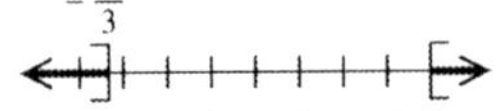

4. **(a)** $|x + 2| < 3$

$$-3 < x + 2 < 3$$
$$-5 < x < 1$$

Check $x = -6$, 0, and 2 in $|x + 2| < 3$.

Check $x = -6$: $|-4| < 3$ *False*
Check $x = 0$: $|2| < 3$ *True*
Check $x = 2$: $|4| < 3$ *False*

Solution set: $(-5, 1)$

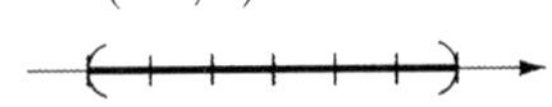

(b) $|3x - 4| \le 11$

$$-11 \le 3x - 4 \le 11$$
$$-7 \le 3x \le 15$$
$$-\tfrac{7}{3} \le x \le 5$$

Check $x = -3$, 0, and 6 in $|3x - 4| \le 11$.

Check $x = -3$: $|-13| \le 11$ *False*
Check $x = 0$: $|-4| \le 11$ *True*
Check $x = 6$: $|14| \le 11$ *False*

Solution set: $\left[-\tfrac{7}{3}, 5\right]$

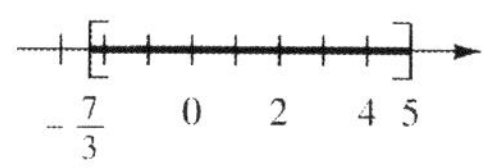

5. **(a)** $|5x + 2| - 9 = -7$
We first *isolate* the absolute value expression, that is, rewrite the equation so that the absolute value expression is alone on one side of the equals sign.
$$|5x + 2| = 2$$

$5x + 2 = 2$ or $5x + 2 = -2$
$5x = 0$ or $5x = -4$
$x = 0$ or $x = -\tfrac{4}{5}$

Check $x = 0$: $|2| - 9 = -7$ *True*
Check $x = -\tfrac{4}{5}$: $|-2| - 9 = -7$ *True*

Solution set: $\left\{-\tfrac{4}{5}, 0\right\}$

(b) $|x + 2| - 3 > 2$
$|x + 2| > 5$ *Isolate.*

$x + 2 > 5$ or $x + 2 < -5$
$x > 3$ or $x < -7$

Solution set: $(-\infty, -7) \cup (3, \infty)$

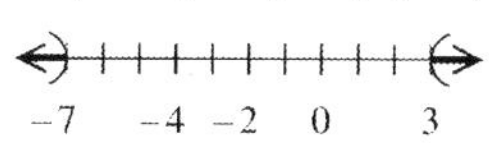

(c) $|3x + 2| + 4 \le 15$
$|3x + 2| \le 11$ *Isolate.*

$$-11 \le 3x + 2 \le 11$$
$$-13 \le 3x \le 9$$
$$-\tfrac{13}{3} \le x \le 3$$

Solution set: $\left[-\tfrac{13}{3}, 3\right]$

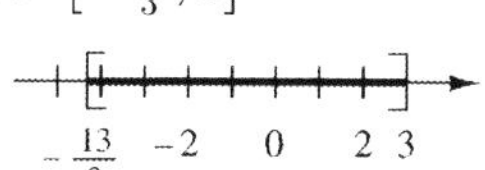

6. **(a)** $|k - 1| = |5k + 7|$

$k - 1 = 5k + 7$ or $k - 1 = -(5k + 7)$
$-8 = 4k$ or $k - 1 = -5k - 7$
$-2 = k$ or $6k = -6$
$k = -1$

Check $k = -2$: $|-3| = |-3|$ *True*
Check $k = -1$: $|-2| = |2|$ *True*

Solution set: $\{-2, -1\}$

(b) $|4r - 1| = |3r + 5|$

$4r - 1 = 3r + 5$ or $4r - 1 = -(3r + 5)$
$r = 6$ or $4r - 1 = -3r - 5$
$7r = -4$
$r = -\tfrac{4}{7}$

Check $r = 6$: $|23| = |23|$ *True*
Check $r = -\tfrac{4}{7}$: $\left|-\tfrac{23}{7}\right| = \left|\tfrac{23}{7}\right|$ *True*

Solution set: $\left\{-\tfrac{4}{7}, 6\right\}$

7. **(a)** $|6x + 7| = -5$

Since the absolute value of an expression can never be negative, there are no solutions for this equation.

Solution set: $\emptyset$

(b) $\left|\tfrac{1}{4}x - 3\right| = 0$
The expression $\tfrac{1}{4}x - 3$ will equal 0 *only* if
$\tfrac{1}{4}x - 3 = 0.$
$x - 12 = 0$ *Multiply by 4.*
$x = 12$

Solution set: $\{12\}$

8. **(a)** $|x| > -1$

The absolute value of a number is never negative. Therefore, the inequality is true for all real numbers.

Solution set: $(-\infty, \infty)$

(b) $|x| < -5$

There is no number whose absolute value is less than a negative number, so this inequality has no solution.

Solution set: $\emptyset$

(c) $|x + 2| \le 0$

The value of $|x + 2|$ will never be less than 0. However, $|x + 2|$ will equal 0 when $x = -2$.

Solution set: $\{-2\}$

3.3 Section Exercises

1. $|x| = 5$ has two solutions, $x = 5$ or $x = -5$. The graph is Choice **E**.

$|x| < 5$ is written $-5 < x < 5$. Notice that -5 and 5 are not included. The graph is Choice **C**, which uses parentheses.

$|x| > 5$ is written $x < -5$ or $x > 5$. The graph is Choice **D**, which uses parentheses.

$|x| \leq 5$ is written $-5 \leq x \leq 5$. This time -5 and 5 are included. The graph is Choice **B**, which uses brackets.

$|x| \geq 5$ is written $x \leq -5$ or $x \geq 5$. The graph is Choice **A**, which uses brackets.

2. $|x| = 9$ has two solutions, $x = 9$ and $x = -9$. The graph is Choice **E**.

$|x| > 9$ is written $x < -9$ or $x > 9$. Notice that -9 and 9 are not included. The graph is Choice **D**, which uses parentheses.

$|x| \geq 9$ is written $x \leq -9$ or $x \geq 9$. This time -9 and 9 are included. The graph is Choice **A**, which uses brackets.

$|x| < 9$ is written $-9 < x < 9$. The graph is Choice **C**, which uses parentheses.

$|x| \leq 9$ is written $-9 \leq x \leq 9$. The graph is Choice **B**, which uses brackets.

3. When solving an absolute value equation or inequality of the form

$$|ax + b| = k, \quad |ax + b| < k, \quad \text{or} \quad |ax + b| > k,$$

where k is a positive number, use *or* for the equality statement and the $>$ statement. Use *and* for the $<$ statement.

4. **(a)** $|ax + b| = k$, $k = 0$
This means the distance from $ax + b$ to 0 is 0, so $ax + b = 0$, which has one solution.

(b) $|ax + b| = k$, $k > 0$
This means the distance from $ax + b$ to 0 is a positive number k, so $ax + b = k$ or $ax + b = -k$. There are two solutions.

(c) $|ax + b| = k$, $k < 0$
This means the distance from $ax + b$ to 0 is a negative number k, which is impossible because distance is always non-negative. There are no solutions.

5. $|x| = 12$

$x = 12$ or $x = -12$

Solution set: $\{-12, 12\}$

6. $|x| = 14$

$x = 14$ or $x = -14$

Solution set: $\{-14, 14\}$

7. $|4x| = 20$

$4x = 20$ or $4x = -20$

$x = 5$ or $x = -5$

Solution set: $\{-5, 5\}$

8. $|5x| = 30$

$5x = 30$ or $5x = -30$

$x = 6$ or $x = -6$

Solution set: $\{-6, 6\}$

9. $|x - 3| = 9$

$x - 3 = 9$ or $x - 3 = -9$

$x = 12$ or $x = -6$

Solution set: $\{-6, 12\}$

10. $|p - 5| = 13$

$p - 5 = 13$ or $p - 5 = -13$

$p = 18$ or $p = -8$

Solution set: $\{-8, 18\}$

11. $|2x + 1| = 9$

$2x + 1 = 9$ or $2x + 1 = -9$

$2x = 8$ or $2x = -10$

$x = 4$ or $x = -5$

Solution set: $\{-5, 4\}$

12. $|2x + 3| = 19$

$2x + 3 = 19$ or $2x + 3 = -19$

$2x = 16$ or $2x = -22$

$x = 8$ or $x = -11$

Solution set: $\{-11, 8\}$

13. $|4r - 5| = 17$

$4r - 5 = 17$ or $4r - 5 = -17$

$4r = 22$ or $4r = -12$

$r = \frac{22}{4} = \frac{11}{2}$ or $r = -3$

Solution set: $\{-3, \frac{11}{2}\}$

14. $|5t - 1| = 21$

$5t - 1 = 21$ or $5t - 1 = -21$

$5t = 22$ or $5t = -20$

$t = \frac{22}{5}$ or $t = -4$

Solution set: $\{-4, \frac{22}{5}\}$

15. $|2x + 5| = 14$

$2x + 5 = 14$ or $2x + 5 = -14$

$2x = 9$ or $2x = -19$

$x = \frac{9}{2}$ or $x = -\frac{19}{2}$

Solution set: $\{-\frac{19}{2}, \frac{9}{2}\}$

16. $|2x-9|=18$

$2x-9=18$ or $2x-9=-18$

$2x=27$ or $2x=-9$

$x=\frac{27}{2}$ or $x=-\frac{9}{2}$

Solution set: $\{-\frac{9}{2}, \frac{27}{2}\}$

17. $\left|\frac{1}{2}x+3\right|=2$

$\frac{1}{2}x+3=2$ or $\frac{1}{2}x+3=-2$

$\frac{1}{2}x=-1$ or $\frac{1}{2}x=-5$

$x=-2$ or $x=-10$

Solution set: $\{-10,-2\}$

18. $\left|\frac{2}{3}q-1\right|=5$

$\frac{2}{3}q-1=5$ or $\frac{2}{3}q-1=-5$

$\frac{2}{3}q=6$ or $\frac{2}{3}q=-4$

$q=\frac{3}{2}(6)$ or $q=\frac{3}{2}(-4)$

$q=9$ or $q=-6$

Solution set: $\{-6,9\}$

19. $\left|1-\frac{3}{4}k\right|=7$

$1-\frac{3}{4}k=7$ or $1-\frac{3}{4}k=-7$

Multiply all sides by 4.

$4-3k=28$ or $4-3k=-28$

$-3k=24$ or $-3k=-32$

$k=-8$ or $k=\frac{32}{3}$

Solution set: $\{-8, \frac{32}{3}\}$

20. $\left|2-\frac{5}{2}m\right|=14$

$2-\frac{5}{2}m=14$ or $2-\frac{5}{2}m=-14$

$-\frac{5}{2}m=12$ or $-\frac{5}{2}m=-16$

$m=(-\frac{2}{5})(12)$ or $m=(-\frac{2}{5})(-16)$

$m=-\frac{24}{5}$ or $m=\frac{32}{5}$

Solution set: $\{-\frac{24}{5}, \frac{32}{5}\}$

21. $|x|>3$

$x>3$ or $x<-3$

Solution set: $(-\infty,-3)\cup(3,\infty)$

22. $|x|>5$

$x<-5$ or $x>5$

Solution set: $(-\infty,-5)\cup(5,\infty)$

23. $|k|\geq 4$

$k\geq 4$ or $k\leq -4$

Solution set: $(-\infty,-4]\cup[4,\infty)$

24. $|r|\geq 6$

$r\leq -6$ or $r\geq 6$

Solution set: $(-\infty,-6]\cup[6,\infty)$

25. $|t+2|>10$

$t+2>10$ or $t+2<-10$

$t>8$ or $t<-12$

Solution set: $(-\infty,-12)\cup(8,\infty)$

26. $|r+5|>20$

$r+5<-20$ or $r+5>20$

$r<-25$ or $r>15$

Solution set: $(-\infty,-25)\cup(15,\infty)$

27. $|3x-1|\geq 8$

$3x-1\geq 8$ or $3x-1\leq -8$

$3x\geq 9$ or $3x\leq -7$

$x\geq 3$ or $x\leq -\frac{7}{3}$

Solution set: $(-\infty,-\frac{7}{3}]\cup[3,\infty)$

28. $|4x+1|\geq 21$

$4x+1\geq 21$ or $4x+1\leq -21$

$4x\geq 20$ or $4x\leq -22$

$x\geq 5$ or $x\leq -\frac{11}{2}$

Solution set: $(-\infty,-\frac{11}{2}]\cup[5,\infty)$

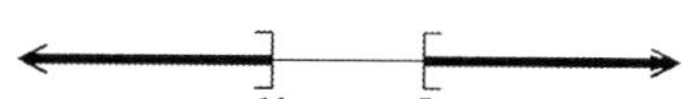

29. $|3-x|>5$

$3-x>5$ or $3-x<-5$

$-x>2$ or $-x<-8$

Multiply by -1,

and reverse the inequality symbols.

$x<-2$ or $x>8$

Solution set: $(-\infty,-2)\cup(8,\infty)$

30. $|5 - x| > 3$

$5 - x > 3$ or $5 - x < -3$

$-x > -2$ or $-x < -8$

$x < 2$ or $x > 8$

Solution set: $(-\infty, 2) \cup (8, \infty)$

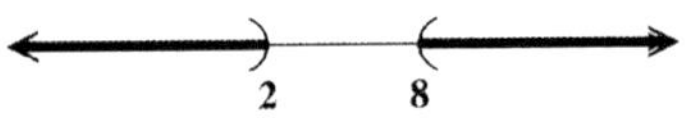

31. **(a)** $|2x + 1| < 9$

The graph of the solution set will be all numbers between -5 and 4, since the absolute value is less than 9.

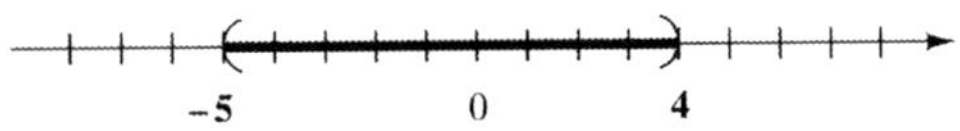

(b) $|2x + 1| > 9$

The graph of the solution set will be all numbers less than -5 or greater than 4, since the absolute value is greater than 9.

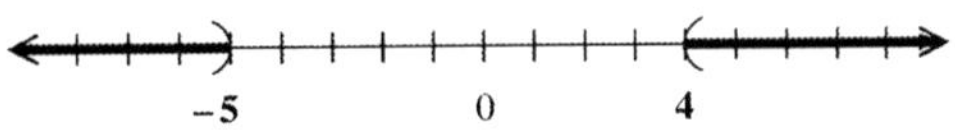

32. **(a)** $|3x - 4| = 5$

The solutions are the numbers at the endpoints in the given graph, $-\frac{1}{3}$ and 3.

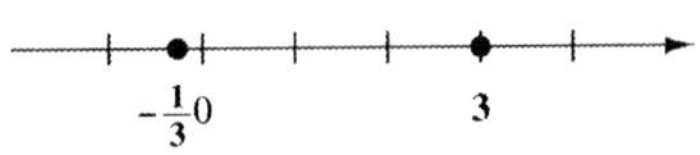

(b) $|3x - 4| > 5$

The solution set is composed of the numbers not in the given graph, not including the endpoints.

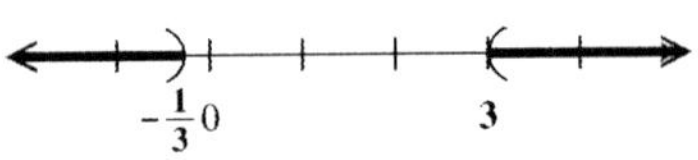

33. $|x| \le 3$

$-3 \le x \le 3$

Solution set: $[-3, 3]$

34. $|x| \le 5$

$-5 \le x \le 5$

Solution set: $[-5, 5]$

35. $|k| < 4$

$-4 < k < 4$

Solution set: $(-4, 4)$

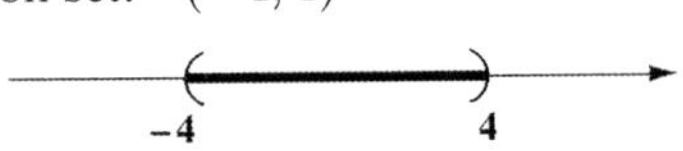

36. $|r| < 6$

$-6 < r < 6$

Solution set: $(-6, 6)$

37. $|t + 2| \le 10$

$-10 \le t + 2 \le 10$

$-12 \le t \le 8$

Solution set: $[-12, 8]$

38. $|r + 5| \le 20$

$-20 \le r + 5 \le 20$

$-25 \le r \le 15$ *Subtract 5.*

Solution set: $[-25, 15]$

39. $|3x - 1| < 8$

$-8 < 3x - 1 < 8$

$-7 < 3x < 9$ *Add 1.*

$-\frac{7}{3} < x < 3$ *Divide by 3.*

Solution set: $(-\frac{7}{3}, 3)$

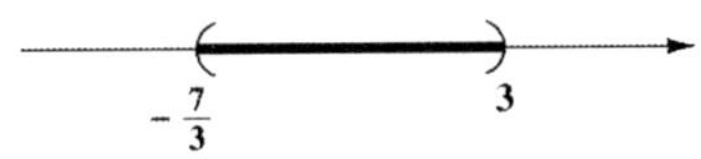

40. $|4x + 1| < 21$

$-21 < 4x + 1 < 21$

$-22 < 4x < 20$

$-\frac{11}{2} < x < 5$

Solution set: $(-\frac{11}{2}, 5)$

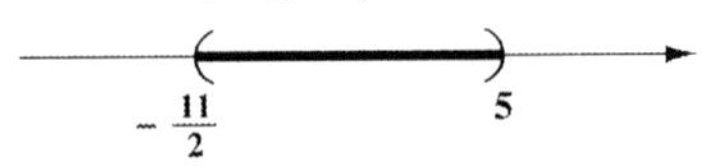

41. $|3 - x| \le 5$

$-5 \le 3 - x \le 5$

$-8 \le -x \le 2$

Multiply each part by -1, and reverse the inequality symbols.

$8 \ge x \ge -2$ or $-2 \le x \le 8$

Solution set: $[-2, 8]$

42. $|5 - x| \le 3$

$-3 \le 5 - x \le 3$

$-8 \le -x \le -2$

Multiply each part by -1, and reverse the inequality symbols.

$8 \ge x \ge 2$ or $2 \le x \le 8$

Solution set: $[2, 8]$

43. $|-4+k| > 9$

$-4+k > 9$ or $-4+k < -9$

$k > 13$ or $k < -5$

Solution set: $(-\infty, -5) \cup (13, \infty)$

44. $|-3+t| > 8$

$-3+t > 8$ or $-3+t < -8$

$t > 11$ or $t < -5$

Solution set: $(-\infty, -5) \cup (11, \infty)$

45. $|7+2z| = 5$

$7+2z = 5$ or $7+2z = -5$

$2z = -2$ or $2z = -12$

$z = -1$ or $z = -6$

Solution set: $\{-6, -1\}$

46. $|9-3p| = 3$

$9-3p = 3$ or $9-3p = -3$

$-3p = -6$ or $-3p = -12$

$p = 2$ or $p = 4$

Solution set: $\{2, 4\}$

47. $|3r-1| \le 11$

$-11 \le 3r - 1 \le 11$

$-10 \le 3r \le 12$

$-\frac{10}{3} \le r \le 4$

Solution set: $\left[-\frac{10}{3}, 4\right]$

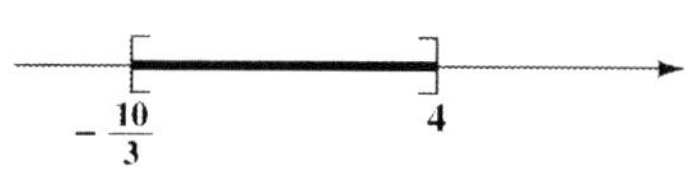

48. $|2s-6| \le 6$

$-6 \le 2s - 6 \le 6$

$0 \le 2s \le 12$

$0 \le s \le 6$

Solution set: $[0, 6]$

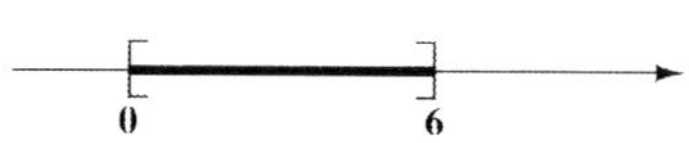

49. $|-6x-6| \le 1$

$-1 \le -6x - 6 \le 1$

$5 \le -6x \le 7$

Divide each side by -6 and reverse the inequality symbols.

$-\frac{5}{6} \ge x \ge -\frac{7}{6}$ or $-\frac{7}{6} \le x \le -\frac{5}{6}$

Solution set: $\left[-\frac{7}{6}, -\frac{5}{6}\right]$

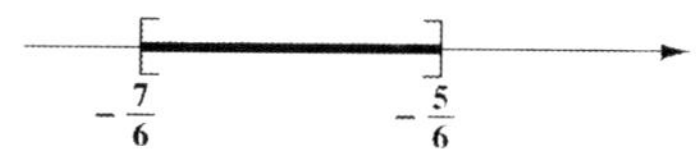

50. $|-2x-6| \le 5$

$-5 \le -2x - 6 \le 5$

$1 \le -2x \le 11$

Divide each side by -2 and reverse the inequality symbols.

$-\frac{1}{2} \ge x \ge -\frac{11}{2}$ or $-\frac{11}{2} \le x \le -\frac{1}{2}$

Solution set: $\left[-\frac{11}{2}, -\frac{1}{2}\right]$

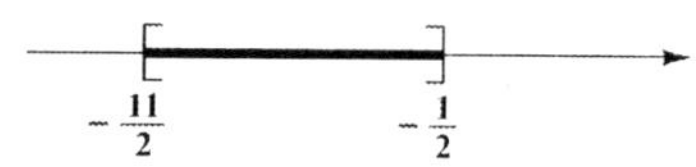

51. $|x| - 1 = 4$

$|x| = 5$

$x = 5$ or $x = -5$

Solution set: $\{-5, 5\}$

52. $|x| + 3 = 10$

$|x| = 7$

$x = 7$ or $x = -7$

Solution set: $\{-7, 7\}$

53. $|x+4| + 1 = 2$

$|x+4| = 1$

$x + 4 = 1$ or $x + 4 = -1$

$x = -3$ or $x = -5$

Solution set: $\{-5, -3\}$

54. $|x+5| - 2 = 12$

$|x+5| = 14$

$x + 5 = 14$ or $x + 5 = -14$

$x = 9$ or $x = -19$

Solution set: $\{-19, 9\}$

55. $|2x+1| + 3 > 8$

$|2x+1| > 5$

$2x + 1 > 5$ or $2x + 1 < -5$

$2x > 4$ or $2x < -6$

$x > 2$ or $x < -3$

Solution set: $(-\infty, -3) \cup (2, \infty)$

56. $|6x - 1| - 2 > 6$

$|6x - 1| > 8$

$6x - 1 > 8$ or $6x - 1 < -8$

$6x > 9$ or $6x < -7$

$x > \frac{3}{2}$ or $x < -\frac{7}{6}$

Solution set: $\left(-\infty, -\frac{7}{6}\right) \cup \left(\frac{3}{2}, \infty\right)$

57. $|x + 5| - 6 \leq -1$

$|x + 5| \leq 5$

$-5 \leq x + 5 \leq 5$

$-10 \leq x \leq 0$

Solution set: $[-10, 0]$

58. $|r - 2| - 3 \leq 4$

$|r - 2| \leq 7$

$-7 \leq r - 2 \leq 7$

$-5 \leq r \leq 9$

Solution set: $[-5, 9]$

59. $|3x + 1| = |2x + 4|$

$3x + 1 = 2x + 4$ or $3x + 1 = -(2x + 4)$

$3x + 1 = -2x - 4$

$5x = -5$

$x = 3$ or $x = -1$

Solution set: $\{-1, 3\}$

60. $|7x + 12| = |x - 8|$

$7x + 12 = x - 8$ or $7x + 12 = -(x - 8)$

$6x = -20$ or $7x + 12 = -x + 8$

$8x = -4$

$x = -\frac{10}{3}$ or $x = -\frac{1}{2}$

Solution set: $\left\{-\frac{10}{3}, -\frac{1}{2}\right\}$

61. $\left|m - \frac{1}{2}\right| = \left|\frac{1}{2}m - 2\right|$

$m - \frac{1}{2} = \frac{1}{2}m - 2$ or $m - \frac{1}{2} = -\left(\frac{1}{2}m - 2\right)$

Multiply by 2. $m - \frac{1}{2} = -\frac{1}{2}m + 2$

$2m - 1 = m - 4$ or $2m - 1 = -m + 4$

$3m = 5$

$m = -3$ or $m = \frac{5}{3}$

Solution set: $\left\{-3, \frac{5}{3}\right\}$

62. $\left|\frac{2}{3}r - 2\right| = \left|\frac{1}{3}r + 3\right|$

$\frac{2}{3}r - 2 = \frac{1}{3}r + 3$ or $\frac{2}{3}r - 2 = -\left(\frac{1}{3}r + 3\right)$

$\frac{1}{3}r = 5$ or $\frac{2}{3}r - 2 = -\frac{1}{3}r - 3$

$r = 15$ or $r = -1$

Solution set: $\{-1, 15\}$

63. $|6x| = |9x + 1|$

$6x = 9x + 1$ or $6x = -(9x + 1)$

$-3x = 1$ or $6x = -9x - 1$

$15x = -1$

$x = -\frac{1}{3}$ or $x = -\frac{1}{15}$

Solution set: $\left\{-\frac{1}{3}, -\frac{1}{15}\right\}$

64. $|13x| = |2x + 1|$

$13x = 2x + 1$ or $13x = -(2x + 1)$

$11x = 1$ or $13x = -2x - 1$

$15x = -1$

$x = \frac{1}{11}$ or $x = -\frac{1}{15}$

Solution set: $\left\{-\frac{1}{15}, \frac{1}{11}\right\}$

65. $|2p - 6| = |2p + 11|$

$2p - 6 = 2p + 11$ or $2p - 6 = -(2p + 11)$

$-6 = 11$ *False* or $2p - 6 = -2p - 11$

$4p = -5$

No solution or $p = -\frac{5}{4}$

Solution set: $\left\{-\frac{5}{4}\right\}$

66. $|3x - 1| = |3x + 9|$

$3x - 1 = 3x + 9$ or $3x - 1 = -(3x + 9)$

$-1 = 9$ *False* or $3x - 1 = -3x - 9$

$6x = -8$

No solution or $x = -\frac{4}{3}$

Solution set: $\left\{-\frac{4}{3}\right\}$

67. $|12t - 3| = -8$

Since the absolute value of an expression can never be negative, there are no solutions for this equation.

Solution set: $\emptyset$

68. $|13w + 1| = -3$

Since the absolute value of an expression can never be negative, there are no solutions for this equation.

Solution set: $\emptyset$

69. $|4x + 1| = 0$

The expression $4x + 1$ will equal 0 *only* for the solution of the equation

$4x + 1 = 0.$

$4x = -1$

$x = \frac{-1}{4}$ or $-\frac{1}{4}$

Solution set: $\left\{-\frac{1}{4}\right\}$

70. $|6r-2|=0$

$$\begin{aligned}6r-2&=0\\6r&=2\\r&=\tfrac{1}{3}\end{aligned}$$

Solution set: $\{\frac{1}{3}\}$

71. $|2q-1|<-6$

There is no number whose absolute value is less than a negative number, so this inequality has no solution.

Solution set: $\emptyset$

72. $|8n+4|<-4$

There is no number whose absolute value is less than a negative number, so this inequality has no solution.

Solution set: $\emptyset$

73. $|x+5|>-9$

Since the absolute value of an expression is always nonnegative (positive or zero), the inequality is true for any real number x.

Solution set: $(-\infty,\infty)$

74. $|x+9|>-3$

Since the absolute value of an expression is always nonnegative (positive or zero), the inequality is true for any real number x.

Solution set: $(-\infty,\infty)$

75. $|7x+3|\le 0$

The absolute value of an expression is always nonnegative (positive or zero), so this inequality is true only when

$$\begin{aligned}7x+3&=0\\7x&=-3\\x&=-\tfrac{3}{7}.\end{aligned}$$

Solution set: $\{-\frac{3}{7}\}$

76. $|4x-1|\le 0$

The absolute value of an expression is always nonnegative (positive or zero), so this inequality is true only when

$$\begin{aligned}4x-1&=0\\4x&=1\\x&=\tfrac{1}{4}.\end{aligned}$$

Solution set: $\{\frac{1}{4}\}$

77. $|5x-2|\ge 0$

The absolute value of an expression is always nonnegative, so the inequality is true for every possible value of x.

Solution set: $(-\infty,\infty)$

78. $|4+7x|\ge 0$

The absolute value of an expression is always nonnegative, so the inequality is true for every possible value of x.

Solution set: $(-\infty,\infty)$

79. $|10z+7|>0$

Since an absolute value of an expression is always nonnegative, there is only one possible value of z that makes this statement false. Solving the equation $10z+7=0$ will give that value of z.

$$\begin{aligned}10z+7&=0\\10z&=-7\\z&=-\tfrac{7}{10}\end{aligned}$$

The solution set of the inequality is

$$\left(-\infty,-\tfrac{7}{10}\right)\cup\left(-\tfrac{7}{10},\infty\right).$$

80. $|4x+1|>0$

Since an absolute value of an expression is always nonnegative, there is only one possible value of x that makes this statement false. Solving the equation $4x+1=0$ will give that value of x.

$$\begin{aligned}4x+1&=0\\4x&=-1\\x&=-\tfrac{1}{4}\end{aligned}$$

The solution set of the inequality is

$$\left(-\infty,-\tfrac{1}{4}\right)\cup\left(-\tfrac{1}{4},\infty\right).$$

81. Let x represent the calcium intake for a specific female. For x to be within 100 mg of 1000 mg, we must have

$$|x-1000|\le 100.$$

$$\begin{aligned}-100\le x-1000&\le 100\\900\le x&\le 1100\end{aligned}$$

82. Let x represent the clotting time for an individual. For x to be within 3.6 seconds of 7.45 seconds, we must have

$$|x-7.45|\le 3.6$$

$$\begin{aligned}-3.6\le x-7.45&\le 3.6\\3.85\le x&\le 11.05\end{aligned}$$

83. Add the given heights with a calculator to get 5359. There are 10 numbers, so divide the sum by 10.

$$\frac{5359}{10} = 535.9$$

The average height is 535.9 ft.

84. $|x - k| < 50$
Substitute 535.9 for k and solve the inequality.

$$|x - 535.9| < 50$$

$$\begin{aligned} -50 < x - 535.9 &< 50 \\ 485.9 < \quad x \quad &< 585.9 \end{aligned}$$

The building with height between 485.9 ft and 585.9 ft is the Hyatt Regency.

85. $|x - k| < 75$
Substitute 535.9 for k and solve the inequality.

$$|x - 535.9| < 75$$

$$\begin{aligned} -75 < x - 535.9 &< 75 \\ 460.9 < \quad x \quad &< 610.9 \end{aligned}$$

The buildings with heights between 460.9 ft and 610.9 ft are Transamerica Tower, Hyatt Regency, and Power and Light Building.

86. **(a)** This would be the opposite of the inequality in Exercise 85, that is,

$$|x - 535.9| \geq 75.$$

(b) $|x - 535.9| \geq 75$

$$\begin{aligned} x - 535.9 \geq 75 \quad &\text{or} \quad x - 535.9 \leq -75 \\ x \geq 610.9 \quad &\text{or} \quad x \leq 460.9 \end{aligned}$$

(c) The buildings that are not within 75 ft of the average have height less than or equal to 460.9 or greater than or equal to 610.9. They are KCTV Tower, One Kansas City Place, City Hall, Federal Office Building, 1201 Walnut Street, Commerce Tower, and City Center Square.

(d) The answer makes sense because it includes all the buildings *not* listed earlier which had heights within 75 ft of the average.

Summary Exercises on Solving Linear and Absolute Value Equations and Inequalities

1.
$$\begin{aligned} 4z + 1 &= 49 \\ 4z &= 48 \\ z &= 12 \end{aligned}$$

Solution set: $\{12\}$

2.
$$|m - 1| = 6$$
$$\begin{aligned} m - 1 = 6 \quad &\text{or} \quad m - 1 = -6 \\ m = 7 \quad &\text{or} \quad m = -5 \end{aligned}$$

Solution set: $\{-5, 7\}$

3.
$$\begin{aligned} 6q - 9 &= 12 + 3q \\ 3q &= 21 \\ q &= 7 \end{aligned}$$

Solution set: $\{7\}$

4.
$$\begin{aligned} 3p + 7 &= 9 + 8p \\ -5p &= 2 \\ p &= -\tfrac{2}{5} \end{aligned}$$

Solution set: $\{-\frac{2}{5}\}$

5. $|a + 3| = -4$

Since the absolute value of an expression is always nonnegative, there is no number that makes this statement true. Therefore, the solution set is $\emptyset$.

6.
$$\begin{aligned} 2m + 1 &\leq m \\ m &\leq -1 \end{aligned}$$

Solution set: $(-\infty, -1]$

7.
$$\begin{aligned} 8r + 2 &\geq 5r \\ 3r &\geq -2 \\ r &\geq -\tfrac{2}{3} \end{aligned}$$

Solution set: $[-\frac{2}{3}, \infty)$

8.
$$\begin{aligned} 4(a - 11) + 3a &= 20a - 31 \\ 4a - 44 + 3a &= 20a - 31 \\ 7a - 44 &= 20a - 31 \\ -13a &= 13 \\ a &= -1 \end{aligned}$$

Solution set: $\{-1\}$

9.
$$\begin{aligned} 2q - 1 &= -7 \\ 2q &= -6 \\ q &= -3 \end{aligned}$$

Solution set: $\{-3\}$

10.
$$\begin{aligned} |3q - 7| - 4 &= 0 \\ |3q - 7| &= 4 \end{aligned}$$
$$\begin{aligned} 3q - 7 = 4 \quad &\text{or} \quad 3q - 7 = -4 \\ 3q = 11 \quad &\text{or} \quad 3q = 3 \\ q = \tfrac{11}{3} \quad &\text{or} \quad q = 1 \end{aligned}$$

Solution set: $\{1, \frac{11}{3}\}$

11.
$$\begin{aligned} 6z - 5 &\leq 3z + 10 \\ 3z &\leq 15 \\ z &\leq 5 \end{aligned}$$

Solution set: $(-\infty, 5]$

12. $|5z - 8| + 9 \geq 7$

$|5z - 8| \geq -2$

The absolute value of an expression is always nonnegative, so the inequality is true for any real number z.

Solution set: $(-\infty, \infty)$

13. $9x - 3(x + 1) = 8x - 7$

$9x - 3x - 3 = 8x - 7$

$6x - 3 = 8x - 7$

$4 = 2x$

$2 = x$

Solution set: $\{2\}$

14. $|x| \geq 8$

$x \geq 8$ or $x \leq -8$

Solution set: $(-\infty, -8] \cup [8, \infty)$

15. $9x - 5 \geq 9x + 3$

$-5 \geq 3$ *False*

This is a false statement, so the inequality is a contradiction.

Solution set: $\emptyset$

16. $13p - 5 > 13p - 8$

$-5 > -8$

This inequality is true for every value of p.

Solution set: $(-\infty, \infty)$

17. $|q| < 5.5$

$-5.5 < q < 5.5$

Solution set: $(-5.5, 5.5)$

18. $4z - 1 = 12 + z$

$3z = 13$

$z = \frac{13}{3}$

Solution set: $\left\{\frac{13}{3}\right\}$

19. $\dfrac{2}{3}x + 8 = \dfrac{1}{4}x$

$8x + 96 = 3x$ *Multiply by 12.*

$5x = -96$

$x = -\frac{96}{5}$

Solution set: $\left\{-\frac{96}{5}\right\}$

20. $-\dfrac{5}{8}x \geq -20$

$-\frac{8}{5}\left(-\frac{5}{8}x\right) \leq -\frac{8}{5}(-20)$

$x \leq 32$

Solution set: $(-\infty, 32]$

21. $\dfrac{1}{4}p < -6$

$4\left(\frac{1}{4}p\right) < 4(-6)$

$p < -24$

Solution set: $(-\infty, -24)$

22. $7z - 3 + 2z = 9z - 8z$

$9z - 3 = z$

$8z = 3$

$z = \frac{3}{8}$

Solution set: $\left\{\frac{3}{8}\right\}$

23. $\dfrac{3}{5}q - \dfrac{1}{10} = 2$

$6q - 1 = 20$ *Multiply by 10.*

$6q = 21$

$q = \frac{21}{6} = \frac{7}{2}$

Solution set: $\left\{\frac{7}{2}\right\}$

24. $|r - 1| < 7$

$-7 < r - 1 < 7$

$-6 < r < 8$

Solution set: $(-6, 8)$

25. $r + 9 + 7r = 4(3 + 2r) - 3$

$8r + 9 = 12 + 8r - 3$

$8r + 9 = 8r + 9$

$0 = 0$ *True*

The last statement is true for any real number r.

Solution set: $(-\infty, \infty)$

26. $6 - 3(2 - p) < 2(1 + p) + 3$

$6 - 6 + 3p < 2 + 2p + 3$

$3p < 5 + 2p$

$p < 5$

Solution set: $(-\infty, 5)$

27. $|2p - 3| > 11$

$2p - 3 > 11$ or $2p - 3 < -11$

$2p > 14$ or $2p < -8$

$p > 7$ or $p < -4$

Solution set: $(-\infty, -4) \cup (7, \infty)$

28. $\dfrac{x}{4} - \dfrac{2x}{3} = -10$

$3x - 8x = -120$ *Multiply by 12.*

$-5x = -120$

$x = 24$

Solution set: $\{24\}$

29. $|5a+1| \le 0$

The expression $|5a+1|$ is never less than 0 since an absolute value expression must be nonnegative. However, $|5a+1| = 0$ if

$$\begin{aligned} 5a+1 &= 0 \\ 5a &= -1 \\ a &= \tfrac{-1}{5} = -\tfrac{1}{5} \end{aligned}$$

Solution set: $\left\{-\frac{1}{5}\right\}$

30. $$\begin{aligned} 5z-(3+z) &\ge 2(3z+1) \\ 5z-3-z &\ge 6z+2 \\ 4z-3 &\ge 6z+2 \\ -2z &\ge 5 \\ z &\le -\tfrac{5}{2} \end{aligned}$$

Solution set: $\left(-\infty, -\frac{5}{2}\right]$

31. $$\begin{aligned} &-2 \le 3x-1 \le 8 \\ &-1 \le 3x \le 9 \\ &-\tfrac{1}{3} \le x \le 3 \end{aligned}$$

Solution set: $\left[-\frac{1}{3}, 3\right]$

32. $$\begin{aligned} &-1 \le 6-x \le 5 \\ &-7 \le -x \le -1 \\ &7 \ge x \ge 1 \quad \text{or} \quad 1 \le x \le 7 \end{aligned}$$

Solution set: $[1, 7]$

33. $|7z-1| = |5z+3|$

$7z-1 = 5z+3$ or $7z-1 = -(5z+3)$

$2z = 4$ or $7z-1 = -5z-3$

$12z = -2$

$z = 2$ or $z = \frac{-2}{12} = -\frac{1}{6}$

Solution set: $\left\{-\frac{1}{6}, 2\right\}$

34. $|p+2| = |p+4|$

$p+2 = p+4$ or $p+2 = -(p+4)$

$2 = 4$ *False* $\quad p+2 = -p-4$

$2p = -6$

No solution or $p = -3$

Solution set: $\{-3\}$

35. $|1-3x| \ge 4$

$1-3x \ge 4$ or $1-3x \le -4$

$-3x \ge 3$ or $-3x \le -5$

$x \le -1$ or $x \ge \frac{5}{3}$

Solution set: $(-\infty, -1] \cup \left[\frac{5}{3}, \infty\right)$

36. $$\begin{aligned} &\frac{1}{2} \le \frac{2}{3}r \le \frac{5}{4} \\ &6 \le 8r \le 15 \quad \textit{Multiply by 12.} \\ &\tfrac{6}{8} \le r \le \tfrac{15}{8} \\ &\tfrac{3}{4} \le r \le \tfrac{15}{8} \end{aligned}$$

Solution set: $\left[\frac{3}{4}, \frac{15}{8}\right]$

37. $$\begin{aligned} -(m+4)+2 &= 3m+8 \\ -m-4+2 &= 3m+8 \\ -m-2 &= 3m+8 \\ -10 &= 4m \\ m &= \tfrac{-10}{4} = -\tfrac{5}{2} \end{aligned}$$

Solution set: $\left\{-\frac{5}{2}\right\}$

38. $$\begin{aligned} \frac{p}{6} - \frac{3p}{5} &= p - 86 \\ 5p - 18p &= 30p - 2580 \quad \textit{Multiply by 30.} \\ -43p &= -2580 \\ p &= \tfrac{-2580}{-43} = 60 \end{aligned}$$

Solution set: $\{60\}$

39. $$\begin{aligned} &-6 \le \frac{3}{2} - x \le 6 \\ &-12 \le 3-2x \le 12 \\ &-15 \le -2x \le 9 \\ &\tfrac{15}{2} \ge x \ge -\tfrac{9}{2} \quad \text{or} \quad -\tfrac{9}{2} \le x \le \tfrac{15}{2} \end{aligned}$$

Solution set: $\left[-\frac{9}{2}, \frac{15}{2}\right]$

40. $$\begin{aligned} &|5-x| < 4 \\ &-4 < 5-x < 4 \\ &-9 < -x < -1 \\ &9 > x > 1 \quad \text{or} \quad 1 < x < 9 \end{aligned}$$

Solution set: $(1, 9)$

41. $|x-1| \ge -6$

The absolute value of an expression is always nonnegative, so the inequality is true for any real number x.

Solution set: $(-\infty, \infty)$

42. $|2r-5| = |r+4|$

$2r-5 = r+4$ or $2r-5 = -(r+4)$

$2r-5 = -r-4$

$3r = 1$

$r = 9$ or $r = \frac{1}{3}$

Solution set: $\left\{\frac{1}{3}, 9\right\}$

43. $$\begin{aligned} 8q-(1-q) &= 3(1+3q)-4 \\ 8q-1+q &= 3+9q-4 \\ 9q-1 &= 9q-1 \quad \textit{True} \end{aligned}$$

This is an identity.

Solution set: $(-\infty, \infty)$

44.
$$\begin{aligned} 8x - (x+3) &= -(2x+1) - 12 \\ 8x - x - 3 &= -2x - 1 - 12 \\ 7x - 3 &= -2x - 13 \\ 9x &= -10 \\ x &= -\tfrac{10}{9} \end{aligned}$$

Solution set: $\left\{-\frac{10}{9}\right\}$

45. $|r-5| = |r+9|$

$r - 5 = r + 9$ or $r - 5 = -(r+9)$

$-5 = 9$ *False* $\quad$ $r - 5 = -r - 9$

$2r = -4$

No solution or $r = -2$

Solution set: $\{-2\}$

46. $|r+2| < -3$

There are no numbers whose absolute value is negative, so this inequality has no solution.

Solution set: $\emptyset$

47.
$$\begin{aligned} 2x + 1 &> 5 &\text{or}\quad 3x + 4 &< 1 \\ 2x &> 4 &\text{or}\quad 3x &< -3 \\ x &> 2 &\text{or}\quad x &< -1 \end{aligned}$$

Solution set: $(-\infty, -1) \cup (2, \infty)$

48.
$$\begin{aligned} 1 - 2x &\ge 5 &\text{and}\quad 7 + 3x &\ge -2 \\ -2x &\ge 4 &\text{and}\quad 3x &\ge -9 \\ x &\le -2 &\text{and}\quad x &\ge -3 \end{aligned}$$

This is an intersection.

Solution set: $[-3, -2]$

Chapter 3 Review Exercises

1. $-\dfrac{2}{3}x < 6$

$-2x < 18$ *Multiply by 3.*

Divide by -2; reverse the inequality symbol.

$x > -9$

Solution set: $(-9, \infty)$

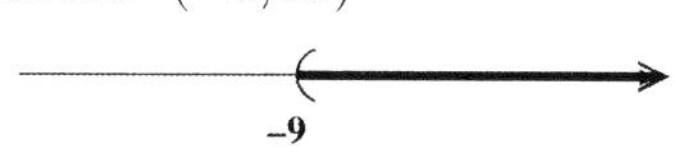

2. $-5x - 4 \ge 11$

$-5x \ge 15$

Divide by -5; reverse the inequality symbol.

$x \le -3$

Solution set: $(-\infty, -3]$

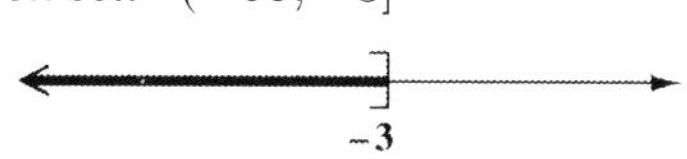

3. $\dfrac{6a+3}{-4} < -3$

Multiply by -4; reverse the inequality symbol.

$$\begin{aligned} 6a + 3 &> 12 \\ 6a &> 9 \\ a &> \tfrac{9}{6} = \tfrac{3}{2} \end{aligned}$$

Solution set: $\left(\frac{3}{2}, \infty\right)$

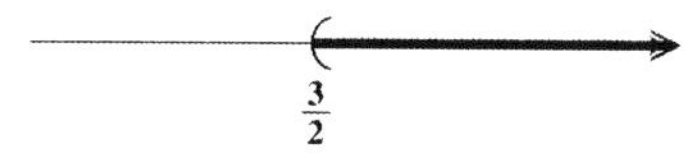

4. $\dfrac{9x+5}{-3} > 3$

Multiply by -3; reverse the inequality symbol.

$$\begin{aligned} 9x + 5 &< -9 \\ 9x &< -14 \\ x &< -\tfrac{14}{9} \end{aligned}$$

Solution set: $\left(-\infty, -\frac{14}{9}\right)$

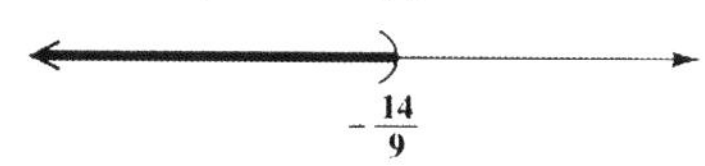

5.
$$\begin{aligned} 5 - (6 - 4t) &\ge 2t - 7 \\ 5 - 6 + 4t &\ge 2t - 7 \\ 4t - 1 &\ge 2t - 7 \\ 2t &\ge -6 \\ t &\ge -3 \end{aligned}$$

Solution set: $[-3, \infty)$

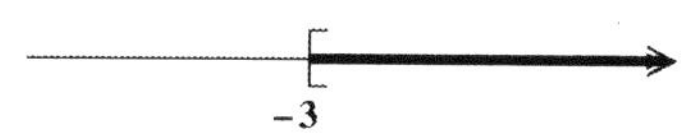

6. $-6 \le 2k \le 24$

$-3 \le k \le 12$

Solution set: $[-3, 12]$

7. $8 \le 3x - 1 < 14$

$9 \le 3x < 15$

$3 \le x < 5$

Solution set: $[3, 5)$

8. $-4 < 3 - 2z < 9$

$-7 < -2z < 6$

Divide all parts by -2; reverse the inequality symbols.

$\frac{7}{2} > z > -3$ or $-3 < z < \frac{7}{2}$

Solution set: $\left(-3, \frac{7}{2}\right)$

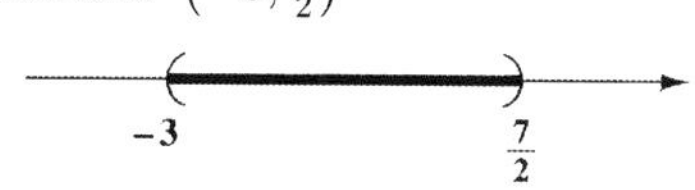

9. Let $x =$ the length of the playground. The width is 22 m, so the perimeter is $2x + 2 \cdot 22$. The perimeter must be no greater than 120 m, so

$$\begin{aligned} 2x + 44 &\leq 120 \\ 2x &\leq 76 \\ x &\leq 38 \end{aligned}$$

The length must be 38 m or less.

10. Let $x =$ the number of miles driven.
The cost of renting from Hughes is \$40 plus the mileage cost of $.15x$, while the cost of renting from Ames is \$48 plus the mileage cost of $.10x$.

$$\begin{aligned} \text{Cost from Hughes} &\leq \text{Cost from Ames} \\ 40 + .15x &\leq 48 + .10x \\ .15x &\leq 8 + .10x \\ .05x &\leq 8 \\ x &\leq 160 \end{aligned}$$

This is the number of miles *per day*, so for 3 days, we would have to travel $3(160) = 480$ miles or less to make the Hughes cost at most as much as the Ames cost.

11. Let $x =$ the student's score on the fifth test.
The average of the five test scores must be at least 70%. The inequality is

$$\begin{aligned} \frac{75 + 79 + 64 + 71 + x}{5} &\geq 70. \\ 75 + 79 + 64 + 71 + x &\geq 350 \\ 289 + x &\geq 350 \\ x &\geq 61 \end{aligned}$$

The student will pass algebra if any score greater than or equal to 61% on the fifth test is achieved.

12. The result, $-8 < -13$, is a false statement. There are no real numbers that make this inequality true. The solution set is $\emptyset$.

For Exercises 13–16, let $A = \{\text{a, b, c, d}\}$, $B = \{\text{a, c, e, f}\}$, and $C = \{\text{a, e, f, g}\}$.

13. $A \cap B = \{\text{a, b, c, d}\} \cap \{\text{a, c, e, f}\} = \{\text{a, c}\}$

14. $A \cap C = \{\text{a, b, c, d}\} \cap \{\text{a, e, f, g}\} = \{\text{a}\}$

15. $B \cup C = \{\text{a, c, e, f}\} \cup \{\text{a, e, f, g}\} = \{\text{a, c, e, f, g}\}$

16. $A \cup C = \{\text{a, b, c, d}\} \cup \{\text{a, e, f, g}\} = \{\text{a, b, c, d, e, f, g}\}$

17. $x > 6$ and $x < 9$

The graph of the solution set will be all numbers which are both greater than 6 and less than 9. The overlap is the numbers between 6 and 9, not including the endpoints.

Solution set: $(6, 9)$

18. $x + 4 > 12$ and $x - 2 < 12$
$x > 8$ and $x < 14$

The graph of the solution set will be all numbers between 8 and 14, not including the endpoints.

Solution set: $(8, 14)$

19. $x > 5$ or $x \leq -3$

The graph of the solution set will be all numbers that are either greater than 5 or less than or equal to -3.

Solution set: $(-\infty, -3] \cup (5, \infty)$

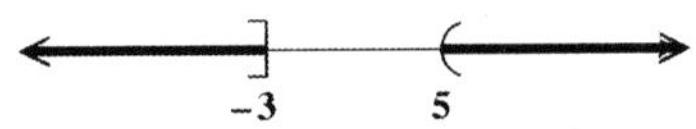

20. $x \geq -2$ or $x < 2$

The graph of the solution set will be all numbers that are either greater than or equal to -2 or less than 2. All real numbers satisfy these criteria.

Solution set: $(-\infty, \infty)$

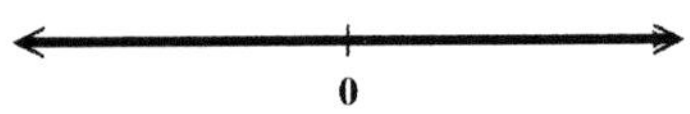

21. $x - 4 > 6$ and $x + 3 \leq 10$
$x > 10$ and $x \leq 7$

The graph of the solution set will be all numbers that are both greater than 10 and less than or equal to 7. There are no real numbers satisfying these criteria.

Solution set: $\emptyset$

22. $-5x + 1 \geq 11$ or $3x + 5 \geq 26$
$-5x \geq 10$ or $3x \geq 21$
$x \leq -2$ or $x \geq 7$

The graph of the solution set will be all numbers that are either less than or equal to -2 or greater than or equal to 7.

Solution set: $(-\infty, -2] \cup [7, \infty)$

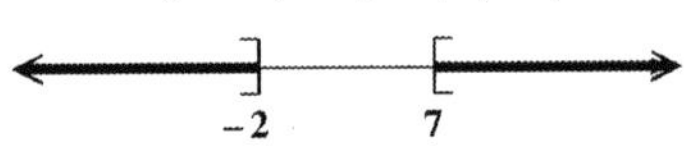

23. $(-3, \infty) \cap (-\infty, 4)$

$(-3, \infty)$ includes all real numbers greater than -3.
$(-\infty, 4)$ includes all real numbers less than 4.
Find the intersection. The numbers common to both sets are greater than -3 and less than 4.

$$-3 < x < 4$$

Solution set: $(-3, 4)$

24. $(-\infty, 6) \cap (-\infty, 2)$

$(-\infty, 6)$ includes all real numbers less than 6.
$(-\infty, 2)$ includes all real numbers less than 2.
Find the intersection. The numbers common to both sets are less than 2.

Solution set: $(-\infty, 2)$

25. $(4, \infty) \cup (9, \infty)$

$(4, \infty)$ includes all real numbers greater than 4.
$(9, \infty)$ includes all real numbers greater than 9.
Find the union. The numbers in the first set, the second set, or in both sets are all the real numbers that are greater than 4.

Solution set: $(4, \infty)$

26. $(1, 2) \cup (1, \infty)$

$(1, 2)$ includes the real numbers between 1 and 2, not including 1 and 2.
$(1, \infty)$ includes all real numbers greater than 1.
Find the union. The numbers in the first set, the second set, or in both sets are all real numbers greater than 1.

Solution set: $(1, \infty)$

27. **(a)** The occupations for which the median earnings for men are less than \$600/week are Sales and office and Production, transportation, and material moving.

The occupations for which the median earnings for women are greater than \$370/week are all the listed occupations except Sales and office.

Thus, the only occupation for which median earnings for men are less than \$600/week *and* for women are greater than \$370/week is Production, transportation, and material moving.

(b) *All* of the listed occupations have median earnings for men greater than \$600 or for women less than \$400.

28. $|x| = 7$
$x = 7$ or $x = -7$

Solution set: $\{-7, 7\}$

29. $|x + 2| = 9$

$x + 2 = 9$ or $x + 2 = -9$
$x = 7$ or $x = -11$

Solution set: $\{-11, 7\}$

30. $|3k - 7| = 8$
$3k - 7 = 8$ or $3k - 7 = -8$
$3k = 15$ or $3k = -1$
$k = 5$ or $k = -\frac{1}{3}$

Solution set: $\left\{-\frac{1}{3}, 5\right\}$

31. $|z - 4| = -12$

Since the absolute value of an expression can never be negative, there are no solutions for this equation.

Solution set: $\emptyset$

32. $|2k - 7| + 4 = 11$
$|2k - 7| = 7$
$2k - 7 = 7$ or $2k - 7 = -7$
$2k = 14$ or $2k = 0$
$k = 7$ or $k = 0$

Solution set: $\{0, 7\}$

33. $|4a + 2| - 7 = -3$
$|4a + 2| = 4$
$4a + 2 = 4$ or $4a + 2 = -4$
$4a = 2$ or $4a = -6$
$a = \frac{2}{4}$ or $a = -\frac{6}{4}$
$a = \frac{1}{2}$ or $a = -\frac{3}{2}$

Solution set: $\left\{-\frac{3}{2}, \frac{1}{2}\right\}$

34. $|3p + 1| = |p + 2|$
$3p + 1 = p + 2$ or $3p + 1 = -(p + 2)$
$2p = 1$ or $3p + 1 = -p - 2$
$4p = -3$
$p = \frac{1}{2}$ or $p = -\frac{3}{4}$

Solution set: $\left\{-\frac{3}{4}, \frac{1}{2}\right\}$

35. $|2m - 1| = |2m + 3|$
$2m - 1 = 2m + 3$ or $2m - 1 = -(2m + 3)$
$-1 = 3$ *False* or $2m - 1 = -2m - 3$
$4m = -2$
No solution or $m = -\frac{2}{4} = -\frac{1}{2}$

Solution set: $\left\{-\frac{1}{2}\right\}$

36. $|x| < 14$
$-14 < x < 14$

Solution set: $(-14, 14)$

37. $|-x+6| \le 7$

$-7 \le -x+6 \le 7$

$-13 \le -x \le 1$

Multiply by -1; reverse the inequality symbols.

$13 \ge x \ge -1$ or $-1 \le x \le 13$

Solution set: $[-1, 13]$

38. $|2p+5| \le 1$

$-1 \le 2p+5 \le 1$

$-6 \le 2p \le -4$

$-3 \le p \le -2$

Solution set: $[-3, -2]$

39. $|x+1| \ge -3$

Since the absolute value of an expression is always nonnegative (positive or zero), the inequality is *true* for any real number x.

Solution set: $(-\infty, \infty)$

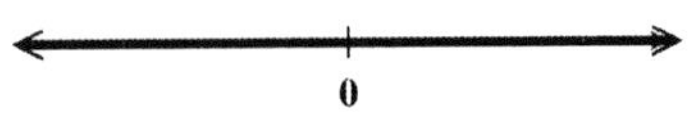

40. $|5r-1| > 9$

$5r-1 > 9$ or $5r-1 < -9$

$5r > 10$ or $5r < -8$

$r > 2$ or $r < -\frac{8}{5}$

Solution set: $\left(-\infty, -\frac{8}{5}\right) \cup (2, \infty)$

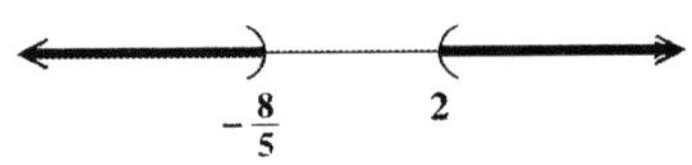

41. $|3x+6| \ge 0$

The absolute value of an expression is always nonnegative, so the inequality is true for any real number x.

Solution set: $(-\infty, \infty)$

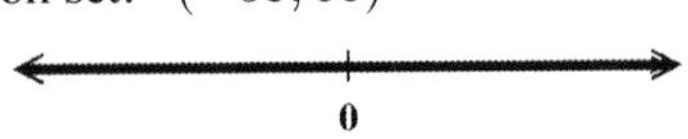

42. **[3.1]** $(7-2x)+3(5-3x) \ge x+8$

$7-2x+15-9x \ge x+8$

$-11x+22 \ge x+8$

$-12x \ge -14$

$x \le \frac{-14}{-12}$

$x \le \frac{7}{6}$

Solution set: $\left(-\infty, \frac{7}{6}\right]$

43. **[3.2]** $x < 5$ and $x \ge -4$

The real numbers that are common to both sets are the numbers greater than or equal to -4 and less than 5.

$$-4 \le x < 5$$

Solution set: $[-4, 5)$

44. **[3.1]** $\frac{3}{4}(a-2) - \frac{1}{3}(5-2a) < -2$

$9(a-2) - 4(5-2a) < -24$ *Multiply by 12.*

$9a - 18 - 20 + 8a < -24$

$17a - 38 < -24$

$17a < 14$

$a < \frac{14}{17}$

Solution set: $\left(-\infty, \frac{14}{17}\right)$

45. **[3.1]** Let $x =$ the employee's earnings during the fifth month. The average of the five months must be at least \$1000.

$$\frac{900+1200+1040+760+x}{5} \ge 1000$$

$$900+1200+1040+760+x \ge 5000$$

$$3900 + x \ge 5000$$

$$x \ge 1100$$

Any amount greater than or equal to \$1100 will qualify the employee for the pension plan.

46. **[3.1]** $-5r \ge -10$

$r \le 2$ *Divide by -5; reverse inequality.*

Solution set: $(-\infty, 2]$

47. **[3.3]** $|7x-2| > 9$

$7x-2 > 9$ or $7x-2 < -9$

$7x > 11$ or $7x < -7$

$x > \frac{11}{7}$ or $x < -1$

Solution set: $(-\infty, -1) \cup \left(\frac{11}{7}, \infty\right)$

48. **[3.3]** $|2x-10| = 20$

$2x-10 = 20$ or $2x-10 = -20$

$2x = 30$ or $2x = -10$

$x = 15$ or $x = -5$

Solution set: $\{-5, 15\}$

49. **[3.3]** $|m+3| \le 13$

$-13 \le m+3 \le 13$

$-16 \le m \le 10$

Solution set: $[-16, 10]$

50. **[3.2]** $x \geq -2$ or $x < 4$

The solution set includes all numbers either greater than or equal to -2 or all numbers less than 4. This is the union and is the set of all real numbers. The solution set is $(-\infty, \infty)$.

51. **[3.3]** $|m - 1| = |2m + 3|$

$$\begin{aligned} m - 1 &= 2m + 3 &\text{or}\quad m - 1 &= -(2m + 3) \\ & & m - 1 &= -2m - 3 \\ & & 3m &= -2 \\ -4 &= m &\text{or}\quad m &= -\tfrac{2}{3} \end{aligned}$$

Solution set: $\left\{-4, -\frac{2}{3}\right\}$

52. **[3.2]** $x > 6$ and $x < 8$

The graph of the solution set is all numbers both greater than 6 *and* less than 8. This is the intersection. The elements common to both sets are the numbers between 6 and 8, not including the endpoints. The solution set is $(6, 8)$.

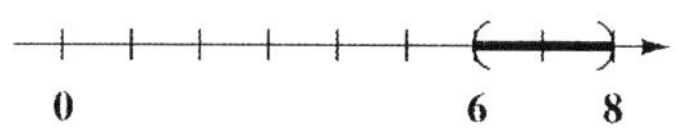

53. **[3.2]** $-5x + 1 \geq 11$ or $3x + 5 \geq 26$

$$\begin{aligned} -5x &\geq 10 &\text{or}\quad 3x &\geq 21 \\ x &\leq -2 &\text{or}\quad x &\geq 7 \end{aligned}$$

The graph of the solution set is all numbers either less than or equal to -2 *or* greater than or equal to 7. This is the union. The solution set is $(-\infty, -2] \cup [7, \infty)$.

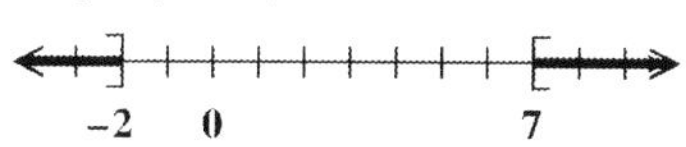

54. **[3.3]** **(a)** $|5x + 3| < k$

If $k < 0$, then $|5x + 3|$ would be less than a negative number. Since the absolute value of an expression is always nonnegative (positive or zero), the solution set is $\emptyset$.

(b) $|5x + 3| > k$

If $k < 0$, then $|5x + 3|$ would be greater than a negative number. Since the absolute value of an expression is always nonnegative (positive or zero), the solution set is the set of all real numbers, $(-\infty, \infty)$.

(c) $|5x + 3| = k$

If $k < 0$, then $|5x + 3|$ would be equal to a negative number. Since the absolute value of an expression is always nonnegative (positive or zero), the solution set is $\emptyset$.

Chapter 3 Test

1. When multiplying or dividing each side of an inequality by a negative number, we must reverse the direction of the inequality symbol.

2.

$$\begin{aligned} 4 - 6(x + 3) &\leq -2 - 3(x + 6) + 3x \\ 4 - 6x - 18 &\leq -2 - 3x - 18 + 3x \\ -6x - 14 &\leq -20 \\ -6x &\leq -6 \end{aligned}$$

Divide each side by -6 and reverse the inequality symbol.

$$x \geq 1$$

Solution set: $[1, \infty)$

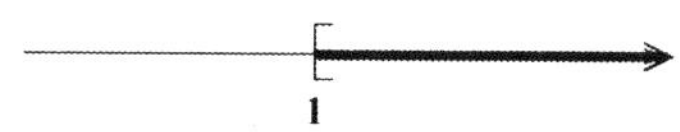

3.

$$-\frac{4}{7}x > -16$$

$$-4x > -112 \qquad \textit{Multiply by 7.}$$

Divide each side by -4 and reverse the inequality symbol.

$$x < 28$$

Solution set: $(-\infty, 28)$

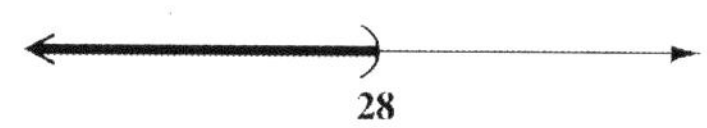

4.

$$-6 \leq \frac{4}{3}x - 2 \leq 2$$

$$\begin{aligned} -18 \leq 4x - 6 \leq 6 &\qquad \textit{Multiply by 3.} \\ -12 \leq 4x \leq 12 &\qquad \textit{Add 6.} \\ -3 \leq x \leq 3 &\qquad \textit{Divide by 4.} \end{aligned}$$

Solution set: $[-3, 3]$

5. Since all the choices have -3 for the coefficient of x, we'll multiply each side of the given inequality by -3, remembering to reverse the direction of the inequality symbol.

$$\begin{aligned} x &< -3 \\ -3(x) &> -3(-3) \\ -3x &> 9 \end{aligned}$$

The answer is choice **C.**

6. Let $x =$ the number of U.S. travelers to Europe (in millions).

(a) From the graph, $x \geq 9$ during 1997–2000.

(b) From the graph, $x < 9$ during 1994–1996.

(c) From the graph, $9 < x < 12$ during 1997–1999.

7. Let $x =$ the score on the fourth test.

$$\frac{83+76+79+x}{4} \geq 80$$
$$\frac{238+x}{4} \geq 80$$
$$238+x \geq 320$$
$$x \geq 82$$

The minimum score must be 82% to guarantee a B.

8. $C = 50x + 5000$; $R = 60x$

$$R \geq C$$
$$60x \geq 50x + 5000$$
$$10x \geq 5000$$
$$x \geq 500$$

The solution set is $[500, \infty)$.

9. **(a)** $A \cap B = \{1,2,5,7\} \cap \{1,5,9,12\}$
$= \{1,5\}$

(b) $A \cup B = \{1,2,5,7\} \cup \{1,5,9,12\}$
$= \{1,2,5,7,9,12\}$

10. $x \leq 2$ and $x \geq 2$

The only number common to *both* inequalities is 2.

Solution set: $\{2\}$

11. $3k \geq 6$ and $k - 4 < 5$
$k \geq 2$ and $k < 9$

The solution set is all numbers both greater than or equal to 2 *and* less than 9. This is the intersection. The numbers common to both sets are between 2 and 9, including 2 but not 9. The solution set is $[2, 9)$.

12. $-4x \leq -24$ or $4x - 2 < 10$
$4x < 12$
$x \geq 6$ or $x < 3$

The solution set is all numbers less than 3 or greater than or equal to 6. This is the union. The solution set is $(-\infty, 3) \cup [6, \infty)$.

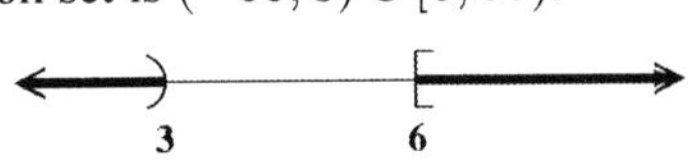

13. $|4x+3| \leq 7$

$$-7 \leq 4x + 3 \leq 7$$
$$-10 \leq 4x \leq 4$$
$$-\tfrac{10}{4} \leq x \leq 1 \quad \text{or} \quad -\tfrac{5}{2} \leq x \leq 1$$

Solution set: $\left[-\frac{5}{2}, 1\right]$

14. $|5 - 6x| > 12$

$5 - 6x > 12$ or $5 - 6x < -12$
$-6x > 7$ or $-6x < -17$
$x < -\frac{7}{6}$ or $x > \frac{17}{6}$

Solution set: $\left(-\infty, -\frac{7}{6}\right) \cup \left(\frac{17}{6}, \infty\right)$

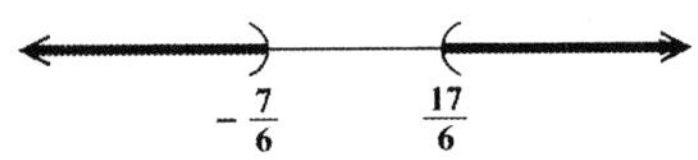

15. $|7 - x| \leq -1$

Since the absolute value of an expression is always nonnegative (positive or zero), the inequality is *false* for any real number x.

Solution set: $\emptyset$

16. $|3k - 2| + 1 = 8$
$|3k - 2| = 7$ *Isolate.*

$3k - 2 = 7$ or $3k - 2 = -7$
$3k = 9$ or $3k = -5$
$k = 3$ or $k = -\frac{5}{3}$

Solution set: $\left\{-\frac{5}{3}, 3\right\}$

17. $|3 - 5x| = |2x + 8|$
$3 - 5x = 2x + 8$ or $3 - 5x = -(2x + 8)$
$-7x = 5$ or $3 - 5x = -2x - 8$
$-3x = -11$
$x = -\frac{5}{7}$ or $x = \frac{11}{3}$

Solution set: $\left\{-\frac{5}{7}, \frac{11}{3}\right\}$

18. $|4x + 3| + 5 = 4$
$|4x + 3| = -1$

Since the absolute value of an expression can never be negative, there are no solutions for this equation.

Solution set: $\emptyset$

Cumulative Review Exercises (Chapters 1–3)

1. $\frac{108}{144} = \frac{3 \cdot 36}{4 \cdot 36} = \frac{3}{4}$

2.
$$\frac{8(7) - 5(6+2)}{3 \cdot 5 + 1} \geq 1 \quad ?$$
$$\frac{8(7) - 5(8)}{3 \cdot 5 + 1} \geq 1 \quad ?$$
$$\frac{56 - 40}{15 + 1} \geq 1 \quad ?$$
$$\frac{16}{16} \geq 1 \quad ?$$
$$1 \geq 1 \quad \textit{True}$$

The inequality is *true*.

3.
$$\begin{aligned} &\frac{5}{6} + \frac{1}{4} - \frac{7}{15} \\ &= \frac{10 \cdot 5}{10 \cdot 6} + \frac{15 \cdot 1}{15 \cdot 4} - \frac{4 \cdot 7}{4 \cdot 15} \quad \textit{LCD} = 60 \\ &= \frac{50}{60} + \frac{15}{60} - \frac{28}{60} \\ &= \frac{37}{60} \end{aligned}$$

4.
$$\begin{aligned} \frac{9}{8} \cdot \frac{16}{3} \div \frac{5}{8} &= \left(\frac{9}{8} \cdot \frac{16}{3}\right) \cdot \frac{8}{5} \\ &= \frac{3}{1} \cdot \frac{16}{1} \cdot \frac{1}{5} \\ &= \frac{3 \cdot 16}{5} \\ &= \frac{48}{5} \end{aligned}$$

5.
$$\begin{aligned} 9 - (-4) + (-2) &= 9 + 4 + (-2) \\ &= 13 + (-2) \\ &= 11 \end{aligned}$$

6. $\frac{-4(9)(-2)}{-3^2} = \frac{-36(-2)}{-(9)} = \frac{72}{-9} = -8$

7.
$$\begin{aligned} &|-7 - 1|(-4) + (-4) \\ &= |-8|(-4) + (-4) \\ &= 8(-4) + (-4) \\ &= -32 + (-4) \\ &= -36 \end{aligned}$$

8. $(-5)^3 = (-5)(-5)(-5) = -125$

9. $\left(\frac{3}{2}\right)^4 = \left(\frac{3}{2}\right)\left(\frac{3}{2}\right)\left(\frac{3}{2}\right)\left(\frac{3}{2}\right) = \frac{81}{16}$

10. Let $x = 2$, $y = -3$, and $z = 4$.
$$\begin{aligned} -2y + 4(x - 3z) &= -2(-3) + 4(2 - 3 \cdot 4) \\ &= 6 + 4(2 - 12) \\ &= 6 + 4(-10) \\ &= 6 - 40 \\ &= -34 \end{aligned}$$

11. Let $x = 2$, $y = -3$, and $z = 4$.
$$\begin{aligned} \frac{3x^2 - y^2}{4z} &= \frac{3(2)^2 - (-3)^2}{4(4)} \\ &= \frac{3(4) - 9}{16} \\ &= \frac{12 - 9}{16} \\ &= \frac{3}{16} \end{aligned}$$

12. $7(k + m) = 7k + 7m$ illustrates the *distributive property* because 7 was distributed to each element within the parentheses.

13. $3 + (5 + 2) = 3 + (2 + 5)$ illustrates the *commutative property* because the order of addition within the parentheses was changed.

14.
$$\begin{aligned} &-4(k + 2) + 3(2k - 1) \\ &= -4k - 8 + 6k - 3 \quad \textit{Distributive property} \\ &= -4k + 6k - 8 - 3 \\ &= 2k - 11 \end{aligned}$$

15.
$$\begin{aligned} 4 - 5(a + 2) &= 3(a + 1) - 1 \\ 4 - 5a - 10 &= 3a + 3 - 1 \\ -5a - 6 &= 3a + 2 \\ -8a - 6 &= 2 \\ -8a &= 8 \\ a &= -1 \end{aligned}$$

Check $a = -1$: $4 - 5 = 0 - 1$ *True*

Solution set: $\{-1\}$

16.
$$\begin{aligned} \frac{2}{3}x + \frac{3}{4}x &= -17 \\ 12\left(\frac{2}{3}x + \frac{3}{4}x\right) &= 12(-17) \quad \textit{Multiply by LCD, 12.} \\ 4(2x) + 3(3x) &= 12(-17) \\ 8x + 9x &= -204 \\ 17x &= -204 \\ x &= -12 \end{aligned}$$

Check $x = -12$: $-8 - 9 = -17$ *True*

Solution set: $\{-12\}$

17.
$$\frac{2x+3}{5} = \frac{x-4}{2}$$
$$10\left(\frac{2x+3}{5}\right) = 10\left(\frac{x-4}{2}\right) \quad \textit{Multiply by LCD, 10.}$$
$$2(2x+3) = 5(x-4)$$
$$4x+6 = 5x-20$$
$$-x+6 = -20$$
$$-x = -26$$
$$x = 26$$

Check $x = 26$: $\frac{55}{5} = \frac{22}{2}$ *True*

Solution set: $\{26\}$

18. $|3m-5| = |m+2|$

$3m-5 = m+2$ or $3m-5 = -(m+2)$
$2m-5 = 2$ or $3m-5 = -m-2$
$2m = 7$ or $4m-5 = -2$
$m = \frac{7}{2}$ or $4m = 3$
$m = \frac{3}{4}$

Check $m = \frac{7}{2}$: $\left|\frac{11}{2}\right| = \left|\frac{11}{2}\right|$ *True*

Check $m = \frac{3}{4}$: $\left|-\frac{11}{4}\right| = \left|\frac{11}{4}\right|$ *True*

Solution set: $\left\{\frac{3}{4}, \frac{7}{2}\right\}$

19. Solve $3x+4y = 24$ for y.
$$4y = 24-3x \quad \textit{Subtract 3x.}$$
$$y = \frac{24-3x}{4} \quad \textit{Divide by 4.}$$

20. Solve $A = P(1+ni)$ for n.
$$A = P + Pni \quad \textit{Distributive property}$$
$$A - P = Pni \quad \textit{Subtract P.}$$
$$\frac{A-P}{Pi} = n \quad \textit{Divide by Pi.}$$
$$\text{or } n = \frac{A-P}{iP}$$

21. $3-2(x+7) \le -x+3$
$3-2x-14 \le -x+3$
$-2x-11 \le -x+3$
$-x \le 14$

Multiply each side by -1 and reverse the inequality symbol.
$$x \ge -14$$

Solution set: $[-14, \infty)$

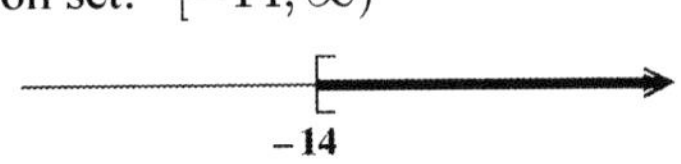

22. $-4 < 5-3x \le 0$
$-9 < -3x \le -5$

Divide each part by -3 and reverse the inequality symbol.
$$3 > x \ge \tfrac{5}{3} \text{ or } \tfrac{5}{3} \le x < 3$$

Solution set: $\left[\frac{5}{3}, 3\right)$

23. $2x+1 > 5$ or $2-x > 2$
$2x > 4$ or $-x > 0$
$x > 2$ or $x < 0$

Solution set: $(-\infty, 0) \cup (2, \infty)$

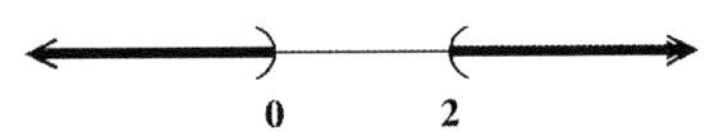

24. $|-7k+3| \ge 4$
$-7k+3 \ge 4$ or $-7k+3 \le -4$
$-7k \ge 1$ $\quad$ $-7k \le -7$
$k \le -\frac{1}{7}$ or $k \ge 1$

Solution set: $\left(-\infty, -\frac{1}{7}\right] \cup [1, \infty)$

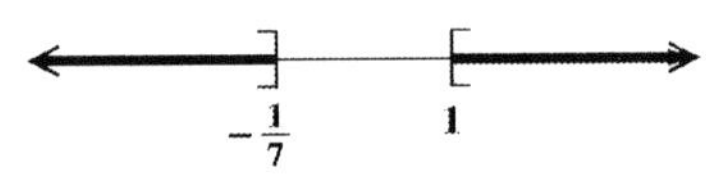

25. Let $x =$ the amount invested at 7% and at 10%.

Rate (as a Decimal)	Principal	Interest
.07	x	$.07x$
.10	x	$.10x$

The total amount invested is $2x$, and the total interest is
$$.1(2x) - 150 = .2x - 150.$$

Interest at 7% + interest at 10% = total interest.
$$.07x + .10x = .2x - 150$$
$$.17x = .20x - 150$$
$$150 = .03x$$
$$5000 = x$$

She invested \$5000 at each rate.

26. Let $x =$ the number of grams of food C. Then $2x$ is the number of grams of food A. There are 5 grams of food B and the total is at most 24 grams.
$$x + 2x + 5 \le 24$$
$$3x \le 19$$
$$x \le \tfrac{19}{3} \text{ or } 6\tfrac{1}{3}$$

He may use at most $6\frac{1}{3}$ grams of food C.

27. Let $x =$ the grade the student must make on the third test.

To find the average of the three tests, add them and divide by 3. This average must be at least 80%.

$$\frac{88+78+x}{3} \geq 80$$
$$\frac{166+x}{3} \geq 80$$
$$166+x \geq 240$$
$$x \geq 74$$

She must score at least 74% on her third test.

28. Let $x =$ the speed of the slower car.

	Rate	Time	Distance
Slower Car	x	4	$4x$
Faster Car	$x+20$	4	$4(x+20)$

The sum of their distances is 400.

$$4x+4(x+20)=400$$
$$4x+4x+80=400$$
$$8x+80=400$$
$$8x=320$$
$$x=40$$

Since $x=40$, $x+20=60$.

The cars' speeds are 40 mph and 60 mph.

29. **(a)** 1990: 1611
2001: 1468

$1611-1468=143$

The number of daily newspapers decreased by 143 between 1990 and 2001.

(b) $\frac{143}{1611} \approx .089$ or 8.9%

The number decreased by approximately 8.9%.

30. Let $x =$ the length of the middle-sized piece. Then $3x =$ the length of the longest piece, and $x-5 =$ the length of the shortest piece.

The total length is 40 cm.

$$x+3x+(x-5)=40$$
$$5x-5=40$$
$$5x=45$$
$$x=9$$

Since $x=9$, $3x=27$, and $x-5=4$.

The three pieces should be cut with lengths 4 cm, 9 cm, and 27 cm.

CHAPTER 4 GRAPHS, LINEAR EQUATIONS, AND FUNCTIONS

4.1 The Rectangular Coordinate System

4.1 Margin Exercises

1. **(a)** To plot $(-4, 2)$, move four units from the origin to the left along the x-axis, and then two units up parallel to the y-axis. The point is located in quadrant II.

   (b) To plot $(3, -2)$, move three units from the origin to the right along the x-axis, and then two units down parallel to the y-axis. The point is located in quadrant IV.

   (c) To plot $(-5, -6)$, move five units from the origin to the left along the x-axis, and then six units down parallel to the y-axis. The point is located in quadrant III.

   (d) To plot $(4, 6)$, move four units from the origin to the right along the x-axis, and then six units up parallel to the y-axis. The point is located in quadrant I.

   (e) To plot $(-3, 0)$, move three units from the origin to the left along the x-axis. Do not move up or down since the y-coordinate is 0. The point belongs to *no quadrant.*

   (f) To plot $(0, -5)$, do not move along the x-axis since the x-coordinate is 0. Move five units from the origin down the y-axis. The point belongs to *no quadrant.*

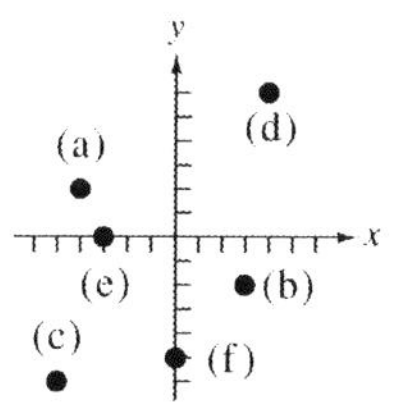

2. **(a)** $3x - 4y = 12$

   To complete the ordered pairs, substitute the given value of x or y in the equation.
   For $(0, \_\_\_\_)$, let $x = 0$.

$$\begin{aligned} 3x - 4y &= 12 \\ 3(0) - 4y &= 12 \\ -4y &= 12 \\ y &= -3 \end{aligned}$$

   The ordered pair is $(0, -3)$.

   For $(\_\_\_\_, 0)$ let $y = 0$.

$$\begin{aligned} 3x - 4y &= 12 \\ 3x - 4(0) &= 12 \\ 3x &= 12 \\ x &= 4 \end{aligned}$$

   The ordered pair is $(4, 0)$.

   For $(\_\_\_\_, -2)$, let $y = -2$.

$$\begin{aligned} 3x - 4y &= 12 \\ 3x - 4(-2) &= 12 \\ 3x + 8 &= 12 \\ 3x &= 4 \\ x &= \tfrac{4}{3} \end{aligned}$$

   The ordered pair is $\left(\frac{4}{3}, -2\right)$.

   For $(-4, \_\_\_\_)$, let $x = -4$.

$$\begin{aligned} 3x - 4y &= 12 \\ 3(-4) - 4y &= 12 \\ -12 - 4y &= 12 \\ -4y &= 24 \\ y &= -6 \end{aligned}$$

   The ordered pair is $(-4, -6)$.

   (b) To find one possible answer, let $x = -6$.

$$\begin{aligned} 3x - 4y &= 12 \\ 3(-6) - 4y &= 12 \\ -18 - 4y &= 12 \\ -4y &= 30 \\ y &= -\tfrac{30}{4} = -\tfrac{15}{2} \end{aligned}$$

   The ordered pair is $\left(-6, -\frac{15}{2}\right)$.

3. Plot the intercepts, $(4, 0)$ and $(0, -3)$ from Margin Exercise 2 and draw the line through them.

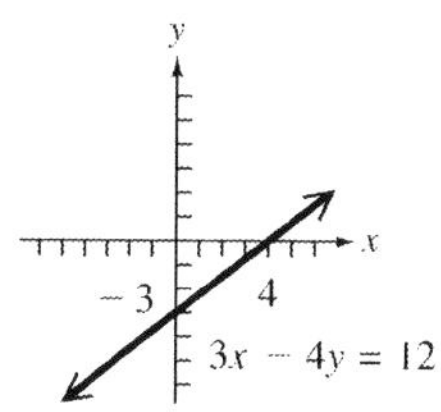

4. $2x - y = 4$
To find the x-intercept, let $y = 0$.

$$\begin{aligned} 2x - y &= 4 \\ 2x - 0 &= 4 \\ 2x &= 4 \\ x &= 2 \end{aligned}$$

The x-intercept is $(2, 0)$.

To find the y-intercept, let $x = 0$.

$$\begin{aligned} 2x - y &= 4 \\ 2(0) - y &= 4 \\ -y &= 4 \\ y &= -4 \end{aligned}$$

The y-intercept is $(0, -4)$.

Plot the intercepts, and draw the line through them.

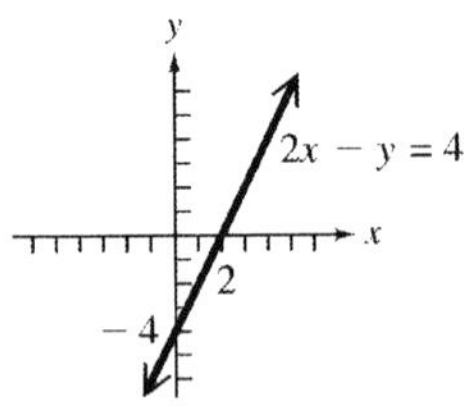

5. $y + 4 = 0$

In standard form, the equation is $0x + y = -4$. Every value of x leads to $y = -4$, so the y-intercept is $(0, -4)$. There is no x-intercept. The graph is the horizontal line through $(0, -4)$.

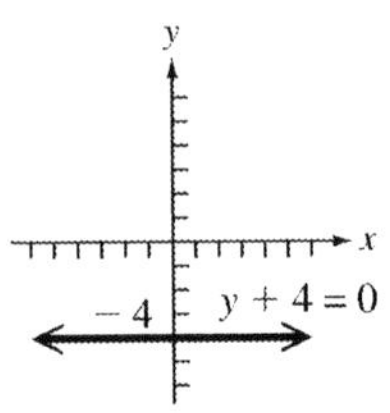

6. $x = 2$

In standard form, the equation is $x + 0y = 2$. Every value of y leads to $x = 2$, the x-intercept is $(2, 0)$. There is no y-intercept. The graph is the vertical line through $(2, 0)$.

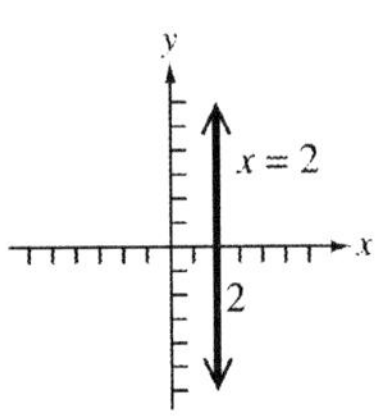

7. $3x - y = 0$
To find the x-intercept, let $y = 0$.

$$\begin{aligned} 3x - 0 &= 0 \\ 3x &= 0 \\ x &= 0 \end{aligned}$$

Since the x-intercept is $(0, 0)$, the y-intercept is also $(0, 0)$.

Find another point. Let $x = 1$.

$$\begin{aligned} 3(1) - y &= 0 \\ 3 - y &= 0 \\ y &= 3 \end{aligned}$$

This gives the ordered pair $(1, 3)$.

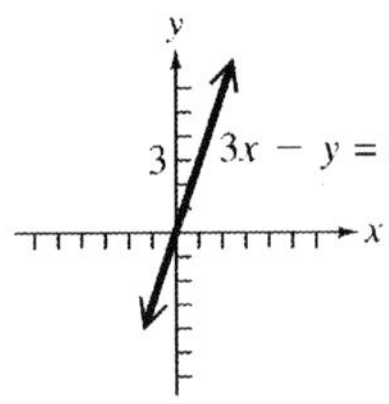

4.1 Section Exercises

1. **(a)** x represents the year; y represents the percent of women in math or computer science professions.

(b) The line goes down from 1990 to 2000. Therefore, from 1990–2000, the percent of women in math or computer science professions decreased.

(c) The highest point on the graph corresponds to the year 1990. Therefore, the percent of women in math or computer science professions reached a maximum in 1990.

(d) The point corresponding to 1980 has a percent of about 27%. Therefore, in 1980 there was about 27% women in math or computer science professions.

2. **(a)** x represents the year; y represents spending in billions of dollars.

(b) The point can be written as $(1999, 1100)$. Therefore, spending in 1999 was about 1100 billion dollars.

(c) The year corresponding to the point nearest 1270 is 2001. Therefore, spending in 2001 was about \$1270 billion.

3. A bar graph could be used in Exercises 1 and 2 for each line graph.

4. Another name for the rectangular coordinate system is the Cartesian coordinate system, named after René Descartes.

5. The point with coordinates $(0, 0)$ is called the origin of a rectangular coordinate system.

6. For any value of x, the point $(x, 0)$ lies on the x-axis.

7. To find the x-intercept of a line, we let y equal 0 and solve for x.

8. The equation $y = 4$ has a horizontal line as its graph.

9. To graph a straight line, we must find a minimum of two points. A third point is sometimes found to check the accuracy of the first two points.

10. Substitute 4 for y in the equation $2x - 3y = 0$ to get $2x - 12 = 0$. Solving for x gives $x = 6$. The point $(6, 4)$ is on the graph of $2x - 3y = 0$.

11. **(a)** The point $(1, 6)$ is located in quadrant I, since the x- and y-coordinates are both positive.

(b) The point $(-4, -2)$ is located in quadrant III, since the x- and y-coordinates are both negative.

(c) The point $(-3, 6)$ is located in quadrant II, since the x-coordinate is negative and the y-coordinate is positive.

(d) The point $(7, -5)$ is located in quadrant IV, since the x-coordinate is positive and the y-coordinate is negative.

(e) The point $(-3, 0)$ is located on the x-axis, so it does not belong to any quadrant.

12. **(a)** The point $(-2, -10)$ is located in quadrant III, since the x- and y-coordinates are both negative.

(b) The point $(4, 8)$ is located in quadrant I, since the x- and y-coordinates are both positive.

(c) The point $(-9, 12)$ is located in quadrant II, since the x-coordinate is negative and the y-coordinate is positive.

(d) The point $(3, -9)$ is located in quadrant IV, since the x-coordinate is positive and the y-coordinate is negative.

(e) The point $(0, -8)$ is located on the y-axis, so it does not belong to any quadrant.

13. **(a)** If $xy > 0$, then both x and y have the same sign.
(x, y) is in quadrant I if x and y are positive.
(x, y) is in quadrant III if x and y are negative.

(b) If $xy < 0$, then x and y have different signs.
(x, y) is in quadrant II if $x < 0$ and $y > 0$.
(x, y) is in quadrant IV if $x > 0$ and $y < 0$.

(c) If $\frac{x}{y} < 0$, then x and y have different signs.
(x, y) is in either quadrant II or IV. (See part (b).)

(d) If $\frac{x}{y} > 0$, then x and y have the same sign.
(x, y) is in either quadrant I or III. (See part (a).)

14. Any point that lies on an axis must have one coordinate that is 0.

For Exercises 15–24, see the rectangular coordinate system after Exercise 24.

15. To plot $(2, 3)$, go two units from zero to the right along the x-axis, and then go three units up parallel to the y-axis.

16. To plot $(-1, 2)$, go 1 unit in the negative direction, that is, left, on the x-axis, then 2 units up.

17. To plot $(-3, -2)$, go three units from zero to the left along the x-axis, and then go two units down parallel to the y-axis.

18. To plot $(1, -4)$, go 1 unit right on the x-axis, then 4 units down.

19. To plot $(0, 5)$, do not move along the x-axis at all since the x-coordinate is 0. Move five units up along the y-axis.

20. To plot $(-2, -4)$, go 2 units left on the x-axis, then 4 units down.

21. To plot $(-2, 4)$, go two units from zero to the left along the x-axis, and then go four units up parallel to the y-axis.

22. To plot $(3, 0)$, go 3 units right on the x-axis, then stop since the y-coordinate is 0.

23. To plot $(-2, 0)$, go two units to the left along the x-axis. Do not move up or down since the y-coordinate is 0.

24. To plot $(3, -3)$, go 3 units right on the x-axis, then 3 units down.

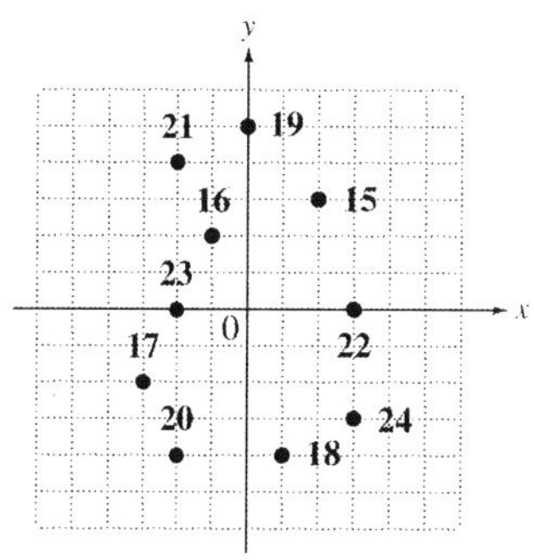

25. $x - y = 3$

For $x = 0$: $x - y = 3$
$0 - y = 3$
$y = -3 \quad (0, -3)$

For $y = 0$: $x - y = 3$
$x - 0 = 3$
$x = 3 \quad (3, 0)$

For $x = 5$: $x - y = 3$
$5 - y = 3$
$-y = -2$
$y = 2 \quad (5, 2)$

For $x = 2$: $x - y = 3$
$2 - y = 3$
$-y = 1$
$y = -1 \quad (2, -1)$

Plot the ordered pairs and draw the line through them.

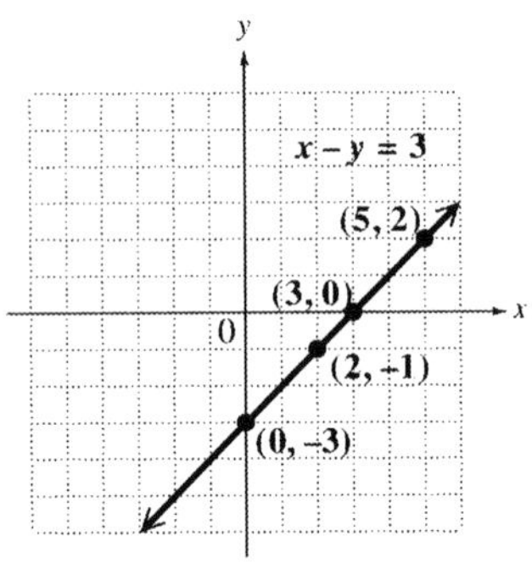

26. $x - y = 5$

For $x = 0$: $0 - y = 5$
$-y = 5$
$y = -5 \quad (0, -5)$

For $y = 0$: $x - 0 = 5$
$x = 5 \quad (5, 0)$

For $x = 1$: $1 - y = 5$
$-y = 4$
$y = -4 \quad (1, -4)$

For $x = 3$: $3 - y = 5$
$-y = 2$
$y = -2 \quad (3, -2)$

Plot the ordered pairs and draw the line through them.

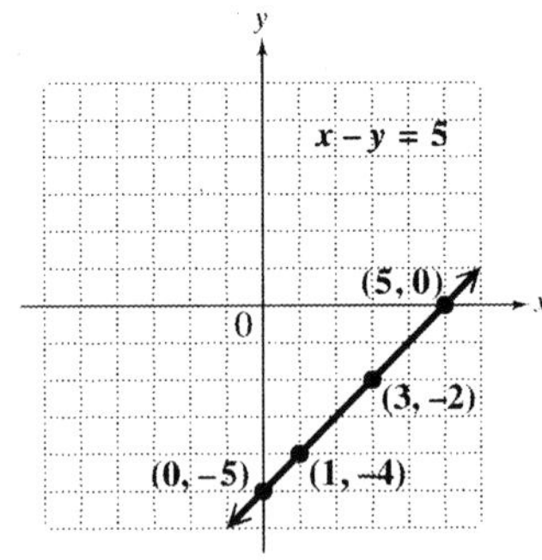

27. $x + 2y = 5$

For $x = 0$: $x + 2y = 5$
$0 + 2y = 5$
$2y = 5$
$y = \frac{5}{2} \quad (0, \frac{5}{2})$

For $y = 0$: $x + 2y = 5$
$x + 2(0) = 5$
$x + 0 = 5$
$x = 5 \quad (5, 0)$

For $x = 2$: $x + 2y = 5$
$2 + 2y = 5$
$2y = 3$
$y = \frac{3}{2} \quad (2, \frac{3}{2})$

For $y = 2$: $x + 2y = 5$
$x + 2(2) = 5$
$x + 4 = 5$
$x = 1 \quad (1, 2)$

Plot the ordered pairs and draw the line through them.

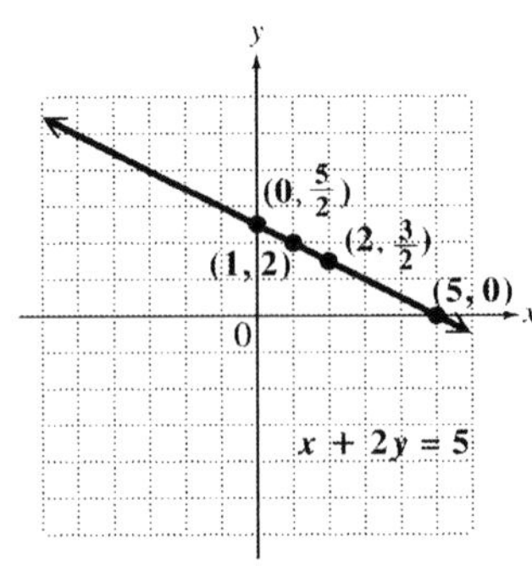

28. $x + 3y = -5$

For $x = 0$: $0 + 3y = -5$
$3y = -5$
$y = -\frac{5}{3} \quad (0, -\frac{5}{3})$

For $y = 0$: $x + 3(0) = -5$
$x = -5 \quad (-5, 0)$

For $x = 1$: $1 + 3y = -5$
$3y = -6$
$y = -2 \quad (1, -2)$

For $y = -1$: $x + 3(-1) = -5$

$$x - 3 = -5$$
$$x = -2 \quad (-2, -1)$$

Plot the ordered pairs and draw the line through them.

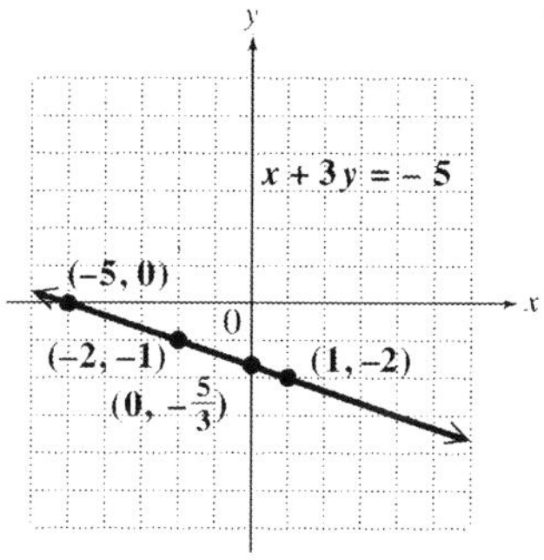

29. $4x - 5y = 20$

For $x = 0$:
$$4x - 5y = 20$$
$$4(0) - 5y = 20$$
$$-5y = 20$$
$$y = -4 \quad (0, -4)$$

For $y = 0$:
$$4x - 5y = 20$$
$$4x - 5(0) = 20$$
$$4x = 20$$
$$x = 5 \quad (5, 0)$$

For $x = 2$:
$$4x - 5y = 20$$
$$4(2) - 5y = 20$$
$$8 - 5y = 20$$
$$-5y = 12$$
$$y = -\tfrac{12}{5} \quad (2, -\tfrac{12}{5})$$

For $y = -3$:
$$4x - 5y = 20$$
$$4x - 5(-3) = 20$$
$$4x + 15 = 20$$
$$4x = 5$$
$$x = \tfrac{5}{4} \quad (\tfrac{5}{4}, -3)$$

Plot the ordered pairs and draw the line through them.

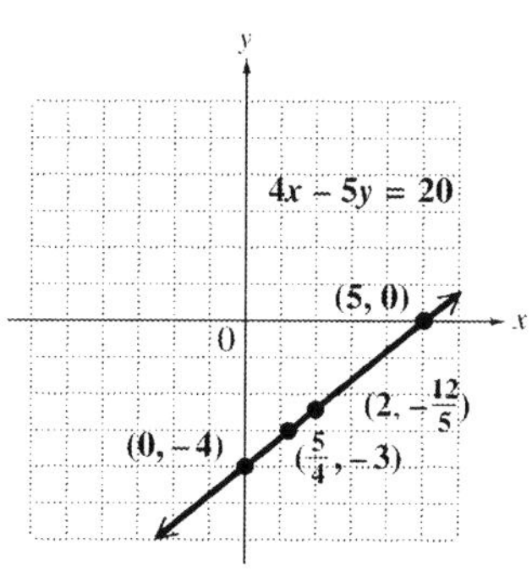

30. $6x - 5y = 30$

For $x = 0$:
$$6(0) - 5y = 30$$
$$-5y = 30$$
$$y = -6 \quad (0, -6)$$

For $y = 0$:
$$6x - 5(0) = 30$$
$$6x = 30$$
$$x = 5 \quad (5, 0)$$

For $x = 3$:
$$6(3) - 5y = 30$$
$$18 - 5y = 30$$
$$-5y = 12$$
$$y = -\tfrac{12}{5} \quad (3, -\tfrac{12}{5})$$

For $y = -2$:
$$6x - 5(-2) = 30$$
$$6x + 10 = 30$$
$$6x = 20$$
$$x = \tfrac{20}{6} = \tfrac{10}{3} \quad (\tfrac{10}{3}, -2)$$

Plot the ordered pairs and draw the line through them.

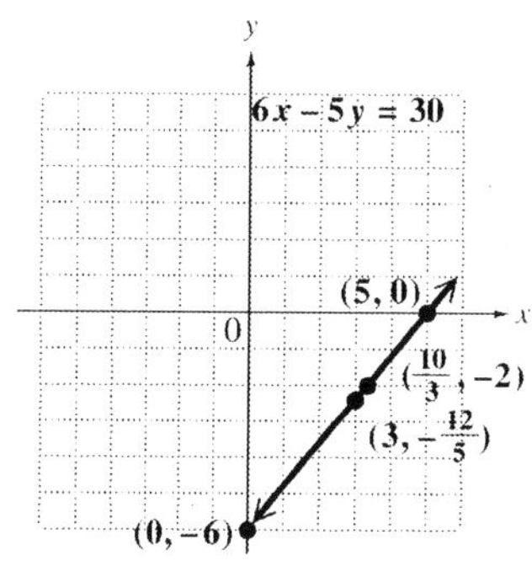

31. $2x + 3y = 12$

To find the x-intercept, let $y = 0$.
$$2x + 3y = 12$$
$$2x + 3(0) = 12$$
$$2x = 12$$
$$x = 6$$

The x-intercept is $(6, 0)$.

To find the y-intercept, let $x = 0$.
$$2x + 3y = 12$$
$$2(0) + 3y = 12$$
$$3y = 12$$
$$y = 4$$

The y-intercept is $(0, 4)$.

Plot the intercepts and draw the line through them.

continued

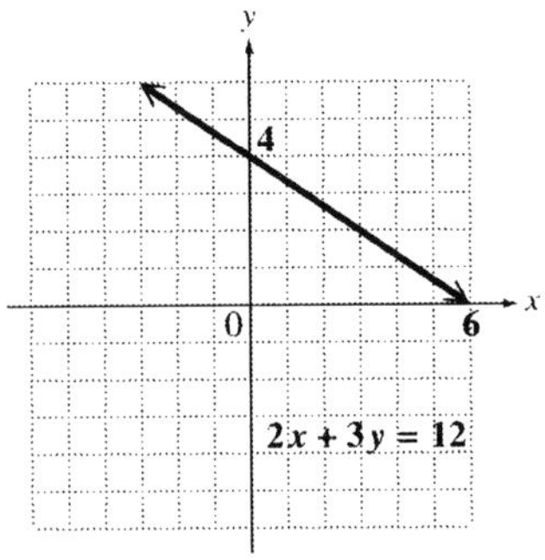

32. $5x + 2y = 10$

To find the x-intercept, let $y = 0$.

$$5x + 2(0) = 10$$
$$5x = 10$$
$$x = 2$$

The x-intercept is $(2, 0)$.

To find the y-intercept, let $x = 0$.

$$5(0) + 2y = 10$$
$$2y = 10$$
$$y = 5$$

The y-intercept is $(0, 5)$.

Plot the intercepts and draw the line through them.

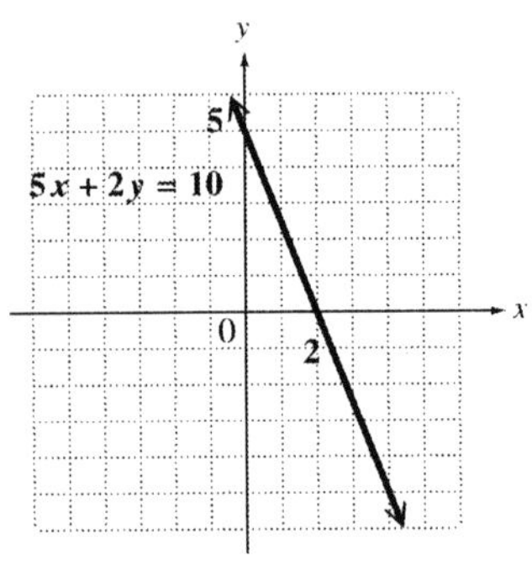

33. $x - 3y = 6$

To find the x-intercept, let $y = 0$.

$$x - 3y = 6$$
$$x - 3(0) = 6$$
$$x - 0 = 6$$
$$x = 6$$

The x-intercept is $(6, 0)$.

To find the y-intercept, let $x = 0$.

$$x - 3y = 6$$
$$0 - 3y = 6$$
$$-3y = 6$$
$$y = -2$$

The y-intercept is $(0, -2)$.

Plot the intercepts and draw the line through them.

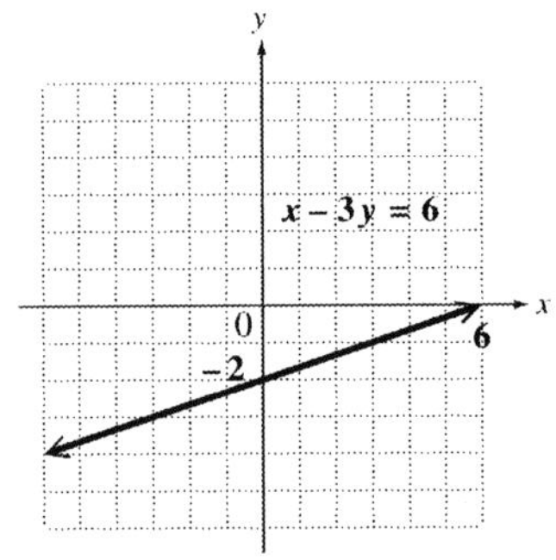

34. $x - 2y = -4$

To find the x-intercept, let $y = 0$.

$$x - 2(0) = -4$$
$$x = -4$$

The x-intercept is $(-4, 0)$.

To find the y-intercept, let $x = 0$.

$$0 - 2y = -4$$
$$-2y = -4$$
$$y = 2$$

The y-intercept is $(0, 2)$.

Plot the intercepts and draw the line through them.

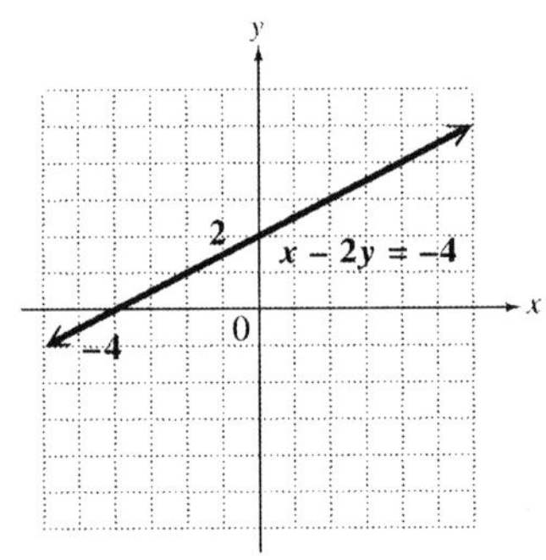

35. $3x - 7y = 9$

To find the x-intercept, let $y = 0$.

$$3x - 7y = 9$$
$$3x - 7(0) = 9$$
$$3x = 9$$
$$x = 3$$

The x-intercept is $(3, 0)$.

To find the y-intercept, let $x = 0$.

$$3x - 7y = 9$$
$$3(0) - 7y = 9$$
$$-7y = 9$$
$$y = -\frac{9}{7}$$

The y-intercept is $\left(0, -\frac{9}{7}\right)$.

Plot the intercepts and draw the line through them.

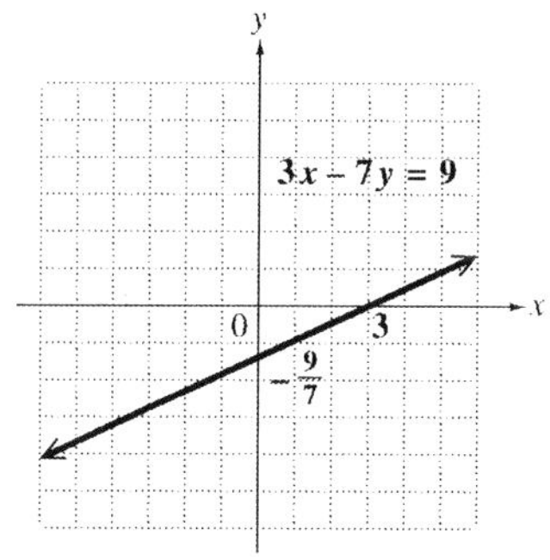

36. $5x + 6y = -10$
To find the x-intercept, let $y = 0$.

$$\begin{aligned} 5x + 6(0) &= -10 \\ 5x &= -10 \\ x &= -2 \end{aligned}$$

The x-intercept is $(-2, 0)$.

To find the y-intercept, let $x = 0$.

$$\begin{aligned} 5(0) + 6y &= -10 \\ 6y &= -10 \\ y &= -\tfrac{10}{6} = -\tfrac{5}{3} \end{aligned}$$

The y-intercept is $\left(0, -\frac{5}{3}\right)$.

Plot the intercepts and draw the line through them.

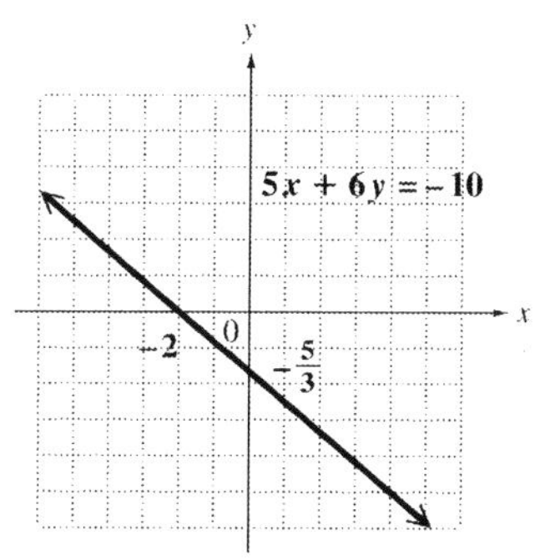

37. $y = 5$
This is a horizontal line. Every point has y-coordinate 5, so no point has y-coordinate 0. There is no x-intercept.
Since every point of the line has y-coordinate 5, the y-intercept is $(0, 5)$. Draw the horizontal line through $(0, 5)$.

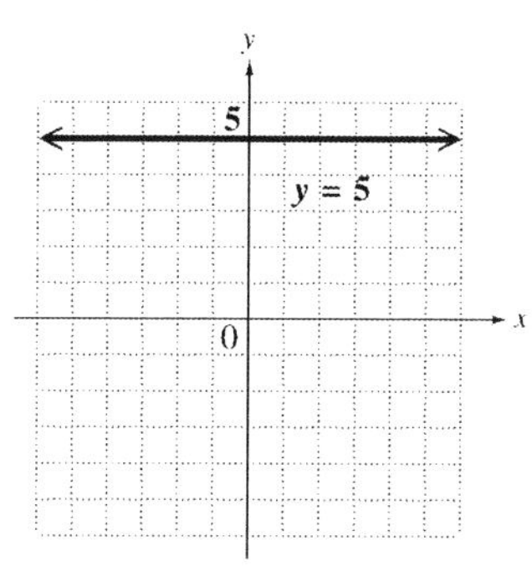

38. $y = -3$
This is a horizontal line. Every point has y-coordinate -3, so no point has y-coordinate 0. There is no x-intercept.
Since every point of the line has y-coordinate -3, the y-intercept is $(0, -3)$. Draw the horizontal line through $(0, -3)$.

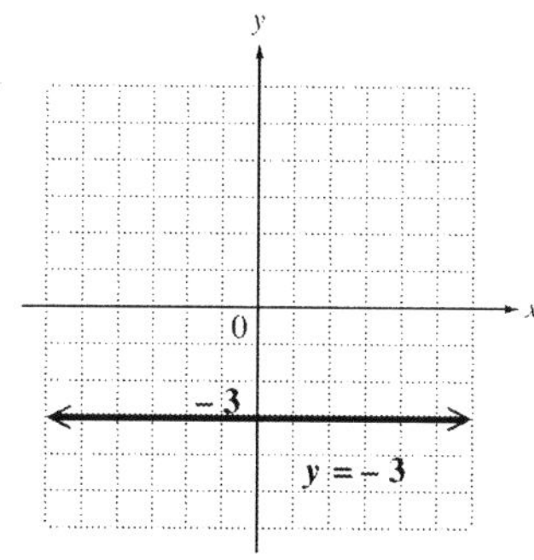

39. $x = 5$
This is a vertical line. Every point has x-coordinate 5, so the x-intercept is $(5, 0)$.
Since every point of the line has x-coordinate 5, no point has x-coordinate 0. There is no y-intercept.
Draw the vertical line through $(5, 0)$.

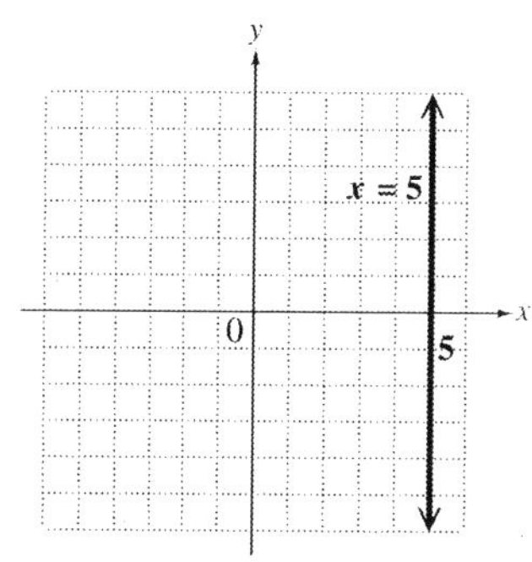

40. $x = -3$
This is a vertical line. Every point has x-coordinate -3, so the x-intercept is $(-3, 0)$.
Since every point of the line has x-coordinate -3, no point has x-coordinate 0. There is no y-intercept. Draw the vertical line through $(-3, 0)$.

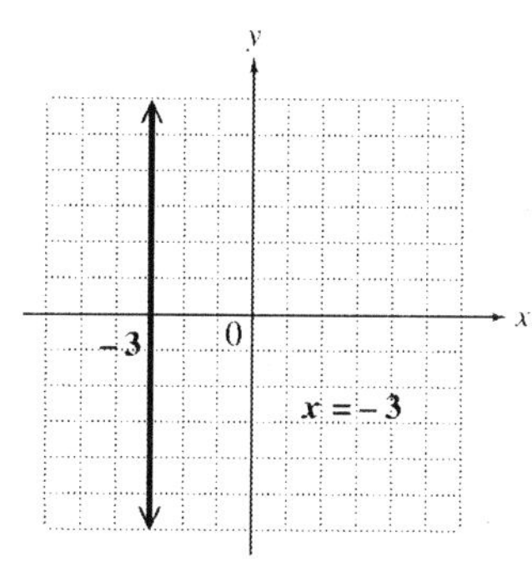

41. $x + 5y = 0$
To find the x-intercept, let $y = 0$.

$$x + 5y = 0$$
$$x + 5(0) = 0$$
$$x = 0$$

The x-intercept is $(0, 0)$, and since $x = 0$, this is also the y-intercept.

Since the intercepts are the same, another point is needed to graph the line. Choose any number for y, say $y = -1$, and solve the equation for x.

$$x + 5y = 0$$
$$x + 5(-1) = 0$$
$$x = 5$$

This gives the ordered pair $(5, -1)$. Plot $(5, -1)$ and $(0, 0)$, and draw the line through them.

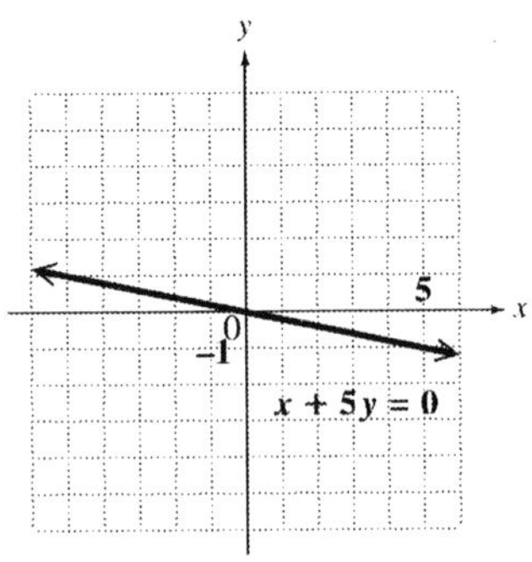

42. $x - 3y = 0$
To find the x-intercept, let $y = 0$.

$$x - 3(0) = 0$$
$$x = 0$$

The x-intercept is $(0, 0)$, and since $x = 0$, this is also the y-intercept.

Since the intercepts are the same, another point is needed to graph the line. Choose any number for y, say $y = 1$, and solve the equation for x.

$$x - 3(1) = 0$$
$$x = 3$$

This gives the ordered pair $(3, 1)$. Plot $(3, 1)$ and $(0, 0)$, and draw the line through them.

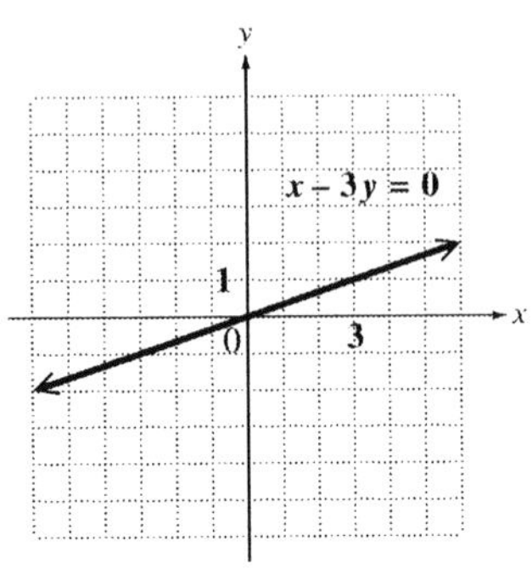

43. $2x = 3y$
If $x = 0$, then $y = 0$, so the x- and y-intercepts are $(0, 0)$. To get another point, let $x = 3$.

$$2(3) = 3y$$
$$2 = y$$

Plot $(3, 2)$ and $(0, 0)$, and draw the line through them.

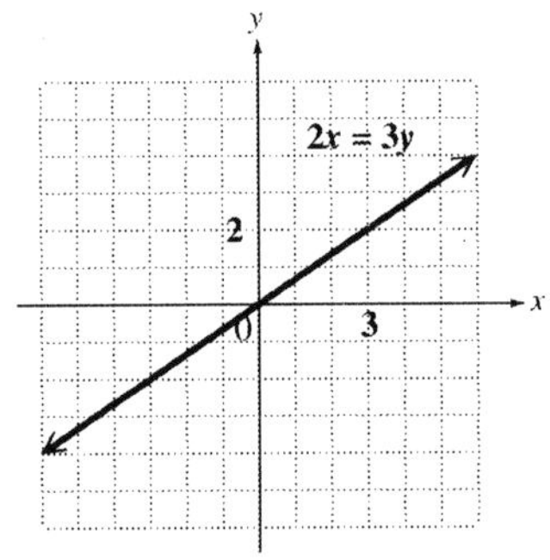

44. $3x = -4y$
If $x = 0$, then $y = 0$, so the x- and y-intercepts are $(0, 0)$. To get another point, let $x = 4$.

$$3(4) = -4y$$
$$-3 = y$$

Plot $(4, -3)$ and $(0, 0)$, and draw the line through them.

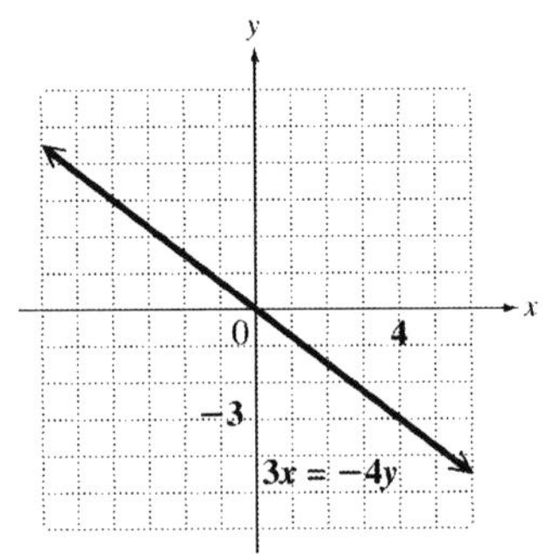

45. R has x-coordinate 6 and y-coordinate -2, so the ordered pair is $(6, -2)$.

46. The midpoint of $\overline{PR}$ has x-coordinate halfway from 4 to 6, which is 5, and y-coordinate -2, so it's the ordered pair $S = (5, -2)$.

47. The midpoint of $\overline{QR}$ has x-coordinate 6 and y-coordinate halfway from -2 to 2, which is 0, so it's the ordered pair $T = (6, 0)$.

48. $M = (x\text{-coordinate of } S, y\text{-coordinate of } T)$
$= (5, 0)$

49. $P = (4, -2)$ and $Q = (6, 2)$.
The average of the x-coordinates of P and Q is

$$\frac{4+6}{2} = \frac{10}{2} = 5.$$

The average of the y-coordinates of P and Q is

$$\frac{-2+2}{2} = \frac{0}{2} = 0.$$

50. The x-coordinate of M is the average of the x-coordinates of P and Q. The y-coordinate of M is the average of the y-coordinates of P and Q.

51. By the midpoint formula, the midpoint of the segment with endpoints $(-8, 4)$ and $(-2, -6)$ is

$$\left(\frac{-8+(-2)}{2}, \frac{4+(-6)}{2}\right) = \left(\frac{-10}{2}, \frac{-2}{2}\right) = (-5, -1).$$

52. By the midpoint formula, the midpoint of the segment with endpoints $(5, 2)$ and $(-1, 8)$ is

$$\left(\frac{5+(-1)}{2}, \frac{2+8}{2}\right) = \left(\frac{4}{2}, \frac{10}{2}\right) = (2, 5).$$

53. By the midpoint formula, the midpoint of the segment with endpoints $(3, -6)$ and $(6, 3)$ is

$$\left(\frac{3+6}{2}, \frac{-6+3}{2}\right) = \left(\frac{9}{2}, \frac{-3}{2}\right) = \left(\frac{9}{2}, -\frac{3}{2}\right).$$

54. By the midpoint formula, the midpoint of the segment with endpoints $(-10, 4)$ and $(7, 1)$ is

$$\left(\frac{-10+7}{2}, \frac{4+1}{2}\right) = \left(\frac{-3}{2}, \frac{5}{2}\right) = \left(-\frac{3}{2}, \frac{5}{2}\right).$$

55. By the midpoint formula, the midpoint of the segment with endpoints $(-9, 3)$ and $(9, 8)$ is

$$\left(\frac{-9+9}{2}, \frac{3+8}{2}\right) = \left(\frac{0}{2}, \frac{11}{2}\right) = \left(0, \frac{11}{2}\right).$$

56. By the midpoint formula, the midpoint of the segment with endpoints $(4, -3)$ and $(-1, 3)$ is

$$\left(\frac{4+(-1)}{2}, \frac{-3+3}{2}\right) = \left(\frac{3}{2}, \frac{0}{2}\right) = \left(\frac{3}{2}, 0\right).$$

4.2 Slope

4.2 Margin Exercises

1. **(a)** rise = vertical change = 2 ft

(b) run = horizontal change = 10 ft

(c) $\text{slope} = \frac{\text{rise}}{\text{run}} = \frac{2 \text{ ft}}{10 \text{ ft}} = \frac{2}{10}$ or $\frac{1}{5}$

2. **(a)** Let $(x_1, y_1) = (-2, 7)$ and $(x_2, y_2) = (4, -3)$
Then

$$m = \frac{y_2 - y_1}{x_2 - x_1} = \frac{-3-7}{4-(-2)} = \frac{-10}{6} = -\frac{5}{3}.$$

The slope is $-\frac{5}{3}$.

(b) Let $(x_1, y_1) = (1, 2)$ and $(x_2, y_2) = (8, 5)$
Then

$$m = \frac{y_2 - y_1}{x_2 - x_1} = \frac{5-2}{8-1} = \frac{3}{7}.$$

The slope is $\frac{3}{7}$.

(c) Let $(x_1, y_1) = (8, -4)$ and $(x_2, y_2) = (3, -2)$
Then

$$m = \frac{y_2 - y_1}{x_2 - x_1} = \frac{-2-(-4)}{3-8} = \frac{2}{-5} = -\frac{2}{5}.$$

The slope is $-\frac{2}{5}$.

3. **(a)** To find the slope of the line with equation

$$2x + y = 6,$$

first find the intercepts. Replace y with 0 to find that the x-intercept is $(3, 0)$; replace x with 0 to find that the y-intercept is $(0, 6)$. The slope is then

$$m = \frac{6-0}{0-3} = \frac{6}{-3} = -2.$$

The slope is -2.

(b) To find the slope of the line with equation

$$3x - 4y = 12,$$

first find the intercepts. The x-intercept is $(4, 0)$, and the y-intercept is $(0, -3)$. The slope is then

$$m = \frac{-3-0}{0-4} = \frac{-3}{-4} = \frac{3}{4}.$$

The slope is $\frac{3}{4}$.

4. **(a)** To find the slope of the line with equation

$$x = -6,$$

select two different points on the line, such as $(-6, 0)$ and $(-6, 3)$, and use the slope formula.

$$m = \frac{3-0}{-6-(-6)} = \frac{3}{0}$$

Since division by zero is undefined, the slope is *undefined*.

(b) To find the slope of the line with equation

$$y + 5 = 0,$$

select two different points on the line, such as $(0, -5)$ and $(2, -5)$, and use the slope formula.

$$m = \frac{-5 - (-5)}{2 - 0} = \frac{0}{2} = 0$$

The slope is 0.

5. Solve the equation for y.

$$\begin{aligned} 2x - 5y &= 8 \\ -5y &= -2x + 8 \quad \textit{Subtract } 2x. \\ y &= \tfrac{2}{5}x - \tfrac{8}{5} \quad \textit{Divide by } -5. \end{aligned}$$

The slope is given by the coefficient of x, so the slope is $\frac{2}{5}$.

6. **(a)** Through $(1, -3)$; $m = -\frac{3}{4}$

Locate the point $(1, -3)$ on the graph. Use the slope formula to find a second point on the line (write $-\frac{3}{4}$ as $\frac{-3}{4}$).

$$m = \frac{\text{change in } y}{\text{change in } x} = \frac{-3}{4}$$

From $(1, -3)$, move *down* 3 units and then 4 units to the *right* to $(5, -6)$. Draw the line through the two points. (Note that we could also think of the slope as $\frac{3}{-4}$, and then move *up* 3 units and *left* 4 units.)

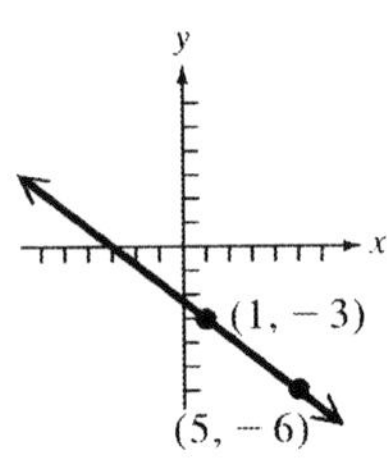

(b) Through $(-1, -4)$; $m = 2$

Locate the point $(-1, -4)$ on the graph. Use the slope formula to find a second point on the line (write 2 as $\frac{2}{1}$).

$$m = \frac{\text{change in } y}{\text{change in } x} = \frac{2}{1}$$

From $(-1, -4)$, move *up* 2 units and then 1 unit to the *right* to $(0, -2)$. Draw the line through the two points.

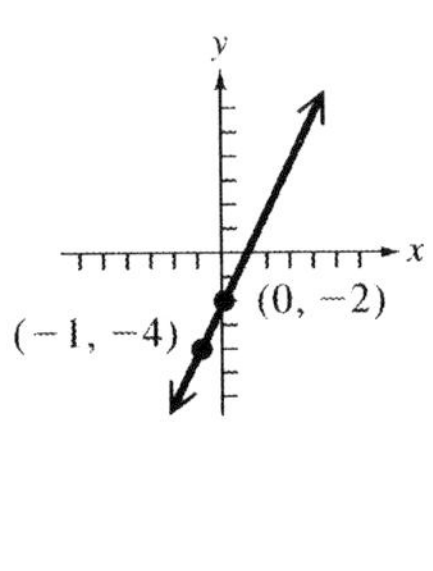

7. **(a)** Find the slope of each line.
The line through $(-1, 2)$ and $(3, 5)$ has slope

$$m_1 = \frac{5 - 2}{3 - (-1)} = \frac{3}{4}.$$

The line through $(4, 7)$ and $(8, 10)$ has slope

$$m_2 = \frac{10 - 7}{8 - 4} = \frac{3}{4}.$$

The slopes are the same, so the lines are *parallel.*

(b) Find the slope of each line.
The line through $(5, -9)$ and $(3, 7)$ has slope

$$m_1 = \frac{7 - (-9)}{3 - 5} = \frac{16}{-2} = -8.$$

The line through $(0, 2)$ and $(8, 3)$ has slope

$$m_2 = \frac{3 - 2}{8 - 0} = \frac{1}{8}.$$

The product of the slopes is

$$(-8)\left(\frac{1}{8}\right) = -1,$$

so the lines are *perpendicular.*

(c) Solve each equation for y.

$$\begin{aligned} 2x - y &= 4 \\ -y &= -2x + 4 \\ y &= 2x - 4 \end{aligned}$$
(so $m_1 = 2$)

$$\begin{aligned} 2x + y &= 6 \\ y &= -2x + 6 \end{aligned}$$
(so $m_2 = -2$)

Since $m_1 \neq m_2$, the lines are not parallel. Since $m_1 m_2 = 2(-2) = -4$, the lines are not perpendicular either. Therefore, the answer is *neither.*

(d) Solve each equation for y.

$$\begin{aligned} 3x + 5y &= 6 \\ 5y &= -3x + 6 \\ y &= -\tfrac{3}{5}x + \tfrac{6}{5} \end{aligned}$$
$\left(\text{so } m_1 = -\frac{3}{5}\right)$

$$\begin{aligned} 5x - 3y &= 2 \\ -3y &= -5x + 2 \\ y &= \tfrac{5}{3}x - \tfrac{2}{3} \end{aligned}$$
$\left(\text{so } m_2 = \frac{5}{3}\right)$

Since $m_1 m_2 = \left(-\frac{3}{5}\right)\left(\frac{5}{3}\right) = -1$, the lines are *perpendicular.*

8. $(x_1, y_1) = (1997, 10)$ and $(x_2, y_2) = (2000, 20.8)$.

$$\begin{aligned} \text{average rate of change} = \frac{y_2 - y_1}{x_2 - x_1} &= \frac{20.8 - 10}{2000 - 1997} \\ &= \frac{10.8}{3} \\ &= 3.6 \end{aligned}$$

The average rate of change is 3.6, which is the same as in Example 8.

9. $(x_1, y_1) = (1997, 36.4)$ and $(x_2, y_2) = (2001, 28.5)$.

$$\begin{aligned}\text{average rate of change} = \frac{y_2 - y_1}{x_2 - x_1} &= \frac{28.5 - 36.4}{2001 - 1997} \\ &= \frac{-7.9}{4} \\ &= -1.975\end{aligned}$$

Thus, the average rate of change from 1997 to 2001 was -1.975% per year.

4.2 Section Exercises

1. $$\begin{aligned}\text{slope} &= \frac{\text{change in vertical position}}{\text{change in horizontal position}} \\ &= \frac{-30 \text{ feet}}{100 \text{ feet}}\end{aligned}$$

Choices **A**, $-.3$, **B**, $-\frac{3}{10}$, and **D**, $-\frac{30}{100}$ are all correct.

2. $$\text{slope} = \frac{\text{change in vertical position}}{\text{change in horizontal position}}$$
$$-.05 = \frac{\text{change in vertical position}}{50 \text{ feet}}$$

So the change in vertical position is $-.05(50 \text{ feet}) = -2.5$ feet, which is a 2.5 ft change in the vertical direction (downward).

3. To get to B from A, we must go up 2 units and move right 1 unit. Thus,

$$\text{slope of } AB = \frac{\text{rise}}{\text{run}} = \frac{2}{1} = 2.$$

4. $\text{slope of } BC = \frac{\text{rise}}{\text{run}} = \frac{0}{-4} = 0$

5. $\text{slope of } CD = \frac{\text{rise}}{\text{run}} = \frac{-7}{0}$, which is undefined.

6. $\text{slope of } DE = \frac{\text{rise}}{\text{run}} = \frac{-1}{3} = -\frac{1}{3}$

7. $m = \frac{6-2}{5-3} = \frac{4}{2} = 2$

8. $m = \frac{5-7}{-4-2} = \frac{-2}{-6} = \frac{1}{3}$

9. $m = \frac{4-(-1)}{-3-(-5)} = \frac{4+1}{-3+5} = \frac{5}{2}$

10. $m = \frac{-6-0}{0-(-3)} = \frac{-6}{3} = -2$

11. $m = \frac{-5-(-5)}{3-2} = \frac{-5+5}{1} = \frac{0}{1} = 0$

12. $m = \frac{7-(-2)}{-3-(-3)} = \frac{7+2}{-3+3} = \frac{9}{0}$, which is undefined.

13. Let $(x_1, y_1) = (-2, -3)$ and $(x_2, y_2) = (-1, 5)$. Then

$$m = \frac{y_2 - y_1}{x_2 - x_1} = \frac{5-(-3)}{-1-(-2)} = \frac{8}{1} = 8.$$

The slope is 8.

14. Let $(x_1, y_1) = (-4, 3)$ and $(x_2, y_2) = (-3, -4)$. Then

$$m = \frac{y_2 - y_1}{x_2 - x_1} = \frac{-4-3}{-3-(-4)} = \frac{-7}{1} = -7.$$

The slope is -7.

15. Let $(x_1, y_1) = (-4, 1)$ and $(x_2, y_2) = (2, 6)$. Then

$$m = \frac{y_2 - y_1}{x_2 - x_1} = \frac{6-1}{2-(-4)} = \frac{5}{6}.$$

The slope is $\frac{5}{6}$.

16. Let $(x_1, y_1) = (-3, -3)$ and $(x_2, y_2) = (5, 6)$. Then

$$m = \frac{y_2 - y_1}{x_2 - x_1} = \frac{6-(-3)}{5-(-3)} = \frac{9}{8}.$$

The slope is $\frac{9}{8}$.

17. Let $(x_1, y_1) = (2, 4)$ and $(x_2, y_2) = (-4, 4)$. Then

$$m = \frac{y_2 - y_1}{x_2 - x_1} = \frac{4-4}{-4-2} = \frac{0}{-6} = 0.$$

The slope is 0.

18. Let $(x_1, y_1) = (-6, 3)$ and $(x_2, y_2) = (2, 3)$. Then

$$m = \frac{y_2 - y_1}{x_2 - x_1} = \frac{3-3}{2-(-6)} = \frac{0}{8} = 0.$$

The slope is 0.

19. The points shown on the line are $(-3, 3)$ and $(-1, -2)$. The slope is

$$m = \frac{-2-3}{-1-(-3)} = \frac{-5}{2} = -\frac{5}{2}.$$

20. The points shown on the line are $(1, -1)$ and $(4, 5)$. The slope is

$$m = \frac{5-(-1)}{4-1} = \frac{6}{3} = 2.$$

21. The points shown on the line are $(3, 3)$ and $(3, -3)$. The slope is

$$m = \frac{-3-3}{3-3} = \frac{-6}{0}, \text{ which is undefined.}$$

22. The points shown on the line are $(-2, 2)$ and $(2, 2)$. The slope is

$$m = \frac{2-2}{2-(-2)} = \frac{0}{4} = 0.$$

The slope is 0.

23. "The line has positive slope" means that the line goes up from left to right. This is line *B*.

24. "The line has negative slope" means that the line goes down from left to right. This is line *C*.

25. "The line has slope 0" means that there is no vertical change; that is, the line is horizontal. This is line *A*.

26. "The line has undefined slope" means that there is no horizontal change; that is, the line is vertical. This is line *D*.

27. To find the slope of the line with equation

$$x + 2y = 4,$$

first find the intercepts. Replace y with 0 to find that the x-intercept is $(4, 0)$; replace x with 0 to find that the y-intercept is $(0, 2)$. The slope is then

$$m = \frac{2-0}{0-4} = -\frac{2}{4} = -\frac{1}{2}.$$

To sketch the graph, plot the intercepts and draw the line through them.

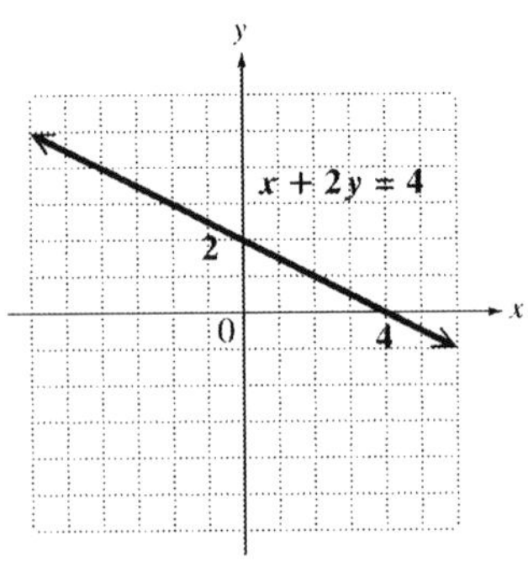

28. To find the slope of the line with equation

$$x + 3y = -6,$$

first find the intercepts. Replace y with 0 to find that the x-intercept is $(-6, 0)$; replace x with 0 to find that the y-intercept is $(0, -2)$. The slope is then

$$m = \frac{-2-0}{0-(-6)} = \frac{-2}{6} = -\frac{1}{3}.$$

To sketch the graph, plot the intercepts and draw the line through them.

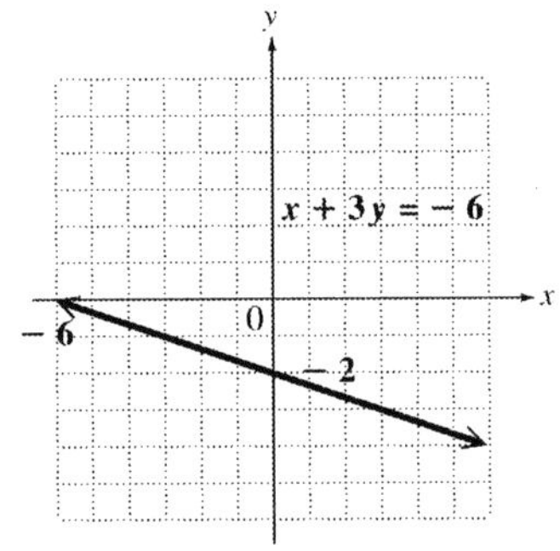

29. To find the slope of the line with equation

$$-x + y = 4,$$

first find the intercepts. Replace y with 0 to find that the x-intercept is $(-4, 0)$; replace x with 0 to find that the y-intercept is $(0, 4)$. The slope is then

$$m = \frac{4-0}{0-(-4)} = \frac{4}{4} = 1.$$

To sketch the graph, plot the intercepts and draw the line through them.

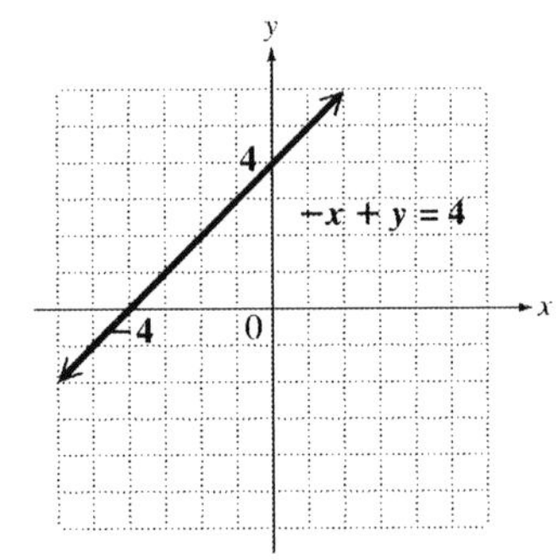

30. To find the slope of the line with equation

$$-x + y = 6,$$

first find the intercepts. Replace y with 0 to find that the x-intercept is $(-6, 0)$; replace x with 0 to find that the y-intercept is $(0, 6)$. The slope is then

$$m = \frac{6-0}{0-(-6)} = \frac{6}{6} = 1.$$

To sketch the graph, plot the intercepts and draw the line through them.

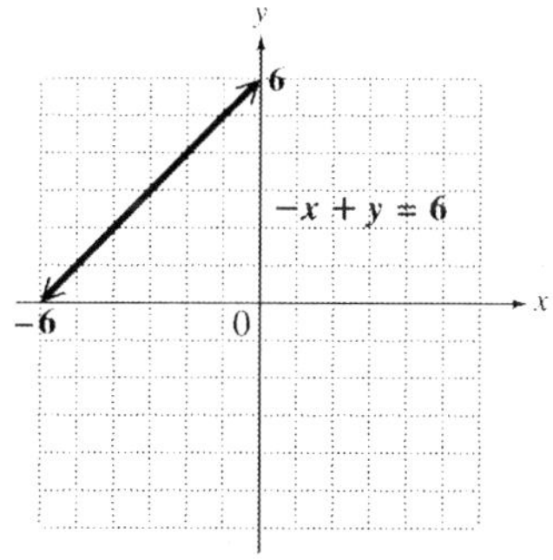

31. To find the slope of the line with equation

$$6x + 5y = 30,$$

first find the intercepts. Replace y with 0 to find that the x-intercept is $(5, 0)$; replace x with 0 to find that the y-intercept is $(0, 6)$. The slope is then

$$m = \frac{6 - 0}{0 - 5} = -\frac{6}{5}.$$

To sketch the graph, plot the intercepts and draw the line through them.

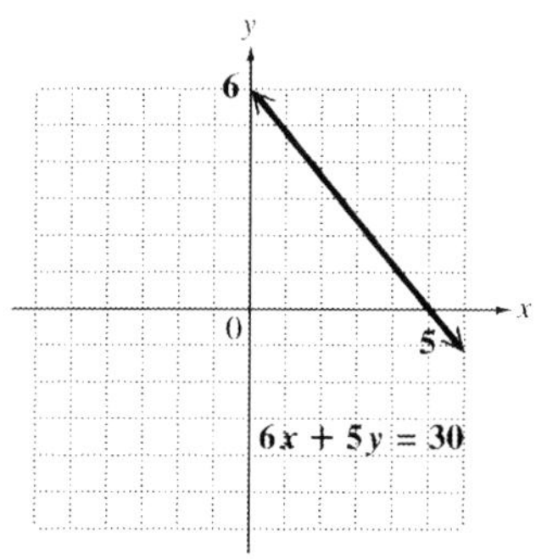

32. To find the slope of the line with equation

$$3x + 4y = 12,$$

first find the intercepts. Replace y with 0 to find that the x-intercept is $(4, 0)$; replace x with 0 to find that the y-intercept is $(0, 3)$. The slope is then

$$m = \frac{3 - 0}{0 - 4} = -\frac{3}{4}.$$

To sketch the graph, plot the intercepts and draw the line through them.

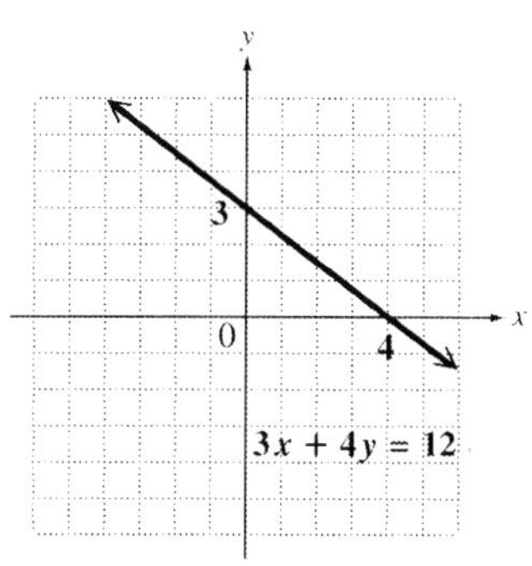

33. The graph of $x + 2 = 0$, or $x = -2$, is a vertical line with x-intercept $(-2, 0)$. The slope of a vertical line is undefined because the denominator equals zero in the slope formula.

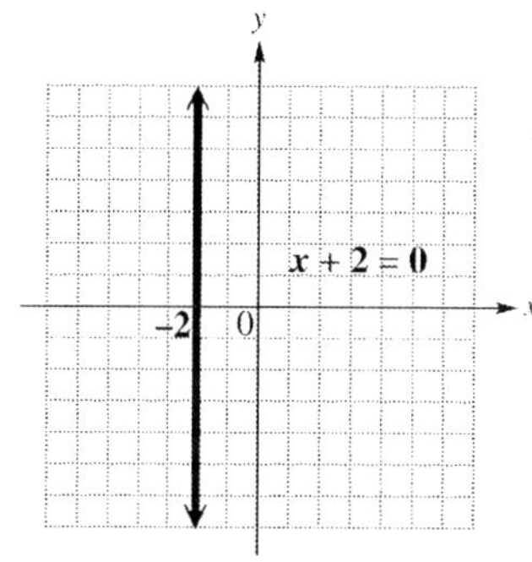

34. The graph of $x - 4 = 0$, or $x = 4$, is a vertical line with x-intercept $(4, 0)$. The slope of a vertical line is undefined because the denominator equals zero in the slope formula.

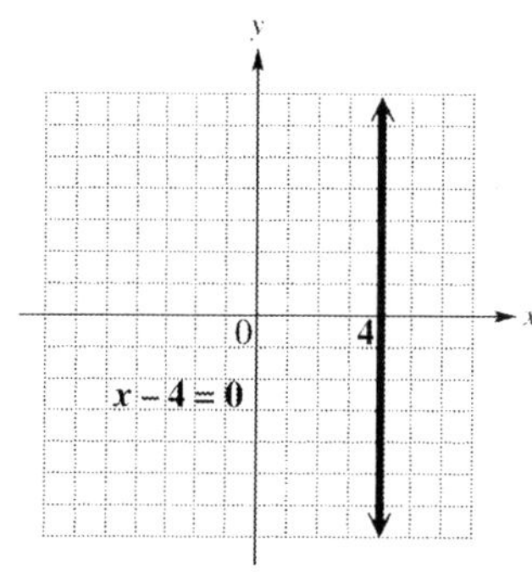

35. In the equation

$$y = 4x,$$

replace x with 0 and then x with 1 to get the ordered pairs $(0, 0)$ and $(1, 4)$, respectively. (There are other possibilities for ordered pairs.) The slope is then

$$m = \frac{4 - 0}{1 - 0} = \frac{4}{1} = 4.$$

To sketch the graph, plot the two points and draw the line through them.

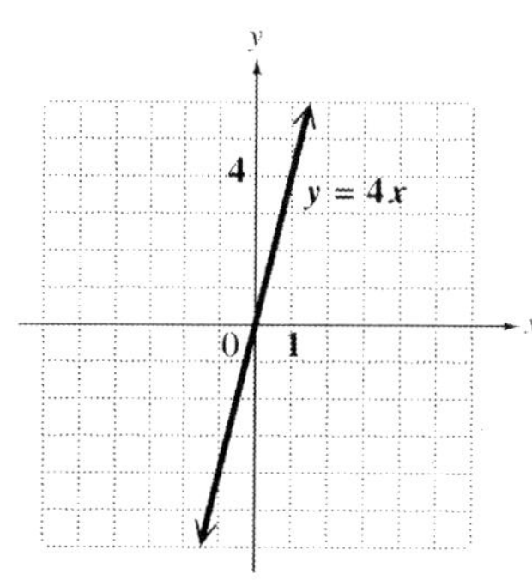

36. In the equation

$$y = -3x,$$

replace x with 0 and then x with 1 to get the ordered pairs $(0, 0)$ and $(1, -3)$, respectively. (There are other possibilities for ordered pairs.) The slope is then

$$m = \frac{-3 - 0}{1 - 0} = \frac{-3}{1} = -3.$$

To sketch the graph, plot the two points and draw the line through them.

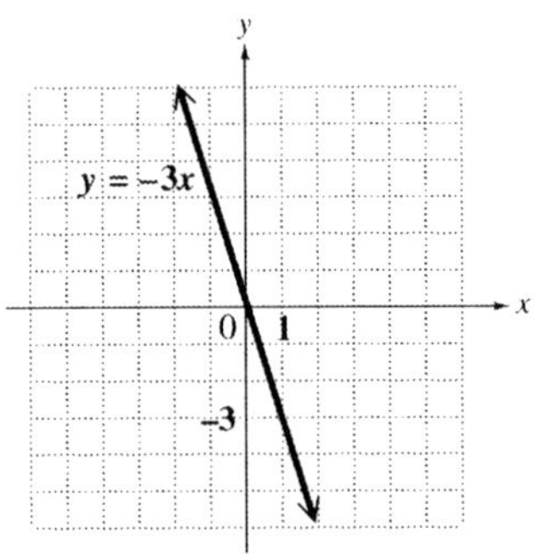

37. $y - 3 = 0 \quad (y = 3)$
The graph of $y = 3$ is the horizontal line with y-intercept $(0, 3)$. The slope of a horizontal line is 0.

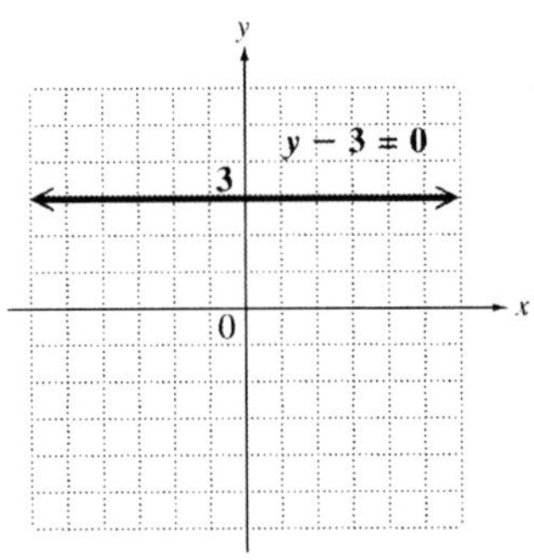

38. $y + 5 = 0 \quad (y = -5)$
The graph of $y = -5$ is the horizontal line with y-intercept $(0, -5)$. The slope of a horizontal line is 0.

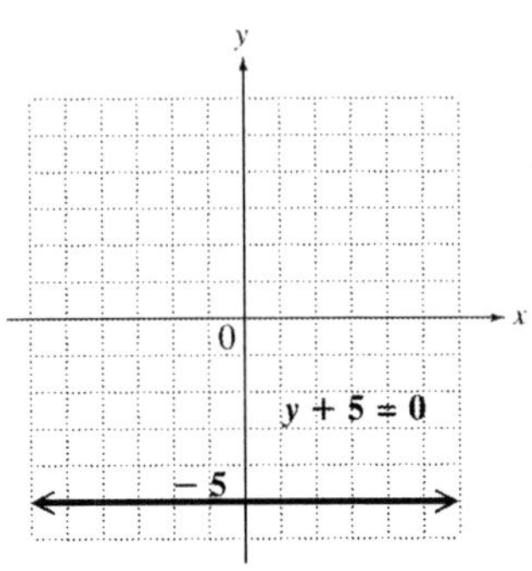

39. To graph the line through $(-4, 2)$ with slope $m = \frac{1}{2}$, locate $(-4, 2)$ on the graph. To find a second point, use the definition of slope.

$$m = \frac{\text{change in } y}{\text{change in } x} = \frac{1}{2}$$

From $(-4, 2)$, go up 1 unit. Then go 2 units to the right to get to $(-2, 3)$. Draw the line through $(-4, 2)$ and $(-2, 3)$.

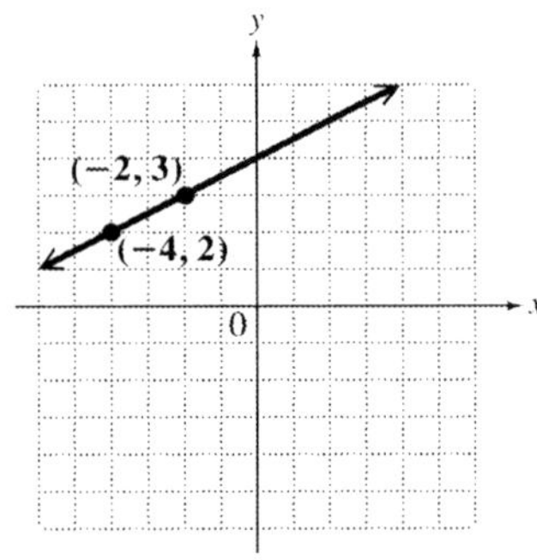

40. To graph the line through $(-2, -3)$ with slope $m = \frac{5}{4}$, locate $(-2, -3)$ on the graph. To find a second point, use the definition of slope.

$$m = \frac{\text{change in } y}{\text{change in } x} = \frac{5}{4}$$

From $(-2, -3)$, go up 5 units. Then go 4 units to the right to get to $(2, 2)$. Draw the line through $(-2, -3)$ and $(2, 2)$.

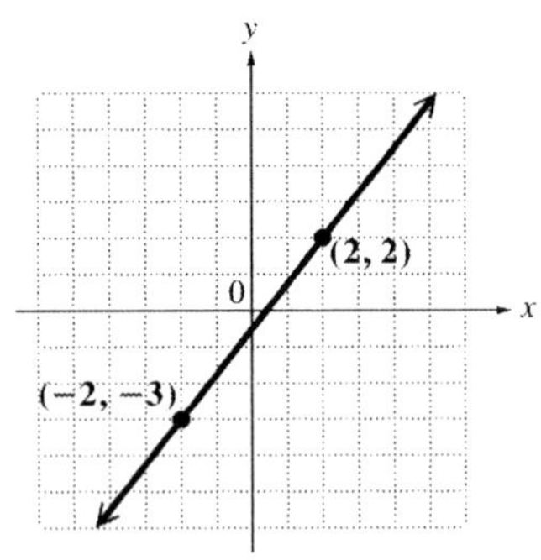

41. To graph the line through $(0, -2)$ with slope $m = -\frac{2}{3}$, locate the point $(0, -2)$ on the graph. To find a second point on the line, use the definition of slope, writing $-\frac{2}{3}$ as $\frac{-2}{3}$.

$$m = \frac{\text{change in } y}{\text{change in } x} = \frac{-2}{3}$$

From $(0, -2)$, move 2 units down and then 3 units to the right to get to $(3, -4)$. Draw the line through $(3, -4)$ and $(0, -2)$. (Note that the slope could also be written as $\frac{2}{-3}$. In this case, move 2 units up and 3 units to the left to get another point on the same line.)

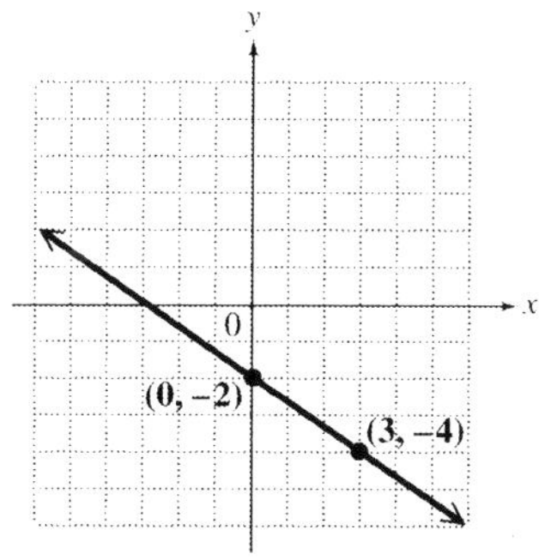

42. To graph the line through $(0, -4)$ with slope $m = -\frac{3}{2}$, locate the point $(0, -4)$ on the graph. To find a second point on the line, use the definition of slope, writing $-\frac{3}{2}$ as $\frac{-3}{2}$.

$$m = \frac{\text{change in } y}{\text{change in } x} = \frac{-3}{2}$$

From $(0, -4)$, move 3 units down and then 2 units to the right to get to $(2, -7)$. Draw the line through $(2, -7)$ and $(0, -4)$. The slope could also be written as $\frac{3}{-2}$. In this case, move 3 units up and 2 units to the left to get to $(-2, -1)$, as shown in the figure.

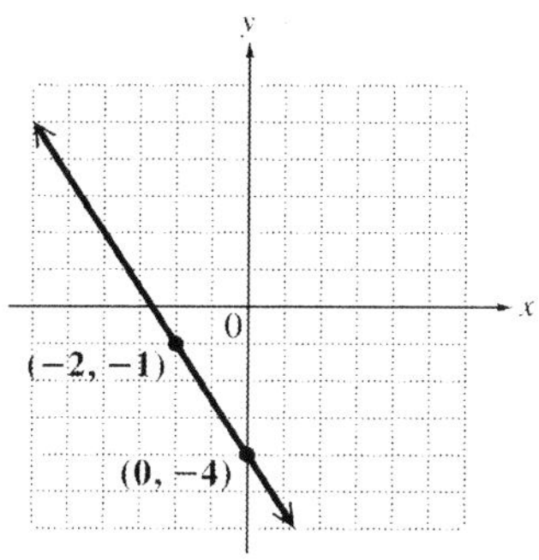

43. Locate $(-1, -2)$. Then use $m = 3 = \frac{3}{1}$ to go 3 units up and 1 unit right to $(0, 1)$.

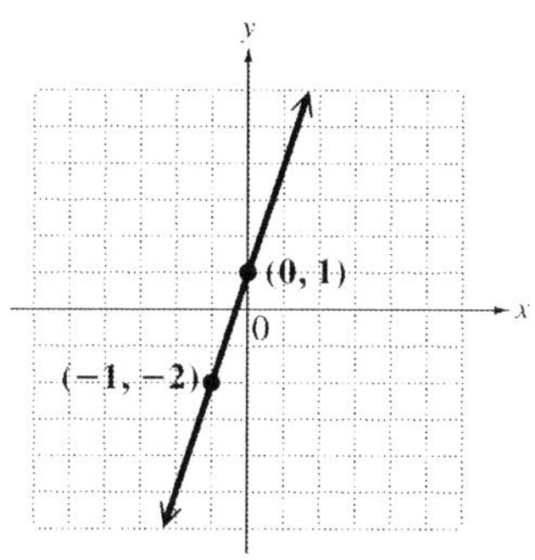

44. Locate $(-2, -4)$. Then use $m = 4 = \frac{4}{1}$ to go 4 units up and 1 unit right to $(-1, 0)$.

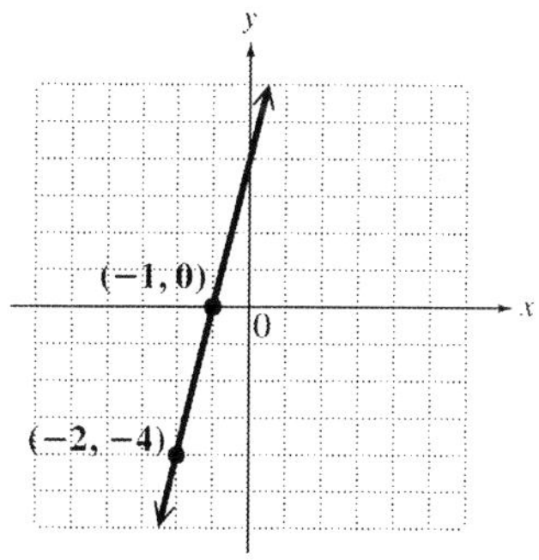

45. The slope of the line through $(4, 6)$ and $(-8, 7)$ is

$$m = \frac{7-6}{-8-4} = \frac{1}{-12} = -\frac{1}{12}.$$

The slope of the line through $(7, 4)$ and $(-5, 5)$ is

$$m = \frac{5-4}{-5-7} = \frac{1}{-12} = -\frac{1}{12}.$$

Since the slopes are equal, the two lines are *parallel.*

46. The slope of the line through $(15, 9)$ and $(12, -7)$ is

$$m = \frac{-7-9}{12-15} = \frac{-16}{-3} = \frac{16}{3}.$$

The slope of the line through $(8, -4)$ and $(5, -20)$ is

$$m = \frac{-20-(-4)}{5-8} = \frac{-16}{-3} = \frac{16}{3}.$$

Since the slopes are equal, the two lines are *parallel.*

47. $2x + 5y = -7$ and $5x - 2y = 1$

Solve the equations for y.

$$5y = -2x - 7 \qquad\qquad -2y = -5x + 1$$
$$y = -\tfrac{2}{5}x - \tfrac{7}{5} \qquad\qquad y = \tfrac{5}{2}x - \tfrac{1}{2}$$

The slopes are the coefficients of x, $-\frac{2}{5}$ and $\frac{5}{2}$. These are negative reciprocals of one another, so the lines are *perpendicular.*

48. $x + 4y = 7$ and $4x - y = 3$

Solve the equations for y.

$$4y = -x + 7 \qquad\qquad -y = -4x + 3$$
$$y = -\tfrac{1}{4}x + \tfrac{7}{4} \qquad\qquad y = 4x - 3$$

The slopes are the coefficients of x, $-\frac{1}{4}$ and 4. These are negative reciprocals of one another, so the lines are *perpendicular.*

49. $2x + y = 6$ and $x - y = 4$
Solve the equations for y.
$y = -2x + 6$ $\quad\quad -y = -x + 4$
$\quad\quad y = x - 4$

The slopes are -2 and 1. The lines are *neither* parallel nor perpendicular.

50. $4x - 3y = 6$ and $3x - 4y = 2$
Solve the equations for y.
$-3y = -4x + 6$ $\quad\quad -4y = -3x + 2$
$y = \frac{4}{3}x - 2$ $\quad\quad y = \frac{3}{4}x - \frac{1}{2}$

The slopes are $\frac{4}{3}$ and $\frac{3}{4}$. The lines are *neither* parallel nor perpendicular.

51. $3x = y$ and $2y - 6x = 5$
The slope of the first line is the coefficient of x, namely 3. Solve the second equation for y.
$2y = 6x + 5$
$y = 3x + \frac{5}{2}$

The slope of the second line is also 3, so the lines are *parallel.*

52. $x = 6$ and $6 - x = 8$
The second equation can be simplified as $x = -2$. Both lines are vertical lines, so they are *parallel.*

53. $2x + 5y = -8$ and $6 + 2x = 5y$
Solve the equations for y.
$5y = -2x - 8$ $\quad\quad 5y = 2x + 6$
$y = -\frac{2}{5}x - \frac{8}{5}$ $\quad\quad y = \frac{2}{5}x + \frac{6}{5}$

The slopes are $-\frac{2}{5}$ and $\frac{2}{5}$. The lines are *neither* parallel nor perpendicular.

54. $4x + y = 0$ and $5x - 8 = 2y$
Solve the equations for y.
$y = -4x$ $\quad\quad \frac{5}{2}x - 4 = y$

The slopes are -4 and $\frac{5}{2}$. The lines are *neither* parallel nor perpendicular.

55. $4x - 3y = 8$ and $4y + 3x = 12$
Solve the equations for y.
$-3y = -4x + 8$ $\quad\quad 4y = -3x + 12$
$y = \frac{4}{3}x - \frac{8}{3}$ $\quad\quad y = -\frac{3}{4}x + 3$

The slopes, $\frac{4}{3}$ and $-\frac{3}{4}$, are negative reciprocals of one another, so the lines are *perpendicular*.

56. $2x = y + 3$ and $2y + x = 3$
Solve the equations for y.
$2x - 3 = y$ $\quad\quad 2y = -x + 3$
$\quad\quad y = -\frac{1}{2}x + \frac{3}{2}$

The slopes, 2 and $-\frac{1}{2}$, are negative reciprocals of one another, so the lines are *perpendicular*.

57. The vertical change is 63 ft, and the horizontal change is $250 - 160 = 90$ ft.
The slope is $\frac{63}{90} = \frac{7}{10}$.

58. Let y be the vertical rise.
Since the slope is the vertical rise divided by the horizontal run,

$$.13 = \frac{y}{150}.$$

Solving for y gives

$$y = .13(150) = 19.5.$$

The vertical rise could be a maximum of 19.5 ft.

59. Use the points $(0, 20)$ and $(4, 4)$.

average rate of change

$$= \frac{\text{change in } y}{\text{change in } x} = \frac{4 - 20}{4 - 0} = \frac{-16}{4} = -4$$

The average rate of change is $-\$4000$ per year, that is, the value of the machine is decreasing \$4000 each year during these years.

60. Use the points $(0, 0)$ and $(4, 200)$.

average rate of change

$$= \frac{\text{change in } y}{\text{change in } x} = \frac{200 - 0}{4 - 0} = \frac{200}{4} = 50$$

The average rate of change is \$50 per month, that is, the amount saved is increasing \$50 each month during these months.

61. We can see that there is no change in the percent of pay raise. Thus, the average rate of change is 0% per year, that is, the percent of pay raise is not changing—it is 3% each year during these years.

62. **(a)** Use $(1996, 25.5)$ and $(2001, 17.3)$.

$$\begin{aligned}\text{average rate of change} &= \frac{17.3 - 25.5}{2001 - 1996} \\ &= -\frac{8.2}{5} \\ &= -1.64\end{aligned}$$

The average rate of change in the number of food stamp recipients is -1.64 million recipients per year during these years.

(b) The negative slope means the numbers of recipients *decreased* by 1.64 million each year.

63. **(a)** Use $(1997, 16)$ and $(2002, 35)$.

$$\begin{aligned}\text{average rate of change} &= \frac{35-16}{2002-1997}\\ &= \frac{19}{5}\\ &= 3.8\end{aligned}$$

The average rate of change in the percent of tax returns filed electronically is 3.8% per yr during these years.

(b) The positive slope means the percent of tax returns filed electronically increased an average of 3.8% each year.

64. Use $(1997, .349)$ and $(2002, 15.5)$.

$$m = \frac{15.5 - .349}{2002-1997} = \frac{15.151}{5} = 3.0302$$

The average rate of change in sales is about 3.03 million per year, that is, the sales of DVD players increased an average of 3.03 million each year from 1997 to 2002.

65. Use $(1997, 500)$ and $(2002, 155)$.

$$m = \frac{155-500}{2002-1997} = \frac{-345}{5} = -69$$

The average rate of change in price is $-\$69$ per year, that is, the price decreased an average of \$69 each year from 1997 to 2002.

66. For $A(3,1)$ and $B(6,2)$, the slope of $\overline{AB}$ is

$$m = \frac{2-1}{6-3} = \frac{1}{3}.$$

67. For $B(6,2)$ and $C(9,3)$, the slope of $\overline{BC}$ is

$$m = \frac{3-2}{9-6} = \frac{1}{3}.$$

68. For $A(3,1)$ and $C(9,3)$, the slope of $\overline{AC}$ is

$$m = \frac{3-1}{9-3} = \frac{2}{6} = \frac{1}{3}.$$

69. The slope of $\overline{AB}$ = slope of $\overline{BC}$
= slope of $\overline{AC}$
$= \frac{1}{3}.$

70. For $A(1,-2)$ and $B(3,-1)$, the slope of $\overline{AB}$ is

$$m = \frac{-1-(-2)}{3-1} = \frac{1}{2}.$$

For $B(3,-1)$ and $C(5,0)$, the slope of $\overline{BC}$ is

$$m = \frac{0-(-1)}{5-3} = \frac{1}{2}.$$

For $A(1,-2)$ and $C(5,0)$, the slope of $\overline{AC}$ is

$$m = \frac{0-(-2)}{5-1} = \frac{2}{4} = \frac{1}{2}.$$

Since the three slopes are the same, the three points are collinear.

71. For $A(0,6)$ and $B(4,-5)$, the slope of $\overline{AB}$ is

$$m = \frac{-5-6}{4-0} = \frac{-11}{4} = -\frac{11}{4}.$$

For $B(4,-5)$ and $C(-2,12)$, the slope of $\overline{BC}$ is

$$m = \frac{12-(-5)}{-2-4} = \frac{17}{-6} = -\frac{17}{6}.$$

Since these two slopes are not the same, the three points are not collinear.

4.3 Linear Equations in Two Variables

4.3 Margin Exercises

1. **(a)** Slope 2; y-intercept $(0,-3)$
Here $m = 2$ and $b = -3$. Substitute these values into the slope-intercept form.

$$\begin{aligned}y &= mx + b\\ y &= 2x + (-3)\\ y &= 2x - 3\end{aligned}$$

(b) Slope $-\frac{2}{3}$; y-intercept $(0,0)$

Here $m = -\frac{2}{3}$ and $b = 0$.

$$\begin{aligned}y &= mx + b\\ y &= -\tfrac{2}{3}x + 0\\ y &= -\tfrac{2}{3}x\end{aligned}$$

2. **(a)** $y = 2x + 3$

Here $m = 2$ and $b = 3$. Plot the y-intercept $(0,3)$. The slope 2 can be interpreted as

$$m = \frac{\text{rise}}{\text{run}} = \frac{\text{change in } y}{\text{change in } x} = \frac{2}{1}.$$

From $(0,3)$, move *up* 2 units and to the *right* 1 unit, and plot a second point at $(1,5)$. Draw a line through the two points.

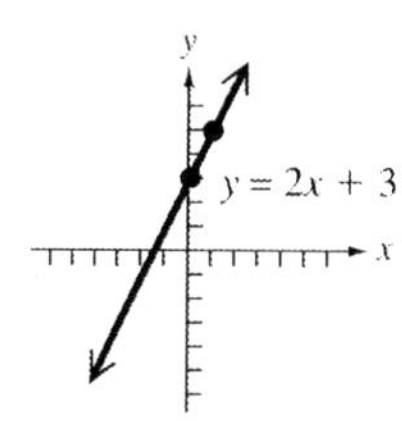

(b) $3x + 4y = 8$

Solve the equation for y.

$$4y = -3x + 8$$
$$y = -\frac{3}{4}x + 2$$

Plot the y-intercept $(0, 2)$. The slope can be interpreted as either $\frac{-3}{4}$ or $\frac{3}{-4}$. Using $\frac{-3}{4}$, move from $(0, 2)$ *down* 3 units and to the *right* 4 units to locate the point $(4, -1)$. Draw a line through the two points.

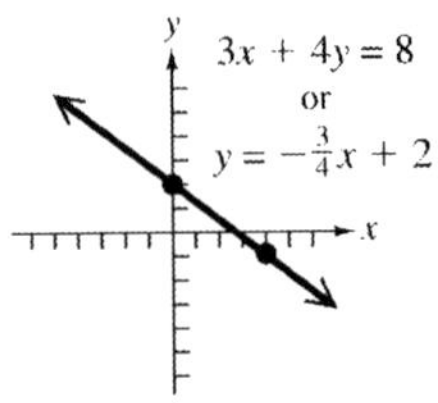

3. **(a)** Through $(-2, 7)$; $m = 3$

Use the point-slope form with $(x_1, y_1) = (-2, 7)$ and $m = 3$.

$$y - y_1 = m(x - x_1)$$
$$y - 7 = 3[x - (-2)]$$
$$y - 7 = 3(x + 2)$$
$$y - 7 = 3x + 6$$
$$y = 3x + 13$$

In standard form, $Ax + By = C$,

$$3x - y = -13.$$

(b) Through $(1, 3)$; $m = -\frac{5}{4}$

$$y - y_1 = m(x - x_1)$$
$$y - 3 = -\frac{5}{4}(x - 1)$$

Multiply by 4 to clear the fraction. Then write the equation in standard form.

$$4y - 12 = -5(x - 1)$$
$$4y - 12 = -5x + 5$$
$$5x + 4y = 17$$

4. First find the slope of the line. Then use the slope and one of the points in the point-slope form.

(a) Through $(-1, 2)$ and $(5, 7)$

$$m = \frac{7 - 2}{5 - (-1)} = \frac{5}{6}$$

Let $(x_1, y_1) = (5, 7)$.

$$y - y_1 = m(x - x_1)$$
$$y - 7 = \frac{5}{6}(x - 5)$$
$$6y - 42 = 5x - 25 \quad \textit{Multiply by 6.}$$
$$5x - 6y = -17 \quad \textit{Standard form}$$

(b) Through $(-2, 6)$ and $(1, 4)$

$$m = \frac{4 - 6}{1 - (-2)} = \frac{-2}{3} = -\frac{2}{3}$$

Let $(x_1, y_1) = (1, 4)$.

$$y - y_1 = m(x - x_1)$$
$$y - 4 = -\frac{2}{3}(x - 1)$$
$$3y - 12 = -2x + 2 \quad \textit{Multiply by 3.}$$
$$2x + 3y = 14 \quad \textit{Standard form}$$

5. **(a)** Through $(8, -2)$; $m = 0$

A line with slope $m = 0$ is horizontal. The horizontal line through the point (c, d) has equation $y = d$. Here $d = -2$, so the equation is $y = -2$.

(b) The vertical line through $(3, 5)$

The vertical line through the point (c, d) has equation $x = c$. Here $c = 3$, so the equation is $x = 3$.

6. **(a)** Through $(-8, 3)$; parallel to the line $2x - 3y = 10$

Find the slope of the given line.

$$2x - 3y = 10$$
$$-3y = -2x + 10$$
$$y = \frac{2}{3}x - \frac{10}{3}$$

The slope is $\frac{2}{3}$, so a line parallel to it also has slope $\frac{2}{3}$. Use $m = \frac{2}{3}$ and $(x_1, y_1) = (-8, 3)$ in the point-slope form.

$$y - y_1 = m(x - x_1)$$
$$y - 3 = \frac{2}{3}[x - (-8)]$$
$$y - 3 = \frac{2}{3}(x + 8)$$
$$y - 3 = \frac{2}{3}x + \frac{16}{3}$$
$$y = \frac{2}{3}x + \frac{16}{3} + \frac{9}{3}$$
$$y = \frac{2}{3}x + \frac{25}{3}$$

(b) Through $(-8, 3)$; perpendicular to $2x - 3y = 10$

The slope of $2x - 3y = 10$ is $\frac{2}{3}$. The negative reciprocal of $\frac{2}{3}$ is $-\frac{3}{2}$, so the slope of the line through $(-8, 3)$ is $-\frac{3}{2}$.

$$y - y_1 = m(x - x_1)$$
$$y - 3 = -\frac{3}{2}[x - (-8)]$$
$$y - 3 = -\frac{3}{2}(x + 8)$$
$$y - 3 = -\frac{3}{2}x - 12$$
$$y = -\frac{3}{2}x - 9$$

7. **(a)** An equation is $y = .1x$.

(b) Since the price you pay is \$.10 per minute plus a flat rate of \$.20, an equation for x minutes is

$$y = .1x + .2.$$

(c) The ordered pair $(15, 1.7)$ indicates that the price of a 15-minute call is \$1.70.

8. **(a)** Use $(0, 38)$ and $(18, 59)$.

$$m = \frac{59-38}{18-0} = \frac{21}{18} = \frac{7}{6} \approx 1.17$$

Use the slope-intercept form.

$$\begin{aligned} y &= mx + b \\ y &= 1.17x + 38 \end{aligned}$$

(b) For 2000, $x = 2000 - 1980 = 20$.

$$\begin{aligned} y &= 1.17x + 38 \\ y &= 1.17(20) + 38 \quad \textit{Let } x = 20. \\ y &= 23.4 + 38 \\ y &= 61.4 \approx 61 \end{aligned}$$

About 61% of mothers of children under 1 yr old participated in the U.S. labor force in 2000.

4.3 Section Exercises

1. Choice **A**, $3x - 2y = 5$, is in the form $Ax + By = C$ with $A \geq 0$ and integers A, B, and C having no common factor (except 1).

2. Choice **C**, $y - 3 = 2(x - 1)$, is in the form $y - y_1 = m(x - x_1)$.

3. Choice **A**, $y = 6x + 2$, is in the form $y = mx + b$.

4. $$\begin{aligned} y + 2 &= -3(x - 4) \\ y + 2 &= -3x + 12 \\ y &= -3x + 10 \end{aligned}$$

5. $$\begin{aligned} y &= -3x + 10 \\ 3x + y &= 10 \quad \textit{Standard form} \end{aligned}$$

6. Solve $10x - 7y = 70$ for y.

$$\begin{aligned} -7y &= -10x + 70 \\ y &= \tfrac{10}{7}x - 10 \end{aligned}$$

7. $y = 2x + 3$
This line is in slope-intercept form with slope $m = 2$ and y-intercept $(0, b) = (0, 3)$. The only graph with positive slope and with a positive y-coordinate of its y-intercept is **A**.

8. $y = -2x + 3$
This line is in slope-intercept form with slope $m = -2$ and y-intercept $(0, b) = (0, 3)$. The only graph with negative slope and with a positive y-coordinate of its y-intercept is **D**.

9. $y = -2x - 3$
This line is in slope-intercept form with slope $m = -2$ and y-intercept $(0, b) = (0, -3)$. The only graph with negative slope and with a negative y-coordinate of its y-intercept is **C**.

10. $y = 2x - 3$
This line has slope $m = 2$ and y-intercept $(0, b) = (0, -3)$. The only graph with positive slope and with a negative y-coordinate of its y-intercept is **F**.

11. $y = 2x$
This line has slope $m = 2$ and y-intercept $(0, b) = (0, 0)$. The only graph with positive slope and with y-intercept $(0, 0)$ is **H**.

12. $y = -2x$
This line has slope $m = -2$ and y-intercept $(0, b) = (0, 0)$. The only graph with negative slope and with y-intercept $(0, 0)$ is **G**.

13. $y = 3$
This line is a horizontal line with y-intercept $(0, 3)$. The only graph that has these characteristics is **B**.

14. $y = -3$
This line is a horizontal line with y-intercept $(0, -3)$. The only graph that has these characteristics is **E**.

15. $m = 5$; $b = 15$
Substitute these values in the slope-intercept form.

$$\begin{aligned} y &= mx + b \\ y &= 5x + 15 \end{aligned}$$

16. $m = -2$; $b = 12$
Substitute these values in the slope-intercept form.

$$\begin{aligned} y &= mx + b \\ y &= -2x + 12 \end{aligned}$$

17. $m = -\frac{2}{3}$; y-intercept $\left(0, \frac{4}{5}\right)$

Here $m = -\frac{2}{3}$ and $b = \frac{4}{5}$. Substitute these values into the slope-intercept form.

$$\begin{aligned} y &= mx + b \\ y &= -\tfrac{2}{3}x + \tfrac{4}{5} \end{aligned}$$

18. $m = -\frac{5}{8}$; y-intercept $\left(0, -\frac{1}{3}\right)$

Here $m = -\frac{5}{8}$ and $b = -\frac{1}{3}$. Substitute these values into the slope-intercept form.

$$\begin{aligned} y &= mx + b \\ y &= -\tfrac{5}{8}x - \tfrac{1}{3} \end{aligned}$$

19. Slope $\frac{2}{5}$; y-intercept $(0, 5)$
Here $m = \frac{2}{5}$ and $b = 5$. Substitute these values in the slope-intercept form.

$$\begin{aligned} y &= mx + b \\ y &= \tfrac{2}{5}x + 5 \end{aligned}$$

20. Slope $-\frac{3}{4}$; y-intercept $(0, 7)$
Here $m = -\frac{3}{4}$ and $b = 7$. Substitute these values in the slope-intercept form.

$$y = mx + b$$
$$y = -\tfrac{3}{4}x + 7$$

21. $-x + y = 2$
(a) Solve for y to get the equation in slope-intercept form.

$$-x + y = 2$$
$$y = x + 2$$

(b) The slope is the coefficient of x, 1.

(c) The y-intercept is the point $(0, b)$, or $(0, 2)$.

(d)

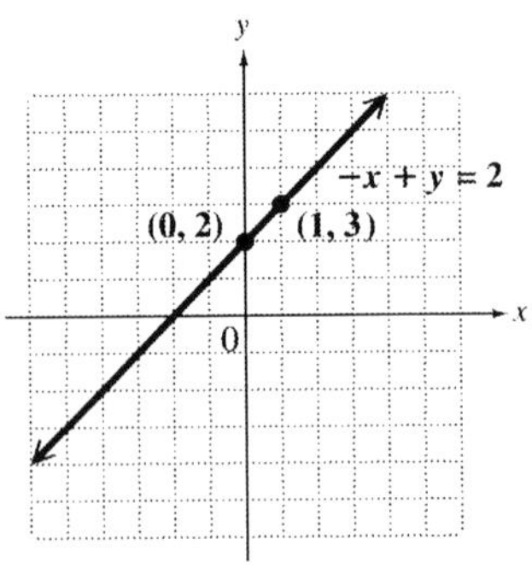

22. $-x + y = 5$
(a) Solve for y to get the equation in slope-intercept form.

$$-x + y = 5$$
$$y = x + 5$$

(b) The slope is the coefficient of x, 1.

(c) The y-intercept is the point $(0, b)$, or $(0, 5)$.

(d)

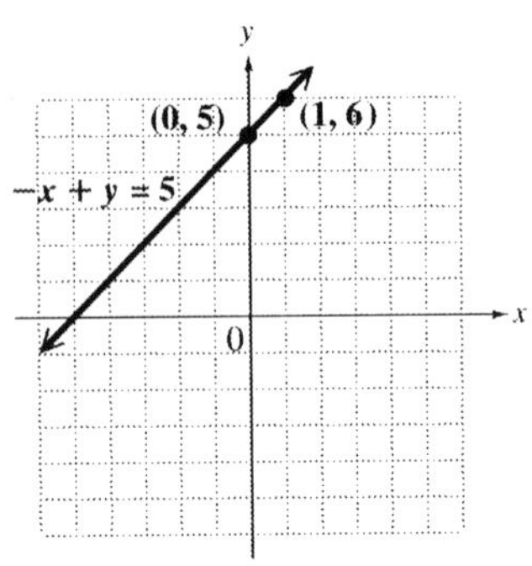

23. $4x - 5y = 20$
(a) Solve for y to get the equation in slope-intercept form.

$$4x - 5y = 20$$
$$-5y = -4x + 20$$
$$y = \tfrac{4}{5}x - 4$$

(b) The slope is the coefficient of x, $\frac{4}{5}$.

(c) The y-intercept is the point $(0, b)$, or $(0, -4)$.

(d)

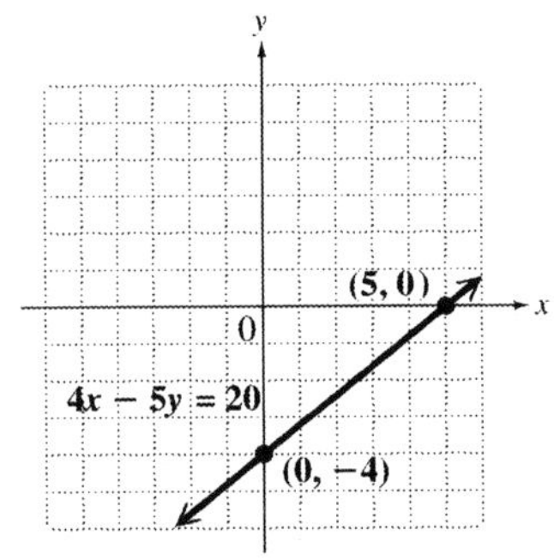

24. $7x - 3y = 3$
(a) Solve for y to get the equation in slope-intercept form.

$$7x - 3y = 3$$
$$-3y = -7x + 3$$
$$y = \tfrac{7}{3}x - 1$$

(b) The slope is the coefficient of x, $\frac{7}{3}$.

(c) The y-intercept is the point $(0, b)$, or $(0, -1)$.

(d)

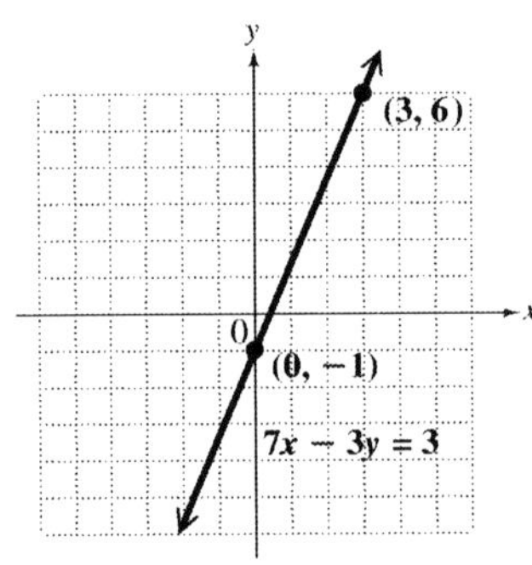

25. $x + 2y = -4$
(a) Solve for y to get the equation in slope-intercept form.

$$x + 2y = -4$$
$$2y = -x - 4$$
$$y = -\tfrac{1}{2}x - 2$$

(b) The slope is the coefficient of x, $-\frac{1}{2}$.

(c) The y-intercept is the point $(0, b)$, or $(0, -2)$.

(d)

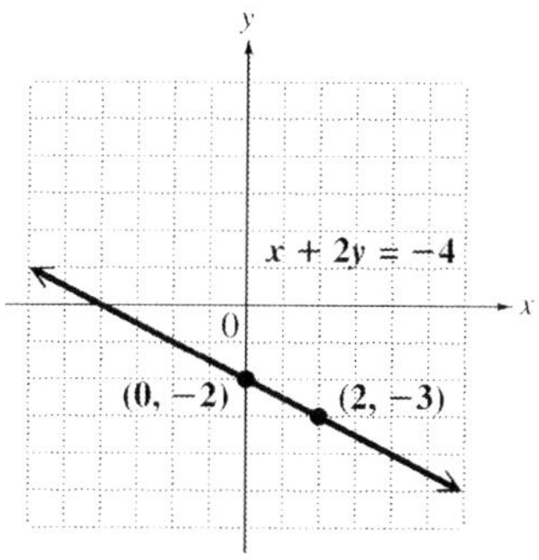

26. $x + 3y = -9$

(a) Solve for y to get the equation in slope-intercept form.

$$\begin{aligned} x + 3y &= -9 \\ 3y &= -x - 9 \\ y &= -\tfrac{1}{3}x - 3 \end{aligned}$$

(b) The slope is the coefficient of x, $-\frac{1}{3}$.

(c) The y-intercept is the point $(0, b)$, or $(0, -3)$.

(d)

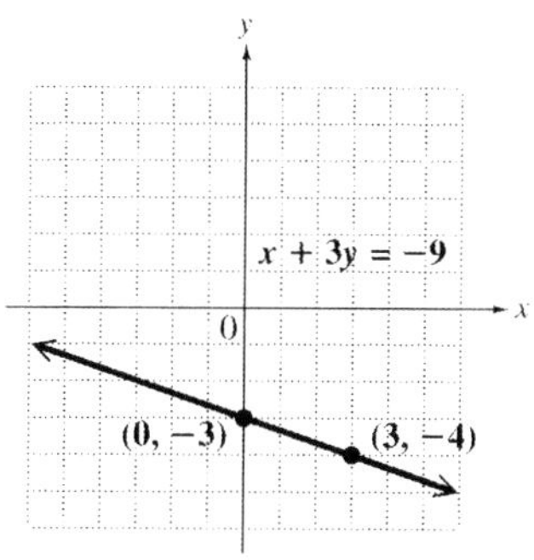

27. Through $(-2, 4)$; $m = -\frac{3}{4}$
Use the point-slope form with $(x_1, y_1) = (-2, 4)$ and $m = -\frac{3}{4}$.

$$\begin{aligned} y - y_1 &= m(x - x_1) \\ y - 4 &= -\tfrac{3}{4}[x - (-2)] \\ 4(y - 4) &= -3(x + 2) \\ 4y - 16 &= -3x - 6 \\ 3x + 4y &= 10 \end{aligned}$$

28. Through $(-1, 6)$; $m = -\frac{5}{6}$
Use the point-slope form with $(x_1, y_1) = (-1, 6)$ and $m = -\frac{5}{6}$.

$$\begin{aligned} y - y_1 &= m(x - x_1) \\ y - 6 &= -\tfrac{5}{6}[x - (-1)] \\ 6(y - 6) &= -5(x + 1) \\ 6y - 36 &= -5x - 5 \\ 5x + 6y &= 31 \end{aligned}$$

29. Through $(5, 8)$; $m = -2$
Use the point-slope form with $(x_1, y_1) = (5, 8)$ and $m = -2$.

$$\begin{aligned} y - y_1 &= m(x - x_1) \\ y - 8 &= -2(x - 5) \\ y - 8 &= -2x + 10 \\ 2x + y &= 18 \end{aligned}$$

30. Through $(12, 10)$; $m = 1$

$$\begin{aligned} y - 10 &= 1(x - 12) \\ y - 10 &= x - 12 \\ -x + y &= -2 \\ x - y &= 2 \end{aligned}$$

31. Through $(-5, 4)$; $m = \frac{1}{2}$
Use the point-slope form with $(x_1, y_1) = (-5, 4)$ and $m = \frac{1}{2}$.

$$\begin{aligned} y - y_1 &= m(x - x_1) \\ y - 4 &= \tfrac{1}{2}[x - (-5)] \\ 2(y - 4) &= 1(x + 5) \\ 2y - 8 &= x + 5 \\ -x + 2y &= 13 \\ x - 2y &= -13 \end{aligned}$$

32. Through $(7, -2)$; $m = \frac{1}{4}$

$$\begin{aligned} y - (-2) &= \tfrac{1}{4}(x - 7) \\ 4(y + 2) &= 1(x - 7) \\ 4y + 8 &= x - 7 \\ -x + 4y &= -15 \\ x - 4y &= 15 \end{aligned}$$

33. Through $(-4, 12)$; horizontal
A horizontal line through the point (c, d) has equation $y = d$. Here $d = 12$, so the equation is $y = 12$.

34. Through $(1, 5)$; horizontal
A horizontal line through the point (c, d) has equation $y = d$. Here $d = 5$, so the equation is $y = 5$.

35. Through $(9, 10)$; undefined slope
A vertical line has undefined slope and equation $x = c$. Since the x-value in $(9, 10)$ is 9, the equation is $x = 9$.

36. Through $(-2, 8)$; 0 slope
A line with 0 slope is a horizontal line in the form $y = d$. Since the y-value in $(-2, 8)$ is 8, the equation of this line is $y = 8$.

37. Through $(.5, .2)$; horizontal
A horizontal line through the point (c, d) has equation $y = d$. Since the y-value in $(.5, .2)$ is .2, the equation of this line is $y = .2$.

38. Through $\left(\frac{5}{8}, \frac{2}{9}\right)$; vertical
A vertical line has the form $x = c$. Since the x-value in $\left(\frac{5}{8}, \frac{2}{9}\right)$ is $\frac{5}{8}$, the equation of this line is $x = \frac{5}{8}$.

39. $(3, 4)$ and $(5, 8)$
Find the slope.

$$m = \frac{8-4}{5-3} = \frac{4}{2} = 2$$

Use the point-slope form with $(x_1, y_1) = (3, 4)$ and $m = 2$.

$$\begin{aligned} y - y_1 &= m(x - x_1) \\ y - 4 &= 2(x - 3) \\ y - 4 &= 2x - 6 \\ -2x + y &= -2 \\ 2x - y &= 2 \end{aligned}$$

40. $(5, -2)$ and $(-3, 14)$
Find the slope.

$$m = \frac{14 - (-2)}{-3 - 5} = \frac{16}{-8} = -2$$

Use the point-slope form with $(x_1, y_1) = (5, -2)$ and $m = -2$.

$$\begin{aligned} y - y_1 &= m(x - x_1) \\ y - (-2) &= -2(x - 5) \\ y + 2 &= -2x + 10 \\ 2x + y &= 8 \end{aligned}$$

41. $(6, 1)$ and $(-2, 5)$
Find the slope.

$$m = \frac{5-1}{-2-6} = \frac{4}{-8} = -\frac{1}{2}$$

Use the point-slope form with $(x_1, y_1) = (6, 1)$ and $m = -\frac{1}{2}$.

$$\begin{aligned} y - y_1 &= m(x - x_1) \\ y - 1 &= -\tfrac{1}{2}(x - 6) \\ 2(y - 1) &= -1(x - 6) \\ 2y - 2 &= -x + 6 \\ x + 2y &= 8 \end{aligned}$$

42. $(-2, 5)$ and $(-8, 1)$
Find the slope.

$$m = \frac{1-5}{-8-(-2)} = \frac{-4}{-6} = \frac{2}{3}$$

Use the point-slope form with $(x_1, y_1) = (-2, 5)$ and $m = \frac{2}{3}$.

$$\begin{aligned} y - 5 &= \tfrac{2}{3}[x - (-2)] \\ 3(y - 5) &= 2(x + 2) \\ 3y - 15 &= 2x + 4 \\ -2x + 3y &= 19 \\ 2x - 3y &= -19 \end{aligned}$$

43. $\left(-\dfrac{2}{5}, \dfrac{2}{5}\right)$ and $\left(\dfrac{4}{3}, \dfrac{2}{3}\right)$
Find the slope.

$$m = \frac{\frac{2}{3} - \frac{2}{5}}{\frac{4}{3} - \left(-\frac{2}{5}\right)} = \frac{\frac{10-6}{15}}{\frac{20+6}{15}}$$

$$= \frac{\frac{4}{15}}{\frac{26}{15}} = \frac{4}{26} = \frac{2}{13}$$

Use the point-slope form with $(x_1, y_1) = \left(-\frac{2}{5}, \frac{2}{5}\right)$ and $m = \frac{2}{13}$.

$$\begin{aligned} y - \tfrac{2}{5} &= \tfrac{2}{13}\left[x - \left(-\tfrac{2}{5}\right)\right] \\ 13\left(y - \tfrac{2}{5}\right) &= 2\left(x + \tfrac{2}{5}\right) \\ 13y - \tfrac{26}{5} &= 2x + \tfrac{4}{5} \\ -2x + 13y &= \tfrac{30}{5} \\ 2x - 13y &= -6 \end{aligned}$$

44. $\left(\dfrac{3}{4}, \dfrac{8}{3}\right)$ and $\left(\dfrac{2}{5}, \dfrac{2}{3}\right)$

$$m = \frac{\frac{2}{3} - \frac{8}{3}}{\frac{2}{5} - \frac{3}{4}} = \frac{\frac{2-8}{3}}{\frac{8-15}{20}}$$

$$= \frac{-\frac{6}{3}}{-\frac{7}{20}} = \frac{2}{\frac{7}{20}} = (2)\left(\frac{20}{7}\right) = \frac{40}{7}$$

Use the point-slope form with $(x_1, y_1) = \left(\frac{3}{4}, \frac{8}{3}\right)$ and $m = \frac{40}{7}$.

$$\begin{aligned} y - \tfrac{8}{3} &= \tfrac{40}{7}\left(x - \tfrac{3}{4}\right) \\ 7\left(y - \tfrac{8}{3}\right) &= 40\left(x - \tfrac{3}{4}\right) \\ 7y - \tfrac{56}{3} &= 40x - 30 \\ -40x + 7y &= \tfrac{56}{3} - \tfrac{90}{3} \\ -40x + 7y &= -\tfrac{34}{3} \\ 120x - 21y &= 34 \end{aligned}$$

45. $(2, 5)$ and $(1, 5)$
Find the slope.

$$m = \frac{5-5}{1-2} = \frac{0}{-1} = 0$$

A line with slope 0 is horizontal. A horizontal line through the point (x, k) has equation $y = k$, where k is the common y-value, so the equation is $y = 5$.

46. $(-2, 2)$ and $(4, 2)$
Find the slope.

$$m = \frac{2-2}{4-(-2)} = \frac{0}{6} = 0$$

A line with slope 0 is horizontal. A horizontal line through the point (x, k) has equation $y = k$, where k is the common y-value, so the equation is $y = 2$.

47. $(7, 6)$ and $(7, -8)$
Find the slope.

$$m = \frac{-8-6}{7-7} = \frac{-14}{0} \quad \textit{Undefined}$$

A line with undefined slope is a vertical line. The equation of a vertical line is $x = k$, where k is the common x-value. So the equation is $x = 7$.

48. $(13, 5)$ and $(13, -1)$
Find the slope.

$$m = \frac{-1-5}{13-13} = \frac{-6}{0} \quad \textit{Undefined}$$

A line with undefined slope is a vertical line. The equation of a vertical line is $x = k$, where k is the common x-value. So the equation is $x = 13$.

49. Through $(7, 2)$; parallel to $3x - y = 8$
Find the slope of the line with equation

$$\begin{aligned} 3x - y &= 8. \\ -y &= -3x + 8 \\ y &= 3x - 8 \end{aligned}$$

The slope is 3, so a line parallel to it also has slope 3. Use $m = 3$ and $(x_1, y_1) = (7, 2)$ in the point-slope form.

$$\begin{aligned} y - y_1 &= m(x - x_1) \\ y - 2 &= 3(x - 7) \\ y - 2 &= 3x - 21 \\ y &= 3x - 19 \end{aligned}$$

50. Through $(4, 1)$; parallel to $2x + 5y = 10$
Find the slope of the line with equation

$$\begin{aligned} 2x + 5y &= 10. \\ 5y &= -2x + 10 \\ y &= -\tfrac{2}{5}x + 2 \end{aligned}$$

The slope is $-\frac{2}{5}$. We are to find the equation of a line parallel to this line so its slope is also $-\frac{2}{5}$. Use the point-slope form.

$$\begin{aligned} y - 1 &= -\tfrac{2}{5}(x - 4) \\ y - 1 &= -\tfrac{2}{5}x + \tfrac{8}{5} \\ y &= -\tfrac{2}{5}x + \tfrac{13}{5} \end{aligned}$$

51. Through $(-2, -2)$; parallel to $-x + 2y = 10$
Find the slope of the line with equation

$$\begin{aligned} -x + 2y &= 10. \\ 2y &= x + 10 \\ y &= \tfrac{1}{2}x + 5 \end{aligned}$$

The slope is $\frac{1}{2}$, so a line parallel to it also has slope $\frac{1}{2}$. Use $m = \frac{1}{2}$ and $(x_1, y_1) = (-2, -2)$ in the point-slope form.

$$\begin{aligned} y - y_1 &= m(x - x_1) \\ y - (-2) &= \tfrac{1}{2}[x - (-2)] \\ y + 2 &= \tfrac{1}{2}(x + 2) \\ y + 2 &= \tfrac{1}{2}x + 1 \\ y &= \tfrac{1}{2}x - 1 \end{aligned}$$

52. Through $(-1, 3)$; parallel to $-x + 3y = 12$

$$\begin{aligned} 3y &= x + 12 \\ y &= \tfrac{1}{3}x + 4 \end{aligned}$$

The slope of the required line is the same as the slope of this line: $\frac{1}{3}$.

$$\begin{aligned} y - 3 &= \tfrac{1}{3}[x - (-1)] \\ y - 3 &= \tfrac{1}{3}(x + 1) \\ y - 3 &= \tfrac{1}{3}x + \tfrac{1}{3} \\ y &= \tfrac{1}{3}x + \tfrac{10}{3} \end{aligned}$$

53. Through $(8, 5)$; perpendicular to $2x - y = 7$
Find the slope of the line with equation

$$\begin{aligned} 2x - y &= 7. \\ -y &= -2x + 7 \\ y &= 2x - 7 \end{aligned}$$

The slope of the line is 2. Therefore, the slope of the line perpendicular to it is $-\frac{1}{2}$ since $2(-\frac{1}{2}) = -1$. Use $m = -\frac{1}{2}$ and $(x_1, y_1) = (8, 5)$ in the point-slope form.

$$\begin{aligned} y - y_1 &= m(x - x_1) \\ y - 5 &= -\tfrac{1}{2}(x - 8) \\ y - 5 &= -\tfrac{1}{2}x + 4 \\ y &= -\tfrac{1}{2}x + 9 \end{aligned}$$

54. Through $(2, -7)$; perpendicular to $5x + 2y = 18$

$$\begin{aligned} 2y &= -5x + 18 \\ y &= -\tfrac{5}{2}x + 9 \end{aligned}$$

The slope of the given line is $-\frac{5}{2}$, so the slope of the required line is the negative reciprocal of $-\frac{5}{2}$, that is, $\frac{2}{5}$.

Use the point-slope form.

$$\begin{aligned} y - (-7) &= \tfrac{2}{5}(x - 2) \\ y + 7 &= \tfrac{2}{5}x - \tfrac{4}{5} \\ y &= \tfrac{2}{5}x - \tfrac{39}{5} \end{aligned}$$

55. Through $(-2, 7)$; perpendicular to $x = 9$
$x = 9$ is a vertical line so a line perpendicular to it will be a horizontal line. It goes through $(-2, 7)$ so its equation is

$$y = 7.$$

56. Through $(8, 4)$; perpendicular to $x = -3$
$x = -3$ is a vertical line so a line perpendicular to it will be a horizontal line. It goes through $(8, 4)$, so its equation is

$$y = 4.$$

57. Distance = (rate)(time), so

$$y = 45x.$$

x	$y = 45x$	Ordered Pair
0	$45(0) = 0$	$(0, 0)$
5	$45(5) = 225$	$(5, 225)$
10	$45(10) = 450$	$(10, 450)$

58. Total cost = (cost/disc)(number of discs), so

$$y = 16x.$$

x	$y = 16x$	Ordered Pair
0	$16(0) = 0$	$(0, 0)$
5	$16(5) = 80$	$(5, 80)$
10	$16(10) = 160$	$(10, 160)$

59. Total cost = (cost/gal)(number of gallons), so

$$y = 2.00x.$$

x	$y = 2.00x$	Ordered Pair
0	$2.00(0) = 0$	$(0, 0)$
5	$2.00(5) = 10.00$	$(5, 10.00)$
10	$2.00(10) = 20.00$	$(10, 20.00)$

60. Total cost = (cost/day)(number of days), so

$$y = 2.50x.$$

x	$y = 2.50x$	Ordered Pair
0	$2.50(0) = 0$	$(0, 0)$
5	$2.50(5) = 12.50$	$(5, 12.50)$
10	$2.50(10) = 25.00$	$(10, 25.00)$

61. **(a)** The fixed cost is \$99, so that is the value of b. The variable cost is \$39, so

$$y = mx + b = 39x + 99.$$

(b) If $x = 5$, $y = 39(5) + 99 = 294$. The ordered pair is $(5, 294)$. The cost of a 5-month membership is \$294.

(c) If $x = 12$, $y = 39(12) + 99 = 567$. The cost of the first year's membership is \$567.

62. **(a)** The fixed cost is \$159, so that is the value of b. The variable cost is \$60, so

$$y = mx + b = 60x + 159.$$

(b) If $x = 5$, $y = 60(5) + 159 = 459$. The ordered pair is $(5, 459)$. The cost of a family membership for a family of six (with five additional members) is \$459.

(c) For a four-person family, $x = 3$, so $y = 60(3) + 159 = 339$. The cost of a family membership for a four-person family is \$339.

63. **(a)** The fixed cost is \$25, so that is the value of b. The variable cost is \$50, so

$$y = mx + b = 50x + 25.$$

(b) If $x = 5$, $y = 50(5) + 25 = 275$. The ordered pair is $(5, 275)$. The cost of the plan for 5 months is \$275.

(c) For a 2-year contract, $x = 24$, so $y = 50(24) + 25 = 1225$. The cost of the plan for 2 years is \$1225.

64. **(a)** The fixed cost is \$19.95 + \$25 = \$44.95, so that is the value of b. The variable cost is \$35, so

$$y = mx + b = 35x + 44.95.$$

(b) If $x = 5$, $y = 35(5) + 44.95 = 219.95$. The ordered pair is $(5, 219.95)$. The cost of the plan for 5 months is \$219.95.

(c) For a 1-year contract, $x = 12$, so $y = 35(12) + 44.95 = 464.95$. The cost of the plan for 1 year is \$464.95.

65. **(a)** The fixed cost is \$50, so that is the value of b. The variable cost is \$.20, so

$$y = mx + b = .20x + 50.$$

(b) If $x = 5$, $y = .20(5) + 50 = 51$. The ordered pair is $(5, 51)$. The charge for driving 5 miles is \$51.

(c)

$$\begin{aligned} 84.60 &= .20x + 50 \qquad \textit{Let } y = 84.60. \\ 34.60 &= .20x \\ x &= \frac{34.60}{.20} = 173 \end{aligned}$$

The car was driven 173 miles.

66. **(a)** The fixed cost is \$30, so that is the value of b. The variable cost is \$6, so

$$y = mx + b = 6x + 30.$$

(b) If $x = 5$, $y = 6(5) + 30 = 60$. The ordered pair is $(5, 60)$. It costs \$60 to rent the saw for 5 days.

(c) $138 = 6x + 30$ *Let y = 138.*

$108 = 6x$

$x = \frac{108}{6} = 18$

The saw is rented for 18 days.

67. **(a)** Use $(0, 9)$ and $(5, 37)$.

$$m = \frac{37 - 9}{5 - 0} = \frac{28}{5} = 5.6$$

The equation is $y = 5.6x + 9$. The slope tells us that the percent of households accessing the Internet by broadband is increasing at a rate of 5.6% per year during the given years.

(b) The year 2006 corresponds to $x = 6$, so $y = 5.6(6) + 9 = 42.6 \approx 43\%$.

68. **(a)** Use $(0, 91)$ and $(5, 63)$.

$$m = \frac{63 - 91}{5 - 0} = \frac{-28}{5} = -5.6$$

The equation is $y = -5.6x + 91$. The slope tells us that the percent of households accessing the Internet by dial-up is decreasing at a rate of 5.6% per year during the given years.

(b) The year 2006 corresponds to $x = 6$, so $y = -5.6(6) + 91 = 57.4 \approx 57\%$.

69. **(a)** Use $(5, 28{,}392)$ and $(10, 27{,}876)$.

$$m = \frac{27{,}876 - 28{,}392}{10 - 5} = \frac{-516}{5} = -103.2$$

Now use the point-slope form.

$$y - 27{,}876 = -103.2(x - 10)$$
$$y - 27{,}876 = -103.2x + 1032$$
$$y = -103.2x + 28{,}908$$

(b) The year 1998 corresponds to $x = 8$, so $y = -103.2(8) + 28{,}908 = 28{,}082.4 \approx 28{,}082$. This value is slightly higher than the actual value of 27,952.

70. **(a)** Use $(5, 22{,}393)$ and $(10, 30{,}436)$.

$$m = \frac{30{,}436 - 22{,}393}{10 - 5} = \frac{8043}{5} = 1608.6$$

Now use the point-slope form.

$$y - 22{,}393 = 1608.6(x - 5)$$
$$y - 22{,}393 = 1608.6x - 8043$$
$$y = 1608.6x + 14{,}350$$

(b) The year 1997 corresponds to $x = 7$, so

$$y = 1608.6(7) + 14{,}350$$
$$= 25{,}610.2 \approx \$25{,}610.$$

This value is a little higher than the actual value of \$25,050.

71. When $C = 0°$, $F = \underline{32}°$, and when $C = 100°$, $F = \underline{212}°$. These are the freezing and boiling temperatures for water.

72. The two points of the form (C, F) would be $(0, 32)$ and $(100, 212)$.

73. $m = \frac{212 - 32}{100 - 0} = \frac{180}{100} = \frac{9}{5}$

74. Let $m = \frac{9}{5}$ and $(x_1, y_1) = (0, 32)$.

$$y - y_1 = m(x - x_1)$$
$$F - 32 = \tfrac{9}{5}(C - 0)$$
$$F - 32 = \tfrac{9}{5}C$$
$$F = \tfrac{9}{5}C + 32$$

75.

$$F = \tfrac{9}{5}C + 32$$
$$F - 32 = \tfrac{9}{5}C$$
$$\tfrac{5}{9}(F - 32) = C$$

76.

$F = \frac{9}{5}C + 32$

$C = \frac{9}{5}C + 32$ *Let F = C.*

$5C = 9C + 160$ *Multiply by 5.*

$0 = 4C + 160$ *Subtract 5C.*

$-160 = 4C$ *Subtract 160.*

$-40 = C$ *Divide by 4.*

The Fahrenheit and Celsius temperatures are equal at $-40°$.

77. $F \approx 2C + 30$

$F \approx 2(15) + 30$ *Let C = 15.*

$= 30 + 30 = 60°$

78. $F = \frac{9}{5}C + 32$

$F = \frac{9}{5}(15) + 32$ *Let C = 15.*

$= 27 + 32$

$= 59°$

$60° - 59° = 1°$

This exact answer is one less than the approximation found in Exercise 77. They differ by $1°$.

79. $F \approx 2C + 30$
$F \approx 2(30) + 30$ *Let C = 30.*
$= 60 + 30$
$= 90°$

$F = \frac{9}{5}C + 32$
$F = \frac{9}{5}(30) + 32$ *Let C = 30.*
$= 54 + 32$
$= 86°$

$90° - 86° = 4°$

The exact answer is four less than the approximation. They differ by 4°.

80. Since $\frac{9}{5}$ is a little less than 2, and 32 is a little more than 30, $\frac{9}{5}C + 32 \approx 2C + 30$.

4.4 Linear Inequalities in Two Variables

4.4 Margin Exercises

1. **(a)** $x + y \leq 4$

Step 1
Graph the line, $x + y = 4$, which has intercepts $(4, 0)$ and $(0, 4)$, as a solid line since the inequality involves " $\leq$ ".

Step 2
Test $(0, 0)$.

$$\begin{aligned} x + y &\leq 4 \\ 0 + 0 &\leq 4 \quad ? \\ 0 &\leq 4 \quad \textit{True} \end{aligned}$$

Step 3
Since the result is true, shade the region that contains $(0, 0)$.

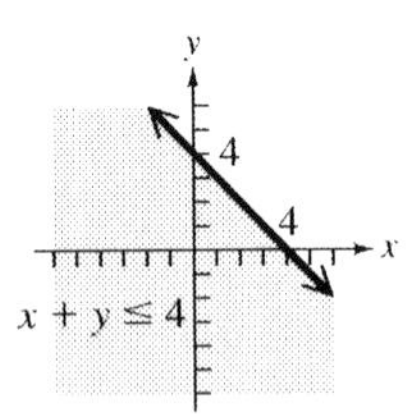

(b) $3x + y \geq 6$

Step 1
Graph the line $3x + y = 6$ as a solid line through $(2, 0)$ and $(0, 6)$.

Step 2
Test $(0, 0)$.

$$\begin{aligned} 3x + y &\geq 6 \\ 3(0) + 0 &\geq 6 \quad ? \\ 0 &\geq 6 \quad \textit{False} \end{aligned}$$

Step 3
Since the result is false, shade the region that does not contain $(0, 0)$.

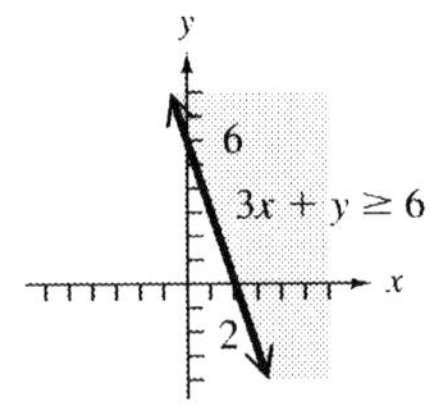

2. **(a)** $x - y > 2$

Solve the inequality for y.

$-y > -x + 2$ *Subtract x.*
$y < x - 2$ *Multiply by −1; change > to <.*

Graph the boundary line, $y = x - 2$ [which has slope 1 and y-intercept $(0, -2)$], as a dashed line because the inequality symbol is $<$. Since the inequality is solved for y and the inequality symbol is $<$, we shade the half-plane below the boundary line.

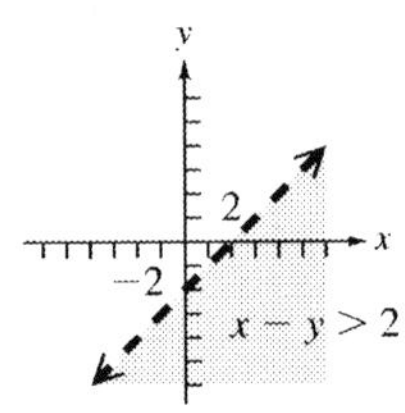

(b) $3x + 4y < 12$

Solve the inequality for y.

$4y < -3x + 12$ *Subtract 3x.*
$y < -\frac{3}{4}x + 3$ *Divide by 4.*

Graph the boundary line, $y = -\frac{3}{4}x + 3$ [which has slope $-\frac{3}{4}$ and y-intercept $(0, 3)$], as a dashed line because the inequality symbol is $<$. Since the inequality is solved for y and the inequality symbol is $<$, we shade the half-plane below the boundary line.

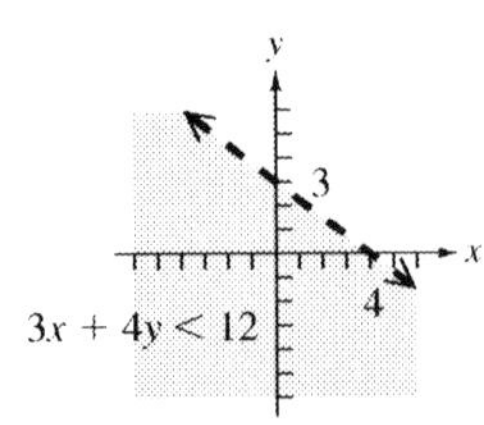

3. $x - y \leq 4$ and $x \geq -2$
Graph $x - y = 4$, which has intercepts $(4, 0)$ and $(0, -4)$, as a solid line since the inequality involves " $\leq$ ". Test $(0, 0)$, which yields $0 \leq 4$, a true statement. Shade the region that includes $(0, 0)$.

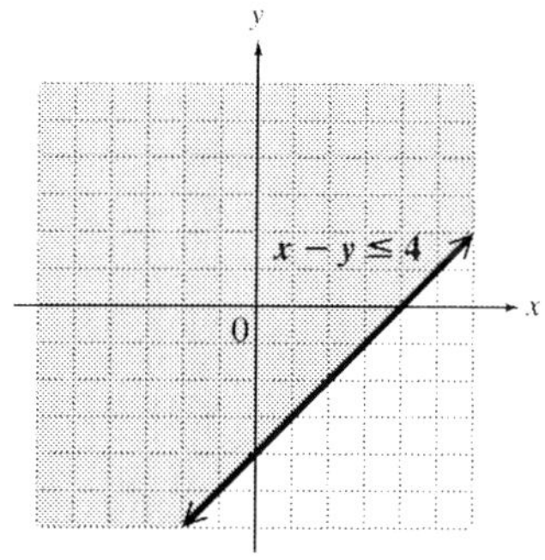

Graph $x = -2$ as a solid vertical line through $(-2, 0)$. Shade the region to the right of $x = -2$.

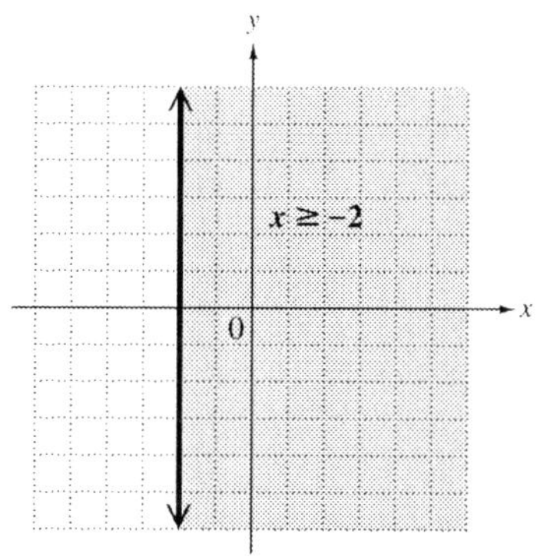

The graph of the intersection is the region common to both graphs.

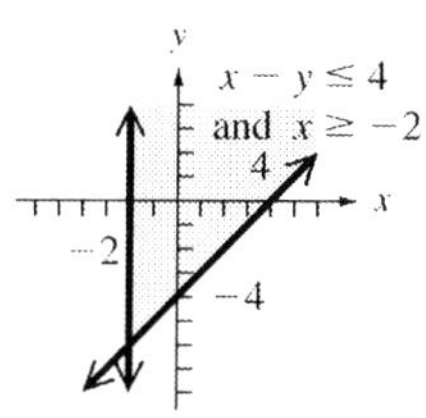

4. $7x - 3y < 21$ or $x > 2$
Graph $7x - 3y = 21$ as a dashed line through its intercepts $(3, 0)$ and $(0, -7)$. Test $(0, 0)$, which yields $0 < 21$, a true statement. Shade the region that includes $(0, 0)$.

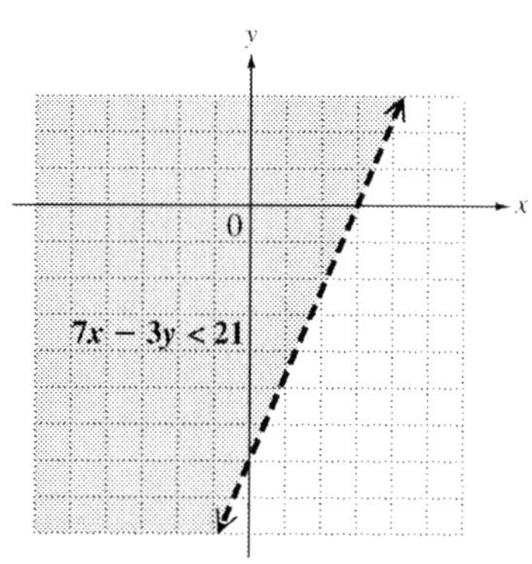

Graph $x = 2$ as a dashed vertical line through $(2, 0)$. Shade the region to the right of $x = 2$.

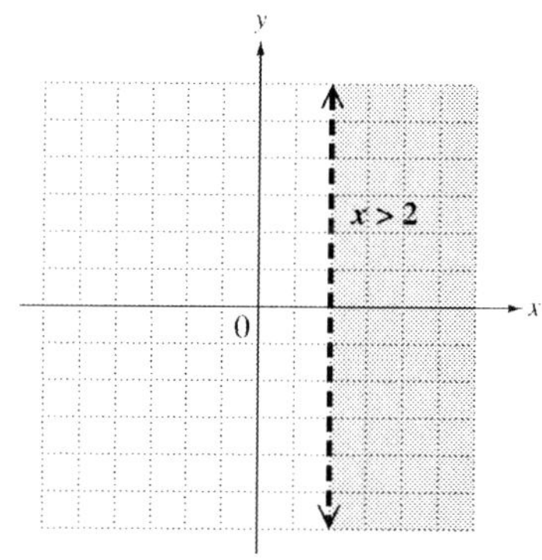

The graph of the union is the region that includes all the points in both graphs.

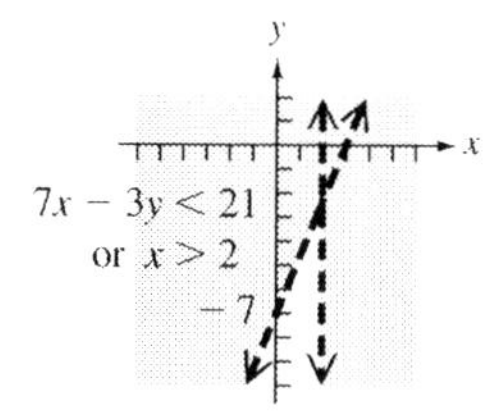

4.4 Section Exercises

1. The boundary of the graph of $y \leq -x + 2$ will be a <u>solid</u> line (since the inequality involves $\leq$), and the shading will be <u>below</u> the line (since the inequality sign is $\leq$ or $<$).

2. The boundary of the graph of $y < -x + 2$ will be a <u>dashed</u> line (since the inequality involves $<$), and the shading will be <u>below</u> the line (since the inequality sign is $\leq$ or $<$).

3. The boundary of the graph of $y > -x + 2$ will be a <u>dashed</u> line (since the inequality involves $>$), and the shading will be <u>above</u> the line (since the inequality sign is $\geq$ or $>$).

4. The boundary of the graph of $y \geq -x + 2$ will be a <u>solid</u> line (since the inequality involves $\geq$), and the shading will be <u>above</u> the line (since the inequality sign is $\geq$ or $>$).

5. The graph of $Ax + By = C$ divides the plane into two regions. In one of these regions, the ordered pairs satisfy $Ax + By < C$; in the other, they satisfy $Ax + By > C$.

6. One method is choosing any test point not on the boundary line and substituting the coordinates into the inequality. If the test point satisfies the inequality, the solution set is the region where it is located; if not, the solution set is the other region. The second method is writing the inequality in the form $y < mx + b$ or $y > mx + b$. The form $y < mx + b$ indicates the solution set is the region *below* the boundary line; $y > mx + b$ indicates the solution set is *above* the boundary line.

7. $x + y \leq 2$
Graph the line $x + y = 2$ by drawing a solid line (since the inequality involves $\leq$) through the intercepts $(2, 0)$ and $(0, 2)$.
Test a point not on this line, such as $(0, 0)$.

$$\begin{aligned} x + y &\leq 2 \\ 0 + 0 &\leq 2 \quad ? \\ 0 &\leq 2 \qquad \textit{True} \end{aligned}$$

Shade that side of the line containing the test point $(0, 0)$.

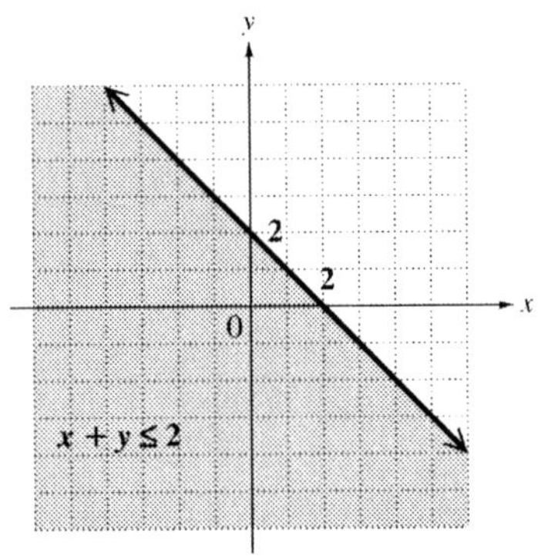

8. $x + y \leq -3$
Graph the line $x + y = -3$ by drawing a solid line (since the inequality involves $\leq$) through the intercepts $(-3, 0)$ and $(0, -3)$.
Test a point not on this line, such as $(0, 0)$.

$$\begin{aligned} x + y &\leq -3 \\ 0 + 0 &\leq -3 \quad ? \\ 0 &\leq -3 \qquad \textit{False} \end{aligned}$$

Shade that side of the line *not* containing the test point $(0, 0)$.

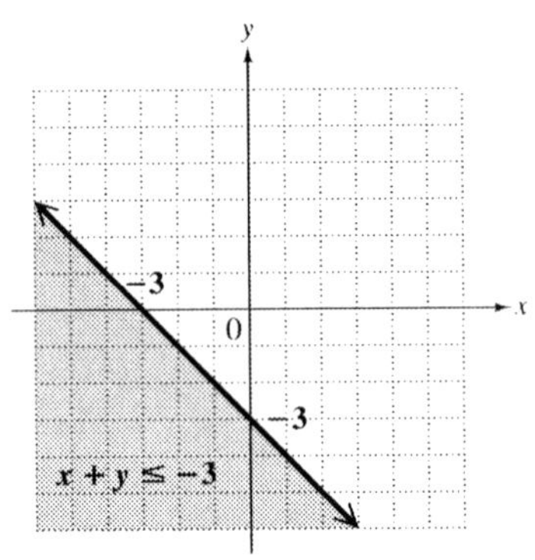

9. $4x - y < 4$
Graph the line $4x - y = 4$ by drawing a dashed line (since the inequality involves $<$) through the intercepts $(1, 0)$ and $(0, -4)$. Instead of using a test point, we will solve the inequality for y.

$$\begin{aligned} -y &< -4x + 4 \\ y &> 4x - 4 \end{aligned}$$

Since we have "$y >$ " in the last inequality, shade the region *above* the boundary line.

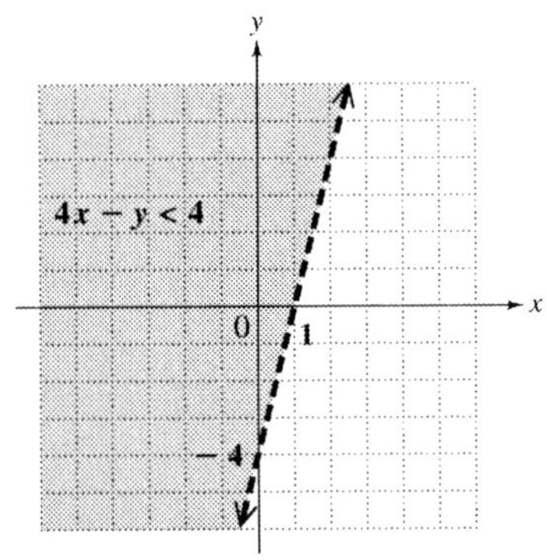

10. $3x - y < 3$
Graph the line $3x - y = 3$ by drawing a dashed line (since the inequality involves $<$) through the intercepts $(1, 0)$ and $(0, -3)$. Instead of using a test point, we will solve the inequality for y.

$$\begin{aligned} -y &< -3x + 3 \\ y &> 3x - 3 \end{aligned}$$

Since we have "$y >$ " in the last inequality, shade the region *above* the boundary line.

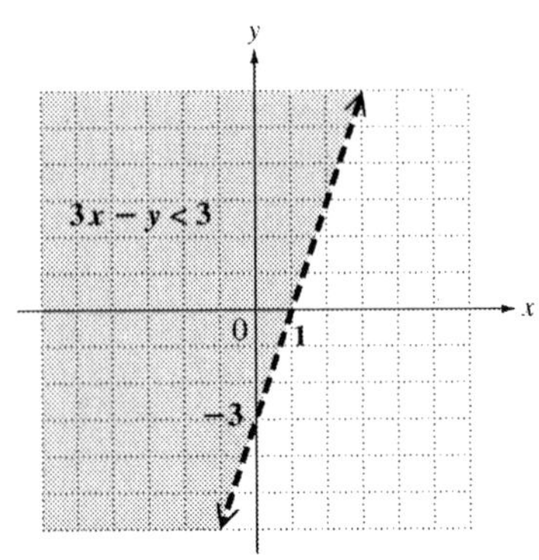

11. $x + 3y \geq -2$

Graph the solid line $x + 3y = -2$ (since the inequality involves $\geq$) through the intercepts $(-2, 0)$ and $\left(0, -\frac{2}{3}\right)$.

Test a point not on this line such as $(0, 0)$.

$$0 + 3(0) \geq -2 \quad ?$$
$$0 \geq -2 \qquad \textit{True}$$

Shade that side of the line containing the test point $(0, 0)$.

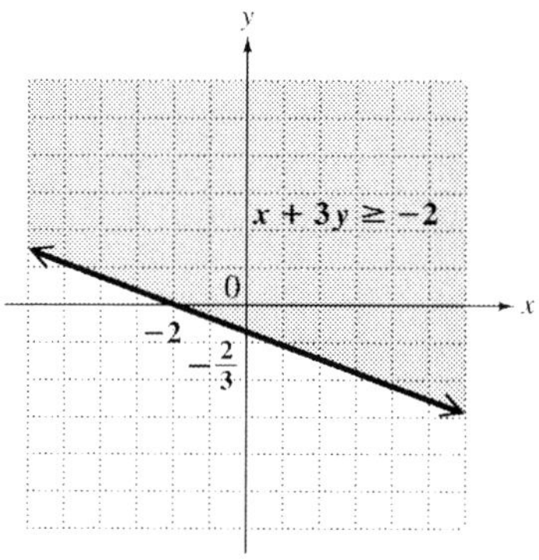

12. $x + 4y \geq -3$

Graph the solid line $x + 4y = -3$ (since the inequality involves $\geq$) through the intercepts $(-3, 0)$ and $\left(0, -\frac{3}{4}\right)$.

Test a point not on this line such as $(0, 0)$.

$$0 + 4(0) \geq -3 \quad ?$$
$$0 \geq -3 \qquad \textit{True}$$

Shade that side of the line containing the test point $(0, 0)$.

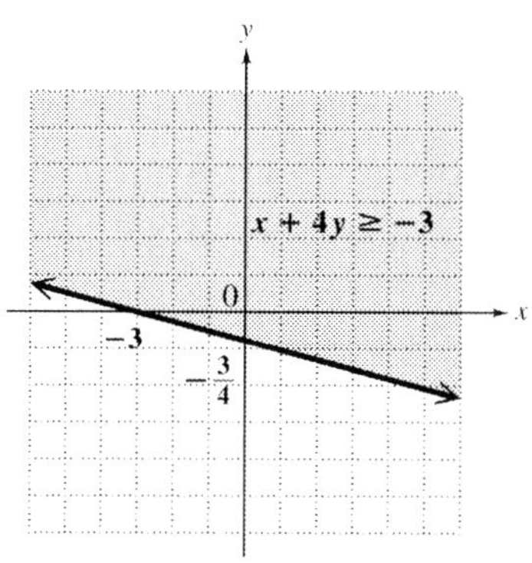

13. $x + y > 0$

Graph the line $x + y = 0$, which includes the points $(0, 0)$ and $(2, -2)$, as a dashed line (since the inequality involves $>$). Solving the inequality for y gives us

$$y > -x,$$

So shade the region above the boundary line.

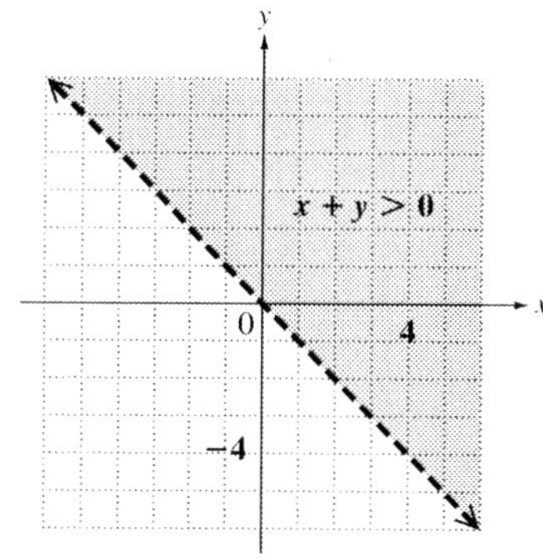

14. $x + 2y > 0$

Graph the line $x + 2y = 0$, which includes the points $(0, 0)$ and $(-4, 2)$, as a dashed line (since the inequality involves $>$). Solving the inequality for y gives us

$$y > -\tfrac{1}{2}x,$$

So shade the region above the boundary line.

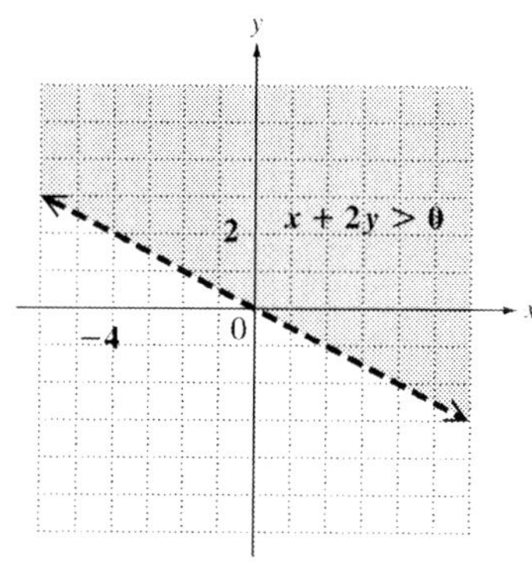

15. $x - 3y \leq 0$

Graph the solid line $x - 3y = 0$ through the points $(0, 0)$ and $(3, 1)$.

Solve the inequality for y.

$$-3y \leq -x$$
$$y \geq \tfrac{1}{3}x$$

Shade the region above the boundary line.

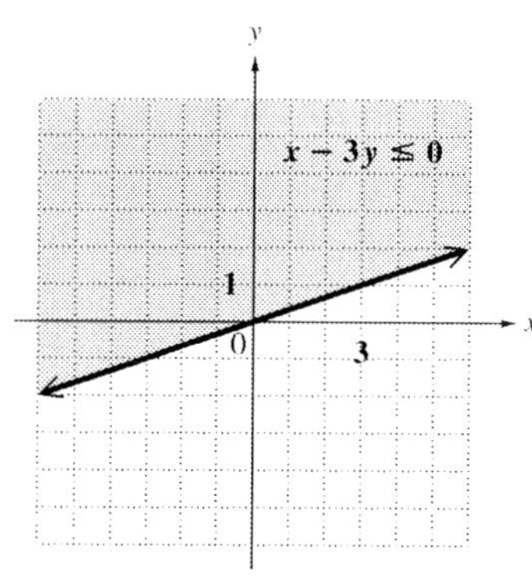

16. $x - 5y \leq 0$

Graph the solid line $x - 5y = 0$ through the points $(0,0)$ and $(5,1)$.

Solve the inequality for y.

$$-5y \leq -x$$
$$y \geq \tfrac{1}{5}x$$

Shade the region above the boundary line.

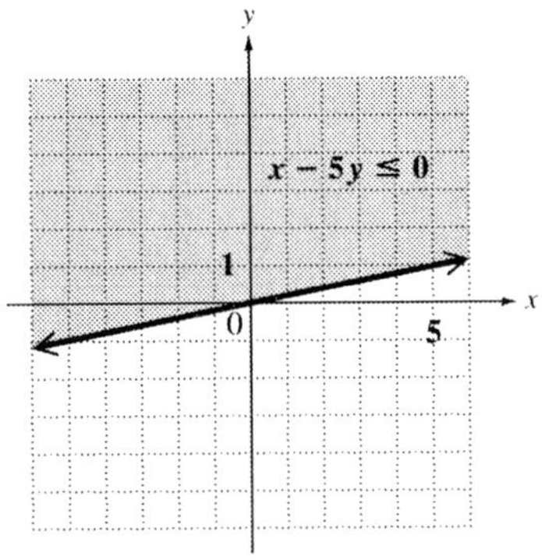

17. $y < x$

Graph the dashed line $y = x$ through $(0,0)$ and $(2,2)$. Since we have "$y <$ " in the inequality, shade the region *below* the boundary line.

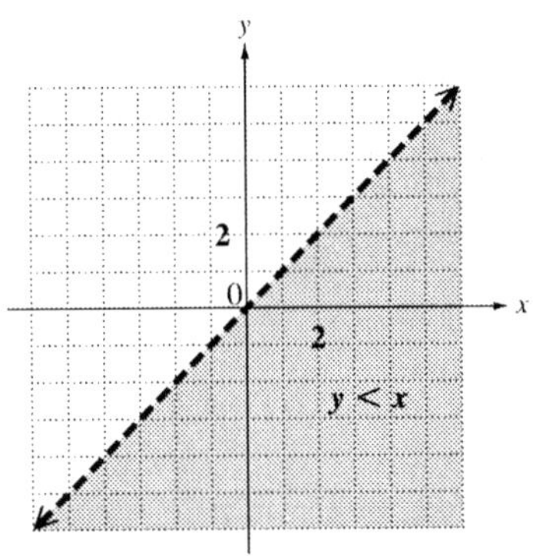

18. $y \leq 4x$

Graph the solid line $y = 4x$ through $(0,0)$ and $(1,4)$. Since we have "$y \leq$ " in the inequality, shade the region *below* the boundary line.

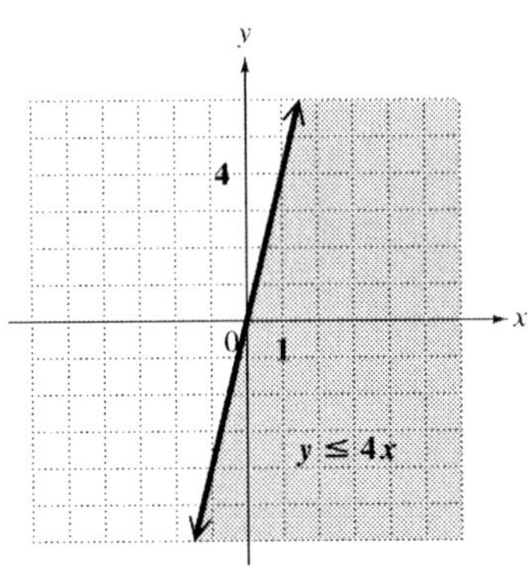

19. $x + y \leq 1$ and $x \geq 1$

Graph the solid line $x + y = 1$ through $(0,1)$ and $(1,0)$. The inequality $x + y \leq 1$ can be written as $y \leq -x + 1$, so shade the region below the boundary line.

Graph the solid vertical line $x = 1$ through $(1,0)$ and shade the region to the right. The required graph is the common shaded area as well as the portions of the lines that bound it.

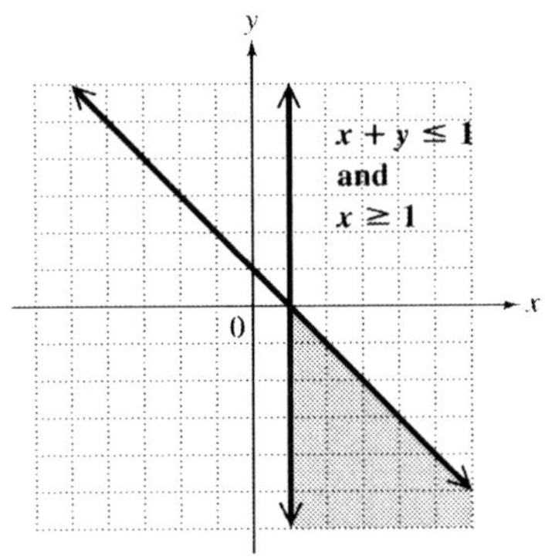

20. $x - y \geq 2$ and $x \geq 3$

Graph $x - y = 2$ as a solid line through $(2,0)$ and $(0,-2)$. Test $(0,0)$.

$$0 - 0 \geq 2 \quad ?$$
$$0 \geq 2 \qquad \textit{False}$$

The graph is the region that does not contain $(0,0)$. Graph $x = 3$ as a solid vertical line through $(3,0)$. The graph of the inequality is the region to the right of the line. Shade the region that includes the overlap of the two graphs.

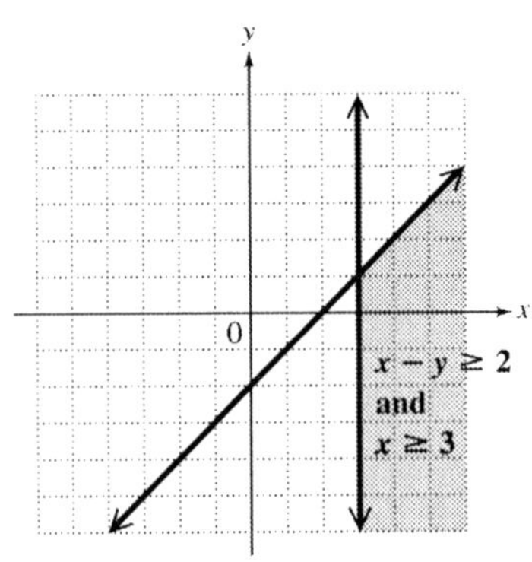

21. $2x - y \geq 2$ and $y < 4$

Graph the solid line $2x - y = 2$ through the intercepts $(1,0)$ and $(0,-2)$. Test $(0,0)$ to get $0 \geq 2$, a false statement. Shade that side of the graph not containing $(0,0)$. To graph $y < 4$ on the same axes, graph the dashed horizontal line through $(0,4)$. Test $(0,0)$ to get $0 < 4$, a true statement. Shade that side of the dashed line containing $(0,0)$.

The word "and" indicates the intersection of the two graphs. The final solution set consists of the region where the two shaded regions overlap.

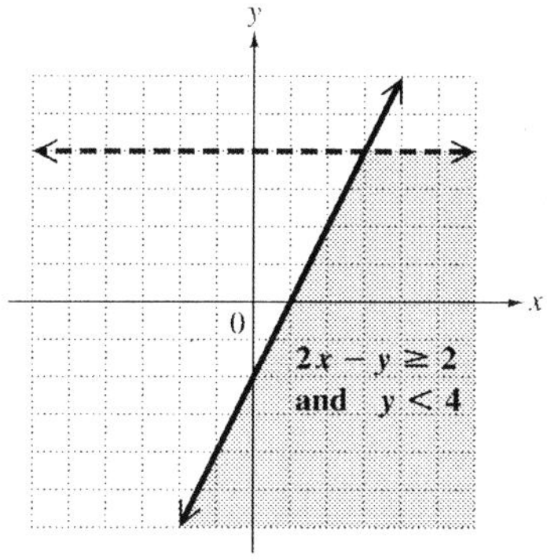

22. $3x - y \geq 3$ and $y < 3$

Graph $3x - y = 3$ as a solid line through $(1, 0)$ and $(0, -3)$. Test $(0, 0)$.

$$3(0) - 0 \geq 3 \quad ?$$
$$0 \geq 3 \qquad \textit{False}$$

The graph is the region that does not contain $(0, 0)$.

Graph $y = 3$ as a dashed horizontal line through $(0, 3)$. The graph of the inequality is the region below the dashed line.

Shade the region that includes the overlap of the two graphs.

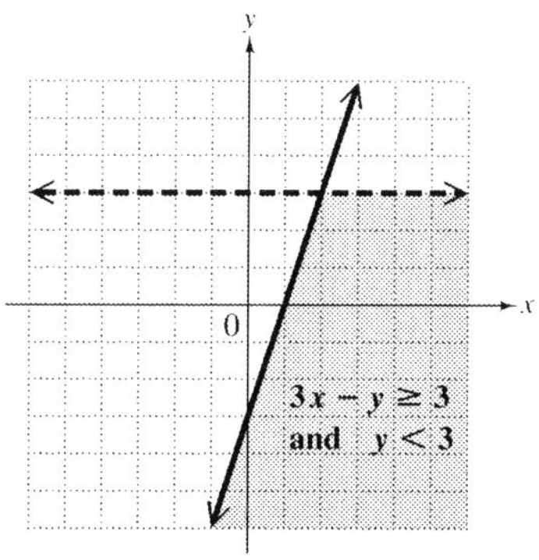

23. $x + y > -5$ and $y < -2$

Graph $x + y = -5$, which has intercepts $(-5, 0)$ and $(0, -5)$, as a dashed line. Test $(0, 0)$, which yields $0 > -5$, a true statement. Shade the region that includes $(0, 0)$.

Graph $y = -2$ as a dashed horizontal line. Shade the region below $y = -2$. The required graph of the intersection is the region common to both graphs.

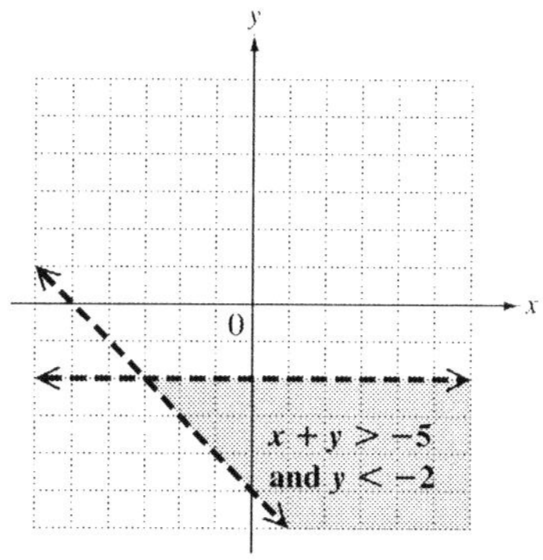

24. $6x - 4y < 10$ and $y > 2$

Graph the dashed line $6x - 4y = 10$ through $\left(\frac{5}{3}, 0\right)$ and $\left(0, -\frac{5}{2}\right)$. Test $(0, 0)$.

$$6(0) - 4(0) < 10 \quad ?$$
$$0 < 10 \qquad \textit{True}$$

The graph includes the region that includes $(0, 0)$. Graph the dashed horizontal line $y = 2$ through $(0, 2)$. The graph includes the region above the line. Shade the region that includes the overlap of the two graphs.

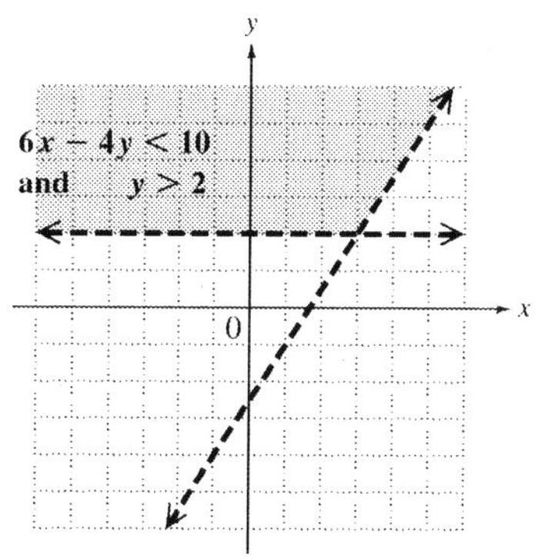

25. $|x| \geq 3$

Rewrite $|x| \geq 3$ as $x \geq 3$ or $x \leq -3$. The graph consists of the region to the left of the solid vertical line $x = -3$ and to the right of the solid vertical line $x = 3$.

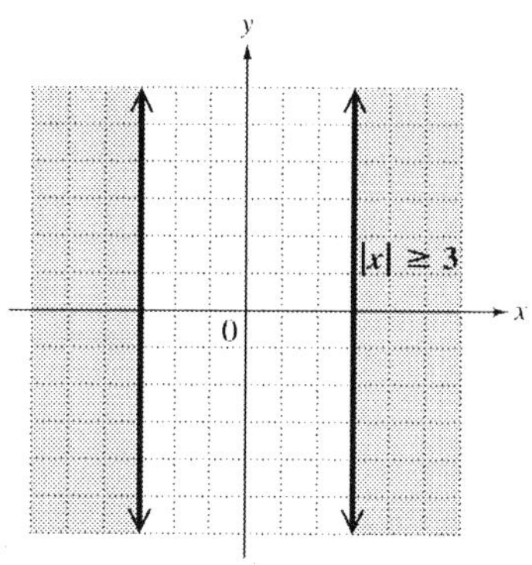

26. $|y| < 5$ can be rewritten as $-5 < y < 5$. The boundaries are the dashed horizontal lines $y = -5$ and $y = 5$. Since y is between -5 and 5, the graph includes all points between the lines.

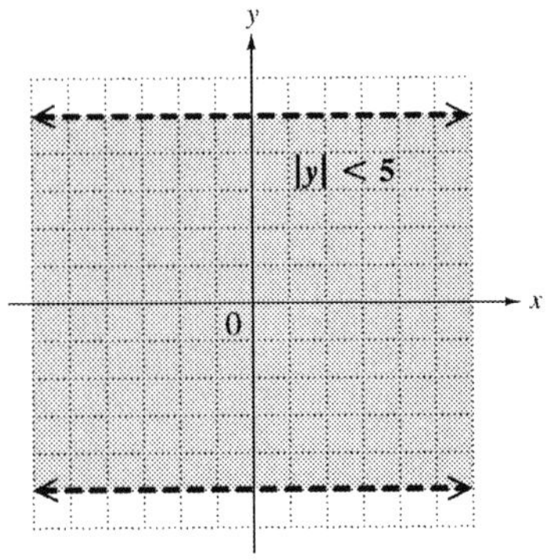

27. $|y + 1| < 2$ can be rewritten as

$$-2 < y + 1 < 2$$
$$-3 < y < 1.$$

The boundaries are the dashed horizontal lines $y = -3$ and $y = 1$. Since y is between -3 and 1, the graph includes all points between the lines.

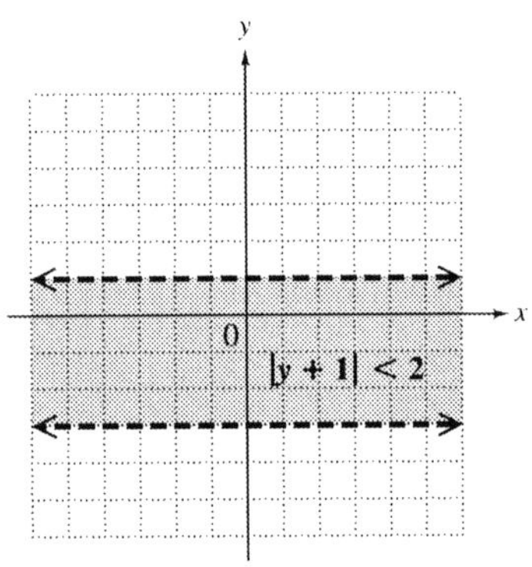

28. $|x - 2| \geq 1$ can be rewritten as

$$x - 2 \geq 1 \quad \text{or} \quad x - 2 \leq -1$$
$$x \geq 3 \quad \text{or} \quad x \leq 1.$$

Graph the vertical boundary lines, $x = 1$ and $x = 3$, as solid lines. Shade to the left of $x = 1$ and to the right of $x = 3$.

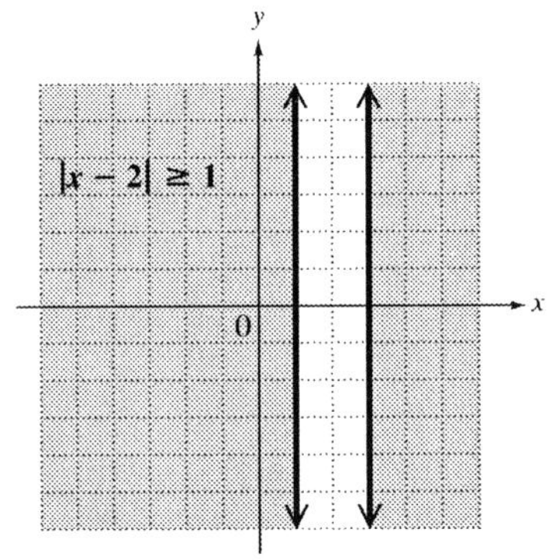

29. $x - y \geq 1$ or $y \geq 2$
Graph the solid line $x - y = 1$, which crosses the y-axis at $(0, -1)$ and the x-axis at $(1, 0)$. Use $(0, 0)$ as a test point, which yields $0 \geq 1$, a false statement. Shade the region that does not include $(0, 0)$.
Now graph the solid line $y = 2$. Since the inequality is $y \geq 2$, shade above this line.
The required graph of the union includes all the shaded regions, that is, all the points that satisfy either inequality.

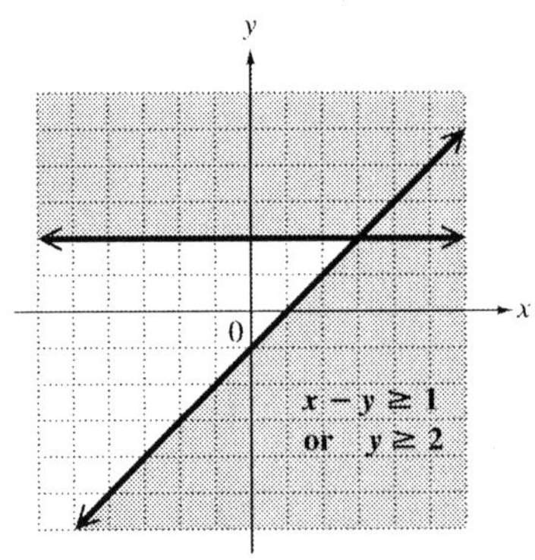

30. $x + y \leq 2$ or $y \geq 3$
Graph the solid line $x + y = 2$ through $(2, 0)$ and $(0, 2)$. Use $(0, 0)$ as a test point, which yields $0 \leq 2$, a true statement. Shade the region that includes $(0, 0)$.
Graph the solid horizontal line $y = 3$ through $(0, 3)$. Shade the region above the line.
The required graph of the union includes all the shaded regions, that is, all the points that satisfy either inequality.

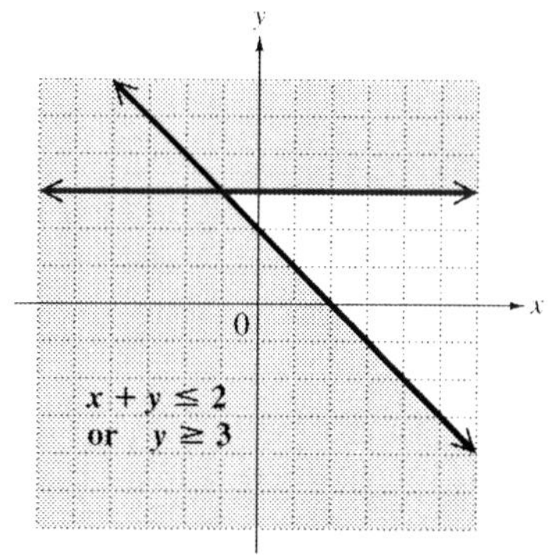

31. $x - 2 > y$ or $x < 1$
Graph $x - 2 = y$, which has intercepts $(2, 0)$ and $(0, -2)$, as a dashed line. Test $(0, 0)$, which yields $-2 > 0$, a false statement. Shade the region that does not include $(0, 0)$.
Graph $x = 1$ as a dashed vertical line. Shade the region to the left of $x = 1$.
The required graph of the union includes all the shaded regions, that is, all the points that satisfy either inequality.

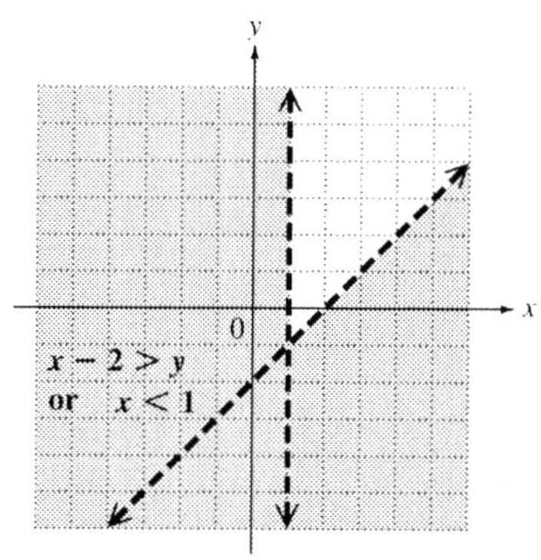

32. $x + 3 < y$ or $x > 3$
Graph the dashed line $x + 3 = y$ through $(-3, 0)$ and $(0, 3)$. Use $(0, 0)$ as a test point, which yields $3 < 0$, a false statement. Shade the region that does not include $(0, 0)$.
Graph the dashed vertical line $x = 3$ through $(3, 0)$. Shade the region to the right of the line.
The required graph of the union includes all the shaded regions, that is, all the points that satisfy either inequality.

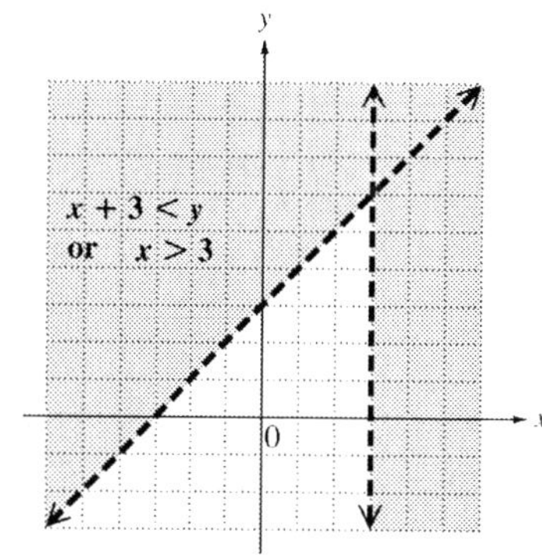

33. $3x + 2y < 6$ or $x - 2y > 2$
Graph $3x + 2y = 6$, which has intercepts $(2, 0)$ and $(0, 3)$, as a dashed line. Test $(0, 0)$, which yields $0 < 6$, a true statement. Shade the region that includes $(0, 0)$.
Graph $x - 2y = 2$, which has intercepts $(2, 0)$ and $(0, -1)$, as a dashed line. Test $(0, 0)$, which yields $0 > 2$, a false statement. Shade the region that does not include $(0, 0)$.
The required graph of the union includes all the shaded regions, that is, all the points that satisfy either inequality.

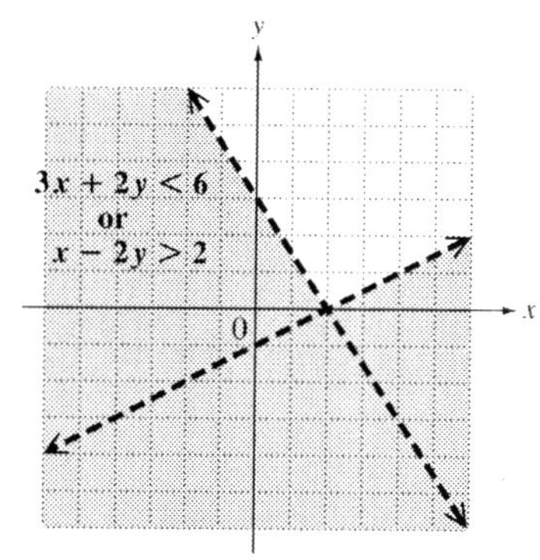

34. $x - y \geq 1$ or $x + y \leq 4$
Graph the solid line $x - y = 1$ through $(1, 0)$ and $(0, -1)$. Test $(0, 0)$, which yields $0 \geq 1$, a false statement. Shade the region that does not include $(0, 0)$.
Graph the solid line $x + y = 4$ through $(4, 0)$ and $(0, 4)$. Test $(0, 0)$, which yields $0 \leq 4$, a true statement. Shade the region that includes $(0, 0)$.
The required graph of the union includes all the shaded regions, that is, all the points that satisfy either inequality.

continued

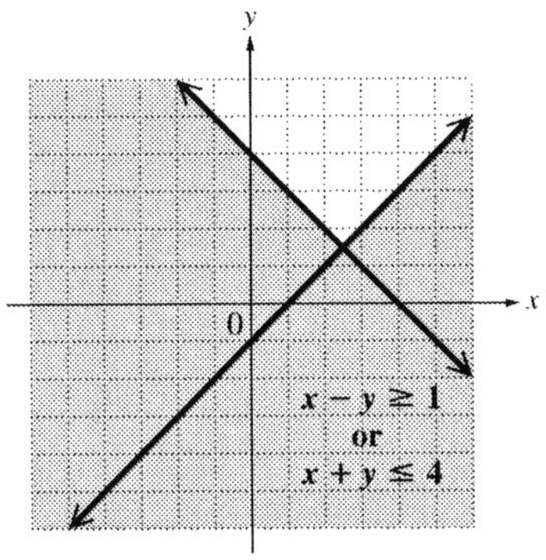

4.5 Introduction to Functions

4.5 Margin Exercises

1. The ordered pair $(40, 320)$ indicates that when you work 40 hours, your paycheck is \$320.

2. **(a)** $\{(0,3),(-1,2),(-1,3)\}$

The last two ordered pairs have the *same* x-value paired with *two different* y-values (-1 is paired with both 2 and 3), so this relation *is not a function.*

(b) $\{(2,-2),(4,-4),(6,-6)\}$

The relation *is a function* because for each different x-value there is exactly one y-value.

(c) $\{(-1,5),(0,5)\}$

The relation *is a function* because for each different x-value there is exactly one y-value. It is acceptable to have different x-values paired with the same y-value.

3. **(a)** $\{(4,0),(4,1),(4,2)\}$

The domain of this relation is the set of x-values, that is, $\{4\}$. The range of this relation is the set of y-values, that is, $\{0,1,2\}$. This relation *is not a function* because the same x-value 4 is paired with more than one y-value.

(b) The domain of this relation is the set of all first components, that is, $\{-1,4,7\}$. The range of this relation is the set of all second components, that is, $\{0,-2,3,7\}$. This relation *is not a function* because the same x-value 4 is paired with more than one y-value.

(c) The domain of this relation is the set of all first components, that is, $\{1995, 1996, 1997, 1998, 1999\}$. The range of this relation is the set of all second components, that is, $\{33{,}786, 44{,}043, 55{,}312, 69{,}209, 86{,}047\}$. This relation *is a function* because for each different first component, there is exactly one second component.

4. **(a)** The domain is the set of x-values, $\{-3,-2,2,3\}$. The range is the set of y-values, $\{-2,-1,2,3\}$.

(b) The arrowheads indicate that the graph extends indefinitely right, as well as up and down. Because there is a smallest x-value, -2, the domain includes all numbers greater than or equal to -2, $[-2,\infty)$. The range includes all real numbers, written $(-\infty,\infty)$.

(c) The arrowheads indicate that the graph extends indefinitely left and right, as well as downward. The domain includes all real numbers, written $(-\infty,\infty)$. Because there is a greatest y-value, 0, the range includes all numbers less than or equal to 0, $(-\infty,0]$.

5. Any vertical line would intersect the graph at most once on graphs **A** and **C**, so graphs **A** and **C** represent functions.

Graph **B** does not represent a function since a vertical line, such as the y-axis, intersects the graph in more than one point.

6. **(a)** $y = 6x + 12$ is a function because each value of x corresponds to exactly one value of y. Its domain is the set of all real numbers, $(-\infty,\infty)$.

(b) $y \le 4x$ is not a function because if $x = 0$, then $y \le 0$. Thus, the x-value 0 corresponds to many y-values. Its domain is the set of all real numbers, $(-\infty,\infty)$.

(c) $y = -\sqrt{3x-2}$ is a function because each value of x corresponds to exactly one value of y. Since the quantity under the radical must be nonnegative, the domain is the set of real numbers that satisfy the condition

$$\begin{aligned} 3x - 2 &\ge 0 \\ x &\ge \tfrac{2}{3}. \end{aligned}$$

Therefore, the domain is $[\frac{2}{3},\infty)$.

(d) $y^2 = 25x$ is not a function. If $x = 1$, for example, $y^2 = 25$ and $y = 5$ or $y = -5$. Since y^2 must be nonnegative, the domain is the set of nonnegative real numbers, $[0,\infty)$.

7. **(a)**
$$\begin{aligned} f(x) &= 6x - 2 \\ f(-3) &= 6(-3) - 2 \\ &= -18 - 2 \\ &= -20 \\ f(p) &= 6p - 2 \\ f(m+1) &= 6(m+1) - 2 \\ &= 6m + 6 - 2 \\ &= 6m + 4 \end{aligned}$$

(b)
$$f(x) = \frac{-3x+5}{2}$$
$$f(-3) = \frac{-3(-3)+5}{2} = \frac{14}{2} = 7$$
$$f(p) = \frac{-3p+5}{2}$$
$$f(m+1) = \frac{-3(m+1)+5}{2} = \frac{-3m-3+5}{2} = \frac{-3m+2}{2}$$

(c)
$$f(x) = \tfrac{1}{6}x - 1$$
$$f(-3) = \tfrac{1}{6}(-3) - 1 = -\tfrac{1}{2} - 1 = -\tfrac{3}{2}$$
$$f(p) = \tfrac{1}{6}p - 1$$
$$f(m+1) = \tfrac{1}{6}(m+1) - 1 = \tfrac{1}{6}m + \tfrac{1}{6} - 1 = \tfrac{1}{6}m - \tfrac{5}{6}$$

8. **(a)**
$$f(x) = -4x - 8$$
$$f(-2) = -4(-2) - 8 \quad \textit{Replace x with } -2.$$
$$= 8 - 8$$
$$= 0$$

(b) $f = \{(0,5), (-1,3), (-2,1)\}$

We want $f(-2)$, the y-value of the ordered pair where $x = -2$. As indicated by the ordered pair $(-2,1)$, when $x = -2$, $y = 1$, so $f(-2) = 1$.

(c) Ordered pairs of a function can be written as $(x, f(x))$. In this case, we have $(-2, f(-2)) = (-2, 4)$, so $f(-2) = 4$.

9. **(a)**
$$y = \sqrt{x+2}$$
$$f(x) = \sqrt{x+2} \quad \textit{Replace y with f(x).}$$
$$f(-1) = \sqrt{-1+2} \quad \textit{Let } x = -1.$$
$$= \sqrt{1} = 1$$

(b) $x^2 - 4y = 3$

Solve for y.
$$-4y = -x^2 + 3$$
$$y = \frac{1}{4}x^2 - \frac{3}{4} \quad \text{or} \quad y = \frac{x^2-3}{4}$$
$$f(x) = \frac{x^2-3}{4} \quad \textit{Replace y with f(x).}$$
$$f(-1) = \frac{(-1)^2-3}{4} = \frac{1-3}{4} = \frac{-2}{4} = -\frac{1}{2}$$

10. **(a)** For $f(x) = \frac{3}{4}x - 2$, the y-intercept is $(0, -2)$ and the slope is $\frac{3}{4}$. Plot the point $(0, -2)$, then move up 3 units and right 4 units to plot another point, namely, $(4, 1)$. Draw a line through the points.

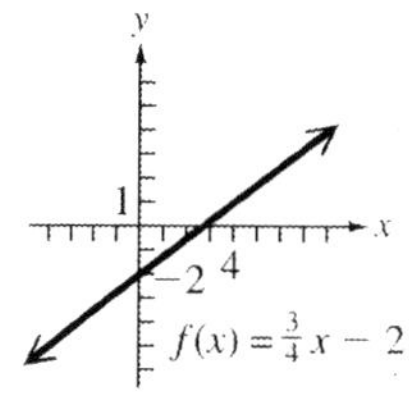

The domain and range are each $(-\infty, \infty)$.

(b) $g(x) = 3$ is a constant function. Its graph is a horizontal line.

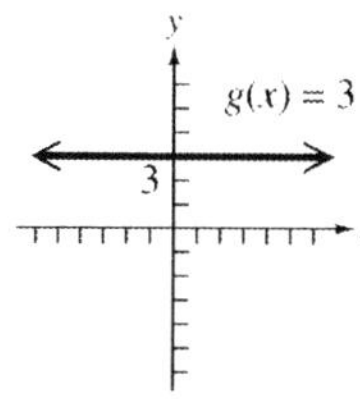

The domain of a constant function is $(-\infty, \infty)$. The range consists of the only y-value, 3, so it is $\{3\}$.

4.5 Section Exercises

1. In an ordered pair of a relation, the first element is the independent variable.

2. A relation that is not a function having domain $\{-3, 2, 6\}$ and range $\{4, 6\}$ must be a relation in which, for some value in the domain, there is more than one value in the range. An example of such a relation is

$$\{(-3,4), (2,4), (2,6), (6,4)\}.$$

Notice that 2, from the domain, is paired with both 4 and 6, from the range. (There are other correct answers possible.)

3. **(a)** A relation is a set of ordered pairs.

(b) The domain is the set of all first components (x-values).

(c) The range is the set of all second components (y-values).

(d) A function is a relation in which each domain element is paired with one and only one range element.

4. The vertical line test is used to determine whether a graph is that of a function. Any vertical line will intersect the graph of a function in at most one point.

5. $\{(5,1),(3,2),(4,9),(7,3)\}$

The relation is a function because for each x-value, there is only one y-value.

The domain, the set of x-values, is $\{5,3,4,7\}$. The range, the set of y-values, is $\{1,2,9,3\}$.

6. $\{(8,0),(5,4),(9,3),(3,9)\}$

The relation is a function because for each x-value, there is only one y-value.

The domain, the set of x-values, is $\{8,5,9,3\}$. The range, the set of y-values, is $\{0,4,3,9\}$.

7. $\{(2,4),(0,2),(2,6)\}$

The relation is not a function since the x-value 2 is paired with two different y-values, 4 and 6.

The domain, the set of x-values, is $\{0,2\}$. The range, the set of y-values, is $\{2,4,6\}$.

8. $\{(9,-2),(-3,5),(9,1)\}$

The relation is not a function since the x-value 9 is paired with two different y-values, -2 and 1.

The domain, the set of x-values, is $\{9,-3\}$. The range, the set of y-values, is $\{-2,5,1\}$.

9. $\{(-3,1),(4,1),(-2,7)\}$

The relation is a function since for each x-value, there is only one y-value.

The domain is the set of x-values: $\{-3,4,-2\}$ The range is the set of y-values: $\{1,7\}$

10. $\{(-12,5),(-10,3),(8,3)\}$

The relation is a function since for each x-value, there is only one y-value.

The domain is the set of x-values: $\{-12,-10,8\}$ The range is the set of y-values: $\{5,3\}$

11. $\{(1,1),(1,-1),(0,0),(2,4),(2,-4)\}$

The relation is not a function since the x-value 1 has two different y-values associated with it, 1 and -1. (A similar statement can be made for $x=2$.)

The domain is the set of x-values: $\{0,1,2\}$. The range is the set of y-values: $\{-4,-1,0,1,4\}$.

12. $\{(2,5),(3,7),(4,9),(5,11)\}$

The relation is a function since for each x-value, there is only one y-value.

The domain is the set of x-values: $\{2,3,4,5\}$. The range is the set of y-values: $\{5,7,9,11\}$.

13. The relation can be described by the set of ordered pairs

$$\{(2,1),(5,1),(11,7),(17,20),(3,20)\}.$$

The relation is a function since for each x-value, there is only one y-value.

The domain is the set of x-values: $\{2,3,5,11,17\}$. The range is the set of y-values: $\{1,7,20\}$.

14. The relation can be described by the set of ordered pairs

$$\{(1,10),(2,15),(2,19),(3,19),(5,27)\}.$$

The relation is not a function since the x-value 2 has two different y-values associated with it, 15 and 19.

The domain is the set of x-values: $\{1,2,3,5\}$. The range is the set of y-values: $\{10,15,19,27\}$.

15. The relation can be described by the set of ordered pairs

$$\{(1,5),(1,2),(1,-1),(1,-4)\}.$$

The relation is not a function since the x-value 1 has four different y-values associated with it, 5, 2, -1, and -4.

The domain is the set of x-values: $\{1\}$. The range is the set of y-values: $\{5,2,-1,-4\}$.

16. The relation can be described by the set of ordered pairs

$$\{(4,-3),(2,-3),(0,-3),(-2,-3)\}.$$

The relation is a function since for each x-value, there is only one y-value.

The domain is the set of x-values: $\{4,2,0,-2\}$. The range is the set of y-values: $\{-3\}$.

17. Using the vertical line test, we find any vertical line will intersect the graph at most once. This indicates that the graph represents a function. This graph extends indefinitely to the left $(-\infty)$ and indefinitely to the right (∞). Therefore, the domain is $(-\infty,\infty)$. This graph extends indefinitely downward $(-\infty)$, and indefinitely upward (∞). Thus, the range is $(-\infty,\infty)$.

18. The relation is not a function since a vertical line may intersect the graph in more than one point. The domain is the set of x-values, $[-2,2]$. The range is the set of y-values, $[-2,2]$.

19. Using the vertical line test, we find any vertical line will intersect the graph at most once. This indicates that the graph represents a function. This graph extends indefinitely to the left $(-\infty)$ and indefinitely to the right (∞). Therefore, the domain is $(-\infty,\infty)$. This graph extends indefinitely downward $(-\infty)$, and reaches a high point at $y=4$. Therefore, the range is $(-\infty,4]$.

20. Since a vertical line may intersect the graph of the relation in more than one point, the relation is not a function. The domain, the x-values of the points on the graph, is $[-4, 4]$. The range, the y-values of the points on the graph, is $[-3, 3]$.

21. Since a vertical line, such as $x = 4$, intersects the graph in two points, the relation is not a function. The domain is $[3, \infty)$, and the range is $(-\infty, \infty)$.

22. Since any vertical line that intersects the graph intersects it in no more than one point, the relation represented by the graph is a function.
The domain is $[-2, 2]$, and the range is $[0, 4]$.

23. $y = x^2$
Each value of x corresponds to one y-value. For example, if $x = 3$, then $y = 3^2 = 9$. Therefore, $y = x^2$ defines y as a function of x.
Since any x-value, positive, negative, or zero, can be squared, the domain is $(-\infty, \infty)$.

24. $y = x^3$ defines y as a function of x because each value, x, in the domain has exactly one value, y, in the range.
Since any x-value, positive, negative, or zero, can be cubed, the domain is $(-\infty, \infty)$.

25. $x = y^6$
The ordered pairs $(64, 2)$ and $(64, -2)$ both satisfy the equation. Since one value of x, 64, corresponds to two values of y, 2 and -2, the relation does not define a function. Because x is equal to the sixth power of y, the values of x must always be nonnegative. The domain is $[0, \infty)$.

26. $x = y^4$
If $x = 16$, y may be either -2 or 2. $x = y^4$ does not define y as a function of x.
All the x-values must be nonnegative, so the domain is $[0, \infty)$.

27. $y = 2x - 6$

For any value of x, there is exactly one value of y, so this equation defines a function. The domain is the set of all real numbers, $(-\infty, \infty)$.

28. $y = -6x + 8$

For any value of x, there is exactly one value of y, so this equation defines a function. The domain is the set of all real numbers, $(-\infty, \infty)$.

29. $x + y < 4$
For a particular x-value, more than one y-value can be selected to satisfy $x + y < 4$. For example, if $x = 2$ and $y = 0$, then

$$2 + 0 < 4. \quad \textit{True}$$

Now, if $x = 2$ and $y = 1$, then

$$2 + 1 < 4. \quad \textit{Also true}$$

Therefore, $x + y < 4$ does not define y as a function of x.
The graph of $x + y < 4$ consists of the shaded region below the dashed line $x + y = 4$, which extends indefinitely from left to right. Therefore, the domain is $(-\infty, \infty)$.

30. $x - y < 3$
Let $x = 1$.

$$\begin{aligned} 1 - y &< 3 \\ -y &< 2 \\ y &> -2 \end{aligned}$$

So for $x = 1$, y may be any number greater than -2.
$x - y < 3$ does not define y as a function of x.
The x-values may be any number. The domain is $(-\infty, \infty)$.

31. $y = \sqrt{x}$
For any value of x, there is exactly one corresponding value for y, so this relation defines a function. Since the radicand must be a nonnegative number, x must always be nonnegative. The domain is $[0, \infty)$.

32. $y = -\sqrt{x}$ defines y as a function of x because each value, x, in the domain has exactly one value, y.
Since $-\sqrt{x}$ must be a real number, $x \geq 0$, so the domain is $[0, \infty)$.

33. $xy = 1$
Rewrite $xy = 1$ as $y = \frac{1}{x}$. Note that x can never equal 0, otherwise the denominator would equal 0. The domain is $(-\infty, 0) \cup (0, \infty)$.
Each nonzero x-value gives exactly one y-value. Therefore, $xy = 1$ defines y as a function of x.

34. $xy = -3$ may be rewritten as $y = -\frac{3}{x}$.
This relation defines y as a function of x because each value of x in the domain has exactly one y-value. Since $\frac{-3}{x}$ must be defined, $x \neq 0$. The domain is $(-\infty, 0) \cup (0, \infty)$.

35. $y = \sqrt{4x + 2}$
To determine the domain of $y = \sqrt{4x + 2}$, recall that the radicand must be nonnegative. Solve the inequality $4x + 2 \geq 0$, which gives us $x \geq -\frac{1}{2}$. Therefore, the domain is $\left[-\frac{1}{2}, \infty\right)$.
Each x-value from the domain produces exactly one y-value. Therefore, $y = \sqrt{4x + 2}$ defines a function.

36. $y = \sqrt{9 - 2x}$ defines y as a function of x because each value of x in the domain has exactly one y-value.
$\sqrt{9 - 2x}$ must be a real number, so

$$\begin{aligned} 9 - 2x &\geq 0 \\ -2x &\geq -9 \\ x &\leq \tfrac{9}{2} \end{aligned}$$

The domain is $\left(-\infty, \frac{9}{2}\right]$.

37. $y = \dfrac{2}{x - 9}$
Given any value of x, y is found by subtracting 9, then dividing the result into 2. This process produces exactly one value of y for each x-value, so the relation represents a function. The domain includes all real numbers except those that make the denominator 0, namely 9. The domain is $(-\infty, 9) \cup (9, \infty)$.

38. $y = \dfrac{-7}{x - 16}$ defines y as a function of x. The domain includes all real numbers except those that make the denominator 0, namely 16. The domain is $(-\infty, 16) \cup (16, \infty)$.

39. **(a)** The independent variable is t, the number of hours, and the possible values are in the set $[0, 100]$. The dependent variable is g, the number of gallons, and the possible values are in the set $[0, 3000]$.

(b) The graph rises for the first 25 hours, so the water level increases for 25 hours. The graph falls for $t = 50$ to $t = 75$, so the water level decreases for 25 hours.

(c) There are 2000 gallons in the pool when $t = 90$.

(d) $g(0)$ is the number of gallons in the pool at time $t = 0$. Here, $g(0) = 0$, which means the pool is empty at time 0.

40. **(a)** The graph passes the vertical line test, so it is the graph of a function.

(b) We start the day at midnight and end the day at midnight. The domain is $[0, 24]$.

(c) At 8 in the morning, it appears that the number of megawatts used is halfway between 1100 and 1300. So 1200 megawatts is a good estimate.

(d) The most electricity was used at 17 hours (5 P.M.) and the least electricity was used at 4 hours (4 A.M.).

41. The amount of income tax you pay depends on your taxable income, so income tax is a function of taxable income.

42. $f(3)$ is the value of the dependent variable when the independent variable is 3—choice **B**.

43.
$$\begin{aligned} f(x) &= -3x + 4 \\ f(0) &= -3(0) + 4 \\ &= 0 + 4 \\ &= 4 \end{aligned}$$

44.
$$\begin{aligned} f(x) &= -3x + 4 \\ f(-3) &= -3(-3) + 4 \\ &= 9 + 4 \\ &= 13 \end{aligned}$$

45.
$$\begin{aligned} g(x) &= -x^2 + 4x + 1 \\ g(-2) &= -(-2)^2 + 4(-2) + 1 \\ &= -(4) - 8 + 1 \\ &= -11 \end{aligned}$$

46.
$$\begin{aligned} g(x) &= -x^2 + 4x + 1 \\ g(10) &= -(10)^2 + 4(10) + 1 \\ &= -100 + 40 + 1 \\ &= -59 \end{aligned}$$

47.
$$\begin{aligned} f(x) &= -3x + 4 \\ f(p) &= -3(p) + 4 \\ &= -3p + 4 \end{aligned}$$

48.
$$\begin{aligned} g(x) &= -x^2 + 4x + 1 \\ g(k) &= -k^2 + 4k + 1 \end{aligned}$$

49.
$$\begin{aligned} f(x) &= -3x + 4 \\ f(-x) &= -3(-x) + 4 \\ &= 3x + 4 \end{aligned}$$

50.
$$\begin{aligned} g(x) &= -x^2 + 4x + 1 \\ g(-x) &= -(-x)^2 + 4(-x) + 1 \\ &= -x^2 - 4x + 1 \end{aligned}$$

51.
$$\begin{aligned} f(x) &= -3x + 4 \\ f(x + 2) &= -3(x + 2) + 4 \\ &= -3x - 6 + 4 \\ &= -3x - 2 \end{aligned}$$

52.
$$\begin{aligned} g(x) &= -x^2 + 4x + 1 \\ g\left(-\frac{1}{x}\right) &= -\left(-\frac{1}{x}\right)^2 + 4\left(-\frac{1}{x}\right) + 1 \\ &= -\frac{1}{x^2} - \frac{4}{x} + 1 \end{aligned}$$

53.
$$\begin{aligned} g(x) &= -x^2 + 4x + 1 \\ g\left(\frac{p}{3}\right) &= -\left(\frac{p}{3}\right)^2 + 4\left(\frac{p}{3}\right) + 1 \\ &= -\frac{p^2}{9} + \frac{4p}{3} + 1 \end{aligned}$$

54.
$$\begin{aligned} f(x) &= -3x + 4 \\ f(3t-2) &= -3(3t-2) + 4 \\ &= -9t + 6 + 4 \\ &= -9t + 10 \end{aligned}$$

55. **(a)** When $x = 2$, $y = 2$, so $f(2) = 2$.

(b) When $x = -1$, $y = 3$, so $f(-1) = 3$.

56. **(a)** When $x = 2$, $y = 5$, so $f(2) = 5$.

(b) When $x = -1$, $y = 11$, so $f(-1) = 11$.

57. **(a)** When $x = 2$, $y = 15$, so $f(2) = 15$.

(b) When $x = -1$, $y = 10$, so $f(-1) = 10$.

58. **(a)** When $x = 2$, $y = 1$, so $f(2) = 1$.

(b) When $x = -1$, $y = 7$, so $f(-1) = 7$.

59. **(a)** The point $(2, 3)$ is on the graph of f, so $f(2) = 3$.

(b) The point $(-1, -3)$ is on the graph of f, so $f(-1) = -3$.

60. **(a)** The point $(2, -3)$ is on the graph of f, so $f(2) = -3$.

(b) The point $(-1, 2)$ is on the graph of f, so $f(-1) = 2$.

61. The equation $2x + y = 4$ has a straight line as its graph. To find y in $(3, y)$, let $x = 3$ in the equation.

$$\begin{aligned} 2x + y &= 4 \\ 2(3) + y &= 4 \\ 6 + y &= 4 \\ y &= -2 \end{aligned}$$

To use functional notation for $2x + y = 4$, solve for y to get

$$y = -2x + 4.$$

Replace y with $f(x)$ to get the linear function

$$\begin{aligned} f(x) &= -2x + 4. \\ f(3) &= -2(3) + 4 = -2 \end{aligned}$$

Because $y = -2$ when $x = 3$, the point $(3, -2)$ lies on the graph of the function.

62. Choice **A** can be written as $f(x) = \frac{1}{4}x - \frac{5}{4}$. It is the only linear function of these choices.

63. **(a)** Solve the equation for y.

$$\begin{aligned} x + 3y &= 12 \\ 3y &= 12 - x \\ y &= \frac{12 - x}{3} \end{aligned}$$

Since $y = f(x)$,

$$f(x) = \frac{12 - x}{3}.$$

(b) $f(3) = \dfrac{12 - 3}{3} = \dfrac{9}{3} = 3$

64. **(a)**
$$\begin{aligned} x - 4y &= 8 \\ -4y &= 8 - x \\ y &= \frac{8 - x}{-4} \\ f(x) &= \frac{8 - x}{-4} \end{aligned}$$

(b) $f(3) = \dfrac{8 - 3}{-4} = \dfrac{5}{-4} = -\dfrac{5}{4}$

65. **(a)** Solve the equation for y.

$$\begin{aligned} y + 2x^2 &= 3 \\ y &= 3 - 2x^2 \end{aligned}$$

Since $y = f(x)$,

$$f(x) = 3 - 2x^2.$$

(b)
$$\begin{aligned} f(3) &= 3 - 2(3)^2 \\ &= 3 - 2(9) \\ &= -15 \end{aligned}$$

66. **(a)**
$$\begin{aligned} y - 3x^2 &= 2 \\ y &= 2 + 3x^2 \\ f(x) &= 2 + 3x^2 \end{aligned}$$

(b)
$$\begin{aligned} f(3) &= 2 + 3(3)^2 \\ &= 2 + 3(9) \\ &= 29 \end{aligned}$$

67. **(a)** Solve the equation for y.

$$\begin{aligned} 4x - 3y &= 8 \\ -3y &= 8 - 4x \\ y &= \frac{8 - 4x}{-3} \end{aligned}$$

Since $y = f(x)$,

$$f(x) = \frac{8 - 4x}{-3}.$$

(b)
$$\begin{aligned} f(3) &= \frac{8 - 4(3)}{-3} \\ &= \frac{8 - 12}{-3} = \frac{-4}{-3} = \frac{4}{3} \end{aligned}$$

68. **(a)**
$$\begin{aligned} -2x + 5y &= 9 \\ 5y &= 9 + 2x \\ y &= \frac{9 + 2x}{5} \\ f(x) &= \frac{9 + 2x}{5} \end{aligned}$$

(b)
$$\begin{aligned} f(3) &= \frac{9 + 2(3)}{5} \\ &= \frac{9+6}{5} = \frac{15}{5} = 3 \end{aligned}$$

69. $f(x) = -2x + 5$
The graph will be a line. The intercepts are $(0, 5)$ and $(\frac{5}{2}, 0)$.
The domain is $(-\infty, \infty)$. The range is $(-\infty, \infty)$.

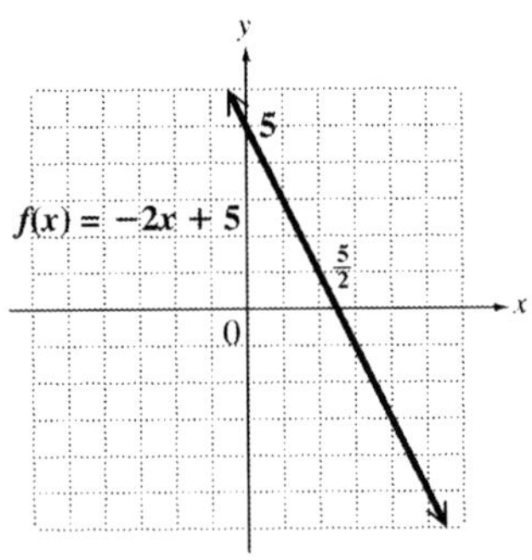

70. $g(x) = 4x - 1$
Using a y-intercept of $(0, -1)$ and a slope of $m = 4 = \frac{4}{1}$, graph the line.
From the graph we see that the domain is $(-\infty, \infty)$. The range is $(-\infty, \infty)$.

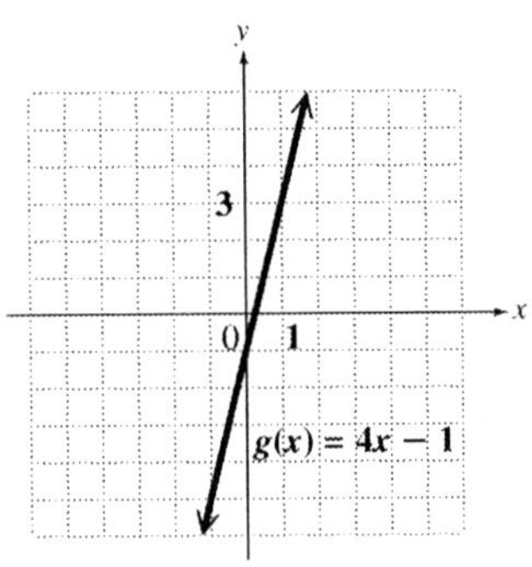

71. $h(x) = \frac{1}{2}x + 2$
The graph will be a line. The intercepts are $(0, 2)$ and $(-4, 0)$.
The domain is $(-\infty, \infty)$. The range is $(-\infty, \infty)$.

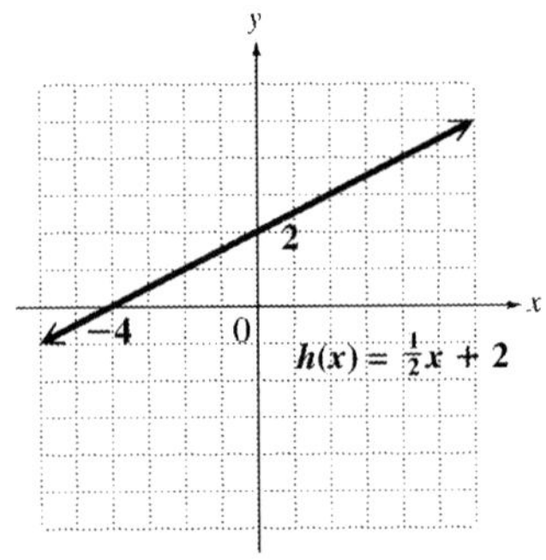

72. $F(x) = -\frac{1}{4}x + 1$
Using a y-intercept of $(0, 1)$ and a slope of $m = -\frac{1}{4}$, we graph the line.
From the graph we see that the domain is $(-\infty, \infty)$. The range is $(-\infty, \infty)$.

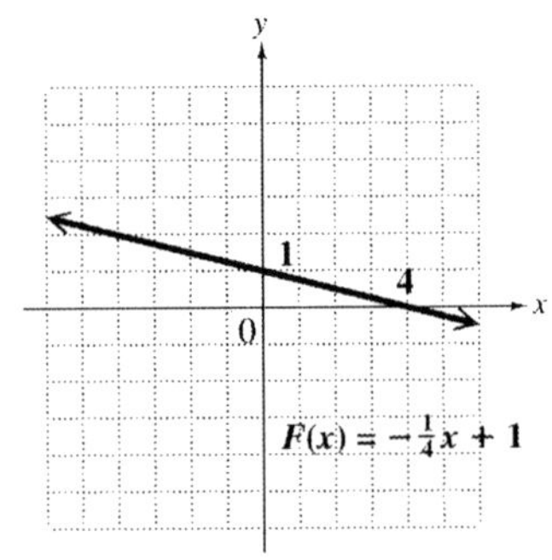

73. $g(x) = -4$
Using a y-intercept of $(0, -4)$ and a slope of $m = 0$, we graph the horizontal line. From the graph we see that the domain is $(-\infty, \infty)$. The range is $\{-4\}$.

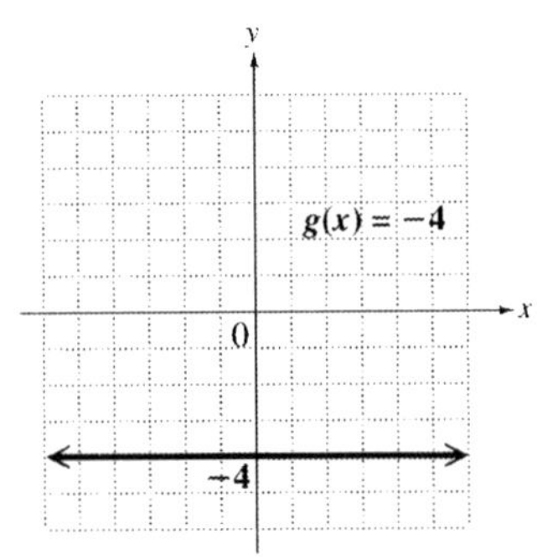

74. $f(x) = 5$
Draw the horizontal line through the point $(0, 5)$. On the horizontal line the value of x can be any real number, so the domain is $(-\infty, \infty)$. The range is $\{5\}$.

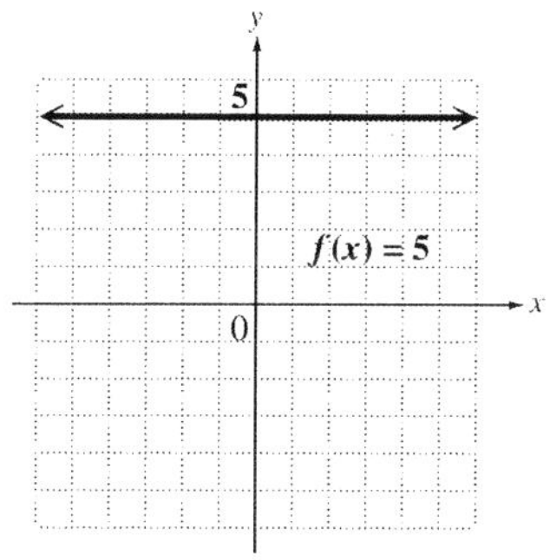

75. **(a)**

x	$f(x)$
0	0
1	\$1.50
2	\$3.00
3	\$4.50

(b) Since the charge equals the cost per mile, \$1.50, times the number of miles, x, $f(x) = \$1.50x$.

(c) To graph $y = f(x)$ for x in $\{0, 1, 2, 3\}$, plot the points $(0, 0)$, $(1, 1.50)$, $(2, 3.00)$, and $(3, 4.50)$ from the chart.

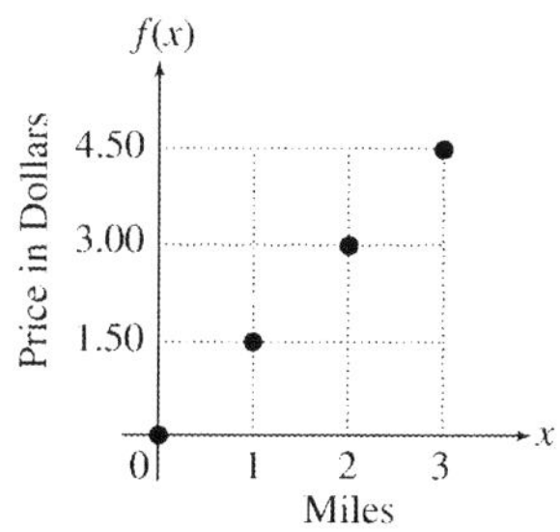

76. **(a)** $f(x) = 2.75x$
$f(3) = 2.75(3)$
$= 8.25$ (dollars)

(b) 3 is the value of the independent variable, which represents a package weight of 3 pounds; $f(3)$ is the value of the dependent variable representing the cost to mail a 3-pound package.

(c) $2.75(5) = \$13.75$, the cost to mail a 5-lb package. Using function notation, we have $f(5) = 13.75$.

77. Since the length of a man's femur is given, use the formula $h(r) = 69.09 + 2.24r$.

$$h(56) = 69.09 + 2.24(56) \quad \textit{Let } r = 56.$$
$$= 194.53$$

The man is 194.53 cm tall.

78. Use the formula $h(t) = 81.69 + 2.39t$.

$$h(40) = 81.69 + 2.39(40) \quad \textit{Let } t = 40.$$
$$= 177.29$$

The man is 177.29 cm tall.

79. Since the length of a woman's femur is given, use the formula $h(r) = 61.41 + 2.32r$.

$$h(50) = 61.41 + 2.32(50) \quad \textit{Let } r = 50.$$
$$= 177.41$$

The woman is 177.41 cm tall.

80. Use the formula $h(t) = 72.57 + 2.53t$.

$$h(36) = 72.57 + 2.53(36) \quad \textit{Let } t = 36.$$
$$= 163.65$$

The woman is 163.65 cm tall.

4.6 Variation

4.6 Margin Exercises

1. **(a)** Let E represent her earnings for d days. E varies directly as d, so

$$E = kd,$$

where k represents Suzanne's daily wage. Let $d = 17$ and $E = 1334.50$.

$$1334.50 = 17k$$
$$78.50 = k$$

Her daily wage is \$78.50. Thus,

$$E = 78.50d.$$

(b) Let d represent the distance traveled in h hours. d varies directly as h, so

$$d = kh,$$

where k represents the speed. Let $d = 100$ and $h = 2$.

$$100 = k \cdot 2$$
$$50 = k$$

Then $d = 50h$.

2. Let c represent the cost of using h kilowatt hours. Use $c = kh$ with $c = 52$ and $h = 800$ to find k.

$$c = kh$$
$$52 = k(800)$$
$$\tfrac{52}{800} = k$$
$$\tfrac{13}{200} = k$$

So $c = \frac{13}{200}h$.

Let $h = 1000$. Find c.

$$c = \tfrac{13}{200}(1000) = 65$$

1000 kilowatt-hours cost \$65.

3. **(a)** Let A represent the area of a circle and r its radius.

A varies directly as r^2, so

$$A = kr^2,$$

for some constant k. Since $A = 28.278$ when $r = 3$, substitute these values in the equation and solve for k.

$$\begin{aligned} A &= kr^2 \\ 28.278 &= k(3)^2 \\ 28.278 &= 9k \\ 3.142 &= k \end{aligned}$$

So $A = 3.142r^2$.

(b) Let $r = 4.1$. Find A.

$$\begin{aligned} A &= 3.142r^2 \\ A &= 3.142(4.1)^2 \\ A &= 52.817 \end{aligned}$$
(to the nearest thousandth)

The area is 52.817 in.^{2}.

4. **(a)** Let V represent the volume, and P the pressure. V varies inversely as P, so

$$V = \frac{k}{P},$$

for some constant k. Since $V = 10$ when $P = 6$, find k.

$$\begin{aligned} V &= \frac{k}{P} \\ 10 &= \frac{k}{6} \\ 60 &= k \end{aligned}$$

So $V = \dfrac{60}{P}$.

(b) Let $P = 12$. Find V.

$$V = \frac{60}{P} = \frac{60}{12} = 5$$

The volume is 5 cm^3.

5. Let V represent the volume, W the width, and L the length. The volume is proportional to the width and length of the box, so

$$V = kWL.$$

Let $W = 2$, $L = 4$, and $V = 12$. Find k.

$$\begin{aligned} 12 &= k(2)(4) \\ 12 &= 8k \\ k &= \tfrac{12}{8} = \tfrac{3}{2} \end{aligned}$$

So $V = \frac{3}{2}WL$.

Now let $L = 5$ and $W = 3$. Find V.

$$\begin{aligned} V &= \tfrac{3}{2}(3)(5) \\ &= \tfrac{45}{2} \quad \text{or} \quad 22.5 \end{aligned}$$

The volume is 22.5 ft^3.

6. Let L represent the maximum load, d the diameter of the cross section, and h the height.

L varies directly as d^4 and inversely as h^2, so

$$L = \frac{kd^4}{h^2},$$

for some constant k. Let $L = 8$ when $h = 9$ and $d = 1$. Find k.

$$\begin{aligned} L &= \frac{kd^4}{h^2} \\ 8 &= \frac{k(1)^4}{9^2} \\ 648 &= k \end{aligned}$$

So $L = \dfrac{648d^4}{h^2}$.

Now let $h = 12$ and $d = \frac{2}{3}$.

$$\begin{aligned} L &= \frac{648\left(\frac{2}{3}\right)^4}{12^2} \\ &= \frac{648\left(\frac{16}{81}\right)}{144} = \frac{128}{144} = \frac{8}{9} \end{aligned}$$

The column can support $\frac{8}{9}$ metric ton.

4.6 Section Exercises

1. The equation $y = \dfrac{3}{x}$ represents *inverse* variation. y varies inversely as x because x is in the denominator.

2. The equation $y = \dfrac{8}{x}$ represents *inverse* variation. y varies inversely as x because x is in the denominator.

3. The equation $y = 10x^2$ represents *direct* variation. The number 10 is the constant of variation, and y varies directly as the square of x.

4. The equation $y = 2x^3$ represents *direct* variation. The number 2 is the constant of variation, and y varies directly as the cube of x.

5. The equation $y = 3xz^4$ represents *joint* variation. y varies directly as x and z^4.

6. The equation $y = 6x^3z^2$ represents *joint* variation. y varies directly as the product of powers of x and z.

7. The equation $y = \frac{4x}{wz}$ represents *combined* variation because it is a combination of direct and inverse variation.

8. The equation $y = \frac{6x}{st}$ represents *combined* variation because it is a combination of direct and inverse variation.

9. "x varies directly as y" means

$$x = ky$$

for some constant k.
Substitute $x = 9$ and $y = 3$ in the equation and solve for k.

$$x = ky$$
$$9 = k(3)$$
$$k = \tfrac{9}{3} = 3$$

So $x = 3y$.
To find x when $y = 12$, substitute 12 for y in the equation.

$$x = 3y = 3(12) = 36$$

10. "x varies directly as y" means

$$x = ky$$

for some constant k.
Let $x = 10$ and $y = 7$ to find k.

$$10 = 7k$$
$$\tfrac{10}{7} = k$$

So $x = \frac{10}{7}y$. Now let $x = 50$.

$$50 = \tfrac{10}{7}y$$
$$y = \tfrac{7}{10}(50) = 35$$

11. "z varies inversely as w" means

$$z = \frac{k}{w}$$

for some constant k. Since $z = 10$ when $w = .5$, substitute these values in the equation and solve for k.

$$z = \frac{k}{w}$$
$$10 = \frac{k}{.5}$$
$$k = 10(.5) = 5$$

So $z = \frac{5}{w}$.
To find z when $w = 8$, substitute 8 for w in the equation.

$$z = \frac{5}{w} = \frac{5}{8} \text{ or } .625$$

12. "t varies inversely as s" means

$$t = \frac{k}{s}$$

for some constant k.
Let $t = 3$ and $s = 5$.

$$3 = \frac{k}{5}$$
$$k = 3(5) = 15$$

So $t = \frac{15}{s}$. Now let $t = 5$.

$$5 = \frac{15}{s}$$
$$5s = 15$$
$$s = 3$$

13. "p varies jointly as q and r^2" means

$$p = kqr^2$$

for some constant k. Given that $p = 200$ when $q = 2$ and $r = 3$, solve for k.

$$p = kqr^2$$
$$200 = k(2)(3)^2$$
$$200 = 18k$$
$$k = \tfrac{200}{18} = \tfrac{100}{9}$$

So $p = \frac{100}{9}qr^2$. Now let $q = 5$ and $r = 2$.

$$p = \tfrac{100}{9}qr^2$$
$$p = \tfrac{100}{9}(5)(2)^2$$
$$= \tfrac{100}{9}(20)$$
$$= \tfrac{2000}{9} \text{ or } 222\tfrac{2}{9}$$

14. "f varies jointly as g^2 and h" means

$$f = kg^2h$$

for some constant k.

Let $f = 50$, $g = 4$, and $h = 2$.

$$50 = k(4^2)(2)$$
$$50 = k(32)$$
$$k = \tfrac{50}{32} = \tfrac{25}{16}$$

So $f = \frac{25}{16}g^2h$. Now let $g = 3$ and $h = 6$.

$$f = \tfrac{25}{16}(3)^2(6)$$
$$= \tfrac{25}{16}(9)(6)$$
$$= \tfrac{1350}{16} = \tfrac{675}{8} \text{ or } 84\tfrac{3}{8}$$

15. For $k > 0$, if y varies directly as x (then $y = kx$), when x increases, y <u>increases</u>, and when x decreases, y <u>decreases</u>.

16. For $k > 0$, if y varies inversely as x $\left(\text{then } y = \frac{k}{x}\right)$, when x increases, then the denominator increases and y <u>decreases</u>, and when x decreases, then the denominator decreases and y <u>increases</u>.

17. If y varies inversely as x, x is in the denominator; however, if y varies directly as x, x is in the numerator. Also, for $k > 0$, with inverse variation, as x increases, y decreases. With direct variation, y increases as x increases.

18. The constant of variation is the coefficient of x. In the graph of the linear equation $y = kx$, k represents the slope of the line.

19. Let $x =$ the number of gallons he bought and let $C =$ the cost.
C varies directly as x, so

$$C = kx,$$

where k represents the cost per gallon.

Since $C = 13.59$ when $x = 8$,

$$\begin{aligned} 13.59 &= k(8) \\ k &= \frac{13.59}{8} = 1.69875 \approx 1.699. \end{aligned}$$

The price per gallon is $\$1.69\frac{9}{10}$.

20. Let $h =$ the number of hours she rode and let $C =$ the cost of the ride.
C varies directly as h, so

$$C = kh,$$

where k represents the price per hour.

Since $C = 50$ when $h = 2.5$,

$$\begin{aligned} 50 &= k(2.5) \\ k &= \frac{50}{2.5} = 20. \end{aligned}$$

The price per hour is \$20.

21. Let $V =$ the volume of the can and let $h =$ the height of the can.
V varies directly as h, so

$$V = kh.$$

Since $V = 300$ when $h = 10.62$,

$$\begin{aligned} 300 &= k(10.62) \\ k &= \frac{300}{10.62} \approx 28.25. \end{aligned}$$

So $V = 28.25h$. Now let $h = 15.92$.

$$\begin{aligned} V &= 28.25h \\ V &= 28.25(15.92) = 449.74 \end{aligned}$$

The volume is about 450 cm^3.

22. Let $y =$ the weight of an object on earth and $x =$ the weight of the object on the moon.
y varies directly as x, so

$$y = kx$$

for some constant k. Since $y = 200$ when $x = 32$, substitute these values in the equation and solve for k.

$$\begin{aligned} y &= kx \\ 200 &= k(32) \\ k &= \tfrac{200}{32} = 6.25 \end{aligned}$$

So $y = 6.25x$.
To find x when $y = 50$, substitute 50 for y in the equation

$$\begin{aligned} y &= 6.25x. \\ 50 &= 6.25x \\ x &= \tfrac{50}{6.25} = 8 \end{aligned}$$

The dog would weigh 8 lb on the moon.

23. Let $d =$ the distance and $t =$ the time.
d varies directly as the square of t, so $d = kt^2$.
Let $d = -576$ and $t = 6$. (You could also use $d = 576$, but the negative sign indicates the direction of the body.)

$$\begin{aligned} -576 &= k(6)^2 \\ -576 &= 36k \\ -16 &= k \end{aligned}$$

So $d = -16t^2$. Now let $t = 4$.

$$d = -16(4)^2 = -256$$

The object fell 256 ft in the first 4 seconds.

24. Let $A =$ the amount of water emptied by a pipe in one hour and $d =$ the diameter of the pipe.
A varies directly as d^2, so

$$A = kd^2$$

for some constant k. Since $A = 200$ when $d = 6$, substitute these values in the equation and solve for k.

$$\begin{aligned} A &= kd^2 \\ 200 &= k(6)^2 \\ 200 &= 36k \\ k &= \tfrac{200}{36} = \tfrac{50}{9} \end{aligned}$$

So $A = \frac{50}{9}d^2$. Now let $d = 12$.

$$\begin{aligned} A &= \tfrac{50}{9}d^2 \\ A &= \tfrac{50}{9}(12)^2 \\ &= \tfrac{50}{9}(144) = 800 \end{aligned}$$

A 12-inch pipe would empty 800 gal of water in one hour.

25. Let C = the current and R = the resistance. C varies inversely as R, so

$$C = \frac{k}{R}$$

for some constant k. Since $C = 20$ when $R = 5$, substitute these values in the equation and solve for k.

$$\begin{aligned} C &= \frac{k}{R} \\ 20 &= \frac{k}{5} \\ k &= 20(5) = 100 \end{aligned}$$

So $C = \dfrac{100}{R}$. Now let $R = 7.5$.

$$C = \frac{100}{7.5} = \frac{1000}{75} = \frac{40}{3} \quad \text{or} \quad 13\frac{1}{3}$$

The current is $13\frac{1}{3}$ amps.

26. The frequency f of a vibrating guitar string varies inversely as its length L, so

$$f = \frac{k}{L}$$

for some constant k. Substitute .65 for L and 4.3 for f.

$$\begin{aligned} 4.3 &= \frac{k}{.65} \\ k &= 4.3(.65) = 2.795 \end{aligned}$$

So $f = \dfrac{2.795}{L}$. Now let $L = .5$.

$$f = \frac{2.795}{.5} = 5.59$$

The frequency would be 5.59 vibrations per second.

27. Let I = the illumination (amount of light) produced by a light source and d = the distance from the source.
I varies inversely as d^2, so

$$I = \frac{k}{d^2}$$

for some constant k. Since $I = 768$ when $d = 1$, substitute these values in the equation and solve for k.

$$\begin{aligned} I &= \frac{k}{d^2} \\ 768 &= \frac{k}{1^2} \\ 768 &= k \end{aligned}$$

So $I = \dfrac{768}{d^2}$. Now let $d = 6$.

$$\begin{aligned} I &= \frac{768}{d^2} \\ I &= \frac{768}{6^2} = \frac{768}{36} = \frac{64}{3} \quad \text{or} \quad 21\frac{1}{3} \end{aligned}$$

The illumination produced by the light source is $21\frac{1}{3}$ foot-candles.

28. Let F = the force and
d = the distance from the center of the earth.

$$F = \frac{k}{d^2}$$

Let $F = 160$ and $d = 4000$.

$$\begin{aligned} 160 &= \frac{k}{4000^2} \\ k &= 160(16{,}000{,}000) \\ &= 2{,}560{,}000{,}000 = 2.56 \times 10^9 \end{aligned}$$

So $F = \dfrac{2.56 \times 10^9}{d^2}$.
Let $d = 6000 = 6 \times 10^3$.

$$\begin{aligned} F &= \frac{2.56 \times 10^9}{(6 \times 10^3)^2} \\ &= \frac{2.56 \times 10^9}{36 \times 10^6} = \frac{256 \times 10^7}{36 \times 10^6} \\ &= \frac{64 \times 10}{9} = \frac{640}{9} \quad \text{or} \quad 71\frac{1}{9} \end{aligned}$$

The force of attraction is $71\frac{1}{9}$ lb.

29. Let I = the simple interest, P the principal, and t the time.
Since I varies jointly as the principal and time, there is a constant k such that $I = kPt$. Find k by replacing I with 280, P with 2000, and t with 4.

$$\begin{aligned} I &= kPt \\ 280 &= k(2000)(4) \\ k &= \tfrac{280}{8000} = .035 \end{aligned}$$

So $I = .035Pt$. Now let $t = 6$.

$$\begin{aligned} I &= .035(2000)(6) \\ &= 420 \end{aligned}$$

The interest would be \$420.

30. Let $c =$ the collision impact, m the mass, and s the speed.
Since c varies jointly as the mass and the square of the speed, there is a constant k such that $c = kms^2$. Find k by replacing c with 6.1, m with 2000, and s with 55.

$$c = kms^2$$
$$6.1 = k(2000)(55)^2$$
$$k = \frac{6.1}{2000(55)^2} \approx 1.008264463 \times 10^{-6}$$

So $c = (1.0083 \times 10^{-6})\, ms^2$. Now let $s = 65$.

$$c = (1.0083 \times 10^{-6})(2000)(65)^2 \approx 8.5$$

The collision impact is about 8.5.

31. Let $F =$ the force, $w =$ the weight of the car, $s =$ the speed, and $r =$ the radius.
The force varies inversely as the radius and jointly as the weight and the square of the speed, so

$$F = \frac{kws^2}{r}.$$

Let $F = 242$, $w = 2000$, $r = 500$, and $s = 30$.

$$242 = \frac{k(2000)(30)^2}{500}$$
$$k = \frac{242(500)}{2000(900)} = \frac{121}{1800}$$

So $F = \dfrac{121ws^2}{1800r}$.

Let $r = 750$, $s = 50$, and $w = 2000$.

$$F = \frac{121(2000)(50)^2}{1800(750)} \approx 448.1$$

Approximately 448.1 lb of force would be needed.

32. Let $V =$ the volume of gas, $p =$ the pressure, and $t =$ the temperature.

Then $V = \dfrac{kt}{p}$.
$V = 1.3$ when $t = 300$ and $p = 18$.

$$1.3 = \frac{k(300)}{18}$$
$$k = \frac{1.3(18)}{300} = .078$$

So $V = \dfrac{.078t}{p}$.
Let $t = 340$ and $p = 24$.

$$V = \frac{.078(340)}{24} = 1.105$$

The volume is 1.105 L.

33. Let $N =$ the number of long distance calls,
$p_1 =$ the population of City 1,
$p_2 =$ the population of City 2,
and $d =$ the distance between them.

$$N = \frac{kp_1p_2}{d}$$

Let $N = 80{,}000$, $p_1 = 70{,}000$, $p_2 = 100{,}000$, and $d = 400$.

$$80{,}000 = \frac{k(70{,}000)(100{,}000)}{400}$$
$$80{,}000 = 17{,}500{,}000k$$
$$k = \frac{80{,}000}{17{,}500{,}000} = \frac{4}{875}$$
$$N = \frac{4}{875}\left(\frac{p_1p_2}{d}\right)$$

Let $p_1 = 50{,}000$, $p_2 = 75{,}000$, and $d = 250$.

$$N = \frac{4}{875}\left(\frac{50{,}000 \cdot 75{,}000}{250}\right) = \frac{480{,}000}{7} \text{ or } 68{,}571\frac{3}{7}$$

Rounded to the nearest hundred, there are approximately 68,600 calls.

34. Use the BMI from Example 6, with $k = 694$.

$$B = \frac{694w}{h^2}$$

Substitute your weight in pounds for w and your height in inches for h to determine your BMI. Answers will vary. A BMI from 19 to 25 is considered desirable.

35. The weight W of a bass varies jointly as its girth G and the square of its length L, so

$$W = kGL^2$$

for some constant k. Substitute 22.7 for W, 21 for G, and 36 for L.

$$22.7 = k(21)(36)^2$$
$$k = \frac{22.7}{27{,}216} \approx .000834$$

So $W = .000834GL^2$. Now let $G = 18$ and $L = 28$.

$$W = .000834GL^2 = .000834(18)(28)^2 \approx 11.8$$

The bass would weigh about 11.8 pounds.

36. The weight W of a trout varies jointly as its length L and the square of its girth G, so

$$W = kLG^2$$

for some constant k. Substitute 10.5 for W, 26 for L, and 18 for G.

$$10.5 = k(26)(18)^2$$
$$k = \frac{10.5}{8424} \approx .001246$$

So $W = .001246LG^2$. Now let $L = 22$ and $G = 15$.

$$\begin{aligned} W &= .001246LG^2 \\ &= .001246(22)(15)^2 \\ &\approx 6.2 \end{aligned}$$

The trout's weight is about 6.2 pounds.

37. The ordered pairs in the form of (gallons, price) are $(0, 0)$ and $(1, 1.75)$.

38. Let $(x_1, y_1) = (0, 0)$ and $(x_2, y_2) = (1, 1.75)$. Then

$$m = \frac{\text{change in } y}{\text{change in } x} = \frac{1.75 - 0}{1 - 0} = 1.75.$$

The slope is 1.75.

39. Since $m = 1.75$ and $b = 0$, the equation is

$$y = 1.75x + 0$$
$$\text{or} \quad y = 1.75x.$$

40. If $f(x) = ax + b$, then $a = 1.75$ and $b = 0$.

41. The value of a, 1.75, is the price in dollars per gallon of gasoline. It is also the slope of the line.

42. Since $f(x) = 1.75x$, it fits the form for a direct variation, that is, $y = kx$. The value of a, 1.75, is the constant of variation $(k = a)$.

Chapter 4 Review Exercises

1. $3x + 2y = 6$

To complete the ordered pairs, substitute the given values for x or y in the equation.

For $(0, \_\_\_\_)$, let $x = 0$.

$$\begin{aligned} 3x + 2y &= 6 \\ 3(0) + 2y &= 6 \\ 2y &= 6 \\ y &= 3 \end{aligned}$$

The ordered pair is $(0, 3)$.

For $(\_\_\_\_, 0)$, let $y = 0$.

$$\begin{aligned} 3x + 2y &= 6 \\ 3x + 2(0) &= 6 \\ 3x &= 6 \\ x &= 2 \end{aligned}$$

The ordered pair is $(2, 0)$.

For $(\_\_\_\_, -2)$, let $y = -2$.

$$\begin{aligned} 3x + 2y &= 6 \\ 3x + 2(-2) &= 6 \\ 3x - 4 &= 6 \\ 3x &= 10 \\ x &= \tfrac{10}{3} \end{aligned}$$

The ordered pair is $\left(\frac{10}{3}, -2\right)$.

Plot the ordered pairs, and draw the line through them.

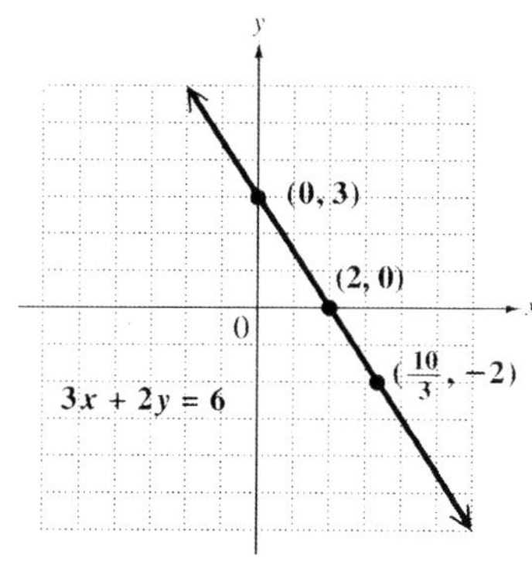

2. $x - y = 6$

To complete the ordered pairs, substitute the given values for x or y in the equation.

For $(2, \_\_\_\_)$, let $x = 2$.

$$\begin{aligned} x - y &= 6 \\ 2 - y &= 6 \\ -y &= 4 \\ y &= -4 \end{aligned}$$

The ordered pair is $(2, -4)$.

For $(\_\_\_\_, -3)$, let $y = -3$.

$$\begin{aligned} x - y &= 6 \\ x - (-3) &= 6 \\ x + 3 &= 6 \\ x &= 3 \end{aligned}$$

The ordered pair is $(3, -3)$.

continued

For (1, ____), let $x = 1$.

$$\begin{aligned} x - y &= 6 \\ 1 - y &= 6 \\ -y &= 5 \\ y &= -5 \end{aligned}$$

The ordered pair is $(1, -5)$.

For (____, -2), let $y = -2$.

$$\begin{aligned} x - y &= 6 \\ x - (-2) &= 6 \\ x + 2 &= 6 \\ x &= 4 \end{aligned}$$

The ordered pair is $(4, -2)$.

Plot the ordered pairs, and draw the line through them.

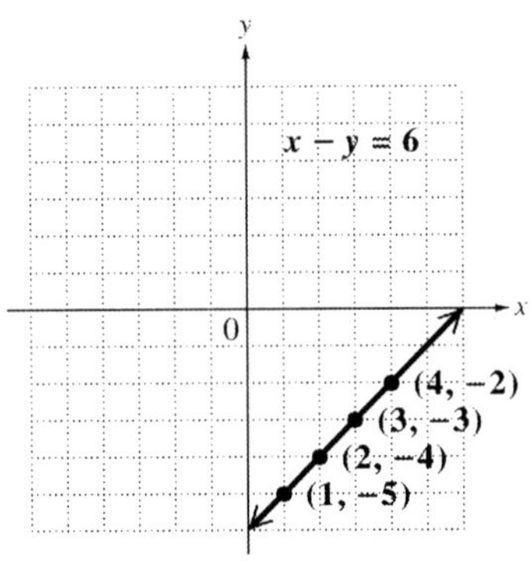

3. $4x + 3y = 12$

To find the x-intercept, let $y = 0$.

$$\begin{aligned} 4x + 3y &= 12 \\ 4x + 3(0) &= 12 \\ 4x &= 12 \\ x &= 3 \end{aligned}$$

The x-intercept is $(3, 0)$.

To find the y-intercept, let $x = 0$.

$$\begin{aligned} 4x + 3y &= 12 \\ 4(0) + 3y &= 12 \\ 3y &= 12 \\ y &= 4 \end{aligned}$$

The y-intercept is $(0, 4)$.

Plot the intercepts and draw the line through them.

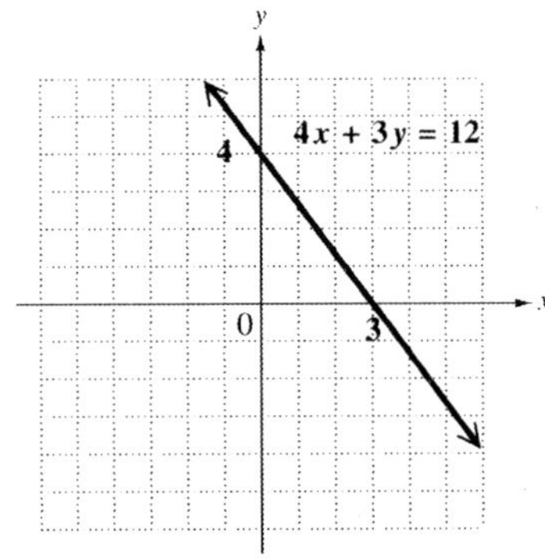

4. $5x + 7y = 15$

To find the x-intercept, let $y = 0$.

$$\begin{aligned} 5x + 7y &= 15 \\ 5x + 7(0) &= 15 \\ 5x &= 15 \\ x &= 3 \end{aligned}$$

The x-intercept is $(3, 0)$.

To find the y-intercept, let $x = 0$.

$$\begin{aligned} 5x + 7y &= 15 \\ 5(0) + 7y &= 15 \\ 7y &= 15 \\ y &= \tfrac{15}{7} \end{aligned}$$

The y-intercept is $\left(0, \frac{15}{7}\right)$.

Plot the intercepts and draw the line through them.

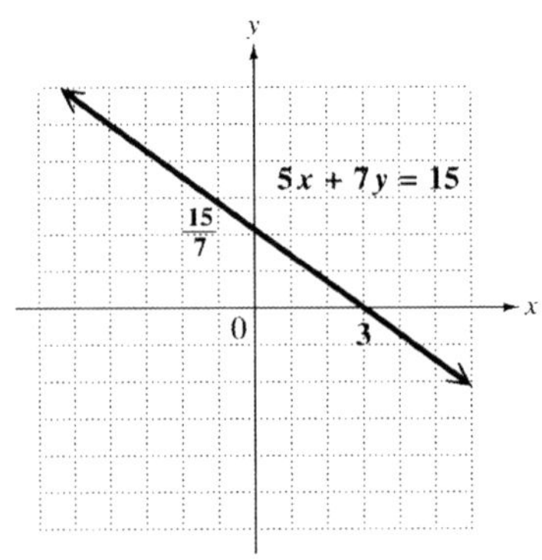

5. Let $(x_1, y_1) = (-1, 2)$ and $(x_2, y_2) = (4, -6)$
Then

$$m = \frac{y_2 - y_1}{x_2 - x_1} = \frac{-6 - 2}{4 - (-1)} = \frac{-8}{5} = -\frac{8}{5}.$$

The slope is $-\frac{8}{5}$.

6. The equation $y = 2x + 3$ is in slope-intercept form, $y = mx + b$, so the slope, m, is 2.

7. $-3x + 4y = 5$

Write the equation in slope-intercept form by solving for y.

$$\begin{aligned} -3x + 4y &= 5 \\ 4y &= 3x + 5 \\ y &= \tfrac{3}{4}x + \tfrac{5}{4} \end{aligned}$$

The slope is $\frac{3}{4}$.

8. The graph of $y = 4$ is the horizontal line with y-intercept $(0, 4)$. The slope of a horizontal line is 0.

9. The line will have the same slope as the line with equation $3y = -2x + 5$ since the two lines are parallel.

$$\begin{aligned} 3y &= -2x + 5 \\ y &= -\tfrac{2}{3}x + \tfrac{5}{3} \end{aligned}$$

The slope of the line parallel to the line with equation $3y = -2x + 5$ is $-\frac{2}{3}$.

10. The slope of the line will be the negative reciprocal of the slope of the line with equation $3x - y = 6$ since the two lines are perpendicular.

$$\begin{aligned} 3x - y &= 6 \\ -y &= -3x + 6 \\ y &= 3x - 6 \end{aligned}$$

The slope of the line with equation $3x - y = 6$ is 3. The negative reciprocal of 3 is $-\frac{1}{3}$.

The slope of a line perpendicular to the line with equation $3x - y = 6$ is $-\frac{1}{3}$.

11. The line goes up from left to right, so it has positive slope.

12. The line goes down from left to right, so it has negative slope.

13. The line is horizontal, so it has 0 slope.

14. The line is vertical, so it has *undefined* slope.

15. To rise 1 foot, we must move 4 feet in the horizontal direction. To rise 3 feet, we must move $3(4) = 12$ feet in the horizontal direction.

16. Use $(1970, 10{,}000)$ and $(2000, 51{,}000)$.

$$\begin{aligned} \text{average rate of change} &= \frac{51{,}000 - 10{,}000}{2000 - 1970} \\ &= \frac{41{,}000}{30} \\ &\approx 1367 \end{aligned}$$

The average rate of change is about \$1367 per year.

17. Slope $\frac{3}{5}$; y-intercept $(0, -8)$

Here, $m = \frac{3}{5}$ and $b = -8$. Substitute these values in the slope-intercept form.

$$\begin{aligned} y &= mx + b \\ y &= \tfrac{3}{5}x - 8 \end{aligned}$$

18. Slope $-\frac{1}{3}$; y-intercept $(0, 5)$

Here, $m = -\frac{1}{3}$ and $b = 5$. Substitute these values in the slope-intercept form.

$$\begin{aligned} y &= mx + b \\ y &= -\tfrac{1}{3}x + 5 \end{aligned}$$

19. Slope 0; y-intercept $(0, 12)$

A line with slope 0 is horizontal. The horizontal line through the point (c, d) has equation $y = d$. Here $d = 12$, so the line has equation $y = 12$.

20. Undefined slope; through $(2, 7)$

A line with undefined slope is vertical. The vertical line through the point (c, d) has equation $x = c$. Here $c = 2$, so the line has equation $x = 2$.

21. Horizontal; through $(-1, 4)$

The horizontal line through the point (c, d) has equation $y = d$. Here $d = 4$, so the line has equation $y = 4$.

22. Vertical; through $(.3, .6)$

The vertical line through the point (c, d) has equation $x = c$. Here $c = .3$, so the line has equation $x = .3$.

23. Through $(2, -5)$ and $(1, 4)$
Find the slope.

$$m = \frac{4 - (-5)}{1 - 2} = \frac{9}{-1} = -9$$

Use the point-slope form with $m = -9$ and $(x_1, y_1) = (2, -5)$.

$$\begin{aligned} y - y_1 &= m(x - x_1) \\ y - (-5) &= -9(x - 2) \\ y + 5 &= -9x + 18 \\ y &= -9x + 13 \end{aligned}$$

24. Through $(-3,-1)$ and $(2,6)$
Find the slope.

$$m=\frac{6-(-1)}{2-(-3)}=\frac{7}{5}$$

Use the point-slope form with $m=\frac{7}{5}$ and $(x_1,y_1)=(2,6)$.

$$\begin{aligned} y-y_1&=m(x-x_1)\\ y-6&=\tfrac{7}{5}(x-2)\\ y-6&=\tfrac{7}{5}x-\tfrac{14}{5}\\ y&=\tfrac{7}{5}x+\tfrac{16}{5}\end{aligned}$$

25. Parallel to $4x-y=3$ and through $(6,-2)$
Writing $4x-y=3$ in slope-intercept form gives us $y=4x-3$, which has slope 4. Lines parallel to it will also have slope 4. The line with slope 4 through $(6,-2)$ is:

$$\begin{aligned} y-y_1&=m(x-x_1)\\ y-(-2)&=4(x-6)\\ y+2&=4x-24\\ y&=4x-26\end{aligned}$$

26. Perpendicular to $2x-5y=7$ and through $(0,1)$
Write the equation in slope-intercept form.

$$\begin{aligned} 2x-5y&=7\\ -5y&=-2x+7\\ y&=\tfrac{2}{5}x-\tfrac{7}{5}\end{aligned}$$

$y=\frac{2}{5}x-\frac{7}{5}$ has slope $\frac{2}{5}$ and is perpendicular to lines with slope $-\frac{5}{2}$.
The line with slope $-\frac{5}{2}$ through $(0,1)$ is

$$\begin{aligned} y-y_1&=m(x-x_1)\\ y-1&=-\tfrac{5}{2}(x-0)\\ y&=-\tfrac{5}{2}x+1\end{aligned}$$

27. **(a)** The fixed cost is \$159, so that is the value of b. The variable cost is \$57, so

$$y=mx+b=57x+159.$$

The cost of a 1-year membership can be found by substituting 12 for x.

$$\begin{aligned} y&=57(12)+159\\ &=684+159=843\end{aligned}$$

The cost is \$843.

(b) As in part (a),

$$\begin{aligned} y&=47x+159.\\ y&=47(12)+159\\ &=564+159=723\end{aligned}$$

The cost is \$723.

28. $3x-2y\le 12$
Graph $3x-2y=12$ as a solid line through $(0,-6)$ and $(4,0)$. Use $(0,0)$ as a test point. Since $(0,0)$ satisfies the inequality, shade the region on the side of the line containing $(0,0)$.

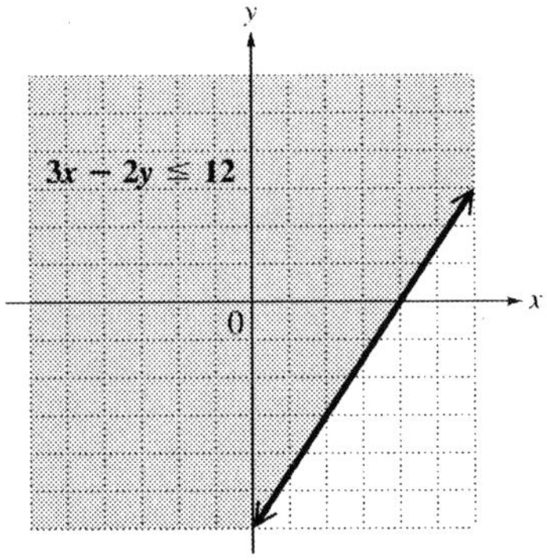

29. $5x-y>6$
Graph $5x-y=6$ as a dashed line through $(0,-6)$ and $\left(\frac{6}{5},0\right)$. Use $(0,0)$ as a test point. Since $(0,0)$ does not satisfy the inequality, shade the region on the side of the line that does not contain $(0,0)$.

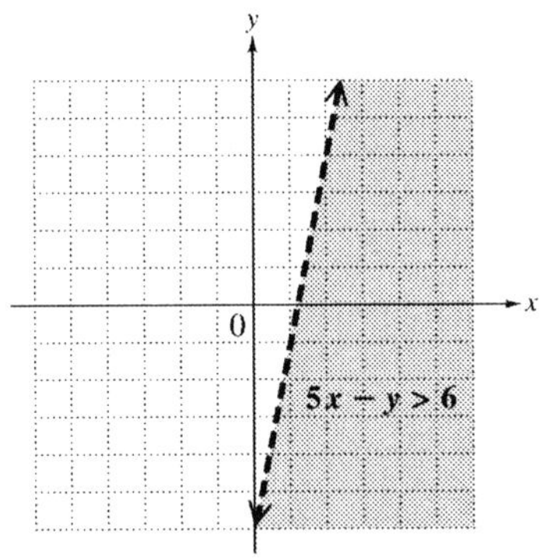

30. $x\ge 2$
Graph $x=2$ as a solid vertical line through $(2,0)$. Shade the region to the right of $x=2$.

$y\ge 2$
Graph $y=2$ as a solid horizontal line through $(0,2)$. Shade the region above $y=2$. The graph of

$$x\ge 2 \quad\text{or}\quad y\ge 2$$

includes all the shaded regions.

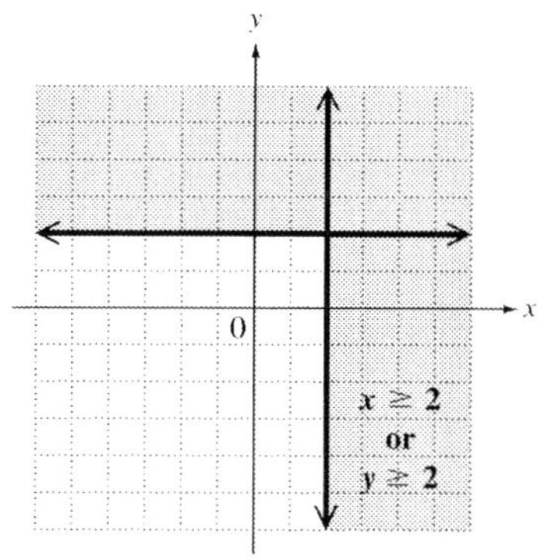

31. $2x + y \le 1$ and $x \ge 2y$
Graph $2x + y = 1$ as a solid line through $\left(\frac{1}{2}, 0\right)$ and $(0, 1)$, and shade the region on the side containing $(0, 0)$ since it satisfies the inequality. Next, graph $x = 2y$ as a solid line through $(0, 0)$ and $(2, 1)$, and shade the region on the side containing $(2, 0)$ since $2 > 2(0)$ or $2 > 0$ is true. The intersection is the region where the graphs overlap.

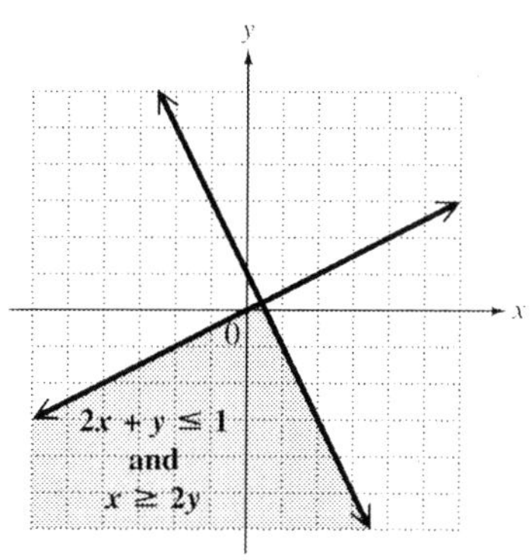

32. $\{(-4, 2), (-4, -2), (1, 5), (1, -5)\}$

The domain, the set of x-values, is $\{-4, 1\}$.
The range, the set of y-values, is $\{2, -2, 5, -5\}$.
Since at least one x-value has more than one y-value, the relation is not a function.

33. The relation can be described by the set of ordered pairs

$$\{(9, 32), (11, 47), (4, 47), (17, 69), (25, 14)\}.$$

The relation is a function since for each x-value, there is only one y-value.

The domain is the set of x-values:

$$\{9, 11, 4, 17, 25\}.$$

The range is the set of y-values: $\{32, 47, 69, 14\}$.

34. The domain, the x-values of the points on the graph, is $[-4, 4]$. The range, the y-values of the points on the graph, is $[0, 2]$. Since any vertical line intersects the graph of the relation in at most one point, the relation is a function.

35. $y = 3x - 3$
For any value of x, there is exactly one value of y, so the equation defines a function, actually a linear function. The domain is the set of all real numbers, $(-\infty, \infty)$.

36. $y < x + 2$
For any value of x, there are many values of y. For example, $(1, 0)$ and $(1, 1)$ are both solutions of the inequality that have the same x-value but different y-values. The inequality does not define a function. The domain is the set of all real numbers, $(-\infty, \infty)$.

37. $y = |x - 4|$
Given any value of x, y is found by subtracting 4, and taking the absolute value of the result. This process produces exactly one value of y for each x-value, so the equation defines a function. The domain is the set of all real numbers, $(-\infty, \infty)$.

38. $y = \sqrt{4x + 7}$
Given any value of x, y is found by multiplying x by 4, adding 7, and taking the square root of the result. This process produces exactly one value of y for each x-value, so the equation defines a function. Since the radicand must be nonnegative,

$$\begin{aligned} 4x + 7 &\ge 0 \\ 4x &\ge -7 \\ x &\ge -\tfrac{7}{4}. \end{aligned}$$

The domain is $\left[-\frac{7}{4}, \infty\right)$.

39. $x = y^2$
The ordered pairs $(4, 2)$ and $(4, -2)$ both satisfy the equation. Since one value of x, 4, corresponds to two values of y, 2 and -2, the equation does not define a function. Because x is equal to the square of y, the values of x must always be nonnegative. The domain is $[0, \infty)$.

40. $y = \dfrac{7}{x - 36}$
Given any value of x, y is found by subtracting 36, then dividing the result into 7. This process produces exactly one value of y for each x-value, so the equation defines a function. The domain includes all real numbers except those that make the denominator 0, namely 36. The domain is $(-\infty, 36) \cup (36, \infty)$.

41. **(a)** For each year, there is exactly one profit associated with the year, so the table defines a function.

(b) The domain is the set of years, that is, $\{1994, 1995, 1996, 1997, 1998, 1999\}$.
The range is the set of profits, that is, $\{40, 60, 80, 130, 180, 200\}$.

(c) Answers will vary. Two possible answers are $(1994, 40)$ and $(1995, 60)$.

(d) $f(1994) = 40$. In 1994, CNBC profits were \$40 million.

(e) Since $f(1998) = 180$, $x = 1998$.

In Exercises 42–45, use

$$f(x) = -2x^2 + 3x - 6.$$

42. $f(0) = -2(0)^2 + 3(0) - 6 = -6$

43. $f(3) = -2(3)^2 + 3(3) - 6$
$= -18 + 9 - 6 = -15$

44. $f(p) = -2p^2 + 3p - 6$

45. $f(-k) = -2(-k)^2 + 3(-k) - 6$
$= -2k^2 - 3k - 6$

46.
$$\begin{aligned} 2x^2 - y &= 0 \\ -y &= -2x^2 \\ y &= 2x^2 \end{aligned}$$

Since $y = f(x)$,
$f(x) = 2x^2$,
and $f(3) = 2(3)^2 = 2(9) = 18$.

47. Solve for y in terms of x.

$$\begin{aligned} 2x - 5y &= 7 \\ 2x - 7 &= 5y \\ \frac{2x - 7}{5} &= y \end{aligned}$$

This is the same as choice **C**,

$$f(x) = \frac{-7 + 2x}{5}.$$

48. If y varies inversely as x, then $y = \frac{k}{x}$, for some constant k. This form fits choice **C**.

49. "m varies inversely as p^2" means

$$m = \frac{k}{p^2}$$

for some constant k. Since $m = 20$ when $p = 2$, substitute these values in the equation and solve for k.

$$\begin{aligned} m &= \frac{k}{p^2} \\ 20 &= \frac{k}{2^2} \\ k &= 20(4) = 80 \end{aligned}$$

So $m = \dfrac{80}{p^2}$. Now let $p = 5$.

$$m = \frac{80}{p^2} = \frac{80}{5^2} = \frac{16}{5}$$

50. Let $v =$ the viewing distance and $e =$ the amount of enlargement.
v varies directly as e, so

$$v = ke$$

for some constant k. Since $v = 250$ when $e = 5$, substitute these values in the equation and solve for k.

$$\begin{aligned} v &= ke \\ 250 &= k(5) \\ 50 &= k \end{aligned}$$

So $v = 50e$. Now let $e = 8.6$.

$$v = 50(8.6) = 430$$

It should be viewed from 430 mm.

51. The volume V of a rectangular box of a given height is proportional to its width W and length L, so

$$V = kWL$$

for some constant k. Substitute 4 for W, 8 for L, and 64 for V.

$$\begin{aligned} 64 &= k(4)(8) \\ k &= \tfrac{64}{32} = 2 \end{aligned}$$

So $V = 2WL$. Now let $W = 3$ and $L = 6$.

$$V = 2(3)(6) = 36$$

The volume is 36 ft^3.

Chapter 4 Test

1. Let $(x_1, y_1) = (6, 4)$ and $(x_2, y_2) = (-4, -1)$. Then

$$m = \frac{y_2 - y_1}{x_2 - x_1} = \frac{-1 - 4}{-4 - 6} = \frac{-5}{-10} = \frac{1}{2}.$$

The slope is $\frac{1}{2}$.

2. To find the slope and y-intercept of

$$3x - 2y = 13,$$

write the equation in slope-intercept form by solving for y.

$$\begin{aligned} 3x - 2y &= 13 \\ -2y &= -3x + 13 \\ y &= \tfrac{3}{2}x - \tfrac{13}{2} \end{aligned}$$

The slope is $\frac{3}{2}$, and the y-intercept is $\left(0, -\frac{13}{2}\right)$. To find the x-intercept, substitute 0 for y in the original equation and solve for x.

$$\begin{aligned} 3x - 2y &= 13 \\ 3x - 2(0) &= 13 \\ 3x &= 13 \\ x &= \tfrac{13}{3} \end{aligned}$$

The x-intercept is $\left(\frac{13}{3}, 0\right)$.

3. The graph of $y = 5$ is the horizontal line with slope 0 and y-intercept $(0, 5)$. There is no x-intercept.

4. The graph of a line with undefined slope is the graph of a vertical line.

5. $4x - 3y = -12$

The x-intercept is $(-3, 0)$ and the y-intercept is $(0, 4)$. To sketch the graph, plot the intercepts and draw the line through them.

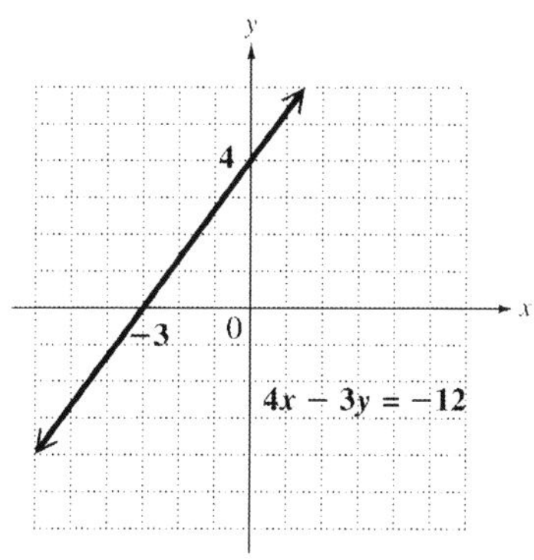

6. $y - 2 = 0$

$y = 2$

The graph of $y = 2$ is the horizontal line with y-intercept $(0, 2)$. There is no x-intercept.

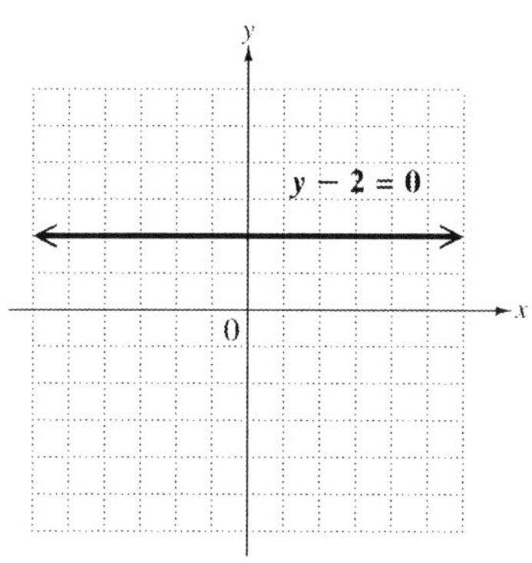

7. $y = -2x$

There is only one intercept, $(0, 0)$. Another ordered pair that satisfies the equation is $(1, -2)$. To sketch the graph, plot the two points and draw the line through them.

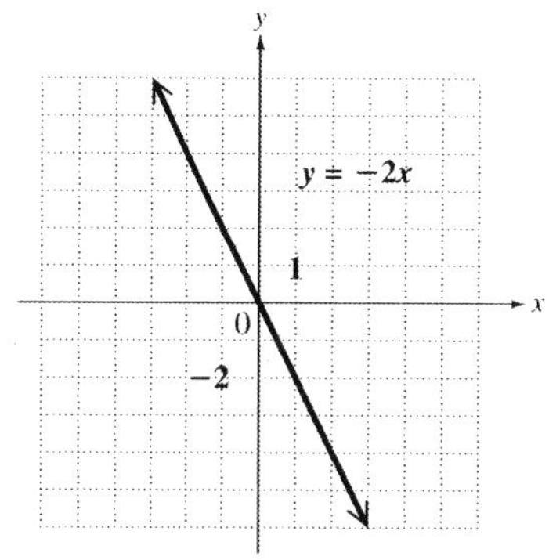

8. $3x - 2y > 6$

Graph the line $3x - 2y = 6$, which has intercepts $(2, 0)$ and $(0, -3)$, as a dashed line since the inequality involves " $>$ ". Test $(0, 0)$, which yields $0 > 6$, a false statement. Shade the region that does not include $(0, 0)$.

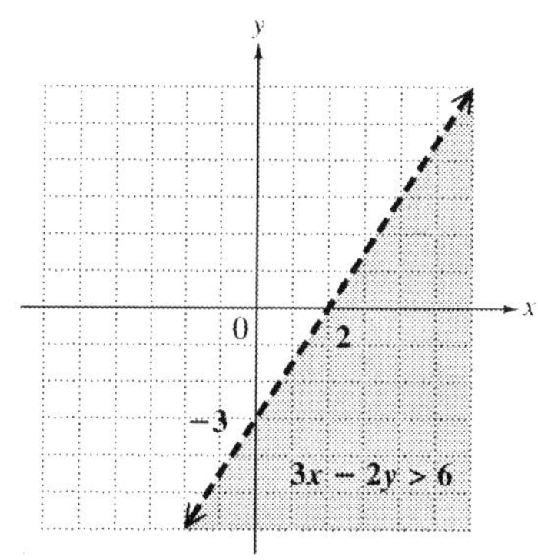

9. Through $(-3, 14)$; horizontal
The horizontal line through the point (c, d) has equation $y = d$. Here $d = 14$, so the line has equation $y = 14$.

10. Through $(4, -1)$; $m = -5$
Let $m = -5$ and $(x_1, y_1) = (4, -1)$ in the point-slope form.

$$\begin{aligned} y - y_1 &= m(x - x_1) \\ y - (-1) &= -5(x - 4) \\ y + 1 &= -5x + 20 \\ y &= -5x + 19 \end{aligned}$$

11. **(a)** Through $(-7, 2)$ and parallel to $3x + 5y = 6$
To find the slope of the line with equation $3x + 5y = 6$, write the equation in slope-intercept form by solving for y.

$$\begin{aligned} 3x + 5y &= 6 \\ 5y &= -3x + 6 \\ y &= -\tfrac{3}{5}x + \tfrac{6}{5} \end{aligned}$$

The slope is $-\frac{3}{5}$, so a line parallel to it also has slope $-\frac{3}{5}$. Let $m = -\frac{3}{5}$ and $(x_1, y_1) = (-7, 2)$ in the point-slope form.

$$\begin{aligned} y - y_1 &= m(x - x_1) \\ y - 2 &= -\tfrac{3}{5}[x - (-7)] \\ y - 2 &= -\tfrac{3}{5}(x + 7) \\ y - 2 &= -\tfrac{3}{5}x - \tfrac{21}{5} \\ y &= -\tfrac{3}{5}x - \tfrac{11}{5} \end{aligned}$$

(b) Through $(-7, 2)$ and perpendicular to $y = 2x$
Since $y = 2x$ is in slope-intercept form $(b = 0)$, the slope, m, of the line with equation $y = 2x$ is 2. A line perpendicular to it has a slope that is the negative reciprocal of 2, that is, $-\frac{1}{2}$. Let $m = -\frac{1}{2}$ and $(x_1, y_1) = (-7, 2)$ in the point-slope form.

$$\begin{aligned} y - y_1 &= m(x - x_1) \\ y - 2 &= -\tfrac{1}{2}(x + 7) \\ y - 2 &= -\tfrac{1}{2}x - \tfrac{7}{2} \\ y &= -\tfrac{1}{2}x - \tfrac{3}{2} \end{aligned}$$

12. Choice **D** is the only graph that passes the vertical line test. Since x can be any value, the domain is $(-\infty, \infty)$. Since $y \geq 0$, the range is $[0, \infty)$.

13. Choice **D** does not define a function, since its domain (input) element 0 is paired with two different range (output) elements, 1 and 2.

The domain is the set of x-values: $\{0, 3, 6\}$
The range is the set of y-values: $\{1, 2, 3\}$

14.
$$\begin{aligned} f(x) &= -x^2 + 2x - 1 \\ f(1) &= -(1)^2 + 2(1) - 1 \\ &= -1 + 2 - 1 \\ &= 0 \\ f(a) &= -a^2 + 2a - 1 \end{aligned}$$

15. The current I is inversely proportional to the resistance R, so

$$I = \frac{k}{R}$$

for some constant k. Let $I = 80$ and $R = 30$. Find k.

$$\begin{aligned} 80 &= \frac{k}{30} \\ k &= 80(30) = 2400 \end{aligned}$$

So $I = \dfrac{2400}{R}$. Now let $R = 12$.

$$I = \frac{2400}{12} = 200$$

The current is 200 amps.

16. The force F of the wind blowing on a vertical surface varies jointly as the area A of the surface and the square of the velocity V, so

$$F = kAV^2$$

for some constant k. Let $F = 50$, $A = 500$, and $V = 40$. Find k.

$$\begin{aligned} 50 &= k(500)(40)^2 \\ k &= \frac{50}{500(1600)} = \frac{1}{16{,}000} \end{aligned}$$

So $F = \frac{1}{16{,}000}AV^2$. Now let $A = 2$ and $V = 80$.

$$F = \frac{1}{16{,}000}(2)(80)^2 = .8$$

The force of the wind is .8 lb.

Cumulative Review Exercises (Chapters 1–4)

1. The absolute value of a negative number is a positive number and the additive inverse of the same negative number is the same positive number. For example, suppose the negative number is -5:

$$|-5| = -(-5) = 5$$
$$\text{and} \quad -(-5) = 5$$

The statement is *always true*.

2. The statement is *always true*; in fact, it is the definition of a rational number.

3. The sum of two negative numbers is another negative number, so the statement is *never true*.

4. The statement is *sometimes true*. For example,

$$3 + (-3) = 0,$$
$$\text{but} \quad 3 + (-1) = 2 \neq 0.$$

5.
$$\begin{aligned} -|-2| - 4 + |-3| + 7 &= -2 - 4 + 3 + 7 \\ &= -6 + 3 + 7 \\ &= -3 + 7 \\ &= 4 \end{aligned}$$

6. $(-.8)^2 = (-.8)(-.8) = .64$

7. $\sqrt{-64}$ is not a real number.

8. The product of two numbers that have the same sign is positive.

$$-\frac{2}{3}\left(-\frac{12}{5}\right)=\frac{2\cdot 12}{3\cdot 5}=\frac{2\cdot 4}{5}=\frac{8}{5}$$

9. $$\begin{aligned}-(-4m+3)&=-(-4m)-3\\&=4m-3\end{aligned}$$

10. $$\begin{aligned}&3x^2-4x+4+9x-x^2\\&=3x^2-x^2-4x+9x+4\\&=2x^2+5x+4\end{aligned}$$

11. $$\begin{aligned}\frac{3\sqrt{16}-(-1)(7)}{4+(-6)}&=\frac{3(4)-(-7)}{-2}\\&=\frac{12+7}{-2}=-\frac{19}{2}\end{aligned}$$

12. $-3<x\leq 5$
This is the set of numbers between -3 and 5, not including -3 (use a parenthesis), but including 5 (use a bracket). In interval notation, the set is $(-3,5]$.

13. $$\sqrt{\frac{-2+4}{-5}}=\sqrt{\frac{2}{-5}}=\sqrt{-\frac{2}{5}}$$

This is not a real number since the number under the radical sign is negative.

For Exercises 14–16, let $p=-4$, $q=-2$, and $r=5$.

14. $$\begin{aligned}-3(2q-3p)&=-3[2(-2)-3(-4)]\\&=-3(-4+12)\\&=-3(8)\\&=-24\end{aligned}$$

15. $$\begin{aligned}|p|^3-|q^3|&=|-4|^3-|(-2)^3|\\&=4^3-|-8|\\&=64-8\\&=56\end{aligned}$$

16. $$\begin{aligned}\frac{\sqrt{r}}{-p+2q}&=\frac{\sqrt{5}}{-(-4)+2(-2)}\\&=\frac{\sqrt{5}}{4-4}\\&=\frac{\sqrt{5}}{0}\end{aligned}$$

This expression is *undefined* since the denominator is zero.

17. $$\begin{aligned}2z-5+3z&=4-(z+2)\\5z-5&=4-z-2\\5z-5&=2-z\\6z&=7\\z&=\tfrac{7}{6}\end{aligned}$$

Solution set: $\{\frac{7}{6}\}$

18. $$\frac{3a-1}{5}+\frac{a+2}{2}=-\frac{3}{10}$$
Multiply each side by the LCD, 10.
$$\begin{aligned}10\left(\frac{3a-1}{5}+\frac{a+2}{2}\right)&=10\left(-\frac{3}{10}\right)\\2(3a-1)+5(a+2)&=-3\\6a-2+5a+10&=-3\\11a+8&=-3\\11a&=-11\\a&=-1\end{aligned}$$

Solution set: $\{-1\}$

19. Solve $V=\frac{1}{3}\pi r^2h$ for h.

$3V=\pi r^2h$ *Multiply by 3.*

$\frac{3V}{\pi r^2}=h$ *Divide by πr^2.*

20. Let $x=$ the time it takes for the planes to be 2100 miles apart.
Make a table. Use the formula $d=rt$.

	r	*t*	*d*
Eastbound Plane	550	x	$550x$
Westbound Plane	500	x	$500x$

The total distance is 2100 miles.

$$\begin{aligned}550x+500x&=2100\\1050x&=2100\\x&=2\end{aligned}$$

It will take 2 hr for the planes to be 2100 mi apart.

21. Let x represent the number of white pills and $2x$ the number of yellow pills.

Make a table.

	Strength (in Units)	**Number of Pills**	**Amount of Medication**
White	3	x	$3x$
Yellow	3	$2x$	$3(2x)$

The total amount of medication must be at least 30 units. Solve the inequality.

$$\begin{aligned}3x+3(2x)&\geq 30\\3x+6x&\geq 30\\9x&\geq 30\\x&\geq \tfrac{30}{9}\\x&\geq \tfrac{10}{3}\ \text{ or }\ 3\tfrac{1}{3}\end{aligned}$$

Since Ms. Bell must take a whole number of pills, she must take at least 4 white pills.

22. Let x denote the side of the original square and $4x$ the perimeter. Now $x+4$ is the side of the new square and $4(x+4)$ is its perimeter.
"The perimeter would be 8 inches less than twice the perimeter of the original square " translates as

$$\begin{aligned} 4(x+4) &= 2(4x) - 8. \\ 4x + 16 &= 8x - 8 \\ 24 &= 4x \\ 6 &= x \end{aligned}$$

The length of a side of the original square is 6 inches.

23. The union of the three solution sets is $(-\infty, \infty)$; that is, the set of all real numbers.

24. $$\begin{aligned} 3 - 2(m+3) &< 4m \\ 3 - 2m - 6 &< 4m \\ -2m - 3 &< 4m \\ -6m &< 3 \end{aligned}$$

Multiply by $-\frac{1}{6}$; reverse the direction of the inequality symbol.

$$\begin{aligned} -\tfrac{1}{6}(-6m) &> -\tfrac{1}{6}(3) \\ m &> -\tfrac{1}{2} \end{aligned}$$

Solution set: $\left(-\frac{1}{2}, \infty\right)$

25. $$\begin{aligned} 2k + 4 < 10 \quad &\text{and} \quad 3k - 1 > 5 \\ 2k < 6 \quad &\text{and} \quad 3k > 6 \\ k < 3 \quad &\text{and} \quad k > 2 \end{aligned}$$

The solution set of these inequalities is the set of numbers that are both less than 3 and greater than 2. This is the set of all numbers between 2 and 3.

Solution set: $(2, 3)$

26. $$\begin{aligned} 2k + 4 > 10 \quad &\text{or} \quad 3k - 1 < 5 \\ 2k > 6 \quad &\text{or} \quad 3k < 6 \\ k > 3 \quad &\text{or} \quad k < 2 \end{aligned}$$

The solution set is the set of numbers that are either greater than 3 or less than 2.

Solution set: $(-\infty, 2) \cup (3, \infty)$

27. $|5x + 3| = 13$

$$\begin{aligned} 5x + 3 = 13 \quad &\text{or} \quad 5x + 3 = -13 \\ 5x = 10 \quad &\text{or} \quad 5x = -16 \\ x = 2 \quad &\text{or} \quad x = -\tfrac{16}{5} \end{aligned}$$

Solution set: $\left\{-\frac{16}{5}, 2\right\}$

28. $|x + 2| < 9$

$-9 < x + 2 < 9$

Subtract 2 from each part of the inequality.

$-11 < x < 7$

Solution set: $(-11, 7)$

29. $|2x - 5| \geq 9$

$$\begin{aligned} 2x - 5 \geq 9 \quad &\text{or} \quad 2x - 5 \leq -9 \\ 2x \geq 14 \quad &\text{or} \quad 2x \leq -4 \\ x \geq 7 \quad &\text{or} \quad x \leq -2 \end{aligned}$$

Solution set: $(-\infty, -2] \cup [7, \infty)$

30. $3x - 4y = 12$

To complete the ordered pairs, substitute the given values for x or y in the equation.

For $(0, \_\_\_\_)$, let $x = 0$.

$$\begin{aligned} 3x - 4y &= 12 \\ 3(0) - 4y &= 12 \\ -4y &= 12 \\ y &= -3 \end{aligned}$$

The ordered pair is $(0, -3)$.

For $(\_\_\_\_, 0)$, let $y = 0$.

$$\begin{aligned} 3x - 4y &= 12 \\ 3x - 4(0) &= 12 \\ 3x &= 12 \\ x &= 4 \end{aligned}$$

The ordered pair is $(4, 0)$.

For $(2, \_\_\_\_)$, let $x = 2$.

$$\begin{aligned} 3x - 4y &= 12 \\ 3(2) - 4y &= 12 \\ 6 - 4y &= 12 \\ -4y &= 6 \\ y &= -\tfrac{6}{4} = -\tfrac{3}{2} \end{aligned}$$

The ordered pair is $\left(2, -\frac{3}{2}\right)$.

31. $-4x + 2y = 8$

To find the x-intercept of the equation, let $y = 0$.

$$\begin{aligned} -4x + 2y &= 8 \\ -4x + 2(0) &= 8 \\ -4x &= 8 \\ x &= -2 \end{aligned}$$

The x-intercept is $(-2, 0)$.

To find the y-intercept of the equation, let $x = 0$.

$$\begin{aligned} -4x + 2y &= 8 \\ -4(0) + 2y &= 8 \\ 2y &= 8 \\ y &= 4 \end{aligned}$$

The y-intercept is $(0, 4)$.

Plot the intercepts and draw the line through them.

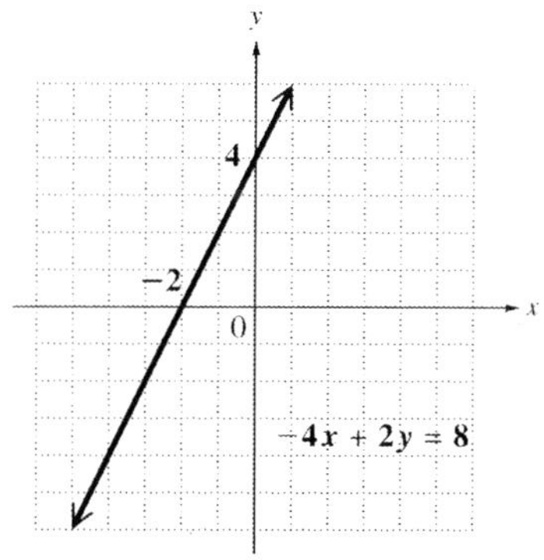

32. Through $(-5, 8)$ and $(-1, 2)$

Let $(x_1, y_1) = (-5, 8)$ and $(x_2, y_2) = (-1, 2)$.

$$m = \frac{y_2 - y_1}{x_2 - x_1} = \frac{2 - 8}{-1 - (-5)} = \frac{-6}{4} = -\frac{3}{2}$$

The slope is $-\frac{3}{2}$.

33. The equation is in the form $y = mx + b$. The slope, m, of the line with equation $y = -\frac{1}{2}x + 5$ is $-\frac{1}{2}$. Thus, the slope of any line parallel to it is $-\frac{1}{2}$.

34. Perpendicular to $4x - 3y = 12$

Solve for y to write the equation in slope-intercept form.

$$\begin{aligned} 4x - 3y &= 12 \\ -3y &= -4x + 12 \\ y &= \tfrac{4}{3}x - 4 \end{aligned}$$

The slope of the given line is $\frac{4}{3}$. Perpendicular lines have slopes that are negative reciprocals of each other. The negative reciprocal of $\frac{4}{3}$ is $-\frac{3}{4}$. The slope of a line perpendicular to the given line is $-\frac{3}{4}$.

35. Slope $-\frac{3}{4}$; y-intercept $(0, -1)$

To write an equation of this line, let $m = -\frac{3}{4}$ and $b = -1$ in the slope-intercept form.

$$\begin{aligned} y &= mx + b \\ y &= -\tfrac{3}{4}x - 1 \end{aligned}$$

36. Horizontal; through $(2, -2)$
A horizontal line through the point (c, d) has equation $y = d$. Here $d = -2$, so the equation of the line is $y = -2$.

37. Through $(4, -3)$ and $(1, 1)$
First find the slope of the line.

$$m = \frac{1 - (-3)}{1 - 4} = \frac{4}{-3} = -\frac{4}{3}$$

Now substitute $(x_1, y_1) = (4, -3)$ and $m = -\frac{4}{3}$ in the point-slope form. Then solve for y.

$$\begin{aligned} y - y_1 &= m(x - x_1) \\ y - (-3) &= -\tfrac{4}{3}(x - 4) \\ y + 3 &= -\tfrac{4}{3}x + \tfrac{16}{3} \\ y &= -\tfrac{4}{3}x + \tfrac{7}{3} \end{aligned}$$

38. $f(x) = -4x + 10$

(a) The variable x can be any real number, so the domain is $(-\infty, \infty)$.

(b) $f(-3) = -4(-3) + 10 = 12 + 10 = 22$

39. From the graph, use the ordered pairs $(1992, 11)$ and $(2000, 95)$.

$$m = \frac{95 - 11}{2000 - 1992} = \frac{84}{8} = 10.5$$

The slope is 10.5.

40. The segment from 1992 to 2000 rises more sharply, so it has a greater slope.

CHAPTER 5 SYSTEMS OF LINEAR EQUATIONS

5.1 Systems of Linear Equations in Two Variables

5.1 Margin Exercises

1. **(a)**
$$\begin{aligned} x - y &= 3 \quad (1) \\ 2x - y &= 4 \quad (2) \end{aligned}$$

When the equations are graphed, the point of intersection appears to be $(1, -2)$. To check, substitute 1 for x and -2 for y in each equation of the system.

$$\begin{aligned} x - y &= 3 \quad (1) \\ 1 - (-2) &= 3\,? \\ 3 &= 3 \quad \textit{True} \end{aligned} \qquad \begin{aligned} 2x - y &= 4 \quad (2) \\ 2(1) - (-2) &= 4\,? \\ 4 &= 4 \quad \textit{True} \end{aligned}$$

Since $(1, -2)$ makes both equations true, the solution set of the system is $\{(1, -2)\}$.

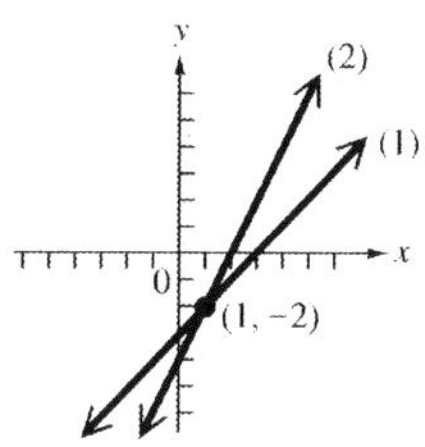

(b)
$$\begin{aligned} 2x + y &= -5 \quad (1) \\ -x + 3y &= 6 \quad (2) \end{aligned}$$

When the equations are graphed, the point of intersection appears to be $(-3, 1)$. To check, substitute -3 for x and 1 for y in each equation of the system.

$$\begin{aligned} 2x + y &= -5 \quad (1) \\ 2(-3) + 1 &= -5\,? \\ -5 &= -5 \quad \textit{True} \end{aligned} \qquad \begin{aligned} -x + 3y &= 6 \quad (2) \\ -(-3) + 3(1) &= 6\,? \\ 6 &= 6 \quad \textit{True} \end{aligned}$$

Since $(-3, 1)$ makes both equations true, the solution set is $\{(-3, 1)\}$.

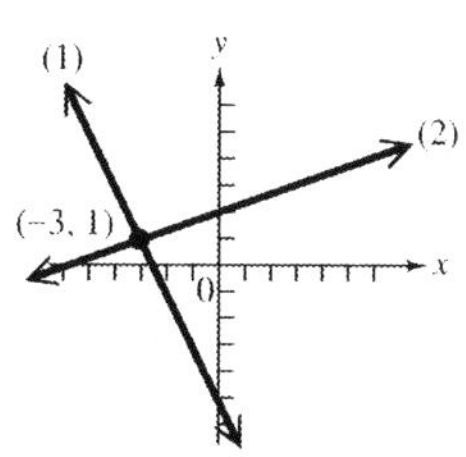

2. **(a)** To determine whether $(-4, 2)$ is a solution of the system

$$\begin{aligned} 2x + y &= -6 \\ x + 3y &= 2, \end{aligned}$$

replace x with -4 and y with 2 in each equation.

$$\begin{aligned} 2x + y &= -6 \quad (1) \\ 2(-4) + 2 &= -6\,? \\ -8 + 2 &= -6\,? \\ -6 &= -6 \quad \textit{True} \end{aligned} \qquad \begin{aligned} x + 3y &= 2 \quad (2) \\ -4 + 3(2) &= 2\,? \\ -4 + 6 &= 2\,? \\ 2 &= 2 \quad \textit{True} \end{aligned}$$

Since $(-4, 2)$ makes both equations true, $(-4, 2)$ is a solution of the system.

(b) To determine whether $(-1, 5)$ is a solution of the system

$$\begin{aligned} 9x - y &= -4 \\ 4x + 3y &= 11, \end{aligned}$$

replace x with -1 and y with 5 in each equation.

$$\begin{aligned} 9x - y &= -4 \\ 9(-1) - 5 &= -4 \quad ? \\ -9 - 5 &= -4 \quad ? \\ -14 &= -4 \qquad \textit{False} \end{aligned}$$

Since $(-1, 5)$ does not satisfy the first equation, it is not necessary to check whether it satisfies the second equation. The ordered pair $(-1, 5)$ is not a solution of the system.

3. **(a)**
$$\begin{aligned} 7x - 2y &= -2 \quad (1) \\ y &= 3x \quad (2) \end{aligned}$$

Since equation (2) is given in terms of y, substitute $3x$ for y in equation (1).

$$\begin{aligned} 7x - 2y &= -2 \quad (1) \\ 7x - 2(3x) &= -2 \\ 7x - 6x &= -2 \\ x &= -2 \end{aligned}$$

Since $y = 3x$ and $x = -2$,

$$y = 3(-2) = -6.$$

Check $(-2, -6)$: $-2 = -2$; $-6 = -6$

Solution set: $\{(-2, -6)\}$

(b)
$$\begin{aligned} 5x - 3y &= -6 \quad (1) \\ x &= 2 - y \quad (2) \end{aligned}$$

Since equation (2) is given in terms of x, substitute $2 - y$ for x in equation (1).

$$\begin{aligned} 5x - 3y &= -6 \quad (1) \\ 5(2 - y) - 3y &= -6 \\ 10 - 5y - 3y &= -6 \\ 10 - 8y &= -6 \\ -8y &= -16 \\ y &= 2 \end{aligned}$$

Since $x = 2 - y$ and $y = 2$,

$$x = 2 - 2 = 0.$$

Check $(0, 2)$: $-6 = -6$; $0 = 0$

Solution set: $\{(0, 2)\}$

4. **(a)** $3x - y = 10$ (1)
$2x + 5y = 1$ (2)

Step 1
To use the substitution method, first solve one of the equations for x or y. Since the coefficient of y in equation (1) is -1, it is easiest to solve for y in this equation.

$$\begin{aligned} 3x - y &= 10 \quad (1) \\ -y &= -3x + 10 \\ y &= 3x - 10 \end{aligned}$$

Step 2
Substitute $3x - 10$ for y in equation (2) and solve for x.

$$\begin{aligned} 2x + 5y &= 1 \quad (2) \\ 2x + 5(3x - 10) &= 1 \end{aligned}$$

Step 3

$$\begin{aligned} 2x + 15x - 50 &= 1 \\ 17x &= 51 \\ x &= 3 \end{aligned}$$

Step 4
Since $y = 3x - 10$ and $x = 3$,

$$y = 3(3) - 10 = 9 - 10 = -1.$$

Step 5
Check $(3, -1)$: $10 = 10$; $1 = 1$

Solution set: $\{(3, -1)\}$

(b) $4x - 5y = -11$ (1)
$x + 2y = 7$ (2)

Step 1
Solve equation (2) for x.

$$\begin{aligned} x + 2y &= 7 \quad (2) \\ x &= 7 - 2y \end{aligned}$$

Step 2
Substitute $7 - 2y$ for x in equation (1).

$$\begin{aligned} 4x - 5y &= -11 \quad (1) \\ 4(7 - 2y) - 5y &= -11 \end{aligned}$$

Step 3

$$\begin{aligned} 28 - 8y - 5y &= -11 \\ 28 - 13y &= -11 \\ -13y &= -39 \\ y &= 3 \end{aligned}$$

Step 4
Since $x = 7 - 2y$ and $y = 3$,

$$x = 7 - 2(3) = 7 - 6 = 1.$$

Step 5
Check $(1, 3)$: $-11 = -11$; $7 = 7$

Solution set: $\{(1, 3)\}$

5. **(a)** $3x - y = -7$ (1)
$2x + y = -3$ (2)

Eliminate y by adding equations (1) and (2).

$$\begin{array}{rcrl} 3x - y &=& -7 & (1) \\ 2x + y &=& -3 & (2) \\ \hline 5x &=& -10 & \textit{Add.} \\ x &=& -2 & \textit{Solve for x.} \end{array}$$

To find y, replace x with -2 in either equation (1) or (2).

$$\begin{aligned} 3x - y &= -7 \quad (1) \\ 3(-2) - y &= -7 \\ -6 - y &= -7 \\ -y &= -1 \\ y &= 1 \end{aligned}$$

Check $(-2, 1)$: $-7 = -7$; $-3 = -3$

Solution set: $\{(-2, 1)\}$

(b) $-2x + 3y = -10$ (1)
$2x + 2y = 5$ (2)

Eliminate x by adding equations (1) and (2).

$$\begin{array}{rcrl} -2x + 3y &=& -10 & (1) \\ 2x + 2y &=& 5 & (2) \\ \hline 5y &=& -5 & \textit{Add.} \\ y &=& -1 & \textit{Solve for y.} \end{array}$$

To find x, replace y with -1 in either equation (1) or (2).

$$\begin{aligned} -2x + 3y &= -10 \quad (1) \\ -2x + 3(-1) &= -10 \\ -2x - 3 &= -10 \\ -2x &= -7 \\ x &= \tfrac{7}{2} \end{aligned}$$

Check $(\frac{7}{2}, -1)$: $-10 = -10$; $5 = 5$

Solution set: $\{(\frac{7}{2}, -1)\}$

6. **(a)** $x + 3y = 8$ (1)
$2x - 5y = -17$ (2)

Step 1
Both equations are in standard form.

Step 2
To eliminate x, multiply equation (1) by -2 and add the result to equation (2).

$$\begin{array}{lrcrl} & -2x - 6y &=& -16 & -2 \times (1) \\ & 2x - 5y &=& -17 & (2) \\ \hline \textit{Step 3} & -11y &=& -33 & \textit{Add.} \\ \textit{Step 4} & y &=& 3 & \textit{Solve for y.} \end{array}$$

Step 5
To find x, substitute 3 for y in equation (1) or (2).

$$\begin{aligned} x + 3y &= 8 \quad (1) \\ x + 3(3) &= 9 \\ x + 9 &= 8 \\ x &= -1 \end{aligned}$$

Step 6
Check $(-1, 3)$: $8 = 8$; $-17 = -17$

Solution set: $\{(-1, 3)\}$

(b) $6x - 2y = -21$ (1)
$-3x + 4y = 36$ (2)

To eliminate x, multiply equation (2) by 2 and add the result to equation (1).

$$\begin{array}{rcrl} 6x - 2y &=& -21 & (1) \\ -6x + 8y &=& 72 & 2 \times (2) \\ \hline 6y &=& 51 & \textit{Add.} \\ y &=& \frac{51}{6} = \frac{17}{2} & \textit{Solve for y.} \end{array}$$

To find x, substitute $\frac{17}{2}$ for y in equation (1) or (2).

$$\begin{aligned} 6x - 2y &= -21 \quad (1) \\ 6x - 2\left(\tfrac{17}{2}\right) &= -21 \\ 6x - 17 &= -21 \\ 6x &= -4 \\ x &= -\tfrac{2}{3} \end{aligned}$$

Check $\left(-\frac{2}{3}, \frac{17}{2}\right)$: $-4 - 17 = -21$; $2 + 34 = 36$

Solution set: $\left\{\left(-\frac{2}{3}, \frac{17}{2}\right)\right\}$

(c) $2x + 3y = 19$ (1)
$3x - 7y = -6$ (2)

To eliminate x, multiply equation (1) by -3 and add the result to equation (2) multiplied by 2.

$$\begin{array}{rcrl} -6x - 9y &=& -57 & -3 \times (1) \\ 6x - 14y &=& -12 & 2 \times (2) \\ \hline -23y &=& -69 & \textit{Add.} \\ y &=& 3 & \textit{Solve for y.} \end{array}$$

To find x, substitute 3 for y in equation (1) or (2).

$$\begin{aligned} 2x + 3y &= 19 \quad (1) \\ 2x + 3(3) &= 19 \\ 2x + 9 &= 19 \\ 2x &= 10 \\ x &= 5 \end{aligned}$$

Check $(5, 3)$: $19 = 19$; $-6 = -6$

Solution set: $\{(5, 3)\}$

7. (a) $\frac{1}{3}x - \frac{1}{2}y = \frac{1}{6}$ (1)
$3x - 2y = 9$ (2)

Multiply equation (1) by the LCD, 6. The new system is

$$\begin{aligned} 2x - 3y &= 1 \quad (3) \\ 3x - 2y &= 9. \quad (2) \end{aligned}$$

Solve equation (2) for x.

$$\begin{aligned} 3x - 2y &= 9 \quad (2) \\ 3x &= 2y + 9 \\ x &= \tfrac{2}{3}y + 3 \end{aligned}$$

Substitute $\frac{2}{3}y + 3$ for x in equation (3) and solve for y.

$$\begin{aligned} 2x - 3y &= 1 \quad (3) \\ 2\left(\frac{2}{3}y + 3\right) - 3y &= 1 \\ \frac{4}{3}y + 6 - 3y &= 1 \\ 4y + 18 - 9y &= 3 \quad \textit{Multiply by 3.} \\ 18 - 5y &= 3 \\ -5y &= -15 \\ y &= 3 \end{aligned}$$

Since $x = \frac{2}{3}y + 3$ and $y = 3$,

$$x = \tfrac{2}{3}(3) + 3 = 2 + 3 = 5.$$

Check $(5, 3)$: $\frac{5}{3} - \frac{3}{2} = \frac{1}{6}$; $9 = 9$

Solution set: $\{(5, 3)\}$

(b) $\frac{x}{5} + \frac{2y}{3} = -\frac{8}{5}$ (1)
$3x - y = 9$ (2)

To eliminate the fractions in equation (1), multiply by 15. The new system is

$$\begin{aligned} 3x + 10y &= -24 \quad (3) \\ 3x - y &= 9. \quad (2) \end{aligned}$$

Solve equation (2) for y.

$$\begin{aligned} 3x - y &= 9 \quad (2) \\ -y &= -3x + 9 \\ y &= 3x - 9 \end{aligned}$$

Substitute $3x - 9$ for y in equation (3) and solve for x.

continued

$$\begin{aligned} 3x + 10y &= -24 \quad (3) \\ 3x + 10(3x - 9) &= -24 \\ 3x + 30x - 90 &= -24 \\ 33x - 90 &= -24 \\ 33x &= 66 \\ x &= 2 \end{aligned}$$

Since $y = 3x - 9$ and $x = 2$,

$$y = 3(2) - 9 = 6 - 9 = -3.$$

Check $(2, -3)$: $\frac{2}{5} - 2 = -\frac{8}{5}$; $9 = 9$

Solution set: $\{(2, -3)\}$

8. $$\begin{aligned} 2x + y &= 6 \quad (1) \\ -8x - 4y &= -24 \quad (2) \end{aligned}$$

Multiply equation (1) by 4, and add the result to equation (2).

$$\begin{array}{rcrl} 8x + 4y &=& 24 & 4 \times (1) \\ -8x - 4y &=& -24 & (2) \\ \hline 0 &=& 0 & \textit{True} \end{array}$$

The equations are dependent.

Solution set: $\{(x, y) \mid 2x + y = 6\}$

Equations (1) and (2) have the same graph.

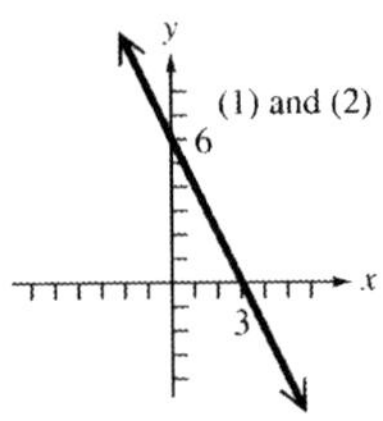

9. $$\begin{aligned} 4x - 3y &= 8 \quad (1) \\ 8x - 6y &= 14 \quad (2) \end{aligned}$$

Multiply equation (1) by -2 and add the result to equation (2).

$$\begin{array}{rcrl} -8x + 6y &=& -16 & -2 \times (1) \\ 8x - 6y &=& 14 & (2) \\ \hline 0 &=& -2 & \textit{False} \end{array}$$

The system is inconsistent.

Solution set: $\emptyset$

The graphs of the equations are parallel lines.

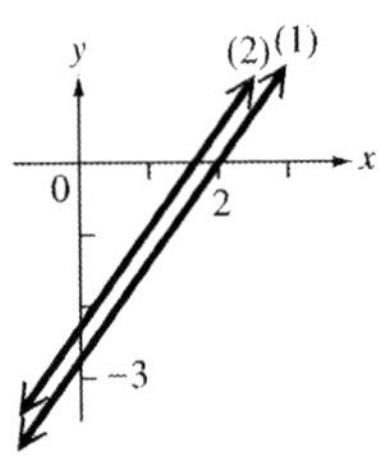

10. Since the equations are equivalent, solve one equation for y.

$$\begin{aligned} 2x - y &= 3 \quad (1) \\ -y &= -2x + 3 \\ y &= 2x - 3 \end{aligned}$$

Both equations are $f(x) = 2x - 3$.

11. $$\begin{aligned} x + 3y &= 4 \quad (1) \\ 3y &= -x + 4 \\ y &= -\tfrac{1}{3}x + \tfrac{4}{3} \\ f(x) &= -\tfrac{1}{3}x + \tfrac{4}{3} \end{aligned}$$

$$\begin{aligned} -2x - 6y &= 3 \quad (2) \\ -6y &= 2x + 3 \\ y &= -\tfrac{1}{3}x - \tfrac{1}{2} \\ f(x) &= -\tfrac{1}{3}x - \tfrac{1}{2} \end{aligned}$$

5.1 Section Exercises

1. If $(3, -6)$ is a solution of a linear system in two variables, then substituting <u>3</u> for x and <u>-6</u> for y leads to true statements in *both* equations.

2. A solution of a system of independent linear equations in two variables is an <u>ordered pair</u>.

3. If the solution process leads to a false statement such as $0 = 5$ when solving a system, the solution set is <u>$\emptyset$</u>.

4. If the solution process leads to a true statement such as $0 = 0$ when solving a system, the system has <u>dependent</u> equations.

5. If the two lines forming a system have the same slope and different y-intercepts, the system has <u>0</u> solutions. (The lines are parallel.)

6. If the two lines forming a system have different slopes, the system has <u>one</u> solution.

7. **D**; The ordered pair solution must be in quadrant IV, since that is where the graphs of the equations intersect.

8. **B**; The ordered pair solution must be on the x-axis, with $x < 0$, since that is where the graphs of the equations intersect.

9. **(a)** $x - y = 0$ implies that $x = y$, so the coordinates of the solution must be equal—this limits our choice to **B** or **C**. Since $x + y = 6$, graph **B** is correct.

   (b) As in part (a), we must have **B** or **C**. Since $x + y = -6$, graph **C** is correct.

   (c) $x + y = 0$ implies that $x = -y$, so the coordinates of the solution must be opposites—this limits our choice to **A** or **D**. Since $x - y = -6$, graph **A** is correct.

(d) As in part (c), we must have **A** or **D**. Since $x - y = 6$, graph **D** is correct.

10. $x + y = 4$
$2x - y = 2$
Graph the line $x + y = 4$ through its intercepts, $(4, 0)$ and $(0, 4)$, and the line $2x - y = 2$ through its intercepts, $(1, 0)$ and $(0, -2)$. The lines appear to intersect at $(2, 2)$.

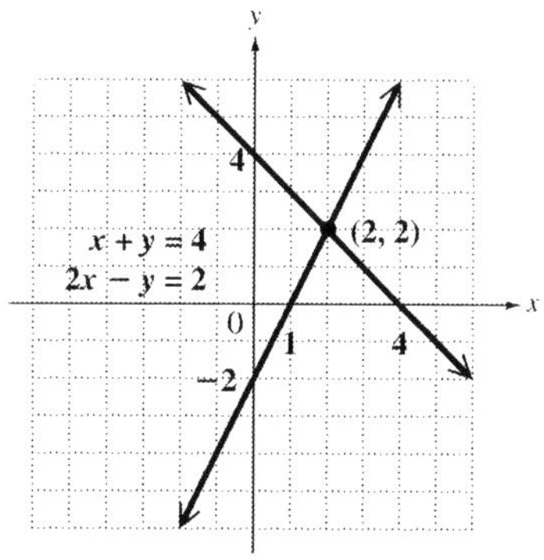

To check, substitute 2 for x and 2 for y in each equation of the system. Since $(2, 2)$ makes both equations true, the solution set of the system is $\{(2, 2)\}$.

11. $x + y = -5$
$-2x + y = 1$
Graph the line $x + y = -5$ through its intercepts, $(-5, 0)$ and $(0, -5)$, and the line $-2x + y = 1$ through its intercepts, $\left(-\frac{1}{2}, 0\right)$ and $(0, 1)$. The lines appear to intersect at $(-2, -3)$.

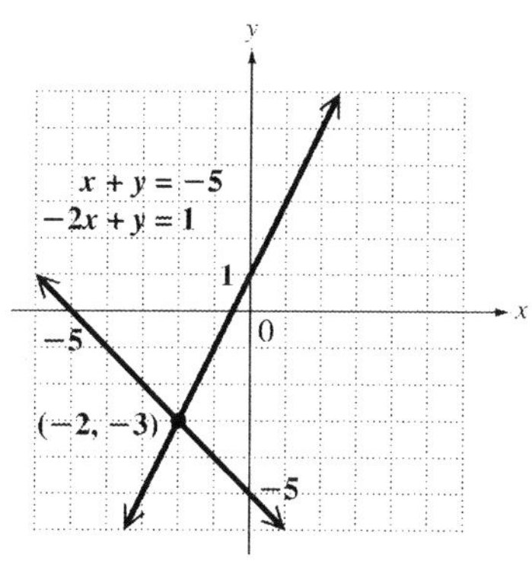

Check this ordered pair in the system.

$$(-2) + (-3) = -5 \quad ?$$
$$-5 = -5 \qquad \textit{True}$$

$$-2(-2) + (-3) = 1 \quad ?$$
$$4 + (-3) = 1 \qquad \textit{True}$$

Solution set: $\{(-2, -3)\}$

12. $x - 4y = -4$
$3x + y = 1$

Graph the line $x - 4y = -4$ through its intercepts, $(-4, 0)$ and $(0, 1)$, and the line $3x + y = 1$ through its intercepts, $\left(\frac{1}{3}, 0\right)$ and $(0, 1)$. The lines have the same y-intercept, $(0, 1)$, so $\{(0, 1)\}$ is the solution set.

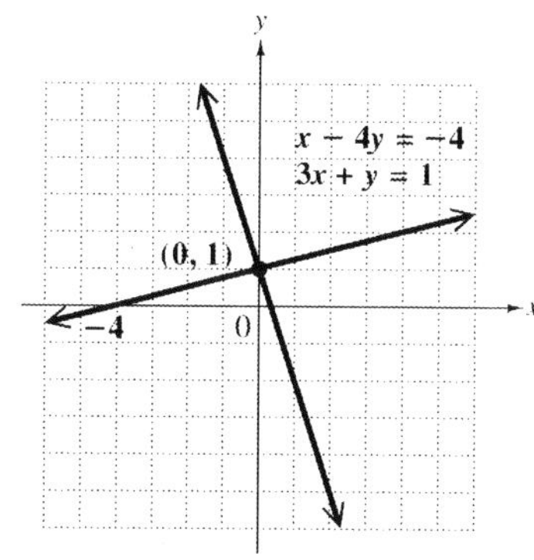

13. To determine whether $(5, 1)$ is a solution of the system

$$x + y = 6$$
$$x - y = 4,$$

replace x with 5 and y with 1 in each equation.

$$x + y = 6$$
$$5 + 1 = 6 \quad ?$$
$$6 = 6 \qquad \textit{True}$$

$$x - y = 4$$
$$5 - 1 = 4 \quad ?$$
$$4 = 4 \qquad \textit{True}$$

Since $(5, 1)$ makes both equations true, $(5, 1)$ is a solution of the system.

14. $x - y = 17$
$x + y = -1$

To decide if $(8, -9)$ is a solution, replace x with 8 and y with -9 in each equation of the system.

$$x - y = 17$$
$$8 - (-9) = 17 \quad ?$$
$$17 = 17 \qquad \textit{True}$$

$$x + y = -1$$
$$8 + (-9) = -1 \quad ?$$
$$-1 = -1 \qquad \textit{True}$$

Since $(8, -9)$ makes both equations true, $(8, -9)$ is a solution of the system.

15. To determine whether $(5, 2)$ is a solution of the system

$$\begin{aligned} 2x - y &= 8 \\ 3x + 2y &= 20, \end{aligned}$$

replace x with 5 and y with 2 in each equation.

$$\begin{aligned} 2x - y &= 8 \\ 2(5) - 2 &= 8 \quad ? \\ 10 - 2 &= 8 \qquad \textit{True} \end{aligned}$$

$$\begin{aligned} 3x + 2y &= 20 \\ 3(5) + 2(2) &= 20 \quad ? \\ 15 + 4 &= 20 \qquad \textit{False} \end{aligned}$$

The ordered pair $(5, 2)$ is *not* a solution of the system since it does not make both equations true.

16. $3x - 5y = -12$
$x - y = 1$
To decide if $(-1, 2)$ is a solution, replace x with -1 and y with 2 in each equation of the system.

$$\begin{aligned} 3(-1) - 5(2) &= -12 \quad ? \\ -3 - 10 &= -12 \quad ? \\ -13 &= -12 \qquad \textit{False} \end{aligned}$$

It is not necessary to check if $(-1, 2)$ satisfies the second equation since it does not satisfy the first equation. Therefore, $(-1, 2)$ is *not* a solution of the system.

17. $4x + y = 6 \quad (1)$
$y = 2x \quad (2)$
Since equation (2) is already solved for y, substitute $2x$ for y in equation (1).

$$\begin{aligned} 4x + y &= 6 \quad (1) \\ 4x + 2x &= 6 \quad \textit{Let } y = 2x. \\ 6x &= 6 \\ x &= 1 \end{aligned}$$

Substitute 1 for x in (2).

$$y = 2(1) = 2$$

The solution $(1, 2)$ checks.

Solution set: $\{(1, 2)\}$

18. $2x - y = 6 \quad (1)$
$y = 5x \quad (2)$
Equation (2) is already solved for y, so substitute $5x$ for y in equation (1).

$$\begin{aligned} 2x - 5x &= 6 \quad \textit{Let } y = 5x. \\ -3x &= 6 \\ x &= -2 \end{aligned}$$

Substitute -2 for x in (2).

$$y = 5(-2) = -10$$

The solution $(-2, -10)$ checks.

Solution set: $\{(-2, -10)\}$

19. $3x - 4y = -22 \quad (1)$
$-3x + y = 0 \quad (2)$
Solve equation (2) for y to get

$$y = 3x. \quad (3)$$

Substitute $3x$ for y in equation (1).

$$\begin{aligned} 3x - 4y &= -22 \quad (1) \\ 3x - 4(3x) &= -22 \quad \textit{Let } y = 3x. \\ 3x - 12x &= -22 \\ -9x &= -22 \\ x &= \tfrac{-22}{-9} = \tfrac{22}{9} \end{aligned}$$

Substitute $\frac{22}{9}$ for x in equation (3) to get

$$y = 3x = 3\left(\tfrac{22}{9}\right) = \tfrac{22}{3}.$$

The solution $\left(\frac{22}{9}, \frac{22}{3}\right)$ checks.

Solution set: $\left\{\left(\frac{22}{9}, \frac{22}{3}\right)\right\}$

20. $-3x + y = -5 \quad (1)$
$x + 2y = 0 \quad (2)$
Solve equation (1) for y.

$$y = -5 + 3x \quad (3)$$

Substitute $-5 + 3x$ for y in (2).

$$\begin{aligned} x + 2y &= 0 \quad (2) \\ x + 2(-5 + 3x) &= 0 \\ x - 10 + 6x &= 0 \\ 7x &= 10 \\ x &= \tfrac{10}{7} \end{aligned}$$

Substitute $\frac{10}{7}$ for x in (3).

$$\begin{aligned} y &= -5 + 3\left(\tfrac{10}{7}\right) \\ &= -5 + \tfrac{30}{7} = -\tfrac{35}{7} + \tfrac{30}{7} = -\tfrac{5}{7} \end{aligned}$$

The solution $\left(\frac{10}{7}, -\frac{5}{7}\right)$ checks.

Solution set: $\left\{\left(\frac{10}{7}, -\frac{5}{7}\right)\right\}$

21. $-x - 4y = -14 \quad (1)$
$2x = y + 1 \quad (2)$
Solve equation (2) for y.

$$\begin{aligned} 2x &= y + 1 \\ 2x - 1 &= y \quad (3) \end{aligned}$$

Substitute $2x - 1$ for y in equation (1) and solve for x.

$$\begin{aligned} -x - 4y &= -14 \quad (1) \\ -x - 4(2x - 1) &= -14 \\ -x - 8x + 4 &= -14 \\ -9x &= -18 \\ x &= 2 \end{aligned}$$

Substitute 2 for x in (3).

$$y = 2(2) - 1 = 3$$

The solution $(2, 3)$ checks.

Solution set: $\{(2, 3)\}$

22. $-3x - 5y = -17 \quad (1)$
$4x = y - 8 \quad (2)$
Solve equation (2) for y.

$$4x + 8 = y \qquad (3)$$

Substitute $4x + 8$ for y in (1).

$$\begin{aligned} -3x - 5(4x + 8) &= -17 \\ -3x - 20x - 40 &= -17 \\ -23x &= 23 \\ x &= -1 \end{aligned}$$

Substitute -1 for x in (3).

$$y = 4(-1) + 8 = 4$$

The solution $(-1, 4)$ checks.

Solution set: $\{(-1, 4)\}$

23. $5x - 4y = 9 \quad (1)$
$3 - 2y = -x \quad (2)$
Solve equation (2) for x.

$$\begin{aligned} 3 - 2y &= -x \qquad (2) \\ 2y - 3 &= x \qquad (3) \quad -1 \times (2) \end{aligned}$$

Substitute $2y - 3$ for x in equation (1).

$$\begin{aligned} 5x - 4y &= 9 \qquad (1) \\ 5(2y - 3) - 4y &= 9 \qquad \textit{Let } x = 2y - 3. \\ 10y - 15 - 4y &= 9 \\ 6y - 15 &= 9 \\ 6y &= 24 \\ y &= 4 \end{aligned}$$

Substitute 4 for y in equation (3).

$$x = 2y - 3 = 2(4) - 3 = 5$$

The solution $(5, 4)$ checks.

Solution set: $\{(5, 4)\}$

24. $6x - y = -9 \quad (1)$
$4 + 7x = -y \quad (2)$
Multiply (2) by -1, which solves the equation for y.

$$-4 - 7x = y \quad (3)$$

Substitute $-4 - 7x$ for y in equation (1).

$$\begin{aligned} 6x - (-4 - 7x) &= -9 \\ 6x + 4 + 7x &= -9 \\ 13x &= -13 \\ x &= -1 \end{aligned}$$

Substitute -1 for x in (3).

$$y = -4 - 7x = -4 - 7(-1) = -4 + 7 = 3$$

The solution $(-1, 3)$ checks.

Solution set: $\{(-1, 3)\}$

25. $x = 3y + 5 \quad (1)$
$x = \frac{3}{2}y \quad (2)$
Both equations are given in terms of x. Choose equation (2), and substitute $\frac{3}{2}y$ for x in equation (1).

$$\begin{aligned} x &= 3y + 5 \qquad (1) \\ \tfrac{3}{2}y &= 3y + 5 \qquad \textit{Let } x = \tfrac{3}{2}y. \\ 3y &= 6y + 10 \qquad \textit{Multiply by 2.} \\ -3y &= 10 \\ y &= -\tfrac{10}{3} \end{aligned}$$

Since $x = \frac{3}{2}y$ and $y = -\frac{10}{3}$,

$$x = \tfrac{3}{2}\left(-\tfrac{10}{3}\right) = -\tfrac{10}{2} = -5.$$

The solution $\left(-5, -\frac{10}{3}\right)$ checks.

Solution set: $\left\{\left(-5, -\frac{10}{3}\right)\right\}$

26. $x = 6y - 2 \quad (1)$
$x = \frac{3}{4}y \quad (2)$
Substitute $\frac{3}{4}y$ for x in (1).

$$\begin{aligned} \tfrac{3}{4}y &= 6y - 2 \\ 3y &= 24y - 8 \qquad \textit{Multiply by 4.} \\ 8 &= 21y \\ \tfrac{8}{21} &= y \end{aligned}$$

Substitute $\frac{8}{21}$ for y in (2).

$$x = \tfrac{3}{4}\left(\tfrac{8}{21}\right) = \tfrac{2}{7}$$

The solution $\left(\frac{2}{7}, \frac{8}{21}\right)$ checks.

Solution set: $\left\{\left(\frac{2}{7}, \frac{8}{21}\right)\right\}$

27. $\frac{1}{2}x + \frac{1}{3}y = 3 \quad (1)$

$y = 3x \quad (2)$

Multiply (1) by its LCD, 6, to eliminate fractions

$$3x + 2y = 18 \quad (3)$$

From (2), we have $y = 3x$, so substitute $3x$ for y in (3).

$$\begin{aligned} 3x + 2(3x) &= 18 \\ 3x + 6x &= 18 \\ 9x &= 18 \\ x &= 2 \end{aligned}$$

Substitute 2 for x in equation (2).

$$y = 3x = 3(2) = 6$$

The solution $(2, 6)$ checks.

Solution set: $\{(2, 6)\}$

28. $\frac{1}{4}x - \frac{1}{5}y = 9 \quad (1)$

$y = 5x \quad (2)$

Multiply (1) by its LCD, 20, to eliminate fractions.

$$5x - 4y = 180 \quad (3)$$

From (2), we have $y = 5x$, so substitute $5x$ for y in (3).

$$\begin{aligned} 5x - 4(5x) &= 180 \\ 5x - 20x &= 180 \\ -15x &= 180 \\ x &= -12 \end{aligned}$$

Substitute -12 for x in equation (2).

$$y = 5x = 5(-12) = -60$$

The solution $(-12, -60)$ checks.

Solution set: $\{(-12, -60)\}$

29. $2x - 5y = 11 \quad (1)$

$3x + y = 8 \quad (2)$

To eliminate y, multiply equation (2) by 5 and add the result to equation (1).

$$\begin{array}{rcrcrl} 2x & - & 5y & = & 11 & (1) \\ 15x & + & 5y & = & 40 & 5 \times (2) \\ \hline 17x & & & = & 51 & \\ & & x & = & 3 & \end{array}$$

To find y, substitute 3 for x in equation (2).

$$\begin{aligned} 3x + y &= 8 \quad (2) \\ 3(3) + y &= 8 \\ 9 + y &= 8 \\ y &= -1 \end{aligned}$$

The ordered pair $(3, -1)$ satisfies both equations, so it checks.

Solution set: $\{(3, -1)\}$

30. $-2x + 3y = 1 \quad (1)$

$-4x + y = -3 \quad (2)$

To eliminate y, multiply equation (2) by -3 and add the result to equation (1).

$$\begin{array}{rcrcrl} -2x & + & 3y & = & 1 & (1) \\ 12x & - & 3y & = & 9 & -3 \times (2) \\ \hline 10x & & & = & 10 & \\ & & x & = & 1 & \end{array}$$

Substitute 1 for x in equation (1) and solve for y.

$$\begin{aligned} -2(1) + 3y &= 1 \\ -2 + 3y &= 1 \\ 3y &= 3 \\ y &= 1 \end{aligned}$$

The ordered pair $(1, 1)$ satisfies both equations, so it checks.

Solution set: $\{(1, 1)\}$

31. $3x + 4y = -6 \quad (1)$

$5x + 3y = 1 \quad (2)$

To eliminate x, multiply equation (1) by 5 and equation (2) by -3. Then add the results.

$$\begin{array}{rcrcrl} 15x & + & 20y & = & -30 & 5 \times (1) \\ -15x & - & 9y & = & -3 & -3 \times (2) \\ \hline & & 11y & = & -33 & \\ & & y & = & -3 & \end{array}$$

To find x, substitute -3 for y in equation (2).

$$\begin{aligned} 5x + 3y &= 1 \quad (2) \\ 5x + 3(-3) &= 1 \\ 5x - 9 &= 1 \\ 5x &= 10 \\ x &= 2 \end{aligned}$$

The ordered pair $(2, -3)$ satisfies both equations, so it checks.

Solution set: $\{(2, -3)\}$

32. $4x + 3y = 1 \quad (1)$
$3x + 2y = 2 \quad (2)$

$$\begin{array}{rcrl} -8x - 6y & = & -2 & -2 \times (1) \\ 9x + 6y & = & 6 & 3 \times (2) \\ \hline x & = & 4 & \end{array}$$

Substitute 4 for x in (1).

$$\begin{aligned} 4(4) + 3y &= 1 \\ 16 + 3y &= 1 \\ 3y &= -15 \\ y &= -5 \end{aligned}$$

The solution $(4, -5)$ checks.

Solution set: $\{(4, -5)\}$

33. $3x + 3y = 0 \quad (1)$
$4x + 2y = 3 \quad (2)$

To eliminate y, multiply equation (1) by 2 and equation (2) by -3. Then add the results.

$$\begin{array}{rcrl} 6x + 6y & = & 0 & 2 \times (1) \\ -12x - 6y & = & -9 & -3 \times (2) \\ \hline -6x & = & -9 & \\ x & = & \frac{-9}{-6} = \frac{3}{2} & \end{array}$$

To find y, substitute $\frac{3}{2}$ for x in equation (1).

$$\begin{aligned} 3x + 3y &= 0 \quad (1) \\ 3\left(\tfrac{3}{2}\right) + 3y &= 0 \\ \tfrac{9}{2} + 3y &= 0 \\ 3y &= -\tfrac{9}{2} \\ y &= \tfrac{1}{3}\left(-\tfrac{9}{2}\right) = -\tfrac{3}{2} \end{aligned}$$

The solution $\left(\frac{3}{2}, -\frac{3}{2}\right)$ checks.

Solution set: $\left\{\left(\frac{3}{2}, -\frac{3}{2}\right)\right\}$

When you get a solution that has non-integer components, it is sometimes more difficult to check the problem than it was to solve it. A graphing calculator can be very helpful in this case. Just store the values for x and y in their respective memory locations, and then type the expressions as shown in the following screen. The results 0 and 3 (the right sides of the equations) indicate that we have found the correct solution.

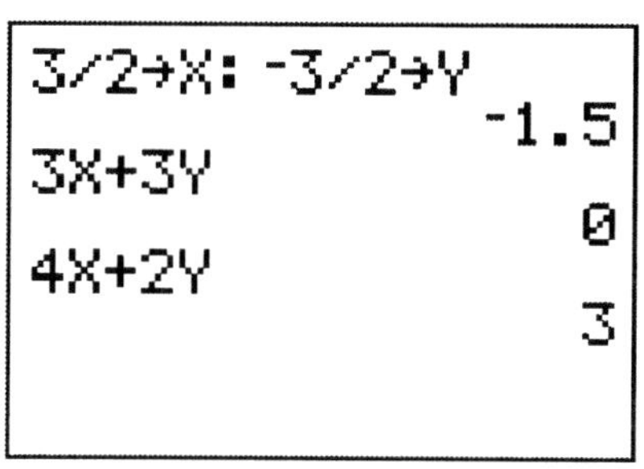

34. $8x + 4y = 0 \quad (1)$
$4x - 2y = 2 \quad (2)$

$$\begin{array}{rcrl} 8x + 4y & = & 0 & (1) \\ 8x - 4y & = & 4 & 2 \times (2) \\ \hline 16x & = & 4 & \\ x & = & \frac{1}{4} & \end{array}$$

Substitute $\frac{1}{4}$ for x in (1).

$$\begin{aligned} 8\left(\tfrac{1}{4}\right) + 4y &= 0 \\ 2 + 4y &= 0 \\ 4y &= -2 \\ y &= -\tfrac{1}{2} \end{aligned}$$

The solution $\left(\frac{1}{4}, -\frac{1}{2}\right)$ checks.

Solution set: $\left\{\left(\frac{1}{4}, -\frac{1}{2}\right)\right\}$

35. $7x + 2y = 6 \quad (1)$
$-14x - 4y = -12 \quad (2)$

To eliminate y, multiply equation (1) by 2 and add the result to equation (2).

$$\begin{array}{rcrl} 14x + 4y & = & 12 & 2 \times (1) \\ -14x - 4y & = & -12 & (2) \\ \hline 0 & = & 0 & \textit{True} \end{array}$$

Multiplying equation (1) by -2 gives equation (2). The equations are dependent, and the solution is the set of all points on the line.

Solution set: $\{(x, y) \mid 7x + 2y = 6\}$

36. $x - 4y = 2 \quad (1)$
$4x - 16y = 8 \quad (2)$

$$\begin{array}{rcrl} -4x + 16y & = & -8 & -4 \times (1) \\ 4x - 16y & = & 8 & (2) \\ \hline 0 & = & 0 & \textit{True} \end{array}$$

The equations are equivalent and have the same line for a graph, so the equations are dependent.

Solution set: $\{(x, y) \mid x - 4y = 2\}$

37. $\dfrac{x}{2} + \dfrac{y}{3} = -\dfrac{1}{3} \quad (1)$
$\dfrac{x}{2} + 2y = -7 \quad (2)$

Eliminate the fractions by multiplying equation (1) by -6 and equation (2) by 6. Then add the results to eliminate x.

$$\begin{array}{rcrl} -3x - 2y & = & 2 & (3) \quad -6 \times (1) \\ 3x + 12y & = & -42 & 6 \times (2) \\ \hline 10y & = & -40 & \\ y & = & -4 & \end{array}$$

continued

To find x, substitute -4 for y in equation (3).

$$\begin{aligned} -3x - 2y &= 2 \quad (3) \\ -3x - 2(-4) &= 2 \\ -3x + 8 &= 2 \\ -3x &= -6 \\ x &= 2 \end{aligned}$$

The solution $(2, -4)$ checks.

Solution set: $\{(2, -4)\}$

38. $\dfrac{x}{5} + y = \dfrac{6}{5}$ (1)

$\dfrac{x}{10} + \dfrac{y}{3} = \dfrac{5}{6}$ (2)

Eliminate the fractions by multiplying each equation by its LCD.

$$\begin{aligned} x + 5y &= 6 \quad (3) \quad 5 \times (1) \\ 3x + 10y &= 25 \quad (4) \quad 30 \times (2) \end{aligned}$$

$$\begin{aligned} -2x - 10y &= -12 \quad -2 \times (3) \\ 3x + 10y &= 25 \quad (4) \\ \hline x &= 13 \end{aligned}$$

Substitute 13 for x in (1).

$$\begin{aligned} \tfrac{13}{5} + y &= \tfrac{6}{5} \\ y &= -\tfrac{7}{5} \end{aligned}$$

The solution $\left(13, -\frac{7}{5}\right)$ checks.

Solution set: $\left\{\left(13, -\frac{7}{5}\right)\right\}$

39. $5x - 5y = 3$ (1)

$x - y = 12$ (2)

To eliminate x, multiply equation (2) by -5 and add the result to equation (1).

$$\begin{aligned} 5x - 5y &= 3 \quad (1) \\ -5x + 5y &= -60 \quad -5 \times (2) \\ \hline 0 &= -57 \quad \textit{False} \end{aligned}$$

The system is *inconsistent*. Since the graphs of the equations are parallel lines, there are no ordered pairs that satisfy both equations.

Solution set: $\emptyset$

40. $2x - 3y = 7$ (1)

$-4x + 6y = 14$ (2)

$$\begin{aligned} 4x - 6y &= 14 \quad 2 \times (1) \\ -4x + 6y &= 14 \quad (2) \\ \hline 0 &= 28 \quad \textit{False} \end{aligned}$$

The system is *inconsistent*. Since the graphs of the equations are parallel lines, there are no ordered pairs that satisfy both equations.

Solution set: $\emptyset$

41. $3x + 7y = 4$ (1)

$6x + 14y = 3$ (2)

Write each equation in slope-intercept form by solving for y.

$$\begin{aligned} 3x + 7y &= 4 \quad (1) \\ 7y &= -3x + 4 \\ y &= -\tfrac{3}{7}x + \tfrac{4}{7} \end{aligned}$$

$$\begin{aligned} 6x + 14y &= 3 \quad (2) \\ 14y &= -6x + 3 \\ y &= -\tfrac{6}{14}x + \tfrac{3}{14} \\ y &= -\tfrac{3}{7}x + \tfrac{3}{14} \end{aligned}$$

Since the equations have the same slope, $-\frac{3}{7}$, but different y-intercepts, $\frac{4}{7}$ and $\frac{3}{14}$, the lines when graphed are parallel. The system is inconsistent and has no solution.

42. $-x + 2y = 8$ in slope-intercept form is

$$\begin{aligned} 2y &= x + 8 \\ y &= \tfrac{1}{2}x + 4. \end{aligned}$$

$4x - 8y = 1$ in slope-intercept form is

$$\begin{aligned} -8y &= -4x + 1 \\ y &= \tfrac{1}{2}x - \tfrac{1}{8}. \end{aligned}$$

The graphs of the equations are parallel lines, so the system has no solution.

43. $2x = -3y + 1$ (1)

$6x = -9y + 3$ (2)

Write each equation in slope-intercept form by solving for y.

$$\begin{aligned} 2x &= -3y + 1 \quad (1) \\ 3y &= -2x + 1 \\ y &= -\tfrac{2}{3}x + \tfrac{1}{3} \end{aligned}$$

$$\begin{aligned} 6x &= -9y + 3 \quad (2) \\ 9y &= -6x + 3 \\ y &= -\tfrac{6}{9}x + \tfrac{3}{9} \\ y &= -\tfrac{2}{3}x + \tfrac{1}{3} \end{aligned}$$

Since both equations are the same, the solution set is all points on the line $y = -\frac{2}{3}x + \frac{1}{3}$. The system has infinitely many solutions.

44. $5x = -2y + 1$ in slope-intercept form is

$$\begin{aligned} 2y &= -5x + 1 \\ y &= -\tfrac{5}{2}x + \tfrac{1}{2}. \end{aligned}$$

$10x = -4y + 2$ in slope-intercept form is

$$\begin{aligned} 4y &= -10x + 2 \\ y &= -\tfrac{5}{2}x + \tfrac{1}{2}. \end{aligned}$$

The graphs of the equations are the same line, so the system has infinitely many solutions.

45. **(a)** $6x - y = 5$ (1)
$y = 11x$ (2)

Use substitution since the second equation is solved for y.

(b) $3x + y = -7$ (1)
$x - y = -5$ (2)

Use elimination since the coefficients of the y-terms are opposites.

(c) $3x - 2y = 0$ (1)
$9x + 8y = 7$ (2)

Use elimination since the equations are in standard form with no coefficients of 1 and -1. Solving by substitution would involve fractions.

46. $6x - y = 5$ (1)
$y = 11x$ (2)

Substitute $11x$ for y in equation (1).

$$\begin{aligned} 6x - 11x &= 5 \\ -5x &= 5 \\ x &= -1 \end{aligned}$$

Substitute -1 for x in equation (2).

$$y = 11(-1) = -11$$

The solution $(-1, -11)$ checks.

Solution set: $\{(-1, -11)\}$

47. $3x + y = -7$ (1)
$x - y = -5$ (2)

Add the equations to eliminate y.

$$\begin{array}{rcrcr} 3x & + & y & = & -7 \\ x & - & y & = & -5 \\ \hline 4x & & & = & -12 \\ & & x & = & -3 \end{array}$$

Substitute -3 for x in (2).

$$\begin{aligned} -3 - y &= -5 \\ -y &= -2 \\ y &= 2 \end{aligned}$$

The solution $(-3, 2)$ checks.

Solution set: $\{(-3, 2)\}$

48. $3x - 2y = 0$ (1)
$9x + 8y = 7$ (2)

Multiply equation (1) by 4 and add the result to equation (2).

$$\begin{array}{rcrcrl} 12x & - & 8y & = & 0 & 4 \times (1) \\ 9x & + & 8y & = & 7 & (2) \\ \hline 21x & & & = & 7 & \\ & & x & = & \frac{1}{3} & \end{array}$$

Substitute $\frac{1}{3}$ for x in equation (1).

$$\begin{aligned} 3\left(\tfrac{1}{3}\right) - 2y &= 0 \\ 1 - 2y &= 0 \\ -2y &= -1 \\ y &= \tfrac{1}{2} \end{aligned}$$

Check $\left(\frac{1}{3}, \frac{1}{2}\right)$: $1 - 1 = 0$; $3 + 4 = 7$

Solution set: $\left\{\left(\frac{1}{3}, \frac{1}{2}\right)\right\}$

49. $2x + 3y = 10$ (1)
$-3x + y = 18$ (2)

Solve equation (2) for y.

$$y = 3x + 18 \quad (3)$$

Substitute $3x + 18$ for y in (1).

$$\begin{aligned} 2x + 3(3x + 18) &= 10 \\ 2x + 9x + 54 &= 10 \\ 11x + 54 &= 10 \\ 11x &= -44 \\ x &= -4 \end{aligned}$$

Substitute -4 for x in (3).

$$y = 3(-4) + 18 = 6$$

The solution $(-4, 6)$ checks.

Solution set: $\{(-4, 6)\}$

50. $3x - 5y = 7$ (1)
$2x + 3y = 30$ (2)

Multiply equation (1) by 3 and equation (2) by 5. Then add.

$$\begin{array}{rcrcrl} 9x & - & 15y & = & 21 & 3 \times (1) \\ 10x & + & 15y & = & 150 & 5 \times (2) \\ \hline 19x & & & = & 171 & \textit{Add.} \\ & & x & = & 9 & \textit{Divide by 19.} \end{array}$$

Substitute 9 for x in equation (2).

$$\begin{aligned} 2(9) + 3y &= 30 \\ 18 + 3y &= 30 \\ 3y &= 12 \\ y &= 4 \end{aligned}$$

The solution $(9, 4)$ checks.

Solution set: $\{(9, 4)\}$

51. $\frac{1}{2}x - \frac{1}{8}y = -\frac{1}{4}$ (1)
$-4x + y = 2$ (2)

Multiply equation (1) by 8 and add to equation (2).

$$\begin{array}{rcrl} 4x - y & = & -2 & (3) \quad 8 \times (1) \\ -4x + y & = & 2 & (2) \\ \hline 0 & = & 0 & \end{array}$$

The equations are dependent.

Solution set: $\{(x, y) \mid 4x - y = -2\}$

52. $\frac{1}{6}x + \frac{1}{3}y = 8$ (1)

$\frac{1}{4}x + \frac{1}{2}y = 12$ (2)

Multiply equation (1) by 6 and equation (2) by -4. Then add.

$$\begin{array}{rcrl} x + 2y & = & 48 & (3) \quad 6 \times (1) \\ -x - 2y & = & -48 & \quad -4 \times (2) \\ \hline 0 & = & 0 & \quad \textit{True} \end{array}$$

The equations are dependent.

Solution set: $\{(x, y) \mid x + 2y = 48\}$

53. $.3x + .2y = .4$ (1)
$.5x + .4y = .7$ (2)

To eliminate y, multiply equation (1) by -20 and equation (2) by 10, then add.

$$\begin{array}{rcrl} -6x - 4y & = & -8 & \quad -20 \times (1) \\ 5x + 4y & = & 7 & (3) \quad 10 \times (2) \\ \hline -x & = & -1 & \\ x & = & 1 & \end{array}$$

Substitute 1 for x in equation (3).

$$5(1) + 4y = 7$$
$$4y = 2$$
$$y = \tfrac{1}{2}$$

Check $(1, \frac{1}{2})$: $.3 + .1 = .4$; $.5 + .2 = .7$

Solution set: $\{(1, \frac{1}{2})\}$

54. $.2x + .5y = 6$ (1)
$.4x + y = 9$ (2)

To eliminate x, multiply equation (1) by 20 and equation (2) by -10. Then add.

$$\begin{array}{rcrl} 4x + 10y & = & 120 & \quad 20 \times (1) \\ -4x - 10y & = & -90 & \quad -10 \times (2) \\ \hline 0 & = & 30 & \quad \textit{False} \end{array}$$

The system is inconsistent.

Solution set: $\emptyset$

55. **(a)** The supply and demand graphs intersect at 4, so supply equals demand at a price of $4 per half-gallon.

(b) At a price of $4 per half-gallon, the supply and demand are both about 300 half-gallons.

(c) At a price of $2 per half-gallon, the supply is 200 half-gallons and the demand is 400 half-gallons.

56. **(a)** The monthly payment for the fixed-rate mortgage is about $650, which is more than the monthly payment for the variable-rate mortgage for years 0 to 6.

(b) The graphs intersect at year 6, so the monthly payments would be the same in year 6, and they are about $650.

57. **(a)** Rising graphs indicate population growth, so Houston, Phoenix, and Dallas will experience population growth.

(b) Philadelphia's graph indicates that it will experience population decline.

(c) In the year 2000, the city populations from least to greatest are Dallas, Phoenix, Philadelphia, and Houston.

(d) The graphs for Dallas and Philadelphia intersect in the year 2010. The population for each city will be about 1.45 million.

(e) The graphs for Houston and Phoenix appear to intersect in the year 2025 with about 2.8 million. This can be represented by the ordered pair (2025, 2.8).

58. **(a)** The graphs of cassette and CD intersect at about (1991, 350). Therefore, in 1991 they both reached the level of about 350 million.

(b) The graphs intersect at (1987, 100 million).

(c) The graph is level (horizontal) between 1988 and 1990, so that is when cassette production remained fairly constant.

(d) CD production generally increased during these years, since the line is increasing from left to right. Therefore, the slope would be positive.

(e) The slope would be negative. Cassette production was generally decreasing during this time.

(f) It could possibly be that downloading from the Internet caused this decrease.

59. The graphs intersect at about 3.5 (years since 2000). So the sales of digital cameras were less than the sales of conventional cameras for 2000, 2001, 2002, and the first half of 2003.

60. The graphs intersect in 2003 at about \$10.5 million.

61.
$$\begin{aligned} 2.5x + y &= 19.4 \quad (1) \\ -1.7x + y &= 4.4 \quad (2) \end{aligned}$$

Solve equation (2) for y.

$$y = 1.7x + 4.4 \quad (3)$$

Substitute $1.7x + 4.4$ for y in equation (1).

$$\begin{aligned} 2.5x + (1.7x + 4.4) &= 19.4 \\ 4.2x + 4.4 &= 19.4 \\ 4.2x &= 15 \\ x &= \tfrac{15}{4.2} = \tfrac{150}{42} = \tfrac{25}{7} \end{aligned}$$

Substitute $\frac{25}{7}$ for x in (3).

$$y = 1.7\left(\tfrac{25}{7}\right) + 4.4 = \tfrac{17}{10}\left(\tfrac{25}{7}\right) + \tfrac{44}{10} \cdot \tfrac{7}{7} = \tfrac{733}{70}$$

Written as an ordered pair, the solution is $\left(\frac{25}{7}, \frac{733}{70} \text{ (million)}\right)$.

62. $\dfrac{25}{7} \approx 3.57$ and $\dfrac{733}{70} \approx 10.47$

During the third year (2003), the sales for both types of cameras were about \$10.5 million.

63.
$$\begin{aligned} 3x + y &= 6 \quad (1) \\ -2x + 3y &= 7 \quad (2) \end{aligned}$$

Multiply equation (1) by -3 and add the result to equation (2).

$$\begin{array}{rcrcrl} -9x & - & 3y & = & -18 & (3) \quad -3 \times (1) \\ -2x & + & 3y & = & 7 & (2) \\ \hline -11x & & & = & -11 & \\ & & x & = & 1 & \end{array}$$

To find y, substitute 1 for x in equation (1).

$$\begin{aligned} 3x + y &= 6 \quad (1) \\ 3(1) + y &= 6 \\ y &= 3 \end{aligned}$$

The solution $(1, 3)$ checks.

Solution set: $\{(1, 3)\}$

64.
$$\begin{aligned} 3x + y &= 6 \\ y &= -3x + 6 \end{aligned}$$

Replace y with $f(x)$.

$$f(x) = -3x + 6$$

Since f is in the form $f(x) = mx + b$, it is a linear function.

65.
$$\begin{aligned} -2x + 3y &= 7 \\ 3y &= 2x + 7 \\ y &= \tfrac{2}{3}x + \tfrac{7}{3} \end{aligned}$$

Replace y with $g(x)$.

$$g(x) = \tfrac{2}{3}x + \tfrac{7}{3}$$

Since g is in the form $g(x) = mx + b$, it is a linear function.

66. Because the graphs of f and g are straight lines that are neither parallel nor coincide, they intersect in exactly <u>one</u> point. The coordinates of the point are (<u>1</u>, <u>3</u>). Using functional notation, this is given by $f(\underline{1}) = \underline{3}$ and $g(\underline{1}) = \underline{3}$.

5.2 Systems of Linear Equations in Three Variables

5.2 Margin Exercises

1. (a)
$$x + 7y - 3z = -14 \quad (2)$$

Let $x = -3$, $y = 1$, $z = 6$.

$$\begin{aligned} -3 + 7(1) - 3(6) &= -14 \quad ? \\ -3 + 7 - 18 &= -14 \quad ? \\ -14 &= -14 \quad \textit{True} \end{aligned}$$

Yes, the solution satisfies equation (2).

(b)
$$2x - 3y + 2z = 3 \quad (3)$$

Let $x = -3$, $y = 1$, $z = 6$.

$$\begin{aligned} 2(-3) - 3(1) + 2(6) &= 3 \quad ? \\ -6 - 3 + 12 &= 3 \quad ? \\ 3 &= 3 \quad \textit{True} \end{aligned}$$

Yes, the solution satisfies equation (3).

2. (a)
$$\begin{aligned} x + y + z &= 2 \quad (1) \\ x - y + 2z &= 2 \quad (2) \\ -x + 2y - z &= 1 \quad (3) \end{aligned}$$

Step 1
Eliminate y by adding equations (1) and (2).

$$\begin{array}{rcrcrcrl} x & + & y & + & z & = & 2 & (1) \\ x & - & y & + & 2z & = & 2 & (2) \\ \hline 2x & & & + & 3z & = & 4 & (4) \end{array}$$

Step 2
To eliminate y again, multiply equation (2) by 2 and add the result to equation (3).

$$\begin{array}{rcrcrcrl} 2x & - & 2y & + & 4z & = & 4 & 2 \times (2) \\ -x & + & 2y & - & z & = & 1 & (3) \\ \hline x & & & + & 3z & = & 5 & (5) \end{array}$$

continued

Step 3
Use equations (4) and (5) to eliminate z. Multiply equation (5) by -1 and add the result to equation (4).

$$\begin{array}{rcrl} -x - 3z &=& -5 & \quad -1 \times (5) \\ 2x + 3z &=& 4 & \quad (4) \\ \hline x &=& -1 & \end{array}$$

Step 4
Substitute -1 for x in equation (5) to find z.

$$\begin{aligned} x + 3z &= 5 && (5) \\ -1 + 3z &= 5 && \textit{Let } x = -1. \\ 3z &= 6 \\ z &= 2 \end{aligned}$$

Step 5
Substitute -1 for x and 2 for z in equation (1) to find y.

$$\begin{aligned} x + y + z &= 2 && (1) \\ -1 + y + 2 &= 2 && \textit{Let } x = -1,\ z = 2. \\ y + 1 &= 2 \\ y &= 1 \end{aligned}$$

Step 6
Check $(-1, 1, 2)$

Equation (1): $-1 + 1 + 2 = 2$ *True*
Equation (2): $-1 - 1 + 4 = 2$ *True*
Equation (3): $1 + 2 - 2 = 1$ *True*

Solution set: $\{(-1, 1, 2)\}$

(b) $2x + y + z = 9$ (1)
$-x - y + z = 1$ (2)
$3x - y + z = 9$ (3)

Eliminate y by adding equations (1) and (2).

$$\begin{array}{rcrl} 2x + y + z &=& 9 & (1) \\ -x - y + z &=& 1 & (2) \\ \hline x + 2z &=& 10 & (4) \end{array}$$

To eliminate y again, add equations (1) and (3).

$$\begin{array}{rcrl} 2x + y + z &=& 9 & (1) \\ 3x - y + z &=& 9 & (3) \\ \hline 5x + 2z &=& 18 & (5) \end{array}$$

Multiply equation (4) by -1 and add to equation (5) to eliminate z.

$$\begin{array}{rcrl} -x - 2z &=& -10 & \quad -1 \times (4) \\ 5x + 2z &=& 18 & \quad (5) \\ \hline 4x &=& 8 & \\ x &=& 2 & \end{array}$$

Substitute 2 for x in equation (4) to find z.

$$\begin{aligned} x + 2z &= 10 && (4) \\ 2 + 2z &= 10 && \textit{Let } x = 2. \\ 2z &= 8 \\ z &= 4 \end{aligned}$$

Substitute 2 for x and 4 for z in equation (1) to find y.

$$\begin{aligned} 2x + y + z &= 9 && (1) \\ 2(2) + y + 4 &= 9 && \textit{Let } x = 2,\ z = 4. \\ 4 + y + 4 &= 9 \\ 8 + y &= 9 \\ y &= 1 \end{aligned}$$

Check $(2, 1, 4)$

Equation (1): $4 + 1 + 4 = 9$ *True*
Equation (2): $-2 - 1 + 4 = 1$ *True*
Equation (3): $6 - 1 + 4 = 9$ *True*

Solution set: $\{(2, 1, 4)\}$

3. **(a)** $x - y = 6$ (1)
$2y + 5z = 1$ (2)
$3x - 4z = 8$ (3)

Since equation (3) is missing y, eliminate y again from equations (1) and (2). Multiply equation (1) by 2 and add the result to equation (2).

$$\begin{array}{rcrl} 2x - 2y &=& 12 & \quad 2 \times (1) \\ 2y + 5z &=& 1 & \quad (2) \\ \hline 2x + 5z &=& 13 & \quad (4) \end{array}$$

Use equation (4) together with equation (3) to eliminate x. Multiply equation (4) by 3 and equation (3) by -2. Then add the results.

$$\begin{array}{rcrl} 6x + 15z &=& 39 & \quad 3 \times (4) \\ -6x + 8z &=& -16 & \quad -2 \times (3) \\ \hline 23z &=& 23 & \\ z &=& 1 & \end{array}$$

Substitute 1 for z in equation (2) to find y.

$$\begin{aligned} 2y + 5z &= 1 && (2) \\ 2y + 5(1) &= 1 && \textit{Let } z = 1. \\ 2y + 5 &= 1 \\ 2y &= -4 \\ y &= -2 \end{aligned}$$

Substitute -2 for y in equation (1) to find x.

$$\begin{aligned} x - y &= 6 && (1) \\ x - (-2) &= 6 && \textit{Let } y = -2. \\ x + 2 &= 6 \\ x &= 4 \end{aligned}$$

Check $(4, -2, 1)$

Equation (1): $4 + 2 = 6$ *True*
Equation (2): $-4 + 5 = 1$ *True*
Equation (3): $12 - 4 = 8$ *True*

Solution set: $\{(4, -2, 1)\}$

(b)
$$\begin{aligned} 5x - y &= 26 \quad (1) \\ 4y + 3z &= -4 \quad (2) \\ x + z &= 5 \quad (3) \end{aligned}$$

Since equation (1) is missing z, eliminate z again from equations (2) and (3). Multiply equation (3) by -3 and add to equation (2).

$$\begin{array}{rrrrl} & 4y & + 3z & = -4 & (2) \\ -3x & & - 3z & = -15 & -3 \times (3) \\ \hline -3x & + 4y & & = -19 & (4) \end{array}$$

Multiply equation (1) by 4 and add to equation (4).

$$\begin{array}{rrrl} 20x & - 4y & = 104 & 4 \times (1) \\ -3x & + 4y & = -19 & (4) \\ \hline 17x & & = 85 & \\ x & & = 5 & \end{array}$$

Substitute 5 for x in equation (1) to find y.

$$\begin{aligned} 5x - y &= 26 \quad (1) \\ 5(5) - y &= 26 \quad \textit{Let } x = 5. \\ 25 - y &= 26 \\ y &= -1 \end{aligned}$$

Substitute 5 for x in equation (3) to find z.

$$\begin{aligned} x + z &= 5 \quad (3) \\ 5 + z &= 5 \quad \textit{Let } x = 5. \\ z &= 0 \end{aligned}$$

Check $(5, -1, 0)$

Equation (1): $25 + 1 = 26$ *True*
Equation (2): $-4 + 0 = -4$ *True*
Equation (3): $5 + 0 = 5$ *True*

Solution set: $\{(5, -1, 0)\}$

4. **(a)**
$$\begin{aligned} 3x - 5y + 2z &= 1 \quad (1) \\ 5x + 8y - z &= 4 \quad (2) \\ -6x + 10y - 4z &= 5 \quad (3) \end{aligned}$$

Multiply equation (1) by 2 and add the result to equation (3).

$$\begin{array}{rrrrl} 6x & - 10y & + 4z & = 2 & 2 \times (1) \\ -6x & + 10y & - 4z & = 5 & (3) \\ \hline & & 0 & = 7 & \textit{False} \end{array}$$

Since a false statement results, the system is inconsistent.

Solution set: $\emptyset$

(b)
$$\begin{aligned} 7x - 9y + 2z &= 0 \quad (1) \\ y + z &= 0 \quad (2) \\ 8x - z &= 0 \quad (3) \end{aligned}$$

Since equation (3) is missing y, eliminate y again from equations (1) and (2). Multiply equation (2) by 9 and add to equation (1).

$$\begin{array}{rrrrl} 7x & - 9y & + 2z & = 0 & (1) \\ & 9y & + 9z & = 0 & 9 \times (2) \\ \hline 7x & & + 11z & = 0 & (4) \end{array}$$

Use equations (3) and (4) to eliminate z. Multiply equation (3) by 11 and add to equation (4).

$$\begin{array}{rrrl} 88x & - 11z & = 0 & 11 \times (3) \\ 7x & + 11z & = 0 & (4) \\ \hline 95x & & = 0 & \\ x & & = 0 & \end{array}$$

Substitute 0 for x in equation (3) to find z.

$$\begin{aligned} 8x - z &= 0 \quad (3) \\ 8(0) - z &= 0 \quad \textit{Let } x = 0. \\ -z &= 0 \\ z &= 0 \end{aligned}$$

Substitute 0 for z in equation (2) to find y.

$$\begin{aligned} y + z &= 0 \quad (2) \\ y + 0 &= 0 \quad \textit{Let } z = 0. \\ y &= 0 \end{aligned}$$

Check The check is trivial since all the terms are equal to 0.

Solution set: $\{(0, 0, 0)\}$

5.
$$\begin{aligned} x - y + z &= 4 \quad (1) \\ -3x + 3y - 3z &= -12 \quad (2) \\ 2x - 2y + 2z &= 8 \quad (3) \end{aligned}$$

Since equation (2) is -3 times equation (1) and equation (3) is 2 times equation (1), the three equations are dependent. All three have the same graph.

Solution set: $\{(x, y, z) \mid x - y + z = 4\}$

5.2 Section Exercises

1. The statement means that when -1 is substituted for x, 2 is substituted for y, and 3 is substituted for z in the three equations, the resulting three statements are true.

2. Substitute 1 for x, 2 for y, and 3 for z in $3x + 2y - z$, which is the left side of each equation.

$$\begin{aligned} 3(1) + 2(2) - (3) &= 3 + 4 - 3 \\ &= 7 - 3 \\ &= 4 \end{aligned}$$

Choice **B** is correct since its right side is 4.

3.
$$\begin{aligned} 2x - 5y + 3z &= -1 \quad (1)\\ x + 4y - 2z &= 9 \quad (2)\\ x - 2y - 4z &= -5 \quad (3)\end{aligned}$$

Eliminate x by adding equation (1) to -2 times equation (2).

$$\begin{array}{rcll} 2x - 5y + 3z &=& -1 & (1)\\ -2x - 8y + 4z &=& -18 & -2\times(2)\\ \hline -13y + 7z &=& -19 & (4)\end{array}$$

Now eliminate x by adding equation (2) to -1 times equation (3).

$$\begin{array}{rcll} x + 4y - 2z &=& 9 & (2)\\ -x + 2y + 4z &=& 5 & -1\times(3)\\ \hline 6y + 2z &=& 14 & (5)\end{array}$$

Use equations (4) and (5) to eliminate z. Multiply equation (4) by -2 and add the result to 7 times equation (5).

$$\begin{array}{rcll} 26y - 14z &=& 38 & -2\times(4)\\ 42y + 14z &=& 98 & 7\times(5)\\ \hline 68y &=& 136 & \\ y &=& 2 & \end{array}$$

Substitute 2 for y in equation (5) to find z.

$$\begin{aligned} 6y + 2z &= 14 \quad (5)\\ 6(2) + 2z &= 14\\ 12 + 2z &= 14\\ 2z &= 2\\ z &= 1\end{aligned}$$

Substitute 2 for y and 1 for z in equation (3) to find x.

$$\begin{aligned} x - 2y - 4z &= -5 \quad (3)\\ x - 2(2) - 4(1) &= -5\\ x - 4 - 4 &= -5\\ x - 8 &= -5\\ x &= 3\end{aligned}$$

The solution $(3, 2, 1)$ checks in all three of the original equations.

Solution set: $\{(3, 2, 1)\}$

4.
$$\begin{aligned} x + 3y - 6z &= 7 \quad (1)\\ 2x - y + z &= 1 \quad (2)\\ x + 2y + 2z &= -1 \quad (3)\end{aligned}$$

Eliminate x by adding -2 times equation (1) to equation (2).

$$\begin{array}{rcll} -2x - 6y + 12z &=& -14 & -2\times(1)\\ 2x - y + z &=& 1 & (2)\\ \hline -7y + 13z &=& -13 & (4)\end{array}$$

Now eliminate x by adding equation (2) to -2 times equation (3).

$$\begin{array}{rcll} 2x - y + z &=& 1 & (2)\\ -2x - 4y - 4z &=& 2 & -2\times(3)\\ \hline -5y - 3z &=& 3 & (5)\end{array}$$

Use equations (4) and (5) to eliminate z. Multiply equation (4) by 3 and add the result to 13 times equation (5).

$$\begin{array}{rcll} -21y + 39z &=& -39 & 3\times(4)\\ -65y - 39z &=& 39 & 13\times(5)\\ \hline -86y &=& 0 & \\ y &=& 0 & \end{array}$$

Substituting 0 for y in equation (4) gives us $13z = -13$, so $z = -1$. Substitute 0 for y and -1 for z in equation (3) to find x.

$$\begin{aligned} x + 2y + 2z &= -1 \quad (3)\\ x + 2(0) + 2(-1) &= -1\\ x - 2 &= -1\\ x &= 1\end{aligned}$$

The solution $(1, 0, -1)$ checks in all three of the original equations.

Solution set: $\{(1, 0, -1)\}$

5.
$$\begin{aligned} 3x + 2y + z &= 8 \quad (1)\\ 2x - 3y + 2z &= -16 \quad (2)\\ x + 4y - z &= 20 \quad (3)\end{aligned}$$

Eliminate z by adding equations (1) and (3).

$$\begin{array}{rcll} 3x + 2y + z &=& 8 & (1)\\ x + 4y - z &=& 20 & (3)\\ \hline 4x + 6y &=& 28 & (4)\end{array}$$

To get another equation without z, multiply equation (3) by 2 and add the result to equation (2).

$$\begin{array}{rcll} 2x + 8y - 2z &=& 40 & 2\times(3)\\ 2x - 3y + 2z &=& -16 & (2)\\ \hline 4x + 5y &=& 24 & (5)\end{array}$$

Use equations (4) and (5) to eliminate x. Multiply equation (4) by -1 and add the result to equation (5).

$$\begin{array}{rcll} -4x - 6y &=& -28 & -1\times(4)\\ 4x + 5y &=& 24 & (5)\\ \hline -y &=& -4 & \\ y &=& 4 & \end{array}$$

Substitute 4 for y in equation (5) to find x.

$$\begin{aligned} 4x + 5y &= 24 \quad (5)\\ 4x + 5(4) &= 24\\ 4x + 20 &= 24\\ 4x &= 4\\ x &= 1\end{aligned}$$

Substitute 1 for x and 4 for y in equation (3) to find z.

$$\begin{aligned} x + 4y - z &= 20 \quad (3) \\ 1 + 4(4) - z &= 20 \\ 1 + 16 - z &= 20 \\ 17 - z &= 20 \\ -z &= 3 \\ z &= -3 \end{aligned}$$

The solution $(1, 4, -3)$ checks in all three of the original equations.

Solution set: $\{(1, 4, -3)\}$

6.
$$\begin{aligned} -3x + y - z &= -10 \quad (1) \\ -4x + 2y + 3z &= -1 \quad (2) \\ 2x + 3y - 2z &= -5 \quad (3) \end{aligned}$$

Multiply (1) by 3 and add to (2).

$$\begin{aligned} -9x + 3y - 3z &= -30 \\ -4x + 2y + 3z &= -1 \quad (2) \\ \hline -13x + 5y &= -31 \quad (4) \end{aligned}$$

Multiply (1) by -2 and add to (3).

$$\begin{aligned} 6x - 2y + 2z &= 20 \\ 2x + 3y - 2z &= -5 \quad (3) \\ \hline 8x + y &= 15 \quad (5) \end{aligned}$$

Multiply (5) by -5 and add to (4).

$$\begin{aligned} -40x - 5y &= -75 \\ -13x + 5y &= -31 \quad (4) \\ \hline -53x &= -106 \\ x &= 2 \end{aligned}$$

Substitute 2 for x in (4).

$$\begin{aligned} -13(2) + 5y &= -31 \\ -26 + 5y &= -31 \\ 5y &= -5 \\ y &= -1 \end{aligned}$$

Substitute 2 for x and -1 for y in (1).

$$\begin{aligned} -3(2) + (-1) - z &= -10 \\ -7 - z &= -10 \\ -z &= -3 \\ z &= 3 \end{aligned}$$

The solution $(2, -1, 3)$ checks in all three of the original equations.

Solution set: $\{(2, -1, 3)\}$

7.
$$\begin{aligned} -x + 2y + 6z &= 2 \quad (1) \\ 3x + 2y + 6z &= 6 \quad (2) \\ x + 4y - 3z &= 1 \quad (3) \end{aligned}$$

Eliminate y and z by adding equation (1) to -1 times equation (2).

$$\begin{aligned} -x + 2y + 6z &= 2 \quad (1) \\ -3x - 2y - 6z &= -6 \qquad -1 \times (2) \\ \hline -4x &= -4 \\ x &= 1 \end{aligned}$$

Eliminate z by adding equation (2) to 2 times equation (3).

$$\begin{aligned} 3x + 2y + 6z &= 6 \quad (2) \\ 2x + 8y - 6z &= 2 \qquad 2 \times (3) \\ \hline 5x + 10y &= 8 \quad (4) \end{aligned}$$

Substitute 1 for x in equation (4).

$$\begin{aligned} 5x + 10y &= 8 \quad (4) \\ 5(1) + 10y &= 8 \\ 10y &= 3 \\ y &= \tfrac{3}{10} \end{aligned}$$

Substitute 1 for x and $\frac{3}{10}$ for y in equation (1).

$$\begin{aligned} -x + 2y + 6z &= 2 \quad (1) \\ -1 + 2\left(\tfrac{3}{10}\right) + 6z &= 2 \\ \tfrac{3}{5} + 6z &= 3 \\ 6z &= \tfrac{12}{5} \\ z &= \tfrac{2}{5} \end{aligned}$$

Solution set: $\left\{\left(1, \frac{3}{10}, \frac{2}{5}\right)\right\}$

8.
$$\begin{aligned} 2x + y + 2z &= 1 \quad (1) \\ x + 2y + z &= 2 \quad (2) \\ x - y - z &= 0 \quad (3) \end{aligned}$$

Add equations (2) and (3) to eliminate z.

$$\begin{aligned} x + 2y + z &= 2 \quad (2) \\ x - y - z &= 0 \quad (3) \\ \hline 2x + y &= 2 \quad (4) \end{aligned}$$

Multiply equation (2) by -2 and add the result to equation (1) to eliminate z again.

$$\begin{aligned} 2x + y + 2z &= 1 \quad (1) \\ -2x - 4y - 2z &= -4 \qquad -2 \times (2) \\ \hline -3y &= -3 \\ y &= 1 \end{aligned}$$

Substitute 1 for y in equation (4) and solve for x.

$$\begin{aligned} 2x + 1 &= 2 \\ 2x &= 1 \\ x &= \tfrac{1}{2} \end{aligned}$$

Substitute $\frac{1}{2}$ for x and 1 for y in equation (2) and solve for z.

$$\begin{aligned} \tfrac{1}{2} + 2(1) + z &= 2 \\ z &= -\tfrac{1}{2} \end{aligned}$$

Solution set: $\left\{\left(\frac{1}{2}, 1, -\frac{1}{2}\right)\right\}$

9.
$$\begin{aligned} 2x + 5y + 2z &= 0 \quad (1)\\ 4x - 7y - 3z &= 1 \quad (2)\\ 3x - 8y - 2z &= -6 \quad (3) \end{aligned}$$

Add equations (1) and (3) to eliminate z.

$$\begin{array}{rcll} 2x + 5y + 2z &=& 0 & (1)\\ 3x - 8y - 2z &=& -6 & (3)\\ \hline 5x - 3y &=& -6 & (4) \end{array}$$

Multiply equation (1) by 3 and equation (2) by 2. Then add the results to eliminate z again.

$$\begin{array}{rcll} 6x + 15y + 6z &=& 0 & 3 \times (1)\\ 8x - 14y - 6z &=& 2 & 2 \times (2)\\ \hline 14x + y &=& 2 & (5) \end{array}$$

Solve the system

$$\begin{aligned} 5x - 3y &= -6 \quad (4)\\ 14x + y &= 2. \quad (5) \end{aligned}$$

Multiply equation (5) by 3 then add this result to (4).

$$\begin{array}{rcll} 5x - 3y &=& -6 & (4)\\ 42x + 3y &=& 6 & 3 \times (5)\\ \hline 47x &=& 0 &\\ x &=& 0 & \end{array}$$

To find y, substitute $x = 0$ into equation (4).

$$\begin{aligned} 5x - 3y &= -6 \quad (4)\\ 5(0) - 3y &= -6\\ y &= 2 \end{aligned}$$

To find z, substitute $x = 0$ and $y = 2$ in equation (1).

$$\begin{aligned} 2x + 5y + 2z &= 0 \quad (1)\\ 2(0) + 5(2) + 2z &= 0\\ 10 + 2z &= 0\\ 2z &= -10\\ z &= -5 \end{aligned}$$

Solution set: $\{(0, 2, -5)\}$

10.
$$\begin{aligned} 5x - 2y + 3z &= -9 \quad (1)\\ 4x + 3y + 5z &= 4 \quad (2)\\ 2x + 4y - 2z &= 14 \quad (3) \end{aligned}$$

Multiply (1) by 3 and (2) by 2; then add.

$$\begin{array}{rcll} 15x - 6y + 9z &=& -27 &\\ 8x + 6y + 10z &=& 8 &\\ \hline 23x + 19z &=& -19 & (4) \end{array}$$

Multiply (1) by 2 and add to (3).

$$\begin{array}{rcll} 10x - 4y + 6z &=& -18 &\\ 2x + 4y - 2z &=& 14 & (3)\\ \hline 12x + 4z &=& -4 & (5) \end{array}$$

Multiply (4) by 4 and (5) by -19; then add.

$$\begin{array}{rcl} 92x + 76z &=& -76\\ -228x - 76z &=& 76\\ \hline -136x &=& 0\\ x &=& 0 \end{array}$$

Substitute 0 for x in equation (4) and solve for z.

$$\begin{aligned} 23(0) + 19z &= -19\\ 0 + 19z &= -19\\ z &= -1 \end{aligned}$$

Substitute 0 for x and -1 for z in (1).

$$\begin{aligned} 5(0) - 2y + 3(-1) &= -9\\ -2y - 3 &= -9\\ -2y &= -6\\ y &= 3 \end{aligned}$$

Solution set: $\{(0, 3, -1)\}$

11.
$$\begin{aligned} x + y - z &= -2 \quad (1)\\ 2x - y + z &= -5 \quad (2)\\ -x + 2y - 3z &= -4 \quad (3) \end{aligned}$$

Eliminate y and z by adding equations (1) and (2).

$$\begin{array}{rcll} x + y - z &=& -2 & (1)\\ 2x - y + z &=& -5 & (2)\\ \hline 3x &=& -7 &\\ x &=& -\frac{7}{3} & \end{array}$$

To get another equation without y, multiply equation (2) by 2 and add the result to equation (3).

$$\begin{array}{rcll} 4x - 2y + 2z &=& -10 & 2 \times (2)\\ -x + 2y - 3z &=& -4 & (3)\\ \hline 3x - z &=& -14 & (4) \end{array}$$

Substitute $-\frac{7}{3}$ for x in equation (4) to find z.

$$\begin{aligned} 3x - z &= -14 \quad (4)\\ 3\left(-\tfrac{7}{3}\right) - z &= -14\\ -7 - z &= -14\\ -z &= -7\\ z &= 7 \end{aligned}$$

Substitute $-\frac{7}{3}$ for x and 7 for z in equation (1) to find y.

$$\begin{aligned} x + y - z &= -2 \quad (1)\\ -\tfrac{7}{3} + y - 7 &= -2\\ -7 + 3y - 21 &= -6 \quad \textit{Multiply by 3.}\\ 3y - 28 &= -6\\ 3y &= 22\\ y &= \tfrac{22}{3} \end{aligned}$$

Solution set: $\left\{\left(-\frac{7}{3}, \frac{22}{3}, 7\right)\right\}$

A calculator check reduces the probability of making any arithmetic errors and is highly recommended. The following screen shows the substitution of the solution for x, y, and z along with the left sides of the three original equations. The evaluation of the three expressions, -2, -5, and -4 (the right sides of the three equations), indicates that we have found the correct solution.

```
-7/3→X:22/3→Y:7→
Z:X+Y-Z
                -2
2X-Y+Z
                -5
-X+2Y-3Z
                -4
```

12.
$$\begin{aligned} x + 2y + 3z &= 1 \quad (1)\\ -x - y + 3z &= 2 \quad (2)\\ -6x + y + z &= -2 \quad (3)\end{aligned}$$

Add (1) and (2).

$$y + 6z = 3 \quad (4)$$

Multiply (1) by 6 and add to (3).

$$\begin{aligned} 6x + 12y + 18z &= 6\\ -6x + y + z &= -2 \quad (3)\\ \hline 13y + 19z &= 4 \quad (5)\end{aligned}$$

Multiply (4) by -13 and add to (5).

$$\begin{aligned} -13y - 78z &= -39\\ 13y + 19z &= 4 \quad (5)\\ \hline -59z &= -35\\ z &= \tfrac{35}{59}\end{aligned}$$

Substitute $\frac{35}{59}$ for z in (4).

$$\begin{aligned} y + 6\left(\tfrac{35}{59}\right) &= 3\\ y + \tfrac{210}{59} &= 3\\ y &= \tfrac{177}{59} - \tfrac{210}{59}\\ y &= -\tfrac{33}{59}\end{aligned}$$

Substitute $-\frac{33}{59}$ for y and $\frac{35}{59}$ for z in (1).

$$\begin{aligned} x + 2\left(-\tfrac{33}{59}\right) + 3\left(\tfrac{35}{59}\right) &= 1\\ x - \tfrac{66}{59} + \tfrac{105}{59} &= 1\\ x &= \tfrac{59}{59} - \tfrac{39}{59}\\ x &= \tfrac{20}{59}\end{aligned}$$

Solution set: $\left\{\left(\frac{20}{59}, -\frac{33}{59}, \frac{35}{59}\right)\right\}$

13.
$$\begin{aligned} 2x - 3y + 2z &= -1 \quad (1)\\ x + 2y + z &= 17 \quad (2)\\ 2y - z &= 7 \quad (3)\end{aligned}$$

Multiply equation (2) by -2, and add the result to equation (1).

$$\begin{aligned} 2x - 3y + 2z &= -1 \quad (1)\\ -2x - 4y - 2z &= -34 \quad -2 \times (2)\\ \hline -7y &= -35\\ y &= 5\end{aligned}$$

To find z, substitute 5 for y in equation (3).

$$\begin{aligned} 2y - z &= 7 \quad (3)\\ 2(5) - z &= 7\\ 10 - z &= 7\\ -z &= -3\\ z &= 3\end{aligned}$$

To find x, substitute $y = 5$ and $z = 3$ into equation (1).

$$\begin{aligned} 2x - 3y + 2z &= -1 \quad (1)\\ 2x - 3(5) + 2(3) &= -1\\ 2x - 9 &= -1\\ 2x &= 8\\ x &= 4\end{aligned}$$

Solution set: $\{(4, 5, 3)\}$

14.
$$\begin{aligned} 2x - y + 3z &= 6 \quad (1)\\ x + 2y - z &= 8 \quad (2)\\ 2y + z &= 1 \quad (3)\end{aligned}$$

Multiply (2) by 3 and add to (1).

$$\begin{aligned} 3x + 6y - 3z &= 24\\ 2x - y + 3z &= 6 \quad (1)\\ \hline 5x + 5y &= 30 \quad (4)\end{aligned}$$

Add (2) and (3).

$$x + 4y = 9 \quad (5)$$

Multiply (5) by -5 and add to (4).

$$\begin{aligned} -5x - 20y &= -45\\ 5x + 5y &= 30 \quad (4)\\ \hline -15y &= -15\\ y &= 1\end{aligned}$$

Substitute 1 for y in equation (5).

$$\begin{aligned} x + 4(1) &= 9\\ x &= 5\end{aligned}$$

Substitute 5 for x and 1 for y in (1).

$$\begin{aligned} 2(5) - 1 + 3z &= 6\\ 10 - 1 + 3z &= 6\\ 3z &= -3\\ z &= -1\end{aligned}$$

Solution set: $\{(5, 1, -1)\}$

15.
$$\begin{aligned} 4x + 2y - 3z &= 6 \quad (1) \\ x - 4y + z &= -4 \quad (2) \\ -x + 2z &= 2 \quad (3) \end{aligned}$$

Equation (3) is missing y. Eliminate y again by multiplying equation (1) by 2 and adding the result to equation (2).

$$\begin{aligned} 8x + 4y - 6z &= 12 \quad 2 \times (1) \\ x - 4y + z &= -4 \quad (2) \\ \hline 9x - 5z &= 8 \quad (4) \end{aligned}$$

Use equations (3) and (4) to eliminate x. Multiply equation (3) by 9 and add the result to equation (4).

$$\begin{aligned} -9x + 18z &= 18 \quad 9 \times (3) \\ 9x - 5z &= 8 \quad (4) \\ \hline 13z &= 26 \\ z &= 2 \end{aligned}$$

Substitute 2 for z in equation (3) to find x.

$$\begin{aligned} -x + 2z &= 2 \quad (3) \\ -x + 2(2) &= 2 \\ -x + 4 &= 2 \\ -x &= -2 \\ x &= 2 \end{aligned}$$

Substitute 2 for x and 2 for z in equation (2) to find y.

$$\begin{aligned} x - 4y + z &= -4 \quad (2) \\ 2 - 4y + 2 &= -4 \\ -4y + 4 &= -4 \\ -4y &= -8 \\ y &= 2 \end{aligned}$$

Solution set: $\{(2, 2, 2)\}$

16.
$$\begin{aligned} 2x + 3y - 4z &= 4 \quad (1) \\ x - 6y + z &= -16 \quad (2) \\ -x + 3z &= 8 \quad (3) \end{aligned}$$

Since (3) has no y-term, eliminate y from (1) and (2). Multiply (1) by 2.

$$\begin{aligned} 4x + 6y - 8z &= 8 \quad 2 \times (1) \\ x - 6y + z &= -16 \quad (2) \\ \hline 5x - 7z &= -8 \quad (4) \end{aligned}$$

Multiply (3) by 5 and add to (4).

$$\begin{aligned} -5x + 15z &= 40 \quad 5 \times (3) \\ 5x - 7z &= -8 \quad (4) \\ \hline 8z &= 32 \\ z &= 4 \end{aligned}$$

Substitute 4 for z in (3).

$$\begin{aligned} -x + 3(4) &= 8 \\ -x + 12 &= 8 \\ -x &= -4 \\ x &= 4 \end{aligned}$$

Substitute 4 for x and 4 for z in (1).

$$\begin{aligned} 2(4) + 3y - 4(4) &= 4 \\ 8 + 3y - 16 &= 4 \\ 3y &= 12 \\ y &= 4 \end{aligned}$$

Solution set: $\{(4, 4, 4)\}$

17.
$$\begin{aligned} 2x + y &= 6 \quad (1) \\ 3y - 2z &= -4 \quad (2) \\ 3x - 5z &= -7 \quad (3) \end{aligned}$$

To eliminate y, multiply equation (1) by -3 and add the result to equation (2).

$$\begin{aligned} -6x - 3y &= -18 \quad -3 \times (1) \\ 3y - 2z &= -4 \quad (2) \\ \hline -6x - 2z &= -22 \quad (4) \end{aligned}$$

Since equation (3) does not have a y-term, we can multiply equation (3) by 2 and add the result to equation (4) to eliminate x and solve for z.

$$\begin{aligned} 6x - 10z &= -14 \quad 2 \times (3) \\ -6x - 2z &= -22 \quad (4) \\ \hline -12z &= -36 \\ z &= 3 \end{aligned}$$

To find x, substitute 3 for z into equation (3).

$$\begin{aligned} 3x - 5z &= -7 \quad (3) \\ 3x - 5(3) &= -7 \\ 3x &= 8 \\ x &= \tfrac{8}{3} \end{aligned}$$

To find y, substitute 3 for z into equation (2).

$$\begin{aligned} 3y - 2z &= -4 \quad (2) \\ 3y - 2(3) &= -4 \\ 3y &= 2 \\ y &= \tfrac{2}{3} \end{aligned}$$

Solution set: $\{(\frac{8}{3}, \frac{2}{3}, 3)\}$

18.
$$\begin{aligned} 4x - 8y &= -7 \quad (1) \\ 4y + z &= 7 \quad (2) \\ -8x + z &= -4 \quad (3) \end{aligned}$$

Multiply (2) by -1.

$$\begin{aligned} -4y - z &= -7 \quad -1 \times (2) \\ -8x + z &= -4 \quad (3) \\ \hline -8x - 4y &= -11 \quad (4) \end{aligned}$$

Multiply (1) by 2.

$$\begin{array}{rcrl} 8x - 16y &=& -14 & 2\times(1) \\ -8x - 4y &=& -11 & (4) \\ \hline -20y &=& -25 & \\ y &=& \frac{5}{4} & \end{array}$$

Substitute $\frac{5}{4}$ for y in (1).

$$\begin{aligned} 4x - 8\left(\tfrac{5}{4}\right) &= -7 \\ 4x - 10 &= -7 \\ 4x &= 3 \\ x &= \tfrac{3}{4} \end{aligned}$$

Substitute $\frac{5}{4}$ for y in (2).

$$\begin{aligned} 4\left(\tfrac{5}{4}\right) + z &= 7 \\ 5 + z &= 7 \\ z &= 2 \end{aligned}$$

Solution set: $\left\{\left(\frac{3}{4}, \frac{5}{4}, 2\right)\right\}$

19. Answers will vary.

(a) One example is two perpendicular walls and the ceiling (or floor) of a room, which meet at one point.

(b) One example is two opposite walls and the floor, which have no points in common. Another is the floors of three different levels of an office building.

(c) One example is the plane through the ceiling of a house, the plane through an outside wall, and the plane through one side of a slanted roof, which all meet in one line; therefore, they intersect in infinitely many points. Another example is three pages of this book, which intersect in the spine.

20. Answers will vary. For example, choose values for x and y such as $x = 1$ and $y = 1$. Substitute these values in the equation to find z.

$$\begin{aligned} x + y + 2z &= 1 \\ 1 + 1 + 2z &= 1 \\ 2z &= -1 \\ z &= -\tfrac{1}{2} \end{aligned}$$

So one ordered triple $\left(1, 1, -\frac{1}{2}\right)$. Using the same method, two other possibilities are $\left(0, 0, \frac{1}{2}\right)$ and $(2, 5, -3)$. The simplest triples are obtained by letting two of the variables equal 0; that is, $\left(0, 0, \frac{1}{2}\right)$, $(1, 0, 0)$, and $(0, 1, 0)$.

21.

$$\begin{array}{rcrl} 2x + 2y - 6z &=& 5 & (1) \\ -3x + y - z &=& -2 & (2) \\ -x - y + 3z &=& 4 & (3) \end{array}$$

Multiply equation (3) by 2 and add the result to equation (1).

$$\begin{array}{rcrl} 2x + 2y - 6z &=& 5 & (1) \\ -2x - 2y + 6z &=& 8 & 2\times(3) \\ \hline 0 &=& 13 & \textit{False} \end{array}$$

Solution set: $\emptyset$; inconsistent system

22.

$$\begin{array}{rcrl} -2x + 5y + z &=& -3 & (1) \\ 5x + 14y - z &=& -11 & (2) \\ 7x + 9y - 2z &=& -5 & (3) \end{array}$$

Add (1) and (2).

$$3x + 19y = -14 \quad (4)$$

Multiply (1) by 2.

$$\begin{array}{rcrl} -4x + 10y + 2z &=& -6 & 2\times(1) \\ 7x + 9y - 2z &=& -5 & (3) \\ \hline 3x + 19y &=& -11 & (5) \end{array}$$

Multiply (4) by -1 and add to (5).

$$\begin{array}{rcrl} -3x - 19y &=& 14 & -1\times(4) \\ 3x + 19y &=& -11 & (5) \\ \hline 0 &=& 3 & \textit{False} \end{array}$$

Solution set: $\emptyset$; inconsistent system

23.

$$\begin{array}{rcrl} -5x + 5y - 20z &=& -40 & (1) \\ x - y + 4z &=& 8 & (2) \\ 3x - 3y + 12z &=& 24 & (3) \end{array}$$

Dividing equation (1) by -5 gives equation (2). Dividing equation (3) by 3 also gives equation (2). The resulting equations are the same, so the three equations are dependent.

Solution set: $\{(x, y, z) \mid x - y + 4z = 8\}$

24.

$$\begin{array}{rcrl} x + 4y - z &=& 3 & (1) \\ -2x - 8y + 2z &=& -6 & (2) \\ 3x + 12y - 3z &=& 9 & (3) \end{array}$$

Dividing equation (3) by 3 gives equation (1). Dividing equation (2) by -2 also gives equation (1). The resulting equations are the same, so the three equations are dependent.

Solution set: $\{(x, y, z) \mid x + 4y - z = 3\}$

25.

$$\begin{array}{rcrl} 2x + y - z &=& 6 & (1) \\ 4x + 2y - 2z &=& 12 & (2) \\ -x - \frac{1}{2}y + \frac{1}{2}z &=& -3 & (3) \end{array}$$

Multiplying equation (1) by 2 gives equation (2). Multiplying equation (3) by -4 also gives equation (2). The resulting equations are the same, so the three equations are dependent.

Solution set: $\{(x, y, z) \mid 2x + y - z = 6\}$

26.
$$\begin{aligned} 2x - 8y + 2z &= -10 && (1)\\ -x + 4y - z &= 5 && (2)\\ \frac{1}{8}x - \frac{1}{2}y + \frac{1}{8}z &= -\frac{5}{8} && (3) \end{aligned}$$

Multiplying equation (2) by -2 gives equation (1). Multiplying equation (3) by 16 also gives equation (1). The resulting equations are the same, so the three equations are dependent.

Solution set: $\{(x, y, z) \mid -x + 4y - z = 5\}$

Note that multiplying equation (2) by -1 gives us $x - 4y + z = -5$.

27.
$$\begin{aligned} x + y - 2z &= 0 && (1)\\ 3x - y + z &= 0 && (2)\\ 4x + 2y - z &= 0 && (3) \end{aligned}$$

Eliminate z by adding equations (2) and (3).

$$\begin{aligned} 3x - y + z &= 0 && (2)\\ 4x + 2y - z &= 0 && (3)\\ \hline 7x + y &= 0 && (4) \end{aligned}$$

To get another equation without z, multiply equation (2) by 2 and add the result to equation (1).

$$\begin{aligned} 6x - 2y + 2z &= 0 && 2 \times (2)\\ x + y - 2z &= 0 && (1)\\ \hline 7x - y &= 0 && (5) \end{aligned}$$

Add equations (4) and (5) to find x.

$$\begin{aligned} 7x + y &= 0 && (4)\\ 7x - y &= 0 && (5)\\ \hline 14x &= 0\\ x &= 0 \end{aligned}$$

Substitute 0 for x in equation (4) to find y.

$$\begin{aligned} 7x + y &= 0 \quad (4)\\ 7(0) + y &= 0\\ 0 + y &= 0\\ y &= 0 \end{aligned}$$

Substitute 0 for x and 0 for y in equation (1) to find z.

$$\begin{aligned} x + y - 2z &= 0 \quad (1)\\ 0 + 0 - 2z &= 0\\ -2z &= 0\\ z &= 0 \end{aligned}$$

Solution set: $\{(0, 0, 0)\}$

28.
$$\begin{aligned} 2x + 3y - z &= 0 && (1)\\ x - 4y + 2z &= 0 && (2)\\ 3x - 5y - z &= 0 && (3) \end{aligned}$$

Multiply (1) by 2 and add to (2).

$$\begin{aligned} 4x + 6y - 2z &= 0 && 2 \times (1)\\ x - 4y + 2z &= 0 && (2)\\ \hline 5x + 2y &= 0 && (4) \end{aligned}$$

Multiply (3) by 2 and add to (2).

$$\begin{aligned} 6x - 10y - 2z &= 0 && 2 \times (3)\\ x - 4y + 2z &= 0 && (2)\\ \hline 7x - 14y &= 0 && (5) \end{aligned}$$

Multiply (4) by 7 and add to (5).

$$\begin{aligned} 35x + 14y &= 0 && 7 \times (4)\\ 7x - 14y &= 0 && (5)\\ \hline 42x &= 0\\ x &= 0 \end{aligned}$$

Substitute 0 for x in equation (4) and solve for y.

$$\begin{aligned} 5(0) + 2y &= 0\\ 2y &= 0\\ y &= 0 \end{aligned}$$

Substitute 0 for x and 0 for y in equation (1). Then solve for z.

$$\begin{aligned} 2(0) + 3(0) - z &= 0\\ -z &= 0\\ z &= 0 \end{aligned}$$

Solution set: $\{(0, 0, 0)\}$

29.
$$\begin{aligned} x + 5y - 2z &= -1 && (1)\\ -2x + 8y + z &= -4 && (2)\\ 3x - y + 5z &= 19 && (3) \end{aligned}$$

Eliminate x by adding (2) to 2 times (1).

$$\begin{aligned} 2x + 10y - 4z &= -2 && 2 \times (1)\\ -2x + 8y + z &= -4 && (2)\\ \hline 18y - 3z &= -6 && (4) \end{aligned}$$

Eliminate x by adding (3) to -3 times (1).

$$\begin{aligned} -3x - 15y + 6z &= 3 && -3 \times (1)\\ 3x - y + 5z &= 19 && (3)\\ \hline -16y + 11z &= 22 && (5) \end{aligned}$$

To eliminate z from equations (4) and (5), we could first divide equation (4) by 3 and then multiply the resulting equation by 11, or we could simply multiply equation (4) by $\frac{11}{3}$ and add it to equation (5).

$$\begin{aligned} 66y - 11z &= -22 && \tfrac{11}{3} \times (4)\\ -16y + 11z &= 22 && (5)\\ \hline 50y &= 0\\ y &= 0 \end{aligned}$$

Substitute 0 for y in equation (5).

$$\begin{aligned} -16y + 11z &= 22 \quad (5) \\ -16(0) + 11z &= 22 \\ 11z &= 22 \\ z &= 2 \end{aligned}$$

Substitute 0 for y and 2 for z in equation (1).

$$\begin{aligned} x + 5y - 2z &= -1 \quad (1) \\ x + 5(0) - 2(2) &= -1 \\ x - 4 &= -1 \\ x &= 3 \end{aligned}$$

Solution set: $\{(3, 0, 2)\}$

30. $$\begin{aligned} x + 3y + z &= 2 \quad (1) \\ 4x + y + 2z &= -4 \quad (2) \\ 5x + 2y + 3z &= -2 \quad (3) \end{aligned}$$

Eliminate x by adding (2) to -4 times (1).

$$\begin{array}{rrrrl} -4x & - 12y & - 4z & = -8 & -4 \times (1) \\ 4x & + y & + 2z & = -4 & (2) \\ \hline & -11y & - 2z & = -12 & (4) \end{array}$$

Eliminate x by adding (3) to -5 times (1).

$$\begin{array}{rrrrl} -5x & - 15y & - 5z & = -10 & -5 \times (1) \\ 5x & + 2y & + 3z & = -2 & (3) \\ \hline & -13y & - 2z & = -12 & (5) \end{array}$$

Eliminate z from (4) and (5).

$$\begin{array}{rrrl} -11y & - 2z & = -12 & (4) \\ 13y & + 2z & = 12 & -1 \times (5) \\ \hline 2y & & = 0 & \\ y & & = 0 & \end{array}$$

Substitute 0 for y in equation (4).

$$\begin{aligned} -11y - 2z &= -12 \quad (4) \\ -11(0) - 2z &= -12 \\ -2z &= -12 \\ z &= 6 \end{aligned}$$

Substitute 0 for y and 6 for z in equation (1).

$$\begin{aligned} x + 3y + z &= 2 \quad (1) \\ x + 3(0) + 6 &= 2 \\ x + 6 &= 2 \\ x &= -4 \end{aligned}$$

Solution set: $\{(-4, 0, 6)\}$

31. $$\begin{aligned} f(1) &= a(1)^2 + b(1) + c \\ &= a + b + c \end{aligned}$$

Since $f(1) = 128$, the first equation is

$$a + b + c = 128.$$

32. $$\begin{aligned} f(1.5) &= a(1.5)^2 + b(1.5) + c \\ &= 2.25a + 1.5b + c \end{aligned}$$

Since $f(1.5) = 140$, the second equation is

$$2.25a + 1.5b + c = 140.$$

33. $$\begin{aligned} f(3) &= a(3)^2 + b(3) + c \\ &= 9a + 3b + c \end{aligned}$$

Since $f(3) = 80$, the third equation is

$$9a + 3b + c = 80.$$

34. Using the three equations from Exercises 31–33, the system is

$$\begin{array}{rrrrl} a & + b & + c & = 128 & (1) \\ 2.25a & + 1.5b & + c & = 140 & (2) \\ 9a & + 3b & + c & = 80. & (3) \end{array}$$

Multiply equation (2) by -1 and add the result to equation (1).

$$\begin{array}{rrrrl} a & + b & + c & = 128 & (1) \\ -2.25a & - 1.5b & - c & = -140 & \\ \hline -1.25a & - .5b & & = -12 & (4) \end{array}$$

Multiply equation (3) by -1 and add the result to equation (1).

$$\begin{array}{rrrrl} a & + b & + c & = 128 & (1) \\ -9a & - 3b & - c & = -80 & \\ \hline -8a & - 2b & & = 48 & (5) \end{array}$$

Use equations (4) and (5) to eliminate b. Multiply equation (4) by -4 and add the result to equation (5).

$$\begin{array}{rrrl} 5a & + 2b & = 48 & -4 \times (4) \\ -8a & - 2b & = 48 & (5) \\ \hline -3a & & = 96 & \\ a & & = -32 & \end{array}$$

To find b, substitute $a = -32$ into equation (5).

$$\begin{aligned} -8a - 2b &= 48 \quad (5) \\ -8(-32) - 2b &= 48 \\ 256 - 2b &= 48 \\ -2b &= -208 \\ b &= 104 \end{aligned}$$

To find c, substitute $a = -32$ and $b = 104$ into equation (1).

$$\begin{aligned} a + b + c &= 128 \quad (1) \\ -32 + 104 + c &= 128 \\ 72 + c &= 128 \\ c &= 56 \end{aligned}$$

Solution set: $\{(-32, 104, 56)\}$

35. If $(a, b, c) = (-32, 104, 56)$, then

$$f(x) = -32x^2 + 104x + 56.$$

36. In the function f written in Exercise 35, the height of the projectile is a function of the time elapsed since it was projected.

37.
$$\begin{aligned} f(x) &= -32x^2 + 104x + 56 \\ f(0) &= -32(0)^2 + 104(0) + 56 \\ &= 56 \end{aligned}$$
The initial height is 56 ft.

38.
$$\begin{aligned} f(1.625) &= -32(1.625)^2 + 104(1.625) + 56 \\ &= -84.5 + 169 + 56 \\ &= 140.5 \end{aligned}$$
The maximum height is 140.5 ft.

5.3 Applications of Systems of Linear Equations

5.3 Margin Exercises

1. *Step 2*
Let $L =$ the length and $W =$ the width of the foundation of the rectangular house.

Step 3
Using the fact that the length is 6 m more than the width and the perimeter formula, we get the following system.

$$\begin{aligned} L &= 6 + W \quad &(1) \\ 48 &= 2L + 2W \quad &(2) \end{aligned}$$

Step 4
Substitute $6 + W$ for L in equation (2).

$$\begin{aligned} 48 &= 2L + 2W \\ 48 &= 2(6 + W) + 2W \\ 48 &= 12 + 2W + 2W \\ 36 &= 4W \\ 9 &= W \end{aligned}$$

Since $L = 6 + W$ and $W = 9$,

$$L = 6 + 9 = 15.$$

Step 5
The length of the house is 15 m and the width is 9 m.

Step 6
Check: 15 m is 6 m more than 9 m and $2(15) + 2(9) = 48$.

2. *Step 2*
Let $x =$ the average cost of a baseball ticket, and $y =$ the average cost of a football ticket.

Step 3
From the given information,

$$\begin{aligned} 3x + 2y &= 159.50 \quad &(1) \\ 2x + y &= 89.66. \quad &(2) \end{aligned}$$

Step 4
Multiply equation (2) by -2 and add to equation (1).

$$\begin{array}{rcrl} 3x + 2y &=& 159.50 & (1) \\ -4x - 2y &=& -179.32 & -2 \times (2) \\ \hline -x &=& -19.82 & \\ x &=& 19.82 & \end{array}$$

Let $x = 19.82$ in equation (2).

$$\begin{aligned} 2(19.82) + y &= 89.66 \\ 39.64 + y &= 89.66 \\ y &= 50.02 \end{aligned}$$

Step 5
The average cost of a baseball ticket is \$19.82 and the average cost of a football ticket is \$50.02.

Step 6
Check: $3(19.82) + 2(50.02) = 159.50$ and $2(19.82) + 50.02 = 89.66$.

3. **(a)** *Step 2*
Let x represent the number of pounds of the \$4 per lb coffee and y represent the number of pounds of the \$8 per lb coffee.

Price per Pound	Number of Pounds	Value of Coffee
\$4	x	$4x$
\$8	y	$8y$
\$5.60	50	$5.6(50) = 280$

Step 3
Write the system from the columns.

$$\begin{aligned} x + y &= 50 \quad &(1) \\ 4x + 8y &= 280 \quad &(2) \end{aligned}$$

Step 4
To eliminate x, multiply equation (1) by -4 and add the result to equation (2).

$$\begin{array}{rcrl} -4x - 4y &=& -200 & -4 \times (1) \\ 4x + 8y &=& 280 & (2) \\ \hline 4y &=& 80 & \\ y &=& 20 & \end{array}$$

Since $y = 20$ and $x + y = 50$, $x = 30$.

Step 5
To mix the coffee, 30 lb of \$4 per lb coffee and 20 lb of \$8 per lb coffee should be used.

Step 6
Check: $30 + 20 = 50$ and $4(30) + 8(20) = 280$.

(b) *Step 2*
Let x = the amount of 40% alcohol solution.
and y = the amount of 80% alcohol solution.

Percent (as a Decimal)	Liters of Solution	Liters of Pure Alcohol
$40\% = .40$	x	$.4x$
$80\% = .80$	y	$.8y$
$50\% = .50$	200	$.50(200) = 100$

Step 3
Write the system from the columns.

$$x + y = 200 \quad (1)$$
$$.4x + .8y = 100 \quad (2)$$

Step 4
To eliminate x, multiply equation (1) by -4 and equation (2) by 10. Then add the results.

$$\begin{array}{rcrl} -4x - 4y &=& -800 & -4 \times (1) \\ 4x + 8y &=& 1000 & 10 \times (2) \\ \hline 4y &=& 200 & \\ y &=& 50 & \end{array}$$

Since $y = 50$ and $x + y = 200$, $x = 150$.

Step 5
The mixture should contain 150 L of 40% solution and 50 L of 80% solution.

Step 6
Check: $150 + 50 = 200$ and $.4(150) + .8(50) = 100$.

4. *Step 2*
Let x = the train's speed and y = the truck's speed.

Make a table. Use $t = \frac{d}{r}$.

	Distance	Rate	Time
Train	600	x	$\frac{600}{x}$
Truck	520	y	$\frac{520}{y}$

Step 3
Since the times are the same,

$$\frac{600}{x} = \frac{520}{y}$$
$$600y = 520x \quad \textit{Multiply by xy.}$$
$$-520x + 600y = 0. \quad (1)$$

Since the train's average speed, x, is 8 mph faster than the truck's speed, y,

$$x = y + 8. \quad (2)$$

Step 4
Substitute $y + 8$ for x in equation (1) to find y.

$$\begin{aligned} -520x + 600y &= 0 \quad (1) \\ -520(y + 8) + 600y &= 0 \\ -520y - 4160 + 600y &= 0 \\ 80y &= 4160 \\ y &= 52 \end{aligned}$$

Since $y = 52$ and $x = y + 8$, $x = 60$.

Step 5
The train's speed is 60 mph, and the truck's speed is 52 mph.

Step 6
Check: 60 mph is 8 mph faster than 52 mph. It would take the train 10 hours to travel 600 miles at 60 mph, which is the same amount of time it would take the truck to travel 520 miles at 52 mph.

5.
$$\begin{aligned} x - 3y &= 0 \quad (1) \\ x - z &= 5 \quad (2) \\ 259x + 299y + 329z &= 6607 \quad (3) \end{aligned}$$

Eliminate z by adding equation (3) to 329 times equation (2).

$$\begin{array}{rcrl} 329x \qquad\quad - 329z &=& 1645 & 329 \times (2) \\ 259x + 299y + 329z &=& 6607 & (3) \\ \hline 588x + 299y \qquad\quad &=& 8252 & (4) \end{array}$$

Eliminate x by adding equation (4) to -588 times equation (1).

$$\begin{array}{rcrl} -588x + 1764y &=& 0 & -588 \times (1) \\ 588x + 299y &=& 8252 & (4) \\ \hline 2063y &=& 8252 & \\ y &=& 4 & \end{array}$$

Substitute 4 for y in equation (1) and solve for x.

$$x - 3(4) = 0$$
$$x = 12$$

Substitute 12 for x in equation (2) and solve for z.

$$12 - z = 5$$
$$7 = z$$

Solution set: $\{(12, 4, 7)\}$

6. Let x = the number of bottles of cheap, at \$8, y = the number of bottles of better, at \$15, and z = the number of bottles of best, at \$32.

There are 10 more bottles of cheap than better, so

$$x = y + 10. \quad (1)$$

There are 3 fewer bottles of best than better, so

$$z = y - 3. \quad (2)$$

continued

The total value is \$589, so

$$8x + 15y + 32z = 589. \quad (3)$$

Substitute $y + 10$ for x and $y - 3$ for z in equation (3) to find y.

$$\begin{aligned} 8(y+10) + 15y + 32(y-3) &= 589 \\ 8y + 80 + 15y + 32y - 96 &= 589 \\ 55y - 16 &= 589 \\ 55y &= 605 \\ y &= 11 \end{aligned}$$

Since $y = 11$,

$$x = y + 10 = 21 \quad \text{and} \quad z = y - 3 = 8.$$

There are 21 bottles of cheap perfume, 11 of better, and 8 of best.

7. Let $x =$ the number of tons of newsprint, $y =$ the number of tons of bond, and $z =$ the number of tons of copy machine paper.

The amount of recycled paper used to make all three items is 4200 tons, so

$$3x + 2y + 2z = 4200. \quad (1)$$

The amount of wood pulp used to make all three items is 5800 tons, so

$$x + 4y + 3z = 5800. \quad (2)$$

The amount of rags used to make all three items is 3900 tons, so

$$3y + 2z = 3900. \quad (3)$$

Solve the system of equations (1), (2), and (3). To eliminate z, multiply equation (3) by -1 and add the result to equation (1).

$$\begin{array}{rrrcrl} & -3y & - 2z & = & -3900 & -1 \times (3) \\ 3x + & 2y & + 2z & = & 4200 & (1) \\ \hline 3x - & y & & = & 300 & (4) \end{array}$$

To eliminate z again, multiply equation (2) by -2 and equation (3) by 3 and add the results.

$$\begin{array}{rrrcrl} -2x - & 8y & - 6z & = & -11{,}600 & -2 \times (2) \\ & 9y & + 6z & = & 11{,}700 & 3 \times (3) \\ \hline -2x + & y & & = & 100 & (5) \end{array}$$

Add equations (4) and (5) to eliminate y.

$$\begin{array}{rrcrl} 3x & - y & = & 300 & (4) \\ -2x & + y & = & 100 & (5) \\ \hline x & & = & 400 & \end{array}$$

Substitute 400 for x in equation (4).

$$\begin{aligned} 3x - y &= 300 \quad (4) \\ 3(400) - y &= 300 \\ 1200 - y &= 300 \\ 900 &= y \end{aligned}$$

Substitute 900 for y in equation (3).

$$\begin{aligned} 3y + 2z &= 3900 \quad (3) \\ 3(900) + 2z &= 3900 \\ 2700 + 2z &= 3900 \\ 2z &= 1200 \\ z &= 600 \end{aligned}$$

The paper mill can make 400 tons of newsprint, 900 tons of bond, and 600 tons of copy machine paper.

5.3 Section Exercises

1. *Step 2*
Let $x =$ the number of games that the Twins won and let $y =$ the number of games that they lost.

Step 3
They played 162 games, so

$$x + y = 162. \quad (1)$$

They won 18 more games than they lost, so

$$x = 18 + y. \quad (2)$$

Step 4
Substitute $18 + y$ for x in (1).

$$\begin{aligned} (18 + y) + y &= 162 \\ 18 + 2y &= 162 \\ 2y &= 144 \\ y &= 72 \end{aligned}$$

Substitute 72 for y in (2).

$$x = 18 + 72 = 90$$

Step 5
The Twins win-loss record was 90 wins and 72 losses.

Step 6
90 is 18 more than 72 and the sum of 90 and 72 is 162.

2. *Step 2*
Let $x =$ the number of games that the Tigers won and let $y =$ the number of games that they lost.

Step 3
They played 162 games, so

$$x + y = 162. \quad (1)$$

They lost 76 more games than they won, so

$$y = 76 + x. \quad (2)$$

Step 4
Substitute $76 + x$ for y in (1).

$$\begin{aligned} x + (76 + x) &= 162 \\ 2x + 76 &= 162 \\ 2x &= 86 \\ x &= 43 \end{aligned}$$

Substitute 43 for x in (2).

$$y = 76 + 43 = 119$$

Step 5
The Tigers win-loss record was 43 wins and 119 losses.

Step 6
119 is 76 more than 43 and the sum of 119 and 43 is 162.

3. Let W = the width of the tennis court and L = the length of the court.

Since the length is 42 ft more than the width,

$$L = W + 42. \quad (1)$$

The perimeter of a rectangle is given by

$$2W + 2L = P.$$

With perimeter $P = 228$ ft,

$$2W + 2L = 228. \quad (2)$$

Substitute $W + 42$ for L in equation (2).

$$\begin{aligned} 2W + 2(W + 42) &= 228 \\ 2W + 2W + 84 &= 228 \\ 4W &= 144 \\ W &= 36 \end{aligned}$$

Substitute $W = 36$ into equation (1).

$$\begin{aligned} L &= W + 42 \quad (1) \\ L &= 36 + 42 \\ &= 78 \end{aligned}$$

The length is 78 ft; the width is 36 ft.

4. Let L = the length of the basketball court and W = the width of the court.

The width was 44 ft less than the length.

$$W = L - 44$$

The perimeter was 288 ft. Use $P = 2L + 2W$.

$$288 = 2L + 2W$$

Rewrite each equation so that the variables are on the left side.

$$\begin{aligned} L - W &= 44 \quad (1) \\ 2L + 2W &= 288 \quad (2) \end{aligned}$$

Multiply (1) by 2 and add to (2).

$$\begin{array}{rcrcrl} 2L & - & 2W & = & 88 & \\ 2L & + & 2W & = & 288 & (2) \\ \hline 4L & & & = & 376 & \\ & & L & = & 94 & \end{array}$$

Since $L = 94$, $W = L - 44 = 94 - 44 = 50$.

The length was 94 ft and the width was 50 ft.

5. Let x = the revenue for Wal-Mart and y = the revenue for ExxonMobil (both in billions of dollars).

The total revenue was \$472 billion.

$$x + y = 472 \quad (1)$$

ExxonMobil's revenue was \$46 billion less than that of Wal-Mart.

$$y = x - 46 \quad (2)$$

Substitute $x - 46$ for y in equation (1).

$$\begin{aligned} x + (x - 46) &= 472 \\ 2x - 46 &= 472 \\ 2x &= 518 \\ x &= 259 \end{aligned}$$

Substitute 259 for x in equation (2).

$$y = x - 46 = 259 - 46 = 213$$

Wal-Mart's revenue was \$259 billion and ExxonMobil's revenue was \$213 billion.

6. Let x = the value of U.S. exports to Canada and y = the value of U.S. exports to Mexico (both in billions of dollars).

Together, exports to these two countries totaled \$258,393 million.

$$x + y = 258{,}393 \quad (1)$$

Exports to Canada were \$63,453 million more than exports to Mexico.

$$x = y + 63{,}453 \quad (2)$$

Substitute $y + 63{,}453$ for x in equation (1).

$$\begin{aligned} (y + 63{,}453) + y &= 258{,}393 \\ 2y + 63{,}453 &= 258{,}393 \\ 2y &= 194{,}940 \\ y &= 97{,}470 \end{aligned}$$

Substitute 97,470 for y in equation (2).

$$x = 97{,}470 + 63{,}453 = 160{,}923$$

The value of the U.S. exports to Canada was \$160,923 million and the value of the U.S. exports to Mexico was \$97,470 million.

7. From the figure in the text, the angles marked y and $3x + 10$ are supplementary, so

$$(3x + 10) + y = 180. \qquad (1)$$

Also, the angles x and y are complementary, so

$$x + y = 90. \qquad (2)$$

Solve equation (2) for y to get

$$y = 90 - x. \qquad (3)$$

Substitute $90 - x$ for y in equation (1).

$$\begin{aligned}(3x + 10) + (90 - x) &= 180\\ 2x + 100 &= 180\\ 2x &= 80\\ x &= 40\end{aligned}$$

Substitute $x = 40$ into equation (3) to get

$$y = 90 - x = 90 - 40 = 50.$$

The angles measure $40°$ and $50°$.

8. Because the angles of $x°$ and $y°$ are supplementary angles,

$$x + y = 180. \qquad (1)$$

Because the angles of $y°$ and $(x - 20)°$ are vertical angles, they are equal.

$$\begin{aligned}x - 20 &= y\\ \text{or} \quad x - y &= 20 \qquad (2)\end{aligned}$$

Solve the following system.

$$\begin{array}{rcrcrl} x & + & y & = & 180 & (1)\\ x & - & y & = & 20 & (2)\\ \hline 2x & & & = & 200 & \\ & & x & = & 100 & \end{array}$$

Substitute 100 for x in (1).

$$\begin{aligned}100 + y &= 180\\ y &= 80\end{aligned}$$

The angles measure $100°$ and $80°$.

9. Let $x =$ the hockey FCI and $y =$ the basketball FCI.

The sum is \$514.91, so

$$x + y = 514.91. \qquad (1)$$

The hockey FCI was \$7.61 less than the basketball FCI, so

$$x = y - 7.61. \qquad (2)$$

From (2), substitute $y - 7.61$ for x in (1).

$$\begin{aligned}(y - 7.61) + y &= 514.91\\ 2y - 7.61 &= 514.91\\ 2y &= 522.52\\ y &= 261.26\end{aligned}$$

From (2),

$$x = y - 7.61 = 261.26 - 7.61 = 253.65.$$

The hockey FCI was \$253.65 and the basketball FCI was \$261.26.

10. Let $x =$ the baseball FCI and $y =$ the football FCI.

The sum is \$457.27, so

$$x + y = 457.27. \qquad (1)$$

The football FCI was \$146.23 more than the baseball FCI, so

$$y = 146.23 + x. \qquad (2)$$

From (2), substitute $146.23 + x$ for y in (1).

$$\begin{aligned}x + (146.23 + x) &= 457.27\\ 2x + 146.23 &= 457.27\\ 2x &= 311.04\\ x &= 155.52\end{aligned}$$

From (2),

$$\begin{aligned}y = 146.23 + x &= 146.23 + 155.52\\ &= 301.75.\end{aligned}$$

The baseball FCI was \$155.52 and the football FCI was \$301.75.

11. Let $x =$ the cost of a single Junior Roast Beef sandwich, and $y =$ the cost of a single Big Montana sandwich.

15 Junior Roast Beef sandwiches and 10 Big Montana sandwiches cost \$64.25.

$$15x + 10y = 64.25 \qquad (1)$$

30 Junior Roast Beef sandwiches and 5 Big Montana sandwiches cost \$65.65.

$$30x + 5y = 65.65 \qquad (2)$$

Multiply equation (1) by -2 and add to equation (2).

$$\begin{array}{rcrcrl} -30x & - & 20y & = & -128.50 & -2 \times (1)\\ 30x & + & 5y & = & 65.65 & (2)\\ \hline & - & 15y & = & -62.85 & \\ & & y & = & 4.19 & \end{array}$$

Substitute 4.19 for y in equation (1).

$$\begin{aligned} 15x + 10y &= 64.25 \\ 15x + 10(4.19) &= 64.25 \\ 15x + 41.90 &= 64.25 \\ 15x &= 22.35 \\ x &= 1.49 \end{aligned}$$

A single Junior Roast Beef sandwich costs \$1.49 and a single Big Montana sandwich costs \$4.19.

12. Let $x =$ the cost per day for Tokyo and $y =$ the cost per day for New York.

Two days in Tokyo and 3 days in New York cost \$2015.

$$2x + 3y = 2015 \quad (1)$$

Four days in Tokyo and 2 days in New York cost \$2490.

$$4x + 2y = 2490 \quad (2)$$

Multiply equation (1) by -2. Add to equation (2).

$$\begin{array}{rcrl} -4x - 6y &=& -4030 & -2 \times (1) \\ 4x + 2y &=& 2490 & (2) \\ \hline -4y &=& -1540 & \\ y &=& 385 & \end{array}$$

Substitute 385 for y in equation (1).

$$\begin{aligned} 2x + 3(385) &= 2015 \\ 2x + 1155 &= 2015 \\ 2x &= 860 \\ x &= 430 \end{aligned}$$

The average cost per day in Tokyo is \$430 whereas that in New York is \$385.

13. Use the formula (rate of percent) • (base amount) = amount (percentage) of pure acid to compute parts (a) – (d).

(a) $.10(60) = 6$ oz

(b) $.25(60) = 15$ oz

(c) $.40(60) = 24$ oz

(d) $.50(60) = 30$ oz

14. Use $I = prt$ with $p = \$5000$ and $t = 1$, so $I = \$5000r$.

(a) $r = 2\% = .02$
$I = \$5000(.02) = \$100$

(b) $r = 3\% = .03$
$I = \$5000(.03) = \$150$

(c) $r = 4\% = .04$
$I = \$5000(.04) = \$200$

(d) $r = 3.5\% = .035$
$I = \$5000(.035) = \$175$

15. The cost is the price per pound, \$.99, times the number of pounds, x, or $\$.99x$.

16. y tickets costing \$8 each cost a total of $\$8y$.

17. Let $x =$ the amount of 25% alcohol solution and
$y =$ the amount of 35% alcohol solution.

Make a table. The percent times the amount of solution gives the amount of pure alcohol in the third column.

Percent (as a Decimal)	Gallons of Solution	Gallons of Pure Alcohol
$25\% = .25$	x	$.25x$
$35\% = .35$	y	$.35y$
$32\% = .32$	20	$.32(20) = 6.4$

The third row gives the total amounts of solution and pure alcohol. From the columns in the table, write a system of equations.

$$\begin{array}{rcll} x + y &=& 20 & (1) \\ .25x + .35y &=& 6.4 & (2) \end{array}$$

Solve the system. Multiply equation (1) by -25 and equation (2) by 100. Then add the results.

$$\begin{array}{rcrl} -25x - 25y &=& -500 & -25 \times (1) \\ 25x + 35y &=& 640 & 100 \times (2) \\ \hline 10y &=& 140 & \\ y &=& 14 & \end{array}$$

Substitute $y = 14$ into equation (1).

$$\begin{aligned} x + y &= 20 \quad (1) \\ x + 14 &= 20 \\ x &= 6 \end{aligned}$$

Mix 6 gal of 25% solution and 14 gal of 35% solution.

18. Let $x =$ the amount of 15% acid
and $y =$ the amount of 33% acid.

Percent (as a Decimal)	Liters of Solution	Liters of Pure Acid
$15\% = .15$	x	$.15x$
$33\% = .33$	y	$.33y$
$21\% = .21$	120	$.21(120) = 25.2$

$$\begin{array}{rcll} x + y &=& 120 & (1) \\ .15x + .33y &=& 25.2 & (2) \end{array}$$

Multiply (2) by 100 to clear the equation of decimals, and multiply (1) by -15.

continued

$$\begin{array}{rcll} -15x - 15y & = & -1800 & -15 \times (1) \\ 15x + 33y & = & 2520 & 100 \times (2) \\ \hline 18y & = & 720 & \\ y & = & \frac{720}{18} = 40 & \end{array}$$

Substitute 40 for y in (1).

$$x + 40 = 120$$
$$x = 120 - 40 = 80$$

80 L of 15% and 40 L of 33% should be mixed.

19. Let $x =$ the amount of pure acid and $y =$ the amount of 10% acid.

Make a table.

Percent (as a Decimal)	Liters of Solution	Liters of Pure Acid
$100\% = 1$	x	$1.00x = x$
$10\% = .10$	y	$.10y$
$20\% = .20$	54	$.20(54) = 10.8$

Solve the following system.

$$\begin{array}{rl} x + y = 54 & (1) \\ x + .10y = 10.8 & (2) \end{array}$$

Multiply equation (2) by 10 to clear the decimals.

$$10x + y = 108 \qquad (3)$$

To eliminate y, multiply equation (1) by -1 and add the result to equation (3).

$$\begin{array}{rcll} -x - y & = & -54 & -1 \times (1) \\ 10x + y & = & 108 & (3) \\ \hline 9x & = & 54 & \\ x & = & 6 & \end{array}$$

Substitute $x = 6$ into equation (1).

$$x + y = 54 \quad (1)$$
$$6 + y = 54$$
$$y = 48$$

Use 6 L of pure acid and 48 L of 10% acid.

20. Let $x =$ the amount of 4% antifreeze and $y =$ the amount of pure antifreeze.

Percent (as a Decimal)	Liters of Solution	Liters of Pure Antifreeze
$4\% = .04$	x	$.04x$
$100\% = 1$	y	y
$20\% = .20$	36	$.2(36) = 7.2$

$$\begin{array}{rl} x + y = 36 & (1) \\ .04x + y = 7.2 & (2) \end{array}$$

Multiply (2) by 100 and (1) by -4.

$$\begin{array}{rcll} -4x - 4y & = & -144 & -4 \times (1) \\ 4x + 100y & = & 720 & 100 \times (2) \\ \hline 96y & = & 576 & \\ y & = & 6 & \end{array}$$

From (1), $x = 30$.
6 L of pure antifreeze should be added to 30 L of the 4% antifreeze.

21. Complete the table.

	Price per Kilogram	Number of Kilograms	Value
Nuts	2.50	x	$2.50x$
Cereal	1.00	y	$1.00y$
Mixture	1.70	30	$1.70(30)$

From the "Number of Kilograms" column,

$$x + y = 30. \quad (1)$$

From the "Value" column,

$$2.50x + 1.00y = 1.70(30) = 51. \quad (2)$$

Solve the system.

$$\begin{array}{rcll} -10x - 10y & = & -300 & -10 \times (1) \\ 25x + 10y & = & 510 & 10 \times (2) \\ \hline 15x & = & 210 & \\ x & = & 14 & \end{array}$$

From (1), $14 + y = 30$, so $y = 16$.
The party mix should be made from 14 kg of nuts and 16 kg of cereal.

22. Complete the table.

	Percent (as a Decimal)	Liters of Drink	Liters of Pure Juice
50% Juice	.50	x	$.50x$
30% Juice	.30	y	$.30y$
Mixture	.45	200	$.45(200)$

From the "Liters of Drink" column,

$$x + y = 200. \quad (1)$$

From the "Liters of Pure Juice" column,

$$.50x + .30y = .45(200) = 90. \quad (2)$$

Multiply (1) by -5 and (2) by 10.

$$\begin{array}{rcll} -5x - 5y & = & -1000 & -5 \times (1) \\ 5x + 3y & = & 900 & 10 \times (2) \\ \hline -2y & = & -100 & \\ y & = & 50 & \end{array}$$

From (1), $x + 50 = 200$, so $x = 150$.
150 L of 50% juice and 50 L of 30% juice should be used.

23. From the "Principal" column in the text,

$$x + y = 3000. \quad (1)$$

From the "Interest" column in the text,

$$.02x + .04y = 100. \quad (2)$$

Multiply equation (2) by 100 to clear the decimals.

$$2x + 4y = 10{,}000 \quad (3)$$

To eliminate x, multiply equation (1) by -2 and add the result to equation (3).

$$\begin{array}{rcrl} -2x - 2y & = & -6000 & -2 \times (1) \\ 2x + 4y & = & 10{,}000 & (3) \\ \hline 2y & = & 4000 & \\ y & = & 2000 & \end{array}$$

From (1), $x + 2000 = 3000$, so $x = 1000$.
\$1000 is invested at 2%, and \$2000 is invested at 4%.

24. Let $x =$ the amount at 4%
and $y =$ the amount at 3%.

Complete the table in the text.

Rate (as a Decimal)	Principal	Interest
.04	x	$.04x$
.03	y	$.03y$
Totals	15,000	550

Use the last two columns to write a system.

$$\begin{array}{rcrl} x + y & = & 15{,}000 & (1) \\ .04x + .03y & = & 550 & (2) \end{array}$$

Multiply (1) by -4 and (2) by 100.

$$\begin{array}{rcrl} -4x - 4y & = & -60{,}000 & -4 \times (1) \\ 4x + 3y & = & 55{,}000 & 100 \times (2) \\ \hline -y & = & -5000 & \\ y & = & 5000 & \end{array}$$

From (1), $x = 10{,}000$.
He should invest \$10,000 at 4% and \$5000 at 3%.

25. **(a)** The speed of the boat going upstream is *decreased* by the speed of the current, so it is $10 - x$ mph.

(b) The speed of the boat going downstream is *increased* by the speed of the current, so it is $10 + x$ mph.

26. In the formula $d = rt$, substitute 25 for r (rate or speed) and y for t (time in hours) to get

$$d = 25y \text{ mi.}$$

27. Let $x =$ speed of the freight train
and $y =$ the speed of the express train.

Make a table. Use $d = rt$.

	r	t	d
Freight Train	x	3	$3x$
Express Train	y	3	$3y$

The trains are 390 km apart, so

$$3x + 3y = 390. \quad (1)$$

The freight train travels 30 km/hr slower than the express train, so

$$x = y - 30. \quad (2)$$

Substitute $y - 30$ for x in equation (1) and solve for y.

$$\begin{aligned} 3x + 3y &= 390 \quad (1) \\ 3(y - 30) + 3y &= 390 \\ 3y - 90 + 3y &= 390 \\ 6y &= 480 \\ y &= 80 \end{aligned}$$

From (2), $x = 80 - 30 = 50$.
The freight train travels at 50 km/hr, while the express train travels at 80 km/hr.

28. Let $x =$ the speed of the train
and $y =$ the speed of the plane.

Make a table. Use $t = \frac{d}{r}$.

	r	t	d
Train	x	$\frac{150}{x}$	150
Plane	y	$\frac{400}{y}$	400

The times are equal, so

$$\frac{150}{x} = \frac{400}{y}. \quad (1)$$

The speed of the plane is 20 km per hour less than 3 times the speed of the train, so

$$3x - 20 = y. \quad (2)$$

Multiply (1) by xy (or use cross products).

$$150y = 400x \quad (3)$$

From (2), substitute $3x - 20$ for y in (3).

$$\begin{aligned} 150(3x - 20) &= 400x \\ 450x - 3000 &= 400x \\ 50x &= 3000 \\ x &= 60 \end{aligned}$$

From (2), $y = 3(60) - 20 = 160$.
The speed of the train is 60 km/hr, and the speed of the plane is 160 km/hr.

29. Let $x =$ the speed of the boat in still water and $y =$ the speed of the current.

Furthermore,

$$\text{rate upstream} = x - y$$
$$\text{and} \quad \text{rate downstream} = x + y.$$

Use these rates and the information in the problem to make a table.

	r	t	d
Upstream	$x - y$	2	36
Downstream	$x + y$	1.5	36

From the table, use the formula $d = rt$ to write a system of equations.

$$36 = 2(x - y)$$
$$36 = 1.5(x + y)$$

Remove the parentheses and move the variables to the left side.

$$2x - 2y = 36 \quad (1)$$
$$1.5x + 1.5y = 36 \quad (2)$$

Solve the system. Multiply equation (1) by -3 and equation (2) by 4. Then add the results.

$$\begin{array}{rcl} -6x + 6y &=& -108 \quad -3 \times (1) \\ 6x + 6y &=& 144 \quad 4 \times (2) \\ \hline 12y &=& 36 \\ y &=& 3 \end{array}$$

Substitute $y = 3$ into equation (1).

$$2x - 2y = 36 \quad (1)$$
$$2x - 2(3) = 36$$
$$2x = 42$$
$$x = 21$$

The speed of the boat is 21 mph, and the speed of the current is 3 mph.

30. Let $x =$ the speed of the plane and $y =$ the speed of the wind. Make a table.

	r	t	d
Against the Wind	$x - y$	3	1650
With the Wind	$x + y$	2	1300

Since $d = rt$, we have the following system.

$$3(x - y) = 1650 \quad (1)$$
$$2(x + y) = 1300 \quad (2)$$

Simplify the system, then add.

$$\begin{array}{rcll} x - y &=& 550 & \quad (1) \div 3 \\ x + y &=& 650 \quad (3) & \quad (2) \div 2 \\ \hline 2x &=& 1200 & \\ x &=& 600 & \end{array}$$

From (3), $600 + y = 650$, so $y = 50$.
The speed of the plane is 600 mph and the speed of the wind is 50 mph.

31. Let $x =$ the revenue from McCartney and $y =$ the revenue from the Rolling Stones (both in millions of dollars).

The total is \$191.2 million, so

$$x + y = 191.2. \quad (1)$$

The Rolling Stones took in \$15.4 million less than McCartney, so

$$y = x - 15.4. \quad (2)$$

Substitute $x - 15.4$ for y in equation (1).

$$x + y = 191.2 \quad (1)$$
$$x + (x - 15.4) = 191.2$$
$$2x = 206.6$$
$$x = 103.3$$

From (2), $y = 103.3 - 15.4 = 87.9$.
McCartney's act generated \$103.3 million and the Rolling Stones act generated \$87.9 million.

32. Complete the table.

	Price per Pound	Number of Pounds	Value
Pecan Clusters	\$3.60	x	\$3.60x
Chocolate Truffles	\$7.20	y	\$7.20y
Valentine Mixture	\$4.95	80	\$4.95(80)

From the "Number of Pounds" column,

$$x + y = 80. \quad (1)$$

From the "Value" column,

$$3.60x + 7.20y = 4.95(80) = 396. \quad (2)$$

Solve the system.

$$\begin{array}{rcll} -36x - 36y &=& -2880 & \quad -36 \times (1) \\ 36x + 72y &=& 3960 & \quad 10 \times (2) \\ \hline 36y &=& 1080 & \\ y &=& 30 & \end{array}$$

From (1), $x + 30 = 80$, so $x = 50$.
She should mix 50 lb of \$3.60/lb pecan clusters with 30 lb of \$7.20/lb chocolate truffles.

33. Let $x =$ the number of general admission tickets and $y =$ the number of student tickets.

Make a table.

Ticket	Number	Value of Tickets
General	x	$5 \cdot x = 5x$
Student	y	$4 \cdot y = 4y$
Totals	184	812

Solve the system.

$$\begin{aligned} x + y &= 184 \quad (1)\\ 5x + 4y &= 812 \quad (2) \end{aligned}$$

To eliminate y, multiply equation (1) by -4 and add the result to equation (2).

$$\begin{array}{rcrcrl} -4x & - & 4y & = & -736 & \quad -4 \times (1)\\ 5x & + & 4y & = & 812 & \quad (2)\\ \hline x & & & = & 76 & \end{array}$$

From (1), $76 + y = 184$, so $y = 108$.

76 general admission tickets and 108 student tickets were sold.

34. Let $x =$ the price of a cappuccino and let $y =$ the price of a house latte.

Two cappuccinos and three house lattes cost \$14.55, so

$$2x + 3y = 14.55. \quad (1)$$

One cappuccino and two house lattes cost \$8.77, so

$$x + 2y = 8.77. \quad (2)$$

Solve the system.

$$\begin{array}{rcrcrl} 2x & + & 3y & = & 14.55 & \quad (1)\\ -2x & - & 4y & = & -17.54 & \quad -2 \times (2)\\ \hline & - & y & = & -2.99 & \\ & & y & = & 2.99 & \end{array}$$

Substitute 2.99 for y in (2).

$$\begin{aligned} x + 2(2.99) &= 8.77\\ x + 5.98 &= 8.77\\ x &= 2.79 \end{aligned}$$

Each cappuccino costs \$2.79 and each house latte costs \$2.99.

35. Let $x =$ the price for a citron and let $y =$ the price for a wood apple.

"9 citrons and 7 fragrant wood apples is 107" gives us

$$9x + 7y = 107. \quad (1)$$

"7 citrons and 9 fragrant wood apples is 101" gives us

$$7x + 9y = 101. \quad (2)$$

Multiply equation (1) by -7 and equation (2) by 9. Then add.

$$\begin{array}{rcrcrl} -63x & - & 49y & = & -749 & \quad -7 \times (1)\\ 63x & + & 81y & = & 909 & \quad 9 \times (2)\\ \hline & & 32y & = & 160 & \\ & & y & = & 5 & \end{array}$$

Substitute 5 for y in equation (1).

$$\begin{aligned} 9x + 7(5) &= 107\\ 9x + 35 &= 107\\ 9x &= 72\\ x &= 8 \end{aligned}$$

The prices are 8 for a citron and 5 for a wood apple.

36. Let $x =$ the top speed of the snow speeder and $y =$ the speed of the wind.

Furthermore,

$$\text{rate into headwind} = x - y$$
$$\text{and} \quad \text{rate with tailwind} = x + y.$$

Use these rates and the information in the problem to make a table.

	r	t	d
Into Headwind	$x - y$	3	4800
With Tailwind	$x + y$	2	4800

From the table, use the formula $d = rt$ to write a system of equations.

$$\begin{aligned} 3(x - y) &= 4800 \quad (1)\\ 2(x + y) &= 4800 \quad (2) \end{aligned}$$

Simplify the system, then add.

$$\begin{array}{rcrcrll} x & - & y & = & 1600 & & (1) \div 3\\ x & + & y & = & 2400 & (3) & (2) \div 2\\ \hline 2x & & & = & 4000 & & \\ & & x & = & 2000 & & \end{array}$$

From (3), $2000 + y = 2400$, so $y = 400$.

The top speed of the snow speeder is 2000 miles per hour and the speed of the wind is 400 miles per hour.

37. Let $x =$ the measure of one angle,
$y =$ the measure of another angle,
and $z =$ the measure of the last angle.

Two equations are given, so

$$\begin{aligned} z &= x + 10 \\ \text{or} \quad -x + z &= 10 \quad (1) \\ \text{and} \quad x + y &= 100. \quad (2) \end{aligned}$$

Since the sum of the measures of the angles of a triangle is $180°$, the third equation of the system is

$$x + y + z = 180. \quad (3)$$

Equation (1) is missing y. To eliminate y again, multiply equation (2) by -1 and add the result to equation (3).

$$\begin{array}{rcl} -x - y &=& -100 \quad -1 \times (2) \\ x + y + z &=& 180 \quad (3) \\ \hline z &=& 80 \end{array}$$

Since $z = 80$,

$$\begin{aligned} -x + z &= 10 \quad (1) \\ -x + 80 &= 10 \\ -x &= -70 \\ x &= 70. \end{aligned}$$

From (2), $70 + y = 100$, so $y = 30$.
The measures of the angles are $70°$, $30°$, and $80°$.

38. The first equation is

$$x = y - 10. \quad (1)$$

The second equation is

$$x = z - 20. \quad (2)$$

Because the sum of the measures of the angles of a triangle is $180°$, we have

$$x + y + z = 180. \quad (3)$$

Solve the system.
Solve (1) for y.

$$x + 10 = y \quad (4)$$

Solve (2) for z.

$$x + 20 = z \quad (5)$$

Substitute $x + 10$ for y and $x + 20$ for z in (3).

$$\begin{aligned} x + (x + 10) + (x + 20) &= 180 \\ 3x + 30 &= 180 \\ 3x &= 150 \\ x &= 50 \end{aligned}$$

From (4), $y = 50 + 10 = 60$.
From (5), $z = 50 + 20 = 70$.
The angle measures are $50°$, $60°$, and $70°$.

39. Let $x =$ the measure of the first angle,
$y =$ the measure of the second angle, and
$z =$ the measure of the third angle.

The sum of the angles in a triangle equals $180°$, so

$$x + y + z = 180. \quad (1)$$

The measure of the second angle is $10°$ more than 3 times that of the first angle, so

$$y = 3x + 10. \quad (2)$$

The third angle is equal to the sum of the other two, so

$$z = x + y. \quad (3)$$

Solve the system. Substitute z for $x + y$ in equation (1).

$$\begin{aligned} (x + y) + z &= 180 \quad (1) \\ z + z &= 180 \\ 2z &= 180 \\ z &= 90 \end{aligned}$$

Substitute 90 for z and $3x + 10$ for y in equation (3).

$$\begin{aligned} z &= x + y \quad (3) \\ 90 &= x + (3x + 10) \\ 80 &= 4x \\ 20 &= x \end{aligned}$$

Substitute $x = 20$ and $z = 90$ into equation (3).

$$\begin{aligned} z &= x + y \quad (3) \\ 90 &= 20 + y \\ 70 &= y \end{aligned}$$

The three angles have measures of $20°$, $70°$, and $90°$.

40. Let $x =$ the measure of the largest angle,
$y =$ the measure of the middle angle,
and $z =$ the measure of the smallest angle.

The measure of the largest angle is $12°$ less than the sum of the measures of the other two.

$$x = y + z - 12 \quad (1)$$

The smallest angle measures $58°$ less than the largest.

$$z = x - 58 \quad (2)$$

The sum of the measures of the angles of a triangle is $180°$.

$$x + y + z = 180 \quad (3)$$

Solve the system of (1), (2), and (3). Rewrite (1) in standard form and add to (3).

$$\begin{array}{rcrcrcrl} x & - & y & - & z & = & -12 & \\ x & + & y & + & z & = & 180 & (3) \\ \hline 2x & & & & & = & 168 & \\ & & & & x & = & 84 & \end{array}$$

From (2), $z = 84 - 58 = 26$.
Substitute 84 for x and 26 for z in (3).

$$\begin{aligned} 84 + y + 26 &= 180 \\ y &= 70 \end{aligned}$$

The largest angle measures 84°, the middle angle measures 70°, and the smallest angle measures 26°.

41. Let $x =$ the length of the longest side,
$y =$ the length of the middle side,
and $z =$ the length of the shortest side.

Perimeter is the sum of the measures of the sides, so

$$x + y + z = 70. \quad (1)$$

The longest side is 4 cm less than the sum of the other sides, so

$$\begin{aligned} x &= y + z - 4 \\ \text{or} \quad x - y - z &= -4. \qquad (2) \end{aligned}$$

Twice the shortest side is 9 cm less than the longest side, so

$$\begin{aligned} 2z &= x - 9 \\ \text{or} \quad -x + 2z &= -9 \qquad (3) \end{aligned}$$

Add equations (1) and (2) to eliminate y and z.

$$\begin{array}{rcrcrcrl} x & + & y & + & z & = & 70 & (1) \\ x & - & y & - & z & = & -4 & (2) \\ \hline 2x & & & & & = & 66 & \\ & & & & x & = & 33 & \end{array}$$

Substitute 33 for x in (3).

$$\begin{aligned} -x + 2z &= -9 \quad (3) \\ -33 + 2z &= -9 \\ 2z &= 24 \\ z &= 12 \end{aligned}$$

Substitute 33 for x and 12 for z in (1).

$$\begin{aligned} x + y + z &= 70 \quad (1) \\ 33 + y + 12 &= 70 \\ y + 45 &= 70 \\ y &= 25 \end{aligned}$$

The shortest side is 12 cm long, the middle side is 25 cm long, and the longest side is 33 cm long.

42. Let $x =$ the length of the longest side,
$y =$ the length of the middle side,
and $z =$ the length of the shortest side.

Since the perimeter is the sum of the three lengths,

$$x + y + z = 56. \quad (1)$$

The longest side measures 4 inches less than the sum of the measures of the other two sides.

$$x = y + z - 4 \quad (2)$$

Three times the shortest side is 4 inches more than the longest side.

$$3z = x + 4 \quad (3)$$

Solve the system of (1), (2), and (3). Rewrite (2) in standard form and add to (1).

$$\begin{array}{rcrcrcrl} x & - & y & - & z & = & -4 & \\ x & + & y & + & z & = & 56 & (1) \\ \hline 2x & & & & & = & 52 & \\ & & & & x & = & 26 & \end{array}$$

Substitute 26 for x in (3).

$$\begin{aligned} 3z &= 26 + 4 \\ 3z &= 30 \\ z &= 10 \end{aligned}$$

Substitute 26 for x and 10 for z in (1).

$$\begin{aligned} 26 + y + 10 &= 56 \\ y + 36 &= 56 \\ y &= 20 \end{aligned}$$

The shortest side is 10 inches, the middle side is 20 inches, and the longest side is 26 inches.

43. Let $W =$ the number of games won,
$L =$ the number of games lost, and
$T =$ the number of games tied.

They won 5 more than they lost, so

$$W = L + 5. \quad (1)$$

They tied 4 fewer than half the number of games they won, so

$$T = \tfrac{1}{2}W - 4. \quad (2)$$

They played a total of 16 games, so

$$W + L + T = 16. \quad (3)$$

Solve equation (1) for L so that L and T are both in terms of W.

$$W = L + 5 \quad \text{gives us} \quad L = W - 5. \quad (4)$$

Substitute $W - 5$ for L and $\frac{1}{2}W - 4$ for T in (3).

continued

$$\begin{aligned} W + (W - 5) + \left(\tfrac{1}{2}W - 4\right) &= 16 \\ 2W + 2W - 10 + W - 8 &= 32 \quad \textit{Multiply by 2.} \\ 5W - 18 &= 32 \\ 5W &= 50 \\ W &= 10 \end{aligned}$$

From (2), $T = \frac{1}{2}(10) - 4 = 1$.
From (4), $L = 10 - 5 = 5$.

The 2002 Steelers won 10 games, lost 5 games, and tied 1 game.

44. Let $x =$ the number of gold medals,
$y =$ the number of silver medals, and
$z =$ the number of bronze medals.

The U.S. earned 3 fewer gold medals than silver, so

$$x = y - 3. \quad (1)$$

The number of bronze medals earned was 15 fewer than twice the number of silver medals, so

$$z = 2y - 15. \quad (2)$$

The total number of medals earned was 34, so

$$x + y + z = 34. \quad (3)$$

Substitute $y - 3$ for x and $2y - 15$ for z in (3).

$$\begin{aligned} (y - 3) + y + (2y - 15) &= 34 \\ 4y - 18 &= 34 \\ 4y &= 52 \\ y &= 13 \end{aligned}$$

From (1), $x = 13 - 3 = 10$.
From (2), $z = 2(13) - 15 = 11$.

The United States earned 10 gold, 13 silver, and 11 bronze medals.

45. Let $x =$ the cost of one jersey,
$y =$ the cost of one bobblehead set, and
$z =$ the cost of one basketball.

From the given information, we can write a system of equations.

$$\begin{aligned} 1x + 2y + 1z &= 302 \quad (1) \\ 2x + 1y + 3z &= 284 \quad (2) \\ 1x + 1y + 1z &= 182 \quad (3) \end{aligned}$$

Eliminate x by adding (2) to -2 times (1).

$$\begin{array}{rcrcrcrl} -2x & - & 4y & - & 2z & = & -604 & -2 \times (1) \\ 2x & + & y & + & 3z & = & 284 & (2) \\ \hline & & -3y & + & z & = & -320 & (4) \end{array}$$

Eliminate x by adding (3) to -1 times (1).

$$\begin{array}{rcrcrcrl} -x & - & 2y & - & z & = & -302 & -1 \times (1) \\ x & + & y & + & z & = & 182 & (3) \\ \hline & & -y & & & = & -120 & \\ & & & & y & = & 120 & \end{array}$$

Substitute 120 for y in equation (4).

$$\begin{aligned} -3y + z &= -320 \quad (4) \\ -3(120) + z &= -320 \\ -360 + z &= -320 \\ z &= 40 \end{aligned}$$

From (3), $x + 120 + 40 = 182$, so $x = 22$.

The price for one jersey is \$22, for one bobblehead set is \$120, and for one basketball is \$40.

46. Let $x =$ the price of each "up close" ticket,
$y =$ the price of each "in the middle" ticket,
and $z =$ the price of each "far out" ticket.

The system of equations is

$$\begin{aligned} x &= y + 10 \quad (1) \\ y &= z + 10 \quad (2) \\ 2x &= 3z + 20. \quad (3) \end{aligned}$$

Substitute $z + 10$ for y in (1).

$$\begin{aligned} x &= (z + 10) + 10 \\ x &= z + 20 \quad (4) \end{aligned}$$

Substitute $z + 20$ for x in (3).

$$\begin{aligned} 2(z + 20) &= 3z + 20 \\ 2z + 40 &= 3z + 20 \\ 20 &= z \end{aligned}$$

From (4), $x = z + 20 = 20 + 20 = 40$.
From (2), $y = z + 10 = 20 + 10 = 30$.
Each "up close" ticket costs \$40, each "in the middle" ticket costs \$30, and each "far out" ticket costs \$20.

47. Let $x =$ the number of type A units,
$y =$ the number of type B units,
and $z =$ the number of type C units.

They must make 10 units more of type C than the total of the other two.

$$z = x + y + 10 \quad (1)$$

They must make twice as many of type B as type A.

$$y = 2x \quad (2)$$

They must produce a total of 490 units per day.

$$x + y + z = 490 \quad (3)$$

Rewrite (1) in standard form and add to (3).

$$\begin{array}{rcrcrcrl} -x & - & y & + & z & = & 10 & \\ x & + & y & + & z & = & 490 & (3) \\ \hline & & & & 2z & = & 500 & \\ & & & & z & = & 250 & \end{array}$$

Substitute 250 for z and $2x$ for y in (3).

$$\begin{aligned} x + 2x + 250 &= 490 \\ 3x &= 240 \\ x &= 80 \end{aligned}$$

From (2), $y = 2x = 2(80) = 160$.
They can make 80 units of type A clamps, 160 units of type B clamps, and 250 units of type C clamps.

48. Let $x =$ the number of cases sent to wholesaler A,
$y =$ the number of cases sent to wholesaler B, and
$z =$ the number of cases sent to wholesaler C.

The total output is 320 cases per day, so

$$x + y + z = 320. \quad (1)$$

The number of cases to A is three times that sent to B, so

$$x = 3y. \quad (2)$$

Wholesaler C gets 160 cases fewer than the sum sent to A and B, so

$$z = x + y - 160. \quad (3)$$

Solve equation (3) for $x + y$ to get

$$x + y = z + 160.$$

Substitute $z + 160$ for $x + y$ in equation (1).

$$\begin{aligned} (x + y) + z &= 320 \quad (1) \\ (z + 160) + z &= 320 \\ 2z &= 160 \\ z &= 80 \end{aligned}$$

Substitute 80 for z and $3y$ for x in equation (3).

$$\begin{aligned} z &= x + y - 160 \quad (3) \\ 80 &= 3y + y - 160 \\ 240 &= 4y \\ 60 &= y \end{aligned}$$

Substitute $y = 60$ into equation (2).

$$x = 3y = 3(60) = 180$$

She should send 180 cases to Wholesaler A, 60 cases to Wholesaler B, and 80 cases to Wholesaler C.

5.4 Solving Systems of Linear Equations by Matrix Methods

5.4 Margin Exercises

1. $x - 2y = 9$
$3x + y = 13$

Write the augmented matrix for this system.

$$\left[\begin{array}{rr|r} 1 & -2 & 9 \\ 3 & 1 & 13 \end{array}\right]$$

$$\left[\begin{array}{rr|r} 1 & -2 & 9 \\ 0 & 7 & -14 \end{array}\right] \quad -3\text{R}_1 + \text{R}_2$$

$$\left[\begin{array}{rr|r} 1 & -2 & 9 \\ 0 & 1 & -2 \end{array}\right] \quad \tfrac{1}{7}\text{R}_2$$

This matrix gives the system

$$\begin{aligned} x - 2y &= 9 \\ y &= -2. \end{aligned}$$

Substitute -2 for y in the first equation.

$$\begin{aligned} x - 2(-2) &= 9 \\ x + 4 &= 9 \\ x &= 5 \end{aligned}$$

Solution set: $\{(5, -2)\}$

2. $2x - y + z = 7$, $x - 3y - z = 7$, $-x + y - 5z = -9$ **OR** $x - 3y - z = 7$, $2x - y + z = 7$, $-x + y - 5z = -9$

Write the augmented matrix for this system.

$$\left[\begin{array}{rrr|r} 1 & -3 & -1 & 7 \\ 2 & -1 & 1 & 7 \\ -1 & 1 & -5 & -9 \end{array}\right]$$

$$\left[\begin{array}{rrr|r} 1 & -3 & -1 & 7 \\ 0 & 5 & 3 & -7 \\ -1 & 1 & -5 & -9 \end{array}\right] \quad -2\text{R}_1 + \text{R}_2$$

$$\left[\begin{array}{rrr|r} 1 & -3 & -1 & 7 \\ 0 & 5 & 3 & -7 \\ 0 & -2 & -6 & -2 \end{array}\right] \quad \text{R}_1 + \text{R}_3$$

$$\left[\begin{array}{rrr|r} 1 & -3 & -1 & 7 \\ 0 & 1 & \frac{3}{5} & -\frac{7}{5} \\ 0 & -2 & -6 & -2 \end{array}\right] \quad \tfrac{1}{5}\text{R}_2$$

$$\left[\begin{array}{rrr|r} 1 & -3 & -1 & 7 \\ 0 & 1 & \frac{3}{5} & -\frac{7}{5} \\ 0 & 0 & -\frac{24}{5} & -\frac{24}{5} \end{array}\right] \quad 2\text{R}_2 + \text{R}_3$$

$$\left[\begin{array}{rrr|r} 1 & -3 & -1 & 7 \\ 0 & 1 & \frac{3}{5} & -\frac{7}{5} \\ 0 & 0 & 1 & 1 \end{array}\right] \quad -\tfrac{5}{24}\text{R}_3$$

continued

This matrix gives the system

$$\begin{aligned} x - 3y - z &= 7 \\ y + \tfrac{3}{5}z &= -\tfrac{7}{5} \\ z &= 1. \end{aligned}$$

Substitute 1 for z in the second equation.

$$\begin{aligned} y + \tfrac{3}{5}(1) &= -\tfrac{7}{5} \\ y &= -2 \end{aligned}$$

Substitute -2 for y and 1 for z in the first equation.

$$\begin{aligned} x - 3(-2) - 1 &= 7 \\ x + 5 &= 7 \\ x &= 2 \end{aligned}$$

Solution set: $\{(2, -2, 1)\}$

3. **(a)** $\begin{aligned} x - y &= 2 \\ -2x + 2y &= 2 \end{aligned}$

Write the augmented matrix for this system.

$$\left[\begin{array}{rr|r} 1 & -1 & 2 \\ -2 & 2 & 2 \end{array}\right]$$

$$\left[\begin{array}{rr|r} 1 & -1 & 2 \\ 0 & 0 & 6 \end{array}\right] \quad 2R_1 + R_2$$

This matrix gives the system

$$\begin{aligned} x - y &= 2 \\ 0 &= 6. \quad \textit{False} \end{aligned}$$

The false statement indicates that the system is inconsistent and has no solution.

Solution set: $\emptyset$

(b) $\begin{aligned} x - y &= 2 \\ -2x + 2y &= -4 \end{aligned}$

Write the augmented matrix for this system.

$$\left[\begin{array}{rr|r} 1 & -1 & 2 \\ -2 & 2 & -4 \end{array}\right]$$

$$\left[\begin{array}{rr|r} 1 & -1 & 2 \\ 0 & 0 & 0 \end{array}\right] \quad 2R_1 + R_2$$

This matrix gives the system

$$\begin{aligned} x - y &= 2 \\ 0 &= 0. \quad \textit{True} \end{aligned}$$

The true statement indicates that the system has dependent equations.

Solution set: $\{(x, y) \mid x - y = 2\}$

5.4 Section Exercises

1. $\begin{bmatrix} -2 & 3 & 1 \\ 0 & 5 & -3 \\ 1 & 4 & 8 \end{bmatrix}$

(a) The elements of the second row are 0, 5, and -3.

(b) The elements of the third column are 1, -3, and 8.

(c) The matrix is square since the number of rows (three) is the same as the number of columns.

(d) The matrix obtained by interchanging the first and third rows is

$$\begin{bmatrix} 1 & 4 & 8 \\ 0 & 5 & -3 \\ -2 & 3 & 1 \end{bmatrix}.$$

(e) The matrix obtained by multiplying the first row by $-\frac{1}{2}$ is

$$\begin{bmatrix} -2(-\frac{1}{2}) & 3(-\frac{1}{2}) & 1(-\frac{1}{2}) \\ 0 & 5 & -3 \\ 1 & 4 & 8 \end{bmatrix} = \begin{bmatrix} 1 & -\frac{3}{2} & -\frac{1}{2} \\ 0 & 5 & -3 \\ 1 & 4 & 8 \end{bmatrix}.$$

(f) The matrix obtained by multiplying the third row by 3 and adding to the first row is

$$\begin{bmatrix} -2+3(1) & 3+3(4) & 1+3(8) \\ 0 & 5 & -3 \\ 1 & 4 & 8 \end{bmatrix} = \begin{bmatrix} 1 & 15 & 25 \\ 0 & 5 & -3 \\ 1 & 4 & 8 \end{bmatrix}.$$

2. **(a)** There are 3 rows and 2 columns, so the dimensions are written as

$$3 \times 2.$$

(b) 2 rows, 3 columns: 2×3

(c) 4 rows, 2 columns: 4×2

3. $\begin{aligned} 4x + 8y &= 44 \\ 2x - y &= -3 \end{aligned}$

$$\left[\begin{array}{rr|r} 4 & 8 & 44 \\ 2 & -1 & -3 \end{array}\right]$$

$$\left[\begin{array}{rr|r} 1 & \underline{2} & \underline{11} \\ 2 & -1 & -3 \end{array}\right] \quad \tfrac{1}{4}R_1$$

$$\left[\begin{array}{rr|r} 1 & 2 & 11 \\ 0 & \underline{-5} & \underline{-25} \end{array}\right] \quad -2R_1 + R_2$$

Note: $\left\{\begin{aligned} -2(2) + (-1) &= -5 \\ -2(11) + (-3) &= -25 \end{aligned}\right.$

$$\left[\begin{array}{rr|r} 1 & 2 & 11 \\ 0 & 1 & \underline{5} \end{array}\right] \quad -\tfrac{1}{5}R_2$$

This represents the system

$$\begin{aligned} x + 2y &= 11 \\ y &= 5. \end{aligned}$$

Substitute $y = 5$ in the first equation.

$$\begin{aligned} x + 2y &= 11 \\ x + 2(5) &= 11 \\ x + 10 &= 11 \\ x &= 1 \end{aligned}$$

Solution set: $\{(1, 5)\}$

4. $2x - 5y = -1$
$3x + \ \ y = \ \ 7$

$$\left[\begin{array}{cc|c} 2 & -5 & -1 \\ 3 & 1 & 7 \end{array}\right]$$

$$\left[\begin{array}{cc|c} 1 & -\frac{5}{2} & -\frac{1}{2} \\ 3 & 1 & 7 \end{array}\right] \quad \tfrac{1}{2}R_1$$

$$\left[\begin{array}{cc|c} 1 & -\frac{5}{2} & -\frac{1}{2} \\ 0 & \frac{17}{2} & \frac{17}{2} \end{array}\right] \quad -3R_1 + R_2$$

$$\left[\begin{array}{cc|c} 1 & -\frac{5}{2} & -\frac{1}{2} \\ 0 & 1 & 1 \end{array}\right] \quad \tfrac{2}{17}R_2$$

This represents the system

$$\begin{aligned} x - \tfrac{5}{2}y &= -\tfrac{1}{2} \\ y &= 1. \end{aligned}$$

Substitute $y = 1$ in the first equation.

$$\begin{aligned} x - \tfrac{5}{2}y &= -\tfrac{1}{2} \\ x - \tfrac{5}{2}(1) &= -\tfrac{1}{2} \\ x - \tfrac{5}{2} &= -\tfrac{1}{2} \\ x &= 2 \end{aligned}$$

Solution set: $\{(2, 1)\}$

5. $x + y = 5$
$x - y = 3$

Write the augmented matrix for this system.

$$\left[\begin{array}{cc|c} 1 & 1 & 5 \\ 1 & -1 & 3 \end{array}\right]$$

$$\left[\begin{array}{cc|c} 1 & 1 & 5 \\ 0 & -2 & -2 \end{array}\right] \quad -1R_1 + R_2$$

$$\left[\begin{array}{cc|c} 1 & 1 & 5 \\ 0 & 1 & 1 \end{array}\right] \quad -\tfrac{1}{2}R_2$$

This matrix gives the system

$$\begin{aligned} x + y &= 5 \\ y &= 1. \end{aligned}$$

Substitute $y = 1$ in the first equation.

$$\begin{aligned} x + y &= 5 \\ x + 1 &= 5 \\ x &= 4 \end{aligned}$$

Solution set: $\{(4, 1)\}$

6. $x + 2y = \ \ 7$
$x - \ \ y = -2$

Write the augmented matrix for this system.

$$\left[\begin{array}{cc|c} 1 & 2 & 7 \\ 1 & -1 & -2 \end{array}\right]$$

$$\left[\begin{array}{cc|c} 1 & 2 & 7 \\ 0 & -3 & -9 \end{array}\right] \quad -1R_1 + R_2$$

$$\left[\begin{array}{cc|c} 1 & 2 & 7 \\ 0 & 1 & 3 \end{array}\right] \quad -\tfrac{1}{3}R_2$$

This augmented matrix leads to the system

$$\begin{aligned} x + 2y &= 7 \\ y &= 3. \end{aligned}$$

Substitute $y = 3$ in the first equation.

$$\begin{aligned} x + 2y &= 7 \\ x + 2(3) &= 7 \\ x + 6 &= 7 \\ x &= 1 \end{aligned}$$

Solution set: $\{(1, 3)\}$

7. $2x + 4y = 6$
$3x - \ \ y = 2$

Write the augmented matrix.

$$\left[\begin{array}{cc|c} 2 & 4 & 6 \\ 3 & -1 & 2 \end{array}\right]$$

The easiest way to get a 1 in the first row, first column position is to multiply the elements in the first row by $\frac{1}{2}$.

$$\left[\begin{array}{cc|c} 1 & 2 & 3 \\ 3 & -1 & 2 \end{array}\right] \quad \tfrac{1}{2}R_1$$

To get a 0 in row two, column 1, we need to subtract 3 from the 3 that is in that position. To do this we will multiply row 1 by -3 and add the result to row 2.

$$\left[\begin{array}{cc|c} 1 & 2 & 3 \\ 0 & -7 & -7 \end{array}\right] \quad -3R_1 + R_2$$

$$\left[\begin{array}{cc|c} 1 & 2 & 3 \\ 0 & 1 & 1 \end{array}\right] \quad -\tfrac{1}{7}R_2$$

continued

This matrix gives the system

$$\begin{aligned} x + 2y &= 3 \\ y &= 1. \end{aligned}$$

Substitute $y = 1$ in the first equation.

$$\begin{aligned} x + 2y &= 3 \\ x + 2(1) &= 3 \\ x &= 1 \end{aligned}$$

Solution set: $\{(1, 1)\}$

8. $4x + 5y = -7$
$x - y = 5$

Write the augmented matrix.

$$\left[\begin{array}{rr|r} 4 & 5 & -7 \\ 1 & -1 & 5 \end{array}\right]$$

$$\left[\begin{array}{rr|r} 1 & -1 & 5 \\ 4 & 5 & -7 \end{array}\right] \quad R_1 \leftrightarrow R_2$$

We use $\leftrightarrow$ to represent the interchanging of 2 rows.

$$\left[\begin{array}{rr|r} 1 & -1 & 5 \\ 0 & 9 & -27 \end{array}\right] \quad -4R_1 + R_2$$

$$\left[\begin{array}{rr|r} 1 & -1 & 5 \\ 0 & 1 & -3 \end{array}\right] \quad \tfrac{1}{9}R_2$$

This matrix gives the system

$$\begin{aligned} x - y &= 5 \\ y &= -3. \end{aligned}$$

Substitute $y = -3$ in the first equation.

$$\begin{aligned} x - (-3) &= 5 \\ x &= 2 \end{aligned}$$

Solution set: $\{(2, -3)\}$

9. $3x + 4y = 13$
$2x - 3y = -14$

Write the augmented matrix.

$$\left[\begin{array}{rr|r} 3 & 4 & 13 \\ 2 & -3 & -14 \end{array}\right]$$

$$\left[\begin{array}{rr|r} 1 & 7 & 27 \\ 2 & -3 & -14 \end{array}\right] \quad -1R_2 + R_1$$

$$\left[\begin{array}{rr|r} 1 & 7 & 27 \\ 0 & -17 & -68 \end{array}\right] \quad -2R_1 + R_2$$

$$\left[\begin{array}{rr|r} 1 & 7 & 27 \\ 0 & 1 & 4 \end{array}\right] \quad -\tfrac{1}{17}R_2$$

This matrix gives the system

$$\begin{aligned} x + 7y &= 27 \\ y &= 4. \end{aligned}$$

Substitute $y = 4$ in the first equation.

$$\begin{aligned} x + 7y &= 27 \\ x + 7(4) &= 27 \\ x + 28 &= 27 \\ x &= -1 \end{aligned}$$

Solution set: $\{(-1, 4)\}$

10. $5x + 2y = 8$
$3x - y = 7$

Write the augmented matrix.

$$\left[\begin{array}{rr|r} 5 & 2 & 8 \\ 3 & -1 & 7 \end{array}\right]$$

$$\left[\begin{array}{rr|r} 5 & 2 & 8 \\ 6 & -2 & 14 \end{array}\right] \quad 2R_2$$

$$\left[\begin{array}{rr|r} 5 & 2 & 8 \\ 1 & -4 & 6 \end{array}\right] \quad -R_1 + R_2$$

$$\left[\begin{array}{rr|r} 1 & -4 & 6 \\ 5 & 2 & 8 \end{array}\right] \quad R_1 \leftrightarrow R_2$$

$$\left[\begin{array}{rr|r} 1 & -4 & 6 \\ 0 & 22 & -22 \end{array}\right] \quad -5R_1 + R_2$$

$$\left[\begin{array}{rr|r} 1 & -4 & 6 \\ 0 & 1 & -1 \end{array}\right] \quad \tfrac{1}{22}R_2$$

This matrix gives the system

$$\begin{aligned} x - 4y &= 6 \\ y &= -1. \end{aligned}$$

Substitute $y = -1$ in the first equation.

$$\begin{aligned} x - 4(-1) &= 6 \\ x + 4 &= 6 \\ x &= 2 \end{aligned}$$

Solution set: $\{(2, -1)\}$

11. $-4x + 12y = 36$
$x - 3y = 9$

Write the augmented matrix.

$$\left[\begin{array}{rr|r} -4 & 12 & 36 \\ 1 & -3 & 9 \end{array}\right]$$

$$\left[\begin{array}{rr|r} -1 & 3 & 9 \\ 1 & -3 & 9 \end{array}\right] \quad \tfrac{1}{4}R_1$$

$$\left[\begin{array}{rr|r} -1 & 3 & 9 \\ 0 & 0 & 18 \end{array}\right] \quad R_1 + R_2$$

The corresponding system is

$$\begin{aligned} -x + 3y &= 9 \\ 0 &= 18 \quad \textit{False} \end{aligned}$$

which is inconsistent and has no solution.

Solution set: $\emptyset$

12. $2x - 4y = 8$
$-3x + 6y = 5$

Write the augmented matrix.

$$\left[\begin{array}{rr|r} 2 & -4 & 8 \\ -3 & 6 & 5 \end{array}\right]$$

$$\left[\begin{array}{rr|r} 1 & -2 & 4 \\ -3 & 6 & 5 \end{array}\right] \quad \tfrac{1}{2}R_1$$

$$\left[\begin{array}{rr|r} 1 & -2 & 4 \\ 0 & 0 & 17 \end{array}\right] \quad 3R_1 + R_2$$

From row 2 $(0 = 17)$ we see that the system is inconsistent.

Solution set: $\emptyset$

13. $2x + y = 4$
$4x + 2y = 8$

Write the augmented matrix.

$$\left[\begin{array}{rr|r} 2 & 1 & 4 \\ 4 & 2 & 8 \end{array}\right]$$

$$\left[\begin{array}{rr|r} 1 & \frac{1}{2} & 2 \\ 4 & 2 & 8 \end{array}\right] \quad \tfrac{1}{2}R_1$$

$$\left[\begin{array}{rr|r} 1 & \frac{1}{2} & 2 \\ 0 & 0 & 0 \end{array}\right] \quad -4R_1 + R_2$$

Row 2, $0 = 0$, indicates that the system has dependent equations.

Solution set: $\{(x, y) \,|\, 2x + y = 4\}$

14. $-3x - 4y = 1$
$6x + 8y = -2$

Write the augmented matrix.

$$\left[\begin{array}{rr|r} -3 & -4 & 1 \\ 6 & 8 & -2 \end{array}\right]$$

$$\left[\begin{array}{rr|r} 1 & \frac{4}{3} & -\frac{1}{3} \\ 6 & 8 & -2 \end{array}\right] \quad -\tfrac{1}{3}R_1$$

$$\left[\begin{array}{rr|r} 1 & \frac{4}{3} & -\frac{1}{3} \\ 0 & 0 & 0 \end{array}\right] \quad -6R_1 + R_2$$

Row 2, $0 = 0$, indicates that the system has dependent equations.

Solution set: $\{(x, y) \,|\, -3x - 4y = 1\}$

15. $\dfrac{1}{2}x + \dfrac{1}{3}y = 0$
$\dfrac{2}{3}x + \dfrac{3}{4}y = 0$

Write the augmented matrix.

$$\left[\begin{array}{rr|r} \frac{1}{2} & \frac{1}{3} & 0 \\ \frac{2}{3} & \frac{3}{4} & 0 \end{array}\right]$$

$$\left[\begin{array}{rr|r} 1 & \frac{2}{3} & 0 \\ \frac{2}{3} & \frac{3}{4} & 0 \end{array}\right] \quad 2R_1$$

$$\left[\begin{array}{rr|r} 1 & \frac{2}{3} & 0 \\ 0 & \frac{11}{36} & 0 \end{array}\right] \quad -\tfrac{2}{3}R_1 + R_2$$

$$\left[\begin{array}{rr|r} 1 & \frac{2}{3} & 0 \\ 0 & 1 & 0 \end{array}\right] \quad \tfrac{36}{11}R_2$$

This matrix gives the system

$$x + \tfrac{2}{3}y = 0$$
$$y = 0.$$

Substitute $y = 0$ in the first equation.

$$x + \tfrac{2}{3}(0) = 0$$
$$x = 0$$

Solution set: $\{(0, 0)\}$

16. $1.2x + .3y = 0$
$2.9x - .6y = 0$

Eliminate decimals first.

$$12x + 3y = 0$$
$$29x - 6y = 0$$

Write the augmented matrix.

$$\left[\begin{array}{rr|r} 12 & 3 & 0 \\ 29 & -6 & 0 \end{array}\right]$$

$$\left[\begin{array}{rr|r} 1 & \frac{1}{4} & 0 \\ 29 & -6 & 0 \end{array}\right] \quad \tfrac{1}{12}R_1$$

$$\left[\begin{array}{rr|r} 1 & \frac{1}{4} & 0 \\ 0 & -\frac{53}{4} & 0 \end{array}\right] \quad -29R_1 + R_2$$

$$\left[\begin{array}{rr|r} 1 & \frac{1}{4} & 0 \\ 0 & 1 & 0 \end{array}\right] \quad -\tfrac{4}{53}R_2$$

This matrix gives the system

$$x + \tfrac{1}{4}y = 0$$
$$y = 0.$$

Substitute $y = 0$ in the first equation.

$$x + \tfrac{1}{4}(0) = 0$$
$$x = 0$$

Solution set: $\{(0, 0)\}$

17.
$$\begin{aligned} x + y - z &= -3 \\ 2x + y + z &= 4 \\ 5x - y + 2z &= 23 \end{aligned}$$

Write the augmented matrix.

$$\left[\begin{array}{rrr|r} 1 & 1 & -1 & -3 \\ 2 & 1 & 1 & 4 \\ 5 & -1 & 2 & 23 \end{array}\right]$$

$$\left[\begin{array}{rrr|r} 1 & 1 & -1 & -3 \\ 0 & -1 & 3 & 10 \\ 0 & -6 & 7 & 38 \end{array}\right] \quad \begin{array}{l} \\ -2R_1 + R_2 \\ -5R_1 + R_3 \end{array}$$

$$\left[\begin{array}{rrr|r} 1 & 1 & -1 & -3 \\ 0 & 1 & -3 & -10 \\ 0 & -6 & 7 & 38 \end{array}\right] \quad -1R_2$$

$$\left[\begin{array}{rrr|r} 1 & 1 & -1 & -3 \\ 0 & 1 & -3 & -10 \\ 0 & 0 & -11 & -22 \end{array}\right] \quad 6R_2 + R_3$$

$$\left[\begin{array}{rrr|r} 1 & 1 & -1 & -3 \\ 0 & 1 & -3 & -10 \\ 0 & 0 & 1 & 2 \end{array}\right] \quad -\tfrac{1}{11}R_3$$

This matrix gives the system

$$\begin{aligned} x + y - z &= -3 \\ y - 3z &= -10 \\ z &= 2. \end{aligned}$$

Substitute $z = 2$ in the second equation.

$$\begin{aligned} y - 3z &= -10 \\ y - 3(2) &= -10 \\ y - 6 &= -10 \\ y &= -4 \end{aligned}$$

Substitute $y = -4$ and $z = 2$ in the first equation.

$$\begin{aligned} x + y - z &= -3 \\ x - 4 - 2 &= -3 \\ x - 6 &= -3 \\ x &= 3 \end{aligned}$$

Solution set: $\{(3, -4, 2)\}$

18.
$$\begin{aligned} 2x + y + 2z &= 11 \\ 2x - y - z &= -3 \\ 3x + 2y + z &= 9 \end{aligned}$$

Write the augmented matrix.

$$\left[\begin{array}{rrr|r} 2 & 1 & 2 & 11 \\ 2 & -1 & -1 & -3 \\ 3 & 2 & 1 & 9 \end{array}\right]$$

$$\left[\begin{array}{rrr|r} 1 & \frac{1}{2} & 1 & \frac{11}{2} \\ 2 & -1 & -1 & -3 \\ 3 & 2 & 1 & 9 \end{array}\right] \quad \tfrac{1}{2}R_1$$

$$\left[\begin{array}{rrr|r} 1 & \frac{1}{2} & 1 & \frac{11}{2} \\ 0 & -2 & -3 & -14 \\ 0 & \frac{1}{2} & -2 & -\frac{15}{2} \end{array}\right] \quad \begin{array}{l} \\ -2R_1 + R_2 \\ -3R_1 + R_3 \end{array}$$

$$\left[\begin{array}{rrr|r} 1 & \frac{1}{2} & 1 & \frac{11}{2} \\ 0 & 1 & \frac{3}{2} & 7 \\ 0 & \frac{1}{2} & -2 & -\frac{15}{2} \end{array}\right] \quad -\tfrac{1}{2}R_2$$

$$\left[\begin{array}{rrr|r} 1 & \frac{1}{2} & 1 & \frac{11}{2} \\ 0 & 1 & \frac{3}{2} & 7 \\ 0 & 0 & -\frac{11}{4} & -11 \end{array}\right] \quad -\tfrac{1}{2}R_2 + R_3$$

$$\left[\begin{array}{rrr|r} 1 & \frac{1}{2} & 1 & \frac{11}{2} \\ 0 & 1 & \frac{3}{2} & 7 \\ 0 & 0 & 1 & 4 \end{array}\right] \quad -\tfrac{4}{11}R_3$$

This matrix gives the system

$$\begin{aligned} x + \tfrac{1}{2}y + z &= \tfrac{11}{2} \\ y + \tfrac{3}{2}z &= 7 \\ z &= 4. \end{aligned}$$

Substitute $z = 4$ in the second equation.

$$\begin{aligned} y + \tfrac{3}{2}z &= 7 \\ y + \tfrac{3}{2}(4) &= 7 \\ y + 6 &= 7 \\ y &= 1 \end{aligned}$$

Substitute $y = 1$ and $z = 4$ in the first equation.

$$\begin{aligned} x + \tfrac{1}{2}y + z &= \tfrac{11}{2} \\ x + \tfrac{1}{2}(1) + 4 &= \tfrac{11}{2} \\ x + \tfrac{9}{2} &= \tfrac{11}{2} \\ x &= 1 \end{aligned}$$

Solution set: $\{(1, 1, 4)\}$

19.
$$\begin{aligned} x + y - 3z &= 1 \\ 2x - y + z &= 9 \\ 3x + y - 4z &= 8 \end{aligned}$$

Write the augmented matrix.

$$\left[\begin{array}{rrr|r} 1 & 1 & -3 & 1 \\ 2 & -1 & 1 & 9 \\ 3 & 1 & -4 & 8 \end{array}\right]$$

$$\left[\begin{array}{rrr|r} 1 & 1 & -3 & 1 \\ 0 & -3 & 7 & 7 \\ 0 & -2 & 5 & 5 \end{array}\right] \quad \begin{array}{l} \\ -2R_1 + R_2 \\ -3R_1 + R_3 \end{array}$$

$$\left[\begin{array}{rrr|r} 1 & 1 & -3 & 1 \\ 0 & 1 & -\frac{7}{3} & -\frac{7}{3} \\ 0 & -2 & 5 & 5 \end{array}\right] \quad -\tfrac{1}{3}R_2$$

$$\left[\begin{array}{rrr|r} 1 & 1 & -3 & 1 \\ 0 & 1 & -\frac{7}{3} & -\frac{7}{3} \\ 0 & 0 & \frac{1}{3} & \frac{1}{3} \end{array}\right] \quad 2R_2 + R_3$$

$$\left[\begin{array}{ccc|c} 1 & 1 & -3 & 1 \\ 0 & 1 & -\frac{7}{3} & -\frac{7}{3} \\ 0 & 0 & 1 & 1 \end{array}\right] \quad 3\text{R}_3$$

This matrix gives the system

$$\begin{aligned} x + y - 3z &= 1 \\ y - \tfrac{7}{3}z &= -\tfrac{7}{3} \\ z &= 1. \end{aligned}$$

Substitute $z = 1$ in the second equation.

$$\begin{aligned} y - \tfrac{7}{3}z &= -\tfrac{7}{3} \\ y - \tfrac{7}{3}(1) &= -\tfrac{7}{3} \\ y &= 0 \end{aligned}$$

Substitute $y = 0$ and $z = 1$ in the first equation.

$$\begin{aligned} x + y - 3z &= 1 \\ x + 0 - 3(1) &= 1 \\ x - 3 &= 1 \\ x &= 4 \end{aligned}$$

Solution set: $\{(4, 0, 1)\}$

20. $$\begin{aligned} 2x + 4y - 3z &= -18 \\ 3x + y - z &= -5 \\ x - 2y + 4z &= 14 \end{aligned}$$

Write the augmented matrix.

$$\left[\begin{array}{ccc|c} 2 & 4 & -3 & -18 \\ 3 & 1 & -1 & -5 \\ 1 & -2 & 4 & 14 \end{array}\right]$$

$$\left[\begin{array}{ccc|c} 0 & 8 & -11 & -46 \\ 0 & 7 & -13 & -47 \\ 1 & -2 & 4 & 14 \end{array}\right] \quad \begin{array}{l} -2\text{R}_3 + \text{R}_1 \\ -3\text{R}_3 + \text{R}_2 \end{array}$$

$$\left[\begin{array}{ccc|c} 0 & 8 & -11 & -46 \\ 0 & 1 & -\frac{13}{7} & -\frac{47}{7} \\ 1 & -2 & 4 & 14 \end{array}\right] \quad \tfrac{1}{7}\text{R}_2$$

$$\left[\begin{array}{ccc|c} 0 & 0 & \frac{27}{7} & \frac{54}{7} \\ 0 & 1 & -\frac{13}{7} & -\frac{47}{7} \\ 1 & -2 & 4 & 14 \end{array}\right] \quad -8\text{R}_2 + \text{R}_1$$

This matrix gives the system

$$\begin{aligned} \tfrac{27}{7}z &= \tfrac{54}{7} \\ y - \tfrac{13}{7}z &= -\tfrac{47}{7} \\ x - 2y + 4z &= 14. \end{aligned}$$

From the first equation,

$$\begin{aligned} \tfrac{27}{7}z &= \tfrac{54}{7} \\ z &= 2. \end{aligned}$$

Substitute $z = 2$ in the second equation.

$$\begin{aligned} y - \tfrac{13}{7}z &= -\tfrac{47}{7} \\ y - \tfrac{13}{7}(2) &= -\tfrac{47}{7} \\ y - \tfrac{26}{7} &= -\tfrac{47}{7} \\ y &= -\tfrac{21}{7} = -3 \end{aligned}$$

Substitute $y = -3$ and $z = 2$ in the third equation.

$$\begin{aligned} x - 2y + 4z &= 14 \\ x - 2(-3) + 4(2) &= 14 \\ x + 6 + 8 &= 14 \\ x + 14 &= 14 \\ x &= 0 \end{aligned}$$

Solution set: $\{(0, -3, 2)\}$

21. $$\begin{aligned} x + y - z &= 6 \\ 2x - y + z &= -9 \\ x - 2y + 3z &= 1 \end{aligned}$$

Write the augmented matrix.

$$\left[\begin{array}{ccc|c} 1 & 1 & -1 & 6 \\ 2 & -1 & 1 & -9 \\ 1 & -2 & 3 & 1 \end{array}\right]$$

$$\left[\begin{array}{ccc|c} 1 & 1 & -1 & 6 \\ 0 & -3 & 3 & -21 \\ 0 & -3 & 4 & -5 \end{array}\right] \quad \begin{array}{l} -2\text{R}_1 + \text{R}_2 \\ -1\text{R}_1 + \text{R}_3 \end{array}$$

$$\left[\begin{array}{ccc|c} 1 & 1 & -1 & 6 \\ 0 & 1 & -1 & 7 \\ 0 & -3 & 4 & -5 \end{array}\right] \quad -\tfrac{1}{3}\text{R}_2$$

$$\left[\begin{array}{ccc|c} 1 & 1 & -1 & 6 \\ 0 & 1 & -1 & 7 \\ 0 & 0 & 1 & 16 \end{array}\right] \quad 3\text{R}_2 + \text{R}_3$$

This matrix gives the system

$$\begin{aligned} x + y - z &= 6 \\ y - z &= 7 \\ z &= 16. \end{aligned}$$

Substitute $z = 16$ in the second equation.

$$\begin{aligned} y - 16 &= 7 \\ y &= 23 \end{aligned}$$

Substitute $y = 23$ and $z = 16$ in the first equation.

$$\begin{aligned} x + 23 - 16 &= 6 \\ x + 7 &= 6 \\ x &= -1 \end{aligned}$$

Solution set: $\{(-1, 23, 16)\}$

22.
$$\begin{aligned} x + 3y - 6z &= 7 \\ 2x - y + 2z &= 0 \\ x + y + 2z &= -1 \end{aligned}$$

Write the augmented matrix.

$$\left[\begin{array}{rrr|r} 1 & 3 & -6 & 7 \\ 2 & -1 & 2 & 0 \\ 1 & 1 & 2 & -1 \end{array}\right]$$

$$\left[\begin{array}{rrr|r} 1 & 3 & -6 & 7 \\ 0 & -7 & 14 & -14 \\ 0 & -2 & 8 & -8 \end{array}\right] \begin{array}{l} \\ -2R_1 + R_2 \\ -1R_1 + R_3 \end{array}$$

$$\left[\begin{array}{rrr|r} 1 & 3 & -6 & 7 \\ 0 & 1 & -2 & 2 \\ 0 & 1 & -4 & 4 \end{array}\right] \begin{array}{l} \\ -\frac{1}{7}R_2 \\ -\frac{1}{2}R_3 \end{array}$$

$$\left[\begin{array}{rrr|r} 1 & 3 & -6 & 7 \\ 0 & 1 & -2 & 2 \\ 0 & 0 & -2 & 2 \end{array}\right] \begin{array}{l} \\ \\ -1R_2 + R_3 \end{array}$$

$$\left[\begin{array}{rrr|r} 1 & 3 & -6 & 7 \\ 0 & 1 & -2 & 2 \\ 0 & 0 & 1 & -1 \end{array}\right] \begin{array}{l} \\ \\ -\frac{1}{2}R_3 \end{array}$$

This matrix gives the system

$$\begin{aligned} x + 3y - 6z &= 7 \\ y - 2z &= 2 \\ z &= -1. \end{aligned}$$

Substitute $z = -1$ in the second equation.

$$\begin{aligned} y - 2z &= 2 \\ y - 2(-1) &= 2 \\ y + 2 &= 2 \\ y &= 0 \end{aligned}$$

Substitute $y = 0$ and $z = -1$ in the first equation.

$$\begin{aligned} x + 3y - 6z &= 7 \\ x + 3(0) - 6(-1) &= 7 \\ x + 6 &= 7 \\ x &= 1 \end{aligned}$$

Solution set: $\{(1, 0, -1)\}$

23.
$$\begin{aligned} x - y &= 1 \\ y - z &= 6 \\ x + z &= -1 \end{aligned}$$

Write the augmented matrix.

$$\left[\begin{array}{rrr|r} 1 & -1 & 0 & 1 \\ 0 & 1 & -1 & 6 \\ 1 & 0 & 1 & -1 \end{array}\right]$$

$$\left[\begin{array}{rrr|r} 1 & -1 & 0 & 1 \\ 0 & 1 & -1 & 6 \\ 0 & 1 & 1 & -2 \end{array}\right] \begin{array}{l} \\ \\ -1R_1 + R_3 \end{array}$$

$$\left[\begin{array}{rrr|r} 1 & -1 & 0 & 1 \\ 0 & 1 & -1 & 6 \\ 0 & 0 & 2 & -8 \end{array}\right] \begin{array}{l} \\ \\ -1R_2 + R_3 \end{array}$$

$$\left[\begin{array}{rrr|r} 1 & -1 & 0 & 1 \\ 0 & 1 & -1 & 6 \\ 0 & 0 & 1 & -4 \end{array}\right] \begin{array}{l} \\ \\ \frac{1}{2}R_3 \end{array}$$

This matrix gives the system

$$\begin{aligned} x - y &= 1 \\ y - z &= 6 \\ z &= -4. \end{aligned}$$

Substitute $z = -4$ in the second equation.

$$\begin{aligned} y - z &= 6 \\ y - (-4) &= 6 \\ y &= 2 \end{aligned}$$

Substitute $y = 2$ in the first equation.

$$\begin{aligned} x - y &= 1 \\ x - 2 &= 1 \\ x &= 3 \end{aligned}$$

Solution set: $\{(3, 2, -4)\}$

24.
$$\begin{aligned} x + y &= 1 \\ 2x - z &= 0 \\ y + 2z &= -2 \end{aligned}$$

Write the augmented matrix.

$$\left[\begin{array}{rrr|r} 1 & 1 & 0 & 1 \\ 2 & 0 & -1 & 0 \\ 0 & 1 & 2 & -2 \end{array}\right]$$

$$\left[\begin{array}{rrr|r} 1 & 1 & 0 & 1 \\ 0 & -2 & -1 & -2 \\ 0 & 1 & 2 & -2 \end{array}\right] \begin{array}{l} \\ -2R_1 + R_2 \\ \\ \end{array}$$

$$\left[\begin{array}{rrr|r} 1 & 1 & 0 & 1 \\ 0 & 0 & 3 & -6 \\ 0 & 1 & 2 & -2 \end{array}\right] \begin{array}{l} \\ 2R_3 + R_2 \\ \\ \end{array}$$

This matrix gives the system

$$\begin{aligned} x + y &= 1 \\ 3z &= -6 \\ y + 2z &= -2. \end{aligned}$$

From the second equation,

$$\begin{aligned} 3z &= -6 \\ z &= -2. \end{aligned}$$

Substitute $z = -2$ in the third equation.

$$\begin{aligned} y + 2z &= -2 \\ y + 2(-2) &= -2 \\ y - 4 &= -2 \\ y &= 2 \end{aligned}$$

Substitute $y = 2$ in the first equation.

$$\begin{aligned} x + y &= 1 \\ x + 2 &= 1 \\ x &= -1 \end{aligned}$$

Solution set: $\{(-1, 2, -2)\}$

25. $$\begin{aligned} 4x + 8y + 4z &= 9 \\ x + 3y + 4z &= 10 \\ 5x + 10y + 5z &= 12 \end{aligned}$$

Write the augmented matrix.

$$\left[\begin{array}{ccc|c} 4 & 8 & 4 & 9 \\ 1 & 3 & 4 & 10 \\ 5 & 10 & 5 & 12 \end{array}\right]$$

$$\left[\begin{array}{ccc|c} 1 & 3 & 4 & 10 \\ 4 & 8 & 4 & 9 \\ 5 & 10 & 5 & 12 \end{array}\right] \quad R_1 \leftrightarrow R_2$$

$$\left[\begin{array}{ccc|c} 1 & 3 & 4 & 10 \\ 0 & -4 & -12 & -31 \\ 0 & -5 & -15 & -38 \end{array}\right] \quad \begin{array}{l} \\ -4R_1 + R_2 \\ -5R_1 + R_3 \end{array}$$

$$\left[\begin{array}{ccc|c} 1 & 3 & 4 & 10 \\ 0 & 1 & 3 & 7 \\ 0 & -5 & -15 & -38 \end{array}\right] \quad -R_3 + R_2$$

$$\left[\begin{array}{ccc|c} 1 & 3 & 4 & 10 \\ 0 & 1 & 3 & 7 \\ 0 & 0 & 0 & -3 \end{array}\right] \quad 5R_2 + R_3$$

From the last row, $0 = -3$, we see that the system is inconsistent.

Solution set: $\emptyset$

26. $$\begin{aligned} x + 2y + 3z &= -2 \\ 2x + 4y + 6z &= -5 \\ x - y + 2z &= 6 \end{aligned}$$

Write the augmented matrix.

$$\left[\begin{array}{ccc|c} 1 & 2 & 3 & -2 \\ 2 & 4 & 6 & -5 \\ 1 & -1 & 2 & 6 \end{array}\right]$$

$$\left[\begin{array}{ccc|c} 1 & 2 & 3 & -2 \\ 0 & 0 & 0 & -1 \\ 0 & -3 & -1 & 8 \end{array}\right] \quad \begin{array}{l} \\ -2R_1 + R_2 \\ -R_1 + R_3 \end{array}$$

From the second row, $0 = -1$, we see that the system is inconsistent.

Solution set: $\emptyset$

27. $$\begin{aligned} x - 2y + z &= 4 \\ 3x - 6y + 3z &= 12 \\ -2x + 4y - 2z &= -8 \end{aligned}$$

Write the augmented matrix.

$$\left[\begin{array}{ccc|c} 1 & -2 & 1 & 4 \\ 3 & -6 & 3 & 12 \\ -2 & 4 & -2 & -8 \end{array}\right]$$

$$\left[\begin{array}{ccc|c} 1 & -2 & 1 & 4 \\ 1 & -2 & 1 & 4 \\ -1 & 2 & -1 & -4 \end{array}\right] \quad \begin{array}{l} \\ \frac{1}{3}R_2 \\ \frac{1}{2}R_3 \end{array}$$

$$\left[\begin{array}{ccc|c} 1 & -2 & 1 & 4 \\ 0 & 0 & 0 & 0 \\ 0 & 0 & 0 & 0 \end{array}\right] \quad \begin{array}{l} \\ -1R_1 + R_2 \\ R_1 + R_3 \end{array}$$

This augmented matrix represents a system of dependent equations.

Solution set: $\{(x, y, z) \mid x - 2y + z = 4\}$

28. $$\begin{aligned} x + 3y + z &= 1 \\ 2x + 6y + 2z &= 2 \\ 3x + 9y + 3z &= 3 \end{aligned}$$

Write the augmented matrix.

$$\left[\begin{array}{ccc|c} 1 & 3 & 1 & 1 \\ 2 & 6 & 2 & 2 \\ 3 & 9 & 3 & 3 \end{array}\right]$$

$$\left[\begin{array}{ccc|c} 1 & 3 & 1 & 1 \\ 0 & 0 & 0 & 0 \\ 0 & 0 & 0 & 0 \end{array}\right] \quad \begin{array}{l} \\ -2R_1 + R_2 \\ -3R_1 + R_3 \end{array}$$

This augmented matrix represents a system of dependent equations.

Solution set: $\{(x, y, z) \mid x + 3y + z = 1\}$

29. $$\begin{aligned} 5x + 3y - z &= 0 \\ 2x - 3y + z &= 0 \\ x + 4y - 2z &= 0 \end{aligned}$$

Write the augmented matrix.

$$\left[\begin{array}{ccc|c} 5 & 3 & -1 & 0 \\ 2 & -3 & 1 & 0 \\ 1 & 4 & -2 & 0 \end{array}\right]$$

$$\left[\begin{array}{ccc|c} 1 & 4 & -2 & 0 \\ 2 & -3 & 1 & 0 \\ 5 & 3 & -1 & 0 \end{array}\right] \quad R_1 \leftrightarrow R_3$$

$$\left[\begin{array}{ccc|c} 1 & 4 & -2 & 0 \\ 0 & -11 & 5 & 0 \\ 0 & -17 & 9 & 0 \end{array}\right] \quad \begin{array}{l} \\ -2R_1 + R_2 \\ -5R_1 + R_2 \end{array}$$

$$\left[\begin{array}{ccc|c} 1 & 4 & -2 & 0 \\ 0 & 1 & -\frac{5}{11} & 0 \\ 0 & -17 & 9 & 0 \end{array}\right] \quad -\frac{1}{11}R_2$$

$$\left[\begin{array}{ccc|c} 1 & 4 & -2 & 0 \\ 0 & 1 & -\frac{5}{11} & 0 \\ 0 & 0 & \frac{14}{11} & 0 \end{array}\right] \quad 17R_2 + R_3$$

$$\left[\begin{array}{ccc|c} 1 & 4 & -2 & 0 \\ 0 & 1 & -\frac{5}{11} & 0 \\ 0 & 0 & 1 & 0 \end{array}\right] \quad \frac{11}{14}R_3$$

continued

This matrix gives the system

$$\begin{aligned} x + 4y - 2z &= 0 \\ y - \tfrac{5}{11}z &= 0 \\ z &= 0. \end{aligned}$$

Substituting $z = 0$ in the second equation gives $y = 0$ and substituting $y = 0$ and $z = 0$ in the first equation gives $x = 0$.

Solution set: $\{(0, 0, 0)\}$

30. $4x + 5y - z = 0$
$7x - 5y + z = 0$
$x + 3y - 2z = 0$

Write the augmented matrix.

$$\left[\begin{array}{rrr|r} 4 & 5 & -1 & 0 \\ 7 & -5 & 1 & 0 \\ 1 & 3 & -2 & 0 \end{array}\right]$$

$$\left[\begin{array}{rrr|r} 1 & 3 & -2 & 0 \\ 7 & -5 & 1 & 0 \\ 4 & 5 & -1 & 0 \end{array}\right] \quad R_1 \leftrightarrow R_3$$

$$\left[\begin{array}{rrr|r} 1 & 3 & -2 & 0 \\ 0 & -26 & 15 & 0 \\ 0 & -7 & 7 & 0 \end{array}\right] \quad \begin{array}{l} \\ -7R_1 + R_2 \\ -4R_1 + R_2 \end{array}$$

$$\left[\begin{array}{rrr|r} 1 & 3 & -2 & 0 \\ 0 & 1 & -\frac{15}{26} & 0 \\ 0 & 1 & -1 & 0 \end{array}\right] \quad \begin{array}{l} \\ -\frac{1}{26}R_2 \\ -\frac{1}{7}R_3 \end{array}$$

$$\left[\begin{array}{rrr|r} 1 & 3 & -2 & 0 \\ 0 & 1 & -\frac{15}{26} & 0 \\ 0 & 0 & -\frac{11}{26} & 0 \end{array}\right] \quad \begin{array}{l} \\ \\ -R_2 + R_3 \end{array}$$

$$\left[\begin{array}{rrr|r} 1 & 3 & -2 & 0 \\ 0 & 1 & -\frac{15}{26} & 0 \\ 0 & 0 & 1 & 0 \end{array}\right] \quad \begin{array}{l} \\ \\ -\frac{26}{11}R_3 \end{array}$$

This matrix gives the system

$$\begin{aligned} x + 3y - 2z &= 0 \\ y - \tfrac{15}{26}z &= 0 \\ z &= 0. \end{aligned}$$

Substituting $z = 0$ in the second equation gives $y = 0$ and substituting $y = 0$ and $z = 0$ in the first equation gives $x = 0$.

Solution set: $\{(0, 0, 0)\}$

Chapter 5 Review Exercises

When solving systems, check each solution in the equations of the original system.

1. $x + 3y = 8 \quad (1)$
$2x - y = 2 \quad (2)$

Graph the two lines. They appear to intersect at the point $(2, 2)$. Check $(2, 2)$ in both equations.

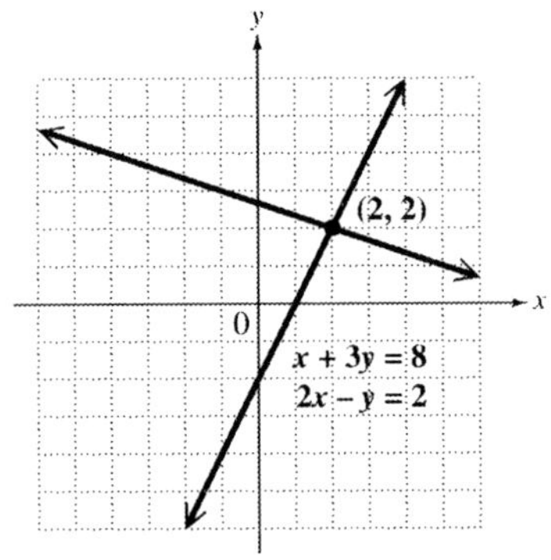

$$\begin{aligned} x + 3y &= 8 && (1) \\ 2 + 3(2) &= 8 \quad ? \\ 8 &= 8 && \textit{True} \end{aligned}$$

$$\begin{aligned} 2x - y &= 2 && (2) \\ 2(2) - 2 &= 2 \quad ? \\ 2 &= 2 && \textit{True} \end{aligned}$$

Solution set: $\{(2, 2)\}$

2. **(a)** The graphs meet between the years of 1978 and 1982. Therefore, the number of men and women reached equal numbers between 1978 and 1982.

(b) The number was just less than 500,000.

3. $3x + y = -4 \quad (1)$
$x = \frac{2}{3}y \quad (2)$

Substitute $\frac{2}{3}y$ for x in equation (1) and solve for y.

$$\begin{aligned} 3x + y &= -4 \quad (1) \\ 3\left(\tfrac{2}{3}y\right) + y &= 4 \\ 2y + y &= -4 \\ 3y &= -4 \\ y &= -\tfrac{4}{3} \end{aligned}$$

Since $x = \frac{2}{3}y$ and $y = -\frac{4}{3}$,

$$x = \tfrac{2}{3}\left(-\tfrac{4}{3}\right) = -\tfrac{8}{9}.$$

Solution set: $\left\{\left(-\frac{8}{9}, -\frac{4}{3}\right)\right\}$

4. $9x - y = -4 \quad (1)$
$y = x + 4 \quad (2)$

Substitute $x + 4$ for y in equation (1) and solve for x.

$$\begin{aligned} 9x - y &= -4 \quad (1) \\ 9x - (x + 4) &= -4 \\ 9x - x - 4 &= -4 \\ 8x &= 0 \\ x &= 0 \end{aligned}$$

Since $x = 0$, $y = x + 4 = 0 + 4 = 4$.

Solution set: $\{(0, 4)\}$

5. $-5x + 2y = -2$ (1)
$x + 6y = 26$ (2)

Solve equation (2) for x.

$$x = 26 - 6y$$

Substitute $26 - 6y$ for x in equation (1).

$$\begin{aligned} -5x + 2y &= -2 \quad (1) \\ -5(26 - 6y) + 2y &= -2 \\ -130 + 30y + 2y &= -2 \\ -130 + 32y &= -2 \\ 32y &= 128 \\ y &= 4 \end{aligned}$$

Since $x = 26 - 6y$ and $y = 4$,

$$x = 26 - 6(4) = 26 - 24 = 2.$$

Solution set: $\{(2, 4)\}$

6. $6x + 5y = 4$ (1)
$-4x + 2y = 8$ (2)

To eliminate x, multiply equation (1) by 2 and equation (2) by 3. Then add the results.

$$\begin{array}{rcrl} 12x + 10y &=& 8 & 2 \times (1) \\ -12x + 6y &=& 24 & 3 \times (2) \\ \hline 16y &=& 32 & \\ y &=& 2 & \end{array}$$

Since $y = 2$,

$$\begin{aligned} -4x + 2y &= 8 \quad (2) \\ -4x + 2(2) &= 8 \\ -4x + 4 &= 8 \\ -4x &= 4 \\ x &= -1. \end{aligned}$$

Solution set: $\{(-1, 2)\}$

7. $\frac{x}{6} + \frac{y}{6} = -\frac{1}{2}$ (1)
$x - y = -9$ (2)

Multiply equation (1) by 6 to clear the fractions. Add the result to equation (2) to eliminate y.

$$\begin{array}{rcrl} x + y &=& -3 & 6 \times (1) \\ x - y &=& -9 & (2) \\ \hline 2x &=& -12 & \\ x &=& -6 & \end{array}$$

Substitute -6 for x in equation (2).

$$\begin{aligned} x - y &= -9 \quad (2) \\ -6 - y &= -9 \\ -y &= -3 \\ y &= 3 \end{aligned}$$

Solution set: $\{(-6, 3)\}$

8. $4x + 5y = 9$ (1)
$3x + 7y = -1$ (2)

To eliminate x, multiply equation (1) by 3 and equation (2) by -4. Then add the results.

$$\begin{array}{rcrl} 12x + 15y &=& 27 & 3 \times (1) \\ -12x - 28y &=& 4 & -4 \times (2) \\ \hline -13y &=& 31 & \\ y &=& -\frac{31}{13} & \end{array}$$

To eliminate y, multiply equation (1) by 7 and equation (2) by -5. Then add the results.

$$\begin{array}{rcrl} 28x + 35y &=& 63 & 7 \times (1) \\ -15x - 35y &=& 5 & -5 \times (2) \\ \hline 13x &=& 68 & \\ x &=& \frac{68}{13} & \end{array}$$

Solution set: $\left\{\left(\frac{68}{13}, -\frac{31}{13}\right)\right\}$

9. $-3x + y = 6$ (1)
$2y = 12 + 6x$ (2)

Since equation (2) can be rewritten as

$$y = 6 + 3x$$
$$\text{or} \quad -3x + y = 6,$$

the two equations are the same, and hence, dependent.

Solution set: $\{(x, y) \mid 3x - y = -6\}$

10. $5x - 4y = 2$ (1)
$-10x + 8y = 7$ (2)

Multiply equation (1) by 2 and add the result to equation (2).

$$\begin{array}{rcrl} 10x - 8y &=& 4 & 2 \times (1) \\ -10x + 8y &=& 7 & (2) \\ \hline 0 &=& 11 & \textit{False} \end{array}$$

Since a false statement results, the system is *inconsistent*. The solution set is $\emptyset$.

11. If the system has a single solution, the graphs of the two linear equations of the system will intersect in one point.

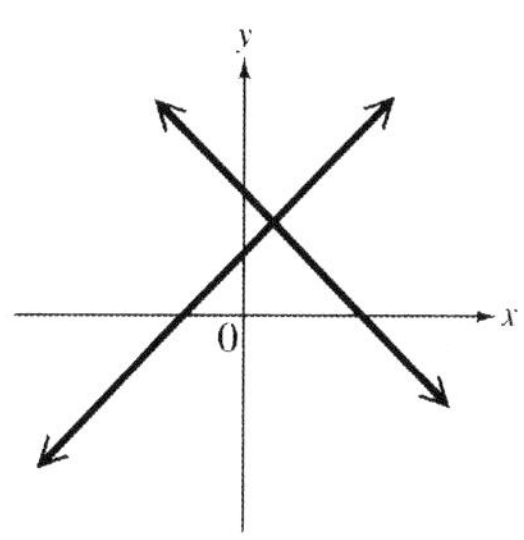

12. If the system has no solution, the graphs of the two linear equations are two parallel lines and do not intersect.

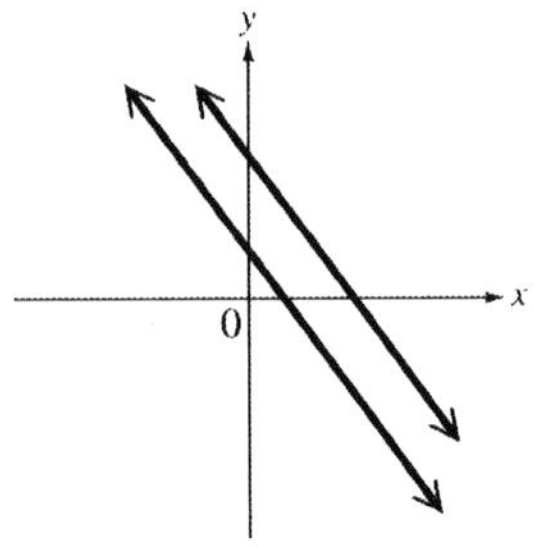

13. If the system has infinitely many solutions, the graphs of the two linear equations are the same line.

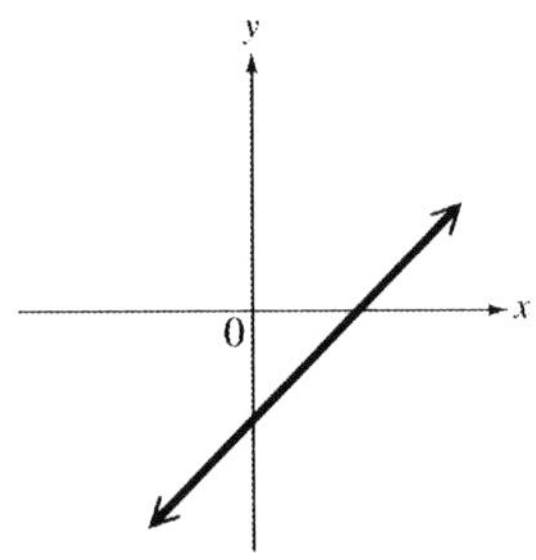

14. $y = 3x + 2$
$y = 3x - 4$

Since both equations are in slope-intercept form, their slopes and y-intercepts can be easily determined. Since they have the same slope but different y-intercepts, they are parallel lines. Thus, there is no point common to both lines, and the system has $\emptyset$ as its solution set.

15.

$$\begin{array}{rcrl} 2x + 3y - z &=& -16 & (1) \\ x + 2y + 2z &=& -3 & (2) \\ -3x + y + z &=& -5 & (3) \end{array}$$

To eliminate z, add equations (1) and (3).

$$\begin{array}{rcrl} 2x + 3y - z &=& -16 & (1) \\ -3x + y + z &=& -5 & (3) \\ \hline -x + 4y &=& -21 & (4) \end{array}$$

To eliminate z again, multiply equation (1) by 2 and add the result to equation (2).

$$\begin{array}{rcrl} 4x + 6y - 2z &=& -32 & 2 \times (1) \\ x + 2y + 2z &=& -3 & (2) \\ \hline 5x + 8y &=& -35 & (5) \end{array}$$

Use equations (4) and (5) to eliminate x. Multiply equation (4) by 5 and add the result to equation (5).

$$\begin{array}{rcrl} -5x + 20y &=& -105 & 5 \times (4) \\ 5x + 8y &=& -35 & (5) \\ \hline 28y &=& -140 & \\ y &=& -5 & \end{array}$$

Substitute -5 for y in equation (4) to find x.

$$\begin{aligned} -x + 4y &= -21 \quad (4) \\ -x + 4(-5) &= -21 \\ -x - 20 &= -21 \\ -x &= -1 \\ x &= 1 \end{aligned}$$

Substitute 1 for x and -5 for y in equation (2) to find z.

$$\begin{aligned} x + 2y + 2z &= -3 \quad (2) \\ 1 + 2(-5) + 2z &= -3 \\ 1 - 10 + 2z &= -3 \\ 2z &= 6 \\ z &= 3 \end{aligned}$$

The solution set is $\{(1, -5, 3)\}$.

16.

$$\begin{array}{rcrl} 3x - y - z &=& -8 & (1) \\ 4x + 2y + 3z &=& 15 & (2) \\ -6x + 2y + 2z &=& 10 & (3) \end{array}$$

To eliminate y, multiply equation (1) by 2 and add the result to equation (3).

$$\begin{array}{rcrl} 6x - 2y - 2z &=& -16 & 2 \times (1) \\ -6x + 2y + 2z &=& 10 & (3) \\ \hline 0 &=& -6 & \textit{False} \end{array}$$

Since a false statement results, equations (1) and (3) have no common solution. The system is *inconsistent*. The solution set is $\emptyset$.

17.

$$\begin{array}{rcrl} 4x - y &=& 2 & (1) \\ 3y + z &=& 9 & (2) \\ x + 2z &=& 7 & (3) \end{array}$$

To eliminate y, multiply equation (1) by 3 and add the result to equation (2).

$$\begin{array}{rcrl} 12x - 3y &=& 6 & 3 \times (1) \\ 3y + z &=& 9 & (2) \\ \hline 12x + z &=& 15 & (4) \end{array}$$

To eliminate z, multiply equation (4) by -2 and add the result to equation (3).

$$\begin{array}{rcrl} -24x - 2z &=& -30 & -2 \times (4) \\ x + 2z &=& 7 & (3) \\ \hline -23x &=& -23 & \\ x &=& 1 & \end{array}$$

Substitute 1 for x in equation (3) to find z.

$$x + 2z = 7 \quad (3)$$
$$1 + 2z = 7$$
$$2z = 6$$
$$z = 3$$

Substitute 1 for x in equation (1) to find y.

$$4x - y = 2 \quad (1)$$
$$4(1) - y = 2$$
$$4 - y = 2$$
$$-y = -2$$
$$y = 2$$

The solution set is $\{(1, 2, 3)\}$.

18. Let x = the width of the rink and y = the length of the rink.

The length is 30 ft longer than twice the width.

$$y = 2x + 30 \quad (1)$$

The perimeter is 570 ft.

$$2x + 2y = 570 \quad (2)$$

Substitute $2x + 30$ for y in equation (2).

$$2x + 2y = 570 \quad (2)$$
$$2x + 2(2x + 30) = 570$$
$$2x + 4x + 60 = 570$$
$$6x = 510$$
$$x = 85$$

From (1), $y = 2(85) + 30 = 200$.

The width is 85 ft and the length is 200 ft.

19. Let x = the number of days she rented the car at the weekday rate, and y = the number of days she rented the car at the weekend rate.

Since the business trip lasted 6 days,

$$x + y = 6. \quad (1)$$

At \$53/day at weekday rates and \$35/day at weekend rates, her total bill was \$264.

$$53x + 35y = 264 \quad (2)$$

Solve the equation (1) for y.

$$y = 6 - x \quad (3)$$

Substitute $6 - x$ for y in equation (2).

$$53x + 35(6 - x) = 264$$
$$53x + 210 - 35x = 264$$
$$18x = 54$$
$$x = 3$$

From (3), $y = 6 - 3 = 3$.

She rented the car for 3 weekdays and 3 weekend days.

20. Let x = the speed of the plane and y = the speed of the wind.

Complete the chart.

	r	t	d
With Wind	$x + y$	1.75	$1.75(x + y)$
Against Wind	$x - y$	2	$2(x - y)$

The distance each way is 560 miles. From the chart,

$$1.75(x + y) = 560.$$

Divide by 1.75.

$$x + y = 320 \quad (1)$$

From the chart,

$$2(x - y) = 560$$
$$x - y = 280. \quad (2)$$

Solve the system by adding equations (1) and (2) to eliminate y.

$$\begin{array}{rcl} x + y &=& 320 \quad (1) \\ x - y &=& 280 \quad (2) \\ \hline 2x &=& 600 \\ x &=& 300 \end{array}$$

From (1), $300 + y = 320$, so $y = 20$.

The speed of the plane was 300 mph, and the speed of the wind was 20 mph.

21. Let x = the amount of \$2-per-pound nuts and y = the amount of \$1-per-pound chocolate candy.

Complete the chart.

	Price per Pound	Number of Pounds	Value
Nuts	2	x	$2x$
Chocolate	1	y	$1y = y$
Mixture	1.30	100	$1.30(100) = 130$

Solve the system formed from the second and third columns.

$$x + y = 100 \quad (1)$$
$$2x + y = 130 \quad (2)$$

Solve equation (1) for y.

$$y = 100 - x \quad (3)$$

Substitute $100 - x$ for y in equation (2).

$$2x + (100 - x) = 130$$
$$x = 30$$

From (3), $y = 100 - 30 = 70$.

She should use 30 lb of \$2-per-pound nuts and 70 lb of \$1-per-pound candy.

22. Let $g =$ the number of vats of green algae to be grown
and $b =$ the number of vats of brown algae to be grown.

Arrange the information in a table. The table shows the amount of each nutrient needed to grow the algae (in kg).

	Amount of Nutrient X Needed	Amount of Nutrient Y Needed
Green Algae	$2g$	$3g$
Brown Algae	$1b$	$2b$
Total Amount of Nutrient Available	15	26

From the last two columns, we have the following system.

$$\begin{aligned} 2g + b &= 15 \quad (1) \\ 3g + 2b &= 26 \quad (2) \end{aligned}$$

To solve, multiply (1) by 2 and (2) by -1 to get

$$\begin{array}{rcrcrl} 4g & + & 2b & = & 30 & \quad 2 \times (1) \\ -3g & - & 2b & = & -26 & \quad -1 \times (2) \\ \hline g & & & = & 4. & \end{array}$$

Substitute $g = 4$ in (1).

$$\begin{aligned} 2(4) + b &= 15 \\ b &= 7 \end{aligned}$$

She should grow 4 vats of green algae and 7 vats of brown algae in order to use all the nutrients.

23. Let $x =$ the measure of the largest angle,
$y =$ the measure of the middle-sized angle,
and $z =$ the measure of the smallest angle.

Since the sum of the measures of the angles of a triangle is 180°,

$$x + y + z = 180. \quad (1)$$

Since the largest angle measures 10° less than the sum of the other two,

$$\begin{aligned} x &= y + z - 10 \\ \text{or} \quad x - y - z &= -10. \quad (2) \end{aligned}$$

Since the measure of the middle-sized angle is the average of the other two,

$$\begin{aligned} y &= \frac{x + z}{2} \\ 2y &= x + z \\ -x + 2y - z &= 0. \quad (3) \end{aligned}$$

Solve the system.

$$\begin{array}{rcrcrcrl} x & + & y & + & z & = & 180 & \quad (1) \\ x & - & y & - & z & = & -10 & \quad (2) \\ -x & + & 2y & - & z & = & 0 & \quad (3) \end{array}$$

Add equations (1) and (2) to find x.

$$\begin{array}{rcrcrcrl} x & + & y & + & z & = & 180 & \quad (1) \\ x & - & y & - & z & = & -10 & \quad (2) \\ \hline 2x & & & & & = & 170 & \\ & & & & x & = & 85 & \end{array}$$

Add equations (1) and (3), to find y.

$$\begin{array}{rcrcrcrl} x & + & y & + & z & = & 180 & \quad (1) \\ -x & + & 2y & - & z & = & 0 & \quad (3) \\ \hline & & 3y & & & = & 180 & \\ & & & & y & = & 60 & \end{array}$$

Substitute 85 for x and 60 for y in equation (1) to find z.

$$\begin{aligned} x + y + z &= 180 \quad (1) \\ 85 + 60 + z &= 180 \\ 145 + z &= 180 \\ z &= 35 \end{aligned}$$

The three angles are 85°, 60°, and 35°.

24. Let $x =$ the number of liters of 8% solution,
$y =$ the number of liters of 10% solution,
and $z =$ the number of liters of 20% solution.

Since the amount of the mixture will be 8 L,

$$x + y + z = 8. \quad (1)$$

Since the final solution will be 12.5% hydrogen peroxide,

$$.08x + .10y + .20z = .125(8).$$

Multiply by 100 to clear the decimals.

$$8x + 10y + 20z = 100 \quad (2)$$

Since the amount of 8% solution used must be 2 L more than the amount of 20% solution,

$$x = z + 2. \quad (3)$$

Solve the system.

$$\begin{array}{rcrcrcrl} x & + & y & + & z & = & 8 & \quad (1) \\ 8x & + & 10y & + & 20z & = & 100 & \quad (2) \\ & & & & x & = & z + 2 & \quad (3) \end{array}$$

Since equation (3) is given in terms of x, substitute $z + 2$ for x in equations (1) and (2).

$$\begin{aligned} x + y + z &= 8 \quad (1) \\ (z + 2) + y + z &= 8 \\ y + 2z &= 6 \quad (4) \end{aligned}$$

$$\begin{aligned} 8x + 10y + 20z &= 100 \quad (2) \\ 8(z+2) + 10y + 20z &= 100 \\ 8z + 16 + 10y + 20z &= 100 \\ 10y + 28z &= 84 \quad (5) \end{aligned}$$

To eliminate y, multiply equation (4) by -10 and add the result to equation (5).

$$\begin{array}{rcrcrl} -10y & - & 20z & = & -60 & -10 \times (4) \\ 10y & + & 28z & = & 84 & (5) \\ \hline & & 8z & = & 24 & \\ & & z & = & 3 & \end{array}$$

From (3), $x = z + 2 = 3 + 2 = 5$.
From (4), $y = 6 - 2z = 6 - 2(3) = 0$.

Mix 5 L of 8% solution, none of 10% solution, and 3 L of 20% solution.

25. Let $x =$ the number of home runs hit by Mantle,
$y =$ the number of home runs hit by Maris,
and $z =$ the number of home runs hit by Blanchard.

They combined for 136 home runs, so

$$x + y + z = 136. \quad (1)$$

Mantle hit 7 fewer than Maris, so

$$x = y - 7. \quad (2)$$

Maris hit 40 more than Blanchard, so

$$y = z + 40 \quad \text{or} \quad z = y - 40. \quad (3)$$

Substitute $y - 7$ for x and $y - 40$ for z in (1).

$$\begin{aligned} x + y + z &= 136 \quad (1) \\ (y - 7) + y + (y - 40) &= 136 \\ 3y - 47 &= 136 \\ 3y &= 183 \\ y &= 61 \end{aligned}$$

From (2), $x = y - 7 = 61 - 7 = 54$.
From (3), $z = y - 40 = 61 - 40 = 21$.

Mantle hit 54 home runs, Maris hit 61 home runs, and Blanchard hit 21 home runs.

26. $2x + 5y = -4$
$4x - y = 14$

Write the augmented matrix.

$$\left[\begin{array}{rr|r} 2 & 5 & -4 \\ 4 & -1 & 14 \end{array}\right]$$

$$\left[\begin{array}{rr|r} 2 & 5 & -4 \\ 0 & -11 & 22 \end{array}\right] \quad -2R_1 + R_2$$

$$\left[\begin{array}{rr|r} 2 & 5 & -4 \\ 0 & 1 & -2 \end{array}\right] \quad -\tfrac{1}{11}R_2$$

This matrix gives the system

$$\begin{aligned} 2x + 5y &= -4 \\ y &= -2. \end{aligned}$$

Substitute $y = -2$ in the first equation.

$$\begin{aligned} 2x + 5y &= -4 \\ 2x + 5(-2) &= -4 \\ 2x - 10 &= -4 \\ 2x &= 6 \\ x &= 3 \end{aligned}$$

Solution set: $\{(3, -2)\}$

27. $6x + 3y = 9$
$-7x + 2y = 17$

Write the augmented matrix.

$$\left[\begin{array}{rr|r} 6 & 3 & 9 \\ -7 & 2 & 17 \end{array}\right]$$

$$\left[\begin{array}{rr|r} 1 & -5 & -26 \\ -7 & 2 & 17 \end{array}\right] \quad -R_1 - R_2$$

$$\left[\begin{array}{rr|r} 1 & -5 & -26 \\ 0 & -33 & -165 \end{array}\right] \quad 7R_1 + R_2$$

$$\left[\begin{array}{rr|r} 1 & -5 & -26 \\ 0 & 1 & 5 \end{array}\right] \quad -\tfrac{1}{33}R_2$$

This matrix gives the system

$$\begin{aligned} x - 5y &= -26 \\ y &= 5. \end{aligned}$$

Substitute $y = 5$ in the first equation.

$$\begin{aligned} x - 5y &= -26 \\ x - 5(5) &= -26 \\ x - 25 &= -26 \\ x &= -1 \end{aligned}$$

Solution set: $\{(-1, 5)\}$

28.

$$\begin{array}{rcrcrcr} x & + & 2y & - & z & = & 1 \\ 3x & + & 4y & + & 2z & = & -2 \\ -2x & - & y & + & z & = & -1 \end{array}$$

$$\left[\begin{array}{rrr|r} 1 & 2 & -1 & 1 \\ 3 & 4 & 2 & -2 \\ -2 & -1 & 1 & -1 \end{array}\right]$$

$$\left[\begin{array}{rrr|r} 1 & 2 & -1 & 1 \\ 0 & -2 & 5 & -5 \\ 0 & 3 & -1 & 1 \end{array}\right] \quad \begin{array}{r} \\ -3R_1 + R_2 \\ 2R_1 + R_3 \end{array}$$

$$\left[\begin{array}{rrr|r} 1 & 2 & -1 & 1 \\ 0 & 1 & 4 & -4 \\ 0 & 3 & -1 & 1 \end{array}\right] \quad R_3 + R_2$$

continued

$$\left[\begin{array}{rrr|r} 1 & 2 & -1 & 1 \\ 0 & 1 & 4 & -4 \\ 0 & 0 & -13 & 13 \end{array}\right] \quad -3R_2 + R_3$$

$$\left[\begin{array}{rrr|r} 1 & 2 & -1 & 1 \\ 0 & 1 & 4 & -4 \\ 0 & 0 & 1 & -1 \end{array}\right] \quad -\tfrac{1}{13}R_3$$

This matrix gives the system

$$\begin{aligned} x + 2y - z &= 1 \\ y + 4z &= -4 \\ z &= -1. \end{aligned}$$

Substitute $z = -1$ in the second equation.

$$\begin{aligned} y + 4z &= -4 \\ y + 4(-1) &= -4 \\ y &= 0 \end{aligned}$$

Substitute $y = 0$ and $z = -1$ in the first equation.

$$\begin{aligned} x + 2y - z &= 1 \\ x + 2(0) - (-1) &= 1 \\ x + 1 &= 1 \\ x &= 0 \end{aligned}$$

Solution set: $\{(0, 0, -1)\}$

29. **[5.1]**
$$\begin{aligned} \frac{2}{3}x + \frac{1}{6}y &= \frac{19}{2} && (1) \\ \frac{1}{3}x - \frac{2}{9}y &= 2 && (2) \end{aligned}$$

Multiply equation (1) by 6 and equation (2) by 9 to clear the fractions.

$$\begin{aligned} 4x + y &= 57 \quad (3) && 6 \times (1) \\ 3x - 2y &= 18 \quad (4) && 9 \times (2) \end{aligned}$$

To eliminate y, multiply equation (3) by 2 and add the result to equation (4).

$$\begin{array}{rcll} 8x + 2y &=& 114 & 2 \times (3) \\ 3x - 2y &=& 18 & (4) \\ \hline 11x &=& 132 & \\ x &=& 12 & \end{array}$$

Substitute 12 for x in equation (3) to find y.

$$\begin{aligned} 4x + y &= 57 \quad (3) \\ 4(12) + y &= 57 \\ 48 + y &= 57 \\ y &= 9 \end{aligned}$$

Solution set: $\{(12, 9)\}$

30. **[5.1]**
$$\begin{aligned} 2x - 5y &= 8 && (1) \\ 3x + 4y &= 10 && (2) \end{aligned}$$

To eliminate y, multiply equation (1) by 4 and equation (2) by 5 and add the results.

$$\begin{array}{rcll} 8x - 20y &=& 32 & 4 \times (1) \\ 15x + 20y &=& 50 & 5 \times (2) \\ \hline 23x &=& 82 & \\ x &=& \frac{82}{23} & \end{array}$$

Instead of substituting to find y, we'll choose different multipliers and eliminate x from the original system.

$$\begin{array}{rcll} 6x - 15y &=& 24 & 3 \times (1) \\ -6x - 8y &=& -20 & -2 \times (2) \\ \hline -23y &=& 4 & \\ y &=& -\frac{4}{23} & \end{array}$$

Solution set: $\left\{\left(\frac{82}{23}, -\frac{4}{23}\right)\right\}$

31. **[5.1]**
$$\begin{aligned} x &= 7y + 10 && (1) \\ 2x + 3y &= 3 && (2) \end{aligned}$$

Since equation (1) is given in terms of x, substitute $7y + 10$ for x in equation (2) and solve for y.

$$\begin{aligned} 2(7y + 10) + 3y &= 3 \\ 14y + 20 + 3y &= 3 \\ 17y &= -17 \\ y &= -1 \end{aligned}$$

From (1), $x = 7(-1) + 10 = 3$.

Solution set: $\{(3, -1)\}$

32. **[5.1]**
$$\begin{aligned} x + 4y &= 17 && (1) \\ -3x + 2y &= -9 && (2) \end{aligned}$$

To eliminate x, multiply equation (1) by 3 and add the result to equation (2).

$$\begin{array}{rcll} 3x + 12y &=& 51 & 3 \times (1) \\ -3x + 2y &=& -9 & (2) \\ \hline 14y &=& 42 & \\ y &=& 3 & \end{array}$$

Substitute 3 for y in equation (1) to find x.

$$\begin{aligned} x + 4y &= 17 \quad (1) \\ x + 4(3) &= 17 \\ x + 12 &= 17 \\ x &= 5 \end{aligned}$$

Solution set: $\{(5, 3)\}$

33. **[5.1]**
$$\begin{aligned} -7x + 3y &= 12 \quad (1)\\ 5x + 2y &= 8 \quad (2) \end{aligned}$$

To eliminate y, multiply equation (1) by 2 and equation (2) by -3. Then add the results.

$$\begin{aligned} -14x + 6y &= 24 \qquad 2 \times (1)\\ -15x - 6y &= -24 \qquad -3 \times (2)\\ \hline -29x &= 0\\ x &= 0 \end{aligned}$$

Substitute 0 for x in equation (1) to find y.

$$\begin{aligned} -7x + 3y &= 12 \quad (1)\\ -7(0) + 3y &= 12\\ 3y &= 12\\ y &= 4 \end{aligned}$$

Solution set: $\{(0, 4)\}$

34. **[5.2]**
$$\begin{aligned} 2x + 5y - z &= 12 \quad (1)\\ -x + y - 4z &= -10 \quad (2)\\ -8x - 20y + 4z &= 31 \quad (3) \end{aligned}$$

Multiply equation (1) by 4 and add the result to equation (3).

$$\begin{aligned} 8x + 20y - 4z &= 48 \qquad 4 \times (1)\\ -8x - 20y + 4z &= 31 \qquad (3)\\ \hline 0 &= 79 \quad \textit{False} \end{aligned}$$

Since a false statement results, the system is *inconsistent*. The solution set is $\emptyset$.

35. **[5.3]** Let x = the number of liters of 5% solution and y = the number of liters of 10% solution.

Percent (as a Decimal)	Liters of Solution	Liters of Pure Acid
.05	x	$.05x$
.20	10	$.20(10) = 2$
.10	y	$.10y$

Solve the system formed from the second and third columns.

$$\begin{aligned} x + 10 &= y \quad (1)\\ .05x + 2 &= .10y \quad (2) \end{aligned}$$

Multiply equation (2) by 100 to clear the decimals.

$$5x + 200 = 10y \quad (3)$$

Substitute $x + 10$ for y in equation (3) and solve for x.

$$\begin{aligned} 5x + 200 &= 10y \quad (3)\\ 5x + 200 &= 10(x + 10)\\ 5x + 200 &= 10x + 100\\ 100 &= 5x\\ 20 &= x \end{aligned}$$

He should use 20 L of 5% solution.

36. **[5.3]** Let x, y, and z denote the number of medals won by the United States, Russia, and China, respectively.

The total number of medals won was 258, so

$$x + y + z = 258. \quad (1)$$

The U.S. won 11 more medals than Russia, so

$$x = y + 11. \quad (2)$$

China won 29 fewer medals than Russia, so

$$z = y - 29. \quad (3)$$

Substitute $y + 11$ for x and $y - 29$ for z in (1).

$$\begin{aligned} (y + 11) + y + (y - 29) &= 258\\ 3y - 18 &= 258\\ 3y &= 276\\ y &= 92 \end{aligned}$$

From (2), $x = 92 + 11 = 103$.
From (3), $z = 92 - 29 = 63$.

The United States won 103 medals, Russia won 92 medals, and China won 63 medals.

37.
$$x^2 + y^2 + ax + by + c = 0$$
Let $x = 2$ and $y = 1$.
$$\begin{aligned} 2^2 + 1^2 + a(2) + b(1) + c &= 0\\ 4 + 1 + 2a + b + c &= 0\\ 2a + b + c &= -5 \end{aligned}$$

38.
$$x^2 + y^2 + ax + by + c = 0$$
Let $x = -1$ and $y = 0$.
$$\begin{aligned} (-1)^2 + 0^2 + a(-1) + b(0) + c &= 0\\ 1 - a + c &= 0\\ -a + c &= -1\\ a - c &= 1 \end{aligned}$$

39.
$$x^2 + y^2 + ax + by + c = 0$$
Let $x = 3$ and $y = 3$.
$$\begin{aligned} 3^2 + 3^2 + a(3) + b(3) + c &= 0\\ 9 + 9 + 3a + 3b + c &= 0\\ 3a + 3b + c &= -18 \end{aligned}$$

40. Use the equations from Exercises 37–39 to form a system of equations.

$$\begin{aligned} 2a + b + c &= -5 \quad (1)\\ a - c &= 1 \quad (2)\\ 3a + 3b + c &= -18 \quad (3) \end{aligned}$$

Add equations (1) and (2).

$$\begin{aligned} 2a + b + c &= -5 \quad (1)\\ a - c &= 1 \quad (2)\\ \hline 3a + b &= -4 \quad (4) \end{aligned}$$

continued

Add equations (2) and (3).

$$\begin{array}{rcrcrcrl} a & & & - & c & = & 1 & (2) \\ 3a & + & 3b & + & c & = & -18 & (3) \\ \hline 4a & + & 3b & & & = & -17 & (5) \end{array}$$

Multiply equation (4) by -3 and add the result to equation (5).

$$\begin{array}{rcrcrl} -9a & - & 3b & = & 12 & -3 \times (4) \\ 4a & + & 3b & = & -17 & (5) \\ \hline -5a & & & = & -5 & \\ & & a & = & 1 & \end{array}$$

Substitute $a = 1$ into equation (4).

$$\begin{aligned} 3a + b &= -4 \quad (4) \\ 3(1) + b &= -4 \\ b &= -7 \end{aligned}$$

Substitute $a = 1$ into equation (2).

$$\begin{aligned} a - c &= 1 \quad (2) \\ 1 - c &= 1 \\ c &= 0 \end{aligned}$$

Since $a = 1$, $b = -7$, and $c = 0$, the equation of the circle is

$$\begin{aligned} x^2 + y^2 + ax + by + c &= 0 \\ x^2 + y^2 + x - 7y &= 0. \end{aligned}$$

41. The graph of the circle is not the graph of a function because it fails the vertical line test.

Chapter 5 Test

1. The graphs intersect to the right of 1997 and to the left of 1998, so the numbers were the same in the year 1997.

2. The graphs intersect about halfway between 100 and 120, so about 110 thousand people of each race were living with AIDS in 1997.

3. When each equation of the system

$$\begin{aligned} x + y &= 7 \\ x - y &= 5 \end{aligned}$$

is graphed, the point of intersection appears to be $(6, 1)$. To check, substitute 6 for x and 1 for y in each of the equations. Since $(6, 1)$ makes both equations true, the solution set of the system is $\{(6, 1)\}$.

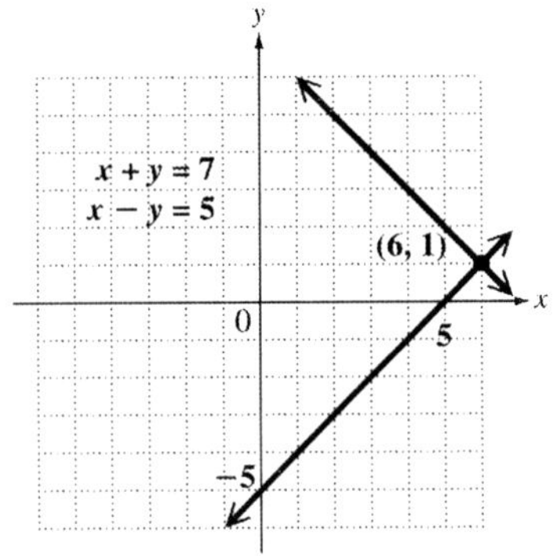

4.
$$\begin{array}{rcrcrl} 2x & - & 3y & = & 24 & (1) \\ & & y & = & -\frac{2}{3}x & (2) \end{array}$$

Since equation (2) is solved for y, substitute $-\frac{2}{3}x$ for y in equation (1) and solve for x.

$$\begin{aligned} 2x - 3y &= 24 \quad (1) \\ 2x - 3\left(-\tfrac{2}{3}x\right) &= 24 \\ 2x + 2x &= 24 \\ 4x &= 24 \\ x &= 6 \end{aligned}$$

From (2), $y = -\frac{2}{3}(6) = -4$.

Solution set: $\{(6, -4)\}$

5.
$$\begin{array}{rrcll} & 12x - 5y & = & 8 & (1) \\ & 3x & = & \frac{5}{4}y + 2 & \\ \text{or} & x & = & \frac{5}{12}y + \frac{2}{3} & (2) \end{array}$$

Substitute $\frac{5}{12}y + \frac{2}{3}$ for x in equation (1) and solve for y.

$$\begin{aligned} 12x - 5y &= 8 \quad (1) \\ 12\left(\tfrac{5}{12}y + \tfrac{2}{3}\right) - 5y &= 8 \\ 5y + 8 - 5y &= 8 \\ 8 &= 8 \quad \textit{True} \end{aligned}$$

Equations (1) and (2) are dependent.

Solution set: $\{(x, y) \mid 12x - 5y = 8\}$

6. $3x - y = -8 \quad (1)$
$2x + 6y = 3 \quad (2)$

Solve equation (1) for y.

$$\begin{aligned} 3x - y &= -8 \\ -y &= -3x - 8 \\ y &= 3x + 8 \end{aligned}$$

Substitute $3x + 8$ for y in equation (2).

$$\begin{aligned} 2x + 6(3x + 8) &= 3 \\ 2x + 18x + 48 &= 3 \\ 20x &= -45 \\ x &= -\tfrac{9}{4} \end{aligned}$$

Since $y = 3x + 8$ and $x = -\frac{9}{4}$,

$$y = 3\left(-\tfrac{9}{4}\right) + 8 = -\tfrac{27}{4} + \tfrac{32}{4} = \tfrac{5}{4}.$$

Solution set: $\left\{\left(-\frac{9}{4}, \frac{5}{4}\right)\right\}$

7. $3x + y = 12 \quad (1)$
$2x - y = 3 \quad (2)$

To eliminate y, add equations (1) and (2).

$$\begin{array}{rcrl} 3x + y &=& 12 & (1) \\ 2x - y &=& 3 & (2) \\ \hline 5x &=& 15 & \\ x &=& 3 & \end{array}$$

Substitute 3 for x in equation (1) to find y.

$$\begin{aligned} 3x + y &= 12 \quad (1) \\ 3(3) + y &= 12 \\ 9 + y &= 12 \\ y &= 3 \end{aligned}$$

Solution set: $\{(3, 3)\}$

8. $-5x + 2y = -4 \quad (1)$
$6x + 3y = -6 \quad (2)$

To eliminate x, multiply equation (1) by 6 and equation (2) by 5. Then add the results.

$$\begin{array}{rcrl} -30x + 12y &=& -24 & 6 \times (1) \\ 30x + 15y &=& -30 & 5 \times (2) \\ \hline 27y &=& -54 & \\ y &=& -2 & \end{array}$$

Substitute -2 for y in equation (1) to find x.

$$\begin{aligned} -5x + 2y &= -4 \quad (1) \\ -5x + 2(-2) &= -4 \\ -5x - 4 &= -4 \\ -5x &= 0 \\ x &= 0 \end{aligned}$$

Solution set: $\{(0, -2)\}$

9. $3x + 4y = 8 \quad (1)$
$8y = 7 - 6x$
or $6x + 8y = 7 \quad (2)$

Multiply equation (1) by -2 and add the result to equation (2).

$$\begin{array}{rcrl} -6x - 8y &=& -16 & -2 \times (1) \\ 6x + 8y &=& 7 & (2) \\ \hline 0 &=& -9 & \textit{False} \end{array}$$

Since a false statement results, the system is *inconsistent*. The solution set is $\emptyset$.

10. $3x + 5y + 3z = 2 \quad (1)$
$6x + 5y + z = 0 \quad (2)$
$3x + 10y - 2z = 6 \quad (3)$

To eliminate x, multiply equation (1) by -1 and add the result to equation (3).

$$\begin{array}{rcrl} -3x - 5y - 3z &=& -2 & -1 \times (1) \\ 3x + 10y - 2z &=& 6 & (3) \\ \hline 5y - 5z &=& 4 & (4) \end{array}$$

To eliminate x again, multiply equation (1) by -2 and add the result to equation (2).

$$\begin{array}{rcrl} -6x - 10y - 6z &=& -4 & -2 \times (1) \\ 6x + 5y + z &=& 0 & (2) \\ \hline -5y - 5z &=& -4 & (5) \end{array}$$

To eliminate y, add equations (4) and (5).

$$\begin{array}{rcrl} 5y - 5z &=& 4 & (4) \\ -5y - 5z &=& -4 & (5) \\ \hline -10z &=& 0 & \\ z &=& 0 & \end{array}$$

Substitute 0 for z in equation (4) to find y.

$$\begin{aligned} 5y - 5z &= 4 \quad (4) \\ 5y - 5(0) &= 4 \\ 5y - 0 &= 4 \\ 5y &= 4 \\ y &= \tfrac{4}{5} \end{aligned}$$

Substitute $\frac{4}{5}$ for y and 0 for z in equation (1) to find x.

$$\begin{aligned} 3x + 5y + 3z &= 2 \quad (1) \\ 3x + 5\left(\tfrac{4}{5}\right) + 3(0) &= 2 \\ 3x + 4 + 0 &= 2 \\ 3x &= -2 \\ x &= -\tfrac{2}{3} \end{aligned}$$

Solution set: $\left\{\left(-\frac{2}{3}, \frac{4}{5}, 0\right)\right\}$

11.
$$\begin{aligned} 4x + y + z &= 11 \quad (1)\\ x - y - z &= 4 \quad (2)\\ y + 2z &= 0 \quad (3) \end{aligned}$$

To eliminate x, multiply equation (2) by -4 and add the result to equation (1).

$$\begin{array}{rcrl} 4x + y + z &=& 11 & (1)\\ -4x + 4y + 4z &=& -16 & -4 \times (2)\\ \hline 5y + 5z &=& -5 & (4) \end{array}$$

To eliminate y, multiply equation (3) by -5 and add the result to equation (4).

$$\begin{array}{rcrl} -5y - 10z &=& 0 & -5 \times (3)\\ 5y + 5z &=& -5 & (4)\\ \hline -5z &=& -5 &\\ z &=& 1 & \end{array}$$

From (3), $y + 2(1) = 0$, so $y = -2$.
From (2), $x - (-2) - 1 = 4$, so $x = 3$.

Solution set: $\{(3, -2, 1)\}$

12. Let $x =$ the gross (in millions of dollars) for *Ocean's Eleven*, and $y =$ the gross (in millions of dollars) for *Runaway Bride*.

Together the movies grossed \$335.5 million, so

$$x + y = 335.5. \quad (1)$$

Runaway Bride grossed \$31.3 million less than *Ocean's Eleven*, so

$$y = x - 31.3. \quad (2)$$

Substitute $x - 31.3$ for y in equation (1).

$$\begin{aligned} x + y &= 335.5 \quad (1)\\ x + (x - 31.3) &= 335.5\\ 2x &= 366.8\\ x &= 183.4 \end{aligned}$$

From (2), $y = 183.4 - 31.3 = 152.1$.

Ocean's Eleven grossed \$183.4 million and *Runaway Bride* grossed \$152.1 million.

13. Let $x =$ the speed of the faster car and $y =$ the speed of the slower car.

Make a table. Use the formula $d = rt$.

	r	t	d
Faster Car	x	3.5	$3.5x$
Slower Car	y	3.5	$3.5y$

Since the fast car travels 30 mph faster than the slow car,

$$x - y = 30. \quad (1)$$

Since the cars travel a total of 420 miles,

$$3.5x + 3.5y = 420.$$

Multiply by 10 to clear the decimals.

$$35x + 35y = 4200 \quad (2)$$

To eliminate y, multiply equation (1) by 35 and add the result to equation (2).

$$\begin{array}{rcrl} 35x - 35y &=& 1050 & 35 \times (1)\\ 35x + 35y &=& 4200 & (2)\\ \hline 70x &=& 5250 &\\ x &=& 75 & \end{array}$$

Substitute 75 for x in equation (1) to find y.

$$\begin{aligned} x - y &= 30 \quad (1)\\ 75 - y &= 30\\ -y &= -45\\ y &= 45 \end{aligned}$$

The faster car is traveling at 75 mph, and the slower car is traveling at 45 mph.

14. Let $x =$ the number of liters of 20% solution and $y =$ the number of liters of 50% solution.

Make a table.

Percent (as a Decimal)	Liters of Solution	Liters of Pure Alcohol
.20	x	$.20x$
.50	y	$.50y$
.40	12	$.40(12) = 4.8$

Since 12 L of the mixture are needed,

$$x + y = 12. \quad (1)$$

Since the amount of pure alcohol in the 20% solution plus the amount of pure alcohol in the 50% solution must equal the amount of alcohol in the mixture,

$$.2x + .5y = 4.8.$$

Multiply by 10 to clear the decimals.

$$2x + 5y = 48 \quad (2)$$

Multiply equation (1) by -2 and add the result to equation (2).

$$\begin{array}{rcrl} -2x - 2y &=& -24 & -2 \times (1)\\ 2x + 5y &=& 48 & (2)\\ \hline 3y &=& 24 &\\ y &=& 8 & \end{array}$$

From (1), $x + 8 = 12$, so $x = 4$.

4 L of 20% solution and 8 L of 50% solution are needed.

15. Let x = the price of an AC adaptor and y = the price of a rechargeable flashlight.

Since 7 AC adaptors and 2 rechargeable flashlights cost \$86,

$$7x + 2y = 86. \quad (1)$$

Since 3 AC adaptors and 4 rechargeable flashlights cost \$84,

$$3x + 4y = 84. \quad (2)$$

Solve the system.

$$\begin{aligned} 7x + 2y &= 86 \quad (1) \\ 3x + 4y &= 84 \quad (2) \end{aligned}$$

To eliminate y, multiply equation (1) by -2 and add the result to equation (2).

$$\begin{aligned} -14x - 4y &= -172 \quad -2 \times (1) \\ 3x + 4y &= 84 \quad (2) \\ \hline -11x &= -88 \\ x &= 8 \end{aligned}$$

Substitute 8 for x in equation (1) to find y.

$$\begin{aligned} 7x + 2y &= 86 \quad (1) \\ 7(8) + 2y &= 86 \\ 56 + 2y &= 86 \\ 2y &= 30 \\ y &= 15 \end{aligned}$$

An AC adaptor costs \$8, and a rechargeable flashlight costs \$15.

16. Let x = the amount of Orange Pekoe, y = the amount of Irish Breakfast, and z = the amount of Earl Grey.

The owner wants 100 oz of tea, so

$$x + y + z = 100. \quad (1)$$

An equation which relates the prices of the tea is

$$.80x + .85y + .95z = .83(100).$$

Multiply by 100 to clear the decimals.

$$80x + 85y + 95z = 8300 \quad (2)$$

The mixture must use twice as much Orange Pekoe as Irish Breakfast, so

$$x = 2y. \quad (3)$$

To eliminate z, multiply equation (1) by -95 and add the result to equation (2).

$$\begin{aligned} -95x - 95y - 95z &= -9500 \quad -95 \times (1) \\ 80x + 85y + 95z &= 8300 \quad (2) \\ \hline -15x - 10y &= -1200 \quad (4) \end{aligned}$$

Divide equation (4) by -5.

$$3x + 2y = 240 \quad (5)$$

Substitute $2y$ for x in equation (5) to find y.

$$\begin{aligned} 3x + 2y &= 240 \quad (5) \\ 3(2y) + 2y &= 240 \\ 8y &= 240 \\ y &= 30 \end{aligned}$$

From (3), $x = 2(30) = 60$.
Substitute 60 for x and 30 for y in equation (1) to find z.

$$\begin{aligned} x + y + z &= 100 \quad (1) \\ 60 + 30 + z &= 100 \\ z &= 10 \end{aligned}$$

He should use 60 oz of Orange Pekoe, 30 oz of Irish Breakfast, and 10 oz of Earl Grey.

17. $$\begin{aligned} 3x + 2y &= 4 \\ 5x + 5y &= 9 \end{aligned}$$

Write the augmented matrix.

$$\left[\begin{array}{cc|c} 3 & 2 & 4 \\ 5 & 5 & 9 \end{array}\right]$$

$$\left[\begin{array}{cc|c} 6 & 4 & 8 \\ 5 & 5 & 9 \end{array}\right] \quad 2R_1$$

$$\left[\begin{array}{cc|c} 1 & -1 & -1 \\ 5 & 5 & 9 \end{array}\right] \quad -R_2 + R_1$$

$$\left[\begin{array}{cc|c} 1 & -1 & -1 \\ 0 & 10 & 14 \end{array}\right] \quad -5R_1 + R_2$$

$$\left[\begin{array}{cc|c} 1 & -1 & -1 \\ 0 & 1 & \frac{7}{5} \end{array}\right] \quad \frac{1}{10}R_2$$

This matrix gives the system

$$\begin{aligned} x - y &= -1 \\ y &= \tfrac{7}{5}. \end{aligned}$$

Substitute $y = \frac{7}{5}$ in the first equation.

$$\begin{aligned} x - \tfrac{7}{5} &= -1 \\ x &= -1 + \tfrac{7}{5} \\ &= -\tfrac{5}{5} + \tfrac{7}{5} = \tfrac{2}{5} \end{aligned}$$

Solution set: $\left\{\left(\frac{2}{5}, \frac{7}{5}\right)\right\}$

18.
$$\begin{aligned} x + 3y + 2z &= 11 \\ 3x + 7y + 4z &= 23 \\ 5x + 3y - 5z &= -14 \end{aligned}$$

Write the augmented matrix.

$$\left[\begin{array}{ccc|c} 1 & 3 & 2 & 11 \\ 3 & 7 & 4 & 23 \\ 5 & 3 & -5 & -14 \end{array}\right]$$

$$\left[\begin{array}{ccc|c} 1 & 3 & 2 & 11 \\ 0 & -2 & -2 & -10 \\ 0 & -12 & -15 & -69 \end{array}\right] \quad \begin{array}{l} \\ -3R_1 + R_2 \\ -5R_1 + R_3 \end{array}$$

$$\left[\begin{array}{ccc|c} 1 & 3 & 2 & 11 \\ 0 & 1 & 1 & 5 \\ 0 & -12 & -15 & -69 \end{array}\right] \quad -\tfrac{1}{2}R_2$$

$$\left[\begin{array}{ccc|c} 1 & 3 & 2 & 11 \\ 0 & 1 & 1 & 5 \\ 0 & 0 & -3 & -9 \end{array}\right] \quad 12R_2 + R_3$$

$$\left[\begin{array}{ccc|c} 1 & 3 & 2 & 11 \\ 0 & 1 & 1 & 5 \\ 0 & 0 & 1 & 3 \end{array}\right] \quad -\tfrac{1}{3}R_3$$

This matrix gives the system

$$\begin{aligned} x + 3y + 2z &= 11 \\ y + z &= 5 \\ z &= 3. \end{aligned}$$

Substitute $z = 3$ in the second equation.

$$\begin{aligned} y + z &= 5 \\ y + 3 &= 5 \\ y &= 2 \end{aligned}$$

Substitute $y = 2$ and $z = 3$ in the first equation.

$$\begin{aligned} x + 3y + 2z &= 11 \\ x + 3(2) + 2(3) &= 11 \\ x + 6 + 6 &= 11 \\ x &= -1 \end{aligned}$$

Solution set: $\{(-1, 2, 3)\}$

Cumulative Review Exercises (Chapters 1–5)

1. $(-3)^4 = (-3)(-3)(-3)(-3) = 81$

2. $-3^4 = -(3)(3)(3)(3) = -81$

3. $-(-3)^4 = -(-3)(-3)(-3)(-3) = -81$

4. $\sqrt{.49} = .7$, since .7 is positive and $(.7)^2 = .49$.

5. $-\sqrt{.49} = -.7$, since $(.7)^2 = .49$ and the negative sign in front of the radical must be applied.

6. $\sqrt{-.49}$ is not a real number because of the negative sign under the radical. No real number squared is negative.

In Exercises 7–9, let $x = -4$, $y = 3$, and $z = 6$.

7.
$$\begin{aligned} |2x| + y^2 - z^3 &= |(2)(-4)| + 3^2 - (6)^3 \\ &= |-8| + 9 - 216 \\ &= 8 + 9 - 216 \\ &= -199 \end{aligned}$$

8.
$$\begin{aligned} -5(x^3 - y^3) &= -5[(-4)^3 - (3)^3] \\ &= -5(-64 - 27) \\ &= -5(-91) \\ &= 455 \end{aligned}$$

9.
$$\begin{aligned} \frac{2x^2 - x + z}{y^2 - z} &= \frac{2(-4)^2 - (-4) + 6}{3^2 - 6} \\ &= \frac{2(16) + 4 + 6}{9 - 6} \\ &= \frac{32 + 4 + 6}{3} \\ &= \frac{42}{3} = 14 \end{aligned}$$

10.
$$\begin{aligned} 7(2x + 3) - 4(2x + 1) &= 2(x + 1) \\ 14x + 21 - 8x - 4 &= 2x + 2 \\ 6x + 17 &= 2x + 2 \\ 4x &= -15 \\ x &= -\tfrac{15}{4} \end{aligned}$$

Solution set: $\{-\frac{15}{4}\}$

11.
$$.04x + .06(x - 1) = 1.04$$

Multiply each side by 100 to clear the decimals.

$$\begin{aligned} 4x + 6(x - 1) &= 104 \\ 4x + 6x - 6 &= 104 \\ 10x - 6 &= 104 \\ 10x &= 110 \\ x &= 11 \end{aligned}$$

Solution set: $\{11\}$

12. To solve $ax + by = cx + d$ for x, get all terms with x alone on one side of the equals sign.

$$\begin{aligned} ax + by &= cx + d \\ ax - cx &= d - by \\ x(a - c) &= d - by \\ x &= \frac{d - by}{a - c} \\ \text{or} \quad x &= \frac{by - d}{c - a} \end{aligned}$$

if the x-terms were put on the right side of the equals sign.

13. $|6x-8|=4$

$$\begin{aligned} 6x-8&=4 \quad\text{or}\quad 6x-8=-4 \\ 6x&=12 \quad\text{or}\quad 6x=4 \\ x&=2 \quad\text{or}\quad x=\tfrac{4}{6}=\tfrac{2}{3} \end{aligned}$$

Solution set: $\{\frac{2}{3},2\}$

14.
$$\frac{2}{3}x+\frac{5}{12}x\le 20$$

Multiply each side by the LCD, 12.

$$\begin{aligned} 12\left(\tfrac{2}{3}x+\tfrac{5}{12}x\right)&\le 12(20) \\ 8x+5x&\le 240 \\ 13x&\le 240 \\ x&\le \tfrac{240}{13} \end{aligned}$$

Solution set: $\left(-\infty,\frac{240}{13}\right]$

15. $|3x+2|\le 4$

$-4\le 3x+2\le 4$

$-6\le 3x\le 2$ *Subtract 2.*

$-2\le x\le \frac{2}{3}$ *Divide by 3.*

Solution set: $\left[-2,\frac{2}{3}\right]$

16. $|12t+7|\ge 0$

The solution set is $(-\infty,\infty)$ since the absolute value of any number is greater than or equal to 0.

17. 80.4% of 2500 is $.804(2500)=2010$
72.5% of 2500 is $.725(2500)=1812.5\approx 1813$

$$\frac{1570}{2500}=.628\text{ or }62.8\%;\quad \frac{1430}{2500}=.572\text{ or }57.2\%$$

Product or Company	Percent	Actual Number
Charmin	80.4%	*2010*
Wheaties	72.5%	*1813*
Budweiser	*62.8%*	1570
State Farm	*57.2%*	1430

18. Let $x=$ the number of nickels,
$x+1=$ the number of dimes, and
$x+6=$ the number of pennies.

The total value is $4.80, so

$$.05x+.10(x+1)+.01(x+6)=4.80.$$

Multiply each side by 100 to clear the decimals.

$$\begin{aligned} 5x+10(x+1)+1(x+6)&=480 \\ 5x+10x+10+x+6&=480 \\ 16x+16&=480 \\ 16x&=464 \\ x&=29 \end{aligned}$$

Since $x=29$, $x+1=29+1=30$, and $x+6=29+6=35$.

There are 35 pennies, 29 nickels, and 30 dimes.

19. Let $x=$ the measure of the equal angles and
$2x-4=$ the measure of the third angle.

The sum of the measures of the angles in a triangle is 180, so

$$\begin{aligned} x+x+(2x-4)&=180 \\ 4x-4&=180 \\ 4x&=184 \\ x&=46. \end{aligned}$$

So, $2x-4=2(46)-4=92-4=88$.

The measures of the angles are 46°, 46°, and 88°.

20. A horizontal line through the point (x,k) has equation $y=k$. Since point A has coordinates $(-2,6)$, $k=6$. The equation of the horizontal line through A is $y=6$.

21. A vertical line through the point (k,y) has equation $x=k$. Since point B has coordinates $(4,-2)$, $k=4$. The equation of the vertical line through B is $x=4$.

22. Let $(x_1,y_1)=(-2,6)$ and $(x_2,y_2)=(4,-2)$. Then,

$$m=\frac{y_2-y_1}{x_2-x_1}=\frac{-2-6}{4-(-2)}=\frac{-8}{6}=-\frac{4}{3}.$$

The slope is $-\frac{4}{3}$.

23. Perpendicular lines have slopes that are negative reciprocals of each other. The slope of line AB is $-\frac{4}{3}$ (from Exercise 22). The negative reciprocal of $-\frac{4}{3}$ is $\frac{3}{4}$, so the slope of a line perpendicular to line AB is $\frac{3}{4}$.

24. Let $m = -\frac{4}{3}$ and $(x_1, y_1) = (4, -2)$ in the point-slope form.

$$\begin{aligned} y - y_1 &= m(x - x_1) \\ y - (-2) &= -\tfrac{4}{3}(x - 4) \\ y + 2 &= -\tfrac{4}{3}x + \tfrac{16}{3} \end{aligned}$$

Multiply by 3 to clear the fractions, and then write the equation in standard form, $Ax + By = C$.

$$\begin{aligned} 3y + 6 &= -4x + 16 \\ 4x + 3y &= 10 \end{aligned}$$

25. Write the equation of the line in slope-intercept form.

$$\begin{aligned} 4x + 3y &= 10 \\ 3y &= -4x + 10 \\ y &= -\tfrac{4}{3}x + \tfrac{10}{3} \end{aligned}$$

Replace y with $f(x)$.

$$f(x) = -\tfrac{4}{3}x + \tfrac{10}{3}$$

26. First locate the point $(-1, -3)$ on a graph. Then use the definition of slope to find a second point on the line.

$$m = \frac{\text{change in } y}{\text{change in } x} = \frac{2}{3}$$

From $(-1, -3)$, move 2 units up and 3 units to the right to get to $(2, -1)$. The line through $(-1, -3)$ and the new point, $(2, -1)$, is the graph.

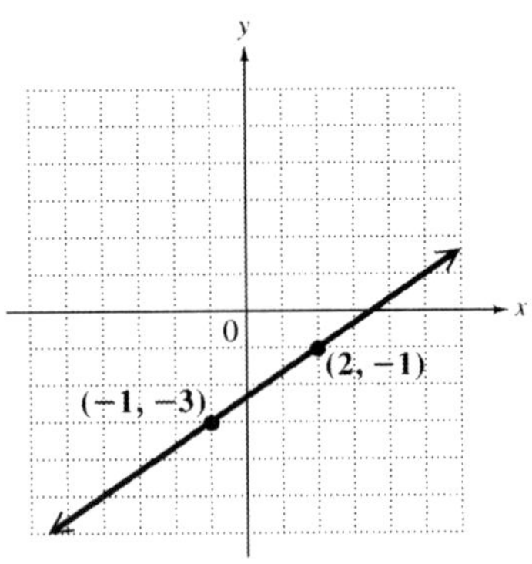

27. $-3x - 2y \le 6$
Graph the line $-3x - 2y = 6$ through its intercepts, $(-2, 0)$ and $(0, -3)$, as a solid line, since the inequality involves $\le$.
To determine the region that belongs to the graph, test $(0, 0)$.

$$\begin{aligned} -3x - 2y &\le 6 \\ -3(0) - 2(0) &\le 6 \quad ? \\ 0 &\le 6 \qquad \textit{True} \end{aligned}$$

Since the result is true, shade the region that includes $(0, 0)$.

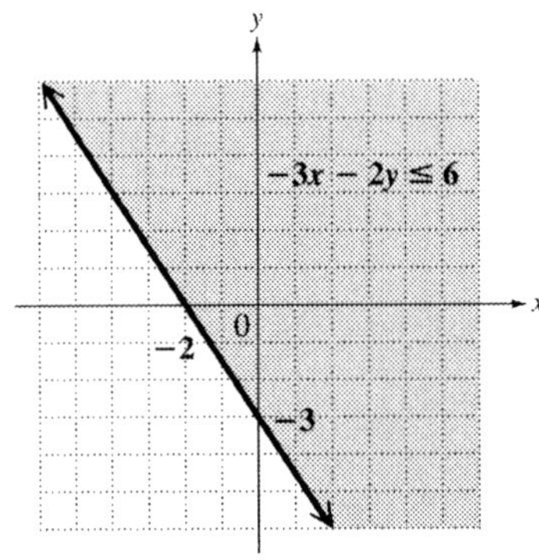

28.

$$\begin{aligned} -2x + 3y &= -15 \quad (1) \\ 4x - y &= 15 \quad (2) \end{aligned}$$

To eliminate x, multiply equation (1) by 2 and add the result to equation (2).

$$\begin{aligned} -4x + 6y &= -30 \qquad 2 \times (1) \\ 4x - y &= 15 \quad (2) \\ \hline 5y &= -15 \\ y &= -3 \end{aligned}$$

Substitute -3 for y in equation (2) to find x.

$$\begin{aligned} 4x - y &= 15 \quad (2) \\ 4x - (-3) &= 15 \\ 4x + 3 &= 15 \\ 4x &= 12 \\ x &= 3 \end{aligned}$$

Solution set: $\{(3, -3)\}$

29.
$$\begin{aligned} x + y + z &= 10 \quad (1)\\ x - y - z &= 0 \quad (2)\\ -x + y - z &= -4 \quad (3)\end{aligned}$$

Add equations (1) and (2) to eliminate y and z. The result is

$$\begin{aligned} 2x &= 10\\ x &= 5.\end{aligned}$$

Add equations (2) and (3) to eliminate x and y. The result is

$$\begin{aligned} -2z &= -4\\ z &= 2.\end{aligned}$$

Substitute 5 for x and 2 for z in equation (1) to find y.

$$\begin{aligned} x + y + z &= 10 \quad (1)\\ 5 + y + 2 &= 10\\ y + 7 &= 10\\ y &= 3\end{aligned}$$

Solution set: $\{(5, 3, 2)\}$

30. Let x = the number of pounds of \$1.20/lb candy, and y = the number of pounds of \$2.40/lb candy.

Price per Pound	Number of Pounds	Value
\$1.20	x	\$1.20x
\$2.40	y	\$2.40y
\$1.65	80	\$1.65(80)

From the last two columns, we have:

$$\begin{aligned} x + y &= 80\\ 1.2x + 2.4y &= 1.65(80)\end{aligned}$$

or

$$\begin{aligned} x &= 80 - y \quad (1)\\ 1.2x + 2.4y &= 132 \quad (2)\end{aligned}$$

Substitute $80 - y$ for x in equation (2).

$$\begin{aligned} 1.2x + 2.4y &= 132 \quad (2)\\ 1.2(80 - y) + 2.4y &= 132\\ 96 - 1.2y + 2.4y &= 132\\ 1.2y &= 36\\ y &= 30\end{aligned}$$

Since $x = 80 - y$ and $y = 30$,

$$x = 80 - 30 = 50.$$

He should mix 50 lb of \$1.20/lb candy with 30 lb of \$2.40/lb candy.

31. Let x = the amount borrowed at 8%, y = the amount borrowed at 9%, and z = the amount borrowed at 10%.

Rate (as a Decimal)	Amount Borrowed	Annual Interest
.08	x	$.08x$
.09	y	$.09y$
.10	z	$.10z$
Totals	25,000	2220

$$\begin{aligned} x + y + z &= 25{,}000\\ z &= \tfrac{1}{2}x + 2000\\ .08x + .09y + .10z &= 2220\end{aligned}$$

or

$$\begin{aligned} x + y + z &= 25{,}000\\ -x + 2z &= 4000\\ 8x + 9y + 10z &= 222{,}000\end{aligned}$$

Write the augmented matrix.

$$\left[\begin{array}{rrr|r} 1 & 1 & 1 & 25{,}000\\ -1 & 0 & 2 & 4000\\ 8 & 9 & 10 & 222{,}000\end{array}\right]$$

$$\left[\begin{array}{rrr|r} 1 & 1 & 1 & 25{,}000\\ 0 & 1 & 3 & 29{,}000\\ 0 & 1 & 2 & 22{,}000\end{array}\right] \quad \begin{array}{l} \\ R_1 + R_2\\ -8R_1 + R_3\end{array}$$

$$\left[\begin{array}{rrr|r} 1 & 1 & 1 & 25{,}000\\ 0 & 1 & 3 & 29{,}000\\ 0 & 0 & -1 & -7000\end{array}\right] \quad \begin{array}{l} \\ \\ -R_2 + R_3\end{array}$$

$$\left[\begin{array}{rrr|r} 1 & 1 & 1 & 25{,}000\\ 0 & 1 & 3 & 29{,}000\\ 0 & 0 & 1 & 7000\end{array}\right] \quad \begin{array}{l} \\ \\ -R_3\end{array}$$

This matrix represents the system

$$\begin{aligned} x + y + z &= 25{,}000\\ y + 3z &= 29{,}000\\ z &= 7000.\end{aligned}$$

Substitute 7000 for z in the second equation.

$$\begin{aligned} y + 3(7000) &= 29{,}000\\ y + 21{,}000 &= 29{,}000\\ y &= 8000\end{aligned}$$

Substitute 8000 for y and 7000 for z in the first equation.

$$\begin{aligned} x + 8000 + 7000 &= 25{,}000\\ x &= 10{,}000\end{aligned}$$

The company borrowed \$10,000 at 8%, \$8000 at 9% and \$7000 at 10%.

32. The lines intersect at $(8, 3000)$, so the cost equals the revenue at $x = 8$ (which is 800 parts). The revenue is \$3000.

33. On the sale of 1100 parts ($x = 11$), the revenue is about \$4100 and the cost is about \$3700.

$$\begin{aligned} \text{Profit} &= \text{Revenue} - \text{Cost} \\ &\approx 4100 - 3700 \\ &= 400 \end{aligned}$$

The profit is about \$400.

CHAPTER 6 EXPONENTS, POLYNOMIALS, AND POLYNOMIAL FUNCTIONS

6.1 Integer Exponents and Scientific Notation

6.1 Margin Exercises

1. Use the product rule for exponents, if possible.

(a) $m^8 \cdot m^6 = m^{8+6} = m^{14}$

(b) $r^7 \cdot r = r^7 \cdot r^1 = r^{7+1} = r^8$

(c) $k^4k^3k^6 = k^{4+3+6} = k^{13}$

(d) $m^5 \cdot p^4$ cannot be simplified further because the bases m and p are not the same. The product rule does not apply.

(e) Use the associative and commutative properties to group constants together and variables together.

$$\begin{aligned}(-4a^3)(6a^2) &= (-4)(6)(a^3a^2)\\ &= -24a^{3+2}\\ &= -24a^5\end{aligned}$$

(f) $$\begin{aligned}(-5p^4)(-9p^5) &= (-5)(-9)(p^4p^5)\\ &= 45p^{4+5}\\ &= 45p^9\end{aligned}$$

2. **(a)** $29^0 = 1$

Any real number (except 0) raised to the power zero is equal to 1.

(b) $(-29)^0 = 1$

(c) $-(-29)^0 = -(1) = -1$

(d) $-29^0 = -(29^0) = -1$

(e) $8^0 - 15^0 = 1 - 1 = 0$

(f) $(-15p^5)^0 = 1$

Since $p \neq 0$, $-15p^5$ will not equal zero. Therefore, $-15p^5$ raised to the power zero will equal 1.

3. **(a)** $6^{-3} = \dfrac{1}{6^3}$

(b) $8^{-1} = \dfrac{1}{8^1} = \dfrac{1}{8}$

(c) $(2x)^{-4} = \dfrac{1}{(2x)^4}, x \neq 0$

(d) $7r^{-6} = 7\left(\dfrac{1}{r^6}\right) = \dfrac{7}{r^6}, r \neq 0$

(e) $-q^{-4} = -\left(\dfrac{1}{q^4}\right) = -\dfrac{1}{q^4}, q \neq 0$

(f) $(-q)^{-4} = \dfrac{1}{(-q)^4}, q \neq 0$

(g) $3^{-1} + 5^{-1} = \dfrac{1}{3} + \dfrac{1}{5} = \dfrac{5}{15} + \dfrac{3}{15} = \dfrac{8}{15}$

(h) $4^{-1} - 2^{-1} = \dfrac{1}{4} - \dfrac{1}{2} = \dfrac{1}{4} - \dfrac{2}{4} = -\dfrac{1}{4}$

4. **(a)** $\dfrac{1}{4^{-3}} = \dfrac{1}{\frac{1}{4^3}} = 1 \div \dfrac{1}{4^3} = 1 \cdot \dfrac{4^3}{1} = 4^3 = 64$

(b) $$\begin{aligned}\frac{3^{-3}}{9^{-1}} &= \frac{\frac{1}{3^3}}{\frac{1}{9}} = \frac{1}{3^3} \div \frac{1}{9} = \frac{1}{3^3} \cdot \frac{9}{1}\\ &= \frac{1}{27} \cdot \frac{9}{1} = \frac{9}{27} = \frac{1}{3}\end{aligned}$$

5. **(a)** $\dfrac{4^8}{4^6} = 4^{8-6} = 4^2$

(b) $\dfrac{x^{12}}{x^3} = x^{12-3} = x^9, x \neq 0$

(c) $\dfrac{r^5}{r^8} = r^{5-8} = r^{-3} = \dfrac{1}{r^3}, r \neq 0$

(d) $\dfrac{2^8}{2^{-4}} = 2^{8-(-4)} = 2^{8+4} = 2^{12}$

(e) $\dfrac{6^{-3}}{6^4} = 6^{-3-4} = 6^{-7} = \dfrac{1}{6^7}$

(f) $\dfrac{8}{8^{-1}} = \dfrac{8^1}{8^{-1}} = 8^{1-(-1)} = 8^{1+1} = 8^2$

(g) $\dfrac{t^{-4}}{t^{-6}} = t^{-4-(-6)} = t^{-4+6} = t^2, t \neq 0$

(h) $\dfrac{x^3}{y^5}, y \neq 0$ cannot be simplified because the quotient rule does not apply. The bases x and y are different.

6. **(a)** $(r^5)^4 = r^{5 \cdot 4} = r^{20}$

(b) $\left(\dfrac{3}{4}\right)^3 = \dfrac{3^3}{4^3} = \dfrac{27}{64}$

(c) $(9x)^3 = 9^3x^3 = 729x^3$

(d) $(5r^6)^3 = 5^3r^{6 \cdot 3} = 5^3r^{18} = 125r^{18}$

(e) $$\begin{aligned}\left(\frac{-3n^4}{m}\right)^3 &= \frac{(-3)^3n^{4 \cdot 3}}{m^3} = \frac{(-3)^3n^{12}}{m^3}\\ &= \frac{-27n^{12}}{m^3}, m \neq 0\end{aligned}$$

7. **(a)** $\left(\frac{3}{4}\right)^{-3} = \left(\frac{4}{3}\right)^{3} = \frac{4^3}{3^3} = \frac{64}{27}$

(b) $\left(\frac{5}{6}\right)^{-2} = \left(\frac{6}{5}\right)^{2} = \frac{6^2}{5^2} = \frac{36}{25}$

8. **(a)** $5^4 \cdot 5^{-6} = 5^{4+(-6)} = 5^{-2} = \frac{1}{5^2}$ or $\frac{1}{25}$

(b) $x^{-4} \cdot x^{-6} \cdot x^8 = x^{-4+(-6)+8} = x^{-2} = \frac{1}{x^2}$

(c) $\left(5^{-3}\right)^{-2} = 5^{(-3)(-2)} = 5^6$

(d) $\left(y^{-2}\right)^7 = y^{(-2)7} = y^{-14} = \frac{1}{y^{14}}$

(e) $\frac{a^{-3}b^5}{a^4b^{-2}} = \frac{a^{-3}}{a^4} \cdot \frac{b^5}{b^{-2}} = a^{-3-4} \cdot b^{5-(-2)}$

$= a^{-7}b^7 = \frac{b^7}{a^7}$

(f) $\left(3^2k^{-4}\right)^{-1} = \left(3^2\right)^{-1} \cdot \left(k^{-4}\right)^{-1} = 3^{-2}k^4$

$= \frac{k^4}{3^2}$ or $\frac{k^4}{9}$

(g) $\left(\frac{2y}{x^3}\right)^2 \left(\frac{4y}{x}\right)^{-1}$

$= \frac{2^2y^2}{x^6} \cdot \frac{4^{-1}y^{-1}}{x^{-1}}$ *Combination of rules*

$= \frac{2^2 4^{-1} y^1}{x^5}$

$= \frac{2^2 y}{4x^5} = \frac{y}{x^5}$

9. **(a)** $400{,}000 = 4_\wedge 00{,}000.$ ← Decimal point

Place a caret after the first nonzero digit, 4. Count 5 places from the decimal point to the caret (understood to be after the last 0). Use a positive exponent on 10 since $400{,}000 > 4$.

$$400{,}000 = 4 \times 10^5$$

(b) $29{,}800{,}000 = 2_\wedge 9{,}800{,}000.$

Count 7 places. Use a positive exponent on 10 since $29{,}800{,}000 > 2.98$.

$$29{,}800{,}000 = 2.98 \times 10^7$$

(c) $-6083 = -6_\wedge 083.$

Count 3 places. Use a positive exponent on 10 since $6083 > 6.083$.

$$-6083 = -6.083 \times 10^3$$

(d) $.00172 = .001_\wedge 72$

Count 3 places. Use a negative exponent on 10 since $.00172 < 1.72$.

$$.00172 = 1.72 \times 10^{-3}$$

(e) $.000\,000\,0503 = .000\,000\,05_\wedge 03$

Count 8 places. Use a negative exponent on 10 since $.000\,000\,0503 < 5.03$.

$$.000\,000\,0503 = 5.03 \times 10^{-8}$$

10. **(a)** $4.98 \times 10^5 = 4.98\,000. = 498{,}000$

Move the decimal 5 places to the right.

(b) $6.8 \times 10^{-7} = .000\,000\,6.8 = .000\,000\,68$

Move the decimal 7 places to the left.

(c) $-5.372 \times 10^0 = -5.372 \times 1 = -5.372$

11.
$$\frac{200{,}000 \times .0003}{.06 \times 4{,}000{,}000} = \frac{2 \times 10^5 \times 3 \times 10^{-4}}{6 \times 10^{-2} \times 4 \times 10^6}$$
$$= \frac{2 \times 3 \times 10^5 \times 10^{-4}}{6 \times 4 \times 10^{-2} \times 10^6}$$
$$= \frac{2 \times 3 \times 10^1}{6 \times 4 \times 10^4}$$
$$= \frac{2 \times 3}{6 \times 4} \times 10^{-3}$$
$$= .25 \times 10^{-3}$$
$$= \left(2.5 \times 10^{-1}\right) \times 10^{-3}$$
$$= 2.5 \times 10^{-4} \text{ or } .00025$$

12. $d = rt$, so

$$t = \frac{d}{r}$$
$$= \frac{9.3 \times 10^7}{3.2 \times 10^3} = \frac{9.3}{3.2} \times 10^{7-3} \approx 2.9 \times 10^4.$$

It would take approximately 2.9×10^4 hours.

6.1 Section Exercises

1. $(ab)^2 = a^2b^2$ by a power rule. Since $a^2b^2 \neq ab^2$, the expression $(ab)^2 = ab^2$ has been simplified incorrectly. The exponent should apply to both a and b.

2. $(5x)^3 = 5^3x^3$ is correct.

3. $\left(\frac{4}{a}\right)^3 = \frac{4^3}{a^3} \quad (a \neq 0)$

Since $\frac{4^3}{a^3} \neq \frac{4^3}{a}$, the expression $\left(\frac{4}{a}\right)^3 = \frac{4^3}{a}$ has been simplified *incorrectly*.

4. $y^2 \cdot y^6 = y^{2+6} = y^8$
Since $y^8 \neq y^{12}$, the expression $y^2 \cdot y^6 = y^{12}$ has been simplified incorrectly.

5. $x^3 \cdot x^4 = x^{3+4} = x^7$ is correct.

6. $xy^0 = x^1 \cdot y^0 = x \cdot 1 = x \quad (y \neq 0)$
Since $y \neq 0$, the expression $xy^0 = 0$ has been simplified incorrectly.

7. $13^4 \cdot 13^8 = 13^{4+8} = 13^{12}$

8. $9^6 \cdot 9^4 = 9^{6+4} = 9^{10}$

9. $x^3 \cdot x^5 \cdot x^9 = x^{3+5+9} = x^{17}$

10. $y^4 \cdot y^5 \cdot y^6 = y^{4+5+6} = y^{15}$

11. $(-3w^5)(9w^3) = (-3)(9)w^{5+3} = -27w^8$

12. $(-5x^2)(3x^4) = (-5)(3)x^{2+4} = -15x^6$

13. $(2x^2y^5)(9xy^3) = (2)(9)x^{2+1}y^{5+3} = 18x^3y^8$

14. $(8s^4t)(3s^3t^5) = 8 \cdot 3s^{4+3}t^{1+5} = 24s^7t^6$

15. $r^2 \cdot s^4$ Because the bases are not the same, the product rule does not apply.

16. $p^3 \cdot q^2$ Because the bases are not the same, the product rule does not apply.

17. **(a)** $9^0 = 1$ **(B)**

(b) $-9^0 = -(9^0) = -(1) = -1$ **(C)**

(c) $(-9)^0 = 1$ **(B)**

(d) $-(-9)^0 = -1$ **(C)**

18. **(a)** $2x^0 = 2(1) = 2$ **(D)**

(b) $-2x^0 = -2(1) = -2$ **(E)**

(c) $(2x)^0 = 1$ **(B)**

(d) $(-2x)^0 = 1$ **(B)**

19. $25^0 = 1$, since $a^0 = 1$ for any nonzero base a.

20. $14^0 = 1$, since $a^0 = 1$ for any nonzero base a.

21. $-7^0 = -(7^0) = -(1) = -1$

22. $-10^0 = -(10^0) = -(1) = -1$

23. $(-15)^0 = 1$, since $a^0 = 1$ for any nonzero base a.

24. $(-20)^0 = 1$, since $a^0 = 1$ for any nonzero base a.

25. $-4^0 - m^0 = -1 - 1 = -2$

26. $-8^0 - k^0 = -1 - 1 = -2$

27. **(a)** $4^{-2} = \frac{1}{4^2} = \frac{1}{16}$ **(B)**

(b) $-4^{-2} = -\frac{1}{4^2} = -\frac{1}{16}$ **(D)**

(c) $(-4)^{-2} = \frac{1}{(-4)^2} = \frac{1}{16}$ **(B)**

(d) $-(-4)^{-2} = -\frac{1}{(-4)^2} = -\frac{1}{16}$ **(D)**

28. **(a)** $5^{-3} = \frac{1}{5^3} = \frac{1}{125}$ **(C)**

(b) $-5^{-3} = -\frac{1}{5^3} = -\frac{1}{125}$ **(D)**

(c) $(-5)^{-3} = \frac{1}{(-5)^3} = \frac{1}{-125} = -\frac{1}{125}$ **(D)**

(d) $-(-5)^{-3} = -\frac{1}{(-5)^3} = -\frac{1}{-125} = \frac{1}{125}$ **(C)**

29. $5^{-4} = \frac{1}{5^4} = \frac{1}{625}$

30. $7^{-2} = \frac{1}{7^2} = \frac{1}{49}$

31. $8^{-1} = \frac{1}{8^1} = \frac{1}{8}$

32. $12^{-1} = \frac{1}{12^1} = \frac{1}{12}$

33. $(4x)^{-2} = \frac{1}{(4x)^2} = \frac{1}{4^2x^2} = \frac{1}{16x^2}$

34. $(5t)^{-3} = \frac{1}{(5t)^3} = \frac{1}{5^3t^3} = \frac{1}{125t^3}$

35. $4x^{-2} = \frac{4}{x^2}$

36. $5t^{-3} = \frac{5}{t^3}$

37. $-a^{-3} = -\frac{1}{a^3}$

38. $-b^{-4} = -\frac{1}{b^4}$

39. $(-a)^{-4} = \frac{1}{(-a)^4} = \frac{1}{a^4}$

40. $(-b)^{-6} = \frac{1}{(-b)^6} = \frac{1}{b^6}$

41. $5^{-1} + 6^{-1} = \frac{1}{5} + \frac{1}{6} = \frac{6}{30} + \frac{5}{30} = \frac{11}{30}$

42. $2^{-1} + 8^{-1} = \frac{1}{2} + \frac{1}{8} = \frac{4}{8} + \frac{1}{8} = \frac{5}{8}$

43. $8^{-1} - 3^{-1} = \frac{1}{8} - \frac{1}{3} = \frac{3}{24} - \frac{8}{24} = -\frac{5}{24}$

44. $6^{-1} - 4^{-1} = \frac{1}{6} - \frac{1}{4} = \frac{2}{12} - \frac{3}{12} = -\frac{1}{12}$

45. $\frac{1}{4^{-2}} = 4^2 = 16$

46. $\frac{1}{3^{-3}} = 3^3 = 27$

47. $\frac{2^{-2}}{3^{-3}} = \frac{3^3}{2^2} = \frac{27}{4}$

48. $\frac{3^{-3}}{2^{-2}} = \frac{2^2}{3^3} = \frac{4}{27}$

49. $\left(\frac{2}{3}\right)^{-3} = \left(\frac{3}{2}\right)^3 = \frac{3^3}{2^3} = \frac{27}{8}$

50. $\left(\frac{3}{2}\right)^{-3} = \left(\frac{2}{3}\right)^3 = \frac{2^3}{3^3} = \frac{8}{27}$

51. $\left(\frac{4}{5}\right)^{-2} = \left(\frac{5}{4}\right)^2 = \frac{5^2}{4^2} = \frac{25}{16}$

52. $\left(\frac{5}{4}\right)^{-2} = \left(\frac{4}{5}\right)^2 = \frac{4^2}{5^2} = \frac{16}{25}$

53. **(a)** $\left(\frac{1}{3}\right)^{-1} = \left(\frac{3}{1}\right)^1 = 3$ **(B)**

(b) $\left(-\frac{1}{3}\right)^{-1} = \left(-\frac{3}{1}\right)^1 = -3$ **(D)**

(c) $-\left(\frac{1}{3}\right)^{-1} = -\left(\frac{3}{1}\right)^1 = -3$ **(D)**

(d) $-\left(-\frac{1}{3}\right)^{-1} = -\left(-\frac{3}{1}\right)^1 = -(-3) = 3$ **(B)**

54. **(a)** $\left(\frac{2}{5}\right)^{-2} = \left(\frac{5}{2}\right)^2 = \frac{25}{4}$ **(A)**

(b) $\left(-\frac{2}{5}\right)^{-2} = \left(-\frac{5}{2}\right)^2 = \frac{25}{4}$ **(A)**

(c) $-\left(\frac{2}{5}\right)^{-2} = -\left(\frac{5}{2}\right)^2 = -\frac{25}{4}$ **(B)**

(d) $-\left(-\frac{2}{5}\right)^{-2} = -\left(-\frac{5}{2}\right)^2 = -\frac{25}{4}$ **(B)**

55. $\frac{4^8}{4^6} = 4^{8-6} = 4^2$ or 16

56. $\frac{5^9}{5^7} = 5^{9-7} = 5^2$ or 25

57. $\frac{x^{12}}{x^8} = x^{12-8} = x^4$

58. $\frac{y^{14}}{y^{10}} = y^{14-10} = y^4$

59. $\frac{r^7}{r^{10}} = r^{7-10} = r^{-3} = \frac{1}{r^3}$

60. $\frac{y^8}{y^{12}} = y^{8-12} = y^{-4} = \frac{1}{y^4}$

61. $\frac{6^4}{6^{-2}} = 6^{4-(-2)} = 6^{4+2} = 6^6$

62. $\frac{7^5}{7^{-3}} = 7^{5-(-3)} = 7^{5+3} = 7^8$

63. $\frac{6^{-3}}{6^7} = 6^{-3-7} = 6^{-10} = \frac{1}{6^{10}}$

64. $\frac{5^{-4}}{5^2} = 5^{-4-2} = 5^{-6} = \frac{1}{5^6}$

65. $\frac{7}{7^{-1}} = 7^{1-(-1)} = 7^2$ or 49

66. $\frac{8}{8^{-1}} = 8^{1-(-1)} = 8^2$ or 64

67. $\frac{r^{-3}}{r^{-6}} = r^{-3-(-6)} = r^{-3+6} = r^3$

68. $\frac{s^{-4}}{s^{-8}} = s^{-4-(-8)} = s^{-4+8} = s^4$

69. $\frac{x^3}{y^2}$ Because the bases are not the same, the quotient rule does not apply.

70. $\frac{y^5}{t^3}$ Because the bases are not the same, the quotient rule does not apply.

71. $(x^3)^6 = x^{3\cdot 6} = x^{18}$

72. $(y^5)^4 = y^{5\cdot 4} = y^{20}$

73. $\left(\frac{3}{5}\right)^3 = \frac{3^3}{5^3} = \frac{27}{125}$

74. $\left(\frac{4}{3}\right)^2 = \frac{4^2}{3^2} = \frac{16}{9}$

75. $(4t)^3 = 4^3t^3 = 64t^3$

76. $(5t)^4 = 5^4t^4 = 625t^4$

77. $(-6x^2)^3 = (-6)^3x^{2\cdot 3} = -216x^6$

78. $(-2x^5)^5 = (-2)^5x^{5\cdot 5} = -32x^{25}$

79. $\left(\frac{-4m^2}{t}\right)^3 = \frac{(-4)^3 m^{2\cdot3}}{t^3} = \frac{-64m^6}{t^3} = -\frac{64m^6}{t^3}$

80. $\left(\frac{-5n^4}{r^2}\right)^3 = \frac{(-5)^3 n^{4\cdot3}}{r^{2\cdot3}} = -\frac{125n^{12}}{r^6}$

81. $3^5 \cdot 3^{-6} = 3^{5+(-6)} = 3^{-1} = \frac{1}{3^1} = \frac{1}{3}$

82. $4^4 \cdot 4^{-6} = 4^{4+(-6)} = 4^{-2} = \frac{1}{4^2}$ or $\frac{1}{16}$

83. $a^{-3}a^2a^{-4} = a^{-3+2+(-4)} = a^{-5} = \frac{1}{a^5}$

84. $k^{-5}k^{-3}k^4 = k^{-5+(-3)+4} = k^{-4} = \frac{1}{k^4}$

85.
$$\begin{aligned}(k^2)^{-3}k^4 &= k^{2(-3)}k^4\\ &= k^{-6}k^4\\ &= k^{-6+4}\\ &= k^{-2}\\ &= \frac{1}{k^2}\end{aligned}$$

86.
$$\begin{aligned}(x^3)^{-4}x^5 &= x^{3(-4)}x^5\\ &= x^{-12}x^5\\ &= x^{-12+5}\\ &= x^{-7}\\ &= \frac{1}{x^7}\end{aligned}$$

87.
$$\begin{aligned}-4r^{-2}(r^4)^2 &= -4r^{-2}(r^8)\\ &= -4r^{-2+8}\\ &= -4r^6\end{aligned}$$

88.
$$\begin{aligned}-2m^{-1}(m^3)^2 &= -2m^{-1}(m^6)\\ &= -2m^{-1+6}\\ &= -2m^5\end{aligned}$$

89.
$$\begin{aligned}(5a^{-1})^4(a^2)^{-3} &= 5^4a^{-1\cdot4}a^{2(-3)}\\ &= 5^4a^{-4}a^{-6}\\ &= 5^4a^{-4-6}\\ &= 5^4a^{-10}\\ &= \frac{5^4}{a^{10}} \text{ or } \frac{625}{a^{10}}\end{aligned}$$

90.
$$\begin{aligned}(3p^{-4})^2(p^3)^{-1} &= 3^2p^{-4\cdot2}p^{3(-1)}\\ &= 3^2p^{-8}p^{-3}\\ &= 3^2p^{-8-3}\\ &= 3^2p^{-11}\\ &= \frac{3^2}{p^{11}} \text{ or } \frac{9}{p^{11}}\end{aligned}$$

91.
$$\begin{aligned}(z^{-4}x^3)^{-1} &= z^{-4(-1)}x^{3(-1)}\\ &= z^4x^{-3}\\ &= \frac{z^4}{x^3}\end{aligned}$$

92.
$$\begin{aligned}(y^{-2}z^4)^{-3} &= y^{-2(-3)}z^{4(-3)}\\ &= y^6z^{-12}\\ &= \frac{y^6}{z^{12}}\end{aligned}$$

93. $\frac{(p^{-2})^3}{5p^4} = \frac{p^{-6}}{5p^4} = \frac{p^{-6-4}}{5} = \frac{p^{-10}}{5} = \frac{1}{5p^{10}}$

94. $\frac{(m^4)^{-1}}{9m^3} = \frac{m^{-4}}{9m^3} = \frac{m^{-4-3}}{9} = \frac{m^{-7}}{9} = \frac{1}{9m^7}$

95.
$$\begin{aligned}\frac{4a^5(a^{-1})^3}{(a^{-2})^{-2}} &= \frac{4a^5a^{-1\cdot3}}{a^{-2(-2)}} = \frac{4a^5a^{-3}}{a^4}\\ &= 4a^{5+(-3)-4}\\ &= 4a^{-2}\\ &= \frac{4}{a^2}\end{aligned}$$

96.
$$\begin{aligned}\frac{12k^{-2}(k^{-3})^{-4}}{6k^5} &= \frac{12k^{-2}k^{12}}{6k^5}\\ &= \frac{12}{6}k^{-2+12-5}\\ &= 2k^5\end{aligned}$$

97. The first step may be the most confusing.

$$\begin{aligned}(-y^{-4})^2 &= (-1\cdot y^{-4})^2\\ &= (-1)^2(y^{-4})^2\\ &= 1\cdot y^{-4(2)} = y^{-8}\end{aligned}$$

It is not necessary to include these steps once the concept of squaring a negative is committed to memory.

$$\begin{aligned}\frac{(-y^{-4})^2}{6(y^{-5})^{-1}} &= \frac{y^{-4(2)}}{6y^{-5(-1)}}\\ &= \frac{y^{-8}}{6y^5}\\ &= \frac{1}{6y^5y^8}\\ &= \frac{1}{6y^{13}}\end{aligned}$$

98. $\dfrac{2(-m^{-1})^{-4}}{9(m^{-3})^2} = \dfrac{2(-1)^{-4}(m^{-1(-4)})}{9m^{-3(2)}}$

$= \dfrac{2m^4}{9(-1)^4m^{-6}}$

$= \dfrac{2m^{4-(-6)}}{9\cdot 1}$

$= \dfrac{2m^{10}}{9}$

99. $\dfrac{(2k)^2m^{-5}}{(km)^{-3}} = \dfrac{2^2k^2m^{-5}}{k^{-3}m^{-3}}$

$= 2^2k^{2-(-3)}m^{-5-(-3)}$

$= 2^2k^5m^{-2}$

$= \dfrac{2^2k^5}{m^2}$ or $\dfrac{4k^5}{m^2}$

100. $\dfrac{(3rs)^{-2}}{3^2r^2s^{-4}} = \dfrac{3^{-2}r^{-2}s^{-2}}{3^2r^2s^{-4}}$

$= 3^{-2-2}r^{-2-2}s^{-2-(-4)}$

$= 3^{-4}r^{-4}s^2$

$= \dfrac{s^2}{3^4r^4}$ or $\dfrac{s^2}{81r^4}$

101. $\left(\dfrac{3k^{-2}}{k^4}\right)^{-1}\cdot\dfrac{2}{k} = \dfrac{(3k^{-2})^{-1}}{(k^4)^{-1}}\cdot\dfrac{2}{k}$

$= \dfrac{3^{-1}k^2\cdot 2}{k^{-4}k^1}$

$= \dfrac{3^{-1}\cdot 2k^2}{k^{-3}}$

$= \dfrac{2k^2k^3}{3}$

$= \dfrac{2k^5}{3}$

102. $\left(\dfrac{7m^{-2}}{m^{-3}}\right)^{-2}\cdot\dfrac{m^3}{4} = \dfrac{7^{-2}m^{-2(-2)}}{m^{-3(-2)}}\cdot\dfrac{m^3}{4}$

$= \dfrac{m^4\cdot m^3}{7^2\cdot 4m^6}$

$= \dfrac{m^{4+3-6}}{7^2\cdot 4}$

$= \dfrac{m}{7^2\cdot 4}$ or $\dfrac{m}{196}$

103. $\left(\dfrac{2p}{q^2}\right)^3\left(\dfrac{3p^4}{q^{-4}}\right)^{-1} = \dfrac{(2p)^3}{(q^2)^3}\cdot\dfrac{(3p^4)^{-1}}{(q^{-4})^{-1}}$

$= \dfrac{2^3p^3}{q^6}\cdot\dfrac{3^{-1}p^{-4}}{q^4}$

$= \dfrac{2^3p^{3+(-4)}}{3^1q^{6+4}}$

$= \dfrac{8p^{-1}}{3q^{10}} = \dfrac{8}{3pq^{10}}$

104. $\left(\dfrac{5z^3}{2a^2}\right)^{-3}\left(\dfrac{8a^{-1}}{15z^{-2}}\right)^{-3}$

$= \dfrac{5^{-3}z^{-9}}{2^{-3}a^{-6}}\cdot\dfrac{8^{-3}a^3}{15^{-3}z^6}$

$= \dfrac{2^3a^6a^315^3}{5^3z^98^3z^6}$

$= \left(\dfrac{15}{5}\right)^3\left(\dfrac{2}{8}\right)^3\dfrac{a^{6+3}}{z^{9+6}}$

$= \dfrac{3^3a^9}{4^3z^{15}} = \dfrac{27a^9}{64z^{15}}$

105. $530 = 5_{\wedge}30.$ ← Decimal point

Count 2 places.

Since the number 5.3 is to be made larger, the exponent on 10 is positive.

$$530 = 5.3\times 10^2$$

106. $1600 = 1_{\wedge}600.$ ← Decimal point

Count 3 places.

Since the number 1.6 is to be made larger, the exponent on 10 is positive.

$$1600 = 1.6\times 10^3$$

107. $.830 = .8_{\wedge}30$

Count 1 place.

Since the number 8.3 is to be made smaller, the exponent on 10 is negative.

$$.830 = 8.3\times 10^{-1}$$

108. $.0072 = .007_{\wedge}2$

Count 3 places.

Since the number 7.2 is to be made smaller, the exponent on 10 is negative.

$$.0072 = 7.2\times 10^{-3}$$

109. $.00000692 = .000006_{\wedge}92$

Count 6 places.

Since the number 6.92 is to be made smaller, the exponent on 10 is negative.

$$.00000692 = 6.92\times 10^{-6}$$

110. $.875 = .8_{\wedge}75$

Count 1 place.

Since the number 8.75 is to be made smaller, the exponent on 10 is negative.

$$.875 = 8.75\times 10^{-1}$$

111. $-38{,}500 = -3_{\wedge}8500.$

Count 4 places.

Since the number 3.85 is to be made larger, the exponent on 10 is positive. Also, affix a negative sign in front of the number.

$$-38{,}500 = -3.85 \times 10^4$$

112. $-976{,}000{,}000 = -9_{\wedge}76000000.$

Count 8 places.

Since the number 9.76 is to be made larger, the exponent on 10 is positive. Also, affix a negative sign in front of the number.

$$-976{,}000{,}000 = -9.76 \times 10^8$$

113. $7.2 \times 10^4 = 7.2000. = 72{,}000$

Move the decimal point 4 places to the *right* because of the *positive* exponent. Attach extra zeros.

114. $8.91 \times 10^2 = 8.91. = 891$

Move the decimal point 2 places to the *right* because of the *positive* exponent.

115. $2.54 \times 10^{-3} = .002.54 = .00254$

Since the exponent is *negative*, move the decimal point 3 places to the *left*.

116. $5.42 \times 10^{-4} = .0005.42 = .000542$

Since the exponent is *negative*, move the decimal point 4 places to the *left*.

117. $-6 \times 10^4 = -6.0000. = -60{,}000$

Move the decimal point 4 places to the *right* because of the *positive* exponent. Attach extra zeros.

118. $-9 \times 10^3 = -9.000. = -9000$

Move the decimal point 3 places to the *right* because of the *positive* exponent. Attach extra zeros.

119. $1.2 \times 10^{-5} = .00001.2 = .000012$

Since the exponent is *negative*, move the decimal point 5 places to the *left*.

120. $2.7 \times 10^{-6} = .000002.7 = .0000027$

Since the exponent is *negative*, move the decimal point 6 places to the *left*.

121.
$$\begin{aligned}\frac{3 \times 10^{-2}}{12 \times 10^3} &= \frac{3 \times 10^{-2}}{1.2 \times 10^4}\\ &= \frac{3}{1.2} \times \frac{10^{-2}}{10^4}\\ &= 2.5 \times 10^{-6}\\ &= .0000025\end{aligned}$$

122.
$$\begin{aligned}\frac{5 \times 10^{-3}}{25 \times 10^2} &= \frac{5}{25} \times 10^{-3-2}\\ &= .2 \times 10^{-5}\\ &= .000002\end{aligned}$$

123.
$$\begin{aligned}\frac{.05 \times 1600}{.0004} &= \frac{5 \times 10^{-2} \times 1.6 \times 10^3}{4 \times 10^{-4}}\\ &= \frac{5(1.6)}{4} \times \frac{10^{-2} \times 10^3}{10^{-4}}\\ &= 2 \times 10^{-2+3-(-4)}\\ &= 2 \times 10^5\\ &= 200{,}000\end{aligned}$$

124.
$$\begin{aligned}\frac{.003 \times 40{,}000}{.00012} &= \frac{3 \times 10^{-3} \times 4 \times 10^4}{1.2 \times 10^{-4}}\\ &= \frac{3 \times 4 \times 10^{-3+4}}{1.2 \times 10^{-4}}\\ &= 10 \times 10^{-3+4-(-4)}\\ &= 10^1 \times 10^5\\ &= 1 \times 10^6\\ &= 1{,}000{,}000\end{aligned}$$

125.
$$\begin{aligned}\$1{,}000{,}000{,}000 &= \$1 \times 10^9\\ \$1{,}000{,}000{,}000{,}000 &= \$1 \times 10^{12}\\ \$2{,}128{,}000{,}000{,}000 &= \$2.128 \times 10^{12}\\ 144{,}419 &= 1.44419 \times 10^5\end{aligned}$$

126.
$$\begin{aligned}1500 &= 1.5 \times 10^3\\ 680{,}000 &= 6.8 \times 10^5\\ 1{,}100{,}000 &= 1.1 \times 10^6\\ 2{,}200{,}000 &= 2.2 \times 10^6\end{aligned}$$

127. (a) 281.4 million
$$\begin{aligned}&= 281{,}400{,}000\\ &= 2.814 \times 10^8\end{aligned}$$

(b) $\$1{,}000{,}000{,}000{,}000 = \$1 \times 10^{12}$

(c) Divide the amount by the number of people to determine how much each person would have to contribute.

$$\begin{aligned}\frac{\$1 \times 10^{12}}{2.814 \times 10^8} &\approx \$.3554 \times 10^4\\ &= \$3554\end{aligned}$$

In 2000, each person in the United States would have had to contribute about $3554 in order to make someone a trillionaire.

128. If x represents the number of operations per second that the normal desktop computer is capable of, then

$$\begin{aligned}15{,}000x &= 3.9 \times 10^8.\\ x &= \frac{3.9 \times 10^8}{1.5 \times 10^4}\\ &= \frac{3.9}{1.5} \times \frac{10^8}{10^4}\\ &= 2.6 \times 10^4\\ &= 26{,}000\end{aligned}$$

In 1998, the normal desktop computer could do about 26,000 operations per second.

129. If x represents the amount that each person should expect to pay, then

$$\begin{aligned}2000x &\approx 8.009 \times 10^7.\\ x &= \frac{8.009 \times 10^7}{2 \times 10^3}\\ &= \frac{8.009}{2} \times \frac{10^7}{10^3}\\ &= 4.0045 \times 10^4\\ &= 40{,}045\end{aligned}$$

Each person should expect to pay about \$40,045.

130. Since $d = rt$, $t = \frac{d}{r}$. Divide the distance traveled by the speed of light.

$$\frac{9 \times 10^{12}}{3 \times 10^{10}} = \frac{9}{3} \times 10^{12-10} = 3 \times 10^2$$

It will take 300 seconds.

131. Use $d = rt$, or $\frac{d}{r} = t$, where $d = 9.3 \times 10^7$ and $r = 2.9 \times 10^3$.

$$\begin{aligned}\frac{9.3 \times 10^7}{2.9 \times 10^3} &= \frac{9.3}{2.9} \times 10^{7-3}\\ &\approx 3.2 \times 10^4\end{aligned}$$

It would take about 3.2×10^4 or 32,000 hours. Note that there are

$$24 \times 365 = 8760$$

hours in one year.

$$\frac{32{,}000 \text{ hours}}{8760 \text{ hours/year}} \approx 3.7 \text{ years}$$

132. First find the number of seconds in a year.

$$\begin{aligned}1 \text{ year} &= 365 \text{ days}\\ &= 365(24) \text{ hr}\\ &= 8760 \text{ hr}\\ &= 8760(60) \text{ min}\\ &= 525{,}600 \text{ min}\\ &= 525{,}600(60) \text{ sec}\\ &= 31{,}536{,}000 \text{ sec}\\ &= 3.1536 \times 10^7 \text{ sec}\end{aligned}$$

Now use $d = rt$ and multiply the rate light travels by the number of seconds in a year.

$$\begin{aligned}1 \text{ light year} &= (1.86 \times 10^5 \text{ mi/sec})\\ &\quad \cdot (3.1536 \times 10^7 \text{ sec})\\ &= 1.86(3.1536) \times 10^{5+7}\\ &\approx 5.87 \times 10^{12} \text{ mi}\end{aligned}$$

There are about 5.87×10^{12} miles in a light year.

133. (a) The distance from Mercury to the sun is 3.6×10^7 mi and the distance from Venus to the sun is 6.7×10^7 mi, so the distance between Mercury and Venus in miles is

$$\begin{aligned}&(6.7 \times 10^7) - (3.6 \times 10^7)\\ &= (6.7 - 3.6) \times 10^7\\ &= 3.1 \times 10^7.\end{aligned}$$

Use $d = rt$, or $\frac{d}{r} = t$, where $d = 3.1 \times 10^7$ and $r = 1.55 \times 10^3$.

$$\begin{aligned}\frac{3.1 \times 10^7}{1.55 \times 10^3} &= \frac{3.1}{1.55} \times 10^{7-3}\\ &= 2 \times 10^4\\ &= 20{,}000\end{aligned}$$

It would take 20,000 hr.

(b) From part (a), it takes 20,000 hr for a spacecraft to travel from Venus to Mercury. Convert this to days (24 hours = 1 day).

$$\begin{aligned}20{,}000 \text{ hr} &= \frac{20{,}000}{24} \text{ days}\\ &\approx 833 \text{ days}\end{aligned}$$

It would take about 833 days.

134. Find the difference.

$$\begin{aligned}&(4.60 \times 10^8) - (4.14 \times 10^8)\\ &= (4.60 - 4.14) \times 10^8\\ &= .46 \times 10^8\\ &= 4.6 \times 10^7\end{aligned}$$

It is 4.6×10^7 m or 46,000,000 m from the center of the moon.

135. Consider $-a^n$ and $(-a)^n$. Let $a = 2$.

If $n = 2$, then
$-a^n = -2^2 = -4$ and $(-a)^n = (-2)^2 = 4$.
If $n = 3$, then
$-a^n = -2^3 = -8$ and $(-a)^n = (-2)^3 = -8$.
If $n = 4$, then
$-a^n = -2^4 = -16$ and $(-a)^n = (-2)^4 = 16$.
If $n = 5$, then
$-a^n = -2^5 = -32$ and $(-a)^n = (-2)^5 = -32$.
If $n = 6$, then
$-a^n = -2^6 = -64$ and $(-a)^n = (-2)^6 = 64$.

Based on these cases, when n is even, the expressions are opposites. When n is odd, the expressions are equal.

136. Your friend multiplied the bases, which is incorrect. Instead, keep the same base and add the exponents.

$$4^5 \cdot 4^2 = 4^{5+2} = 4^7$$

137. Write the fraction as its reciprocal raised to the opposite of the negative power.

138. To raise a power to a power, keep the same base and multiply the exponents.

6.2 Adding and Subtracting Polynomials

6.2 Margin Exercises

1. **(a)** In the term $-9m^5$, the coefficient is -9.

(b) In $12y^2x$, the coefficient is 12.

(c) In x, the coefficient is 1 since $x = 1 \cdot x$.

(d) In $-y$, the coefficient is -1 since $-y = -1 \cdot y$.

(e) In $\frac{z}{4}$, the coefficient is $\frac{1}{4}$ since $\frac{z}{4} = \frac{1}{4}z$.

2. **(a)** $-4 + 9y + y^3$ is written in descending powers as $y^3 + 9y - 4$.

(b) $-3z^4 + 2z^3 + z^5 - 6z$ is written in descending powers as $z^5 - 3z^4 + 2z^3 - 6z$.

(c) $-12m^{10} + 8m^9 + 10m^{12}$ is written in descending powers as $10m^{12} - 12m^{10} + 8m^9$.

3. **(a)** $12m^4 - 6m^2$ is a binomial since it has two terms.

(b) $-6y^3 + 2y^2 - 8y$ is a trinomial since it has three terms.

(c) $3a^5$ is a monomial since it has one term.

(d) $-2k^{10} + 2k^9 - 8k^5 + 2k$ has four terms, so it is none of the choices.

4. The degree of a polynomial is the highest degree of any of the terms.

(a) $9y^4 + 8y^3 - 6$ has degree 4 since the largest exponent on y is 4.

(b) $-12m^7 + 11m^3 + m^9$ has degree 9.

(c) $-2k$ has degree 1 since $-2k = -2k^1$.

(d) 10 has degree 0 since $10 = 10x^0$.

(e) $3mn^2 + 2m^3n$ has degree 4.

Since the degree of $3mn^2$, or $3m^1n^2$, is $1 + 2 = 3$, and the degree of $2m^3n$, or $2m^3n^1$, is $3 + 1 = 4$, the degree of the polynomial is the greatest degree, 4.

5. **(a)** $11x + 12x - 7x - 3x$
$= (11 + 12 - 7 - 3)x = 13x$

(b) $11p^5 + 4p^5 - 6p^3 + 8p^3$
$= (11 + 4)p^5 + (-6 + 8)p^3$
$= 15p^5 + 2p^3$

(c) $2y^2z^4 + 3y^4 + 5y^4 - 9y^4z^2$
$= 2y^2z^4 + (3 + 5)y^4 - 9y^4z^2$
$= 2y^2z^4 + 8y^4 - 9y^4z^2$

6. **(a)** Add using the horizontal method.

$(12y^2 - 7y + 9) + (-4y^2 - 11y + 5)$
$= 12y^2 - 4y^2 - 7y - 11y + 9 + 5$
$= 8y^2 - 18y + 14$

Add using the vertical method.

$$\begin{array}{rcrcr} 12y^2 & - & 7y & + & 9 \\ -4y^2 & - & 11y & + & 5 \\ \hline 8y^2 & - & 18y & + & 14 \end{array}$$

(b) Add using the vertical method.

$$\begin{array}{rcrcr} -6r^5 & + & 2r^3 & - & r^2 \\ 8r^5 & - & 2r^3 & + & 5r^2 \\ \hline 2r^5 & + & 0r^3 & + & 4r^2 \end{array}$$

Add using the horizontal method.

$(-6r^5 + 2r^3 - r^2) + (8r^5 - 2r^3 + 5r^2)$
$= -6r^5 + 8r^5 + 2r^3 - 2r^3 - r^2 + 5r^2$
$= 2r^5 + 4r^2$

7. **(a)** Subtract using the horizontal method.

$$\begin{aligned}&(6y^3 - 9y^2 + 8) - (2y^3 + y^2 + 5)\\&= 6y^3 - 9y^2 + 8 - 2y^3 - y^2 - 5\\&= 6y^3 - 2y^3 - 9y^2 - y^2 + 8 - 5\\&= 4y^3 - 10y^2 + 3\end{aligned}$$

Subtract using the vertical method.

$$\begin{array}{rcrcr} 6y^3 & - & 9y^2 & + & 8 \\ \underline{2y^3} & \underline{+} & \underline{y^2} & \underline{+} & \underline{5} \end{array}$$

Change all the signs in the second polynomial, and add.

$$\begin{array}{rcrcrl} 6y^3 & - & 9y^2 & + & 8 & \\ -2y^3 & - & y^2 & - & 5 & \textit{Change all signs.} \\ \hline 4y^3 & - & 10y^2 & + & 3 & \textit{Add in columns.} \end{array}$$

(b) Subtract using the vertical method.

$$\begin{array}{rcrcr} 6y^3 & - & 2y^2 & + & 5y \\ \underline{-2y^3} & \underline{+} & \underline{8y^2} & \underline{-} & \underline{11y} \end{array}$$

Change all the signs in the second polynomial, and add.

$$\begin{array}{rcrcrl} 6y^3 & - & 2y^2 & + & 5y & \\ 2y^3 & - & 8y^2 & + & 11y & \textit{Change all signs.} \\ \hline 8y^3 & - & 10y^2 & + & 16y & \textit{Add in columns.} \end{array}$$

Subtract using the horizontal method.

$$\begin{aligned}&(6y^3 - 2y^2 + 5y) - (-2y^3 + 8y^2 - 11y)\\&= 6y^3 - 2y^2 + 5y + 2y^3 - 8y^2 + 11y\\&= 6y^3 + 2y^3 - 2y^2 - 8y^2 + 5y + 11y\\&= 8y^3 - 10y^2 + 16y\end{aligned}$$

6.2 Section Exercises

1. In $2x^3 + x - 3x^2$, the degrees of the terms are 3, 1, and 2 from left to right. Therefore, this polynomial is written in neither ascending nor descending powers.

2. $3x^5 + x^4 - 2x^3 + x$ is written in descending powers because the degrees of the terms decrease from left to right.

3. The polynomial $4p^3 - 8p^5 + p^7$ is written in ascending powers, since the exponents are increasing from left to right.

4. $q^2 + 3q^4 - 2q + 1$ is written in neither descending nor ascending powers because the degrees of the terms neither decrease nor increase from left to right.

5. In $-m^3 + 5m^2 + 3m + 10$, the degrees of the terms are 3, 2, 1, and 0 from left to right. Therefore, this polynomial is written in descending powers.

6. $4 - x + 3x^2$ is written in ascending powers. The term 4 can be thought of as $4x^0$ and $-x$ as $-x^1$. The degrees of the terms increase from left to right.

7. In $7z$, the coefficient is 7 and, since $7z = 7z^1$, the degree is 1.

8. In $3r$, the coefficient is 3 and, since $3r = 3r^1$, the degree is 1.

9. In $-15p^2$, the coefficient is -15 and the degree is 2.

10. In $-27k^3$, the coefficient is -27 and the degree is 3.

11. In x^4, since $x^4 = 1x^4$, the coefficient is 1 and the degree is 4.

12. In y^6, since $y^6 = 1y^6$, the coefficient is 1 and the degree is 6.

13. In $-mn^5$, the coefficient is -1, since $-mn^5 = -1mn^5$, and the degree is 6, since the sum of the exponents on m and n is $1 + 5 = 6$.

14. In $-a^5b$, since $-a^5b = -1a^5b$, the coefficient is -1 and, since $b = b^1$ and $5 + 1 = 6$, the degree is 6.

15. 24 is one term, so it's a *monomial*. 24 is a nonzero constant, so it has degree zero.

16. 5 is one term, so it's a *monomial*. 5 is a nonzero constant, so it has degree zero.

17. $7m - 21$ has two terms, so it's a *binomial*. The exponent on m is 1, so $7m - 21$ has degree 1.

18. $-x^2 + 3x^5$ is a *binomial* of degree 5.

19. $2r^3 + 3r^2 + 5r$ has three terms, so it's a *trinomial*. The greatest exponent is 3, so the degree is 3.

20. $5z^2 - 5z + 7$ has three terms, so it's a *trinomial*. The greatest exponent is 2, so the degree is 2.

21. $-6p^4q - 3p^3q^2 + 2pq^3 - q^4$ has four terms, so it is classified as *none of these*. The greatest sum of exponents on any term is 5, so the polynomial has degree 5.

22. $8s^3t - 3s^2t^2 + 2st^3 + 9$ is *none of these* and has degree 4.

23. $5z^4 + 3z^4 = (5 + 3)z^4 = 8z^4$

24. $8r^5 - 2r^5 = (8 - 2)r^5 = 6r^5$

25. $-m^3 + 2m^3 + 6m^3 = (-1 + 2 + 6)m^3 = 7m^3$

26. $3p^4 + 5p^4 - 2p^4 = (3 + 5 - 2)p^4 = 6p^4$

27. $x + x + x + x + x$
$= (1 + 1 + 1 + 1 + 1)x$
$= 5x$

28. $z - z - z + z = (1 - 1 - 1 + 1)z$
$= 0z = 0$

29. $m^4 - 3m^2 + m$ is *already simplified* since there are no like terms to be combined.

30. $5a^5 + 2a^4 - 9a^3$ is *already simplified* since there are no like terms to be combined.

31. $y^2 + 7y - 4y^2 = (y^2 - 4y^2) + 7y$
$= (1 - 4)y^2 + 7y$
$= -3y^2 + 7y$

32. $2c^2 - 4 + 8 - c^2 = (2c^2 - c^2) - 4 + 8$
$= (2 - 1)c^2 + 4$
$= c^2 + 4$

33. $2k + 3k^2 + 5k^2 - 7$
$= (3k^2 + 5k^2) + 2k - 7$
$= (3 + 5)k^2 + 2k - 7$
$= 8k^2 + 2k - 7$

34. $4x^2 + 2x - 6x^2 - 6$
$= (4x^2 - 6x^2) + 2x - 6$
$= (4 - 6)x^2 + 2x - 6$
$= -2x^2 + 2x - 6$

35. $n^4 - 2n^3 + n^2 - 3n^4 + n^3$
$= n^4 - 3n^4 - 2n^3 + n^3 + n^2$
$= (1 - 3)n^4 + (-2 + 1)n^3 + n^2$
$= -2n^4 - n^3 + n^2$

36. $2q^3 + 3q^2 - 4q - q^3 + 5q^2$
$= (2q^3 - q^3) + (3q^2 + 5q^2) - 4q$
$= (2 - 1)q^3 + (3 + 5)q^2 - 4q$
$= q^3 + 8q^2 - 4q$

37.
$$\begin{array}{r} -12p^2 + 4p - 1 \\ 3p^2 + 7p - 8 \\ \hline -9p^2 + 11p - 9 \end{array}$$
Add vertically.

38.
$$\begin{array}{r} -6y^3 + 8y + 5 \\ 9y^3 + 4y - 6 \\ \hline 3y^3 + 12y - 1 \end{array}$$
Add vertically.

39.
$$\begin{array}{r} 12a + 15 \\ 7a - 3 \\ \hline \end{array}$$

To subtract, change all the signs in the second polynomial, and add.

$$\begin{array}{r} 12a + 15 \\ -7a + 3 \\ \hline 5a + 18 \end{array}$$

40.
$$\begin{array}{r} -3b + 6 \\ 2b - 8 \\ \hline \end{array}$$

To subtract, change all the signs in the second polynomial, and add.

$$\begin{array}{r} -3b + 6 \\ -2b + 8 \\ \hline -5b + 14 \end{array}$$

41.
$$\begin{array}{r} 6m^2 - 11m + 5 \\ -8m^2 + 2m - 1 \\ \hline \end{array}$$

To subtract, change all the signs in the second polynomial, and add.

$$\begin{array}{r} 6m^2 - 11m + 5 \\ 8m^2 - 2m + 1 \\ \hline 14m^2 - 13m + 6 \end{array}$$

42.
$$\begin{array}{r} -4z^2 + 2z - 1 \\ 3z^2 - 5z + 2 \\ \hline \end{array}$$

To subtract, change all the signs in the second polynomial, and add.

$$\begin{array}{r} -4z^2 + 2z - 1 \\ -3z^2 + 5z - 2 \\ \hline -7z^2 + 7z - 3 \end{array}$$

43. Add column by column to obtain the result on the bottom line.

$$\begin{array}{r} 12z^2 - 11z + 8 \\ 5z^2 + 16z - 2 \\ -4z^2 + 5z - 9 \\ \hline 13z^2 + 10z - 3 \end{array}$$

44. Add in columns.

$$\begin{array}{r} -6m^3 + 2m^2 + 5m \\ 8m^3 + 4m^2 - 6m \\ -3m^3 + 2m^2 - 7m \\ \hline -m^3 + 8m^2 - 8m \end{array}$$

45.
$$\begin{array}{r} 6y^3 - 9y^2 + 8 \\ 4y^3 + 2y^2 + 5y \\ \hline 10y^3 - 7y^2 + 5y + 8 \end{array}$$

46.
$$\begin{array}{r} -7r^8 + 2r^6 - r^5 \\ 3r^6 \qquad + 5 \\ \hline -7r^8 + 5r^6 - r^5 + 5 \end{array}$$

47.
$$\begin{array}{r} -5a^4 \qquad + 8a^2 - 9 \\ 6a^3 - a^2 + 2 \\ \hline \end{array}$$

To subtract, change all the signs in the second polynomial, and add.

$$\begin{array}{r} -5a^4 \qquad + 8a^2 - 9 \\ -6a^3 + a^2 - 2 \\ \hline -5a^4 - 6a^3 + 9a^2 - 11 \end{array}$$

48.
$$\begin{array}{r} -2m^3 + 8m^2 \qquad \\ m^4 - m^3 \qquad + 2m \\ \hline \end{array}$$

Change signs; add in columns.

$$\begin{array}{r} -2m^3 + 8m^2 \qquad \\ -m^4 + m^3 \qquad - 2m \\ \hline -m^4 - m^3 + 8m^2 - 2m \end{array}$$

49. $(3r + 8) - (2r - 5)$

Change all signs in the second polynomial and add.

$$\begin{aligned} &= (3r + 8) + (-2r + 5) \\ &= 3r + 8 - 2r + 5 \\ &= 3r - 2r + 8 + 5 \\ &= r + 13 \end{aligned}$$

50. $(2d + 7) - (3d - 1)$

Change all signs in the second polynomial and add.

$$\begin{aligned} &= (2d + 7) + (-3d + 1) \\ &= 2d - 3d + 7 + 1 \\ &= -d + 8 \end{aligned}$$

51. $(5x^2 + 7x - 4) + (3x^2 - 6x + 2)$

$$\begin{aligned} &= 5x^2 + 3x^2 + 7x - 6x - 4 + 2 \\ &= 8x^2 + x - 2 \end{aligned}$$

52. $(4k^3 + k^2 + k) + (2k^3 - 4k^2 - 3k)$

$$\begin{aligned} &= 4k^3 + 2k^3 + k^2 - 4k^2 + k - 3k \\ &= 6k^3 - 3k^2 - 2k \end{aligned}$$

53. $(2a^2 + 3a - 1) - (4a^2 + 5a + 6)$

$$\begin{aligned} &= (2a^2 + 3a - 1) + (-4a^2 - 5a - 6) \\ &= 2a^2 - 4a^2 + 3a - 5a - 1 - 6 \\ &= -2a^2 - 2a - 7 \end{aligned}$$

54. $(q^4 - 2q^2 + 10) - (3q^4 + 5q^2 - 5)$

$$\begin{aligned} &= (q^4 - 2q^2 + 10) + (-3q^4 - 5q^2 + 5) \\ &= q^4 - 3q^4 - 2q^2 - 5q^2 + 10 + 5 \\ &= -2q^4 - 7q^2 + 15 \end{aligned}$$

55. $(z^5 + 3z^2 + 2z) - (4z^5 + 2z^2 - 5z)$

$$\begin{aligned} &= z^5 + 3z^2 + 2z - 4z^5 - 2z^2 + 5z \\ &= z^5 - 4z^5 + 3z^2 - 2z^2 + 2z + 5z \\ &= -3z^5 + z^2 + 7z \end{aligned}$$

56. $(5t^3 - 3t^2 + 2t) - (4t^3 + 2t^2 + 3t)$

$$\begin{aligned} &= 5t^3 - 3t^2 + 2t - 4t^3 - 2t^2 - 3t \\ &= 5t^3 - 4t^3 - 3t^2 - 2t^2 + 2t - 3t \\ &= t^3 - 5t^2 - t \end{aligned}$$

6.3 Polynomial Functions

6.3 Margin Exercises

1. $f(x) = -x^2 + 5x - 11$

(a)
$$\begin{aligned} f(1) &= -1^2 + 5(1) - 11 \\ &= -1 + 5 - 11 \\ &= -7 \end{aligned}$$

(b)
$$\begin{aligned} f(-4) &= -(-4)^2 + 5(-4) - 11 \\ &= -16 - 20 - 11 \\ &= -47 \end{aligned}$$

(c)
$$\begin{aligned} f(0) &= -0^2 + 5(0) - 11 \\ &= 0 + 0 - 11 \\ &= -11 \end{aligned}$$

2. $P(x) = .808x^2 + 2.625x + .502$

$$\begin{aligned} P(6) &= .808(6)^2 + 2.625(6) + .502 \quad \textit{Let } x = 6. \\ &= 45.34 \end{aligned}$$

Thus, in 2006 about 45.34 million households are expected to pay at least one bill on-line each month.

3. $f(x) = 3x^2 + 8x - 6$, $g(x) = -4x^2 + 4x - 8$

(a) $(f + g)(x)$

$$\begin{aligned} &= f(x) + g(x) \\ &= (3x^2 + 8x - 6) + (-4x^2 + 4x - 8) \\ &= 3x^2 + 8x - 6 - 4x^2 + 4x - 8 \\ &= -x^2 + 12x - 14 \end{aligned}$$

(b) $(f - g)(x)$

$$\begin{aligned} &= f(x) - g(x) \\ &= (3x^2 + 8x - 6) - (-4x^2 + 4x - 8) \\ &= 3x^2 + 8x - 6 + 4x^2 - 4x + 8 \\ &= 7x^2 + 4x + 2 \end{aligned}$$

4. **(a)** $(f+g)(x)$
$= f(x) + g(x)$
$= (18x^2 - 24x) + (3x)$
$= 18x^2 - 21x$
$(f+g)(-1)$
$= 18(-1)^2 - 21(-1)$
$= 18 + 21$
$= 39$

(b) $(f-g)(x)$
$= f(x) - g(x)$
$= (18x^2 - 24x) - (3x)$
$= 18x^2 - 27x$
$(f-g)(1)$
$= 18(1)^2 - 27(1)$
$= 18 - 27$
$= -9$

5.

x	$f(x) = -2x^2$
-2	-8
-1	-2
0	0
1	-2
2	-8

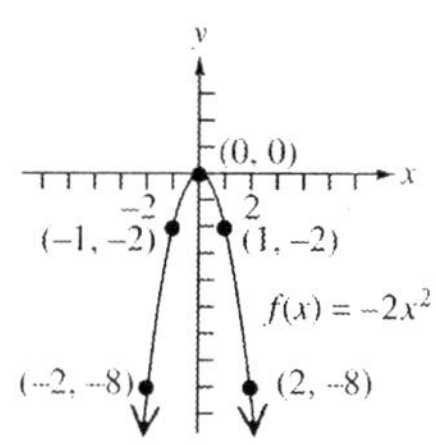

Any value of x can be used, so the domain is $(-\infty, \infty)$. The maximum y-value is 0 and there is no minimum y-value, so the range is $(-\infty, 0]$.

6.3 Section Exercises

1. $f(x) = 6x - 4$

(a) $f(-1) = 6(-1) - 4$
$= -6 - 4 = -10$

(b) $f(2) = 6(2) - 4$
$= 12 - 4 = 8$

2. $f(x) = -2x + 5$

(a) $f(-1) = -2(-1) + 5$
$= 2 + 5 = 7$

(b) $f(2) = -2(2) + 5$
$= -4 + 5 = 1$

3. $f(x) = x^2 - 3x + 4$

(a) $f(-1) = (-1)^2 - 3(-1) + 4$
$= 1 + 3 + 4 = 8$

(b) $f(2) = (2)^2 - 3(2) + 4$
$= 4 - 6 + 4 = 2$

4. $f(x) = 3x^2 + x - 5$

(a) $f(-1) = 3(-1)^2 + (-1) - 5$
$= 3 - 1 - 5 = -3$

(b) $f(2) = 3(2)^2 + (2) - 5$
$= 12 + 2 - 5 = 9$

5. $f(x) = 5x^4 - 3x^2 + 6$

(a) $f(-1) = 5(-1)^4 - 3(-1)^2 + 6$
$= 5 \cdot 1 - 3 \cdot 1 + 6$
$= 5 - 3 + 6 = 8$

(b) $f(2) = 5(2)^4 - 3(2)^2 + 6$
$= 5 \cdot 16 - 3 \cdot 4 + 6$
$= 80 - 12 + 6 = 74$

6. $f(x) = -4x^4 + 2x^2 - 1$

(a) $f(-1) = -4(-1)^4 + 2(-1)^2 - 1$
$= -4 \cdot 1 + 2 \cdot 1 - 1$
$= -4 + 2 - 1 = -3$

(b) $f(2) = -4(2)^4 + 2(2)^2 - 1$
$= -4 \cdot 16 + 2 \cdot 4 - 1$
$= -64 + 8 - 1 = -57$

7. $f(x) = -x^2 + 2x^3 - 8$

(a) $f(-1) = -(-1)^2 + 2(-1)^3 - 8$
$= -(1) + 2(-1) - 8$
$= -1 - 2 - 8 = -11$

(b) $f(2) = -(2)^2 + 2(2)^3 - 8$
$= -(4) + 2 \cdot 8 - 8$
$= -4 + 16 - 8 = 4$

8. $f(x) = -x^2 - x^3 + 11x$

(a) $f(-1) = -(-1)^2 - (-1)^3 + 11(-1)$
$= -(1) - (-1) - 11$
$= -1 + 1 - 11 = -11$

(b) $f(2) = -(2)^2 - (2)^3 + 11(2)$
$= -(4) - (8) + 22$
$= -4 - 8 + 22 = 10$

9. $f(x) = -2.1852x^2 + 236.28x + 15{,}191$

(a) $x = 1980 - 1980 = 0$

$f(0) = -2.1852(0)^2 + 236.28(0) + 15{,}191$
$= 15{,}191$ airports

(b) $x = 1990 - 1980 = 10$

$$\begin{aligned} f(10) &= -2.1852(10)^2 + 236.28(10) + 15{,}191 \\ &= 17{,}335.28 \approx 17{,}335 \text{ airports} \end{aligned}$$

(c) $x = 1999 - 1980 = 19$

$$\begin{aligned} f(19) &= -2.1852(19)^2 + 236.28(19) + 15{,}191 \\ &= 18{,}891.4628 \approx 18{,}891 \text{ airports} \end{aligned}$$

10. $f(x) = -.066x^2 + 1.12x + 28.3$

(a) $x = 1990 - 1990 = 0$

$$\begin{aligned} f(0) &= -.066(0)^2 + 1.12(0) + 28.3 \\ &= 28.3 \quad (28.3\%) \end{aligned}$$

(b) $x = 1995 - 1990 = 5$

$$\begin{aligned} f(5) &= -.066(5) + 1.12(5) + 28.3 \\ &= 32.25 \quad (\approx 32.3\%) \end{aligned}$$

(c) $x = 2000 - 1990 = 10$

$$\begin{aligned} f(10) &= -.066(10)^2 + 1.12(10) + 28.3 \\ &= 32.9 \quad (32.9\%) \end{aligned}$$

11. $P(x) = -.31x^3 + 5.8x^2 - 15x + 9$

(a) $x = 1990 - 1990 = 0$

$$\begin{aligned} P(0) &= -.31(0)^3 + 5.8(0)^2 - 15(0) + 9 \\ &= 9 \text{ million} \end{aligned}$$

(b) $x = 1996 - 1990 = 6$

$$\begin{aligned} P(6) &= -.31(6)^3 + 5.8(6)^2 - 15(6) + 9 \\ &= 60.84 \approx 61 \text{ million} \end{aligned}$$

(c) $x = 1999 - 1990 = 9$

$$\begin{aligned} P(9) &= -.31(9)^3 + 5.8(9)^2 - 15(9) + 9 \\ &= 117.81 \approx 118 \text{ million} \end{aligned}$$

12. $P(x) = -.0325x^3 + 4.524x^2 + 102.5x + 5128$

(a) $x = 1980 - 1980 = 0$

$P(0) = 5128$ cases

(b) $x = 1990 - 1980 = 10$

$$\begin{aligned} P(10) &= -.0325(10)^3 + 4.524(10)^2 \\ &\quad + 102.5(10) + 5128 \\ &= 6572.9 \approx 6573 \text{ cases} \end{aligned}$$

(c) $x = 2000 - 1980 = 20$

$$\begin{aligned} P(20) &= -.0325(20)^3 + 4.524(20)^2 \\ &\quad + 102.5(20) + 5128 \\ &= 8727.6 \approx 8728 \text{ cases} \end{aligned}$$

13. (a)
$$\begin{aligned} (f+g)(x) &= f(x) + g(x) \\ &= (5x - 10) + (3x + 7) \\ &= 8x - 3 \end{aligned}$$

(b)
$$\begin{aligned} (f-g)(x) &= f(x) - g(x) \\ &= (5x - 10) - (3x + 7) \\ &= (5x - 10) + (-3x - 7) \\ &= 2x - 17 \end{aligned}$$

14. (a)
$$\begin{aligned} (f+g)(x) &= f(x) + g(x) \\ &= (-4x + 1) + (6x + 2) \\ &= 2x + 3 \end{aligned}$$

(b)
$$\begin{aligned} (f-g)(x) &= f(x) - g(x) \\ &= (-4x + 1) - (6x + 2) \\ &= (-4x + 1) + (-6x - 2) \\ &= -10x - 1 \end{aligned}$$

15. (a)
$$\begin{aligned} &(f+g)(x) \\ &= f(x) + g(x) \\ &= (4x^2 + 8x - 3) + (-5x^2 + 4x - 9) \\ &= -x^2 + 12x - 12 \end{aligned}$$

(b)
$$\begin{aligned} &(f-g)(x) \\ &= f(x) - g(x) \\ &= (4x^2 + 8x - 3) - (-5x^2 + 4x - 9) \\ &= (4x^2 + 8x - 3) + (5x^2 - 4x + 9) \\ &= 9x^2 + 4x + 6 \end{aligned}$$

16. (a)
$$\begin{aligned} &(f+g)(x) \\ &= f(x) + g(x) \\ &= (3x^2 - 9x + 10) + (-4x^2 + 2x + 12) \\ &= -x^2 - 7x + 22 \end{aligned}$$

(b)
$$\begin{aligned} &(f-g)(x) \\ &= f(x) - g(x) \\ &= (3x^2 - 9x + 10) - (-4x^2 + 2x + 12) \\ &= (3x^2 - 9x + 10) + (4x^2 - 2x - 12) \\ &= 7x^2 - 11x - 2 \end{aligned}$$

For Exercises 17–28, let $f(x) = x^2 - 9$, $g(x) = 2x$, and $h(x) = x - 3$.

17.
$$\begin{aligned} (f+g)(x) &= f(x) + g(x) \\ &= \left(x^2 - 9\right) + (2x) \\ &= x^2 + 2x - 9 \end{aligned}$$

18.
$$\begin{aligned} (f-g)(x) &= f(x) - g(x) \\ &= \left(x^2 - 9\right) - (2x) \\ &= x^2 - 2x - 9 \end{aligned}$$

19.
$$\begin{aligned} (f+g)(3) &= f(3) + g(3) \\ &= \left(3^2 - 9\right) + 2(3) \\ &= 0 + 6 = 6 \end{aligned}$$

Alternatively, we could evaluate the polynomial in Exercise 17, $x^2 + 2x - 9$, using $x = 3$.

20. $$\begin{aligned}(f-g)(-3) &= f(-3)-g(-3)\\ &= \left[(-3)^2-9\right]-[2(-3)]\\ &= (9-9)-(-6)\\ &= 0+6=6\end{aligned}$$

Alternatively, we could evaluate the polynomial in Exercise 18, x^2-2x-9, using $x=-3$.

21. $$\begin{aligned}(f-h)(x) &= f(x)-h(x)\\ &= \left(x^2-9\right)-(x-3)\\ &= x^2-9-x+3\\ &= x^2-x-6\end{aligned}$$

22. $$\begin{aligned}(f+h)(x) &= f(x)+h(x)\\ &= \left(x^2-9\right)+(x-3)\\ &= x^2+x-12\end{aligned}$$

23. $$\begin{aligned}(f-h)(-3) &= f(-3)-h(-3)\\ &= \left[(-3)^2-9\right]-[(-3)-3]\\ &= (9-9)-(-6)\\ &= 0+6=6\end{aligned}$$

Alternatively, we could evaluate the polynomial in Exercise 21, x^2-x-6, using $x=-3$.

24. $$\begin{aligned}(f+h)(-2) &= f(-2)+h(-2)\\ &= \left[(-2)^2-9\right]+[(-2)-3]\\ &= 4-9+(-5)\\ &= -5-5=-10\end{aligned}$$

Alternatively, we could evaluate the polynomial in Exercise 22, x^2+x-12, using $x=-2$.

25. $$\begin{aligned}(g+h)(-10) &= g(-10)+h(-10)\\ &= 2(-10)+[(-10)-3]\\ &= -20+(-13)\\ &= -33\end{aligned}$$

26. $$\begin{aligned}(g-h)(10) &= g(10)-h(10)\\ &= 2(10)-(10-3)\\ &= 20-7\\ &= 13\end{aligned}$$

27. $$\begin{aligned}(g-h)(-3) &= g(-3)-h(-3)\\ &= 2(-3)-[(-3)-3]\\ &= -6-(-6)\\ &= -6+6=0\end{aligned}$$

28. $$\begin{aligned}(g+h)\left(\tfrac{1}{3}\right) &= g\left(\tfrac{1}{3}\right)+h\left(\tfrac{1}{3}\right)\\ &= 2\left(\tfrac{1}{3}\right)+\left(\tfrac{1}{3}-3\right)\\ &= \tfrac{2}{3}+\left(-\tfrac{8}{3}\right)\\ &= -\tfrac{6}{3}=-2\end{aligned}$$

29.

x	$f(x)=-2x+1$
-2	$-2(-2)+1=5$
-1	$-2(-1)+1=3$
0	$-2(0)+1=1$
1	$-2(1)+1=-1$
2	$-2(2)+1=-3$

This is a linear function, so plot the points and draw a line through them.

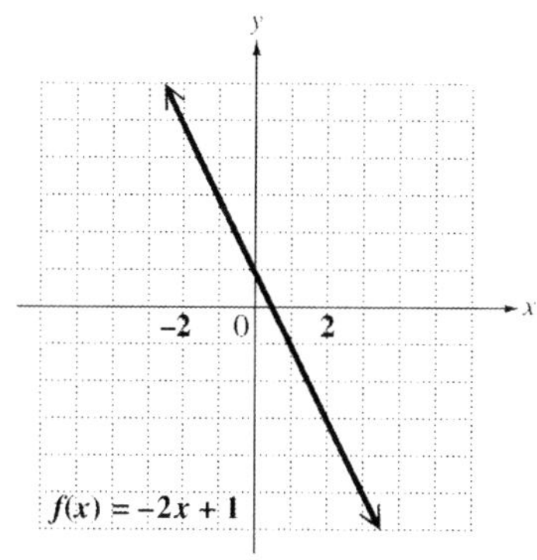

Any x-value can be used, so the domain is $(-\infty,\infty)$. From the graph, we see that any y-value can be obtained from the function, so the range is $(-\infty,\infty)$.

30.

x	$f(x)=3x+2$
-2	$3(-2)+2=-4$
-1	$3(-1)+2=-1$
0	$3(0)+2=2$
1	$3(1)+2=5$
2	$3(2)+2=8$

This is a linear function, so plot the points and draw a line through them.

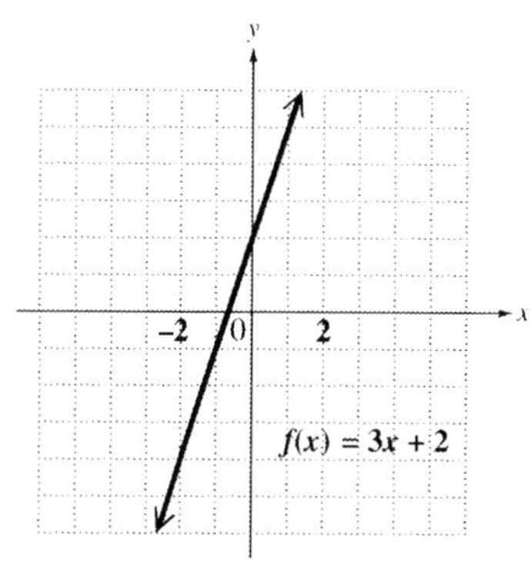

Any x-value can be used, so the domain is $(-\infty,\infty)$. From the graph, we see that any y-value can be obtained from the function, so the range is $(-\infty,\infty)$.

31.

x	$f(x) = -3x^2$
-2	$-3(-2)^2 = -12$
-1	$-3(-1)^2 = -3$
0	$-3(0)^2 = 0$
1	$-3(1)^2 = -3$
2	$-3(2)^2 = -12$

Since the greatest exponent is 2, the graph of f is a parabola.

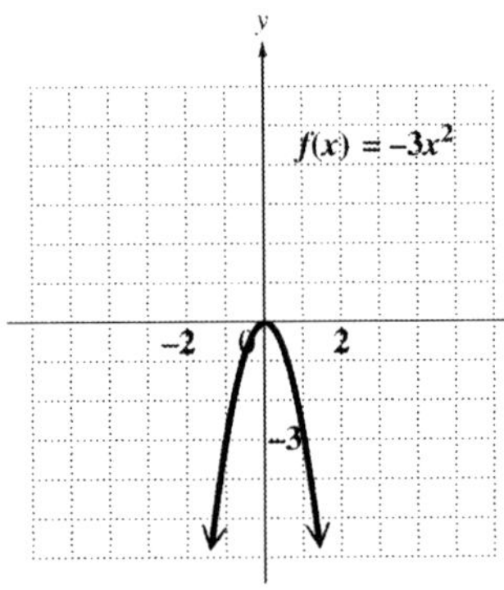

Any x-value can be used, so the domain is $(-\infty, \infty)$. From the graph, we see that the y-values are at most 0, so the range is $(-\infty, 0]$.

32.

x	$f(x) = \frac{1}{2}x^2$
-2	$\frac{1}{2}(-2)^2 = 2$
-1	$\frac{1}{2}(-1)^2 = \frac{1}{2}$
0	$\frac{1}{2}(0)^2 = 0$
1	$\frac{1}{2}(1)^2 = \frac{1}{2}$
2	$\frac{1}{2}(2)^2 = 2$

Since the greatest exponent is 2, the graph of f is a parabola.

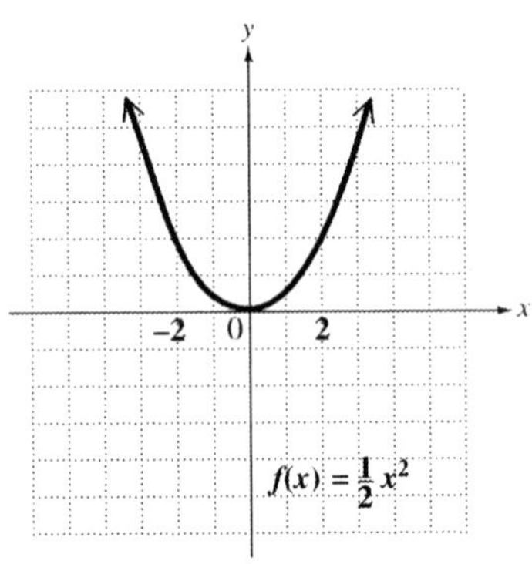

Any x-value can be used, so the domain is $(-\infty, \infty)$. From the graph, we see that the y-values are at least 0, so the range is $[0, \infty)$.

33.

x	$f(x) = x^3 + 1$
-2	$(-2)^3 + 1 = -7$
-1	$(-1)^3 + 1 = 0$
0	$(0)^3 + 1 = 1$
1	$(1)^3 + 1 = 2$
2	$(2)^3 + 1 = 9$

The greatest exponent is 3, so the graph of f is s-shaped.

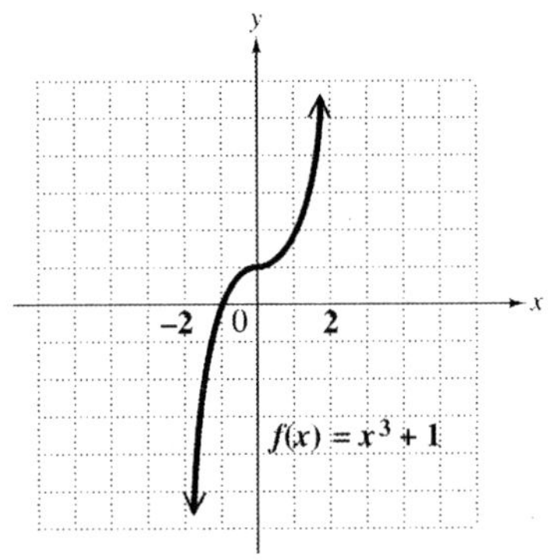

Any x-value can be used, so the domain is $(-\infty, \infty)$. From the graph, we see that any y-value can be obtained from the function, so the range is $(-\infty, \infty)$.

34.

x	$f(x) = -x^3 + 2$
-2	$-(-2)^3 + 2 = 10$
-1	$-(-1)^3 + 2 = 3$
0	$-(0)^3 + 2 = 2$
1	$-(1)^3 + 2 = 1$
2	$-(2)^3 + 2 = -6$

The greatest exponent is 3, so the graph of f is s-shaped.

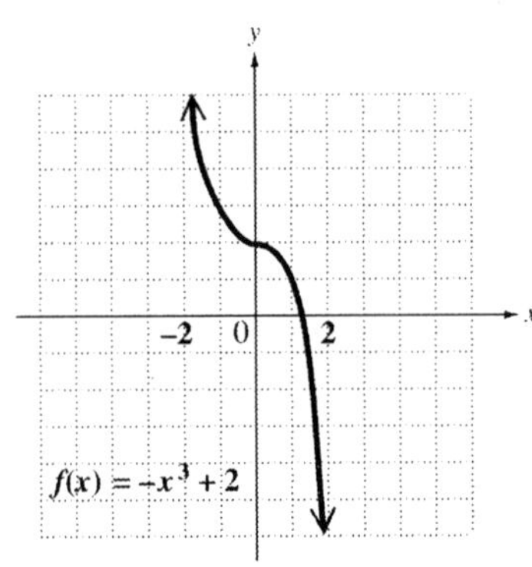

Any x-value can be used, so the domain is $(-\infty, \infty)$. From the graph, we see that any y-value can be obtained from the function, so the range is $(-\infty, \infty)$.

6.4 Multiplying Polynomials

6.4 Margin Exercises

1. **(a)** $-6m^5(2m^4) = (-6)(2)m^5 \cdot m^4$
$= -12m^{5+4}$
$= -12m^9$

(b) $(8k^3y)(9ky^3) = (8)(9)k^3 \cdot k^1 \cdot y^1 \cdot y^3$
$= 72k^{3+1}y^{1+3}$
$= 72k^4y^4$

2. **(a)** $-2r(9r-5) = -2r(9r) - 2r(-5)$
$= -18r^2 + 10r$

(b) $3p^2(5p^3 + 2p^2 - 7)$
$= 3p^2(5p^3) + 3p^2(2p^2) + 3p^2(-7)$
$= 15p^5 + 6p^4 - 21p^2$

(c) $(4a-5)(3a+6)$
$= (4a-5)(3a) + (4a-5)(6)$
$= (4a)(3a) + (-5)(3a) + (4a)(6) + (-5)(6)$
$= 12a^2 - 15a + 24a - 30$
$= 12a^2 + 9a - 30$

(d) $3x^3(x+4)(x-6)$
$= 3x^3[(x+4)(x) + (x+4)(-6)]$
$= 3x^3[x^2 + 4x - 6x - 24]$
$= 3x^3(x^2 - 2x - 24)$
$= 3x^5 - 6x^4 - 72x^3$

3. **(a)**

$$\begin{array}{rrrl} & 2m & -\ 5 & \\ & 3m & +\ 4 & \\ \hline & 8m & -\ 20 & \leftarrow 4(2m-5) \\ 6m^2 & -\ 15m & & \leftarrow 3m(2m-5) \\ \hline 6m^2 & -\ 7m & -\ 20 & \text{Combine like terms.} \end{array}$$

(b)

$$\begin{array}{rrrrr} & 5a^3 & -\ 6a^2 & +\ 2a & -\ 3 \\ & & & 2a & -\ 5 \\ \hline & -\ 25a^3 & +\ 30a^2 & -\ 10a & +\ 15 \\ 10a^4 & -\ 12a^3 & +\ 4a^2 & -\ 6a & \\ \hline 10a^4 & -\ 37a^3 & +\ 34a^2 & -\ 16a & +\ 15 \end{array}$$

4. **(a)** $(3z+2)(z+1)$
F O I L
$= 3z \cdot z + 3z \cdot 1 + 2 \cdot z + 2 \cdot 1$
$= 3z^2 + 3z + 2z + 2$
$= 3z^2 + 5z + 2$

(b) $(5r-3)(2r-5)$
F O I L
$= 10r^2 - 25r - 6r + 15$
$= 10r^2 - 31r + 15$

(c) $(4p+5q)(3p-2q)$
F O I L
$= 12p^2 - 8pq + 15pq - 10q^2$
$= 12p^2 + 7pq - 10q^2$

(d) $(4y-z)(2y+3z)$
F O I L
$= 8y^2 + 12yz - 2yz - 3z^2$
$= 8y^2 + 10yz - 3z^2$

(e) $(8r+1)(8r-1)$
F O I L
$= 64r^2 - 8r + 8r - 1$
$= 64r^2 - 1$

5. Use $(x+y)(x-y) = x^2 - y^2$.

(a) $(m+5)(m-5) = m^2 - 5^2$
$= m^2 - 25$

(b) $(x-4y)(x+4y) = x^2 - (4y)^2$
$= x^2 - 4^2y^2$
$= x^2 - 16y^2$

(c) $(7m-2n)(7m+2n) = (7m)^2 - (2n)^2$
$= 49m^2 - 4n^2$

(d) $4y^2(y+7)(y-7) = 4y^2(y^2 - 49)$
$= 4y^4 - 196y^2$

6. Use $(x+y)^2 = x^2 + 2xy + y^2$
or $(x-y)^2 = x^2 - 2xy + y^2$.

(a) $(a+2)^2 = a^2 + 2 \cdot a \cdot 2 + 2^2$
$= a^2 + 4a + 4$

(b) $(2m-5)^2 = (2m)^2 - 2(2m)(5) + 5^2$
$= 4m^2 - 20m + 25$

(c) $(y+6z)^2 = y^2 + 2(y)(6z) + (6z)^2$
$= y^2 + 12yz + 36z^2$

(d) $(3k-2n)^2 = (3k)^2 - 2(3k)(2n) + (2n)^2$
$= 9k^2 - 12kn + 4n^2$

7. **(a)** $[(m-2n)-3] \cdot [(m-2n)+3]$
This is the product of a sum and a difference.
$= (m-2n)^2 - 3^2$
The first term is the square of a binomial.
$= m^2 - 2(m)(2n) + (2n)^2 - 3^2$
$= m^2 - 4mn + 4n^2 - 9$

(b) $[(k-5h)+2]^2$

This is the square of a binomial, with $(k-5h)$ as the first term.

$= (k-5h)^2 + 2(k-5h)(2) + 2^2$
$= k^2 - 2(k)(5h) + (5h)^2 + 4(k-5h) + 4$
$= k^2 - 10kh + 25h^2 + 4k - 20h + 4$

(c) $(p+2q)^3$

$= (p+2q)^2(p+2q)$
$= (p^2 + 4pq + 4q^2)(p+2q)$ *Square p + 2q.*
$= p^3 + 4p^2q + 4pq^2 + 2p^2q + 8pq^2 + 8q^3$
$= p^3 + 6p^2q + 12pq^2 + 8q^3$

(d) $(x+2)^4$

$= (x+2)^2(x+2)^2$
$= (x^2+4x+4)(x^2+4x+4)$
$= x^4 + 4x^3 + 4x^2 + 4x^3 + 16x^2 + 16x + 4x^2 + 16x + 16$
$= x^4 + 8x^3 + 24x^2 + 32x + 16$

8. $f(x) = 2x+7, \quad g(x) = x^2 - 4$

$(fg)(x) = f(x) \cdot g(x)$
$= (2x+7)(x^2-4)$
$= 2x^3 - 8x + 7x^2 - 28$
$= 2x^3 + 7x^2 - 8x - 28$

$(fg)(2) = 2(2)^3 + 7(2)^2 - 8(2) - 28$
$= 16 + 28 - 16 - 28$
$= 0$

6.4 Section Exercises

1. $-8m^3(3m^2) = -8(3)m^{3+2} = -24m^5$

2. $4p^2(-5p^4) = 4(-5)p^{2+4} = -20p^6$

3. $3x(-2x+5) = 3x(-2x) + 3x(5) = -6x^2 + 15x$

4. $5y(-6y-1) = 5y(-6y) + 5y(-1) = -30y^2 - 5y$

5. $-q^3(2+3q) = -q^3(2) - q^3(3q) = -2q^3 - 3q^4$

6. $-3a^4(4-a) = -3a^4(4) - 3a^4(-a)$
$= -12a^4 + 3a^5$

7. $6k^2(3k^2+2k+1)$
$= 6k^2(3k^2) + 6k^2(2k) + 6k^2(1)$
$= 18k^4 + 12k^3 + 6k^2$

8. $5r^3(2r^2-3r-4)$
$= 5r^3(2r^2) + 5r^3(-3r) + 5r^3(-4)$
$= 10r^5 - 15r^4 - 20r^3$

9. $(2m+3)(3m^2-4m-1)$
$= 2m(3m^2-4m-1) + 3(3m^2-4m-1)$
$= 2m(3m^2) + 2m(-4m) + 2m(-1) + 3(3m^2) + 3(-4m) + 3(-1)$
$= 6m^3 - 8m^2 - 2m + 9m^2 - 12m - 3$
$= 6m^3 - 8m^2 + 9m^2 - 2m - 12m - 3$
$= 6m^3 + m^2 - 14m - 3$

10. $(4z-2)(z^2+3z+5)$
$= 4z(z^2+3z+5) - 2(z^2+3z+5)$
$= 4z(z^2) + 4z(3z) + 4z(5) - 2(z^2) - 2(3z) - 2(5)$
$= 4z^3 + 12z^2 + 20z - 2z^2 - 6z - 10$
$= 4z^3 + 10z^2 + 14z - 10$

11. $4x^3(x-3)(x+2)$
$= 4x^3[(x-3)(x) + (x-3)(2)]$
$= 4x^3[x^2 - 3x + 2x - 6]$
$= 4x^3(x^2 - x - 6)$
$= 4x^3(x^2) + 4x^3(-x) + 4x^3(-6)$
$= 4x^5 - 4x^4 - 24x^3$

12. $2y^5(y-8)(y+2)$
$= 2y^5[(y-8)(y) + (y-8)(2)]$
$= 2y^5[y^2 - 8y + 2y - 16]$
$= 2y^5(y^2 - 6y - 16)$
$= 2y^5(y^2) + 2y^5(-6y) + 2y^5(-16)$
$= 2y^7 - 12y^6 - 32y^5$

13. $(2y+3)(3y-4)$

$$\begin{array}{rrrl} & 2y & +\ 3 & \\ & 3y & -\ 4 & \\ \hline & -\ 8y & -\ 12 & \leftarrow -4(2y+3) \\ 6y^2 & +\ 9y & & \leftarrow 3y(2y+3) \\ \hline 6y^2 & +\ y & -\ 12 & \textit{Combine like terms.} \end{array}$$

14. $(5m-3)(2m+6)$

$$\begin{array}{rrrl} & 5m & -\ 3 & \\ & 2m & +\ 6 & \\ \hline & 30m & -\ 18 & \leftarrow 6(5m-3) \\ 10m^2 & -\ 6m & & \leftarrow 2m(5m-3) \\ \hline 10m^2 & +\ 24m & -\ 18 & \textit{Combine like terms.} \end{array}$$

15.

$$\begin{array}{rrr} & 5m & -\ 3n \\ & 5m & +\ 3n \\ \hline & 15mn & -\ 9n^2 \\ 25m^2 & -\ 15mn & \\ \hline 25m^2 & & -\ 9n^2 \end{array}$$

16.
$$\begin{array}{rr} & 2k + 6q \\ & 2k - 6q \\ \hline & -12kq - 36q^2 \\ 4k^2 + 12kq & \\ \hline 4k^2 & - 36q^2 \end{array}$$

17.
$$\begin{array}{r} -b^2 + 3b + 3 \\ 2b + 4 \\ \hline -4b^2 + 12b + 12 \\ -2b^3 + 6b^2 + 6b \\ \hline -2b^3 + 2b^2 + 18b + 12 \end{array}$$

18.
$$\begin{array}{r} -r^2 - 4r + 8 \\ 3r - 2 \\ \hline 2r^2 + 8r - 16 \\ -3r^3 - 12r^2 + 24r \\ \hline -3r^3 - 10r^2 + 32r - 16 \end{array}$$

19.
$$\begin{array}{r} 2z^3 - 5z^2 + 8z - 1 \\ 4z + 3 \\ \hline 6z^3 - 15z^2 + 24z - 3 \\ 8z^4 - 20z^3 + 32z^2 - 4z \\ \hline 8z^4 - 14z^3 + 17z^2 + 20z - 3 \end{array}$$

20.
$$\begin{array}{r} 3z^4 - 2z^3 + z - 5 \\ 2z - 5 \\ \hline -15z^4 + 10z^3 - 5z + 25 \\ 6z^5 - 4z^4 + 2z^2 - 10z \\ \hline 6z^5 - 19z^4 + 10z^3 + 2z^2 - 15z + 25 \end{array}$$

21.
$$\begin{array}{r} 2p^2 + 3p + 6 \\ 3p^2 - 4p - 1 \\ \hline -2p^2 - 3p - 6 \\ -8p^3 - 12p^2 - 24p \\ 6p^4 + 9p^3 + 18p^2 \\ \hline 6p^4 + p^3 + 4p^2 - 27p - 6 \end{array}$$

22.
$$\begin{array}{r} 5y^2 - 2y + 4 \\ 2y^2 + y + 3 \\ \hline 15y^2 - 6y + 12 \\ 5y^3 - 2y^2 + 4y \\ 10y^4 - 4y^3 + 8y^2 \\ \hline 10y^4 + y^3 + 21y^2 - 2y + 12 \end{array}$$

23. $(m+5)(m-8)$

F O I L

$= m^2 - 8m + 5m - 40$

$= m^2 - 3m - 40$

24. $(p-6)(p+4)$

F O I L

$= p^2 + 4p - 6p - 24$

$= p^2 - 2p - 24$

25. $(4k+3)(3k-2)$

F O I L

$= 12k^2 - 8k + 9k - 6$

$= 12k^2 + k - 6$

26. $(5w+2)(2w+5)$

F O I L

$= 10w^2 + 25w + 4w + 10$

$= 10w^2 + 29w + 10$

27. $(z-w)(3z+4w)$

F O I L

$= 3z^2 + 4zw - 3zw - 4w^2$

$= 3z^2 + zw - 4w^2$

28. $(s+t)(2s-5t)$

F O I L

$= 2s^2 - 5st + 2st - 5t^2$

$= 2s^2 - 3st - 5t^2$

29. $(6c-d)(2c+3d)$

F O I L

$= 12c^2 + 18cd - 2cd - 3d^2$

$= 12c^2 + 16cd - 3d^2$

30. $(2m-n)(3m+5n)$

F O I L

$= 6m^2 + 10mn - 3mn - 5n^2$

$= 6m^2 + 7mn - 5n^2$

31. $(.2x+1.3)(.5x-.1)$

F O I L

$= .1x^2 - .02x + .65x - .13$

$= .1x^2 + .63x - .13$

32. $(.5y-.4)(.1y+2.1)$

F O I L

$= .05y^2 + 1.05y - .04y - .84$

$= .05y^2 + 1.01y - .84$

33. $\left(3r + \frac{1}{4}y\right)(r-2y)$

F O I L

$= 3r^2 - 6ry + \frac{1}{4}ry - \frac{1}{2}y^2$

$= 3r^2 - \frac{23}{4}ry - \frac{1}{2}y^2$

34. $\left(5w - \frac{2}{3}z\right)(w+5z)$

F O I L

$= 5w^2 + 25wz - \frac{2}{3}wz - \frac{10}{3}z^2$

$= 5w^2 + \frac{73}{3}wz - \frac{10}{3}z^2$

35. The product of two binomials is the sum of the product of the first terms, the product of the outer terms, the product of the inner terms, and the product of the last terms.

36. The product is not a trinomial because, using the FOIL method, the inner and outer products are additive inverses and add to 0, leaving only two terms.

37. Use the formula for the product of the sum and difference of two terms.

$$\begin{aligned}(2p-3)(2p+3) &= (2p)^2-(3)^2\\ &= 4p^2-9\end{aligned}$$

38. $$\begin{aligned}(3x-8)(3x+8) &= (3x)^2-(8)^2\\ &= 9x^2-64\end{aligned}$$

39. $$\begin{aligned}(5m-1)(5m+1) &= (5m)^2-(1)^2\\ &= 25m^2-1\end{aligned}$$

40. $$\begin{aligned}(6y+3)(6y-3) &= (6y)^2-(3)^2\\ &= 36y^2-9\end{aligned}$$

41. $$\begin{aligned}(3a+2c)(3a-2c) &= (3a)^2-(2c)^2\\ &= 9a^2-4c^2\end{aligned}$$

42. $$\begin{aligned}(5r-4s)(5r+4s) &= (5r)^2-(4s)^2\\ &= 25r^2-16s^2\end{aligned}$$

43. $$\begin{aligned}\left(4x-\frac{2}{3}\right)\left(4x+\frac{2}{3}\right) &= (4x)^2-\left(\frac{2}{3}\right)^2\\ &= 16x^2-\frac{4}{9}\end{aligned}$$

44. $$\begin{aligned}\left(3t+\frac{5}{4}\right)\left(3t-\frac{5}{4}\right) &= (3t)^2-\left(\frac{5}{4}\right)^2\\ &= 9t^2-\frac{25}{16}\end{aligned}$$

45. $$\begin{aligned}\left(4m+7n^2\right)\left(4m-7n^2\right) &= (4m)^2-\left(7n^2\right)^2\\ &= 16m^2-49n^4\end{aligned}$$

46. $$\begin{aligned}\left(2k^2+6h\right)\left(2k^2-6h\right) &= \left(2k^2\right)^2-(6h)^2\\ &= 4k^4-36h^2\end{aligned}$$

47. $$\begin{aligned}5y^3(y+2)(y-2) &= 5y^3\left(y^2-4\right)\\ &= 5y^5-20y^3\end{aligned}$$

48. $$\begin{aligned}3x^3(x-4)(x+4) &= 3x^3\left(x^2-16\right)\\ &= 3x^5-48x^3\end{aligned}$$

49. Use the formula for the square of a binomial.

$$\begin{aligned}(y-5)^2 &= y^2-2(y)(5)+5^2\\ &= y^2-10y+25\end{aligned}$$

50. $$\begin{aligned}(a-3)^2 &= a^2-2(a)(3)+3^2\\ &= a^2-6a+9\end{aligned}$$

51. $$\begin{aligned}(2p+7)^2 &= (2p)^2+2(2p)(7)+7^2\\ &= 4p^2+28p+49\end{aligned}$$

52. $$\begin{aligned}(3z+8)^2 &= (3z)^2+2(3z)(8)+8^2\\ &= 9z^2+48z+64\end{aligned}$$

53. $$\begin{aligned}(4n-3m)^2 &= (4n)^2-2(4n)(3m)+(3m)^2\\ &= 16n^2-24nm+9m^2\end{aligned}$$

54. $$\begin{aligned}(5r-7s)^2 &= (5r)^2-2(5r)(7s)+(7s)^2\\ &= 25r^2-70rs+49s^2\end{aligned}$$

55. $$\begin{aligned}\left(k-\frac{5}{7}p\right)^2 &= k^2-2(k)\left(\frac{5}{7}p\right)+\left(\frac{5}{7}p\right)^2\\ &= k^2-\tfrac{10}{7}kp+\tfrac{25}{49}p^2\end{aligned}$$

56. $$\begin{aligned}\left(q-\frac{3}{4}r\right)^2 &= q^2-2(q)\left(\frac{3}{4}r\right)+\left(\frac{3}{4}r\right)^2\\ &= q^2-\tfrac{3}{2}qr+\tfrac{9}{16}r^2\end{aligned}$$

57. $(x+y)^2=x^2+2xy+y^2$. The expression x^2+y^2 is missing the $2xy$ term, so $(x+y)^2\neq x^2+y^2$ in general.

58. To find the product $101\cdot 99$ using the special product rule

$$(x+y)(x-y)=x^2-y^2,$$

let $x=100$ and $y=1$. Then

$$\begin{aligned}101\cdot 99 &= (100+1)(100-1)\\ &= 100^2-1^2\\ &= 10{,}000-1\\ &= 9999.\end{aligned}$$

59. $$\begin{aligned}&[(5x+1)+6y]^2\\ &= (5x+1)^2+2(5x+1)(6y)+(6y)^2 \quad \textit{Square of a binomial}\\ &= (5x+1)^2+12y(5x+1)+36y^2\\ &= \left[(5x)^2+2(5x)(1)+1^2\right]\\ &\quad +60xy+12y+36y^2 \quad \textit{Square of a binomial}\\ &= 25x^2+10x+1+60xy+12y+36y^2\end{aligned}$$

60. $$\begin{aligned}&[(3m-2)+p]^2\\ &= (3m-2)^2+2(3m-2)p+p^2\\ &= (3m)^2-2(3m)(2)+2^2\\ &\quad +2p(3m-2)+p^2\\ &= 9m^2-12m+4+6mp-4p+p^2\end{aligned}$$

61. $[(2a+b)-3][(2a+b)+3]$
$= (2a+b)^2 - 3^2$
Product of the sum and difference of two terms
$= \left[(2a)^2 + 2(2a)(b) + b^2\right] - 9$
Square of a binomial
$= 4a^2 + 4ab + b^2 - 9$

62. $[(m+p)+5][(m+p)-5]$
$= (m+p)^2 - 5^2$
$= m^2 + 2mp + p^2 - 25$

63. $[(2h-k)+j][(2h-k)-j]$
$= (2h-k)^2 - j^2$
$= (2h)^2 - 2(2h)(k) + k^2 - j^2$
$= 4h^2 - 4hk + k^2 - j^2$

64. $[(3m-y)+z][(3m-y)-z]$
$= (3m-y)^2 - z^2$
$= (3m)^2 - 2(3m)(y) + y^2 - z^2$
$= 9m^2 - 6my + y^2 - z^2$

65. $(x+2)^3 = (x+2)^2(x+2)$
$= \left[x^2 + 2(x)(2) + 2^2\right](x+2)$
$= \left(x^2 + 4x + 4\right)(x+2)$

$$\begin{array}{r} x^2 + 4x + 4 \\ x + 2 \\ \hline 2x^2 + 8x + 8 \\ x^3 + 4x^2 + 4x \quad\;\; \\ \hline x^3 + 6x^2 + 12x + 8 \end{array}$$

66. $(z-3)^3 = (z-3)^2(z-3)$
$= \left[z^2 - 2(3)(z) + 9\right](z-3)$
$= \left(z^2 - 6z + 9\right)(z-3)$

$$\begin{array}{r} z^2 - 6z + 9 \\ z - 3 \\ \hline -3z^2 + 18z - 27 \\ z^3 - 6z^2 + 9z \quad\;\;\;\; \\ \hline z^3 - 9z^2 + 27z - 27 \end{array}$$

67. $(5r-s)^3$
$= (5r-s)^2(5r-s)$
$= \left(25r^2 - 10rs + s^2\right)(5r-s)$
$= 125r^3 - 50r^2s + 5rs^2 - 25r^2s + 10rs^2 - s^3$
$= 125r^3 - 75r^2s + 15rs^2 - s^3$

68. $(x+3y)^3$
$= (x+3y)^2(x+3y)$
$= \left(x^2 + 6xy + 9y^2\right)(x+3y)$
$= x^3 + 6x^2y + 9xy^2 + 3x^2y + 18xy^2 + 27y^3$
$= x^3 + 9x^2y + 27xy^2 + 27y^3$

69. $(q-2)^4 = (q-2)^2(q-2)^2$
$= \left(q^2 - 4q + 4\right)\left(q^2 - 4q + 4\right)$

$$\begin{array}{r} q^2 - 4q + 4 \\ q^2 - 4q + 4 \\ \hline 4q^2 - 16q + 16 \\ -4q^3 + 16q^2 - 16q \quad\;\;\;\; \\ q^4 - 4q^3 + 4q^2 \quad\quad\quad\quad\quad\;\; \\ \hline q^4 - 8q^3 + 24q^2 - 32q + 16 \end{array}$$

70. $(m-p)^4$
$= (m-p)^2(m-p)^2$
$= \left(m^2 - 2mp + p^2\right)\left(m^2 - 2mp + p^2\right)$
$= m^4 - 2m^3p + m^2p^2 - 2m^3p + 4m^2p^2$
$\quad - 2mp^3 + m^2p^2 - 2mp^3 + p^4$
$= m^4 - 4m^3p + 6m^2p^2 - 4mp^3 + p^4$

71. The length of each side of the entire square is a. To find the length of each side of the blue square, subtract b, the length of a side of the green rectangle. The result is $a - b$.

72. The formula for the area A of a square with side s is $A = s^2$. Since $s = a - b$, the formula for the area of the blue square would be $A = (a-b)^2$.

73. The formula for the area of a rectangle is $A = LW$. The green rectangle has length $a - b$ and width b, so each green rectangle has an area of $\underline{(a-b)b \text{ or } ab - b^2}$.

Since there are two green rectangles, the total area in green is $\underline{2(ab - b^2) \text{ or } 2ab - 2b^2}$.

74. The length of each side of the yellow square is b, so the yellow square has an area of $\underline{b^2}$.

75. The area of the entire colored region is $\underline{a^2}$, because each side of the entire colored region has length $\underline{a}$.

76. Using the results from Exercises 71–75, the area of the blue square equals

$$a^2 - \left(2ab - 2b^2\right) - b^2 = a^2 - 2ab + b^2.$$

77. Both expressions for the area of the blue square must be equal to each other; that is, $(a-b)^2$ from Exercise 72 and $a^2 - 2ab + b^2$ from Exercise 76.

78. From Exercise 72, the area of the blue square is $(a-b)^2$. From Exercise 76, the area of the blue square is $a^2 - 2ab + b^2$. Since these expressions must be equal

$$(a-b)^2 = a^2 - 2ab + b^2.$$

This reinforces the special product for the square of a binomial difference.

79. $f(x) = 2x,\ g(x) = 5x - 1$

$$\begin{aligned}(fg)(x) &= f(x) \cdot g(x)\\ &= 2x(5x-1)\\ &= 10x^2 - 2x\end{aligned}$$

80. $f(x) = 3x,\ g(x) = 6x - 8$

$$\begin{aligned}(fg)(x) &= f(x) \cdot g(x)\\ &= 3x(6x-8)\\ &= 18x^2 - 24x\end{aligned}$$

81. $f(x) = x + 1,\ g(x) = 2x - 3$

$$\begin{aligned}(fg)(x) &= f(x) \cdot g(x)\\ &= (x+1)(2x-3)\\ &= 2x^2 - 3x + 2x - 3\\ &= 2x^2 - x - 3\end{aligned}$$

82. $f(x) = x - 7,\ g(x) = 4x + 5$

$$\begin{aligned}(fg)(x) &= f(x) \cdot g(x)\\ &= (x-7)(4x+5)\\ &= 4x^2 + 5x - 28x - 35\\ &= 4x^2 - 23x - 35\end{aligned}$$

83. $f(x) = 2x - 3,\ g(x) = 4x^2 + 6x + 9$

$$\begin{aligned}(fg)(x) &= f(x) \cdot g(x)\\ &= (2x-3)(4x^2+6x+9)\end{aligned}$$

Multiply vertically.

$$\begin{array}{rrrr} & 4x^2 & +\ 6x & +\ 9\\ & & 2x & -\ 3\\ \hline & -\ 12x^2 & -\ 18x & -\ 27\\ 8x^3 & +\ 12x^2 & +\ 18x & \\ \hline 8x^3 & & & -\ 27\end{array}$$

84. $f(x) = 3x + 4,\ g(x) = 9x^2 - 12x + 16$

$$\begin{aligned}(fg)(x) &= f(x) \cdot g(x)\\ &= (3x+4)(9x^2-12x+16)\end{aligned}$$

Multiply vertically.

$$\begin{array}{rrrr} & 9x^2 & -\ 12x & +\ 16\\ & & 3x & +\ 4\\ \hline & 36x^2 & -\ 48x & +\ 64\\ 27x^3 & -\ 36x^2 & +\ 48x & \\ \hline 27x^3 & & & +\ 64\end{array}$$

85.
$$\begin{aligned}(fg)(2) &= f(2) \cdot g(2)\\ &= (2^2-9)[2(2)]\\ &= -5 \cdot 4 = -20\end{aligned}$$

86.
$$\begin{aligned}(fh)(1) &= f(1) \cdot h(1)\\ &= (1^2-9) \cdot (1-3)\\ &= -8(-2) = 16\end{aligned}$$

87.
$$\begin{aligned}(fh)(-1) &= f(-1) \cdot h(-1)\\ &= \left[(-1)^2-9\right] \cdot [(-1)-3]\\ &= -8(-4) = 32\end{aligned}$$

88.
$$\begin{aligned}(gh)(-3) &= g(-3) \cdot h(-3)\\ &= [2(-3)] \cdot [(-3)-3]\\ &= -6(-6) = 36\end{aligned}$$

89.
$$\begin{aligned}(fg)(-2) &= f(-2) \cdot g(-2)\\ &= \left[(-2)^2-9\right] \cdot [2(-2)]\\ &= (4-9) \cdot (-4)\\ &= -5(-4) = 20\end{aligned}$$

90. Substitute 2 for x and 3 for y.

$$(x+y)^2 = (2+3)^2 = 5^2 = 25$$

Again, substitute 2 for x and 3 for y.

$$x^2 + y^2 = 2^2 + 3^2 = 4 + 9 = 13$$

Since $25 \neq 13$, $(x+y)^2 \neq x^2 + y^2$.

The correct product is

$$(x+y)^2 = x^2 + 2xy + y^2.$$

91. Substitute 2 for x and 3 for y.

$$(x+y)^3 = (2+3)^3 = 5^3 = 125$$

Again, substitute 2 for x and 3 for y.

$$x^3 + y^3 = 2^3 + 3^3 = 8 + 27 = 35$$

Since $125 \neq 35$, $(x+y)^3 \neq x^3 + y^3$.

The correct product is

$$\begin{aligned}(x+y)^3 &= (x+y)^2(x+y)\\ &= \left(x^2+2xy+y^2\right)(x+y)\\ &= x^3 + 2x^2y + xy^2 + x^2y + 2xy^2 + y^3\\ &= x^3 + 3x^2y + 3xy^2 + y^3.\end{aligned}$$

92. Let $x = 2$, $y = 3$.

$$\begin{aligned}(x+y)^4 &= (2+3)^4 = 5^4 = 625\\ x^4 + y^4 &= 2^4 + 3^4 = 16 + 81 = 97\end{aligned}$$

Since $625 \neq 97$, $(x+y)^4 \neq x^4 + y^4$.

$$\begin{aligned}(x+y)^4 &= \left[(x+y)^2\right]^2\\ &= \left(x^2+2xy+y^2\right)^2\end{aligned}$$

$$\begin{array}{rrrrr} & & x^2 & +\ 2xy & +\ y^2\\ & & x^2 & +\ 2xy & +\ y^2\\ \hline & & x^2y^2 & +\ 2xy^3 & +\ y^4\\ & 2x^3y & +\ 4x^2y^2 & +\ 2xy^3 & \\ x^4 & +\ 2x^3y & +\ x^2y^2 & & \\ \hline x^4 & +\ 4x^3y & +\ 6x^2y^2 & +\ 4xy^3 & +\ y^4\end{array}$$

6.5 Dividing Polynomials

6.5 Margin Exercises

1. (a) $\dfrac{12p+30}{6}=\dfrac{12p}{6}+\dfrac{30}{6}=2p+5$

(b) $\dfrac{9y^3-4y^2+8y}{2y^2}=\dfrac{9y^3}{2y^2}-\dfrac{4y^2}{2y^2}+\dfrac{8y}{2y^2}$
$=\dfrac{9y}{2}-2+\dfrac{4}{y}$

(c) $\dfrac{8a^2b^2-20ab^3}{4a^3b}=\dfrac{8a^2b^2}{4a^3b}-\dfrac{20ab^3}{4a^3b}$
$=\dfrac{2b}{a}-\dfrac{5b^2}{a^2}$

2. (a) $\dfrac{2r^2+r-21}{r-3}$

$$\begin{array}{r|l}
 & 2r + 7 \\ \hline
r-3 & 2r^2 + r - 21 \\
 & \underline{2r^2 - 6r} \\
 & \quad 7r - 21 \\
 & \quad \underline{7r - 21} \\
 & \qquad 0
\end{array}$$

First step: $\dfrac{2r^2}{r}=2r$

Check: $(r-3)(2r+7)=2r^2+r-21$

Answer: $2r+7$

(b) $\dfrac{2k^2+17k+30}{2k+5}$

$$\begin{array}{r|l}
 & k + 6 \\ \hline
2k+5 & 2k^2 + 17k + 30 \\
 & \underline{2k^2 + 5k} \\
 & \quad 12k + 30 \\
 & \quad \underline{12k + 30} \\
 & \qquad 0
\end{array}$$

First step: $\dfrac{2k^2}{2k}=k$

Check: $(2k+5)(k+6)=2k^2+17k+30$

Answer: $k+6$

3. Add a term with 0 coefficient as a placeholder for the missing k^2-term.

$$\begin{array}{r|l}
 & 3k^2 + 6k + 21 \\ \hline
k-2 & 3k^3 + 0k^2 + 9k - 14 \\
 & \underline{3k^3 - 6k^2} \\
 & \quad 6k^2 + 9k \\
 & \quad \underline{6k^2 - 12k} \\
 & \qquad 21k - 14 \\
 & \qquad \underline{21k - 42} \\
 & \qquad\quad 28
\end{array}$$

Remainder

Check: $(k-2)(3k^2+6k+21)+28$
$=(3k^3+9k-42)+28$
$=3k^3+9k-14$

Answer: $3k^2+6k+21+\dfrac{28}{k-2}$

4.

$$\begin{array}{r|l}
 & p^2 + \frac{5}{2}p + 2 \\ \hline
2p+2 & 2p^3 + 7p^2 + 9p + 2 \\
 & \underline{2p^3 + 2p^2} \\
 & \quad 5p^2 + 9p \\
 & \quad \underline{5p^2 + 5p} \\
 & \qquad 4p + 2 \\
 & \qquad \underline{4p + 4} \\
 & \qquad\quad -2
\end{array}$$

Remainder

First step: $\dfrac{2p^3}{2p}=p^2$

Note: $\dfrac{5p^2}{2p}=\dfrac{5p}{2}$

Check: $(2p+2)\left(p^2+\dfrac{5}{2}p+2\right)-2$
$=(2p^3+7p^2+9p+4)-2$
$=2p^3+7p^2+9p+2$

Answer: $p^2+\dfrac{5}{2}p+2+\dfrac{-2}{2p+2}$

5. (a) $\dfrac{3r^5-15r^4-2r^3+19r^2-7}{3r^2-2}$

Each polynomial has a missing r-term.

$$\begin{array}{r|l}
 & r^3 - 5r^2 \qquad + 3 \\ \hline
3r^2+0r-2 & 3r^5 - 15r^4 - 2r^3 + 19r^2 + 0r - 7 \\
 & \underline{3r^5 + 0r^4 - 2r^3} \\
 & \quad -15r^4 + 0r^3 + 19r^2 \\
 & \quad \underline{-15r^4 + 0r^3 + 10r^2} \\
 & \qquad 9r^2 + 0r - 7 \\
 & \qquad \underline{9r^2 + 0r - 6} \\
 & \qquad\quad -1
\end{array}$$

Remainder

Answer: $r^3-5r^2+3+\dfrac{-1}{3r^2-2}$

(b) $\dfrac{4x^4-7x^2+x+5}{2x^2-x}$

The numerator has a missing x^3-term.

$$\begin{array}{r|l}
 & 2x^2 + x - 3 \\ \hline
2x^2-x & 4x^4 + 0x^3 - 7x^2 + x + 5 \\
 & \underline{4x^4 - 2x^3} \\
 & \quad 2x^3 - 7x^2 \\
 & \quad \underline{2x^3 - x^2} \\
 & \qquad -6x^2 + x \\
 & \qquad \underline{-6x^2 + 3x} \\
 & \qquad\quad -2x + 5
\end{array}$$

Remainder

Answer: $2x^2+x-3+\dfrac{-2x+5}{2x^2-x}$

6. $f(x) = 2x^2 + 17x + 30,\ g(x) = 2x + 5$

$$\left(\frac{f}{g}\right)(x) = \frac{f(x)}{g(x)} = \frac{2x^2 + 17x + 30}{2x + 5}$$

This quotient, found in Margin Exercise 2(b), with x replacing k, is $x + 6$. Thus, $\left(\frac{f}{g}\right)(x) = x + 6$, provided the denominator, $2x + 5$, is *not* equal to zero; that is, $x \neq \frac{5}{2}$.

Using the above result, $\left(\frac{f}{g}\right)(x) = x + 6$, we have

$$\left(\frac{f}{g}\right)(-1) = (-1) + 6 = 5.$$

6.5 Section Exercises

1. $$\frac{15x^3 - 10x^2 + 5}{5} = \frac{15x^3}{5} - \frac{10x^2}{5} + \frac{5}{5} = 3x^3 - 2x^2 + 1$$

2. $$\frac{27m^4 - 18m^3 + 9m}{9} = \frac{27m^4}{9} - \frac{18m^3}{9} + \frac{9m}{9} = 3m^4 - 2m^3 + m$$

3. $$\frac{9y^2 + 12y - 15}{3y} = \frac{9y^2}{3y} + \frac{12y}{3y} - \frac{15}{3y} = 3y + 4 - \frac{5}{y}$$

4. $$\frac{80r^2 - 40r + 10}{10r} = \frac{80r^2}{10r} - \frac{40r}{10r} + \frac{10}{10r} = 8r - 4 + \frac{1}{r}$$

5. $$\frac{15m^3 + 25m^2 + 30m}{5m^2} = \frac{15m^3}{5m^2} + \frac{25m^2}{5m^2} + \frac{30m}{5m^2} = 3m + 5 + \frac{6}{m}$$

6. $$\frac{64x^3 - 72x^2 + 12x}{8x^3} = \frac{64x^3}{8x^3} - \frac{72x^2}{8x^3} + \frac{12x}{8x^3} = 8 - \frac{9}{x} + \frac{3}{2x^2}$$

7. $$\frac{14m^2n^2 - 21mn^3 + 28m^2n}{14m^2n} = \frac{14m^2n^2}{14m^2n} - \frac{21mn^3}{14m^2n} + \frac{28m^2n}{14m^2n} = n - \frac{3n^2}{2m} + 2$$

8. $$\frac{24h^2k + 56hk^2 - 28hk}{16h^2k^2} = \frac{24h^2k}{16h^2k^2} + \frac{56hk^2}{16h^2k^2} - \frac{28hk}{16h^2k^2} = \frac{3}{2k} + \frac{7}{2h} - \frac{7}{4hk}$$

9. $$\frac{y^2 + 3y - 18}{y + 6}$$

$$\begin{array}{r|l} & y - 3 \\ y + 6 & y^2 + 3y - 18 \\ & \underline{y^2 + 6y} \\ & \quad -3y - 18 \\ & \quad \underline{-3y - 18} \\ & \qquad\quad 0 \end{array}$$

Answer: $y - 3$

10. $$\frac{q^2 + 4q - 32}{q - 4}$$

$$\begin{array}{r|l} & q + 8 \\ q - 4 & q^2 + 4q - 32 \\ & \underline{q^2 - 4q} \\ & \quad 8q - 32 \\ & \quad \underline{8q - 32} \\ & \qquad\quad 0 \end{array}$$

Answer: $q + 8$

11. $$\frac{3t^2 + 17t + 10}{3t + 2}$$

$$\begin{array}{r|l} & t + 5 \\ 3t + 2 & 3t^2 + 17t + 10 \\ & \underline{3t^2 + 2t} \\ & \quad 15t + 10 \\ & \quad \underline{15t + 10} \\ & \qquad\quad 0 \end{array}$$

Answer: $t + 5$

12. $$\frac{2k^2 - 3k - 20}{2k + 5}$$

$$\begin{array}{r|l} & k - 4 \\ 2k + 5 & 2k^2 - 3k - 20 \\ & \underline{2k^2 + 5k} \\ & \quad -8k - 20 \\ & \quad \underline{-8k - 20} \\ & \qquad\quad 0 \end{array}$$

Answer: $k - 4$

13. $$\frac{p^2 + 2p + 20}{p + 6}$$

$$\begin{array}{r|l} & p - 4 \\ p + 6 & p^2 + 2p + 20 \\ & \underline{p^2 + 6p} \\ & \quad -4p + 20 \\ & \quad \underline{-4p - 24} \\ & \qquad\quad 44 \end{array}$$

Remainder

Answer: $p - 4 + \frac{44}{p + 6}$

14. $\dfrac{x^2+11x+16}{x+8}$

$$\begin{array}{r} x+3 \\ x+8\,\overline{)\,x^2+11x+16} \\ \underline{x^2+8x} \\ 3x+16 \\ \underline{3x+24} \\ -8 \end{array}$$

Remainder

Answer: $x+3+\dfrac{-8}{x+8}$

15. $(2z^3-5z^2+6z-15)\div(2z-5)$

$$\begin{array}{r} z^2+3 \\ 2z-5\,\overline{)\,2z^3-5z^2+6z-15} \\ \underline{2z^3-5z^2} \\ 6z-15 \\ \underline{6z-15} \\ 0 \end{array}$$

Answer: z^2+3

16. $(3p^3+p^2+18p+6)\div(3p+1)$

$$\begin{array}{r} p^2+6 \\ 3p+1\,\overline{)\,3p^3+p^2+18p+6} \\ \underline{3p^3+p^2} \\ 18p+6 \\ \underline{18p+6} \\ 0 \end{array}$$

Answer: p^2+6

17. $(4x^3+9x^2-10x+3)\div(4x+1)$

$$\begin{array}{r} x^2+2x-3 \\ 4x+1\,\overline{)\,4x^3+9x^2-10x+3} \\ \underline{4x^3+x^2} \\ 8x^2-10x \\ \underline{8x^2+2x} \\ -12x+3 \\ \underline{-12x-3} \\ 6 \end{array}$$

Remainder

Answer: $x^2+2x-3+\dfrac{6}{4x+1}$

18. $(10z^3-26z^2+17z-13)\div(5z-3)$

$$\begin{array}{r} 2z^2-4z+1 \\ 5z-3\,\overline{)\,10z^3-26z^2+17z-13} \\ \underline{10z^3-6z^2} \\ -20z^2+17z \\ \underline{-20z^2+12z} \\ 5z-13 \\ \underline{5z-3} \\ -10 \end{array}$$

Remainder

Answer: $2z^2-4z+1+\dfrac{-10}{5z-3}$

19. $\dfrac{14x+6x^3-15-19x^2}{3x^2-2x+4}$

Rewrite as $\dfrac{6x^3-19x^2+14x-15}{3x^2-2x+4}$.

$$\begin{array}{r} 2x-5 \\ 3x^2-2x+4\,\overline{)\,6x^3-19x^2+14x-15} \\ \underline{6x^3-4x^2+8x} \\ -15x^2+6x-15 \\ \underline{-15x^2+10x-20} \\ -4x+5 \end{array}$$

Remainder

Answer: $2x-5+\dfrac{-4x+5}{3x^2-2x+4}$

20. $\dfrac{37m-18m^2-13+8m^3}{2m^2-3m+6}$

Rewrite as $\dfrac{8m^3-18m^2+37m-13}{2m^2-3m+6}$.

$$\begin{array}{r} 4m-3 \\ 2m^2-3m+6\,\overline{)\,8m^3-18m^2+37m-13} \\ \underline{8m^3-12m^2+24m} \\ -6m^2+13m-13 \\ \underline{-6m^2+9m-18} \\ 4m+5 \end{array}$$

Remainder

Answer: $4m-3+\dfrac{4m+5}{2m^2-3m+6}$

21. $(3x^3-x+4)\div(x-2)$

$$\begin{array}{r} 3x^2+6x+11 \\ x-2\,\overline{)\,3x^3+0x^2-x+4} \\ \underline{3x^3-6x^2} \\ 6x^2-x \\ \underline{6x^2-12x} \\ 11x+4 \\ \underline{11x-22} \\ 26 \end{array}$$

Answer: $3x^2+6x+11+\dfrac{26}{x-2}$

22. $(4x^3 - 3x - 2) \div (x + 1)$

$$
\begin{array}{r|l}
 & 4x^2 - 4x + 1 \\ \cline{2-2}
x + 1 & 4x^3 + 0x^2 - 3x - 2 \\
 & \underline{4x^3 + 4x^2} \\
 & \quad -4x^2 - 3x \\
 & \quad \underline{-4x^2 - 4x} \\
 & \qquad\qquad x - 2 \\
 & \qquad\qquad \underline{x + 1} \\
 & \qquad\qquad\quad -3
\end{array}
$$

Answer: $4x^2 - 4x + 1 + \dfrac{-3}{x + 1}$

23. $(2x^3 - 11x^2 + 28) \div (x - 5)$

$$
\begin{array}{r|l}
 & 2x^2 - x - 5 \\ \cline{2-2}
x - 5 & 2x^3 - 11x^2 + 0x + 28 \\
 & \underline{2x^3 - 10x^2} \\
 & \quad -x^2 + 0x \\
 & \quad \underline{-x^2 + 5x} \\
 & \qquad\qquad -5x + 28 \\
 & \qquad\qquad \underline{-5x + 25} \\
 & \qquad\qquad\qquad 3
\end{array}
$$

Answer: $2x^2 - x - 5 + \dfrac{3}{x - 5}$

24. $(3x^3 - 4x + 2) \div (x - 1)$

$$
\begin{array}{r|l}
 & 3x^2 + 3x - 1 \\ \cline{2-2}
x - 1 & 3x^3 + 0x^2 - 4x + 2 \\
 & \underline{3x^3 - 3x^2} \\
 & \quad 3x^2 - 4x \\
 & \quad \underline{3x^2 - 3x} \\
 & \qquad\qquad -x + 2 \\
 & \qquad\qquad \underline{-x + 1} \\
 & \qquad\qquad\qquad 1
\end{array}
$$

Answer: $3x^2 + 3x - 1 + \dfrac{1}{x - 1}$

25. $\dfrac{4k^4 + 6k^3 + 3k - 1}{2k^2 + 1}$

$$
\begin{array}{r|l}
 & 2k^2 + 3k - 1 \\ \cline{2-2}
2k^2 + 1 & 4k^4 + 6k^3 + 0k^2 + 3k - 1 \\
 & \underline{4k^4 \qquad\quad + 2k^2} \\
 & \quad 6k^3 - 2k^2 + 3k \\
 & \quad \underline{6k^3 \qquad\quad + 3k} \\
 & \qquad\qquad -2k^2 \qquad - 1 \\
 & \qquad\qquad \underline{-2k^2 \qquad - 1} \\
 & \qquad\qquad\qquad\qquad 0
\end{array}
$$

Answer: $2k^2 + 3k - 1$

26. $\dfrac{6y^4 + 9y^3 + 10y^2 + 6y + 4}{3y^2 + 2}$

$$
\begin{array}{r|l}
 & 2y^2 + 3y + 2 \\ \cline{2-2}
3y^2 + 2 & 6y^4 + 9y^3 + 10y^2 + 6y + 4 \\
 & \underline{6y^4 \qquad\quad + 4y^2} \\
 & \quad 9y^3 + 6y^2 + 6y \\
 & \quad \underline{9y^3 \qquad\quad + 6y} \\
 & \qquad\qquad 6y^2 \qquad + 4 \\
 & \qquad\qquad \underline{6y^2 \qquad + 4} \\
 & \qquad\qquad\qquad\qquad 0
\end{array}
$$

Answer: $2y^2 + 3y + 2$

27. $(9z^4 - 13z^3 + 23z^2 - 10z + 8) \div (z^2 - z + 2)$

$$
\begin{array}{r|l}
 & 9z^2 - 4z + 1 \\ \cline{2-2}
z^2 - z + 2 & 9z^4 - 13z^3 + 23z^2 - 10z + 8 \\
 & \underline{9z^4 - 9z^3 + 18z^2} \\
 & \quad -4z^3 + 5z^2 - 10z \\
 & \quad \underline{-4z^3 + 4z^2 - 8z} \\
 & \qquad\qquad z^2 - 2z + 8 \\
 & \qquad\qquad \underline{z^2 - z + 2} \\
 & \qquad\qquad\quad -z + 6
\end{array}
$$

Answer: $9z^2 - 4z + 1 + \dfrac{-z + 6}{z^2 - z + 2}$

28. $(2q^4 + 5q^3 - 11q^2 + 11q - 20) \div (2q^2 - q + 2)$

$$
\begin{array}{r|l}
 & q^2 + 3q - 5 \\ \cline{2-2}
2q^2 - q + 2 & 2q^4 + 5q^3 - 11q^2 + 11q - 20 \\
 & \underline{2q^4 - q^3 + 2q^2} \\
 & \quad 6q^3 - 13q^2 + 11q \\
 & \quad \underline{6q^3 - 3q^2 + 6q} \\
 & \qquad\qquad -10q^2 + 5q - 20 \\
 & \qquad\qquad \underline{-10q^2 + 5q - 10} \\
 & \qquad\qquad\qquad\qquad -10
\end{array}
$$

Answer: $q^2 + 3q - 5 + \dfrac{-10}{2q^2 - q + 2}$

29. $\left(2x^2 - \frac{7}{3}x - 1\right) \div (3x + 1)$

To start: $\dfrac{2x^2}{3x} = \dfrac{2}{3}x$

$$
\begin{array}{r|l}
 & \frac{2}{3}x - 1 \\ \cline{2-2}
3x + 1 & 2x^2 - \frac{7}{3}x - 1 \\
 & \underline{2x^2 + \frac{2}{3}x} \\
 & \qquad -3x - 1 \\
 & \qquad \underline{-3x - 1} \\
 & \qquad\qquad\quad 0
\end{array}
$$

Answer: $\frac{2}{3}x - 1$

30. $\left(m^2 + \frac{7}{2}m + 3\right) \div (2m + 3)$

To start: $\dfrac{m^2}{2m} = \dfrac{1}{2}m$

$$\begin{array}{r}
\frac{1}{2}m + 1 \\
2m + 3 \overline{\big) m^2 + \frac{7}{2}m + 3} \\
\underline{m^2 + \frac{3}{2}m } \\
2m + 3 \\
\underline{2m + 3} \\
0
\end{array}$$

Answer: $\frac{1}{2}m + 1$

31. $\left(3a^2 - \frac{23}{4}a - 5\right) \div (4a + 3)$

$$\begin{array}{r}
\frac{3}{4}a - 2 \\
4a + 3 \overline{\big) 3a^2 - \frac{23}{4}a - 5} \\
\underline{3a^2 + \frac{9}{4}a } \\
-8a - 5 \\
\underline{-8a - 6} \\
1
\end{array}$$

Remainder

Answer: $\dfrac{3}{4}a - 2 + \dfrac{1}{4a + 3}$

32. $\left(3q^2 + \frac{19}{5}q - 3\right) \div (5q - 2)$

$$\begin{array}{r}
\frac{3}{5}q + 1 \\
5q - 2 \overline{\big) 3q^2 + \frac{19}{5}q - 3} \\
\underline{3q^2 - \frac{6}{5}q } \\
5q - 3 \\
\underline{5q - 2} \\
-1
\end{array}$$

Remainder

Answer: $\dfrac{3}{5}q + 1 + \dfrac{-1}{5q - 2}$

33. $f(x) = 10x^2 - 2x,\ g(x) = 2x$

$$\begin{aligned}
\left(\frac{f}{g}\right)(x) = \frac{f(x)}{g(x)} &= \frac{10x^2 - 2x}{2x} \\
&= \frac{10x^2}{2x} - \frac{2x}{2x} \\
&= 5x - 1
\end{aligned}$$

The x-values that are not in the domain of the quotient function g are found by solving $g(x) = 0$.

$$\begin{aligned}
2x &= 0 \\
x &= 0
\end{aligned}$$

34. $f(x) = 18x^2 - 24x,\ g(x) = 3x$

$$\begin{aligned}
\left(\frac{f}{g}\right)(x) = \frac{f(x)}{g(x)} &= \frac{18x^2 - 24x}{3x} \\
&= \frac{18x^2}{3x} - \frac{24x}{3x} \\
&= 6x - 8
\end{aligned}$$

The x-values that are not in the domain of the quotient function g are found by solving $g(x) = 0$.

$$\begin{aligned}
3x &= 0 \\
x &= 0
\end{aligned}$$

35. $f(x) = 2x^2 - x - 3,\ g(x) = x + 1$

$$\left(\frac{f}{g}\right)(x) = \frac{f(x)}{g(x)} = \frac{2x^2 - x - 3}{x + 1}$$

$$\begin{array}{r}
2x - 3 \\
x + 1 \overline{\big) 2x^2 - x - 3} \\
\underline{2x^2 + 2x } \\
-3x - 3 \\
\underline{-3x - 3} \\
0
\end{array}$$

Quotient: $2x - 3$

$$\begin{aligned}
g(x) &= 0 \\
x + 1 &= 0 \\
x &= -1
\end{aligned}$$

36. $f(x) = 4x^2 - 23x - 35,\ g(x) = x - 7$

$$\left(\frac{f}{g}\right)(x) = \frac{f(x)}{g(x)} = \frac{4x^2 - 23x - 35}{x - 7}$$

$$\begin{array}{r}
4x + 5 \\
x - 7 \overline{\big) 4x^2 - 23x - 35} \\
\underline{4x^2 - 28x } \\
5x - 35 \\
\underline{5x - 35} \\
0
\end{array}$$

Quotient: $4x + 5$

$$\begin{aligned}
g(x) &= 0 \\
x - 7 &= 0 \\
x &= 7
\end{aligned}$$

37. $f(x) = 8x^3 - 27$, $g(x) = 2x - 3$

$$\left(\frac{f}{g}\right)(x) = \frac{f(x)}{g(x)} = \frac{8x^3 - 27}{2x - 3}$$

$$\begin{array}{r|l}
 & 4x^2 + 6x + 9 \\
\hline
2x - 3 & 8x^3 + 0x^2 + 0x - 27 \\
 & \underline{8x^3 - 12x^2} \\
 & 12x^2 + 0x \\
 & \underline{12x^2 - 18x} \\
 & 18x - 27 \\
 & \underline{18x - 27} \\
 & 0
\end{array}$$

Quotient: $4x^2 + 6x + 9$

$$\begin{aligned} g(x) &= 0 \\ 2x - 3 &= 0 \\ 2x &= 3 \\ x &= \tfrac{3}{2} \end{aligned}$$

38. $f(x) = 27x^3 + 64$, $g(x) = 3x + 4$

$$\left(\frac{f}{g}\right)(x) = \frac{f(x)}{g(x)} = \frac{27x^3 + 64}{3x + 4}$$

$$\begin{array}{r|l}
 & 9x^2 - 12x + 16 \\
\hline
3x + 4 & 27x^3 + 0x^2 + 0x + 64 \\
 & \underline{27x^3 + 36x^2} \\
 & -36x^2 + 0x \\
 & \underline{-36x^2 - 48x} \\
 & 48x + 64 \\
 & \underline{48x + 64} \\
 & 0
\end{array}$$

Quotient: $9x^2 - 12x + 16$

$$\begin{aligned} g(x) &= 0 \\ 3x + 4 &= 0 \\ 3x &= -4 \\ x &= -\tfrac{4}{3} \end{aligned}$$

For Exercises 39–46, let $f(x) = x^2 - 9$, $g(x) = 2x$, and $h(x) = x - 3$.

39. $\left(\frac{f}{g}\right)(x) = \frac{f(x)}{g(x)} = \frac{x^2 - 9}{2x}$

We must exclude any values of x that make the denominator equal to zero, that is, $x \neq 0$.

40. $\left(\frac{f}{h}\right)(x) = \frac{f(x)}{h(x)} = \frac{x^2 - 9}{x - 3}$

$$\begin{array}{r|l}
 & x + 3 \\
\hline
x - 3 & x^2 + 0x - 9 \\
 & \underline{x^2 - 3x} \\
 & 3x - 9 \\
 & \underline{3x - 9} \\
 & 0
\end{array}$$

Thus, $\left(\frac{f}{h}\right)(x) = x + 3$ and $x \neq 3$.

41. $\left(\frac{f}{g}\right)(2) = \frac{f(2)}{g(2)} = \frac{2^2 - 9}{2(2)} = \frac{-5}{4} = -\frac{5}{4}$

42. $\left(\frac{f}{h}\right)(1) = \frac{f(1)}{h(1)} = \frac{1^2 - 9}{1 - 3} = \frac{-8}{-2} = 4$

43. $\left(\frac{h}{g}\right)(x) = \frac{h(x)}{g(x)} = \frac{x - 3}{2x}, x \neq 0$

44. $\left(\frac{f}{h}\right)(-3) = \frac{f(-3)}{h(-3)} = \frac{(-3)^2 - 9}{(-3) - 3} = \frac{0}{-6} = 0$

45. $\left(\frac{h}{g}\right)(3) = \frac{h(3)}{g(3)} = \frac{(3) - 3}{2(3)} = \frac{0}{6} = 0$

46. $\left(\frac{f}{g}\right)(-1) = \frac{f(-1)}{g(-1)} = \frac{(-1)^2 - 9}{2(-1)} = \frac{-8}{-2} = 4$

47. The volume of a box is the product of the height, length, and width. Use the formula $V = LWH$.

$$\begin{aligned} V &= LWH \\ \frac{V}{LH} &= W \end{aligned}$$

Here, $L \cdot H = (p + 4)p = p^2 + 4p$, so

$$W = \frac{V}{LH} = \frac{2p^3 + 15p^2 + 28p}{p^2 + 4p}.$$

$$\begin{array}{r|l}
 & 2p + 7 \\
\hline
p^2 + 4p & 2p^3 + 15p^2 + 28p \\
 & \underline{2p^3 + 8p^2} \\
 & 7p^2 + 28p \\
 & \underline{7p^2 + 28p} \\
 & 0
\end{array}$$

The width is $2p + 7$ ft.

48. Use $r = \frac{d}{t}$ with $d = 2m^3 + 15m^2 + 13m - 63$ and $t = 2m + 9$.

$$\begin{array}{r|l}
 & m^2 + 3m - 7 \\
\hline
2m + 9 & 2m^3 + 15m^2 + 13m - 63 \\
 & \underline{2m^3 + 9m^2} \\
 & 6m^2 + 13m \\
 & \underline{6m^2 + 27m} \\
 & -14m - 63 \\
 & \underline{-14m - 63} \\
 & 0
\end{array}$$

The rate is $\left(m^2 + 3m - 7\right)$ km/hr.

Chapter 6 Review Exercises

1. $4^3 = 4 \cdot 4 \cdot 4 = 64$

2. $\left(\frac{1}{3}\right)^4 = \frac{1}{3} \cdot \frac{1}{3} \cdot \frac{1}{3} \cdot \frac{1}{3} = \frac{1}{81}$

3. $(-5)^3 = (-5)(-5)(-5) = -125$

4. $\frac{2}{(-3)^{-2}} = \frac{2}{\frac{1}{(-3)^2}}$
$= 2 \cdot (-3)^2$
$= 2 \cdot (-3)(-3)$
$= 18$

5. $\left(\frac{2}{3}\right)^{-4} = \left(\frac{3}{2}\right)^4$
$= \frac{3}{2} \cdot \frac{3}{2} \cdot \frac{3}{2} \cdot \frac{3}{2}$
$= \frac{81}{16}$

6. $\left(\frac{5}{4}\right)^{-2} = \left(\frac{4}{5}\right)^2 = \frac{4}{5} \cdot \frac{4}{5} = \frac{16}{25}$

7. $5^{-1} + 6^{-1} = \frac{1}{5} + \frac{1}{6} = \frac{6}{30} + \frac{5}{30} = \frac{11}{30}$

8. $-3^0 + 3^0 = -1 + 1 = 0$

9. $\left(-3x^4y^3\right)\left(4x^{-2}y^5\right) = -3(4)x^4x^{-2}y^3y^5$
$= -12x^{4+(-2)}y^{3+5}$
$= -12x^2y^8$

10. $\frac{6m^{-4}n^3}{-3mn^2} = -2m^{-4-1}n^{3-2}$
$= -2m^{-5}n^1$
$= -\frac{2n}{m^5}$ or $\frac{-2n}{m^5}$

11. $\frac{(5p^{-2}q)(4p^5q^{-3})}{2p^{-5}q^5} = \frac{20p^{-2+5}q^{1+(-3)}}{2p^{-5}q^5}$
$= \frac{10p^3q^{-2}}{p^{-5}q^5}$
$= 10p^{3-(-5)}q^{-2-5}$
$= 10p^8q^{-7}$
$= \frac{10p^8}{q^7}$

12. $\frac{x^{-2}y^{-4}}{x^{-4}y^{-2}} = x^{-2-(-4)}y^{-4-(-2)}$
$= x^{-2+4}y^{-4+2}$
$= x^2y^{-2}$
$= \frac{x^2}{y^2}$

13. $\left(3^{-4}\right)^2 = 3^{(-4)\cdot 2} = 3^{-8} = \frac{1}{3^8}$

14. $\left(x^{-4}\right)^{-2} = x^{-4(-2)} = x^8$

15. $\left(xy^{-3}\right)^{-2} = x^{1(-2)}y^{(-3)(-2)}$
$= x^{-2}y^6$
$= \frac{1}{x^2} \cdot y^6 = \frac{y^6}{x^2}$

16. $\left(z^{-3}\right)^3 z^{-6} = z^{-9}z^{-6}$
$= z^{-9+(-6)}$
$= z^{-15} = \frac{1}{z^{15}}$

17. $\left(5m^{-3}\right)^2\left(m^4\right)^{-3} = 5^2m^{-6}m^{-12}$
$= 25m^{-6+(-12)}$
$= 25m^{-18} = \frac{25}{m^{18}}$

18. $\frac{(3r)^2r^4}{r^{-2}r^{-3}}\left(9r^{-3}\right)^{-2} = \frac{3^2r^2r^49^{-2}r^6}{r^{-2}r^{-3}}$
$= \frac{9^19^{-2}r^{2+4+6}}{r^{-2+(-3)}}$
$= \frac{9^{-1}r^{12}}{r^{-5}}$
$= \frac{r^{12-(-5)}}{9} = \frac{r^{17}}{9}$

19. $\left(\frac{5z^{-3}}{z^{-1}}\right)\frac{5}{z^2} = \frac{25z^{-3}}{z^{-1+2}}$
$= \frac{25}{z^3z^1}$
$= \frac{25}{z^{3+1}} = \frac{25}{z^4}$

20. $\left(\frac{6m^{-4}}{m^{-9}}\right)^{-1}\left(\frac{m^{-2}}{16}\right) = \frac{6^{-1}m^4}{m^9} \cdot \frac{m^{-2}}{16}$
$= \frac{1}{6 \cdot 16}m^{4+(-2)-9}$
$= \frac{1}{96}m^{-7} = \frac{1}{96m^7}$

21. $\left(\frac{3r^5}{5r^{-3}}\right)^{-2}\left(\frac{9r^{-1}}{2r^{-5}}\right)^3$
$= \frac{3^{-2}r^{-10}}{5^{-2}r^6} \cdot \frac{9^3r^{-3}}{2^3r^{-15}}$
$= \frac{5^2}{3^2}r^{-10-6} \cdot \frac{9^3}{2^3}r^{-3-(-15)}$
$= \frac{25}{9}r^{-16} \cdot \frac{729}{8}r^{12}$
$= \frac{(25)(729)}{(9)(8)}r^{-16+12}$
$= \frac{2025}{8}r^{-4} = \frac{2025}{8r^4}$

22. $13{,}450 = 1_{\wedge}3450.$

Place a caret to the right of the first nonzero digit. Count 4 places.

Since the number 1.345 is to be made larger, the exponent on 10 is positive.

$$13{,}450 = 1.345 \times 10^4$$

23. $.000\,000\,0765 = .0\,0\,0\,0\,0\,0\,0\,7{\wedge}65$

Count 8 places

Since the number 7.65 is to be made smaller, the exponent on 10 is negative.

$$.000\,000\,0765 = 7.65 \times 10^{-8}$$

24. $.138 = .1{\wedge}38$

Count 1 place.

Since the number 1.38 is to be made smaller, the exponent on 10 is negative.

$$.138 = 1.38 \times 10^{-1}$$

25. $281{,}400{,}000 = 2.814 \times 10^8$

$$50{,}454 = 5.0454 \times 10^4$$

$$100 = 1 \times 10^2$$

26. $1.21 \times 10^6 = 1{,}210{,}000$

Move the decimal point 6 places to the right because the exponent is positive. Attach extra zeros.

27. $5.8 \times 10^{-3} = .0058$

Move the decimal point 3 places to the left because the exponent is negative.

28. $$\frac{16 \times 10^4}{8 \times 10^8} = \frac{16}{8} \times 10^{4-8} = 2 \times 10^{-4} \text{ or } .0002$$

29. $$\frac{6 \times 10^{-2}}{4 \times 10^{-5}} = \frac{6}{4} \times 10^{-2-(-5)} = 1.5 \times 10^3 \text{ or } 1500$$

30. $$\frac{.000\,000\,0164}{.0004} = \frac{1.64 \times 10^{-8}}{4 \times 10^{-4}} = \frac{1.64}{4} \times 10^{-8-(-4)} = .41 \times 10^{-4} = 4.1 \times 10^{-5} \text{ or } .000041$$

31. $$\frac{.0009 \times 12{,}000{,}000}{400{,}000} = \frac{9 \times 10^{-4} \times 1.2 \times 10^7}{4 \times 10^5} = \frac{9 \times 1.2}{4} \times \frac{10^{-4} \times 10^7}{10^5} = \frac{10.8}{4} \times 10^{-4+7-5} = 2.7 \times 10^{-2} \text{ or } .027$$

32. The coefficient of $14p^5$ is 14.

33. The coefficient of $-z$ is -1.

34. $9k + 11k^3 - 3k^2$

(a) In descending powers of k, the polynomial is

$$11k^3 - 3k^2 + 9k.$$

(b) The polynomial is a trinomial since it has three terms.

(c) The degree of the polynomial is 3 since the highest power of k is 3.

35. $14m^6 + 9m^7$

(a) In descending powers of m, the polynomial is

$$9m^7 + 14m^6.$$

(b) The polynomial is a binomial since it has two terms.

(c) The degree of the polynomial is 7 since the highest power of m is 7.

36. $-7q^5r^3$

(a) The polynomial is already written in descending powers.

(b) The polynomial is a monomial since it has just one term.

(c) The degree is $5 + 3 = 8$, the sum of the exponents of this term.

37. One example of a polynomial in the variable x that has degree 5, is lacking a third-degree term, and is written in descending powers of the variable is

$$x^5 + 2x^4 - x^2 + x + 2.$$

38. Add by columns.

$$\begin{array}{rcrcr} 3x^2 & - & 5x & + & 6 \\ -4x^2 & + & 2x & - & 5 \\ \hline -1x^2 & - & 3x & + & 1 \end{array} \text{ or } -x^2 - 3x + 1$$

39. Subtract.

$$\begin{array}{rcrcrcr} -5y^3 & & & + & 8y & - & 3 \\ & & 4y^2 & + & 2y & + & 9 \\ \hline \end{array}$$

Change the signs in the second polynomial and add.

$$\begin{array}{rcrcrcr} -5y^3 & & & + & 8y & - & 3 \\ & - & 4y^2 & - & 2y & - & 9 \\ \hline -5y^3 & - & 4y^2 & + & 6y & - & 12 \end{array}$$

40. $$\begin{aligned} &(4a^3 - 9a + 15) - (-2a^3 + 4a^2 + 7a) \\ &= 4a^3 - 9a + 15 + 2a^3 - 4a^2 - 7a \\ &= 4a^3 + 2a^3 - 4a^2 - 9a - 7a + 15 \\ &= 6a^3 - 4a^2 - 16a + 15 \end{aligned}$$

41. $(3y^2 + 2y - 1) + (5y^2 - 11y + 6)$
$= 3y^2 + 5y^2 + 2y - 11y - 1 + 6$
$= 8y^2 - 9y + 5$

42. To find the perimeter, add the measures of the three sides.

$(4x^2 + 2) + (6x^2 + 5x + 2) + (2x^2 + 3x + 1)$
$= 4x^2 + 6x^2 + 2x^2 + 5x + 3x + 2 + 2 + 1$
$= 12x^2 + 8x + 5$

The perimeter is $12x^2 + 8x + 5$.

43. $f(x) = -2x^2 + 5x + 7$

(a) $f(-2) = -2(-2)^2 + 5(-2) + 7$
$= -2(4) - 10 + 7$
$= -8 - 10 + 7 = -11$

(b) $f(3) = -2(3)^2 + 5(3) + 7$
$= -2(9) + 15 + 7$
$= -18 + 15 + 7 = 4$

44. **(a)** $(f + g)(x) = f(x) + g(x)$
$= (2x + 3) + (5x^2 - 3x + 2)$
$= 5x^2 + 2x - 3x + 3 + 2$
$= 5x^2 - x + 5$

(b) $(f - g)(x) = f(x) - g(x)$
$= (2x + 3) - (5x^2 - 3x + 2)$
$= 2x + 3 - 5x^2 + 3x - 2$
$= -5x^2 + 5x + 1$

(c) $(f + g)(-1)$
$= f(-1) + g(-1)$
$= [2(-1) + 3] + [5(-1)^2 - 3(-1) + 2]$
$= [1] + [10]$
$= 11$

Alternatively, use the result from part (a).

(d) $(f - g)(-1) = f(-1) - g(-1)$
$= 1 - 10$ *from part (c)*
$= -9$

Alternatively, use the result from part (b).

45. $f(x) = .241x^2 + 3.26x + 30.0$

(a) $x = 1990 - 1990 = 0$

$f(0) = .241(0)^2 + 3.26(0) + 30.0$
$= 30$ million

(b) $x = 1995 - 1990 = 5$

$f(5) = .241(5)^2 + 3.26(5) + 30.0$
$= 52.325$ million

(c) $x = 2000 - 1990 = 10$

$f(10) = .241(10)^2 + 3.26(10) + 30.0$
$= 86.7$ million

46. $y = f(x) = -2x + 5$

This is a linear function. The y-intercept is $(0, 5)$. To find another point, let $x = 2$.

$y = -2(2) + 5$ *Let $x = 2$.*
$= -4 + 5$
$= 1$

Plot the points $(0, 5)$ and $(2, 1)$, and draw a line through them.

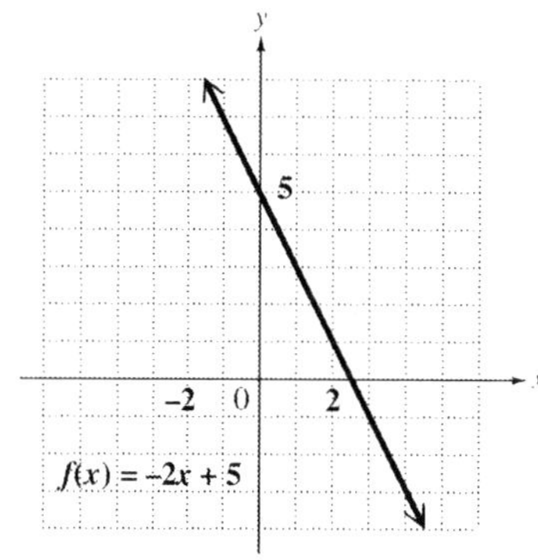

47.

x	$f(x) = x^2 - 6$
-2	$(-2)^2 - 6 = -2$
-1	$(-1)^2 - 6 = -5$
0	$(0)^2 - 6 = -6$
1	$(1)^2 - 6 = -5$
2	$(2)^2 - 6 = -2$

Since the greatest exponent is 2, the graph of f is a parabola.

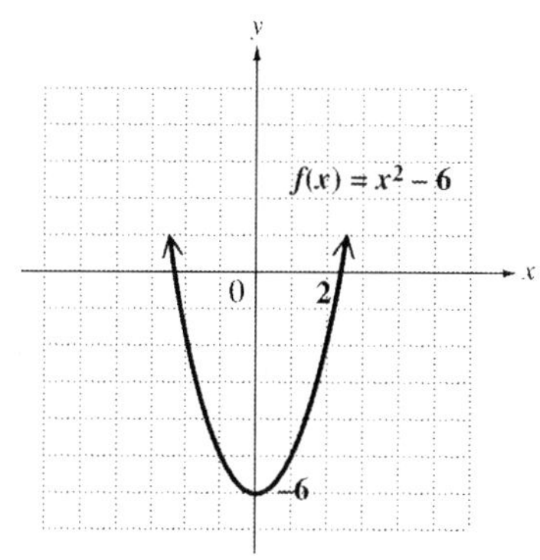

48.

x	$f(x) = -x^3 + 1$
-2	$-(-2)^3 + 1 = 9$
-1	$-(-1)^3 + 1 = 2$
0	$-(0)^3 + 1 = 1$
1	$-(1)^3 + 1 = 0$
2	$-(2)^3 + 1 = -7$

The greatest exponent is 3, so the graph of f is s-shaped.

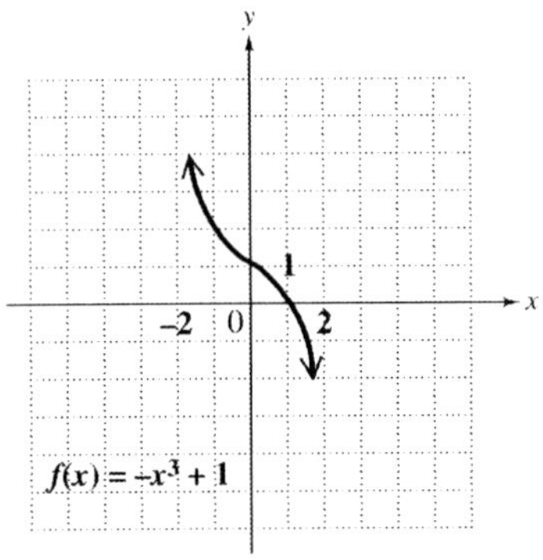

49. $-6k(2k^2 + 7) = -6k(2k^2) - 6k(7)$
$= -12k^3 - 42k$

50. $(7y - 8)(2y + 3)$
F O I L
$= 14y^2 + 21y - 16y - 24$
$= 14y^2 + 5y - 24$

51. $(3w - 2t)(2w - 3t)$
F O I L
$= 6w^2 - 9wt - 4wt + 6t^2$
$= 6w^2 - 13wt + 6t^2$

52. $(2p^2 + 6p)(5p^2 - 4)$
F O I L
$= 10p^4 - 8p^2 + 30p^3 - 24p$
$= 10p^4 + 30p^3 - 8p^2 - 24p$

53. $(3z^3 - 2z^2 + 4z - 1)(3z - 2)$
$= (3z^3 - 2z^2 + 4z - 1)(3z)$
$+ (3z^3 - 2z^2 + 4z - 1)(-2)$
$= 9z^4 - 6z^3 + 12z^2 - 3z$
$- 6z^3 + 4z^2 - 8z + 2$
$= 9z^4 - 6z^3 - 6z^3 + 12z^2 + 4z^2$
$- 3z - 8z + 2$
$= 9z^4 - 12z^3 + 16z^2 - 11z + 2$

54. $(6r^2 - 1)(6r^2 + 1) = (6r^2)^2 - 1^2$
$= 36r^4 - 1$

55. $\left(z + \frac{3}{5}\right)\left(z - \frac{3}{5}\right) = z^2 - \left(\frac{3}{5}\right)^2$
$= z^2 - \frac{9}{25}$

56. $(4m + 3)^2 = (4m)^2 + 2(4m)(3) + 3^2$
$= 16m^2 + 24m + 9$

57. $(2x + 5)^3 = (2x + 5)(2x + 5)^2$
$= (2x + 5)(4x^2 + 20x + 25)$
$= 8x^3 + 40x^2 + 50x + 20x^2 + 100x$
$+ 125$
$= 8x^3 + 60x^2 + 150x + 125$

58. $\frac{4y^3 - 12y^2 + 5y}{4y} = \frac{4y^3}{4y} - \frac{12y^2}{4y} + \frac{5y}{4y}$
$= y^2 - 3y + \frac{5}{4}$

59. $\frac{2p^3 + 9p^2 + 27}{2p - 3}$

$$
\begin{array}{r|l}
 & p^2 + 6p + 9 \\
\hline
2p - 3 & 2p^3 + 9p^2 + 0p + 27 \\
 & \underline{2p^3 - 3p^2} \\
 & \quad 12p^2 + 0p \\
 & \quad \underline{12p^2 - 18p} \\
 & \qquad 18p + 27 \\
 & \qquad \underline{18p - 27} \\
 & \qquad\qquad 54
\end{array}
$$

Remainder

Answer: $p^2 + 6p + 9 + \frac{54}{2p - 3}$

60. $\frac{5p^4 + 15p^3 - 33p^2 - 9p + 18}{5p^2 - 3}$

$$
\begin{array}{r|l}
 & p^2 + 3p - 6 \\
\hline
5p^2 - 3 & 5p^4 + 15p^3 - 33p^2 - 9p + 18 \\
 & \underline{5p^4 \qquad\quad - 3p^2} \\
 & \quad 15p^3 - 30p^2 - 9p \\
 & \quad \underline{15p^3 \qquad\quad - 9p} \\
 & \qquad -30p^2 \qquad + 18 \\
 & \qquad \underline{-30p^2 \qquad + 18} \\
 & \qquad\qquad 0
\end{array}
$$

Answer: $p^2 + 3p - 6$

61. **[6.1]** **(a)** $4^{-2} = \frac{1}{4^2} = \frac{1}{16}$ **(A)**

(b) $-4^2 = -(4^2) = -16$ **(G)**

(c) $4^0 = 1$ **(C)**

(d) $(-4)^0 = 1$ **(C)**

(e) $(-4)^{-2} = \frac{1}{(-4)^2} = \frac{1}{16}$ **(A)**

(f) $-4^0 = -(4^0) = -1$ **(E)**

(g) $-4^0 + 4^0 = -1 + 1 = 0$ **(B)**

(h) $-4^0 - 4^0 = (-1) - 1 = -2$ **(H)**

(i) $4^{-2} + 4^{-1} = \frac{1}{4^2} + \frac{1}{4} = \frac{1}{16} + \frac{4}{16} = \frac{5}{16}$ **(F)**

62. **[6.1]** The density D is the population P divided by the area A.

$$D = \frac{P}{A}$$

We want to find the area, so solve the formula for A.

$$\begin{aligned} DA &= P \\ A &= \frac{P}{D} \\ &= \frac{3.92 \times 10^5}{400} \\ &= \frac{3.92 \times 10^5}{4 \times 10^2} \\ &= .98 \times 10^{5-2} \\ &= .98 \times 10^3 = 980 \end{aligned}$$

The area is 980 mi^2.

63. **[6.4]** $(4x + 1)(2x - 3)$

F O I L

$$\begin{aligned} &= 8x^2 - 12x + 2x - 3 \\ &= 8x^2 - 10x - 3 \end{aligned}$$

64. **[6.1]** $\frac{6^{-1}y^3(y^2)^{-2}}{6y^{-4}(y^{-1})} = \frac{y^3y^{-4}}{6^1 \cdot 6y^{-4}y^{-1}}$

$$\begin{aligned} &= \frac{y^{3+(-4)}}{36y^{-4+(-1)}} = \frac{y^{-1}}{36y^{-5}} \\ &= \frac{y^{-1-(-5)}}{36} = \frac{y^4}{36} \end{aligned}$$

65. **[6.1]** $(y^6)^{-5}(2y^{-3})^{-4} = y^{-30}(2)^{-4}y^{12}$

$$\begin{aligned} &= \frac{y^{-30+12}}{2^4} \\ &= \frac{y^{-18}}{2^4} = \frac{1}{16y^{18}} \end{aligned}$$

66. **[6.4]** $(2x - 9)^2 = (2x)^2 - 2(2x)(9) + 9^2$

$$= 4x^2 - 36x + 81$$

67. **[6.5]** $\frac{20y^3x^3 + 15y^4x + 25yx^4}{10yx^2}$

$$\begin{aligned} &= \frac{20y^3x^3}{10yx^2} + \frac{15y^4x}{10yx^2} + \frac{25yx^4}{10yx^2} \\ &= 2y^2x + \frac{3y^3}{2x} + \frac{5x^2}{2} \end{aligned}$$

68. **[6.4]** $7p^5(3p^4 + p^3 + 2p^2)$

$$\begin{aligned} &= 7p^5(3p^4) + 7p^5(p^3) + 7p^5(2p^2) \\ &= 21p^9 + 7p^8 + 14p^7 \end{aligned}$$

69. **[6.1]** $\frac{(-z^{-2})^3}{5(z^{-3})^{-1}} = \frac{(-1)^3z^{-2(3)}}{5z^{-3(-1)}}$

$$\begin{aligned} &= \frac{-z^{-6}}{5z^3} \\ &= \frac{-z^{-6-3}}{5} \\ &= \frac{-z^{-9}}{5} = -\frac{1}{5z^9} \end{aligned}$$

70. **[6.5]** $\frac{x^3 + 7x^2 + 7x - 15}{x + 5}$

$$\begin{array}{r} x^2 + 2x - 3 \\ x + 5 \overline{) x^3 + 7x^2 + 7x - 15} \\ \underline{x^3 + 5x^2 \qquad\qquad} \\ 2x^2 + 7x \qquad \\ \underline{2x^2 + 10x \qquad} \\ -3x - 15 \\ \underline{-3x - 15} \\ 0 \end{array}$$

Answer: $x^2 + 2x - 3$

71. **[6.2]** $(-5 + 11w) + (6 + 5w) + (-15 - 8w^2)$

$$\begin{aligned} &= -5 + 6 - 15 + 11w + 5w - 8w^2 \\ &= -14 + 16w - 8w^2 \end{aligned}$$

72. **[6.2]** $(2k - 1) - (3k^2 - 2k + 6)$

$$\begin{aligned} &= 2k - 1 - 3k^2 + 2k - 6 \\ &= -3k^2 + 2k + 2k - 1 - 6 \\ &= -3k^2 + 4k - 7 \end{aligned}$$

Chapter 6 Test

1. **(a)** $7^{-2} = \frac{1}{7^2} = \frac{1}{49}$ **(C)**

(b) $7^0 = 1$ **(A)**

(c) $-7^0 = -(1) = -1$ **(D)**

(d) $(-7)^0 = 1$ **(A)**

(e) $-7^2 = -49$ **(E)**

(f) $7^{-1}+2^{-1}=\frac{1}{7}+\frac{1}{2}$
$=\frac{2}{14}+\frac{7}{14}=\frac{9}{14}$ **(F)**

(g) $(7+2)^{-1}=9^{-1}=\frac{1}{9}$ **(B)**

(h) $\frac{7^{-1}}{2^{-1}}=\frac{2^1}{7^1}=\frac{2}{7}$ **(G)**

(i) $(-7)^{-2}=\frac{1}{(-7)^2}=\frac{1}{49}$ **(C)**

2. $\left(3x^{-2}y^3\right)^{-2}\left(4x^3y^{-4}\right)$
$=3^{-2}x^{-2(-2)}y^{3(-2)}4x^3y^{-4}$
$=3^{-2}x^4y^{-6}4x^3y^{-4}$
$=\frac{4x^{4+3}y^{-6-4}}{3^2}$
$=\frac{4x^7y^{-10}}{9}=\frac{4x^7}{9y^{10}}$

3. $\frac{36r^{-4}(r^2)^{-3}}{6r^4}=\frac{36r^{-4}r^{2(-3)}}{6r^4}$
$=\frac{6r^{-4}r^{-6}}{r^4}$
$=\frac{6r^{-10}}{r^4}$
$=\frac{6}{r^4r^{10}}=\frac{6}{r^{14}}$

4. $\left(\frac{4p^2}{q^4}\right)^3\left(\frac{6p^8}{q^{-8}}\right)^{-2}=\frac{4^3p^6}{q^{12}}\cdot\frac{6^{-2}p^{-16}}{q^{16}}$
$=\frac{4^3p^{-10}}{6^2q^{28}}$
$=\frac{64}{36p^{10}q^{28}}=\frac{16}{9p^{10}q^{28}}$

5. $\left(-2x^4y^{-3}\right)^0\left(-4x^{-3}y^{-8}\right)^2$
$=1(-4)^2x^{-6}y^{-16}$
$=\frac{16}{x^6y^{16}}$

6. $9.1\times10^{-7}=.000\,000\,91$
Move the decimal point 7 places to the left because the exponent is negative.

7. $\frac{2{,}500{,}000\times.00003}{.05\times5{,}000{,}000}$
$=\frac{2.5\times10^6\times3\times10^{-5}}{5\times10^{-2}\times5\times10^6}$
$=\frac{7.5\times10^1}{25\times10^4}$
$=.3\times10^{1-4}$
$=.3\times10^{-3}$
$=3\times10^{-4}$ or $.0003$

8. **(a)** $f(x)=-2x^2+5x-6$
$f(4)=-2(4)^2+5(4)-6$
$=-2\cdot16+20-6$
$=-32+20-6=-18$

(b) $(f+g)(x)=f(x)+g(x)$
$=\left(-2x^2+5x-6\right)+(7x-3)$
$=-2x^2+12x-9$

(c) $(f-g)(x)=f(x)-g(x)$
$=\left(-2x^2+5x-6\right)-(7x-3)$
$=-2x^2+5x-6-7x+3$
$=-2x^2-2x-3$

(d) Using the answer in part (c), we have
$(f-g)(-2)=-2(-2)^2-2(-2)-3$
$=-8+4-3=-7.$

9.

x	$f(x)=-2x^2+3$
-2	$-2(-2)^2+3=-5$
-1	$-2(-1)^2+3=1$
0	$-2(0)^2+3=3$
1	$-2(1)^2+3=1$
2	$-2(2)^2+3=-5$

Since the greatest exponent is 2, the graph of f is a parabola.

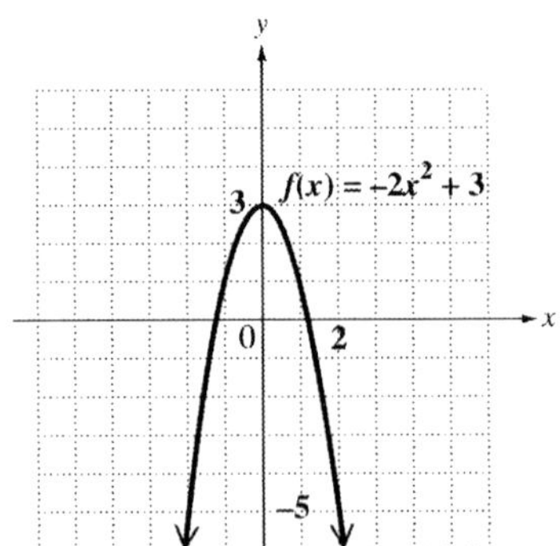

10. $f(x)=-.197x^2+21.7x+615$

$f(x)$ represents the number of medical doctors, in thousands, in the United States.

	Year	x	$f(x)$
(a)	1990	0	615
(b)	1996	6	$738.108\approx738$
(c)	2000	10	$812.3\approx812$

11. $\left(4x^3-3x^2+2x-5\right)$
$-\left(3x^3+11x+8\right)+\left(x^2-x\right)$
$=4x^3-3x^2+2x-5-3x^3-11x$
$-8+x^2-x$
$=x^3-2x^2-10x-13$

12. $(5x-3)(2x+1)$

F O I L

$= 10x^2 + 5x - 6x - 3$
$= 10x^2 - x - 3$

13. $(2m-5)(3m^2+4m-5)$
$= 2m(3m^2+4m-5)$
$\quad + (-5)(3m^2+4m-5)$
$= 6m^3 + 8m^2 - 10m$
$\quad - 15m^2 - 20m + 25$
$= 6m^3 - 7m^2 - 30m + 25$

14. $(6x+y)(6x-y) = (6x)^2 - y^2$
$= 36x^2 - y^2$

15. $(3k+q)^2 = (3k)^2 + 2(3k)(q) + q^2$
$= 9k^2 + 6kq + q^2$

16. $[2y+(3z-x)][2y-(3z-x)]$
$= (2y)^2 - (3z-x)^2$
$= 4y^2 - (9z^2 - 6zx + x^2)$
$= 4y^2 - 9z^2 + 6zx - x^2$

17. $\dfrac{16p^3 - 32p^2 + 24p}{4p^2}$
$= \dfrac{16p^3}{4p^2} - \dfrac{32p^2}{4p^2} + \dfrac{24p}{4p^2}$
$= 4p - 8 + \dfrac{6}{p}$

18. $(x^3 + 3x^2 - 4) \div (x-1)$

Insert $0x$ for the missing x-term.

$$\begin{array}{r|l} & x^2 + 4x + 4 \\ \hline x - 1 & x^3 + 3x^2 + 0x - 4 \\ & \underline{x^3 - x^2} \\ & \quad 4x^2 + 0x \\ & \quad \underline{4x^2 - 4x} \\ & \qquad 4x - 4 \\ & \qquad \underline{4x - 4} \\ & \qquad\quad 0 \end{array}$$

Answer: $x^2 + 4x + 4$

19. **(a)** $(fg)(x) = f(x) \cdot g(x)$
$= (x^2+3x+2)(x+1)$
$= (x^2+3x+2)(x)$
$\quad + (x^2+3x+2)(1)$
$= x^3 + 3x^2 + 2x + x^2 + 3x + 2$
$= x^3 + 4x^2 + 5x + 2$

(b) $(fg)(-2) = f(-2) \cdot g(-2)$
$= [(-2)^2 + 3(-2) + 2] \cdot [(-2) + 1]$
$= [4 - 6 + 2] \cdot [-1]$
$= 0(-1) = 0$

Alternatively, we could have substituted -2 for x into our answer from part (a).

20. **(a)** $\left(\dfrac{f}{g}\right)(x) = \dfrac{f(x)}{g(x)} = \dfrac{x^2+3x+2}{x+1}$

$$\begin{array}{r|l} & x + 2 \\ \hline x + 1 & x^2 + 3x + 2 \\ & \underline{x^2 + x} \\ & \quad 2x + 2 \\ & \quad \underline{2x + 2} \\ & \qquad 0 \end{array}$$

Thus, $\left(\dfrac{f}{g}\right)(x) = x + 2$ if $x + 1 \neq 0$, that is, $x \neq -1$.

(b) Using our answer from part (a),

$$\left(\frac{f}{g}\right)(-2) = (-2) + 2 = 0.$$

Cumulative Review Exercises (Chapters 1–6)

1. 34 is a natural number, so it is also a whole number, an integer, a rational number, and a real number. **A, B, C, D, F**
2. 0 is a whole number, so it is also an integer, a rational number, and a real number. **B, C, D, F**
3. 2.16 is a rational number, so it is also a real number. **D, F**
4. $-\sqrt{36} = -6$ is an integer, so it is also a rational number and a real number. **C, D, F**
5. $\sqrt{13}$ is an irrational number, so it is also a real number. **E, F**
6. $-\frac{4}{5}$ is a rational number, so it is also a real number. **D, F**
7. $9 \cdot 4 - 16 \div 4 = (9 \cdot 4) - (16 \div 4) = 36 - 4 = 32$
8. $-|8 - 13| - |-4| + |-9| = -|-5| - 4 + 9$
$= -5 - 4 + 9$
$= -9 + 9 = 0$

9.
$$\begin{aligned} -5(8-2z)+4(7-z) &= 7(8+z)-3 \\ -40+10z+28-4z &= 56+7z-3 \quad \textit{Distributive property} \\ 6z-12 &= 7z+53 \quad \textit{Combine like terms} \\ -65 &= z \quad \textit{Subtract 6z; 53} \end{aligned}$$

Thus, the solution set is $\{-65\}$.

10.
$$\begin{aligned} 3(x+2)-5(x+2) &= -2x-4 \\ 3x+6-5x-10 &= -2x-4 \\ -2x-4 &= -2x-4 \end{aligned}$$

The last statement is true for all real numbers, so the solution set is $(-\infty, \infty)$.

11.
$$\begin{aligned} 2(m+5)-3m+1 &> 5 \\ 2m+10-3m+1 &> 5 \\ -m+11 &> 5 \\ -m &> -6 \\ m &< 6 \end{aligned}$$

Solution set: $(-\infty, 6)$

12. $|3x-1|=2$

$$\begin{aligned} 3x-1=2 \quad &\text{or} \quad 3x-1=-2 \\ 3x=3 \quad &\text{or} \quad 3x=-1 \\ x=1 \quad &\text{or} \quad x=-\tfrac{1}{3} \end{aligned}$$

Solution set: $\left\{-\frac{1}{3}, 1\right\}$

13. $|3z+1| \geq 7$

$$\begin{aligned} 3z+1 \geq 7 \quad &\text{or} \quad 3z+1 \leq -7 \\ 3z \geq 6 \quad &\text{or} \quad 3z \leq -8 \\ z \geq 2 \quad &\text{or} \quad z \leq -\tfrac{8}{3} \end{aligned}$$

Solution set: $\left(-\infty, -\frac{8}{3}\right] \cup [2, \infty)$

14. Personal computer: $\dfrac{480}{1500} = 32\%$

Pacemaker:

26% of $1500 = .26(1500) = 390$

Wireless communication:

18% of $1500 = .18(1500) = 270$

Television: $\dfrac{150}{1500} = 10\%$

15. The sum of the measures of the angles of any triangle is $180°$, so

$$(x+15)+(6x+10)+(x-5)=180.$$

Solve this equation.

$$\begin{aligned} 8x+20 &= 180 \\ 8x &= 160 \\ x &= 20 \end{aligned}$$

Substitute 20 for x to find the measures of the angles.

$$\begin{aligned} x-5 &= 20-5 = 15 \\ x+15 &= 20+15 = 35 \\ 6x+10 &= 6(20)+10 = 130 \end{aligned}$$

The measures of the angles of the triangle are $15°$, $35°$, and $130°$.

16. Through $(-4, 5)$ and $(2, -3)$

Use the definition of slope with $x_1 = -4$, $y_1 = 5$, $x_2 = 2$, and $y_2 = -3$.

$$\begin{aligned} m = \frac{y_2 - y_1}{x_2 - x_1} &= \frac{-3-5}{2-(-4)} \\ &= \frac{-8}{6} = -\frac{4}{3} \end{aligned}$$

Use the point-slope form of a line.

$$\begin{aligned} y - y_1 &= m(x - x_1) \\ y - 5 &= -\tfrac{4}{3}[x-(-4)] \\ 3y - 15 &= -4(x+4) \quad \textit{Multiply by 3.} \\ 3y - 15 &= -4x - 16 \\ 4x + 3y &= -1 \quad \textit{Standard form} \end{aligned}$$

17. $-3x+4y=12$

If $y=0$, $x=-4$, so the x-intercept is $(-4, 0)$.

If $x=0$, $y=3$, so the y-intercept is $(0, 3)$.

Draw a line through these intercepts. A third point may be used as a check.

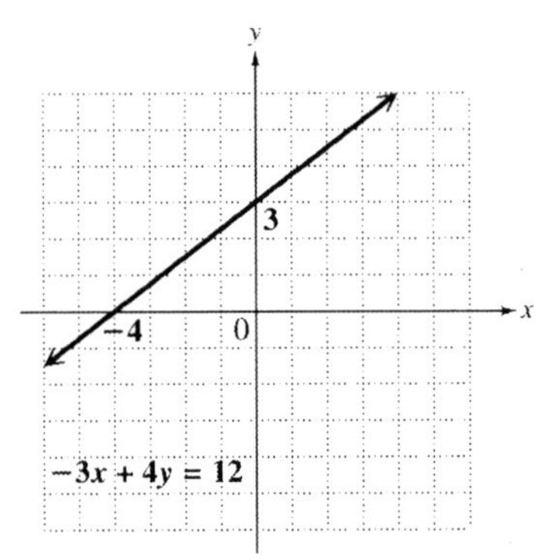

18. $y \leq 2x - 6$

Graph the boundary, $y = 2x - 6$, as a solid line through the intercepts $(3, 0)$ and $(0, -6)$. A third point such as $(1, -4)$ can be used as a check. Using $(0, 0)$ as a test point results in the false inequality $0 \leq -6$, so shade the region *not* containing the origin. This is the region below the line. The solid line shows that the boundary is part of the graph.

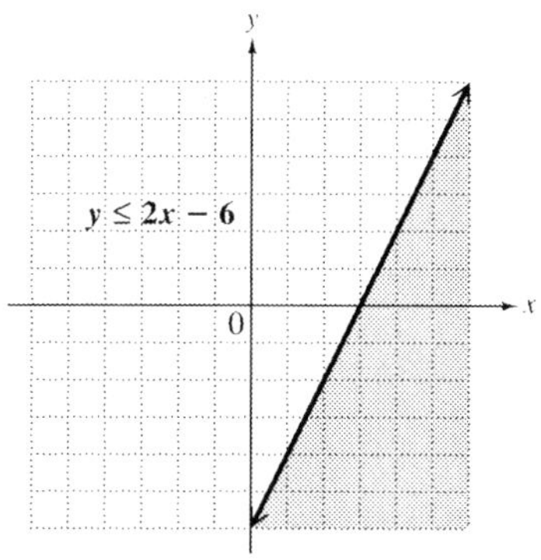

19. $3x + 2y < 0$

Graph the boundary, $3x + 2y = 0$, as a dashed line through $(0, 0)$, $(-2, 3)$, and $(2, -3)$. Choose a test point not on the line. Using $(1, 1)$ results in the false statement $5 < 0$, so shade the region *not* containing $(1, 1)$. This is the region below the line. The dashed line shows that the boundary is not part of the graph.

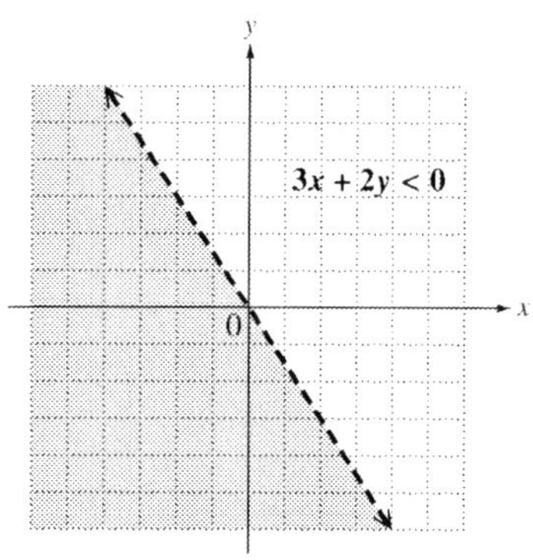

20. **(a)** $m = \dfrac{118{,}916 - 93{,}865}{10 - 0} = \dfrac{25{,}051}{10} = 2505.1$

The average rate of change is 2505.1 twin births per year, that is, the number of twin births increased an average of 2505.1 per year.

(b) $y = mx + b$

$y = 2505.1x + 93{,}865$

(c) If $x = 2002 - 1990 = 12$, then

$$\begin{aligned} y &= 2505.1(12) + 93{,}865 \\ &= 123{,}926.2 \end{aligned}$$

The model predicts about 123,926 twin births in 2002.

21. $\{(-4, -2), (-1, 0), (2, 0), (5, 2)\}$

The domain is the set of first components, that is, $\{-4, -1, 2, 5\}$.
The range is the set of second components, that is, $\{-2, 0, 2\}$.
The relation is a function since each first component is paired with a unique second component.

22. $3x - 4y = 1 \quad (1)$
$2x + 3y = 12 \quad (2)$

To eliminate y, multiply equation (1) by 3 and equation (2) by 4. Then add the results.

$$\begin{array}{rcrcll} 9x & - & 12y & = & 3 & 3 \times (1) \\ 8x & + & 12y & = & 48 & 4 \times (2) \\ \hline 17x & & & = & 51 & \\ & & x & = & 3 & \end{array}$$

Substitute $x = 3$ into equation (1).

$$\begin{aligned} 3x - 4y &= 1 \quad (1) \\ 3(3) - 4y &= 1 \\ 9 - 4y &= 1 \\ -4y &= -8 \\ y &= 2 \end{aligned}$$

Solution set: $\{(3, 2)\}$

23. $3x - 2y = 4 \quad (1)$
$-6x + 4y = 7 \quad (2)$

Multiply equation (1) by 2 and add the result to equation (2).

$$\begin{array}{rcrcll} 6x & - & 4y & = & 8 & 2 \times (1) \\ -6x & + & 4y & = & 7 & (2) \\ \hline & & 0 & = & 15 & \textit{False} \end{array}$$

Since a false statement results, the system is *inconsistent*. The solution set is $\emptyset$.

24.
$$\begin{aligned} x+3y-6z &= 7 \quad (1)\\ 2x-y+z &= 1 \quad (2)\\ x+2y+2z &= -1 \quad (3) \end{aligned}$$

To eliminate x, multiply equation (1) by -2 and add the result to equation (2).

$$\begin{array}{rcrcrcrl} -2x & - & 6y & + & 12z & = & -14 & \quad -2\times(1)\\ 2x & - & y & + & z & = & 1 & \quad (2)\\ \hline & - & 7y & + & 13z & = & -13 & \quad (4) \end{array}$$

To eliminate x again, multiply equation (3) by -2 and add the result to equation (2).

$$\begin{array}{rcrcrcrl} 2x & - & y & + & z & = & 1 & \quad (2)\\ -2x & - & 4y & - & 4z & = & 2 & \quad -2\times(3)\\ \hline & - & 5y & - & 3z & = & 3 & \quad (5) \end{array}$$

Use equations (4) and (5) to eliminate z. Multiply equation (4) by 3 and add the result to 13 times equation (5).

$$\begin{array}{rcrcrl} -21y & + & 39z & = & -39 & \quad 3\times(4)\\ -65y & - & 39z & = & 39 & \quad 13\times(5)\\ \hline -86y & & & = & 0 & \\ y & & & = & 0 & \end{array}$$

From (5) with $y=0$, $-3z=3$, so $z=-1$.
From (3) with $y=0$ and $z=-1$, $x-2=-1$, so $x=1$.

Solution set: $\{(1,0,-1)\}$

25. The length L of the rectangular flag measured 12 feet more than its width W, so

$$L = W + 12. \quad (1)$$

The perimeter is 144 feet.

$$P = 2L + 2W \quad (2)$$

Substitute $W+12$ for L into equation (2).

$$\begin{aligned} 144 &= 2(W+12)+2W\\ 144 &= 2W+24+2W\\ 120 &= 4W\\ 30 &= W \end{aligned}$$

From (1), $L = 30 + 12 = 42$.

The length is 42 feet and the width is 30 feet.

26. $$\left(\frac{2m^3n}{p^2}\right)^3 = \frac{2^3(m^3)^3n^3}{(p^2)^3} = \frac{8m^9n^3}{p^6}$$

27. $$\frac{x^{-6}y^3z^{-1}}{x^7y^{-4}z} = \frac{y^4y^3}{x^6x^7z^1z} = \frac{y^7}{x^{13}z^2}$$

28.
$$\begin{aligned} (2m^{-2}n^3)^{-3} &= 2^{-3}(m^{-2})^{-3}(n^3)^{-3}\\ &= 2^{-3}m^{(-2)(-3)}n^{3(-3)}\\ &= 2^{-3}m^6n^{-9}\\ &= \frac{m^6}{2^3n^9} = \frac{m^6}{8n^9} \end{aligned}$$

29.
$$\begin{aligned} &2(3x^2-8x+1)-4(x^2-3x-9)\\ &= 6x^2-16x+2-4x^2+12x+36 && \textit{Distributive property}\\ &= (6x^2-4x^2)+(-16x+12x)+(2+36) && \textit{Combine like terms}\\ &= 2x^2-4x+38 \end{aligned}$$

30. $(3x+2y)(5x-y)$

F O I L

$$\begin{aligned} &= 3x(5x)+3x(-y)+2y(5x)+2y(-y)\\ &= 15x^2-3xy+10xy-2y^2\\ &= 15x^2+7xy-2y^2 \end{aligned}$$

31.
$$\begin{aligned} (8m+5n)(8m-5n) &= (8m)^2-(5n)^2\\ &= 64m^2-25n^2 \end{aligned}$$

32. $$\frac{m^3-3m^2+5m-3}{m-1}$$

$$\begin{array}{r|l} & m^2 - 2m + 3\\ \hline m-1 & m^3 - 3m^2 + 5m - 3\\ & m^3 - m^2\\ \hline & -2m^2 + 5m\\ & -2m^2 + 2m\\ \hline & 3m - 3\\ & 3m - 3\\ \hline & 0 \end{array}$$

The remainder is 0. The answer is the quotient,

$$m^2 - 2m + 3.$$

CHAPTER 7 FACTORING

7.1 Greatest Common Factors; Factoring by Grouping

7.1 Margin Exercises

1. **(a)** $7k + 28 = 7 \cdot k + 7 \cdot 4 \quad GCF = 7$
$= 7(k + 4)$

Check: $7(k + 4) = 7(k) + 7(4)$
$= 7k + 28$

(b) $32m + 24 = 8 \cdot 4m + 8 \cdot 3 \quad GCF = 8$
$= 8(4m + 3)$

Check: $8(4m + 3) = 8(4m) + 8(3)$
$= 32m + 24$

(c) $8a - 9$ *cannot be factored* because $8a$ and 9 do not have a common factor other than 1.

(d) $5z + 5 = 5 \cdot z + 5 \cdot 1 \quad GCF = 5$
$= 5(z + 1)$

Check: $5(z + 1) = 5(z) + 5(1)$
$= 5z + 5$

2. **(a)** $16y^4 + 8y^3$

The numerical part of the GCF is 8, the largest number that divides into both 16 and 8. For the variable parts, y^4 and y^3, use the least exponent that appears on y; here the least exponent is 3. The GCF is $8y^3$.

$$16y^4 + 8y^3 = 8y^3(2y) + 8y^3(1) = 8y^3(2y + 1)$$

(b) $14p^2 - 9p^3 + 6p^4$

The least exponent is 2. The GCF is p^2.

$$14p^2 - 9p^3 + 6p^4 = p^2(14) + p^2(-9p) + p^2(6p^2) = p^2(14 - 9p + 6p^2)$$

(c) $15z^2 + 45z^5 - 60z^6 = 15z^2(1 + 3z^3 - 4z^4)$

Check:
$$15z^2(1 + 3z^3 - 4z^4) = 15z^2(1) + 15z^2(3z^3) + 15z^2(-4z^4) = 15z^2 + 45z^5 - 60z^6$$

(d) $4x^2z - 2xz + 8z^2 = 2z(2x^2 - x + 4z)$

Check: $2z(2x^2 - x + 4z)$
$= 2z(2x^2) + 2z(-x) + 2z(4z)$
$= 4x^2z - 2xz + 8z^2$

(e) $12y^5x^2 + 8y^3x^3$

The numerical part of the GCF is 4. The least exponent on x is 2, and the least exponent on y is 3. Therefore, the GCF is $4y^3x^2$.

$$12y^5x^2 + 8y^3x^3 = 4y^3x^2(3y^2 + 2x)$$

(f) $5m^4x^3 + 15m^5x^6 - 20m^4x^6$

The numerical part of the GCF is 5. The least exponent on m is 4, and the least exponent on x is 3. Therefore, the GCF is $5m^4x^3$.

$$5m^4x^3 + 15m^5x^6 - 20m^4x^6 = 5m^4x^3(1 + 3mx^3 - 4x^3)$$

3. **(a)** $(a + 2)(a - 3) + (a + 2)(a + 6)$

The greatest common factor here is $(a + 2)$.

$$= (a + 2)[(a - 3) + (a + 6)] = (a + 2)(a - 3 + a + 6) = (a + 2)(2a + 3)$$

(b) $(y - 1)(y + 3) - (y - 1)(y + 4)$
$= (y - 1)[(y + 3) - (y + 4)]$
$= (y - 1)(y + 3 - y - 4)$
$= (y - 1)(-1)$ or $-y + 1$

(c) $k^2(a + 5b) + m^2(a + 5b)$
$= (a + 5b)(k^2 + m^2) \qquad GCF = a + 5b$

(d) $r^2(y + 6) + r^2(y + 3)$
$= r^2[(y + 6) + (y + 3)] \quad GCF = r^2$
$= r^2(y + 6 + y + 3)$
$= r^2(2y + 9)$

4. **(a)** $-k^2 + 3k$

Factor with GCF $= k$.

$$-k^2 + 3k = k(-k) + k(3) = k(-k + 3)$$

Factor with GCF $= -k$.

$$-k^2 + 3k = -k(k) + (-k)(-3) = -k(k - 3)$$

(b) $-6r^3 - 5r^2 + 14r$

Factor with GCF $= r$.

$$-6r^3 - 5r^2 + 14r = r(-6r^2) + r(-5r) + r(14) = r(-6r^2 - 5r + 14)$$

Factor with GCF $= -r$.

$$-6r^3 - 5r^2 + 14r = -r(6r^2) + (-r)(5r) + (-r)(-14) = -r(6r^2 + 5r - 14)$$

5. $6p - 6q + rp - rq = (6p - 6q) + (rp - rq)$
$= 6(p - q) + r(p - q)$
$= (p - q)(6 + r)$

6. $xy - 2y - 4x + 8 = (xy - 2y) - (4x - 8)$
$= y(x - 2) - 4(x - 2)$
$= (x - 2)(y - 4)$

7. $2xy + 3y + 2x + 3 = (2xy + 3y) + (2x + 3)$
$= y(2x + 3) + 1(2x + 3)$
$= (2x + 3)(y + 1)$

8. **(a)** $mn + 6 + 2n + 3m$

Rearrange the terms so that there is a common factor in the first two terms and a common factor in the last two terms.

$= mn + 2n + 6 + 3m$

Group the first two terms and the last two terms.

$= (mn + 2n) + (6 + 3m)$
$= n(m + 2) + 3(2 + m)$
$= n(m + 2) + 3(m + 2)$
$= (m + 2)(n + 3)$

(b) $4y - zx + yx - 4z$
$= yx - zx + 4y - 4z$
$= (yx - zx) + (4y - 4z)$
$= x(y - z) + 4(y - z)$
$= (y - z)(x + 4)$

7.1 Section Exercises

1. Answers will vary. A sample answer follows: To factor a polynomial means to write it as the product of two or more polynomials.

2. The first step in factoring a polynomial is to find the greatest common factor and factor it out.

3. $9m^3, 3m^2, 15m$

The numerical part of the GCF is 3. The least exponent on m is 1. Thus, the GCF is $3m$.

4. $4a^2, 6ab, 2a^3$

The numerical part of the GCF is 2. The least exponent on a is 1. Thus, the GCF is $2a$.

5. $6m(r + t)^2, 3p(r + t)^4$

The numerical part of the GCF is 3. The least exponent on $(r + t)$ is 2. Thus, the GCF is $3(r + t)^2$.

6. $7z^2(m + n)^4, 9z^3(m + n)^5$

The least exponent on z is 2 and the least exponent on $(m + n)$ is 4. Thus, the GCF is $z^2(m + n)^4$.

7. $6x^3y^4 - 12x^5y^2 + 24x^4y^8$

The numerical part of the GCF is 6. The least exponent that appears on x is 3, while 2 is the least exponent on y. Thus, $6x^3y^2$ is the GCF. Therefore,

$$6x^3y^4 - 12x^5y^2 + 24x^4y^8 = 6x^3y^2\left(y^2 - 2x^2 + 4xy^6\right).$$

Choice **A** is the correct response.

8. Yes, the teacher was justified in her grading because the polynomial was not factored *completely*. The second factor can be factored further as $2(xy^2 - 2)$. Factored completely,

$$4x^2y^5 - 8xy^3 = 4xy^3(xy^2 - 2).$$

9. $10x - 30 = 10 \cdot x - 10 \cdot 3 \quad GCF = 10$
$= 10(x - 3)$

10. $15y - 60 = 15 \cdot y - 15 \cdot 4 \quad GCF = 15$
$= 15(y - 4)$

11. $8s + 16t = 8(s) + 8(2t) \quad GCF = 8$
$= 8(s + 2t)$

12. $35p + 70q = 35(p) + 35(2q) \quad GCF = 35$
$= 35(p + 2q)$

13. $6 + 12r = 6(1) + 6(2r) \quad GCF = 6$
$= 6(1 + 2r)$

14. $9 + 18m = 9(1) + 9(2m) \quad GCF = 9$
$= 9(1 + 2m)$

15. $8k^3 + 24k = 8k \cdot k^2 + 8k \cdot 3$
$= 8k(k^2 + 3)$

16. $9z^4 + 27z = 9z \cdot z^3 + 9z \cdot 3$
$= 9z\left(z^3 + 3\right)$

17. $3xy - 5xy^2 = xy \cdot 3 - xy \cdot 5y$
$= xy(3 - 5y)$

18. $5h^2j + 7hj = hj \cdot 5h + hj \cdot 7$
$= hj(5h + 7)$

19. $-4p^3q^4 - 2p^2q^5 = -2p^2q^4 \cdot 2p - 2p^2q^4 \cdot q$
$= -2p^2q^4(2p + q)$

20. $-3z^5w^2 - 18z^3w^4 = -3z^3w^2\left(z^2 + 6w^2\right)$

21. $21x^5 + 35x^4 - 14x^3$
$= 7x^3\left(3x^2\right) + 7x^3(5x) + 7x^3(-2)$
$= 7x^3\left(3x^2 + 5x - 2\right)$

22. $18k^3 - 36k^4 + 48k^5$
$= 6k^3(3) + 6k^3(-6k) + 6k^3\left(8k^2\right)$
$= 6k^3\left(3 - 6k + 8k^2\right)$

23. $15a^2c^3 - 25ac^2 + 5a^2c$
$= 5ac(3ac^2) + 5ac(-5c) + 5ac(a)$
$= 5ac(3ac^2 - 5c + a)$

24. $15y^3z^3 + 27y^2z^4 - 36yz^5$
$= 3yz^3(5y^2) + 3yz^3(9yz) + 3yz^3(-12z^2)$
$= 3yz^3(5y^2 + 9yz - 12z^2)$

25. $-27m^3p^5 + 5r^4s^3 - 8x^5z^4$ *cannot be factored* because there is no common factor (except ± 1).

26. $-50r^4t^2 + 81x^3y^3 - 49p^2q^4$ *cannot be factored* because there is no common factor (except ± 1).

27. $(m-4)(m+2) + (m-4)(m+3)$
The GCF is $(m-4)$.
$= (m-4)[(m+2) + (m+3)]$
$= (m-4)(m+2+m+3)$
$= (m-4)(2m+5)$

28. $(z-5)(z+7) + (z-5)(z+9)$
$= (z-5)[(z+7) + (z+9)]$
$= (z-5)(2z+16)$
$= 2(z-5)(z+8)$

29. $(2z-1)(z+6) - (2z-1)(z-5)$
$= (2z-1)[(z+6) - (z-5)]$
$= (2z-1)[z+6-z+5]$
$= (2z-1)(11)$
$= 11(2z-1)$

30. $(3x+2)(x-4) - (3x+2)(x+8)$
$= (3x+2)[(x-4) - (x+8)]$
$= (3x+2)(-12)$ or $-12(3x+2)$

31. $5(2-x)^3 - (2-x)^4 + 4(2-x)^2$
$= (2-x)^2[5(2-x) - (2-x)^2 + 4]$
Factor out $(2-x)^2$.
$= (2-x)^2[10 - 5x - (4 - 4x + x^2) + 4]$
$= (2-x)^2[10 - 5x - 4 + 4x - x^2 + 4]$
Clear parentheses inside brackets.
$= (2-x)^2(10 - x - x^2)$

32. $3(5-x)^4 + 2(5-x)^3 - (5-x)^2$
$= (5-x)^2[3(5-x)^2 + 2(5-x) - 1]$
$= (5-x)^2$
$\cdot [3(25 - 10x + x^2) + 10 - 2x - 1]$
$= (5-x)^2(75 - 30x + 3x^2 + 10 - 2x - 1)$
$= (5-x)^2(3x^2 - 32x + 84)$

33. $-r^3 + 3r^2 + 5r$

Factor out r.

$$-r^3 + 3r^2 + 5r = r(-r^2 + 3r + 5)$$

Factor out $-r$.

$$-r^3 + 3r^2 + 5r = -r(r^2 - 3r - 5)$$

34. $-t^4 + 8t^3 - 12t$

Factor out t.

$$-t^4 + 8t^3 - 12t = t(-t^3 + 8t^2 - 12)$$

Factor out $-t$.

$$-t^4 + 8t^3 - 12t = -t(t^3 - 8t^2 + 12)$$

35. $-12s^5 + 48s^4$

Factor out $12s^4$.

$$-12s^5 + 48s^4 = 12s^4(-s + 4)$$

Factor out $-12s^4$.

$$-12s^5 + 48s^4 = -12s^4(s - 4)$$

36. $-16y^4 + 64y^3$

Factor out $16y^3$.

$$-16y^4 + 64y^3 = 16y^3(-y + 4)$$

Factor out $-16y^3$.

$$-16y^4 + 64y^3 = -16y^3(y - 4)$$

37. $-2x^5 + 6x^3 + 4x^2$

Factor out $2x^2$.

$$-2x^5 + 6x^3 + 4x^2 = 2x^2(-x^3 + 3x + 2)$$

Factor out $-2x^2$.

$$-2x^5 + 6x^3 + 4x^2 = -2x^2(x^3 - 3x - 2)$$

38. $-5a^3 + 10a^4 - 15a^5$

Factor out $5a^3$.

$$-5a^3 + 10a^4 - 15a^5 = 5a^3(-1 + 2a - 3a^2)$$

Factor out $-5a^3$.

$$-5a^3 + 10a^4 - 15a^5 = -5a^3(1 - 2a + 3a^2)$$

39. $mx + 3qx + my + 3qy$
$= (mx + 3qx) + (my + 3qy)$
$= x(m + 3q) + y(m + 3q)$
$= (m + 3q)(x + y)$

40. $2k + 2h + jk + jh$
$= (2k + 2h) + (jk + jh)$
$= 2(k + h) + j(k + h)$
$= (k + h)(2 + j)$

41. $10m + 2n + 5mk + nk$
$= (10m + 2n) + (5mk + nk)$
$= 2(5m + n) + k(5m + n)$
$= (5m + n)(2 + k)$

42. $3ma + 3mb + 2ab + 2b^2$
$= (3ma + 3mb) + (2ab + 2b^2)$
$= 3m(a + b) + 2b(a + b)$
$= (a + b)(3m + 2b)$

43. $4 - 2q - 6p + 3pq$
$= (4 - 2q) + (-6p + 3pq)$
$= 2(2 - q) - 3p(2 - q)$
$= (2 - q)(2 - 3p)$

44. $20 + 5m + 12n + 3mn$
$= (20 + 5m) + (12n + 3mn)$
$= 5(4 + m) + 3n(4 + m)$
$= (4 + m)(5 + 3n)$

45. $p^2 - 4zq + pq - 4pz$
$= (p^2 + pq) + (-4zq - 4pz)$
$= p(p + q) - 4z(q + p)$
$= (p + q)(p - 4z)$

46. $r^2 - 9tw + 3rw - 3rt$
$= r^2 + 3rw - 3rt - 9tw$
$= (r^2 + 3rw) + (-3rt - 9tw)$
$= r(r + 3w) - 3t(r + 3w)$
$= (r + 3w)(r - 3t)$

47. $6y^2 + 9y + 4xy + 6x$
$= (6y^2 + 9y) + (4xy + 6x)$
$= 3y(2y + 3) + 2x(2y + 3)$
$= (2y + 3)(3y + 2x)$

48. $8xy + 12x^2 + 10y + 15x$
$= (8xy + 12x^2) + (10y + 15x)$
$= 4x(2y + 3x) + 5(2y + 3x)$
$= (2y + 3x)(4x + 5)$

49. $m^3 + 4m^2 - 6m - 24$
$= (m^3 + 4m^2) + (-6m - 24)$
$= m^2(m + 4) - 6(m + 4)$
$= (m + 4)(m^2 - 6)$

50. $2a^3 + a^2 - 14a - 7$
$= (2a^3 + a^2) + (-14a - 7)$
$= a^2(2a + 1) - 7(2a + 1)$
$= (2a + 1)(a^2 - 7)$

51. $-3a^3 - 3ab^2 + 2a^2b + 2b^3$
$= (-3a^3 - 3ab^2) + (2a^2b + 2b^3)$
$= -3a(a^2 + b^2) + 2b(a^2 + b^2)$
$= (a^2 + b^2)(-3a + 2b)$

52. $-16m^3 + 4m^2p^2 - 4mp + p^3$
$= (-16m^3 + 4m^2p^2) + (-4mp + p^3)$
$= 4m^2(-4m + p^2) + p(-4m + p^2)$
$= (-4m + p^2)(4m^2 + p)$

53. $4 + xy - 2y - 2x$
$= xy - 2x - 2y + 4$
$= (xy - 2x) + (-2y + 4)$
$= x(y - 2) - 2(y - 2)$
$= (y - 2)(x - 2)$

54. $2ab^2 - 4 - 8b^2 + a$
$= 2ab^2 + a - 8b^2 - 4$
$= (2ab^2 + a) + (-8b^2 - 4)$
$= a(2b^2 + 1) - 4(2b^2 + 1)$
$= (2b^2 + 1)(a - 4)$

55. $8 + 9y^4 - 6y^3 - 12y$
$= 9y^4 - 6y^3 - 12y + 8$
$= (9y^4 - 6y^3) + (-12y + 8)$
$= 3y^3(3y - 2) - 4(3y - 2)$
$= (3y - 2)(3y^3 - 4)$

56. $x^3y^2 - 3 - 3y^2 + x^3$
$= x^3y^2 - 3y^2 + x^3 - 3$
$= (x^3y^2 - 3y^2) + (x^3 - 3)$
$= y^2(x^3 - 3) + 1(x^3 - 3)$
$= (x^3 - 3)(y^2 + 1)$

57. $3m^{-5} + m^{-3}$

Factor out m^{-5} since -5 is the smaller exponent.

$= m^{-5}(3) + m^{-5}(m^{-3-(-5)})$
$= m^{-5}(3 + m^2)$ or $\dfrac{3 + m^2}{m^5}$

58. $k^{-2} + 2k^{-4}$

Factor out k^{-4} since -4 is the smaller exponent.

$= k^{-4}(k^{-2-(-4)}) + k^{-4}(2)$
$= k^{-4}(k^2 + 2)$ or $\dfrac{k^2 + 2}{k^4}$

59. $3p^{-3} + 2p^{-2}$

Factor out p^{-3} since -3 is the smaller exponent.

$= p^{-3}(3) + p^{-3}(2p^{-2-(-3)})$
$= p^{-3}(3 + 2p)$ or $\dfrac{3 + 2p}{p^3}$

7.2 Factoring Trinomials

7.2 Margin Exercises

1. **(a)** $p^2 + 6p + 5$

Step 1 Find pairs of numbers whose product is 5.	*Step 2* Write sums of those numbers.
$5(1)$	$5 + 1 = 6 \leftarrow$
$-5(-1)$	$-5 + (-1) = -6$

The coefficient of the middle term is 6, so the required numbers are 1 and 5. The factored form is

$$p^2 + 6p + 5 = (p + 1)(p + 5).$$

Check: $(p + 1)(p + 5) = p^2 + 6p + 5$

(b) $a^2 + 9a + 20$

Step 1 Find pairs of numbers whose product is 20.	*Step 2* Write sums of those numbers.
$20(1)$	$20 + 1 = 21$
$-20(-1)$	$-20 + (-1) = -21$
$10(2)$	$10 + 2 = 12$
$-10(-2)$	$-10 + (-2) = -12$
$5(4)$	$5 + 4 = 9 \leftarrow$
$-5(-4)$	$-5 + (-4) = -9$

The coefficient of the middle term is 9, so the required numbers are 5 and 4. The factored form is

$$a^2 + 9a + 20 = (a + 5)(a + 4).$$

Check: $(a + 5)(a + 4) = a^2 + 9a + 20$

(c) $k^2 - k - 6$

Step 1 Find pairs of numbers whose product is -6.	*Step 2* Write sums of those numbers.
$-6(1)$	$-6 + 1 = -5$
$6(-1)$	$6 + (-1) = 5$
$-3(2)$	$-3 + 2 = -1 \leftarrow$
$3(-2)$	$3 + (-2) = 1$

The coefficient of the middle term is -1, so the required numbers are -3 and 2. The factored form is

$$k^2 - k - 6 = (k - 3)(k + 2).$$

Check: $(k - 3)(k + 2) = k^2 - k - 6$

(d) $b^2 - 7b + 10$

Step 1 Find pairs of numbers whose product is 10.	*Step 2* Write sums of those numbers.
$10(1)$	$10 + 1 = 11$
$-10(-1)$	$-10 + (-1) = -11$
$5(2)$	$5 + 2 = 7$
$-5(-2)$	$-5 + (-2) = -7 \leftarrow$

The coefficient of the middle term is -7, so the required numbers are -5 and -2. The factored form is

$$b^2 - 7b + 10 = (b - 5)(b - 2).$$

Check: $(b - 5)(b - 2) = b^2 - 7b + 10$

(e) $y^2 - 8y + 6$

Step 1 Find pairs of numbers whose product is 6.	*Step 2* Write sums of those numbers.
$6(1)$	$6 + 1 = 7$
$-6(-1)$	$-6 + (-1) = -7$
$3(2)$	$3 + 2 = 5$
$-3(-2)$	$-3 + (-2) = -5$

The coefficient of the middle term is -8, and none of the sums equal -8. Thus, $y^2 - 8y + 6$ cannot be factored with integer coefficients and is *prime*. (Note that we have assumed that the GCF has already been factored out.)

2. **(a)** $m^2 + 2mn - 8n^2$

Look for two expressions whose product is $-8n^2$ and whose sum is $2n$. The quantities $4n$ and $-2n$ have the necessary product and sum, so

$$m^2 + 2mn - 8n^2 = (m + 4n)(m - 2n).$$

Check:

$(m + 4n)(m - 2n)$

$= m^2 - 2mn + 4mn - 8n^2$ *FOIL*

$= m^2 + 2mn - 8n^2$ *Original polynomial*

(b) $z^2 - 7zx + 9x^2$

Look for two expressions whose product is $9x^2$ and whose sum is $-7x$. There are no such quantities. Therefore, the trinomial cannot be factored and is *prime.*

3. $5m^4 - 5m^3 - 100m^2$

Start by factoring out the GCF, $5m^2$.

$5m^4 - 5m^3 - 100m^2 = 5m^2(m^2 - m - 20)$

To factor $m^2 - m - 20$, look for two integers whose product is -20 and whose sum is -1. The necessary integers are -5 and 4. Remember to write the common factor $5m^2$ as part of the answer.

$5m^4 - 5m^3 - 100m^2 = 5m^2(m-5)(m+4)$

4. **(a)** $3y^2 - 11y - 4$

The product ac is $3(-4) = -12$. Look for two integers whose product is -12 and whose sum is -11. The necessary integers are -12 and 1. Write $-11y$ as $-12y + y$ and then factor by grouping.

$$\begin{aligned} 3y^2 - 11y - 4 &= 3y^2 - 12y + y - 4 \\ &= \left(3y^2 - 12y\right) + (y - 4) \\ &= 3y(y-4) + 1(y-4) \\ &= (y-4)(3y+1) \end{aligned}$$

(b) $6k^2 - 19k + 10$

The product ac is $6(10) = 60$. Look for two integers whose product is 60 and whose sum is -19. The necessary integers are -15 and -4. Write $-19k$ as $-15k - 4k$ and then factor by grouping.

$$\begin{aligned} 6k^2 - 19k + 10 &= 6k^2 - 15k - 4k + 10 \\ &= \left(6k^2 - 15k\right) + (-4k + 10) \\ &= 3k(2k-5) - 2(2k-5) \\ &= (2k-5)(3k-2) \end{aligned}$$

5. By trial and error, the following are factored.

(a) $10x^2 + 17x + 3 = (5x+1)(2x+3)$

(b) $16y^2 - 34y - 15 = (8y+3)(2y-5)$

(c) $8t^2 - 13t + 5 = (8t-5)(t-1)$

6. **(a)** $7p^2 + 15pq + 2q^2$

Try some possibilities using $7p$ and p along with $2q$ and q. Use FOIL to check.

$(7p+2q)(p+q) = 7p^2 + 9pq + 2q^2$ *No*
$(7p+q)(p+2q) = 7p^2 + 15pq + 2q^2$ *Yes*

The correct factoring is

$$7p^2 + 15pq + 2q^2 = (7p+q)(p+2q).$$

(b) $6m^2 + 7mn - 5n^2$

Try some possibilities.

$(6m+n)(m-5n) = 6m^2 - 29mn - 5n^2$ *No*
$(6m-5n)(m+n) = 6m^2 + mn - 5n^2$ *No*
$(3m+n)(2m-5n) = 6m^2 - 13mn - 5n^2$ *No*
$(3m+5n)(2m-n) = 6m^2 + 7mn - 5n^2$ *Yes*

The correct factoring is

$$6m^2 + 7mn - 5n^2 = (3m+5n)(2m-n).$$

(c) $12z^2 - 5zy - 2y^2$

Try some possibilities.

$(12z+y)(z-2y) = 12z^2 - 23zy - 2y^2$ *No*
$(3z-2y)(4z+y) = 12z^2 - 5zy - 2y^2$ *Yes*

The correct factoring is

$$12z^2 - 5zy - 2y^2 = (3z-2y)(4z+y).$$

(d) $8m^2 + 18mx - 5x^2$

Try some possibilities.

$(8m+x)(m-5x) = 8m^2 - 39mx - 5x^2$ *No*
$(8m+5x)(m-x) = 8m^2 - 3mx - 5x^2$ *No*
$(4m+x)(2m-5x) = 8m^2 - 18mx - 5x^2$ *No*
$(4m-x)(2m+5x) = 8m^2 + 18mx - 5x^2$ *Yes*

The correct factoring is

$$8m^2 + 18mx - 5x^2 = (4m-x)(2m+5x).$$

7. **(a)** $-6r^2 + 13r + 5$

First factor out -1, then proceed.

$$\begin{aligned} -6r^2 + 13r + 5 &= -1\left(6r^2 - 13r - 5\right) \\ &= -1(2r-5)(3r+1) \\ &= -(2r-5)(3r+1) \end{aligned}$$

(b) $-8x^2 + 10x - 3$

First factor out -1, then proceed.

$$\begin{aligned} -8x^2 + 10x - 3 &= -1\left(8x^2 - 10x + 3\right) \\ &= -1(4x-3)(2x-1) \\ &= -(4x-3)(2x-1) \end{aligned}$$

8. **(a)** $2m^3 - 4m^2 - 6m$

First, factor out the GCF, $2m$.

$$= 2m(m^2 - 2m - 3)$$

Look for two integers whose product is -3 and whose sum is -2. The integers are 1 and -3.

$$= 2m(m+1)(m-3)$$

(b) $12r^4 + 6r^3 - 90r^2$

First, factor out the GCF, $6r^2$.

$$= 6r^2(2r^2 + r - 15)$$

Look for two integers whose product is $2(-15) = -30$ and whose sum is 1. The integers are 6 and -5. Write r as $6r - 5r$.

$$\begin{aligned} &= 6r^2(2r^2 + 6r - 5r - 15) \\ &= 6r^2[(2r^2 + 6r) + (-5r - 15)] \\ &= 6r^2[(2r(r+3) - 5(r+3)] \\ &= 6r^2(r+3)(2r-5) \end{aligned}$$

(c) $30y^5 - 55y^4 - 50y^3$

First, factor out the GCF, $5y^3$.

$$= 5y^3(6y^2 - 11y - 10)$$

Look for two integers whose product is $6(-10) = -60$ and whose sum is -11. The integers are 4 and -15. Write $-11y$ as $4y - 15y$.

$$\begin{aligned} &= 5y^3(6y^2 + 4y - 15y - 10) \\ &= 5y^3[(6y^2 + 4y) + (-15y - 10)] \\ &= 5y^3[2y(3y+2) - 5(3y+2)] \\ &= 5y^3(3y+2)(2y-5) \end{aligned}$$

9. **(a)** $6(a-1)^2 + (a-1) - 2$

$$\begin{aligned} &= 6x^2 + x - 2 && \textit{Let } x = a - 1. \\ &= (2x-1)(3x+2) \end{aligned}$$

Now replace x with $a - 1$.

$$\begin{aligned} &6(a-1)^2 + (a-1) - 2 \\ &= [2(a-1) - 1][3(a-1) + 2] \\ &= (2a - 2 - 1)(3a - 3 + 2) \\ &= (2a-3)(3a-1) \end{aligned}$$

(b) $8(z+5)^2 - 2(z+5) - 3$

$$\begin{aligned} &= 8x^2 - 2x - 3 && \textit{Let } x = z + 5. \\ &= (2x+1)(4x-3) \end{aligned}$$

Now replace x with $z + 5$.

$$\begin{aligned} &8(z+5)^2 - 2(z+5) - 3 \\ &= [2(z+5) + 1][4(z+5) - 3] \\ &= (2z + 10 + 1)(4z + 20 - 3) \\ &= (2z+11)(4z+17) \end{aligned}$$

(c) $15(m-4)^2 - 11(m-4) + 2$

$$\begin{aligned} &= 15x^2 - 11x + 2 && \textit{Let } x = m - 4. \\ &= (5x-2)(3x-1) \end{aligned}$$

Now replace x with $m - 4$.

$$\begin{aligned} &15(m-4)^2 - 11(m-4) + 2 \\ &= [5(m-4) - 2][3(m-4) - 1] \\ &= (5m - 20 - 2)(3m - 12 - 1) \\ &= (5m-22)(3m-13) \end{aligned}$$

10. **(a)** $y^4 + y^2 - 6$

$$\begin{aligned} &= (y^2)^2 + y^2 - 6 \\ &= x^2 + x - 6 && \textit{Let } x = y^2. \\ &= (x-2)(x+3) && \textit{Factor.} \\ &= (y^2-2)(y^2+3) && x = y^2 \end{aligned}$$

(b) $2p^4 + 7p^2 - 15$

$$\begin{aligned} &= 2(p^2)^2 + 7p^2 - 15 \\ &= 2x^2 + 7x - 15 && \textit{Let } x = p^2. \\ &= (2x-3)(x+5) && \textit{Factor.} \\ &= (2p^2-3)(p^2+5) && x = p^2 \end{aligned}$$

(c) $6r^4 - 13r^2 + 5$

$$\begin{aligned} &= 6(r^2)^2 - 13r^2 + 5 \\ &= 6x^2 - 13x + 5 && \textit{Let } x = r^2. \\ &= (3x-5)(2x-1) && \textit{Factor.} \\ &= (3r^2-5)(2r^2-1) && x = r^2 \end{aligned}$$

7.2 Section Exercises

1. **D** is not valid.

$$(8x)(4x) = 32x^2 \neq 12x^2$$

2. **A** is the completely factored form.

$$\begin{aligned} x^4(2x+1)(x-3) &= x^4(2x^2 - 5x - 3) \\ &= 2x^6 - 5x^5 - 3x^4 \end{aligned}$$

3. **C** is the completely factored form.

$$\begin{aligned} 4(x+2)(x-3) &= 4(x^2 - x - 6) \\ &= 4x^2 - 4x - 24 \end{aligned}$$

4. **B** is not a factored form.

$$\begin{aligned} (-x-10)(x+6) &= -x^2 - 16x - 60 \\ &\neq -x^2 + 16x - 60 \end{aligned}$$

5. To factor $y^2 + 7y - 30$, we need two integer factors whose sum is 7 (coefficient of the middle term) and whose product is -30 (the last term). Since $-3 + 10 = 7$ and $-3 \cdot 10 = -30$, we have

$$y^2 + 7y - 30 = (y-3)(y+10).$$

6. To factor $z^2 + 2z - 24$, find two integers whose product is -24 and whose sum is 2. The numbers are 6 and -4.

$$z^2 + 2z - 24 = (z + 6)(z - 4)$$

7. $p^2 - p - 56$

Look for two integers whose product is -56 and whose sum is -1. Since the sum is negative, we'll restrict the products in the following table to those that have negative sums.

Products	Sums
$-56 \cdot 1$	-55
$-28 \cdot 2$	-26
$-14 \cdot 4$	-10
$-8 \cdot 7$	$-1 \leftarrow$

Thus, $p^2 - p - 56 = (p - 8)(p + 7)$.

8. $k^2 - 11k + 30$

Two integer factors whose product is 30 and whose sum is -11 are -6 and -5.

$$k^2 - 11k + 30 = (k - 6)(k - 5)$$

9. $-m^2 + 16m - 60$

First factor -1 out of the trinomial to get a trinomial whose coefficient of the squared term is 1.

$$-m^2 + 16m - 60 = -(m^2 - 16m + 60)$$

To factor $m^2 - 16m + 60$, look for two integers whose product is 60 and whose sum is -16. Since $-10 \cdot (-6) = 60$ and $-10 + (-6) = -16$, we have

$$m^2 - 16m + 60 = (m - 10)(m - 6).$$

Thus, $-m^2 + 16m - 60 = -(m - 10)(m - 6)$.

10. $-p^2 + 6p + 27$

First factor -1 out of the trinomial to get a trinomial whose coefficient of the squared term is 1.

$$-p^2 + 6p + 27 = -(p^2 - 6p - 27)$$

To factor $p^2 - 6p - 27$, look for two integers whose product is -27 and whose sum is -6. The integers are 3 and -9.

Thus, $-p^2 + 6p + 27 = -(p + 3)(p - 9)$.

11. $a^2 - 2ab - 35b^2$

Two integer factors whose product is -35 and whose sum is -2 are 5 and -7.

$$a^2 - 2ab - 35b^2 = (a + 5b)(a - 7b)$$

12. $z^2 + 8zw + 15w^2$

Two integer factors whose product is 15 and whose sum is 8 are 3 and 5.

$$z^2 + 8zw + 15w^2 = (z + 3w)(z + 5w)$$

13. $y^2 - 3yq - 15q^2$

There are no integer factors of -15 that add up to -3, so this trinomial is *prime.*

14. $k^2 - 11hk + 28h^2$

Two integer factors whose product is 28 and whose sum is -11 are -4 and -7.

$$k^2 - 11hk + 28h^2 = (k - 4h)(k - 7h)$$

15. $x^2y^2 + 11xy + 18$

Two integer factors whose product is 18 and whose sum is 11 are 9 and 2.

$$x^2y^2 + 11xy + 18 = (xy + 9)(xy + 2)$$

16. $p^2q^2 - 5pq - 18$

There are no integer factors of -18 that add up to -5, so this trinomial is *prime.*

17. $-6m^2 - 13m + 15 = -1(6m^2 + 13m - 15)$

We'll try to factor $6m^2 + 13m - 15$.
Multiply the first and last coefficients to get $6(-15) = -90$.
Two integer factors whose product is -90 and whose sum is 13 are -5 and 18.
Rewrite the trinomial in a form that can be factored by grouping.

$$\begin{aligned} &6m^2 + (+13m) - 15 \\ &= 6m^2 + (-5m + 18m) - 15 \\ &= (6m^2 - 5m) + (18m - 15) \\ &= m(6m - 5) + 3(6m - 5) \\ &= (6m - 5)(m + 3) \end{aligned}$$

Thus, the final factored form is

$$-1(6m - 5)(m + 3).$$

Note: These exercises can be worked using the alternative method of repeated combinations and FOIL or the grouping method.

18. $-15y^2 + 17y + 18 = -1(15y^2 - 17y - 18)$

We'll try to factor $15y^2 - 17y - 18$.
Two integer factors whose product is $(15)(-18) = -270$ and whose sum is -17 are 10 and -27.
Rewrite the trinomial in a form that can be factored by grouping.

$= 15y^2 + 10y - 27y - 18$
$= 5y(3y + 2) - 9(3y + 2)$
$= (3y + 2)(5y - 9)$

Thus, the final factored form is

$$-1(3y + 2)(5y - 9).$$

19. $10x^2 + 3x - 18$
Two integer factors whose product is $(10)(-18) = -180$ and whose sum is 3 are 15 and -12.
Rewrite the trinomial in a form that can be factored by grouping.

$10x^2 + 3x - 18$
$= 10x^2 + 15x - 12x - 18$
$= 5x(2x + 3) - 6(2x + 3)$
$= (2x + 3)(5x - 6)$

20. $8k^2 + 34k + 35$
Two integer factors whose product is $(8)(35) = 280$ and whose sum is 34 are 14 and 20.
Rewrite the trinomial in a form that can be factored by grouping.

$8k^2 + 34k + 35$
$= 8k^2 + 14k + 20k + 35$
$= 2k(4k + 7) + 5(4k + 7)$
$= (4k + 7)(2k + 5)$

21. $20k^2 + 47k + 24$
Two integer factors whose product is $(20)(24) = 480$ and whose sum is 47 are 15 and 32.
Rewrite the trinomial in a form that can be factored by grouping.

$= 20k^2 + 15k + 32k + 24$
$= 5k(4k + 3) + 8(4k + 3)$
$= (4k + 3)(5k + 8)$

22. $27z^2 + 42z - 5$
Two integer factors whose product is $(27)(-5) = -135$ and whose sum is 42 are -3 and 45.

$= 27z^2 - 3z + 45z - 5$
$= 3z(9z - 1) + 5(9z - 1)$
$= (9z - 1)(3z + 5)$

23. $15a^2 - 22ab + 8b^2$
Two integer factors whose product is $(15)(8) = 120$ and whose sum is -22 are -10 and -12.
Rewrite the trinomial in a form that can be factored by grouping.

$= 15a^2 - 10ab - 12ab + 8b^2$
$= 5a(3a - 2b) - 4b(3a - 2b)$
$= (3a - 2b)(5a - 4b)$

24. $15p^2 + 24pq + 8q^2$
There are no integer factors of $(15)(8) = 120$ that add up to 24, so this trinomial is *prime*.

25. $36m^2 - 60m + 25$
Use the alternative method and write $36m^2$ as $6m \cdot 6m$ and 25 as $5 \cdot 5$. Use these factors in the binomial factors to obtain

$$36m^2 - 60m + 25 = (6m - 5)(6m - 5)$$
$$= (6m - 5)^2.$$

26. $25r^2 - 90r + 81$
Use the alternative method and write $25r^2$ as $5r \cdot 5r$ and 81 as $9 \cdot 9$. Use these factors in the binomial factors to obtain

$$25r^2 - 90r + 81 = (5r - 9)(5r - 9)$$
$$= (5r - 9)^2.$$

27. $40x^2 + xy + 6y^2$
There are no integer factors of $(40)(6) = 240$ that add up to 1, so this trinomial is *prime*.

28. $14c^2 - 17cd - 6d^2$
Two integer factors whose product is $(14)(-6) = -84$ and whose sum is -17 are 4 and -21.

$= 14c^2 + 4cd - 21cd - 6d^2$
$= 2c(7c + 2d) - 3d(7c + 2d)$
$= (7c + 2d)(2c - 3d)$

29. $6x^2z^2 + 5xz - 4$
Two integer factors whose product is $(6)(-4) = -24$ and whose sum is 5 are 8 and -3.
Rewrite the trinomial in a form that can be factored by grouping.

$= 6x^2z^2 + 8xz - 3xz - 4$
$= 2xz(3xz + 4) - 1(3xz + 4)$
$= (3xz + 4)(2xz - 1)$

30. $8m^2n^2 - 10mn + 3$
$= (4mn - 3)(2mn - 1)$
by the alternative method.

31. $24x^2 + 42x + 15$

Always factor out the GCF first.

$= 3(8x^2 + 14x + 5)$

Now factor $8x^2 + 14x + 5$ by the alternative method.

$$8x^2 + 14x + 5 = (4x + 5)(2x + 1)$$

The final factored form is

$$3(4x + 5)(2x + 1).$$

32. $36x^2 + 18x - 4$
$= 2(18x^2 + 9x - 2)$
$= 2(6x - 1)(3x + 2)$

33. $-15a^2 - 70a + 120$
$= -5(3a^2 + 14a - 24)$
$= -5(a + 6)(3a - 4)$

34. $-12a^2 - 10a + 42$
$= -2(6a^2 + 5a - 21)$
$= -2(3a + 7)(2a - 3)$

35. $-11x^3 + 110x^2 - 264x$
$= -11x(x^2 - 10x + 24)$
$= -11x(x - 6)(x - 4)$

36. $-9k^3 - 36k^2 + 189k$
$= -9k(k^2 + 4k - 21)$
$= -9k(k + 7)(k - 3)$

37. $2x^3y^3 - 48x^2y^4 + 288xy^5$
$= 2xy^3(x^2 - 24xy + 144y^2)$
$= 2xy^3(x - 12y)(x - 12y)$
or $2xy^3(x - 12y)^2$

38. $6m^3n^2 - 24m^2n^3 - 30mn^4$
$= 6mn^2(m^2 - 4mn - 5n^2)$
$= 6mn^2(m - 5n)(m + n)$

39. $10(k + 1)^2 - 7(k + 1) + 1$

Let $x = k + 1$ to obtain

$$10x^2 - 7x + 1 = (5x - 1)(2x - 1).$$

Replace x with $k + 1$.

$10(k + 1)^2 - 7(k + 1) + 1$
$= [5(k + 1) - 1][2(k + 1) - 1]$
$= (5k + 5 - 1)(2k + 2 - 1)$
$= (5k + 4)(2k + 1)$

40. $4(m - 5)^2 - 4(m - 5) - 15$

Let $x = m - 5$ to obtain

$$4x^2 - 4x - 15 = (2x - 5)(2x + 3).$$

Replace x with $m - 5$.

$4(m - 5)^2 - 4(m - 5) - 15$
$= [2(m - 5) - 5][2(m - 5) + 3]$
$= (2m - 10 - 5)(2m - 10 + 3)$
$= (2m - 15)(2m - 7)$

41. $3(m + p)^2 - 7(m + p) - 20$

Let $x = m + p$ to obtain

$$3x^2 - 7x - 20 = (3x + 5)(x - 4).$$

Replace x with $m + p$.

$3(m + p)^2 - 7(m + p) - 20$
$= [3(m + p) + 5][(m + p) - 4]$
$= (3m + 3p + 5)(m + p - 4)$

42. $4(x - y)^2 - 23(x - y) - 6$
$= 4m^2 - 23m - 6$ *Let $m = x - y$.*
$= (4m + 1)(m - 6)$
$= [4(x - y) + 1][(x - y) - 6]$ *Resubstitute.*
$= (4x - 4y + 1)(x - y - 6)$

43. $a^2(a + b)^2 - ab(a + b)^2 - 6b^2(a + b)^2$

Factor out the GCF, $(a + b)^2$.

$= (a + b)^2(a^2 - ab - 6b^2)$

Factor the trinomial.

$= (a + b)^2(a - 3b)(a + 2b)$

44. $m^2(m - p) + mp(m - p) - 2p^2(m - p)$

Factor out the GCF, $m - p$.

$= (m - p)(m^2 + mp - 2p^2)$

Factor the trinomial.

$= (m - p)(m - p)(m + 2p)$
$= (m - p)^2(m + 2p)$

45. In $2x^4 - 9x^2 - 18$, let $y = x^2$ to obtain

$$2y^2 - 9y - 18 = (2y + 3)(y - 6).$$

Replace y with x^2.

$$2x^4 - 9x^2 - 18 = (2x^2 + 3)(x^2 - 6)$$

46. In $6z^4 + z^2 - 1$, let $x = z^2$ to obtain

$$6x^2 + x - 1 = (3x - 1)(2x + 1).$$

Replace x with z^2.

$$6z^4 + z^2 - 1 = (3z^2 - 1)(2z^2 + 1)$$

47. In $16x^4 + 16x^2 + 3$, let $m = x^2$ to obtain

$$16m^2 + 16m + 3 = (4m + 3)(4m + 1).$$

Replace m with x^2.

$$16x^4 + 16x^2 + 3 = (4x^2 + 3)(4x^2 + 1)$$

48. $9r^4 + 9r^2 + 2 = (3r^2 + 1)(3r^2 + 2)$

49. In $12p^6 - 32p^3r + 5r^2$, let $x = p^3$ to obtain

$$12x^2 - 32xr + 5r^2 = (6x - r)(2x - 5r).$$

Replace x with p^3.

$$12p^6 - 32p^3r + 5r^2 = \left(6p^3 - r\right)\left(2p^3 - 5r\right)$$

50. $2y^6 + 7xy^3 + 6x^2$

$$\begin{aligned} &= 2m^2 + 7xm + 6x^2 \qquad \textit{Let } m = y^3. \\ &= (2m + 3x)(m + 2x) \\ &= \left(2y^3 + 3x\right)\left(y^3 + 2x\right) \quad \textit{Resubstitute.} \end{aligned}$$

51. No, 2 is not a factor of 45 since 45 is an odd number.

52. The positive factors of 45 are 1 and 45, 3 and 15, and 5 and 9. No, 2 is not a factor of any of these factors.

53. No, 5 is not a factor of $10x^2 + 29x + 10$ since it is not a factor of 29.

54. $10x^2 + 29x + 10$

Look for two integers whose product is $10(10) = 100$ and whose sum is 29. The integers are 4 and 25. Write $29x$ as $4x + 25x$.

$$\begin{aligned} 10x^2 + 29x + 10 &= 10x^2 + 4x + 25x + 10 \\ &= \left(10x^2 + 4x\right) + (25x + 10) \\ &= 2x(5x + 2) + 5(5x + 2) \\ &= (5x + 2)(2x + 5) \end{aligned}$$

Since 5 is not a factor of 2, 5 is not a factor of $5x + 2$ or $2x + 5$.

55. Since k is odd, 2 is not a factor of $2x^2 + kx + 8$, and because 2 is a factor of $2x + 4$, the binomial $2x + 4$ cannot be a factor.

56. $3y + 15$ cannot be a factor of $12y^2 - 11y - 15$ because 3 is a factor of $3y + 15$, but 3 is not a factor of $12y^2 - 11y - 15$.

7.3 Special Factoring

7.3 Margin Exercises

1. Use the difference of squares formula,

$$x^2 - y^2 = (x + y)(x - y).$$

(a) $$\begin{aligned} 2x^2 - 18 &= 2\left(x^2 - 9\right) \qquad GCF = 2 \\ &= 2\left(x^2 - 3^2\right) \\ &= 2(x + 3)(x - 3) \end{aligned}$$

(b) $$\begin{aligned} 9a^2 - 16b^2 &= (3a)^2 - (4b)^2 \\ &= (3a + 4b)(3a - 4b) \end{aligned}$$

(c) $(m + 3)^2 - 49z^2$

$$\begin{aligned} &= (m + 3)^2 - (7z)^2 \\ &= (m + 3 + 7z)(m + 3 - 7z) \end{aligned}$$

(d) $$\begin{aligned} y^4 - 16 &= \left(y^2\right)^2 - 4^2 \\ &= \left(y^2 + 4\right)\left(y^2 - 4\right)) \\ &= \left(y^2 + 4\right)\left(y^2 - 2^2\right) \\ &= \left(y^2 + 4\right)(y + 2)(y - 2) \end{aligned}$$

2. **(a)** $z^2 + 12z + 36 = z^2 + 12z + 6^2$

If this is a perfect square trinomial, the middle term must be

$$2(z)(6) = 12z.$$

Since this is true, $z^2 + 12z + 36$ is a perfect square trinomial.

(b) $2x^2 - 4x + 4$

This is not a perfect square trinomial since $2x^2$ is not a perfect square.

(c) $9a^2 + 12ab + 16b^2 = (3a)^2 + 12ab + (4b)^2$

If this is a perfect square trinomial, the middle term must be

$$2(3a)(4b) = 24ab.$$

Since the middle term is $12ab$, the trinomial is not a perfect square trinomial.

3. **(a)** $$\begin{aligned} 49z^2 - 14zk + k^2 &= (7z)^2 - 14zk + k^2 \\ &= (7z - k)^2 \end{aligned}$$

Check: $2(7z)(-k) = -14zk$, which is the middle term.

Thus, $49z^2 - 14zk + k^2 = (7z - k)^2$.

(b) $$\begin{aligned} 9a^2 + 48ab + 64b^2 &= (3a)^2 + 48ab + (8b)^2 \\ &= (3a + 8b)^2 \end{aligned}$$

Check: $2(3a)(8b) = 48ab$, which is the middle term.

Thus, $9a^2 + 48ab + 64b^2 = (3a + 8b)^2$.

(c) $(k + m)^2 - 12(k + m) + 36$

$$\begin{aligned} &= [(k + m) - 6]^2 \\ &= (k + m - 6)^2 \end{aligned}$$

Check: $2(k + m)(-6) = -12(k + m)$, which is the middle term.

Thus, $(k + m)^2 - 12(k + m) + 36$
$= (k + m - 6)^2$.

(d) $x^2 - 2x + 1 - y^2$

Group the first three terms.

$= (x^2 - 2x + 1) - y^2$

Factor the perfect square trinomial.

$= (x - 1)^2 - y^2$

This is the difference of two squares.

$$= [(x-1) + y][(x-1) - y]$$
$$= (x - 1 + y)(x - 1 - y)$$

4. Use the difference of cubes formula,

$$x^3 - y^3 = (x - y)(x^2 + xy + y^2).$$

(a) $x^3 - 1000 = x^3 - 10^3$
$= (x - 10)(x^2 + 10x + 10^2)$
$= (x - 10)(x^2 + 10x + 100)$

(b) $8k^3 - y^3 = (2k)^3 - y^3$
$= (2k - y)\left[(2k)^2 + (2k)(y) + y^2\right]$
$= (2k - y)(4k^2 + 2ky + y^2)$

(c) $27m^3 - 64$
$= (3m)^3 - 4^3$
$= (3m - 4)\left[(3m)^2 + (3m)(4) + 4^2\right]$
$= (3m - 4)(9m^2 + 12m + 16)$

5. Use the sum of cubes formula,

$$x^3 + y^3 = (x + y)(x^2 - xy + y^2).$$

(a) $8p^3 + 125 = (2p)^3 + 5^3$
$= (2p + 5)\left[(2p)^2 - (2p)(5) + 5^2\right]$
$= (2p + 5)(4p^2 - 10p + 25)$

(b) $27m^3 + 125n^3$
$= (3m)^3 + (5n)^3$
$= (3m + 5n)\left[(3m)^2 - (3m)(5n) + (5n)^2\right]$
$= (3m + 5n)(9m^2 - 15mn + 25n^2)$

(c) $2x^3 + 2000$
$= 2(x^3 + 1000)$ *GCF = 2*
$= 2(x^3 + 10^3)$
$= 2(x + 10)(x^2 - 10x + 10^2)$
$= 2(x + 10)(x^2 - 10x + 100)$

7.3 Section Exercises

1. **A.** Yes, 64 and m^2 are squares.

B. No, $2x^2$ is not a square.

C. No, $k^2 + 9$ is a *sum* of squares.

D. Yes, $4z^2$ and 49 are squares.

So, the binomials that are differences of squares are **A** and **D**.

2. **A.** $64 + y^3$ is the sum of cubes, $4^3 + y^3$.

B. $125 - p^6$ is the difference of cubes, $5^3 - (p^2)^3$.

C. $9x^3 + 125$ is neither the sum nor the difference of cubes since 9 is not a perfect cube.

D. $(x + y)^3 - 1$ is the difference of cubes, $(x + y)^3 - 1^3$.

Choices **A**, **B**, and **D** are the sums or differences of cubes.

3. **A.** Since $x^2 - 8x - 16$ has a negative third term, it is not a perfect square trinomial.

B. $4m^2 + 20m + 25 = (2m)^2 + 20m + 5^2$

$2(2m)(5) = 20m$, the middle term. Therefore, this trinomial is a perfect square.

C. $9z^4 + 30z^2 + 25 = (3z^2)^2 + 30z^2 + 5^2$

$2(3z^2)(5) = 30z^2$, the middle term. Therefore, this trinomial is a perfect square.

D. $25a^2 - 45a + 81 = (5a)^2 - 45a + (-9)^2$
$2(5a)(-9) = -90a$

This is not the middle term, so the trinomial is not a perfect square.

So, the perfect square trinomials are **B** and **C**.

4. In Exercise 1, choices **A** and **D** are differences of squares. In Exercise 2, choices **A**, **B**, and **D** are sums or differences of cubes. In Exercise 3, choices **B** and **C** are perfect square trinomials. Therefore, **1A**, **1D**, **2A**, **2B**, **2D**, **3B**, and **3C** can be factored using the methods of this section.

5. The sum of two squares can be factored only if the binomial has a common factor. For example,

$$9x^2 + 81 = 9(x^2 + 9).$$

6. **(a)** $8 + t^3 = 2^3 + t^3 = (2 \underline{+} t)(4 \underline{-} 2t \underline{+} t^2)$

(b) $z^3 - 1 = z^3 - 1^3 = (z \underline{-} 1)(z^2 \underline{+} z \underline{+} 1)$

7. $p^2 - 16 = p^2 - 4^2$
$= (p + 4)(p - 4)$

8. $k^2 - 9 = k^2 - 3^2$
$= (k+3)(k-3)$

9. $25x^2 - 4 = (5x)^2 - 2^2$
$= (5x+2)(5x-2)$

10. $36m^2 - 25 = (6m)^2 - 5^2$
$= (6m+5)(6m-5)$

11. $18a^2 - 98b^2 = 2(9a^2 - 49b^2)$
$= 2[(3a)^2 - (7b)^2]$
$= 2[(3a+7b)(3a-7b)]$
$= 2(3a+7b)(3a-7b)$

12. $32c^2 - 98d^2 = 2(16c^2 - 49d^2)$
$= 2[(4c)^2 - (7d)^2]$
$= 2[(4c+7d)(4c-7d)]$
$= 2(4c+7d)(4c-7d)$

13. $64m^4 - 4y^4$
$= 4(16m^4 - y^4)$
$= 4\left[(4m^2)^2 - (y^2)^2\right]$
Factor the difference of squares.
$= 4(4m^2 + y^2)(4m^2 - y^2)$
$= 4(4m^2 + y^2)[(2m)^2 - y^2]$
Factor the difference of squares again.
$= 4(4m^2 + y^2)(2m+y)(2m-y)$

14. $243x^4 - 3t^4$
$= 3(81x^4 - t^4)$
$= 3\left[(9x^2)^2 - (t^2)^2\right]$
$= 3(9x^2 + t^2)(9x^2 - t^2)$
$= 3(9x^2 + t^2)[(3x)^2 - t^2]$
$= 3(9x^2 + t^2)(3x+t)(3x-t)$

15. $(y+z)^2 - 81$
$= (y+z)^2 - 9^2$
$= [(y+z)+9][(y+z)-9]$
$= (y+z+9)(y+z-9)$

16. $(h+k)^2 - 9 = (h+k)^2 - 3^2$
$= (h+k+3)(h+k-3)$

17. $16 - (x+3y)^2$
$= 4^2 - z^2$ *Let $z = (x+3y)$.*
$= (4+z)(4-z)$
Substitute $x + 3y$ for z.
$= [4 + (x+3y)][4 - (x+3y)]$
$= (4+x+3y)(4-x-3y)$

18. $64 - (r+2t)^2$
$= 8^2 - (r+2t)^2$
$= [8 + (r+2t)][8 - (r+2t)]$
$= (8+r+2t)(8-r-2t)$

19. $(p+q)^2 - (p-q)^2$
$= [(p+q) + (p-q)]$
$\cdot [(p+q) - (p-q)]$
$= (p+q+p-q)(p+q-p+q)$
$= 2p(2q)$
$= 4pq$

20. $(a+b)^2 - (a-b)^2$
$= [(a+b) + (a-b)]$
$\cdot [(a+b) - (a-b)]$
$= (a+b+a-b)(a+b-a+b)$
$= 2a(2b)$
$= 4ab$

21. $k^2 - 6k + 9 = (k)^2 - 2(k)(3) + 3^2$
$= (k-3)^2$

22. $x^2 + 10x + 25 = x^2 + 2(x)(5) + 5^2$
$= (x+5)^2$

23. $4z^2 + 4zw + w^2$
$= (2z)^2 + 2(2z)(w) + w^2$
$= (2z+w)^2$

24. $9y^2 + 6yz + z^2 = (3y)^2 + 2(3y)(z) + z^2$
$= (3y+z)^2$

25. $16m^2 - 8m + 1 - n^2$
Group the first three terms.
$= (16m^2 - 8m + 1) - n^2$
$= [(4m)^2 - 2(4m)(1) + 1^2] - n^2$
$= (4m-1)^2 - n^2$
$= [(4m-1)+n][(4m-1)-n]$
$= (4m-1+n)(4m-1-n)$

26. $25c^2 - 20c + 4 - d^2$
$= [(5c)^2 - 2(5c)(2) + (2)^2] - d^2$
$= (5c-2)^2 - d^2$
$= (5c-2+d)(5c-2-d)$

27. $4r^2 - 12r + 9 - s^2$
Group the first three terms.
$= (4r^2 - 12r + 9) - s^2$
$= [(2r)^2 - 2(2r)(3) + 3^2] - s^2$
$= (2r-3)^2 - s^2$
$= [(2r-3)+s][(2r-3)-s]$
$= (2r-3+s)(2r-3-s)$

28. $9a^2 - 24a + 16 - b^2$
$= [(3a)^2 - 2(3a)(4) + (4)^2] - b^2$
$= (3a-4)^2 - b^2$
$= (3a-4+b)(3a-4-b)$

29. $x^2 - y^2 + 2y - 1$

Group the last three terms.

$$= x^2 - (y^2 - 2y + 1)$$
$$= x^2 - (y-1)^2$$
$$= [x + (y-1)][x - (y-1)]$$
$$= (x + y - 1)(x - y + 1)$$

30. $-k^2 - h^2 + 2kh + 4$

$$= -(k^2 + h^2 - 2kh - 4)$$
$$= -[(k^2 - 2kh + h^2) - 2^2]$$
$$= -[(k-h)^2 - 2^2]$$
$$= -(k - h + 2)(k - h - 2)$$
$$= (2 + k - h)(-1)(k - h - 2)$$
$$= (2 + k - h)(2 - k + h)$$

31. $98m^2 + 84mn + 18n^2$

$$= 2(49m^2 + 42mn + 9n^2)$$
$$= 2[(7m)^2 + 2(7m)(3n) + (3n)^2]$$
$$= 2(7m + 3n)^2$$

32. $80z^2 - 40zw + 5w^2$

$$= 5(16z^2 - 8zw + w^2)$$
$$= 5[(4z)^2 - 2(4z)(w) + (w)^2]$$
$$= 5(4z - w)^2$$

33. $(p+q)^2 + 2(p+q) + 1$

$$= x^2 + 2x + 1 \qquad \textit{Let } x = p + q.$$
$$= (x+1)^2$$
$$= (p + q + 1)^2 \qquad \textit{Resubstitute.}$$

34. $(x+y)^2 + 6(x+y) + 9$

$$= (x+y)^2 + 2(x+y)(3) + 3^2$$
$$= (x + y + 3)^2$$

35. $(a-b)^2 + 8(a-b) + 16$

$$= (a-b)^2 + 2(a-b)(4) + 4^2$$
$$= [(a-b) + 4]^2$$
$$= (a - b + 4)^2$$

36. $(m-n)^2 + 4(m-n) + 4$

$$= (m-n)^2 + 2(m-n)(2) + 2^2$$
$$= (m - n + 2)^2$$

37. $8x^3 - y^3$

$$= (2x)^3 - y^3$$
$$= [2x - y][(2x)^2 + (2x)(y) + y^2]$$
$$= (2x - y)(4x^2 + 2xy + y^2)$$

38. $z^3 + 125p^3$

$$= z^3 + (5p)^3$$
$$= (z + 5p)[z^2 - z(5p) + (5p)^2]$$
$$= (z + 5p)(z^2 - 5zp + 25p^2)$$

39. $64g^3 + 27h^3$

$$= (4g)^3 + (3h)^3$$
$$= (4g + 3h)[(4g)^2 - (4g)(3h) + (3h)^2]$$
$$= (4g + 3h)(16g^2 - 12gh + 9h^2)$$

40. $27a^3 - 8b^3$

$$= (3a)^3 - (2b)^3$$
$$= (3a - 2b)[(3a)^2 + (3a)(2b) + (2b)^2]$$
$$= (3a - 2b)(9a^2 + 6ab + 4b^2)$$

41. $24n^3 + 81p^3$

$$= 3(8n^3 + 27p^3)$$
$$= 3[(2n)^3 + (3p)^3]$$
$$= 3[2n + 3p][(2n)^2 - (2n)(3p) + (3p)^2]$$
$$= 3(2n + 3p)(4n^2 - 6np + 9p^2)$$

42. $250x^3 - 16y^3$

$$= 2(125x^3 - 8y^3)$$
$$= 2[(5x)^3 - (2y)^3]$$
$$= 2(5x - 2y)[(5x)^2 + (5x)(2y) + (2y)^2]$$
$$= 2(5x - 2y)(25x^2 + 10xy + 4y^2)$$

43. $(y+z)^3 - 64$

$$= (y+z)^3 - 4^3$$
$$= [(y+z) - 4][(y+z)^2 + (y+z)(4) + 4^2]$$
$$= (y + z - 4)(y^2 + 2yz + z^2 + 4y + 4z + 16)$$

44. $(p-q)^3 + 125$

$$= (p-q)^3 + 5^3$$
$$= [(p-q) + 5] \cdot [(p-q)^2 - (p-q)(5) + 5^2]$$
$$= (p - q + 5)$$
$$\cdot (p^2 - 2pq + q^2 - 5p + 5q + 25)$$

45. $m^6 - 125 = (m^2)^3 - (5)^3$

$$= [m^2 - 5]\left[(m^2)^2 + (m^2)(5) + 5^2\right]$$
$$= (m^2 - 5)(m^4 + 5m^2 + 25)$$

46. $x^6 + 729 = (x^2)^3 + 9^3$

$$= (x^2 + 9)\left[(x^2)^2 - (x^2)(9) + 9^2\right]$$
$$= (x^2 + 9)(x^4 - 9x^2 + 81)$$

47. $k^6 + (k+3)^3$

$$= (k^2)^3 + (k+3)^3$$
$$= [k^2 + (k+3)] \cdot \left[(k^2)^2 - k^2(k+3) + (k+3)^2\right]$$
$$= (k^2 + k + 3) \cdot (k^4 - k^3 - 3k^2 + k^2 + 6k + 9)$$
$$= (k^2 + k + 3)(k^4 - k^3 - 2k^2 + 6k + 9)$$

48. $(a+b)^3-(a-b)^3$
$= [(a+b)-(a-b)]$
$\cdot\left[(a+b)^2+(a+b)(a-b)+(a-b)^2\right]$
$= (a+b-a+b)$
$\cdot\left[\left(a^2+2ab+b^2\right)+\left(a^2-b^2\right)+\left(a^2-2ab+b^2\right)\right]$
$= 2b\left(3a^2+b^2\right)$

49. x^6-y^6
$= \left(x^3\right)^2-\left(y^3\right)^2$
$= \left(x^3+y^3\right)\left(x^3-y^3\right)$
$= \left[(x+y)\left(x^2-xy+y^2\right)\right]$
$\cdot\left[(x-y)\left(x^2+xy+y^2\right)\right]$
$= (x+y)\left(x^2-xy+y^2\right)(x-y)$
$\cdot\left(x^2+xy+y^2\right)$

50. $x^6-y^6 = (x-y)(x+y)$
$\cdot\ \underline{\left(x^2+xy+y^2\right)\left(x^2-xy+y^2\right)}$

51. x^6-y^6
$= \left(x^2\right)^3-\left(y^2\right)^3$
$= \left(x^2-y^2\right)\left(x^4+x^2y^2+y^4\right)$
$= (x+y)(x-y)\left(x^4+x^2y^2+y^4\right)$

52. $x^6-y^6 = (x-y)(x+y)$
$\cdot\ \underline{\left(x^4+x^2y^2+y^4\right)}$

53. The product written on the blank in Exercise 50 must equal the product written on the blank in Exercise 52. To verify this, multiply the two factors written in Exercise 50.

$$\begin{aligned}&\left(x^2+xy+y^2\right)\left(x^2-xy+y^2\right)\\&= x^2\left(x^2-xy+y^2\right)\\&\quad+xy\left(x^2-xy+y^2\right)\\&\quad+y^2\left(x^2-xy+y^2\right)\\&= x^4-x^3y+x^2y^2+x^3y-x^2y^2\\&\quad+xy^3+x^2y^2-xy^3+y^4\\&= x^4+x^2y^2+y^4\end{aligned}$$

They are equal.

54. Start by factoring as the difference of squares since doing so resulted in the complete factorization more directly.

Summary Exercises on Factoring

1. $100a^2-9b^2$
$= (10a)^2-(3b)^2$ *Difference of squares*
$= (10a+3b)(10a-3b)$

2. $10r^2+13r-3$

Two integer factors whose product is $(10)(-3)=-30$ and whose sum is 13 are -2 and 15.

$= 10r^2-2r+15r-3$
$= 2r(5r-1)+3(5r-1)$
$= (5r-1)(2r+3)$

3. $18p^5-24p^3+12p^6 = 6p^3(3p^2-4+2p^3)$

4. $15x^2-20x = 5x(3x-4)$

5. $x^2+2x-35$

Look for two integers whose product is -35 and whose sum is 2. The integers are 7 and -5.

$$x^2+2x-35=(x+7)(x-5)$$

6. $9-a^2+2ab-b^2$
Group the last three terms.
$= 9-\left(a^2-2ab+b^2\right)$
$= 3^2-(a-b)^2$
$= [3+(a-b)][3-(a-b)]$
$= (3+a-b)(3-a+b)$

7. $225p^2+256$ is the *sum* of squares and cannot be factored. The binomial is prime.

8. x^3-1000
$= (x)^3-10^3$ *Difference of cubes*
$= (x-10)\left(x^2+10x+100\right)$

9. $6b^2-17b-3$

Two integer factors whose product is $(6)(-3)=-18$ and whose sum is -17 are -18 and 1.

$= 6b^2-18b+b-3$
$= 6b(b-3)+1(b-3)$
$= (b-3)(6b+1)$

10. $k^2-6k+16$ cannot be factored. The trinomial is *prime.*

11. $18m^3n+3m^2n^2-6mn^3$
$= 3mn\left(6m^2+mn-2n^2\right)$
Factor the trinomial.
$= 3mn(3m+2n)(2m-n)$

12. $6t^2+19tu-77u^2$

Two integer factors whose product is $(6)(-77)=-462$ and whose sum is 19 are -14 and 33.

$= 6t^2-14tu+33tu-77u^2$
$= 2t(3t-7u)+11u(3t-7u)$
$= (3t-7u)(2t+11u)$

13. $2p^2 + 11pq + 15q^2$

Look for two integers whose product is $2(15) = 30$ and whose sum is 11. The integers are 6 and 5. Write $11pq$ as $6pq + 5pq$ and factor by grouping.

$$\begin{aligned} &= 2p^2 + 6pq + 5pq + 15q^2 \\ &= (2p^2 + 6pq) + (5pq + 15q^2) \\ &= 2p(p + 3q) + 5q(p + 3q) \\ &= (p + 3q)(2p + 5q) \end{aligned}$$

14. $9m^2 - 45m + 18m^3$

Factor out the common factor $9m$.

$$= 9m(m - 5 + 2m^2) \quad \text{or} \quad 9m(2m^2 + m - 5)$$

There is no pair of integers with a product of -10 and a sum of 1, so this expression cannot be factored further.

15. $4k^2 + 28kr + 49r^2$

$$\begin{aligned} &= (2k)^2 + 2(2k)(7r) + (7r)^2 \quad \textit{Perfect square trinomial} \\ &= (2k + 7r)^2 \end{aligned}$$

16. $54m^3 - 2000$

Factor out the GCF, 2.

$2(27m^3 - 1000)$

$$\begin{aligned} &= 2[(3m)^3 - 10^3] \quad \textit{Difference of cubes} \\ &= 2(3m - 10) \cdot [(3m)^2 + (3m)(10) + 10^2] \\ &= 2(3m - 10)(9m^2 + 30m + 100) \end{aligned}$$

17. $mn - 2n + 5m - 10$

$$\begin{aligned} &= n(m - 2) + 5(m - 2) \\ &= (m - 2)(n + 5) \end{aligned}$$

18. $9m^2 - 30mn + 25n^2 - p^2$

$$\begin{aligned} &= (9m^2 - 30mn + 25n^2) - p^2 \\ &= [(3m)^2 - 2(3m)(5n) + (5n)^2] - p^2 \\ &= (3m - 5n)^2 - p^2 \\ &= [(3m - 5n) + p][(3m - 5n) - p] \\ &= (3m - 5n + p)(3m - 5n - p) \end{aligned}$$

19. $x^3 + 3x^2 - 9x - 27$

Factor by grouping.

$$\begin{aligned} &= (x^3 + 3x^2) + (-9x - 27) \\ &= x^2(x + 3) - 9(x + 3) \\ &= (x + 3)(x^2 - 9) \end{aligned}$$

Since $x^2 - 9$ is the difference of two squares, $x^2 - 3^2$, factor it as $(x + 3)(x - 3)$.

$$\begin{aligned} &= (x + 3)(x + 3)(x - 3) \\ &= (x + 3)^2(x - 3) \end{aligned}$$

20. $56k^3 - 875$

$$\begin{aligned} &= 7(8k^3 - 125) \\ &= 7[(2k)^3 - 5^3] \\ &= 7(2k - 5)[(2k)^2 + (2k)(5) + 5^2] \\ &= 7(2k - 5)(4k^2 + 10k + 25) \end{aligned}$$

21. $9r^2 + 100 = (3r)^2 + 10^2$

There is no GCF and the sum of squares cannot be factored. This binomial is *prime.*

22. $8p^3 - 125$

Factor as the difference of two cubes.

$$\begin{aligned} &= (2p)^3 - 5^3 \\ &= (2p - 5)[(2p)^2 + (2p)(5) + 5^2] \\ &= (2p - 5)(4p^2 + 10p + 25) \end{aligned}$$

23. $6k^2 - k - 1$

Multiply $6(-1) = -6$. The numbers -3 and 2 have the product -6 and the sum -1.

$$\begin{aligned} 6k^2 - k - 1 &= 6k^2 - 3k + 2k - 1 \\ &= (6k^2 - 3k) + (2k - 1) \\ &= 3k(2k - 1) + 1(2k - 1) \\ &= (2k - 1)(3k + 1) \end{aligned}$$

24. $27m^2 + 144mn + 192n^2$

$$\begin{aligned} &= 3(9m^2 + 48mn + 64n^2) \\ &= 3[(3m)^2 + 2(3m)(8n) + (8n)^2] \\ &= 3(3m + 8n)^2 \end{aligned}$$

25. $x^4 - 625$

$$\begin{aligned} &= (x^2)^2 - 25^2 \quad \textit{Difference of squares} \\ &= (x^2 + 25)(x^2 - 25) \\ &= (x^2 + 25)(x + 5)(x - 5) \quad \textit{Difference of squares again} \end{aligned}$$

26. $125m^6 + 216$

$$\begin{aligned} &= (5m^2)^3 + 6^3 \\ &= (5m^2 + 6)\left[(5m^2)^2 - (5m^2)(6) + 6^2\right] \\ &= (5m^2 + 6)(25m^4 - 30m^2 + 36) \end{aligned}$$

27. $ab + 6b + ac + 6c$

$$\begin{aligned} &= (ab + 6b) + (ac + 6c) \quad \textit{Group.} \\ &= b(a + 6) + c(a + 6) \\ &= (a + 6)(b + c) \end{aligned}$$

28. $p^3 + 64$

$= p^3 + 4^3$ *Sum of cubes*

$= (p+4)(p^2 - 4p + 4^2)$

$= (p+4)(p^2 - 4p + 16)$

29. $4y^2 - 8y = 4y(y-2)$

30. $6a^4 - 11a^2 - 10 = 6(a^2)^2 - 11a^2 - 10$

Since 4 and -15 have a product of -60 and a sum of -11, we can write

$6(a^2)^2 - 11a^2 - 10$

$= 6(a^2)^2 + 4a^2 - 15a^2 - 10$

$= 2a^2(3a^2+2) - 5(3a^2+2)$

$= (3a^2+2)(2a^2-5).$

31. $14z^2 - 3zk - 2k^2$

Two integer factors whose product is $(14)(-2) = -28$ and whose sum is -3 are 4 and -7.

$= 14z^2 + 4zk - 7zk - 2k^2$

$= 2z(7z+2k) - k(7z+2k)$

$= (7z+2k)(2z-k)$

32. $12z^3 - 6z^2 + 18z$

Factor out the GCF, $6z$.

$$6z(2z^2 - z + 3)$$

Further factoring is not possible. There is no pair of integers whose product is 6 and whose sum is -1.

33. $256b^2 - 400c^2$

$= 16(16b^2 - 25c^2)$

$= 16[(4b)^2 - (5c)^2]$ *Difference of squares*

$= 16(4b+5c)(4b-5c)$

34. $z^2 - zp + 20p^2$

There are no expressions that have a product of $20p^2$ and a sum of $-p$. The trinomial cannot be factored and is *prime.*

35. $1000z^3 + 512$

$= 8(125z^3 + 64)$

$= 8[(5z)^3 + 4^3]$ *Sum of cubes*

$= 8[5z+4][(5z)^2 - (5z)(4) + 4^2]$

$= 8(5z+4)(25z^2 - 20z + 16)$

36. $64m^2 - 25n^2$

$= (8m)^2 - (5n)^2$ *Difference of squares*

$= (8m+5n)(8m-5n)$

37. $10r^2 + 23rs - 5s^2$

Two integer factors whose product is $(10)(-5) = -50$ and whose sum is 23 are 25 and -2.

$= 10r^2 + 25rs - 2rs - 5s^2$

$= 5r(2r+5s) - s(2r+5s)$

$= (2r+5s)(5r-s)$

38. $12k^2 - 17kq - 5q^2$

Two integer factors whose product is $(12)(-5) = -60$ and whose sum is -17 are 3 and -20.

$= 12k^2 + 3kq - 20kq - 5q^2$

$= 3k(4k+q) - 5q(4k+q)$

$= (4k+q)(3k-5q)$

39. $32x^2 + 16x^3 - 24x^5 = 8x^2(4 + 2x - 3x^3)$

40. $48k^4 - 243$

$= 3(16k^4 - 81)$

$= 3[(4k^2)^2 - 9^2]$

$= 3(4k^2+9)(4k^2-9)$

$= 3(4k^2+9)[(2k)^2 - 3^2]$

$= 3(4k^2+9)(2k+3)(2k-3)$

41. $14x^2 - 25xq - 25q^2$

Two integer factors whose product is $(14)(-25) = -350$ and whose sum is -25 are -35 and 10.

$= 14x^2 - 35xq + 10xq - 25q^2$

$= 7x(2x-5q) + 5q(2x-5q)$

$= (2x-5q)(7x+5q)$

42. $5p^2 - 10p = 5p(p-2)$

43. $y^2 + 3y - 10 = (y+5)(y-2)$

44. $b^2 - 7ba - 18a^2$

Multiply $1(-18a^2) = -18a^2$. The expressions $-9a$ and $2a$ have the product $-18a^2$ and the sum $-7a$.

$b^2 - 7ba - 18a^2 = b^2 - 9ba + 2ba - 18a^2$

$= (b^2 - 9ba) + (2ba - 18a^2)$

$= b(b-9a) + 2a(b-9a)$

$= (b-9a)(b+2a)$

45. $2a^3 + 6a^2 - 4a = 2a(a^2 + 3a - 2)$

Since -2 factors only as $1(-2)$ and $(-1)2$, and $1 + (-2) = -1 \neq 3$ and $-1 + 2 = 1 \neq 3$, we know that $a^2 + 3a - 2$ cannot be factored further.

46. $12m^2rx + 4mnrx + 40n^2rx$

Factor out the GCF, $4rx$.

$= 4rx(3m^2 + mn + 10n^2)$

If we try to find two integer factors whose product is $3(10) = 30$ and whose sum is 1, we will be unsuccessful. Thus, $3m^2 + mn + 10n^2$ is prime, and the final factored form is listed above.

47. $18p^2 + 53pr - 35r^2$

Two integer factors whose product is $(18)(-35) = -630$ and whose sum is 53 are -10 and 63.

$= 18p^2 - 10pr + 63pr - 35r^2$
$= 2p(9p - 5r) + 7r(9p - 5r)$
$= (9p - 5r)(2p + 7r)$

48. $21a^2 - 5ab - 4b^2$

Look for two integers whose product is $21(-4) = -84$ and whose sum is -5. The integers are 7 and -12. Write $-5ab$ as $7ab - 12ab$.

$= 21a^2 + 7ab - 12ab - 4b^2$
$= (21a^2 + 7ab) + (-12ab - 4b^2)$
$= 7a(3a + b) - 4b(3a + b)$
$= (3a + b)(7a - 4b)$

49. $(x - 2y)^2 - 4 = (x - 2y)^2 - 2^2$
$= [(x - 2y) + 2][(x - 2y) - 2]$
$= (x - 2y + 2)(x - 2y - 2)$

50. $(3m - n)^2 - 25 = (3m - n)^2 - 5^2$
$= [(3m - n) + 5][(3m - n) - 5]$
$= (3m - n + 5)(3m - n - 5)$

51. $(5r + 2s)^2 - 6(5r + 2s) + 9$
$= x^2 - 6x + 9$ *Let $x = 5r + 2s$.*
$= (x - 3)^2$
$= [(5r + 2s) - 3]^2$ *Resubstitute.*
$= (5r + 2s - 3)^2$

52. $(p + 8q)^2 - 10(p + 8q) + 25$
$= x^2 - 10x + 25$ *Let $x = p + 8q$.*
$= (x - 5)^2$
$= [(p + 8q) - 5]^2$ *Resubstitute.*
$= (p + 8q - 5)^2$

53. $z^4 - 9z^2 + 20$
$= x^2 - 9x + 20$ *Let $x = z^2$.*
$= (x - 5)(x - 4)$
$= (z^2 - 5)(z^2 - 4)$ *Resubstitute.*
$= (z^2 - 5)(z + 2)(z - 2)$

54. $21m^4 - 32m^2 - 5$
$= 21x^2 - 32x - 5$ *Let $x = m^2$.*

Two integer factors whose product is $(21)(-5) = -105$ and whose sum is -32 are -35 and 3.

$= 21x^2 - 35x + 3x - 5$
$= 7x(3x - 5) + 1(3x - 5)$
$= (3x - 5)(7x + 1)$
$= (3m^2 - 5)(7m^2 + 1)$ *Resubstitute.*

7.4 Solving Equations by Factoring

7.4 Margin Exercises

1. **(a)** $(3x + 5)(x + 1) = 0$

$3x + 5 = 0$ or $x + 1 = 0$
$3x = -5$ or $x = -1$
$x = -\frac{5}{3}$

Check $x = -\frac{5}{3}$: $0(-\frac{2}{3}) = 0$ *True*
Check $x = -1$: $2(0) = 0$ *True*

Solution set: $\{-\frac{5}{3}, -1\}$

(b) $(3x + 11)(5x - 2) = 0$

$3x + 11 = 0$ or $5x - 2 = 0$
$3x = -11$ or $5x = 2$
$x = -\frac{11}{3}$ or $x = \frac{2}{5}$

Check $x = -\frac{11}{3}$: $0(-\frac{61}{3}) = 0$ *True*
Check $x = \frac{2}{5}$: $\frac{61}{5}(0) = 0$ *True*

Solution set: $\{-\frac{11}{3}, \frac{2}{5}\}$

2. **(a)** *Step 1 Standard form*

$3x^2 - x = 4$
$3x^2 - x - 4 = 0$

Step 2 Factor.
$(3x - 4)(x + 1) = 0$

Step 3 Zero-factor property
$3x - 4 = 0$ or $x + 1 = 0$

Step 4 Solve each equation.
$3x = 4$ or $x = -1$
$x = \frac{4}{3}$

Step 5
Check each solution in the original equation.

Check $x = \frac{4}{3}$: $\frac{16}{3} - \frac{4}{3} = 4$ *True*
Check $x = -1$: $3 - (-1) = 4$ *True*

Solution set: $\{-1, \frac{4}{3}\}$

(b)
$$25x^2 = -20x - 4$$
$$25x^2 + 20x + 4 = 0 \quad \textit{Standard form}$$
$$(5x+2)^2 = 0 \quad \textit{Factor.}$$
$$5x + 2 = 0 \quad \textit{Zero-factor property}$$
$$5x = -2$$
$$x = -\tfrac{2}{5}$$

Check $x = -\frac{2}{5}$: $4 = 8 - 4$ *True*

Solution set: $\{-\frac{2}{5}\}$

3. **(a)**
$$x^2 = -12x$$
$$x^2 + 12x = 0$$
$$x(x+12) = 0$$

$$x = 0 \quad \text{or} \quad x + 12 = 0$$
$$x = -12$$

Check $x = 0$: $0 = 0$ *True*
Check $x = -12$: $144 = 144$ *True*

Solution set: $\{-12, 0\}$

(b)
$$t^2 - 16 = 0$$
$$(t+4)(t-4) = 0$$

$$t + 4 = 0 \quad \text{or} \quad t - 4 = 0$$
$$t = -4 \quad \text{or} \quad t = 4$$

Check $t = \pm 4$: $16 - 16 = 0$ *True*

Solution set: $\{-4, 4\}$

4.
$$(x+6)(x-2) = -8 + x$$
$$x^2 + 4x - 12 = -8 + x$$
$$x^2 + 3x - 4 = 0$$
$$(x-1)(x+4) = 0$$

$$x - 1 = 0 \quad \text{or} \quad x + 4 = 0$$
$$x = 1 \quad \text{or} \quad x = -4$$

Check $x = 1$: $7(-1) = -7$ *True*
Check $x = -4$: $2(-6) = -12$ *True*

Solution set: $\{-4, 1\}$

5.
$$3x^3 + x^2 = 4x$$
$$3x^3 + x^2 - 4x = 0$$
$$x\left(3x^2 + x - 4\right) = 0$$
$$x(x-1)(3x+4) = 0$$

$$x = 0 \quad \text{or} \quad x - 1 = 0 \quad \text{or} \quad 3x + 4 = 0$$
$$x = 1 \quad \text{or} \quad x = -\tfrac{4}{3}$$

Check $x = 0$: $0 = 0$ *True*
Check $x = 1$: $4 = 4$ *True*
Check $x = -\frac{4}{3}$: $-\frac{64}{9} + \frac{16}{9} = -\frac{16}{3}$ *True*

Solution set: $\left\{-\frac{4}{3}, 0, 1\right\}$

6. Let $x =$ the length of the deck. Then $\frac{1}{2}x - 1 =$ the width of the deck.

Use the formula $LW = A$, where $L = x$, $W = \frac{1}{2}x - 1$, and $A = 60$.

$$x\left(\tfrac{1}{2}x - 1\right) = 60$$
$$\tfrac{1}{2}x^2 - x = 60$$
$$x^2 - 2x = 120 \quad \textit{Multiply by 2.}$$
$$x^2 - 2x - 120 = 0$$
$$(x-12)(x+10) = 0$$

$$x - 12 = 0 \quad \text{or} \quad x + 10 = 0$$
$$x = 12 \quad \text{or} \quad x = -10$$

The deck cannot have a negative length, so reject $x = -10$. The length of the deck is 12 m and the width is $\frac{1}{2}(12) - 1 = 5$ m.

7.
$$h(t) = -16t^2 + 128t$$
$$256 = -16t^2 + 128t$$
$$16t^2 - 128t + 256 = 0$$
$$t^2 - 8t + 16 = 0 \quad \textit{Divide by 16.}$$
$$(t-4)^2 = 0 \quad \textit{Factor.}$$
$$t - 4 = 0 \quad \textit{Zero-factor property}$$
$$t = 4$$

It will reach a height of 256 ft in 4 seconds.

7.4 Section Exercises

1. First rewrite the equation so one side is 0. Factor the other side and set each factor equal to 0. The solutions of these linear equations are solutions of the quadratic equation.

2. Choices **A**, **B**, and **C** are in factored form and are in proper form for using the zero-factor property. Choice **D** is not because the polynomial is not in factored form. The correct answer is choice **D**.

In the exercises in this section, check all solutions to the equations by substituting them back in the original equations.

3. $(x+10)(x-5) = 0$

$$x + 10 = 0 \quad \text{or} \quad x - 5 = 0$$
$$x = -10 \quad \text{or} \quad x = 5$$

Check $x = -10$: $(0)(-15) = 0$ *True*
Check $x = 5$: $(15)(0) = 0$ *True*

Solution set: $\{-10, 5\}$

4. $(x+7)(x+3) = 0$

$$x + 7 = 0 \quad \text{or} \quad x + 3 = 0$$
$$x = -7 \quad \text{or} \quad x = -3$$

Solution set: $\{-7, -3\}$

5. $(3k+8)(2k-5)=0$

$3k+8=0$ or $2k-5=0$
$3k=-8$ or $2k=5$
$k=-\frac{8}{3}$ or $k=\frac{5}{2}$

Solution set: $\{-\frac{8}{3},\frac{5}{2}\}$

6. $(2q+5)(3q-4)=0$

$2q+5=0$ or $3q-4=0$
$2q=-5$ or $3q=4$
$q=-\frac{5}{2}$ or $q=\frac{4}{3}$

Solution set: $\{-\frac{5}{2},\frac{4}{3}\}$

7. $m^2-3m-10=0$
$(m+2)(m-5)=0$
Use the zero-factor property.
$m+2=0$ or $m-5=0$
$m=-2$ or $m=5$

Solution set: $\{-2,5\}$

8. $x^2+x-12=0$
$(x+4)(x-3)=0$

$x+4=0$ or $x-3=0$
$x=-4$ or $x=3$

Solution set: $\{-4,3\}$

9. $z^2+9z+18=0$
$(z+6)(z+3)=0$

$z+6=0$ or $z+3=0$
$z=-6$ or $z=-3$

Solution set: $\{-6,-3\}$

10. $x^2-18x+80=0$
$(x-8)(x-10)=0$

$x-8=0$ or $x-10=0$
$x=8$ or $x=10$

Solution set: $\{8,10\}$

11. $2x^2=7x+4$
Get 0 on one side.
$2x^2-7x-4=0$
$(2x+1)(x-4)=0$

$2x+1=0$ or $x-4=0$
$2x=-1$ or $x=4$
$x=-\frac{1}{2}$

Solution set: $\{-\frac{1}{2},4\}$

12. $2x^2=3-x$
$2x^2+x-3=0$
$(2x+3)(x-1)=0$

$2x+3=0$ or $x-1=0$
$2x=-3$ or $x=1$
$x=-\frac{3}{2}$

Solution set: $\{-\frac{3}{2},1\}$

13. $15k^2-7k=4$
$15k^2-7k-4=0$
$(3k+1)(5k-4)=0$

$3k+1=0$ or $5k-4=0$
$3k=-1$ or $5k=4$
$k=-\frac{1}{3}$ or $k=\frac{4}{5}$

Solution set: $\{-\frac{1}{3},\frac{4}{5}\}$

14. $12x^2+4x=5$
$12x^2+4x-5=0$
$(6x+5)(2x-1)=0$
$6x+5=0$ or $2x-1=0$
$6x=-5$ or $2x=1$
$x=-\frac{5}{6}$ or $x=\frac{1}{2}$

Solution set: $\{-\frac{5}{6},\frac{1}{2}\}$

15. $16x^2+24x=-9$
$16x^2+24x+9=0$
$(4x+3)^2=0$
$4x+3=0$
$4x=-3$
$x=-\frac{3}{4}$

Solution set: $\{-\frac{3}{4}\}$

16. $49x^2+14x=-1$
$49x^2+14x+1=0$
$(7x+1)^2=0$
$7x+1=0$
$7x=-1$
$x=-\frac{1}{7}$

Solution set: $\{-\frac{1}{7}\}$

17. $(5x+1)(x+3)=-2(5x+1)$
$(5x+1)(x+3)+2(5x+1)=0$
$(5x+1)[(x+3)+2]=0$
$(5x+1)(x+5)=0$

$5x+1=0$ or $x+5=0$
$5x=-1$ or $x=-5$
$x=-\frac{1}{5}$

Solution set: $\{-5,-\frac{1}{5}\}$

18. $(3x+1)(x-3) = 2+3(x+5)$

$$3x^2 - 8x - 3 = 2 + 3x + 15$$
$$3x^2 - 11x - 20 = 0$$
$$(3x+4)(x-5) = 0$$

$3x+4=0$ or $x-5=0$

$3x=-4$ or $x=5$

$x=-\frac{4}{3}$

Solution set: $\{-\frac{4}{3}, 5\}$

19. $6m^2 - 36m = 0$

$$6m(m-6)=0$$

$6m=0$ or $m-6=0$

$m=0$ or $m=6$

Solution set: $\{0,6\}$

20. $-3m^2 + 27m = 0$

$$-3m(m-9)=0$$

$-3m=0$ or $m-9=0$

$m=0$ or $m=9$

Solution set: $\{0,9\}$

21. $-3m^2 + 27 = 0$

$$-3(m^2-9)=0$$
$$-3(m+3)(m-3)=0$$

Note that the leading -3 does not affect the solution set of the equation.

$m+3=0$ or $m-3=0$

$m=-3$ or $m=3$

Solution set: $\{-3,3\}$

22. $-2x^2+8=0$

$$-2(x^2-4)=0$$
$$-2(x+2)(x-2)=0$$

$x+2=0$ or $x-2=0$

$x=-2$ or $x=2$

Solution set: $\{-2,2\}$

23. $4p^2-16=0$

$$4(p^2-4)=0$$
$$4(p+2)(p-2)=0$$

$p+2=0$ or $p-2=0$

$p=-2$ or $p=2$

Solution set: $\{-2,2\}$

24. $9x^2-81=0$

$$9(x^2-9)=0$$
$$9(x+3)(x-3)=0$$

$x+3=0$ or $x-3=0$

$x=-3$ or $x=3$

Solution set: $\{-3,3\}$

25. $(x-3)(x+5)=-7$

Multiply the factors, and then add 7 on both sides of the equation to get 0 on the right.

$$x^2+5x-3x-15=-7$$
$$x^2+2x-8=0$$

Now factor the polynomial.

$$(x+4)(x-2)=0$$

$x+4=0$ or $x-2=0$

$x=-4$ or $x=2$

Solution set: $\{-4,2\}$

26. $(x+8)(x-2)=-21$

Multiply the factors, and then add 21 on both sides of the equation to get 0 on the right.

$$x^2-2x+8x-16=-21$$
$$x^2+6x+5=0$$

Now factor the polynomial.

$$(x+5)(x+1)=0$$

$x+5=0$ or $x+1=0$

$x=-5$ or $x=-1$

Solution set: $\{-5,-1\}$

27. $(2x+1)(x-3)=6x+3$

$$2x^2-6x+x-3=6x+3$$
$$2x^2-5x-3=6x+3$$
$$2x^2-11x-6=0$$
$$(2x+1)(x-6)=0$$

$2x+1=0$ or $x-6=0$

$2x=-1$ or $x=6$

$x=-\frac{1}{2}$

Solution set: $\{-\frac{1}{2},6\}$

28. $(3x+2)(x-3)=7x-1$

$$3x^2-7x-6=7x-1$$
$$3x^2-14x-5=0$$
$$(3x+1)(x-5)=0$$

$3x+1=0$ or $x-5=0$

$3x=-1$ or $x=5$

$x=-\frac{1}{3}$

Solution set: $\{-\frac{1}{3},5\}$

29. $(x+3)(x-6)=(2x+2)(x-6)$

$$x^2-3x-18=2x^2-10x-12$$
$$0=x^2-7x+6$$
$$0=(x-1)(x-6)$$

$x-1=0$ or $x-6=0$

$x=1$ or $x=6$

Solution set: $\{1,6\}$

30. $$\begin{aligned}(2x+1)(x+5)&=(x+11)(x+3)\\ 2x^2+11x+5&=x^2+14x+33\\ x^2-3x-28&=0\\ (x+4)(x-7)&=0\end{aligned}$$

$x+4=0$ or $x-7=0$
$x=-4$ or $x=7$

Solution set: $\{-4, 7\}$

31. $$\begin{aligned}2x^3-9x^2-5x&=0\\ x(2x^2-9x-5)&=0\\ x(2x+1)(x-5)&=0\end{aligned}$$

$x=0$ or $2x+1=0$ or $x-5=0$
$2x=-1$ or $x=5$
$x=-\frac{1}{2}$

Solution set: $\{-\frac{1}{2}, 0, 5\}$

32. $$\begin{aligned}6x^3-13x^2-5x&=0\\ x(6x^2-13x-5)&=0\\ x(3x+1)(2x-5)&=0\end{aligned}$$

$x=0$ or $3x+1=0$ or $2x-5=0$
$3x=-1$ or $2x=5$
$x=-\frac{1}{3}$ or $x=\frac{5}{2}$

Solution set: $\{-\frac{1}{3}, 0, \frac{5}{2}\}$

33. $$\begin{aligned}9t^3&=16t\\ 9t^3-16t&=0\\ t(9t^2-16)&=0\\ t(3t+4)(3t-4)&=0\end{aligned}$$

$t=0$ or $3t+4=0$ or $3t-4=0$
$3t=-4$ or $3t=4$
$t=-\frac{4}{3}$ or $t=\frac{4}{3}$

Solution set: $\{-\frac{4}{3}, 0, \frac{4}{3}\}$

34. $$\begin{aligned}25x^3&=64x\\ 25x^3-64x&=0\\ x(25x^2-64)&=0\\ x(5x+8)(5x-8)&=0\end{aligned}$$

$x=0$ or $5x+8=0$ or $5x-8=0$
$5x=-8$ or $5x=8$
$x=-\frac{8}{5}$ or $x=\frac{8}{5}$

Solution set: $\{-\frac{8}{5}, 0, \frac{8}{5}\}$

35. $$2r^3+5r^2-2r-5=0$$
Factor by grouping.
$$\begin{aligned}(2r^3-2r)+(5r^2-5)&=0\\ 2r(r^2-1)+5(r^2-1)&=0\\ (r^2-1)(2r+5)&=0\\ (r+1)(r-1)(2r+5)&=0\end{aligned}$$

$r+1=0$ or $r-1=0$ or $2r+5=0$
$r=-1$ or $r=1$ or $r=-\frac{5}{2}$

Solution set: $\{-\frac{5}{2}, -1, 1\}$

36. $$2p^3+p^2-98p-49=0$$
Factor by grouping.
$$\begin{aligned}(2p^3-98p)+(p^2-49)&=0\\ 2p(p^2-49)+1(p^2-49)&=0\\ (p^2-49)(2p+1)&=0\\ (p+7)(p-7)(2p+1)&=0\end{aligned}$$

$p+7=0$ or $p-7=0$ or $2p+1=0$
$p=-7$ or $p=7$ or $2p=-1$
$p=-\frac{1}{2}$

Solution set: $\{-7, -\frac{1}{2}, 7\}$

37. By dividing each side by a variable expression, she "lost" the solution 0.

38. An equation will have 0 in its solution set if the factor x appears in each term. Thus, the correct choice is **D**.

39. Let $x =$ the width of the garden.
Then $x + 4 =$ the length of the garden.

The area of the rectangular-shaped garden is 320 ft^2, so use the formula $A = LW$ and substitute 320 for A, $x + 4$ for L, and x for W.

$$\begin{aligned}A&=LW\\ 320&=(x+4)x\\ 320&=x^2+4x\\ 0&=x^2+4x-320\\ 0&=(x-16)(x+20)\end{aligned}$$

$x-16=0$ or $x+20=0$
$x=16$ or $x=-20$

A rectangle cannot have a width that is a negative measure, so reject -20 as a solution. The only possible solution is 16.

The width of the garden is 16 ft, and the length is $16 + 4 = 20$ ft.

40. Let $x =$ the length of the side of the square painting. Then $x - 2 =$ the length of the side of the square mirror.

The formula for the area of a square is $A = s^2$, so the area of the painting is x^2 and the area of the mirror is $(x - 2)^2$. The difference between their areas is 32 square feet.

$$\begin{aligned} x^2 - (x-2)^2 &= 32 \\ x^2 - \left(x^2 - 4x + 4\right) &= 32 \\ x^2 - x^2 + 4x - 4 &= 32 \\ 4x - 4 &= 32 \\ 4x &= 36 \\ x &= 9 \end{aligned}$$

The length of the side of the painting is 9 feet. The length of the side of the mirror is $9 - 2 = 7$ feet.

41. Let $h =$ the length of the height.
Then $h - 3 =$ the length of the base.

The area of the triangle is $44\ \text{m}^2$, so use the formula $A = \frac{1}{2}bh$ and substitute 44 for A and $h - 3$ for b.

$$\begin{aligned} A &= \tfrac{1}{2}bh \\ 44 &= \tfrac{1}{2}(h-3)h \\ 88 &= (h-3)h \qquad \textit{Multiply by 2.} \\ 88 &= h^2 - 3h \\ 0 &= h^2 - 3h - 88 \\ 0 &= (h-11)(h+8) \end{aligned}$$

$$h - 11 = 0 \quad \text{or} \quad h + 8 = 0$$
$$h = 11 \quad \text{or} \quad h = -8$$

A triangle cannot have a height that is negative, so reject -8 as a solution. The only possible solution is 11.
The height of the triangle is 11 m, and the base is $11 - 3 = 8$ m.

42. Let $x =$ the height of the parallelogram.
Then $x + 7 =$ the base of the parallelogram.

The area is $60\ \text{ft}^2$, so use the formula $A = bh$ and substitute 60 for A, $x + 7$ for b, and x for h.

$$\begin{aligned} A &= bh \\ 60 &= (x+7)x \\ 60 &= x^2 + 7x \\ 0 &= x^2 + 7x - 60 \\ 0 &= (x+12)(x-5) \end{aligned}$$

$$x + 12 = 0 \quad \text{or} \quad x - 5 = 0$$
$$x = -12 \quad \text{or} \quad x = 5$$

A parallelogram cannot have a height that is negative, so reject -12 as a solution. The only possible solution is 5.
The height of the parallelogram is 5 ft and the base is $5 + 7 = 12$ ft.

43. Let $L =$ the length of the rectangular area and $W =$ the width.

Use the formula for perimeter, $P = 2L + 2W$, and solve for W in terms of L. The perimeter is 300 ft.

$$\begin{aligned} 300 &= 2L + 2W \\ 300 - 2L &= 2W \\ 150 - L &= W \end{aligned}$$

Now use the formula for area, $A = LW$, substitute 5000 for A, and solve for L.

$$\begin{aligned} 5000 &= L(150 - L) \\ 5000 &= 150L - L^2 \\ L^2 - 150L + 5000 &= 0 \\ (L-50)(L-100) &= 0 \end{aligned}$$

$$L - 50 = 0 \quad \text{or} \quad L - 100 = 0$$
$$L = 50 \quad \text{or} \quad L = 100$$

When $L = 50$, $W = 150 - 50 = 100$.
When $L = 100$, $W = 150 - 100 = 50$.
The dimensions should be 50 ft by 100 ft.

44. Let $x =$ the width. Then $x + 200 =$ the length.

The area is $30{,}000\ \text{ft}^2$, so use the formula $A = LW$ and substitute 30,000 for A, $x + 200$ for L, and x for W.

$$\begin{aligned} LW &= A \\ (x+200)x &= 30{,}000 \\ x^2 + 200x &= 30{,}000 \\ x^2 + 200x - 30{,}000 &= 0 \\ (x-100)(x+300) &= 0 \end{aligned}$$

$$x - 100 = 0 \quad \text{or} \quad x + 300 = 0$$
$$x = 100 \quad \text{or} \quad x = -300$$

The width cannot be a negative measure, so reject -300 as a solution. The only possible solution is 100.
The width is 100 ft and the length is $100 + 200 = 300$ ft.

45. Let $w =$ the width of the cardboard.
Then $w + 6 =$ the length of the cardboard.

If squares that measure 2 inches are cut from each corner of the cardboard, then the width becomes $w - 4$ and the length becomes $(w + 6) - 4 = w + 2$. Use the formula $V = LWH$ and substitute 110 for V, $w + 2$ for L, $w - 4$ for W, and 2 for H.

continued

$$\begin{aligned} V &= LWH \\ 110 &= (w+2)(w-4)2 \\ 110 &= (w^2-2w-8)2 \\ 55 &= w^2-2w-8 \\ 0 &= w^2-2w-63 \\ 0 &= (w-9)(w+7) \end{aligned}$$

$$w-9=0 \quad \text{or} \quad w+7=0$$
$$w=9 \quad \text{or} \quad w=-7$$

A box cannot have a negative width, so reject -7 as a solution. The only possible solution is 9. The piece of cardboard has width 9 inches and length $9+6=15$ inches.

46. Let $x =$ the length of the side of the base of the box.

The surface area is 161.

$$\begin{aligned} S(x) &= x^2+16x \\ 161 &= x^2+16x \\ 0 &= x^2+16x-161 \\ 0 &= (x+23)(x-7) \end{aligned}$$

$$x+23=0 \quad \text{or} \quad x-7=0$$
$$x=-23 \quad \text{or} \quad x=7$$

The box cannot have a length that is a negative measure, so reject -23 as a solution. The only possible solution is 7.
Each side of the base of the box measures 7 in.

47. From Example 7,

$$\begin{aligned} h(t) &= -16t^2+128t. \\ 240 &= -16t^2+128t && \textit{Let } h(t)=240. \\ 16t^2-128t+240 &= 0 && \textit{Standard form} \\ t^2-8t+15 &= 0 && \textit{Divide by 16.} \\ (t-3)(t-5) &= 0 \end{aligned}$$

$$t-3=0 \quad \text{or} \quad t-5=0$$
$$t=3 \quad \text{or} \quad t=5$$

The height of the rocket will be 240 feet after 3 seconds (on the way up) and after 5 seconds (on the way down).

$$\begin{aligned} 112 &= -16t^2+128t && \textit{Let } h(t)=112. \\ 16t^2-128t+112 &= 0 && \textit{Standard form} \\ t^2-8t+7 &= 0 && \textit{Divide by 16.} \\ (t-1)(t-7) &= 0 \end{aligned}$$

$$t-1=0 \quad \text{or} \quad t-7=0$$
$$t=1 \quad \text{or} \quad t=7$$

The height of the rocket will be 112 feet after 1 second (on the way up) and after 7 seconds (on the way down).

48. When the object hits the ground, its height is given by $f(t)=0$. Let $f(t)=0$ in the given equation, and solve for t.

$$\begin{aligned} -16t^2+64t+80 &= f(t) \\ -16t^2+64t+80 &= 0 \\ t^2-4t-5 &= 0 && \textit{Divide by } -16. \\ (t-5)(t+1) &= 0 \end{aligned}$$

$$t-5=0 \quad \text{or} \quad t+1=0$$
$$t=5 \quad \text{or} \quad t=-1$$

Time cannot be negative, so reject -1 as a solution. The only possible solution is 5.
The object will hit the ground 5 seconds after it is thrown.

49. Use $f(t)=-16t^2+576$ with $f(t)=0$.

$$\begin{aligned} 0 &= -16t^2+576 \\ 0 &= t^2-36 && \textit{Divide by } -16. \\ 0 &= (t+6)(t-6) \end{aligned}$$

$$t+6=0 \quad \text{or} \quad t-6=0$$
$$t=-6 \quad \text{or} \quad t=6$$

Time cannot be negative, so reject -6 as a solution. The only possible solution is 6.
The rock will hit the ground after 6 seconds.

50. Use $f(t)=-16t^2+625$ with $f(t)=0$.

$$\begin{aligned} 0 &= -16t^2+625 \\ 0 &= 16t^2-625 && \textit{Multiply by } -1. \\ 0 &= (4t+25)(4t-25) \end{aligned}$$

$$4t+25=0 \quad \text{or} \quad 4t-25=0$$
$$4t=-25 \qquad\qquad 4t=25$$
$$t=-\tfrac{25}{4} \quad \text{or} \quad t=\tfrac{25}{4}$$

Time cannot be negative, so reject $-\frac{25}{4}$ as a solution. The only possible solution is $\frac{25}{4}$ or $6\frac{1}{4}$.
The ball will hit the ground after $6\frac{1}{4}$ seconds.

Chapter 7 Review Exercises

1. $21y^2+35y=7y(3y+5)$

2. $12q^2b+8qb^2-20q^3b^2$
$=4qb(3q+2b-5q^2b)$

3. $(x+3)(4x-1)-(x+3)(3x+2)$

The GCF is $(x+3)$.

$$\begin{aligned} &= (x+3)[(4x-1)-(3x+2)] \\ &= (x+3)(4x-1-3x-2) \\ &= (x+3)(x-3) \end{aligned}$$

4. $(z+1)(z-4)+(z+1)(2z+3)$
$= (z+1)[(z-4)+(2z+3)]$
$= (z+1)(3z-1)$

5. $4m+nq+mn+4q$

Rearrange the terms.

$= 4m+4q+mn+nq$
$= 4(m+q)+n(m+q)$
$= (m+q)(4+n)$

6. $x^2+5y+5x+xy$
$= x^2+xy+5x+5y$
$= x(x+y)+5(x+y)$
$= (x+y)(x+5)$

7. $2m+6-am-3a$
$= 2(m+3)-a(m+3)$
$= (m+3)(2-a)$

8. $2am-2bm-ap+bp$
$= (2am-2bm)+(-ap+bp)$
$= 2m(a-b)-p(a-b)$
$= (a-b)(2m-p)$

9. $3p^2-p-4$

Two integer factors whose product is $(3)(-4)=-12$ and whose sum is -1 are -4 and 3.

$= 3p^2-4p+3p-4$
$= p(3p-4)+1(3p-4)$
$= (3p-4)(p+1)$

10. $12r^2-5r-3$

Two integer factors whose product is $(12)(-3)=-36$ and whose sum is -5 are -9 and 4.

$= 12r^2-9r+4r-3$
$= 3r(4r-3)+1(4r-3)$
$= (4r-3)(3r+1)$

11. $10m^2+37m+30$

Two integer factors whose product is $(10)(30)=300$ and whose sum is 37 are 12 and 25.

$= 10m^2+12m+25m+30$
$= 2m(5m+6)+5(5m+6)$
$= (5m+6)(2m+5)$

12. $10k^2-11kh+3h^2$

Two integer factors whose product is $(10)(3)=30$ and whose sum is -11 are -6 and -5.

$= 10k^2-6kh-5kh+3h^2$
$= 2k(5k-3h)-h(5k-3h)$
$= (5k-3h)(2k-h)$

13. $9x^2+4xy-2y^2$

There are no integers that have a product of $9(-2)=-18$ and a sum of 4. Therefore, the trinomial cannot be factored and is *prime*.

14. $24x-2x^2-2x^3$
$= 2x(12-x-x^2)$
$= 2x(4+x)(3-x)$

15. $2k^4-5k^2-3$
$= 2(k^2)^2-5k^2-3$
$= (2k^2+1)(k^2-3)$

16. $p^2(p+2)^2+p(p+2)^2-6(p+2)^2$
Factor out $(p+2)^2$.
$= (p+2)^2(p^2+p-6)$
$= (p+2)^2(p+3)(p-2)$

17. $16x^2-25=(4x)^2-5^2$ *Difference of squares*
$= (4x+5)(4x-5)$

18. $9t^2-49=(3t)^2-7^2$
$= (3t+7)(3t-7)$

19. $x^2+14x+49=x^2+2(x)(7)+7^2$
Perfect square trinomial
$= (x+7)^2$

20. $9k^2-12k+4=(3k)^2-2(3k)(2)+2^2$
$= (3k-2)^2$

21. $r^3+27=r^3+3^3$ *Sum of cubes*
$= (r+3)(r^2-3r+9)$

22. $125x^3-1=(5x)^3-1^3$ *Difference of cubes*
$= (5x-1)(25x^2+5x+1)$

23. $m^6-1=(m^3)^2-1^2$
Difference of squares
$= (m^3+1)(m^3-1)$
$= (m^3+1^3)(m^3-1^3)$
Sum of cubes; Difference of cubes
$= (m+1)(m^2-m+1)$
$\cdot (m-1)(m^2+m+1)$

24. $x^8 - 1 = (x^4)^2 - 1^2$

Difference of squares

$= (x^4 + 1)(x^4 - 1)$

$= (x^4 + 1)\left[(x^2)^2 - 1^2\right]$

Difference of squares again

$= (x^4 + 1)(x^2 + 1)(x^2 - 1)$

Difference of squares again

$= (x^4 + 1)(x^2 + 1)(x + 1)(x - 1)$

25. $x^2 + 6x + 9 - 25y^2$

$= (x^2 + 6x + 9) - 25y^2$

$= [x^2 + 2(x)(3) + 3^2] - 25y^2$

Perfect square trinomial

$= (x + 3)^2 - (5y)^2$

Difference of squares

$= [(x + 3) + 5y][(x + 3) - 5y]$

$= (x + 3 + 5y)(x + 3 - 5y)$

26. $(x + 1)(5x + 2) = 0$

$x + 1 = 0$ or $5x + 2 = 0$

$x = -1$ or $x = -\frac{2}{5}$

Solution set: $\left\{-1, -\frac{2}{5}\right\}$

27. $p^2 - 5p + 6 = 0$

$(p - 2)(p - 3) = 0$

$p - 2 = 0$ or $p - 3 = 0$

$p = 2$ or $p = 3$

Solution set: $\{2, 3\}$

28. $6z^2 = 5z + 50$

$6z^2 - 5z - 50 = 0$

$(3z - 10)(2z + 5) = 0$

$3z - 10 = 0$ or $2z + 5 = 0$

$3z = 10$ or $2z = -5$

$z = \frac{10}{3}$ or $z = -\frac{5}{2}$

Solution set: $\left\{-\frac{5}{2}, \frac{10}{3}\right\}$

29. $6r^2 + 7r = 3$

$6r^2 + 7r - 3 = 0$

$(2r + 3)(3r - 1) = 0$

$2r + 3 = 0$ or $3r - 1 = 0$

$2r = -3$ or $3r = 1$

$r = -\frac{3}{2}$ or $r = \frac{1}{3}$

Solution set: $\left\{-\frac{3}{2}, \frac{1}{3}\right\}$

30. $-4m^2 + 36 = 0$

$m^2 - 9 = 0$ *Divide by −4.*

$(m + 3)(m - 3) = 0$

$m + 3 = 0$ or $m - 3 = 0$

$m = -3$ or $m = 3$

Solution set: $\{-3, 3\}$

31. $6x^2 + 9x = 0$

$3x(2x + 3) = 0$

$3x = 0$ or $2x + 3 = 0$

$x = 0$ or $2x = -3$

$x = -\frac{3}{2}$

Solution set: $\left\{-\frac{3}{2}, 0\right\}$

32. $(2x + 1)(x - 2) = -3$

$2x^2 - 3x - 2 = -3$

$2x^2 - 3x + 1 = 0$

$(2x - 1)(x - 1) = 0$

$2x - 1 = 0$ or $x - 1 = 0$

$2x = 1$ or $x = 1$

$x = \frac{1}{2}$

Solution set: $\left\{\frac{1}{2}, 1\right\}$

33. $x^2 - 8x + 16 = 0$

$(x - 4)(x - 4) = 0$

$x - 4 = 0$ or $x - 4 = 0$

$x = 4$ or $x = 4$

Solution set: $\{4\}$

34. $2x^3 - x^2 - 28x = 0$

$x(2x^2 - x - 28) = 0$

$x(2x + 7)(x - 4) = 0$

$x = 0$ or $2x + 7 = 0$ or $x - 4 = 0$

$2x = -7$ or $x = 4$

$x = -\frac{7}{2}$

Solution set: $\left\{-\frac{7}{2}, 0, 4\right\}$

35. Let x be the length of the shorter side. Then the length of the longer side will be $2x + 1$. The area is 10.5 ft^2. Use the formula for area of a triangle, $A = \frac{1}{2}bh$.

$\frac{1}{2}(2x + 1)(x) = 10.5$

$x(2x + 1) = 21$ *Multiply by 2.*

$2x^2 + x = 21$

$2x^2 + x - 21 = 0$

$(2x + 7)(x - 3) = 0$

$$\begin{aligned} 2x+7=0 \quad &\text{or} \quad x-3=0 \\ 2x=-7 \quad &\text{or} \quad x=3 \\ x=-\tfrac{7}{2} \quad & \end{aligned}$$

The side cannot have a negative length, so reject $x=-\frac{7}{2}$.
The length of the shorter side is 3 ft.

36. Let w be the width of the lot. Then $w+20$ will be the length of the lot. The area is 2400 ft^2. Use the formula $LW=A$.

$$\begin{aligned} (w+20)w &= 2400 \\ w^2+20w &= 2400 \\ w^2+20w-2400 &= 0 \\ (w-40)(w+60) &= 0 \end{aligned}$$

$$\begin{aligned} w-40=0 \quad &\text{or} \quad w+60=0 \\ w=40 \quad &\text{or} \quad w=-60 \end{aligned}$$

The lot cannot have a negative width, so reject $w=-60$.
The width of the lot is 40 ft, and the length is $40+20=60$ ft.

37. The height is 0 when the rock returns to the ground.

$$\begin{aligned} f(t) &= -16t^2+256t \\ 0 &= -16t^2+256t \\ 0 &= -16t(t-16) \end{aligned}$$

$$\begin{aligned} -16t=0 \quad &\text{or} \quad t-16=0 \\ t=0 \quad &\text{or} \quad t=16 \end{aligned}$$

The rock is on the ground when $t=0$. It will return to the ground again after 16 seconds.

38.
$$\begin{aligned} f(t) &= -16t^2+256t \\ 240 &= -16t^2+256t \qquad \textit{Let } f(t)=240. \\ 0 &= -16t^2+256t-240 \\ 0 &= -16\left(t^2-16t+15\right) \\ 0 &= -16(t-15)(t-1) \end{aligned}$$

$$\begin{aligned} t-15=0 \quad &\text{or} \quad t-1=0 \\ t=15 \quad &\text{or} \quad t=1 \end{aligned}$$

The rock will be 240 ft above the ground after 1 second and again after 15 seconds.

39. The question in Exercise 38 has two answers because the rock will be 240 ft above the ground after 1 second on the way up and again after 15 seconds on the way back down.

40. **[7.2]** $30a+am-am^2$
$$\begin{aligned} &= a\left(30+m-m^2\right) \\ &= a(6-m)(5+m) \end{aligned}$$

41. **[7.3]** $8-a^3$
$$\begin{aligned} &= 2^3-a^3 \\ &= (2-a)\left(2^2+2a+a^2\right) \\ &= (2-a)\left(4+2a+a^2\right) \end{aligned}$$

42. **[7.2]** $9x^2+13xy-3y^2$

There are no integers that have a product of $9(-3)=-27$ and a sum of 13. Therefore, the trinomial cannot be factored and is *prime*.

43. **[7.1]** $15y^3+20y^2=5y^2(3y+4)$

44. **[7.4]** $5x^2-17x-12=0$
$$(5x+3)(x-4)=0$$

$$\begin{aligned} 5x+3=0 \quad &\text{or} \quad x-4=0 \\ 5x=-3 \quad &\text{or} \quad x=4 \\ x=-\tfrac{3}{5} \quad & \end{aligned}$$

Solution set: $\left\{-\frac{3}{5},4\right\}$

45. **[7.4]**
$$\begin{aligned} x^3-x &= 0 \\ x\left(x^2-1\right) &= 0 \\ x(x+1)(x-1) &= 0 \end{aligned}$$

$$\begin{aligned} x=0 \quad \text{or} \quad x+1=0 \quad &\text{or} \quad x-1=0 \\ x=-1 \quad &\text{or} \quad x=1 \end{aligned}$$

Solution set: $\{-1,0,1\}$

46. **[7.4]** Let x be the width of the floor. Then $x+85$ will be the length of the floor. The area is 2750 ft^2. Use the formula $LW=A$.

$$\begin{aligned} (x+85)x &= 2750 \\ x^2+85x &= 2750 \\ x^2+85x-2750 &= 0 \\ (x+110)(x-25) &= 0 \end{aligned}$$

$$\begin{aligned} x+110=0 \quad &\text{or} \quad x-25=0 \\ x=-110 \quad &\text{or} \quad x=25 \end{aligned}$$

The floor cannot have a negative width, so reject $x=-110$.
The width of the floor was 25 feet and the length was $25+85=110$ feet.

Chapter 7 Test

1. $11z^2-44z=11z(z-4)$

2. $10x^2y^5-5x^2y^3-25x^5y^3$
Factor out the GCF, $5x^2y^3$.
$$= 5x^2y^3\left(2y^2-1-5x^3\right)$$

3. $3x+by+bx+3y$
$$\begin{aligned} &= 3x+3y+bx+by \\ &= 3(x+y)+b(x+y) \\ &= (x+y)(3+b) \end{aligned}$$

4. $-2x^2 - x + 36 = -1(2x^2 + x - 36)$

Two integer factors whose product is $(2)(-36) = -72$ and whose sum is 1 are 9 and -8.

$$\begin{aligned} 2x^2 + x - 36 &= 2x^2 + 9x - 8x - 36 \\ &= x(2x+9) - 4(2x+9) \\ &= (2x+9)(x-4) \end{aligned}$$

Thus, the final factored form is

$$-(2x+9)(x-4).$$

5. $6x^2 + 11x - 35$

Two integer factors whose product is $(6)(-35) = -210$ and whose sum is 11 are 21 and -10.

$$\begin{aligned} &= 6x^2 + 21x - 10x - 35 \\ &= 3x(2x+7) - 5(2x+7) \\ &= (2x+7)(3x-5) \end{aligned}$$

6. $4p^2 + 3pq - q^2$

Two integer factors whose product is $(4)(-1) = -4$ and whose sum is 3 are 4 and -1.

$$\begin{aligned} &= 4p^2 + 4pq - pq - q^2 \\ &= 4p(p+q) - q(p+q) \\ &= (p+q)(4p-q) \end{aligned}$$

7. $16a^2 + 40ab + 25b^2$

$$\begin{aligned} &= (4a)^2 + 2(4a)(5b) + (5b)^2 \\ &= (4a+5b)^2 \end{aligned}$$

8. $x^2 + 2x + 1 - 4z^2$

$$\begin{aligned} &= \left(x^2 + 2x + 1\right) - 4z^2 \\ &= (x+1)^2 - (2z)^2 \\ &= [(x+1) + 2z][(x+1) - 2z] \\ &= (x+1+2z)(x+1-2z) \end{aligned}$$

9. $a^3 + 2a^2 - ab^2 - 2b^2$

$$\begin{aligned} &= a^2(a+2) - b^2(a+2) \\ &= (a+2)\left(a^2 - b^2\right) \\ &= (a+2)(a+b)(a-b) \end{aligned}$$

10. $$\begin{aligned} 9k^2 - 121j^2 &= (3k)^2 - (11j)^2 \\ &= (3k+11j)(3k-11j) \end{aligned}$$

11. $$\begin{aligned} y^3 - 216 &= y^3 - 6^3 \\ &= (y-6)\left(y^2 + 6y + 6^2\right) \\ &= (y-6)\left(y^2 + 6y + 36\right) \end{aligned}$$

12. $6k^4 - k^2 - 35 = 6\left(k^2\right)^2 - k^2 - 35$

Two integer factors whose product is $(6)(-35) = -210$ and whose sum is -1 are -15 and 14.

$$\begin{aligned} &= 6\left(k^2\right)^2 - 15k^2 + 14k^2 - 35 \\ &= 3k^2\left(2k^2 - 5\right) + 7\left(2k^2 - 5\right) \\ &= \left(2k^2 - 5\right)\left(3k^2 + 7\right) \end{aligned}$$

13. $27x^6 + 1$

$$\begin{aligned} &= \left(3x^2\right)^3 + (1)^3 \\ &= \left(3x^2 + 1\right)\left[\left(3x^2\right)^2 - \left(3x^2\right)(1) + 1^2\right] \\ &= \left(3x^2 + 1\right)\left(9x^4 - 3x^2 + 1\right) \end{aligned}$$

14. $$\begin{aligned} -x^2 + x + 30 &= -1\left(x^2 - x - 30\right) \\ &= -1(x+5)(x-6) \\ &= -(x+5)(x-6) \end{aligned}$$

15. $\left(t^2 + 3\right)^2 + 4\left(t^2 + 3\right) - 5$

$$\begin{aligned} &= \left[\left(t^2 + 3\right) + 5\right]\left[\left(t^2 + 3\right) - 1\right] \\ &= \left(t^2 + 3 + 5\right)\left(t^2 + 3 - 1\right) \\ &= \left(t^2 + 8\right)\left(t^2 + 2\right) \end{aligned}$$

16. It is not in factored form because there are two terms: $(x^2 + 2y)p$ and $3(x^2 + 2y)$. The common factor is $x^2 + 2y$, and the factored form is $(x^2 + 2y)(p + 3)$.

17. **A** $(3 - x)(x + 4)$

$$\begin{aligned} &= 3x + 12 - x^2 - 4x \\ &= -x^2 - x + 12 \end{aligned}$$

B $-(x - 3)(x + 4)$

$$\begin{aligned} &= -\left(x^2 + 4x - 3x - 12\right) \\ &= -\left(x^2 + x - 12\right) \\ &= -x^2 - x + 12 \end{aligned}$$

C $(-x + 3)(x + 4)$

$$\begin{aligned} &= -x^2 - 4x + 3x + 12 \\ &= -x^2 - x + 12 \end{aligned}$$

D $(x - 3)(-x + 4)$

$$\begin{aligned} &= -x^2 + 4x + 3x - 12 \\ &= -x^2 + 7x - 12 \end{aligned}$$

Therefore, only **D** is *not* a factored form of $-x^2 - x + 12$.

18. $$\begin{aligned} 3x^2 + 8x &= -4 \\ 3x^2 + 8x + 4 &= 0 \\ (x+2)(3x+2) &= 0 \end{aligned}$$

$$\begin{aligned} x + 2 &= 0 \quad \text{or} \quad & 3x + 2 &= 0 \\ x &= -2 \quad \text{or} \quad & 3x &= -2 \\ & & x &= -\tfrac{2}{3} \end{aligned}$$

Solution set: $\left\{-2, -\frac{2}{3}\right\}$

19.
$$\begin{aligned} 3x^2 - 5x &= 0 \\ x(3x-5) &= 0 \end{aligned}$$
$$x = 0 \quad \text{or} \quad 3x - 5 = 0$$
$$3x = 5$$
$$x = \tfrac{5}{3}$$

Solution set: $\left\{0, \frac{5}{3}\right\}$

20.
$$\begin{aligned} 5m(m-1) &= 2(1-m) \\ 5m^2 - 5m &= 2 - 2m \\ 5m^2 - 3m - 2 &= 0 \\ (5m+2)(m-1) &= 0 \end{aligned}$$
$$5m + 2 = 0 \quad \text{or} \quad m - 1 = 0$$
$$5m = -2 \quad \text{or} \quad m = 1$$
$$m = -\tfrac{2}{5}$$

Solution set: $\left\{-\frac{2}{5}, 1\right\}$

21. Using $A = LW$, substitute 40 for A, $x + 7$ for L, and $2x + 3$ for W.
$$\begin{aligned} A &= LW \\ 40 &= (x+7)(2x+3) \\ 40 &= 2x^2 + 3x + 14x + 21 \\ 40 &= 2x^2 + 17x + 21 \\ 0 &= 2x^2 + 17x - 19 \\ 0 &= (2x+19)(x-1) \end{aligned}$$
$$2x + 19 = 0 \quad \text{or} \quad x - 1 = 0$$
$$2x = -19 \quad \text{or} \quad x = 1$$
$$x = -\tfrac{19}{2}$$

Length and width will be negative if $x = -\frac{19}{2}$, so reject it as a possible solution. Since $x = 1$, $x + 7 = 1 + 7 = 8$ and $2x + 3 = 2(1) + 3 = 5$.

The length is 8 inches, and the width is 5 inches.

22. Substitute 128 for $f(t)$ in the equation.
$$\begin{aligned} f(t) &= -16t^2 + 96t \\ 128 &= -16t^2 + 96t \\ 16t^2 - 96t + 128 &= 0 \\ 16\left(t^2 - 6t + 8\right) &= 0 \\ 16(t-4)(t-2) &= 0 \end{aligned}$$
$$t - 4 = 0 \quad \text{or} \quad t - 2 = 0$$
$$t = 4 \quad \text{or} \quad t = 2$$

The ball is 128 ft high at 2 seconds (on the way up) and again at 4 seconds (on the way down).

Cumulative Review Exercises (Chapters 1–7)

1. $-2(m-3) = -2(m) - 2(-3) = -2m + 6$

2. $-(-4m+3) = -(-4m) - (3) = 4m - 3$

3.
$$\begin{aligned} &3x^2 - 4x + 4 + 9x - x^2 \\ &= 3x^2 - x^2 - 4x + 9x + 4 \\ &= 2x^2 + 5x + 4 \end{aligned}$$

For Exercises 4–7, let $p = -4$, $q = -2$, and $r = 5$.

4.
$$\begin{aligned} -3(2q - 3p) &= -3[2(-2) - 3(-4)] \\ &= -3(-4 + 12) \\ &= -3(8) = -24 \end{aligned}$$

5.
$$\begin{aligned} 8r^2 + q^2 &= 8(5)^2 + (-2)^2 \\ &= 8(25) + 4 \\ &= 200 + 4 = 204 \end{aligned}$$

6.
$$\begin{aligned} \frac{\sqrt{r}}{-p + 2q} &= \frac{\sqrt{5}}{-(-4) + 2(-2)} \\ &= \frac{\sqrt{5}}{4 - 4} = \frac{\sqrt{5}}{0} \end{aligned}$$
This is *undefined* since the denominator is zero.

7.
$$\begin{aligned} \frac{rp + 6r^2}{p^2 + q - 1} &= \frac{5(-4) + 6(5)^2}{(-4)^2 + (-2) - 1} \\ &= \frac{-20 + 6(25)}{16 - 2 - 1} \\ &= \frac{-20 + 150}{13} \\ &= \frac{130}{13} = 10 \end{aligned}$$

8.
$$\begin{aligned} 2z - 5 + 3z &= 4 - (z + 2) \\ 5z - 5 &= 4 - z - 2 \\ 5z - 5 &= 2 - z \\ 6z &= 7 \\ z &= \tfrac{7}{6} \end{aligned}$$

Solution set: $\left\{\frac{7}{6}\right\}$

9.
$$\begin{aligned} \frac{3a - 1}{5} + \frac{a + 2}{2} &= -\frac{3}{10} \\ 2(3a - 1) + 5(a + 2) &= -3 \qquad \textit{Multiply by 10.} \\ 6a - 2 + 5a + 10 &= -3 \\ 11a + 8 &= -3 \\ 11a &= -11 \\ a &= -1 \end{aligned}$$

Solution set: $\{-1\}$

10.
$$\begin{aligned} -\tfrac{4}{3}d &\geq -5 \\ -\tfrac{3}{4}\left(-\tfrac{4}{3}d\right) &\leq -\tfrac{3}{4}(-5) \\ d &\leq \tfrac{15}{4} \end{aligned}$$

Solution set: $\left(-\infty, \frac{15}{4}\right]$

11. $3 - 2(m+3) < 4m$
$3 - 2m - 6 < 4m$
$-2m - 3 < 4m$
$-6m < 3$
$m > -\frac{3}{6}$ or $-\frac{1}{2}$

Solution set: $\left(-\frac{1}{2}, \infty\right)$

12. $2k + 4 < 10$ and $3k - 1 > 5$
$2k < 6$ and $3k > 6$
$k < 3$ and $k > 2$

The overlap of these inequalities is the set of all numbers between 2 and 3.

Solution set: $(2, 3)$

13. $2k + 4 > 10$ or $3k - 1 < 5$
$2k > 6$ or $3k < 6$
$k > 3$ or $k < 2$

The solution set is the set of numbers that are either greater than 3 or less than 2.

Solution set: $(-\infty, 2) \cup (3, \infty)$

14. $|5x + 3| - 10 = 3$
$|5x + 3| = 13$

$5x + 3 = 13$ or $5x + 3 = -13$
$5x = 10$ or $5x = -16$
$x = 2$ or $x = -\frac{16}{5}$

Solution set: $\left\{-\frac{16}{5}, 2\right\}$

15. $|x + 2| < 9$
$-9 < x + 2 < 9$
$-11 < x < 7$

Solution set: $(-11, 7)$

16. $|2y - 5| \geq 9$

$2y - 5 \geq 9$ or $2y - 5 \leq -9$
$2y \geq 14$ or $2y \leq -4$
$y \geq 7$ or $y \leq -2$

Solution set: $(-\infty, -2] \cup [7, \infty)$

17. Solve $V = lwh$ for h.

$$\frac{V}{lw} = \frac{lwh}{lw} \quad \textit{Divide by lw.}$$
$$\frac{V}{lw} = h$$

18. Let x be the time it takes for the planes to be 2100 mi apart. Use the formula $d = rt$ to complete the table.

Plane	r	t	d
Eastbound	550	x	$550x$
Westbound	500	x	$500x$

The total distance is 2100 mi.

$$550x + 500x = 2100$$
$$1050x = 2100$$
$$x = 2$$

It will take 2 hours for the planes to be 2100 miles apart.

19. $4x + 2y = -8$

Draw a line through the x- and y-intercepts, $(-2, 0)$ and $(0, -4)$, respectively.

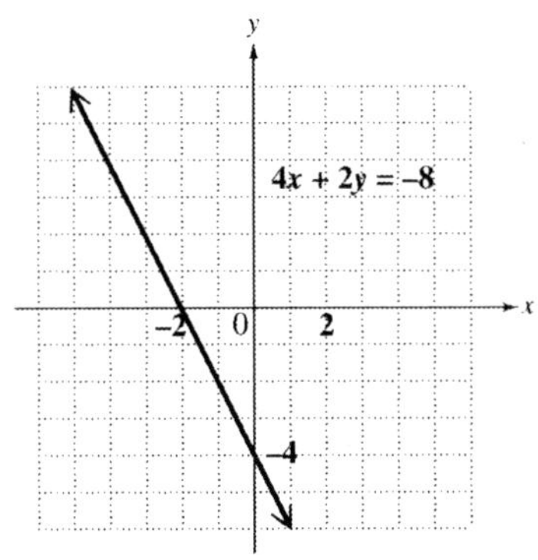

20. The slope m of the line through $(-4, 8)$ and $(-2, 6)$ is

$$m = \frac{6 - 8}{-2 - (-4)} = \frac{-2}{2} = -1.$$

21. $y = -3$ is an equation of a horizontal line. Its slope is 0.

22. $f(x) = 2x + 7$
$f(-4) = 2(-4) + 7$
$= -8 + 7 = -1$

23. To find the x-intercept of the graph of $f(x) = 2x + 7$, let $f(x) = 0$ (which is the same as letting $y = 0$) and solve for x.

$$0 = 2x + 7$$
$$-7 = 2x$$
$$-\frac{7}{2} = x$$

The x-intercept is $\left(-\frac{7}{2}, 0\right)$.

24. $f(0) = 7$, so the y-intercept is $(0, 7)$.

25. $3x - 2y = -7 \quad (1)$
$2x + 3y = 17 \quad (2)$
To eliminate y, multiply (1) by 3 and (2) by 2, and then add the resulting equations.

$$\begin{array}{rcrl} 9x - 6y & = & -21 & 3 \times (1) \\ 4x + 6y & = & 34 & 2 \times (2) \\ \hline 13x & = & 13 & \\ x & = & 1 & \end{array}$$

Substitute 1 for x in (1).

$$\begin{aligned} 3x - 2y &= -7 \quad (1) \\ 3(1) - 2y &= -7 \\ -2y &= -10 \\ y &= 5 \end{aligned}$$

Solution set: $\{(1, 5)\}$

26. $2x + 3y - 6z = 5 \quad (1)$
$8x - y + 3z = 7 \quad (2)$
$3x + 4y - 3z = 7 \quad (3)$

To eliminate z, add (2) and (3).

$$\begin{array}{rcrl} 8x - y + 3z & = & 7 & (2) \\ 3x + 4y - 3z & = & 7 & (3) \\ \hline 11x + 3y & = & 14 & (4) \end{array}$$

To eliminate z again, multiply (2) by 2 and add the result to (1).

$$\begin{array}{rcrl} 2x + 3y - 6z & = & 5 & (1) \\ 16x - 2y + 6z & = & 14 & 2 \times (2) \\ \hline 18x + y & = & 19 & (5) \end{array}$$

To eliminate y, multiply (5) by -3 and add the result to (4).

$$\begin{array}{rcrl} 11x + 3y & = & 14 & (4) \\ -54x - 3y & = & -57 & -3 \times (5) \\ \hline -43x & = & -43 & \\ x & = & 1 & \end{array}$$

From (5), $18(1) + y = 19$, so $y = 1$.
To find z, let $x = 1$ and $y = 1$ in (1).

$$\begin{aligned} 2x + 3y - 6z &= 5 \quad (1) \\ 2(1) + 3(1) - 6z &= 5 \\ 5 - 6z &= 5 \\ -6z &= 0 \\ z &= 0 \end{aligned}$$

Solution set: $\{(1, 1, 0)\}$

27. $\left(3x^2y^{-1}\right)^{-2}\left(2x^{-3}y\right)^{-1}$
$= 3^{-2}x^{-4}y^2 \cdot 2^{-1}x^3y^{-1}$
$= 3^{-2}2^{-1}x^{-1}y$
$= \dfrac{y}{3^2 \cdot 2x} = \dfrac{y}{18x}$

28. $\dfrac{5m^{-2}y^3}{3m^{-3}y^{-1}} = \dfrac{5}{3} \cdot \dfrac{m^{-2}}{m^{-3}} \cdot \dfrac{y^3}{y^{-1}}$
$= \dfrac{5}{3}m^{-2-(-3)}y^{3-(-1)}$
$= \dfrac{5}{3}m^1y^4$ or $\dfrac{5my^4}{3}$

29. $\left(3x^3 + 4x^2 - 7\right) - \left(2x^3 - 8x^2 + 3x\right)$
$= 3x^3 + 4x^2 - 7 - 2x^3 + 8x^2 - 3x$
$= x^3 + 12x^2 - 3x - 7$

30. $(7x + 3y)^2 = (7x)^2 + 2(7x)(3y) + (3y)^2$
$= 49x^2 + 42xy + 9y^2$

31. $(2p + 3)\left(5p^2 - 4p - 8\right)$
$= 10p^3 - 8p^2 - 16p + 15p^2 - 12p - 24$
$= 10p^3 + 7p^2 - 28p - 24$

32. $16w^2 + 50wz - 21z^2$
Two integer factors whose product is $(16)(-21) = -336$ and whose sum is 50 are 56 and -6.

$= 16w^2 + 56wz - 6wz - 21z^2$
$= 8w(2w + 7z) - 3z(2w + 7z)$
$= (2w + 7z)(8w - 3z)$

33. $4x^2 - 4x + 1 - y^2$
Group the first three terms.
$= \left(4x^2 - 4x + 1\right) - y^2$
$= (2x - 1)^2 - y^2$
$= [(2x - 1) + y][(2x - 1) - y]$
$= (2x - 1 + y)(2x - 1 - y)$

34. $4y^2 - 36y + 81$
$= (2y)^2 - 2(2y)(9) + 9^2$
$= (2y - 9)^2$

35. $100x^4 - 81 = \left(10x^2\right)^2 - 9^2$
$= \left(10x^2 + 9\right)\left(10x^2 - 9\right)$

36. $8p^3 + 27$
$= (2p)^3 + 3^3$
$= (2p + 3)\left[(2p)^2 - (2p)(3) + 3^2\right]$
$= (2p + 3)\left(4p^2 - 6p + 9\right)$

37. $(p + 4)(2p + 3)(p - 1) = 0$

$p + 4 = 0$ or $2p + 3 = 0$ or $p - 1 = 0$
$p = -4$ or $p = -\frac{3}{2}$ or $p = 1$

Solution set: $\left\{-4, -\frac{3}{2}, 1\right\}$

38.

$$\begin{aligned} 9q^2 &= 6q - 1 \\ 9q^2 - 6q + 1 &= 0 \\ (3q - 1)^2 &= 0 \\ 3q - 1 &= 0 \\ 3q &= 1 \\ q &= \tfrac{1}{3} \end{aligned}$$

Solution set: $\left\{\frac{1}{3}\right\}$

39. Let x be the length of the base. Then $x + 3$ will be the height. The area is 14 square feet. Use the formula $A = \frac{1}{2}bh$, and substitute 14 for A, x for b, and $x + 3$ for h.

$$\begin{aligned} \tfrac{1}{2}bh &= A \\ \tfrac{1}{2}(x)(x+3) &= 14 \\ x(x+3) &= 28 \quad \textit{Multiply by 2.} \\ x^2 + 3x &= 28 \\ x^2 + 3x - 28 &= 0 \\ (x+7)(x-4) &= 0 \end{aligned}$$

$$x + 7 = 0 \quad \text{or} \quad x - 4 = 0$$
$$x = -7 \quad \text{or} \quad x = 4$$

The length cannot be negative, so reject -7 as a solution. The only possible solution is 4. The base is 4 feet long.

40. Let x be the distance between the longer sides. (This is actually the width.) Then $x + 2$ will be the length of the longer side. The area of the rectangle is 288 in^2. Use the formula $LW = A$. Substitute 288 for A, $x + 2$ for L, and x for W.

$$\begin{aligned} (x+2)x &= 288 \\ x^2 + 2x &= 288 \\ x^2 + 2x - 288 &= 0 \\ (x+18)(x-16) &= 0 \end{aligned}$$

$$x + 18 = 0 \quad \text{or} \quad x - 16 = 0$$
$$x = -18 \quad \text{or} \quad x = 16$$

The distance cannot be negative, so reject -18 as a solution. The only possible solution is 16.
The distance between the longer sides is 16 inches, and the length of the longer sides is $16 + 2 = 18$ inches.

CHAPTER 8 RATIONAL EXPRESSIONS AND FUNCTIONS

8.1 Rational Expressions and Functions; Multiplying and Dividing

8.1 Margin Exercises

1. **(a)** $f(x) = \dfrac{x+4}{x-6}$

Set the denominator equal to zero, and solve the equation.

$$\begin{aligned} x - 6 &= 0 \\ x &= 0 \end{aligned}$$

The number 6 makes the function undefined, so it is not in the domain of f.

Domain: $\{x \mid x \neq 6\}$

(b) $f(x) = \dfrac{x+6}{x^2 - x - 6}$

Set the denominator equal to zero, and solve the equation.

$$\begin{aligned} x^2 - x - 6 &= 0 \\ (x-3)(x+2) &= 0 \\ x - 3 = 0 \quad \text{or} \quad x + 2 &= 0 \\ x = 3 \quad \text{or} \quad x &= -2 \end{aligned}$$

Both 3 and -2 make the function undefined.

Domain: $\{x \mid x \neq -2, 3\}$

(c) $f(x) = \dfrac{3+2x}{5}$

The denominator, 5, can never be 0, so the domain includes all real numbers, $(-\infty, \infty)$.

(d) $f(x) = \dfrac{2}{x^2+1}$

Set the denominator equal to zero, and solve the equation.

$$\begin{aligned} x^2 + 1 &= 0 \\ x^2 &= -1 \end{aligned}$$

There is no real number whose square is -1, so the domain includes all real numbers, $(-\infty, \infty)$.

2. **(a)** $\dfrac{y^2+2y-3}{y^2-3y+2} = \dfrac{(y+3)(y-1)}{(y-2)(y-1)}$ *Factor.*

$= \dfrac{y+3}{y-2} \cdot 1$

$= \dfrac{y+3}{y-2}$ *Lowest terms*

(b) $\dfrac{3y+9}{y^2-9} = \dfrac{3(y+3)}{(y-3)(y+3)} = \dfrac{3}{y-3}$

(c) $\dfrac{y+2}{y^2+4}$

The denominator cannot be factored, so this expression cannot be simplified further and is in lowest terms.

(d) $\dfrac{1+p^3}{1+p}$

$= \dfrac{(1+p)(1-p+p^2)}{1+p}$ *Factor the sum of cubes.*

$= 1 - p + p^2$ *Lowest terms*

(e) $\dfrac{3x+3y+rx+ry}{5x+5y-rx-ry}$

$= \dfrac{(3x+3y)+(rx+ry)}{(5x+5y)-(rx+ry)}$ *Group terms.*

$= \dfrac{3(x+y)+r(x+y)}{5(x+y)-r(x+y)}$ *Factor within groups.*

$= \dfrac{(x+y)(3+r)}{(x+y)(5-r)}$ *Factor by grouping.*

$= \dfrac{3+r}{5-r}$ *Lowest terms*

3. **(a)** $\dfrac{y-2}{2-y} = \dfrac{y-2}{-1(y-2)} = \dfrac{1}{-1} = -1$

(b) $\dfrac{8-b}{8+b}$

Since this expression cannot be simplified further, it is in lowest terms.

(c)

$\dfrac{p-2}{4-p^2} = \dfrac{p-2}{(2-p)(2+p)}$ *Factor.*

$= \dfrac{p-2}{-1(p-2)(2+p)}$

$= \dfrac{-1}{2+p}$ *Fundamental property*

4. **(a)** $\dfrac{2r+4}{5r} \cdot \dfrac{3r}{5r+10} = \dfrac{2(r+2)}{5r} \cdot \dfrac{3r}{5(r+2)}$

$= \dfrac{2 \cdot 3}{5 \cdot 5} = \dfrac{6}{25}$

(b) $\dfrac{c^2+2c}{c^2-4} \cdot \dfrac{c^2-4c+4}{c^2-c}$

$= \dfrac{c(c+2)}{(c-2)(c+2)} \cdot \dfrac{(c-2)(c-2)}{c(c-1)}$

$= \dfrac{c-2}{c-1}$

(c) $\frac{m^2-16}{m+2} \cdot \frac{1}{m+4}$

$= \frac{(m-4)(m+4)}{m+2} \cdot \frac{1}{m+4}$

$= \frac{m-4}{m+2}$

(d) $\frac{x-3}{x^2+2x-15} \cdot \frac{x^2-25}{x^2+3x-40}$

$= \frac{x-3}{(x+5)(x-3)} \cdot \frac{(x-5)(x+5)}{(x+8)(x-5)}$

$= \frac{1}{x+8}$

5. To find the reciprocal, interchange the numerator and denominator of the rational expression.

	Expression	Reciprocal
(a)	$\frac{-3}{r}$	$\frac{r}{-3}$
(b)	$\frac{7}{y+8}$	$\frac{y+8}{7}$
(c)	$\frac{a^2+7a}{2a-1}$	$\frac{2a-1}{a^2+7a}$
(d)	$\frac{0}{-5}$	Undefined; there is no reciprocal for 0.

6. In each case, multiply by the reciprocal of the divisor.

(a) $\frac{16k^2}{5} \div \frac{3k}{10} = \frac{16k^2}{5} \cdot \frac{10}{3k}$

$= \frac{16k^2}{5} \cdot \frac{2 \cdot 5}{3k} = \frac{32k}{3}$

(b) $\frac{5p+2}{6} \div \frac{15p+6}{5} = \frac{5p+2}{6} \cdot \frac{5}{15p+6}$

$= \frac{5p+2}{6} \cdot \frac{5}{3(5p+2)}$

$= \frac{5}{18}$

(c) $\frac{y^2-2y-3}{y^2+4y+4} \div \frac{y^2-1}{y^2+y-2}$

$= \frac{y^2-2y-3}{y^2+4y+4} \cdot \frac{y^2+y-2}{y^2-1}$

$= \frac{(y-3)(y+1)}{(y+2)(y+2)} \cdot \frac{(y+2)(y-1)}{(y+1)(y-1)}$

$= \frac{y-3}{y+2}$

8.1 Section Exercises

1. $\frac{x-3}{x+4} = \frac{(-1)(x-3)}{(-1)(x+4)}$

$= \frac{-x+3}{-x-4}$ or $\frac{3-x}{-x-4}$ **(C)**

2. $\frac{x+3}{x-4} = \frac{(-1)(x+3)}{(-1)(x-4)}$

$= \frac{-x-3}{-x+4}$ or $\frac{-x-3}{4-x}$ **(A)**

3. $\frac{x-3}{x-4} = \frac{(-1)(x-3)}{(-1)(x-4)}$

$= \frac{-x+3}{-x+4}$ **(D)**

4. $\frac{x+3}{x+4} = \frac{(-1)(x+3)}{(-1)(x+4)}$

$= \frac{-x-3}{-x-4}$ **(B)**

5. $\frac{3-x}{x+4} = \frac{(-1)(3-x)}{(-1)(x+4)}$

$= \frac{-3+x}{-x-4}$ or $\frac{x-3}{-x-4}$ **(E)**

6. $\frac{x+3}{4-x} = \frac{(-1)(x+3)}{(-1)(4-x)}$

$= \frac{-x-3}{-4+x}$ or $\frac{-x-3}{x-4}$ **(F)**

7. Replacing x with 2 makes the denominator 0 and the value of the expression undefined. To find the values excluded from the domain, set the denominator equal to 0 and solve the equation. All solutions of the equation are excluded from the domain.

8. There is no number that makes the denominator 0. In fact, the denominator is greater than or equal to 3 for any replacement of x. Thus, the domain includes all real numbers.

9. $f(x) = \frac{x}{x-7}$

Set the denominator equal to zero, and solve the equation.

$$x - 7 = 0$$
$$x = 7$$

The number 7 makes the rational expression undefined, so 7 is not in the domain of the function. In set notation, the domain is $\{x \mid x \neq 7\}$.

10. $f(x) = \dfrac{x}{x+3}$ is undefined when the denominator $x + 3 = 0$. Solving this equation gives $x = -3$, so $x = -3$ makes the rational expression undefined and -3 is not in the domain of the function. In set notation, the domain is $\{x \mid x \neq -3\}$.

11. $f(x) = \dfrac{6x-5}{7x+1}$

Set the denominator equal to zero, and solve the equation.

$$\begin{aligned} 7x + 1 &= 0 \\ 7x &= -1 \\ x &= -\tfrac{1}{7} \end{aligned}$$

The number $-\frac{1}{7}$ makes the rational expression undefined, so $-\frac{1}{7}$ is not in the domain of the function. In set notation, the domain is $\{x \mid x \neq -\frac{1}{7}\}$.

12. $f(x) = \dfrac{8x-3}{2x+7}$ is undefined when the denominator $2x + 7 = 0$.

$$\begin{aligned} 2x + 7 &= 0 \\ 2x &= -7 \\ x &= -\tfrac{7}{2} \end{aligned}$$

In set notation, the domain is $\{x \mid x \neq -\frac{7}{2}\}$.

13. $f(x) = \dfrac{12x+3}{x}$

Set the denominator equal to zero and solve.

$$x = 0$$

The number 0 makes the rational expression undefined, so 0 is not in the domain of the function. In set notation, the domain is $\{x \mid x \neq 0\}$.

14. $f(x) = \dfrac{9x+8}{x}$ is undefined when the denominator $x = 0$. In set notation, the domain is $\{x \mid x \neq 0\}$.

15. $f(x) = \dfrac{3x+1}{2x^2+x-6}$

Set the denominator equal to zero and solve.

$$\begin{aligned} 2x^2 + x - 6 &= 0 \\ (x+2)(2x-3) &= 0 \end{aligned}$$

$$\begin{aligned} x + 2 &= 0 \quad \text{or} \quad & 2x - 3 &= 0 \\ x &= -2 \quad \text{or} \quad & 2x &= 3 \\ & & x &= \tfrac{3}{2} \end{aligned}$$

The numbers -2 and $\frac{3}{2}$ are not in the domain of the function. In set notation, the domain is $\{x \mid x \neq -2, \frac{3}{2}\}$.

16. $f(x) = \dfrac{2x+4}{3x^2+11x-42}$ is undefined when the denominator is equal to 0.

$$\begin{aligned} 3x^2 + 11x - 42 &= 0 \\ (x+6)(3x-7) &= 0 \end{aligned}$$

$$\begin{aligned} x + 6 &= 0 \quad \text{or} \quad & 3x - 7 &= 0 \\ x &= -6 \quad \text{or} \quad & 3x &= 7 \\ & & x &= \tfrac{7}{3} \end{aligned}$$

In set notation, the domain is $\{x \mid x \neq -6, \frac{7}{3}\}$.

17. $f(x) = \dfrac{x+2}{14}$

The denominator is never zero, so all numbers are in the domain of the function. In set notation, the domain is $(-\infty, \infty)$.

18. $f(x) = \dfrac{x-9}{26}$ is never undefined, so all numbers are in the domain of the function. In set notation, the domain is $(-\infty, \infty)$.

19. $f(x) = \dfrac{2x^2-3x+4}{3x^2+8}$

Set the denominator equal to zero and solve.

$$\begin{aligned} 3x^2 + 8 &= 0 \\ 3x^2 &= -8 \\ x^2 &= -\tfrac{8}{3} \end{aligned}$$

The square of any real number x is positive or zero, so this equation has no solution. There are no real numbers which make this rational expression undefined, so all numbers are in the domain of the function. In set notation, the domain is $(-\infty, \infty)$.

20. $f(x) = \dfrac{9x^2-8x+3}{4x^2+1}$

Set the denominator equal to zero and solve.

$$\begin{aligned} 4x^2 + 1 &= 0 \\ 4x^2 &= -1 \\ x^2 &= -\tfrac{1}{4} \end{aligned}$$

The square of any real number x is positive or zero, so this equation has no solution. There are no real numbers which make this rational expression undefined, so all numbers are in the domain of the function. In set notation, the domain is $(-\infty, \infty)$.

21. **(a)** $\dfrac{x^2+4x}{x+4}$

The two terms in the numerator are x^2 and $4x$. The two terms in the denominator are x and 4.

(b) First factor the numerator, getting $x(x+4)$, then divide the numerator and denominator by the common factor of $x+4$ to get $\frac{x}{1}$ or x.

$$\begin{aligned}\frac{x^2+4x}{x+4} &= \frac{x(x+4)}{x+4}\\ &= x \cdot 1 = x\end{aligned}$$

22. **A.** The numerator and denominator of

$$\frac{x^2+2}{x^2}$$

have no common factors, so the expression cannot be simplified.

B. The numerator and denominator of

$$\frac{x^2+2}{2}$$

have no common factors. The expression is simplified.

C. The numerator and denominator of

$$\frac{x^2+y^2}{y^2}$$

have no common factors. The expression is simplified.

D. $x^2-5x = x(x-5)$, so

$$\begin{aligned}\frac{x^2-5x}{x} &= \frac{x(x-5)}{x}\\ &= 1(x-5)\\ &= x-5.\end{aligned}$$

This expression, **D**, is the only one that can be simplified.

23. **A.** $\dfrac{3-x}{x-4} = \dfrac{-1(x-3)}{-1(4-x)} = \dfrac{x-3}{4-x}$

B. $\dfrac{x+3}{4+x}$ cannot be transformed to equal $\dfrac{x-3}{4-x}$.

C. $-\dfrac{3-x}{4-x} = \dfrac{-(3-x)}{4-x} = \dfrac{x-3}{4-x}$

D. $-\dfrac{x-3}{x-4} = \dfrac{x-3}{-(x-4)} = \dfrac{x-3}{4-x}$

Only the expression in **B** is *not* equivalent to $\dfrac{x-3}{4-x}$.

24. **A.** $\dfrac{2x+3}{2x-3}$ cannot be simplified.

B. $\dfrac{2x-3}{3-2x} = \dfrac{2x-3}{-(2x-3)} = -1$

C. $\dfrac{2x+3}{3+2x} = \dfrac{2x+3}{2x+3} = 1$

D. $\dfrac{2x+3}{-2x-3} = \dfrac{2x+3}{-(2x+3)} = -1$

The correct choices are **B** and **D**.

25. $\dfrac{x^2(x+1)}{x(x+1)} = \dfrac{x}{1} \cdot \dfrac{x(x+1)}{x(x+1)} = \dfrac{x}{1} \cdot 1 = x$

26. $\dfrac{y^3(y-4)}{y^2(y-4)} = \dfrac{y}{1} \cdot \dfrac{y^2(y-4)}{y^2(y-4)} = \dfrac{y}{1} \cdot 1 = y$

27. $\dfrac{(x+4)(x-3)}{(x+5)(x+4)} = \dfrac{x-3}{x+5} \cdot \dfrac{x+4}{x+4} = \dfrac{x-3}{x+5}$

28. $\dfrac{(2x+7)(x-1)}{(2x+3)(2x+7)} = \dfrac{x-1}{2x+3} \cdot \dfrac{2x+7}{2x+7} = \dfrac{x-1}{2x+3}$

29. $$\begin{aligned}\frac{4x(x+3)}{8x^2(x-3)} &= \frac{(x+3) \cdot 4x}{2x(x-3) \cdot 4x}\\ &= \frac{x+3}{2x(x-3)}\end{aligned}$$

30. $$\begin{aligned}\frac{5y^2(y+8)}{15y(y-8)} &= \frac{5y \cdot y(y+8)}{3 \cdot 5y(y-8)}\\ &= \frac{y(y+8)}{3(y-8)}\end{aligned}$$

31. $\dfrac{3x+7}{3}$ Since the numerator and denominator have no common factors, the expression is already in lowest terms.

32. $\dfrac{4x-9}{4}$ Since the numerator and denominator have no common factors, the expression is already in lowest terms.

33. $\dfrac{6m+18}{7m+21} = \dfrac{6(m+3)}{7(m+3)} = \dfrac{6}{7}$

34. $\dfrac{5r-20}{3r-12} = \dfrac{5(r-4)}{3(r-4)} = \dfrac{5}{3}$

35. $\dfrac{3z^2+z}{18z+6} = \dfrac{z(3z+1)}{6(3z+1)} = \dfrac{z}{6}$

36. $\dfrac{2x^2-5x}{16x-40} = \dfrac{x(2x-5)}{8(2x-5)} = \dfrac{x}{8}$

37. $\dfrac{2t+6}{t^2-9} = \dfrac{2(t+3)}{(t-3)(t+3)} = \dfrac{2}{t-3}$

38. $\dfrac{5s-25}{s^2-25} = \dfrac{5(s-5)}{(s+5)(s-5)} = \dfrac{5}{s+5}$

39. $\frac{x^2+2x-15}{x^2+6x+5} = \frac{(x+5)(x-3)}{(x+5)(x+1)}$
$= \frac{x-3}{x+1}$

40. $\frac{y^2-5y-14}{y^2+y-2} = \frac{(y-7)(y+2)}{(y-1)(y+2)}$
$= \frac{y-7}{y-1}$

41. $\frac{8x^2-10x-3}{8x^2-6x-9} = \frac{(4x+1)(2x-3)}{(4x+3)(2x-3)}$
$= \frac{4x+1}{4x+3}$

42. $\frac{12x^2-4x-5}{8x^2-6x-5} = \frac{(6x-5)(2x+1)}{(4x-5)(2x+1)}$
$= \frac{6x-5}{4x-5}$

43. $\frac{a^3+b^3}{a+b} = \frac{(a+b)(a^2-ab+b^2)}{a+b}$
$= a^2-ab+b^2$

44. $\frac{r^3-s^3}{r-s} = \frac{(r-s)(r^2+rs+s^2)}{r-s}$
$= r^2+rs+s^2$

45. $\frac{2c^2+2cd-60d^2}{2c^2-12cd+10d^2}$
$= \frac{2(c^2+cd-30d^2)}{2(c^2-6cd+5d^2)}$
$= \frac{2(c+6d)(c-5d)}{2(c-d)(c-5d)}$
$= \frac{c+6d}{c-d}$

46. $\frac{3s^2-9st-54t^2}{3s^2-6st-72t^2}$
$= \frac{3(s^2-3st-18t^2)}{3(s^2-2st-24t^2)}$
$= \frac{3(s-6t)(s+3t)}{3(s-6t)(s+4t)}$
$= \frac{s+3t}{s+4t}$

47. $\frac{ac-ad+bc-bd}{ac-ad-bc+bd}$
$= \frac{a(c-d)+b(c-d)}{a(c-d)-b(c-d)}$
$= \frac{(c-d)(a+b)}{(c-d)(a-b)}$ *Factor by grouping.*
$= \frac{a+b}{a-b}$

48. $\frac{2xy+2xw+y+w}{2xy+y-2xw-w}$
$= \frac{2x(y+w)+1(y+w)}{y(2x+1)-w(2x+1)}$
$= \frac{(y+w)(2x+1)}{(2x+1)(y-w)}$
$= \frac{y+w}{y-w}$

49. $\frac{7-b}{b-7} = \frac{-1(b-7)}{b-7} = -1$

50. $\frac{r-13}{13-r} = \frac{r-13}{-1(r-13)} = -1$

In Exercises 51–56, there are other acceptable ways to express each answer.

51. $\frac{x^2-y^2}{y-x} = \frac{(x-y)(x+y)}{y-x}$
$= \frac{-1(y-x)(x+y)}{y-x}$
$= -(x+y)$

52. $\frac{m^2-n^2}{n-m} = \frac{(m+n)(m-n)}{-1(m-n)}$
$= \frac{m+n}{-1} = -(m+n)$

53. $\frac{(a-3)(x+y)}{(3-a)(x-y)} = \frac{(a-3)(x+y)}{-1(a-3)(x-y)}$
$= \frac{x+y}{-1(x-y)}$
$= -\frac{x+y}{x-y}$

54. $\frac{(8-p)(x+2)}{(p-8)(x-2)} = \frac{-(p-8)(x+2)}{(p-8)(x-2)}$
$= -\frac{x+2}{x-2}$

55. $\frac{5k-10}{20-10k} = \frac{-5(2-k)}{10(2-k)} = \frac{-5}{10} = -\frac{1}{2}$

56. $\frac{7x-21}{63-21x} = \frac{7(x-3)}{-21(x-3)}$
$= \frac{7(x-3)}{-3 \cdot 7(x-3)} = -\frac{1}{3}$

57. $\frac{a^2-b^2}{a^2+b^2} = \frac{(a+b)(a-b)}{a^2+b^2}$
The numerator and denominator have no common factors except 1, so the original expression is already in lowest terms.

58. $\frac{p^2+q^2}{p^2-q^2}=\frac{p^2+q^2}{(p+q)(p-q)}$

The numerator and denominator have no common factors except 1, so the original expression is already in lowest terms.

59. $\frac{(x+2)(x+1)}{(x+3)(x-2)}\cdot\frac{(x+3)(x+4)}{(x+2)(x+1)}$

$=\frac{(x+1)(x+2)(x+3)}{(x+1)(x+2)(x+3)}\cdot\frac{x+4}{x-2}=\frac{x+4}{x-2}$

60. $\frac{(x+3)(x-4)}{(x-4)(x+2)}\cdot\frac{(x+5)(x-6)}{(x+3)(x-6)}$

$=\frac{(x+3)(x-4)(x-6)}{(x+3)(x-4)(x-6)}\cdot\frac{x+5}{x+2}=\frac{x+5}{x+2}$

61. $\frac{(2x+3)(x-4)}{(x+8)(x-4)}\div\frac{(x-4)(x+2)}{(x-4)(x+8)}$

$=\frac{2x+3}{x+8}\div\frac{x+2}{x+8}$

$=\frac{2x+3}{x+8}\cdot\frac{x+8}{x+2}$

$=\frac{2x+3}{x+2}$

62. $\frac{(6x+5)(x-3)}{(x+9)(x-1)}\div\frac{(x-3)(2x+7)}{(x-1)(x+9)}$

$=\frac{(6x+5)(x-3)}{(x+9)(x-1)}\cdot\frac{(x-1)(x+9)}{(x-3)(2x+7)}$

$=\frac{(6x+5)(x-3)(x-1)(x+9)}{(x+9)(x-1)(x-3)(2x+7)}$

$=\frac{6x+5}{2x+7}$

63. $\frac{7t+7}{-6}\div\frac{4t+4}{15}$

$=\frac{7t+7}{-6}\cdot\frac{15}{4t+4}$

$=\frac{7(t+1)}{-2\cdot 3}\cdot\frac{3\cdot 5}{4(t+1)}$

$=-\frac{35}{8}$

64. $\frac{8z-16}{-20}\div\frac{3z-6}{40}=\frac{8(z-2)}{-20}\cdot\frac{40}{3(z-2)}$

$=\frac{8(z-2)\cdot 2\cdot 20}{-20\cdot 3(z-2)}$

$=-\frac{16}{3}$

65. $\frac{4x}{8x+4}\cdot\frac{14x+7}{6}=\frac{4x\cdot 7(2x+1)}{4(2x+1)\cdot 6}$

$=\frac{7x}{6}$

66. $\frac{12x-20}{5x}\cdot\frac{6}{9x-15}=\frac{4(3x-5)\cdot 2\cdot 3}{5x\cdot 3(3x-5)}$

$=\frac{8}{5x}$

For Exercises 67–70, there are several other ways to express the answer.

67. $\frac{p^2-25}{4p}\cdot\frac{2}{5-p}=\frac{(p+5)(p-5)2}{2\cdot 2p(-1)(p-5)}$

$=\frac{p+5}{(2p)(-1)}=-\frac{p+5}{2p}$

68. $\frac{a^2-1}{4a}\cdot\frac{2}{1-a}=\frac{(a-1)(a+1)\cdot 2}{4a\cdot(1-a)}$

$=\frac{2(-1)(1-a)(a+1)}{4a(1-a)}$

$=\frac{-2(a+1)}{4a}=-\frac{a+1}{2a}$

69. $\frac{m^2-49}{m+1}\div\frac{7-m}{m}$

$=\frac{(m-7)(m+7)}{m+1}\cdot\frac{m}{7-m}$

$=\frac{(-1)(7-m)(m+7)}{m+1}\cdot\frac{m}{7-m}$

$=\frac{-m(m+7)}{m+1}$

70. $\frac{k^2-4}{3k^2}\div\frac{2-k}{11k}$

$=\frac{(k+2)(k-2)}{3k^2}\cdot\frac{11k}{-(k-2)}$

$=\frac{-11(k+2)}{3k}$

71. $\frac{12x-10y}{3x+2y}\cdot\frac{6x+4y}{10y-12x}$

$=\frac{2(6x-5y)\cdot 2(3x+2y)}{(3x+2y)\cdot 2(5y-6x)}$

$=\frac{2(-1)(5y-6x)}{(5y-6x)}=-2$

72. $\frac{9s-12t}{2s+2t}\cdot\frac{3s+3t}{4t-3s}$

$=\frac{3(3s-4t)}{2(s+t)}\cdot\frac{3(s+t)}{-(3s-4t)}=-\frac{9}{2}$

73. $\frac{x^2-25}{x^2+x-20}\cdot\frac{x^2+7x+12}{x^2-2x-15}$

$=\frac{(x-5)(x+5)}{(x+5)(x-4)}\cdot\frac{(x+3)(x+4)}{(x-5)(x+3)}$

$=\frac{x+4}{x-4}$

74. $\dfrac{t^2-49}{t^2+4t-21} \cdot \dfrac{t^2+8t+15}{t^2-2t-35}$

$= \dfrac{(t+7)(t-7)}{(t-3)(t+7)} \cdot \dfrac{(t+3)(t+5)}{(t+5)(t-7)}$

$= \dfrac{t+3}{t-3}$

75. $\dfrac{6x^2+5xy-6y^2}{12x^2-11xy+2y^2} \div \dfrac{4x^2-12xy+9y^2}{8x^2-14xy+3y^2}$

$= \dfrac{(3x-2y)(2x+3y)}{(3x-2y)(4x-y)} \cdot \dfrac{(4x-y)(2x-3y)}{(2x-3y)(2x-3y)}$

$= \dfrac{2x+3y}{2x-3y}$

76. $\dfrac{8a^2-6ab-9b^2}{6a^2-5ab-6b^2} \div \dfrac{4a^2+11ab+6b^2}{9a^2+12ab+4b^2}$

$= \dfrac{(2a-3b)(4a+3b)}{(2a-3b)(3a+2b)} \cdot \dfrac{(3a+2b)(3a+2b)}{(4a+3b)(a+2b)}$

$= \dfrac{3a+2b}{a+2b}$

77. $\dfrac{3k^2+17kp+10p^2}{6k^2+13kp-5p^2} \div \dfrac{6k^2+kp-2p^2}{6k^2-5kp+p^2}$

$= \dfrac{(3k+2p)(k+5p)}{(3k-p)(2k+5p)} \cdot \dfrac{(3k-p)(2k-p)}{(3k+2p)(2k-p)}$

$= \dfrac{k+5p}{2k+5p}$

78. $\dfrac{16c^2+24cd+9d^2}{16c^2-16cd+3d^2} \div \dfrac{16c^2-9d^2}{16c^2-24cd+9d^2}$

$= \dfrac{(4c+3d)(4c+3d)}{(4c-3d)(4c-d)} \cdot \dfrac{(4c-3d)(4c-3d)}{(4c+3d)(4c-3d)}$

$= \dfrac{4c+3d}{4c-d}$

79. $\left(\dfrac{6k^2-13k-5}{k^2+7k} \div \dfrac{2k-5}{k^3+6k^2-7k}\right) \cdot \dfrac{k^2-5k+6}{3k^2-8k-3}$

Factor k from the denominator of the divisor; multiply by the reciprocal.

$= \left[\dfrac{6k^2-13k-5}{k^2+7k} \cdot \dfrac{k(k^2+6k-7)}{2k-5}\right] \cdot \dfrac{k^2-5k+6}{3k^2-8k-3}$

$= \left[\dfrac{(3k+1)(2k-5)}{k(k+7)} \cdot \dfrac{k(k+7)(k-1)}{2k-5}\right] \cdot \dfrac{(k-2)(k-3)}{(3k+1)(k-3)}$

$= (k-1)(k-2)$

80. $\left(\dfrac{2x^3+3x^2-2x}{3x-15} \div \dfrac{2x^3-x^2}{x^2-3x-10}\right) \cdot \dfrac{5x^2-10x}{3x^2+12x+12}$

$= \left[\dfrac{x(2x^2+3x-2)}{3(x-5)} \cdot \dfrac{(x-5)(x+2)}{x^2(2x-1)}\right] \cdot \dfrac{5x(x-2)}{3(x^2+4x+4)}$

$= \left[\dfrac{x(2x-1)(x+2)}{3(x-5)} \cdot \dfrac{(x-5)(x+2)}{x^2(2x-1)}\right] \cdot \dfrac{5x(x-2)}{3(x+2)(x+2)}$

$= \left[\dfrac{(x+2)(x+2)}{3x}\right] \cdot \dfrac{5x(x-2)}{3(x+2)(x+2)}$

$= \dfrac{5(x-2)}{9}$

8.2 Adding and Subtracting Rational Expressions

8.2 Margin Exercises

1. **(a)** $\dfrac{3m}{8} + \dfrac{5n}{8} = \dfrac{3m+5n}{8}$

(b) $\dfrac{7}{3a} + \dfrac{10}{3a} = \dfrac{7+10}{3a} = \dfrac{17}{3a}$

(c) $\dfrac{2}{y^2} - \dfrac{5}{y^2} = \dfrac{2-5}{y^2} = \dfrac{-3}{y^2} = -\dfrac{3}{y^2}$

(d) $\dfrac{a}{a+b} + \dfrac{b}{a+b} = \dfrac{a+b}{a+b} = 1$

(e) $\dfrac{2y-1}{y^2+y-2} - \dfrac{y}{y^2+y-2} = \dfrac{(2y-1)-y}{y^2+y-2}$

$= \dfrac{y-1}{(y+2)(y-1)}$

$= \dfrac{1}{y+2}$

2. **(a)** $5k^3s, 10ks^4$

Factor the denominators.

$5k^3s = 5 \cdot k^3 \cdot s$
$10ks^4 = 2 \cdot 5 \cdot k \cdot s^4$

The least common denominator (LCD) is the product of all the different factors, with each factor raised to the greatest power in any denominator.

$\text{LCD} = 2 \cdot 5 \cdot k^3 \cdot s^4$
$= 10k^3s^4$

(b) $3-x, 9-x^2$

Factor the denominators.

$$3-x=3-x$$
$$9-x^2=(3+x)(3-x)$$

The least common denominator (LCD) is the product of all the different factors, with each factor raised to the greatest power in any denominator.

$\text{LCD}=(3+x)(3-x)$

(c) $z, z+6$

Each denominator is already factored.

$\text{LCD}=z(z+6)$

(d) $2y^2-3y-2, 2y^2+3y+1$

Factor the denominators.

$$2y^2-3y-2=(2y+1)(y-2)$$
$$2y^2+3y+1=(2y+1)(y+1)$$

$\text{LCD}=(y-2)(2y+1)(y+1)$

(e) $x^2-2x+1, x^2-4x+3, 4x-4$

Factor the denominators.

$$x^2-2x+1=(x-1)(x-1)$$
$$x^2-4x+3=(x-1)(x-3)$$
$$4x-4=4(x-1)$$

$\text{LCD}=4(x-1)^2(x-3)$

3. **(a)** $\frac{6}{7}+\frac{1}{5}$

The LCD of 7 and 5 is 35. Multiply $\frac{6}{7}$ by $\frac{5}{5}$ and $\frac{1}{5}$ by $\frac{7}{7}$ so that each fraction has denominator 35. Then add the numerators.

$$\begin{aligned}\frac{6}{7}+\frac{1}{5}&=\frac{6\cdot5}{7\cdot5}+\frac{1\cdot7}{5\cdot7}\\&=\frac{30}{35}+\frac{7}{35}\\&=\frac{30+7}{35}\\&=\frac{37}{35}\end{aligned}$$

(b) $\frac{8}{3k}-\frac{2}{9k}$

The LCD for $3k$ and $9k$ is $9k$. To write $\frac{8}{3k}$ with a denominator of $9k$, multiply by $\frac{3}{3}$.

$$\begin{aligned}\frac{8}{3k}-\frac{2}{9k}&=\frac{8\cdot3}{3k\cdot3}-\frac{2}{9k}\\&=\frac{24}{9k}-\frac{2}{9k}\\&=\frac{24-2}{9k}\\&=\frac{22}{9k}\end{aligned}$$

(c) $\frac{2}{y}-\frac{1}{y+4}$

The LCD is $y(y+4)$. Rewrite each rational expression with this denominator.

$$\begin{aligned}\frac{2}{y}-\frac{1}{y+4}&=\frac{2(y+4)}{y(y+4)}-\frac{(1)y}{(y+4)y}\\&=\frac{2(y+4)}{y(y+4)}-\frac{y}{y(y+4)}\\&=\frac{2y+8-y}{y(y+4)}\\&=\frac{y+8}{y(y+4)}\end{aligned}$$

4. **(a)** $\frac{5x+7}{2x+7}-\frac{-x-14}{2x+7}$

The denominator is already the same for both rational expressions, so write them as a single expression. Be careful to apply the subtraction sign to both terms of the second expression.

$$\begin{aligned}&=\frac{5x+7-(-x-14)}{2x+7}\\&=\frac{5x+7+x+14}{2x+7}\\&=\frac{6x+21}{2x+7}\\&=\frac{3(2x+7)}{2x+7}\\&=3\end{aligned}$$

(b) $\frac{2}{r-2}-\frac{r}{r-1}$

The LCD is $(r-2)(r-1)$.

$$\begin{aligned}&=\frac{2(r-1)}{(r-2)(r-1)}-\frac{r(r-2)}{(r-1)(r-2)}\\&=\frac{2r-2-(r^2-2r)}{(r-2)(r-1)}\\&=\frac{2r-2-r^2+2r}{(r-2)(r-1)}\\&=\frac{-r^2+4r-2}{(r-2)(r-1)}\end{aligned}$$

5. **(a)** $\dfrac{8}{x-4}+\dfrac{2}{4-x}$

To get a common denominator of $x-4$, multiply both the numerator and denominator of the second expression by -1.

$$=\frac{8}{x-4}+\frac{2(-1)}{(4-x)(-1)}$$
$$=\frac{8}{x-4}+\frac{-2}{x-4}$$
$$=\frac{8+(-2)}{x-4}$$
$$=\frac{6}{x-4}$$

An equivalent answer is

$$\frac{-1(6)}{-1(x-4)}=\frac{-6}{4-x}.$$

(b) $\dfrac{9}{2x-9}-\dfrac{4}{9-2x}$

$$=\frac{9}{2x-9}-\frac{4(-1)}{(9-2x)(-1)}$$
$$=\frac{9}{2x-9}-\frac{-4}{2x-9}$$
$$=\frac{9-(-4)}{2x-9}$$
$$=\frac{13}{2x-9}$$

An equivalent answer is

$$\frac{-1(13)}{-1(2x-9)}=\frac{-13}{9-2x}.$$

6. $\dfrac{4}{x-5}+\dfrac{-2}{x}-\dfrac{10}{x^2-5x}$

$$=\frac{4}{x-5}+\frac{-2}{x}-\frac{10}{x(x-5)}$$

Factor denominators.

$$=\frac{4x}{(x-5)x}+\frac{-2(x-5)}{x(x-5)}-\frac{10}{x(x-5)}$$

LCD = x(x − 5)

$$=\frac{4x+(-2)(x-5)-10}{x(x-5)}$$
$$=\frac{4x-2x+10-10}{x(x-5)}$$
$$=\frac{2x}{x(x-5)}$$
$$=\frac{2}{x-5}$$

7. $\dfrac{-a}{a^2+3a-4}-\dfrac{4a}{a^2+7a+12}$

$$=\frac{-a}{(a+4)(a-1)}-\frac{4a}{(a+3)(a+4)}$$

The LCD is $(a+4)(a-1)(a+3)$.

$$=\frac{-a(a+3)}{(a+4)(a-1)(a+3)}$$
$$-\frac{4a(a-1)}{(a+3)(a+4)(a-1)}$$
$$=\frac{-a(a+3)-4a(a-1)}{(a+4)(a-1)(a+3)}$$
$$=\frac{-a^2-3a-4a^2+4a}{(a+4)(a-1)(a+3)}$$
$$=\frac{-5a^2+a}{(a+4)(a-1)(a+3)}$$

8. $\dfrac{4}{p^2-6p+9}+\dfrac{1}{p^2+2p-15}$

$$=\frac{4}{(p-3)(p-3)}+\frac{1}{(p+5)(p-3)}$$

The LCD is $(p-3)^2(p+5)$.

$$=\frac{4(p+5)}{(p-3)^2(p+5)}+\frac{1(p-3)}{(p+5)(p-3)^2}$$
$$=\frac{4(p+5)+1(p-3)}{(p-3)^2(p+5)}$$
$$=\frac{4p+20+p-3}{(p-3)^2(p+5)}$$
$$=\frac{5p+17}{(p-3)^2(p+5)}$$

8.2 Section Exercises

1. To add or subtract rational expressions that have a common denominator, first add or subtract the numerators. Then place the result over the common denominator. Write the answer in lowest terms.

2. To add or subtract rational expressions that have different denominators, first write each expression as an equivalent expression with the least common denominator. Then follow the procedure described in the answer to Exercise 1.

3. $\dfrac{7}{t}+\dfrac{2}{t}=\dfrac{7+2}{t}=\dfrac{9}{t}$

4. $\dfrac{5}{r}+\dfrac{9}{r}=\dfrac{5+9}{r}=\dfrac{14}{r}$

5. $\dfrac{11}{5x}-\dfrac{1}{5x}=\dfrac{11-1}{5x}=\dfrac{10}{5x}=\dfrac{2}{x}$

6. $\dfrac{7}{4y}-\dfrac{3}{4y}=\dfrac{7-3}{4y}=\dfrac{4}{4y}=\dfrac{1}{y}$

7. $\frac{5x+4}{6x+5}+\frac{x+1}{6x+5}$
$=\frac{5x+4+x+1}{6x+5}$
$=\frac{6x+5}{6x+5}=1$

8. $\frac{6y+12}{4y+3}+\frac{2y-6}{4y+3}=\frac{6y+12+2y-6}{4y+3}$
$=\frac{8y+6}{4y+3}$
$=\frac{2(4y+3)}{4y+3}=2$

9. $\frac{x^2}{x+5}-\frac{25}{x+5}=\frac{x^2-25}{x+5}$
$=\frac{(x+5)(x-5)}{x+5}$
$=x-5$

10. $\frac{y^2}{y+6}-\frac{36}{y+6}=\frac{y^2-36}{y+6}$
$=\frac{(y+6)(y-6)}{y+6}$
$=y-6$

11. $\frac{4}{p^2+7p+12}+\frac{p}{p^2+7p+12}$
$=\frac{4+p}{p^2+7p+12}$
$=\frac{p+4}{(p+3)(p+4)}$
$=\frac{1}{p+3}$

12. $\frac{5}{x^2+x-20}+\frac{x}{x^2+x-20}$
$=\frac{5+x}{x^2+x-20}$
$=\frac{5+x}{(x+5)(x-4)}$
$=\frac{1}{x-4}$

13. $\frac{a^3}{a^2+ab+b^2}-\frac{b^3}{a^2+ab+b^2}$
$=\frac{a^3-b^3}{a^2+ab+b^2}$
$=\frac{(a-b)(a^2+ab+b^2)}{a^2+ab+b^2}$
$=a-b$

14. $\frac{p^3}{p^2-pq+q^2}+\frac{q^3}{p^2-pq+q^2}$
$=\frac{p^3+q^3}{p^2-pq+q^2}$
$=\frac{(p+q)(p^2-pq+q^2)}{p^2-pq+q^2}$
$=p+q$

15. $18x^2y^3$, $24x^4y^5$

Factor each denominator.

$$18x^2y^3=2\cdot3\cdot3\cdot x^2\cdot y^3$$
$$=2\cdot3^2\cdot x^2\cdot y^3$$
$$24x^4y^5=2\cdot2\cdot2\cdot3\cdot x^4\cdot y^5$$
$$=2^3\cdot3\cdot x^4\cdot y^5$$

The least common denominator (LCD) is the product of all the different factors, with each factor raised to the greatest power in any denominator.

$$\text{LCD}=2^3\cdot3^2\cdot x^4\cdot y^5$$
$$=8\cdot9\cdot x^4\cdot y^5$$
$$=72x^4y^5$$

The LCD is $72x^4y^5$.

16. $24a^3b^4$, $18a^5b^2$

$$24a^3b^4=2^3\cdot3\cdot a^3\cdot b^4$$
$$18a^5b^2=2\cdot3^2\cdot a^5\cdot b^2$$

The least common denominator (LCD) is the product of all the different factors, with each factor raised to the greatest power in any denominator.

$\text{LCD}=2^3\cdot3^2\cdot a^5\cdot b^4=72a^5b^4$

17. $z-2$, z

Both $z-2$ and z have only 1 and themselves for factors.

$\text{LCD}=z(z-2)$

18. $k+3$, k

Both $k+3$ and k have only 1 and themselves for factors.

$\text{LCD}=k(k+3)$

19. $2y+8,\ y+4$

Factor each denominator.

$$2y+8=2(y+4)$$

The second denominator, $y+4$, is already factored. The LCD is

$$2(y+4).$$

20. $3r-21,\ r-7$

Factor $3r-21$ as $3(r-7)$.
$r-7$ cannot be factored further.

$\text{LCD}=3(r-7)$

21. $x^2-81,\ x^2+18x+81$

Factor each denominator.

$$\begin{aligned} x^2-81&=(x+9)(x-9)\\ x^2+18x+81&=(x+9)(x+9) \end{aligned}$$

$\text{LCD}=(x+9)^2(x-9)$

22. $y^2-16,\ y^2-8y+16$

Factor each denominator.

$$\begin{aligned} y^2-16&=(y+4)(y-4)\\ y^2-8y+16&=(y-4)(y-4) \end{aligned}$$

$\text{LCD}=(y-4)^2(y+4)$

23. $m+n,\ m-n,\ m^2-n^2$

Both $m+n$ and $m-n$ have only 1 and themselves for factors, while $m^2-n^2=(m+n)(m-n)$.

$\text{LCD}=(m+n)(m-n)$

24. $r+s,\ r-s,\ r^2-s^2$

Both $r+s$ and $r-s$ have only 1 and themselves for factors, while $r^2-s^2=(r+s)(r-s)$.

$\text{LCD}=(r+s)(r-s)$

25. $x^2-3x-4,\ x+x^2$

Factor each denominator.

$$\begin{aligned} x^2-3x-4&=(x-4)(x+1)\\ x+x^2&=x(1+x)=x(x+1) \end{aligned}$$

The LCD is $x(x-4)(x+1)$.

26. $y^2-8y+12,\ y^2-6y$

Factor each denominator.

$$\begin{aligned} y^2-8y+12&=(y-2)(y-6)\\ y^2-6y&=y(y-6) \end{aligned}$$

$\text{LCD}=y(y-2)(y-6)$

27. $2t^2+7t-15,\ t^2+3t-10$

Factor each denominator.

$$\begin{aligned} 2t^2+7t-15&=(2t-3)(t+5)\\ t^2+3t-10&=(t+5)(t-2) \end{aligned}$$

The LCD is $(2t-3)(t+5)(t-2)$.

28. $s^2-3s-4,\ 3s^2+s-2$

Factor each denominator.

$$\begin{aligned} s^2-3s-4&=(s+1)(s-4)\\ 3s^2+s-2&=(3s-2)(s+1) \end{aligned}$$

$\text{LCD}=(s+1)(s-4)(3s-2)$

29. $2y+6,\ y^2-9,\ y$

Factor each denominator.

$$\begin{aligned} 2y+6&=2(y+3)\\ y^2-9&=(y+3)(y-3) \end{aligned}$$

Remember the factor y from the third denominator. The LCD is

$$2y(y+3)(y-3).$$

30. $9x+18,\ x^2-4,\ x$

Factor each denominator.

$$\begin{aligned} 9x+18&=9(x+2)\\ x^2-4&=(x+2)(x-2) \end{aligned}$$

Remember the factor x from the third denominator.

$\text{LCD}=9x(x+2)(x-2)$

31. Yes, they could both be correct because the expressions are equivalent. Multiplying $\frac{3}{5-y}$ by 1 in the form $\frac{-1}{-1}$ gives $\frac{-3}{y-5}$.

32. The first step is incorrect. The third term in the numerator should be $+1$, because the subtraction sign should be distributed to both $4x$ and -1. The correct solution follows.

$$\begin{aligned} \frac{x}{x+2}-\frac{4x-1}{x+2}&=\frac{x-(4x-1)}{x+2}\\ &=\frac{x-4x+1}{x+2}\\ &=\frac{-3x+1}{x+2} \end{aligned}$$

33. $\frac{8}{t}+\frac{7}{3t}$ The LCD is $3t$.

$$\frac{8}{t}+\frac{7}{3t}=\frac{8\cdot 3}{t\cdot 3}+\frac{7}{3t}$$
$$=\frac{24+7}{3t}=\frac{31}{3t}$$

34. $\frac{5}{x}+\frac{9}{4x}$ The LCD is $4x$.

$$=\frac{5\cdot 4}{x\cdot 4}+\frac{9}{4x}$$
$$=\frac{20+9}{4x}=\frac{29}{4x}$$

35. $\frac{5}{12x^2y}-\frac{11}{6xy}$ The LCD is $12x^2y$.

$$\frac{5}{12x^2y}-\frac{11}{6xy}=\frac{5}{12x^2y}-\frac{11\cdot 2x}{6xy\cdot 2x}$$
$$=\frac{5}{12x^2y}-\frac{22x}{12x^2y}$$
$$=\frac{5-22x}{12x^2y}$$

36. $\frac{7}{18a^3b^2}-\frac{2}{9ab}$ LCD $=18a^3b^2$

$$=\frac{7}{18a^3b^2}-\frac{2\cdot 2a^2b}{9ab\cdot 2a^2b}$$
$$=\frac{7-4a^2b}{18a^3b^2}$$

37. $\frac{1}{x-1}-\frac{1}{x}$ LCD $=x(x-1)$

$$\frac{1}{x-1}-\frac{1}{x}=\frac{1\cdot x}{(x-1)x}-\frac{1\cdot(x-1)}{x(x-1)}$$
$$=\frac{x-(x-1)}{x(x-1)}$$
$$=\frac{x-x+1}{x(x-1)}$$
$$=\frac{1}{x(x-1)}$$

38. $\frac{3}{x-3}-\frac{1}{x}$ LCD $=x(x-3)$

$$=\frac{3\cdot x}{(x-3)x}-\frac{1\cdot(x-3)}{x(x-3)}$$
$$=\frac{3x-x+3}{x(x-3)}=\frac{2x+3}{x(x-3)}$$

39. $\frac{3a}{a+1}+\frac{2a}{a-3}$ LCD $=(a+1)(a-3)$

$$=\frac{3a(a-3)}{(a+1)(a-3)}+\frac{2a(a+1)}{(a-3)(a+1)}$$
$$=\frac{3a(a-3)+2a(a+1)}{(a+1)(a-3)}$$
$$=\frac{3a^2-9a+2a^2+2a}{(a+1)(a-3)}$$
$$=\frac{5a^2-7a}{(a+1)(a-3)}$$

40. $\frac{2x}{x+4}+\frac{3x}{x-7}$ LCD $=(x+4)(x-7)$

$$=\frac{2x(x-7)}{(x+4)(x-7)}+\frac{3x(x+4)}{(x-7)(x+4)}$$
$$=\frac{2x^2-14x+3x^2+12x}{(x+4)(x-7)}$$
$$=\frac{5x^2-2x}{(x+4)(x-7)}$$

41. $\frac{17y+3}{9y+7}-\frac{-10y-18}{9y+7}$

$$=\frac{17y+3-(-10y-18)}{9y+7}$$
$$=\frac{17y+3+10y+18}{9y+7}$$
$$=\frac{27y+21}{9y+7}$$
$$=\frac{3(9y+7)}{9y+7}$$
$$=3$$

42. $\frac{7x+8}{3x+2}-\frac{x+4}{3x+2}=\frac{7x+8-(x+4)}{3x+2}$

$$=\frac{7x+8-x-4}{3x+2}$$
$$=\frac{6x+4}{3x+2}$$
$$=\frac{2(3x+2)}{3x+2}$$
$$=2$$

43. $\frac{2}{4-x}+\frac{5}{x-4}$

To get a common denominator of $x-4$, multiply both the numerator and denominator of the first expression by -1.

$$=\frac{(2)(-1)}{(4-x)(-1)}+\frac{5}{x-4}$$
$$=\frac{-2}{x-4}+\frac{5}{x-4}$$
$$=\frac{-2+5}{x-4}$$
$$=\frac{3}{x-4}$$

If you chose $4-x$ for the LCD, then you should have obtained the equivalent answer, $\frac{-3}{4-x}$.

44. $\dfrac{3}{2-t}+\dfrac{1}{t-2}$

To get a common denominator of $t-2$, multiply both the numerator and denominator of the first expression by -1.

$$=\frac{(3)(-1)}{(2-t)(-1)}+\frac{1}{t-2}$$
$$=\frac{-3}{t-2}+\frac{1}{t-2}$$
$$=\frac{-2}{t-2} \text{ or } \frac{2}{2-t}$$

45. $\dfrac{w}{w-z}-\dfrac{z}{z-w}$

$w-z$ and $z-w$ are opposites, so factor out -1 from $z-w$ to get a common denominator.

$$=\frac{w}{w-z}-\frac{z}{-1(w-z)}$$
$$=\frac{w}{w-z}+\frac{z}{w-z}$$
$$=\frac{w+z}{w-z} \text{ or } \frac{-w-z}{z-w}$$

46. $\dfrac{a}{a-b}-\dfrac{b}{b-a}$

$a-b$ and $b-a$ are opposites, so factor out -1 from $b-a$ to get a common denominator.

$$=\frac{a}{a-b}-\frac{b}{-1(a-b)}$$
$$=\frac{a}{a-b}+\frac{b}{a-b}$$
$$=\frac{a+b}{a-b} \text{ or } \frac{-a-b}{b-a}$$

47. $\dfrac{5}{12+4x}-\dfrac{7}{9+3x}=\dfrac{5}{4(3+x)}-\dfrac{7}{3(3+x)}$

The LCD is $12(3+x)$.

$$=\frac{5\cdot 3}{4(3+x)\cdot 3}-\frac{7\cdot 4}{3(3+x)\cdot 4}$$
$$=\frac{15-28}{12(3+x)}$$
$$=\frac{-13}{12(3+x)}$$

48. $\dfrac{3}{10x+15}-\dfrac{8}{12x+18}$

$$=\frac{3}{5(2x+3)}-\frac{8}{6(2x+3)}$$

The LCD is $5\cdot 6(2x+3)$.

$$=\frac{3\cdot 6}{5\cdot 6(2x+3)}-\frac{8\cdot 5}{5\cdot 6(2x+3)}$$
$$=\frac{18-40}{30(2x+3)}$$
$$=\frac{-22}{30(2x+3)}=\frac{-11}{15(2x+3)}$$

49. $\dfrac{4x}{x-1}-\dfrac{2}{x+1}-\dfrac{4}{x^2-1}$

$x^2-1=(x+1)(x-1)$, the LCD.

$$\frac{4x}{x-1}-\frac{2}{x+1}-\frac{4}{x^2-1}$$
$$=\frac{4x(x+1)}{(x-1)(x+1)}-\frac{2(x-1)}{(x+1)(x-1)}-\frac{4}{(x+1)(x-1)}$$
$$=\frac{4x(x+1)-2(x-1)-4}{(x+1)(x-1)}$$
$$=\frac{4x^2+4x-2x+2-4}{(x-1)(x+1)}$$
$$=\frac{4x^2+2x-2}{(x-1)(x+1)}$$
$$=\frac{2(2x^2+x-1)}{(x-1)(x+1)}$$
$$=\frac{2(2x-1)(x+1)}{(x-1)(x+1)}$$
$$=\frac{2(2x-1)}{x-1}$$

50. $\dfrac{4}{x+3}-\dfrac{x}{x-3}-\dfrac{18}{x^2-9}$

$x^2-9=(x+3)(x-3)$, the LCD.

$$=\frac{4(x-3)}{(x+3)(x-3)}-\frac{x(x+3)}{(x-3)(x+3)}-\frac{18}{(x+3)(x-3)}$$
$$=\frac{4x-12-x^2-3x-18}{(x+3)(x-3)}$$
$$=\frac{-x^2+x-30}{(x+3)(x-3)}$$

51. $\dfrac{15}{y^2+3y}+\dfrac{2}{y}+\dfrac{5}{y+3}$

$y^2+3y=y(y+3)$, the LCD.

$$=\frac{15}{y(y+3)}+\frac{2(y+3)}{y(y+3)}+\frac{5y}{(y+3)y}$$
$$=\frac{15+2(y+3)+5y}{y(y+3)}$$
$$=\frac{15+2y+6+5y}{y(y+3)}$$
$$=\frac{7y+21}{y(y+3)}$$
$$=\frac{7(y+3)}{y(y+3)}=\frac{7}{y}$$

52. $\dfrac{7}{t-2}-\dfrac{6}{t^2-2t}-\dfrac{3}{t}$

$t^2-2t=t(t-2)$, the LCD.

$$=\frac{7t}{t(t-2)}-\frac{6}{t(t-2)}-\frac{3(t-2)}{t(t-2)}$$
$$=\frac{7t-6-3t+6}{t(t-2)}$$
$$=\frac{4t}{t(t-2)}$$
$$=\frac{4}{t-2}$$

53. $\dfrac{5}{x-2}+\dfrac{1}{x}+\dfrac{2}{x^2-2x}$

$x^2-2x=x(x-2)$, the LCD.

$$=\frac{5x}{(x-2)x}+\frac{1(x-2)}{x(x-2)}+\frac{2}{x(x-2)}$$
$$=\frac{5x+x-2+2}{x(x-2)}$$
$$=\frac{6x}{x(x-2)}=\frac{6}{x-2}$$

54. $\dfrac{5x}{x-3}+\dfrac{2}{x}+\dfrac{6}{x^2-3x}$

$x^2-3x=x(x-3)$, the LCD.

$$=\frac{5x\cdot x}{(x-3)x}+\frac{2(x-3)}{x(x-3)}+\frac{6}{x(x-3)}$$
$$=\frac{5x^2+2x-6+6}{x(x-3)}$$
$$=\frac{5x^2+2x}{x(x-3)}$$
$$=\frac{x(5x+2)}{x(x-3)}=\frac{5x+2}{x-3}$$

55. $\dfrac{3x}{x+1}+\dfrac{4}{x-1}-\dfrac{6}{x^2-1}$

$x^2-1=(x+1)(x-1)$, the LCD.

$$=\frac{3x(x-1)}{(x+1)(x-1)}+\frac{4(x+1)}{(x-1)(x+1)}-\frac{6}{(x+1)(x-1)}$$
$$=\frac{3x(x-1)+4(x+1)-6}{(x+1)(x-1)}$$
$$=\frac{3x^2-3x+4x+4-6}{(x+1)(x-1)}$$
$$=\frac{3x^2+x-2}{(x+1)(x-1)}$$
$$=\frac{(3x-2)(x+1)}{(x+1)(x-1)}$$
$$=\frac{3x-2}{x-1}$$

56. $\dfrac{5x}{x+3}+\dfrac{x+2}{x}-\dfrac{6}{x^2+3x}$

$x^2+3x=x(x+3)$, the LCD.

$$=\frac{5x\cdot x}{(x+3)x}+\frac{(x+2)(x+3)}{x(x+3)}-\frac{6}{x(x+3)}$$
$$=\frac{5x^2+x^2+5x+6-6}{x(x+3)}$$
$$=\frac{6x^2+5x}{x(x+3)}=\frac{x(6x+5)}{x(x+3)}$$
$$=\frac{6x+5}{x+3}$$

57. $\dfrac{4}{x+1}+\dfrac{1}{x^2-x+1}-\dfrac{12}{x^3+1}$

$x^3+1=(x+1)(x^2-x+1)$, the LCD.

$$=\frac{4(x^2-x+1)}{(x+1)(x^2-x+1)}+\frac{1\cdot(x+1)}{(x^2-x+1)(x+1)}-\frac{12}{(x+1)(x^2-x+1)}$$
$$=\frac{4(x^2-x+1)+(x+1)-12}{(x+1)(x^2-x+1)}$$
$$=\frac{4x^2-4x+4+x+1-12}{(x+1)(x^2-x+1)}$$
$$=\frac{4x^2-3x-7}{(x+1)(x^2-x+1)}$$
$$=\frac{(4x-7)(x+1)}{(x+1)(x^2-x+1)}$$
$$=\frac{4x-7}{x^2-x+1}$$

58. $\dfrac{5}{x+2}+\dfrac{2}{x^2-2x+4}-\dfrac{60}{x^3+8}$

$x^3+8=(x+2)(x^2-2x+4)$, the LCD.

$$=\frac{5(x^2-2x+4)}{(x+2)(x^2-2x+4)}+\frac{2(x+2)}{(x+2)(x^2-2x+4)}-\frac{60}{(x+2)(x^2-2x+4)}$$
$$=\frac{5x^2-10x+20+2x+4-60}{(x+2)(x^2-2x+4)}$$
$$=\frac{5x^2-8x-36}{(x+2)(x^2-2x+4)}$$
$$=\frac{(5x-18)(x+2)}{(x+2)(x^2-2x+4)}$$
$$=\frac{5x-18}{x^2-2x+4}$$

59. $\dfrac{2x+4}{x+3}+\dfrac{3}{x}-\dfrac{6}{x^2+3x}$

$x^2+3x=x(x+3)$, the LCD.

$$=\frac{(2x+4)x}{(x+3)x}+\frac{3(x+3)}{x(x+3)}-\frac{6}{x(x+3)}$$
$$=\frac{(2x+4)x+3(x+3)-6}{x(x+3)}$$
$$=\frac{2x^2+4x+3x+9-6}{x(x+3)}$$
$$=\frac{2x^2+7x+3}{x(x+3)}$$
$$=\frac{(2x+1)(x+3)}{x(x+3)}=\frac{2x+1}{x}$$

60. $\dfrac{4x+1}{x+5}-\dfrac{2}{x}+\dfrac{10}{x^2+5x}$

$x^2+5x=x(x+5)$, the LCD.

$$=\frac{x(4x+1)}{x(x+5)}-\frac{2(x+5)}{x(x+5)}+\frac{10}{x(x+5)}$$
$$=\frac{4x^2+x-2x-10+10}{x(x+5)}$$
$$=\frac{4x^2-x}{x(x+5)}=\frac{x(4x-1)}{x(x+5)}$$
$$=\frac{4x-1}{x+5}$$

61. $\dfrac{3}{x^2-5x+6}-\dfrac{2}{x^2-4x+4}$

$$=\frac{3}{(x-2)(x-3)}-\frac{2}{(x-2)(x-2)}$$
$$=\frac{3(x-2)}{(x-2)(x-3)(x-2)}$$
$$-\frac{2(x-3)}{(x-2)(x-2)(x-3)}$$

LCD $=(x-2)^2(x-3)$

$$=\frac{3x-6-2x+6}{(x-2)^2(x-3)}$$
$$=\frac{x}{(x-2)^2(x-3)}$$

62. $\dfrac{2}{m^2-4m+4}+\dfrac{3}{m^2+m-6}$

$$=\frac{2}{(m-2)^2}+\frac{3}{(m+3)(m-2)}$$
$$=\frac{2(m+3)}{(m-2)^2(m+3)}+\frac{3(m-2)}{(m+3)(m-2)^2}$$

LCD $=(m+3)(m-2)^2$

$$=\frac{2m+6+3m-6}{(m-2)^2(m+3)}$$
$$=\frac{5m}{(m-2)^2(m+3)}$$

63. $\dfrac{3}{x^2+4x+4}+\dfrac{7}{x^2+5x+6}$

$$=\frac{3}{(x+2)^2}+\frac{7}{(x+2)(x+3)}$$

LCD $=(x+2)^2(x+3)$

$$=\frac{3(x+3)}{(x+2)^2(x+3)}+\frac{7(x+2)}{(x+2)^2(x+3)}$$
$$=\frac{3x+9+7x+14}{(x+2)^2(x+3)}$$
$$=\frac{10x+23}{(x+2)^2(x+3)}$$

64. $\dfrac{5}{x^2+6x+9}-\dfrac{2}{x^2+4x+3}$

$$=\frac{5}{(x+3)^2}-\frac{2}{(x+1)(x+3)}$$

LCD $=(x+3)^2(x+1)$

$$=\frac{5(x+1)}{(x+3)^2(x+1)}-\frac{2(x+3)}{(x+3)^2(x+1)}$$
$$=\frac{5x+5-2x-6}{(x+3)^2(x+1)}$$
$$=\frac{3x-1}{(x+3)^2(x+1)}$$

65. $\dfrac{5x}{x^2+xy-2y^2}-\dfrac{3x}{x^2+5xy-6y^2}$

Factor each denominator.

$$x^2+xy-2y^2=(x+2y)(x-y)$$
$$x^2+5xy-6y^2=(x+6y)(x-y)$$

The LCD is $(x+2y)(x-y)(x+6y)$.

$$\frac{5x}{(x+2y)(x-y)}-\frac{3x}{(x+6y)(x-y)}$$
$$=\frac{(5x)(x+6y)}{(x+2y)(x-y)(x+6y)}$$
$$-\frac{(3x)(x+2y)}{(x+6y)(x-y)(x+2y)}$$
$$=\frac{(5x)(x+6y)-(3x)(x+2y)}{(x+6y)(x-y)(x+2y)}$$
$$=\frac{5x^2+30xy-(3x^2+6xy)}{(x+2y)(x-y)(x+6y)}$$
$$=\frac{2x^2+24xy}{(x+2y)(x-y)(x+6y)}$$
$$=\frac{2x(x+12y)}{(x+2y)(x-y)(x+6y)}$$

66. $\dfrac{6x}{6x^2+5xy-4y^2}-\dfrac{2y}{9x^2-16y^2}$

$=\dfrac{6x}{(3x+4y)(2x-y)}$

$-\dfrac{2y}{(3x+4y)(3x-4y)}$

The LCD is $(3x+4y)(2x-y)(3x-4y)$.

$=\dfrac{(6x)(3x-4y)}{(3x+4y)(2x-y)(3x-4y)}$

$-\dfrac{(2y)(2x-y)}{(3x+4y)(3x-4y)(2x-y)}$

$=\dfrac{18x^2-24xy-(4xy-2y^2)}{(3x+4y)(2x-y)(3x-4y)}$

$=\dfrac{18x^2-28xy+2y^2}{(3x+4y)(2x-y)(3x-4y)}$

$=\dfrac{2(9x^2-14xy+y^2)}{(3x+4y)(2x-y)(3x-4y)}$

67. $c(x)=\dfrac{1010}{49(101-x)}-\dfrac{10}{49}$

$=\dfrac{1010}{49(101-x)}-\dfrac{10(101-x)}{49(101-x)}$

$=\dfrac{1010-1010+10x}{49(101-x)}$

$=\dfrac{10x}{49(101-x)}$

68. $c(95)=\dfrac{10(95)}{49(101-95)}$

$=\dfrac{950}{294}\approx 3.23$

It would cost approximately 3.23 thousand dollars to win 95 points.

69. $\dfrac{3}{7}+\dfrac{5}{9}-\dfrac{6}{63}$

The LCD is $7(9)=63$.

$=\dfrac{3\cdot 9}{7\cdot 9}+\dfrac{5\cdot 7}{9\cdot 7}-\dfrac{6}{63}$

$=\dfrac{27+35-6}{63}$

$=\dfrac{56}{63}=\dfrac{8}{9}$

70. From Example 6,

$$\frac{3}{x-2}+\frac{5}{x}-\frac{6}{x^2-2x}.$$

Substitute 9 for x.

$$\frac{3}{9-2}+\frac{5}{9}-\frac{6}{9^2-2(9)}=\frac{3}{7}+\frac{5}{9}-\frac{6}{63}$$

The problems in Exercises 69 and 70 are the same.

71. From Exercise 69,

$$\frac{3}{7}+\frac{5}{9}-\frac{6}{63}=\frac{8}{9}.$$

From Example 6, the answer is $\frac{8}{x}$. If we substitute 9 for x, the answer becomes $\frac{8}{9}$. The answers agree.

72. Answers will vary. For example, suppose the last name is Sosa so that $x=4$. The problem in Example 6 becomes

$$\frac{3}{4-2}+\frac{5}{4}-\frac{6}{4^2-2(4)}=\frac{3}{2}+\frac{5}{4}-\frac{6}{8}.$$

The predicted answer is

$$\frac{8}{x},\ \text{or}\ \frac{8}{4}=2.$$

Perform the operations to verify our prediction.

$$\frac{3}{2}+\frac{5}{4}-\frac{6}{8}=\frac{3\cdot 4}{2\cdot 4}+\frac{5\cdot 2}{4\cdot 2}-\frac{6}{8}$$

$$=\frac{12+10-6}{8}$$

$$=\frac{16}{8}=2$$

The prediction is correct.

73. If $x=2$, then the problem from Example 6,

$$\frac{3}{x-2}+\frac{5}{x}-\frac{6}{x^2-2x},$$

becomes

$$\frac{3}{2-2}+\frac{5}{2}-\frac{6}{2^2-2(2)}=\frac{3}{0}+\frac{5}{2}-\frac{6}{0}.$$

Thus, if $x=2$, then

$$\frac{3}{x-2}\ \text{and}\ \frac{6}{x^2-2x}$$

are undefined.

74. If $x=0$, then

$$\frac{5}{x}=\frac{5}{0}\ \text{and}\ \frac{6}{x^2-2x}=\frac{6}{0},$$

which are undefined. Therefore, 0 is not allowed as a value of x.

8.3 Complex Fractions

8.3 Margin Exercises

1. **(a)** $\dfrac{\frac{a+2}{5a}}{\frac{a-3}{7a}}$

Both the numerator and denominator are already simplified.

$$= \frac{a+2}{5a} \div \frac{a-3}{7a}$$

$$= \frac{a+2}{5a} \cdot \frac{7a}{a-3} \quad \textit{Multiply by the reciprocal of the divisor.}$$

$$= \frac{7(a+2)}{5(a-3)}$$

(b) $\dfrac{2+\frac{1}{k}}{2-\frac{1}{k}}$

$$= \frac{\frac{2k}{k}+\frac{1}{k}}{\frac{2k}{k}-\frac{1}{k}} \quad \textit{Simplify the numerator and denominator.}$$

$$= \frac{\frac{2k+1}{k}}{\frac{2k-1}{k}}$$

$$= \frac{2k+1}{k} \cdot \frac{k}{2k-1} \quad \textit{Multiply by the reciprocal of the denominator.}$$

$$= \frac{2k+1}{2k-1}$$

(c) $\dfrac{\frac{r^2-4}{4}}{1+\frac{2}{r}}$

$$= \frac{\frac{r^2-4}{4}}{\frac{r+2}{r}} \quad \textit{Simplify the denominator.}$$

$$= \frac{r^2-4}{4} \cdot \frac{r}{r+2} \quad \textit{Multiply by the reciprocal of the denominator.}$$

$$= \frac{(r-2)(r+2)}{4} \cdot \frac{r}{r+2}$$

$$= \frac{r(r-2)}{4}$$

2. **(a)** $\dfrac{\frac{5}{y}+6}{\frac{8}{3y}-1}$

The LCD of $\frac{5}{y}$ and $\frac{8}{3y}$ is $3y$. Multiply the numerator and denominator by $3y$.

$$= \frac{\left(\frac{5}{y}+6\right)\cdot 3y}{\left(\frac{8}{3y}-1\right)\cdot 3y}$$

$$= \frac{\frac{5}{y}\cdot 3y + 6\cdot 3y}{\frac{8}{3y}\cdot 3y - 1\cdot 3y} \quad \textit{Distributive property}$$

$$= \frac{15+18y}{8-3y} \quad \textit{Simplify.}$$

(b) $\dfrac{\frac{1}{y}+\frac{1}{y-1}}{\frac{1}{y}-\frac{2}{y-1}}$

Multiply the numerator and denominator by $y(y-1)$, the LCD of all the fractions.

$$= \frac{\left(\frac{1}{y}+\frac{1}{y-1}\right)\cdot y(y-1)}{\left(\frac{1}{y}-\frac{2}{y-1}\right)\cdot y(y-1)}$$

$$= \frac{\frac{1}{y}\cdot y(y-1) + \frac{1}{y-1}\cdot y(y-1)}{\frac{1}{y}\cdot y(y-1) - \frac{2}{y-1}\cdot y(y-1)}$$

$$= \frac{(y-1)+y}{(y-1)-2y}$$

$$= \frac{2y-1}{-y-1}$$

An equivalent answer is

$$\frac{(-1)(2y-1)}{(-1)(-y-1)} = \frac{1-2y}{y+1}.$$

3. **(a) Method 1**

$$\frac{\frac{5}{y+2}}{\frac{-3}{y^2-4}} = \frac{\frac{5}{y+2}}{\frac{-3}{(y+2)(y-2)}}$$

$$= \frac{5}{y+2} \div \frac{-3}{(y+2)(y-2)}$$

$$= \frac{5}{y+2} \cdot \frac{(y+2)(y-2)}{-3}$$

$$= \frac{5(y-2)}{-3}$$

continued

Method 2

$$\frac{\frac{5}{y+2}}{\frac{-3}{y^2-4}} = \frac{\frac{5}{y+2}}{\frac{-3}{(y+2)(y-2)}}$$

The LCD of the numerator and denominator is $(y+2)(y-2)$.

$$= \frac{\frac{5}{y+2}\cdot (y+2)(y-2)}{\frac{-3}{(y+2)(y-2)}\cdot (y+2)(y-2)}$$

$$= \frac{5(y-2)}{-3}$$

(b) Method 1

$$\frac{\frac{1}{a}-\frac{1}{b}}{\frac{1}{a^2}-\frac{1}{b^2}} = \frac{\frac{b}{ab}-\frac{a}{ab}}{\frac{b^2}{a^2b^2}-\frac{a^2}{a^2b^2}} \quad \begin{array}{l} LCD = ab \\ LCD = a^2b^2 \end{array}$$

$$= \frac{\frac{b-a}{ab}}{\frac{b^2-a^2}{a^2b^2}}$$

$$= \frac{\frac{b-a}{ab}}{\frac{(b+a)(b-a)}{a^2b^2}}$$

$$= \frac{b-a}{ab} \div \frac{(b+a)(b-a)}{a^2b^2}$$

$$= \frac{b-a}{ab}\cdot\frac{a^2b^2}{(b+a)(b-a)}$$

$$= \frac{ab}{b+a}$$

Method 2

$$\frac{\frac{1}{a}-\frac{1}{b}}{\frac{1}{a^2}-\frac{1}{b^2}}$$

The LCD of the numerator and the denominator is a^2b^2.

$$= \frac{\left(\frac{1}{a}-\frac{1}{b}\right)\cdot a^2b^2}{\left(\frac{1}{a^2}-\frac{1}{b^2}\right)\cdot a^2b^2}$$

$$= \frac{ab^2-a^2b}{b^2-a^2}$$

$$= \frac{ab(b-a)}{(b+a)(b-a)}$$

$$= \frac{ab}{b+a}$$

4. (a) $$\frac{r^{-2}-s^{-1}}{4r^{-1}+s^{-2}} = \frac{\frac{1}{r^2}-\frac{1}{s}}{\frac{4}{r}+\frac{1}{s^2}} \qquad LCD = r^2s^2$$

$$= \frac{r^2s^2\left(\frac{1}{r^2}-\frac{1}{s}\right)}{r^2s^2\left(\frac{4}{r}+\frac{1}{s^2}\right)}$$

$$= \frac{r^2s^2\cdot\frac{1}{r^2} - r^2s^2\cdot\frac{1}{s}}{r^2s^2\cdot\frac{4}{r}+r^2s^2\cdot\frac{1}{s^2}}$$

$$= \frac{s^2-r^2s}{4rs^2+r^2}$$

(b) $$\frac{b^{-4}}{b^{-5}+2} = \frac{\frac{1}{b^4}}{\frac{1}{b^5}+2} \qquad LCD = b^5$$

$$= \frac{b^5\left(\frac{1}{b^4}\right)}{b^5\left(\frac{1}{b^5}+2\right)}$$

$$= \frac{b^5\cdot\frac{1}{b^4}}{b^5\cdot\frac{1}{b^5}+b^5\cdot 2}$$

$$= \frac{b}{1+2b^5}$$

8.3 Section Exercises

1. Answers will vary. Begin by simplifying the numerator to a single fraction. Then simplify the denominator to a single fraction. Write as a division problem, and multiply by the reciprocal of the denominator. Simplify the result, if possible.

2. Multiplying the numerator and denominator of a complex fraction by the same nonzero expression is justified by the *identity property for multiplication.*

3. $$\frac{\frac{12}{x-1}}{\frac{6}{x}} = \frac{12}{x-1} \div \frac{6}{x}$$

Multiply by the reciprocal of the divisor.

$$= \frac{12}{x-1}\cdot\frac{x}{6}$$

$$= \frac{2x}{x-1}$$

4. $$\frac{\frac{24}{t+4}}{\frac{6}{t}} = \frac{24}{t+4} \div \frac{6}{t}$$

Multiply by the reciprocal of the divisor.

$$= \frac{24}{t+4} \cdot \frac{t}{6}$$
$$= \frac{4t}{t+4}$$

5. $$\frac{\frac{k+1}{2k}}{\frac{3k-1}{4k}} = \frac{k+1}{2k} \cdot \frac{4k}{3k-1}$$
$$= \frac{4k(k+1)}{2k(3k-1)}$$
$$= \frac{2(k+1)}{3k-1}$$

6. $$\frac{\frac{1-r}{4r}}{\frac{-1-r}{8r}} = \frac{1-r}{4r} \cdot \frac{8r}{-1-r}$$
$$= \frac{2(1-r)}{-1-r}$$

7. $$\frac{\frac{4z^2x^4}{9}}{\frac{12x^2z^5}{15}} = \frac{\frac{4z^2x^4}{9}}{\frac{4x^2z^5}{5}}$$
$$= \frac{4z^2x^4}{9} \div \frac{4x^2z^5}{5}$$
$$= \frac{4z^2x^4}{9} \cdot \frac{5}{4x^2z^5}$$
$$= \frac{5z^2x^4}{9x^2z^5} = \frac{5x^2}{9z^3}$$

8. $$\frac{\frac{3y^2x^3}{8}}{\frac{9y^3x^4}{16}} = \frac{3y^2x^3}{8} \cdot \frac{16}{9y^3x^4}$$
$$= \frac{3 \cdot 2y^2x^3}{3 \cdot 3y^3x^4} = \frac{2}{3yx}$$

9. $$\frac{\frac{1}{x}+1}{-\frac{1}{x}+1}$$

Multiply the numerator and denominator by x, the LCD of all the fractions.

$$= \frac{x\left(\frac{1}{x}+1\right)}{x\left(-\frac{1}{x}+1\right)}$$
$$= \frac{x \cdot \frac{1}{x} + x \cdot 1}{x\left(-\frac{1}{x}\right) + x \cdot 1}$$
$$= \frac{1+x}{-1+x}$$

10. $$\frac{\frac{2}{k}-1}{\frac{2}{k}+1} = \frac{\frac{2-k}{k}}{\frac{2+k}{k}}$$
$$= \frac{2-k}{k} \cdot \frac{k}{2+k}$$
$$= \frac{2-k}{2+k}$$

11. $$\frac{\frac{3}{x}+\frac{3}{y}}{\frac{3}{x}-\frac{3}{y}}$$

Multiply the numerator and denominator by xy, the LCD of all the fractions.

$$= \frac{\left(\frac{3}{x}+\frac{3}{y}\right)xy}{\left(\frac{3}{x}-\frac{3}{y}\right)xy}$$
$$= \frac{\frac{3}{x} \cdot xy + \frac{3}{y} \cdot xy}{\frac{3}{x} \cdot xy - \frac{3}{y} \cdot xy}$$
$$= \frac{3y+3x}{3y-3x}$$
$$= \frac{3(y+x)}{3(y-x)}$$
$$= \frac{y+x}{y-x}$$

12. $$\frac{\frac{4}{t}-\frac{4}{s}}{\frac{4}{t}+\frac{4}{s}} = \frac{\frac{4s-4t}{ts}}{\frac{4s+4t}{ts}}$$
$$= \frac{4(s-t)}{ts} \cdot \frac{ts}{4(s+t)}$$
$$= \frac{s-t}{s+t}$$

13. $$\frac{\frac{8x-24y}{10}}{\frac{x-3y}{5x}} = \frac{8x-24y}{10} \cdot \frac{5x}{x-3y}$$
$$= \frac{8(x-3y)5x}{10(x-3y)}$$
$$= \frac{40x}{10} = 4x$$

14. $\dfrac{\dfrac{10x-5y}{12}}{\dfrac{2x-y}{6y}} = \dfrac{\dfrac{5(2x-y)}{12}}{\dfrac{2x-y}{6y}}$

$= \dfrac{5(2x-y)}{12} \cdot \dfrac{6y}{2x-y}$

$= \dfrac{5y}{2}$

15. $\dfrac{\dfrac{x^2-16y^2}{xy}}{\dfrac{1}{y}-\dfrac{4}{x}}$

Multiply the numerator and denominator by xy, the LCD of all the fractions.

$= \dfrac{\left(\dfrac{x^2-16y^2}{xy}\right)xy}{\left(\dfrac{1}{y}-\dfrac{4}{x}\right)xy}$

$= \dfrac{x^2-16y^2}{\dfrac{1}{y} \cdot xy - \dfrac{4}{x} \cdot xy}$

$= \dfrac{x^2-16y^2}{x-4y}$

$= \dfrac{(x+4y)(x-4y)}{x-4y}$

$= x+4y$

16. $\dfrac{\dfrac{2}{s}-\dfrac{3}{t}}{\dfrac{4t^2-9s^2}{st}}$

Multiply the numerator and denominator by st, the LCD of all the fractions.

$= \dfrac{st\left(\dfrac{2}{s}-\dfrac{3}{t}\right)}{st\left(\dfrac{4t^2-9s^2}{st}\right)}$

$= \dfrac{2t-3s}{4t^2-9s^2}$

$= \dfrac{2t-3s}{(2t+3s)(2t-3s)}$

$= \dfrac{1}{2t+3s}$

17. $\dfrac{y-\dfrac{y-3}{3}}{\dfrac{4}{9}+\dfrac{2}{3y}}$

Multiply the numerator and denominator by $9y$, the LCD of all the fractions.

$= \dfrac{9y\left(y-\dfrac{y-3}{3}\right)}{9y\left(\dfrac{4}{9}+\dfrac{2}{3y}\right)}$

$= \dfrac{9y^2-3y(y-3)}{4y+6}$

$= \dfrac{9y^2-3y^2+9y}{4y+6}$

$= \dfrac{6y^2+9y}{4y+6}$

$= \dfrac{3y(2y+3)}{2(2y+3)} = \dfrac{3y}{2}$

18. $\dfrac{p-\dfrac{p+2}{4}}{\dfrac{3}{4}-\dfrac{5}{2p}}$ $\qquad$ *LCD* $= 4p$

$= \dfrac{4p\left(p-\dfrac{p+2}{4}\right)}{4p\left(\dfrac{3}{4}-\dfrac{5}{2p}\right)}$

$= \dfrac{4p^2-p^2-2p}{3p-10}$

$= \dfrac{3p^2-2p}{3p-10}$

19. $\dfrac{\dfrac{x+2}{x}+\dfrac{1}{x+2}}{\dfrac{5}{x}+\dfrac{x}{x+2}}$

Multiply the numerator and denominator by $x(x+2)$, the LCD of all the fractions.

$= \dfrac{x(x+2)\left(\dfrac{x+2}{x}+\dfrac{1}{x+2}\right)}{x(x+2)\left(\dfrac{5}{x}+\dfrac{x}{x+2}\right)}$

$= \dfrac{x(x+2)\left(\dfrac{x+2}{x}\right)+x(x+2)\left(\dfrac{1}{x+2}\right)}{x(x+2)\left(\dfrac{5}{x}\right)+x(x+2)\left(\dfrac{x}{x+2}\right)}$

$= \dfrac{(x+2)(x+2)+x}{5(x+2)+x^2}$

$= \dfrac{x^2+4x+4+x}{5x+10+x^2}$

$= \dfrac{x^2+5x+4}{x^2+5x+10}$

20. $$\frac{\frac{y+3}{y}-\frac{4}{y-1}}{\frac{y}{y-1}+\frac{1}{y}} \qquad LCD = y(y-1)$$

$$= \frac{y(y-1)\left(\frac{y+3}{y}-\frac{4}{y-1}\right)}{y(y-1)\left(\frac{y}{y-1}+\frac{1}{y}\right)}$$

$$= \frac{(y-1)(y+3)-4y}{y^2+y-1}$$

$$= \frac{y^2+2y-3-4y}{y^2+y-1}$$

$$= \frac{y^2-2y-3}{y^2+y-1}$$

21. To add the fractions in the numerator, use the LCD, $m(m-1)$.

$$\frac{4}{m}+\frac{m+2}{m-1} = \frac{4(m-1)}{m(m-1)}+\frac{m(m+2)}{m(m-1)}$$

$$= \frac{4m-4+m^2+2m}{m(m-1)}$$

$$= \frac{m^2+6m-4}{m(m-1)}$$

22. To subtract the fractions in the denominator, use the same LCD, $m(m-1)$.

$$\frac{m+2}{m}-\frac{2}{m-1}$$

$$= \frac{(m+2)(m-1)}{m(m-1)}-\frac{m\cdot 2}{m(m-1)}$$

$$= \frac{m^2+m-2-2m}{m(m-1)}$$

$$= \frac{m^2-m-2}{m(m-1)}$$

23. Exercise 21 answer ↓ Exercise 22 answer ↓

$$\frac{m^2+6m-4}{m(m-1)} \div \frac{m^2-m-2}{m(m-1)}$$

Multiply by the reciprocal.

$$= \frac{m^2+6m-4}{m(m-1)}\cdot\frac{m(m-1)}{m^2-m-2}$$

$$= \frac{m^2+6m-4}{m^2-m-2}$$

24. The LCD of all the denominators in the complex fraction is $m(m-1)$.

25. $$\frac{\left(\frac{4}{m}+\frac{m+2}{m-1}\right)\cdot m(m-1)}{\left(\frac{m+2}{m}-\frac{2}{m-1}\right)\cdot m(m-1)}$$

$$= \frac{4(m-1)+m(m+2)}{(m+2)(m-1)-2m}$$

$$= \frac{4m-4+m^2+2m}{m^2+m-2-2m}$$

$$= \frac{m^2+6m-4}{m^2-m-2}$$

26. Answers will vary. Because of the complicated nature of the numerator and denominator of the complex fraction, using Method 1 takes much longer to simplify the complex fraction. Method 2 is a simpler, more direct means of simplifying and is most likely the preferred method.

27. $$\frac{1}{x^{-2}+y^{-2}} = \frac{1}{\frac{1}{x^2}+\frac{1}{y^2}}$$

$$= \frac{x^2y^2(1)}{x^2y^2\left(\frac{1}{x^2}+\frac{1}{y^2}\right)} \qquad LCD = x^2y^2$$

$$= \frac{x^2y^2}{y^2+x^2}$$

28. $$\frac{1}{p^{-2}-q^{-2}} = \frac{1}{\frac{1}{p^2}-\frac{1}{q^2}}$$

$$= \frac{p^2q^2(1)}{p^2q^2\left(\frac{1}{p^2}-\frac{1}{q^2}\right)}$$

$$= \frac{p^2q^2}{q^2-p^2} \text{ or } \frac{p^2q^2}{(q+p)(q-p)}$$

29. $$\frac{x^{-2}+y^{-2}}{x^{-1}+y^{-1}}$$

$$= \frac{\frac{1}{x^2}+\frac{1}{y^2}}{\frac{1}{x}+\frac{1}{y}}$$

Multiply the numerator and denominator by x^2y^2, the LCD of all the fractions.

$$= \frac{x^2y^2\left(\frac{1}{x^2}+\frac{1}{y^2}\right)}{x^2y^2\left(\frac{1}{x}+\frac{1}{y}\right)}$$

$$= \frac{x^2y^2\cdot\frac{1}{x^2}+x^2y^2\cdot\frac{1}{y^2}}{x^2y^2\cdot\frac{1}{x}+x^2y^2\cdot\frac{1}{y}}$$

$$= \frac{y^2+x^2}{xy^2+x^2y} \text{ or } \frac{y^2+x^2}{xy(y+x)}$$

30. $$\frac{x^{-1}-y^{-1}}{x^{-2}-y^{-2}}=\frac{\frac{1}{x}-\frac{1}{y}}{\frac{1}{x^2}-\frac{1}{y^2}}$$
$$=\frac{x^2y^2\left(\frac{1}{x}-\frac{1}{y}\right)}{x^2y^2\left(\frac{1}{x^2}-\frac{1}{y^2}\right)}$$
$$=\frac{xy^2-x^2y}{y^2-x^2}$$
$$=\frac{xy(y-x)}{(y+x)(y-x)}$$
$$=\frac{xy}{y+x}$$

31. $$\frac{x^{-1}+2y^{-1}}{2y+4x}=\frac{\frac{1}{x}+\frac{2}{y}}{2y+4x}$$

Multiply the numerator and denominator by xy, the LCD of all the fractions.

$$=\frac{xy\left(\frac{1}{x}+\frac{2}{y}\right)}{xy(2y+4x)}$$
$$=\frac{y+2x}{2xy(y+2x)}$$
$$=\frac{1}{2xy}$$

32. $$\frac{a^{-2}-4b^{-2}}{3b-6a}=\frac{\frac{1}{a^2}-\frac{4}{b^2}}{3b-6a}$$

Multiply the numerator and denominator by a^2b^2, the LCD of all the fractions.

$$=\frac{a^2b^2\left(\frac{1}{a^2}-\frac{4}{b^2}\right)}{a^2b^2(3b-6a)}$$
$$=\frac{b^2-4a^2}{3a^2b^2(b-2a)}$$
$$=\frac{(b+2a)(b-2a)}{3a^2b^2(b-2a)}$$
$$=\frac{b+2a}{3a^2b^2}$$

8.4 Equations with Rational Expressions and Graphs

8.4 Margin Exercises

1. **(a)** $\frac{3}{x}+\frac{1}{2}=\frac{5}{6x}$

Find the domain of each term.

For $\frac{3}{x}$, $x\neq 0$, so the domain is $\{x\,|\,x\neq 0\}$.

The domain of $\frac{1}{2}$ is $(-\infty,\infty)$.

For $\frac{5}{6x}$, $6x\neq 0$, so the domain is $\{x\,|\,x\neq 0\}$.

The intersection of these three domains is $\{x\,|\,x\neq 0\}$.

(b) $\frac{4}{x-5}-\frac{2}{x+5}=\frac{1}{x^2-25}$

Find the domain of each term.

For $\frac{4}{x-5}$, $x-5\neq 0$, so the domain is $\{x\,|\,x\neq 5\}$.

For $\frac{2}{x+5}$, $x+5\neq 0$, so the domain is $\{x\,|\,x\neq -5\}$.

For $\frac{1}{x^2-25}$, $x^2-25=(x+5)(x-5)\neq 0$, so the domain is $\{x\,|\,x\neq \pm 5\}$.

The intersection of these three domains is $\{x\,|\,x\neq \pm 5\}$.

2. $-\frac{3}{20}+\frac{2}{x}=\frac{5}{4x}$

The domain is $\{x\,|\,x\neq 0\}$.

Multiply each side by the LCD, $20x$.

$$20x\left(-\frac{3}{20}+\frac{2}{x}\right)=20x\left(\frac{5}{4x}\right)$$
$$20x\left(-\frac{3}{20}\right)+20x\left(\frac{2}{x}\right)=20x\left(\frac{5}{4x}\right)$$
$$-3x+40=25$$
$$-3x=-15$$
$$x=5$$

Check $x=5$: $-\frac{3}{20}+\frac{8}{20}=\frac{5}{20}$ *True*

Solution set: $\{5\}$

3. **(a)** $\frac{3}{x+1} = \frac{1}{x-1} - \frac{2}{x^2-1}$

The domain is $\{x \mid x \neq \pm 1\}$.

Multiply each side by the LCD, $(x+1)(x-1) = x^2 - 1$.

$$(x^2-1)\left(\frac{3}{x+1}\right) = (x^2-1)\left(\frac{1}{x-1} - \frac{2}{x^2-1}\right)$$

$$(x^2-1)\left(\frac{3}{x+1}\right) = (x^2-1)\left(\frac{1}{x-1}\right) - (x^2-1)\left(\frac{2}{x^2-1}\right)$$

$$3(x-1) = x+1-2$$
$$3x-3 = x-1$$
$$2x = 2$$
$$x = 1$$

Since 1 is not in the domain, the solution set is $\emptyset$.

(b) $\frac{1}{x-3} + \frac{1}{x+3} = \frac{6}{x^2-9}$

The domain is $\{x \mid x \neq \pm 3\}$.

Multiply each side by the LCD, $(x+3)(x-3) = x^2 - 9$.

$$(x^2-9)\left(\frac{1}{x-3} + \frac{1}{x+3}\right) = (x^2-9)\left(\frac{6}{x^2-9}\right)$$

$$(x^2-9)\left(\frac{1}{x-3}\right) + (x^2-9)\left(\frac{1}{x+3}\right) = (x^2-9)\left(\frac{6}{x^2-9}\right)$$

$$x+3+x-3 = 6$$
$$2x = 6$$
$$x = 3$$

Since 3 is not in the domain, the solution set is $\emptyset$.

4. $\frac{4}{x^2+x-6} - \frac{1}{x^2-4} = \frac{2}{x^2+5x+6}$

Factor the denominators and determine the domain.

$$x^2+x-6 = (x+3)(x-2), \text{ so } x \neq -3, 2.$$
$$x^2-4 = (x+2)(x-2), \text{ so } x \neq \pm 2.$$
$$x^2+5x+6 = (x+3)(x+2), \text{ so } x \neq -3, -2.$$

The domain is $\{x \mid x \neq -3, \pm 2\}$.

Multiply each side by the LCD, $(x+3)(x+2)(x-2)$.

$$(x+3)(x+2)(x-2)\left[\frac{4}{(x+3)(x-2)} - \frac{1}{(x+2)(x-2)}\right] = (x+3)(x+2)(x-2)\frac{2}{(x+3)(x+2)}$$

$$4(x+2) - (x+3) = 2(x-2)$$
$$4x+8-x-3 = 2x-4$$
$$3x+5 = 2x-4$$
$$x = -9$$

The solution set is $\{-9\}$.

5. $\frac{1}{x+4} + \frac{x}{x-4} = \frac{-8}{x^2-16}$

The domain is $\{x \mid x \neq \pm 4\}$.

Multiply each side by the LCD, $(x+4)(x-4) = x^2 - 16$.

$$(x^2-16)\left(\frac{1}{x+4} + \frac{x}{x-4}\right) = (x^2-16)\left(\frac{-8}{x^2-16}\right)$$

$$(x^2-16)\frac{1}{x+4} + (x^2-16)\frac{x}{x-4} = (x^2-16)\frac{-8}{x^2-16}$$

$$x-4+x(x+4) = -8$$
$$x-4+x^2+4x = -8$$
$$x^2+5x+4 = 0$$
$$(x+4)(x+1) = 0$$

$$x+4 = 0 \quad \text{or} \quad x+1 = 0$$
$$x = -4 \quad \text{or} \quad x = -1$$

Because -4 is not in the domain of the equation, it is not a solution.

Check $x = -1$: $\frac{1}{3} + \frac{1}{5} = \frac{8}{15}$ *True*

Solution set: $\{-1\}$

6. **(a)** $f(x) = -\frac{1}{x}$

x	-10	-1	$-\frac{1}{10}$	$\frac{1}{10}$	1	10
y	$\frac{1}{10}$	1	10	-10	-1	$-\frac{1}{10}$

Since $x \neq 0$, the equation of the vertical asymptote is $x = 0$. As the values of x get larger, the values of y get closer to 0, so $y = 0$ is the equation of the horizontal asymptote.

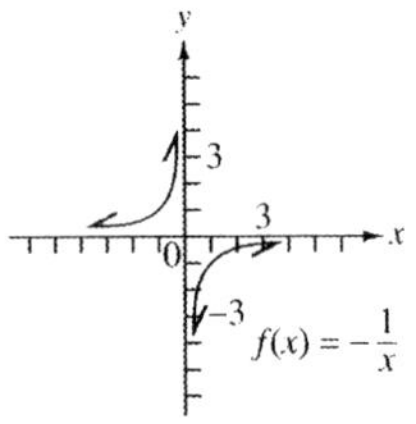

(b) $f(x) = \frac{2}{x+3}$

x	-13	-4	-3.1	-2.9	-2	7
y	$-\frac{1}{5}$	-2	-20	20	2	$\frac{1}{5}$

Set the denominator equal to zero and solve for x.

$$x + 3 = 0$$
$$x = -3$$

Since $x \neq -3$, the equation of the vertical asymptote is $x = -3$. As the values of x get larger, the values of y get closer to 0, so $y = 0$ is the equation of the horizontal asymptote.

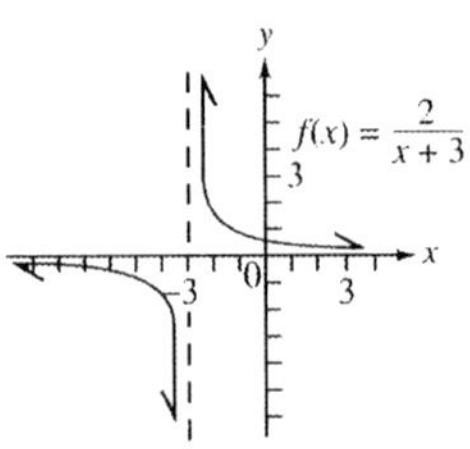

8.4 Section Exercises

1. **(a)** $\frac{1}{x+1} - \frac{1}{x-2} = 0$

Set each denominator equal to 0 and solve.

$$x + 1 = 0 \quad \text{or} \quad x - 2 = 0$$
$$x = -1 \quad \text{or} \quad x = 2$$

Solutions of -1 and 2 would be rejected since these values would make a denominator of the original equation equal to 0.

(b) The domain is $\{x \mid x \neq -1, 2\}$.

2. **(a)** $\frac{3}{x+4} - \frac{2}{x-9} = 0$

Set each denominator equal to 0 and solve.

$$x + 4 = 0 \quad \text{or} \quad x - 9 = 0$$
$$x = -4 \quad \text{or} \quad x = 9$$

So -4 and 9 would have to be rejected as potential solutions.

(b) The domain is $\{x \mid x \neq -4, 9\}$.

3. **(a)** $\frac{5}{3x+5} - \frac{1}{x} = \frac{1}{2x+3}$

Set each denominator equal to 0 and solve.

$$3x + 5 = 0 \quad \text{or} \quad x = 0 \quad \text{or} \quad 2x + 3 = 0$$
$$3x = -5 \qquad\qquad 2x = -3$$
$$x = -\tfrac{5}{3} \qquad\qquad x = -\tfrac{3}{2}$$

So $-\frac{5}{3}$, 0, and $-\frac{3}{2}$ would have to be rejected as potential solutions.

(b) The domain is $\left\{x \mid x \neq -\frac{5}{3}, 0, -\frac{3}{2}\right\}$.

4. **(a)** $\frac{6}{4x+7} - \frac{3}{x} = \frac{5}{6x-13}$

Set each denominator equal to 0 and solve.

$$4x + 7 = 0 \quad \text{or} \quad x = 0 \quad \text{or} \quad 6x - 13 = 0$$
$$4x = -7 \qquad\qquad 6x = 13$$
$$x = -\tfrac{7}{4} \qquad\qquad x = \tfrac{13}{6}$$

So $-\frac{7}{4}$, 0, and $\frac{13}{6}$ would have to be rejected as potential solutions.

(b) The domain is $\left\{x \mid x \neq -\frac{7}{4}, 0, \frac{13}{6}\right\}$.

5. **(a)** $\frac{1}{3x} + \frac{1}{2x} = \frac{x}{3}$

Only 0 would make any of the denominators equal to 0, so 0 would have to be rejected as a potential solution.

(b) The domain is $\{x \mid x \neq 0\}$.

6. **(a)** $\frac{5}{6x} - \frac{8}{2x} = \frac{x}{4}$

Only 0 would make any of the denominators equal to 0, so 0 would have to be rejected as a potential solution.

(b) The domain is $\{x \mid x \neq 0\}$.

7. **(a)** $\dfrac{3x+1}{x-4}=\dfrac{6x+5}{2x-7}$

Set each denominator equal to 0 and solve.

$$\begin{aligned} x-4&=0 \quad\text{or}\quad & 2x-7&=0\\ x&=4 \quad\text{or}\quad & 2x&=7\\ & & x&=\tfrac{7}{2} \end{aligned}$$

So 4 and $\frac{7}{2}$ would have to be rejected as potential solutions.

(b) The domain is $\{x \mid x \neq 4, \frac{7}{2}\}$.

8. **(a)** $\dfrac{4x-1}{2x+3}=\dfrac{12x-25}{6x-2}$

Set each denominator equal to 0 and solve.

$$\begin{aligned} 2x+3&=0 \quad\text{or}\quad & 6x-2&=0\\ 2x&=-3 \quad\text{or}\quad & 6x&=2\\ x&=-\tfrac{3}{2} \quad\text{or}\quad & x&=\tfrac{2}{6}=\tfrac{1}{3} \end{aligned}$$

So $-\frac{3}{2}$ and $\frac{1}{3}$ would have to be rejected as potential solutions.

(b) The domain is $\{x \mid x \neq -\frac{3}{2}, \frac{1}{3}\}$.

9. **(a)** $\dfrac{2}{x^2-x}+\dfrac{1}{x+3}=\dfrac{4}{x-2}$

$x^2-x=x(x-1)$ is 0 if $x=0$ or $x=1$.
$x+3$ is 0 if $x=-3$.
$x-2$ is 0 if $x=2$.

So 0, 1, -3, and 2 would have to be rejected as potential solutions.

(b) The domain is $\{x \mid x \neq 0, 1, -3, 2\}$.

10. Since there are no variables in denominators, the domain includes all real numbers, and no potential solutions must be excluded.

In Exercises 11–37, check each potential solution in the original equation.

11. $\dfrac{-5}{2x}+\dfrac{3}{4x}=\dfrac{-7}{4}$

Multiply by the LCD, $4x$. $(x \neq 0)$

$$\begin{aligned} 4x\left(\frac{-5}{2x}+\frac{3}{4x}\right)&=4x\left(\frac{-7}{4}\right)\\ 2(-5)+1(3)&=x(-7)\\ -10+3&=-7x\\ -7&=-7x\\ 1&=x \end{aligned}$$

Check $x=1$: $-\frac{5}{2}+\frac{3}{4}=-\frac{7}{4}$ *True*

Solution set: $\{1\}$

12. $\dfrac{6}{5x}-\dfrac{2}{3x}=\dfrac{-8}{45}$

Multiply by the LCD, $45x$. $(x \neq 0)$

$$\begin{aligned} 45x\left(\frac{6}{5x}-\frac{2}{3x}\right)&=45x\left(\frac{-8}{45}\right)\\ 54-30&=-8x\\ 24&=-8x\\ -3&=x \end{aligned}$$

Solution set: $\{-3\}$

13. $x-\dfrac{24}{x}=-2$

Multiply by the LCD, x. $(x \neq 0)$

$$\begin{aligned} x\left(x-\frac{24}{x}\right)&=-2\cdot x\\ x^2-24&=-2x\\ x^2+2x-24&=0\\ (x+6)(x-4)&=0 \end{aligned}$$

$$\begin{aligned} x+6&=0 \quad\text{or}\quad & x-4&=0\\ x&=-6 \quad\text{or}\quad & x&=4 \end{aligned}$$

Check $x=-6$: $-6+4=-2$ *True*
Check $x=4$: $4-6=-2$ *True*

Solution set: $\{-6, 4\}$

14. $p+\dfrac{15}{p}=-8$

Multiply by the LCD, p. $(p \neq 0)$

$$\begin{aligned} p^2+15&=-8p\\ p^2+8p+15&=0\\ (p+5)(p+3)&=0 \end{aligned}$$

$$\begin{aligned} p+5&=0 \quad\text{or}\quad & p+3&=0\\ p&=-5 \quad\text{or}\quad & p&=-3 \end{aligned}$$

Solution set: $\{-5, -3\}$

15. $\dfrac{x-4}{x+6}=\dfrac{2x+3}{2x-1}$

Multiply by the LCD, $(x+6)(2x-1)$.
Note that $x \neq -6$ and $x \neq \frac{1}{2}$.

$$\begin{aligned} (x+6)(2x-1)\left(\frac{x-4}{x+6}\right)&=(x+6)(2x-1)\left(\frac{2x+3}{2x-1}\right)\\ (2x-1)(x-4)&=(x+6)(2x+3)\\ 2x^2-9x+4&=2x^2+15x+18\\ -24x&=14\\ x&=\frac{14}{-24}=-\frac{7}{12} \end{aligned}$$

A calculator check is suggested.

continued

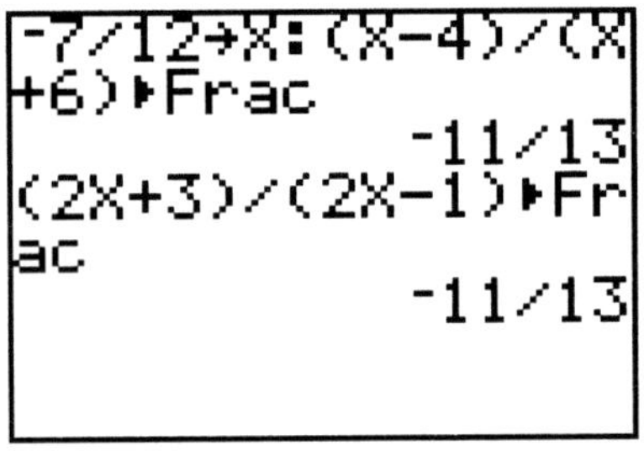

Check $x = -\frac{7}{12}$: $-\frac{11}{13} = -\frac{11}{13}$ *True*

Solution set: $\left\{-\frac{7}{12}\right\}$

16. $$\frac{5x-8}{x+2} = \frac{5x-1}{x+3}$$

Multiply by the LCD, $(x+2)(x+3)$.
Note that $x \neq -2$ and $x \neq -3$.

$$(x+2)(x+3)\left(\frac{5x-8}{x+2}\right) = (x+2)(x+3)\left(\frac{5x-1}{x+3}\right)$$
$$(x+3)(5x-8) = (x+2)(5x-1)$$
$$5x^2+7x-24 = 5x^2+9x-2$$
$$-22 = 2x$$
$$-11 = x$$

Solution set: $\{-11\}$

17. $$\frac{3x+1}{x-4} = \frac{6x+5}{2x-7}$$

Multiply by the LCD, $(x-4)(2x-7)$.
Note that $x \neq 4$ and $x \neq \frac{7}{2}$.

$$(x-4)(2x-7)\left(\frac{3x+1}{x-4}\right) = (x-4)(2x-7)\left(\frac{6x+5}{2x-7}\right)$$
$$(2x-7)(3x+1) = (x-4)(6x+5)$$
$$6x^2-19x-7 = 6x^2-19x-20$$
$$-7 = -20 \quad \textit{False}$$

The false statement indicates that the original equation has no solution.

Solution set: $\emptyset$

18. $$\frac{4x-1}{2x+3} = \frac{12x-25}{6x-2}$$

Multiply by the LCD, $(2x+3)(6x-2)$.
Note that $x \neq -\frac{3}{2}$ and $x \neq \frac{1}{3}$.

$$(6x-2)(4x-1) = (2x+3)(12x-25)$$
$$24x^2-14x+2 = 24x^2-14x-75$$
$$2 = -75 \quad \textit{False}$$

The false statement indicates that the original equation has no solution.

Solution set: $\emptyset$

19. $$\frac{1}{y-1} + \frac{5}{12} = \frac{-2}{3y-3}$$
$$\frac{1}{y-1} + \frac{5}{12} = \frac{-2}{3(y-1)}$$

Multiply by the LCD, $12(y-1)$. $(y \neq 1)$

$$12(y-1)\left(\frac{1}{y-1} + \frac{5}{12}\right) = 12(y-1)\left(\frac{-2}{3(y-1)}\right)$$
$$12 + 5(y-1) = -8$$
$$12 + 5y - 5 = -8$$
$$5y + 7 = -8$$
$$5y = -15$$
$$y = -3$$

Check $y = -3$: $-\frac{1}{4} + \frac{5}{12} = \frac{1}{6}$ *True*

Solution set: $\{-3\}$

20. $$\frac{4}{m+2} - \frac{11}{9} = \frac{1}{3m+6}$$
$$\frac{4}{m+2} - \frac{11}{9} = \frac{1}{3(m+2)}$$

Multiply by the LCD, $9(m+2)$. $(m \neq -2)$

$$9(m+2)\left(\frac{4}{m+2} - \frac{11}{9}\right) = 9(m+2)\left[\frac{1}{3(m+2)}\right]$$
$$36 - 11(m+2) = 3 \cdot 1$$
$$36 - 11m - 22 = 3$$
$$-11m = -11$$
$$m = 1$$

Check $m = 1$: $\frac{4}{3} - \frac{11}{9} = \frac{1}{9}$ *True*

Solution set: $\{1\}$

21. $$\frac{-2}{3t-6} - \frac{1}{36} = \frac{-3}{4t-8}$$
$$\frac{-2}{3(t-2)} - \frac{1}{36} = \frac{-3}{4(t-2)}$$

Multiply by the LCD, $36(t-2)$. $(t \neq 2)$

$$36(t-2)\left(\frac{-2}{3(t-2)} - \frac{1}{36}\right) = 36(t-2)\left(\frac{-3}{4(t-2)}\right)$$
$$12(-2) - 1(t-2) = 9(-3)$$
$$-24 - t + 2 = -27$$
$$-t - 22 = -27$$
$$-t = -5$$
$$t = 5$$

Check $t = 5$: $-\frac{2}{9} - \frac{1}{36} = \frac{-3}{12}$ *True*

Solution set: $\{5\}$

22. $$\frac{3}{4m+2}=\frac{17}{2}-\frac{7}{2m+1}$$
$$\frac{3}{2(2m+1)}=\frac{17}{2}-\frac{7}{2m+1}$$
Multiply by the LCD, $2(2m+1)$. $(m \neq -\frac{1}{2})$
$$2(2m+1)\left(\frac{3}{2(2m+1)}\right)=2(2m+1)\left(\frac{17}{2}-\frac{7}{2m+1}\right)$$
$$\begin{aligned}3&=17(2m+1)-7(2)\\3&=34m+17-14\\3&=34m+3\\0&=34m\\0&=m\end{aligned}$$

Check $m=0$: $\frac{3}{2}=\frac{17}{2}-7$ *True*

Solution set: $\{0\}$

23. $$\frac{3}{k+2}-\frac{2}{k^2-4}=\frac{1}{k-2}$$
$$\frac{3}{k+2}-\frac{2}{(k+2)(k-2)}=\frac{1}{k-2}$$
Multiply by the LCD, $(k+2)(k-2)$.
$(k \neq -2, 2)$
$$(k+2)(k-2)\left(\frac{3}{k+2}-\frac{2}{(k+2)(k-2)}\right)=(k+2)(k-2)\left(\frac{1}{k-2}\right)$$
$$\begin{aligned}3(k-2)-2&=k+2\\3k-6-2&=k+2\\3k-8&=k+2\\2k&=10\\k&=5\end{aligned}$$
Check $k=5$: $\frac{3}{7}-\frac{2}{21}=\frac{1}{3}$ *True*

Solution set: $\{5\}$

24. $$\frac{3}{x-2}+\frac{21}{x^2-4}=\frac{14}{x+2}$$
$$\frac{3}{x-2}+\frac{21}{(x+2)(x-2)}=\frac{14}{x+2}$$
Multiply by the LCD, $(x+2)(x-2)$.
$(x \neq -2, 2)$
$$\begin{aligned}3(x+2)+21&=14(x-2)\\3x+6+21&=14x-28\\3x+27&=14x-28\\55&=11x\\5&=x\end{aligned}$$
Check $x=5$: $1+1=2$ *True*

Solution set: $\{5\}$

25. $$\frac{1}{y+2}+\frac{3}{y+7}=\frac{5}{y^2+9y+14}$$
$$\frac{1}{y+2}+\frac{3}{y+7}=\frac{5}{(y+2)(y+7)}$$
Multiply by the LCD, $(y+2)(y+7)$.
$(y \neq -2, -7)$
$$(y+2)(y+7)\left(\frac{1}{y+2}+\frac{3}{y+7}\right)=(y+2)(y+7)\left(\frac{5}{(y+2)(y+7)}\right)$$
$$\begin{aligned}(y+7)+3(y+2)&=5\\y+7+3y+6&=5\\4y+13&=5\\4y&=-8\\y&=-2\end{aligned}$$

But y cannot equal -2 because that would make the denominator $y+2$ equal to 0. Since division by 0 is undefined, the equation has no solution.

Solution set: $\emptyset$

26. $$\frac{1}{t+3}+\frac{4}{t+5}=\frac{2}{t^2+8t+15}$$
$$\frac{1}{t+3}+\frac{4}{t+5}=\frac{2}{(t+5)(t+3)}$$
Multiply by the LCD, $(t+5)(t+3)$.
$(t \neq -5, -3)$
$$\begin{aligned}1(t+5)+4(t+3)&=2\\t+5+4t+12&=2\\5t+17&=2\\5t&=-15\\t&=-3\end{aligned}$$
But t cannot equal -3 because that would make the denominator $t+3$ equal to 0. Since division by 0 is undefined, the equation has no solution.
Solution set: $\emptyset$

27. $$\frac{9}{x}+\frac{4}{6x-3}=\frac{2}{6x-3}$$
Multiply by the LCD, $x(6x-3)$. $(x \neq 0, \frac{1}{2})$
$$x(6x-3)\left(\frac{9}{x}+\frac{4}{6x-3}\right)=x(6x-3)\left(\frac{2}{6x-3}\right)$$
$$\begin{aligned}9(6x-3)+4x&=2x\\54x-27+4x&=2x\\56x&=27\\x&=\tfrac{27}{56}\end{aligned}$$

Check $x=\frac{27}{56}$: $\frac{56}{3}+\left(-\frac{112}{3}\right)=-\frac{56}{3}$ *True*

Solution set: $\left\{\frac{27}{56}\right\}$

28. $$\frac{5}{n}+\frac{4}{6-3n}=\frac{2n}{6-3n}$$

Multiply by the LCD, $n(6-3n)$. $(n \neq 0, 2)$

$$5(6-3n)+4n=2n(n)$$
$$30-15n+4n=2n^2$$
$$0=2n^2+11n-30$$
$$0=(2n+15)(n-2)$$

$2n+15=0$ or $n-2=0$
$2n=-15$ or $n=2$
$n=-\frac{15}{2}$

Since $n \neq 2$, we only need to check $-\frac{15}{2}$.

Check $n=-\frac{15}{2}$: $-\frac{2}{3}+\frac{8}{57}=-\frac{10}{19}$ *True*

Solution set: $\{-\frac{15}{2}\}$

29. $$\frac{6}{w+3}+\frac{-7}{w-5}=\frac{-48}{w^2-2w-15}$$
$$\frac{6}{w+3}+\frac{-7}{w-5}=\frac{-48}{(w+3)(w-5)}$$

Multiply by the LCD, $(w+3)(w-5)$. $(w \neq -3, 5)$

$$(w+3)(w-5)\left(\frac{6}{w+3}+\frac{-7}{w-5}\right)=(w+3)(w-5)\left[\frac{-48}{(w+3)(w-5)}\right]$$
$$6(w-5)-7(w+3)=-48$$
$$6w-30-7w-21=-48$$
$$-w-51=-48$$
$$-w=3$$
$$w=-3$$

But w cannot equal -3 because that would make the denominator $w+3$ equal to 0. Since division by 0 is undefined, the equation has no solution.

Solution set: $\emptyset$

30. $$\frac{2}{r-5}+\frac{3}{2r+1}=\frac{22}{2r^2-9r-5}$$
$$\frac{2}{r-5}+\frac{3}{2r+1}=\frac{22}{(2r+1)(r-5)}$$

Multiply by the LCD, $(2r+1)(r-5)$. $\left(r \neq -\frac{1}{2}, 5\right)$

$$2(2r+1)+3(r-5)=22$$
$$4r+2+3r-15=22$$
$$7r-13=22$$
$$7r=35$$
$$r=5$$

But $r \neq 5$.

Solution set: $\emptyset$

31. $$\frac{x}{x-3}+\frac{4}{x+3}=\frac{18}{x^2-9}$$
$$\frac{x}{x-3}+\frac{4}{x+3}=\frac{18}{(x-3)(x+3)}$$

Multiply by the LCD, $(x-3)(x+3)$. $(x \neq 3, -3)$

$$(x-3)(x+3)\left(\frac{x}{x-3}+\frac{4}{x+3}\right)=(x-3)(x+3)\left(\frac{18}{(x-3)(x+3)}\right)$$
$$x(x+3)+4(x-3)=18$$
$$x^2+3x+4x-12=18$$
$$x^2+7x-30=0$$
$$(x-3)(x+10)=0$$

$x-3=0$ or $x+10=0$
$x=3$ or $x=-10$

But $x \neq 3$ since a denominator of 0 results. The only solution to check is -10.

Check $x=-10$: $\frac{10}{13}+\left(-\frac{4}{7}\right)=\frac{18}{91}$ *True*

Solution set: $\{-10\}$

32. $$\frac{2x}{x-3}+\frac{4}{x+3}=\frac{-24}{x^2-9}$$
$$\frac{2x}{x-3}+\frac{4}{x+3}=\frac{-24}{(x+3)(x-3)}$$

Multiply by the LCD, $(x+3)(x-3)$. $(x \neq -3, 3)$

$$2x(x+3)+4(x-3)=-24$$
$$2x^2+6x+4x-12=-24$$
$$2x^2+10x-12=-24$$
$$2x^2+10x+12=0$$
$$x^2+5x+6=0 \quad \textit{Divide by 2.}$$
$$(x+3)(x+2)=0$$

$x+3=0$ or $x+2=0$
$x=-3$ or $x=-2$

But x cannot equal -3, so we only need to check -2.

Check $x=-2$: $\frac{4}{5}+4=\frac{24}{5}$ *True*

Solution set: $\{-2\}$

33. $$\frac{6}{x-4}+\frac{5}{x}=\frac{-20}{x^2-4x}$$
$$\frac{6}{x-4}+\frac{5}{x}=\frac{-20}{x(x-4)}$$

Multiply by the LCD, $x(x-4)$. $(x \neq 0, 4)$

$$x(x-4)\left(\frac{6}{x-4}+\frac{5}{x}\right)=x(x-4)\left(\frac{-20}{x(x-4)}\right)$$

$$6x + 5(x - 4) = -20$$
$$6x + 5x - 20 = -20$$
$$11x = 0$$
$$x = 0$$

Since 0 cannot appear in the denominator, this equation has no solution.

Solution set: $\emptyset$

34. $$\frac{7}{x-4} + \frac{3}{x} = \frac{-12}{x^2 - 4x}$$
$$\frac{7}{x-4} + \frac{3}{x} = \frac{-12}{x(x-4)}$$

Multiply by the LCD, $x(x-4)$. $(x \neq 0, 4)$

$$x(x-4)\left(\frac{7}{x-4} + \frac{3}{x}\right) = x(x-4)\left(\frac{-12}{x(x-4)}\right)$$
$$7x + 3(x-4) = -12$$
$$7x + 3x - 12 = -12$$
$$10x = 0$$
$$x = 0$$

But $x \neq 0$.

Solution set: $\emptyset$

35. $$\frac{2}{4x+7} + \frac{x}{3} = \frac{6}{12x+21}$$
$$\frac{2}{4x+7} + \frac{x}{3} = \frac{6}{3(4x+7)}$$

Multiply by the LCD, $3(4x+7)$. $(x \neq -\frac{7}{4})$

$$2(3) + x(4x+7) = 6$$
$$6 + 4x^2 + 7x = 6$$
$$4x^2 + 7x = 0$$
$$x(4x+7) = 0$$
$$x = 0 \quad \text{or} \quad 4x + 7 = 0$$
$$4x = -7$$
$$x = -\tfrac{7}{4}$$

But x cannot equal $-\frac{7}{4}$, so we only need to check 0.

Check $x = 0$: $\frac{2}{7} + 0 = \frac{6}{21}$ *True*

Solution set: $\{0\}$

36. $$\frac{5x+14}{x^2-9} = \frac{-2x^2-5x+2}{x^2-9} + \frac{2x+4}{x-3}$$
$$\frac{5x+14}{(x+3)(x-3)} = \frac{-2x^2-5x+2}{(x+3)(x-3)} + \frac{2x+4}{x-3}$$

Multiply by the LCD, $(x+3)(x-3)$. $(x \neq -3, 3)$

$$5x + 14 = -2x^2 - 5x + 2 + (2x+4)(x+3)$$
$$5x + 14 = -2x^2 - 5x + 2 + 2x^2 + 10x + 12$$
$$5x + 14 = 5x + 14 \quad \textit{True}$$

This equation is true for every real number value of x, but we have already determined that $x \neq -3$ or $x \neq 3$. So every real number except -3 and 3 is a solution.

Solution set: $\{x \mid x \neq -3, 3\}$ or $(-\infty, -3) \cup (-3, 3) \cup (3, \infty)$

37. $$\frac{4x-7}{4x^2-9} = \frac{-2x^2+5x-4}{4x^2-9} + \frac{x+1}{2x+3}$$
$$\frac{4x-7}{(2x+3)(2x-3)} = \frac{-2x^2+5x-4}{(2x+3)(2x-3)} + \frac{x+1}{2x+3}$$

Multiply by the LCD, $(2x+3)(2x-3)$. $(x \neq -\frac{3}{2}, \frac{3}{2})$

$$4x - 7 = -2x^2 + 5x - 4 + (2x-3)(x+1)$$
$$4x - 7 = -2x^2 + 5x - 4 + 2x^2 - x - 3$$
$$4x - 7 = 4x - 7 \quad \textit{True}$$

This equation is true for every real number value of x, but we have already determined that $x \neq -\frac{3}{2}$ or $x \neq \frac{3}{2}$. So every real number except $-\frac{3}{2}$ and $\frac{3}{2}$ is a solution.

Solution set: $\{x \mid x \neq -\frac{3}{2}, \frac{3}{2}\}$ or $\left(-\infty, -\frac{3}{2}\right) \cup \left(-\frac{3}{2}, \frac{3}{2}\right) \cup \left(\frac{3}{2}, \infty\right)$

38. "Solve" refers to finding the solution set of an equation. What appears here is not an equation, but an expression. "Solve" should be replaced by "Simplify" or "Add."

39. $f(x) = \dfrac{2}{x}$ is not defined when $x = 0$, so an equation of the vertical asymptote is $x = 0$. As the values of x get larger, the values of y get closer to 0, so $y = 0$ is the equation of the horizontal asymptote.

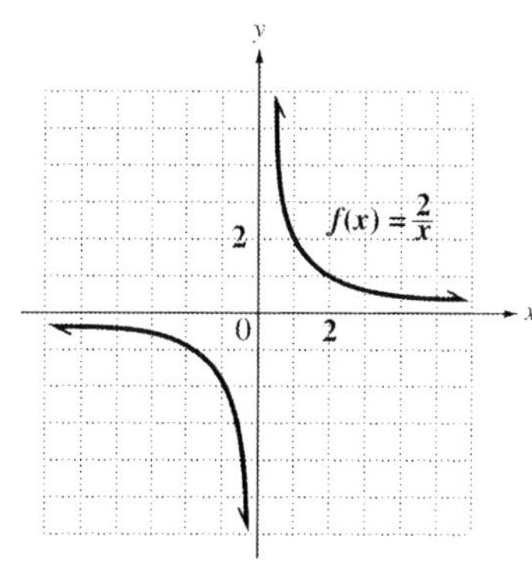

40. $f(x) = \dfrac{3}{x}$ is not defined when $x = 0$, so an equation of the vertical asymptote is $x = 0$. As the values of x get larger, the values of y get closer to 0, so $y = 0$ is the equation of the horizontal asymptote.

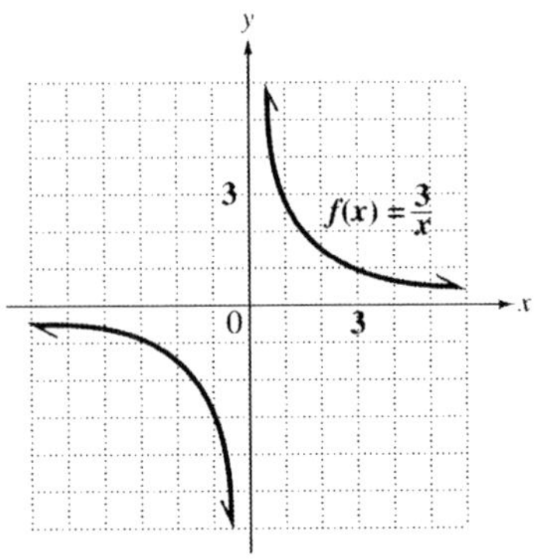

41. $f(x) = \dfrac{1}{x-2}$ is not defined when $x = 2$, so an equation of the vertical asymptote is $x = 2$. As the values of x get larger, the values of y get closer to 0, so $y = 0$ is the equation of the horizontal asymptote.

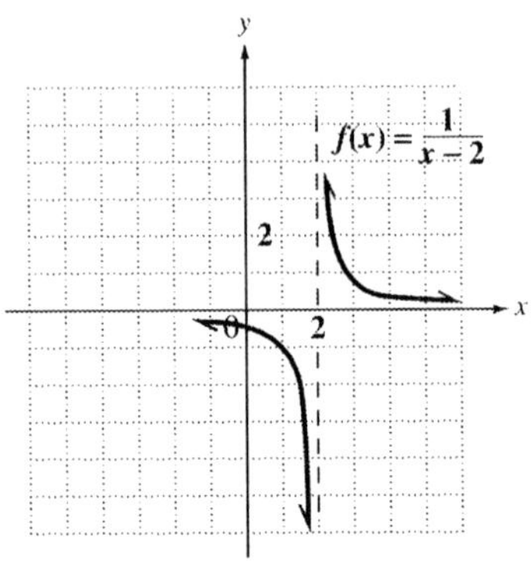

42. $f(x) = \dfrac{1}{x+2}$ is not defined when $x = -2$, so an equation of the vertical asymptote is $x = -2$. As the values of x get larger, the values of y get closer to 0, so $y = 0$ is the equation of the horizontal asymptote.

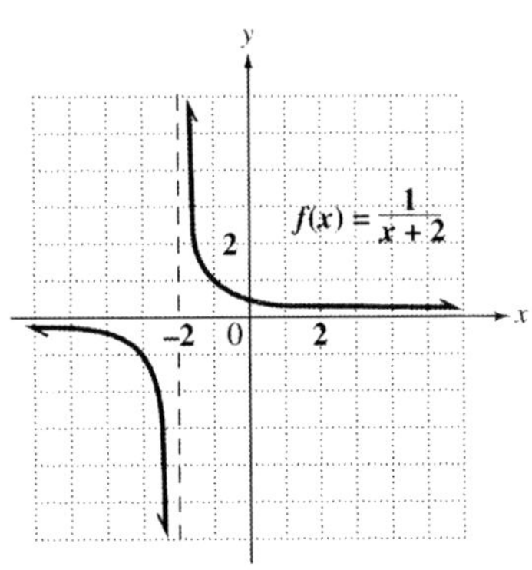

43. $w(x) = \dfrac{x^2}{2(1-x)}$

(a) $w(.1) = \dfrac{(.1)^2}{2(1-.1)}$

$= \dfrac{.01}{2(.9)} \approx .006$

To the nearest tenth, $w(.1)$ is 0.

(b) $w(.8) = \dfrac{(.8)^2}{2(1-.8)}$

$= \dfrac{.64}{2(.2)} = 1.6$

(c) $w(.9) = \dfrac{(.9)^2}{2(1-.9)}$

$= \dfrac{.81}{2(.1)} = 4.05 \approx 4.1$

(d) Based on the answers in (a), (b), and (c), we see that as the traffic intensity increases, the waiting time also increases.

44. $p(x) = \dfrac{x-1}{x}$

(a) $p(10) = \dfrac{10-1}{10} = \dfrac{9}{10} = .9$ or 90%

(b) $80\% = .8$, so let $p(x) = .8$.

$$.8 = \frac{x-1}{x}$$
$$.8x = x - 1$$
$$-.2x = -1$$
$$x = \frac{-1}{-.2} = 5$$

(c) The incidence rate x cannot equal 0 because 0 is not in the domain of this function.

45. Substituting -1 for x gives a true statement, $\frac{4}{3} = \frac{4}{3}$. Substituting -2 for x leads to 0 in the first and third denominators.

46. If $A = 2$, and $B = 1$, then

$$C = -2AB = -2(2)(1) = -4.$$

The equation determined by A, B, and C is

$$\frac{2}{x+1} + \frac{x}{x-1} = \frac{-4}{x^2-1}.$$
$$\frac{2}{x+1} + \frac{x}{x-1} = \frac{-4}{(x+1)(x-1)}$$

Multiply by the LCD, $(x+1)(x-1)$. $(x \neq \pm 1)$

$$(x+1)(x-1)\left(\frac{2}{x+1}+\frac{x}{x-1}\right) = (x+1)(x-1)\left(\frac{-4}{(x+1)(x-1)}\right)$$

$$2(x-1)+x(x+1) = -4$$
$$2x-2+x^2+x = -4$$
$$x^2+3x+2 = 0$$
$$(x+2)(x+1) = 0$$

$$x+2=0 \quad \text{or} \quad x+1=0$$
$$x=-2 \quad \text{or} \quad x=-1$$

But $x \neq -1$, so it must be rejected.

Check $x=-2$: $-2+\frac{2}{3}=-\frac{4}{3}$ *True*

Solution set: $\{-2\}$

47. If $A=4$ and $B=-3$, then

$$C=-2AB=-2(4)(-3)=24.$$

The equation determined by A, B, and C is

$$\frac{4}{x-3}+\frac{x}{x+3}=\frac{24}{x^2-9}.$$
$$\frac{4}{x-3}+\frac{x}{x+3}=\frac{24}{(x+3)(x-3)}$$

Multiply by the LCD, $(x+3)(x-3)$. $(x \neq \pm 3)$

$$(x+3)(x-3)\left(\frac{4}{x-3}+\frac{x}{x+3}\right) = (x+3)(x-3)\left(\frac{24}{(x+3)(x-3)}\right)$$

$$4(x+3)+x(x-3) = 24$$
$$4x+12+x^2-3x = 24$$
$$x^2+x-12 = 0$$
$$(x+4)(x-3) = 0$$

$$x+4=0 \quad \text{or} \quad x-3=0$$
$$x=-4 \quad \text{or} \quad x=3$$

But $x \neq 3$, so it must be rejected.

Check $x=-4$: $-\frac{4}{7}+4=\frac{24}{7}$ *True*

Solution set: $\{-4\}$

48. Answers will vary. However, in every case, $-B$ will be the rejected solution, and $\{-A\}$ will be the solution set.

Summary Exercises on Rational Expressions and Equations

1. $\frac{x}{2}-\frac{x}{4}=5$

There is an equals sign, so this is an *equation.*

$$4\left(\frac{x}{2}-\frac{x}{4}\right)=4(5) \quad \textit{Multiply by 4.}$$
$$2x-x=20$$
$$x=20$$

Check $x=20$: $10-5=5$ *True*

Solution set: $\{20\}$

2. $\frac{4x-20}{x^2-25}\cdot\frac{(x+5)^2}{10}$

No equals sign appears so this is an *expression.*

$$=\frac{4(x-5)}{(x+5)(x-5)}\cdot\frac{(x+5)(x+5)}{10}$$
$$=\frac{2(x+5)}{5}$$

3. $\frac{6}{7x}-\frac{4}{x}$

No equals sign appears so this is an *expression.*

$$=\frac{6}{7x}-\frac{4\cdot 7}{x\cdot 7} \qquad LCD=7x$$
$$=\frac{6-28}{7x}=\frac{-22}{7x} \text{ or } -\frac{22}{7x}$$

4. $\dfrac{\frac{1}{x}+\frac{1}{y}}{\frac{1}{x}-\frac{1}{y}}$

No equals sign appears so this is an *expression.* Multiply the numerator and denominator by the LCD of all the fractions, xy.

$$=\frac{xy\left(\frac{1}{x}+\frac{1}{y}\right)}{xy\left(\frac{1}{x}-\frac{1}{y}\right)}=\frac{y+x}{y-x}$$

5. $\frac{5}{7t}=\frac{52}{7}-\frac{3}{t}$

There is an equals sign, so this is an *equation.* Multiply each side by the LCD, $7t$. $(t \neq 0)$

$$7t\left(\frac{5}{7t}\right)=7t\left(\frac{52}{7}-\frac{3}{t}\right)$$
$$5=52t-3(7)$$
$$5=52t-21$$
$$26=52t$$
$$t=\tfrac{26}{52}=\tfrac{1}{2}$$

Check $t=\frac{1}{2}$: $\frac{10}{7}=\frac{52}{7}-6$ *True*

Solution set: $\left\{\frac{1}{2}\right\}$

6. $\dfrac{x-5}{3}+\dfrac{1}{3}=\dfrac{x-2}{5}$

There is an equals sign, so this is an *equation.* Multiply each side by the LCD, 15.

$$15\left(\frac{x-5}{3}+\frac{1}{3}\right)=15\left(\frac{x-2}{5}\right)$$
$$5(x-5)+5=3(x-2)$$
$$5x-25+5=3x-6$$
$$5x-20=3x-6$$
$$2x=14$$
$$x=7$$

Check $x=7$: $\frac{2}{3}+\frac{1}{3}=1$ *True*

Solution set: $\{7\}$

7. $\dfrac{7}{6x}+\dfrac{5}{8x}$

No equals sign appears so this is an *expression.*

$$=\frac{7}{3\cdot 2x}+\frac{5}{4\cdot 2x}$$
$$=\frac{7(4)}{4\cdot 3\cdot 2x}+\frac{5(3)}{3\cdot 4\cdot 2x} \qquad \textit{LCD} = 24x$$
$$=\frac{28+15}{3\cdot 4\cdot 2x}=\frac{43}{24x}$$

8. $\dfrac{4}{x}-\dfrac{8}{x+1}=0$

There is an equals sign, so this is an *equation.* Multiply by the LCD, $x(x+1)$. $(x\neq -1,0)$

$$x(x+1)\left(\frac{4}{x}-\frac{8}{x+1}\right)=x(x+1)\cdot 0$$
$$4(x+1)-8x=0$$
$$4x+4-8x=0$$
$$-4x=-4$$
$$x=1$$

Check $x=1$: $4-4=0$ *True*

Solution set: $\{1\}$

9. $\dfrac{\dfrac{6}{x+1}-\dfrac{1}{x}}{\dfrac{2}{x}-\dfrac{4}{x+1}}$

No equals sign appears so this is an *expression.* Multiply the numerator and denominator by the LCD of all the fractions, $x(x+1)$.

$$=\frac{6(x)-1(x+1)}{2(x+1)-4(x)}$$
$$=\frac{6x-x-1}{2x+2-4x}$$
$$=\frac{5x-1}{-2x+2} \text{ or } \frac{5x-1}{-2(x-1)}$$

10. $\dfrac{8}{r+2}-\dfrac{7}{4r+8}$

No equals sign appears so this is an *expression.*

$$=\frac{8}{r+2}-\frac{7}{4(r+2)}$$
$$=\frac{4(8)}{4(r+2)}-\frac{7}{4(r+2)} \qquad \textit{LCD} = 4(r+2)$$
$$=\frac{32}{4(r+2)}-\frac{7}{4(r+2)}$$
$$=\frac{25}{4(r+2)}$$

11. $\dfrac{x}{x+y}+\dfrac{2y}{x-y}$

No equals sign appears so this is an *expression.*

$$=\frac{x(x-y)}{(x+y)(x-y)}+\frac{2y(x+y)}{(x+y)(x-y)}$$
$$\textit{LCD} = (x+y)(x-y)$$
$$=\frac{x^2-xy+2xy+2y^2}{(x+y)(x-y)}$$
$$=\frac{x^2+xy+2y^2}{(x+y)(x-y)}$$

12. $\dfrac{3p^2-6p}{p+5}\div\dfrac{p^2-4}{8p+40}$

No equals sign appears so this is an *expression.*

$$=\frac{3p(p-2)}{p+5}\cdot\frac{8(p+5)}{(p+2)(p-2)}$$
$$=\frac{24p(p-2)(p+5)}{(p+2)(p-2)(p+5)}$$
$$=\frac{24p}{p+2}$$

13. $\dfrac{x-2}{9}\cdot\dfrac{5}{8-4x}$

No equals sign appears so this is an *expression.*

$$=\frac{x-2}{9}\cdot\frac{5}{-4(x-2)}$$
$$=\frac{5}{-36}=-\frac{5}{36}$$

14. $\dfrac{a-4}{3}+\dfrac{11}{6}=\dfrac{a+1}{2}$

There is an equals sign, so this is an *equation.* Multiply each side by the LCD, 6.

$$6\left(\frac{a-4}{3}+\frac{11}{6}\right)=6\left(\frac{a+1}{2}\right)$$
$$2(a-4)+11=3(a+1)$$
$$2a-8+11=3a+3$$
$$2a+3=3a+3$$
$$0=a$$

Check $a=0$: $-\frac{4}{3}+\frac{11}{6}=\frac{1}{2}$ *True*

Solution set: $\{0\}$

15. $\dfrac{b^2+b-6}{b^2+2b-8}\cdot\dfrac{b^2+8b+16}{3b+12}$

No equals sign appears so this is an *expression.*

$$=\frac{(b+3)(b-2)}{(b+4)(b-2)}\cdot\frac{(b+4)(b+4)}{3(b+4)}$$
$$=\frac{b+3}{3}$$

16. $\dfrac{10z^2-5z}{3z^3-6z^2}\div\dfrac{2z^2+5z-3}{z^2+z-6}$

No equals sign appears so this is an *expression.*

$$=\frac{5z(2z-1)}{3z^2(z-2)}\cdot\frac{(z+3)(z-2)}{(2z-1)(z+3)}$$
$$=\frac{5z(2z-1)(z+3)(z-2)}{3z^2(2z-1)(z+3)(z-2)}$$
$$=\frac{5}{3z}$$

17. $\dfrac{5}{x^2-2x}-\dfrac{3}{x^2-4}$

No equals sign appears so this is an *expression.*

$$=\frac{5}{x(x-2)}-\frac{3}{(x+2)(x-2)}$$
$$=\frac{5(x+2)}{x(x-2)(x+2)}-\frac{3x}{x(x-2)(x+2)}$$
$$=\frac{5x+10-3x}{x(x-2)(x+2)}$$
$$=\frac{2x+10}{x(x-2)(x+2)}$$

18. $\dfrac{6}{t+1}+\dfrac{4}{5t+5}=\dfrac{34}{15}$

$$\frac{6}{t+1}+\frac{4}{5(t+1)}=\frac{34}{15}$$

There is an equals sign, so this is an *equation.*
Multiply by the LCD, $15(t+1)$. $(t\neq -1)$

$$15(t+1)\left(\frac{6}{t+1}+\frac{4}{5(t+1)}\right)=15(t+1)\left(\frac{34}{15}\right)$$
$$90+12=34(t+1)$$
$$102=34t+34$$
$$68=34t$$
$$2=t$$

Check $t=2$: $2+\frac{4}{15}=\frac{34}{15}$ *True*
Solution set: $\{2\}$

19. $\dfrac{\frac{5}{x}-\frac{3}{y}}{\frac{9x^2-25y^2}{x^2y}}$

No equals sign appears so this is an *expression.*
Multiply the numerator and denominator by the LCD of all the fractions, x^2y.

$$=\frac{x^2y\left(\frac{5}{x}-\frac{3}{y}\right)}{x^2y\left(\frac{9x^2-25y^2}{x^2y}\right)}$$
$$=\frac{5xy-3x^2}{9x^2-25y^2}$$
$$=\frac{-x(3x-5y)}{(3x+5y)(3x-5y)}$$
$$=\frac{-x}{(3x+5y)}$$

20. $\dfrac{-2}{a^2+2a-3}-\dfrac{5}{3-3a}=\dfrac{4}{3a+9}$

$$\frac{-2}{(a+3)(a-1)}-\frac{5}{3(1-a)}=\frac{4}{3(a+3)}$$
$$\frac{-2}{(a+3)(a-1)}+\frac{5}{3(a-1)}=\frac{4}{3(a+3)}$$

There is an equals sign, so this is an *equation.*
Multiply by the LCD, $3(a+3)(a-1)$.
$(a\neq -3, 1)$

$$-2\cdot 3+5(a+3)=4(a-1)$$
$$-6+5a+15=4a-4$$
$$5a+9=4a-4$$
$$a=-13$$

Check $a=-13$: $-\frac{1}{70}-\frac{5}{42}=-\frac{2}{15}$ *True*
Solution set: $\{-13\}$

21. $\dfrac{4y^2-13y+3}{2y^2-9y+9}\div\dfrac{4y^2+11y-3}{6y^2-5y-6}$

No equals sign appears so this is an *expression.*

$$=\frac{(4y-1)(y-3)}{(2y-3)(y-3)}\cdot\frac{(2y-3)(3y+2)}{(4y-1)(y+3)}$$
$$=\frac{3y+2}{y+3}$$

22. $\dfrac{8}{3k+9}-\dfrac{8}{15}=\dfrac{2}{5k+15}$

$$\frac{8}{3(k+3)}-\frac{8}{15}=\frac{2}{5(k+3)}$$

There is an equals sign, so this is an *equation.*
Multiply by the LCD, $15(k+3)$. $(k\neq -3)$

$$5(8)-8(k+3)=3(2)$$
$$40-8k-24=6$$
$$-8k+16=6$$
$$-8k=-10$$
$$k=\tfrac{-10}{-8}=\tfrac{5}{4}$$

Check $k=\frac{5}{4}$: $\frac{32}{51}-\frac{8}{15}=\frac{8}{85}$ *True*
Solution set: $\{\frac{5}{4}\}$

23. $\dfrac{3r}{r-2} = 1 + \dfrac{6}{r-2}$

There is an equals sign, so this is an *equation.*

Multiply by the LCD, $r-2$. $(r \neq 2)$

$$3r = r - 2 + 6$$
$$2r = 4$$
$$r = 2$$

But $r \neq 2$.

Solution set: $\emptyset$

24. $\dfrac{6z^2-5z-6}{6z^2+5z-6} \cdot \dfrac{12z^2-17z+6}{12z^2-z-6}$

No equals sign appears so this is an *expression.*

$$= \frac{(2z-3)(3z+2)}{(2z+3)(3z-2)} \cdot \frac{(3z-2)(4z-3)}{(3z+2)(4z-3)}$$
$$= \frac{2z-3}{2z+3}$$

25. $\dfrac{-1}{3-x} - \dfrac{2}{x-3}$

No equals sign appears so this is an *expression.*

$$= \frac{-1}{-(x-3)} - \frac{2}{x-3}$$
$$= \frac{1}{x-3} - \frac{2}{x-3}$$
$$= \frac{-1}{x-3} \text{ or } \frac{-1}{-(3-x)} = \frac{1}{3-x}$$

26. $\dfrac{\frac{t}{4} - \frac{1}{t}}{1 + \frac{t+4}{t}}$

No equals sign appears so this is an *expression.* Multiply the numerator and denominator by the LCD of all the fractions, $4t$.

$$= \frac{4t\left(\frac{t}{4} - \frac{1}{t}\right)}{4t\left(1 + \frac{t+4}{t}\right)} = \frac{t^2-4}{4t+4(t+4)}$$
$$= \frac{t^2-4}{4t+4t+16}$$
$$= \frac{t^2-4}{8t+16}$$
$$= \frac{(t+2)(t-2)}{8(t+2)} = \frac{t-2}{8}$$

27. $\dfrac{2}{y+1} - \dfrac{3}{y^2-y-2} = \dfrac{3}{y-2}$

$\dfrac{2}{y+1} - \dfrac{3}{(y-2)(y+1)} = \dfrac{3}{y-2}$

There is an equals sign, so this is an *equation.*

Multiply by the LCD, $(y-2)(y+1)$.

$(y \neq -1, 2)$

$$2(y-2) - 3 = 3(y+1)$$
$$2y - 4 - 3 = 3y + 3$$
$$2y - 7 = 3y + 3$$
$$-10 = y$$

Check $y = -10$: $-\frac{2}{9} - \frac{1}{36} = -\frac{1}{4}$ *True*

Solution set: $\{-10\}$

28. $\dfrac{7}{2x^2-8x} + \dfrac{3}{x^2-16}$

No equals sign appears so this is an *expression.*

$$= \frac{7}{2x(x-4)} + \frac{3}{(x-4)(x+4)}$$
$$= \frac{7(x+4)}{2x(x-4)(x+4)} + \frac{(3)2x}{(x-4)(x+4)2x}$$

LCD=2x(x-4)(x+4)

$$= \frac{7(x+4)}{2x(x-4)(x+4)} + \frac{6x}{2x(x-4)(x+4)}$$
$$= \frac{7x+28+6x}{2x(x-4)(x+4)}$$
$$= \frac{13x+28}{2x(x-4)(x+4)}$$

29. $\dfrac{3}{y-3} - \dfrac{3}{y^2-5y+6} = \dfrac{2}{y-2}$

$\dfrac{3}{y-3} - \dfrac{3}{(y-3)(y-2)} = \dfrac{2}{y-2}$

There is an equals sign, so this is an *equation.*

Multiply by the LCD, $(y-3)(y-2)$. $(y \neq 2, 3)$

$$3(y-2) - 3 = 2(y-3)$$
$$3y - 6 - 3 = 2y - 6$$
$$3y - 9 = 2y - 6$$
$$y = 3$$

But $y \neq 3$.

Solution set: $\emptyset$

30. $\dfrac{2k + \frac{5}{k-1}}{3k - \frac{2}{k}}$

No equals sign appears so this is an *expression.* Multiply the numerator and denominator by the LCD of all the fractions, $k(k-1)$.

$$= \frac{k(k-1)\left(2k + \frac{5}{k-1}\right)}{k(k-1)\left(3k - \frac{2}{k}\right)}$$

$$= \frac{2k^2(k-1) + 5k}{3k^2(k-1) - 2(k-1)}$$

$$= \frac{k[2k(k-1)+5]}{(k-1)(3k^2-2)} \quad \textit{Factor out k and k − 1.}$$

$$= \frac{k(2k^2-2k+5)}{(k-1)(3k^2-2)}$$

8.5 Applications of Rational Expressions

8.5 Margin Exercises

1. **(a)** $\frac{1}{f} = \frac{1}{p} + \frac{1}{q}$

$\frac{1}{15} = \frac{1}{p} + \frac{1}{25}$ *Let f = 15, q = 25.*

Multiply by the LCD, $75p$.

$$75p \cdot \frac{1}{15} = 75p\left(\frac{1}{p} + \frac{1}{25}\right)$$
$$75p \cdot \frac{1}{15} = 75p\left(\frac{1}{p}\right) + 75p\left(\frac{1}{25}\right)$$
$$5p = 75 + 3p$$
$$2p = 75$$
$$p = \tfrac{75}{2}$$

(b) $\frac{1}{f} = \frac{1}{p} + \frac{1}{q}$

$\frac{1}{f} = \frac{1}{6} + \frac{1}{9}$ *Let p = 6, q = 9.*

Multiply by the LCD, $18f$.

$$18f \cdot \frac{1}{f} = 18f\left(\frac{1}{6} + \frac{1}{9}\right)$$
$$18f \cdot \frac{1}{f} = 18f\left(\frac{1}{6}\right) + 18f\left(\frac{1}{9}\right)$$
$$18 = 3f + 2f$$
$$18 = 5f$$
$$\tfrac{18}{5} = f$$

(c) $\frac{1}{f} = \frac{1}{p} + \frac{1}{q}$

$\frac{1}{12} = \frac{1}{16} + \frac{1}{q}$ *Let f = 12, p = 16.*

Multiply by the LCD, $48q$.

$$48q \cdot \frac{1}{12} = 48q\left(\frac{1}{16} + \frac{1}{q}\right)$$
$$48q \cdot \frac{1}{12} = 48q\left(\frac{1}{16}\right) + 48q\left(\frac{1}{q}\right)$$
$$4q = 3q + 48$$
$$q = 48$$

2. Solve $\frac{3}{p} + \frac{3}{q} = \frac{5}{r}$ for q.

Multiply by the LCD, pqr.

$$pqr\left(\frac{3}{p} + \frac{3}{q}\right) = pqr\left(\frac{5}{r}\right)$$
$$pqr\left(\frac{3}{p}\right) + pqr\left(\frac{3}{q}\right) = pqr\left(\frac{5}{r}\right)$$
$$3qr + 3pr = 5pq$$

Get the q terms on one side.

$$3qr - 5pq = -3pr$$
$$q(3r - 5p) = -3pr \quad \textit{Factor out q.}$$
$$q = \frac{-3pr}{3r-5p} \text{ or } \frac{3pr}{5p-3r}$$

3. Solve $A = \frac{Rr}{R+r}$ for R.

Multiply by the LCD, $R + r$.

$$A(R+r) = \frac{Rr}{R+r}(R+r)$$
$$A(R+r) = Rr$$
$$AR + Ar = Rr$$

Get the R terms on one side.

$$AR - Rr = -Ar$$
$$R(A - r) = -Ar \quad \textit{Factor out R.}$$
$$R = \frac{-Ar}{A-r} \text{ or } \frac{Ar}{r-A}$$

4. *Step 2*
Let x = the number (in millions) who had no health insurance.

Step 3
Set up a proportion.

$$\frac{11.6}{100} = \frac{x}{73{,}500{,}000}$$

Step 4
To solve the equation, multiply by the LCD.

$$73{,}500{,}000\left(\frac{11.6}{100}\right) = 73{,}500{,}000\left(\frac{x}{73{,}500{,}000}\right)$$
$$735{,}000(11.6) = x$$
$$x = 8{,}526{,}000$$

continued

Step 5
There were 8,526,000 children under 18 years of age in the United States with no health insurance in 2002.

Step 6
The ratio of 8,526,000 to 73,500,000 equals $\frac{11.6}{100}$.

5. *Step 2*
Let $x =$ the average family income.

Step 3
Set up a proportion.

$$\frac{3665}{x} = \frac{8.2}{100}$$

Step 4
Multiply by the LCD, $100x$.

$$100x\left(\frac{3665}{x}\right) = 100x\left(\frac{8.2}{100}\right)$$
$$366{,}500 = 8.2x$$
$$x = \frac{366{,}500}{8.2} \approx 44{,}695$$

Step 5
The average family income was \$44,695.

Step 6
The ratio of 3665 to 44,695 is approximately equal to $\frac{8.2}{100}$.

6. **(a)** Let $x =$ the speed of the plane in still air.

Use $d = rt$, or $t = \frac{d}{r}$, to complete the table.

	d	r	t
Against Wind	100	$x - 20$	$\frac{100}{x-20}$
With Wind	120	$x + 20$	$\frac{120}{x+20}$

(b) Because the time against the wind equals the time with the wind,

$$\frac{100}{x-20} = \frac{120}{x+20}.$$

Multiply by the LCD, $(x-20)(x+20)$.

$$\frac{100}{x-20}(x-20)(x+20)$$
$$= \frac{120}{x+20}(x-20)(x+20)$$
$$100(x+20) = 120(x-20)$$
$$100x + 2000 = 120x - 2400$$
$$4400 = 20x$$
$$220 = x$$

The speed of the plane is 220 mph in still air.

7. Let $x =$ the distance at reduced speed.

Use $d = rt$, or $t = \frac{d}{r}$, to complete the table.

	d	r	t
Normal Speed	$300 - x$	55	$\frac{300-x}{55}$
Reduced Speed	x	15	$\frac{x}{15}$

Now write an equation.

Time on the freeway	plus	time at reduced speed	equals 6 hr.
↓	↓	↓	↓
$\frac{300-x}{55}$	$+$	$\frac{x}{15}$	$= 6$

Multiply by the LCD, 165.

$$165\left(\frac{300-x}{55} + \frac{x}{15}\right) = 165(6)$$
$$3(300-x) + 11x = 990$$
$$900 - 3x + 11x = 990$$
$$8x = 90$$
$$x = \frac{90}{8} \text{ or } 11\tfrac{1}{4}$$

She drove $11\frac{1}{4}$ mi at reduced speed.

8. **(a)** Let $x =$ the time it will take them working together.

Make a table.

	Rate	Time Working Together	Fractional Part of the Job Done
Stan	$\frac{1}{45}$	x	$\frac{1}{45}x$
Deb	$\frac{1}{30}$	x	$\frac{1}{30}x$

Part done by Stan	plus	Part done by Deb	equals	1 whole job.
↓	↓	↓	↓	↓
$\frac{x}{45}$	$+$	$\frac{x}{30}$	$=$	1

Multiply by the LCD, 90.

$$90\left(\frac{1}{45}x + \frac{1}{30}x\right) = 90 \cdot 1$$
$$2x + 3x = 90$$
$$5x = 90$$
$$x = 18$$

It will take them 18 min working together.

(b) Let $x =$ the time for Deb to do the job alone.

Make a table.

	Rate	Time Working Together	Fractional Part of the Job Done
Stan	$\frac{1}{35}$	15	$\frac{15}{35}$
Deb	$\frac{1}{x}$	15	$\frac{15}{x}$

Since together they complete one job, the sum of the fractional parts should be 1.

$$\frac{15}{35}+\frac{15}{x}=1$$
$$\frac{3}{7}+\frac{15}{x}=1$$

Multiply by the LCD, $7x$.

$$7x\left(\frac{3}{7}+\frac{15}{x}\right)=7x\cdot 1$$
$$3x+105=7x$$
$$105=4x$$
$$x=\tfrac{105}{4} \text{ or } 26\tfrac{1}{4}$$

Deb can do the dishes alone in $26\frac{1}{4}$ min.

8.5 Section Exercises

1. **A.** $b=\frac{p}{r}$ is the same as $p=br$.

B. $r=\frac{b}{p}$ is the same as $b=pr$.

C. $b=\frac{r}{p}$ is the same as $r=bp$.

D. $p=\frac{r}{b}$ is the same as $r=bp$.

Choice **A** is correct.

2. **A.** $H=\frac{LW}{V}$ is the same as $HV=LW$.

B. $L=\frac{V}{WH}$ is the same as $LWH=V$.

C. $L=\frac{WH}{V}$ is the same as $LV=WH$.

D. $W=\frac{H}{VL}$ is the same as $WVL=H$.

Choice **B** is correct.

3. **A.** $a=mF$ is the same as $m=\frac{a}{F}$.

B. $F=\frac{m}{a}$ is the same as $m=Fa$.

C. $F=\frac{a}{m}$ is the same as $Fm=a$, which is the same as $m=\frac{a}{F}$.

D. $F=ma$ is the same as $m=\frac{F}{a}$.

Choice **D** is correct.

4. **A.** $R=\frac{I}{E}$ is the same as $RE=I$.

B. $R=IE$ is the same as $I=\frac{R}{E}$.

C. $E=\frac{I}{R}$ is the same as $ER=I$.

D. $E=RI$ is the same as $\frac{E}{R}=I$.

Choice **D** is correct.

5. $$\frac{PV}{T}=\frac{pv}{t}$$

Suppose that $T=300$, $t=350$, $V=9$, $P=50$, $v=8$. Find p.

$$\frac{50\cdot 9}{300}=\frac{p\cdot 8}{350}$$
$$p=\frac{50\cdot 9\cdot 350}{300\cdot 8}$$
$$=65.625$$

6. $$\frac{1}{a}=\frac{1}{b}+\frac{1}{c}$$

Let $a=8$ and $c=12$. Find b.

$$\frac{1}{8}=\frac{1}{b}+\frac{1}{12}$$

Multiply each side by the LCD, $24b$.

$$24b\left(\frac{1}{8}\right)=24b\left(\frac{1}{b}+\frac{1}{12}\right)$$
$$3b=24+2b$$
$$b=24$$

7. $$c=\frac{100b}{L}$$

Let $c=80$, $b=5$. Find L.

$$80=\frac{100\cdot 5}{L}$$
$$80L=500 \qquad \textit{Multiply by } L.$$
$$L=\tfrac{500}{80}=\tfrac{25}{4} \text{ or } 6.25$$

8. Solve $F = \frac{GMm}{d^2}$ for M.

$$Fd^2 = GMm \qquad \textit{Multiply by } d^2.$$
$$\frac{Fd^2}{Gm} = M \qquad \textit{Divide by Gm.}$$

Substitute 10 for F, 6.67×10^{-11} for G, 1 for m, and 3×10^{-6} for d.

$$M = \frac{Fd^2}{Gm}$$
$$M = \frac{(10)(3 \times 10^{-6})^2}{(6.67 \times 10^{-11})(1)}$$
$$\approx 1.349$$

9. Solve $F = \frac{GMm}{d^2}$ for G.

$$Fd^2 = GMm \qquad \textit{Multiply by } d^2.$$
$$\frac{Fd^2}{Mm} = G \qquad \textit{Divide by } Mm.$$

10. Solve $F = \frac{GMm}{d^2}$ for M.

$$Fd^2 = GMm \qquad \textit{Multiply by } d^2.$$
$$\frac{Fd^2}{Gm} = M \qquad \textit{Divide by Gm.}$$

11. Solve $\frac{1}{a} = \frac{1}{b} + \frac{1}{c}$ for a.

Multiply each side by the LCD, abc.

$$abc\left(\frac{1}{a}\right) = abc\left(\frac{1}{b} + \frac{1}{c}\right)$$
$$bc = ac + ab$$
$$bc = a(c + b) \qquad \textit{Factor out a.}$$
$$\frac{bc}{c + b} = a \qquad \textit{Divide by } c + b.$$

12. Solve $\frac{1}{a} = \frac{1}{b} + \frac{1}{c}$ for b.

Multiply each side by the LCD, abc.

$$abc\left(\frac{1}{a}\right) = abc\left(\frac{1}{b} + \frac{1}{c}\right)$$
$$bc = ac + ab$$
$$bc - ab = ac$$
$$b(c - a) = ac$$
$$b = \frac{ac}{c - a}$$
$$\text{or} \quad b = \frac{-ac}{a - c}$$

13. Solve $\frac{PV}{T} = \frac{pv}{t}$ for v.

$$PVt = pvT \qquad \textit{Multiply by Tt.}$$
$$\frac{PVt}{pT} = v \qquad \textit{Divide by pT.}$$

14. Solve $\frac{PV}{T} = \frac{pv}{t}$ for T.

$$PVt = pvT \qquad \textit{Multiply by Tt.}$$
$$\frac{PVt}{pv} = T \qquad \textit{Divide by pv.}$$

15. Solve $I = \frac{nE}{R + nr}$ for r.

$$I(R + nr) = nE$$
$$IR + Inr = nE$$
$$Inr = nE - IR$$
$$r = \frac{nE - IR}{In}$$

16. Solve $a = \frac{V - v}{t}$ for V.

$$at = V - v$$
$$at + v = V$$

17. Solve $A = \frac{1}{2}h(B + b)$ for b.

$$2A = h(B + b) \qquad \textit{Multiply by 2.}$$
$$\frac{2A}{h} = B + b \qquad \textit{Divide by } h.$$
$$\frac{2A}{h} - B = b \quad \text{or} \quad b = \frac{2A - Bh}{h}$$

18. Solve $S = \frac{n}{2}(a + \ell)d$ for n.

$$2S = n(a + \ell)d$$
$$\frac{2S}{(a + \ell)d} = n$$

19. $\frac{E}{e} = \frac{R + r}{r}$ for r

$$Er = e(R + r) \qquad \textit{Multiply by er.}$$
$$Er = eR + er$$
$$Er - er = eR \qquad \textit{Subtract er.}$$
$$r(E - e) = eR$$
$$r = \frac{eR}{E - e} \qquad \textit{Divide by } E - e.$$

20. Solve $y = \frac{x + z}{a - x}$ for x.

$$(a - x)y = x + z$$
$$ay - xy = x + z$$
$$ay - z = x + xy$$
$$ay - z = x(1 + y)$$
$$\frac{ay - z}{1 + y} = x$$

21. To solve the equation $m = \frac{ab}{a - b}$ for a, the first step is to multiply each side of the equation by the LCD, $a - b$.

22. $rp - rq = p + q$

The first step is to factor out r on the left side.

$$r(p - q) = p + q$$

23. Let $x =$ the number of girls in the class.
Write and solve a proportion.

$$\frac{3}{4} = \frac{x}{20}$$

Multiply by the LCD, 20.

$$20\left(\frac{3}{4}\right) = 20\left(\frac{x}{20}\right)$$
$$15 = x$$

There are 15 girls and $20 - 15 = 5$ boys in the class.

24. Let $x =$ the sales tax on a purchase of \$6.00.
Write a proportion using ratios of sales tax to price of purchases.

$$\frac{x}{6.00} = \frac{.12}{1.50}$$

Multiply by the LCD, 6.

$$6\left(\frac{x}{6.00}\right) = 6\left(\frac{.12}{1.50}\right)$$
$$x = .48$$

The sales tax on \$6.00 is \$.48.

25.
$$\text{Marin's rate} = \frac{1 \text{ job}}{\text{time to complete 1 job}} = \frac{1 \text{ job}}{2 \text{ hours}} = \tfrac{1}{2} \text{ job per hour}$$

26. Using the formula $d = rt$, we get

$$d = 50(14) = 700.$$

The distance from Atlanta to Detroit is 700 miles.

27. 1996: $\frac{4.8}{13.3} \approx .361$ (least)

1997: $\frac{7.7}{16.0} \approx .481$

1998: $\frac{7.3}{13.8} \approx .529$

1999: $\frac{6.2}{11.6} \approx .534$

2000: $\frac{8.8}{15.9} \approx .553$ (greatest)

The ratio of truck accidents to car accidents was greatest in 2000.

28. From Exercise 27, the ratio of truck accidents to car accidents was least in 1996.

29. 1996: $\frac{13.3}{4.8} \approx 2.77$

1997: $\frac{16.0}{7.7} \approx 2.08$

1998: $\frac{13.8}{7.3} \approx 1.89$

1999: $\frac{11.6}{6.2} \approx 1.87$

2000: $\frac{15.9}{8.8} \approx 1.81$

The ratio of car accidents to truck accidents was closest to 3 to 1 in 1996.

30. From Exercise 29, the ratio of car accidents to truck accidents was less than 2 to 1 in 1998, 1999, and 2000.

31. Let $x =$ the number of fast food restaurants in the United States in 2003.

Write and solve a proportion.

$$\frac{x}{844{,}000} = \frac{1}{5}$$
$$844{,}000\left(\frac{x}{844{,}000}\right) = 844{,}000\left(\frac{1}{5}\right)$$
$$x = 168{,}800$$

There were 168,800 fast food restaurants.

32. Let $x =$ the number of teachers.

Write a proportion.

$$\frac{1}{14} = \frac{x}{554}$$

Multiply each side by $554 \cdot 14$.

$$554 \cdot 14 \cdot \frac{1}{14} = 554 \cdot 14 \cdot \frac{x}{554}$$
$$554 = 14x$$
$$x = \tfrac{554}{14} \approx 40$$

There would be 40 teachers.

33. Let $x =$ the number of fish in the lake.
Write and solve a proportion.

$$\frac{\text{total in lake}}{\text{tagged in lake}} = \frac{\text{total in sample}}{\text{tagged in sample}}$$
$$\frac{x}{500} = \frac{400}{8}$$
$$\frac{x}{500} = 50$$
$$x = 500(50) = 25{,}000$$

There are approximately 25,000 fish in the lake.

34. In Exercise 33, change 8 to 10.

$$\begin{aligned}\frac{x}{500} &= \frac{400}{10}\\ \frac{x}{500} &= 40\\ x &= 500(40)\\ &= 20{,}000\end{aligned}$$

The estimated fish population would be 20,000.

35. Let $x =$ the amount earned on 75 shares of stock.

Write and solve a proportion.

$$\frac{191.50}{50} = \frac{x}{75}$$

Multiply each side by the LCD, 150.

$$\begin{aligned}150\left(\frac{191.50}{50}\right) &= 150\left(\frac{x}{75}\right)\\ 3(191.50) &= 2x\\ x &= \frac{574.50}{2} = 287.25\end{aligned}$$

75 shares would have earned

$$\$287.25 - \$191.50 = \$95.75$$

more than 50 shares.

36. Let $x =$ the total income on a \$260,000 investment.

Write and solve a proportion.

$$\begin{aligned}\frac{x}{260{,}000} &= \frac{22{,}950}{100{,}000}\\ 260{,}000\left(\frac{x}{260{,}000}\right) &= 260{,}000\left(\frac{22{,}950}{100{,}000}\right)\\ x &= 59{,}670\end{aligned}$$

There would have been

$$\$59{,}670 - \$22{,}950 = \$36{,}720$$

more income produced.

37. Let x represent the amount to administer in milliliters.

$$\frac{100 \text{ mg}}{2 \text{ mL}} = \frac{120 \text{ mg}}{x \text{ mL}}$$

Multiply each side by the LCD, $2x$.

$$\begin{aligned}2x\left(\frac{100}{2}\right) &= 2x\left(\frac{120}{x}\right)\\ 100x &= 240\\ x &= 2.4\end{aligned}$$

The correct dose is 2.4 mL.

38. Let x represent the amount to administer in milliliters.

$$\frac{20 \text{ mg}}{10 \text{ mL}} = \frac{1.5 \text{ mg}}{x \text{ mL}}$$

Multiply each side by the LCD, $10x$.

$$\begin{aligned}10x\left(\frac{20}{10}\right) &= 10x\left(\frac{1.5}{x}\right)\\ 20x &= 15\\ x &= \tfrac{15}{20} = .75\end{aligned}$$

The correct dose is .75 mL.

39. Since $\frac{2}{3} = \frac{4}{6} = \frac{6}{9}$, use the proportion

$$\begin{aligned}\frac{2}{3} &= \frac{2x+1}{2x+5}.\\ 2(2x+5) &= 3(2x+1)\\ 4x+10 &= 6x+3\\ 7 &= 2x\\ \tfrac{7}{2} &= x\end{aligned}$$

Since $x = \frac{7}{2}$,

$$AC = 2x + 1 = 2\left(\tfrac{7}{2}\right) + 1 = 8$$

and $DF = 2x + 5 = 2\left(\tfrac{7}{2}\right) + 5 = 12.$

40. Write a proportion.

$$\begin{aligned}\frac{4}{3} &= \frac{5y-2}{2y+2}\\ 4(2y+2) &= 3(5y-2)\\ 8y+8 &= 15y-6\\ 14 &= 7y\\ 2 &= y\end{aligned}$$

Also note that

$$\begin{aligned}\frac{MP}{QT} &= \frac{4}{3}\\ \frac{MP}{2y} &= \frac{4}{3}\\ MP &= \frac{4(2y)}{3} = \frac{8y}{3}.\end{aligned}$$

Now, since $y = 2$,

side $NP = 5y - 2 = 5(2) - 2 = 8$,

side $MP = \dfrac{8y}{3} = \dfrac{8(2)}{3} = \dfrac{16}{3}$ or $5\dfrac{1}{3}$,

side $RT = 2y + 2 = 2(2) + 2 = 6$, and

side $QT = 2y = 2(2) = 4$.

Thus, the length of the two longest sides of triangle MNP are 8 and $5\frac{1}{3}$, and of triangle QRT are 6 and 4.

41. *Step 2*
Let x represent the speed of the current of the river. The boat goes 12 mph, so the downstream speed is $12 + x$ and the upstream speed is $12 - x$.

Use $t = \frac{d}{r}$ and make a table.

	Distance	Rate	Time
Downstream	10	$12 + x$	$\frac{10}{12 + x}$
Upstream	6	$12 - x$	$\frac{6}{12 - x}$

Step 3
Because the time upstream equals the time downstream,

$$\frac{6}{12 - x} = \frac{10}{12 + x}$$

Step 4
Multiply by the LCD, $(12 - x)(12 + x)$.

$$(12 - x)(12 + x)\left(\frac{6}{12 - x}\right)$$
$$= (12 - x)(12 + x)\left(\frac{10}{12 + x}\right)$$
$$6(12 + x) = 10(12 - x)$$
$$72 + 6x = 120 - 10x$$
$$16x = 48$$
$$x = 3$$

Step 5
The speed of the current of the river is 3 mph.

Step 6
Check: The rate downstream is $12 + 3 = 15$ mph, so she can go 10 miles in $\frac{10}{15} = \frac{2}{3}$ hour. The rate upstream is $12 - 3 = 9$ mph, so she can go 6 miles in $\frac{6}{9} = \frac{2}{3}$ hour. The times are the same, as required.

42. Let $x =$ the rate of the current.

Make a table. Use $t = \frac{d}{r}$.

	Distance	Rate	Time
Downstream	12	$15 + x$	$\frac{12}{15 + x}$
Upstream	8	$15 - x$	$\frac{8}{15 - x}$

Set the times for each journey equal to each other and solve for x.

$$\frac{12}{15 + x} = \frac{8}{15 - x}$$

Multiply by the LCD, $(15 + x)(15 - x)$.

$$(15 + x)(15 - x)\left(\frac{12}{15 + x}\right) = (15 + x)(15 - x)\left(\frac{8}{15 - x}\right)$$
$$12(15 - x) = 8(15 + x)$$
$$3(15 - x) = 2(15 + x) \quad \textit{Divide by 4.}$$
$$45 - 3x = 30 + 2x$$
$$15 = 5x$$
$$3 = x$$

The rate of the current is 3 mph.

43. *Step 2*
Find the distance from Tulsa to Detroit.
Let x represent that distance.

Complete the table.

	d	***r***	***t***
Actual Trip	x	50	$\frac{x}{50}$
Alternative Trip	x	60	$\frac{x}{60}$

Step 3
At 60 mph, his time at 50 mph would be decreased 3 hr.

$$\frac{x}{60} = \frac{x}{50} - 3$$

Step 4
Multiply by 300.

$$5x = 6x - 900$$
$$900 = x$$

Step 5
The distance from Tulsa to Detroit is 900 miles.

Step 6
Check: 900 miles at 50 mph takes $\frac{900}{50}$ or 18 hours; 900 miles at 60 mph takes $\frac{900}{60}$ or 15 hours; $15 = 18 - 3$ as required.

44. Let $x =$ the distance between his office and home. Make a table using the information in the problem and the formula $t = \frac{d}{r}$.

	d	***r***	***t***
Bike	x	12	$\frac{x}{12}$
Car	x	36	$\frac{x}{36}$

From the problem, the equation is stated in words:
Car driving time = Bike riding time less $\frac{1}{4}$ hour.

continued

Use the car time and the bike time given in the table to write the equation.

$$\frac{x}{36} = \frac{x}{12} - \frac{1}{4}$$
$$36\left(\frac{x}{36}\right) = 36\left(\frac{x}{12}\right) - 36\left(\frac{1}{4}\right)$$
$$x = 3x - 9$$
$$-2x = -9$$
$$x = 4\tfrac{1}{2} \text{ miles}$$

His office is $4\frac{1}{2}$ miles from home.

45. *Step 2*
Let $x =$ the distance from San Francisco to the secret rendezvous.

Make a table.

	d	r	t
First Trip	x	200	$\frac{x}{200}$
Return Trip	x	300	$\frac{x}{300}$

Step 3

Time there	plus	time back	equals	4 hr.
$\frac{x}{200}$	$+$	$\frac{x}{300}$	$=$	4

Step 4
Multiply by the LCD, 600.

$$600\left(\frac{x}{200} + \frac{x}{300}\right) = 600(4)$$
$$3x + 2x = 2400$$
$$5x = 2400$$
$$x = 480$$

Step 5
The distance is 480 miles.

Step 6
Check: 480 miles at 200 mph takes $\frac{480}{200}$ or 2.4 hours; 480 miles at 300 mph takes $\frac{480}{300}$ or 1.6 hours. The total time is 4 hours, as required.

46. *Step 2*
Let $x =$ the distance to the fishing hole.

Make a table.

	d	r	t
Old Highway	x	30	$\frac{x}{30}$
Interstate	x	50	$\frac{x}{50}$

Step 3
The time on the interstate is 2 hr less than the time on the old highway.

$$\frac{x}{50} = \frac{x}{30} - 2$$

Step 4
Multiply by 150.

$$3x = 5x - 300$$
$$300 = 2x$$
$$150 = x$$

Step 5
The distance to the fishing hole is 150 miles.

Step 6
Check: 150 miles at 30 mph takes $\frac{150}{30}$ or 5 hours; 150 miles at 50 mph takes $\frac{150}{50}$ or 3 hours. The interstate time is 2 hours less than the highway time, as required.

47. *Step 2*
Let $x =$ the distance on the first part of the trip.

Make a table.

	d	r	t
First Part	x	60	$\frac{x}{60}$
Second Part	$x + 10$	50	$\frac{x+10}{50}$

Step 3
From the problem, the equation is stated in words: (Note that 30 min $= \frac{1}{2}$ hr.) Time for the second part $=$ time for the first part $+ \frac{1}{2}$. Use the times given in the table to write the equation.

$$\frac{x+10}{50} = \frac{x}{60} + \frac{1}{2}$$

Step 4
Multiply by the LCD, 300.

$$300\left(\frac{x+10}{50}\right) = 300\left(\frac{x}{60} + \frac{1}{2}\right)$$
$$6(x + 10) = 5x + 150$$
$$6x + 60 = 5x + 150$$
$$x = 90$$

Step 5
The distance for both parts of the trip is given by

$$x + (x + 10) = 90 + (90 + 10) = 190.$$

The distance is 190 miles.

Step 6

Check: 90 miles at 60 mph takes $\frac{90}{60}$ or $1\frac{1}{2}$ hours; 100 miles at 50 mph takes 2 hours. The second part of the trip takes $\frac{1}{2}$ hour more than the first part, as required.

48. Let $x =$ the distance they drove on the highway. Then $x - 100 =$ the distance they drove in Houston.

	d	r	t
Highway	x	60	$\frac{x}{60}$
In Houston	$x - 100$	30	$\frac{x-100}{30}$

They spent 50 min or $\frac{5}{6}$ hr more on the highway than in Houston.

$$\frac{x}{60} = \frac{x-100}{30} + \frac{5}{6}$$
$$x = 2(x - 100) + 10(5) \quad \textit{Multiply by 60.}$$
$$x = 2x - 200 + 50$$
$$150 = x$$

The distance they drove on the highway was 150 mi and the distance in Houston was $150 - 100 = 50$ mi. Their total distance was $150 + 50 = 200$ mi.

49. Let $x =$ the time it would take them working together.

Complete the table.

Worker	Rate	Time Working Together	Fractional Part of the Job Done
Butch	$\frac{1}{15}$	x	$\frac{1}{15}x$
Peggy	$\frac{1}{12}$	x	$\frac{1}{12}x$

Part done by Butch	+	part done by Peggy	=	1 whole job.
$\frac{1}{15}x$	+	$\frac{1}{12}x$	=	1

Multiply by the LCD, 60.

$$60\left(\frac{1}{15}x + \frac{1}{12}x\right) = 60 \cdot 1$$
$$4x + 5x = 60$$
$$9x = 60$$
$$x = \frac{60}{9} = \frac{20}{3} \text{ or } 6\tfrac{2}{3}$$

Together they could do the job in $\frac{20}{3}$ or $6\frac{2}{3}$ minutes.

50. Let $x =$ the time it would take them working together.

Complete the table.

Worker	Rate	Time Working Together	Fractional Part of the Job Done
Lou	$\frac{1}{8}$	x	$\frac{1}{8}x$
Janet	$\frac{1}{5}$	x	$\frac{1}{5}x$

Part done by Lou	+	part done by Janet	=	1 whole job.
$\frac{1}{8}x$	+	$\frac{1}{5}x$	=	1

Multiply by the LCD, 40.

$$40\left(\frac{1}{8}x + \frac{1}{5}x\right) = 40 \cdot 1$$
$$5x + 8x = 40$$
$$13x = 40$$
$$x = \frac{40}{13} \text{ or } 3\tfrac{1}{13}$$

Together they could groom the dogs in $\frac{40}{13}$ or $3\frac{1}{13}$ hr.

51. Let $x =$ the time it would take Jason working alone.

Worker	Rate	Time Working Together	Fractional Part of the Job Done
Ron	$\frac{1}{6}$	4	$\frac{1}{6}(4) = \frac{2}{3}$
Jason	$\frac{1}{x}$	4	$\frac{1}{x}(4) = \frac{4}{x}$

Part done by Ron	+	part done by Jason	=	1 whole job.
$\frac{2}{3}$	+	$\frac{4}{x}$	=	1

Multiply by the LCD, $3x$.

$$2x + 12 = 3x$$
$$12 = x$$

It would take Jason 12 hours to do the job alone.

52. Let $x =$ the time for Sandi to do the job alone. Make a table.

Worker	Rate	Time Working Together	Fractional Part of the Job Done
Cary	$\frac{1}{7}$	5	$\frac{1}{7}(5) = \frac{5}{7}$
Sandi	$\frac{1}{x}$	5	$\frac{1}{x}(5) = \frac{5}{x}$

Part done by Cary + part done by Sandi = 1 whole job.

$$\frac{5}{7} + \frac{5}{x} = 1$$

Multiply by the LCD, $7x$.

$$7x\left(\frac{5}{7} + \frac{5}{x}\right) = 7x \cdot 1$$
$$5x + 35 = 7x$$
$$-2x = -35$$
$$x = \tfrac{35}{2} \text{ or } 17\tfrac{1}{2}$$

Sandi can do the job alone in $\frac{35}{2}$ or $17\frac{1}{2}$ hours.

Another solution: Cary will do $\frac{5}{7}$ of the job, leaving $\frac{2}{7}$ of the job for Sandi.

$$A = rt$$
$$\frac{2}{7} = \frac{1}{x} \cdot 5$$
$$x = 5 \cdot \tfrac{7}{2} = \tfrac{35}{2}$$

53. Let $x =$ the time it will take to fill the vat if both pipes are open.

	Rate	Time to Fill the Vat	Fractional Part of the Job Done
Inlet Pipe	$\frac{1}{10}$	x	$\frac{1}{10}x$
Outlet Pipe	$-\frac{1}{20}$	x	$-\frac{1}{20}x$

Notice that the rate of the outlet pipe is negative because it will empty the vat, not fill it.

Part done with the inlet pipe open + Part done with the outlet pipe open = 1 whole job.

$$\tfrac{1}{10}x + \left(-\tfrac{1}{20}x\right) = 1$$

Multiply by the LCD, 20.

$$2x - x = 20$$
$$x = 20$$

It will take 20 hours to fill the vat.

54. Let $x =$ the time it will take to fill the barrel if both pipes are open.

	Rate	Time to Fill Together	Fractional Part of the Job Done
Inlet Pipe	$\frac{1}{9}$	x	$\frac{1}{9}x$
Outlet Pipe	$-\frac{1}{12}$	x	$-\frac{1}{12}x$

Notice that the rate of the outlet pipe is negative because it will empty the barrel, not fill it.

Part done with the inlet pipe open + Part done with the outlet pipe open = 1 whole job.

$$\tfrac{1}{9}x + \left(-\tfrac{1}{12}x\right) = 1$$

Multiply by the LCD, 36.

$$36\left(\tfrac{1}{9}x - \tfrac{1}{12}x\right) = 36 \cdot 1$$
$$4x - 3x = 36$$
$$x = 36$$

It will take 36 hours to fill the barrel.

55. Let $x =$ the time from Mimi's arrival home to the time the place is a shambles.

	Rate	Time to Mess up House	Fractional Part of the Job Done
Hortense and Mort	$-\frac{1}{7}$	x	$-\frac{1}{7}x$
Mimi	$\frac{1}{2}$	x	$\frac{1}{2}x$

Notice that Hortense and Mort's rate is negative since they are opposing the messing up by cleaning the house.

Part done by Hortense and Mort + Part done by Mimi = 1 whole job of messing up.

$$-\tfrac{1}{7}x + \tfrac{1}{2}x = 1$$

Multiply by the LCD, 14.

$$-2x + 7x = 14$$
$$5x = 14$$
$$x = \tfrac{14}{5} \text{ or } 2\tfrac{4}{5}$$

It would take $\frac{14}{5}$ or $2\frac{4}{5}$ hours after Mimi got home for the house to be a shambles.

56. Let $x =$ the time it will take to fill the pond if both pipes are open.

	Rate	Time to Fill Together	Fractional Part of the Job Done
Inlet Pipe	$\frac{1}{60}$	x	$\frac{1}{60}x$
Outlet Pipe	$-\frac{1}{80}$	x	$-\frac{1}{80}x$

Notice that the rate of the outlet pipe is negative because it will empty the pond, not fill it.

Part done with the inlet pipe open	+	Part done with the outlet pipe open	=	1 whole job.
$\frac{1}{60}x$	+	$\left(-\frac{1}{80}x\right)$	=	1

$$\frac{1}{60}x - \frac{1}{80}x = 1$$

Multiply by the LCD, 240.

$$\begin{aligned} 240\left(\tfrac{1}{60}x - \tfrac{1}{80}x\right) &= 240 \cdot 1 \\ 4x - 3x &= 240 \\ x &= 240 \end{aligned}$$

It will take 240 minutes (4 hours) to fill the pond.

Chapter 8 Review Exercises

1. **(a)** $f(x) = \dfrac{-7}{3x+18}$

Set the denominator equal to zero and solve.

$$\begin{aligned} 3x + 18 &= 0 \\ 3x &= -18 \\ x &= -6 \end{aligned}$$

The number -6 makes the expression undefined, so it is excluded from the domain.

(b) The domain is $\{x \mid x \neq -6\}$.

2. **(a)** $f(x) = \dfrac{5x+17}{x^2-7x+10}$

Set the denominator equal to zero and solve.

$$\begin{aligned} x^2 - 7x + 10 &= 0 \\ (x-5)(x-2) &= 0 \end{aligned}$$

$$\begin{aligned} x - 5 = 0 \quad &\text{or} \quad x - 2 = 0 \\ x = 5 \quad &\text{or} \quad x = 2 \end{aligned}$$

The numbers 2 and 5 make the expression undefined, so they are excluded from the domain.

(b) The domain is $\{x \mid x \neq 2, 5\}$.

3. **(a)** $f(x) = \dfrac{9}{x^2-18x+81}$

Set the denominator equal to zero and solve.

$$\begin{aligned} x^2 - 18x + 81 &= 0 \\ (x-9)^2 &= 0 \\ x - 9 &= 0 \\ x &= 9 \end{aligned}$$

The number 9 makes the expression undefined, so it is excluded from the domain.

(b) The domain is $\{x \mid x \neq 9\}$.

4. $\dfrac{12x^2+6x}{24x+12} = \dfrac{6x(2x+1)}{12(2x+1)} = \dfrac{x}{2}$

5. $$\begin{aligned} \frac{25m^2-n^2}{25m^2-10mn+n^2} &= \frac{(5m+n)(5m-n)}{(5m-n)(5m-n)} \\ &= \frac{5m+n}{5m-n} \end{aligned}$$

6. $$\begin{aligned} \frac{r-2}{4-r^2} &= \frac{r-2}{(2+r)(2-r)} \\ &= \frac{(-1)(2-r)}{(2+r)(2-r)} \\ &= \frac{-1}{2+r} \end{aligned}$$

7. The reciprocal of a rational expression is another rational expression such that the two rational expressions have a product of 1.

8. $$\begin{aligned} &\frac{(2y+3)^2}{5y} \cdot \frac{15y^3}{4y^2-9} \\ &= \frac{15y^3(2y+3)^2}{5y(2y+3)(2y-3)} \\ &= \frac{3y^2(2y+3)}{2y-3} \end{aligned}$$

9. $$\begin{aligned} &\frac{w^2-16}{w} \cdot \frac{3}{4-w} \\ &= \frac{(w-4)(w+4)}{w} \cdot \frac{3}{4-w} \\ &= \frac{(-1)(4-w)(w+4)}{w} \cdot \frac{3}{4-w} \\ &= \frac{-3(w+4)}{w} \end{aligned}$$

10. $$\begin{aligned} &\frac{z^2-z-6}{z-6} \div \frac{z^2+2z-15}{z^2-6z} \\ &= \frac{z^2-z-6}{z-6} \cdot \frac{z^2-6z}{z^2+2z-15} \\ &= \frac{(z-3)(z+2)}{z-6} \cdot \frac{z(z-6)}{(z-3)(z+5)} \\ &= \frac{z(z+2)}{z+5} \end{aligned}$$

11. $\dfrac{m^3-n^3}{m^2-n^2} \div \dfrac{m^2+mn+n^2}{m+n}$

Multiply by the reciprocal.

$$= \frac{m^3-n^3}{m^2-n^2} \cdot \frac{m+n}{m^2+mn+n^2}$$
$$= \frac{(m-n)(m^2+mn+n^2)}{(m-n)(m+n)} \cdot \frac{m+n}{m^2+mn+n^2}$$
$$= 1$$

12. $32b^3,\ 24b^5$

Factor each denominator.

$$32b^3 = 2\cdot 2\cdot 2\cdot 2\cdot 2\cdot b^3 = 2^5\cdot b^3$$
$$24b^5 = 2\cdot 2\cdot 2\cdot 3\cdot b^5 = 2^3\cdot 3\cdot b^5$$

$\text{LCD} = 2^5\cdot 3\cdot b^5 = 96b^5$

13. $9r^2,\ 3r+1$

Factor each denominator.

$$9r^2 = 3^2\cdot r^2$$

The second denominator is already in factored form. The LCD is

$$3^2\cdot r^2\cdot(3r+1) \quad \text{or} \quad 9r^2(3r+1).$$

14. $6x^2+13x-5,\ 9x^2+9x-4$

Factor each denominator.

$$6x^2+13x-5 = (3x-1)(2x+5)$$
$$9x^2+9x-4 = (3x-1)(3x+4)$$

The LCD is $(3x-1)(2x+5)(3x+4)$.

15. $\dfrac{8}{z} - \dfrac{3}{2z^2}$

$$= \frac{8\cdot 2z}{z\cdot 2z} - \frac{3}{2z^2} \qquad LCD = 2z^2$$
$$= \frac{16z}{2z^2} - \frac{3}{2z^2}$$
$$= \frac{16z-3}{2z^2}$$

16. $\dfrac{5y+13}{y+1} - \dfrac{1-7y}{y+1}$

$$= \frac{5y+13-(1-7y)}{y+1}$$
$$= \frac{5y+13-1+7y}{y+1}$$
$$= \frac{12y+12}{y+1}$$
$$= \frac{12(y+1)}{y+1} = 12$$

17. $\dfrac{6}{5a+10} + \dfrac{7}{6a+12}$

$$= \frac{6}{5(a+2)} + \frac{7}{6(a+2)}$$

The LCD is $30(a+2)$.

$$= \frac{6\cdot 6}{5(a+2)\cdot 6} + \frac{7\cdot 5}{6(a+2)\cdot 5}$$
$$= \frac{36}{30(a+2)} + \frac{35}{30(a+2)}$$
$$= \frac{36+35}{30(a+2)} = \frac{71}{30(a+2)}$$

18. $\dfrac{3r}{10r^2-3rs-s^2} + \dfrac{2r}{2r^2+rs-s^2}$

$$= \frac{3r}{(5r+s)(2r-s)} + \frac{2r}{(2r-s)(r+s)}$$

The LCD is $(5r+s)(2r-s)(r+s)$.

$$= \frac{3r(r+s)}{(5r+s)(2r-s)(r+s)}$$
$$+ \frac{2r(5r+s)}{(2r-s)(r+s)(5r+s)}$$
$$= \frac{3r^2+3rs+10r^2+2rs}{(5r+s)(2r-s)(r+s)}$$
$$= \frac{13r^2+5rs}{(5r+s)(2r-s)(r+s)}$$

19. $\dfrac{\frac{3}{t}+2}{\frac{4}{t}-7}$

Multiply the numerator and denominator by the LCD of all the fractions, t.

$$= \frac{t\left(\frac{3}{t}+2\right)}{t\left(\frac{4}{t}-7\right)} = \frac{3+2t}{4-7t}$$

20. $\dfrac{\frac{2}{m-3n}}{\frac{1}{3n-m}} = \dfrac{2}{m-3n} \div \dfrac{1}{3n-m}$

$$= \frac{2}{m-3n} \cdot \frac{3n-m}{1}$$
$$= \frac{2}{m-3n} \cdot \frac{-1(m-3n)}{1}$$
$$= \frac{2\cdot(-1)}{1} = -2$$

21. $\dfrac{\frac{3}{p}-\frac{2}{q}}{\frac{9q^2-4p^2}{qp}}$

Multiply the numerator and denominator by the LCD of all the fractions, qp.

$$= \frac{qp\left(\frac{3}{p}-\frac{2}{q}\right)}{qp\left(\frac{9q^2-4p^2}{qp}\right)}$$
$$= \frac{3q-2p}{9q^2-4p^2}$$
$$= \frac{3q-2p}{(3q+2p)(3q-2p)}$$
$$= \frac{1}{3q+2p}$$

22. $\dfrac{x^{-2}-y^{-2}}{x^{-1}-y^{-1}} = \dfrac{\frac{1}{x^2}-\frac{1}{y^2}}{\frac{1}{x}-\frac{1}{y}}$

Multiply the numerator and denominator by the LCD of all the fractions, x^2y^2.

$$= \frac{x^2y^2\left(\frac{1}{x^2}-\frac{1}{y^2}\right)}{x^2y^2\left(\frac{1}{x}-\frac{1}{y}\right)}$$
$$= \frac{y^2-x^2}{xy^2-x^2y}$$
$$= \frac{(y+x)(y-x)}{xy(y-x)}$$
$$= \frac{y+x}{xy}$$

23. $$\frac{1}{t+4}+\frac{1}{2}=\frac{3}{2t+8}$$
$$\frac{1}{t+4}+\frac{1}{2}=\frac{3}{2(t+4)}$$

Multiply by the LCD, $2(t+4)$. $(t \neq -4)$

$$2(t+4)\left(\frac{1}{t+4}+\frac{1}{2}\right) = 2(t+4)\left(\frac{3}{2(t+4)}\right)$$
$$2+(t+4)=3$$
$$t+6=3$$
$$t=-3$$

Check $t=-3$: $1+\frac{1}{2}=\frac{3}{2}$ *True*

Solution set: $\{-3\}$

24. $$\frac{-5m}{m+1}+\frac{m}{3m+3}=\frac{56}{6m+6}$$
$$\frac{-5m}{m+1}+\frac{m}{3(m+1)}=\frac{56}{6(m+1)}$$
$$\frac{-5m}{m+1}+\frac{m}{3(m+1)}=\frac{28}{3(m+1)}$$

Multiply by the LCD, $3(m+1)$. $(m \neq -1)$

$$3(m+1)\left(\frac{-5m}{m+1}+\frac{m}{3(m+1)}\right) = 3(m+1)\left(\frac{28}{3(m+1)}\right)$$
$$-15m+m=28$$
$$-14m=28$$
$$m=-2$$

Check $m=-2$: $-10+\frac{2}{3}=-\frac{56}{6}$ *True*

Solution set: $\{-2\}$

25. $$\frac{2}{k-1}-\frac{4k+1}{k^2-1}=\frac{-1}{k+1}$$
$$\frac{2}{k-1}-\frac{4k+1}{(k+1)(k-1)}=\frac{-1}{k+1}$$

Multiply by the LCD, $(k+1)(k-1)$. $(k \neq \pm 1)$

$$(k+1)(k-1)\left(\frac{2}{k-1}-\frac{4k+1}{(k+1)(k-1)}\right) = (k+1)(k-1)\left(\frac{-1}{k+1}\right)$$
$$2(k+1)-(4k+1)=-1(k-1)$$
$$2k+2-4k-1=-k+1$$
$$-2k+1=-k+1$$
$$0=k$$

Check $k=0$: $-2+1=-1$ *True*

Solution set: $\{0\}$

26. $$\frac{5}{x+2}+\frac{3}{x+3}=\frac{x}{x^2+5x+6}$$
$$\frac{5}{x+2}+\frac{3}{x+3}=\frac{x}{(x+2)(x+3)}$$

Multiply by the LCD, $(x+2)(x+3)$. $(x \neq -3, -2)$

$$(x+2)(x+3)\left(\frac{5}{x+2}+\frac{3}{x+3}\right) = (x+2)(x+3)\left(\frac{x}{(x+2)(x+3)}\right)$$
$$5(x+3)+3(x+2)=x$$
$$5x+15+3x+6=x$$
$$8x+21=x$$
$$7x=-21$$
$$x=-3$$

Substituting -3 in the original equation results in division by 0, so -3 is not a solution.

Solution set: $\emptyset$

27. Although her algebra was correct, 3 is not a solution because it is not in the domain of the equation. Thus, $\emptyset$ is correct.

28. In simplifying the expression, we are combining terms to get a single fraction with a denominator of $6x$. In solving the equation, we are finding a value for x that makes the equation true.

29. The graph in choice **C** has a vertical asymptote. Its equation is $x = 0$. The equation of its horizontal asymptote is $y = 0$.

30. $\frac{1}{A} = \frac{1}{B} + \frac{1}{C}$

Let $B = 30$ and $C = 10$. Find A.

$\frac{1}{A} = \frac{1}{30} + \frac{1}{10}$

To solve for A, multiply each side by the LCD, $30A$.

$$30A\left(\frac{1}{A}\right) = 30A\left(\frac{1}{30} + \frac{1}{10}\right)$$
$$30 = A + 3A$$
$$30 = 4A$$
$$A = \tfrac{30}{4} = \tfrac{15}{2}$$

31. Solve $F = \frac{GMm}{d^2}$ for m.

$Fd^2 = GMm$ *Multiply by d^2.*

$\frac{Fd^2}{GM} = m$ *Divide by GM.*

32. Solve $\mu = \frac{Mv}{M+m}$ for M.

$\mu(M + m) = Mv$ *Multiply by $M + m$.*

$$\mu M + \mu m = Mv$$
$$\mu m = Mv - \mu M$$
$$m\mu = M(v - \mu)$$
$$M = \frac{m\mu}{v - \mu}$$

33. Let $x =$ the number of passenger-kilometers per day provided by high-speed trains.

Write a proportion.

$$\frac{x}{15{,}000} = \frac{23{,}200}{58{,}000}$$
$$x = \frac{23{,}200(15{,}000)}{58{,}000}$$
$$= 6000$$

34. Let $x =$ the speed of the boat in still water.

Use $d = rt$, or $t = \frac{d}{r}$, to make a table.

	Distance	Rate	Time
Downstream	40	$x + 4$	$\frac{40}{x+4}$
Upstream	24	$x - 4$	$\frac{24}{x-4}$

Because the times are equal,

$$\frac{40}{x+4} = \frac{24}{x-4}.$$

Multiply by the LCD, $(x+4)(x-4)$. $(x \neq -4, 4)$

$$(x+4)(x-4)\left(\frac{40}{x+4}\right) = (x+4)(x-4)\left(\frac{24}{x-4}\right)$$
$$40(x-4) = 24(x+4)$$
$$40x - 160 = 24x + 96$$
$$16x = 256$$
$$x = 16$$

The speed of the boat in still water is 16 km/hr.

35. Let $x =$ the time it takes to fill the sink with both taps open.

Make a table.

	Rate	Time Working Together	Fractional Part of the Job Done
Cold	$\frac{1}{8}$	x	$\frac{x}{8}$
Hot	$\frac{1}{12}$	x	$\frac{x}{12}$

Part done by cold	plus	part done by hot	equals	1 whole job.
$\frac{x}{8}$	$+$	$\frac{x}{12}$	$=$	1

Multiply by the LCD, 24.

$$24\left(\frac{x}{8} + \frac{x}{12}\right) = 24 \cdot 1$$
$$3x + 2x = 24$$
$$5x = 24$$
$$x = \tfrac{24}{5} \text{ or } 4\tfrac{4}{5}$$

The sink will be filled in $\frac{24}{5}$ or $4\frac{4}{5}$ minutes.

36. Let x = the time to do the job working together. Make a table.

Worker	Rate	Time Working Together	Fractional Part of the Job Done
Dona	$\frac{1}{9}$	x	$\frac{x}{9}$
Jay	$\frac{1}{6}$	x	$\frac{x}{6}$

Part done by Dona plus part done by Jay equals 1 whole job.

$$\frac{x}{9} + \frac{x}{6} = 1$$

Multiply by the LCD, 36.

$$\begin{aligned} 36\left(\frac{x}{9}+\frac{x}{6}\right) &= 36 \cdot 1 \\ 4x + 6x &= 36 \\ 10x &= 36 \\ x &= \tfrac{36}{10} \\ x &= \tfrac{18}{5} \text{ or } 3\tfrac{3}{5} \end{aligned}$$

Working together, they can do the job in $\frac{18}{5}$ or $3\frac{3}{5}$ hours.

37. [8.1] $\dfrac{x+2y}{x^2-4y^2} = \dfrac{x+2y}{(x+2y)(x-2y)} = \dfrac{1}{x-2y}$

38. [8.1] $\dfrac{x^2+2x-15}{x^2-x-6} = \dfrac{(x+5)(x-3)}{(x-3)(x+2)} = \dfrac{x+5}{x+2}$

39. [8.2] $\dfrac{2}{m} + \dfrac{5}{3m^2}$

The LCD is $3m^2$.

$$= \frac{2 \cdot 3m}{m \cdot 3m} + \frac{5}{3m^2} = \frac{6m}{3m^2} + \frac{5}{3m^2} = \frac{6m+5}{3m^2}$$

40. [8.1]

$$\frac{k^2-6k+9}{1-216k^3} \cdot \frac{6k^2+17k-3}{9-k^2}$$

Factor $1-216k^3$ as the difference of cubes, $1^3-(6k)^3$.

$$= \frac{(k-3)(k-3)}{(1-6k)(1+6k+36k^2)} \cdot \frac{(6k-1)(k+3)}{(3-k)(3+k)}$$

$$= \frac{(k-3)(k-3)}{(-1)(6k-1)(1+6k+36k^2)} \cdot \frac{(6k-1)(k+3)}{(-1)(k-3)(k+3)}$$

$$= \frac{k-3}{1+6k+36k^2} \text{ or } \frac{k-3}{36k^2+6k+1}$$

41. [8.3]

$$\frac{\frac{-3}{x}+\frac{x}{2}}{1+\frac{x+1}{x}}$$

Multiply the numerator and denominator by the LCD of all the fractions, $2x$.

$$= \frac{2x\left(\frac{-3}{x}+\frac{x}{2}\right)}{2x\left(1+\frac{x+1}{x}\right)} = \frac{-6+x^2}{2x+2(x+1)} = \frac{x^2-6}{2x+2x+2} = \frac{x^2-6}{4x+2} = \frac{x^2-6}{2(2x+1)}$$

42. [8.1] $\dfrac{9x^2+46x+5}{3x^2-2x-1} \div \dfrac{x^2+11x+30}{x^3+5x^2-6x}$

Multiply by the reciprocal.

$$= \frac{9x^2+46x+5}{3x^2-2x-1} \cdot \frac{x(x^2+5x-6)}{x^2+11x+30} = \frac{(9x+1)(x+5)}{(3x+1)(x-1)} \cdot \frac{x(x+6)(x-1)}{(x+6)(x+5)} = \frac{x(9x+1)}{3x+1}$$

43. [8.3] $\dfrac{\frac{3}{x}-5}{6+\frac{1}{x}}$

Multiply the numerator and denominator by the LCD of all the fractions, x.

$$= \frac{x\left(\frac{3}{x}-5\right)}{x\left(6+\frac{1}{x}\right)} = \frac{3-5x}{6x+1}$$

44. **[8.2]** $\dfrac{9}{3-x}-\dfrac{2}{x-3}$

$$=\frac{9}{3-x}-\frac{2(-1)}{(x-3)(-1)}$$
$$=\frac{9}{3-x}-\frac{-2}{3-x}$$
$$=\frac{9-(-2)}{3-x}$$
$$=\frac{11}{3-x}\text{ or }\frac{-11}{x-3}$$

45. **[8.1]** $\dfrac{4y+16}{30}\div\dfrac{2y+8}{5}$

Multiply by the reciprocal.

$$=\frac{4y+16}{30}\cdot\frac{5}{2y+8}$$
$$=\frac{4(y+4)}{30}\cdot\frac{5}{2(y+4)}$$
$$=\frac{4\cdot 5}{2\cdot 30}=\frac{2}{6}=\frac{1}{3}$$

46. **[8.3]** $\dfrac{t^{-2}+s^{-2}}{t^{-1}-s^{-1}}=\dfrac{\frac{1}{t^2}+\frac{1}{s^2}}{\frac{1}{t}-\frac{1}{s}}$

Multiply the numerator and denominator by the LCD of all the fractions, t^2s^2.

$$=\frac{t^2s^2\left(\frac{1}{t^2}+\frac{1}{s^2}\right)}{t^2s^2\left(\frac{1}{t}-\frac{1}{s}\right)}$$
$$=\frac{s^2+t^2}{ts^2-t^2s}$$
$$=\frac{s^2+t^2}{st(s-t)}$$

47. **[8.2]**

$$\frac{4a}{a^2-ab-2b^2}-\frac{6b-a}{a^2+4ab+3b^2}$$
$$=\frac{4a}{(a-2b)(a+b)}-\frac{6b-a}{(a+3b)(a+b)}$$

The LCD is $(a+3b)(a-2b)(a+b)$.

$$=\frac{4a(a+3b)}{(a-2b)(a+b)(a+3b)}-\frac{(6b-a)(a-2b)}{(a+3b)(a+b)(a-2b)}$$
$$=\frac{4a(a+3b)-(6b-a)(a-2b)}{(a+3b)(a+b)(a-2b)}$$
$$=\frac{4a^2+12ab-(6ab-12b^2-a^2+2ab)}{(a+3b)(a+b)(a-2b)}$$
$$=\frac{4a^2+12ab-6ab+12b^2+a^2-2ab}{(a+3b)(a+b)(a-2b)}$$
$$=\frac{5a^2+4ab+12b^2}{(a+3b)(a+b)(a-2b)}$$

48. **[8.2]** $\dfrac{a}{b}+\dfrac{b}{c}+\dfrac{c}{d}$

The LCD is bcd.

$$=\frac{a\cdot cd}{b\cdot cd}+\frac{b\cdot bd}{c\cdot bd}+\frac{c\cdot bc}{d\cdot bc}$$
$$=\frac{acd+b^2d+bc^2}{bcd}$$

49. **[8.4]**

$$\frac{x+3}{x^2-5x+4}-\frac{1}{x}=\frac{2}{x^2-4x}$$
$$\frac{x+3}{(x-4)(x-1)}-\frac{1}{x}=\frac{2}{x(x-4)}$$

Multiply by the LCD, $x(x-4)(x-1)$. $(x\neq 0,1,4)$

$$x(x-4)(x-1)\left(\frac{x+3}{(x-4)(x-1)}-\frac{1}{x}\right)=x(x-4)(x-1)\cdot\left(\frac{2}{x(x-4)}\right)$$
$$x(x+3)-(x-4)(x-1)=2(x-1)$$
$$x^2+3x-(x^2-5x+4)=2x-2$$
$$x^2+3x-x^2+5x-4=2x-2$$
$$8x-4=2x-2$$
$$6x=2$$
$$x=\tfrac{1}{3}$$

Check $x=\frac{1}{3}$: $\frac{15}{11}-3=-\frac{18}{11}$ *True*

Solution set: $\left\{\frac{1}{3}\right\}$

50. **[8.4]** Solve $A=\dfrac{Rr}{R+r}$ for r.

$$A(R+r)=Rr$$
$$AR+Ar=Rr$$
$$AR=Rr-Ar$$
$$AR=(R-A)r$$
$$\frac{AR}{R-A}=r\text{ or }r=\frac{-AR}{A-R}$$

51. **[8.4]** $1-\dfrac{5}{r}=\dfrac{-4}{r^2}$

Multiply by the LCD, r^2. $(r\neq 0)$

$$r^2\left(1-\frac{5}{r}\right)=r^2\left(\frac{-4}{r^2}\right)$$
$$r^2-5r=-4$$
$$r^2-5r+4=0$$
$$(r-4)(r-1)=0$$

$$r-4=0 \quad \text{or} \quad r-1=0$$
$$r=4 \quad \text{or} \quad r=1$$

Check $r=1$: $1-5=-4$ *True*

Check $r=4$: $1-\frac{5}{4}=-\frac{1}{4}$ *True*

Solution set: $\{1,4\}$

52. **[8.4]**
$$\frac{3x}{x-4}+\frac{2}{x}=\frac{48}{x^2-4x}$$
$$\frac{3x}{x-4}+\frac{2}{x}=\frac{48}{x(x-4)}$$

Multiply by the LCD, $x(x-4)$. $(x \neq 0, 4)$

$$x(x-4)\left(\frac{3x}{x-4}+\frac{2}{x}\right)=x(x-4)\left(\frac{48}{x(x-4)}\right)$$
$$3x^2+2(x-4)=48$$
$$3x^2+2x-8=48$$
$$3x^2+2x-56=0$$
$$(3x+14)(x-4)=0$$
$$3x+14=0 \quad \text{or} \quad x-4=0$$
$$3x=-14 \quad \text{or} \quad x=4$$
$$x=-\frac{14}{3}$$

The number 4 is not allowed as a solution because substituting it in the original equation results in division by 0.

Check $x=-\frac{14}{3}$: $\frac{21}{13}-\frac{3}{7}=\frac{108}{91}$ *True*

Solution set: $\left\{-\frac{14}{3}\right\}$

53. **[8.5]** $a(d)=\dfrac{337}{d}$

(a) $a(40.5)=\dfrac{337}{40.5}$ *Let d = 40.5.*
$$\approx 8.32$$

(b) $7.51=\dfrac{337}{d}$ *Let a = 7.51.*
$$7.51d=337$$
$$d=\frac{337}{7.51}\approx 44.9$$

54. **[8.5]** Let $x=$ the time to fill the tub working together. Make a table.

	Rate	Time Working Together	Fractional Part of the Job Done
Hot	$\frac{1}{20}$	x	$\frac{x}{20}$
Cold	$\frac{1}{15}$	x	$\frac{x}{15}$

Part done by hot	plus	part done by cold	equals	1 whole job.
$\frac{x}{20}$	$+$	$\frac{x}{15}$	$=$	1

Multiply by the LCD, 60.

$$60\left(\frac{x}{20}+\frac{x}{15}\right)=60\cdot 1$$
$$3x+4x=60$$
$$7x=60$$
$$x=\frac{60}{7} \quad \text{or} \quad 8\tfrac{4}{7}$$

Working together, the tub can be filled in $\frac{60}{7}$ or $8\frac{4}{7}$ minutes.

Chapter 8 Test

1. $f(x)=\dfrac{x+3}{3x^2+2x-8}$

Set the denominator equal to zero and solve.

$$3x^2+2x-8=0$$
$$(3x-4)(x+2)=0$$
$$3x-4=0 \quad \text{or} \quad x+2=0$$
$$3x=4 \quad \text{or} \quad x=-2$$
$$x=\frac{4}{3}$$

The numbers -2 and $\frac{4}{3}$ make the rational expression undefined and are excluded from the domain of f, which can be written in set notation as $\{x\,|\,x \neq -2, \frac{4}{3}\}$.

2.
$$\frac{6x^2-13x-5}{9x^3-x}=\frac{(3x+1)(2x-5)}{x(9x^2-1)}$$
$$=\frac{(3x+1)(2x-5)}{x(3x+1)(3x-1)}$$
$$=\frac{2x-5}{x(3x-1)}$$

3.
$$\frac{(x+3)^2}{4}\cdot\frac{6}{2x+6}=\frac{(x+3)^2}{4}\cdot\frac{2\cdot 3}{2(x+3)}$$
$$=\frac{3(x+3)}{4}$$

4.
$$\frac{y^2-16}{y^2-25}\cdot\frac{y^2+2y-15}{y^2-7y+12}$$
$$=\frac{(y+4)(y-4)}{(y+5)(y-5)}\cdot\frac{(y+5)(y-3)}{(y-4)(y-3)}$$
$$=\frac{y+4}{y-5}$$

5.
$$\frac{x^2-9}{x^3+3x^2}\div\frac{x^2+x-12}{x^3+9x^2+20x}$$

Multiply by the reciprocal.

$$=\frac{x^2-9}{x^3+3x^2}\cdot\frac{x(x^2+9x+20)}{x^2+x-12}$$
$$=\frac{(x+3)(x-3)}{x^2(x+3)}\cdot\frac{x(x+5)(x+4)}{(x+4)(x-3)}$$
$$=\frac{x+5}{x}$$

6. $t^2 + t - 6, t^2 + 3t, t^2$

Factor each denominator.

$$t^2 + t - 6 = (t+3)(t-2)$$
$$t^2 + 3t = t(t+3)$$

The third denominator is already in factored form. The LCD is

$$t^2(t+3)(t-2).$$

7. $\frac{7}{6t^2} - \frac{1}{3t}$

The LCD is $6t^2$.

$$= \frac{7}{6t^2} - \frac{1 \cdot 2t}{3t \cdot 2t}$$
$$= \frac{7}{6t^2} - \frac{2t}{6t^2} = \frac{7-2t}{6t^2}$$

8. $\frac{9}{x-7} + \frac{4}{x+7}$

The LCD is $(x-7)(x+7)$.

$$= \frac{9(x+7)}{(x-7)(x+7)} + \frac{4(x-7)}{(x+7)(x-7)}$$
$$= \frac{9(x+7)+4(x-7)}{(x-7)(x+7)}$$
$$= \frac{9x+63+4x-28}{(x-7)(x+7)}$$
$$= \frac{13x+35}{(x-7)(x+7)}$$

9. $\frac{6}{x+4} + \frac{1}{x+2} - \frac{3x}{x^2+6x+8}$

$$= \frac{6}{x+4} + \frac{1}{x+2} - \frac{3x}{(x+4)(x+2)}$$

The LCD is $(x+4)(x+2)$.

$$= \frac{6(x+2)}{(x+4)(x+2)} + \frac{1(x+4)}{(x+2)(x+4)}$$
$$- \frac{3x}{(x+4)(x+2)}$$
$$= \frac{6(x+2)+x+4-3x}{(x+4)(x+2)}$$
$$= \frac{6x+12+x+4-3x}{(x+4)(x+2)}$$
$$= \frac{4x+16}{(x+4)(x+2)}$$
$$= \frac{4(x+4)}{(x+4)(x+2)} = \frac{4}{x+2}$$

10. $\dfrac{\frac{12}{r+4}}{\frac{11}{6r+24}} = \frac{12}{r+4} \div \frac{11}{6r+24}$

Multiply by the reciprocal.

$$= \frac{12}{r+4} \cdot \frac{6r+24}{11}$$
$$= \frac{12}{r+4} \cdot \frac{6(r+4)}{11} = \frac{72}{11}$$

11. $\dfrac{\frac{1}{a} - \frac{1}{b}}{\frac{a}{b} - \frac{b}{a}}$

Multiply the numerator and the denominator by the LCD of all the fractions, ab.

$$= \frac{ab\left(\frac{1}{a} - \frac{1}{b}\right)}{ab\left(\frac{a}{b} - \frac{b}{a}\right)} = \frac{b-a}{a^2-b^2}$$
$$= \frac{b-a}{(a-b)(a+b)} = \frac{(-1)(a-b)}{(a-b)(a+b)}$$
$$= \frac{-1}{a+b} \text{ or } -\frac{1}{a+b}$$

12. $\dfrac{\frac{2}{x^2} + \frac{1}{y^2}}{\frac{1}{x} - \frac{1}{y}}$

Multiply the numerator and denominator by the LCD of all the fractions, x^2y^2.

$$= \frac{x^2y^2\left(\frac{2}{x^2} + \frac{1}{y^2}\right)}{x^2y^2\left(\frac{1}{x} - \frac{1}{y}\right)}$$
$$= \frac{2y^2+x^2}{xy^2-x^2y}$$
$$= \frac{2y^2+x^2}{xy(y-x)}$$

13. **(a)** Simplify this expression.

$$\frac{2x}{3} + \frac{x}{4} - \frac{11}{2}$$

The LCD is 12.

$$= \frac{2x \cdot 4}{3 \cdot 4} + \frac{x \cdot 3}{4 \cdot 3} - \frac{11 \cdot 6}{2 \cdot 6}$$
$$= \frac{8x}{12} + \frac{3x}{12} - \frac{66}{12}$$
$$= \frac{8x+3x-66}{12}$$
$$= \frac{11x-66}{12} = \frac{11(x-6)}{12}$$

(b) Solve this equation.

$$\frac{2x}{3}+\frac{x}{4}=\frac{11}{2}$$

Multiply by the LCD, 12.

$$12\left(\frac{2x}{3}+\frac{x}{4}\right)=12\left(\frac{11}{2}\right)$$
$$8x+3x=66$$
$$11x=66$$
$$x=6$$

Check $x=6$: $4+\frac{3}{2}=\frac{11}{2}$ *True*

Solution set: $\{6\}$

14. $$\frac{1}{x}-\frac{4}{3x}=\frac{1}{x-2}$$

Multiply by the LCD, $3x(x-2)$. $(x\neq 0,2)$

$$3x(x-2)\left(\frac{1}{x}-\frac{4}{3x}\right)=3x(x-2)\left(\frac{1}{x-2}\right)$$
$$3(x-2)-4(x-2)=3x$$
$$3x-6-4x+8=3x$$
$$-x+2=3x$$
$$-4x=-2$$
$$x=\tfrac{1}{2}$$

Check $x=\frac{1}{2}$: $2-\frac{8}{3}=-\frac{2}{3}$ *True*

Solution set: $\left\{\frac{1}{2}\right\}$

15. $$\frac{y}{y+2}-\frac{1}{y-2}=\frac{8}{y^2-4}$$
$$\frac{y}{y+2}-\frac{1}{y-2}=\frac{8}{(y+2)(y-2)}$$

Multiply by the LCD, $(y+2)(y-2)$. $(y\neq -2,2)$

$$(y+2)(y-2)\left(\frac{y}{y+2}-\frac{1}{y-2}\right)=(y+2)(y-2)\left(\frac{8}{(y+2)(y-2)}\right)$$
$$y(y-2)-1(y+2)=8$$
$$y^2-2y-y-2=8$$
$$y^2-3y-10=0$$
$$(y-5)(y+2)=0$$

$$y-5=0 \quad\text{or}\quad y+2=0$$
$$y=5 \quad\text{or}\quad y=-2$$

The number -2 is not allowed as a solution because substituting it in the original equation results in division by 0.

Check $y=5$: $\frac{5}{7}-\frac{1}{3}=\frac{8}{21}$ *True*

Solution set: $\{5\}$

16. A solution cannot make a denominator 0. This is illustrated in Exercise 15 where -2 is not allowed as a solution for this reason.

17. $f(x)=\dfrac{-2}{x+1}$ is not defined when $x=-1$, so an equation of the vertical asymptote is $x=-1$. As the values of x get larger, the values of y get closer to 0, so $y=0$ is the equation of the horizontal asymptote.

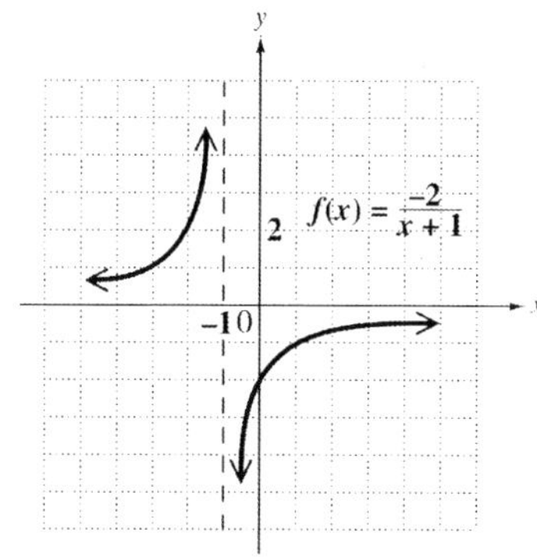

18. Solve $S=\frac{n}{2}(a+\ell)$ for ℓ.

$$2S=n(a+\ell) \qquad \textit{Multiply by 2.}$$
$$\frac{2S}{n}=a+\ell \qquad \textit{Divide by } n.$$
$$\frac{2S}{n}-a=\ell \quad\text{or}\quad \ell=\frac{2S-na}{n}$$

19. Let $x=$ the time to do the job working together. Make a table.

Worker	Rate	Time Working Together	Fractional Part of the Job Done
Wayne	$\frac{1}{9}$	x	$\frac{x}{9}$
Susan	$\frac{1}{5}$	x	$\frac{x}{5}$

Part done by Wayne	plus	part done by Susan	equals	1 whole job.
$\frac{x}{9}$	$+$	$\frac{x}{5}$	$=$	1

Multiply by the LCD, 45.

$$45\left(\frac{x}{9}+\frac{x}{5}\right)=45\cdot 1$$
$$5x+9x=45$$
$$14x=45$$
$$x=\tfrac{45}{14}\ \text{or}\ 3\tfrac{3}{14}$$

Working together, they can do the job in $\frac{45}{14}$ or $3\frac{3}{14}$ hours.

20. Let $x =$ the speed of the boat in still water. Make a table. Use $d = rt$, or $t = \frac{d}{r}$.

	d	r	t
Downstream	36	$x+3$	$\dfrac{36}{x+3}$
Upstream	24	$x-3$	$\dfrac{24}{x-3}$

Because the times are equal,

$$\frac{36}{x+3} = \frac{24}{x-3}.$$

Multiply by the LCD, $(x+3)(x-3)$. $(x \neq -3, 3)$

$$(x+3)(x-3)\left(\frac{36}{x+3}\right) = (x+3)(x-3)\left(\frac{24}{x-3}\right)$$
$$36(x-3) = 24(x+3)$$
$$36x - 108 = 24x + 72$$
$$12x = 180$$
$$x = 15$$

The speed of the boat in still water is 15 mph.

21. Let $x =$ the number of fish in Lake Linda. Write a proportion.

$$\frac{x}{600} = \frac{800}{10}$$

Multiply by the LCD, 600.

$$600\left(\frac{x}{600}\right) = 600\left(\frac{800}{10}\right)$$
$$x = 60 \cdot 800$$
$$x = 48{,}000$$

There are about 48,000 fish in the lake.

22. $g(x) = \dfrac{5x}{2+x}$

(a) $3 = \dfrac{5x}{2+x}$ *Let g(x) = 3.*

$$3(2+x) = 5x$$
$$6 + 3x = 5x$$
$$6 = 2x$$
$$x = 3 \text{ units}$$

3 units of food will produce a growth rate of 3 units of growth per unit of food.

(b) If no food is available, then $x = 0$.

$$g(0) = \frac{5(0)}{2+0} = \frac{0}{2} = 0$$

The growth rate is 0.

Cumulative Review Exercises (Chapters 1–8)

In Exercises 1–2, $x = -4$, $y = 3$, and $z = 6$.

1. $|2x| + 3y - z^3$
$$= |(2)(-4)| + 3(3) - (6)^3$$
$$= |-8| + 3(3) - 216$$
$$= 8 + 9 - 216 = -199$$

2. $\dfrac{x(2x-1)}{3y-z} = \dfrac{-4[2(-4)-1]}{3(3)-6}$
$$= \frac{-4[-8-1]}{9-6}$$
$$= \frac{-4(-9)}{3} = 12$$

3. $7(2x+3) - 4(2x+1) = 2(x+1)$
$$14x + 21 - 8x - 4 = 2x + 2$$
$$6x + 17 = 2x + 2$$
$$4x = -15$$
$$x = -\tfrac{15}{4}$$

Solution set: $\left\{-\frac{15}{4}\right\}$

4. $|6x - 8| - 4 = 0$
$$|6x \quad 8| = 4$$

$6x - 8 = 4$ or $6x - 8 = -4$
$6x = 12$ or $6x = 4$
$x = 2$ or $x = \frac{4}{6} = \frac{2}{3}$

Solution set: $\left\{\frac{2}{3}, 2\right\}$

5. Solve $ax + by = cx + d$ for x. Get the x-terms on one side.

$$ax - cx = d - by$$
$$x(a-c) = d - by$$
$$x = \frac{d-by}{a-c}$$
$$\text{or } x = \frac{by-d}{c-a}$$

6. $\dfrac{2}{3}x + \dfrac{5}{12}x \leq 20$

Multiply each side by the LCD, 12.

$$12\left(\tfrac{2}{3}x + \tfrac{5}{12}x\right) \leq 12(20)$$
$$8x + 5x \leq 240$$
$$13x \leq 240$$
$$x \leq \tfrac{240}{13}$$

Solution set: $\left(-\infty, \frac{240}{13}\right]$

7. $|3x+2| \geq 4$

$$\begin{aligned} 3x+2 &\geq 4 \quad \text{or} \quad 3x+2 \leq -4 \\ 3x &\geq 2 \quad \text{or} \quad 3x \leq -6 \\ x &\geq \tfrac{2}{3} \quad \text{or} \quad x \leq -2 \end{aligned}$$

Solution set: $(-\infty, -2] \cup \left[\frac{2}{3}, \infty\right)$

8. Let x = the amount of money invested at 4% and $2x$ = the amount of money invested at 3%.

Use $I = prt$ with the time, t, equal to 1 yr.

$$\begin{aligned} .04x + .03(2x) &= 400 \\ .04x + .06x &= 400 \\ .10x &= 400 \end{aligned}$$

Multiply by 10 to clear the decimal.

$$x = 4000$$

Since $x = 4000$, $2x = 2(4000) = 8000$.

He invested \$4000 at 4% and \$8000 at 3%.

9. Let h = the height of the triangle.
Use the formula $A = \frac{1}{2}bh$. Here, $A = 42$ and $b = 14$.

$$\begin{aligned} A &= \tfrac{1}{2}bh \\ 42 &= \tfrac{1}{2}(14)h \\ 42 &= 7h \\ 6 &= h \end{aligned}$$

The height is 6 meters.

10. $-4x + 2y = 8$
To find the x-intercept, let $y = 0$.

$$\begin{aligned} -4x + 2(0) &= 8 \\ -4x &= 8 \\ x &= -2 \end{aligned}$$

The x-intercept is $(-2, 0)$.
To find the y-intercept, let $x = 0$.

$$\begin{aligned} -4(0) + 2y &= 8 \\ 2y &= 8 \\ y &= 4 \end{aligned}$$

The y-intercept is $(0, 4)$. Plot the intercepts, and draw the line through them.

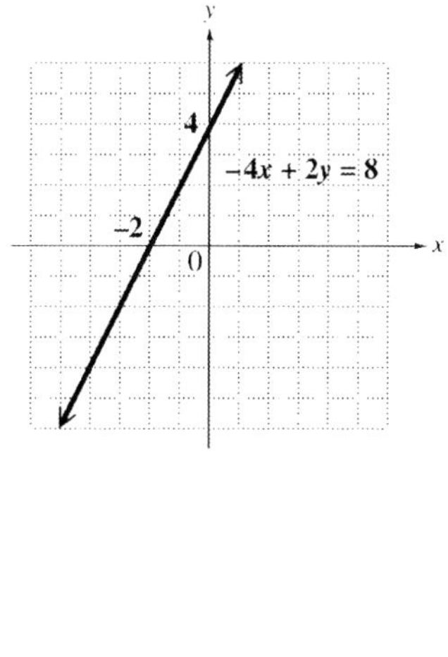

11. Through $(-5, 8)$ and $(-1, 2)$
Let $(x_1, y_1) = (-5, 8)$ and $(x_2, y_2) = (-1, 2)$. Then,

$$m = \frac{y_2 - y_1}{x_2 - x_1} = \frac{2-8}{-1-(-5)} = \frac{-6}{4} = -\frac{3}{2}.$$

The slope is $-\frac{3}{2}$.

12. Perpendicular to $4x - 3y = 12$

Solve for y to write the equation in slope-intercept form and find the slope.

$$\begin{aligned} 4x - 3y &= 12 \\ -3y &= -4x + 12 \\ y &= \tfrac{4}{3}x - 4 \end{aligned}$$

The slope is $\frac{4}{3}$. Perpendicular lines have slopes that are negative reciprocals of each other. The negative reciprocal of $\frac{4}{3}$ is $-\frac{3}{4}$. The slope of a line perpendicular to the given line is $-\frac{3}{4}$.

13. Use $(x_1, y_1) = (-5, 8)$ and $m = -\frac{3}{2}$ in the point-slope form.

$$\begin{aligned} y - y_1 &= m(x - x_1) \\ y - 8 &= -\tfrac{3}{2}[x - (-5)] \\ y - 8 &= -\tfrac{3}{2}(x + 5) \\ y - 8 &= -\tfrac{3}{2}x - \tfrac{15}{2} \\ y &= -\tfrac{3}{2}x + \tfrac{1}{2} \end{aligned}$$

14. $2x + 5y > 10$
Graph the line $2x + 5y = 10$ by drawing a dashed line (since the inequality involves $>$) through the intercepts $(5, 0)$ and $(0, 2)$.
Test a point not on this line, such as $(0, 0)$.

$$\begin{aligned} 2x + 5y &> 10 \\ 2(0) + 5(0) &> 10 \quad ? \\ 0 &> 10 \qquad \textit{False} \end{aligned}$$

Shade the side of the line not containing $(0, 0)$.

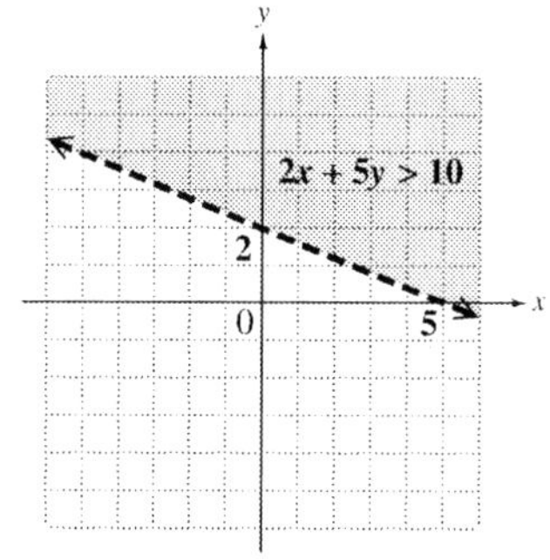

15. $x - y \geq 3$ and $3x + 4y \leq 12$
Graph the solid line $x - y = 3$ through $(3, 0)$ and $(0, -3)$. The inequality $x - y \geq 3$ can be written as $y \leq x - 3$, so shade the region below the boundary line.
Graph the solid line $3x + 4y = 12$ through $(4, 0)$ and $(0, 3)$. The inequality $3x + 4y \leq 12$ can be written as $y \leq -\frac{3}{4}x + 3$, so shade the region below the boundary line.
The required graph is the common shaded area as well as the portions of the lines that bound it.

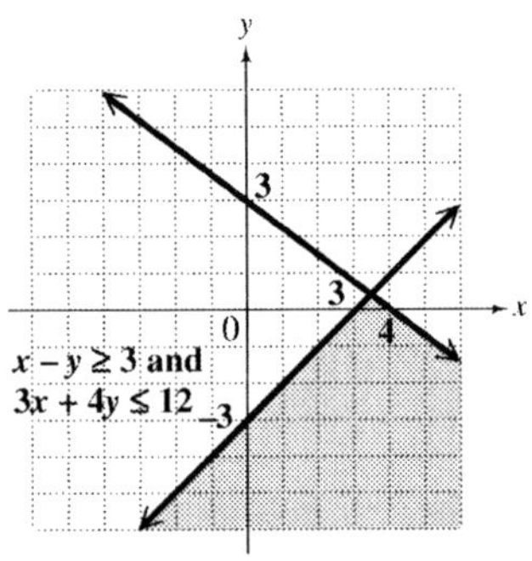

16. Since no year is repeated, the relation defines a function.
The domain is

$$\{1990, 1992, 1994, 1996, 1998, 2000\}.$$

The range is $\{1.25, 1.61, 1.80, 1.21, 1.94, 2.26\}$.

17. The relation is not a function since a vertical line may intersect the graph in more than one point. The domain is the set of x-values, $[-2, \infty)$. The range is the set of y-values, $(-\infty, \infty)$.

18. $y = -\sqrt{x+2}$
The relation is a function since each input value corresponds to exactly one output value. The radicand must be nonnegative; that is, $x + 2 \geq 0$, or $x \geq -2$. The domain is $[-2, \infty)$. The values of $\sqrt{x+2}$ are nonnegative, so the values of $-\sqrt{x+2}$ are nonpositive. Thus, the range is $(-\infty, 0]$.

19. **(a)** Solve the equation for y.

$$\begin{aligned} 5x - 3y &= 8 \\ 5x - 8 &= 3y \\ \frac{5x-8}{3} &= y \end{aligned}$$

So $f(x) = \dfrac{5x-8}{3}$ or $f(x) = \dfrac{5}{3}x - \dfrac{8}{3}$.

(b) $f(1) = \dfrac{5(1)-8}{3} = \dfrac{-3}{3} = -1$

20.
$$\begin{aligned} f(x) &= 3x + 6 \\ f(x+3) &= 3(x+3) + 6 \\ &= 3x + 9 + 6 \\ &= 3x + 15 \end{aligned}$$

21.
$$\begin{aligned} 4x - y &= -7 \quad (1) \\ 5x + 2y &= 1 \quad (2) \end{aligned}$$

To eliminate y, multiply equation (1) by 2 and add the result to (2).

$$\begin{array}{rcrcrl} 8x & - & 2y & = & -14 & 2 \times (1) \\ 5x & + & 2y & = & 1 & (2) \\ \hline 13x & & & = & -13 & \\ & & x & = & -1 & \end{array}$$

To find y, substitute -1 for x in (1).

$$\begin{aligned} 4x - y &= -7 \quad (1) \\ 4(-1) - y &= -7 \\ -4 - y &= -7 \\ -y &= -3 \\ y &= 3 \end{aligned}$$

Solution set: $\{(-1, 3)\}$

22.
$$\begin{aligned} x + y - 2z &= -1 \quad (1) \\ 2x - y + z &= -6 \quad (2) \\ 3x + 2y - 3z &= -3 \quad (3) \end{aligned}$$

Add (1) and (2) to eliminate y.

$$\begin{array}{rcrcrcrl} x & + & y & - & 2z & = & -1 & (1) \\ 2x & - & y & + & z & = & -6 & (2) \\ \hline 3x & & & - & z & = & -7 & (4) \end{array}$$

Multiply (2) by 2 and add it to (3) to eliminate y.

$$\begin{array}{rcrcrcrl} 4x & - & 2y & + & 2z & = & -12 & 2 \times (2) \\ 3x & + & 2y & - & 3z & = & -3 & (3) \\ \hline 7x & & & - & z & = & -15 & (5) \end{array}$$

Multiply (4) by -1 and add to (5) to eliminate z.

$$\begin{array}{rcrcrl} -3x & + & z & = & 7 & -1 \times (4) \\ 7x & - & z & = & -15 & (5) \\ \hline 4x & & & = & -8 & \\ & & x & = & -2 & \end{array}$$

Substitute -2 for x into (4) and solve for z.

$$\begin{aligned} 3(-2) - z &= -7 \\ -6 - z &= -7 \\ -z &= -1 \\ z &= 1 \end{aligned}$$

Substitute -2 for x and 1 for z into (1) and solve for y.

$$\begin{aligned} -2+y-2(1) &= -1 \\ -2+y-2 &= -1 \\ y-4 &= -1 \\ y &= 3 \end{aligned}$$

Solution set: $\{(-2,3,1)\}$

23. $$\begin{aligned} x+2y+z &= 5 \quad (1) \\ x-y+z &= 3 \quad (2) \\ 2x+4y+2z &= 11 \quad (3) \end{aligned}$$

Multiply (1) by -1 and add to (2).

$$\begin{array}{rcl} -x - 2y - z &=& -5 \qquad -1\times(1) \\ x - y + z &=& 3 \quad (2) \\ \hline -3y &=& -2 \\ y &=& \frac{2}{3} \end{array}$$

Substitute $\frac{2}{3}$ for y in (1) and (3), then add the resulting equations.

$$\begin{aligned} x+2\left(\tfrac{2}{3}\right)+z &= 5 \qquad \text{Let } y=\tfrac{2}{3} \text{ in (1).} \\ x+\tfrac{4}{3}+z &= 5 \\ x+z &= \tfrac{11}{3} \qquad (4) \end{aligned}$$

$$\begin{aligned} 2x+4\left(\tfrac{2}{3}\right)+2z &= 11 \qquad \text{Let } y=\tfrac{2}{3} \text{ in (3).} \\ 2x+\tfrac{8}{3}+2z &= 11 \\ 2x+2z &= \tfrac{25}{3} \qquad (5) \end{aligned}$$

Multiply (4) by -2 and add to (5).

$$\begin{array}{rcl} -2x - 2z &=& -\frac{22}{3} \qquad -2\times(4) \\ 2x + 2z &=& \frac{25}{3} \quad (5) \\ \hline 0 &=& 1 \qquad \textit{False} \end{array}$$

Since this statement is false, the solution is $\emptyset$.

24. Let $x =$ the number of career points for Swoopes. Then $x+799 =$ the number of career points for Leslie. Together they have 6737 points, so

$$\begin{aligned} x+(x+799) &= 6737. \\ 2x+799 &= 6737 \\ 2x &= 5938 \\ x &= 2969 \end{aligned}$$

Since $x=2969$, $x+799=3768$. As of June 2004, Swoopes had 2969 career points and Leslie had 3768 career points.

25. $$\left(\frac{a^{-3}b^4}{a^2b^{-1}}\right)^{-2} = \left(\frac{b^4b^1}{a^2a^3}\right)^{-2} = \left(\frac{b^5}{a^5}\right)^{-2} = \left(\frac{a^5}{b^5}\right)^2 = \frac{a^{10}}{b^{10}}$$

26. $$\begin{aligned} &\left(\frac{m^{-4}n^2}{m^2n^{-3}}\right)\cdot\left(\frac{m^5n^{-1}}{m^{-2}n^5}\right) \\ &= \left(\frac{n^2n^3}{m^2m^4}\right)\cdot\left(\frac{m^5m^2}{n^1n^5}\right) \\ &= \left(\frac{n^5}{m^6}\right)\cdot\left(\frac{m^7}{n^6}\right) \\ &= \frac{n^5m^7}{n^6m^6} = \frac{m}{n} \end{aligned}$$

27. $$\begin{aligned} &(3y^2-2y+6)-(-y^2+5y+12) \\ &= 3y^2-2y+6+y^2-5y-12 \\ &= 4y^2-7y-6 \end{aligned}$$

28. $-6x^4(x^2-3x+2) = -6x^6+18x^5-12x^4$

29. $$\begin{aligned} (4f+3)(3f-1) &= 12f^2-4f+9f-3 \\ &= 12f^2+5f-3 \end{aligned}$$

30. $(7t^3+8)(7t^3-8)$

This is the product of the sum and difference of two terms.

$$\begin{aligned} (7t^3+8)(7t^3-8) &= \left(7t^3\right)^2-8^2 \\ &= 49t^6-64 \end{aligned}$$

31. $\left(\frac{1}{4}x+5\right)^2$

Use the formula for the square of a binomial, $(a+b)^2=a^2+2ab+b^2$.

$$\begin{aligned} \left(\tfrac{1}{4}x+5\right)^2 &= \left(\tfrac{1}{4}x\right)^2+2\left(\tfrac{1}{4}x\right)(5)+5^2 \\ &= \tfrac{1}{16}x^2+\tfrac{5}{2}x+25 \end{aligned}$$

32. $(3x^3+13x^2-17x-7)\div(3x+1)$

$$\begin{array}{r|l} & x^2 + 4x - 7 \\ \hline 3x+1 & 3x^3 + 13x^2 - 17x - 7 \\ & 3x^3 + x^2 \\ \hline & 12x^2 - 17x \\ & 12x^2 + 4x \\ \hline & -21x - 7 \\ & -21x - 7 \\ \hline & 0 \end{array}$$

Answer: x^2+4x-7

33. **(a)** $$\begin{aligned} (f+g)(x) &= f(x)+g(x) \\ &= \left(x^2+2x-3\right) \\ &\quad + \left(2x^3-3x^2+4x-1\right) \\ &= 2x^3-2x^2+6x-4 \end{aligned}$$

(b) $(g-f)(x) = g(x) - f(x)$
$$= (2x^3 - 3x^2 + 4x - 1) - (x^2 + 2x - 3)$$
$$= 2x^3 - 3x^2 + 4x - 1 - x^2 - 2x + 3$$
$$= 2x^3 - 4x^2 + 2x + 2$$

(c) Use part (a).
$$(f+g)(-1) = 2(-1)^3 - 2(-1)^2 + 6(-1) - 4$$
$$= 2(-1) - 2(1) + 6(-1) - 4$$
$$= -2 - 2 - 6 - 4$$
$$= -14$$

34. $2x^2 - 13x - 45 = (2x+5)(x-9)$

35. $100t^4 - 25 = 25(4t^4 - 1)$
$$= 25\left[(2t^2)^2 - 1^2\right]$$
$$= 25(2t^2+1)(2t^2-1)$$

36. Use the sum of cubes formula,
$$x^3 + y^3 = (x+y)(x^2 - xy + y^2).$$
$$8p^3 + 125 = (2p)^3 + 5^3$$
$$= (2p+5)\left[(2p)^2 - (2p)(5) + 5^2\right]$$
$$= (2p+5)(4p^2 - 10p + 25)$$

37.
$$3x^2 + 4x = 7$$
$$3x^2 + 4x - 7 = 0$$
$$(3x+7)(x-1) = 0$$

$3x + 7 = 0$ or $x - 1 = 0$
$3x = -7$ or $x = 1$
$x = -\frac{7}{3}$

Solution set: $\{-\frac{7}{3}, 1\}$

38. $$\frac{y^2-16}{y^2-8y+16} = \frac{(y+4)(y-4)}{(y-4)(y-4)} = \frac{y+4}{y-4}$$

39. $$\frac{8x^2-18}{8x^2+4x-12} = \frac{2(4x^2-9)}{4(2x^2+x-3)} = \frac{2(2x+3)(2x-3)}{4(2x+3)(x-1)} = \frac{2x-3}{2(x-1)}$$

40. $$\frac{2a^2}{a+b} \cdot \frac{a-b}{4a} = \frac{2a^2(a-b)}{4a(a+b)} = \frac{a(a-b)}{2(a+b)}$$

41. $$\frac{x+4}{x-2} + \frac{2x-10}{x-2} = \frac{x+4+2x-10}{x-2} = \frac{3x-6}{x-2} = \frac{3(x-2)}{x-2} = 3$$

42. $$\frac{2x}{2x-1} + \frac{4}{2x+1} + \frac{8}{4x^2-1}$$
$$= \frac{2x}{2x-1} + \frac{4}{2x+1} + \frac{8}{(2x+1)(2x-1)}$$

The LCD is $(2x+1)(2x-1)$.
$$= \frac{2x(2x+1)}{(2x-1)(2x+1)} + \frac{4(2x-1)}{(2x+1)(2x-1)} + \frac{8}{(2x+1)(2x-1)}$$
$$= \frac{2x(2x+1) + 4(2x-1) + 8}{(2x+1)(2x-1)}$$
$$= \frac{4x^2 + 2x + 8x - 4 + 8}{(2x+1)(2x-1)}$$
$$= \frac{4x^2 + 10x + 4}{(2x+1)(2x-1)}$$
$$= \frac{2(2x^2 + 5x + 2)}{(2x+1)(2x-1)}$$
$$= \frac{2(2x+1)(x+2)}{(2x+1)(2x-1)}$$
$$= \frac{2(x+2)}{2x-1}$$

43. $$\frac{-3x}{x+1} + \frac{4x+1}{x} = \frac{-3}{x^2+x}$$
$$\frac{-3x}{x+1} + \frac{4x+1}{x} = \frac{-3}{x(x+1)}$$

Multiply by the LCD, $x(x+1)$. $(x \neq -1, 0)$
$$x(x+1)\left(\frac{-3x}{x+1} + \frac{4x+1}{x}\right) = x(x+1)\left(\frac{-3}{x(x+1)}\right)$$
$$x(-3x) + (x+1)(4x+1) = -3$$
$$-3x^2 + 4x^2 + x + 4x + 1 = -3$$
$$x^2 + 5x + 4 = 0$$
$$(x+4)(x+1) = 0$$

$x + 4 = 0$ or $x + 1 = 0$
$x = -4$ or $x = -1$

The number -1 is not allowed as a solution because substituting it in the original equation results in division by 0.

Check $x = -4$: $-4 + \frac{15}{4} = -\frac{1}{4}$ *True*

Solution set: $\{-4\}$

44. Solve $\frac{1}{f} = \frac{1}{p} + \frac{1}{q}$ for q.

Multiply by the LCD, fpq.

$$fpq\left(\frac{1}{f}\right) = fpq\left(\frac{1}{p} + \frac{1}{q}\right)$$
$$pq = fq + fp$$

Get all terms with q on one side.

$$pq - fq = fp$$
$$q(p - f) = fp$$
$$q = \frac{fp}{p - f}$$
$$\text{or} \quad q = \frac{-fp}{f - p}$$

45. Let $x =$ the speed of the plane in still air. Use $d = rt$, or $t = \frac{d}{r}$, to make a table.

	d	r	t
Against the Wind	200	$x - 30$	$\frac{200}{x - 30}$
With the Wind	300	$x + 30$	$\frac{300}{x + 30}$

The times are the same, so

$$\frac{200}{x - 30} = \frac{300}{x + 30}.$$

Multiply by the LCD, $(x - 30)(x + 30)$.

$$(x - 30)(x + 30)\left(\frac{200}{x - 30}\right) = (x - 30)(x + 30)\left(\frac{300}{x + 30}\right)$$
$$200(x + 30) = 300(x - 30)$$
$$200x + 6000 = 300x - 9000$$
$$-100x = -15{,}000$$
$$x = 150$$

The speed of the plane in still air is 150 mph.

46. Let $x =$ the time to complete the job working together.

Make a table.

	Rate	**Time Working Together**	**Fractional Part of the Job Done**
Machine A	$\frac{1}{2}$	x	$\frac{x}{2}$
Machine B	$\frac{1}{3}$	x	$\frac{x}{3}$

Part done by A	plus	part done by B	equals	1 whole job.
$\frac{x}{2}$	$+$	$\frac{x}{3}$	$=$	1

Multiply each side by the LCD, 6.

$$6\left(\frac{x}{2} + \frac{x}{3}\right) = 6 \cdot 1$$
$$3x + 2x = 6$$
$$5x = 6$$
$$x = \tfrac{6}{5} \quad \text{or} \quad 1\tfrac{1}{5}$$

Working together, the machines can complete the job in $\frac{6}{5}$ or $1\frac{1}{5}$ hours.

CHAPTER 9 ROOTS, RADICALS, AND ROOT FUNCTIONS

9.1 Radical Expressions and Graphs

9.1 Margin Exercises

1. **(a)** $\sqrt[3]{8} = 2$, because $2^3 = 8$.

(b) $\sqrt[3]{1000} = 10$, because $10^3 = 1000$.

(c) $\sqrt[4]{81} = 3$, because $3^4 = 81$.

(d) $\sqrt[6]{64} = 2$, because $2^6 = 64$.

2. **(a)** $\sqrt{4} = 2$, because $2^2 = 4$.

(b) $\sqrt[3]{27} = 3$, because $3^3 = 27$.

(c) $\sqrt{36} = 6$, because $6^2 = 36$, so $-\sqrt{36} = -6$. The negative sign outside the radical makes the answer negative.

(d) $\sqrt[4]{625} = 5$, because $5^4 = 625$.

(e) $\sqrt[5]{-32} = -2$, because $(-2)^5 = -32$.

(f) $\sqrt[4]{-16}$ is *not a real number*, since the index, 4, is even and the radicand, -16, is negative.

3. **(a)** $f(x) = \sqrt{x} + 2$

For the radicand to be nonnegative, we must have $x \geq 0$. Therefore, the domain is $[0, \infty)$. The function values are at least 2 since $\sqrt{x} \geq 0$ and we add 2 to $\sqrt{x}$, so the range is $[2, \infty)$.

x	$f(x) = \sqrt{x} + 2$
0	$\sqrt{0} + 2 = 2$
1	$\sqrt{1} + 2 = 3$
4	$\sqrt{4} + 2 = 4$
9	$\sqrt{9} + 2 = 5$

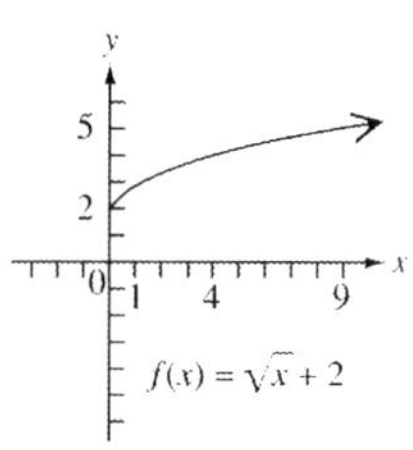

(b) $f(x) = \sqrt[3]{x - 1}$

Since we can take the cube root of any real number, the domain is $(-\infty, \infty)$.
The result of a cube root can be any real number, so the range is $(-\infty, \infty)$.

x	$f(x) = \sqrt[3]{x - 1}$
-7	$\sqrt[3]{-7 - 1} = -2$
0	$\sqrt[3]{0 - 1} = -1$
1	$\sqrt[3]{1 - 1} = 0$
2	$\sqrt[3]{2 - 1} = 1$
9	$\sqrt[3]{9 - 1} = 2$

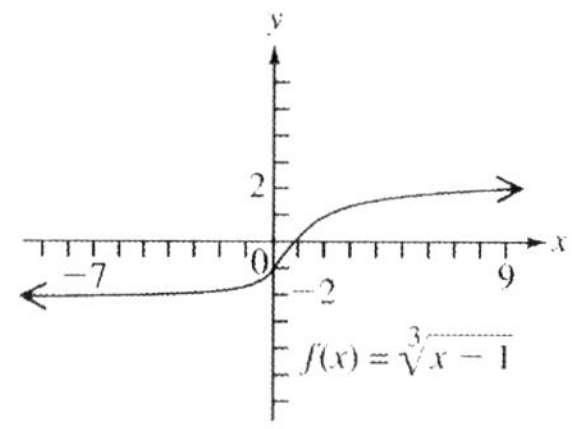

4. **(a)** $\sqrt{49} = \sqrt{7^2} = |7| = 7$

(b) $-\sqrt{\frac{36}{25}} = -\sqrt{\left(\frac{6}{5}\right)^2} = -\left|\frac{6}{5}\right| = -\frac{6}{5}$

(c) $\sqrt{(-6)^2} = |-6| = 6$

(d) $\sqrt{r^2} = |r|$

5. **(a)** $\sqrt[6]{64} = \sqrt[6]{2^6} = |2| = 2$ (n is even)

(b) $-\sqrt[4]{16} = -\sqrt[4]{2^4} = -|2| = -2$ (n is even)

(c) $\sqrt[3]{\frac{216}{125}} = \sqrt[3]{\left(\frac{6}{5}\right)^3} = \frac{6}{5}$ (n is odd)

(d) $\sqrt[5]{-243} = \sqrt[5]{(-3)^5} = -3$ (n is odd)

(e) $\sqrt[6]{(-p)^6} = |-p| = |p|$ (n is even)

(f) $-\sqrt[6]{y^{24}} = -\sqrt[6]{(y^4)^6} = -|y^4| = -y^4$ (n is even)
No absolute value bars are needed here because y^4 is nonnegative for any real number value of y.

6. Use a calculator, and round to three decimal places.

(a) $\sqrt{17} \approx 4.123$

(b) $-\sqrt{362} \approx -19.026$

(c) $\sqrt[3]{9482} \approx 21.166$

(d) $\sqrt[4]{6825} \approx 9.089$

7.
$$f = \frac{1}{2\pi\sqrt{LC}}$$
$$= \frac{1}{2\pi\sqrt{(6 \times 10^{-5})(4 \times 10^{-9})}}$$
$$\approx 324{,}874$$
or about 325,000 cycles per second.

9.1 Section Exercises

1. $-\sqrt{16} = -(4) = -4$ **(E)**

2. $\sqrt{-16}$ is not a real number (the index, 2, is even and the radicand, -16, is negative). **(F)**

3. $\sqrt[3]{-27} = -3$ because $(-3)^3 = -27$. **(D)**

4. $\sqrt[5]{-32} = -2$ because $(-2)^5 = -32$. **(B)**

5. $\sqrt[4]{81} = 3$ because $3^4 = 81$. **(A)**

6. $\sqrt[3]{8} = 2$ because $2^3 = 8$. **(C)**

7. $\sqrt{123.5} \approx \sqrt{121} = 11$ because $11^2 = 121$. **(C)**

8. $\sqrt{67.8} \approx \sqrt{64} = 8$ because $8^2 = 64$. **(B)**

9. The length $\sqrt{98}$ is closer to $\sqrt{100} = 10$ than to $\sqrt{81} = 9$. The width $\sqrt{26}$ is closer to $\sqrt{25} = 5$ than to $\sqrt{36} = 6$. Use the estimates $L = 10$ and $W = 5$ in $A = LW$ to find an estimate of the area.

$$A \approx 10 \cdot 5 = 50$$

Choice **C** is the best estimate.

10. The length $\sqrt{98}$ is closer to $\sqrt{100} = 10$ than to $\sqrt{81} = 9$. The width $\sqrt{26}$ is closer to $\sqrt{25} = 5$ than to $\sqrt{36} = 6$. Use the estimates $L = 10$ and $W = 5$ in $P = 2L + 2W$ to find an estimate of the perimeter.

$$P \approx 2(10) + 2(5) = 30$$

Choice **D** is the best estimate.

11. **(a)** If $a > 0$, then $-a < 0$ and $\sqrt{-a}$ is not a real number. Thus, the given expression, $-\sqrt{-a}$, is *not a real number.*

(b) If $a < 0$, then $-a > 0$, so $\sqrt{-a}$ is a positive number and $-\sqrt{-a}$ is a *negative* number.

(c) If $a = 0$, then $-a = 0$, so $\sqrt{-a} = \sqrt{0} = 0$.

12. **(a)** If n is odd and a is *positive*, then $\sqrt[n]{a}$ is positive.

(b) If n is odd and a is *negative*, then $\sqrt[n]{a}$ is negative.

(c) If n is odd and a is 0, then $\sqrt[n]{a}$ is 0.

13. $\sqrt{81} = 9$, because $9^2 = 81$, so $-\sqrt{81} = -9$.

14. $\sqrt{121} = 11$, because $11^2 = 121$, so $-\sqrt{121} = -11$.

15. $\sqrt[3]{216} = 6$, because $6^3 = 216$.

16. $\sqrt[3]{343} = 7$, because $7^3 = 343$.

17. $\sqrt[3]{-64} = -4$, because $(-4)^3 = -64$.

18. $\sqrt[3]{-125} = -5$, because $(-5)^3 = -125$.

19. $\sqrt[3]{512} = 8$, because $8^3 = 512$, so $-\sqrt[3]{512} = -8$.

20. $\sqrt[3]{1000} = 10$, because $10^3 = 1000$, so $-\sqrt[3]{1000} = -10$.

21. $\sqrt[4]{1296} = 6$, because $6^4 = 1296$.

22. $\sqrt[4]{625} = 5$, because $5^4 = 625$.

23. $\sqrt[4]{81} = 3$, because $3^4 = 81$, so $-\sqrt[4]{81} = -3$.

24. $\sqrt[4]{256} = 4$, because $4^4 = 256$, so $-\sqrt[4]{256} = -4$.

25. $\sqrt[4]{-16}$ is not a real number since no real number to the fourth power equals -16. Any real number raised to the fourth power is 0 or positive.

26. $\sqrt[4]{-81}$ is not a real number since no real number to the fourth power equals -81.

27. $\sqrt[6]{(-2)^6} = \sqrt[6]{64} = 2$, because $2^6 = 64$.

28. $\sqrt[6]{(-4)^6} = \sqrt[6]{4096} = 4$, because $4^6 = 4096$.

29. $\sqrt[5]{(-9)^5} = \sqrt[5]{-59{,}049} = -9$, because $(-9)^5 = -59{,}049$.

30. $\sqrt[5]{(-8)^5} = \sqrt[5]{-32{,}768} = -8$, because $(-8)^5 = -32{,}768$.

31. $\sqrt{\frac{64}{81}} = \frac{8}{9}$, because $\left(\frac{8}{9}\right)^2 = \frac{64}{81}$.

32. $\sqrt{\frac{100}{9}} = \frac{10}{3}$, because $\left(\frac{10}{3}\right)^2 = \frac{100}{9}$.

33. $\sqrt[3]{\frac{8}{27}} = \frac{2}{3}$, because $\left(\frac{2}{3}\right)^3 = \frac{8}{27}$.

34. $\sqrt[4]{\frac{81}{16}} = \frac{3}{2}$, because $\left(\frac{3}{2}\right)^4 = \frac{81}{16}$.

35. $\sqrt[6]{\frac{1}{64}} = \frac{1}{2}$, because $\left(\frac{1}{2}\right)^6 = \frac{1}{64}$.

36. $\sqrt[5]{\frac{1}{32}} = \frac{1}{2}$, because $\left(\frac{1}{2}\right)^5 = \frac{1}{32}$.

37. $f(x) = \sqrt{x+3}$
For the radicand to be nonnegative, we must have

$$x + 3 \geq 0 \quad \text{or} \quad x \geq -3.$$

Thus, the domain is $[-3, \infty)$.
The function values are positive or zero (the result of the radical), so the range is $[0, \infty)$.

x	$f(x) = \sqrt{x+3}$
-3	$\sqrt{-3+3} = 0$
-2	$\sqrt{-2+3} = 1$
1	$\sqrt{1+3} = 2$

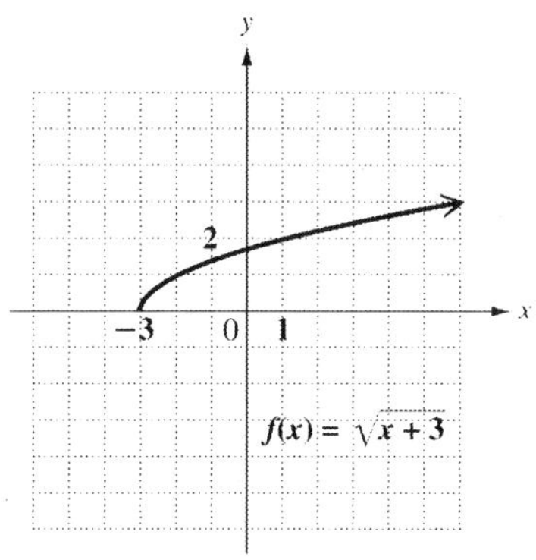

38. $f(x) = \sqrt{x-5}$
For the radicand to be nonnegative, we must have

$$x - 5 \geq 0 \quad \text{or} \quad x \geq 5.$$

Thus, the domain is $[5, \infty)$.
The function values are positive or zero (the result of the radical), so the range is $[0, \infty)$.

x	$f(x) = \sqrt{x-5}$
5	$\sqrt{5-5} = 0$
6	$\sqrt{6-5} = 1$
9	$\sqrt{9-5} = 2$

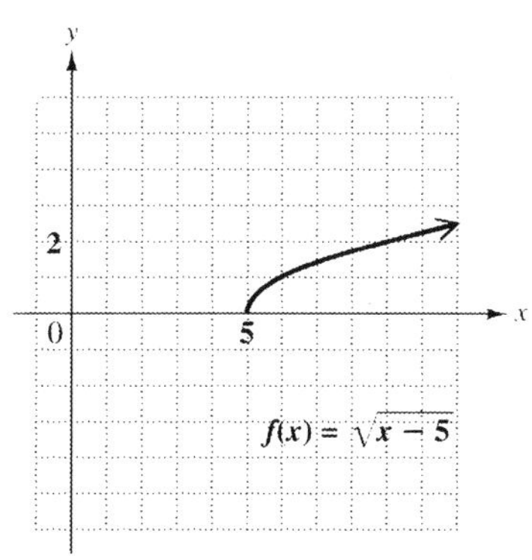

39. $f(x) = \sqrt{x} - 2$
For the radicand to be nonnegative, we must have

$$x \geq 0.$$

Note that the "-2" does not affect the domain, which is $[0, \infty)$.
The result of the radical is positive or zero, but the function values are 2 less than those values, so the range is $[-2, \infty)$.

x	$f(x) = \sqrt{x} - 2$
0	$\sqrt{0} - 2 = -2$
1	$\sqrt{1} - 2 = -1$
4	$\sqrt{4} - 2 = 0$

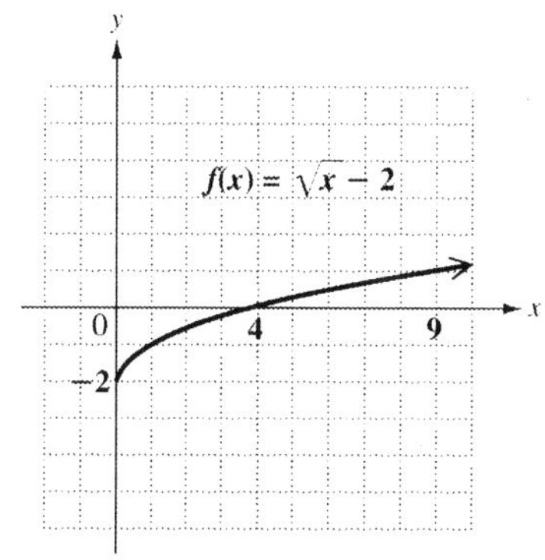

40. $f(x) = \sqrt{x} + 4$
For the radicand to be nonnegative, we must have

$$x \geq 0.$$

Note that the "$+4$" does not affect the domain, which is $[0, \infty)$.
The result of the radical is positive or zero, but the function values are 4 more than those values, so the range is $[4, \infty)$.

x	$f(x) = \sqrt{x} + 4$
0	$\sqrt{0} + 4 = 4$
1	$\sqrt{1} + 4 = 5$
4	$\sqrt{4} + 4 = 6$

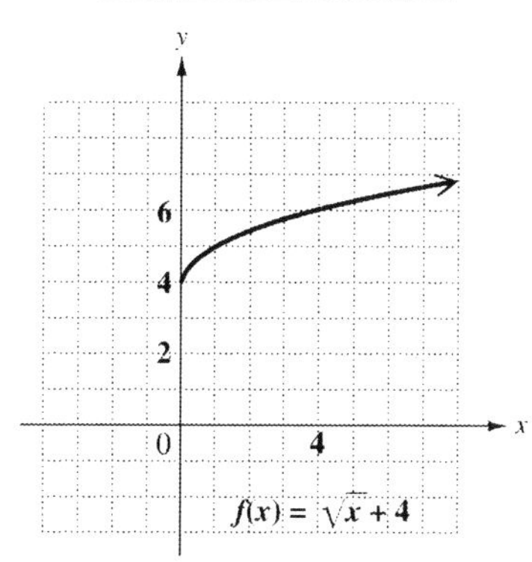

41. $f(x) = \sqrt[3]{x} - 3$

Since we can take the cube root of any real number, the domain is $(-\infty, \infty)$.
The result of a cube root can be any real number, so the range is $(-\infty, \infty)$. (The "-3" does not affect the range.)

x	$f(x) = \sqrt[3]{x} - 3$
-8	$\sqrt[3]{-8} - 3 = -5$
-1	$\sqrt[3]{-1} - 3 = -4$
0	$\sqrt[3]{0} - 3 = -3$
1	$\sqrt[3]{1} - 3 = -2$
8	$\sqrt[3]{8} - 3 = -1$

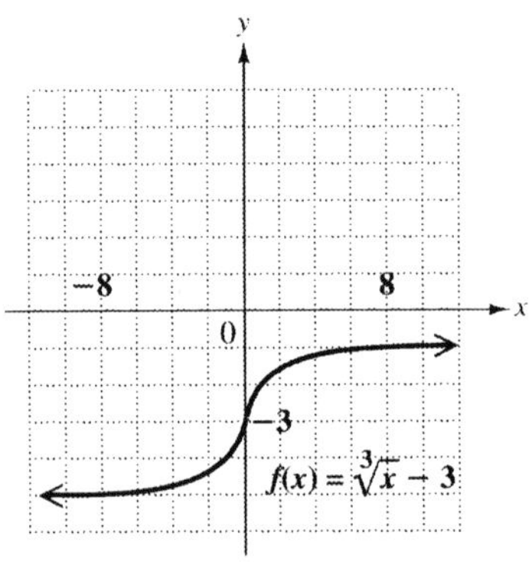

42. $f(x) = \sqrt[3]{x} + 1$

Since we can take the cube root of any real number, the domain is $(-\infty, \infty)$.
The result of a cube root can be any real number, so the range is $(-\infty, \infty)$. (The "$+1$" does not affect the range.)

x	$f(x) = \sqrt[3]{x} + 1$
-8	$\sqrt[3]{-8} + 1 = -1$
-1	$\sqrt[3]{-1} + 1 = 0$
0	$\sqrt[3]{0} + 1 = 1$
1	$\sqrt[3]{1} + 1 = 2$
8	$\sqrt[3]{8} + 1 = 3$

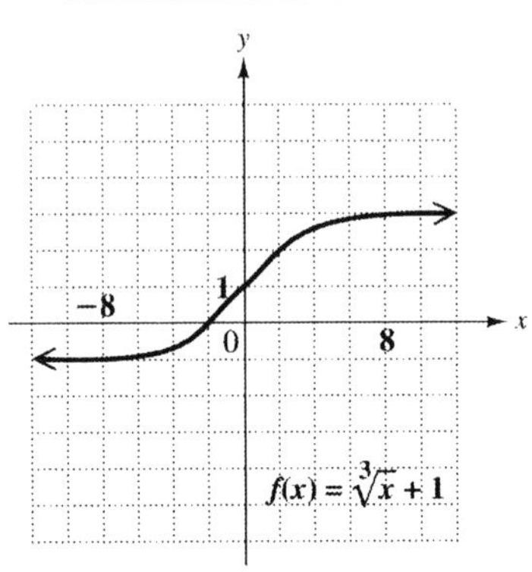

43. Since the index is even, $\sqrt{x^2} = |x|$.

44. Since the index is even, $\sqrt{x^2} = |x|$. Thus,
$-\sqrt{x^2} = -\left(\sqrt{x^2}\right) = -(|x|) = -|x|$.

45. Since the index is odd, $\sqrt[3]{x^3} = x$.

46. Since the index is odd, $\sqrt[3]{x^3} = x$. Thus,
$-\sqrt[3]{x^3} = -\left(\sqrt[3]{x^3}\right) = -(x) = -x$.

47. $\sqrt[6]{x^{30}} = \sqrt[6]{(x^5)^6} = |x^5|$ or $x^4|x|$

48. $\sqrt[4]{k^{20}} = \sqrt[4]{(k^5)^4} = |k^5|$ (4 is even)

49. $\sqrt{x^6} = \sqrt{(x^3)^2} = |x^3|$ or $x^2|x|$

50. $\sqrt[4]{x^{12}} = \sqrt[4]{(x^3)^4} = |x^3|$ or $x^2|x|$

51. $\sqrt{9483} \approx 97.381$

52. $\sqrt{6825} \approx 82.614$

53. $\sqrt{284.361} \approx 16.863$

54. $\sqrt{846.104} \approx 29.088$

55. $\sqrt[4]{19.4481} = 2.1$

56. $\sqrt[4]{39.0625} = 2.5$

57. $\sqrt[3]{3.375} = 1.5$

58. $\sqrt[3]{238.328} = 6.2$

59. $\sqrt{16} = 4$, because $4^2 = 16$. The principal root of 16 is 4, and the negative root of 16 is -4.

60. The principal square root is the positive root. The principal square root of $\sqrt{16}$ is 4.

61. $\sqrt{16} = 4$ and $-\sqrt{16} = -(4) = -4$.

62.
$$x^2 = 16$$
$$x^2 - 16 = 0$$
$$(x+4)(x-4) = 0$$
$$x + 4 = 0 \quad \text{or} \quad x - 4 = 0$$
$$x = -4 \quad \text{or} \quad x = 4$$

Solution set: $\{-4, 4\}$

63. $\pm\sqrt{16}$ means both the positive (principal) and negative square roots of 16, that is, $\sqrt{16} = 4$ and $-\sqrt{16} = -4$.

64. $\sqrt{x^2}$ represents the *nonnegative* square root. $\sqrt{x^2}$ is simplified as $|x|$ because the index is even and x could be a negative number. For example, if $x = -2$ then $\sqrt{x^2} = \sqrt{(-2)^2} = \sqrt{4} = 2 \neq x$. However, $|x| = |-2| = 2$.

65. $f = \dfrac{1}{2\pi\sqrt{LC}}$

$= \dfrac{1}{2\pi\sqrt{(7.237 \times 10^{-5})(2.5 \times 10^{-10})}}$

$\approx 1{,}183{,}235$

or about 1,183,000 cycles per second.

66. Use $h = 12.3\sqrt[3]{T}$ with $T = 216$ to find h in inches.

$$h = 12.3\sqrt[3]{216} = 12.3(6) = 73.8 \approx 74 \text{ inches}$$

In feet, $h = \frac{74}{12} \approx 6.17$ ft. To the nearest tenth of a foot this is 6.2 feet.

67. Since $H = 44 + 6 = 50$ ft, substitute 50 for H in the formula.

$$D = \sqrt{2H} = \sqrt{2 \cdot 50} = \sqrt{100} = 10$$

She will be able to see about 10 miles.

68. Use $t = 2\pi\sqrt{\dfrac{L}{g}}$ with $g = 32$ and $L = 2$.

$$t = 2\pi\sqrt{\tfrac{2}{32}} = 2\pi\sqrt{\tfrac{1}{16}} = 2\pi\left(\tfrac{1}{4}\right) = \tfrac{1}{2}\pi \approx 1.57$$

The time t is about 1.6 seconds.

69. Let $a = 850$, $b = 925$, and $c = 1300$.
First find the semiperimeter s.

$$\begin{aligned} s &= \tfrac{1}{2}(a + b + c) \\ &= \tfrac{1}{2}(850 + 925 + 1300) \\ &= \tfrac{3075}{2} = 1537.5 \end{aligned}$$

Now find the area A using Heron's formula.

$$\begin{aligned} A &= \sqrt{s(s-a)(s-b)(s-c)} \\ &= \sqrt{1537.5(687.5)(612.5)(237.5)} \\ &\approx 392{,}128.8 \end{aligned}$$

The area of the Bermuda Triangle is about 392,000 square miles.

70. Let $a = b = 246.75$ and $c = 438.14$.

$$\begin{aligned} s &= \tfrac{1}{2}(a + b + c) \\ &= \tfrac{1}{2}(246.75 + 246.75 + 438.14) \\ &= \tfrac{931.64}{2} = 465.82 \\ A &= \sqrt{s(s-a)(s-b)(s-c)} \\ &= \sqrt{465.82(219.07)^2(27.68)} \\ &\approx 24{,}875.7 \end{aligned}$$

The area of this enclosure is about 24,900 square feet.

71. $I = \sqrt{\dfrac{2P}{L}}$

$= \sqrt{\dfrac{2(120)}{80}} = \sqrt{3} \approx 1.732$ amps

72. $I = \sqrt{\dfrac{2P}{L}}$

$= \sqrt{\dfrac{2(100)}{40}} = \sqrt{5} \approx 2.236$ amps

9.2 Rational Exponents

9.2 Margin Exercises

1. **(a)** $8^{1/3} = \sqrt[3]{8} = 2$

(b) $9^{1/2} = \sqrt{9} = 3$

(c) $-81^{1/4} = -\sqrt[4]{81} = -3$

(d) $(-16)^{1/4} = \sqrt[4]{-16}$ is *not a real number* because the radicand, -16, is negative and the index, 4, is even.

(e) $64^{1/3} = \sqrt[3]{64} = 4$

(f) $\left(\dfrac{1}{32}\right)^{1/5} = \sqrt[5]{\dfrac{1}{32}} = \dfrac{1}{2}$

2. **(a)** $64^{2/3} = \left(64^{1/3}\right)^2 = 4^2 = 16$

(b) $100^{3/2} = \left(100^{1/2}\right)^3 = 10^3 = 1000$

(c) $-16^{3/4} = -\left(16^{3/4}\right) = -\left(16^{1/4}\right)^3 = -(2)^3 = -8$

(d) $(-16)^{3/4}$ is *not a real number,* since $(-16)^{1/4}$ is not a real number.

3. **(a)** $36^{-3/2} = \dfrac{1}{36^{3/2}} = \dfrac{1}{\left(\sqrt{36}\right)^3} = \dfrac{1}{6^3} = \dfrac{1}{216}$

(b) $32^{-4/5} = \dfrac{1}{32^{4/5}} = \dfrac{1}{\left(\sqrt[5]{32}\right)^4} = \dfrac{1}{2^4} = \dfrac{1}{16}$

(c) $\left(\dfrac{4}{9}\right)^{-5/2} = \left(\dfrac{9}{4}\right)^{5/2} = \left(\sqrt{\dfrac{9}{4}}\right)^5 = \left(\dfrac{3}{2}\right)^5 = \dfrac{243}{32}$

4. **(a)** $5^{2/3} = \left(\sqrt[3]{5}\right)^2$

(b) $4k^{3/5} = 4\left(k^{3/5}\right) = 4\left(\sqrt[5]{k}\right)^3$

(c) $(7r)^{4/3} = \left(\sqrt[3]{7r}\right)^4$

(d) $(m^3 + n^3)^{1/3} = \sqrt[3]{m^3 + n^3}$

5. (a) $\sqrt{y^{10}} = y^{10/2} = y^5$

(b) $\sqrt[3]{27y^9} = (27y^9)^{1/3} = 27^{1/3}y^{9/3} = 3y^3$

(c) $\sqrt[4]{t^4} = t^{4/4} = t^1 = t$ (t is positive)

6. (a) $11^{3/4} \cdot 11^{5/4} = 11^{3/4+5/4} = 11^{8/4}$
$= 11^2$ or 121

(b) $\dfrac{7^{3/4}}{7^{7/4}} = 7^{3/4-7/4} = 7^{-4/4} = 7^{-1} = \dfrac{1}{7}$

(c) $\dfrac{9^{2/3}(x^{1/3})^4}{9^{-1/3}} = 9^{2/3-(-1/3)}x^{4/3}$
$= 9^{2/3+1/3}x^{4/3}$
$= 9^1x^{4/3}$
$= 9x^{4/3}$

(d) $\left(\dfrac{a^3b^{-4}}{a^{-2}b^{1/5}}\right)^{-1/2} = \dfrac{(a^3)^{-1/2}(b^{-4})^{-1/2}}{(a^{-2})^{-1/2}(b^{1/5})^{-1/2}}$
$= \dfrac{a^{-3/2}b^2}{a^1b^{-1/10}}$
$= a^{-3/2-1}b^{2-(-1/10)}$
$= a^{-5/2}b^{21/10}$
$= \dfrac{b^{21/10}}{a^{5/2}}$

(e) $a^{2/3}(a^{7/3} + a^{1/3}) = a^{2/3} \cdot a^{7/3} + a^{2/3} \cdot a^{1/3}$
$= a^{2/3+7/3} + a^{2/3+1/3}$
$= a^{9/3} + a^{3/3}$
$= a^3 + a$

7. (a) $\sqrt[5]{m^3} \cdot \sqrt{m} = m^{3/5} \cdot m^{1/2}$
$= m^{3/5+1/2}$
$= m^{6/10+5/10}$
$= m^{11/10}$

(b) $\dfrac{\sqrt[3]{p^5}}{\sqrt{p^3}} = \dfrac{p^{5/3}}{p^{3/2}} = p^{5/3-3/2} = p^{10/6-9/6} = p^{1/6}$

(c) $\sqrt[4]{\sqrt[3]{x}} = \sqrt[4]{x^{1/3}} = (x^{1/3})^{1/4}$
$= x^{(1/3)(1/4)} = x^{1/12}$

9.2 Section Exercises

1. $2^{1/2} = \sqrt[2]{2^1} = \sqrt{2}$ **(C)**

2. $(-27)^{1/3} = \sqrt[3]{-27} = -3$ **(E)**

3. $-16^{1/2} = -\sqrt{16} = -(4) = -4$ **(A)**

4. $(-16)^{1/2} = \sqrt{-16}$ is not a real number **(J)**

5. $(-32)^{1/5} = \sqrt[5]{-32} = -2$ **(H)**

6. $(-32)^{2/5} = \left[(-32)^{1/5}\right]^2 = (-2)^2 = 4$ **(G)**

7. $4^{3/2} = (4^{1/2})^3 = 2^3 = 8$ **(B)**

8. $6^{2/4} = 6^{1/2} = \sqrt{6}$ **(F)**

9. $-6^{2/4} = -(6^{1/2}) = -\sqrt{6}$ **(D)**

10. $36^{.5} = 36^{1/2} = \sqrt{36} = 6$ **(I)**

11. $169^{1/2} = \sqrt{169} = 13$

12. $121^{1/2} = \sqrt{121} = 11$

13. $729^{1/3} = \sqrt[3]{729} = 9$

14. $512^{1/3} = \sqrt[3]{512} = 8$

15. $16^{1/4} = \sqrt[4]{16} = 2$

16. $625^{1/4} = \sqrt[4]{625} = 5$

17. $\left(\dfrac{64}{81}\right)^{1/2} = \sqrt{\dfrac{64}{81}} = \dfrac{8}{9}$

18. $\left(\dfrac{8}{27}\right)^{1/3} = \sqrt[3]{\dfrac{8}{27}} = \dfrac{2}{3}$

19. $(-27)^{1/3} = \sqrt[3]{-27} = -3$

20. $(-32)^{1/5} = \sqrt[5]{-32} = -2$

21. $100^{3/2} = (100^{1/2})^3 = 10^3 = 1000$

22. $64^{3/2} = (64^{1/2})^3 = 8^3 = 512$

23. $-16^{5/2} = -(16^{1/2})^5 = -(4)^5 = -1024$

24. $-32^{3/5} = -(32^{1/5})^3 = -(2)^3 = -8$

25. $(-144)^{1/2} = \sqrt{-144}$ is not a real number because no real number squared equals -144.

26. $(-36)^{1/2}$ is not a real number.

27. $64^{-3/2} = \dfrac{1}{64^{3/2}} = \dfrac{1}{(64^{1/2})^3} = \dfrac{1}{8^3} = \dfrac{1}{512}$

28. $81^{-3/2} = \dfrac{1}{81^{3/2}} = \dfrac{1}{(81^{1/2})^3} = \dfrac{1}{9^3} = \dfrac{1}{729}$

29. $\left(-\dfrac{8}{27}\right)^{-2/3} = \left(-\dfrac{27}{8}\right)^{2/3}$
$= \left(\sqrt[3]{-\dfrac{27}{8}}\right)^2$
$= \left(-\dfrac{3}{2}\right)^2 = \dfrac{9}{4}$

30. $\left(-\frac{64}{125}\right)^{-2/3} = \left(-\frac{125}{64}\right)^{2/3}$
$= \left(\sqrt[3]{-\frac{125}{64}}\right)^2$
$= \left(-\frac{5}{4}\right)^2 = \frac{25}{16}$

31. $(-64)^{1/2} = \sqrt{-64}$ is an even root of a negative number. No real number squared will give -64. On the other hand, $-64^{1/2} = -\sqrt{64} = -8$, which is a real number. ($-64^{1/2}$ is the opposite of $64^{1/2}$.)

32. Because both $a^{1/n}$ and $\sqrt[n]{a}$ are equal to a when raised to the nth power, they are defined to be equal to each other.

33. $12^{1/2} = \sqrt{12}$

34. $3^{1/2} = \sqrt{3}$

35. $8^{3/4} = (8^{1/4})^3 = (\sqrt[4]{8})^3$

36. $7^{2/3} = (7^{1/3})^2 = (\sqrt[3]{7})^2$

37. $(9q)^{5/8} - (2x)^{2/3}$
$= [(9q)^{1/8}]^5 - [(2x)^{1/3}]^2$
$= (\sqrt[8]{9q})^5 - (\sqrt[3]{2x})^2$

38. $(3p)^{3/4} + (4x)^{1/3}$
$= [(3p)^{1/4}]^3 + (4x)^{1/3}$
$= (\sqrt[4]{3p})^3 + \sqrt[3]{4x}$

39. $(2m)^{-3/2} = \frac{1}{(2m)^{3/2}}$
$= \frac{1}{[(2m)^{1/2}]^3}$
$= \frac{1}{(\sqrt{2m})^3}$

40. $(5y)^{-3/5} = \frac{1}{(5y)^{3/5}}$
$= \frac{1}{[(5y)^{1/5}]^3}$
$= \frac{1}{(\sqrt[5]{5y})^3}$

41. $(2y+x)^{2/3} = [(2y+x)^{1/3}]^2$
$= (\sqrt[3]{2y+x})^2$

42. $(r+2z)^{3/2} = [(r+2z)^{1/2}]^3$
$= (\sqrt{r+2z})^3$

43. $(3m^4 + 2k^2)^{-2/3} = \frac{1}{(3m^4+2k^2)^{2/3}}$
$= \frac{1}{\left[(3m^4+2k^2)^{1/3}\right]^2}$
$= \frac{1}{(\sqrt[3]{3m^4+2k^2})^2}$

44. $(5x^2 + 3z^3)^{-5/6} = \frac{1}{(5x^2+3z^3)^{5/6}}$
$= \frac{1}{\left[(5x^2+3z^3)^{1/6}\right]^5}$
$= \frac{1}{(\sqrt[6]{5x^2+3z^3})^5}$

45. We are to show that, in general,

$$\sqrt{a^2+b^2} \neq a+b.$$

When $a = 3$ and $b = 4$,

$$\sqrt{a^2+b^2} = \sqrt{3^2+4^2} = \sqrt{9+16} = \sqrt{25} = 5,$$

but

$$a + b = 3 + 4 = 7.$$

Since $5 \neq 7$, $\sqrt{a^2+b^2} \neq a+b$.

46. Although the expression may be true when $a = 1$ and $b = 0$, in Exercise 45 it was shown that

$$\sqrt{a^2+b^2} \neq a+b$$

when $a = 3$ and $b = 4$. In order for an expression to be true, it must be true for *all* values of a, b, and exponent n. Although the expression may be true for $a = 1$ and $b = 0$, this does not mean it is always true.

47. $\sqrt{2^{12}} = (2^{12})^{1/2} = 2^{12/2} = 2^6 = 64$

48. $\sqrt{5^{10}} = (5^{10})^{1/2} = 5^{10/2} = 5^5 = 3125$

49. $\sqrt[3]{4^9} = 4^{9/3} = 4^3 = 64$

50. $\sqrt[4]{6^8} = 6^{8/4} = 6^2 = 36$

51. $\sqrt{x^{20}} = x^{20/2} = x^{10}$

52. $\sqrt{r^{50}} = r^{50/2} = r^{25}$

53. $\sqrt[3]{x} \cdot \sqrt{x} = x^{1/3} \cdot x^{1/2} = x^{1/3+1/2} = x^{2/6+3/6}$
$= x^{5/6} = \sqrt[6]{x^5}$

54. $\sqrt[4]{y} \cdot \sqrt[5]{y^2} = y^{1/4} \cdot y^{2/5}$
$= y^{1/4+2/5} = y^{5/20+8/20}$
$= y^{13/20} = \sqrt[20]{y^{13}}$

55. $\frac{\sqrt[3]{t^4}}{\sqrt[5]{t^4}} = \frac{t^{4/3}}{t^{4/5}} = \frac{t^{20/15}}{t^{12/15}} = t^{20/15-12/15}$
$= t^{8/15} = \sqrt[15]{t^8}$

56. $\frac{\sqrt[4]{w^3}}{\sqrt[6]{w}} = \frac{w^{3/4}}{w^{1/6}} = \frac{w^{9/12}}{w^{2/12}} = w^{9/12-2/12}$
$= w^{7/12} = \sqrt[12]{w^7}$

57. $3^{1/2} \cdot 3^{3/2} = 3^{1/2+3/2} = 3^{4/2} = 3^2 = 9$

58. $6^{4/3} \cdot 6^{2/3} = 6^{4/3+2/3} = 6^{6/3} = 6^2 = 36$

59. $\frac{64^{5/3}}{64^{4/3}} = 64^{5/3-4/3} = 64^{1/3} = \sqrt[3]{64} = 4$

60. $\frac{125^{7/3}}{125^{5/3}} = 125^{7/3-5/3}$
$= 125^{2/3} = \left(125^{1/3}\right)^2$
$= 5^2 = 25$

61. $y^{7/3} \cdot y^{-4/3} = y^{7/3+(-4/3)} = y^{3/3} = y$

62. $r^{-8/9} \cdot r^{17/9} = r^{-8/9+17/9}$
$= r^{9/9} = r^1 = r$

63. $\frac{k^{1/3}}{k^{2/3} \cdot k^{-1}} = \frac{k^{1/3}}{k^{-1/3}} = k^{1/3-(-1/3)} = k^{2/3}$

64. $\frac{z^{3/4}}{z^{5/4} \cdot z^{-2}} = \frac{z^{3/4}}{z^{-3/4}} = z^{3/4-(-3/4)}$
$= z^{6/4} = z^{3/2}$

65. $\frac{\left(x^{1/4}y^{2/5}\right)^{20}}{x^2} = \frac{x^{(1/4) \cdot 20}y^{(2/5) \cdot 20}}{x^2}$
$= \frac{x^5y^8}{x^2}$
$= x^3y^8$

66. $\frac{\left(r^{1/5}s^{2/3}\right)^{15}}{r^2} = \frac{r^{(1/5) \cdot 15}s^{(2/3) \cdot 15}}{r^2}$
$= \frac{r^3s^{10}}{r^2}$
$= rs^{10}$

67. $\frac{\left(x^{2/3}\right)^2}{\left(x^2\right)^{7/3}} = \frac{x^{4/3}}{x^{14/3}} = x^{4/3-14/3}$
$= x^{-10/3} = \frac{1}{x^{10/3}}$

68. $\frac{\left(p^3\right)^{1/4}}{\left(p^{5/4}\right)^2} = \frac{p^{3/4}}{p^{10/4}} = p^{3/4-10/4}$
$= p^{-7/4} = \frac{1}{p^{7/4}}$

69. $\frac{m^{3/4}n^{-1/4}}{\left(m^2n\right)^{1/2}} = \frac{m^{3/4}n^{-1/4}}{m^1n^{1/2}}$
$= m^{3/4-1}n^{-1/4-1/2}$
$= m^{3/4-4/4}n^{-1/4-2/4}$
$= m^{-1/4}n^{-3/4}$
$= \frac{1}{m^{1/4}n^{3/4}}$

70. $\frac{\left(a^2b^5\right)^{-1/4}}{\left(a^{-3}b^2\right)^{1/6}} = \frac{a^{-1/2}b^{-5/4}}{a^{-1/2}b^{1/3}}$
$= a^{-1/2-(-1/2)}b^{-5/4-1/3}$
$= a^0b^{-15/12-4/12}$
$= 1 \cdot b^{-19/12}$
$= \frac{1}{b^{19/12}}$

71. $\frac{p^{1/5}p^{7/10}p^{1/2}}{\left(p^3\right)^{-1/5}} = \frac{p^{2/10+7/10+5/10}}{p^{-3/5}}$
$= \frac{p^{14/10}}{p^{-6/10}}$
$= p^{14/10-(-6/10)}$
$= p^{20/10} = p^2$

72. $\frac{z^{1/3}z^{-2/3}z^{1/6}}{\left(z^{-1/6}\right)^3} = \frac{z^{2/6-4/6+1/6}}{z^{-3/6}}$
$= \frac{z^{-1/6}}{z^{-3/6}}$
$= z^{-1/6-(-3/6)}$
$= z^{2/6} = z^{1/3}$

73. $\left(\frac{b^{-3/2}}{c^{-5/3}}\right)^2 \left(b^{-1/4}c^{-1/3}\right)^{-1}$
$= \left(\frac{c^{5/3}}{b^{3/2}}\right)^2 \left(b^{1/4}c^{1/3}\right)$
$= \frac{c^{10/3}}{b^3}\left(b^{1/4}c^{1/3}\right)$
$= \frac{c^{10/3}b^{1/4}c^{1/3}}{b^3}$
$= c^{10/3+1/3}b^{1/4-3}$
$= c^{11/3}b^{-11/4}$
$= \frac{c^{11/3}}{b^{11/4}}$

74. $\left(\dfrac{m^{-2/3}}{a^{-3/4}}\right)^4 \left(m^{-3/8}a^{1/4}\right)^{-2}$
$= \dfrac{m^{-8/3}}{a^{-3}} \cdot m^{3/4}a^{-1/2}$
$= m^{-8/3+3/4}a^{-1/2-(-3)}$
$= m^{-32/12+9/12}a^{-1/2+6/2}$
$= m^{-23/12}a^{5/2}$
$= \dfrac{a^{5/2}}{m^{23/12}}$

75. $\left(\dfrac{p^{-1/4}q^{-3/2}}{3^{-1}p^{-2}q^{-2/3}}\right)^{-2}$
$= \dfrac{p^{1/2}q^3}{3^2p^4q^{4/3}}$
$= \dfrac{p^{1/2-4}q^{3-4/3}}{9}$
$= \dfrac{p^{-7/2}q^{5/3}}{9}$
$= \dfrac{q^{5/3}}{9p^{7/2}}$

76. $\left(\dfrac{2^{-2}w^{-3/4}x^{-5/8}}{w^{3/4}x^{-1/2}}\right)^{-3}$
$= \dfrac{2^6w^{9/4}x^{15/8}}{w^{-9/4}x^{3/2}}$
$= 64w^{9/4-(-9/4)}x^{15/8-3/2}$
$= 64w^{9/2}x^{3/8}$

77. $p^{2/3}\left(p^{1/3}+2p^{4/3}\right)$
$= p^{2/3}p^{1/3}+p^{2/3}\left(2p^{4/3}\right)$
$= p^{2/3+1/3}+2p^{2/3+4/3}$
$= p^{3/3}+2p^{6/3}$
$= p^1+2p^2$
$= p+2p^2$

78. $z^{5/8}\left(3z^{5/8}+5z^{11/8}\right)$
$= z^{5/8}\left(3z^{5/8}\right)+z^{5/8}\left(5z^{11/8}\right)$
$= 3z^{5/8+5/8}+5z^{5/8+11/8}$
$= 3z^{10/8}+5z^{16/8}$
$= 3z^{5/4}+5z^2$

79. $k^{1/4}\left(k^{3/2}-k^{1/2}\right)$
$= k^{1/4+3/2}-k^{1/4+1/2}$
$= k^{1/4+6/4}-k^{1/4+2/4}$
$= k^{7/4}-k^{3/4}$

80. $r^{3/5}\left(r^{1/2}+r^{3/4}\right)$
$= r^{3/5+1/2}+r^{3/5+3/4}$
$= r^{11/10}+r^{27/20}$

81. $6a^{7/4}\left(a^{-7/4}+3a^{-3/4}\right)$
$= 6a^{7/4+(-7/4)}+18a^{7/4+(-3/4)}$
$= 6a^0+18a^{4/4}$
$= 6(1)+18a^1 = 6+18a$

82. $4m^{5/3}\left(m^{-2/3}-4m^{-5/3}\right)$
$= 4m^{5/3+(-2/3)}-4\cdot 4m^{5/3+(-5/3)}$
$= 4m^{3/3}-16m^0$
$= 4m^1-16(1) = 4m-16$

83. $5m^{-2/3}\left(m^{2/3}+m^{-7/3}\right)$
$= 5m^{-2/3}\cdot m^{2/3}+5m^{-2/3}\cdot m^{-7/3}$
$= 5m^{-2/3+2/3}+5m^{-2/3+(-7/3)}$
$= 5m^0+5m^{-9/3}$
$= 5\cdot 1+5m^{-3}$
$= 5+\dfrac{5}{m^3}$

84. $\sqrt[5]{x^3}\cdot\sqrt[4]{x} = x^{3/5}\cdot x^{1/4}$
$= x^{3/5+1/4}$
$= x^{12/20+5/20}$
$= x^{17/20}$

85. $\sqrt[6]{y^5}\cdot\sqrt[3]{y^2} = y^{5/6}\cdot y^{2/3}$
$= y^{5/6+2/3}$
$= y^{5/6+4/6}$
$= y^{9/6} = y^{3/2}$

86. $\dfrac{\sqrt{x^5}}{\sqrt{x^8}} = \dfrac{\left(x^5\right)^{1/2}}{\left(x^8\right)^{1/2}}$
$= \dfrac{x^{5/2}}{x^{8/2}}$
$= x^{5/2-8/2}$
$= x^{-3/2} = \dfrac{1}{x^{3/2}}$

87. $\dfrac{\sqrt[3]{k^5}}{\sqrt[3]{k^7}} = \dfrac{\left(k^5\right)^{1/3}}{\left(k^7\right)^{1/3}}$
$= \dfrac{k^{5(1/3)}}{k^{7(1/3)}}$
$= \dfrac{k^{5/3}}{k^{7/3}}$
$= k^{5/3-7/3}$
$= k^{-2/3} = \dfrac{1}{k^{2/3}}$

88. $\sqrt{y}\cdot\sqrt[3]{yz} = y^{1/2}\cdot(yz)^{1/3}$
$= y^{1/2}y^{1/3}z^{1/3}$
$= y^{1/2+1/3}z^{1/3}$
$= y^{3/6+2/6}z^{1/3}$
$= y^{5/6}z^{1/3}$

89. $\sqrt[3]{xz} \cdot \sqrt{z} = (xz)^{1/3} \cdot z^{1/2}$
$= x^{1/3} z^{1/3} z^{1/2}$
$= x^{1/3} z^{1/3+1/2}$
$= x^{1/3} z^{5/6}$

90. $\sqrt[4]{\sqrt[3]{m}} = \sqrt[4]{m^{1/3}} = \left(m^{1/3}\right)^{1/4} = m^{1/12}$

91. $\sqrt[3]{\sqrt{k}} = \sqrt[3]{k^{1/2}} = \left(k^{1/2}\right)^{1/3} = k^{1/6}$

92. $\sqrt{\sqrt[3]{\sqrt[4]{x}}} = \sqrt{\sqrt[3]{x^{1/4}}}$
$= \sqrt{\left(x^{1/4}\right)^{1/3}}$
$= \left(x^{1/12}\right)^{1/2}$
$= x^{1/24}$

93. $\sqrt[3]{\sqrt[5]{\sqrt{y}}} = \sqrt[3]{\sqrt[5]{y^{1/2}}}$
$= \sqrt[3]{\left(y^{1/2}\right)^{1/5}}$
$= \left(y^{1/10}\right)^{1/3}$
$= y^{1/30}$

94. $\sqrt{y^{5/4}} = \left(y^{5/4}\right)^{1/2} = y^{(5/4)(1/2)} = y^{5/8}$

95. $\sqrt[3]{x^{5/9}} = \left(x^{5/9}\right)^{1/3} = x^{(5/9)(1/3)} = x^{5/27}$

96. Use $T(D) = .07D^{3/2}$ with $D = 16$.

$T(16) = .07(16)^{3/2} = .07\left(16^{1/2}\right)^3$
$= .07(4)^3 = .07(64) = 4.48$

To the nearest tenth of an hour, time T is 4.5 hours.

97. Use $h(T) = (1860.867T)^{1/3}$ with $T = 200$.

$h(200) = [1860.867(200)]^{1/3}$
$\approx 71.93 \approx 72$ inches

The height is 6.0 feet (to the nearest tenth of a foot).

98. $3x^{-1/2} - 4x^{1/2} = x^{-1/2}\left(3 - 4x^{(1/2)-(-1/2)}\right)$
$= x^{-1/2}\left(3 - 4x^1\right)$
$= x^{-1/2}(3 - 4x)$

Note that we are *subtracting* exponents.

99. $m^3 - 3m^{5/2} = m^{5/2}\left(m^{3-(5/2)} - 3\right)$
$= m^{5/2}\left(m^{1/2} - 3\right)$

100. $4t^{-1/2} + 7t^{3/2} = t^{-1/2}\left(4 + 7t^{(3/2)-(-1/2)}\right)$
$= t^{-1/2}\left(4 + 7t^2\right)$

101. $8x^{2/3} + 5x^{-1/3} = x^{-1/3}\left(8x^{(2/3)-(-1/3)} + 5\right)$
$= x^{-1/3}\left(8x^1 + 5\right)$
$= x^{-1/3}(8x + 5)$

102. $4p - p^{3/4} = p^{3/4}\left(4p^{1-(3/4)} - 1\right)$
$= p^{3/4}\left(4p^{1/4} - 1\right)$

103. $2m^{1/8} - m^{5/8} = m^{1/8}\left(2 - m^{(5/8)-(1/8)}\right)$
$= m^{1/8}\left(2 - m^{4/8}\right)$
$= m^{1/8}\left(2 - m^{1/2}\right)$

9.3 Simplifying Radical Expressions

9.3 Margin Exercises

1. **(a)** $\sqrt{5} \cdot \sqrt{13} = \sqrt{5 \cdot 13} = \sqrt{65}$

(b) $\sqrt{10y} \cdot \sqrt{3k} = \sqrt{10y \cdot 3k} = \sqrt{30yk}$

(c) $\sqrt{\frac{5}{a}} \cdot \sqrt{\frac{11}{z}} = \sqrt{\frac{5}{a} \cdot \frac{11}{z}} = \sqrt{\frac{55}{az}}$

2. **(a)** $\sqrt[3]{2} \cdot \sqrt[3]{7} = \sqrt[3]{2 \cdot 7} = \sqrt[3]{14}$

(b) $\sqrt[6]{8r^2} \cdot \sqrt[6]{2r^3} = \sqrt[6]{8r^2 \cdot 2r^3} = \sqrt[6]{16r^5}$

(c) $\sqrt[5]{9y^2x} \cdot \sqrt[5]{8xy^2} = \sqrt[5]{9y^2x \cdot 8xy^2} = \sqrt[5]{72y^4x^2}$

(d) $\sqrt{7} \cdot \sqrt[3]{5}$ cannot be simplified using the product rule for radicals, because the indexes (2 and 3) are different.

3. **(a)** $\sqrt{\frac{100}{81}} = \frac{\sqrt{100}}{\sqrt{81}} = \frac{10}{9}$

(b) $\sqrt{\frac{11}{25}} = \frac{\sqrt{11}}{\sqrt{25}} = \frac{\sqrt{11}}{5}$

(c) $\sqrt[3]{\frac{18}{125}} = \frac{\sqrt[3]{18}}{\sqrt[3]{125}} = \frac{\sqrt[3]{18}}{5}$

(d) $\sqrt{\frac{y^8}{16}} = \frac{\sqrt{y^8}}{\sqrt{16}} = \frac{y^4}{4}$

(e) $\sqrt[3]{\frac{x^2}{r^{12}}} = \frac{\sqrt[3]{x^2}}{\sqrt[3]{r^{12}}} = \frac{\sqrt[3]{x^2}}{r^4}$

4. **(a)** $\sqrt{32} = \sqrt{16 \cdot 2} = \sqrt{16} \cdot \sqrt{2} = 4\sqrt{2}$

(b) $\sqrt{45} = \sqrt{9 \cdot 5} = \sqrt{9} \cdot \sqrt{5} = 3\sqrt{5}$

(c) $\sqrt{300} = \sqrt{100 \cdot 3} = \sqrt{100} \cdot \sqrt{3} = 10\sqrt{3}$

(d) $\sqrt{35}$

No perfect square (other than 1) divides into 35, so $\sqrt{35}$ *cannot be simplified further.*

(e) $-\sqrt[3]{54} = -\sqrt[3]{27 \cdot 2} = -\sqrt[3]{27} \cdot \sqrt[3]{2} = -3\sqrt[3]{2}$

(f) $\sqrt[4]{243} = \sqrt[4]{81 \cdot 3} = \sqrt[4]{81} \cdot \sqrt[4]{3} = 3\sqrt[4]{3}$

5. **(a)** $\sqrt{25p^7} = \sqrt{25p^6 \cdot p}$
$= \sqrt{25p^6} \cdot \sqrt{p} = 5p^3\sqrt{p}$

(b) $\sqrt{72y^3x} = \sqrt{36y^2 \cdot 2yx}$
$= \sqrt{36y^2} \cdot \sqrt{2yx}$
$= 6y\sqrt{2yx}$

(c) $\sqrt[3]{y^7x^5z^6} = \sqrt[3]{y^6x^3z^6 \cdot yx^2}$
$= \sqrt[3]{y^6x^3z^6} \cdot \sqrt[3]{yx^2}$
$= y^2xz^2\sqrt[3]{yx^2}$

(d) $-\sqrt[4]{32a^5b^7} = -\sqrt[4]{16a^4b^4 \cdot 2ab^3}$
$= -\sqrt[4]{16a^4b^4} \cdot \sqrt[4]{2ab^3}$
$= -2ab\sqrt[4]{2ab^3}$

6. **(a)** $\sqrt[12]{2^3} = 2^{3/12} = 2^{1/4} = \sqrt[4]{2}$

(b) $\sqrt[6]{t^2} = t^{2/6} = t^{1/3} = \sqrt[3]{t}$

7. $\sqrt{5} \cdot \sqrt[3]{4}$

The least common index of 2 and 3 is 6. Write each radical as a sixth root.

$$\sqrt{5} = 5^{1/2} = 5^{3/6} = \sqrt[6]{5^3} = \sqrt[6]{125}$$
$$\sqrt[3]{4} = 4^{1/3} = 4^{2/6} = \sqrt[6]{4^2} = \sqrt[6]{16}$$

Therefore,

$$\sqrt{5} \cdot \sqrt[3]{4} = \sqrt[6]{125} \cdot \sqrt[6]{16} = \sqrt[6]{2000}.$$

8. **(a)** Substitute 14 for a and 8 for b in the Pythagorean formula to find c.

$$c^2 = a^2 + b^2$$
$$c^2 = 14^2 + 8^2$$
$$c^2 = 260$$
$$c = \sqrt{260}$$
$$c = \sqrt{4 \cdot 65}$$
$$c = \sqrt{4} \cdot \sqrt{65}$$
$$c = 2\sqrt{65}$$

The length of the hypotenuse is $2\sqrt{65}$.

(b) Substitute 4 for a and 6 for c to find b.

$$a^2 + b^2 = c^2$$
$$b^2 = c^2 - a^2$$
$$b^2 = 6^2 - 4^2$$
$$b^2 = 20$$
$$b = \sqrt{20}$$
$$b = \sqrt{4 \cdot 5}$$
$$b = \sqrt{4} \cdot \sqrt{5}$$
$$b = 2\sqrt{5}$$

The length of the unknown side is $2\sqrt{5}$.

9. Use the distance formula,

$$d = \sqrt{(x_2 - x_1)^2 + (y_2 - y_1)^2}.$$

(a) $(x_1, y_1) = (2, -1)$ and $(x_2, y_2) = (5, 3)$

$$d = \sqrt{(5-2)^2 + [3-(-1)]^2}$$
$$= \sqrt{3^2 + 4^2}$$
$$= \sqrt{9 + 16}$$
$$= \sqrt{25}$$
$$= 5$$

(b) $(x_1, y_1) = (-3, 2)$ and $(x_2, y_2) = (0, -4)$

$$d = \sqrt{[0-(-3)]^2 + (-4-2)^2}$$
$$= \sqrt{3^2 + (-6)^2}$$
$$= \sqrt{9 + 36}$$
$$= \sqrt{45}$$
$$= \sqrt{9} \cdot \sqrt{5}$$
$$= 3\sqrt{5}$$

9.3 Section Exercises

1. Does $2\sqrt{12} = \sqrt{48}$?

$2\sqrt{12} = 2\sqrt{4 \cdot 3}$
$= 2\sqrt{4} \cdot \sqrt{3} = 2 \cdot 2 \cdot \sqrt{3} = 4\sqrt{3}$
$\sqrt{48} = \sqrt{16 \cdot 3} = \sqrt{16} \cdot \sqrt{3} = 4\sqrt{3}$

The calculator approximation for each expression is 6.92820323. The statement is true.

2. Does $\sqrt{72} = 2\sqrt{18}$?

$\sqrt{72} = \sqrt{36 \cdot 2} = \sqrt{36}\sqrt{2} = 6\sqrt{2}$
$2\sqrt{18} = 2\sqrt{9 \cdot 2} = 2\sqrt{9}\sqrt{2} = 2 \cdot 3\sqrt{2} = 6\sqrt{2}$

The calculator approximation for each expression is 8.485281374. The statement is true.

3. Does $3\sqrt{8} = 2\sqrt{18}$?

$3\sqrt{8} = 3\sqrt{4 \cdot 2} = 3\sqrt{4}\sqrt{2} = 3 \cdot 2\sqrt{2} = 6\sqrt{2}$
$2\sqrt{18} = 2\sqrt{9 \cdot 2} = 2\sqrt{9}\sqrt{2} = 2 \cdot 3\sqrt{2} = 6\sqrt{2}$

The calculator approximation for each expression is 8.485281374. The statement is true.

4. Does $5\sqrt{72} = 6\sqrt{50}$?

$5\sqrt{72} = 5\sqrt{36 \cdot 2} = 5\sqrt{36}\sqrt{2}$
$= 5 \cdot 6\sqrt{2} = 30\sqrt{2}$
$6\sqrt{50} = 6\sqrt{25 \cdot 2} = 6\sqrt{25}\sqrt{2}$
$= 6 \cdot 5\sqrt{2} = 30\sqrt{2}$

The calculator approximation for each expression is 42.42640687. The statement is true.

5. $\sqrt[3]{x} \cdot \sqrt[3]{x}$ is not equal to x because there are only two factors of $\sqrt[3]{x}$, and we would need three factors of $\sqrt[3]{x}$ to equal x.

$$\sqrt[3]{x} \cdot \sqrt[3]{x} = \left(\sqrt[3]{x}\right)^2 \quad \text{or} \quad \sqrt[3]{x^2}$$

6. Because there are just two factors, $\sqrt[4]{x} \cdot \sqrt[4]{x} = \left(\sqrt[4]{x}\right)^2 = x^{2/4} = x^{1/2} = \sqrt{x}$, for $x \geq 0$.

7. **A.** $.5 = \frac{1}{2}$ so $\sqrt{.5} = \sqrt{\frac{1}{2}}$

B. $\frac{2}{4} = \frac{1}{2}$ so $\sqrt{\frac{2}{4}} = \sqrt{\frac{1}{2}}$

C. $\frac{3}{6} = \frac{1}{2}$ so $\sqrt{\frac{3}{6}} = \sqrt{\frac{1}{2}}$

D. $\frac{\sqrt{4}}{\sqrt{16}} = \sqrt{\frac{4}{16}} = \sqrt{\frac{1}{4}} \neq \sqrt{\frac{1}{2}}$

Choice **D** is not equal to $\sqrt{\frac{1}{2}}$.

8. $\pi \approx 3.14159265$

$\sqrt[4]{\frac{2143}{22}} \approx 3.14159265$

The radical is not actually equal to π. They are only approximations that agree in the first few decimal places.

9. $\sqrt{5} \cdot \sqrt{6} = \sqrt{5 \cdot 6} = \sqrt{30}$

10. $\sqrt{10} \cdot \sqrt{3} = \sqrt{10 \cdot 3} = \sqrt{30}$

11. $\sqrt{14} \cdot \sqrt{x} = \sqrt{14 \cdot x} = \sqrt{14x}$

12. $\sqrt{23} \cdot \sqrt{t} = \sqrt{23 \cdot t} = \sqrt{23t}$

13. $\sqrt{14} \cdot \sqrt{3pqr} = \sqrt{14 \cdot 3pqr} = \sqrt{42pqr}$

14. $\sqrt{7} \cdot \sqrt{5xt} = \sqrt{7 \cdot 5xt} = \sqrt{35xt}$

15. $\sqrt[3]{7x} \cdot \sqrt[3]{2y} = \sqrt[3]{7x \cdot 2y} = \sqrt[3]{14xy}$

16. $\sqrt[3]{9x} \cdot \sqrt[3]{4y} = \sqrt[3]{9x \cdot 4y} = \sqrt[3]{36xy}$

17. $\sqrt[4]{11} \cdot \sqrt[4]{3} = \sqrt[4]{11 \cdot 3} = \sqrt[4]{33}$

18. $\sqrt[4]{6} \cdot \sqrt[4]{9} = \sqrt[4]{6 \cdot 9} = \sqrt[4]{54}$

19. $\sqrt[4]{2x} \cdot \sqrt[4]{3y^2} = \sqrt[4]{2x \cdot 3y^2} = \sqrt[4]{6xy^2}$

20. $\sqrt[4]{3y^2} \cdot \sqrt[4]{6yz} = \sqrt[4]{3y^2 \cdot 6yz} = \sqrt[4]{18y^3z}$

21. $\sqrt[3]{7} \cdot \sqrt[4]{3}$ cannot be multiplied using the product rule, because the indexes (3 and 4) are different.

22. $\sqrt[5]{8} \cdot \sqrt[6]{12}$ cannot be multiplied using the product rule, because the indexes (5 and 6) are different.

23. $\sqrt{\frac{64}{121}} = \frac{\sqrt{64}}{\sqrt{121}} = \frac{8}{11}$

24. $\sqrt{\frac{16}{49}} = \frac{\sqrt{16}}{\sqrt{49}} = \frac{4}{7}$

25. $\sqrt{\frac{3}{25}} = \frac{\sqrt{3}}{\sqrt{25}} = \frac{\sqrt{3}}{5}$

26. $\sqrt{\frac{13}{49}} = \frac{\sqrt{13}}{\sqrt{49}} = \frac{\sqrt{13}}{7}$

27. $\sqrt{\frac{x}{25}} = \frac{\sqrt{x}}{\sqrt{25}} = \frac{\sqrt{x}}{5}$

28. $\sqrt{\frac{k}{100}} = \frac{\sqrt{k}}{\sqrt{100}} = \frac{\sqrt{k}}{10}$

29. $\sqrt{\frac{p^6}{81}} = \frac{\sqrt{p^6}}{\sqrt{81}} = \frac{\sqrt{(p^3)^2}}{9} = \frac{p^3}{9}$

30. $\sqrt{\frac{w^{10}}{36}} = \frac{\sqrt{w^{10}}}{\sqrt{36}} = \frac{\sqrt{(w^5)^2}}{6} = \frac{w^5}{6}$

31. $\sqrt[3]{\frac{27}{64}} = \frac{\sqrt[3]{27}}{\sqrt[3]{64}} = \frac{3}{4}$

32. $\sqrt[3]{\frac{216}{125}} = \frac{\sqrt[3]{216}}{\sqrt[3]{125}} = \frac{6}{5}$

33. $\sqrt[3]{-\frac{r^2}{8}} = \frac{\sqrt[3]{r^2}}{\sqrt[3]{-8}} = \frac{\sqrt[3]{r^2}}{-2} = -\frac{\sqrt[3]{r^2}}{2}$

34. $\sqrt[3]{-\frac{t}{125}} = \frac{\sqrt[3]{t}}{\sqrt[3]{-125}} = \frac{\sqrt[3]{t}}{-5} = -\frac{\sqrt[3]{t}}{5}$

35. $-\sqrt[4]{\frac{81}{x^4}} = -\frac{\sqrt[4]{3^4}}{\sqrt[4]{x^4}} = -\frac{3}{x}$

36. $-\sqrt[4]{\frac{625}{y^4}} = -\frac{\sqrt[4]{5^4}}{\sqrt[4]{y^4}} = -\frac{5}{y}$

37. $\sqrt[5]{\frac{1}{x^{15}}} = \frac{\sqrt[5]{1}}{\sqrt[5]{(x^3)^5}} = \frac{1}{x^3}$

38. $\sqrt[5]{\frac{32}{y^{20}}} = \frac{\sqrt[5]{32}}{\sqrt[5]{(y^4)^5}} = \frac{2}{y^4}$

39. $\sqrt{12} = \sqrt{4 \cdot 3} = \sqrt{4} \cdot \sqrt{3} = 2\sqrt{3}$

40. $\sqrt{18} = \sqrt{9 \cdot 2} = \sqrt{9} \cdot \sqrt{2} = 3\sqrt{2}$

41. $\sqrt{288} = \sqrt{144 \cdot 2} = \sqrt{144} \cdot \sqrt{2} = 12\sqrt{2}$

42. $\sqrt{72} = \sqrt{36 \cdot 2} = \sqrt{36} \cdot \sqrt{2} = 6\sqrt{2}$

43. $-\sqrt{32} = -\sqrt{16 \cdot 2} = -\sqrt{16} \cdot \sqrt{2} = -4\sqrt{2}$

44. $-\sqrt{48} = -\sqrt{16 \cdot 3} = -\sqrt{16} \cdot \sqrt{3} = -4\sqrt{3}$

45. $-\sqrt{28} = -\sqrt{4 \cdot 7} = -\sqrt{4} \cdot \sqrt{7} = -2\sqrt{7}$

46. $-\sqrt{24} = -\sqrt{4 \cdot 6} = -\sqrt{4} \cdot \sqrt{6} = -2\sqrt{6}$

47. $\sqrt{-300}$ is not a real number (even index, negative radicand).

48. $\sqrt{-150}$ is not a real number (even index, negative radicand).

49. $\sqrt[3]{128} = \sqrt[3]{64 \cdot 2} = \sqrt[3]{64} \cdot \sqrt[3]{2} = 4\sqrt[3]{2}$

50. $\sqrt[3]{24} = \sqrt[3]{8 \cdot 3} = \sqrt[3]{8} \cdot \sqrt[3]{3} = 2\sqrt[3]{3}$

51. $\sqrt[3]{-16} = \sqrt[3]{-8 \cdot 2} = \sqrt[3]{-8} \cdot \sqrt[3]{2} = -2\sqrt[3]{2}$

52. $\sqrt[3]{-250} = \sqrt[3]{-125 \cdot 2} = \sqrt[3]{-125} \cdot \sqrt[3]{2} = -5\sqrt[3]{2}$

53. $\sqrt[3]{40} = \sqrt[3]{8 \cdot 5} = \sqrt[3]{8} \cdot \sqrt[3]{5} = 2\sqrt[3]{5}$

54. $\sqrt[3]{375} = \sqrt[3]{125 \cdot 3} = \sqrt[3]{125} \cdot \sqrt[3]{3} = 5\sqrt[3]{3}$

55. $-\sqrt[4]{512} = -\sqrt[4]{256 \cdot 2} = -\sqrt[4]{4^4} \cdot \sqrt[4]{2} = -4\sqrt[4]{2}$

56. $-\sqrt[4]{1250} = -\sqrt[4]{625 \cdot 2} = -\sqrt[4]{5^4} \cdot \sqrt[4]{2} = -5\sqrt[4]{2}$

57. $\sqrt[5]{64} = \sqrt[5]{32 \cdot 2} = \sqrt[5]{2^5} \cdot \sqrt[5]{2} = 2\sqrt[5]{2}$

58. $\sqrt[5]{128} = \sqrt[5]{32 \cdot 4} = \sqrt[5]{2^5} \cdot \sqrt[5]{4} = 2\sqrt[5]{4}$

59. His reasoning was incorrect. The radicand 14 must be written as a product of two factors (not a sum of two terms) where one of the two factors is a perfect cube.

60. $\sqrt[3]{k^4}$ is not a simplified radical because the radicand k^4 has a factor, k, raised to a power, 4, that is greater than the index, 3.

$$\sqrt[3]{k^4} = \sqrt[3]{k^3 \cdot k} = \sqrt[3]{k^3} \cdot \sqrt[3]{k} = k\sqrt[3]{k}$$

61. $\sqrt{72k^2} = \sqrt{36k^2 \cdot 2} = \sqrt{36k^2} \cdot \sqrt{2} = 6k\sqrt{2}$

62. $\sqrt{18m^2} = \sqrt{9m^2 \cdot 2} = \sqrt{9m^2} \cdot \sqrt{2} = 3m\sqrt{2}$

63. $\sqrt[3]{\dfrac{81}{64}} = \dfrac{\sqrt[3]{81}}{\sqrt[3]{64}} = \dfrac{\sqrt[3]{27 \cdot 3}}{4} = \dfrac{\sqrt[3]{27} \cdot \sqrt[3]{3}}{4} = \dfrac{3\sqrt[3]{3}}{4}$

64. $\sqrt[3]{\dfrac{32}{216}} = \dfrac{\sqrt[3]{2^5}}{\sqrt[3]{6^3}} = \dfrac{\sqrt[3]{2^3 \cdot 2^2}}{\sqrt[3]{6^3}} = \dfrac{2\sqrt[3]{4}}{6} = \dfrac{\sqrt[3]{4}}{3}$

65. $\sqrt{121x^6} = \sqrt{(11x^3)^2} = 11x^3$

66. $\sqrt{256z^{12}} = \sqrt{(16z^6)^2} = 16z^6$

67. $-\sqrt[3]{27t^{12}} = -\sqrt[3]{(3t^4)^3} = -3t^4$

68. $-\sqrt[3]{64y^{18}} = -\sqrt[3]{(4y^6)^3} = -4y^6$

69. $-\sqrt{100m^8z^4} = -\sqrt{(10m^4z^2)^2} = -10m^4z^2$

70. $-\sqrt{25t^6s^{20}} = -\sqrt{(5t^3s^{10})^2} = -5t^3s^{10}$

71. $-\sqrt[3]{-125a^6b^9c^{12}} = -\sqrt[3]{(-5a^2b^3c^4)^3}$
$= -(-5a^2b^3c^4)$
$= 5a^2b^3c^4$

72. $-\sqrt[3]{-216y^{15}x^6z^3} = -\sqrt[3]{(-6y^5x^2z)^3}$
$= -(-6y^5x^2z)$
$= 6y^5x^2z$

73. $\sqrt[4]{\dfrac{1}{16}r^8t^{20}} = \sqrt[4]{\left(\dfrac{1}{2}r^2t^5\right)^4} = \dfrac{1}{2}r^2t^5$

74. $\sqrt[4]{\dfrac{81}{256}t^{12}u^8} = \sqrt[4]{\left(\dfrac{3}{4}t^3u^2\right)^4} = \dfrac{3}{4}t^3u^2$

75. $\sqrt{50x^3} = \sqrt{25x^2 \cdot 2x} = \sqrt{(5x)^2} \cdot \sqrt{2x}$
$= 5x\sqrt{2x}$

76. $\sqrt{300z^3} = \sqrt{100z^2 \cdot 3z}$
$= \sqrt{(10z)^2} \cdot \sqrt{3z} = 10z\sqrt{3z}$

77. $-\sqrt{500r^{11}} = -\sqrt{100r^{10} \cdot 5r}$
$= -\sqrt{(10r^5)^2} \cdot \sqrt{5r} = -10r^5\sqrt{5r}$

78. $-\sqrt{200p^{13}} = -\sqrt{100p^{12} \cdot 2p}$
$= -\sqrt{(10p^6)^2 \cdot 2p} = -10p^6\sqrt{2p}$

79. $\sqrt{13x^7y^8} = \sqrt{(x^6y^8)(13x)}$
$= \sqrt{x^6y^8} \cdot \sqrt{13x}$
$= x^3y^4\sqrt{13x}$

80. $\sqrt{23k^9p^{14}} = \sqrt{(k^8p^{14})(23k)}$
$= k^4p^7\sqrt{23k}$

81. $\sqrt[3]{8z^6w^9} = \sqrt[3]{(2z^2w^3)^3} = 2z^2w^3$

82. $\sqrt[3]{64a^{15}b^{12}} = \sqrt[3]{(4a^5b^4)^3} = 4a^5b^4$

83. $\sqrt[3]{-16z^5t^7} = \sqrt[3]{(-2^3z^3t^6)(2z^2t)}$
$= -2zt^2\sqrt[3]{2z^2t}$

84. $\sqrt[3]{-81m^4n^{10}} = \sqrt[3]{(-3^3m^3n^9)(3mn)}$
$= -3mn^3\sqrt[3]{3mn}$

85. $\sqrt[4]{81x^{12}y^{16}} = \sqrt[4]{(3x^3y^4)^4} = 3x^3y^4$

86. $\sqrt[4]{81t^8u^{28}} = \sqrt[4]{(3t^2u^7)^4} = 3t^2u^7$

87. $-\sqrt[4]{162r^{15}s^{10}} = -\sqrt[4]{81r^{12}s^8(2r^3s^2)}$
$= -\sqrt[4]{81r^{12}s^8}\cdot\sqrt[4]{2r^3s^2}$
$= -3r^3s^2\sqrt[4]{2r^3s^2}$

88. $-\sqrt[4]{32k^5m^{10}} = -\sqrt[4]{16k^4m^8(2km^2)}$
$= -2km^2\sqrt[4]{2km^2}$

89. $\sqrt{\frac{y^{11}}{36}} = \frac{\sqrt{y^{11}}}{\sqrt{36}} = \frac{\sqrt{y^{10}\cdot y}}{6} = \frac{y^5\sqrt{y}}{6}$

90. $\sqrt{\frac{v^{13}}{49}} = \frac{\sqrt{v^{12}\cdot v}}{\sqrt{49}} = \frac{v^6\sqrt{v}}{7}$

91. $\sqrt[3]{\frac{x^{16}}{27}} = \frac{\sqrt[3]{x^{15}\cdot x}}{\sqrt[3]{27}} = \frac{x^5\sqrt[3]{x}}{3}$

92. $\sqrt[3]{\frac{y^{17}}{125}} = \frac{\sqrt[3]{y^{15}\cdot y^2}}{\sqrt[3]{125}} = \frac{y^5\sqrt[3]{y^2}}{5}$

93. $\sqrt[4]{48^2} = 48^{2/4} = 48^{1/2} = \sqrt{48} = \sqrt{16\cdot 3}$
$= \sqrt{16}\cdot\sqrt{3} = 4\sqrt{3}$

94. $\sqrt[4]{50^2} = 50^{2/4} = 50^{1/2} = \sqrt{50}$
$= \sqrt{25\cdot 2} = 5\sqrt{2}$

95. $\sqrt[4]{25} = 25^{1/4} = (5^2)^{1/4} = 5^{2/4} = 5^{1/2} = \sqrt{5}$

96. $\sqrt[6]{8} = 8^{1/6} = (2^3)^{1/6} = 2^{3/6} = 2^{1/2} = \sqrt{2}$

97. $\sqrt[10]{x^{25}} = x^{25/10} = x^{5/2} = \sqrt{x^5}$
$= \sqrt{x^4\cdot x} = x^2\sqrt{x}$

98. $\sqrt[12]{x^{44}} = x^{44/12} = x^{11/3} = \sqrt[3]{x^{11}}$
$= \sqrt[3]{x^9\cdot x^2} = x^3\sqrt[3]{x^2}$

99. $\sqrt[3]{4}\cdot\sqrt{3}$

The least common index of 3 and 2 is 6. Write each radical as a sixth root.

$$\sqrt[3]{4} = 4^{1/3} = 4^{2/6} = \sqrt[6]{4^2} = \sqrt[6]{16}$$
$$\sqrt{3} = 3^{1/2} = 3^{3/6} = \sqrt[6]{3^3} = \sqrt[6]{27}$$

Therefore,

$$\sqrt[3]{4}\cdot\sqrt{3} = \sqrt[6]{16}\cdot\sqrt[6]{27} = \sqrt[6]{16\cdot 27} = \sqrt[6]{432}.$$

100. $\sqrt[3]{5}\cdot\sqrt{6}$

The least common index of 3 and 2 is 6. Write each radical as a sixth root.

$$\sqrt[3]{5} = 5^{1/3} = 5^{2/6} = \sqrt[6]{5^2} = \sqrt[6]{25}$$
$$\sqrt{6} = 6^{1/2} = 6^{3/6} = \sqrt[6]{6^3} = \sqrt[6]{216}$$

So $\sqrt[3]{5}\cdot\sqrt{6} = \sqrt[6]{25}\cdot\sqrt[6]{216} = \sqrt[6]{5400}$.

101. $\sqrt[4]{3}\cdot\sqrt[3]{4}$

The least common index of 4 and 3 is 12. Write each radical as a twelfth root.

$$\sqrt[4]{3} = 3^{1/4} = 3^{3/12} = \sqrt[12]{3^3} = \sqrt[12]{27}$$
$$\sqrt[3]{4} = 4^{1/3} = 4^{4/12} = \sqrt[12]{4^4} = \sqrt[12]{256}$$

Therefore,

$$\sqrt[4]{3}\cdot\sqrt[3]{4} = \sqrt[12]{27}\cdot\sqrt[12]{256} = \sqrt[12]{27\cdot 256} = \sqrt[12]{6912}.$$

102. $\sqrt[5]{7}\cdot\sqrt[7]{5}$

The least common index of 5 and 7 is 35. Write each radical as a thirty-fifth root.

$$\sqrt[5]{7} = 7^{1/5} = 7^{7/35} = \sqrt[35]{7^7}$$
$$\sqrt[7]{5} = 5^{1/7} = 5^{5/35} = \sqrt[35]{5^5}$$

So $\sqrt[5]{7}\cdot\sqrt[7]{5} = \sqrt[35]{7^7\cdot 5^5} = \sqrt[35]{2{,}573{,}571{,}875}$.

103. $\sqrt{x}\cdot\sqrt[3]{x}$

$$\sqrt{x} = x^{1/2} = x^{3/6} = \sqrt[6]{x^3}$$
$$\sqrt[3]{x} = x^{1/3} = x^{2/6} = \sqrt[6]{x^2}$$

So $\sqrt{x}\cdot\sqrt[3]{x} = \sqrt[6]{x^3}\cdot\sqrt[6]{x^2} = \sqrt[6]{x^3\cdot x^2} = \sqrt[6]{x^5}$.

104. $\sqrt[3]{x}\cdot\sqrt[4]{x}$

$$\sqrt[3]{x} = x^{1/3} = x^{4/12} = \sqrt[12]{x^4}$$
$$\sqrt[4]{x} = x^{1/4} = x^{3/12} = \sqrt[12]{x^3}$$

So $\sqrt[3]{x}\cdot\sqrt[4]{x} = \sqrt[12]{x^4\cdot x^3} = \sqrt[12]{x^7}$.

105. Substitute 3 for a and 4 for b in the Pythagorean formula to find c.

$$c^2 = a^2 + b^2$$
$$c = \sqrt{a^2+b^2} = \sqrt{3^2+4^2} = \sqrt{9+16} = \sqrt{25} = 5$$

The length of the hypotenuse is 5.

106. Substitute 5 for a and 13 for c in the Pythagorean formula to find b.

$$a^2 + b^2 = c^2$$
$$b = \sqrt{c^2 - a^2} = \sqrt{13^2 - 5^2}$$
$$= \sqrt{169 - 25} = \sqrt{144} = 12$$

The length of the unknown leg is 12.

107. Substitute 12 for c and 4 for a in the Pythagorean formula to find b.

$$a^2 + b^2 = c^2$$
$$b = \sqrt{c^2 - a^2} = \sqrt{12^2 - 4^2}$$
$$= \sqrt{144 - 16} = \sqrt{128}$$
$$= \sqrt{64}\sqrt{2} = 8\sqrt{2}$$

The length of the unknown leg is $8\sqrt{2}$.

108. Substitute 6 for a and 9 for c in the Pythagorean formula to find b.

$$a^2 + b^2 = c^2$$
$$b = \sqrt{c^2 - a^2} = \sqrt{9^2 - 6^2}$$
$$= \sqrt{81 - 36} = \sqrt{45}$$
$$= \sqrt{9}\sqrt{5} = 3\sqrt{5}$$

The length of the unknown leg is $3\sqrt{5}$.

In Exercises 109–120, use the distance formula,

$$d = \sqrt{(x_2 - x_1)^2 + (y_2 - y_1)^2}.$$

109. $(6, 13)$ and $(1, 1)$

$$d = \sqrt{(1 - 6)^2 + (1 - 13)^2}$$
$$= \sqrt{(-5)^2 + (-12)^2} = \sqrt{25 + 144}$$
$$= \sqrt{169} = 13$$

110. $(8, 13)$ and $(2, 5)$

$$d = \sqrt{(2 - 8)^2 + (5 - 13)^2}$$
$$= \sqrt{(-6)^2 + (-8)^2} = \sqrt{36 + 64}$$
$$= \sqrt{100} = 10$$

111. $(-6, 5)$ and $(3, -4)$

$$d = \sqrt{[3 - (-6)]^2 + (-4 - 5)^2}$$
$$= \sqrt{9^2 + (-9)^2} = \sqrt{81 + 81}$$
$$= \sqrt{162} = \sqrt{81} \cdot \sqrt{2} = 9\sqrt{2}$$

112. $(-1, 5)$ and $(-7, 7)$

$$d = \sqrt{[-7 - (-1)]^2 + (7 - 5)^2}$$
$$= \sqrt{(-6)^2 + 2^2} = \sqrt{36 + 4}$$
$$= \sqrt{40} = \sqrt{4} \cdot \sqrt{10} = 2\sqrt{10}$$

113. $(-8, 2)$ and $(-4, 1)$

$$d = \sqrt{[-4 - (-8)]^2 + (1 - 2)^2}$$
$$= \sqrt{4^2 + (-1)^2}$$
$$= \sqrt{16 + 1} = \sqrt{17}$$

114. $(-1, 2)$ and $(5, 3)$

$$d = \sqrt{[5 - (-1)]^2 + (3 - 2)^2}$$
$$= \sqrt{6^2 + 1^2}$$
$$= \sqrt{36 + 1} = \sqrt{37}$$

115. $(4.7, 2.3)$ and $(1.7, -1.7)$

$$d = \sqrt{(1.7 - 4.7)^2 + (-1.7 - 2.3)^2}$$
$$= \sqrt{(-3)^2 + (-4)^2} = \sqrt{9 + 16}$$
$$= \sqrt{25} = 5$$

116. $(-2.9, 18.2)$ and $(2.1, 6.2)$

$$d = \sqrt{[2.1 - (-2.9)]^2 + (6.2 - 18.2)^2}$$
$$= \sqrt{5^2 + (-12)^2} = \sqrt{25 + 144}$$
$$= \sqrt{169} = 13$$

117. $(\sqrt{2}, \sqrt{6})$ and $(-2\sqrt{2}, 4\sqrt{6})$

$$d = \sqrt{\left(-2\sqrt{2} - \sqrt{2}\right)^2 + \left(4\sqrt{6} - \sqrt{6}\right)^2}$$
$$= \sqrt{\left(-3\sqrt{2}\right)^2 + \left(3\sqrt{6}\right)^2}$$
$$= \sqrt{9 \cdot 2 + 9 \cdot 6} = \sqrt{18 + 54}$$
$$= \sqrt{72} = \sqrt{36} \cdot \sqrt{2} = 6\sqrt{2}$$

118. $(\sqrt{7}, 9\sqrt{3})$ and $(-\sqrt{7}, 4\sqrt{3})$

$$d = \sqrt{\left(-\sqrt{7} - \sqrt{7}\right)^2 + \left(4\sqrt{3} - 9\sqrt{3}\right)^2}$$
$$= \sqrt{\left(-2\sqrt{7}\right)^2 + \left(-5\sqrt{3}\right)^2}$$
$$= \sqrt{4 \cdot 7 + 25 \cdot 3}$$
$$= \sqrt{28 + 75} = \sqrt{103}$$

119. $(x + y, y)$ and $(x - y, x)$

$$d = \sqrt{[(x - y) - (x + y)]^2 + (x - y)^2}$$
$$= \sqrt{(-2y)^2 + (x - y)^2}$$
$$= \sqrt{4y^2 + x^2 - 2xy + y^2}$$
$$= \sqrt{5y^2 - 2xy + x^2}$$

120. $(c, c - d)$ and $(d, c + d)$

$$d = \sqrt{(d - c)^2 + [(c + d) - (c - d)]^2}$$
$$= \sqrt{(d - c)^2 + (2d)^2}$$
$$= \sqrt{d^2 - 2cd + c^2 + 4d^2}$$
$$= \sqrt{c^2 - 2cd + 5d^2}$$

121. Substitute 21.7 for a and 16 for b in the Pythagorean formula to find c.

$$c^2 = a^2 + b^2 = (21.7)^2 + 16^2 = 726.89$$
$$c = \sqrt{726.89} \approx 26.96$$

To the nearest tenth of an inch, the diagonal of the screen is 27.0 inches.

122. Use $D = \sqrt{L^2 + W^2 + H^2}$ with $L = 4, W = 2,$ and $H = 3$ to find D.

$$\begin{aligned} D &= \sqrt{4^2 + 2^2 + 3^2} \\ &= \sqrt{16 + 4 + 9} = \sqrt{29} \approx 5.4 \end{aligned}$$

The length of the diagonal is $\sqrt{29}$ feet or about 5.4 feet.

123. Substitute 282 for E, 100 for R, 264 for L, and 120π for ω in the formula to find I.

$$\begin{aligned} I &= \frac{E}{\sqrt{R^2 + \omega^2 L^2}} \\ &= \frac{282}{\sqrt{100^2 + (120\pi)^2(264)^2}} \\ &\approx \frac{282}{99{,}526} \approx .003 \end{aligned}$$

124. Use $f_1 = f_2\sqrt{\dfrac{F_1}{F_2}}$ with $F_1 = 300$, $F_2 = 60$, and $f_2 = 260$ to find f_1.

$$f_1 = 260\sqrt{\frac{300}{60}} = 260\sqrt{5} \approx 581$$

125. $$\begin{aligned} d &= 1.224\sqrt{h} \\ &= 1.224\sqrt{156} \approx 15.3 \text{ miles} \end{aligned}$$

9.4 Adding and Subtracting Radical Expressions

9.4 Margin Exercises

1. **(a)** $3\sqrt{5} + 7\sqrt{5} = (3 + 7)\sqrt{5} = 10\sqrt{5}$

(b) $$\begin{aligned} &2\sqrt{11} - \sqrt{11} + 3\sqrt{44} \\ &= 2\sqrt{11} - 1\sqrt{11} + 3\sqrt{4}\cdot\sqrt{11} \\ &= 2\sqrt{11} - 1\sqrt{11} + 3\cdot 2\cdot\sqrt{11} \\ &= (2 - 1 + 6)\sqrt{11} \\ &= 7\sqrt{11} \end{aligned}$$

(c) $$\begin{aligned} &5\sqrt{12y} + 6\sqrt{75y},\ y \geq 0 \\ &= 5\sqrt{4}\cdot\sqrt{3y} + 6\sqrt{25}\cdot\sqrt{3y} \\ &= 5\cdot 2\sqrt{3y} + 6\cdot 5\sqrt{3y} \\ &= 10\sqrt{3y} + 30\sqrt{3y} \\ &= (10 + 30)\sqrt{3y} \\ &= 40\sqrt{3y} \end{aligned}$$

(d) $$\begin{aligned} &3\sqrt{8} - 6\sqrt{50} + 2\sqrt{200} \\ &= 3\sqrt{4}\cdot\sqrt{2} - 6\sqrt{25}\cdot\sqrt{2} + 2\sqrt{100}\cdot\sqrt{2} \\ &= 3\cdot 2\sqrt{2} - 6\cdot 5\sqrt{2} + 2\cdot 10\sqrt{2} \\ &= 6\sqrt{2} - 30\sqrt{2} + 20\sqrt{2} \\ &= (6 - 30 + 20)\sqrt{2} \\ &= -4\sqrt{2} \end{aligned}$$

(e) $9\sqrt{5} - 4\sqrt{10}$

Here the radicals differ and are already simplified, so this expression *cannot be simplified further.*

2. **(a)** $$\begin{aligned} 7\sqrt[3]{81} + 3\sqrt[3]{24} &= 7\sqrt[3]{27\cdot 3} + 3\sqrt[3]{8\cdot 3} \\ &= 7\sqrt[3]{27}\cdot\sqrt[3]{3} + 3\sqrt[3]{8}\cdot\sqrt[3]{3} \\ &= 7\cdot 3\sqrt[3]{3} + 3\cdot 2\sqrt[3]{3} \\ &= 21\sqrt[3]{3} + 6\sqrt[3]{3} \\ &= 27\sqrt[3]{3} \end{aligned}$$

(b) $$\begin{aligned} &-2\sqrt[4]{32} - 7\sqrt[4]{162} \\ &= -2\sqrt[4]{16\cdot 2} - 7\sqrt[4]{81\cdot 2} \\ &= -2\cdot 2\sqrt[4]{2} - 7\cdot 3\sqrt[4]{2} \\ &= -4\sqrt[4]{2} - 21\sqrt[4]{2} \\ &= -25\sqrt[4]{2} \end{aligned}$$

(c) $$\begin{aligned} \sqrt[3]{p^4q^7} - \sqrt[3]{64pq} &= \sqrt[3]{p^3q^6\cdot pq} - \sqrt[3]{64\cdot pq} \\ &= pq^2\sqrt[3]{pq} - 4\sqrt[3]{pq} \\ &= (pq^2 - 4)\sqrt[3]{pq} \end{aligned}$$

3. $$\begin{aligned} \sqrt{\frac{80}{y^4}} + \sqrt{\frac{81}{y^{10}}} &= \frac{\sqrt{16\cdot 5}}{\sqrt{y^4}} + \frac{\sqrt{81}}{\sqrt{y^{10}}} \\ &= \frac{4\sqrt{5}}{y^2} + \frac{9}{y^5} \\ &= \frac{4y^3\sqrt{5}}{y^5} + \frac{9}{y^5} \\ &= \frac{4y^3\sqrt{5} + 9}{y^5} \end{aligned}$$

9.4 Section Exercises

1. Only choice **B** has like radical factors, so it can be simplified without first simplifying the individual radical expressions.

$$3\sqrt{6} + 9\sqrt{6} = 12\sqrt{6}$$

2. Let $a = 1$ and $b = 64$.

(a) $$\begin{aligned} \sqrt{a} + \sqrt{b} &= \sqrt{1} + \sqrt{64} \\ &= 1 + 8 = 9 \\ \sqrt{a + b} &= \sqrt{1 + 64} \\ &= \sqrt{65} \end{aligned}$$

$9 \neq \sqrt{65}$, so $\sqrt{a} + \sqrt{b} \neq \sqrt{a + b}$.

(b) $\sqrt[3]{a} + \sqrt[3]{b} = \sqrt[3]{1} + \sqrt[3]{64}$
$= 1 + 4 = 5$
$\sqrt[3]{a+b} = \sqrt[3]{1+64}$
$= \sqrt[3]{65}$
$5 \neq \sqrt[3]{65}$, so $\sqrt[3]{a} + \sqrt[3]{b} \neq \sqrt[3]{a+b}$.

(c) In general, $\sqrt[n]{a} + \sqrt[n]{b} \neq \sqrt[n]{a+b}$, based on the observations in parts (a) and (b) of this exercise.

3. $\sqrt{64} + \sqrt[3]{125} + \sqrt[4]{16} = \sqrt{8^2} + \sqrt[3]{5^3} + \sqrt[4]{2^4}$
$= 8 + 5 + 2 = 15$
This sum can be found easily since each radicand has a whole number power corresponding to the index of the radical; that is, each radical expression simplifies to a whole number.

4. $28 - 4\sqrt{2} \neq 24\sqrt{2}$ because you cannot subtract $28 - 4$ first. Multiplication must take place before subtraction, according to the rules for order of operations.

5. Simplify each radical and subtract.

$$\sqrt{36} - \sqrt{100} = 6 - 10 = -4$$

6. Simplify each radical and subtract.

$$\sqrt{25} - \sqrt{81} = 5 - 9 = -4$$

7. $-2\sqrt{48} + 3\sqrt{75}$
$= -2\sqrt{16 \cdot 3} + 3\sqrt{25 \cdot 3}$
$= -2 \cdot 4\sqrt{3} + 3 \cdot 5\sqrt{3}$
$= -8\sqrt{3} + 15\sqrt{3} = 7\sqrt{3}$

8. $4\sqrt{32} - 2\sqrt{8}$
$= 4\sqrt{16}\sqrt{2} - 2\sqrt{4}\sqrt{2}$
$= 4 \cdot 4 \cdot \sqrt{2} - 2 \cdot 2\sqrt{2}$
$= 16\sqrt{2} - 4\sqrt{2} = 12\sqrt{2}$

9. $\sqrt[3]{16} + 4\sqrt[3]{54}$
$= \sqrt[3]{8 \cdot 2} + 4\sqrt[3]{27 \cdot 2}$
$= \sqrt[3]{8}\sqrt[3]{2} + 4\sqrt[3]{27}\sqrt[3]{2}$
$= 2\sqrt[3]{2} + 4 \cdot 3\sqrt[3]{2}$
$= 2\sqrt[3]{2} + 12\sqrt[3]{2} = 14\sqrt[3]{2}$

10. $3\sqrt[3]{24} - 2\sqrt[3]{192}$
$= 3\sqrt[3]{8 \cdot 3} - 2\sqrt[3]{64 \cdot 3}$
$= 3 \cdot 2\sqrt[3]{3} - 2 \cdot 4\sqrt[3]{3}$
$= 6\sqrt[3]{3} - 8\sqrt[3]{3} = -2\sqrt[3]{3}$

11. $\sqrt[4]{32} + 3\sqrt[4]{2}$
$= \sqrt[4]{16 \cdot 2} + 3\sqrt[4]{2}$
$= \sqrt[4]{16}\sqrt[4]{2} + 3\sqrt[4]{2}$
$= 2\sqrt[4]{2} + 3\sqrt[4]{2} = 5\sqrt[4]{2}$

12. $\sqrt[4]{405} - 2\sqrt[4]{5}$
$= \sqrt[4]{81 \cdot 5} - 2\sqrt[4]{5}$
$= 3\sqrt[4]{5} - 2\sqrt[4]{5} = \sqrt[4]{5}$

13. $6\sqrt{18} - \sqrt{32} + 2\sqrt{50}$
$= 6\sqrt{9 \cdot 2} - \sqrt{16 \cdot 2} + 2\sqrt{25 \cdot 2}$
$= 6 \cdot 3\sqrt{2} - 4\sqrt{2} + 2 \cdot 5\sqrt{2}$
$= 18\sqrt{2} - 4\sqrt{2} + 10\sqrt{2}$
$= 24\sqrt{2}$

14. $5\sqrt{8} + 3\sqrt{72} - 3\sqrt{50}$
$= 5\sqrt{4}\sqrt{2} + 3\sqrt{36}\sqrt{2} - 3\sqrt{25}\sqrt{2}$
$= 5 \cdot 2\sqrt{2} + 3 \cdot 6\sqrt{2} - 3 \cdot 5\sqrt{2}$
$= 10\sqrt{2} + 18\sqrt{2} - 15\sqrt{2}$
$= 13\sqrt{2}$

15. $5\sqrt{6} + 2\sqrt{10}$

The radicals differ and are already simplified, so the expression cannot be simplified further.

16. $3\sqrt{11} - 5\sqrt{13}$

The radicals differ and are already simplified, so the expression cannot be simplified further.

17. $2\sqrt{5} + 3\sqrt{20} + 4\sqrt{45}$
$= 2\sqrt{5} + 3\sqrt{4 \cdot 5} + 4\sqrt{9 \cdot 5}$
$= 2\sqrt{5} + 3 \cdot 2\sqrt{5} + 4 \cdot 3\sqrt{5}$
$= 2\sqrt{5} + 6\sqrt{5} + 12\sqrt{5}$
$= 20\sqrt{5}$

18. $5\sqrt{54} - 2\sqrt{24} - 2\sqrt{96}$
$= 5\sqrt{9}\sqrt{6} - 2\sqrt{4}\sqrt{6} - 2\sqrt{16}\sqrt{6}$
$= 5 \cdot 3 \cdot \sqrt{6} - 2 \cdot 2 \cdot \sqrt{6} - 2 \cdot 4 \cdot \sqrt{6}$
$= 15\sqrt{6} - 4\sqrt{6} - 8\sqrt{6}$
$= 3\sqrt{6}$

19. $8\sqrt{2x} - \sqrt{8x} + \sqrt{72x}$
$= 8\sqrt{2x} - \sqrt{4 \cdot 2x} + \sqrt{36 \cdot 2x}$
$= 8\sqrt{2x} - 2\sqrt{2x} + 6\sqrt{2x}$
$= 12\sqrt{2x}$

20. $4\sqrt{18k} - \sqrt{72k} + \sqrt{50k}$
$= 4\sqrt{9}\sqrt{2k} - \sqrt{36}\sqrt{2k} + \sqrt{25}\sqrt{2k}$
$= 4 \cdot 3\sqrt{2k} - 6\sqrt{2k} + 5\sqrt{2k}$
$= 12\sqrt{2k} - 6\sqrt{2k} + 5\sqrt{2k}$
$= 11\sqrt{2k}$

21. $3\sqrt{72m^2} - 5\sqrt{32m^2}$
$= 3\sqrt{36m^2 \cdot 2} - 5\sqrt{16m^2 \cdot 2}$
$= 3 \cdot 6m\sqrt{2} - 5 \cdot 4m\sqrt{2}$
$= 18m\sqrt{2} - 20m\sqrt{2}$
$= -2m\sqrt{2}$

22. $9\sqrt{27p^2} - 14\sqrt{108p^2}$
$= 9\sqrt{9p^2}\sqrt{3} - 14\sqrt{36p^2}\sqrt{3}$
$= 9 \cdot 3p\sqrt{3} - 14 \cdot 6p\sqrt{3}$
$= 27p\sqrt{3} - 84p\sqrt{3}$
$= -57p\sqrt{3}$

23. $-\sqrt[3]{54} + 2\sqrt[3]{16} = -\sqrt[3]{27 \cdot 2} + 2\sqrt[3]{8 \cdot 2}$
$= -3\sqrt[3]{2} + 2 \cdot 2\sqrt[3]{2}$
$= -3\sqrt[3]{2} + 4\sqrt[3]{2}$
$= \sqrt[3]{2}$

24. $15\sqrt[3]{81} - 4\sqrt[3]{24} = 15\sqrt[3]{27}\sqrt[3]{3} - 4\sqrt[3]{8}\sqrt[3]{3}$
$= 15 \cdot 3\sqrt[3]{3} - 4 \cdot 2\sqrt[3]{3}$
$= 45\sqrt[3]{3} - 8\sqrt[3]{3}$
$= 37\sqrt[3]{3}$

25. $2\sqrt[3]{27x} - 2\sqrt[3]{8x}$
$= 2\sqrt[3]{27 \cdot x} - 2\sqrt[3]{8 \cdot x}$
$= 2 \cdot 3\sqrt[3]{x} - 2 \cdot 2\sqrt[3]{x}$
$= 6\sqrt[3]{x} - 4\sqrt[3]{x}$
$= 2\sqrt[3]{x}$

26. $6\sqrt[3]{128m} + 3\sqrt[3]{16m}$
$= 6\sqrt[3]{64} \cdot \sqrt[3]{2m} + 3\sqrt[3]{8} \cdot \sqrt[3]{2m}$
$= 6 \cdot 4 \cdot \sqrt[3]{2m} + 3 \cdot 2 \cdot \sqrt[3]{2m}$
$= 24\sqrt[3]{2m} + 6\sqrt[3]{2m}$
$= 30\sqrt[3]{2m}$

27. $\sqrt[3]{x^2y} - \sqrt[3]{8x^2y} = \sqrt[3]{x^2y} - \sqrt[3]{8}\sqrt[3]{x^2y}$
$= 1\sqrt[3]{x^2y} - 2\sqrt[3]{x^2y}$
$= (1-2)\sqrt[3]{x^2y}$
$= -\sqrt[3]{x^2y}$

28. $3\sqrt[3]{x^2y^2} - 2\sqrt[3]{64x^2y^2}$
$= 3\sqrt[3]{x^2y^2} - 2 \cdot \sqrt[3]{64} \cdot \sqrt[3]{x^2y^2}$
$= 3\sqrt[3]{x^2y^2} - 2 \cdot 4 \cdot \sqrt[3]{x^2y^2}$
$= (3-8)\sqrt[3]{x^2y^2} = -5\sqrt[3]{x^2y^2}$

29. $3x\sqrt[3]{xy^2} - 2\sqrt[3]{8x^4y^2}$
$= 3x\sqrt[3]{xy^2} - 2\sqrt[3]{8x^3} \cdot \sqrt[3]{xy^2}$
$= 3x\sqrt[3]{xy^2} - 2 \cdot 2x \cdot \sqrt[3]{xy^2}$
$= (3x - 4x)\sqrt[3]{xy^2} = -x\sqrt[3]{xy^2}$

30. $6q^2\sqrt[3]{5q} - 2q\sqrt[3]{40q^4}$
$= 6q^2\sqrt[3]{5q} - 2q \cdot \sqrt[3]{8q^3} \cdot \sqrt[3]{5q}$
$= 6q^2\sqrt[3]{5q} - 2q \cdot 2q \cdot \sqrt[3]{5q}$
$= \left(6q^2 - 4q^2\right)\sqrt[3]{5q} = 2q^2\sqrt[3]{5q}$

31. $5\sqrt[4]{32} + 3\sqrt[4]{162} = 5\sqrt[4]{16 \cdot 2} + 3\sqrt[4]{81 \cdot 2}$
$= 5 \cdot 2\sqrt[4]{2} + 3 \cdot 3\sqrt[4]{2}$
$= 10\sqrt[4]{2} + 9\sqrt[4]{2}$
$= 19\sqrt[4]{2}$

32. $2\sqrt[4]{512} + 4\sqrt[4]{32}$
$= 2\sqrt[4]{256} \cdot \sqrt[4]{2} + 4\sqrt[4]{16} \cdot \sqrt[4]{2}$
$= 2 \cdot 4\sqrt[4]{2} + 4 \cdot 2 \cdot \sqrt[4]{2}$
$= 8\sqrt[4]{2} + 8\sqrt[4]{2}$
$= 16\sqrt[4]{2}$

33. $3\sqrt[4]{x^5y} - 2x\sqrt[4]{xy}$
$= 3\sqrt[4]{x^4 \cdot xy} - 2x\sqrt[4]{xy}$
$= 3x\sqrt[4]{xy} - 2x\sqrt[4]{xy}$
$= (3x - 2x)\sqrt[4]{xy} = x\sqrt[4]{xy}$

34. $2\sqrt[4]{m^9p^6} - 3m^2p\sqrt[4]{mp^2}$
$= 2\sqrt[4]{m^8p^4} \cdot \sqrt[4]{mp^2} - 3m^2p\sqrt[4]{mp^2}$
$= 2m^2p\sqrt[4]{mp^2} - 3m^2p\sqrt[4]{mp^2}$
$= \left(2m^2p - 3m^2p\right)\sqrt[4]{mp^2} = -m^2p\sqrt[4]{mp^2}$

35. $2\sqrt[4]{32a^3} + 5\sqrt[4]{2a^3}$
$= 2\sqrt[4]{16} \cdot \sqrt[4]{2a^3} + 5\sqrt[4]{2a^3}$
$= 2 \cdot 2 \cdot \sqrt[4]{2a^3} + 5\sqrt[4]{2a^3}$
$= (4+5)\sqrt[4]{2a^3} = 9\sqrt[4]{2a^3}$

36. $-\sqrt[4]{16r} + 5\sqrt[4]{r}$
$= -\sqrt[4]{16} \cdot \sqrt[4]{r} + 5\sqrt[4]{r}$
$= -2\sqrt[4]{r} + 5\sqrt[4]{r}$
$= (-2+5)\sqrt[4]{r} = 3\sqrt[4]{r}$

37. $$\frac{2\sqrt{5}}{3}+\frac{\sqrt{5}}{6}=\frac{4\sqrt{5}}{6}+\frac{1\sqrt{5}}{6}=\frac{4\sqrt{5}+1\sqrt{5}}{6}=\frac{5\sqrt{5}}{6}$$

38. $$\frac{4\sqrt{3}}{3}+\frac{2\sqrt{3}}{9}=\frac{12\sqrt{3}}{9}+\frac{2\sqrt{3}}{9}=\frac{12\sqrt{3}+2\sqrt{3}}{9}=\frac{14\sqrt{3}}{9}$$

39. $$\sqrt{\frac{8}{9}}+\sqrt{\frac{18}{36}}=\frac{\sqrt{8}}{\sqrt{9}}+\frac{\sqrt{18}}{\sqrt{36}}=\frac{\sqrt{4}\sqrt{2}}{3}+\frac{\sqrt{9}\sqrt{2}}{6}=\frac{2\sqrt{2}}{3}+\frac{3\sqrt{2}}{6}=\frac{4\sqrt{2}}{6}+\frac{3\sqrt{2}}{6}=\frac{4\sqrt{2}+3\sqrt{2}}{6}=\frac{7\sqrt{2}}{6}$$

40. $$\sqrt{\frac{12}{16}}+\sqrt{\frac{48}{64}}=\frac{\sqrt{12}}{\sqrt{16}}+\frac{\sqrt{48}}{\sqrt{64}}=\frac{\sqrt{4}\sqrt{3}}{4}+\frac{\sqrt{16}\sqrt{3}}{8}=\frac{2\sqrt{3}}{4}+\frac{4\sqrt{3}}{8}=\frac{1}{2}\sqrt{3}+\frac{1}{2}\sqrt{3}=\sqrt{3}$$

41. $$\frac{\sqrt{32}}{3}+\frac{2\sqrt{2}}{3}-\frac{\sqrt{2}}{\sqrt{9}}=\frac{\sqrt{16}\sqrt{2}}{3}+\frac{2\sqrt{2}}{3}-\frac{\sqrt{2}}{3}=\frac{4\sqrt{2}+2\sqrt{2}-\sqrt{2}}{3}=\frac{5\sqrt{2}}{3}$$

42. $$\frac{\sqrt{27}}{2}-\frac{3\sqrt{3}}{2}+\frac{\sqrt{3}}{\sqrt{4}}=\frac{\sqrt{9}\sqrt{3}}{2}-\frac{3\sqrt{3}}{2}+\frac{\sqrt{3}}{2}=\frac{3\sqrt{3}}{2}-\frac{3\sqrt{3}}{2}+\frac{\sqrt{3}}{2}=\frac{\sqrt{3}}{2}$$

43. $$3\sqrt{\frac{50}{9}}+8\frac{\sqrt{2}}{\sqrt{8}}=3\frac{\sqrt{50}}{\sqrt{9}}+8\frac{\sqrt{2}}{2\sqrt{2}}=3\cdot\frac{5\sqrt{2}}{3}+8\cdot\frac{1}{2}=5\sqrt{2}+4$$

44. $$9\sqrt{\frac{48}{25}}-2\frac{\sqrt{2}}{\sqrt{98}}=9\frac{\sqrt{48}}{\sqrt{25}}-2\frac{\sqrt{2}}{\sqrt{49}\cdot\sqrt{2}}=9\cdot\frac{4\sqrt{3}}{5}-2\cdot\frac{1}{7}=\frac{36\sqrt{3}\cdot 7}{5\cdot 7}-\frac{2\cdot 5}{7\cdot 5}=\frac{252\sqrt{3}-10}{35}$$

45. $$\sqrt{\frac{25}{x^8}}-\sqrt{\frac{9}{x^6}}=\frac{\sqrt{25}}{\sqrt{x^8}}-\frac{\sqrt{9}}{\sqrt{x^6}}=\frac{5}{x^4}-\frac{3}{x^3}=\frac{5}{x^4}-\frac{3\cdot x}{x^3\cdot x}\quad LCD=x^4$$
$$=\frac{5-3x}{x^4}$$

46. $$\sqrt{\frac{100}{y^4}}+\sqrt{\frac{81}{y^{10}}}=\frac{\sqrt{100}}{\sqrt{y^4}}+\frac{\sqrt{81}}{\sqrt{y^{10}}}=\frac{10}{y^2}+\frac{9}{y^5}=\frac{10\cdot y^3}{y^2\cdot y^3}+\frac{9}{y^5}\quad LCD=y^5$$
$$=\frac{10y^3+9}{y^5}$$

47. $$3\sqrt[3]{\frac{m^5}{27}}-2m\sqrt[3]{\frac{m^2}{64}}=\frac{3\sqrt[3]{m^5}}{\sqrt[3]{27}}-\frac{2m\sqrt[3]{m^2}}{\sqrt[3]{64}}=\frac{3\sqrt[3]{m^3}\sqrt[3]{m^2}}{3}-\frac{2m\sqrt[3]{m^2}}{4}=\frac{m\sqrt[3]{m^2}}{1}-\frac{m\sqrt[3]{m^2}}{2}=\frac{2m\sqrt[3]{m^2}-m\sqrt[3]{m^2}}{2}=\frac{m\sqrt[3]{m^2}}{2}$$

48. $$2a\sqrt[4]{\frac{a}{16}}-5a\sqrt[4]{\frac{a}{81}}=\frac{2a\sqrt[4]{a}}{\sqrt[4]{16}}-\frac{5a\sqrt[4]{a}}{\sqrt[4]{81}}=\frac{2a\sqrt[4]{a}}{2}-\frac{5a\sqrt[4]{a}}{3}=\frac{a\sqrt[4]{a}}{1}-\frac{5a\sqrt[4]{a}}{3}=\frac{3a\sqrt[4]{a}-5a\sqrt[4]{a}}{3}=-\frac{2a\sqrt[4]{a}}{3}$$

49. $3\sqrt[3]{\frac{2}{x^6}}-4\sqrt[3]{\frac{5}{x^9}}=3\frac{\sqrt[3]{2}}{\sqrt[3]{x^6}}-4\frac{\sqrt[3]{5}}{\sqrt[3]{x^9}}$

$$=3\frac{\sqrt[3]{2}}{x^2}-4\frac{\sqrt[3]{5}}{x^3}$$

$$=\frac{3\cdot x\cdot\sqrt[3]{2}}{x^2\cdot x}-\frac{4\sqrt[3]{5}}{x^3}$$

$LCD = x^3$

$$=\frac{3x\sqrt[3]{2}-4\sqrt[3]{5}}{x^3}$$

50. $-4\sqrt[3]{\frac{4}{t^9}}+3\sqrt[3]{\frac{9}{t^{12}}}=-4\frac{\sqrt[3]{4}}{\sqrt[3]{t^9}}+3\frac{\sqrt[3]{9}}{\sqrt[3]{t^{12}}}$

$$=-4\frac{\sqrt[3]{4}}{t^3}+3\frac{\sqrt[3]{9}}{t^4}$$

$$=\frac{-4\cdot t\cdot\sqrt[3]{4}}{t^3\cdot t}+\frac{3\sqrt[3]{9}}{t^4}$$

$$=\frac{-4t\sqrt[3]{4}+3\sqrt[3]{9}}{t^4}$$

51. The perimeter, P, of a triangle is the sum of the measures of the sides.

$$P=3\sqrt{20}+2\sqrt{45}+\sqrt{75}$$
$$=3\sqrt{4\cdot5}+2\sqrt{9\cdot5}+\sqrt{25\cdot3}$$
$$=3\cdot2\sqrt{5}+2\cdot3\sqrt{5}+5\sqrt{3}$$
$$=6\sqrt{5}+6\sqrt{5}+5\sqrt{3}$$
$$=12\sqrt{5}+5\sqrt{3}$$

The perimeter is $12\sqrt{5}+5\sqrt{3}$ inches.

52. Use $P=2L+2W$ with $L=\sqrt{192}$ and $W=\sqrt{48}$.

$$P=2\sqrt{192}+2\sqrt{48}$$
$$=2\sqrt{64}\cdot\sqrt{3}+2\sqrt{16}\cdot\sqrt{3}$$
$$=2\cdot8\sqrt{3}+2\cdot4\sqrt{3}$$
$$=16\sqrt{3}+8\sqrt{3}$$
$$=24\sqrt{3}$$

The perimeter is $24\sqrt{3}$ meters.

53. To find the perimeter, add the lengths of the sides.

$$4\sqrt{18}+\sqrt{108}+2\sqrt{72}+3\sqrt{12}$$
$$=4\sqrt{9}\sqrt{2}+\sqrt{36}\sqrt{3}+2\sqrt{36}\sqrt{2}+3\sqrt{4}\sqrt{3}$$
$$=4\cdot3\sqrt{2}+6\sqrt{3}+2\cdot6\sqrt{2}+3\cdot2\sqrt{3}$$
$$=12\sqrt{2}+6\sqrt{3}+12\sqrt{2}+6\sqrt{3}$$
$$=24\sqrt{2}+12\sqrt{3}$$

The perimeter is $24\sqrt{2}+12\sqrt{3}$ inches.

54. Use $A=\frac{1}{2}h(B+b)$ with $h=\sqrt{24}$, $B=\sqrt{288}$, and $b=\sqrt{72}$.

$$A=\tfrac{1}{2}\sqrt{24}\left(\sqrt{288}+\sqrt{72}\right)$$
$$=\frac{\sqrt{4}\cdot\sqrt{6}}{2}\left(\sqrt{144}\cdot\sqrt{2}+\sqrt{36}\cdot\sqrt{2}\right)$$
$$=\sqrt{6}\left(12\sqrt{2}+6\sqrt{2}\right)=\sqrt{6}\left(18\sqrt{2}\right)$$
$$=18\sqrt{12}=18\cdot2\sqrt{3}=36\sqrt{3}$$

The area is $36\sqrt{3}$ square inches.

9.5 Multiplying and Dividing Radical Expressions

9.5 Margin Exercises

1. Use the FOIL method and multiply as two binomials.

(a) $(2+\sqrt{3})(1+\sqrt{5})$

F O I L

$$=2\cdot1+2\sqrt{5}+\sqrt{3}\cdot1+\sqrt{3}\cdot\sqrt{5}$$
$$=2+2\sqrt{5}+\sqrt{3}+\sqrt{15}$$

(b) $(2\sqrt{3}+\sqrt{5})(\sqrt{6}-3\sqrt{5})$

$$=2\sqrt{3}\cdot\sqrt{6}-2\sqrt{3}\cdot3\sqrt{5}+\sqrt{5}\cdot\sqrt{6}$$
$$-\sqrt{5}\cdot3\sqrt{5}$$
$$=2\sqrt{18}-6\sqrt{15}+\sqrt{30}-3(5)$$
$$=2\sqrt{9}\cdot\sqrt{2}-6\sqrt{15}+\sqrt{30}-15$$
$$=6\sqrt{2}-6\sqrt{15}+\sqrt{30}-15$$

(c) $(4+\sqrt{3})(4-\sqrt{3})$

$$=4\cdot4-4\sqrt{3}+4\sqrt{3}-\sqrt{3}\cdot\sqrt{3}$$
$$=16-3$$
$$=13$$

(d) $(\sqrt{6}-\sqrt{5})^2$

$$=(\sqrt{6}-\sqrt{5})(\sqrt{6}-\sqrt{5})$$
$$=\sqrt{6}\cdot\sqrt{6}-\sqrt{6}\cdot\sqrt{5}-\sqrt{5}\cdot\sqrt{6}$$
$$+\sqrt{5}\cdot\sqrt{5}$$
$$=6-\sqrt{30}-\sqrt{30}+5$$
$$=11-2\sqrt{30}$$

(e) $(4+\sqrt[3]{7})(4-\sqrt[3]{7})$

$$=4\cdot4-4\sqrt[3]{7}+4\sqrt[3]{7}-\sqrt[3]{7}\cdot\sqrt[3]{7}$$
$$=16-\sqrt[3]{49}$$

(f) $\left(\sqrt{p}+\sqrt{2}\right)\left(\sqrt{p}-\sqrt{2}\right)$

$$=\left(\sqrt{p}\right)^2-\left(\sqrt{2}\right)^2$$
$$=p-2,\quad p\ge0$$

2. **(a)** $\dfrac{8}{\sqrt{3}} = \dfrac{8 \cdot \sqrt{3}}{\sqrt{3} \cdot \sqrt{3}} = \dfrac{8\sqrt{3}}{3}$

(b) $\dfrac{\sqrt{3}}{\sqrt{7}} = \dfrac{\sqrt{3} \cdot \sqrt{7}}{\sqrt{7} \cdot \sqrt{7}} = \dfrac{\sqrt{21}}{7}$

(c) $\dfrac{3}{\sqrt{48}} = \dfrac{3}{\sqrt{16 \cdot 3}} = \dfrac{3}{4\sqrt{3}}$

$= \dfrac{3 \cdot \sqrt{3}}{4\sqrt{3} \cdot \sqrt{3}} = \dfrac{3\sqrt{3}}{4 \cdot 3}$

$= \dfrac{\sqrt{3}}{4}$

(d) $\dfrac{-16}{\sqrt{32}} = -\dfrac{16}{\sqrt{16 \cdot 2}} = -\dfrac{16}{4\sqrt{2}}$

$= -\dfrac{4}{\sqrt{2}} = -\dfrac{4 \cdot \sqrt{2}}{\sqrt{2} \cdot \sqrt{2}}$

$= -\dfrac{4\sqrt{2}}{2} = -2\sqrt{2}$

3. **(a)** $\sqrt{\dfrac{8}{45}} = \dfrac{\sqrt{8}}{\sqrt{45}} = \dfrac{\sqrt{4 \cdot 2}}{\sqrt{9 \cdot 5}} = \dfrac{2\sqrt{2}}{3\sqrt{5}}$

$= \dfrac{2\sqrt{2} \cdot \sqrt{5}}{3\sqrt{5} \cdot \sqrt{5}} = \dfrac{2\sqrt{10}}{3 \cdot 5} = \dfrac{2\sqrt{10}}{15}$

(b) $\sqrt{\dfrac{72}{y}} = \dfrac{\sqrt{72} \cdot \sqrt{y}}{\sqrt{y} \cdot \sqrt{y}}$

$= \dfrac{\sqrt{72y}}{y} = \dfrac{\sqrt{36 \cdot 2y}}{y} = \dfrac{6\sqrt{2y}}{y}$

(c) $\sqrt{\dfrac{200k^6}{y^7}} = \dfrac{\sqrt{100k^6 \cdot 2}}{\sqrt{y^6 \cdot y}}$

$= \dfrac{10k^3\sqrt{2}}{y^3\sqrt{y}}$

$= \dfrac{10k^3\sqrt{2} \cdot \sqrt{y}}{y^3\sqrt{y} \cdot \sqrt{y}}$

$= \dfrac{10k^3\sqrt{2y}}{y^3 \cdot y}$

$= \dfrac{10k^3\sqrt{2y}}{y^4}$

4. **(a)** $\sqrt[3]{\dfrac{15}{32}} = \dfrac{\sqrt[3]{15}}{\sqrt[3]{32}} = \dfrac{\sqrt[3]{15}}{\sqrt[3]{8} \cdot \sqrt[3]{4}} = \dfrac{\sqrt[3]{15}}{2\sqrt[3]{4}}$

$= \dfrac{\sqrt[3]{15} \cdot \sqrt[3]{2}}{2\sqrt[3]{4} \cdot \sqrt[3]{2}} = \dfrac{\sqrt[3]{30}}{2\sqrt[3]{8}}$

$= \dfrac{\sqrt[3]{30}}{2 \cdot 2} = \dfrac{\sqrt[3]{30}}{4}$

(b) $\sqrt[3]{\dfrac{m^{12}}{n}} = \dfrac{\sqrt[3]{m^{12}}}{\sqrt[3]{n}}$

$= \dfrac{m^4 \cdot \sqrt[3]{n^2}}{\sqrt[3]{n} \cdot \sqrt[3]{n^2}}$

$= \dfrac{m^4\sqrt[3]{n^2}}{n} \quad (n \neq 0)$

(c) $\sqrt[4]{\dfrac{6y}{w^2}} = \dfrac{\sqrt[4]{6y}}{\sqrt[4]{w^2}} \cdot \dfrac{\sqrt[4]{w^2}}{\sqrt[4]{w^2}}$

$= \dfrac{\sqrt[4]{6yw^2}}{\sqrt[4]{w^4}}$

$= \dfrac{\sqrt[4]{6yw^2}}{w} \quad (y \geq 0, w \neq 0)$

5. **(a)** $\dfrac{-4}{\sqrt{5}+2}$

Multiply both the numerator and denominator by the conjugate of the denominator, $\sqrt{5} - 2$.

$= \dfrac{-4\left(\sqrt{5}-2\right)}{\left(\sqrt{5}+2\right)\left(\sqrt{5}-2\right)}$

$= \dfrac{-4\left(\sqrt{5}-2\right)}{5-4}$

$= -4\left(\sqrt{5}-2\right)$

(b) $\dfrac{15}{\sqrt{7}+\sqrt{2}} = \dfrac{15\left(\sqrt{7}-\sqrt{2}\right)}{\left(\sqrt{7}+\sqrt{2}\right)\left(\sqrt{7}-\sqrt{2}\right)}$

$= \dfrac{15\left(\sqrt{7}-\sqrt{2}\right)}{7-2}$

$= \dfrac{15\left(\sqrt{7}-\sqrt{2}\right)}{5}$

$= 3\left(\sqrt{7}-\sqrt{2}\right)$

(c) $\dfrac{\sqrt{3}+\sqrt{5}}{\sqrt{2}-\sqrt{7}}$

$= \dfrac{\left(\sqrt{3}+\sqrt{5}\right)\left(\sqrt{2}+\sqrt{7}\right)}{\left(\sqrt{2}-\sqrt{7}\right)\left(\sqrt{2}+\sqrt{7}\right)}$

$= \dfrac{\sqrt{6}+\sqrt{21}+\sqrt{10}+\sqrt{35}}{2-7}$

$= \dfrac{-\left(\sqrt{6}+\sqrt{21}+\sqrt{10}+\sqrt{35}\right)}{5}$

(d) $\dfrac{2}{\sqrt{k}+\sqrt{z}} = \dfrac{2\left(\sqrt{k}-\sqrt{z}\right)}{\left(\sqrt{k}+\sqrt{z}\right)\left(\sqrt{k}-\sqrt{z}\right)}$
$= \dfrac{2\left(\sqrt{k}-\sqrt{z}\right)}{k-z}$
$(k \neq z,\, k > 0, z > 0)$

6. **(a)** $\dfrac{15-5\sqrt{3}}{5} = \dfrac{5\left(3-\sqrt{3}\right)}{5} = 3-\sqrt{3}$

(b) $\dfrac{24-36\sqrt{7}}{16} = \dfrac{4\left(6-9\sqrt{7}\right)}{16} = \dfrac{6-9\sqrt{7}}{4}$

9.5 Section Exercises

1. $(x+\sqrt{y})(x-\sqrt{y})$
$= x^2 - x\sqrt{y} + x\sqrt{y} - (\sqrt{y})^2$
$= x^2 - y$ **(E)**

2. $(\sqrt{x}+y)(\sqrt{x}-y)$
$= (\sqrt{x})^2 - \sqrt{x}y + \sqrt{x}y - y^2$
$= x - y^2$ **(C)**

3. $(\sqrt{x}+\sqrt{y})(\sqrt{x}-\sqrt{y})$
$= (\sqrt{x})^2 - (\sqrt{y})^2$
$= x - y$ **(A)**

4. $(\sqrt{x}+\sqrt{y})^2$
$= (\sqrt{x})^2 + 2\sqrt{x}\sqrt{y} + (\sqrt{y})^2$
$= x + 2\sqrt{xy} + y$ **(F)**

5. $(\sqrt{x}-\sqrt{y})^2$
$= (\sqrt{x})^2 - 2\sqrt{x}\sqrt{y} + (\sqrt{y})^2$
$= x - 2\sqrt{xy} + y$ **(D)**

6. $(\sqrt{x}+y)^2$
$= (\sqrt{x})^2 + 2\sqrt{x}\cdot y + y^2$
$= x + 2y\sqrt{x} + y^2$ **(B)**

7. $\sqrt{3}\left(\sqrt{12}-4\right) = \sqrt{3}\cdot\sqrt{12} - \sqrt{3}\cdot 4$
$= \sqrt{3\cdot 12} - 4\sqrt{3}$
$= \sqrt{36} - 4\sqrt{3}$
$= 6 - 4\sqrt{3}$

8. $\sqrt{5}\left(\sqrt{125}-6\right) = \sqrt{5}\cdot\sqrt{125} - \sqrt{5}\cdot 6$
$= \sqrt{5\cdot 125} - 6\sqrt{5}$
$= \sqrt{625} - 6\sqrt{5}$
$= 25 - 6\sqrt{5}$

9. $\sqrt{2}\left(\sqrt{18}-\sqrt{3}\right) = \sqrt{2}\cdot\sqrt{18} - \sqrt{2}\cdot\sqrt{3}$
$= \sqrt{36} - \sqrt{6}$
$= 6 - \sqrt{6}$

10. $\sqrt{5}\left(\sqrt{15}+\sqrt{5}\right) = \sqrt{5}\cdot\sqrt{15} + \sqrt{5}\cdot\sqrt{5}$
$= \sqrt{75} + 5$
$= 5\sqrt{3} + 5$

11. $(\sqrt{6}+2)(\sqrt{6}-2)$
$= \sqrt{6}\cdot\sqrt{6} - 2\sqrt{6} + 2\sqrt{6} - 2\cdot 2$
$= 6 - 4$
$= 2$

12. $(\sqrt{7}+8)(\sqrt{7}-8)$
$= \sqrt{7}\cdot\sqrt{7} - 8\cdot\sqrt{7} + 8\cdot\sqrt{7} - 8\cdot 8$
$= 7 - 64$
$= -57$

13. $(\sqrt{12}-\sqrt{3})(\sqrt{12}+\sqrt{3}) = (\sqrt{12})^2 - (\sqrt{3})^2$
$= 12 - 3$
$= 9$

14. $\left(\sqrt{18}+\sqrt{8}\right)\left(\sqrt{18}-\sqrt{8}\right)$
$= (\sqrt{18})^2 - (\sqrt{8})^2$
$= 18 - 8$
$= 10$

15. $(\sqrt{3}+2)(\sqrt{6}-5)$
$= \sqrt{3}\cdot\sqrt{6} - 5\sqrt{3} + 2\sqrt{6} - 10$
$= \sqrt{18} - 5\sqrt{3} + 2\sqrt{6} - 10$
$= \sqrt{9\cdot 2} - 5\sqrt{3} + 2\sqrt{6} - 10$
$= 3\sqrt{2} - 5\sqrt{3} + 2\sqrt{6} - 10$

16. $(\sqrt{7}+1)(\sqrt{2}-4) = \sqrt{14} - 4\sqrt{7} + \sqrt{2} - 4$

17. $(\sqrt{3x}+2)(\sqrt{3x}-2) = (\sqrt{3x})^2 - 2^2$
$= 3x - 4$

18. $(\sqrt{6y}-4)(\sqrt{6y}+4) = (\sqrt{6y})^2 - 4^2$
$= 6y - 16$

19. $(2\sqrt{x}+\sqrt{y})(2\sqrt{x}-\sqrt{y})$
$= (2\sqrt{x})^2 - (\sqrt{y})^2$
$= 2^2(\sqrt{x})^2 - y$
$= 4x - y$

20. $(\sqrt{p}+5\sqrt{s})(\sqrt{p}-5\sqrt{s})$
$= (\sqrt{p})^2 - (5\sqrt{s})^2$
$= p - 25s$

21. $(4\sqrt{x}+3)^2 = (4\sqrt{x})^2 + 2(4\sqrt{x})(3) + 3^2$
$= 16x + 24\sqrt{x} + 9$

22. $\left(5\sqrt{p}-6\right)^2 = \left(5\sqrt{p}\right)^2 + 2\left(5\sqrt{p}\right)(-6) + (-6)^2$
$= 25p - 60\sqrt{p} + 36$

23. $\left(9-\sqrt[3]{2}\right)\left(9+\sqrt[3]{2}\right) = 9^2 - \left(\sqrt[3]{2}\right)^2$
$= 81 - \sqrt[3]{4}$

24. $\left(7+\sqrt[3]{6}\right)\left(7-\sqrt[3]{6}\right) = 7^2 - \left(\sqrt[3]{6}\right)^2$
$= 49 - \sqrt[3]{36}$

25. $6 - 4\sqrt{3}$ is not equal to $2\sqrt{3}$ because 6 and $4\sqrt{3}$ are not like terms, so they cannot be combined.

26. In rationalizing the denominator in the radical expression $\frac{1}{\sqrt{2}}$, we multiply both the numerator and denominator by $\sqrt{2}$ because this process is the same as multiplying the expression by

$$\frac{\sqrt{2}}{\sqrt{2}} = 1.$$

This procedure is justified by the *identity property for multiplication.*

27. $\dfrac{7}{\sqrt{7}} = \dfrac{7\cdot\sqrt{7}}{\sqrt{7}\cdot\sqrt{7}} = \dfrac{7\sqrt{7}}{7} = \sqrt{7}$

28. $\dfrac{11}{\sqrt{11}} = \dfrac{11\cdot\sqrt{11}}{\sqrt{11}\cdot\sqrt{11}} = \dfrac{11\sqrt{11}}{11} = \sqrt{11}$

29. $\dfrac{15}{\sqrt{3}} = \dfrac{15\cdot\sqrt{3}}{\sqrt{3}\cdot\sqrt{3}} = \dfrac{15\sqrt{3}}{3} = 5\sqrt{3}$

30. $\dfrac{12}{\sqrt{6}} = \dfrac{12\cdot\sqrt{6}}{\sqrt{6}\cdot\sqrt{6}} = \dfrac{12\sqrt{6}}{6} = 2\sqrt{6}$

31. $\dfrac{\sqrt{3}}{\sqrt{2}} = \dfrac{\sqrt{3}\cdot\sqrt{2}}{\sqrt{2}\cdot\sqrt{2}} = \dfrac{\sqrt{6}}{2}$

32. $\dfrac{\sqrt{7}}{\sqrt{6}} = \dfrac{\sqrt{7}\cdot\sqrt{6}}{\sqrt{6}\cdot\sqrt{6}} = \dfrac{\sqrt{42}}{6}$

33. $\dfrac{9\sqrt{3}}{\sqrt{5}} = \dfrac{9\sqrt{3}\cdot\sqrt{5}}{\sqrt{5}\cdot\sqrt{5}} = \dfrac{9\sqrt{15}}{5}$

34. $\dfrac{3\sqrt{2}}{\sqrt{11}} = \dfrac{3\sqrt{2}\cdot\sqrt{11}}{\sqrt{11}\cdot\sqrt{11}} = \dfrac{3\sqrt{22}}{11}$

35. $\dfrac{-6}{\sqrt{18}} = \dfrac{-6}{\sqrt{9\cdot 2}} = \dfrac{-6}{3\sqrt{2}} = \dfrac{-2}{\sqrt{2}} = \dfrac{-2\cdot\sqrt{2}}{\sqrt{2}\cdot\sqrt{2}}$
$= \dfrac{-2\sqrt{2}}{2} = -\sqrt{2}$

36. $\dfrac{-5}{\sqrt{24}} = \dfrac{-5}{2\sqrt{6}} = \dfrac{-5\cdot\sqrt{6}}{2\sqrt{6}\cdot\sqrt{6}}$
$= \dfrac{-5\sqrt{6}}{2\cdot 6} = -\dfrac{5\sqrt{6}}{12}$

37. $\dfrac{-8\sqrt{3}}{\sqrt{k}} = \dfrac{-8\sqrt{3}\cdot\sqrt{k}}{\sqrt{k}\cdot\sqrt{k}} = \dfrac{-8\sqrt{3k}}{k}$

38. $\dfrac{-4\sqrt{13}}{\sqrt{m}} = \dfrac{-4\sqrt{13}\cdot\sqrt{m}}{\sqrt{m}\cdot\sqrt{m}} = \dfrac{-4\sqrt{13m}}{m}$

39. $\dfrac{6\sqrt{3y}}{\sqrt{y^3}} = \dfrac{6\sqrt{3y}\cdot\sqrt{y}}{\sqrt{y^3}\cdot\sqrt{y}}$
$= \dfrac{6\sqrt{3y^2}}{\sqrt{y^4}}$
$= \dfrac{6y\sqrt{3}}{y^2}$
$= \dfrac{6\sqrt{3}}{y}$

40. $\dfrac{-8\sqrt{5y}}{\sqrt{y^5}} = \dfrac{-8\sqrt{5y}\cdot\sqrt{y}}{\sqrt{y^5}\cdot\sqrt{y}}$
$= \dfrac{-8\sqrt{5y^2}}{\sqrt{y^6}}$
$= \dfrac{-8y\sqrt{5}}{y^3}$
$= -\dfrac{8\sqrt{5}}{y^2}$

41. Multiply the numerator and denominator by $\sqrt{y}$:

$$\frac{6\sqrt{3y}}{\sqrt{y^3}} = \frac{6\sqrt{3y}\cdot\sqrt{y}}{\sqrt{y^3}\cdot\sqrt{y}} = \frac{6\sqrt{3y^2}}{\sqrt{y^4}} = \frac{6y\sqrt{3}}{y^2} = \frac{6\sqrt{3}}{y}.$$

Multiply the numerator and denominator by $\sqrt{y^3}$:

$$\frac{6\sqrt{3y}}{\sqrt{y^3}} = \frac{6\sqrt{3y}\cdot\sqrt{y^3}}{\sqrt{y^3}\cdot\sqrt{y^3}} = \frac{6\sqrt{3y^4}}{y^3} = \frac{6y^2\sqrt{3}}{y^3} = \frac{6\sqrt{3}}{y}.$$

Answers will vary. A sample answer follows:

Both methods lead to the same result, $\dfrac{6\sqrt{3}}{y}$, but multiplying the numerator and denominator by $\sqrt{y}$ produces this result more directly, with less simplification required.

42. Multiply the numerator and the denominator by $\sqrt[3]{2}$:

$$\frac{1}{\sqrt[3]{2}} = \frac{1 \cdot \sqrt[3]{2}}{\sqrt[3]{2} \cdot \sqrt[3]{2}} = \frac{\sqrt[3]{2}}{\sqrt[3]{4}}.$$

The denominator is not yet rationalized. Multiply both the numerator and the denominator by $\sqrt[3]{4}$:

$$\frac{1}{\sqrt[3]{2}} = \frac{1 \cdot \sqrt[3]{4}}{\sqrt[3]{2} \cdot \sqrt[3]{4}} = \frac{\sqrt[3]{4}}{\sqrt[3]{8}} = \frac{\sqrt[3]{4}}{2}.$$

To rationalize a cube root, three factors of the quantity under the radical sign are needed. We must multiply by $\sqrt[3]{2^2}$ or $\sqrt[3]{4}$ to rationalize $\sqrt[3]{2}$.

43. $\sqrt{\frac{7}{2}} = \frac{\sqrt{7}}{\sqrt{2}} = \frac{\sqrt{7} \cdot \sqrt{2}}{\sqrt{2} \cdot \sqrt{2}} = \frac{\sqrt{14}}{2}$

44. $\sqrt{\frac{10}{3}} = \frac{\sqrt{10}}{\sqrt{3}} = \frac{\sqrt{10} \cdot \sqrt{3}}{\sqrt{3} \cdot \sqrt{3}} = \frac{\sqrt{30}}{3}$

45. $-\sqrt{\frac{7}{50}} = -\frac{\sqrt{7}}{\sqrt{25 \cdot 2}} = -\frac{\sqrt{7}}{5\sqrt{2}}$

$= -\frac{\sqrt{7} \cdot \sqrt{2}}{5\sqrt{2} \cdot \sqrt{2}} = -\frac{\sqrt{14}}{5 \cdot 2} = -\frac{\sqrt{14}}{10}$

46. $-\sqrt{\frac{13}{75}} = -\frac{\sqrt{13}}{\sqrt{75}} = -\frac{\sqrt{13}}{\sqrt{25 \cdot 3}} = -\frac{\sqrt{13}}{5\sqrt{3}}$

$= -\frac{\sqrt{13} \cdot \sqrt{3}}{5\sqrt{3} \cdot \sqrt{3}} = -\frac{\sqrt{39}}{5 \cdot 3} = -\frac{\sqrt{39}}{15}$

47. $\sqrt{\frac{24}{x}} = \frac{\sqrt{24}}{\sqrt{x}} = \frac{\sqrt{4 \cdot 6}}{\sqrt{x}} = \frac{2\sqrt{6}}{\sqrt{x}}$

$= \frac{2\sqrt{6} \cdot \sqrt{x}}{\sqrt{x} \cdot \sqrt{x}} = \frac{2\sqrt{6x}}{x}$

48. $\sqrt{\frac{52}{y}} = \frac{\sqrt{52}}{\sqrt{y}} = \frac{2\sqrt{13}}{\sqrt{y}}$

$= \frac{2\sqrt{13} \cdot \sqrt{y}}{\sqrt{y} \cdot \sqrt{y}} = \frac{2\sqrt{13y}}{y}$

49. $-\sqrt{\frac{98r^3}{s}} = -\frac{\sqrt{98r^3}}{\sqrt{s}} = -\frac{\sqrt{49r^2 \cdot 2r}}{\sqrt{s}}$

$= -\frac{7r\sqrt{2r}}{\sqrt{s}} = -\frac{7r\sqrt{2r} \cdot \sqrt{s}}{\sqrt{s} \cdot \sqrt{s}}$

$= -\frac{7r\sqrt{2rs}}{s}$

50. $-\sqrt{\frac{150m^5}{n}} = -\frac{\sqrt{150m^5}}{\sqrt{n}} = -\frac{\sqrt{25m^4 \cdot 6m}}{\sqrt{n}}$

$= -\frac{5m^2\sqrt{6m}}{\sqrt{n}} = -\frac{5m^2\sqrt{6m} \cdot \sqrt{n}}{\sqrt{n} \cdot \sqrt{n}}$

$= -\frac{5m^2\sqrt{6mn}}{n}$

51. $\sqrt{\frac{288x^7}{y^9}} = \frac{\sqrt{288x^7}}{\sqrt{y^9}} = \frac{\sqrt{144x^6 \cdot 2x}}{\sqrt{y^8 \cdot y}}$

$= \frac{12x^3\sqrt{2x}}{y^4\sqrt{y}} = \frac{12x^3\sqrt{2x} \cdot \sqrt{y}}{y^4\sqrt{y} \cdot \sqrt{y}}$

$= \frac{12x^3\sqrt{2xy}}{y^4 \cdot y} = \frac{12x^3\sqrt{2xy}}{y^5}$

52. $\sqrt{\frac{242t^9}{u^{11}}} = \frac{\sqrt{242t^9}}{\sqrt{u^{11}}} = \frac{\sqrt{121t^8 \cdot 2t}}{\sqrt{u^{10} \cdot u}}$

$= \frac{11t^4\sqrt{2t}}{u^5\sqrt{u}} = \frac{11t^4\sqrt{2t} \cdot \sqrt{u}}{u^5\sqrt{u} \cdot \sqrt{u}}$

$= \frac{11t^4\sqrt{2tu}}{u^6}$

53. $\sqrt[3]{\frac{2}{3}} = \frac{\sqrt[3]{2} \cdot \sqrt[3]{9}}{\sqrt[3]{3} \cdot \sqrt[3]{9}} = \frac{\sqrt[3]{18}}{\sqrt[3]{27}} = \frac{\sqrt[3]{18}}{3}$

54. $\sqrt[3]{\frac{4}{5}} = \frac{\sqrt[3]{4}}{\sqrt[3]{5}} = \frac{\sqrt[3]{4} \cdot \sqrt[3]{25}}{\sqrt[3]{5} \cdot \sqrt[3]{25}} = \frac{\sqrt[3]{100}}{\sqrt[3]{125}} = \frac{\sqrt[3]{100}}{5}$

55. $\sqrt[3]{\frac{4}{9}} = \frac{\sqrt[3]{4}}{\sqrt[3]{9}} = \frac{\sqrt[3]{4}}{\sqrt[3]{3^2}} = \frac{\sqrt[3]{4} \cdot \sqrt[3]{3}}{\sqrt[3]{3^2} \cdot \sqrt[3]{3}}$

$= \frac{\sqrt[3]{12}}{\sqrt[3]{3^3}} = \frac{\sqrt[3]{12}}{3}$

56. $\sqrt[3]{\frac{5}{16}} = \frac{\sqrt[3]{5}}{\sqrt[3]{16}} = \frac{\sqrt[3]{5}}{\sqrt[3]{8 \cdot 2}} = \frac{\sqrt[3]{5}}{2\sqrt[3]{2}}$

$= \frac{\sqrt[3]{5} \cdot \sqrt[3]{4}}{2\sqrt[3]{2} \cdot \sqrt[3]{4}} = \frac{\sqrt[3]{20}}{2\sqrt[3]{8}} = \frac{\sqrt[3]{20}}{2 \cdot 2} = \frac{\sqrt[3]{20}}{4}$

57. $-\sqrt[3]{\frac{2p}{r^2}} = -\frac{\sqrt[3]{2p}}{\sqrt[3]{r^2}} = -\frac{\sqrt[3]{2p} \cdot \sqrt[3]{r}}{\sqrt[3]{r^2} \cdot \sqrt[3]{r}}$

$= -\frac{\sqrt[3]{2pr}}{\sqrt[3]{r^3}} = -\frac{\sqrt[3]{2pr}}{r}$

58. $-\sqrt[3]{\frac{6x}{y^2}} = -\frac{\sqrt[3]{6x}}{\sqrt[3]{y^2}} = -\frac{\sqrt[3]{6x} \cdot \sqrt[3]{y}}{\sqrt[3]{y^2} \cdot \sqrt[3]{y}}$

$= -\frac{\sqrt[3]{6xy}}{\sqrt[3]{y^3}} = -\frac{\sqrt[3]{6xy}}{y}$

59. $\sqrt[4]{\frac{16}{x}} = \frac{\sqrt[4]{16}}{\sqrt[4]{x}} = \frac{2}{\sqrt[4]{x}} = \frac{2\cdot\sqrt[4]{x^3}}{\sqrt[4]{x}\cdot\sqrt[4]{x^3}}$

$= \frac{2\sqrt[4]{x^3}}{\sqrt[4]{x^4}} = \frac{2\sqrt[4]{x^3}}{x}$

60. $\sqrt[4]{\frac{81}{y}} = \frac{\sqrt[4]{81}}{\sqrt[4]{y}} = \frac{3}{\sqrt[4]{y}} = \frac{3\cdot\sqrt[4]{y^3}}{\sqrt[4]{y}\cdot\sqrt[4]{y^3}}$

$= \frac{3\sqrt[4]{y^3}}{\sqrt[4]{y^4}} = \frac{3\sqrt[4]{y^3}}{y}$

61. To rationalize the denominator of the expression $\frac{2}{4+\sqrt{3}}$, we must multiply the numerator and denominator by $4-\sqrt{3}$. The denominator becomes $(4+\sqrt{3})(4-\sqrt{3}) = 16-3 = 13$, which is a rational number.

62. No, because $(4+\sqrt{3})(4+\sqrt{3}) = 16+8\sqrt{3}+3 = 19+8\sqrt{3}$, which is not a rational number.

63. $\frac{2}{4+\sqrt{3}}$

Multiply both the numerator and denominator by the conjugate of the denominator, $4-\sqrt{3}$.

$= \frac{2(4-\sqrt{3})}{(4+\sqrt{3})(4-\sqrt{3})}$

$= \frac{2(4-\sqrt{3})}{16-3}$

$= \frac{2(4-\sqrt{3})}{13}$

64. $\frac{6}{5+\sqrt{2}}$

Multiply both the numerator and denominator by the conjugate of the denominator, $5-\sqrt{2}$.

$= \frac{6(5-\sqrt{2})}{(5+\sqrt{2})(5-\sqrt{2})}$

$= \frac{6(5-\sqrt{2})}{25-2}$

$= \frac{6(5-\sqrt{2})}{23}$

65. $\frac{6}{\sqrt{5}+\sqrt{3}}$

Multiply both the numerator and denominator by the conjugate of the denominator, $\sqrt{5}-\sqrt{3}$.

$= \frac{6(\sqrt{5}-\sqrt{3})}{(\sqrt{5}+\sqrt{3})(\sqrt{5}-\sqrt{3})}$

$= \frac{6(\sqrt{5}-\sqrt{3})}{(\sqrt{5})^2-(\sqrt{3})^2}$

$= \frac{6(\sqrt{5}-\sqrt{3})}{5-3}$

$= \frac{6(\sqrt{5}-\sqrt{3})}{2}$

$= 3(\sqrt{5}-\sqrt{3})$

66. $\frac{12}{\sqrt{6}+\sqrt{3}}$

Multiply both the numerator and denominator by the conjugate of the denominator, $\sqrt{6}-\sqrt{3}$.

$= \frac{12(\sqrt{6}-\sqrt{3})}{(\sqrt{6}+\sqrt{3})(\sqrt{6}-\sqrt{3})}$

$= \frac{12(\sqrt{6}-\sqrt{3})}{(\sqrt{6})^2-(\sqrt{3})^2}$

$= \frac{12(\sqrt{6}-\sqrt{3})}{6-3}$

$= \frac{12(\sqrt{6}-\sqrt{3})}{3}$

$= 4(\sqrt{6}-\sqrt{3})$

67. $\frac{-4}{\sqrt{3}-\sqrt{7}}$

Multiply both the numerator and denominator by the conjugate of the denominator, $\sqrt{3}+\sqrt{7}$.

$= \frac{-4(\sqrt{3}+\sqrt{7})}{(\sqrt{3}-\sqrt{7})(\sqrt{3}+\sqrt{7})}$

$= \frac{-4(\sqrt{3}+\sqrt{7})}{3-7}$

$= \frac{-4(\sqrt{3}+\sqrt{7})}{-4}$

$= \sqrt{3}+\sqrt{7}$

68. $\dfrac{-3}{\sqrt{2}+\sqrt{5}}$

Multiply both the numerator and denominator by the conjugate of the denominator, $\sqrt{2}-\sqrt{5}$.

$$= \frac{-3\left(\sqrt{2}-\sqrt{5}\right)}{(\sqrt{2}+\sqrt{5})(\sqrt{2}-\sqrt{5})}$$
$$= \frac{-3\left(\sqrt{2}-\sqrt{5}\right)}{2-5}$$
$$= \frac{-3\left(\sqrt{2}-\sqrt{5}\right)}{-3}$$
$$= \sqrt{2}-\sqrt{5}$$

69. $\dfrac{1-\sqrt{2}}{\sqrt{7}+\sqrt{6}}$

Multiply both the numerator and denominator by the conjugate of the denominator, $\sqrt{7}-\sqrt{6}$.

$$= \frac{(1-\sqrt{2})(\sqrt{7}-\sqrt{6})}{(\sqrt{7}+\sqrt{6})(\sqrt{7}-\sqrt{6})}$$
$$= \frac{\sqrt{7}-\sqrt{6}-\sqrt{14}+\sqrt{12}}{7-6}$$
$$= \frac{\sqrt{7}-\sqrt{6}-\sqrt{14}+\sqrt{4\cdot 3}}{1}$$
$$= \sqrt{7}-\sqrt{6}-\sqrt{14}+2\sqrt{3}$$

70. $\dfrac{-1-\sqrt{3}}{\sqrt{6}+\sqrt{5}}$

Multiply both the numerator and denominator by the conjugate of the denominator, $\sqrt{6}-\sqrt{5}$.

$$= \frac{(-1-\sqrt{3})(\sqrt{6}-\sqrt{5})}{(\sqrt{6}+\sqrt{5})(\sqrt{6}-\sqrt{5})}$$
$$= \frac{-\sqrt{6}+\sqrt{5}-\sqrt{18}+\sqrt{15}}{6-5}$$
$$= \frac{-\sqrt{6}+\sqrt{5}-\sqrt{9\cdot 2}+\sqrt{15}}{1}$$
$$= -\sqrt{6}+\sqrt{5}-3\sqrt{2}+\sqrt{15}$$

71. $\dfrac{4\sqrt{x}}{\sqrt{x}-2\sqrt{y}}$

Multiply both the numerator and denominator by the conjugate of the denominator, $\sqrt{x}+2\sqrt{y}$.

$$= \frac{4\sqrt{x}(\sqrt{x}+2\sqrt{y})}{(\sqrt{x}-2\sqrt{y})(\sqrt{x}+2\sqrt{y})}$$
$$= \frac{4\sqrt{x}(\sqrt{x}+2\sqrt{y})}{x-4y}$$

72. $\dfrac{5\sqrt{r}}{3\sqrt{r}+\sqrt{s}}$

Multiply both the numerator and denominator by the conjugate of the denominator, $3\sqrt{r}-\sqrt{s}$.

$$= \frac{5\sqrt{r}(3\sqrt{r}-\sqrt{s})}{(3\sqrt{r}+\sqrt{s})(3\sqrt{r}-\sqrt{s})}$$
$$= \frac{5\sqrt{r}(3\sqrt{r}-\sqrt{s})}{9r-s}$$

73. $\dfrac{\sqrt{x}-\sqrt{y}}{\sqrt{2x}+\sqrt{3y}}$

Multiply both the numerator and denominator by the conjugate of the denominator, $\sqrt{2x}-\sqrt{3y}$.

$$= \frac{\sqrt{x}-\sqrt{y}}{\sqrt{2x}+\sqrt{3y}}\cdot\frac{\sqrt{2x}-\sqrt{3y}}{\sqrt{2x}-\sqrt{3y}}$$
$$= \frac{\sqrt{x}\cdot\sqrt{2x}-\sqrt{x}\cdot\sqrt{3y}-\sqrt{y}\cdot\sqrt{2x}+\sqrt{y}\cdot\sqrt{3y}}{\left(\sqrt{2x}\right)^2-\left(\sqrt{3y}\right)^2}$$
$$= \frac{\sqrt{2x^2}-\sqrt{3xy}-\sqrt{2xy}+\sqrt{3y^2}}{2x-3y}$$
$$= \frac{x\sqrt{2}-\sqrt{3xy}-\sqrt{2xy}+y\sqrt{3}}{2x-3y}$$

74. $\dfrac{\sqrt{a}+\sqrt{b}}{\sqrt{5a}-\sqrt{2b}}$

Multiply both the numerator and denominator by the conjugate of the denominator, $\sqrt{5a}+\sqrt{2b}$.

$$= \frac{\sqrt{a}+\sqrt{b}}{\sqrt{5a}-\sqrt{2b}}\cdot\frac{\sqrt{5a}+\sqrt{2b}}{\sqrt{5a}+\sqrt{2b}}$$
$$= \frac{\sqrt{a}\cdot\sqrt{5a}+\sqrt{a}\cdot\sqrt{2b}+\sqrt{b}\cdot\sqrt{5a}+\sqrt{b}\cdot\sqrt{2b}}{\left(\sqrt{5a}\right)^2-\left(\sqrt{2b}\right)^2}$$
$$= \frac{\sqrt{5a^2}+\sqrt{2ab}+\sqrt{5ab}+\sqrt{2b^2}}{5a-2b}$$
$$= \frac{a\sqrt{5}+\sqrt{2ab}+\sqrt{5ab}+b\sqrt{2}}{5a-2b}$$

75. Let $a=\dfrac{\sqrt{6}-\sqrt{2}}{4}$ and $b=\dfrac{\sqrt{2-\sqrt{3}}}{2}$.
Since the numerators and denominators are all positive, a and b are both positive.

$$a^2=\left(\frac{\sqrt{6}-\sqrt{2}}{4}\right)^2$$
$$= \frac{(\sqrt{6}-\sqrt{2})(\sqrt{6}-\sqrt{2})}{4^2}$$
$$= \frac{6-2\sqrt{12}+2}{16}=\frac{8-2\sqrt{12}}{16}$$

$$= \frac{8 - 2\sqrt{4 \cdot 3}}{16} = \frac{8 - 4\sqrt{3}}{16}$$
$$= \frac{4(2 - \sqrt{3})}{4 \cdot 4} = \frac{2 - \sqrt{3}}{4}$$
$$b^2 = \left(\frac{\sqrt{2 - \sqrt{3}}}{2}\right)^2$$
$$= \frac{\sqrt{2 - \sqrt{3}} \cdot \sqrt{2 - \sqrt{3}}}{2^2}$$
$$= \frac{2 - \sqrt{3}}{4}$$

Since $a^2 = b^2$ and a and b are both positive, $a = b$.

76. Each original expression is approximately equal to .2588190451.

77. $$\frac{25 + 10\sqrt{6}}{20} = \frac{5\left(5 + 2\sqrt{6}\right)}{5 \cdot 4} = \frac{5 + 2\sqrt{6}}{4}$$

78. $$\frac{12 - 6\sqrt{2}}{24} = \frac{6\left(2 - \sqrt{2}\right)}{6 \cdot 4} = \frac{2 - \sqrt{2}}{4}$$

79. $$\frac{16 + 4\sqrt{8}}{12} = \frac{16 + 4(2\sqrt{2})}{12} = \frac{16 + 8\sqrt{2}}{12} = \frac{4(4 + 2\sqrt{2})}{4 \cdot 3} = \frac{4 + 2\sqrt{2}}{3}$$

80. $$\frac{12 + 9\sqrt{72}}{18} = \frac{12 + 9 \cdot 6\sqrt{2}}{18} = \frac{6(2 + 9\sqrt{2})}{6 \cdot 3} = \frac{2 + 9\sqrt{2}}{3}$$

81. $$\frac{6x + \sqrt{24x^3}}{3x}$$
$$= \frac{6x + \sqrt{4x^2 \cdot 6x}}{3x} = \frac{6x + 2x\sqrt{6x}}{3x}$$
$$= \frac{x(6 + 2\sqrt{6x})}{3x} = \frac{6 + 2\sqrt{6x}}{3}$$

82. $$\frac{11y + \sqrt{242y^5}}{22y}$$
$$= \frac{11y + \sqrt{121y^4 \cdot 2y}}{22y} = \frac{11y + 11y^2\sqrt{2y}}{22y}$$
$$= \frac{11y(1 + y\sqrt{2y})}{11y \cdot 2} = \frac{1 + y\sqrt{2y}}{2}$$

83. $$\frac{8\sqrt{5} - 1}{6} = \frac{(8\sqrt{5} - 1)(8\sqrt{5} + 1)}{6(8\sqrt{5} + 1)}$$
$$= \frac{(8\sqrt{5})^2 - 1^2}{6(8\sqrt{5} + 1)}$$
$$= \frac{64(5) - 1}{6(8\sqrt{5} + 1)}$$
$$= \frac{319}{6(8\sqrt{5} + 1)}$$

84. $$\frac{3\sqrt{a} + \sqrt{b}}{\sqrt{b} - \sqrt{a}} = \frac{(3\sqrt{a} + \sqrt{b})(3\sqrt{a} - \sqrt{b})}{(\sqrt{b} - \sqrt{a})(3\sqrt{a} - \sqrt{b})}$$
$$= \frac{(3\sqrt{a})^2 - (\sqrt{b})^2}{(\sqrt{b} - \sqrt{a})(3\sqrt{a} - \sqrt{b})}$$
$$= \frac{9a - b}{(\sqrt{b} - \sqrt{a})(3\sqrt{a} - \sqrt{b})}$$

85. $$\frac{3\sqrt{a} + \sqrt{b}}{\sqrt{b} - \sqrt{a}} = \frac{(3\sqrt{a} + \sqrt{b})(\sqrt{b} + \sqrt{a})}{(\sqrt{b} - \sqrt{a})(\sqrt{b} + \sqrt{a})}$$
$$= \frac{(3\sqrt{a} + \sqrt{b})(\sqrt{b} + \sqrt{a})}{(\sqrt{b})^2 - (\sqrt{a})^2}$$
$$= \frac{(3\sqrt{a} + \sqrt{b})(\sqrt{b} + \sqrt{a})}{b - a}$$

86. In Exercise 84, we multiplied the numerator and denominator by the conjugate of the numerator, while in Exercise 85 we multiplied by the conjugate of the denominator.

Summary Exercises on Operations with Radicals and Rational Exponents

1. $$6\sqrt{10} - 12\sqrt{10} = (6 - 12)\sqrt{10} = -6\sqrt{10}$$

2. $$\sqrt{7}(\sqrt{7} - \sqrt{2}) = \sqrt{7} \cdot \sqrt{7} - \sqrt{7} \cdot \sqrt{2} = 7 - \sqrt{14}$$

3. $$(1 - \sqrt{3})(2 + \sqrt{6})$$
F O I L
$$= 2 + \sqrt{6} - 2\sqrt{3} - \sqrt{18}$$
$$= 2 + \sqrt{6} - 2\sqrt{3} - \sqrt{9} \cdot \sqrt{2}$$
$$= 2 + \sqrt{6} - 2\sqrt{3} - 3\sqrt{2}$$

4. $$\sqrt{50} - \sqrt{98} + \sqrt{72}$$
$$= \sqrt{25} \cdot \sqrt{2} - \sqrt{49} \cdot \sqrt{2} + \sqrt{36} \cdot \sqrt{2}$$
$$= 5\sqrt{2} - 7\sqrt{2} + 6\sqrt{2}$$
$$= (5 - 7 + 6)\sqrt{2}$$
$$= 4\sqrt{2}$$

5. $(3\sqrt{5}+2\sqrt{7})^2$
$= (3\sqrt{5})^2 + 2(3\sqrt{5})(2\sqrt{7}) + (2\sqrt{7})^2$
$= 9\cdot 5 + 12\sqrt{35} + 4\cdot 7$
$= 45 + 12\sqrt{35} + 28$
$= 73 + 12\sqrt{35}$

6. $\dfrac{-3}{\sqrt{6}} = \dfrac{-3}{\sqrt{6}}\cdot\dfrac{\sqrt{6}}{\sqrt{6}}$
$= \dfrac{-3\sqrt{6}}{6}$
$= \dfrac{-\sqrt{6}}{2}$

7. $\dfrac{8}{\sqrt{7}+\sqrt{5}} = \dfrac{8}{\sqrt{7}+\sqrt{5}}\cdot\dfrac{\sqrt{7}-\sqrt{5}}{\sqrt{7}-\sqrt{5}}$
$= \dfrac{8(\sqrt{7}-\sqrt{5})}{7-5}$
$= \dfrac{8(\sqrt{7}-\sqrt{5})}{2}$
$= 4(\sqrt{7}-\sqrt{5})$

8. $\sqrt[3]{16x^2} - \sqrt[3]{54x^2} + \sqrt[3]{128x^2}$
$= \sqrt[3]{8}\cdot\sqrt[3]{2x^2} - \sqrt[3]{27}\cdot\sqrt[3]{2x^2} + \sqrt[3]{64}\cdot\sqrt[3]{2x^2}$
$= 2\sqrt[3]{2x^2} - 3\sqrt[3]{2x^2} + 4\sqrt[3]{2x^2}$
$= (2-3+4)\sqrt[3]{2x^2}$
$= 3\sqrt[3]{2x^2}$

9. $\dfrac{1-\sqrt{2}}{1+\sqrt{2}} = \dfrac{1-\sqrt{2}}{1+\sqrt{2}}\cdot\dfrac{1-\sqrt{2}}{1-\sqrt{2}}$
$= \dfrac{(1-\sqrt{2})^2}{1^2-(\sqrt{2})^2}$
$= \dfrac{1-2\sqrt{2}+2}{1-2}$
$= \dfrac{3-2\sqrt{2}}{-1}$
$= -3+2\sqrt{2}$

10. $(1-\sqrt[3]{3})(1+\sqrt[3]{3}+\sqrt[3]{9})$
$= 1+\sqrt[3]{3}+\sqrt[3]{9}-\sqrt[3]{3}-\sqrt[3]{9}-\sqrt[3]{27}$
$= 1-\sqrt[3]{27}$
$= 1-3 = -2$

11. $(\sqrt{5}+7)(\sqrt{5}-7) = (\sqrt{5})^2 - 7^2$
$= 5-49$
$= -44$

12. $\dfrac{1}{\sqrt{x}-\sqrt{5}} = \dfrac{1}{\sqrt{x}-\sqrt{5}}\cdot\dfrac{\sqrt{x}+\sqrt{5}}{\sqrt{x}+\sqrt{5}}$
$= \dfrac{1(\sqrt{x}+\sqrt{5})}{(\sqrt{x})^2-(\sqrt{5})^2}$
$= \dfrac{\sqrt{x}+\sqrt{5}}{x-5} \quad (x \neq 5)$

13. $\sqrt[3]{8a^3b^5c^9} = \sqrt[3]{8a^3b^3c^9}\cdot\sqrt[3]{b^2}$
$= 2abc^3\sqrt[3]{b^2}$

14. $\dfrac{15}{\sqrt[3]{9}} = \dfrac{15}{\sqrt[3]{9}}\cdot\dfrac{\sqrt[3]{3}}{\sqrt[3]{3}}$
$= \dfrac{15\sqrt[3]{3}}{\sqrt[3]{27}}$
$= \dfrac{15\sqrt[3]{3}}{3} = 5\sqrt[3]{3}$

15. $\dfrac{3}{\sqrt{5}+2} = \dfrac{3}{\sqrt{5}+2}\cdot\dfrac{\sqrt{5}-2}{\sqrt{5}-2}$
$= \dfrac{3(\sqrt{5}-2)}{5-4}$
$= 3(\sqrt{5}-2)$

16. $\sqrt{\dfrac{3}{5x}} = \dfrac{\sqrt{3}}{\sqrt{5x}}\cdot\dfrac{\sqrt{5x}}{\sqrt{5x}}$
$= \dfrac{\sqrt{15x}}{5x}$

17. $\dfrac{16\sqrt{3}}{5\sqrt{12}} = \dfrac{16\sqrt{3}}{5\cdot\sqrt{4}\cdot\sqrt{3}}$
$= \dfrac{16}{5\cdot 2}$
$= \dfrac{8}{5}$

18. $\dfrac{2\sqrt{25}}{8\sqrt{50}} = \dfrac{2\cdot 5}{8\cdot\sqrt{25}\cdot\sqrt{2}}\cdot\dfrac{\sqrt{2}}{\sqrt{2}}$
$= \dfrac{2\cdot 5\cdot\sqrt{2}}{8\cdot 5\cdot 2}$
$= \dfrac{\sqrt{2}}{8}$

19. $\dfrac{-10}{\sqrt[3]{10}} = \dfrac{-10}{\sqrt[3]{10}}\cdot\dfrac{\sqrt[3]{100}}{\sqrt[3]{100}}$
$= \dfrac{-10\sqrt[3]{100}}{\sqrt[3]{1000}}$
$= \dfrac{-10\sqrt[3]{100}}{10} = -\sqrt[3]{100}$

20. $\dfrac{\sqrt{6}+\sqrt{5}}{\sqrt{6}-\sqrt{5}} = \dfrac{\sqrt{6}+\sqrt{5}}{\sqrt{6}-\sqrt{5}} \cdot \dfrac{\sqrt{6}+\sqrt{5}}{\sqrt{6}+\sqrt{5}}$

$= \dfrac{6+2\cdot\sqrt{6}\cdot\sqrt{5}+5}{6-5}$

$= \dfrac{11+2\sqrt{30}}{1} = 11+2\sqrt{30}$

21. $\sqrt{12x}-\sqrt{75x} = \sqrt{4}\cdot\sqrt{3x}-\sqrt{25}\cdot\sqrt{3x}$

$= 2\sqrt{3x}-5\sqrt{3x}$

$= -3\sqrt{3x}$

22. $(5-3\sqrt{3})^2 = 5^2 - 2\cdot 5\cdot 3\sqrt{3} + (3\sqrt{3})^2$

$= 25 - 30\sqrt{3} + 9\cdot 3$

$= 25 - 30\sqrt{3} + 27$

$= 52 - 30\sqrt{3}$

23. $(\sqrt{74}-\sqrt{73})(\sqrt{74}+\sqrt{73}) = 74 - 73 = 1$

24. $\sqrt[3]{\dfrac{13}{81}} = \dfrac{\sqrt[3]{13}}{\sqrt[3]{81}}$

$= \dfrac{\sqrt[3]{13}}{\sqrt[3]{27}\cdot\sqrt[3]{3}} \cdot \dfrac{\sqrt[3]{9}}{\sqrt[3]{9}}$

$= \dfrac{\sqrt[3]{13\cdot 9}}{3\cdot\sqrt[3]{27}}$

$= \dfrac{\sqrt[3]{117}}{3\cdot 3} = \dfrac{\sqrt[3]{117}}{9}$

25. $-t^2\sqrt[4]{t} + 3\sqrt[4]{t^9} - t\sqrt[4]{t^5}$

$= -t^2\sqrt[4]{t} + 3\sqrt[4]{t^8}\cdot\sqrt[4]{t} - t\cdot\sqrt[4]{t^4}\cdot\sqrt[4]{t}$

$= -t^2\sqrt[4]{t} + 3t^2\sqrt[4]{t} - t\cdot t\cdot\sqrt[4]{t}$

$= -t^2\sqrt[4]{t} + 3t^2\sqrt[4]{t} - t^2\sqrt[4]{t}$

$= (-t^2 + 3t^2 - t^2)\sqrt[4]{t} = t^2\sqrt[4]{t}$

26. $\dfrac{\sqrt{3}+\sqrt{7}}{\sqrt{6}-\sqrt{5}} = \dfrac{\sqrt{3}+\sqrt{7}}{\sqrt{6}-\sqrt{5}} \cdot \dfrac{\sqrt{6}+\sqrt{5}}{\sqrt{6}+\sqrt{5}}$

$= \dfrac{\sqrt{18}+\sqrt{15}+\sqrt{42}+\sqrt{35}}{(\sqrt{6})^2-(\sqrt{5})^2}$

$= \dfrac{\sqrt{9\cdot 2}+\sqrt{15}+\sqrt{42}+\sqrt{35}}{6-5}$

$= 3\sqrt{2}+\sqrt{15}+\sqrt{42}+\sqrt{35}$

27. $\dfrac{6}{\sqrt[4]{3}} = \dfrac{6}{\sqrt[4]{3}} \cdot \dfrac{\sqrt[4]{3^3}}{\sqrt[4]{3^3}}$

$= \dfrac{6\sqrt[4]{27}}{3} = 2\sqrt[4]{27}$

28. $\dfrac{1}{1-\sqrt[3]{3}}$

The denominator is the difference of *cube* roots, not the difference of *square* roots, so multiplying the numerator and denominator by $1+\sqrt[3]{3}$ will not help simplify the expression. We must use the difference of cubes factoring formula,

$$x^3 - y^3 = (x-y)(x^2+xy+y^2),$$

and recognize that $1-\sqrt[3]{3}$ plays the role of the factor $(x-y)$. Thus, we must multiply the numerator and denominator by the factor (x^2+xy+y^2), which is

$$(1)^2 + (1)\sqrt[3]{3} + (\sqrt[3]{3})^2 = 1 + \sqrt[3]{3} + \sqrt[3]{9},$$

to get $x^3 - y^3$.

$\dfrac{1}{1-\sqrt[3]{3}} \cdot \dfrac{1+\sqrt[3]{3}+\sqrt[3]{9}}{1+\sqrt[3]{3}+\sqrt[3]{9}} = \dfrac{1+\sqrt[3]{3}+\sqrt[3]{9}}{(1)^3-(\sqrt[3]{3})^3}$

$= \dfrac{1+\sqrt[3]{3}+\sqrt[3]{9}}{1-3}$

$= \dfrac{1+\sqrt[3]{3}+\sqrt[3]{9}}{-2}$

Note: You may have recognized this product from Exercise 10.

29. $\sqrt[3]{\dfrac{x^2y}{x^{-3}y^4}} = \sqrt[3]{x^{2-(-3)}y^{1-4}}$

$= \sqrt[3]{x^5y^{-3}}$

$= \dfrac{\sqrt[3]{x^5}}{\sqrt[3]{y^3}}$

$= \dfrac{\sqrt[3]{x^3\cdot x^2}}{y} = \dfrac{x\sqrt[3]{x^2}}{y}$

30. $\sqrt{12}-\sqrt{108}-\sqrt[3]{27} = \sqrt{4\cdot 3}-\sqrt{36\cdot 3}-3$

$= 2\sqrt{3}-6\sqrt{3}-3$

$= -4\sqrt{3}-3$

31. $\dfrac{x^{-2/3}y^{4/5}}{x^{-5/3}y^{-2/5}} = x^{-2/3-(-5/3)}y^{4/5-(-2/5)}$

$= x^{3/3}y^{6/5}$

$= xy^{6/5}$

32. $\left(\dfrac{x^{3/4}y^{2/3}}{x^{1/3}y^{5/8}}\right)^{24} = \dfrac{(x^{3/4})^{24}(y^{2/3})^{24}}{(x^{1/3})^{24}(y^{5/8})^{24}}$

$= \dfrac{x^{18}y^{16}}{x^8y^{15}}$

$= x^{18-8}y^{16-15}$

$= x^{10}y$

33. $$(125x^3)^{-2/3} = \frac{1}{(125x^3)^{2/3}} = \frac{1}{\left(\sqrt[3]{125x^3}\right)^2} = \frac{1}{(5x)^2} = \frac{1}{25x^2}$$

34. $$(3x^{-2/3}y^{1/2})(-2x^{5/8}y^{-1/3})$$
$$= 3(-2)x^{-2/3+5/8}y^{1/2+(-1/3)}$$
$$= -6x^{-16/24+15/24}y^{3/6-2/6}$$
$$= -6x^{-1/24}y^{1/6}$$
$$= \frac{-6y^{1/6}}{x^{1/24}}$$

35. $$\frac{4^{1/2}+3^{1/2}}{4^{1/2}-3^{1/2}} = \frac{2+\sqrt{3}}{2-\sqrt{3}}$$
$$= \frac{2+\sqrt{3}}{2-\sqrt{3}} \cdot \frac{2+\sqrt{3}}{2+\sqrt{3}}$$
$$= \frac{2^2+2(2)(\sqrt{3})+(\sqrt{3})^2}{2^2-(\sqrt{3})^2}$$
$$= \frac{4+4\sqrt{3}+3}{4-3}$$
$$= 7+4\sqrt{3}$$

36. $$(\sqrt{6}-\sqrt{5})^2(\sqrt{6}+\sqrt{5})^2$$
$$= [(\sqrt{6}-\sqrt{5})(\sqrt{6}+\sqrt{5})]^2 \quad a^2b^2 = (ab)^2$$
$$= \left[(\sqrt{6})^2-(\sqrt{5})^2\right]^2$$
$$= [6-5]^2$$
$$= 1^2 = 1$$

9.6 Solving Equations with Radicals

9.6 Margin Exercises

1. **(a)** $\sqrt{r} = 3$

Square both sides.

$$(\sqrt{r})^2 = 3^2$$
$$r = 9$$

To check, substitute the potential solution in the *original* equation.

$$\sqrt{r} = 3$$
$$\sqrt{9} = 3 \quad ? \quad \textit{Let } r = 9.$$
$$3 = 3 \quad \textit{True}$$

Solution set: $\{9\}$

(b) $\sqrt{5x+1} = 4$

Square both sides.

$$(\sqrt{5x+1})^2 = 4^2$$
$$5x+1 = 16$$
$$5x = 15$$
$$x = 3$$

Check 3 in the original equation.

$$\sqrt{5x+1} = 4$$
$$\sqrt{5 \cdot 3+1} = 4 \quad ? \quad \textit{Let } x = 3.$$
$$\sqrt{16} = 4 \quad ?$$
$$4 = 4 \quad \textit{True}$$

Solution set: $\{3\}$

2. **(a)** $\sqrt{x}+4 = -3$

Step 1

Isolate the radical on one side of the equation.

$$\sqrt{x} = -7$$

The equation has no solution because the square root of a real number must be nonnegative.

Solution set: $\emptyset$

(b) $\sqrt{x-9}-3 = 0$

Step 1

Isolate the radical on one side.

$$\sqrt{x-9} = 3$$

Step 2

Square both sides.

$$(\sqrt{x-9})^2 = 3^2$$

Step 3

$$x-9 = 9$$
$$x = 18$$

Step 4

Check 18 in the original equation.

$$\sqrt{x-9}-3 = 0$$
$$\sqrt{18-9}-3 = 0 \quad ?$$
$$\sqrt{9}-3 = 0 \quad ?$$
$$3-3 = 0 \quad ?$$
$$0 = 0 \quad \textit{True}$$

The potential solution, 18, checks.

Solution set: $\{18\}$

3. **(a)** $\sqrt{3x-5} = x-1$

Square both sides.

$$\left(\sqrt{3x-5}\right)^2 = (x-1)^2$$
$$3x-5 = x^2-2x+1$$
$$0 = x^2-5x+6$$
$$0 = (x-3)(x-2)$$
$$x-3=0 \quad \text{or} \quad x-2=0$$
$$x=3 \quad \text{or} \quad x=2$$

Check $x=3$: $\sqrt{4}=2$ *True*

Check $x=2$: $\sqrt{1}=1$ *True*

Solution set: $\{2,3\}$

(b) $x+1 = \sqrt{-2x-2}$

Square both sides.

$$(x+1)^2 = \left(\sqrt{-2x-2}\right)^2$$
$$x^2+2x+1 = -2x-2$$
$$x^2+4x+3=0$$
$$(x+3)(x+1)=0$$
$$x+3=0 \quad \text{or} \quad x+1=0$$
$$x=-3 \quad \text{or} \quad x=-1$$

Check $x=-3$: $-2=\sqrt{4}$ *False*

Check $x=-1$: $0=\sqrt{0}$ *True*

Solution set: $\{-1\}$

4. $\sqrt{4x^2+2x-3} = 2x+7$

Square both sides.

$$\left(\sqrt{4x^2+2x-3}\right)^2 = (2x+7)^2$$
$$4x^2+2x-3 = 4x^2+28x+49$$
$$-52 = 26x$$
$$-2 = x$$

Check $x=-2$: $\sqrt{9}=3$ *True*

Solution set: $\{-2\}$

5. **(a) Check** $x=15$:

$$\sqrt{5x+6}+\sqrt{3x+4} = 2$$
$$\sqrt{81}+\sqrt{49} = 2 \quad ? \quad \textit{Let } x = 15.$$
$$9+7=2 \quad ?$$
$$16=2 \quad \textit{False}$$

Thus, 15 is not a solution.

(b) $\sqrt{x+1}-\sqrt{x-4}=1$

Get one radical on each side of the equals sign.

$$\sqrt{x+1} = 1+\sqrt{x-4}$$

Square both sides.

$$\left(\sqrt{x+1}\right)^2 = \left(1+\sqrt{x-4}\right)^2$$
$$x+1 = 1+2\sqrt{x-4}+(x-4)$$

Isolate the remaining radical.

$$x+1 = 2\sqrt{x-4}+x-3$$
$$4 = 2\sqrt{x-4}$$
$$2 = \sqrt{x-4}$$

Square both sides again.

$$2^2 = \left(\sqrt{x-4}\right)^2$$
$$4 = x-4$$
$$8 = x$$

Check $x=8$: $\sqrt{9}-\sqrt{4}=1$ *True*

Solution set: $\{8\}$

6. **(a)** $\sqrt[3]{x^2+3x+12} = \sqrt[3]{x^2}$

Cube both sides.

$$\left(\sqrt[3]{x^2+3x+12}\right)^3 = \left(\sqrt[3]{x^2}\right)^3$$
$$x^2+3x+12 = x^2$$
$$3x+12=0$$
$$3x=-12$$
$$x=-4$$

Check $x=-4$: $\sqrt[3]{16}=\sqrt[3]{16}$ *True*

Solution set: $\{-4\}$

(b) $\sqrt[4]{2x+5}+1=0$

$$\sqrt[4]{2x+5}=-1$$

Since no real number has a principal fourth root that is negative, the equation has no solution.

Solution set: $\emptyset$

9.6 Section Exercises

1. $\sqrt{3x+18}=x$

(a) Check $x=6$.

Note: Sometimes we will use the notation " $\stackrel{?}{=}$ " when we are checking to determine if one side of an equation is equal to the other side.

$$\sqrt{3(6)+18} \stackrel{?}{=} 6$$
$$\sqrt{18+18} \stackrel{?}{=} 6$$
$$\sqrt{36} \stackrel{?}{=} 6$$
$$6=6 \quad \textit{True}$$

The number 6 is a solution.

(b) Check $x = -3$.

$\sqrt{3(-3)+18} = -3$ is a false statement since the principal square root of a number is nonnegative. The number -3 is not a solution.

2. $\sqrt{3x-3} = x-1$

(a) Check $x = 1$.

$$\begin{aligned} \sqrt{3(1)-3} &\stackrel{?}{=} 1-1 \\ \sqrt{0} &\stackrel{?}{=} 0 \\ 0 &= 0 \qquad \textit{True} \end{aligned}$$

The number 1 is a solution.

(b) Check $x = 4$.

$$\begin{aligned} \sqrt{3(4)-3} &\stackrel{?}{=} 4-1 \\ \sqrt{9} &\stackrel{?}{=} 3 \\ 3 &= 3 \qquad \textit{True} \end{aligned}$$

The number 4 is a solution.

3. $\sqrt{x+2} = \sqrt{9x-2} - 2\sqrt{x-1}$

(a) Check $x = 2$.

$$\begin{aligned} \sqrt{2+2} &\stackrel{?}{=} \sqrt{9(2)-2} - 2\sqrt{2-1} \\ \sqrt{4} &\stackrel{?}{=} \sqrt{16} - 2\sqrt{1} \\ 2 &\stackrel{?}{=} 4-2 \\ 2 &= 2 \quad \textit{True} \end{aligned}$$

The number 2 is a solution.

(b) Check $x = 7$.

$$\begin{aligned} \sqrt{7+2} &\stackrel{?}{=} \sqrt{9(7)-2} - 2\sqrt{7-1} \\ \sqrt{9} &\stackrel{?}{=} \sqrt{61} - 2\sqrt{6} \\ 3 &= \sqrt{61} - 2\sqrt{6} \quad \textit{False} \end{aligned}$$

The number 7 is not a solution.

4. $\sqrt{8x-3} = 2x$

(a) Check $x = \frac{3}{2}$.

$$\begin{aligned} \sqrt{8\left(\tfrac{3}{2}\right)-3} &\stackrel{?}{=} 2\left(\tfrac{3}{2}\right) \\ \sqrt{9} &\stackrel{?}{=} 3 \\ 3 &= 3 \quad \textit{True} \end{aligned}$$

The number $\frac{3}{2}$ is a solution.

(b) Check $x = \frac{1}{2}$.

$$\begin{aligned} \sqrt{8\left(\tfrac{1}{2}\right)-3} &\stackrel{?}{=} 2\left(\tfrac{1}{2}\right) \\ \sqrt{1} &\stackrel{?}{=} 1 \\ 1 &= 1 \quad \textit{True} \end{aligned}$$

The number $\frac{1}{2}$ is a solution.

5. $\sqrt{9} = 3$, not -3. There is no solution of $\sqrt{x} = -3$ since the value of a principal square root cannot equal a negative number.

6. $\sqrt{3x+18} = x$ does not have a negative solution because the expression on the left cannot be negative so x on the right cannot be negative.

7. $\sqrt{x-2} = 3$

Square both sides.

$$\begin{aligned} \left(\sqrt{x-2}\right)^2 &= 3^2 \\ x-2 &= 9 \\ x &= 11 \end{aligned}$$

Check the potential solution, 11.

Check $x = 11$: $\sqrt{9} = 3$ *True*

Solution set: $\{11\}$

8. $\sqrt{x+1} = 7$

Square both sides.

$$\begin{aligned} \left(\sqrt{x+1}\right)^2 &= 7^2 \\ x+1 &= 49 \\ x &= 48 \end{aligned}$$

Check the potential solution, 48.

Check $x = 48$: $\sqrt{49} = 7$ *True*

Solution set: $\{48\}$

9. $\sqrt{6x-1} = 1$

Square both sides.

$$\begin{aligned} \left(\sqrt{6x-1}\right)^2 &= 1^2 \\ 6x-1 &= 1 \\ 6x &= 2 \\ x &= \tfrac{2}{6} = \tfrac{1}{3} \end{aligned}$$

Check the potential solution, $\frac{1}{3}$.

Check $x = \frac{1}{3}$: $\sqrt{1} = 1$ *True*

Solution set: $\left\{\frac{1}{3}\right\}$

10. $\sqrt{7x-3} = 5$

Square both sides.

$$\begin{aligned} \left(\sqrt{7x-3}\right)^2 &= 5^2 \\ 7x-3 &= 25 \\ 7x &= 28 \\ x &= 4 \end{aligned}$$

Check the potential solution, 4.

Check $x = 4$: $\sqrt{25} = 5$ *True*

Solution set: $\{4\}$

11. $\sqrt{4x+3}+1=0$
Isolate the radical.
$$\sqrt{4x+3}=-1$$
This equation has no solution because $\sqrt{4x+3}$ cannot be negative.

Solution set: $\emptyset$

12. $\sqrt{5x-3}+2=0$
Isolate the radical.
$$\sqrt{5x-3}=-2$$
This equation has no solution because $\sqrt{5x-3}$ cannot be negative.

Solution set: $\emptyset$

13. $\sqrt{3k+1}-4=0$
Isolate the radical.
$$\sqrt{3k+1}=4$$
Square both sides.
$$\left(\sqrt{3k+1}\right)^2=4^2$$
$$3k+1=16$$
$$3k=15$$
$$k=5$$
Check $k=5$: $\sqrt{16}-4=0$ *True*

Solution set: $\{5\}$

14. $\sqrt{5z+1}-11=0$
Isolate the radical.
$$\sqrt{5z+1}=11$$
Square both sides.
$$\left(\sqrt{5z+1}\right)^2=11^2$$
$$5z+1=121$$
$$5z=120$$
$$z=24$$
Check $z=24$: $\sqrt{121}-11=0$ *True*

Solution set: $\{24\}$

15. $4-\sqrt{x-2}=0$
$$4=\sqrt{x-2} \quad \textit{Isolate}$$
$$4^2=\left(\sqrt{x-2}\right)^2 \quad \textit{Square}$$
$$16=x-2$$
$$18=x$$
Check $x=18$: $4-\sqrt{16}=0$ *True*

Solution set: $\{18\}$

16. $9-\sqrt{4k+1}=0$
$$9=\sqrt{4k+1} \quad \textit{Isolate}$$
$$9^2=\left(\sqrt{4k+1}\right)^2 \quad \textit{Square}$$
$$81=4k+1$$
$$80=4k$$
$$20=k$$
Check $k=20$: $9-\sqrt{81}=0$ *True*

Solution set: $\{20\}$

17. $\sqrt{9a-4}=\sqrt{8a+1}$
$$\left(\sqrt{9a-4}\right)^2=\left(\sqrt{8a+1}\right)^2 \quad \textit{Square}$$
$$9a-4=8a+1$$
$$a=5$$
Check $a=5$: $\sqrt{41}=\sqrt{41}$ *True*

Solution set: $\{5\}$

18. $\sqrt{4p-2}=\sqrt{3p+5}$
$$\left(\sqrt{4p-2}\right)^2=\left(\sqrt{3p+5}\right)^2 \quad \textit{Square}$$
$$4p-2=3p+5$$
$$p=7$$
Check $p=7$: $\sqrt{26}=\sqrt{26}$ *True*

Solution set: $\{7\}$

19. $2\sqrt{x}=\sqrt{3x+4}$
$$\left(2\sqrt{x}\right)^2=\left(\sqrt{3x+4}\right)^2 \quad \textit{Square}$$
$$4x=3x+4$$
$$x=4$$
Check $x=4$: $4=\sqrt{16}$ *True*

Solution set: $\{4\}$

20. $2\sqrt{m}=\sqrt{5m-16}$
$$\left(2\sqrt{m}\right)^2=\left(\sqrt{5m-16}\right)^2 \quad \textit{Square}$$
$$4m=5m-16$$
$$16=m$$
Check $m=16$: $8=\sqrt{64}$ *True*

Solution set: $\{16\}$

21. $3\sqrt{z-1}=2\sqrt{2z+2}$
$$\left(3\sqrt{z-1}\right)^2=\left(2\sqrt{2z+2}\right)^2 \quad \textit{Square}$$
$$9(z-1)=4(2z+2)$$
$$9z-9=8z+8$$
$$z=17$$
Check $z=17$: $3(4)=2(6)$ *True*

Solution set: $\{17\}$

22.
$$5\sqrt{4x+1}=3\sqrt{10x+25}$$
$$\left(5\sqrt{4x+1}\right)^2=\left(3\sqrt{10x+25}\right)^2 \quad \textit{Square}$$
$$25(4x+1)=9(10x+25)$$
$$100x+25=90x+225$$
$$10x=200$$
$$x=20$$

Check $x=20$: $5(9)=3(15)$ *True*

Solution set: $\{20\}$

23.
$$k=\sqrt{k^2+4k-20}$$
$$k^2=\left(\sqrt{k^2+4k-20}\right)^2 \quad \textit{Square}$$
$$k^2=k^2+4k-20$$
$$20=4k$$
$$5=k$$

Check $k=5$: $5=\sqrt{25}$ *True*

Solution set: $\{5\}$

24.
$$p=\sqrt{p^2-3p+18}$$
$$p^2=\left(\sqrt{p^2-3p+18}\right)^2 \quad \textit{Square}$$
$$p^2=p^2-3p+18$$
$$3p=18$$
$$p=6$$

Check $p=6$: $6=\sqrt{36}$ *True*

Solution set: $\{6\}$

25.
$$x=\sqrt{x^2+3x+9}$$
$$x^2=\left(\sqrt{x^2+3x+9}\right)^2 \quad \textit{Square}$$
$$x^2=x^2+3x+9$$
$$-3x=9$$
$$x=-3$$

Substituting -3 for x makes the left side of the original equation negative, but the right side is nonnegative, so the solution set is $\emptyset$.

26.
$$z=\sqrt{z^2-4z-8}$$
$$z^2=\left(\sqrt{z^2-4z-8}\right)^2 \quad \textit{Square}$$
$$z^2=z^2-4z-8$$
$$4z=-8$$
$$z=-2$$

Substituting -2 for z makes the left side of the original equation negative, but the right side is nonnegative, so the solution set is $\emptyset$.

27.
$$\sqrt{9-x}=x+3$$
$$\left(\sqrt{9-x}\right)^2=(x+3)^2 \quad \textit{Square}$$
$$9-x=x^2+6x+9$$
$$0=x^2+7x$$
$$0=x(x+7)$$
$$x=0 \quad \text{or} \quad x+7=0$$
$$x=-7$$

Check $x=-7$: $\sqrt{16}=-4$ *False*

Check $x=0$: $\sqrt{9}=3$ *True*

Solution set: $\{0\}$

28.
$$\sqrt{5-x}=x+1$$
$$\left(\sqrt{5-x}\right)^2=(x+1)^2 \quad \textit{Square}$$
$$5-x=x^2+2x+1$$
$$0=x^2+3x-4$$
$$0=(x+4)(x-1)$$
$$x+4=0 \quad \text{or} \quad x-1=0$$
$$x=-4 \quad \text{or} \quad x=1$$

Check $x=-4$: $\sqrt{9}=-3$ *False*

Check $x=1$: $\sqrt{4}=2$ *True*

Solution set: $\{1\}$

29.
$$\sqrt{k^2+2k+9}=k+3$$
$$\left(\sqrt{k^2+2k+9}\right)^2=(k+3)^2 \quad \textit{Square}$$
$$k^2+2k+9=k^2+6k+9$$
$$0=4k$$
$$0=k$$

Check $k=0$: $\sqrt{9}=3$ *True*

Solution set: $\{0\}$

30.
$$\sqrt{x^2-3x+3}=x-1$$
$$\left(\sqrt{x^2-3x+3}\right)^2=(x-1)^2 \quad \textit{Square}$$
$$x^2-3x+3=x^2-2x+1$$
$$2=x$$

Check $x=2$: $\sqrt{1}=1$ *True*

Solution set: $\{2\}$

31.
$$\sqrt{r^2+9r+3}=-r$$
$$\left(\sqrt{r^2+9r+3}\right)^2=(-r)^2 \quad \textit{Square}$$
$$r^2+9r+3=r^2$$
$$9r=-3$$
$$r=-\tfrac{1}{3}$$

Check $r=-\frac{1}{3}$: $\sqrt{\frac{1}{9}}=\frac{1}{3}$ *True*

Solution set: $\left\{-\frac{1}{3}\right\}$

32. $$\sqrt{p^2-15p+15}=p-5$$
$$\left(\sqrt{p^2-15p+15}\right)^2=(p-5)^2 \quad \textit{Square}$$
$$p^2-15p+15=p^2-10p+25$$
$$-5p=10$$
$$p=-2$$

Substituting -2 for p makes the right side of the original equation negative, but the left side is nonnegative, so the solution set is $\emptyset$.

33. $$\sqrt{z^2+12z-4}+4-z=0$$
$$\sqrt{z^2+12z-4}=z-4 \quad \textit{Isolate}$$
$$\left(\sqrt{z^2+12z-4}\right)^2=(z-4)^2 \quad \textit{Square}$$
$$z^2+12z-4=z^2-8z+16$$
$$20z=20$$
$$z=1$$

Substituting 1 for z makes the left side of the original equation positive, but the right side is zero, so the solution set is $\emptyset$.

34. $$\sqrt{m^2+3m+12}-m-2=0$$
$$\sqrt{m^2+3m+12}=m+2 \quad \textit{Isolate}$$
$$\left(\sqrt{m^2+3m+12}\right)^2=(m+2)^2 \quad \textit{Square}$$
$$m^2+3m+12=m^2+4m+4$$
$$8=m$$

Check $m=8$: $\sqrt{100}-8-2=0$ *True*

Solution set: $\{8\}$

35. $\sqrt{3x+4}=8-x$

$(8-x)^2$ equals $64-16x+x^2$, not $64+x^2$. The first step should be

$$3x+4=64-16x+x^2$$
$$0=x^2-19x+60$$
$$0=(x-4)(x-15)$$

$$x-4=0 \quad \text{or} \quad x-15=0$$
$$x=4 \quad \text{or} \quad x=15$$

Check $x=4$: $\sqrt{16}=8-4$ *True*
Check $x=15$: $\sqrt{49}=8-15$ *False*

Solution set: $\{4\}$

36. In solving $\sqrt{5x+6}-\sqrt{x+3}=3$, the incorrect solution attempted to use the power rule by squaring both sides. But

$$\left(\sqrt{5x+6}-\sqrt{x+3}\right)^2$$
$$=5x+6-2\sqrt{(5x+6)(x+3)}+x+3,$$

not $(5x+6)+(x+3)$.
The correct method is to add $\sqrt{x+3}$ to both sides and then square both sides. After simplifying the result, it will be necessary to isolate the radical expression on one side, and then square both sides again, as follows.

$$\sqrt{5x+6}=3+\sqrt{x+3}$$
$$\left(\sqrt{5x+6}\right)^2=\left(3+\sqrt{x+3}\right)^2$$
$$5x+6=9+6\sqrt{x+3}+x+3$$
$$4x-6=6\sqrt{x+3}$$
$$2x-3=3\sqrt{x+3}$$
$$(2x-3)^2=\left(3\sqrt{x+3}\right)^2$$
$$4x^2-12x+9=9(x+3)$$
$$4x^2-12x+9=9x+27$$
$$4x^2-21x-18=0$$
$$(4x+3)(x-6)=0$$

$$4x+3=0 \quad \text{or} \quad x-6=0$$
$$x=-\tfrac{3}{4} \quad \text{or} \quad x=6$$

Check $x=-\frac{3}{4}$: $\sqrt{\frac{9}{4}}-\sqrt{\frac{9}{4}}=3$ *False*

Check $x=6$: $\sqrt{36}-\sqrt{9}=3$ *True*

Solution set: $\{6\}$

37. $\sqrt[3]{2x+5}=\sqrt[3]{6x+1}$

Cube both sides.

$$\left(\sqrt[3]{2x+5}\right)^3=\left(\sqrt[3]{6x+1}\right)^3$$
$$2x+5=6x+1$$
$$4=4x$$
$$1=x$$

Check $x=1$: $\sqrt[3]{7}=\sqrt[3]{7}$ *True*
Solution set: $\{1\}$

38. $$\sqrt[3]{p-1}=2$$
$$\left(\sqrt[3]{p-1}\right)^3=2^3 \quad \textit{Cube}$$
$$p-1=8$$
$$p=9$$

Check $p=9$: $\sqrt[3]{8}=2$ *True*
Solution set: $\{9\}$

39. $$\sqrt[3]{a^2+5a+1}=\sqrt[3]{a^2+4a}$$
$$\left(\sqrt[3]{a^2+5a+1}\right)^3=\left(\sqrt[3]{a^2+4a}\right)^3 \quad \textit{Cube}$$
$$a^2+5a+1=a^2+4a$$
$$a=-1$$

Check $a=-1$: $\sqrt[3]{-3}=\sqrt[3]{-3}$ *True*
Solution set: $\{-1\}$

40. $\sqrt[3]{r^2+2r+8} = \sqrt[3]{r^2}$

$\left(\sqrt[3]{r^2+2r+8}\right)^3 = \left(\sqrt[3]{r^2}\right)^3$ *Cube*

$r^2+2r+8 = r^2$

$2r = -8$

$r = -4$

Check $r=-4$: $\sqrt[3]{16} = \sqrt[3]{16}$ *True*

Solution set: $\{-4\}$

41. $\sqrt[3]{2m-1} = \sqrt[3]{m+13}$

$\left(\sqrt[3]{2m-1}\right)^3 = \left(\sqrt[3]{m+13}\right)^3$ *Cube*

$2m-1 = m+13$

$m = 14$

Check $m=14$: $\sqrt[3]{27} = \sqrt[3]{27}$ *True*

Solution set: $\{14\}$

42. $\sqrt[3]{2k-11} - \sqrt[3]{5k+1} = 0$

Get one radical on each side of the equals sign.

$\sqrt[3]{2k-11} = \sqrt[3]{5k+1}$

$\left(\sqrt[3]{2k-11}\right)^3 = \left(\sqrt[3]{5k+1}\right)^3$ *Cube*

$2k-11 = 5k+1$

$-12 = 3k$

$-4 = k$

Check $k=-4$: $\sqrt[3]{-19} - \sqrt[3]{-19} = 0$ *True*

Solution set: $\{-4\}$

43. $\sqrt[4]{a+8} = \sqrt[4]{2a}$

Raise each side to the fourth power.

$\left(\sqrt[4]{a+8}\right)^4 = \left(\sqrt[4]{2a}\right)^4$

$a+8 = 2a$

$8 = a$

Check $a=8$: $\sqrt[4]{16} = \sqrt[4]{16}$ *True*

Solution set: $\{8\}$

44. $\sqrt[4]{z+11} = \sqrt[4]{2z+6}$

Raise each side to the fourth power.

$\left(\sqrt[4]{z+11}\right)^4 = \left(\sqrt[4]{2z+6}\right)^4$

$z+11 = 2z+6$

$5 = z$

Check $z=5$: $\sqrt[4]{16} = \sqrt[4]{16}$ *True*

Solution set: $\{5\}$

45. $\sqrt[3]{x-8}+2 = 0$

$\sqrt[3]{x-8} = -2$ *Isolate*

$\left(\sqrt[3]{x-8}\right)^3 = (-2)^3$ *Cube*

$x-8 = -8$

$x = 0$

Check $x=0$: $\sqrt[3]{-8}+2 = 0$ *True*

Solution set: $\{0\}$

46. $\sqrt[3]{r+1}+1 = 0$

$\sqrt[3]{r+1} = -1$ *Isolate*

$\left(\sqrt[3]{r+1}\right)^3 = (-1)^3$ *Cube*

$r+1 = -1$

$r = -2$

Check $r=-2$: $\sqrt[3]{-1}+1 = 0$ *True*

Solution set: $\{-2\}$

47. $\sqrt[4]{2k-5}+4 = 0$

$\sqrt[4]{2k-5} = -4$ *Isolate*

This equation has no solution because $\sqrt[4]{2k-5}$ cannot be negative.

Solution set: $\emptyset$

48. $\sqrt[4]{8z-3}+2 = 0$

$\sqrt[4]{8z-3} = -2$ *Isolate*

This equation has no solution because $\sqrt[4]{8z-3}$ cannot be negative.

Solution set: $\emptyset$

49. $\sqrt{k+2} - \sqrt{k-3} = 1$

Get one radical on each side of the equals sign.

$\sqrt{k+2} = 1+\sqrt{k-3}$

$\left(\sqrt{k+2}\right)^2 = \left(1+\sqrt{k-3}\right)^2$ *Square*

$k+2 = 1+2\sqrt{k-3}+k-3$

$4 = 2\sqrt{k-3}$ *Isolate*

$2 = \sqrt{k-3}$ *Divide by 2.*

$2^2 = \left(\sqrt{k-3}\right)^2$ *Square again*

$4 = k-3$

$7 = k$

Check $k=7$: $\sqrt{9} - \sqrt{4} = 1$ *True*

Solution set: $\{7\}$

50. $\sqrt{r+6}-\sqrt{r-2}=2$

Get one radical on each side of the equals sign.

$$\sqrt{r+6}=2+\sqrt{r-2}$$
$$\left(\sqrt{r+6}\right)^2=\left(2+\sqrt{r-2}\right)^2 \quad \textit{Square}$$
$$r+6=4+4\sqrt{r-2}+r-2$$
$$4=4\sqrt{r-2} \quad \textit{Isolate}$$
$$1=\sqrt{r-2} \quad \textit{Divide by 4.}$$
$$1^2=\left(\sqrt{r-2}\right)^2 \quad \textit{Square again}$$
$$1=r-2$$
$$3=r$$

Check $r=3$: $\sqrt{9}-\sqrt{1}=2$ *True*

Solution set: $\{3\}$

51. $\sqrt{2r+11}-\sqrt{5r+1}=-1$

Get one radical on each side of the equals sign.

$$\sqrt{2r+11}=-1+\sqrt{5r+1}$$

Square both sides.

$$\left(\sqrt{2r+11}\right)^2=\left(-1+\sqrt{5r+1}\right)^2$$
$$2r+11=1-2\sqrt{5r+1}+5r+1$$

Isolate the remaining radical.

$$2\sqrt{5r+1}=3r-9$$

Square both sides again.

$$\left(2\sqrt{5r+1}\right)^2=(3r-9)^2$$
$$4(5r+1)=9r^2-54r+81$$
$$20r+4=9r^2-54r+81$$
$$0=9r^2-74r+77$$
$$0=(9r-11)(r-7)$$

$9r-11=0$ or $r-7=0$

$r=\frac{11}{9}$ or $r=7$

Check $r=\frac{11}{9}$: $\frac{11}{3}-\frac{8}{3}=-1$ *False*

Check $r=7$: $5-6=-1$ *True*

Solution set: $\{7\}$

52. $\sqrt{3x-2}-\sqrt{x+3}=1$

Get one radical on each side of the equals sign.

$$\sqrt{3x-2}=1+\sqrt{x+3}$$

Square both sides.

$$\left(\sqrt{3x-2}\right)^2=\left(1+\sqrt{x+3}\right)^2$$
$$3x-2=1+2\sqrt{x+3}+x+3$$

Isolate the remaining radical.

$$2x-6=2\sqrt{x+3}$$
$$x-3=\sqrt{x+3}$$

Square both sides again.

$$(x-3)^2=\left(\sqrt{x+3}\right)^2$$
$$x^2-6x+9=x+3$$
$$x^2-7x+6=0$$
$$(x-1)(x-6)=0$$

$x-1=0$ or $x-6=0$

$x=1$ or $x=6$

Check $x=1$: $\sqrt{1}-\sqrt{4}=1$ *False*

Check $x=6$: $\sqrt{16}-\sqrt{9}=1$ *True*

Solution set: $\{6\}$

53. $\sqrt{3p+4}-\sqrt{2p-4}=2$

Get one radical on each side of the equals sign.

$$\sqrt{3p+4}=2+\sqrt{2p-4}$$

Square both sides.

$$\left(\sqrt{3p+4}\right)^2=\left(2+\sqrt{2p-4}\right)^2$$
$$3p+4=4+4\sqrt{2p-4}+2p-4$$

Isolate the remaining radical.

$$p+4=4\sqrt{2p-4}$$

Square both sides again.

$$(p+4)^2=\left(4\sqrt{2p-4}\right)^2$$
$$p^2+8p+16=16(2p-4)$$
$$p^2+8p+16=32p-64$$
$$p^2-24p+80=0$$
$$(p-4)(p-20)=0$$

$p-4=0$ or $p-20=0$

$p=4$ or $p=20$

Check $p=4$: $\sqrt{16}-\sqrt{4}=2$ *True*

Check $p=20$: $\sqrt{64}-\sqrt{36}=2$ *True*

Solution set: $\{4, 20\}$

54. $\sqrt{4x+5}-\sqrt{2x+2}=1$

Get one radical on each side of the equals sign.

$$\sqrt{4x+5}=1+\sqrt{2x+2}$$

Square both sides.

$$\left(\sqrt{4x+5}\right)^2=\left(1+\sqrt{2x+2}\right)^2$$
$$4x+5=1+2\sqrt{2x+2}+2x+2$$

Isolate the remaining radical.

$$2x+2=2\sqrt{2x+2}$$
$$x+1=\sqrt{2x+2}$$

Square both sides again.

$$(x+1)^2=\left(\sqrt{2x+2}\right)^2$$
$$x^2+2x+1=2x+2$$
$$x^2-1=0$$
$$(x+1)(x-1)=0$$

continued

$x+1=0$ or $x-1=0$
$x=-1$ or $x=1$

Check $x=-1$: $\sqrt{1}-0=1$ *True*
Check $x=1$: $\sqrt{9}-\sqrt{4}=1$ *True*
Solution set: $\{-1,1\}$

55. $$\sqrt{3-3p}-3=\sqrt{3p+2}$$
Square both sides.
$$\left(\sqrt{3-3p}-3\right)^2=\left(\sqrt{3p+2}\right)^2$$
$$3-3p-6\sqrt{3-3p}+9=3p+2$$
Isolate the remaining radical.
$$-6\sqrt{3-3p}=6p-10$$
$$-3\sqrt{3-3p}=3p-5$$
Square both sides again.
$$\left(-3\sqrt{3-3p}\right)^2=(3p-5)^2$$
$$9(3-3p)=9p^2-30p+25$$
$$27-27p=9p^2-30p+25$$
$$0=9p^2-3p-2$$
$$0=(3p+1)(3p-2)$$

$3p+1=0$ or $3p-2=0$
$p=-\frac{1}{3}$ or $p=\frac{2}{3}$

Check $p=-\frac{1}{3}$: $\sqrt{4}-3=\sqrt{1}$ *False*
Check $p=\frac{2}{3}$: $\sqrt{1}-3=\sqrt{4}$ *False*
Solution set: $\emptyset$

56. $$\sqrt{4x+7}-4=\sqrt{4x-1}$$
Square both sides.
$$\left(\sqrt{4x+7}-4\right)^2=\left(\sqrt{4x-1}\right)^2$$
$$4x+7-8\sqrt{4x+7}+16=4x-1$$
Isolate the remaining radical.
$$-8\sqrt{4x+7}=-24$$
$$\sqrt{4x+7}=3$$
Square both sides again.
$$\left(\sqrt{4x+7}\right)^2=3^2$$
$$4x+7=9$$
$$4x=2$$
$$x=\tfrac{1}{2}$$

Check $x=\frac{1}{2}$: $\sqrt{9}-4=\sqrt{1}$ *False*
Solution set: $\emptyset$

57. $$\sqrt{2\sqrt{x+11}}=\sqrt{4x+2}$$
$$2\sqrt{x+11}=4x+2 \qquad \textit{Square}$$
$$\left(2\sqrt{x+11}\right)^2=(4x+2)^2 \qquad \textit{Square again}$$
$$4(x+11)=16x^2+16x+4$$
$$4x+44=16x^2+16x+4$$
$$0=16x^2+12x-40$$
$$0=4x^2+3x-10$$
$$0=(x+2)(4x-5)$$

$x+2=0$ or $4x-5=0$
$x=-2$ or $x=\frac{5}{4}$

Check $x=-2$: $\sqrt{6}=\sqrt{-6}$ *False*
Check $x=\frac{5}{4}$: $\sqrt{7}=\sqrt{7}$ *True*
Solution set: $\left\{\frac{5}{4}\right\}$

58. $$\sqrt{1+\sqrt{24-10x}}=\sqrt{3x+5}$$
$$1+\sqrt{24-10x}=3x+5 \qquad \textit{Square}$$
Isolate the remaining radical.
$$\sqrt{24-10x}=3x+4$$
$$\left(\sqrt{24-10x}\right)^2=(3x+4)^2 \qquad \textit{Square}$$
$$24-10x=9x^2+24x+16$$
$$0=9x^2+34x-8$$
$$0=(x+4)(9x-2)$$

$x+4=0$ or $9x-2=0$
$x=-4$ or $x=\frac{2}{9}$

Check $x=-4$: $\sqrt{9}=\sqrt{-7}$ *False*
Check $x=\frac{2}{9}$: $\sqrt{1+\frac{14}{3}}=\sqrt{\frac{17}{3}}$ *True*
Solution set: $\left\{\frac{2}{9}\right\}$

59. $$(2x-9)^{1/2}=2+(x-8)^{1/2}$$
$$\sqrt{2x-9}=2+\sqrt{x-8}$$
$$\left(\sqrt{2x-9}\right)^2=\left(2+\sqrt{x-8}\right)^2$$
$$2x-9=4+4\sqrt{x-8}+x-8$$
$$x-5=4\sqrt{x-8}$$
$$(x-5)^2=\left(4\sqrt{x-8}\right)^2$$
$$x^2-10x+25=16(x-8)$$
$$x^2-10x+25=16x-128$$
$$x^2-26x+153=0$$
$$(x-9)(x-17)=0$$
$x=9$ or $x=17$

Check $x=9$: $9^{1/2}\stackrel{?}{=}2+1^{1/2}$
$3=2+1$ *True*

Check $x=17$: $25^{1/2}\stackrel{?}{=}2+9^{1/2}$
$5=2+3$ *True*

Solution set: $\{9,17\}$

60.
$$\begin{aligned}(3w+7)^{1/2} &= 1+(w+2)^{1/2}\\ \sqrt{3w+7} &= 1+\sqrt{w+2}\\ \left(\sqrt{3w+7}\right)^2 &= \left(1+\sqrt{w+2}\right)^2\\ 3w+7 &= 1+2\sqrt{w+2}+w+2\\ 2w+4 &= 2\sqrt{w+2}\\ w+2 &= \sqrt{w+2}\\ (w+2)^2 &= \left(\sqrt{w+2}\right)^2\\ w^2+4w+4 &= w+2\\ w^2+3w+2 &= 0\\ (w+2)(w+1) &= 0\end{aligned}$$

$w=-2$ or $w=-1$

Check $w=-2$: $1^{1/2} \stackrel{?}{=} 1+0^{1/2}$

$1=1$ *True*

Check $w=-1$: $4^{1/2} \stackrel{?}{=} 1+1^{1/2}$

$2=1+1$ *True*

Solution set: $\{-2,-1\}$

61.
$$\begin{aligned}(2w-1)^{2/3}-w^{1/3} &= 0\\ \sqrt[3]{(2w-1)^2} &= \sqrt[3]{w}\\ \left[\sqrt[3]{(2w-1)^2}\right]^3 &= \left(\sqrt[3]{w}\right)^3\\ (2w-1)^2 &= w\\ 4w^2-4w+1 &= w\\ 4w^2-5w+1 &= 0\\ (4w-1)(w-1) &= 0\end{aligned}$$

$4w-1=0$ or $w-1=0$

$w=\frac{1}{4}$ or $w=1$

Check $w=\frac{1}{4}$: $\left(-\frac{1}{2}\right)^{2/3}-\left(\frac{1}{4}\right)^{1/3} \stackrel{?}{=} 0$

$\left(\frac{1}{4}\right)^{1/3}-\left(\frac{1}{4}\right)^{1/3}=0$ *True*

Check $w=1$: $1-1=0$ *True*

Solution set: $\left\{\frac{1}{4},1\right\}$

62.
$$\begin{aligned}\left(x^2-2x\right)^{1/3}-x^{1/3} &= 0\\ \sqrt[3]{x^2-2x} &= \sqrt[3]{x}\\ \left(\sqrt[3]{x^2-2x}\right)^3 &= \left(\sqrt[3]{x}\right)^3\\ x^2-2x &= x\\ x^2-3x &= 0\\ x(x-3) &= 0\end{aligned}$$

$x=0$ or $x=3$

Check $x=0$: $0^{1/3}-0^{1/3}=0$ *True*

Check $x=3$: $3^{1/3}-3^{1/3}=0$ *True*

Solution set: $\{0,3\}$

63. Solve $V=\sqrt{\dfrac{2K}{m}}$ for K.

$$\begin{aligned}(V)^2 &= \left(\sqrt{\frac{2K}{m}}\right)^2 && \textit{Square}\\ V^2 &= \frac{2K}{m} && \textit{Multiply by } \frac{m}{2}.\\ \frac{V^2m}{2} &= K\end{aligned}$$

64. Solve $V=\sqrt{\dfrac{2K}{m}}$ for m.

$$\begin{aligned}V^2 &= \frac{2K}{m}\\ V^2m &= 2K\\ m &= \frac{2K}{V^2}\end{aligned}$$

65. Solve $f=\dfrac{1}{2\pi\sqrt{LC}}$ for L.

$$\begin{aligned}2\pi f\sqrt{LC} &= 1\\ \left(2\pi f\sqrt{LC}\right)^2 &= 1^2\\ 4\pi^2f^2LC &= 1\\ L &= \frac{1}{4\pi^2f^2C}\end{aligned}$$

66. Solve $r=\sqrt{\dfrac{Mm}{F}}$ for F.

$$\begin{aligned}(r)^2 &= \left(\sqrt{\frac{Mm}{F}}\right)^2\\ r^2 &= \frac{Mm}{F}\\ Fr^2 &= Mm\\ F &= \frac{Mm}{r^2}\end{aligned}$$

67. Solve $N=\dfrac{1}{2\pi}\sqrt{\dfrac{a}{r}}$ for r.

$$\begin{aligned}2\pi N &= \sqrt{\frac{a}{r}}\\ (2\pi N)^2 &= \left(\sqrt{\frac{a}{r}}\right)^2\\ 4\pi^2N^2 &= \frac{a}{r}\\ 4\pi^2N^2r &= a\\ r &= \frac{a}{4\pi^2N^2}\end{aligned}$$

68.
$$\begin{aligned}r &= \frac{9.8}{4\pi^2(.063)^2}\\ &\approx 62.5 \text{ meters}\end{aligned}$$

9.7 Complex Numbers

9.7 Margin Exercises

1. **(a)** $\sqrt{-16} = i\sqrt{16} = 4i$

   (b) $-\sqrt{-81} = -i\sqrt{81} = -9i$

   (c) $\sqrt{-7} = i\sqrt{7}$

   (d) $\sqrt{-32} = i\sqrt{32} = i\sqrt{16} \cdot \sqrt{2} = 4i\sqrt{2}$

2. **(a)** $\sqrt{-7} \cdot \sqrt{-7} = i\sqrt{7} \cdot i\sqrt{7}$
   $= i^2\left(\sqrt{7}\right)^2$
   $= -1(7)$
   $= -7$

   (b) $\sqrt{-5} \cdot \sqrt{-10} = i\sqrt{5} \cdot i\sqrt{10}$
   $= i^2\sqrt{50}$
   $= i^2\sqrt{25 \cdot 2}$
   $= -1\left(5\sqrt{2}\right)$
   $= -5\sqrt{2}$

   (c) $\sqrt{-15} \cdot \sqrt{2} = i\sqrt{15} \cdot \sqrt{2} = i\sqrt{30}$

3. **(a)** $\dfrac{\sqrt{-32}}{\sqrt{-2}} = \dfrac{i\sqrt{32}}{i\sqrt{2}} = \sqrt{\dfrac{32}{2}} = \sqrt{16} = 4$

   (b) $\dfrac{\sqrt{-27}}{\sqrt{-3}} = \dfrac{i\sqrt{27}}{i\sqrt{3}} = \sqrt{\dfrac{27}{3}} = \sqrt{9} = 3$

   (c) $\dfrac{\sqrt{-40}}{\sqrt{10}} = \dfrac{i\sqrt{40}}{\sqrt{10}} = i\sqrt{\dfrac{40}{10}} = i\sqrt{4} = 2i$

4. To add complex numbers, add the real parts and add the imaginary parts.

   (a) $(4 + 6i) + (-3 + 5i)$
   $= [4 + (-3)] + (6 + 5)i$
   $= 1 + 11i$

   (b) $(-1 + 8i) + (9 - 3i)$
   $= (-1 + 9) + (8 - 3)i$
   $= 8 + 5i$

5. To subtract complex numbers, subtract the real parts and subtract the imaginary parts.

   (a) $(7 + 3i) - (4 + 2i)$
   $= (7 - 4) + (3 - 2)i$
   $= 3 + i$

   (b) $(-6 - i) - (-5 - 4i)$
   $= (-6 + 5) + (-1 + 4)i$
   $= -1 + 3i$

   (c) $8 - (3 - 2i) = (8 - 3) + 2i$
   $= 5 + 2i$

6. **(a)** $6i(4 + 3i) = 6i(4) + 6i(3i)$
   $= 24i + 18i^2$
   $= 24i + 18(-1)$
   $= -18 + 24i$

   (b) $(6 - 4i)(2 + 4i)$
   F O I L
   $= 6(2) + 6(4i) + 2(-4i) + 4i(-4i)$
   $= 12 + 24i - 8i - 16i^2$
   $= 12 + 16i - 16(-1)$
   $= 12 + 16i + 16$
   $= 28 + 16i$

   (c) $(3 - 2i)(3 + 2i)$
   F O I L
   $= 3(3) + 3(2i) + (-2i)(3) + (-2i)(2i)$
   $= 9 + 6i - 6i - 4i^2$
   $= 9 - 4(-1)$
   $= 9 + 4$
   $= 13$

7. In each case, multiply the numerator and denominator by the conjugate of the denominator.

   (a) $\dfrac{2 + i}{3 - i} = \dfrac{(2 + i)(3 + i)}{(3 - i)(3 + i)}$
   $= \dfrac{6 + 2i + 3i + i^2}{3^2 - i^2}$
   $= \dfrac{6 + 2i + 3i - 1}{9 - (-1)}$
   $= \dfrac{5 + 5i}{10}$
   $= \dfrac{5(1 + i)}{10}$
   $= \dfrac{1 + i}{2}$
   $= \dfrac{1}{2} + \dfrac{1}{2}i$

   (b) $\dfrac{8 - 4i}{1 - i} = \dfrac{(8 - 4i)(1 + i)}{(1 - i)(1 + i)}$
   $= \dfrac{8 + 8i - 4i - 4i^2}{1^2 + 1^2}$
   $= \dfrac{8 + 4i + 4}{1 + 1}$ *$i^2 = -1$*
   $= \dfrac{12 + 4i}{2}$
   $= \dfrac{2(6 + 2i)}{2}$
   $= 6 + 2i$

(c) $\frac{5}{3-2i} = \frac{5(3+2i)}{(3-2i)(3+2i)}$
$= \frac{5(3+2i)}{3^2+2^2}$
$= \frac{5(3+2i)}{9+4}$
$= \frac{5(3+2i)}{13}$
$= \frac{15+10i}{13}$
$= \frac{15}{13} + \frac{10}{13}i$

(d) $\frac{5-i}{i}$

The conjugate of i is $-i$.

$= \frac{(5-i)(-i)}{i(-i)}$
$= \frac{-5i+i^2}{-i^2}$
$= \frac{-5i+(-1)}{-(-1)}$
$= \frac{-5i-1}{1}$
$= -1-5i$

8. (a) $i^{21} = i^{20} \cdot i = (i^4)^5 \cdot i = 1^5 \cdot i = i$

(b) $i^{36} = (i^4)^9 = 1^9 = 1$

(c) $i^{50} = i^{48} \cdot i^2 = (i^4)^{12} \cdot i^2 = 1^{12} \cdot (-1) = -1$

(d) $i^{-9} = \frac{1}{i^9} = \frac{1}{i^8 \cdot i} = \frac{1}{(i^4)^2 \cdot i} = \frac{1}{1^2 \cdot i} = \frac{1}{i}$

Now multiply by $-i$, the conjugate of i.

$\frac{1}{i} = \frac{1(-i)}{i(-i)} = \frac{-i}{-i^2} = \frac{-i}{-(-1)} = \frac{-i}{1} = -i$

9.7 Section Exercises

1. $\sqrt{-1} = i$

2. $-i^2 = -(i^2) = -(-1) = 1$

3. $\frac{1}{i} = \frac{1}{i} \cdot \frac{-i}{-i}$ *$-i$ is the conjugate of i*
$= \frac{-i}{-i^2} = \frac{i}{i^2} = \frac{i}{-1} = -i$

4. $(-i)^2 = i^2 = -1$

5. Since any real number can be written as $a+bi$, where $b=0$, every real number is also a complex number.

6. Not every complex number is a real number. For example, any number $a+bi$, $b \neq 0$, such as $3+7i$ is a complex number that is not a real number.

7. $\sqrt{-169} = i\sqrt{169} = 13i$

8. $\sqrt{-225} = i\sqrt{225} = 15i$

9. $-\sqrt{-144} = -i\sqrt{144} = -12i$

10. $-\sqrt{-196} = -i\sqrt{196} = -14i$

11. $\sqrt{-5} = i\sqrt{5}$

12. $\sqrt{-21} = i\sqrt{21}$

13. $\sqrt{-48} = i\sqrt{48} = i\sqrt{16 \cdot 3} = 4i\sqrt{3}$

14. $\sqrt{-96} = i\sqrt{96} = i\sqrt{16 \cdot 6} = 4i\sqrt{6}$

15. $\sqrt{-15} \cdot \sqrt{-15} = i\sqrt{15} \cdot i\sqrt{15}$
$= i^2(\sqrt{15})^2$
$= -1(15) = -15$

16. $\sqrt{-19} \cdot \sqrt{-19} = i\sqrt{19} \cdot i\sqrt{19}$
$= i^2(\sqrt{19})^2$
$= -1(19) = -19$

17. $\sqrt{-4} \cdot \sqrt{-25} = i\sqrt{4} \cdot i\sqrt{25} = 2i \cdot 5i = 10i^2$
$= 10(-1) = -10$

18. $\sqrt{-9} \cdot \sqrt{-81} = i\sqrt{9} \cdot i\sqrt{81} = 3i \cdot 9i = 27i^2$
$= 27(-1) = -27$

19. $\frac{\sqrt{-300}}{\sqrt{-100}} = \frac{i\sqrt{300}}{i\sqrt{100}} = \sqrt{\frac{300}{100}} = \sqrt{3}$

20. $\frac{\sqrt{-40}}{\sqrt{-10}} = \frac{i\sqrt{40}}{i\sqrt{10}} = \sqrt{\frac{40}{10}} = \sqrt{4} = 2$

21. $\frac{\sqrt{-75}}{\sqrt{3}} = \frac{i\sqrt{75}}{\sqrt{3}} = i\sqrt{\frac{75}{3}} = i\sqrt{25} = 5i$

22. $\frac{\sqrt{-160}}{\sqrt{10}} = \frac{i\sqrt{160}}{\sqrt{10}} = i\sqrt{\frac{160}{10}} = i\sqrt{16} = 4i$

23. $(3+2i) + (-4+5i)$
$= [(3+(-4)] + (2+5)i$
$= -1+7i$

24. $(7+15i) + (-11+14i)$
$= (7-11) + (15+14)i$
$= -4+29i$

25. $(5-i) + (-5+i)$
$= (5-5) + (-1+1)i$
$= 0$

26. $(-2+6i)+(2-6i)$
$=(-2+2)+(6-6)i$
$=0$

27. $(4+i)-(-3-2i)$
$=[(4-(-3)]+[(1-(-2)]i$
$=7+3i$

28. $(9+i)-(3+2i)$
$=(9-3)+(1-2)i$
$=6-i$

29. $(-3-4i)-(-1-4i)$
$=[-3-(-1)]+[-4-(-4)]i$
$=-2$

30. $(-2-3i)-(-5-3i)$
$=[-2-(-5)]+[(-3)-(-3)]i$
$=3$

31. $(-4+11i)+(-2-4i)+(7+6i)$
$=(-4-2+7)+(11-4+6)i$
$=1+13i$

32. $(-1+i)+(2+5i)+(3+2i)$
$=(-1+2+3)+(1+5+2)i$
$=4+8i$

33. $[(7+3i)-(4-2i)]+(3+i)$
Work inside the brackets first.
$=[(7-4)+(3+2)i]+(3+i)$
$=(3+5i)+(3+i)$
$=(3+3)+(5+1)i$
$=6+6i$

34. $[(7+2i)+(-4-i)]-(2+5i)$
$=[(7-4)+(2-1)i]-(2+5i)$
$=(3+i)-(2+5i)$
$=(3-2)+(1-5)i$
$=1-4i$

35. If $a-c=b$, then $b+c=a$.
So, $(4+2i)-(3+i)=1+i$ implies that

$$(1+i)+(3+i)=\underline{4+2i}.$$

36. Because $\dfrac{-5}{2-i}=-2-i$, using the definition of division, we can check this to find that $(-2-i)(2-i)=\underline{-5}$.

37. $(3i)(27i)=81i^2=81(-1)=-81$

38. $(5i)(125i)=625i^2=625(-1)=-625$

39. $(-8i)(-2i)=16i^2=16(-1)=-16$

40. $(-32i)(-2i)=64i^2=64(-1)=-64$

41. $5i(-6+2i)=(5i)(-6)+(5i)(2i)$
$=-30i+10i^2$
$=-30i+10(-1)$
$=-10-30i$

42. $3i(4+9i)=(3i)(4)+(3i)(9i)$
$=12i+27i^2$
$=12i+27(-1)$
$=-27+12i$

43. $(4+3i)(1-2i)$
F O I L
$=(4)(1)+4(-2i)+(3i)(1)+(3i)(-2i)$
$=4-8i+3i-6i^2$
$=4-5i-6(-1)$
$=4-5i+6=10-5i$

44. $(7-2i)(3+i)$
F O I L
$=7(3)+7(i)+(-2i)(3)+(-2i)(i)$
$=21+7i-6i-2i^2$
$=21+i-2(-1)$
$=21+i+2=23+i$

45. $(4+5i)^2=4^2+2(4)(5i)+(5i)^2$
$=16+40i+25i^2$
$=16+40i+25(-1)$
$=16+40i-25$
$=-9+40i$

46. $(3+2i)^2=3^2+2(3)(2i)+(2i)^2$
$=9+12i+4i^2$
$=9+12i+4(-1)$
$=9+12i-4$
$=5+12i$

47. $(12+3i)(12-3i)$
$=12^2-(3i)^2=144-9i^2$
$=144-9(-1)=144+9=153$

48. $(6+7i)(6-7i)$
$=6^2-(7i)^2=36-49i^2$
$=36-49(-1)=36+49=85$

49. **(a)** The conjugate of $a+bi$ is $a-bi$.

(b) $(a+bi)(a-bi)$
$=a^2-abi+abi-b^2i^2$
$=a^2-b^2(-1)$
$=a^2+b^2$,
which is always a real number.

50. To find the quotient,

$$\frac{-1+5i}{3+2i},$$

multiply both the numerator and denominator by the complex conjugate of the denominator, $3-2i$. Then simplify the result by multiplying in the numerator and denominator and combining like terms. Write the quotient in the form $a+bi$.

51. $\dfrac{2}{1-i}$

Multiply the numerator and the denominator by the conjugate of the denominator, $1+i$.

$$=\frac{2(1+i)}{(1-i)(1+i)}=\frac{2(1+i)}{1^2-i^2}$$
$$=\frac{2(1+i)}{1-(-1)}=\frac{2(1+i)}{2}=1+i$$

52. $\dfrac{29}{5+2i}$

Multiply the numerator and the denominator by the conjugate of the denominator, $5-2i$.

$$=\frac{29(5-2i)}{(5+2i)(5-2i)}=\frac{29(5-2i)}{5^2+2^2}$$
$$=\frac{29(5-2i)}{29}=5-2i$$

53. $\dfrac{-7+4i}{3+2i}$

Multiply the numerator and the denominator by the conjugate of the denominator, $3-2i$.

$$=\frac{(-7+4i)(3-2i)}{(3+2i)(3-2i)}$$

In the denominator, we make use of the fact that $(a+bi)(a-bi)=a^2+b^2$.

$$=\frac{-21+14i+12i+8}{3^2+2^2}$$
$$=\frac{-13+26i}{13}=\frac{13(-1+2i)}{13}=-1+2i$$

54. $$\frac{-38-8i}{7+3i}=\frac{-2(19+4i)(7-3i)}{(7+3i)(7-3i)}$$
$$=\frac{-2(133-57i+28i+12)}{7^2+3^2}$$
$$=\frac{-2(145-29i)}{49+9}=\frac{-2(29)(5-i)}{58}$$
$$=-(5-i)=-5+i$$

55. $\dfrac{8i}{2+2i}$

Write in lowest terms.

$$=\frac{2\cdot 4i}{2(1+i)}=\frac{4i}{1+i}$$

Multiply the numerator and the denominator by the conjugate of the denominator, $1-i$.

$$=\frac{4i(1-i)}{(1+i)(1-i)}=\frac{4(i-i^2)}{1^2+1^2}$$
$$=\frac{4(i+1)}{2}=2(i+1)=2+2i$$

56. $$\frac{-8i}{1+i}=\frac{-8i(1-i)}{(1+i)(1-i)}$$
$$=\frac{-8(i-i^2)}{1^2+1^2}=\frac{-8(i+1)}{2}$$
$$=-4(i+1)=-4-4i$$

57. $\dfrac{2-3i}{2+3i}$

Multiply the numerator and the denominator by the conjugate of the denominator, $2-3i$.

$$=\frac{(2-3i)(2-3i)}{(2+3i)(2-3i)}=\frac{2^2-2(2)(3i)+(3i)^2}{2^2+3^2}$$
$$=\frac{4-12i+9i^2}{4+9}=\frac{4-12i-9}{13}$$
$$=\frac{-5-12i}{13}=-\frac{5}{13}-\frac{12}{13}i$$

58. $$\frac{-1+5i}{3+2i}=\frac{(-1+5i)(3-2i)}{(3+2i)(3-2i)}$$
$$=\frac{-3+2i+15i-10i^2}{3^2+2^2}$$
$$=\frac{-3+10+17i}{13}$$
$$=\frac{7}{13}+\frac{17}{13}i$$

59. **(a)** $(x+2)+(3x-1)$
$$=(1+3)x+(2-1)$$
$$=4x+1$$

(b) $(1+2i)+(3-i)$
$$=(1+3)+(2-1)i$$
$$=4+i$$

60. **(a)** $(x+2)-(3x-1)$
$$=(x-3x)+[2-(-1)]$$
$$=(1-3)x+(2+1)$$
$$=-2x+3$$

(b) $(1+2i)-(3-i)$
$$=(1-3)+[2-(-1)]i$$
$$=-2+3i$$

61. **(a)** $(x+2)(3x-1)$
$= 3x^2 - x + 6x - 2$
$= 3x^2 + 5x - 2$

(b) $(1+2i)(3-i)$
$= 3 - i + 6i - 2i^2$
$= 3 + 5i - 2(-1)$
$= 3 + 5i + 2$
$= 5 + 5i$

62. **(a)** $\dfrac{\sqrt{3}-1}{1+\sqrt{2}} = \dfrac{(\sqrt{3}-1)(1-\sqrt{2})}{(1+\sqrt{2})(1-\sqrt{2})}$
$= \dfrac{\sqrt{3}-\sqrt{6}-1+\sqrt{2}}{1^2-(\sqrt{2})^2}$
$= \dfrac{\sqrt{3}-\sqrt{6}-1+\sqrt{2}}{1-2}$
$= \dfrac{\sqrt{3}-\sqrt{6}-1+\sqrt{2}}{-1}$
$= -(\sqrt{3}-\sqrt{6}-1+\sqrt{2})$
$= -\sqrt{3}+\sqrt{6}+1-\sqrt{2}$

(b) $\dfrac{3-i}{1+2i} = \dfrac{(3-i)(1-2i)}{(1+2i)(1-2i)}$
$= \dfrac{3-6i-i+2i^2}{1-(2i)^2}$
$= \dfrac{3-7i+2(-1)}{1-4(-1)}$
$= \dfrac{3-7i-2}{1+4}$
$= \dfrac{1-7i}{5}$
$= \frac{1}{5} - \frac{7}{5}i$

63. In parts (a) and (b) of Exercises 59 and 60, real and imaginary parts are added, just like coefficients of similar terms in the binomials, and the answers correspond. In Exercise 61, introducing $i^2 = -1$ when a product is found leads to answers that do not correspond.

64. In parts (a) and (b) of Exercises 59 and 60, real and imaginary parts are added, just like coefficients of similar terms in the binomials, and the answers correspond. In Exercise 62, introducing $i^2 = -1$ when performing the division leads to answers that do not correspond.

65. The reciprocal of $5-4i$ is $\dfrac{1}{5-4i}$.

$\dfrac{1}{5-4i} = \dfrac{1\cdot(5+4i)}{(5-4i)(5+4i)}$
$= \dfrac{5+4i}{5^2+4^2} = \dfrac{5+4i}{25+16}$
$= \dfrac{5+4i}{41} = \dfrac{5}{41}+\dfrac{4}{41}i$

66. $(4-3i)^{-1} = \dfrac{1}{(4-3i)^1} = \dfrac{1}{4-3i}$
$\dfrac{1}{4-3i} = \dfrac{4+3i}{(4-3i)(4+3i)}$
$= \dfrac{4+3i}{4^2+3^2} = \dfrac{4+3i}{16+9}$
$= \dfrac{4+3i}{25} = \dfrac{4}{25}+\dfrac{3}{25}i$

67. $i^{18} = i^{16}\cdot i^2 = (i^4)^4\cdot i^2$
$= 1^4\cdot(-1) = 1\cdot(-1) = -1$

68. $i^{26} = i^{24}\cdot i^2$
$= (i^4)^6\cdot i^2 = 1^6(-1) = -1$

69. $i^{89} = i^{88}\cdot i = (i^4)^{22}\cdot i = 1^{22}\cdot i$
$= 1\cdot i = i$

70. $i^{45} = i^{44}\cdot i = (i^4)^{11}\cdot i = 1^{11}\cdot i = i$

71. $i^{96} = (i^4)^{24} = 1^{24} = 1$

72. $i^{48} = (i^4)^{12} = 1^{12} = 1$

73. $i^{-5} = \dfrac{1}{i^5} = \dfrac{1}{i^4\cdot i} = \dfrac{1}{1\cdot i} = \dfrac{1}{i}$

From Exercise 3, $\dfrac{1}{i} = -i$.

74. $i^{-17} = \dfrac{1}{i^{17}} = \dfrac{1}{i^{16}\cdot i} = \dfrac{1}{(i^4)^4\cdot i} = \dfrac{1}{1^4\cdot i} = \dfrac{1}{i}$

From Exercise 3, $\dfrac{1}{i} = -i$.

75. Since $i^{20} = (i^4)^5 = 1^5 = 1$, the student multiplied by 1, which is justified by the identity property for multiplication.

76. $i^{12} = (i^4)^3 = 1^3 = 1$, so by the identity property for multiplication, the two products must be equal.

77. $I = \dfrac{E}{R+(X_L-X_c)i}$

Substitute $2+3i$ for E, 5 for R, 4 for X_L, and 3 for X_c.

$I = \dfrac{2+3i}{5+(4-3)i} = \dfrac{2+3i}{5+i}$
$= \dfrac{(2+3i)(5-i)}{(5+i)(5-i)} = \dfrac{10-2i+15i-3i^2}{5^2+1^2}$
$= \dfrac{10+3+13i}{25+1} = \dfrac{13+13i}{26}$
$= \dfrac{13(1+i)}{13\cdot 2} = \dfrac{1+i}{2} = \dfrac{1}{2}+\dfrac{1}{2}i$

78. Use $I = \dfrac{E}{R + (X_L - X_c)i}$ to find E if $I = 1 - i$, $R = 2$, $X_L = 3$, and $X_c = 1$.

$$\begin{aligned} 1 - i &= \frac{E}{2 + (3-1)i} \\ 1 - i &= \frac{E}{2 + 2i} \\ (2+2i)(1-i) &= E \\ 2 - 2i + 2i - 2i^2 &= E \\ 2 + 0i + 2 &= E \\ 4 &= E \end{aligned}$$

79. To check that $1 + 5i$ is a solution of the equation, substitute $1 + 5i$ for x.

$$\begin{aligned} x^2 - 2x + 26 &= 0 \\ (1+5i)^2 - 2(1+5i) + 26 &\stackrel{?}{=} 0 \\ \left(1 + 10i + 25i^2\right) - 2 - 10i + 26 &\stackrel{?}{=} 0 \\ 1 + 10i - 25 - 2 - 10i + 26 &\stackrel{?}{=} 0 \\ (1 - 25 - 2 + 26) + (10 - 10)i &\stackrel{?}{=} 0 \\ 0 &= 0 \quad \textit{True} \end{aligned}$$

Thus, $1 + 5i$ is a solution of

$$x^2 - 2x + 26 = 0.$$

80. Check $3 + 2i$ in $x^2 - 6x + 13 = 0$.

$$\begin{aligned} (3+2i)^2 - 6(3+2i) + 13 &\stackrel{?}{=} 0 \\ \left(9 + 12i + 4i^2\right) - 18 - 12i + 13 &\stackrel{?}{=} 0 \\ (9 + 12i - 4) - 18 - 12i + 13 &\stackrel{?}{=} 0 \\ (9 - 4 - 18 + 13) + (12 - 12)i &\stackrel{?}{=} 0 \\ 0 &= 0 \quad \textit{True} \end{aligned}$$

Thus, $3 + 2i$ is a solution of

$$x^2 - 6x + 13 = 0.$$

Chapter 9 Review Exercises

1. $\sqrt{1764} = 42$, because $42^2 = 1764$.

2. $\sqrt{289} = 17$, because $17^2 = 289$, so $-\sqrt{289} = -17$.

3. $-\sqrt{-841}$ is *not a real number*, since the square of a real number is nonnegative.

4. $\sqrt[3]{216} = 6$, because $6^3 = 216$.

5. $\sqrt[5]{-32} = -2$, because $(-2)^5 = -32$.

6. $\sqrt{x^2} = |x|$

7. $\sqrt[3]{x^3} = x$

8. $\sqrt[4]{x^{20}} = \sqrt[4]{(x^5)^4} = |x^5| \quad \text{or} \quad x^4|x|$.

9. $f(x) = \sqrt{x-1}$
For the radicand to be nonnegative, we must have

$$x - 1 \geq 0 \quad \text{or} \quad x \geq 1.$$

Thus, the domain is $[1, \infty)$.
The function values are positive or zero (the result of the radical), so the range is $[0, \infty)$.

x	$f(x) = \sqrt{x-1}$
1	$\sqrt{1-1} = 0$
2	$\sqrt{2-1} = 1$
5	$\sqrt{5-1} = 2$

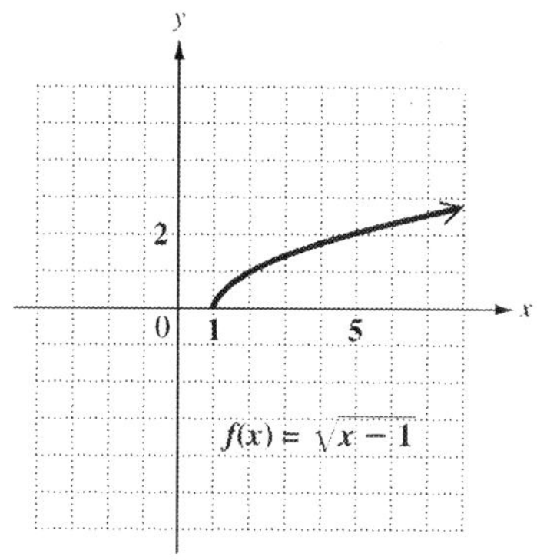

10. $f(x) = \sqrt[3]{x} + 4$
Since we can take the cube root of any real number, the domain is $(-\infty, \infty)$.
The result of a cube root can be any real number, so the range is $(-\infty, \infty)$. (The "+4" does not affect that range.)

x	$f(x) = \sqrt[3]{x} + 4$
-8	$\sqrt[3]{-8} + 4 = 2$
-1	$\sqrt[3]{-1} + 4 = 3$
0	$\sqrt[3]{0} + 4 = 4$
1	$\sqrt[3]{1} + 4 = 5$
8	$\sqrt[3]{8} + 4 = 6$

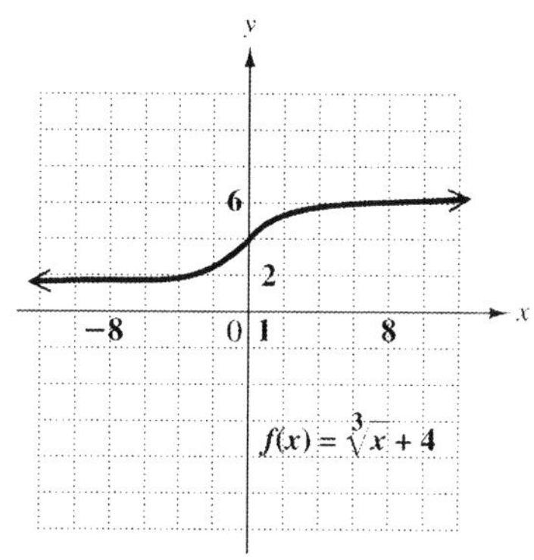

11. $\sqrt[n]{a}$ is not a real number if n is even and a is negative.

12. If a is negative and n is even, then $a^{1/n}$ *is not a real number*. An example is $(-4)^{1/2}$, not a real number.

13. $\sqrt{40} \approx 6.325$

14. $\sqrt{77} \approx 8.775$

15. $\sqrt{310} \approx 17.607$

16. Substitute 3 for L and 32 for g.

$$t = 2\pi\sqrt{\frac{L}{g}} = 2\pi\sqrt{\frac{3}{32}} \approx 1.9$$

The time to complete one swing is approximately 1.9 sec.

17. Let $a = 11$, $b = 13$, and $c = 20$. First find the semiperimeter s.

$$\begin{aligned} s &= \tfrac{1}{2}(a+b+c) \\ &= \tfrac{1}{2}(11+13+20) \\ &= \tfrac{44}{2} = 22 \end{aligned}$$

Now find the area A using Heron's formula.

$$\begin{aligned} A &= \sqrt{s(s-a)(s-b)(s-c)} \\ &= \sqrt{22(22-11)(22-13)(22-20)} \\ &= \sqrt{22(11)(9)(2)} \\ &= 66 \end{aligned}$$

The area of the triangle is 66 in.^{2}.

18. $49^{1/2} = \sqrt{49} = 7$

19. $-8^{1/3} = -\sqrt[3]{8} = -2$

20. $(-16)^{1/4} = \sqrt[4]{-16}$ is *not a real number* because the radicand, -16, is negative and the index, 4, is even.

21. The expression with fractional exponents, $a^{m/n}$, is equivalent to the radical expression, $\sqrt[n]{a^m}$. The denominator of the exponent is the index of the radical. For example, $8^{2/3} = (8^{1/3})^2 = 2^2 = 4$, and $\sqrt[3]{8^2} = \sqrt[3]{64} = 4$.

22. $16^{5/4} = (16^{1/4})^5 = 2^5 = 32$

23. $-8^{2/3} = -(8^{1/3})^2 = -2^2 = -4$

24. $$\begin{aligned} -\left(\frac{36}{25}\right)^{3/2} &= -\left(\sqrt{\frac{36}{25}}\right)^3 \\ &= -\left(\frac{6}{5}\right)^3 = -\frac{216}{125} \end{aligned}$$

25. $$\begin{aligned} \left(-\frac{1}{8}\right)^{-5/3} &= (-8)^{5/3} = \left[(-8)^{1/3}\right]^5 \\ &= (-2)^5 = -32 \end{aligned}$$

26. $$\begin{aligned} \left(\frac{81}{10{,}000}\right)^{-3/4} &= \left(\frac{10{,}000}{81}\right)^{3/4} \\ &= \left(\sqrt[4]{\frac{10{,}000}{81}}\right)^3 \\ &= \left(\frac{10}{3}\right)^3 = \frac{1000}{27} \end{aligned}$$

27. $7^{1/3} \cdot 7^{5/3} = 7^{1/3+5/3} = 7^{6/3} = 7^2 = 49$

28. $\dfrac{96^{2/3}}{96^{-1/3}} = 96^{2/3-(-1/3)} = 96^1 = 96$

29. $$\begin{aligned} \frac{k^{2/3}k^{-1/2}k^{3/4}}{2(k^2)^{-1/4}} &= \frac{k^{2/3-1/2+3/4}}{2k^{-2/4}} \\ &= \frac{k^{8/12-6/12+9/12}}{2k^{-6/12}} \\ &= \frac{k^{11/12}}{2k^{-6/12}} \\ &= \frac{k^{17/12}}{2} \end{aligned}$$

30. $2^{4/5} = (2^4)^{1/5} = \sqrt[5]{2^4}$ or $\sqrt[5]{16}$

31. $\sqrt{3^{18}} = 3^{18/2} = 3^9$

32. $$\begin{aligned} \sqrt{7^9} &= 7^{9/2} = 7^{8/2+1/2} \\ &= 7^{4+1/2} = 7^4 \cdot 7^{1/2} = 7^4\sqrt{7} \end{aligned}$$

33. $$\begin{aligned} \sqrt[3]{m^5} \cdot \sqrt[3]{m^8} &= (m^5)^{1/3}(m^8)^{1/3} \\ &= m^{5/3}m^{8/3} = m^{5/3+8/3} \\ &= m^{13/3} = m^{12/3}m^{1/3} \\ &= m^4\sqrt[3]{m} \end{aligned}$$

34. $$\begin{aligned} \sqrt[4]{k^2} \cdot \sqrt[4]{k^7} &= k^{2/4} \cdot k^{7/4} \\ &= k^{2/4+7/4} \\ &= k^{9/4} \\ &= k^{2+1/4} \\ &= k^2 \cdot k^{1/4} \\ &= k^2\sqrt[4]{k} \end{aligned}$$

35. $$\begin{aligned} \sqrt[3]{\sqrt{m}} &= (\sqrt{m})^{1/3} = (m^{1/2})^{1/3} \\ &= m^{(1/2)(1/3)} = m^{1/6} = \sqrt[6]{m} \end{aligned}$$

36. $$\begin{aligned} \sqrt[4]{16y^5} &= (16y^5)^{1/4} = 16^{1/4}y^{5/4} \\ &= 2y^{4/4}y^{1/4} = 2y\sqrt[4]{y} \end{aligned}$$

37. $\sqrt[5]{y}\cdot\sqrt[3]{y} = y^{1/5}\cdot y^{1/3}$
$= y^{1/5+1/3}$
$= y^{3/15+5/15}$
$= y^{8/15}$
$= \sqrt[15]{y^8}$

38. $\dfrac{\sqrt[3]{y^2}}{\sqrt[4]{y}} = \dfrac{y^{2/3}}{y^{1/4}} = y^{2/3-1/4} = y^{8/12-3/12}$
$= y^{5/12} = \sqrt[12]{y^5}$

39. $\sqrt{6}\cdot\sqrt{11} = \sqrt{6\cdot 11} = \sqrt{66}$

40. $\sqrt{5}\cdot\sqrt{r} = \sqrt{5\cdot r} = \sqrt{5r}$

41. $\sqrt[3]{6}\cdot\sqrt[3]{5} = \sqrt[3]{6\cdot 5} = \sqrt[3]{30}$

42. $\sqrt[4]{7}\cdot\sqrt[4]{3} = \sqrt[4]{7\cdot 3} = \sqrt[4]{21}$

43. $\sqrt{20} = \sqrt{4\cdot 5} = \sqrt{4}\sqrt{5} = 2\sqrt{5}$

44. $-\sqrt{125} = -\sqrt{25\cdot 5} = -5\sqrt{5}$

45. $\sqrt[3]{-108x^4y} = \sqrt[3]{-1\cdot 27\cdot x^3\cdot 4xy}$
$= -3x\sqrt[3]{4xy}$

46. $\sqrt[3]{64p^4q^6} = \sqrt[3]{64p^3q^6\cdot p} = 4pq^2\sqrt[3]{p}$

47. $\sqrt{\dfrac{49}{81}} = \dfrac{\sqrt{49}}{\sqrt{81}} = \dfrac{7}{9}$

48. $\sqrt{\dfrac{y^3}{144}} = \dfrac{\sqrt{y^3}}{\sqrt{144}} = \dfrac{\sqrt{y^2\cdot y}}{12} = \dfrac{y\sqrt{y}}{12}$

49. $\sqrt[3]{\dfrac{m^{15}}{27}} = \dfrac{\sqrt[3]{m^{15}}}{\sqrt[3]{27}} = \dfrac{\sqrt[3]{(m^5)^3}}{\sqrt[3]{3^3}} = \dfrac{m^5}{3}$

50. $\sqrt[3]{\dfrac{r^2}{8}} = \dfrac{\sqrt[3]{r^2}}{\sqrt[3]{8}} = \dfrac{\sqrt[3]{r^2}}{2}$

51. $\dfrac{\sqrt[3]{2^4}}{\sqrt[4]{32}} = \dfrac{\sqrt[3]{2^3}\sqrt[3]{2}}{\sqrt[4]{16}\sqrt[4]{2}} = \dfrac{2\cdot 2^{1/3}}{2\cdot 2^{1/4}}$
$= 2^{1/3-1/4}$
$= 2^{4/12-3/12}$
$= 2^{1/12}$
$= \sqrt[12]{2}$

52. $\dfrac{\sqrt{x}}{\sqrt[5]{x}} = \dfrac{x^{1/2}}{x^{1/5}} = x^{1/2-1/5} = x^{5/10-2/10} = x^{3/10}$
$= \sqrt[10]{x^3}$

53. $(x_1, y_1) = (2, 7)$ and $(x_2, y_2) = (-1, -4)$
$d = \sqrt{(x_2 - x_1)^2 + (y_2 - y_1)^2}$
$= \sqrt{(-1-2)^2 + (-4-7)^2}$
$= \sqrt{(-3)^2 + (-11)^2}$
$= \sqrt{9 + 121}$
$= \sqrt{130}$

54. $(x_1, y_1) = (-3, -5)$ and $(x_2, y_2) = (4, -3)$
$d = \sqrt{(x_2 - x_1)^2 + (y_2 - y_1)^2}$
$= \sqrt{[4-(-3)]^2 + [-3-(-5)]^2}$
$= \sqrt{7^2 + 2^2}$
$= \sqrt{49 + 4}$
$= \sqrt{53}$

55. $2\sqrt{8} - 3\sqrt{50} = 2\sqrt{4\cdot 2} - 3\sqrt{25\cdot 2}$
$= 2\cdot 2\sqrt{2} - 3\cdot 5\sqrt{2}$
$= 4\sqrt{2} - 15\sqrt{2} = -11\sqrt{2}$

56. $8\sqrt{80} - 3\sqrt{45} = 8\sqrt{16\cdot 5} - 3\sqrt{9\cdot 5}$
$= 8\cdot 4\sqrt{5} - 3\cdot 3\sqrt{5}$
$= 32\sqrt{5} - 9\sqrt{5} = 23\sqrt{5}$

57. $-\sqrt{27y} + 2\sqrt{75y} = -\sqrt{9\cdot 3y} + 2\sqrt{25\cdot 3y}$
$= -3\sqrt{3y} + 2\cdot 5\sqrt{3y}$
$= -3\sqrt{3y} + 10\sqrt{3y}$
$= 7\sqrt{3y}$

58. $2\sqrt{54m^3} + 5\sqrt{96m^3}$
$= 2\sqrt{9m^2\cdot 6m} + 5\sqrt{16m^2\cdot 6m}$
$= 2\cdot 3m\sqrt{6m} + 5\cdot 4m\sqrt{6m}$
$= 6m\sqrt{6m} + 20m\sqrt{6m} = 26m\sqrt{6m}$

59. $3\sqrt[3]{54} + 5\sqrt[3]{16} = 3\sqrt[3]{27\cdot 2} + 5\sqrt[3]{8\cdot 2}$
$= 3\cdot 3\sqrt[3]{2} + 5\cdot 2\sqrt[3]{2}$
$= 9\sqrt[3]{2} + 10\sqrt[3]{2} = 19\sqrt[3]{2}$

60. $-6\sqrt[4]{32} + \sqrt[4]{512} = -6\sqrt[4]{16\cdot 2} + \sqrt[4]{256\cdot 2}$
$= -6\cdot 2\sqrt[4]{2} + 4\sqrt[4]{2}$
$= -12\sqrt[4]{2} + 4\sqrt[4]{2} = -8\sqrt[4]{2}$

61. $(\sqrt{3}+1)(\sqrt{3}-2) = 3 - 2\sqrt{3} + \sqrt{3} - 2$
$= 1 - \sqrt{3}$

62. $(\sqrt{7}+\sqrt{5})(\sqrt{7}-\sqrt{5}) = (\sqrt{7})^2 - (\sqrt{5})^2$
$= 7 - 5 = 2$

63. $(3\sqrt{2}+1)(2\sqrt{2}-3)$
$= 6\cdot 2 - 9\sqrt{2} + 2\sqrt{2} - 3$
$= 12 - 7\sqrt{2} - 3 = 9 - 7\sqrt{2}$

64. $(\sqrt{11}+3\sqrt{5})(\sqrt{11}+5\sqrt{5})$
$= 11 + 5\sqrt{55} + 3\sqrt{55} + 15 \cdot 5$
$= 11 + 8\sqrt{55} + 75$
$= 86 + 8\sqrt{55}$

65. $(\sqrt{13}-\sqrt{2})^2$
$= (\sqrt{13})^2 - 2 \cdot \sqrt{13} \cdot \sqrt{2} + (\sqrt{2})^2$
$= 13 - 2\sqrt{26} + 2 = 15 - 2\sqrt{26}$

66. $(\sqrt{5}-\sqrt{7})^2$
$= (\sqrt{5})^2 - 2 \cdot \sqrt{5} \cdot \sqrt{7} + (\sqrt{7})^2$
$= 5 - 2\sqrt{35} + 7$
$= 12 - 2\sqrt{35}$

67. $\dfrac{-6\sqrt{3}}{\sqrt{2}} = \dfrac{-6\sqrt{3} \cdot \sqrt{2}}{\sqrt{2} \cdot \sqrt{2}} = \dfrac{-6\sqrt{6}}{2} = -3\sqrt{6}$

68. $\dfrac{3\sqrt{7p}}{\sqrt{y}} = \dfrac{3\sqrt{7p} \cdot \sqrt{y}}{\sqrt{y} \cdot \sqrt{y}} = \dfrac{3\sqrt{7py}}{y}$

69. $-\sqrt[3]{\dfrac{9}{25}} = -\dfrac{\sqrt[3]{9}}{\sqrt[3]{5^2}} = -\dfrac{\sqrt[3]{9} \cdot \sqrt[3]{5}}{\sqrt[3]{5^2} \cdot \sqrt[3]{5}}$
$= -\dfrac{\sqrt[3]{45}}{\sqrt[3]{5^3}} = -\dfrac{\sqrt[3]{45}}{5}$

70. $\sqrt[3]{\dfrac{108m^3}{n^5}} = \dfrac{\sqrt[3]{108m^3}}{\sqrt[3]{n^5}} = \dfrac{\sqrt[3]{27m^3 \cdot 4}}{\sqrt[3]{n^3 \cdot n^2}}$
$= \dfrac{3m\sqrt[3]{4}}{n\sqrt[3]{n^2}} = \dfrac{3m\sqrt[3]{4} \cdot \sqrt[3]{n}}{n\sqrt[3]{n^2} \cdot \sqrt[3]{n}}$
$= \dfrac{3m\sqrt[3]{4n}}{n \cdot n} = \dfrac{3m\sqrt[3]{4n}}{n^2}$

71. $\dfrac{1}{\sqrt{2}+\sqrt{7}}$

Multiply the numerator and denominator by the conjugate of the denominator, $\sqrt{2}-\sqrt{7}$.

$= \dfrac{1(\sqrt{2}-\sqrt{7})}{(\sqrt{2}+\sqrt{7})(\sqrt{2}-\sqrt{7})}$
$= \dfrac{\sqrt{2}-\sqrt{7}}{2-7} = \dfrac{\sqrt{2}-\sqrt{7}}{-5}$

72. $\dfrac{-5}{\sqrt{6}-\sqrt{3}}$

Multiply the numerator and denominator by the conjugate of the denominator, $\sqrt{6}+\sqrt{3}$.

$= \dfrac{-5(\sqrt{6}+\sqrt{3})}{(\sqrt{6}-\sqrt{3})(\sqrt{6}+\sqrt{3})}$
$= \dfrac{-5(\sqrt{6}+\sqrt{3})}{6-3}$
$= \dfrac{-5(\sqrt{6}+\sqrt{3})}{3}$

73. $\sqrt{8x+9} = 5$
$(\sqrt{8x+9})^2 = 5^2$ *Square*
$8x + 9 = 25$
$8x = 16$
$x = 2$

Check $x = 2$: $\sqrt{25} = 5$ *True*

Solution set: $\{2\}$

74. $\sqrt{2z-3} - 3 = 0$
$\sqrt{2z-3} = 3$ *Isolate*
$(\sqrt{2z-3})^2 = 3^2$ *Square*
$2z - 3 = 9$
$2z = 12$
$z = 6$

Check $z = 6$: $\sqrt{9} - 3 = 0$ *True*

Solution set: $\{6\}$

75. $\sqrt{3m+1} = -1$

This equation has no solution because $\sqrt{3m+1}$ cannot be negative.

Solution set: $\emptyset$

76. $\sqrt{7z+1} = z+1$
$(\sqrt{7z+1})^2 = (z+1)^2$ *Square*
$7z + 1 = z^2 + 2z + 1$
$0 = z^2 - 5z$
$0 = z(z-5)$

$z = 0$ or $z = 5$

Check $z = 0$: $\sqrt{1} = 1$ *True*

Check $z = 5$: $\sqrt{36} = 6$ *True*

Solution set: $\{0, 5\}$

77. $3\sqrt{m} = \sqrt{10m-9}$
$(3\sqrt{m})^2 = (\sqrt{10m-9})^2$ *Square*
$9m = 10m - 9$
$9 = m$

Check $m = 9$: $3\sqrt{9} = \sqrt{81}$ *True*

Solution set: $\{9\}$

78. $\sqrt{p^2+3p+7} = p+2$
$\left(\sqrt{p^2+3p+7}\right)^2 = (p+2)^2$ *Square*
$p^2 + 3p + 7 = p^2 + 4p + 4$
$3 = p$

Check $p = 3$: $\sqrt{25} = 5$ *True*

Solution set: $\{3\}$

79. $\sqrt{x+2}-\sqrt{x-3}=1$

Get one radical on each side of the equals sign.

$$\sqrt{x+2}=1+\sqrt{x-3}$$

Square both sides.

$$\begin{aligned}(\sqrt{x+2})^2&=(1+\sqrt{x-3})^2\\x+2&=1+2\sqrt{x-3}+x-3\\4&=2\sqrt{x-3}\\2&=\sqrt{x-3}\end{aligned}$$

Square both sides again.

$$\begin{aligned}2^2&=(\sqrt{x-3})^2\\4&=x-3\\7&=x\end{aligned}$$

Check $x=7$: $\sqrt{9}-\sqrt{4}=1$ *True*

Solution set: $\{7\}$

80. $\sqrt[3]{5m-1}=\sqrt[3]{3m-2}$

Cube both sides.

$$\begin{aligned}(\sqrt[3]{5m-1})^3&=(\sqrt[3]{3m-2})^3\\5m-1&=3m-2\\2m&=-1\\m&=-\tfrac{1}{2}\end{aligned}$$

Check $m=-\frac{1}{2}$: $\sqrt[3]{-\frac{7}{2}}=\sqrt[3]{-\frac{7}{2}}$ *True*

Solution set: $\{-\frac{1}{2}\}$

81. $\sqrt[4]{x+6}=\sqrt[4]{2x}$

Raise both sides to the fourth power.

$$\begin{aligned}(\sqrt[4]{x+6})^4&=(\sqrt[4]{2x})^4\\x+6&=2x\\6&=x\end{aligned}$$

Check $x=6$: $\sqrt[4]{12}=\sqrt[4]{12}$ *True*

Solution set: $\{6\}$

82. $\sqrt{-25}=i\sqrt{25}=5i$

83. $\sqrt{-200}=i\sqrt{100\cdot 2}=10i\sqrt{2}$

84. $\sqrt{-160}=i\sqrt{16\cdot 10}=4i\sqrt{10}$

85. $(-2+5i)+(-8-7i)$

$$\begin{aligned}&=[-2+(-8)]+[5+(-7)]i\\&=-10-2i\end{aligned}$$

86. $(5+4i)-(-9-3i)$

$$\begin{aligned}&=[(5-(-9)]+[(4-(-3)]i\\&=14+7i\end{aligned}$$

87. $\sqrt{-5}\cdot\sqrt{-7}=i\sqrt{5}\cdot i\sqrt{7}=i^2\sqrt{35}$

$$=-1\left(\sqrt{35}\right)=-\sqrt{35}$$

88. $\sqrt{-25}\cdot\sqrt{-81}=i\sqrt{25}\cdot i\sqrt{81}$

$$\begin{aligned}&=5i\cdot 9i=45i^2\\&=45(-1)=-45\end{aligned}$$

89. $\dfrac{\sqrt{-72}}{\sqrt{-8}}=\dfrac{i\sqrt{72}}{i\sqrt{8}}=\sqrt{\dfrac{72}{8}}=\sqrt{9}=3$

90. $(2+3i)(1-i)=2-2i+3i-3i^2$

$$\begin{aligned}&=2+i-3(-1)\\&=2+i+3\\&=5+i\end{aligned}$$

91. $(6-2i)^2=6^2-2\cdot 6\cdot 2i+(2i)^2$

$$\begin{aligned}&=36-24i+4i^2\\&=36-24i+4(-1)\\&=36-24i-4\\&=32-24i\end{aligned}$$

92. $\dfrac{3-i}{2+i}$

Multiply the numerator and denominator by the conjugate of the denominator, $2-i$.

$$\begin{aligned}&=\frac{(3-i)(2-i)}{(2+i)(2-i)}\\&=\frac{6-3i-2i+i^2}{4-i^2}\\&=\frac{6-5i-1}{4-(-1)}=\frac{5-5i}{5}\\&=\frac{5(1-i)}{5}=1-i\end{aligned}$$

93. $\dfrac{5+14i}{2+3i}$

Multiply the numerator and denominator by the conjugate of the denominator, $2-3i$.

$$\begin{aligned}&=\frac{(5+14i)(2-3i)}{(2+3i)(2-3i)}\\&=\frac{10-15i+28i-42i^2}{2^2+3^2}\\&=\frac{10+13i-42(-1)}{4+9}=\frac{52+13i}{13}\\&=\frac{13(4+i)}{13}=4+i\end{aligned}$$

94. $i^{11}=i^8\cdot i^3=\left(i^4\right)^2\cdot i^2\cdot i$

$$=1^2\cdot(-1)\cdot i=-i$$

95. $i^{52}=\left(i^4\right)^{13}=1^{13}=1$

96. $i^{-13}=i^{-13}\cdot i^{16}$, since $i^{16}=\left(i^4\right)^4=1^4=1$. So, $i^{-13}=i^{-13}\cdot i^{16}=i^3=i^2\cdot i=-i$.

97. **[9.1]** $-\sqrt{169a^2b^4}=-\sqrt{(13ab^2)^2}=-13ab^2$

98. [9.2] $1000^{-2/3} = \frac{1}{1000^{2/3}} = \frac{1}{(1000^{1/3})^2} = \frac{1}{10^2} = \frac{1}{100}$

99. [9.2] $\frac{y^{-1/3} \cdot y^{5/6}}{y} = \frac{y^{-2/6+5/6}}{y} = \frac{y^{3/6}}{y} = \frac{y^{1/2}}{y} = y^{1/2-1} = y^{-1/2} = \frac{1}{y^{1/2}}$

100. [9.2] $\frac{z^{-1/4}x^{1/2}}{z^{1/2}x^{-1/4}} = z^{-1/4-1/2} \cdot x^{1/2-(-1/4)} = z^{-1/4-2/4} \cdot x^{2/4+1/4} = z^{-3/4} \cdot x^{3/4} = \frac{x^{3/4}}{z^{3/4}}$

101. [9.3] $\sqrt[4]{k^{24}} = k^{24/4} = k^6$

102. [9.3] $\sqrt[3]{54z^9t^8} = \sqrt[3]{27z^9t^6 \cdot 2t^2} = 3z^3t^2\sqrt[3]{2t^2}$

103. [9.4] $-5\sqrt{18} + 12\sqrt{72}$
$= -5\sqrt{9 \cdot 2} + 12\sqrt{36 \cdot 2}$
$= -5 \cdot 3\sqrt{2} + 12 \cdot 6\sqrt{2}$
$= -15\sqrt{2} + 72\sqrt{2} = 57\sqrt{2}$

104. [9.4] $8\sqrt[3]{x^3y^2} - 2x\sqrt[3]{y^2} = 8x\sqrt[3]{y^2} - 2x\sqrt[3]{y^2} = 6x\sqrt[3]{y^2}$

105. [9.5] $(\sqrt{5} - \sqrt{3})(\sqrt{7} + \sqrt{3})$
$= \sqrt{35} + \sqrt{15} - \sqrt{21} - 3$

106. [9.5] $\frac{-1}{\sqrt{12}} = \frac{-1}{\sqrt{4 \cdot 3}} = \frac{-1}{2\sqrt{3}} = \frac{-1 \cdot \sqrt{3}}{2\sqrt{3} \cdot \sqrt{3}} = \frac{-\sqrt{3}}{2 \cdot 3} = \frac{-\sqrt{3}}{6}$

107. [9.5] $\sqrt[3]{\frac{12}{25}} = \frac{\sqrt[3]{12}}{\sqrt[3]{25}} = \frac{\sqrt[3]{12}}{\sqrt[3]{5^2}} = \frac{\sqrt[3]{12} \cdot \sqrt[3]{5}}{\sqrt[3]{5^2} \cdot \sqrt[3]{5}} = \frac{\sqrt[3]{60}}{\sqrt[3]{5^3}} = \frac{\sqrt[3]{60}}{5}$

108. [9.5] $\frac{2\sqrt{z}}{\sqrt{z} - 2}$

Multiply the numerator and denominator by the conjugate of the denominator, $\sqrt{z} + 2$.

$= \frac{2\sqrt{z}(\sqrt{z} + 2)}{(\sqrt{z} - 2)(\sqrt{z} + 2)} = \frac{2\sqrt{z}(\sqrt{z} + 2)}{z - 4}$

109. [9.7] $\sqrt{-49} = i\sqrt{49} = 7i$

110. [9.7] $(4 - 9i) + (-1 + 2i)$
$= (4 - 1) + (-9 + 2)i$
$= 3 - 7i$

111. [9.7] $\frac{\sqrt{50}}{\sqrt{-2}} = \frac{\sqrt{25 \cdot 2}}{i\sqrt{2}} = \frac{5\sqrt{2}}{i\sqrt{2}} = \frac{5}{i}$

The conjugate of i is $-i$.

$\frac{5(-i)}{i(-i)} = \frac{-5i}{-i^2} = \frac{-5i}{-(-1)} = -5i$

112. [9.6]
$\sqrt{x + 4} = x - 2$
$(\sqrt{x + 4})^2 = (x - 2)^2$ *Square*
$x + 4 = x^2 - 4x + 4$
$0 = x^2 - 5x$
$0 = x(x - 5)$

$x = 0$ or $x = 5$

Check $x = 0$: $\sqrt{4} = -2$ *False*

Check $x = 5$: $\sqrt{9} = 3$ *True*

Solution set: $\{5\}$

113. [9.6] $\sqrt{6 + 2x} - 1 = \sqrt{7 - 2x}$

Square both sides.

$(\sqrt{6 + 2x} - 1)^2 = (\sqrt{7 - 2x})^2$
$6 + 2x - 2\sqrt{6 + 2x} + 1 = 7 - 2x$
$4x = 2\sqrt{6 + 2x}$
$2x = \sqrt{6 + 2x}$

Square both sides again.

$(2x)^2 = (\sqrt{6 + 2x})^2$
$4x^2 = 6 + 2x$
$4x^2 - 2x - 6 = 0$
$2x^2 - x - 3 = 0$
$(2x - 3)(x + 1) = 0$

$2x - 3 = 0$ or $x + 1 = 0$
$x = \frac{3}{2}$ or $x = -1$

Check $x = \frac{3}{2}$: $\sqrt{9} - 1 = \sqrt{4}$ *True*

Check $x = -1$: $\sqrt{4} - 1 = \sqrt{9}$ *False*

Solution set: $\{\frac{3}{2}\}$

114. **[9.6]** Substitute 12 for L and 9 for W.

$$\begin{aligned} L &= \sqrt{H^2+W^2} \\ 12 &= \sqrt{H^2+9^2} \\ 12^2 &= \left(\sqrt{H^2+81}\right)^2 \\ 144 &= H^2+81 \\ 63 &= H^2 \\ \sqrt{63} &= H \end{aligned}$$

To the nearest tenth of a foot, the height is approximately 7.9 feet.

115. **[9.4]** Add the measures of the sides.

$$\begin{aligned} P &= a+b+c \\ P &= 2\sqrt{27}+\sqrt{108}+\sqrt{50} \\ &= 2\sqrt{9\cdot 3}+\sqrt{36\cdot 3}+\sqrt{25\cdot 2} \\ &= 2\cdot 3\sqrt{3}+6\sqrt{3}+5\sqrt{2} \\ &= 6\sqrt{3}+6\sqrt{3}+5\sqrt{2} \\ &= 12\sqrt{3}+5\sqrt{2} \end{aligned}$$

The perimeter is $12\sqrt{3}+5\sqrt{2}$ feet.

Chapter 9 Test

1. $\sqrt{841}=29$, because $29^2=841$, so $-\sqrt{841}=-29$.

2. $125^{1/3}=\sqrt[3]{125}=5$, because $5^3=125$.

3. $\sqrt{146.25}\approx\sqrt{144}=12$, because $12^2=144$, so choice **C** is the best estimate.

4. $\sqrt{146.25}\approx 12.09$

5. $f(x)=\sqrt{x+6}$
For the radicand to be nonnegative, we must have

$$x+6\ge 0 \quad \text{or} \quad x\ge -6.$$

Thus, the domain is $[-6,\infty)$.
The function values are positive or zero (the result of the radical), so the range is $[0,\infty)$.

x	$f(x)=\sqrt{x+6}$
-6	$\sqrt{-6+6}=0$
-5	$\sqrt{-5+6}=1$
-2	$\sqrt{-2+6}=2$

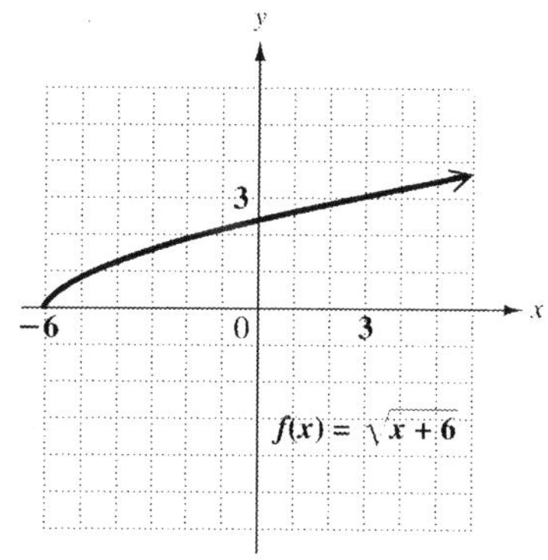

6. $(-64)^{-4/3}=\dfrac{1}{(-64)^{4/3}}=\dfrac{1}{\left(\sqrt[3]{-64}\right)^4}$

$$=\frac{1}{(-4)^4}=\frac{1}{256}$$

7. $\dfrac{3^{2/5}x^{-1/4}y^{2/5}}{3^{-8/5}x^{7/4}y^{1/10}}$

$$\begin{aligned} &= 3^{2/5-(-8/5)}x^{-1/4-7/4}y^{2/5-1/10} \\ &= 3^{10/5}x^{-8/4}y^{4/10-1/10} \\ &= 3^2x^{-2}y^{3/10}=\frac{9y^{3/10}}{x^2} \end{aligned}$$

8. $\sqrt{54x^5y^6}=\sqrt{9x^4y^6\cdot 6x}=3x^2y^3\sqrt{6x}$

9. $\sqrt[4]{32a^7b^{13}}=\sqrt[4]{16a^4b^{12}\cdot 2a^3b}$

$$=2ab^3\sqrt[4]{2a^3b}$$

10. $\sqrt{2}\cdot\sqrt[3]{5}=2^{1/2}\cdot 5^{1/3}=2^{3/6}\cdot 5^{2/6}$

$$\begin{aligned} &= \left(2^3\cdot 5^2\right)^{1/6}=\sqrt[6]{2^3\cdot 5^2} \\ &= \sqrt[6]{8\cdot 25}=\sqrt[6]{200} \end{aligned}$$

11. $3\sqrt{20}-5\sqrt{80}+4\sqrt{500}$

$$\begin{aligned} &= 3\sqrt{4\cdot 5}-5\sqrt{16\cdot 5}+4\sqrt{100\cdot 5} \\ &= 3\cdot 2\sqrt{5}-5\cdot 4\sqrt{5}+4\cdot 10\sqrt{5} \\ &= 6\sqrt{5}-20\sqrt{5}+40\sqrt{5}=26\sqrt{5} \end{aligned}$$

12. $(7\sqrt{5}+4)(2\sqrt{5}-1)$

$$\begin{aligned} &= 14\cdot 5-7\sqrt{5}+8\sqrt{5}-4 \\ &= 70+\sqrt{5}-4=66+\sqrt{5} \end{aligned}$$

13. $\dfrac{-4}{\sqrt{7}+\sqrt{5}}$

Multiply the numerator and denominator by the conjugate of the denominator, $\sqrt{7}-\sqrt{5}$.

$$\begin{aligned} &= \frac{-4\left(\sqrt{7}-\sqrt{5}\right)}{\left(\sqrt{7}+\sqrt{5}\right)\left(\sqrt{7}-\sqrt{5}\right)} \\ &= \frac{-4\left(\sqrt{7}-\sqrt{5}\right)}{7-5} \\ &= \frac{-4\left(\sqrt{7}-\sqrt{5}\right)}{2}=-2\left(\sqrt{7}-\sqrt{5}\right) \end{aligned}$$

14. $\dfrac{-5}{\sqrt{40}} = \dfrac{-5}{\sqrt{4 \cdot 10}} = \dfrac{-5}{2\sqrt{10}} = \dfrac{-5 \cdot \sqrt{10}}{2\sqrt{10} \cdot \sqrt{10}}$

$= \dfrac{-5\sqrt{10}}{2 \cdot 10} = \dfrac{-5\sqrt{10}}{20} = \dfrac{-\sqrt{10}}{4}$

15. $\dfrac{2}{\sqrt[3]{5}} = \dfrac{2 \cdot \sqrt[3]{5^2}}{\sqrt[3]{5}\sqrt[3]{5^2}} = \dfrac{2\sqrt[3]{25}}{5}$

16. The distance d between $(x_1, y_1) = (-3, 8)$ and $(x_2, y_2) = (2, 7)$ is

$$\begin{aligned} d &= \sqrt{(x_2 - x_1)^2 + (y_2 - y_1)^2} \\ &= \sqrt{[2 - (-3)]^2 + (7 - 8)^2} \\ &= \sqrt{5^2 + (-1)^2} \\ &= \sqrt{26}. \end{aligned}$$

17.
$$\begin{aligned} a^2 + b^2 &= c^2 \\ 12^2 + b^2 &= 17^2 \qquad \textit{Let } a = 12,\ c = 17. \\ 144 + b^2 &= 289 \\ b^2 &= 145 \\ b &= \sqrt{145} \end{aligned}$$

18.
$$\begin{aligned} \sqrt[3]{5x} &= \sqrt[3]{2x - 3} \\ \left(\sqrt[3]{5x}\right)^3 &= \left(\sqrt[3]{2x - 3}\right)^3 \qquad \textit{Cube} \\ 5x &= 2x - 3 \\ 3x &= -3 \\ x &= -1 \end{aligned}$$

Check $x = -1$: $\sqrt[3]{-5} = \sqrt[3]{-5}$ *True*

Solution set: $\{-1\}$

19.
$$\begin{aligned} \sqrt{7 - x} + 5 &= x \\ \sqrt{7 - x} &= x - 5 \\ \left(\sqrt{7 - x}\right)^2 &= (x - 5)^2 \qquad \textit{Square} \\ 7 - x &= x^2 - 10x + 25 \\ 0 &= x^2 - 9x + 18 \\ 0 &= (x - 3)(x - 6) \end{aligned}$$

$x - 3 = 0$ or $x - 6 = 0$

$x = 3$ or $x = 6$

Check $x = 3$: $\sqrt{4} + 5 = 3$ *False*

Check $x = 6$: $\sqrt{1} + 5 = 6$ *True*

Solution set: $\{6\}$

20.
$$\begin{aligned} \sqrt{x + 4} - \sqrt{1 - x} &= -1 \\ \sqrt{x + 4} &= -1 + \sqrt{1 - x} \\ \left(\sqrt{x + 4}\right)^2 &= \left(-1 + \sqrt{1 - x}\right)^2 \\ x + 4 &= 1 - 2\sqrt{1 - x} + 1 - x \\ 2x + 2 &= -2\sqrt{1 - x} \\ x + 1 &= -\sqrt{1 - x} \\ (x + 1)^2 &= \left(-\sqrt{1 - x}\right)^2 \\ x^2 + 2x + 1 &= 1 - x \\ x^2 + 3x &= 0 \\ x(x + 3) &= 0 \end{aligned}$$

$x = 0$ or $x + 3 = 0$

$x = -3$

Check $x = -3$: $1 - 2 = -1$ *True*

Check $x = 0$: $2 - 1 = -1$ *False*

Solution set: $\{-3\}$

21.
$$\begin{aligned} &(-2 + 5i) - (3 + 6i) - 7i \\ &= (-2 - 3) + (5 - 6 - 7)i \\ &= -5 - 8i \end{aligned}$$

22. $(-4 + 2i)(3 - i)$

F O I L

$$\begin{aligned} &= -4(3) + (-4)(-i) + 2i(3) + 2i(-i) \\ &= -12 + 4i + 6i - 2i^2 \\ &= -12 + 10i + 2 \\ &= -10 + 10i \end{aligned}$$

23. $\dfrac{7 + i}{1 - i}$

Multiply the numerator and denominator by the conjugate of the denominator, $1 + i$.

$$\begin{aligned} &= \frac{(7 + i)(1 + i)}{(1 - i)(1 + i)} \\ &= \frac{7 + 7i + i + i^2}{1 - i^2} \\ &= \frac{7 + 8i - 1}{1 - (-1)} \\ &= \frac{6 + 8i}{2} = \frac{2(3 + 4i)}{2} = 3 + 4i \end{aligned}$$

24.
$$\begin{aligned} i^{35} &= i^{32} \cdot i^3 = \left(i^4\right)^8 \cdot i^3 \\ &= 1^8 \cdot i^2 \cdot i = 1(-1) \cdot i = -i \end{aligned}$$

Cumulative Review Exercises (Chapters 1–9)

1.
$$\begin{aligned} 7 - (4 + 3t) + 2t &= -6(t - 2) - 5 \\ 7 - 4 - 3t + 2t &= -6t + 12 - 5 \\ 3 - t &= -6t + 7 \\ 5t &= 4 \\ t &= \tfrac{4}{5} \end{aligned}$$

Solution set: $\left\{\frac{4}{5}\right\}$

2. $|6x-9| = |-4x+2|$

$$\begin{array}{rcl} 6x-9=-4x+2 & \text{or} & 6x-9=-(-4x+2) \\ 10x=11 & \text{or} & 6x-9=4x-2 \\ & & 2x=7 \\ x=\frac{11}{10} & \text{or} & x=\frac{7}{2} \end{array}$$

Solution set: $\left\{\frac{11}{10}, \frac{7}{2}\right\}$

3.
$$\begin{aligned} -5-3(x-2) &< 11-2(x+2) \\ -5-3x+6 &< 11-2x-4 \\ 1-3x &< 7-2x \\ -x &< 6 \end{aligned}$$

Multiply by -1; reverse the inequality.

$$x > -6$$

Solution set: $(-6, \infty)$

4.
$$\begin{array}{lcl} 1+4x>5 & \text{and} & -2x>-6 \\ 4x>4 & & \\ x>1 & \text{and} & x<3 \end{array}$$

Solution set: $(1, 3)$

5.
$$\begin{aligned} &-2 < 1-3x < 7 \\ &-3 < -3x < 6 \\ &1 > x > -2 \quad \text{or} \quad -2 < x < 1 \end{aligned}$$

Solution set: $(-2, 1)$

6. To write an equation of the line through $(-4, 6)$ and $(7, -6)$, first find the slope.

$$m = \frac{-6-6}{7-(-4)} = \frac{-12}{11} = -\frac{12}{11}$$

Now use the point-slope form with $(x_1, y_1) = (-4, 6)$ and $m = -\frac{12}{11}$.

$$\begin{aligned} y-y_1 &= m(x-x_1) \\ y-6 &= -\tfrac{12}{11}[x-(-4)] \\ 11y-66 &= -12(x+4) \qquad \textit{Multiply by 11.} \\ 11y-66 &= -12x-48 \\ 12x+11y &= 18 \end{aligned}$$

7. $2x+3y=8$ and $6y=4x+16$

Find the slope of each line by writing them in slope-intercept form.

$$\begin{aligned} 2x+3y &= 8 \\ 3y &= -2x+8 \\ y &= -\tfrac{2}{3}x+\tfrac{8}{3} \end{aligned}$$

The slope of the first line is $-\frac{2}{3}$.

$$\begin{aligned} 6y &= 4x+16 \\ y &= \tfrac{4}{6}x+\tfrac{16}{6} \\ y &= \tfrac{2}{3}x+\tfrac{8}{3} \end{aligned}$$

The slope of this line is $\frac{2}{3}$.

The slopes are not the same, so the lines are not parallel. The slopes are not negative reciprocals of one another, so the lines are not perpendicular. Therefore, the answer is **C**.

8. $f(x) = -3x+6$ or $y = -3x+6$

(a) From the equation, the y-intercept is $(0, 6)$.

(b) Let $y=0$ to find the x-intercept.

$$\begin{aligned} 0 &= -3x+6 \\ 3x &= 6 \\ x &= 2 \end{aligned}$$

The x-intercept is $(2, 0)$.

9. Let $c =$ the cost per item and $n =$ the number of items manufactured.

c varies inversely as n, so $c = \frac{k}{n}$ for some constant k. Substitute $c = 200$ and $n = 1500$ in the equation and solve for k.

$$\begin{aligned} c &= \frac{k}{n} \\ 200 &= \frac{k}{1500} \\ 300{,}000 &= k \end{aligned}$$

So, $c = \frac{300{,}000}{n}$. When $n = 2500$,

$$c = \frac{300{,}000}{2500} = 120.$$

It will cost \$120 per widget.

10. To graph $-2x+y<-6$, graph the line $-2x+y=-6$, which has intercepts $(3, 0)$ and $(0, -6)$, as a dashed line since the inequality involves " $<$ ". Test $(0, 0)$, which yields $0 < -6$, a false statement. Shade the region that does not include $(0, 0)$.

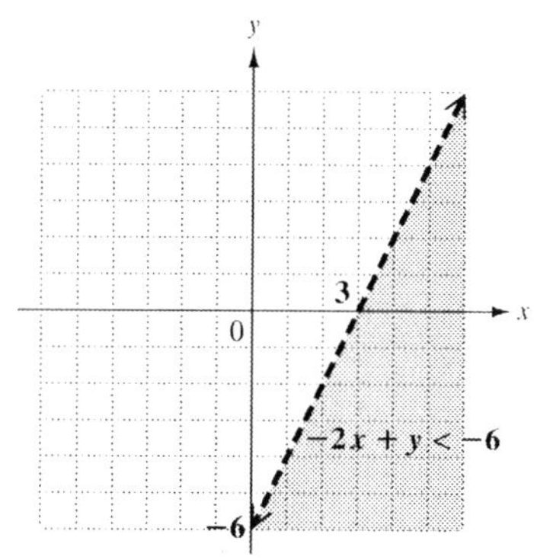

11. The two angles have the same measure, so

$$\begin{aligned} 10x - 70 &= 7x - 25 \\ 3x &= 45 \\ x &= 15. \end{aligned}$$

Since $x = 15$, $10x - 70 = 10(15) - 70$
$= 150 - 70 = 80.$

Each angle measures $80°$.

12. $3x - y = 23$ (1)
$2x + 3y = 8$ (2)
To eliminate y, multiply equation (1) by 3 and add the result to equation (2).

$$\begin{array}{rcrl} 9x - 3y &=& 69 & 3 \times (1) \\ 2x + 3y &=& 8 & (2) \\ \hline 11x &=& 77 & \\ x &=& 7 & \end{array}$$

Substitute 7 for x in (2).

$$\begin{aligned} 2(7) + 3y &= 8 \\ 14 + 3y &= 8 \\ 3y &= -6 \\ y &= -2 \end{aligned}$$

Solution set: $\{(7, -2)\}$

13. $5x + 2y = 7$ (1)
$10x + 4y = 12$ (2)

Multiplying equation (1) by -2 and adding the result to equation (2) yields the false statement $0 = -2$. The system is inconsistent.

Solution set: $\emptyset$

14. $2x + y - z = 5$ (1)
$3x + 2y + z = 8$ (2)
$4x + 2y - 2z = 10$ (3)

Multiplying equation (1) by -2 and adding the result to equation (3) yields the true statement $0 = 0$. Equations (1) and (3) are dependent. Since the two equations that are left include *three* variables, it is not possible to find a unique solution. Therefore, the system has an *infinite number of solutions.*

15. Let $x =$ the number of 2-ounce letters and $y =$ the number of 3-ounce letters.

$5x + 3y = 5.49$ (1)
$3x + 5y = 5.95$ (2)

To eliminate x, multiply (1) by -3 and (2) by 5 and add the results.

$$\begin{array}{rcrl} -15x - 9y &=& -16.47 & -3 \times (1) \\ 15x + 25y &=& 29.75 & 5 \times (2) \\ \hline 16y &=& 13.28 & \\ y &=& .83 & \end{array}$$

Substitute $y = .83$ in (1).

$$\begin{aligned} 5x + 3y &= 5.49 \quad (1) \\ 5x + 3(.83) &= 5.49 \\ 5x + 2.49 &= 5.49 \\ 5x &= 3.00 \\ x &= .60 \end{aligned}$$

The 2004 postage rate for a 2-ounce letter was \$.60 and for a 3-ounce letter, \$.83.

16. $(3k^3 - 5k^2 + 8k - 2) - (4k^3 + 11k + 7) + (2k^2 - 5k)$
$= 3k^3 - 4k^3 - 5k^2 + 2k^2 + 8k - 11k - 5k - 2 - 7$
$= -k^3 - 3k^2 - 8k - 9$

17. $(8x - 7)(x + 3)$
$= 8x^2 + 24x - 7x - 21$
$= 8x^2 + 17x - 21$

18. $\dfrac{8z^3 - 16z^2 + 24z}{8z^2} = \dfrac{8z^3}{8z^2} - \dfrac{16z^2}{8z^2} + \dfrac{24z}{8z^2}$
$= z - 2 + \dfrac{3}{z}$

19. $\dfrac{6y^4 - 3y^3 + 5y^2 + 6y - 9}{2y + 1}$

$$\begin{array}{r} 3y^3 - 3y^2 + 4y + 1 \\ 2y + 1 \overline{\smash{)}\, 6y^4 - 3y^3 + 5y^2 + 6y - 9} \\ \underline{6y^4 + 3y^3} \qquad\qquad\qquad \\ -6y^3 + 5y^2 \qquad\qquad \\ \underline{-6y^3 - 3y^2} \qquad\qquad \\ 8y^2 + 6y \qquad \\ \underline{8y^2 + 4y} \qquad \\ 2y - 9 \\ \underline{2y + 1} \\ -10 \end{array}$$

The answer is

$$3y^3 - 3y^2 + 4y + 1 + \frac{-10}{2y + 1}.$$

20. $2p^2 - 5pq + 3q^2 = (2p - 3q)(p - q)$

21. $18k^4 + 9k^2 - 20 = (3k^2 + 4)(6k^2 - 5)$

22. $x^3 + 512 = x^3 + 8^3$
$= (x + 8)(x^2 - 8x + 64)$

23. $\dfrac{y^2+y-12}{y^3+9y^2+20y} \div \dfrac{y^2-9}{y^3+3y^2}$

$= \dfrac{y^2+y-12}{y(y^2+9y+20)} \cdot \dfrac{y^3+3y^2}{y^2-9}$

$= \dfrac{(y+4)(y-3)}{y(y+4)(y+5)} \cdot \dfrac{y^2(y+3)}{(y+3)(y-3)}$

$= \dfrac{y}{y+5}$

24. $\dfrac{1}{x+y} + \dfrac{3}{x-y}$ *The LCD is* $(x+y)(x-y)$.

$= \dfrac{1(x-y)}{(x+y)(x-y)} + \dfrac{3(x+y)}{(x-y)(x+y)}$

$= \dfrac{(x-y)+3(x+y)}{(x+y)(x-y)}$

$= \dfrac{x-y+3x+3y}{(x+y)(x-y)}$

$= \dfrac{4x+2y}{(x+y)(x-y)}$

25. $\dfrac{\dfrac{-6}{x-2}}{\dfrac{8}{3x-6}} = \dfrac{-6}{x-2} \div \dfrac{8}{3x-6}$

$= \dfrac{-6}{x-2} \cdot \dfrac{3x-6}{8}$

$= \dfrac{-6}{x-2} \cdot \dfrac{3(x-2)}{8}$

$= \dfrac{-2 \cdot 3 \cdot 3}{2 \cdot 4} = -\dfrac{9}{4}$

26. $\dfrac{\dfrac{1}{a}-\dfrac{1}{b}}{\dfrac{a}{b}-\dfrac{b}{a}}$ *The LCD in both numerator and denominator is ab.*

$= \dfrac{\dfrac{b-a}{ab}}{\dfrac{a^2-b^2}{ab}}$

$= \dfrac{b-a}{ab} \div \dfrac{a^2-b^2}{ab}$

$= \dfrac{b-a}{ab} \cdot \dfrac{ab}{a^2-b^2}$

$= \dfrac{b-a}{a^2-b^2}$

$= \dfrac{-(a-b)}{(a-b)(a+b)} = \dfrac{-1}{a+b}$ or $-\dfrac{1}{a+b}$

27. $2x^2+11x+15=0$

$(x+3)(2x+5)=0$

$x+3=0$ or $2x+5=0$

$x=-3$ or $x=-\frac{5}{2}$

Solution set: $\left\{-3, -\frac{5}{2}\right\}$

28. $5t(t-1)=2(1-t)$

$5t^2-5t=2-2t$

$5t^2-3t-2=0$

$(5t+2)(t-1)=0$

$5t+2=0$ or $t-1=0$

$t=-\frac{2}{5}$ or $t=1$

Solution set: $\left\{-\frac{2}{5}, 1\right\}$

29. $27^{-5/3} = \dfrac{1}{27^{5/3}} = \dfrac{1}{(27^{1/3})^5} = \dfrac{1}{3^5} = \dfrac{1}{243}$

30. $\dfrac{x^{-2/3}}{x^{-3/4}} = x^{-2/3-(-3/4)} = x^{-8/12+9/12} = x^{1/12}$

31. $8\sqrt{20}+3\sqrt{80}-2\sqrt{500}$

$= 8\sqrt{4\cdot 5}+3\sqrt{16\cdot 5}-2\sqrt{100\cdot 5}$

$= 8\cdot 2\sqrt{5}+3\cdot 4\sqrt{5}-2\cdot 10\sqrt{5}$

$= 16\sqrt{5}+12\sqrt{5}-20\sqrt{5} = 8\sqrt{5}$

32. $\dfrac{-9}{\sqrt{80}} = \dfrac{-9}{\sqrt{16\cdot 5}} = \dfrac{-9}{4\sqrt{5}}$

$= \dfrac{-9}{4\sqrt{5}} \cdot \dfrac{\sqrt{5}}{\sqrt{5}} = \dfrac{-9\sqrt{5}}{20}$

33. $\dfrac{4}{\sqrt{6}-\sqrt{5}} = \dfrac{4}{\sqrt{6}-\sqrt{5}} \cdot \dfrac{\sqrt{6}+\sqrt{5}}{\sqrt{6}+\sqrt{5}}$

$= \dfrac{4(\sqrt{6}+\sqrt{5})}{6-5} = 4(\sqrt{6}+\sqrt{5})$

34. $\dfrac{12}{\sqrt[3]{2}} = \dfrac{12}{\sqrt[3]{2}} \cdot \dfrac{\sqrt[3]{4}}{\sqrt[3]{4}} = \dfrac{12\sqrt[3]{4}}{\sqrt[3]{8}} = \dfrac{12\sqrt[3]{4}}{2} = 6\sqrt[3]{4}$

35. The distance d between $(x_1, y_1) = (-4, 4)$ and $(x_2, y_2) = (-2, 9)$ is

$d = \sqrt{(x_2-x_1)^2+(y_2-y_1)^2}$

$= \sqrt{[-2-(-4)]^2+(9-4)^2}$

$= \sqrt{2^2+5^2}$

$= \sqrt{4+25}$

$= \sqrt{29}$.

36. $\sqrt{8x-4}-\sqrt{7x+2}=0$

$\sqrt{8x-4}=\sqrt{7x+2}$

$8x-4=7x+2$ *Square*

$x=6$

Check $x=6$: $\sqrt{44}-\sqrt{44}=0$ *True*

Solution set: $\{6\}$

37. Let $x =$ the speed of the boat in still water. Make a table. Use $d = rt$, or $t = \frac{d}{r}$.

Direction	Distance	Rate	Time
Downstream	36	$x+3$	$\frac{36}{x+3}$
Upstream	24	$x-3$	$\frac{24}{x-3}$

Both times are the same.

$$\frac{36}{x+3} = \frac{24}{x-3}$$

Multiply each side by the LCD, $(x-3)(x+3)$.

$$\begin{aligned} 36(x-3) &= 24(x+3) \\ 36x - 108 &= 24x + 72 \\ 12x &= 180 \\ x &= 15 \end{aligned}$$

The speed of the boat in still water is 15 mph.

38. Let $x =$ the amount of pure alcohol.

Solution	Strength	Amount	Pure Alcohol
#1	100%	x	$1 \cdot x = x$
#2	18%	40	$.18(40) = 7.2$
Mixture	22%	$x+40$	$.22(x+40)$

The last column gives the equation.

$$\begin{aligned} x + 7.2 &= .22(x+40) \\ x + 7.2 &= .22x + 8.8 \\ .78x &= 1.6 \\ x &= \frac{1.6}{.78} = \frac{160}{78} = \frac{80}{39} \text{ or } 2\frac{2}{39} \end{aligned}$$

The required amount is $\frac{80}{39}$ or $2\frac{2}{39}$ L.

39. Let $x =$ the number of dimes. Then $29 - x =$ the number of quarters.

The total value of the coins is \$4.70, so

$$.10x + .25(29 - x) = 4.70.$$

Multiply by 100 to clear the decimals.

$$\begin{aligned} 10x + 25(29 - x) &= 470 \\ 10x + 725 - 25x &= 470 \\ -15x &= -255 \\ x &= 17 \end{aligned}$$

Since $x = 17$, $29 - x = 29 - 17 = 12$.

There are 17 dimes and 12 quarters.

40. Let $x =$ Chuck's speed. Then $x + 4 =$ Brenda's speed.

Make a table. Use $d = rt$, or $t = \frac{d}{r}$.

	Distance	Rate	Time
Chuck	24	x	$\frac{24}{x}$
Brenda	48	$x+4$	$\frac{48}{x+4}$

The times are the same.

$$\frac{24}{x} = \frac{48}{x+4}$$

Multiply each side by the LCD, $x(x+4)$.

$$\begin{aligned} x(x+4)\left(\frac{24}{x}\right) &= x(x+4)\left(\frac{48}{x+4}\right) \\ 24(x+4) &= 48x \\ 24x + 96 &= 48x \\ -24x &= -96 \\ x &= 4 \end{aligned}$$

Chuck's speed is 4 mph and Brenda's speed is $x + 4 = 4 + 4 = 8$ mph.

CHAPTER 10 QUADRATIC EQUATIONS, INEQUALITIES, AND FUNCTIONS

10.1 The Square Root Property and Completing the Square

10.1 Margin Exercises

1. **(a)** **B** and **C** are quadratic equations because they are equations with a squared term and no terms of higher degree.

   (b) Equation **B**, $x^2 - 8x + 16 = 0$, is in standard form, $ax^2 + bx + c = 0$.

2. **(a)**
$$x^2 + 3x + 2 = 0$$
$$(x+1)(x+2) = 0$$
$$x + 1 = 0 \quad \text{or} \quad x + 2 = 0$$
$$x = -1 \quad \text{or} \quad x = -2$$

   Solution set: $\{-2, -1\}$

   (b)
$$3m^2 = 3 - 8m$$
$$3m^2 + 8m - 3 = 0$$
$$(m+3)(3m-1) = 0$$
$$m + 3 = 0 \quad \text{or} \quad 3m - 1 = 0$$
$$m = -3 \quad \text{or} \quad m = \tfrac{1}{3}$$

   Solution set: $\{-3, \frac{1}{3}\}$

3. **(a)** $m^2 = 64 \qquad (64 = 8^2)$

   By the square root property,
$$m = 8 \quad \text{or} \quad m = -8.$$

   Solution set: $\{-8, 8\}$

   (b) $p^2 = 7$

   By the square root property,
$$p = \sqrt{7} \quad \text{or} \quad p = -\sqrt{7}.$$

   Solution set: $\left\{-\sqrt{7}, \sqrt{7}\right\}$

   (c) $3x^2 - 54 = 0$
$$3x^2 = 54$$
$$x^2 = 18$$

   By the square root property,
$$x = \sqrt{18} \quad \text{or} \quad x = -\sqrt{18}$$
$$x = 3\sqrt{2} \quad \text{or} \quad x = -3\sqrt{2}$$

   Solution set: $\left\{3\sqrt{2}, -3\sqrt{2}\right\}$

4.
$$d = 16t^2$$
$$4 = 16t^2 \qquad \textit{Let } d = 4.$$
$$\tfrac{1}{4} = t^2$$

   By the square root property,
$$t = \tfrac{1}{2} \quad \text{or} \quad t = -\tfrac{1}{2}.$$

   Since time cannot be negative, we discard the negative solution. Therefore, .5 sec elapses between the dropping of the coin and the shot.

5. **(a)** $(x-3)^2 = 25$

   By the square root property,
$$x - 3 = 5 \quad \text{or} \quad x - 3 = -5$$
$$x = 8 \quad \text{or} \quad x = -2.$$

   Solution set: $\{-2, 8\}$

   (b) $(3k+1)^2 = 2$
$$3k + 1 = \sqrt{2} \quad \text{or} \quad 3k + 1 = -\sqrt{2}$$
$$3k = -1 + \sqrt{2} \quad \text{or} \quad 3k = -1 - \sqrt{2}$$
$$k = \frac{-1 + \sqrt{2}}{3} \quad \text{or} \quad k = \frac{-1 - \sqrt{2}}{3}$$

   Solution set: $\left\{\dfrac{-1+\sqrt{2}}{3}, \dfrac{-1-\sqrt{2}}{3}\right\}$

   (c) $(2r+3)^2 = 8$
$$2r + 3 = \sqrt{8} \quad \text{or} \quad 2r + 3 = -\sqrt{8}$$
$$2r + 3 = 2\sqrt{2} \quad \text{or} \quad 2r + 3 = -2\sqrt{2}$$
$$2r = -3 + 2\sqrt{2} \quad \text{or} \quad 2r = -3 - 2\sqrt{2}$$
$$r = \frac{-3 + 2\sqrt{2}}{2} \quad \text{or} \quad r = \frac{-3 - 2\sqrt{2}}{2}$$

   Solution set: $\left\{\dfrac{-3+2\sqrt{2}}{2}, \dfrac{-3-2\sqrt{2}}{2}\right\}$

6. **(a)** $x^2 + 4x + \underline{4}$ since $\left[\frac{1}{2}(4)\right]^2 = 2^2 = 4$

   (b) $t^2 - 2t + \underline{1}$ since $\left[\frac{1}{2}(-2)\right]^2 = (-1)^2 = 1$

   (c) $m^2 + 5m + \underline{\frac{25}{4}}$ since $\left[\frac{1}{2}(5)\right]^2 = \left(\frac{5}{2}\right)^2 = \frac{25}{4}$

   (d) $x^2 - \frac{2}{3}x + \underline{\frac{1}{9}}$ since $\left[\frac{1}{2}\left(-\frac{2}{3}\right)\right]^2 = \left(-\frac{1}{3}\right)^2 = \frac{1}{9}$

7. $n^2 + 6n + 4 = 0$
$$n^2 + 6n = -4$$

Complete the square. Take half of 6, the coefficient of n, and square the result.

$$\left[\tfrac{1}{2}(6)\right]^2 = 3^2 = 9$$

Add 9 to each side of the equation.

$$n^2 + 6n + 9 = -4 + 9$$
$$(n+3)^2 = 5$$

Use the square root property.

$$n + 3 = \sqrt{5} \quad \text{or} \quad n + 3 = -\sqrt{5}$$
$$n = -3 + \sqrt{5} \quad \text{or} \quad n = -3 - \sqrt{5}$$

Solution set: $\left\{-3+\sqrt{5}, -3-\sqrt{5}\right\}$

8. **(a)** $x^2 + 2x - 10 = 0$
$$x^2 + 2x = 10$$

Complete the square.

$$\left[\tfrac{1}{2}(2)\right]^2 = 1^2 = 1$$

Add 1 to each side.

$$x^2 + 2x + 1 = 10 + 1$$
$$(x+1)^2 = 11$$

Use the square root property.

$$x + 1 = \sqrt{11} \quad \text{or} \quad x + 1 = -\sqrt{11}$$
$$x = -1 + \sqrt{11} \quad \text{or} \quad x = -1 - \sqrt{11}$$

Solution set: $\left\{-1+\sqrt{11}, -1-\sqrt{11}\right\}$

(b) $r^2 + 3r - 1 = 0$
$$r^2 + 3r = 1$$

Complete the square.

$$\left[\tfrac{1}{2}(3)\right]^2 = \left(\tfrac{3}{2}\right)^2 = \tfrac{9}{4}$$

Add $\frac{9}{4}$ to each side.

$$r^2 + 3r + \tfrac{9}{4} = 1 + \tfrac{9}{4}$$
$$\left(r + \tfrac{3}{2}\right)^2 = \tfrac{13}{4}$$

Use the square root property.

$$r + \frac{3}{2} = \sqrt{\frac{13}{4}} \quad \text{or} \quad r + \frac{3}{2} = -\sqrt{\frac{13}{4}}$$
$$r + \frac{3}{2} = \frac{\sqrt{13}}{2} \quad \text{or} \quad r + \frac{3}{2} = -\frac{\sqrt{13}}{2}$$
$$r = \frac{-3+\sqrt{13}}{2} \quad \text{or} \quad r = \frac{-3-\sqrt{13}}{2}$$

Solution set: $\left\{\dfrac{-3+\sqrt{13}}{2}, \dfrac{-3-\sqrt{13}}{2}\right\}$

9. **(a)** $2r^2 - 4r + 1 = 0$
$$2r^2 - 4r = -1$$
$$r^2 - 2r = -\tfrac{1}{2}$$

Complete the square.

$$\left[\tfrac{1}{2}(-2)\right]^2 = (-1)^2 = 1$$

Add 1 to each side.

$$r^2 - 2r + 1 = -\tfrac{1}{2} + 1$$
$$(r-1)^2 = \tfrac{1}{2}$$

Use the square root property.

$$r - 1 = \sqrt{\frac{1}{2}} \quad \text{or} \quad r - 1 = -\sqrt{\frac{1}{2}}$$
$$r - 1 = \frac{\sqrt{2}}{2} \quad \text{or} \quad r - 1 = -\frac{\sqrt{2}}{2}$$
$$r = 1 + \frac{\sqrt{2}}{2} \quad \text{or} \quad r = 1 - \frac{\sqrt{2}}{2}$$
$$r = \frac{2+\sqrt{2}}{2} \quad \text{or} \quad r = \frac{2-\sqrt{2}}{2}$$

Solution set: $\left\{\dfrac{2+\sqrt{2}}{2}, \dfrac{2-\sqrt{2}}{2}\right\}$

(b) $3z^2 - 6z - 2 = 0$
$$3z^2 - 6z = 2$$
$$z^2 - 2z = \tfrac{2}{3}$$

Complete the square.

$$\left[\tfrac{1}{2}(-2)\right]^2 = (-1)^2 = 1$$

Add 1 to each side.

$$z^2 - 2z + 1 = \tfrac{2}{3} + 1$$
$$(z-1)^2 = \tfrac{5}{3}$$

Use the square root property.

$$z - 1 = \sqrt{\frac{5}{3}} \quad \text{or} \quad z - 1 = -\sqrt{\frac{5}{3}}$$
$$z - 1 = \frac{\sqrt{15}}{3} \quad \text{or} \quad z - 1 = -\frac{\sqrt{15}}{3}$$
$$z = 1 + \frac{\sqrt{15}}{3} \quad \text{or} \quad z = 1 - \frac{\sqrt{15}}{3}$$
$$z = \frac{3+\sqrt{15}}{3} \quad \text{or} \quad z = \frac{3-\sqrt{15}}{3}$$

Solution set: $\left\{\dfrac{3+\sqrt{15}}{3}, \dfrac{3-\sqrt{15}}{3}\right\}$

(c) $8x^2 - 4x - 2 = 0$

$$8x^2 - 4x = 2$$
$$x^2 - \tfrac{1}{2}x = \tfrac{1}{4}$$

Complete the square.

$$\left[\tfrac{1}{2}\left(-\tfrac{1}{2}\right)\right]^2 = \left(-\tfrac{1}{4}\right)^2 = \tfrac{1}{16}$$

Add $\frac{1}{16}$ to each side.

$$x^2 - \tfrac{1}{2}x + \tfrac{1}{16} = \tfrac{1}{4} + \tfrac{1}{16}$$
$$\left(x - \tfrac{1}{4}\right)^2 = \tfrac{5}{16}$$

$$x - \frac{1}{4} = \sqrt{\frac{5}{16}} \quad \text{or} \quad x - \frac{1}{4} = -\sqrt{\frac{5}{16}}$$
$$x - \frac{1}{4} = \frac{\sqrt{5}}{4} \quad \text{or} \quad x - \frac{1}{4} = -\frac{\sqrt{5}}{4}$$
$$x = \frac{1 + \sqrt{5}}{4} \quad \text{or} \quad x = \frac{1 - \sqrt{5}}{4}$$

Solution set: $\left\{\frac{1 + \sqrt{5}}{4}, \frac{1 - \sqrt{5}}{4}\right\}$

10. **(a)** $x^2 = -17$

$$x = \sqrt{-17} \quad \text{or} \quad x = -\sqrt{-17}$$
$$x = i\sqrt{17} \quad \text{or} \quad x = -i\sqrt{17}$$

Solution set: $\left\{i\sqrt{17}, -i\sqrt{17}\right\}$

(b) $(k + 5)^2 = -100$

$$k + 5 = \sqrt{-100} \quad \text{or} \quad k + 5 = -\sqrt{-100}$$
$$k + 5 = 10i \quad \text{or} \quad k + 5 = -10i$$
$$k = -5 + 10i \quad \text{or} \quad k = -5 - 10i$$

Solution set: $\{-5 + 10i, -5 - 10i\}$

(c) $5t^2 - 15t + 12 = 0$

$$5t^2 - 15t = -12$$
$$t^2 - 3t = -\tfrac{12}{5}$$

Complete the square.

$$\left[\tfrac{1}{2}(-3)\right]^2 = \left(-\tfrac{3}{2}\right)^2 = \tfrac{9}{4}$$

Add $\frac{9}{4}$ to each side.

$$t^2 - 3t + \tfrac{9}{4} = -\tfrac{12}{5} + \tfrac{9}{4}$$
$$\left(t - \tfrac{3}{2}\right)^2 = -\tfrac{3}{20}$$

$$t - \frac{3}{2} = \sqrt{-\frac{3}{20}} \quad \text{or} \quad t - \frac{3}{2} = -\sqrt{-\frac{3}{20}}$$
$$t - \frac{3}{2} = \frac{i\sqrt{3}}{\sqrt{20}} \cdot \frac{\sqrt{5}}{\sqrt{5}} \quad \text{or} \quad t - \frac{3}{2} = \frac{-i\sqrt{3}}{\sqrt{20}} \cdot \frac{\sqrt{5}}{\sqrt{5}}$$
$$t - \frac{3}{2} = \frac{i\sqrt{15}}{10} \quad \text{or} \quad t - \frac{3}{2} = \frac{-i\sqrt{15}}{10}$$
$$t = \frac{3}{2} + \frac{i\sqrt{15}}{10} \quad \text{or} \quad t = \frac{3}{2} - \frac{i\sqrt{15}}{10}$$
$$t = \frac{15 + i\sqrt{15}}{10} \quad \text{or} \quad t = \frac{15 - i\sqrt{15}}{10}$$

Solution set: $\left\{\frac{15 + i\sqrt{15}}{10}, \frac{15 - i\sqrt{15}}{10}\right\}$

10.1 Section Exercises

1. By the square root property, if $x^2 = 16$, then

$$x = +\sqrt{16} \quad \text{or} \quad x = -\sqrt{16}.$$

Thus, the equation is also true for $x = -4$.

2. Some quadratic equations cannot be easily factored.

3. **(a)** A quadratic equation in standard form has a second-degree polynomial in decreasing powers equal to 0.

(b) The zero-factor property states that if a product equals 0, then at least one of the factors equals 0.

(c) The square root property states that if the square of a quantity equals a number, then the quantity equals the positive or negative square root of the number.

4. The zero-factor property requires a product equal to 0. In this solution, the first step should have been to rewrite the equation with 0 on one side.

5. $x^2 = 81$

$$x = 9 \quad \text{or} \quad x = -9$$

Solution set: $\{9, -9\}$

6. $z^2 = 225$

$$z = \sqrt{225} \quad \text{or} \quad z = -\sqrt{225}$$
$$z = 15 \quad \text{or} \quad z = -15$$

Solution set: $\{15, -15\}$

7. $t^2 = 17$

$$t = \sqrt{17} \quad \text{or} \quad t = -\sqrt{17}$$

Solution set: $\left\{\sqrt{17}, -\sqrt{17}\right\}$

8. $k^2 = 19$

$$k = \sqrt{19} \quad \text{or} \quad k = -\sqrt{19}$$

Solution set: $\left\{\sqrt{19}, -\sqrt{19}\right\}$

9. $m^2 = 32$

$m = \sqrt{32}$ or $m = -\sqrt{32}$

$m = 4\sqrt{2}$ or $m = -4\sqrt{2}$

Solution set: $\{4\sqrt{2}, -4\sqrt{2}\}$

10. $x^2 = 54$

$x = \sqrt{54}$ or $x = -\sqrt{54}$

$x = 3\sqrt{6}$ or $x = -3\sqrt{6}$

Solution set: $\{3\sqrt{6}, -3\sqrt{6}\}$

11. $t^2 - 20 = 0$

$t^2 = 20$

$t = \sqrt{20}$ or $t = -\sqrt{20}$

$t = 2\sqrt{5}$ or $t = -2\sqrt{5}$

Solution set: $\{2\sqrt{5}, -2\sqrt{5}\}$

12. $p^2 - 50 = 0$

$p^2 = 50$

$p = \sqrt{50}$ or $p = -\sqrt{50}$

$p = 5\sqrt{2}$ or $p = -5\sqrt{2}$

Solution set: $\{5\sqrt{2}, -5\sqrt{2}\}$

13. $3n^2 - 72 = 0$

$3n^2 = 72$

$n = 24$

$n = \sqrt{24}$ or $n = -\sqrt{24}$

$n = 2\sqrt{6}$ or $n = -2\sqrt{6}$

Solution set: $\{2\sqrt{6}, -2\sqrt{6}\}$

14. $5z^2 - 200 = 0$

$5z^2 = 200$

$z^2 = 40$

$z = \sqrt{40}$ or $z = -\sqrt{40}$

$z = 2\sqrt{10}$ or $z = -2\sqrt{10}$

Solution set: $\{2\sqrt{10}, -2\sqrt{10}\}$

15. $(x+2)^2 = 25$

$x + 2 = \sqrt{25}$ or $x + 2 = -\sqrt{25}$

$x + 2 = 5$ or $x + 2 = -5$

$x = 3$ or $x = -7$

Solution set: $\{-7, 3\}$

16. $(t+8)^2 = 9$

$t + 8 = \sqrt{9}$ or $t + 8 = -\sqrt{9}$

$t + 8 = 3$ or $t + 8 = -3$

$t = -5$ or $t = -11$

Solution set: $\{-5, -11\}$

17. $(x-4)^2 = 3$

$x - 4 = \sqrt{3}$ or $x - 4 = -\sqrt{3}$

$x = 4 + \sqrt{3}$ or $x = 4 - \sqrt{3}$

Solution set: $\{4 + \sqrt{3}, 4 - \sqrt{3}\}$

18. $(x+3)^2 = 11$

$x + 3 = \sqrt{11}$ or $x + 3 = -\sqrt{11}$

$x = -3 + \sqrt{11}$ or $x = -3 - \sqrt{11}$

Solution set: $\{-3 + \sqrt{11}, -3 - \sqrt{11}\}$

19. $(t+5)^2 = 48$

$t + 5 = \sqrt{48}$ or $t + 5 = -\sqrt{48}$

$t + 5 = 4\sqrt{3}$ or $t + 5 = -4\sqrt{3}$

$t = -5 + 4\sqrt{3}$ or $t = -5 - 4\sqrt{3}$

Solution set: $\{-5 + 4\sqrt{3}, -5 - 4\sqrt{3}\}$

20. $(m-6)^2 = 27$

$m - 6 = \sqrt{27}$ or $m - 6 = -\sqrt{27}$

$m - 6 = 3\sqrt{3}$ or $m - 6 = -3\sqrt{3}$

$m = 6 + 3\sqrt{3}$ or $m = 6 - 3\sqrt{3}$

Solution set: $\{6 + 3\sqrt{3}, 6 - 3\sqrt{3}\}$

21. $(3k-1)^2 = 7$

$3k - 1 = \sqrt{7}$ or $3k - 1 = -\sqrt{7}$

$3k = 1 + \sqrt{7}$ or $3k = 1 - \sqrt{7}$

$k = \dfrac{1+\sqrt{7}}{3}$ or $k = \dfrac{1-\sqrt{7}}{3}$

Solution set: $\left\{\dfrac{1+\sqrt{7}}{3}, \dfrac{1-\sqrt{7}}{3}\right\}$

22. $(2x+4)^2 = 10$

$2x + 4 = \sqrt{10}$ or $2x + 4 = -\sqrt{10}$

$2x = -4 + \sqrt{10}$ or $2x = -4 - \sqrt{10}$

$x = \dfrac{-4+\sqrt{10}}{2}$ or $x = \dfrac{-4-\sqrt{10}}{2}$

Solution set: $\left\{\dfrac{-4+\sqrt{10}}{2}, \dfrac{-4-\sqrt{10}}{2}\right\}$

23. $(4p+1)^2 = 24$

$4p + 1 = \sqrt{24}$ or $4p + 1 = -\sqrt{24}$

$4p + 1 = 2\sqrt{6}$ or $4p + 1 = -2\sqrt{6}$

$4p = -1 + 2\sqrt{6}$ or $4p = -1 - 2\sqrt{6}$

$p = \dfrac{-1+2\sqrt{6}}{4}$ or $p = \dfrac{-1-2\sqrt{6}}{4}$

Solution set: $\left\{\dfrac{-1+2\sqrt{6}}{4}, \dfrac{-1-2\sqrt{6}}{4}\right\}$

24. $(5k-2)^2 = 12$

$$5k-2=\sqrt{12} \quad \text{or} \quad 5k-2=-\sqrt{12}$$
$$5k=2+\sqrt{12} \quad \text{or} \quad 5k=2-\sqrt{12}$$
$$k=\frac{2+\sqrt{12}}{5} \quad \text{or} \quad k=\frac{2-\sqrt{12}}{5}$$
$$k=\frac{2+2\sqrt{3}}{5} \quad \text{or} \quad k=\frac{2-2\sqrt{3}}{5}$$

Solution set: $\left\{\frac{2+2\sqrt{3}}{5}, \frac{2-2\sqrt{3}}{5}\right\}$

25.
$$d=16t^2$$
$$500=16t^2 \qquad \textit{Let } d = 500.$$
$$t^2=\frac{500}{16}=31.25$$
$$t=\sqrt{31.25} \quad \text{or} \quad t=-\sqrt{31.25}$$
$$t\approx 5.6 \quad \text{or} \quad t\approx -5.6$$

Since time cannot be negative, -5.6 is discarded. Therefore, it takes the object about 5.6 seconds to fall 500 feet.

26.
$$d=16t^2$$
$$630=16t^2 \qquad \textit{Let } d = 630.$$
$$t^2=\frac{630}{16}=39.375$$
$$t=\sqrt{39.375} \quad \text{or} \quad t=-\sqrt{39.375}$$
$$t\approx 6.3 \quad \text{or} \quad t\approx -6.3$$

Since time cannot be negative, -6.3 is discarded. Therefore, it takes the object about 6.3 seconds to fall 630 feet.

27. To solve $(2x+1)^2=5$, we use the square root property and find that $2x+1=\sqrt{5}$ or $2x+1=-\sqrt{5}$. Then we solve for x in both equations.
To solve $x^2+4x=12$, we find that

$$x^2+4x+4=12+4$$
$$(x+2)^2=16$$

by completing the square.

28. The coefficient of the second-degree term is 1 in $x^2+4x=20$, so the first step of the solution process can be skipped. The other equation requires first dividing each side by 5 and then working with fractions.

29. **(a)** We need to add the square of half the coefficient of x to get a perfect square trinomial.

$$\tfrac{1}{2}(6)=3 \quad \text{and} \quad 3^2=9$$

Add 9 to x^2+6x to get a perfect square trinomial.

(b) $\frac{1}{2}(14)=7$ and $7^2=49$

(c) $\frac{1}{2}(-12)=-6$ and $(-6)^2=36$

(d) $\frac{1}{2}(3)=\frac{3}{2}$ and $\left(\frac{3}{2}\right)^2=\frac{9}{4}$

(e) $\frac{1}{2}(-9)=-\frac{9}{2}$ and $\left(-\frac{9}{2}\right)^2=\frac{81}{4}$

(f) $\frac{1}{2}\left(-\frac{1}{2}\right)=-\frac{1}{4}$ and $\left(-\frac{1}{4}\right)^2=\frac{1}{16}$

30. To solve $2x^2+8x=9$ by completing the square, the first step would be to divide both sides by 2 so the coefficient of x^2 would be 1.

31.
$$x^2+4x-2=0$$
$$x^2+4x=2$$
$$\left[\tfrac{1}{2}(4)\right]^2=2^2=4$$

32.
$$t^2+2t-1=0$$
$$t^2+2t=1$$
$$\left[\tfrac{1}{2}(2)\right]^2=1^2=1$$

33.
$$x^2+10x+18=0$$
$$x^2+10x=-18$$
$$\left[\tfrac{1}{2}(10)\right]^2=5^2=25$$

34.
$$x^2+8x+11=0$$
$$x^2+8x=-11$$
$$\left[\tfrac{1}{2}(8)\right]^2=4^2=16$$

35.
$$3w^2-w-24=0$$
$$w^2-\tfrac{1}{3}w-8=0 \qquad \textit{Divide by 3.}$$
$$w^2-\tfrac{1}{3}w=8$$
$$\left[\tfrac{1}{2}\left(-\tfrac{1}{3}\right)\right]^2=\left(-\tfrac{1}{6}\right)^2=\tfrac{1}{36}$$

36.
$$4z^2-z-39=0$$
$$z^2-\tfrac{1}{4}z-\tfrac{39}{4}=0 \qquad \textit{Divide by 4.}$$
$$z^2-\tfrac{1}{4}z=\tfrac{39}{4}$$
$$\left[\tfrac{1}{2}\left(-\tfrac{1}{4}\right)\right]^2=\left(-\tfrac{1}{8}\right)^2=\tfrac{1}{64}$$

37. $x^2-2x-24=0$
Get the variable terms alone on the left side.

$$x^2-2x=24$$

Complete the square by taking half of -2, the coefficient of x, and squaring the result.

$$\left[\tfrac{1}{2}(-2)\right]^2=(-1)^2=1$$

Add 1 to each side.

$$x^2-2x+1=24+1$$

Factor the left side.

$$(x-1)^2=25$$

continued

Use the square root property.

$x-1=\sqrt{25}$ or $x-1=-\sqrt{25}$
$x-1=5$ or $x-1=-5$
$x=6$ or $x=-4$

Solution set: $\{-4, 6\}$

38. $m^2-4m-32=0$
Isolate terms involving m on the left side.

$$m^2-4m=32$$

Complete the square.

$$\left[\tfrac{1}{2}(-4)\right]^2=(-2)^2=4$$

Add 4 to both sides.

$$m^2-4m+4=32+4$$
$$(m-2)^2=36$$

$m-2=\sqrt{36}$ or $m-2=-\sqrt{36}$
$m-2=6$ or $m-2=-6$
$m=8$ or $m=-4$

Solution set: $\{-4, 8\}$

39. $x^2+4x-2=0$
$x^2+4x=2$
$x^2+4x+4=2+4$ $\quad\left[\tfrac{1}{2}(4)\right]^2=4$
$(x+2)^2=6$

$x+2=\sqrt{6}$ or $x+2=-\sqrt{6}$
$x=-2+\sqrt{6}$ or $x=-2-\sqrt{6}$

Solution set: $\left\{-2+\sqrt{6}, -2-\sqrt{6}\right\}$

40. $t^2+2t-1=0$
$t^2+2t=1$
$t^2+2t+1=1+1$ $\quad\left[\tfrac{1}{2}(2)\right]^2=1$
$(t+1)^2=2$

$t+1=\sqrt{2}$ or $t+1=-\sqrt{2}$
$t=-1+\sqrt{2}$ or $t=-1-\sqrt{2}$

Solution set: $\left\{-1+\sqrt{2}, -1-\sqrt{2}\right\}$

41. $x^2+10x+18=0$
$x^2+10x=-18$
$x^2+10x+25=-18+25$ $\quad\left[\tfrac{1}{2}(10)\right]^2=25$
$(x+5)^2=7$

$x+5=\sqrt{7}$ or $x+5=-\sqrt{7}$
$x=-5+\sqrt{7}$ or $x=-5-\sqrt{7}$

Solution set: $\left\{-5+\sqrt{7}, -5-\sqrt{7}\right\}$

42. $x^2+8x+11=0$
$x^2+8x=-11$
$x^2+8x+16=-11+16$ $\quad\left[\tfrac{1}{2}(8)\right]^2=16$
$(x+4)^2=5$

$x+4=\sqrt{5}$ or $x+4=-\sqrt{5}$
$x=-4+\sqrt{5}$ or $x=-4-\sqrt{5}$

Solution set: $\left\{-4+\sqrt{5}, -4-\sqrt{5}\right\}$

43. $3w^2-w=24$
$w^2-\frac{1}{3}w=8$ *Divide by 3.*
By Exercise 35, add $\frac{1}{36}$ to each side.
$w^2-\frac{1}{3}w+\frac{1}{36}=8+\frac{1}{36}$
$\left(w-\frac{1}{6}\right)^2=\frac{288}{36}+\frac{1}{36}$
$\left(w-\frac{1}{6}\right)^2=\frac{289}{36}$

$w-\frac{1}{6}=\sqrt{\frac{289}{36}}$ or $w-\frac{1}{6}=-\sqrt{\frac{289}{36}}$
$w=\frac{1}{6}+\frac{\sqrt{289}}{\sqrt{36}}$ or $w=\frac{1}{6}-\frac{\sqrt{289}}{\sqrt{36}}$
$w=\frac{1}{6}+\frac{17}{6}$ or $w=\frac{1}{6}-\frac{17}{6}$
$w=\frac{18}{6}$ or $w=-\frac{16}{6}$
$w=3$ or $w=-\frac{8}{3}$

Solution set: $\left\{-\frac{8}{3}, 3\right\}$

44. $4z^2-z=39$
$z^2-\frac{1}{4}z=\frac{39}{4}$ *Divide by 4.*
By Exercise 36, add $\frac{1}{64}$ to each side.
$z^2-\frac{1}{4}z+\frac{1}{64}=\frac{39}{4}+\frac{1}{64}$
$\left(z-\frac{1}{8}\right)^2=\frac{624}{64}+\frac{1}{64}$
$\left(z-\frac{1}{8}\right)^2=\frac{625}{64}$

$z-\frac{1}{8}=\sqrt{\frac{625}{64}}$ or $z-\frac{1}{8}=-\sqrt{\frac{625}{64}}$
$z-\frac{1}{8}=\frac{25}{8}$ or $z-\frac{1}{8}=-\frac{25}{8}$
$z=\frac{26}{8}=\frac{13}{4}$ or $z=\frac{-24}{8}=-3$

Solution set: $\left\{-3, \frac{13}{4}\right\}$

45. $2k^2+5k-2=0$

$2k^2+5k=2$

$k^2+\frac{5}{2}k=1$ *Divide by 2.*

Complete the square.

$$\left[\frac{1}{2}\left(\frac{5}{2}\right)\right]^2=\left(\frac{5}{4}\right)^2=\frac{25}{16}$$

Add $\frac{25}{16}$ to each side.

$k^2+\frac{5}{2}k+\frac{25}{16}=1+\frac{25}{16}$

$\left(k+\frac{5}{4}\right)^2=\frac{41}{16}$

$k+\frac{5}{4}=\sqrt{\frac{41}{16}}$ or $k+\frac{5}{4}=-\sqrt{\frac{41}{16}}$

$k=-\frac{5}{4}+\frac{\sqrt{41}}{4}$ or $k=-\frac{5}{4}-\frac{\sqrt{41}}{4}$

$k=\frac{-5+\sqrt{41}}{4}$ or $k=\frac{-5-\sqrt{41}}{4}$

Solution set: $\left\{\frac{-5+\sqrt{41}}{4},\frac{-5-\sqrt{41}}{4}\right\}$

46. $3r^2+2r-2=0$

$3r^2+2r=2$

$r^2+\frac{2}{3}r=\frac{2}{3}$ *Divide by 3.*

Complete the square.

$$\left[\frac{1}{2}\left(\frac{2}{3}\right)\right]^2=\left(\frac{1}{3}\right)^2=\frac{1}{9}$$

Add $\frac{1}{9}$ to each side.

$r^2+\frac{2}{3}r+\frac{1}{9}=\frac{2}{3}+\frac{1}{9}$

$\left(r+\frac{1}{3}\right)^2=\frac{6+1}{9}=\frac{7}{9}$

$r+\frac{1}{3}=\sqrt{\frac{7}{9}}$ or $r+\frac{1}{3}=-\sqrt{\frac{7}{9}}$

$r+\frac{1}{3}=\frac{\sqrt{7}}{3}$ or $r+\frac{1}{3}=-\frac{\sqrt{7}}{3}$

$r=-\frac{1}{3}+\frac{\sqrt{7}}{3}$ or $r=-\frac{1}{3}-\frac{\sqrt{7}}{3}$

$r=\frac{-1+\sqrt{7}}{3}$ or $r=\frac{-1-\sqrt{7}}{3}$

Solution set: $\left\{\frac{-1+\sqrt{7}}{3},\frac{-1-\sqrt{7}}{3}\right\}$

47. $5x^2-10x+2=0$

$5x^2-10x=-2$

$x^2-2x=-\frac{2}{5}$ *Divide by 5.*

Complete the square.

$$\left[\frac{1}{2}(-2)\right]^2=(-1)^2=1$$

Add 1 to each side.

$x^2-2x+1=-\frac{2}{5}+1$

$(x-1)^2=\frac{3}{5}$

$x-1=\sqrt{\frac{3}{5}}$ or $x-1=-\sqrt{\frac{3}{5}}$

$x-1=\frac{\sqrt{3}}{\sqrt{5}}\cdot\frac{\sqrt{5}}{\sqrt{5}}$ or $x-1=-\frac{\sqrt{3}}{\sqrt{5}}\cdot\frac{\sqrt{5}}{\sqrt{5}}$

$x=\frac{5}{5}+\frac{\sqrt{15}}{5}$ or $x=\frac{5}{5}-\frac{\sqrt{15}}{5}$

$x=\frac{5+\sqrt{15}}{5}$ or $x=\frac{5-\sqrt{15}}{5}$

Solution set: $\left\{\frac{5+\sqrt{15}}{5},\frac{5-\sqrt{15}}{5}\right\}$

48. $2x^2-16x+25=0$

$2x^2-16x=-25$

$x^2-8x=\frac{-25}{2}$ *Divide by 2.*

Complete the square.

$$\left[\frac{1}{2}(-8)\right]^2=(-4)^2=16$$

Add 16 to each side.

$x^2-8x+16=\frac{-25}{2}+16$

$(x-4)^2=\frac{-25+32}{2}=\frac{7}{2}$

$x-4=\sqrt{\frac{7}{2}}$ or $x-4=-\sqrt{\frac{7}{2}}$

$x-4=\frac{\sqrt{7}}{\sqrt{2}}\cdot\frac{\sqrt{2}}{\sqrt{2}}$ or $x-4=-\frac{\sqrt{7}}{\sqrt{2}}\cdot\frac{\sqrt{2}}{\sqrt{2}}$

$x=\frac{8}{2}+\frac{\sqrt{14}}{2}$ or $x=\frac{8}{2}-\frac{\sqrt{14}}{2}$

$x=\frac{8+\sqrt{14}}{2}$ or $x=\frac{8-\sqrt{14}}{2}$

Solution set: $\left\{\frac{8+\sqrt{14}}{2},\frac{8-\sqrt{14}}{2}\right\}$

49. $9x^2-24x=-13$

$x^2-\frac{24}{9}x=\frac{-13}{9}$ *Divide by 9.*

$x^2-\frac{8}{3}x=\frac{-13}{9}$

Complete the square.

$$\left[\frac{1}{2}\left(-\frac{8}{3}\right)\right]^2=\left(-\frac{4}{3}\right)^2=\frac{16}{9}$$

Add $\frac{16}{9}$ to each side.

$x^2-\frac{8}{3}x+\frac{16}{9}=\frac{-13}{9}+\frac{16}{9}$

$\left(x-\frac{4}{3}\right)^2=\frac{3}{9}$

$x-\frac{4}{3}=\sqrt{\frac{3}{9}}$ or $x-\frac{4}{3}=-\sqrt{\frac{3}{9}}$

$x=\frac{4}{3}+\frac{\sqrt{3}}{3}$ or $x=\frac{4}{3}-\frac{\sqrt{3}}{3}$

$x=\frac{4+\sqrt{3}}{3}$ or $x=\frac{4-\sqrt{3}}{3}$

Solution set: $\left\{\frac{4+\sqrt{3}}{3},\frac{4-\sqrt{3}}{3}\right\}$

50. $25n^2 - 20n = 1$

$n^2 - \frac{4}{5}n = \frac{1}{25}$ *Divide by 25.*

Complete the square.

$$\left[\tfrac{1}{2}\left(-\tfrac{4}{5}\right)\right]^2 = \left(-\tfrac{2}{5}\right)^2 = \tfrac{4}{25}$$

Add $\frac{4}{25}$ to each side.

$n^2 - \frac{4}{5}n + \frac{4}{25} = \frac{1}{25} + \frac{4}{25}$

$\left(n - \frac{2}{5}\right)^2 = \frac{5}{25}$

$n - \frac{2}{5} = \sqrt{\frac{5}{25}}$ or $n - \frac{2}{5} = -\sqrt{\frac{5}{25}}$

$n - \frac{2}{5} = \frac{\sqrt{5}}{5}$ or $n - \frac{2}{5} = -\frac{\sqrt{5}}{5}$

$n = \frac{2+\sqrt{5}}{5}$ or $n = \frac{2-\sqrt{5}}{5}$

Solution set: $\left\{\frac{2+\sqrt{5}}{5}, \frac{2-\sqrt{5}}{5}\right\}$

51. $z^2 - \frac{4}{3}z = -\frac{1}{9}$

Complete the square.

$$\left[\tfrac{1}{2}\left(-\tfrac{4}{3}\right)\right]^2 = \left(-\tfrac{2}{3}\right)^2 = \tfrac{4}{9}$$

Add $\frac{4}{9}$ to each side.

$z^2 - \frac{4}{3}z + \frac{4}{9} = -\frac{1}{9} + \frac{4}{9}$

$\left(z - \frac{2}{3}\right)^2 = \frac{3}{9}$

$z - \frac{2}{3} = \sqrt{\frac{3}{9}}$ or $z - \frac{2}{3} = -\sqrt{\frac{3}{9}}$

$z = \frac{2}{3} + \frac{\sqrt{3}}{3}$ or $z = \frac{2}{3} - \frac{\sqrt{3}}{3}$

$z = \frac{2+\sqrt{3}}{3}$ or $z = \frac{2-\sqrt{3}}{3}$

Solution set: $\left\{\frac{2+\sqrt{3}}{3}, \frac{2-\sqrt{3}}{3}\right\}$

52. $p^2 - \frac{8}{3}p = -1$

Complete the square.

$$\left[\tfrac{1}{2}\left(-\tfrac{8}{3}\right)\right]^2 = \left(-\tfrac{4}{3}\right)^2 = \tfrac{16}{9}$$

Add $\frac{16}{9}$ to each side.

$p^2 - \frac{8}{3}p + \frac{16}{9} = -1 + \frac{16}{9}$

$\left(p - \frac{4}{3}\right)^2 = \frac{7}{9}$

$p - \frac{4}{3} = \sqrt{\frac{7}{9}}$ or $p - \frac{4}{3} = -\sqrt{\frac{7}{9}}$

$p = \frac{4}{3} + \frac{\sqrt{7}}{3}$ or $p = \frac{4}{3} - \frac{\sqrt{7}}{3}$

$p = \frac{4+\sqrt{7}}{3}$ or $p = \frac{4-\sqrt{7}}{3}$

Solution set: $\left\{\frac{4+\sqrt{7}}{3}, \frac{4-\sqrt{7}}{3}\right\}$

53. $.1x^2 - .2x - .1 = 0$

Multiply each side by 10 to clear the decimals.

$x^2 - 2x - 1 = 0$

$x^2 - 2x = 1$

Complete the square.

$$\left[\tfrac{1}{2}(-2)\right]^2 = (-1)^2 = 1$$

Add 1 to each side.

$x^2 - 2x + 1 = 1 + 1$

$(x-1)^2 = 2$

$x - 1 = \sqrt{2}$ or $x - 1 = -\sqrt{2}$

$x = 1 + \sqrt{2}$ or $x = 1 - \sqrt{2}$

Solution set: $\left\{1+\sqrt{2}, 1-\sqrt{2}\right\}$

54. $.1p^2 - .4p + .1 = 0$

Multiply each side by 10 to clear the decimals.

$p^2 - 4p + 1 = 0$

$p^2 - 4p = -1$

Complete the square.

$$\left[\tfrac{1}{2}(-4)\right]^2 = (-2)^2 = 4$$

Add 4 to each side.

$p^2 - 4p + 4 = -1 + 4$

$(p-2)^2 = 3$

$p - 2 = \sqrt{3}$ or $p - 2 = -\sqrt{3}$

$p = 2 + \sqrt{3}$ or $p = 2 - \sqrt{3}$

Solution set: $\left\{2+\sqrt{3}, 2-\sqrt{3}\right\}$

55. $x^2 = -12$

$x = \sqrt{-12}$ or $x = -\sqrt{-12}$

$x = i\sqrt{12}$ or $x = -i\sqrt{12}$

$x = 2i\sqrt{3}$ or $x = -2i\sqrt{3}$

Solution set: $\left\{2i\sqrt{3}, -2i\sqrt{3}\right\}$

56. $x^2 = -18$

$x = \sqrt{-18}$ or $x = -\sqrt{-18}$

$x = i\sqrt{18}$ or $x = -i\sqrt{18}$

$x = 3i\sqrt{2}$ or $x = -3i\sqrt{2}$

Solution set: $\left\{3i\sqrt{2}, -3i\sqrt{2}\right\}$

57. $(r-5)^2 = -3$

$r - 5 = \sqrt{-3}$ or $r - 5 = -\sqrt{-3}$

$r = 5 + i\sqrt{3}$ or $r = 5 - i\sqrt{3}$

Solution set: $\left\{5+i\sqrt{3}, 5-i\sqrt{3}\right\}$

58. $(t+6)^2 = -5$

$t+6 = \sqrt{-5}$ or $t+6 = -\sqrt{-5}$

$t = -6 + i\sqrt{5}$ or $t = -6 - i\sqrt{5}$

Solution set: $\left\{-6+i\sqrt{5}, -6-i\sqrt{5}\right\}$

59. $(6k-1)^2 = -8$

$6k-1 = \sqrt{-8}$ or $6k-1 = -\sqrt{-8}$

$6k-1 = i\sqrt{8}$ or $6k-1 = -i\sqrt{8}$

$6k-1 = 2i\sqrt{2}$ or $6k-1 = -2i\sqrt{2}$

$6k = 1+2i\sqrt{2}$ or $6k = 1-2i\sqrt{2}$

$k = \dfrac{1+2i\sqrt{2}}{6}$ or $k = \dfrac{1-2i\sqrt{2}}{6}$

Solution set: $\left\{\dfrac{1+2i\sqrt{2}}{6}, \dfrac{1-2i\sqrt{2}}{6}\right\}$

60. $(4m-7)^2 = -27$

$4m-7 = \sqrt{-27}$ or $4m-7 = -\sqrt{-27}$

$4m-7 = 3i\sqrt{3}$ or $4m-7 = -3i\sqrt{3}$

$4m = 7+3i\sqrt{3}$ or $4m = 7-3i\sqrt{3}$

$m = \dfrac{7+3i\sqrt{3}}{4}$ or $m = \dfrac{7-3i\sqrt{3}}{4}$

Solution set: $\left\{\dfrac{7+3i\sqrt{3}}{4}, \dfrac{7-3i\sqrt{3}}{4}\right\}$

61. $m^2 + 4m + 13 = 0$

$m^2 + 4m = -13$

Complete the square.

$\left[\frac{1}{2}(4)\right]^2 = 2^2 = 4$

Add 4 to each side.

$m^2 + 4m + 4 = -13 + 4$

$(m+2)^2 = -9$

$m+2 = \sqrt{-9}$ or $m+2 = -\sqrt{-9}$

$m = -2+3i$ or $m = -2-3i$

Solution set: $\{-2+3i, -2-3i\}$

62. $t^2 + 6t + 10 = 0$

$t^2 + 6t = -10$

$\left[\frac{1}{2}(6)\right]^2 = 3^2 = 9$

$t^2 + 6t + 9 = -10 + 9$

$(t+3)^2 = -1$

$t+3 = \sqrt{-1}$ or $t+3 = -\sqrt{-1}$

$t+3 = i$ or $t+3 = -i$

$t = -3+i$ or $t = -3-i$

Solution set: $\{-3+i, -3-i\}$

63. $3r^2 + 4r + 4 = 0$

$3r^2 + 4r = -4$

$r^2 + \frac{4}{3}r = \frac{-4}{3}$ *Divide by 3.*

Complete the square.

$\left[\frac{1}{2}\left(\frac{4}{3}\right)\right]^2 = \left(\frac{2}{3}\right)^2 = \frac{4}{9}$

Add $\frac{4}{9}$ to each side.

$r^2 + \frac{4}{3}r + \frac{4}{9} = \frac{-4}{3} + \frac{4}{9}$

$\left(r+\frac{2}{3}\right)^2 = \frac{-8}{9}$

$r + \dfrac{2}{3} = \dfrac{\sqrt{-8}}{\sqrt{9}}$ or $r + \dfrac{2}{3} = -\dfrac{\sqrt{-8}}{\sqrt{9}}$

$r = -\dfrac{2}{3} + \dfrac{2i\sqrt{2}}{3}$ or $r = -\dfrac{2}{3} - \dfrac{2i\sqrt{2}}{3}$

$r = \dfrac{-2+2i\sqrt{2}}{3}$ or $r = \dfrac{-2-2i\sqrt{2}}{3}$

Solution set: $\left\{\dfrac{-2+2i\sqrt{2}}{3}, \dfrac{-2-2i\sqrt{2}}{3}\right\}$

64. $4x^2 + 5x + 5 = 0$

$4x^2 + 5x = -5$

$x^2 + \frac{5}{4}x = -\frac{5}{4}$ *Divide by 4.*

$\left[\frac{1}{2}\left(\frac{5}{4}\right)\right]^2 = \left(\frac{5}{8}\right)^2 = \frac{25}{64}$

$x^2 + \frac{5}{4}x + \frac{25}{64} = -\frac{5}{4} + \frac{25}{64}$

$\left(x+\frac{5}{8}\right)^2 = \frac{-80+25}{64} = \frac{-55}{64}$

$x + \dfrac{5}{8} = \sqrt{\dfrac{-55}{64}}$ or $x + \dfrac{5}{8} = -\sqrt{\dfrac{-55}{64}}$

$x + \dfrac{5}{8} = \dfrac{i\sqrt{55}}{8}$ or $x + \dfrac{5}{8} = \dfrac{-i\sqrt{55}}{8}$

$x = \dfrac{-5+i\sqrt{55}}{8}$ or $x = \dfrac{-5-i\sqrt{55}}{8}$

Solution set: $\left\{\dfrac{-5+i\sqrt{55}}{8}, \dfrac{-5-i\sqrt{55}}{8}\right\}$

65. $-m^2 - 6m - 12 = 0$

Multiply each side by -1.

$m^2 + 6m + 12 = 0$

$m^2 + 6m = -12$

Complete the square.

$\left[\frac{1}{2}(6)\right]^2 = 3^2 = 9$

Add 9 to each side.

$m^2 + 6m + 9 = -12 + 9$

$(m+3)^2 = -3$

$m+3 = \sqrt{-3}$ or $m+3 = -\sqrt{-3}$

$m = -3+i\sqrt{3}$ or $m = -3-i\sqrt{3}$

Solution set: $\left\{-3+i\sqrt{3}, -3-i\sqrt{3}\right\}$

66. $-k^2 - 5k - 10 = 0$

Multiply each side by -1.

$$k^2 + 5k + 10 = 0$$
$$k^2 + 5k = -10$$
$$\left[\tfrac{1}{2}(5)\right]^2 = \left(\tfrac{5}{2}\right)^2 = \tfrac{25}{4}$$
$$k^2 + 5k + \tfrac{25}{4} = -10 + \tfrac{25}{4}$$
$$\left(k + \tfrac{5}{2}\right)^2 = \tfrac{-15}{4}$$

$$k + \frac{5}{2} = \sqrt{\frac{-15}{4}} \quad \text{or} \quad k + \frac{5}{2} = -\sqrt{\frac{-15}{4}}$$
$$k = -\frac{5}{2} + \frac{i\sqrt{15}}{2} \quad \text{or} \quad k = -\frac{5}{2} - \frac{i\sqrt{15}}{2}$$
$$k = \frac{-5 + i\sqrt{15}}{2} \quad \text{or} \quad k = \frac{-5 - i\sqrt{15}}{2}$$

Solution set: $\left\{\frac{-5 + i\sqrt{15}}{2}, \frac{-5 - i\sqrt{15}}{2}\right\}$

67. The area of the original square is $x \cdot x$, or x^2.

68. Each rectangular strip has length x and width 1, so each strip has an area of $x \cdot 1$, or x.

69. From Exercise 68, the area of a rectangular strip is x. The area of 6 rectangular strips is $6x$.

70. These are 1 by 1 squares, so each has an area of $1 \cdot 1$, or 1.

71. There are 9 small squares, each with area 1 (from Exercise 70), so the total area is $9 \cdot 1$, or 9.

72. The area of the larger square is $(x + 3)^2$. Using the results from Exercises 67–71,

$$(x + 3)^2 = x^2 + 6x + 9.$$

10.2 The Quadratic Formula

10.2 Margin Exercises

1. **(a)** $-3q^2 + 9q - 4 = 0$

Here $a = -3$, $b = 9$, and $c = -4$.

(b) $3x^2 = 6x + 2$

$$3x^2 - 6x - 2 = 0$$

Here $a = 3$, $b = -6$, and $c = -2$.

2. $4x^2 - 11x - 3 = 0$

Here $a = 4$, $b = -11$, and $c = -3$.

$$x = \frac{-b \pm \sqrt{b^2 - 4ac}}{2a}$$
$$x = \frac{-(-11) \pm \sqrt{(-11)^2 - 4(4)(-3)}}{2(4)}$$
$$x = \frac{11 \pm \sqrt{121 + 48}}{8}$$
$$x = \frac{11 \pm \sqrt{169}}{8}$$
$$x = \frac{11 \pm 13}{8}$$

$$x = \frac{11 + 13}{8} = \frac{24}{8} = 3$$
$$\text{or} \quad x = \frac{11 - 13}{8} = \frac{-2}{8} = -\frac{1}{4}$$

Solution set: $\left\{-\frac{1}{4}, 3\right\}$

3. **(a)** $6x^2 + 4x - 1 = 0$

Here $a = 6$, $b = 4$, and $c = -1$.

$$x = \frac{-b \pm \sqrt{b^2 - 4ac}}{2a}$$
$$= \frac{-4 \pm \sqrt{4^2 - 4(6)(-1)}}{2(6)}$$
$$= \frac{-4 \pm \sqrt{16 + 24}}{12}$$
$$= \frac{-4 \pm \sqrt{40}}{12}$$
$$= \frac{-4 \pm 2\sqrt{10}}{12}$$
$$= \frac{2\left(-2 \pm \sqrt{10}\right)}{12}$$
$$= \frac{-2 \pm \sqrt{10}}{6}$$

Solution set: $\left\{\frac{-2 + \sqrt{10}}{6}, \frac{-2 - \sqrt{10}}{6}\right\}$

(b) $2k^2 + 19 = 14k$

$$2k^2 - 14k + 19 = 0$$

Here $a = 2$, $b = -14$, and $c = 19$.

$$k = \frac{-b \pm \sqrt{b^2 - 4ac}}{2a}$$
$$= \frac{-(-14) \pm \sqrt{(-14)^2 - 4(2)(19)}}{2(2)}$$
$$= \frac{14 \pm \sqrt{196 - 152}}{4}$$
$$= \frac{14 \pm \sqrt{44}}{4}$$
$$= \frac{14 \pm 2\sqrt{11}}{4}$$
$$= \frac{2\left(7 \pm \sqrt{11}\right)}{4}$$
$$= \frac{7 \pm \sqrt{11}}{2}$$

Solution set: $\left\{\frac{7 + \sqrt{11}}{2}, \frac{7 - \sqrt{11}}{2}\right\}$

4. **(a)** $x^2 + x + 1 = 0$

Here $a = 1,\ b = 1,$ and $c = 1.$

$$\begin{aligned} x &= \frac{-b \pm \sqrt{b^2 - 4ac}}{2a} \\ &= \frac{-1 \pm \sqrt{1^2 - 4(1)(1)}}{2(1)} \\ &= \frac{-1 \pm \sqrt{1-4}}{2} \\ &= \frac{-1 \pm \sqrt{-3}}{2} \\ &= \frac{-1 \pm i\sqrt{3}}{2} \end{aligned}$$

Solution set: $\left\{\frac{-1 + i\sqrt{3}}{2}, \frac{-1 - i\sqrt{3}}{2}\right\}$

(b)
$$\begin{aligned} (z+2)(z-6) &= -17 \\ z^2 + 2z - 6z - 12 &= -17 \\ z^2 - 4z - 12 &= -17 \\ z^2 - 4z + 5 &= 0 \end{aligned}$$

Here $a = 1,\ b = -4,$ and $c = 5.$

$$\begin{aligned} z &= \frac{-b \pm \sqrt{b^2 - 4ac}}{2a} \\ &= \frac{-(-4) \pm \sqrt{(-4)^2 - 4(1)(5)}}{2(1)} \\ &= \frac{4 \pm \sqrt{16 - 20}}{2} \\ &= \frac{4 \pm \sqrt{-4}}{2} \\ &= \frac{4 \pm 2i}{2} \\ &= \frac{2(2 \pm i)}{2} \\ &= 2 \pm i \end{aligned}$$

Solution set: $\{2 + i, 2 - i\}$

5. **(a)**
$$\begin{aligned} 2x^2 + 3x &= 4 \\ 2x^2 + 3x - 4 &= 0 \end{aligned}$$

Here $a = 2,\ b = 3,$ and $c = -4,$ so the discriminant is

$$\begin{aligned} b^2 - 4ac &= 3^2 - 4(2)(-4) \\ &= 9 + 32 \\ &= 41. \end{aligned}$$

The discriminant is positive but not a perfect square and a, b, and c are integers, so there are two different irrational solutions.

(b) $2x^2 + 3x + 4 = 0$

Here $a = 2,\ b = 3,$ and $c = 4,$ so the discriminant is

$$\begin{aligned} b^2 - 4ac &= 3^2 - 4(2)(4) \\ &= 9 - 32 \\ &= -23. \end{aligned}$$

The discriminant is negative and a, b, and c are integers, so there are two different nonreal complex solutions.

(c) $x^2 + 20x + 100 = 0$

Here $a = 1,\ b = 20,$ and $c = 100,$ so the discriminant is

$$\begin{aligned} b^2 - 4ac &= 20^2 - 4(1)(100) \\ &= 400 - 400 \\ &= 0. \end{aligned}$$

The discriminant is zero, so there is one rational solution.

(d)
$$\begin{aligned} 15k^2 + 11k &= 14 \\ 15k^2 + 11k - 14 &= 0 \end{aligned}$$

Here $a = 15,\ b = 11,$ and $c = -14,$ so the discriminant is

$$\begin{aligned} b^2 - 4ac &= 11^2 - 4(15)(-14) \\ &= 121 + 840 \\ &= 961 \quad \text{or} \quad 31^2. \end{aligned}$$

The discriminant is a perfect square and a, b, and c are integers, so there are two different rational solutions.

(e) If the discriminant is a perfect square (including 0), then the equation can be solved by factoring, so the answer is (c) and (d).

10.2 Section Exercises

1. If the expression $\frac{-b \pm \sqrt{b^2 - 4ac}}{2a}$ were written as two terms, we would have $-\frac{b}{2a} \pm \frac{\sqrt{b^2 - 4ac}}{2a}$. Therefore,

$$\frac{-b \pm \sqrt{b^2 - 4ac}}{2a} \neq -b \pm \frac{\sqrt{b^2 - 4ac}}{2a}.$$

The student was incorrect, since the fraction bar should extend under the term $-b$.

2. The last step is wrong. Because 5 is not a common factor in the numerator, the fraction cannot be simplified. The solutions are $\frac{5 \pm \sqrt{5}}{10}$.

3. $m^2 - 8m + 15 = 0$

Here $a = 1$, $b = -8$, and $c = 15$.

$$m = \frac{-b \pm \sqrt{b^2 - 4ac}}{2a}$$
$$m = \frac{-(-8) \pm \sqrt{(-8)^2 - 4(1)(15)}}{2(1)}$$
$$= \frac{8 \pm \sqrt{64 - 60}}{2}$$
$$= \frac{8 \pm \sqrt{4}}{2} = \frac{8 \pm 2}{2}$$
$$m = \frac{8+2}{2} = \frac{10}{2} = 5$$
$$\text{or } m = \frac{8-2}{2} = \frac{6}{2} = 3$$

Solution set: $\{3, 5\}$

4. $x^2 + 3x - 28 = 0$

Here $a = 1$, $b = 3$, and $c = -28$.

$$x = \frac{-b \pm \sqrt{b^2 - 4ac}}{2a}$$
$$x = \frac{-3 \pm \sqrt{3^2 - 4(1)(-28)}}{2(1)}$$
$$= \frac{-3 \pm \sqrt{9 + 112}}{2}$$
$$= \frac{-3 \pm \sqrt{121}}{2} = \frac{-3 \pm 11}{2}$$
$$x = \frac{8}{2} = 4 \text{ or } x = \frac{-14}{2} = -7$$

Solution set: $\{-7, 4\}$

5. $2k^2 + 4k + 1 = 0$

Here $a = 2$, $b = 4$, and $c = 1$.

$$k = \frac{-b \pm \sqrt{b^2 - 4ac}}{2a}$$
$$k = \frac{-4 \pm \sqrt{4^2 - 4(2)(1)}}{2(2)}$$
$$= \frac{-4 \pm \sqrt{16 - 8}}{4}$$
$$= \frac{-4 \pm \sqrt{8}}{4} = \frac{-4 \pm 2\sqrt{2}}{4}$$
$$= \frac{2\left(-2 \pm \sqrt{2}\right)}{2 \cdot 2} = \frac{-2 \pm \sqrt{2}}{2}$$

Solution set: $\left\{\frac{-2+\sqrt{2}}{2}, \frac{-2-\sqrt{2}}{2}\right\}$

6. $2w^2 + 3w - 1 = 0$

Here $a = 2$, $b = 3$, and $c = -1$.

$$w = \frac{-b \pm \sqrt{b^2 - 4ac}}{2a}$$
$$w = \frac{-3 \pm \sqrt{3^2 - 4(2)(-1)}}{2(2)}$$
$$= \frac{-3 \pm \sqrt{9 + 8}}{4} = \frac{-3 \pm \sqrt{17}}{4}$$

Solution set: $\left\{\frac{-3+\sqrt{17}}{4}, \frac{-3-\sqrt{17}}{4}\right\}$

7. $2x^2 - 2x = 1$

$2x^2 - 2x - 1 = 0$

Here $a = 2$, $b = -2$, and $c = -1$.

$$x = \frac{-b \pm \sqrt{b^2 - 4ac}}{2a}$$
$$x = \frac{-(-2) \pm \sqrt{(-2)^2 - 4(2)(-1)}}{2(2)}$$
$$= \frac{2 \pm \sqrt{4 + 8}}{4} = \frac{2 \pm \sqrt{12}}{4}$$
$$= \frac{2 \pm 2\sqrt{3}}{4} = \frac{2\left(1 \pm \sqrt{3}\right)}{2 \cdot 2}$$
$$= \frac{1 \pm \sqrt{3}}{2}$$

Solution set: $\left\{\frac{1+\sqrt{3}}{2}, \frac{1-\sqrt{3}}{2}\right\}$

8. $9t^2 + 6t = 1$

$9t^2 + 6t - 1 = 0$

Here $a = 9$, $b = 6$, and $c = -1$.

$$t = \frac{-b \pm \sqrt{b^2 - 4ac}}{2a}$$
$$t = \frac{-6 \pm \sqrt{6^2 - 4(9)(-1)}}{2(9)}$$
$$= \frac{-6 \pm \sqrt{36 + 36}}{18}$$
$$= \frac{-6 \pm \sqrt{72}}{18} = \frac{-6 \pm 6\sqrt{2}}{18}$$
$$= \frac{6\left(-1 \pm \sqrt{2}\right)}{6 \cdot 3} = \frac{-1 \pm \sqrt{2}}{3}$$

Solution set: $\left\{\frac{-1+\sqrt{2}}{3}, \frac{-1-\sqrt{2}}{3}\right\}$

9. $x^2 + 18 = 10x$

$x^2 - 10x + 18 = 0$

Here $a = 1$, $b = -10$, and $c = 18$.

$$x = \frac{-b \pm \sqrt{b^2 - 4ac}}{2a}$$

$$x = \frac{-(-10) \pm \sqrt{(-10)^2 - 4(1)(18)}}{2(1)}$$

$$= \frac{10 \pm \sqrt{100 - 72}}{2} = \frac{10 \pm \sqrt{28}}{2}$$

$$= \frac{10 \pm 2\sqrt{7}}{2} = \frac{2\left(5 \pm \sqrt{7}\right)}{2}$$

$$= 5 \pm \sqrt{7}$$

Solution set: $\left\{5 + \sqrt{7}, 5 - \sqrt{7}\right\}$

10. $x^2 - 4 = 2x$

$x^2 - 2x - 4 = 0$

Here $a = 1$, $b = -2$, and $c = -4$.

$$x = \frac{-b \pm \sqrt{b^2 - 4ac}}{2a}$$

$$x = \frac{-(-2) \pm \sqrt{(-2)^2 - 4(1)(-4)}}{2(1)}$$

$$= \frac{2 \pm \sqrt{4 + 16}}{2} = \frac{2 \pm \sqrt{20}}{2}$$

$$= \frac{2 \pm 2\sqrt{5}}{2} = 1 \pm \sqrt{5}$$

Solution set: $\left\{1 + \sqrt{5}, 1 - \sqrt{5}\right\}$

11. $4k^2 + 4k - 1 = 0$

Here $a = 4$, $b = 4$, and $c = -1$.

$$k = \frac{-b \pm \sqrt{b^2 - 4ac}}{2a}$$

$$k = \frac{-4 \pm \sqrt{4^2 - 4(4)(-1)}}{2(4)}$$

$$= \frac{-4 \pm \sqrt{16 + 16}}{8} = \frac{-4 \pm \sqrt{32}}{8}$$

$$= \frac{-4 \pm 4\sqrt{2}}{8} = \frac{4\left(-1 \pm \sqrt{2}\right)}{2 \cdot 4}$$

$$= \frac{-1 \pm \sqrt{2}}{2}$$

Solution set: $\left\{\frac{-1 + \sqrt{2}}{2}, \frac{-1 - \sqrt{2}}{2}\right\}$

12. $4r^2 - 4r - 19 = 0$

Here $a = 4$, $b = -4$, and $c = -19$.

$$r = \frac{-b \pm \sqrt{b^2 - 4ac}}{2a}$$

$$r = \frac{-(-4) \pm \sqrt{(-4)^2 - 4(4)(-19)}}{2(4)}$$

$$= \frac{4 \pm \sqrt{16 + 304}}{8} = \frac{4 \pm \sqrt{320}}{8}$$

$$= \frac{4 \pm 8\sqrt{5}}{8} = \frac{4\left(1 \pm 2\sqrt{5}\right)}{2 \cdot 4}$$

$$= \frac{1 \pm 2\sqrt{5}}{2}$$

Solution set: $\left\{\frac{1 + 2\sqrt{5}}{2}, \frac{1 - 2\sqrt{5}}{2}\right\}$

13. $2 - 2x = 3x^2$

$0 = 3x^2 + 2x - 2$

Here $a = 3$, $b = 2$, and $c = -2$.

$$x = \frac{-b \pm \sqrt{b^2 - 4ac}}{2a}$$

$$x = \frac{-2 \pm \sqrt{2^2 - 4(3)(-2)}}{2(3)}$$

$$= \frac{-2 \pm \sqrt{4 + 24}}{6} = \frac{-2 \pm \sqrt{28}}{6}$$

$$= \frac{-2 \pm 2\sqrt{7}}{6} = \frac{2\left(-1 \pm \sqrt{7}\right)}{2 \cdot 3}$$

$$= \frac{-1 \pm \sqrt{7}}{3}$$

Solution set: $\left\{\frac{-1 + \sqrt{7}}{3}, \frac{-1 - \sqrt{7}}{3}\right\}$

14. $26r - 2 = 3r^2$

$0 = 3r^2 - 26r + 2$

Here $a = 3$, $b = -26$, and $c = 2$.

$$r = \frac{-b \pm \sqrt{b^2 - 4ac}}{2a}$$

$$r = \frac{-(-26) \pm \sqrt{(-26)^2 - 4(3)(2)}}{2(3)}$$

$$= \frac{26 \pm \sqrt{676 - 24}}{6} = \frac{26 \pm \sqrt{652}}{6}$$

$$= \frac{26 \pm 2\sqrt{163}}{6} = \frac{13 \pm \sqrt{163}}{3}$$

Solution set: $\left\{\frac{13 + \sqrt{163}}{3}, \frac{13 - \sqrt{163}}{3}\right\}$

15. $\frac{x^2}{4} - \frac{x}{2} = 1$

$\frac{x^2}{4} - \frac{x}{2} - 1 = 0$

$x^2 - 2x - 4 = 0$ *Multiply by 4.*

Here $a = 1$, $b = -2$, and $c = -4$.

$$x = \frac{-b \pm \sqrt{b^2 - 4ac}}{2a}$$

$$x = \frac{-(-2) \pm \sqrt{(-2)^2 - 4(1)(-4)}}{2(1)}$$

$$= \frac{2 \pm \sqrt{4 + 16}}{2} = \frac{2 \pm \sqrt{20}}{2}$$

$$= \frac{2 \pm 2\sqrt{5}}{2} = 1 \pm \sqrt{5}$$

Solution set: $\left\{1 + \sqrt{5}, 1 - \sqrt{5}\right\}$

16. $p^2 + \frac{p}{3} = \frac{1}{6}$

$p^2 + \frac{p}{3} - \frac{1}{6} = 0$

$6p^2 + 2p - 1 = 0$ *Multiply by 6.*

Here $a = 6$, $b = 2$, and $c = -1$.

$$p = \frac{-b \pm \sqrt{b^2 - 4ac}}{2a}$$

$$p = \frac{-2 \pm \sqrt{2^2 - 4(6)(-1)}}{2(6)}$$

$$= \frac{-2 \pm \sqrt{4 + 24}}{12}$$

$$= \frac{-2 \pm \sqrt{28}}{12} = \frac{-2 \pm 2\sqrt{7}}{12}$$

$$= \frac{-1 \pm \sqrt{7}}{6}$$

Solution set: $\left\{\frac{-1 + \sqrt{7}}{6}, \frac{-1 - \sqrt{7}}{6}\right\}$

17. $-2t(t + 2) = -3$

$-2t^2 - 4t = -3$

$-2t^2 - 4t + 3 = 0$

Here $a = -2$, $b = -4$, and $c = 3$.

$$t = \frac{-b \pm \sqrt{b^2 - 4ac}}{2a}$$

$$t = \frac{-(-4) \pm \sqrt{(-4)^2 - 4(-2)(3)}}{2(-2)}$$

$$= \frac{4 \pm \sqrt{16 + 24}}{-4} = \frac{4 \pm \sqrt{40}}{-4}$$

$$= \frac{4 \pm 2\sqrt{10}}{-4} = \frac{2\left(2 \pm \sqrt{10}\right)}{-2 \cdot 2}$$

$$= \frac{2 \pm \sqrt{10}}{-2} \cdot \frac{-1}{-1} = \frac{-2 \mp \sqrt{10}}{2}$$

$$= \frac{-2 \pm \sqrt{10}}{2}$$

Solution set: $\left\{\frac{-2 + \sqrt{10}}{2}, \frac{-2 - \sqrt{10}}{2}\right\}$

18. $-3x(x + 2) = -4$

$-3x^2 - 6x = -4$

$0 = 3x^2 + 6x - 4$

Here $a = 3$, $b = 6$, and $c = -4$.

$$x = \frac{-b \pm \sqrt{b^2 - 4ac}}{2a}$$

$$x = \frac{-6 \pm \sqrt{6^2 - 4(3)(-4)}}{2(3)}$$

$$= \frac{-6 \pm \sqrt{36 + 48}}{6} = \frac{-6 \pm \sqrt{84}}{6}$$

$$= \frac{-6 \pm 2\sqrt{21}}{6} = \frac{2\left(-3 \pm \sqrt{21}\right)}{2 \cdot 3}$$

$$= \frac{-3 \pm \sqrt{21}}{3}$$

Solution set: $\left\{\frac{-3 + \sqrt{21}}{3}, \frac{-3 - \sqrt{21}}{3}\right\}$

19. $(r - 3)(r + 5) = 2$

$r^2 + 2r - 15 = 2$

$r^2 + 2r - 17 = 0$

Here $a = 1$, $b = 2$, and $c = -17$.

$$r = \frac{-b \pm \sqrt{b^2 - 4ac}}{2a}$$

$$r = \frac{-2 \pm \sqrt{2^2 - 4(1)(-17)}}{2(1)}$$

$$= \frac{-2 \pm \sqrt{4 + 68}}{2} = \frac{-2 \pm \sqrt{72}}{2}$$

$$= \frac{-2 \pm 6\sqrt{2}}{2} = \frac{2\left(-1 \pm 3\sqrt{2}\right)}{2}$$

$$= -1 \pm 3\sqrt{2}$$

Solution set: $\left\{-1 + 3\sqrt{2}, -1 - 3\sqrt{2}\right\}$

20. $(k+1)(k-7)=1$

$$k^2-6k-7=1$$
$$k^2-6k-8=0$$

Here $a=1$, $b=-6$, and $c=-8$.

$$k=\frac{-b\pm\sqrt{b^2-4ac}}{2a}$$
$$k=\frac{-(-6)\pm\sqrt{(-6)^2-4(1)(-8)}}{2(1)}$$
$$=\frac{6\pm\sqrt{36+32}}{2}=\frac{6\pm\sqrt{68}}{2}$$
$$=\frac{6\pm2\sqrt{17}}{2}=3\pm\sqrt{17}$$

Solution set: $\left\{3+\sqrt{17},3-\sqrt{17}\right\}$

21. $x^2-3x+17=0$

Here $a=1$, $b=-3$, and $c=17$.

$$x=\frac{-b\pm\sqrt{b^2-4ac}}{2a}$$
$$x=\frac{-(-3)\pm\sqrt{(-3)^2-4(1)(17)}}{2(1)}$$
$$=\frac{3\pm\sqrt{9-68}}{2}$$
$$=\frac{3\pm\sqrt{-59}}{2}$$
$$=\frac{3\pm i\sqrt{59}}{2}$$

Solution set: $\left\{\frac{3+i\sqrt{59}}{2},\frac{3-i\sqrt{59}}{2}\right\}$

22. $x^2-5x+20=0$

Here $a=1$, $b=-5$, and $c=20$.

$$x=\frac{-b\pm\sqrt{b^2-4ac}}{2a}$$
$$x=\frac{-(-5)\pm\sqrt{(-5)^2-4(1)(20)}}{2(1)}$$
$$=\frac{5\pm\sqrt{25-80}}{2}$$
$$=\frac{5\pm\sqrt{-55}}{2}$$
$$=\frac{5\pm i\sqrt{55}}{2}$$

Solution set: $\left\{\frac{5+i\sqrt{55}}{2},\frac{5-i\sqrt{55}}{2}\right\}$

23. $r^2-6r+14=0$

Here $a=1$, $b=-6$, and $c=14$.

$$r=\frac{-b\pm\sqrt{b^2-4ac}}{2a}$$
$$r=\frac{-(-6)\pm\sqrt{(-6)^2-4(1)(14)}}{2(1)}$$
$$=\frac{6\pm\sqrt{36-56}}{2}$$
$$=\frac{6\pm\sqrt{-20}}{2}=\frac{6\pm2i\sqrt{5}}{2}$$
$$=\frac{2\left(3\pm i\sqrt{5}\right)}{2}=3\pm i\sqrt{5}$$

Solution set: $\left\{3+i\sqrt{5},3-i\sqrt{5}\right\}$

24. $t^2+4t+11=0$

Here $a=1$, $b=4$, and $c=11$.

$$t=\frac{-b\pm\sqrt{b^2-4ac}}{2a}$$
$$t=\frac{-4\pm\sqrt{4^2-4(1)(11)}}{2(1)}$$
$$=\frac{-4\pm\sqrt{16-44}}{2}=\frac{-4\pm\sqrt{-28}}{2}$$
$$=\frac{-4\pm2i\sqrt{7}}{2}=-2\pm i\sqrt{7}$$

Solution set: $\left\{-2+i\sqrt{7},-2-i\sqrt{7}\right\}$

25. $4x^2-4x=-7$

$$4x^2-4x+7=0$$

Here $a=4$, $b=-4$, and $c=7$.

$$x=\frac{-b\pm\sqrt{b^2-4ac}}{2a}$$
$$x=\frac{-(-4)\pm\sqrt{(-4)^2-4(4)(7)}}{2(4)}$$
$$=\frac{4\pm\sqrt{16-112}}{8}=\frac{4\pm\sqrt{-96}}{8}$$
$$=\frac{4\pm4i\sqrt{6}}{8}=\frac{4\left(1\pm i\sqrt{6}\right)}{2\bullet4}$$
$$=\frac{1\pm i\sqrt{6}}{2}$$

Solution set: $\left\{\frac{1+i\sqrt{6}}{2},\frac{1-i\sqrt{6}}{2}\right\}$

26. $9x^2 - 6x = -7$

$9x^2 - 6x + 7 = 0$

Here $a = 9$, $b = -6$, and $c = 7$.

$$x = \frac{-b \pm \sqrt{b^2 - 4ac}}{2a}$$
$$x = \frac{-(-6) \pm \sqrt{(-6)^2 - 4(9)(7)}}{2(9)}$$
$$= \frac{6 \pm \sqrt{36 - 252}}{18} = \frac{6 \pm \sqrt{-216}}{18}$$
$$= \frac{6 \pm 6i\sqrt{6}}{18} = \frac{6\left(1 \pm i\sqrt{6}\right)}{3 \cdot 6}$$
$$= \frac{1 \pm i\sqrt{6}}{3}$$

Solution set: $\left\{\frac{1 + i\sqrt{6}}{3}, \frac{1 - i\sqrt{6}}{3}\right\}$

27. $x(3x + 4) = -2$

$3x^2 + 4x = -2$

$3x^2 + 4x + 2 = 0$

Here $a = 3$, $b = 4$, and $c = 2$.

$$x = \frac{-b \pm \sqrt{b^2 - 4ac}}{2a}$$
$$x = \frac{-4 \pm \sqrt{4^2 - 4(3)(2)}}{2(3)}$$
$$= \frac{-4 \pm \sqrt{16 - 24}}{6} = \frac{-4 \pm \sqrt{-8}}{6}$$
$$= \frac{-4 \pm 2i\sqrt{2}}{6} = \frac{2\left(-2 \pm i\sqrt{2}\right)}{2 \cdot 3}$$
$$= \frac{-2 \pm i\sqrt{2}}{3}$$

Solution set: $\left\{\frac{-2 + i\sqrt{2}}{3}, \frac{-2 - i\sqrt{2}}{3}\right\}$

28. $p(2p + 3) = -2$

$2p^2 + 3p + 2 = 0$

Here $a = 2$, $b = 3$, and $c = 2$.

$$p = \frac{-b \pm \sqrt{b^2 - 4ac}}{2a}$$
$$p = \frac{-3 \pm \sqrt{3^2 - 4(2)(2)}}{2(2)}$$
$$= \frac{-3 \pm \sqrt{9 - 16}}{4} = \frac{-3 \pm \sqrt{-7}}{4}$$
$$= \frac{-3 \pm i\sqrt{7}}{4}$$

Solution set: $\left\{\frac{-3 + i\sqrt{7}}{4}, \frac{-3 - i\sqrt{7}}{4}\right\}$

29. $25x^2 + 70x + 49 = 0$

Here $a = 25$, $b = 70$, and $c = 49$, so the discriminant is

$$\begin{aligned} b^2 - 4ac &= 70^2 - 4(25)(49) \\ &= 4900 - 4900 \\ &= 0. \end{aligned}$$

Since the discriminant is 0, the quantity under the radical in the quadratic formula is 0, and there is only one rational solution. The answer is **B**.

30. $4k^2 - 28k + 49 = 0$

Here $a = 4$, $b = -28$, and $c = 49$, so the discriminant is

$$\begin{aligned} b^2 - 4ac &= (-28)^2 - 4(4)(49) \\ &= 784 - 784 \\ &= 0. \end{aligned}$$

Since the discriminant is 0, the quantity under the radical in the quadratic formula is 0, and there is only one rational solution. The answer is **B**.

31. $x^2 + 4x + 2 = 0$

Here $a = 1$, $b = 4$, and $c = 2$, so the discriminant is

$$\begin{aligned} b^2 - 4ac &= 4^2 - 4(1)(2) \\ &= 16 - 8 \\ &= 8. \end{aligned}$$

Since the discriminant is positive, but not a perfect square, there are two distinct irrational number solutions. The answer is **C**.

32. $9x^2 - 12x - 1 = 0$

Here $a = 9$, $b = -12$, and $c = -1$, so the discriminant is

$$\begin{aligned} b^2 - 4ac &= (-12)^2 - 4(9)(-1) \\ &= 144 + 36 \\ &= 180. \end{aligned}$$

Since the discriminant is positive, but not a perfect square, there are two distinct irrational number solutions. The answer is **C**.

33. $3x^2 = 5x + 2$

$3x^2 - 5x - 2 = 0$

Here $a = 3$, $b = -5$, and $c = -2$, so the discriminant is

$$\begin{aligned} b^2 - 4ac &= (-5)^2 - 4(3)(-2) \\ &= 25 + 24 \\ &= 49. \end{aligned}$$

Since the discriminant is a perfect square, there are two distinct rational solutions. The answer is **A**.

34. $$4x^2 = 4x + 3$$
$$4x^2 - 4x - 3 = 0$$

Here $a = 4$, $b = -4$, and $c = -3$, so the discriminant is

$$\begin{aligned} b^2 - 4ac &= (-4)^2 - 4(4)(-3) \\ &= 16 + 48 \\ &= 64. \end{aligned}$$

Since the discriminant is a perfect square, there are two distinct rational solutions. The answer is **A**.

35. $3m^2 - 10m + 15 = 0$

Here $a = 3$, $b = -10$, and $c = 15$, so the discriminant is

$$\begin{aligned} b^2 - 4ac &= (-10)^2 - 4(3)(15) \\ &= 100 - 180 \\ &= -80. \end{aligned}$$

Since the discriminant is negative, there are two distinct nonreal complex number solutions. The answer is **D**.

36. $18x^2 + 60x + 82 = 0$

Here $a = 18$, $b = 60$, and $c = 82$, so the discriminant is

$$\begin{aligned} b^2 - 4ac &= 60^2 - 4(18)(82) \\ &= 3600 - 5904 \\ &= -2304. \end{aligned}$$

Since the discriminant is negative, there are two distinct nonreal complex number solutions. The answer is **D**.

37. The equations in Exercises 29, 30, 33, and 34 can be solved by factoring because the discriminant is a perfect square.

38. **(a)** $25x^2 + 70x + 49 = 0$
$$(5x + 7)^2 = 0$$
$$5x + 7 = 0$$
$$x = -\tfrac{7}{5}$$

Solution set: $\{-\frac{7}{5}\}$

(b) $$3x^2 = 5x + 2$$
$$3x^2 - 5x - 2 = 0$$
$$(3x + 1)(x - 2) = 0$$

$3x + 1 = 0$ or $x - 2 = 0$

$x = -\frac{1}{3}$ or $x = 2$

Solution set: $\{-\frac{1}{3}, 2\}$

10.3 Equations Quadratic in Form

10.3 Margin Exercises

1. **(a)** $\dfrac{5}{m} + \dfrac{12}{m^2} = 2$

Clear fractions by multiplying each term by the LCD, m^2.

$$m^2\left(\frac{5}{m}\right) + m^2\left(\frac{12}{m^2}\right) = m^2(2)$$
$$5m + 12 = 2m^2$$
$$-2m^2 + 5m + 12 = 0$$
$$2m^2 - 5m - 12 = 0$$
$$(2m + 3)(m - 4) = 0$$

$2m + 3 = 0$ or $m - 4 = 0$

$m = -\frac{3}{2}$ or $m = 4$

Check $m = -\frac{3}{2}$: $-\frac{10}{3} + \frac{16}{3} = 2$ *True*

Check $m = 4$: $\frac{5}{4} + \frac{3}{4} = 2$ *True*

Solution set: $\{-\frac{3}{2}, 4\}$

(b) $\dfrac{2}{x} + \dfrac{1}{x - 2} = \dfrac{5}{3}$

Multiply each side by the LCD, $3x(x - 2)$.

$$3x(x - 2)\left(\frac{2}{x} + \frac{1}{x - 2}\right) = 3x(x - 2)\left(\frac{5}{3}\right)$$
$$6(x - 2) + 3x = 5x(x - 2)$$
$$6x - 12 + 3x = 5x^2 - 10x$$
$$-5x^2 + 19x - 12 = 0$$
$$5x^2 - 19x + 12 = 0$$
$$(5x - 4)(x - 3) = 0$$

$5x - 4 = 0$ or $x - 3 = 0$

$x = \frac{4}{5}$ or $x = 3$

Check $x = \frac{4}{5}$: $\frac{5}{2} - \frac{5}{6} = \frac{5}{3}$ *True*

Check $x = 3$: $\frac{2}{3} + 1 = \frac{5}{3}$ *True*

Solution set: $\{\frac{4}{5}, 3\}$

(c) $\dfrac{4}{m - 1} + 9 = -\dfrac{7}{m}$

Multiply by the LCD, $m(m - 1)$.

$$m(m - 1)\left(\frac{4}{m - 1} + 9\right) = m(m - 1)\left(-\frac{7}{m}\right)$$
$$4m + 9m(m - 1) = -7(m - 1)$$
$$4m + 9m^2 - 9m = -7m + 7$$
$$9m^2 + 2m - 7 = 0$$
$$(9m - 7)(m + 1) = 0$$

$9m - 7 = 0$ or $m + 1 = 0$

$m = \frac{7}{9}$ or $m = -1$

continued

Check $m = \frac{7}{9}$: $-18 + 9 = -9$ *True*
Check $m = -1$: $-2 + 9 = 7$ *True*

Solution set: $\left\{-1, \frac{7}{9}\right\}$

2. **(a)** Let $x =$ the speed of the boat in still water.

Since the speed of the current is 5 mph, the speed upriver (*against* the current) is $x - 5$ and the speed downriver (*with* the current) is $x + 5$.

Kerrie went 15 mi each way. Complete the table. Use $d = rt$, or $t = \frac{d}{r}$.

	d	*r*	*t*
Up	15	$x-5$	$\frac{15}{x-5}$
Down	15	$x+5$	$\frac{15}{x+5}$

(b) To solve, use the fact that she can make the entire trip in 4 hr; that is, the time going upriver added to the time going downriver is 4.

$$\frac{15}{x-5} + \frac{15}{x+5} = 4$$

Multiply each term by the LCD, $(x-5)(x+5)$.

$$\frac{15}{x-5}(x-5)(x+5) + \frac{15}{x+5}(x-5)(x+5) = 4(x-5)(x+5)$$
$$15(x+5) + 15(x-5) = 4\left(x^2 - 25\right)$$
$$15x + 75 + 15x - 75 = 4x^2 - 100$$
$$30x = 4x^2 - 100$$
$$0 = 4x^2 - 30x - 100$$
$$0 = 2x^2 - 15x - 50$$
$$0 = (2x+5)(x-10)$$

$2x + 5 = 0$ or $x - 10 = 0$
$x = -\frac{5}{2}$ or $x = 10$

The speed must be nonnegative, so $-\frac{5}{2}$ is not a solution. The speed of the boat is 10 mph.

(c) Let $x =$ the speed Ken can row.

Make a table. Use $t = \frac{d}{r}$.

	d	*r*	*t*
Up	5	$x-3$	$\frac{5}{x-3}$
Down	5	$x+3$	$\frac{5}{x+3}$

The time going upriver added to the time going downriver is $1\frac{3}{4}$ or $\frac{7}{4}$ hr.

$$\frac{5}{x-3} + \frac{5}{x+3} = \frac{7}{4}$$

Multiply each term by the LCD, $4(x-3)(x+3)$.

$$4(x-3)(x+3)\frac{5}{x-3} + 4(x-3)(x+3)\frac{5}{x+3} = 4(x-3)(x+3)\left(\frac{7}{4}\right)$$
$$20(x+3) + 20(x-3) = 7(x-3)(x+3)$$
$$20x + 60 + 20x - 60 = 7\left(x^2 - 9\right)$$
$$40x = 7x^2 - 63$$
$$0 = 7x^2 - 40x - 63$$
$$0 = (7x+9)(x-7)$$

$7x + 9 = 0$ or $x - 7 = 0$
$x = -\frac{9}{7}$ or $x = 7$

Speed can't be negative, so Ken rows at the speed of 7 mph.

3. **(a)** Let $x =$ Jaime's time alone. Then $x - 2 =$ Carlos' time alone.

Make a table.

	Rate	**Time Working Together**	**Fractional Part of the Job Done**
Carlos	$\frac{1}{x-2}$	2	$\frac{2}{x-2}$
Jaime	$\frac{1}{x}$	2	$\frac{2}{x}$

Since together they complete 1 whole job,

$$\frac{2}{x-2} + \frac{2}{x} = 1.$$

Multiply each side by the LCD, $x(x-2)$.

$$x(x-2)\left(\frac{2}{x-2} + \frac{2}{x}\right) = x(x-2)(1)$$
$$2x + 2(x-2) = x(x-2)$$
$$2x + 2x - 4 = x^2 - 2x$$
$$0 = x^2 - 6x + 4$$

Use the quadratic formula.

Here $a = 1$, $b = -6$, and $c = 4$.

$$x = \frac{-b \pm \sqrt{b^2 - 4ac}}{2a}$$
$$= \frac{-(-6) \pm \sqrt{(-6)^2 - 4(1)(4)}}{2(1)}$$
$$= \frac{6 \pm \sqrt{36 - 16}}{2}$$
$$= \frac{6 \pm \sqrt{20}}{2} = \frac{6 \pm 2\sqrt{5}}{2}$$
$$= \frac{2\left(3 \pm \sqrt{5}\right)}{2} = 3 \pm \sqrt{5}$$

$x = 3 + \sqrt{5} \approx 5.2$ or $x = 3 - \sqrt{5} \approx .8$

Jaime's time cannot be .8, since Carlos' time would then be $.8 - 2$ or -1.2. So, Jaime's time alone is 5.2 hr, and Carlos' time alone is 3.2 hr.

(b) Let $x =$ the slow chef's time alone. Then $x - 2 =$ the fast chef's time alone.

Make a table.

	Rate	Time Working Together	Fractional Part of the Job Done
Slow	$\frac{1}{x}$	5	$\frac{5}{x}$
Fast	$\frac{1}{x-2}$	5	$\frac{5}{x-2}$

Since together they complete 1 whole job,

$$\frac{5}{x} + \frac{5}{x-2} = 1.$$

Multiply each term by the LCD, $x(x-2)$.

$$x(x-2)\left(\frac{5}{x}\right) + x(x-2)\left(\frac{5}{x-2}\right) = x(x-2)(1)$$
$$5(x-2) + 5x = x(x-2)$$
$$5x - 10 + 5x = x^2 - 2x$$
$$0 = x^2 - 12x + 10$$

Here $a = 1$, $b = -12$, and $c = 10$.

$$x = \frac{-b \pm \sqrt{b^2 - 4ac}}{2a}$$
$$= \frac{-(-12) \pm \sqrt{(-12)^2 - 4(1)(10)}}{2(1)}$$
$$= \frac{12 \pm \sqrt{144 - 40}}{2}$$
$$= \frac{12 \pm \sqrt{104}}{2} = \frac{12 \pm 2\sqrt{26}}{2}$$
$$= \frac{2\left(6 \pm \sqrt{26}\right)}{2} = 6 \pm \sqrt{26}$$

$x = 6 + \sqrt{26} \approx 11.1$ or $x = 6 - \sqrt{26} \approx .9$

The slow chef's time cannot be .9 since the fast chef's time would then be $.9 - 2$ or -1.1. So the slow chef's time working alone is 11.1 hr and the fast chef's time working alone is $11.1 - 2 = 9.1$ hr.

4. **(a)**

$$x = \sqrt{7x - 10}$$
$$x^2 = \left(\sqrt{7x-10}\right)^2 \quad \textit{Square}$$
$$x^2 = 7x - 10$$
$$x^2 - 7x + 10 = 0$$
$$(x-2)(x-5) = 0$$
$$x - 2 = 0 \quad \text{or} \quad x - 5 = 0$$
$$x = 2 \quad \text{or} \quad x = 5$$

Check $x = 2$: $2 = \sqrt{4}$ *True*

Check $x = 5$: $5 = \sqrt{25}$ *True*

Solution set: $\{2, 5\}$

(b)

$$2x = \sqrt{x} + 1$$
$$2x - 1 = \sqrt{x} \quad \textit{Isolate}$$
$$(2x-1)^2 = \left(\sqrt{x}\right)^2 \quad \textit{Square}$$
$$4x^2 - 4x + 1 = x$$
$$4x^2 - 5x + 1 = 0$$
$$(4x-1)(x-1) = 0$$
$$4x - 1 = 0 \quad \text{or} \quad x - 1 = 0$$
$$x = \tfrac{1}{4} \quad \text{or} \quad x = 1$$

Check $x = \frac{1}{4}$: $\frac{1}{2} = \frac{1}{2} + 1$ *False*

Check $x = 1$: $2 = \sqrt{1} + 1$ *True*

Solution set: $\{1\}$

5. **(a)** $m^4 - 10m^2 + 9 = 0$

Let $y = m^2$, so $y^2 = \left(m^2\right)^2 = m^4$.

$$y^2 - 10y + 9 = 0$$
$$(y-9)(y-1) = 0$$
$$y - 9 = 0 \quad \text{or} \quad y - 1 = 0$$
$$y = 9 \quad \text{or} \quad y = 1$$

To find m, substitute m^2 for y.

$$m^2 = 9 \quad \text{or} \quad m^2 = 1$$
$$m = \pm 3 \quad \text{or} \quad m = \pm 1$$

Check $m = \pm 3$: $81 - 90 + 9 = 0$ *True*

Check $m = \pm 1$: $1 - 10 + 9 = 0$ *True*

Solution set: $\{-3, -1, 1, 3\}$

(b) $9k^4 - 37k^2 + 4 = 0$

Let $y = k^2$, so $y^2 = (k^2)^2 = k^4$.

$$9y^2 - 37y + 4 = 0$$
$$(y-4)(9y-1) = 0$$
$$y - 4 = 0 \quad \text{or} \quad 9y - 1 = 0$$
$$y = 4 \quad \text{or} \quad y = \tfrac{1}{9}$$

To find k, substitute k^2 for y.

$$k^2 = 4 \quad \text{or} \quad k^2 = \tfrac{1}{9}$$
$$k = \pm 2 \quad \text{or} \quad k = \pm \tfrac{1}{3}$$

continued

Check $k = \pm 2$: $144 - 148 + 4 = 0$ *True*

Check $k = \pm \frac{1}{3}$: $\frac{1}{9} - \frac{37}{9} + 4 = 0$ *True*

Solution set: $\left\{-2, -\frac{1}{3}, \frac{1}{3}, 2\right\}$

(c) $x^4 - 4x^2 = -2$

$x^4 - 4x^2 + 2 = 0$

Let $y = x^2$, so $\left(x^2\right)^2 = x^4$.

$y^2 - 4y + 2 = 0$

Use the quadratic formula with $a = 1, b = -4$, and $c = 2$.

$$\begin{aligned} y &= \frac{-b \pm \sqrt{b^2 - 4ac}}{2a} \\ &= \frac{-(-4) \pm \sqrt{(-4)^2 - 4(1)(2)}}{2(1)} \\ &= \frac{4 \pm \sqrt{16 - 8}}{2} \\ &= \frac{4 \pm \sqrt{8}}{2} = \frac{4 \pm 2\sqrt{2}}{2} \\ &= \frac{2\left(2 \pm \sqrt{2}\right)}{2} = 2 \pm \sqrt{2} \end{aligned}$$

To find x, substitute x^2 for y.

$x^2 = 2 \pm \sqrt{2}$

$x = \pm\sqrt{2 \pm \sqrt{2}}$

Check $x = \pm\sqrt{2 + \sqrt{2}}$:

$\left(2 + \sqrt{2}\right)^2 - 4\left(2 + \sqrt{2}\right) = -2$?

$4 + 4\sqrt{2} + 2 - 8 - 4\sqrt{2} = -2$?

$-2 = -2$ *True*

Check $x = \pm\sqrt{2 - \sqrt{2}}$:

$\left(2 - \sqrt{2}\right)^2 - 4\left(2 - \sqrt{2}\right) = -2$?

$4 - 4\sqrt{2} + 2 - 8 + 4\sqrt{2} = -2$?

$-2 = -2$ *True*

The solution set contains four numbers:

$\left\{\sqrt{2 + \sqrt{2}}, -\sqrt{2 + \sqrt{2}}, \sqrt{2 - \sqrt{2}}, -\sqrt{2 - \sqrt{2}}\right\}$

6. **(a)** $5(r + 3)^2 + 9(r + 3) = 2$

Let $x = r + 3$. The equation becomes:

$5x^2 + 9x = 2$

$5x^2 + 9x - 2 = 0$

$(5x - 1)(x + 2) = 0$

$5x - 1 = 0$ or $x + 2 = 0$

$x = \frac{1}{5}$ or $x = -2$

To find r, substitute $r + 3$ for x.

$r + 3 = \frac{1}{5}$ or $r + 3 = -2$

$r = -\frac{14}{5}$ or $r = -5$

Check $r = -\frac{14}{5}$: $\frac{1}{5} + \frac{9}{5} = 2$ *True*

Check $r = -5$: $20 - 18 = 2$ *True*

Solution set: $\left\{-5, -\frac{14}{5}\right\}$

(b) $4m^{2/3} = 3m^{1/3} + 1$

Let $x = m^{1/3}$, so $x^2 = \left(m^{1/3}\right)^2 = m^{2/3}$.

$4x^2 = 3x + 1$

$4x^2 - 3x - 1 = 0$

$(4x + 1)(x - 1) = 0$

$4x + 1 = 0$ or $x - 1 = 0$

$x = -\frac{1}{4}$ or $x = 1$

To find m, substitute $m^{1/3}$ for x.

$m^{1/3} = -\frac{1}{4}$ or $m^{1/3} = 1$

Cube each side of each equation.

$\left(m^{1/3}\right)^3 = \left(-\frac{1}{4}\right)^3$ or $\left(m^{1/3}\right)^3 = 1^3$

$m = -\frac{1}{64}$ or $m = 1$

Check $m = -\frac{1}{64}$: $\frac{1}{4} = -\frac{3}{4} + 1$ *True*

Check $m = 1$: $4 = 3 + 1$ *True*

Solution set: $\left\{-\frac{1}{64}, 1\right\}$

10.3 Section Exercises

1. $\frac{14}{x} = x - 5$

This is a rational equation, so multiply both sides by the LCD, x.

2. $\sqrt{1 + x} + x = 5$

This is a radical equation, so isolate the radical term on one side.

3. $\left(r^2 + r\right)^2 - 8\left(r^2 + r\right) + 12 = 0$

This is quadratic in form, so substitute a variable for $r^2 + r$.

4. $3t = \sqrt{16 - 10t}$

This is a radical equation, so square both sides to eliminate the radical.

5. The potential solution -1 does not check. The solution set is $\{4\}$.

6. The solutions stated are the values of u, $u = \frac{1}{2}$ or $u = 1$. We must give the values of the original variable, m. Substitute $m - 1$ for u.

$$u = \tfrac{1}{2} \quad \text{or} \quad u = 1$$
$$m - 1 = \tfrac{1}{2} \qquad m - 1 = 1$$
$$m = \tfrac{3}{2} \quad \text{or} \quad m = 2$$

Solution set: $\left\{\frac{3}{2}, 2\right\}$

7. $$1 - \frac{3}{x} - \frac{28}{x^2} = 0$$

Multiply by the LCD, x^2.

$$x^2(1) - x^2\left(\frac{3}{x}\right) - x^2\left(\frac{28}{x^2}\right) = x^2 \bullet 0$$
$$x^2 - 3x - 28 = 0$$
$$(x+4)(x-7) = 0$$
$$x + 4 = 0 \quad \text{or} \quad x - 7 = 0$$
$$x = -4 \quad \text{or} \quad x = 7$$

Check $x = -4$: $1 + \frac{3}{4} - \frac{7}{4} = 0$ *True*

Check $x = 7$: $1 - \frac{3}{7} - \frac{4}{7} = 0$ *True*

Solution set: $\{-4, 7\}$

8. $$4 - \frac{7}{r} - \frac{2}{r^2} = 0$$

$4r^2 - 7r - 2 = 0$ *Multiply by r^2.*

$$(4r+1)(r-2) = 0$$
$$4r + 1 = 0 \quad \text{or} \quad r - 2 = 0$$
$$r = -\tfrac{1}{4} \quad \text{or} \quad r = 2$$

Check $r = -\frac{1}{4}$: $4 + 28 - 32 = 0$ *True*

Check $r = 2$: $4 - \frac{7}{2} - \frac{1}{2} = 0$ *True*

Solution set: $\left\{-\frac{1}{4}, 2\right\}$

9. $$3 - \frac{1}{t} = \frac{2}{t^2}$$

Multiply each term by the LCD, t^2.

$$t^2(3) - t^2\left(\frac{1}{t}\right) = t^2\left(\frac{2}{t^2}\right)$$
$$3t^2 - t = 2$$
$$3t^2 - t - 2 = 0$$
$$(3t+2)(t-1) = 0$$
$$3t + 2 = 0 \quad \text{or} \quad t - 1 = 0$$
$$t = -\tfrac{2}{3} \quad \text{or} \quad t = 1$$

Check $t = -\frac{2}{3}$: $3 + \frac{3}{2} = \frac{9}{2}$ *True*

Check $t = 1$: $3 - 1 = 2$ *True*

Solution set: $\left\{-\frac{2}{3}, 1\right\}$

10. $$1 + \frac{2}{k} = \frac{3}{k^2}$$

Multiply by the LCD, k^2.

$$k^2\left(1 + \frac{2}{k}\right) = k^2\left(\frac{3}{k^2}\right)$$
$$k^2 + 2k = 3$$
$$k^2 + 2k - 3 = 0$$
$$(k+3)(k-1) = 0$$
$$k + 3 = 0 \quad \text{or} \quad k - 1 = 0$$
$$k = -3 \quad \text{or} \quad k = 1$$

Check $k = -3$: $1 - \frac{2}{3} = \frac{1}{3}$ *True*

Check $k = 1$: $1 + 2 = 3$ *True*

Solution set: $\{-3, 1\}$

11. $$\frac{1}{x} + \frac{2}{x+2} = \frac{17}{35}$$

Multiply by the LCD, $35x(x+2)$.

$$35x(x+2)\left(\frac{1}{x}\right) + 35x(x+2)\left(\frac{2}{x+2}\right) = 35x(x+2)\left(\frac{17}{35}\right)$$
$$35(x+2) + 35x(2) = 17x(x+2)$$
$$35x + 70 + 70x = 17x^2 + 34x$$
$$70 + 105x = 17x^2 + 34x$$
$$0 = 17x^2 - 71x - 70$$
$$0 = (17x+14)(x-5)$$
$$17x + 14 = 0 \quad \text{or} \quad x - 5 = 0$$
$$x = -\tfrac{14}{17} \quad \text{or} \quad x = 5$$

Check $x = -\frac{14}{17}$: $-\frac{17}{14} + \frac{17}{10} = \frac{17}{35}$ *True*

Check $x = 5$: $\frac{1}{5} + \frac{2}{7} = \frac{17}{35}$ *True*

Solution set: $\left\{-\frac{14}{17}, 5\right\}$

12. $$\frac{2}{m} + \frac{3}{m+9} = \frac{11}{4}$$

Multiply by the LCD, $4m(m+9)$.

$$4(m+9)(2) + 4m(3) = 11m(m+9)$$
$$8m + 72 + 12m = 11m^2 + 99m$$
$$0 = 11m^2 + 79m - 72$$
$$0 = (m+8)(11m-9)$$
$$m + 8 = 0 \quad \text{or} \quad 11m - 9 = 0$$
$$m = -8 \quad \text{or} \quad m = \tfrac{9}{11}$$

Check $m = -8$: $-\frac{1}{4} + 3 = \frac{11}{4}$ *True*

Check $m = \frac{9}{11}$: $\frac{22}{9} + \frac{11}{36} = \frac{11}{4}$ *True*

Solution set: $\left\{-8, \frac{9}{11}\right\}$

13. $\frac{2}{x+1}+\frac{3}{x+2}=\frac{7}{2}$

Multiply by the LCD, $2(x+1)(x+2)$.

$$2(x+1)(x+2)\left(\frac{2}{x+1}+\frac{3}{x+2}\right) = 2(x+1)(x+2)\left(\frac{7}{2}\right)$$

$$2(x+2)(2)+2(x+1)(3) = (x+1)(x+2)(7)$$
$$4x+8+6x+6=(x^2+3x+2)(7)$$
$$10x+14=7x^2+21x+14$$
$$7x^2+11x=0$$
$$x(7x+11)=0$$

$x=0$ or $7x+11=0$
$x=-\frac{11}{7}$

Check $x=-\frac{11}{7}$: $-\frac{7}{2}+7=\frac{7}{2}$ *True*

Check $x=0$: $2+\frac{3}{2}=\frac{7}{2}$ *True*

Solution set: $\{-\frac{11}{7},0\}$

14. $\frac{4}{3-p}+\frac{2}{5-p}=\frac{26}{15}$

Multiply by the LCD, $15(3-p)(5-p)$.

$$15(5-p)(4)+15(3-p)(2) = 26(3-p)(5-p)$$
$$60(5-p)+30(3-p)=26(15-8p+p^2)$$
$$300-60p+90-30p=390-208p+26p^2$$
$$26p^2-118p=0$$
$$2p(13p-59)=0$$

$2p=0$ or $13p-59=0$
$p=0$ or $p=\frac{59}{13}$

Check $p=0$: $\frac{4}{3}+\frac{2}{5}=\frac{26}{15}$ *True*

Check $p=\frac{59}{13}$: $-\frac{13}{5}+\frac{13}{3}=\frac{26}{15}$ *True*

Solution set: $\{0,\frac{59}{13}\}$

15. $\frac{3}{2x}-\frac{1}{2(x+2)}=1$

Multiply by the LCD, $2x(x+2)$.

$$2x(x+2)\left(\frac{3}{2x}-\frac{1}{2(x+2)}\right) = 2x(x+2)\cdot 1$$
$$3(x+2)-x(1)=2x(x+2)$$
$$3x+6-x=2x^2+4x$$
$$0=2x^2+2x-6$$
$$0=x^2+x-3$$

Use $a=1$, $b=1$, $c=-3$ in the quadratic formula.

$$x=\frac{-b\pm\sqrt{b^2-4ac}}{2a}$$
$$x=\frac{-1\pm\sqrt{1^2-4(1)(-3)}}{2(1)}$$
$$=\frac{-1\pm\sqrt{1+12}}{2}=\frac{-1\pm\sqrt{13}}{2}$$

Use a calculator to check both potential solutions. Both solutions check.

Solution set: $\left\{\frac{-1+\sqrt{13}}{2},\frac{-1-\sqrt{13}}{2}\right\}$

16. $\frac{4}{3x}-\frac{1}{2(x+1)}=1$

Multiply by the LCD, $6x(x+1)$.

$$2(x+1)(4)-3x(1)=6x(x+1)(1)$$
$$8x+8-3x=6x^2+6x$$
$$6x^2+x-8=0$$

Use $a=6$, $b=1$, and $c=-8$ in the quadratic formula.

$$x=\frac{-b\pm\sqrt{b^2-4ac}}{2a}$$
$$x=\frac{-1\pm\sqrt{193}}{2(6)}=\frac{-1\pm\sqrt{193}}{12}$$

Use a calculator to check both potential solutions. Both solutions check.

Solution set: $\left\{\frac{-1+\sqrt{193}}{12},\frac{-1-\sqrt{193}}{12}\right\}$

17. $\frac{6}{p}=2+\frac{p}{p+1}$

Multiply each term by the LCD, $p(p+1)$.

$$6(p+1)=2p(p+1)+p\cdot p$$
$$6p+6=2p^2+2p+p^2$$
$$0=3p^2-4p-6$$

Use $a=3$, $b=-4$, and $c=-6$ in the quadratic formula.

$$p=\frac{-b\pm\sqrt{b^2-4ac}}{2a}$$
$$p=\frac{-(-4)\pm\sqrt{(-4)^2-4(3)(-6)}}{2(3)}$$
$$=\frac{4\pm\sqrt{16+72}}{2(3)}=\frac{4\pm\sqrt{88}}{2(3)}$$
$$=\frac{4\pm2\sqrt{22}}{2(3)}=\frac{2\pm\sqrt{22}}{3}$$

Use a calculator to check both potential solutions. Both solutions check.

Solution set: $\left\{\frac{2+\sqrt{22}}{3}, \frac{2-\sqrt{22}}{3}\right\}$

18. $\frac{k}{2-k}+\frac{2}{k}=5$

Multiply by the LCD, $k(2-k)$.

$k \cdot k+2(2-k)=5k(2-k)$

$k^2+4-2k=10k-5k^2$

$6k^2-12k+4=0$

$3k^2-6k+2=0$

Use $a=3$, $b=-6$, and $c=2$ in the quadratic formula.

$$k=\frac{-b \pm \sqrt{b^2-4ac}}{2a}$$

$$k=\frac{-(-6) \pm \sqrt{(-6)^2-4(3)(2)}}{2(3)}$$

$$=\frac{6 \pm \sqrt{36-24}}{2(3)}=\frac{6 \pm \sqrt{12}}{2(3)}$$

$$=\frac{6 \pm 2\sqrt{3}}{2(3)}=\frac{3 \pm \sqrt{3}}{3}$$

Use a calculator to check both potential solutions. Both solutions check.

Solution set: $\left\{\frac{3+\sqrt{3}}{3}, \frac{3-\sqrt{3}}{3}\right\}$

19. Rate in still water: 20 mph
Rate of current: t mph

(a) When the boat travels upstream, the current works against the rate of the boat in still water, so the rate is $(20-t)$ mph.

(b) When the boat travels downstream, the current works with the rate of the boat in still water, so the rate is $(20+t)$ mph.

20. The grader's rate $=\frac{1 \text{ job}}{\text{time to finish}}$

$=\frac{1}{m}$ job per hour.

21. Let x = rate of the boat in still water.
With the speed of the current at 15 mph, then
$x-15$ = rate going upstream and
$x+15$ = rate going downstream.
Complete a table using the information in the problem, the rates given above, and the formula $d=rt$ or $t=\frac{d}{r}$.

	d	r	t
Upstream	4	$x-15$	$\frac{4}{x-15}$
Downstream	16	$x+15$	$\frac{16}{x+15}$

The time, 48 min, is written as $\frac{48}{60}=\frac{4}{5}$ hr. The time upstream plus the time downstream equals $\frac{4}{5}$. So, from the table, the equation is written as $\frac{4}{x-15}+\frac{16}{x+15}=\frac{4}{5}$.

Multiply by the LCD, $5(x-15)(x+15)$.

$$5(x-15)(x+15)\left(\frac{4}{x-15}+\frac{16}{x+15}\right)$$
$$=5(x-15)(x+15) \cdot \frac{4}{5}$$

$20(x+15)+80(x-15)$
$=4(x-15)(x+15)$

$20x+300+80x-1200$
$=4(x^2-225)$

$100x-900=4x^2-900$

$0=4x^2-100x$

$0=4x(x-25)$

$4x=0$ or $x-25=0$

$x=0$ or $x=25$

Reject $x=0$ mph as a possible boat speed. Yoshiaki's boat had a top speed of 25 mph.

22. Let x represent the speed of the plane in still air. Then the plane's speed with the wind is $x+15$ mph.
Similarly, the plane's speed against the wind is $x-15$ mph. Make a table. Use $t=\frac{d}{r}$.

	d	r	t
With the Wind	810	$x+15$	$\frac{810}{x+15}$
Against the Wind	720	$x-15$	$\frac{720}{x-15}$

Since the sum of the times with and against the wind is 6 hr,

$\frac{810}{x+15}+\frac{720}{x-15}=6.$

Multiply by the LCD, $(x+15)(x-15)$.

$810(x-15)+720(x+15)$
$=6(x^2-225)$

$810x-12{,}150+720x+10{,}800$
$=6x^2-1350$

$1530x-1350=6x^2-1350$

$6x^2-1530x=0$

$6x(x-255)=0$

continued

$6x = 0$ or $x - 255 = 0$
$x = 0$ or $x = 255$

Reject $x = 0$ for this problem. The plane's speed is 255 mph.

23. Let $x =$ Harry's average speed.
Then $x - 20 =$ Yoshi's average speed.

	d	r	t
Harry	300	x	$\frac{300}{x}$
Yoshi	300	$x - 20$	$\frac{300}{x-20}$

It takes Harry $1\frac{1}{4}$ or $\frac{5}{4}$ hours less time than Yoshi.

$$\frac{300}{x} = \frac{300}{x-20} - \frac{5}{4}$$

Multiply by the LCD, $4x(x - 20)$.

$$4x(x-20)\left(\frac{300}{x}\right) = 4x(x-20)\left(\frac{300}{x-20} - \frac{5}{4}\right)$$
$$1200(x - 20) = 4x(300) - x(x - 20) \cdot 5$$
$$1200x - 24{,}000 = 1200x - 5x^2 + 100x$$
$$5x^2 - 100x - 24{,}000 = 0$$
$$x^2 - 20x - 4800 = 0$$
$$(x - 80)(x + 60) = 0$$

$x - 80 = 0$ or $x + 60 = 0$
$x = 80$ or $x = -60$

Reject $x = -60$. Harry's average speed is 80 km/hr.

24. Let $x =$ Rico's speed from Jackson to Lodi.
Then $x + 10 =$ his speed from Lodi to Manteca.

Make a table. Use $t = \frac{d}{r}$.

	d	r	t
Jackson to Lodi	40	x	$\frac{40}{x}$
Lodi to Manteca	40	$x + 10$	$\frac{40}{x+10}$

Driving time for the entire trip was 88 minutes, or $\frac{88}{60} = \frac{22}{15}$ hours.

$$\frac{40}{x} + \frac{40}{x+10} = \frac{22}{15}$$

Multiply by the LCD, $15x(x + 10)$.

$$15x(x+10)\left(\frac{40}{x} + \frac{40}{x+10}\right)$$
$$= 15x(x+10)\left(\frac{22}{15}\right)$$

$$600(x + 10) + 600x = 22x(x + 10)$$
$$600x + 6000 + 600x = 22x^2 + 220x$$
$$0 = 22x^2 - 980x - 6000$$
$$0 = 11x^2 - 490x - 3000$$
$$0 = (11x + 60)(x - 50)$$

$11x + 60 = 0$ or $x - 50 = 0$
$x = -\frac{60}{11}$ or $x = 50$

Reject $-\frac{60}{11}$ since speed cannot be negative. Rico's speed from Jackson to Lodi is 50 mph.

25. Let x be the time in hours required for the faster person to cut the lawn. Then the slower person requires $x + 1$ hours.
Complete the chart.

	Rate	Time Working Together	Fractional Part of the Job Done
Faster Worker	$\frac{1}{x}$	2	$\frac{2}{x}$
Slower Worker	$\frac{1}{x+1}$	2	$\frac{2}{x+1}$

Part done by faster person + Part done by slower person = one whole job.

$$\frac{2}{x} + \frac{2}{x+1} = 1$$

Multiply each side by the LCD, $x(x + 1)$.

$$x(x+1)\left(\frac{2}{x} + \frac{2}{x+1}\right) = x(x+1) \cdot 1$$
$$2(x + 1) + 2x = x^2 + x$$
$$2x + 2 + 2x = x^2 + x$$
$$0 = x^2 - 3x - 2$$

Solve for x using the quadratic formula with $a = 1$, $b = -3$, and $c = -2$.

$$x = \frac{-(-3) \pm \sqrt{(-3)^2 - 4(1)(-2)}}{2(1)}$$
$$= \frac{3 \pm \sqrt{9+8}}{2} = \frac{3 \pm \sqrt{17}}{2}$$

$x = \frac{3 + \sqrt{17}}{2}$ or $x = \frac{3 - \sqrt{17}}{2}$
$x \approx 3.6$ or $x \approx -.6$

Discard $-.6$ as a solution since time cannot be negative.
It would take the faster person approximately 3.6 hours.

26. Let x be the time in hours required for the new worker (slower person) working alone to clean the building. It would take the faster person $x - 2$ hours working alone.
Complete the chart.

	Rate	Time Working Together	Fractional Part of the Job Done
Faster Worker	$\frac{1}{x-2}$	5	$\frac{5}{x-2}$
Slower Person	$\frac{1}{x}$	5	$\frac{5}{x}$

Part done by faster person + Part done by slower person = one whole job.

$$\frac{5}{x-2} + \frac{5}{x} = 1$$

Multiply each side by the LCD, $x(x-2)$.

$$x(x-2)\left(\frac{5}{x-2} + \frac{5}{x}\right) = x(x-2) \cdot 1$$
$$5x + 5(x-2) = x^2 - 2x$$
$$5x + 5x - 10 = x^2 - 2x$$
$$0 = x^2 - 12x + 10$$

Solve for x using the quadratic formula with $a = 1$, $b = -12$, and $c = 10$.

$$x = \frac{-(-12) \pm \sqrt{(-12)^2 - 4(1)(10)}}{2(1)}$$
$$= \frac{12 \pm \sqrt{104}}{2}$$

$$x = \frac{12 + \sqrt{104}}{2} \quad \text{or} \quad x = \frac{12 - \sqrt{104}}{2}$$
$$x \approx 11.1 \quad \text{or} \quad x \approx .9$$

The solution .9 makes no sense in the problem since a negative time results for the faster person. Therefore, the new worker (slower person) would require about 11.1 hours.

27. Let $x =$ the number of minutes it takes for the cold water tap alone to fill the washer.
$x + 9 =$ the number of minutes it takes for the hot water tap alone to fill the washer.

Working together, both taps can fill the washer in 6 minutes. Complete a chart using the above information.

Tap	Rate	Time	Fractional Part of Washer Filled
Cold	$\frac{1}{x}$	6	$\frac{6}{x}$
Hot	$\frac{1}{x+9}$	6	$\frac{6}{x+9}$

Since together the hot and cold taps fill one washer, the sum of their fractional parts is 1; that is,

$$\frac{6}{x} + \frac{6}{x+9} = 1.$$

Multiply by the LCD, $x(x+9)$.

$$x(x+9)\left(\frac{6}{x} + \frac{6}{x+9}\right) = x(x+9) \cdot 1$$
$$6(x+9) + 6x = x(x+9)$$
$$6x + 54 + 6x = x^2 + 9x$$
$$0 = x^2 - 3x - 54$$
$$0 = (x-9)(x+6)$$

$$x - 9 = 0 \quad \text{or} \quad x + 6 = 0$$
$$x = 9 \quad \text{or} \quad x = -6$$

Reject -6 as a possible time. The cold water tap can fill the washer in 9 minutes.

28. Let $x =$ the number of hours it takes for the faster pipe alone to fill the tank.
$x + 3 =$ the number of hours it takes for the slower pipe alone to fill the tank.

Working together, both pipes can fill the tank in 2 hours. Make a chart.

Pipe	Rate	Time	Fractional Part of Tank Filled
Faster	$\frac{1}{x}$	2	$\frac{2}{x}$
Slower	$\frac{1}{x+3}$	2	$\frac{2}{x+3}$

Since together the faster and slower pipes fill one tank, the sum of their fractional parts is 1; that is,

$$\frac{2}{x} + \frac{2}{x+3} = 1.$$

Multiply by the LCD, $x(x+3)$.

$$2(x+3) + 2(x) = x(x+3)$$
$$2x + 6 + 2x = x^2 + 3x$$
$$0 = x^2 - x - 6$$
$$0 = (x-3)(x+2)$$

$$x - 3 = 0 \quad \text{or} \quad x + 2 = 0$$
$$x = 3 \quad \text{or} \quad x = -2$$

Reject -2. The faster pipe takes 3 hours to fill the tank alone and the slower pipe takes 6 hours to fill the tank alone.

29.

$$z = \sqrt{5z-4}$$
$$(z)^2 = \left(\sqrt{5z-4}\right)^2$$
$$z^2 = 5z-4$$
$$z^2 - 5z + 4 = 0$$
$$(z-1)(z-4) = 0$$

$z - 1 = 0$ or $z - 4 = 0$
$z = 1$ or $z = 4$

Check $z = 1$: $1 = \sqrt{1}$ *True*

Check $z = 4$: $4 = \sqrt{16}$ *True*

Solution set: $\{1, 4\}$

30.

$$x = \sqrt{9x-14}$$
$$x^2 = \left(\sqrt{9x-14}\right)^2 \quad \textit{Square}$$
$$x^2 = 9x - 14$$
$$x^2 - 9x + 14 = 0$$
$$(x-2)(x-7) = 0$$

$x - 2 = 0$ or $x - 7 = 0$
$x = 2$ or $x = 7$

Check $x = 2$: $2 = \sqrt{4}$ *True*

Check $x = 7$: $7 = \sqrt{49}$ *True*

Solution set: $\{2, 7\}$

31.

$$2x = \sqrt{11x+3}$$
$$(2x)^2 = \left(\sqrt{11x+3}\right)^2$$
$$4x^2 = 11x + 3$$
$$4x^2 - 11x - 3 = 0$$
$$(4x+1)(x-3) = 0$$

$4x + 1 = 0$ or $x - 3 = 0$
$x = -\frac{1}{4}$ or $x = 3$

Check $x = -\frac{1}{4}$: $-\frac{1}{2} = \sqrt{\frac{1}{4}}$ *False*

Check $x = 3$: $6 = \sqrt{36}$ *True*

Solution set: $\{3\}$

32.

$$4x = \sqrt{6x+1}$$
$$(4x)^2 = \left(\sqrt{6x+1}\right)^2$$
$$16x^2 = 6x + 1$$
$$16x^2 - 6x - 1 = 0$$
$$(8x+1)(2x-1) = 0$$

$8x + 1 = 0$ or $2x - 1 = 0$
$x = -\frac{1}{8}$ or $x = \frac{1}{2}$

Check $x = -\frac{1}{8}$: $-\frac{1}{2} = \sqrt{\frac{1}{4}}$ *False*

Check $x = \frac{1}{2}$: $2 = \sqrt{4}$ *True*

Solution set: $\left\{\frac{1}{2}\right\}$

33.

$$3x = \sqrt{16-10x}$$
$$(3x)^2 = \left(\sqrt{16-10x}\right)^2$$
$$9x^2 = 16 - 10x$$
$$9x^2 + 10x - 16 = 0$$
$$(9x-8)(x+2) = 0$$

$9x - 8 = 0$ or $x + 2 = 0$
$x = \frac{8}{9}$ or $x = -2$

Check $x = \frac{8}{9}$: $\frac{8}{3} = \sqrt{\frac{64}{9}}$ *True*

Check $x = -2$: $-6 = \sqrt{36}$ *False*

Solution set: $\left\{\frac{8}{9}\right\}$

34.

$$4t = \sqrt{8t+3}$$
$$(4t)^2 = \left(\sqrt{8t+3}\right)^2$$
$$16t^2 = 8t + 3$$
$$16t^2 - 8t - 3 = 0$$
$$(4t-3)(4t+1) = 0$$

$4t - 3 = 0$ or $4t + 1 = 0$
$t = \frac{3}{4}$ or $t = -\frac{1}{4}$

Check $t = \frac{3}{4}$: $3 = \sqrt{9}$ *True*

Check $t = -\frac{1}{4}$: $-1 = \sqrt{1}$ *False*

Solution set: $\left\{\frac{3}{4}\right\}$

35.

$$p - 2\sqrt{p} = 8$$
$$p - 8 = 2\sqrt{p}$$
$$(p-8)^2 = \left(2\sqrt{p}\right)^2$$
$$p^2 - 16p + 64 = 4p$$
$$p^2 - 20p + 64 = 0$$
$$(p-4)(p-16) = 0$$

$p - 4 = 0$ or $p - 16 = 0$
$p = 4$ or $p = 16$

Check $p = 4$: $4 - 4 = 8$ *False*

Check $p = 16$: $16 - 8 = 8$ *True*

Solution set: $\{16\}$

36.

$$k + \sqrt{k} = 12$$
$$\sqrt{k} = 12 - k$$
$$\left(\sqrt{k}\right)^2 = (12-k)^2$$
$$k = 144 - 24k + k^2$$
$$k^2 - 25k + 144 = 0$$
$$(k-9)(k-16) = 0$$

$k - 9 = 0$ or $k - 16 = 0$
$k = 9$ or $k = 16$

Check $k = 9$: $9 + 3 = 12$ *True*

Check $k = 16$: $16 + 4 = 12$ *False*

Solution set: $\{9\}$

37.
$$m = \sqrt{\frac{6 - 13m}{5}}$$
$$m^2 = \frac{6 - 13m}{5}$$
$$5m^2 = 6 - 13m$$
$$5m^2 + 13m - 6 = 0$$
$$(5m - 2)(m + 3) = 0$$

$5m - 2 = 0$ or $m + 3 = 0$

$m = \frac{2}{5}$ or $m = -3$

Check $m = \frac{2}{5}$: $\frac{2}{5} = \sqrt{\frac{4}{25}}$ *True*

Check $m = -3$: $-3 = \sqrt{9}$ *False*

Solution set: $\{\frac{2}{5}\}$

38.
$$r = \sqrt{\frac{20 - 19r}{6}}$$
$$r^2 = \frac{20 - 19r}{6}$$
$$6r^2 = 20 - 19r$$
$$6r^2 + 19r - 20 = 0$$
$$(r + 4)(6r - 5) = 0$$

$r + 4 = 0$ or $6r - 5 = 0$

$r = -4$ or $r = \frac{5}{6}$

Check $r = -4$: $-4 = \sqrt{16}$ *False*

Check $r = \frac{5}{6}$: $\frac{5}{6} = \sqrt{\frac{25}{36}}$ *True*

Solution set: $\{\frac{5}{6}\}$

39. $t^4 - 18t^2 + 81 = 0$

Let $u = t^2$, so $u^2 = t^4$. The equation becomes
$$u^2 - 18u + 81 = 0.$$
$$(u - 9)^2 = 0$$
$$u - 9 = 0$$
$$u = 9$$

To find t, substitute t^2 for u.
$$t^2 = 9$$

$t = 3$ or $t = -3$

Check $t = \pm 3$: $81 - 162 + 81 = 0$ *True*

Solution set: $\{-3, 3\}$

40. $x^4 - 8x^2 + 16 = 0$

Let $u = x^2$, so $u^2 = x^4$.
$$u^2 - 8u + 16 = 0$$
$$(u - 4)^2 = 0$$
$$u - 4 = 0$$
$$u = 4$$
$$x^2 = 4$$
$$x = \pm 2$$

Check $x = \pm 2$: $16 - 32 + 16 = 0$ *True*

Solution set: $\{-2, 2\}$

41. $4k^4 - 13k^2 + 9 = 0$

Let $u = k^2$ and $u^2 = k^4$ to get
$$4u^2 - 13u + 9 = 0$$
$$(4u - 9)(u - 1) = 0$$

$4u - 9 = 0$ or $u - 1 = 0$

$u = \frac{9}{4}$ or $u = 1.$

To find k, substitute k^2 for u.

$k^2 = \frac{9}{4}$ or $k^2 = 1$

$k = \pm \frac{3}{2}$ or $k = \pm 1$

Check $k = \pm \frac{3}{2}$: $\frac{81}{4} - \frac{117}{4} + 9 = 0$ *True*

Check $k = \pm 1$: $4 - 13 + 9 = 0$ *True*

Solution set: $\{-\frac{3}{2}, -1, 1, \frac{3}{2}\}$

42. $9x^4 - 25x^2 + 16 = 0$

Let $u = x^2$, so $u^2 = x^4$.
$$9u^2 - 25u + 16 = 0$$
$$(u - 1)(9u - 16) = 0$$

$u - 1 = 0$ or $9u - 16 = 0$

$u = 1$ or $u = \frac{16}{9}$

$x^2 = 1$ or $x^2 = \frac{16}{9}$

$x = \pm\sqrt{1}$ or $x = \pm\sqrt{\frac{16}{9}}$

$x = \pm 1$ or $x = \pm\frac{4}{3}$

Check $x = \pm 1$: $9 - 25 + 16 = 0$ *True*

Check $x = \pm \frac{4}{3}$: $\frac{256}{9} - \frac{400}{9} + 16 = 0$ *True*

Solution set: $\{-\frac{4}{3}, -1, 1, \frac{4}{3}\}$

43. $x^4 + 48 = 16x^2$

$x^4 - 16x^2 + 48 = 0$

Let $u = x^2$, so $u^2 = x^4$. The equation becomes

$$u^2 - 16u + 48 = 0.$$
$$(u-4)(u-12) = 0$$
$$u - 4 = 0 \quad \text{or} \quad u - 12 = 0$$
$$u = 4 \quad \text{or} \quad u = 12$$

To find x substitute x^2 for u.

$$x^2 = 4 \quad \text{or} \quad x^2 = 12$$
$$x = \pm\sqrt{4} \quad \text{or} \quad x = \pm\sqrt{12}$$
$$x = \pm 2 \quad \text{or} \quad x = \pm 2\sqrt{3}$$

Check $x = \pm 2$: $16 + 48 = 64$ *True*

Check $x = \pm 2\sqrt{3}$: $144 + 48 = 192$ *True*

Solution set: $\left\{-2\sqrt{3}, -2, 2, 2\sqrt{3}\right\}$

44. $z^4 = 17z^2 - 72$

$z^4 - 17z^2 + 72 = 0$

Let $u = z^2$, so $u^2 = z^4$.

$$u^2 - 17u + 72 = 0$$
$$(u-8)(u-9) = 0$$
$$u = 8 \quad \text{or} \quad u = 9$$

To find z, substitute z^2 for u.

$$z^2 = 8 \quad \text{or} \quad z^2 = 9$$
$$z = \pm 2\sqrt{2} \quad \text{or} \quad z = \pm 3$$

Check $z = \pm 3$: $81 = 153 - 72$ *True*

Check $z = \pm 2\sqrt{2}$: $64 = 136 - 72$ *True*

Solution set: $\left\{-3, -2\sqrt{2}, 2\sqrt{2}, 3\right\}$

45. $2x^4 - 9x^2 = -2$

$2x^4 - 9x^2 + 2 = 0$

Let $u = x^2$, so $u^2 = x^4$.

$2u^2 - 9u + 2 = 0$

Use $a = 2$, $b = -9$, and $c = 2$ in the quadratic formula.

$$u = \frac{-b \pm \sqrt{b^2 - 4ac}}{2a}$$
$$u = \frac{-(-9) \pm \sqrt{(-9)^2 - 4(2)(2)}}{2(2)}$$
$$= \frac{9 \pm \sqrt{81 - 16}}{4}$$
$$= \frac{9 \pm \sqrt{65}}{4}$$

To find x, substitute x^2 for u.

$$x^2 = \frac{9 \pm \sqrt{65}}{4}$$
$$x = \pm\sqrt{\frac{9 \pm \sqrt{65}}{4}}$$
$$= \pm\frac{\sqrt{9 \pm \sqrt{65}}}{2}$$

Note: the last expression represents four numbers. All four potential solutions check.

Solution set: $\left\{\frac{\sqrt{9+\sqrt{65}}}{2}, -\frac{\sqrt{9+\sqrt{65}}}{2}, \frac{\sqrt{9-\sqrt{65}}}{2}, -\frac{\sqrt{9-\sqrt{65}}}{2}\right\}$

46. $8x^4 + 1 = 11x^2$

$8x^4 - 11x^2 + 1 = 0$

Let $u = x^2$, so $u^2 = x^4$.

$8u^2 - 11u + 1 = 0$

Use $a = 8$, $b = -11$, and $c = 1$ in the quadratic formula.

$$u = \frac{-b \pm \sqrt{b^2 - 4ac}}{2a}$$
$$u = \frac{-(-11) \pm \sqrt{(-11)^2 - 4(8)(1)}}{2(8)}$$
$$= \frac{11 \pm \sqrt{121 - 32}}{16}$$
$$= \frac{11 \pm \sqrt{89}}{16}$$

To find x, substitute x^2 for u.

$$x^2 = \frac{11 \pm \sqrt{89}}{16}$$
$$x = \pm\sqrt{\frac{11 \pm \sqrt{89}}{16}}$$
$$= \pm\frac{\sqrt{11 \pm \sqrt{89}}}{4}$$

All four potential solutions check.

Solution set: $\left\{\frac{\sqrt{11+\sqrt{89}}}{4}, -\frac{\sqrt{11+\sqrt{89}}}{4}, \frac{\sqrt{11-\sqrt{89}}}{4}, -\frac{\sqrt{11-\sqrt{89}}}{4}\right\}$

47. $(x+3)^2+5(x+3)+6=0$

Let $u=x+3$, so $u^2=(x+3)^2$.

$$u^2+5u+6=0$$
$$(u+3)(u+2)=0$$

$u+3=0$ or $u+2=0$

$u=-3$ or $u=-2$

To find x, substitute $x+3$ for u.

$x+3=-3$ or $x+3=-2$

$x=-6$ or $x=-5$

Check $x=-6$: $9-15+6=0$ *True*

Check $x=-5$: $4-10+6=0$ *True*

Solution set: $\{-6,-5\}$

48. $(k-4)^2+(k-4)-20=0$

Let $x=k-4$, so $x^2=(k-4)^2$.

$$x^2+x-20=0$$
$$(x+5)(x-4)=0$$

$x+5=0$ or $x-4=0$

$x=-5$ or $x=4$

$k-4=-5$ or $k-4=4$

$k=-1$ or $k=8$

Check $k=-1$: $25-5-20=0$ *True*

Check $k=8$: $16+4-20=0$ *True*

Solution set: $\{-1,8\}$

49. $(t+5)^2+6=7(t+5)$

Let $u=t+5$, so $u^2=(t+5)^2$.

$$u^2+6=7u$$
$$u^2-7u+6=0$$
$$(u-6)(u-1)=0$$

$u-6=0$ or $u-1=0$

$u=6$ or $u=1$

To find t, substitute $t+5$ for u.

$t+5=6$ or $t+5=1$

$t=1$ or $t=-4$

Check $t=1$: $36+6=42$ *True*

Check $t=-4$: $1+6=7$ *True*

Solution set: $\{-4,1\}$

50. $3(m+4)^2-8=2(m+4)$

Let $x=m+4$, so $x^2=(m+4)^2$.

$$3x^2-8=2x$$
$$3x^2-2x-8=0$$
$$(3x+4)(x-2)=0$$

$3x+4=0$ or $x-2=0$

$x=-\frac{4}{3}$ or $x=2$

$m+4=-\frac{4}{3}$ or $m+4=2$

$m=-\frac{16}{3}$ or $m=-2$

Check $m=-\frac{16}{3}$: $\frac{16}{3}-8=-\frac{8}{3}$ *True*

Check $m=-2$: $12-8=4$ *True*

Solution set: $\{-\frac{16}{3},-2\}$

51. $$2+\frac{5}{3k-1}=\frac{-2}{(3k-1)^2}$$

Let $u=3k-1$, so $u^2=(3k-1)^2$.

$$2+\frac{5}{u}=-\frac{2}{u^2}$$

Multiply by the LCD, u^2.

$$u^2\left(2+\frac{5}{u}\right)=u^2\left(-\frac{2}{u^2}\right)$$
$$2u^2+5u=-2$$
$$2u^2+5u+2=0$$
$$(2u+1)(u+2)=0$$

$2u+1=0$ or $u+2=0$

$u=-\frac{1}{2}$ or $u=-2$

To find k, substitute $3k-1$ for u.

$3k-1=-\frac{1}{2}$ or $3k-1=-2$

$3k=\frac{1}{2}$ or $3k=-1$

$k=\frac{1}{6}$ or $k=-\frac{1}{3}$

Check $k=\frac{1}{6}$: $2-10=-8$ *True*

Check $k=-\frac{1}{3}$: $2-\frac{5}{2}=-\frac{1}{2}$ *True*

Solution set: $\{-\frac{1}{3},\frac{1}{6}\}$

52. $$3-\frac{7}{2p+2}=\frac{6}{(2p+2)^2}$$

Let $x=2p+2$, so $x^2=(2p+2)^2$.

$$3-\frac{7}{x}=\frac{6}{x^2}$$

Multiply by the LCD, x^2.

$$3x^2-7x=6$$
$$3x^2-7x-6=0$$
$$(3x+2)(x-3)=0$$

$3x+2=0$ or $x-3=0$

$x=-\frac{2}{3}$ or $x=3$

$2p+2=-\frac{2}{3}$ or $2p+2=3$

$2p=-\frac{8}{3}$ or $2p=1$

$p=-\frac{4}{3}$ or $p=\frac{1}{2}$

Check $p=-\frac{4}{3}$: $3+\frac{21}{2}=\frac{27}{2}$ *True*

Check $p=\frac{1}{2}$: $3-\frac{7}{3}=\frac{2}{3}$ *True*

Solution set: $\{-\frac{4}{3},\frac{1}{2}\}$

53. $x^{2/3} + x^{1/3} - 2 = 0$

Let $u = x^{1/3}$, so $u^2 = x^{2/3}$.

$u^2 + u - 2 = 0$

$(u + 2)(u - 1) = 0$

$u + 2 = 0$ or $u - 1 = 0$

$u = -2$ or $u = 1$

To find x, substitute $x^{1/3}$ for u.

$x^{1/3} = -2$ or $x^{1/3} = 1$

Cube both sides of each equation.

$(x^{1/3})^3 = (-2)^3$ or $(x^{1/3})^3 = 1^3$

$x = -8$ or $x = 1$

Check $x = -8$: $4 - 2 - 2 = 0$ *True*

Check $x = 1$: $1 + 1 - 2 = 0$ *True*

Solution set: $\{-8, 1\}$

54. $x^{2/3} - 2x^{1/3} - 3 = 0$

Let $u = x^{1/3}$, so $u^2 = x^{2/3}$.

$u^2 - 2u - 3 = 0$

$(u - 3)(u + 1) = 0$

$u - 3 = 0$ or $u + 1 = 0$

$u = 3$ or $u = -1$

$x^{1/3} = 3$ or $x^{1/3} = -1$

$(x^{1/3})^3 = 3^3$ or $(x^{1/3})^3 = (-1)^3$

$x = 27$ or $x = -1$

Check $x = -1$: $1 + 2 - 3 = 0$ *True*

Check $x = 27$: $9 - 6 - 3 = 0$ *True*

Solution set: $\{-1, 27\}$

55. $r^{2/3} + r^{1/3} - 12 = 0$

Let $u = r^{1/3}$, so $u^2 = r^{2/3}$. The equation becomes

$u^2 + u - 12 = 0.$

$(u + 4)(u - 3) = 0$

$u + 4 = 0$ or $u - 3 = 0$

$u = -4$ or $u = 3$

To find r, substitute $r^{1/3}$ for u.

$r^{1/3} = -4$ or $r^{1/3} = 3$

$(r^{1/3})^3 = (-4)^3$ or $(r^{1/3})^3 = 3^3$

$r = -64$ or $r = 27$

Check $r = -64$: $16 - 4 - 12 = 0$ *True*

Check $r = 27$: $9 + 3 - 12 = 0$ *True*

Solution set: $\{-64, 27\}$

56. $3x^{2/3} - x^{1/3} - 24 = 0$

Let $y = x^{1/3}$, so $y^2 = x^{2/3}$.

$3y^2 - y - 24 = 0$

$(3y + 8)(y - 3) = 0$

$3y + 8 = 0$ or $y - 3 = 0$

$y = -\frac{8}{3}$ $\quad y = 3$

$x^{1/3} = -\frac{8}{3}$ or $x^{1/3} = 3$

$(x^{1/3})^3 = \left(-\frac{8}{3}\right)^3$ $\quad (x^{1/3})^3 = 3^3$

$x = -\frac{512}{27}$ $\quad x = 27$

Check $x = -\frac{512}{27}$: $\frac{64}{3} + \frac{8}{3} - 24 = 0$ *True*

Check $x = 27$: $27 - 3 - 24 = 0$ *True*

Solution set: $\left\{-\frac{512}{27}, 27\right\}$

57. $2\left(1 + \sqrt{r}\right)^2 = 13\left(1 + \sqrt{r}\right) - 6$

Let $u = 1 + \sqrt{r}$.

$2u^2 = 13u - 6$

$2u^2 - 13u + 6 = 0$

$(2u - 1)(u - 6) = 0$

$2u - 1 = 0$ or $u - 6 = 0$

$u = \frac{1}{2}$ or $u = 6$

Replace u with $1 + \sqrt{r}$.

$1 + \sqrt{r} = \frac{1}{2}$ or $1 + \sqrt{r} = 6$

$\sqrt{r} = -\frac{1}{2}$ or $\sqrt{r} = 5$

Not possible, $\quad r = 25$

since $\sqrt{r} \geq 0$.

Check $r = 25$: $72 = 78 - 6$ *True*

Solution set: $\{25\}$

58. $\left(k^2 + k\right)^2 + 12 = 8\left(k^2 + k\right)$

Let $x = k^2 + k$, so $x^2 = \left(k^2 + k\right)^2$.

$x^2 + 12 = 8x$

$x^2 - 8x + 12 = 0$

$(x - 2)(x - 6) = 0$

$x - 2 = 0$ or $x - 6 = 0$

$x = 2$ $\quad x = 6$

Replace x with $k^2 + k$.

$k^2 + k = 2$ or $k^2 + k = 6$

$k^2 + k - 2 = 0$ $\quad k^2 + k - 6 = 0$

$(k + 2)(k - 1) = 0$ $\quad (k + 3)(k - 2) = 0$

$k + 2 = 0$ or $k - 1 = 0$

$k = -2$ or $k = 1$

$$k + 3 = 0 \quad \text{or} \quad k - 2 = 0$$
$$k = -3 \quad \text{or} \quad k = 2$$

Check $k = -3$: $36 + 12 = 48$ *True*

Check $k = -2$: $4 + 12 = 16$ *True*

Check $k = 1$: $4 + 12 = 16$ *True*

Check $k = 2$: $36 + 12 = 48$ *True*

Solution set: $\{-3, -2, 1, 2\}$

For Exercises 59–64, use the equation

$$\frac{x^2}{(x-3)^2} + \frac{3x}{x-3} - 4 = 0.$$

59. Substituting 3 for x would cause both denominators to be 0, and division by 0 is undefined.

60.
$$(x-3)^2\left(\frac{x^2}{(x-3)^2}\right) + (x-3)^2\left(\frac{3x}{x-3}\right) - (x-3)^2 \bullet 4 = (x-3)^2 \bullet 0$$
$$x^2 + 3x(x-3) - 4(x^2 - 6x + 9) = 0$$
$$x^2 + 3x^2 - 9x - 4x^2 + 24x - 36 = 0$$
$$15x - 36 = 0$$
$$15x = 36$$
$$x = \tfrac{36}{15} = \tfrac{12}{5}$$

The solution is $\frac{12}{5}$.

61.
$$\frac{x^2}{(x-3)^2} + \frac{3x}{x-3} - 4 = 0$$
$$\left(\frac{x}{x-3}\right)^2 + 3\left(\frac{x}{x-3}\right) - 4 = 0$$

62. If a fraction is equal to 1, then the numerator must be equal to the denominator. But the numerator can never equal the denominator, since the denominator is 3 less than the numerator.

63. From Exercise 61,

$$\left(\frac{x}{x-3}\right)^2 + 3\left(\frac{x}{x-3}\right) - 4 = 0.$$

Let $t = \frac{x}{x-3}$, so $t^2 = \left(\frac{x}{x-3}\right)^2$.

$$t^2 + 3t - 4 = 0$$
$$(t-1)(t+4) = 0$$
$$t - 1 = 0 \quad \text{or} \quad t + 4 = 0$$
$$t = 1 \quad \text{or} \quad t = -4$$

To find x, substitute $\frac{x}{x-3}$ for t.

$$\frac{x}{x-3} = 1 \quad \text{or} \quad \frac{x}{x-3} = -4$$

The equation $\frac{x}{x-3} = 1$ has no solution since there is no value of x for which $x = x - 3$. (See Exercise 62.) Therefore, $t = 1$ is impossible.

$$\frac{x}{x-3} = -4$$
$$x = -4(x-3)$$
$$x = -4x + 12$$
$$5x = 12$$
$$x = \tfrac{12}{5}$$

Solution set: $\left\{\frac{12}{5}\right\}$

64. $x^2(x-3)^{-2} + 3x(x-3)^{-1} - 4 = 0$

Let $s = (x-3)^{-1}$, so $s^2 = (x-3)^{-2}$.

$$x^2s^2 + 3xs - 4 = 0$$
$$(xs + 4)(xs - 1) = 0$$
$$xs + 4 = 0 \quad \text{or} \quad xs - 1 = 0$$
$$s = -\frac{4}{x} \quad \text{or} \quad s = \frac{1}{x}$$

To find x, substitute $\frac{1}{x-3}$ for s.

$$\frac{1}{x-3} = -\frac{4}{x} \quad \text{or} \quad \frac{1}{x-3} = \frac{1}{x}$$

$$x = -4(x-3)$$
$$x = -4x + 12 \qquad\qquad x = x - 3$$
$$5x = 12 \qquad\qquad 0 = -3 \quad \textit{False}$$
$$x = \frac{12}{5}$$

Thus, $s = \frac{1}{x}$ is impossible.

Solution set: $\left\{\frac{12}{5}\right\}$

Summary Exercises on Solving Quadratic Equations

1. $(2x + 3)^2 = 4$

Since the equation has the form $(ax + b)^2 = c$, use the *square root property*.

2.
$$4x^2 - 3x = 1$$
$$4x^2 - 3x - 1 = 0$$

The discriminant is

$$b^2 - 4ac = (-3)^2 - 4(4)(-1)$$
$$= 9 + 16 = 25.$$

Since the discriminant is a perfect square, use *factoring*.

3. $z^2 + 5z - 8 = 0$

The discriminant is

$$\begin{aligned} b^2 - 4ac &= 5^2 - 4(1)(-8) \\ &= 25 + 32 = 57. \end{aligned}$$

Since the discriminant is not a perfect square, use the *quadratic formula.*

4. $2k^2 + 3k = 1$

$2k^2 + 3k - 1 = 0$

The discriminant is

$$\begin{aligned} b^2 - 4ac &= 3^2 - 4(2)(-1) \\ &= 9 + 8 = 17. \end{aligned}$$

Since the discriminant is not a perfect square, use the *quadratic formula.*

5. $3m^2 = 2 - 5m$

$3m^2 + 5m - 2 = 0$

The discriminant is

$$\begin{aligned} b^2 - 4ac &= 5^2 - 4(3)(-2) \\ &= 25 + 24 = 49. \end{aligned}$$

Since the discriminant is a perfect square, use *factoring.*

6. $p^2 = 5$

Since the equation has the form $x^2 = c$, use the *square root property.*

7. $p^2 = 47$

$p = \sqrt{47}$ or $p = -\sqrt{47}$

Solution set: $\{\sqrt{47}, -\sqrt{47}\}$

8. $6x^2 - x - 15 = 0$

$(2x + 3)(3x - 5) = 0$

$2x + 3 = 0$ or $3x - 5 = 0$

$x = -\frac{3}{2}$ or $x = \frac{5}{3}$

Solution set: $\{-\frac{3}{2}, \frac{5}{3}\}$

9. $n^2 + 8n + 6 = 0$

$n^2 + 8n = -6$

$n^2 + 8n + 16 = -6 + 16$ $\quad [\frac{1}{2}(8)]^2 = 16$

$(n + 4)^2 = 10$

$n + 4 = \sqrt{10}$ or $n + 4 = -\sqrt{10}$

$n = -4 + \sqrt{10}$ or $n = -4 - \sqrt{10}$

Solution set: $\{-4 + \sqrt{10}, -4 - \sqrt{10}\}$

10. $(x - 4)^2 = 49$

$x - 4 = \sqrt{49}$ or $x - 4 = -\sqrt{49}$

$x = 4 + 7$ or $x = 4 - 7$

$x = 11$ or $x = -3$

Solution set: $\{-3, 11\}$

11. $\frac{9}{m} + \frac{5}{m^2} = 2$

Multiply by the LCD, m^2.

$9m + 5 = 2m^2$

$0 = 2m^2 - 9m - 5$

$0 = (2m + 1)(m - 5)$

$2m + 1 = 0$ or $m - 5 = 0$

$m = -\frac{1}{2}$ or $m = 5$

Solution set: $\{-\frac{1}{2}, 5\}$

12. $3m^2 = 3 - 8m$

$3m^2 + 8m - 3 = 0$

$(m + 3)(3m - 1) = 0$

$m + 3 = 0$ or $3m - 1 = 0$

$m = -3$ or $m = \frac{1}{3}$

Solution set: $\{-3, \frac{1}{3}\}$

13. $3x^2 - 9x + 4 = 0$

Use $a = 3$, $b = -9$, and $c = 4$ in the quadratic formula.

$$\begin{aligned} x &= \frac{-b \pm \sqrt{b^2 - 4ac}}{2a} \\ x &= \frac{-(-9) \pm \sqrt{(-9)^2 - 4(3)(4)}}{2(3)} \\ &= \frac{9 \pm \sqrt{81 - 48}}{6} \\ &= \frac{9 \pm \sqrt{33}}{6} \end{aligned}$$

Solution set: $\left\{\frac{9 + \sqrt{33}}{6}, \frac{9 - \sqrt{33}}{6}\right\}$

14. $x^2 = -12$

$x = \sqrt{-12}$ or $x = -\sqrt{-12}$

$x = i\sqrt{12}$ $\quad x = -i\sqrt{12}$

$x = 2i\sqrt{3}$ or $x = -2i\sqrt{3}$

Solution set: $\{2i\sqrt{3}, -2i\sqrt{3}\}$

15. $x\sqrt{2} = \sqrt{5x - 2}$

$(x\sqrt{2})^2 = (\sqrt{5x - 2})^2$

$x^2 \cdot 2 = 5x - 2$

$2x^2 - 5x + 2 = 0$

$(2x - 1)(x - 2) = 0$

$$2x - 1 = 0 \quad \text{or} \quad x - 2 = 0$$
$$x = \tfrac{1}{2} \qquad\qquad x = 2$$

Check $x = \frac{1}{2}$: $\frac{1}{2}\sqrt{2} = \sqrt{\frac{1}{2}}$ *True*

Check $x = 2$: $2\sqrt{2} = \sqrt{8}$ *True*

Solution set: $\{\frac{1}{2}, 2\}$

16. $12x^4 - 11x^2 + 2 = 0$

Let $u = x^2$, so $u^2 = x^4$.

$$12u^2 - 11u + 2 = 0$$
$$(4u - 1)(3u - 2) = 0$$
$$4u - 1 = 0 \quad \text{or} \quad 3u - 2 = 0$$
$$u = \tfrac{1}{4} \quad \text{or} \quad u = \tfrac{2}{3}$$
$$x^2 = \tfrac{1}{4} \quad \text{or} \quad x^2 = \tfrac{2}{3}$$
$$x = \pm\sqrt{\tfrac{1}{4}} \quad \text{or} \quad x = \pm\sqrt{\tfrac{2}{3}}$$
$$x = \pm\frac{1}{2} \quad \text{or} \quad x = \pm\frac{\sqrt{2}}{\sqrt{3}} \cdot \frac{\sqrt{3}}{\sqrt{3}}$$
$$= \pm\frac{\sqrt{6}}{3}$$

Solution set: $\left\{-\frac{\sqrt{6}}{3}, -\frac{1}{2}, \frac{1}{2}, \frac{\sqrt{6}}{3}\right\}$

17. $(2k + 5)^2 = 12$

$$2k + 5 = \sqrt{12} \quad \text{or} \quad 2k + 5 = -\sqrt{12}$$
$$2k = -5 + 2\sqrt{3} \quad \text{or} \quad 2k = -5 - 2\sqrt{3}$$
$$k = \frac{-5 + 2\sqrt{3}}{2} \quad \text{or} \quad k = \frac{-5 - 2\sqrt{3}}{2}$$

Solution set: $\left\{\frac{-5 + 2\sqrt{3}}{2}, \frac{-5 - 2\sqrt{3}}{2}\right\}$

18. $$\frac{2}{x} + \frac{1}{x - 2} - \frac{5}{3} = 0$$

Multiply by the LCD, $3x(x - 2)$.

$$2 \cdot 3(x - 2) + 1 \cdot 3x - 5 \cdot x(x - 2) = 0$$
$$6x - 12 + 3x - 5x^2 + 10x = 0$$
$$-5x^2 + 19x - 12 = 0$$
$$5x^2 - 19x + 12 = 0$$
$$(5x - 4)(x - 3) = 0$$

$$5x - 4 = 0 \quad \text{or} \quad x - 3 = 0$$
$$x = \tfrac{4}{5} \quad \text{or} \quad x = 3$$

Check $x = \frac{4}{5}$: $\frac{5}{2} - \frac{5}{6} - \frac{5}{3} = 0$ *True*

Check $x = 3$: $\frac{2}{3} + 1 - \frac{5}{3} = 0$ *True*

Solution set: $\{\frac{4}{5}, 3\}$

19. $$t^4 + 14 = 9t^2$$
$$t^4 - 9t^2 + 14 = 0$$
$$(t^2 - 2)(t^2 - 7) = 0$$
$$t^2 - 2 = 0 \quad \text{or} \quad t^2 - 7 = 0$$
$$t^2 = 2 \qquad\qquad t^2 = 7$$
$$t = \pm\sqrt{2} \quad \text{or} \quad t = \pm\sqrt{7}$$

Solution set: $\{-\sqrt{7}, -\sqrt{2}, \sqrt{2}, \sqrt{7}\}$

20. $$2x^2 + 4x = 5$$
$$2x^2 + 4x - 5 = 0$$

Use $a = 2$, $b = 4$, and $c = -5$ in the quadratic formula.

$$x = \frac{-b \pm \sqrt{b^2 - 4ac}}{2a}$$
$$x = \frac{-4 \pm \sqrt{4^2 - 4(2)(-5)}}{2(2)}$$
$$= \frac{-4 \pm \sqrt{16 + 40}}{2(2)} = \frac{-4 \pm \sqrt{56}}{2(2)}$$
$$= \frac{-4 \pm 2\sqrt{14}}{2(2)} = \frac{-2 \pm \sqrt{14}}{2}$$

Solution set: $\left\{\frac{-2 + \sqrt{14}}{2}, \frac{-2 - \sqrt{14}}{2}\right\}$

21. $z^2 + z + 2 = 0$

Use $a = 1$, $b = 1$, and $c = 2$ in the quadratic formula.

$$z = \frac{-b \pm \sqrt{b^2 - 4ac}}{2a}$$
$$z = \frac{-1 \pm \sqrt{1^2 - 4(1)(2)}}{2(1)}$$
$$= \frac{-1 \pm \sqrt{1 - 8}}{2} = \frac{-1 \pm \sqrt{-7}}{2}$$
$$= \frac{-1 \pm i\sqrt{7}}{2}$$

Solution set: $\left\{\frac{-1 + i\sqrt{7}}{2}, \frac{-1 - i\sqrt{7}}{2}\right\}$

22. $$x^4 - 8x^2 = -1$$
$$x^4 - 8x^2 + 1 = 0$$

Let $u = x^2$, so $u^2 = x^4$.

$$u^2 - 8u + 1 = 0$$

Use $a = 1$, $b = -8$, and $c = 1$ in the quadratic formula.

continued

$$u = \frac{-b \pm \sqrt{b^2 - 4ac}}{2a}$$

$$u = \frac{-(-8) \pm \sqrt{(-8)^2 - 4(1)(1)}}{2(1)}$$

$$= \frac{8 \pm \sqrt{64 - 4}}{2} = \frac{8 \pm \sqrt{60}}{2}$$

$$= \frac{8 \pm 2\sqrt{15}}{2} = 4 \pm \sqrt{15}$$

$$x^2 = 4 \pm \sqrt{15}$$

$$x = \pm\sqrt{4 \pm \sqrt{15}}$$

Solution set: $\left\{\sqrt{4+\sqrt{15}}, -\sqrt{4+\sqrt{15}}, \sqrt{4-\sqrt{15}}, -\sqrt{4-\sqrt{15}}\right\}$

23.
$$4t^2 - 12t + 9 = 0$$
$$(2t-3)(2t-3) = 0$$
$$(2t-3)^2 = 0$$
$$2t - 3 = 0$$
$$t = \tfrac{3}{2}$$

Solution set: $\left\{\frac{3}{2}\right\}$

24.
$$x\sqrt{3} = \sqrt{2-x}$$
$$\left(x\sqrt{3}\right)^2 = \left(\sqrt{2-x}\right)^2$$
$$x^2 \cdot 3 = 2 - x$$
$$3x^2 + x - 2 = 0$$
$$(3x-2)(x+1) = 0$$

$3x - 2 = 0$ or $x + 1 = 0$
$x = \frac{2}{3}$ or $x = -1$

Check $x = \frac{2}{3}$: $\frac{2}{3}\sqrt{3} = \sqrt{\frac{4}{3}}$ *True*

Check $x = -1$: $-\sqrt{3} = \sqrt{3}$ *False*

Solution set: $\left\{\frac{2}{3}\right\}$

25.
$$r^2 - 72 = 0$$
$$r^2 = 72$$
$$r = \pm\sqrt{72} = \pm 6\sqrt{2}$$

Solution set: $\left\{6\sqrt{2}, -6\sqrt{2}\right\}$

26.
$$-3x^2 + 4x = -4$$
$$0 = 3x^2 - 4x - 4$$
$$0 = (3x+2)(x-2)$$

$3x + 2 = 0$ or $x - 2 = 0$
$x = -\frac{2}{3}$ or $x = 2$

Solution set: $\left\{-\frac{2}{3}, 2\right\}$

27.
$$x^2 - 5x - 36 = 0$$
$$(x+4)(x-9) = 0$$

$x + 4 = 0$ or $x - 9 = 0$
$x = -4$ or $x = 9$

Solution set: $\{-4, 9\}$

28. $w^2 = 169$

$w = \sqrt{169}$ or $w = -\sqrt{169}$
$w = 13$ or $w = -13$

Solution set: $\{13, -13\}$

29.
$$3p^2 = 6p - 4$$
$$3p^2 - 6p + 4 = 0$$

Use $a = 3$, $b = -6$, and $c = 4$ in the quadratic formula.

$$p = \frac{-b \pm \sqrt{b^2 - 4ac}}{2a}$$

$$p = \frac{-(-6) \pm \sqrt{(-6)^2 - 4(3)(4)}}{2(3)}$$

$$= \frac{6 \pm \sqrt{36 - 48}}{2(3)} = \frac{6 \pm \sqrt{-12}}{2(3)}$$

$$= \frac{6 \pm 2i\sqrt{3}}{2(3)}$$

$$= \frac{3 \pm i\sqrt{3}}{3}$$

Solution set: $\left\{\frac{3+i\sqrt{3}}{3}, \frac{3-i\sqrt{3}}{3}\right\}$

30.
$$z = \sqrt{\frac{5z+3}{2}}$$
$$(z)^2 = \left(\sqrt{\frac{5z+3}{2}}\right)^2$$
$$z^2 = \frac{5z+3}{2}$$
$$2z^2 = 5z + 3$$
$$2z^2 - 5z - 3 = 0$$
$$(2z+1)(z-3) = 0$$

$2z + 1 = 0$ or $z - 3 = 0$
$z = -\frac{1}{2}$ or $z = 3$

Check $z = -\frac{1}{2}$: $-\frac{1}{2} = \frac{1}{2}$ *False*
Check $z = 3$: $3 = 3$ *True*

Solution set: $\{3\}$

31. $2(3k-1)^2+5(3k-1)=-2$
Let $u=3k-1$.
$$2u^2+5u+2=0$$
$$(2u+1)(u+2)=0$$
$2u+1=0$ or $u+2=0$
$u=-\frac{1}{2}$ $\quad u=-2$
$3k-1=-\frac{1}{2}$ or $3k-1=-2$
$3k=\frac{1}{2}$ $\quad 3k=-1$
$k=\frac{1}{6}$ $\quad k=-\frac{1}{3}$

Solution set: $\left\{-\frac{1}{3},\frac{1}{6}\right\}$

32. $\frac{4}{r^2}+3=\frac{1}{r}$
Multiply by the LCD, r^2.
$4+3r^2=r$
$3r^2-r+4=0$
Use $a=3$, $b=-1$, and $c=4$ in the quadratic formula.
$$r=\frac{-b\pm\sqrt{b^2-4ac}}{2a}$$
$$r=\frac{-(-1)\pm\sqrt{(-1)^2-4(3)(4)}}{2(3)}$$
$$=\frac{1\pm\sqrt{1-48}}{6}=\frac{1\pm\sqrt{-47}}{6}$$
$$=\frac{1\pm i\sqrt{47}}{6}$$

Solution set: $\left\{\frac{1+i\sqrt{47}}{6},\frac{1-i\sqrt{47}}{6}\right\}$

33. $x-\sqrt{15-2x}=0$
$x=\sqrt{15-2x}$ *Isolate*
$x^2=\left(\sqrt{15-2x}\right)^2$ *Square*
$x^2=15-2x$
$x^2+2x-15=0$
$(x+5)(x-3)=0$
$x+5=0$ or $x-3=0$
$x=-5$ or $x=3$

Check $x=-5$: $-5-\sqrt{25}=0$ *False*
Check $x=3$: $3-\sqrt{9}=0$ *True*

Solution set: $\{3\}$

34. $3=\frac{1}{t+2}+\frac{2}{(t+2)^2}$
Let $x=\frac{1}{t+2}$, so $x^2=\frac{1}{(t+2)^2}$.
$3=x+2x^2$
$0=2x^2+x-3$
$0=(2x+3)(x-1)$

$2x+3=0$ or $x-1=0$
$x=-\frac{3}{2}$ or $x=1$
$\frac{1}{t+2}=-\frac{3}{2}$ or $\frac{1}{t+2}=1$
$1(2)=-3(t+2)$ or $1=1(t+2)$
$2=-3t-6$ or $1=t+2$
$8=-3t$
$-\frac{8}{3}=t$ or $-1=t$

Check $t=-\frac{8}{3}$: $3=-\frac{3}{2}+\frac{9}{2}$ *True*
Check $t=-1$: $3=1+2$ *True*

Solution set: $\left\{-\frac{8}{3},-1\right\}$

35. $4k^4+5k^2+1=0$
Let $x=k^2$, so $x^2=k^4$.
$$4x^2+5x+1=0$$
$$(4x+1)(x+1)=0$$
$4x+1=0$ or $x+1=0$
$x=-\frac{1}{4}$ or $x=-1$
$k^2=-\frac{1}{4}$ or $k^2=-1$
$k=\pm\sqrt{-\frac{1}{4}}$ or $k=\pm\sqrt{-1}$
$k=\pm\frac{1}{2}i$ or $k=\pm i$

Check $k=\pm\frac{1}{2}i$: $\frac{1}{4}-\frac{5}{4}+1=0$ *True*
Check $k=\pm i$: $4-5+1=0$ *True*

Solution set: $\left\{-i,i,-\frac{1}{2}i,\frac{1}{2}i\right\}$

36. $(x+1)^{2/3}-(x+1)^{1/3}=2$
Let $u=(x+1)^{1/3}$, so $u^2=(x+1)^{2/3}$.
$$u^2-u=2$$
$$u^2-u-2=0$$
$$(u-2)(u+1)=0$$
$u-2=0$ or $u+1=0$
$u=2$ or $u=-1$
$(x+1)^{1/3}=2$ or $(x+1)^{1/3}=-1$
$\left[(x+1)^{1/3}\right]^3=2^3$ or $\left[(x+1)^{1/3}\right]^3=(-1)^3$
$x+1=8$ or $x+1=-1$
$x=7$ or $x=-2$

Check $x=7$: $4-2=2$ *True*
Check $x=-2$: $1-(-1)=2$ *True*

Solution set: $\{-2,7\}$

10.4 Formulas and Further Applications

10.4 Margin Exercises

1. **(a)** Solve $A = \pi r^2$ for r.

$$\frac{A}{\pi} = r^2 \quad \textit{Divide by } \pi.$$

$$r = \pm\sqrt{\frac{A}{\pi}} \quad \textit{Square root property}$$

$$r = \frac{\pm\sqrt{A}}{\sqrt{\pi}} \cdot \frac{\sqrt{\pi}}{\sqrt{\pi}} \quad \textit{Rationalize the denominator.}$$

$$r = \frac{\pm\sqrt{A\pi}}{\pi}$$

(b) Solve $s = 30\sqrt{\frac{a}{p}}$ for a.

$$s^2 = 900 \cdot \frac{a}{p} \quad \textit{Square both sides.}$$

$$ps^2 = 900a \quad \textit{Multiply by p.}$$

$$\frac{ps^2}{900} = a \quad \textit{Divide by 900.}$$

2. Solve $2t^2 - 5t + k = 0$ for t.

Use $a = 2$, $b = -5$, and $c = k$ in the quadratic formula.

$$t = \frac{-b \pm \sqrt{b^2 - 4ac}}{2a}$$

$$t = \frac{-(-5) \pm \sqrt{(-5)^2 - 4(2)k}}{2(2)}$$

$$t = \frac{5 \pm \sqrt{25 - 8k}}{4}$$

The solutions are

$$t = \frac{5 + \sqrt{25 - 8k}}{4} \text{ and } t = \frac{5 - \sqrt{25 - 8k}}{4}.$$

3. *Step 2*
Let $x =$ the distance to the top of the ladder from the ground. Then $x - 7 =$ the distance to the bottom of the ladder from the house.

Step 3
The wall of the house is perpendicular to the ground, so this is a right triangle. Use the Pythagorean formula.

$$a^2 + b^2 = c^2$$
$$x^2 + (x - 7)^2 = 13^2$$

Step 4

$$x^2 + x^2 - 14x + 49 = 169$$
$$2x^2 - 14x - 120 = 0$$
$$x^2 - 7x - 60 = 0$$
$$(x - 12)(x + 5) = 0$$

$$x - 12 = 0 \quad \text{or} \quad x + 5 = 0$$
$$x = 12 \quad \text{or} \quad x = -5$$

Step 5
Since x represents a length, it must be positive, so reject -5. Since $x = 12$, $x - 7 = 12 - 7 = 5$. The bottom of the ladder is 5 ft from the house.

Step 6
Since $5^2 + 12^2 = 13^2$ ft and 5 ft is 7 ft less than 12 ft, the distances are correct.

4. *Step 2*
Let $x =$ the width of the grass strip.

Step 3
The width of the large rectangle is $20 + 2x$, and the length is $40 + 2x$.

Area of rectangle − area of pool = area of grass.

$$(20 + 2x)(40 + 2x) - 20(40) = 700$$

Step 4

$$800 + 120x + 4x^2 - 800 = 700$$
$$4x^2 + 120x - 700 = 0$$
$$x^2 + 30x - 175 = 0$$
$$(x + 35)(x - 5) = 0$$

$$x + 35 = 0 \quad \text{or} \quad x - 5 = 0$$
$$x = -35 \quad \text{or} \quad x = 5$$

Step 5
The width cannot be -35, so the grass strip should be 5 ft wide.

Step 6
If $x = 5$, then the area of the large rectangle is

$$(40 + 2 \cdot 5)(20 + 2 \cdot 5) = 50 \cdot 30 = 1500 \text{ ft}^2.$$

The area of the pool is $40 \cdot 20 = 800$ ft^2. So, the area of the grass strip is $1500 - 800 = 700$ ft^2, as required. The answer is correct.

5.

$$s(t) = -16t^2 + 64t$$
$$32 = -16t^2 + 64t \quad \textit{Let } s(t) = 32.$$
$$16t^2 - 64t + 32 = 0$$
$$t^2 - 4t + 2 = 0 \quad \textit{Divide by 16.}$$

Use $a = 1$, $b = -4$, and $c = 2$ in the quadratic formula.

$$t = \frac{-b \pm \sqrt{b^2 - 4ac}}{2a}$$
$$= \frac{-(-4) \pm \sqrt{(-4)^2 - 4(1)(2)}}{2(1)}$$
$$= \frac{4 \pm \sqrt{16-8}}{2} = \frac{4 \pm \sqrt{8}}{2}$$
$$= \frac{4 \pm 2\sqrt{2}}{2} = 2 \pm \sqrt{2}$$
$$t = 2 + \sqrt{2} \approx 3.4 \quad \text{or} \quad t = 2 - \sqrt{2} \approx .6$$

The ball will be at a height of 32 ft at about .6 sec and 3.4 sec.

6. $$x = \frac{28.6 \pm \sqrt{(-28.6)^2 - 4(3.37)(-17)}}{2(3.37)}$$
$$= \frac{28.6 \pm \sqrt{1047.12}}{6.74}$$
$$x \approx 9.044 \approx 9.0 \quad \text{or} \quad x \approx -.558 \approx -.6$$

The positive solution, $x \approx 9.0$, is valid for this problem.

10.4 Section Exercises

1. The first step in solving a formula like $gw^2 = 2r$ for w is to divide both sides by g. This isolates w^2 and allows us to apply the square root property.

2. We must recognize that a formula like
$$gw^2 = kw + 24$$
is quadratic in w. So the first step is to write the formula in standard form (with 0 on one side, in decreasing powers of w). This allows us to apply the quadratic formula to solve for w.

3. Since the triangle is a right triangle, use the Pythagorean formula with legs m and n and hypotenuse p.
$$m^2 + n^2 = p^2$$
$$m^2 = p^2 - n^2$$
$$m = \sqrt{p^2 - n^2}$$
Only the positive square root is given since m represents the side of a triangle.

4. m is the length of the longest side (hypotenuse) of a right triangle and n and p are the lengths of the shorter sides, so, by the Pythagorean formula,
$$m^2 = n^2 + p^2.$$
Solve for m using the square root property.
$$m = \pm\sqrt{n^2 + p^2}$$
Because m represents a length, m must be positive.
$$m = \sqrt{n^2 + p^2}$$

5. Solve $d = kt^2$ for t.

$kt^2 = d$

$t^2 = \frac{d}{k}$ *Divide by k.*

$t = \pm\sqrt{\frac{d}{k}}$ *Use square root property.*

$= \frac{\pm\sqrt{d}}{\sqrt{k}} \cdot \frac{\sqrt{k}}{\sqrt{k}}$ *Rationalize denominator.*

$t = \frac{\pm\sqrt{dk}}{k}$ *Simplify*

6. Solve $s = kwd^2$ for d.

$kwd^2 = s$

$d^2 = \frac{s}{kw}$ *Divide by kw.*

$d = \pm\sqrt{\frac{s}{kw}}$ *Use square root property.*

$= \pm\sqrt{\frac{s}{kw}} \cdot \frac{\sqrt{kw}}{\sqrt{kw}}$ *Rationalize denominator.*

$d = \frac{\pm\sqrt{skw}}{kw}$ *Simplify*

7. Solve $I = \frac{ks}{d^2}$ for d.

$Id^2 = ks$ *Multiply by d^2.*

$d^2 = \frac{ks}{I}$ *Divide by I.*

$d = \pm\sqrt{\frac{ks}{I}}$ *Use square root property.*

$= \pm\frac{\sqrt{ks}}{\sqrt{I}} \cdot \frac{\sqrt{I}}{\sqrt{I}}$ *Rationalize denominator.*

$d = \frac{\pm\sqrt{ksI}}{I}$ *Simplify*

8. Solve $R = \frac{k}{d^2}$ for d.

$Rd^2 = k$ *Multiply by d^2.*

$d^2 = \frac{k}{R}$ *Divide by R.*

$d = \pm\sqrt{\frac{k}{R}}$ *Use square root property.*

$= \pm\frac{\sqrt{k}}{\sqrt{R}} \cdot \frac{\sqrt{R}}{\sqrt{R}}$ *Rationalize denominator.*

$d = \frac{\pm\sqrt{kR}}{R}$ *Simplify*

9. Solve $F = \dfrac{kA}{v^2}$ for v.

$v^2 F = kA$ *Multiply by v^2.*

$v^2 = \dfrac{kA}{F}$ *Divide by F.*

$v = \pm\sqrt{\dfrac{kA}{F}}$ *Use square root property.*

$= \dfrac{\pm\sqrt{kA}}{\sqrt{F}} \cdot \dfrac{\sqrt{F}}{\sqrt{F}}$ *Rationalize denominator.*

$v = \dfrac{\pm\sqrt{kAF}}{F}$ *Simplify*

10. Solve $L = \dfrac{kd^4}{h^2}$ for h.

$h^2 L = kd^4$ *Multiply by h^2.*

$h^2 = \dfrac{kd^4}{L}$ *Divide by L.*

$h = \pm\sqrt{\dfrac{kd^4}{L}}$ *Use square root property.*

$= \pm\dfrac{\sqrt{kd^4 L}}{L}$ *Rationalize denominator.*

$h = \dfrac{\pm d^2\sqrt{kL}}{L}$ *Simplify*

11. Solve $V = \dfrac{1}{3}\pi r^2 h$ for r.

$3V = \pi r^2 h$ *Multiply by 3.*

$\dfrac{3V}{\pi h} = r^2$ *Divide by πh.*

$r = \pm\sqrt{\dfrac{3V}{\pi h}}$ *Use square root property.*

$= \dfrac{\pm\sqrt{3V}\cdot\sqrt{\pi h}}{\sqrt{\pi h}\cdot\sqrt{\pi h}}$ *Rationalize denominator.*

$r = \dfrac{\pm\sqrt{3\pi V h}}{\pi h}$ *Simplify*

12. Solve $V = \pi\left(r^2 + R^2\right)h$ for r.

$$V = \pi r^2 h + \pi R^2 h$$
$$\pi r^2 h = V - \pi R^2 h$$
$$r^2 = \frac{V - \pi R^2 h}{\pi h}$$
$$r = \pm\sqrt{\frac{V - \pi R^2 h}{\pi h}}$$
$$= \frac{\pm\sqrt{V - \pi R^2 h}}{\sqrt{\pi h}} \cdot \frac{\sqrt{\pi h}}{\sqrt{\pi h}}$$
$$r = \frac{\pm\sqrt{V\pi h - \pi^2 R^2 h^2}}{\pi h}$$

13. Solve $At^2 + Bt = -C$ for t.

$$At^2 + Bt + C = 0$$

Use the quadratic formula.

$$t = \frac{-B \pm \sqrt{B^2 - 4AC}}{2A}$$

14. Solve $S = 2\pi rh + \pi r^2$ for r.

$(\pi)r^2 + (2\pi h)r - S = 0$

Use the quadratic formula with $a = \pi$, $b = 2\pi h$, and $c = -S$.

$$r = \frac{-2\pi h \pm \sqrt{4\pi^2 h^2 - 4(\pi)(-S)}}{2\pi}$$
$$= \frac{-2\pi h \pm \sqrt{4(\pi^2 h^2 + \pi S)}}{2\pi}$$
$$= \frac{-2\pi h \pm 2\sqrt{\pi^2 h^2 + \pi S}}{2\pi}$$
$$r = \frac{-\pi h \pm \sqrt{\pi^2 h^2 + \pi S}}{\pi}$$

15. Solve $D = \sqrt{kh}$ for h.

$D^2 = kh$ *Square both sides.*

$\dfrac{D^2}{k} = h$ *Divide by k.*

16. Solve $F = \dfrac{k}{\sqrt{d}}$ for d.

$\sqrt{d}F = k$ *Multiply by $\sqrt{d}$.*

$\sqrt{d} = \dfrac{k}{F}$ *Divide by F.*

$d = \dfrac{k^2}{F^2}$ *Square both sides.*

17. Solve $p = \sqrt{\dfrac{k\ell}{g}}$ for ℓ.

$p^2 = \dfrac{k\ell}{g}$ *Square both sides.*

$p^2 g = k\ell$ *Multiply by g.*

$\dfrac{p^2 g}{k} = \ell$ *Divide by k.*

18. Solve $p = \sqrt{\dfrac{k\ell}{g}}$ for g.

$p^2 = \dfrac{k\ell}{g}$ *Square both sides.*

$gp^2 = k\ell$ *Multiply by g.*

$g = \dfrac{k\ell}{p^2}$ *Divide by p^2.*

19. Apply the Pythagorean formula.

$$(5m)^2 = (2m)^2 + (2m+3)^2$$
$$25m^2 = 4m^2 + 4m^2 + 12m + 9$$
$$17m^2 - 12m - 9 = 0$$

Here $a = 17$, $b = -12$, and $c = -9$.

$$m = \frac{-(-12) \pm \sqrt{(-12)^2 - 4(17)(-9)}}{2(17)}$$
$$= \frac{12 \pm \sqrt{144 + 612}}{34} = \frac{12 \pm \sqrt{756}}{34}$$
$$m = \frac{12 + \sqrt{756}}{34} \approx 1.16 \text{ or}$$
$$m = \frac{12 - \sqrt{756}}{34} \approx -.46$$

Reject the negative solution.
If $m = 1.16$, then

$$5m = 5(1.16) = 5.80,$$
$$2m = 2(1.16) = 2.32, \text{ and}$$
$$2m + 3 = 2(1.16) + 3 = 5.32.$$

The lengths of the sides of the triangle are approximately 2.3, 5.3, and 5.8.

20. Apply the Pythagorean formula.

$$(x+4)^2 = x^2 + (x+1)^2$$
$$x^2 + 8x + 16 = x^2 + x^2 + 2x + 1$$
$$0 = x^2 - 6x - 15$$

Here $a = 1$, $b = -6$, and $c = -15$.

$$x = \frac{-(-6) \pm \sqrt{(-6)^2 - 4(1)(-15)}}{2(1)}$$
$$= \frac{6 \pm \sqrt{36 + 60}}{2} = \frac{6 \pm \sqrt{96}}{2}$$
$$x = \frac{6 + \sqrt{96}}{2} \approx 7.9 \text{ or}$$
$$x = \frac{6 - \sqrt{96}}{2} \approx -1.9$$

Reject the negative solution.
If $x = 7.9$, then

$$x + 4 = 11.9 \quad \text{and} \quad x + 1 = 8.9.$$

The lengths of the sides of the triangle are approximately 7.9, 8.9, and 11.9.

21. Let $x =$ the distance traveled by the eastbound ship. Then $x + 70 =$ the distance traveled by the southbound ship.

Since the ships are traveling at right angles to one another, the distance d between them can be found using the Pythagorean formula.

$$c^2 = a^2 + b^2$$
$$d^2 = x^2 + (x+70)^2$$

Let $d = 170$, and solve for x.

$$170^2 = x^2 + (x+70)^2$$
$$28{,}900 = x^2 + x^2 + 140x + 4900$$
$$0 = 2x^2 + 140x - 24{,}000$$
$$0 = x^2 + 70x - 12{,}000$$
$$0 = (x + 150)(x - 80)$$

$$x + 150 = 0 \quad \text{or} \quad x - 80 = 0$$
$$x = -150 \quad \text{or} \quad x = 80$$

Distance cannot be negative, so reject -150. If $x = 80$, then $x + 70 = 150$. The eastbound ship traveled 80 miles, and the southbound ship traveled 150 miles.

22. Let $x =$ the horizontal distance of the kite from her hand. Then $30 + x =$ the vertical distance, or the height of the kite.

Use the Pythagorean formula.

$$x^2 + (30 + x)^2 = 150^2$$
$$x^2 + 900 + 60x + x^2 = 22{,}500$$
$$2x^2 + 60x - 21{,}600 = 0$$
$$x^2 + 30x - 10{,}800 = 0$$
$$(x + 120)(x - 90) = 0$$

$$x + 120 = 0 \quad \text{or} \quad x - 90 = 0$$
$$x = -120 \qquad x = 90$$

Discard the negative solution. The horizontal distance is 90 feet, so the height of the kite is $30 + 90 = 120$ feet.

23. Let $x =$ the width of the uncovered strip of flooring.

From the problem,
(length of the rug) • (width of the rug) $= 234$.

The rug is centered in the room a distance x from the walls (width of the strip x), so the length of the rug

$=$ length of the room $- 2 \cdot$ (width of the strip)
$= 20 - 2x$,

and the width of the rug

$=$ width of the room $- 2 \cdot$ (width of the strip)
$= 15 - 2x$.

continued

The equation

(length of rug) • (width of rug) = 234

becomes

$$(20 - 2x)(15 - 2x) = 234.$$
$$300 - 70x + 4x^2 = 234$$
$$4x^2 - 70x + 66 = 0$$

Divide by 2.

$$2x^2 - 35x + 33 = 0$$
$$(2x - 33)(x - 1) = 0$$

$2x - 33 = 0$ or $x - 1 = 0$

$x = \frac{33}{2}$ or $x = 1$

Reject $\frac{33}{2} = 16\frac{1}{2}$ since $16\frac{1}{2}$ is wider than the room itself. The width of the uncovered strip is 1 foot.

24. Let x = the width of the border. Then the width of the pool and the border is $30 + 2x$ ft and the length of the pool and border is $40 + 2x$ ft. Since the area of the pool is $30 \cdot 40 = 1200$ ft^2, we can write an equation using the total area of the pool and the border.

The area of the pool and border	is	the area of the pool	plus	the area of the border.
$(30 + 2x)(40 + 2x)$	$=$	1200	$+$	296

$$1200 + 140x + 4x^2 = 1496$$
$$4x^2 + 140x - 296 = 0$$
$$x^2 + 35x - 74 = 0$$
$$(x - 2)(x + 37) = 0$$

$x - 2 = 0$ or $x + 37 = 0$

$x = 2$ or $x = -37$

Discard the negative solution. The strip can be 2 feet wide.

25. Let x be the width of the sheet metal. Then the length is $2x - 4$.

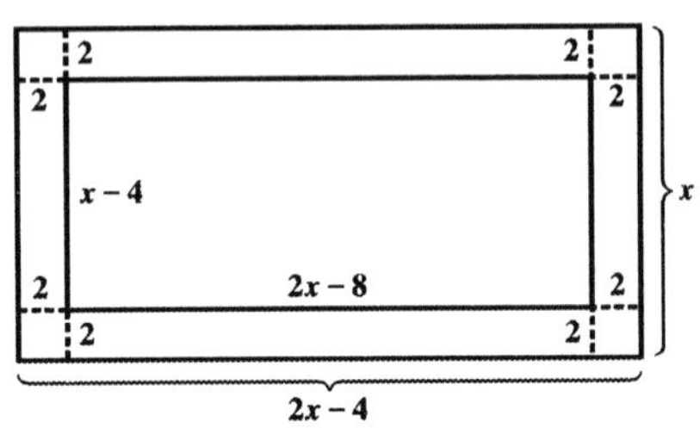

By cutting out 2-inch squares from each corner we get a rectangle with width $x - 4$ and length $(2x - 4) - 4 = 2x - 8$. The uncovered box then has height 2 inches, length $2x - 8$ inches, and width $x - 4$ inches.
Use the formula $V = LWH$ or $V = HLW$.

$$256 = 2(2x - 8)(x - 4)$$
$$256 = 4(x - 4)(x - 4) \quad \textit{Factor out 2.}$$
$$64 = (x - 4)^2 \quad \textit{Divide by 4.}$$

Use the square root property.

$\pm 8 = x - 4$

$x - 4 = 8$ or $x - 4 = -8$

$x = 12$ or $x = -4$

Since x represents width, discard the negative solution.
The width is 12 inches, and the length is $2(12) - 4 = 20$ inches.

26. Let x = the width of the piece of sheet metal. Then $x + 2$ = the length.

The width of the box is $x - 6$, since two pieces 3 inches on a side are cut from the corners of the sheet to make the box. Likewise, the length of the box is $x + 2 - 6$, or $x - 4$. The height of the box is 3 inches. Use $V = LWH$ to write the equation.

$$765 = (x - 4)(x - 6)(3)$$
$$255 = (x - 4)(x - 6) \quad \textit{Divide by 3.}$$
$$255 = x^2 - 10x + 24$$
$$0 = x^2 - 10x - 231$$
$$0 = (x - 21)(x + 11)$$

$x - 21 = 0$ or $x + 11 = 0$

$x = 21$ or $x = -11$

Discard the negative solution. The dimensions of the original piece of metal are 21 inches by $21 + 2 = 23$ inches.

27. Let $s = 213$ in the equation.

$$s = -16t^2 + 128t$$
$$213 = -16t^2 + 128t$$
$$0 = -16t^2 + 128t - 213$$

Here $a = -16$, $b = 128$, and $c = -213$.

$$t = \frac{-b \pm \sqrt{b^2 - 4ac}}{2a}$$
$$t = \frac{-128 \pm \sqrt{128^2 - 4(-16)(-213)}}{2(-16)}$$
$$= \frac{-128 \pm \sqrt{16{,}384 - 13{,}632}}{-32}$$
$$= \frac{-128 \pm \sqrt{2752}}{-32}$$
$$t = \frac{-128 + \sqrt{2752}}{-32} \approx 2.4 \text{ or}$$
$$t = \frac{-128 - \sqrt{2752}}{-32} \approx 5.6$$

The ball will be 213 feet from the ground after 2.4 seconds and again after 5.6 seconds.

28. Let $s = 550$ in $s = -16t^2 + 208t$.

$$550 = -16t^2 + 208t$$
$$16t^2 - 208t + 550 = 0$$
$$8t^2 - 104t + 275 = 0 \quad \textit{Divide by 2.}$$

Here $a = 8$, $b = -104$, and $c = 275$.

$$t = \frac{-b \pm \sqrt{b^2 - 4ac}}{2a}$$
$$t = \frac{-(-104) \pm \sqrt{(-104)^2 - 4(8)(275)}}{2(8)}$$
$$= \frac{104 \pm \sqrt{10{,}816 - 8800}}{16}$$
$$= \frac{104 \pm \sqrt{2016}}{16}$$
$$t = \frac{104 + \sqrt{2016}}{16} \quad \text{or} \quad t = \frac{104 - \sqrt{2016}}{16}$$
$$t \approx 9.3 \quad \text{or} \quad t \approx 3.7$$

The rocket will be 550 feet from the ground after 3.7 seconds and again after 9.3 seconds.

29. $D(t) = 13t^2 - 100t$

$$180 = 13t^2 - 100t \qquad \textit{Let } D(t) = 180.$$
$$0 = 13t^2 - 100t - 180$$

Here $a = 13$, $b = -100$, and $c = -180$.

$$t = \frac{-b \pm \sqrt{b^2 - 4ac}}{2a}$$
$$t = \frac{-(-100) \pm \sqrt{(-100)^2 - 4(13)(-180)}}{2(13)}$$
$$= \frac{100 \pm \sqrt{10{,}000 + 9360}}{26}$$
$$= \frac{100 \pm \sqrt{19{,}360}}{26}$$
$$t = \frac{100 + \sqrt{19{,}360}}{26} \approx 9.2 \quad \text{or}$$
$$t = \frac{100 - \sqrt{19{,}360}}{26} \approx -1.5$$

Discard the negative solution. The car will skid 180 feet in approximately 9.2 seconds.

30. $D(t) = 13t^2 - 73t$

$$218 = 13t^2 - 73t \qquad \textit{Let } D(t) = 218.$$
$$0 = 13t^2 - 73t - 218$$

Here $a = 13$, $b = -73$, and $c = -218$.

$$t = \frac{-b \pm \sqrt{b^2 - 4ac}}{2a}$$
$$t = \frac{-(-73) \pm \sqrt{(-73)^2 - 4(13)(-218)}}{2(13)}$$
$$= \frac{73 \pm \sqrt{5329 + 11{,}336}}{26}$$
$$= \frac{73 \pm \sqrt{16{,}665}}{26}$$
$$t = \frac{73 + \sqrt{16{,}665}}{26} \approx 7.8 \quad \text{or}$$
$$t = \frac{73 - \sqrt{16{,}665}}{26} \approx -2.2$$

Discard the negative solution. The car will skid 218 feet in approximately 7.8 seconds.

31. $s(t) = -16t^2 + 160t$

$$400 = -16t^2 + 160t \qquad \textit{Let } s(t) = 400.$$
$$0 = -16t^2 + 160t - 400$$
$$0 = t^2 - 10t + 25 \qquad \textit{Divide by } -16.$$
$$0 = (t-5)(t-5)$$
$$0 = (t-5)^2$$
$$0 = t - 5$$
$$5 = t$$

The rock reaches a height of 400 feet after 5 seconds. This is its maximum height since this is the only time it reaches 400 feet.

32.

$$s(t) = -16t^2 + 160t$$
$$425 = -16t^2 + 160t \qquad \textit{Let } s(t) = 425.$$
$$16t^2 - 160t + 425 = 0$$

Here $a = 16$, $b = -160$, and $c = 425$.

$$t = \frac{-b \pm \sqrt{b^2 - 4ac}}{2a}$$
$$t = \frac{-(-160) \pm \sqrt{(-160)^2 - 4(16)(425)}}{2(16)}$$
$$= \frac{160 \pm \sqrt{25{,}600 - 27{,}200}}{32}$$
$$= \frac{160 \pm \sqrt{1600}}{32}$$

The negative discriminant indicates that the solutions are nonreal complex numbers. This fact indicates that the rock never reaches 425 feet. Notice that in Exercise 31 the height of 400 feet is the maximum height.

33. Supply and demand are equal when

$$3p - 200 = \frac{3200}{p}.$$

Solve for p.

$$3p^2 - 200p = 3200$$
$$3p^2 - 200p - 3200 = 0$$

Use the quadratic formula with $a = 3$, $b = -200$, and $c = -3200$.

$$p = \frac{-(-200) \pm \sqrt{(-200)^2 - 4(3)(-3200)}}{2(3)}$$
$$= \frac{200 \pm \sqrt{40{,}000 + 38{,}400}}{6}$$
$$= \frac{200 \pm \sqrt{78{,}400}}{6} = \frac{200 \pm 280}{6}$$

$p = \frac{480}{6} = 80$ or $p = \frac{-80}{6} = -\frac{40}{3}$

Discard the negative solution. The supply and demand are equal when the price is 80 cents or \$.80.

34. Supply and demand are equal when

$$5P - 1 = \frac{700}{P}.$$

Solve for P.

$$P(5P - 1) = P\left(\frac{700}{P}\right)$$
$$5P^2 - P = 700$$
$$5P^2 - P - 700 = 0$$

Here $a = 5$, $b = -1$, and $c = -700$.

$$P = \frac{-(-1) \pm \sqrt{(-1)^2 - 4(5)(-700)}}{2(5)}$$
$$= \frac{1 \pm \sqrt{1 + 14{,}000}}{10} = \frac{1 \pm \sqrt{14{,}001}}{10}$$
$$P = \frac{1 + \sqrt{14{,}001}}{10} \approx 11.93 \text{ or}$$
$$P = \frac{1 - \sqrt{14{,}001}}{10} \approx -11.73$$

Discard the negative solution since P represents dollars. The supply equals the demand at a price of \$11.93.

35. **(a)** 2.4 million

(b) Use $f(x) = .016x^2 + .124x + .787$ with $x = 7$.

$$f(7) = .016(7)^2 + .124(7) + .787$$
$$= 2.439 \approx 2.4 \text{ million}$$

This answer agrees with part (a).

36. **(a)** Use $f(x) = .016x^2 + .124x + .787$ with $x = 10$.

$$f(10) = .016(10)^2 + .124(10) + .787$$
$$= 3.627$$

Sales in 2000 were about 3.6 million.

(b) Yes; Sales were about $\frac{2.9 \text{ million}}{10 \text{ months}} \approx .3$ million each month, so at this rate, sales for 2000 would be about $(.3)(12) = 3.6$ million.

37. Use $f(x) = .016x^2 + .124x + .787$ with $f(x) = 2$.

$$2 = .016x^2 + .124x + .787$$

Multiply each side by 1000.

$$2000 = 16x^2 + 124x + 787$$
$$0 = 16x^2 + 124x - 1213$$

Here $a = 16$, $b = 124$, and $c = -1213$.

$$x = \frac{-b \pm \sqrt{b^2 - 4ac}}{2a}$$
$$x = \frac{-124 \pm \sqrt{124^2 - 4(16)(-1213)}}{2(16)}$$
$$= \frac{-124 \pm \sqrt{15{,}376 + 77{,}632}}{32}$$
$$= \frac{-124 \pm \sqrt{93{,}008}}{32}$$
$$x = \frac{-124 + \sqrt{93{,}008}}{32} \approx 5.7 \quad \text{or}$$
$$x = \frac{-124 - \sqrt{93{,}008}}{32} \approx -13.4$$

Since time cannot be negative, sales reached 2 million in 1995 ($x \approx 5$, rounded down). The graph indicates that sales reached 2 million in 1996.

38. Use $f(x) = .016x^2 + .124x + .787$ with $f(x) = 3$.

$$3 = .016x^2 + .124x + .787$$

Multiply each side by 1000.

$$3000 = 16x^2 + 124x + 787$$
$$0 = 16x^2 + 124x - 2213$$

Here $a = 16$, $b = 124$, and $c = -2213$.

$$x = \frac{-b \pm \sqrt{b^2 - 4ac}}{2a}$$
$$x = \frac{-124 \pm \sqrt{124^2 - 4(16)(-2213)}}{2(16)}$$
$$= \frac{-124 \pm \sqrt{15{,}376 + 141{,}632}}{32}$$
$$= \frac{-124 \pm \sqrt{157{,}008}}{32}$$

$$x = \frac{-124 + \sqrt{157{,}008}}{32} \approx 8.5 \quad \text{or}$$
$$x = \frac{-124 - \sqrt{157{,}008}}{32} \approx -16.3$$

Since time cannot be negative, sales reached 3 million in 1998 ($x \approx 8$, rounded down). The graph indicates that sales reached 3 million in 1999.

39. Let F denote the Froude number. Solve

$$F = \frac{v^2}{g\ell}$$

for v.

$$v^2 = Fg\ell$$
$$v = \pm\sqrt{Fg\ell}$$

v is positive, so

$$v = \sqrt{Fg\ell}.$$

For the rhinoceros, $\ell = 1.2$ and $F = 2.57$.

$$v = \sqrt{(2.57)(9.8)(1.2)} \approx 5.5$$

or 5.5 meters per second.

40. See Exercise 39. For the triceratops, $\ell = 2.8$ and $F = .16$.

$$v = \sqrt{(.16)(9.8)(2.8)} \approx 2.1$$

or 2.1 meters per second.

41. Write a proportion.

$$\frac{x-4}{3x-19} = \frac{4}{x-3}$$

Multiply by the LCD, $(3x-19)(x-3)$.

$$(3x-19)(x-3)\left(\frac{x-4}{3x-19}\right) = (3x-19)(x-3)\left(\frac{4}{x-3}\right)$$
$$(x-3)(x-4) = (3x-19)4$$
$$x^2 - 7x + 12 = 12x - 76$$
$$x^2 - 19x + 88 = 0$$
$$(x-8)(x-11) = 0$$

$$x - 8 = 0 \quad \text{or} \quad x - 11 = 0$$
$$x = 8 \quad \text{or} \quad x = 11$$

If $x = 8$, then

$$3x - 19 = 3(8) - 19 = 5.$$

If $x = 11$, then

$$3x - 19 = 3(11) - 19 = 14.$$

Thus, $AC = 5$ or $AC = 14$.

42. Write a proportion.

$$\frac{x-5}{3} = \frac{3x-11}{x+3}$$
$$(x-5)(x+3) = 3(3x-11)$$
$$x^2 - 2x - 15 = 9x - 33$$
$$x^2 - 11x + 18 = 0$$
$$(x-2)(x-9) = 0$$

$$x - 2 = 0 \quad \text{or} \quad x - 9 = 0$$
$$x = 2 \quad \text{or} \quad x = 9$$

If $x = 2$, both $x - 5$ and $3x - 11$ are negative so reject this solution.
If $x = 9$, then

$$RQ = x - 5 = 9 - 5 = 4.$$

10.5 Graphs of Quadratic Functions

10.5 Margin Exercises

1. (a)

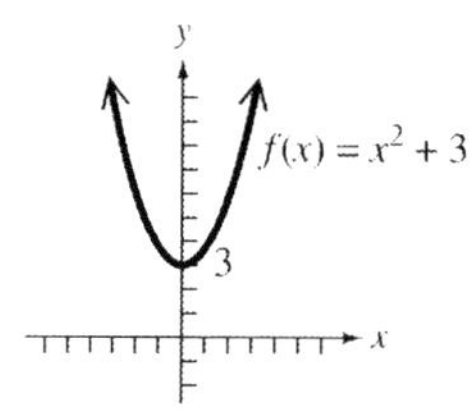

The graph of $f(x) = x^2 + 3$ has the same shape as the graph of $f(x) = x^2$. Here $k = 3$, so the graph is shifted up 3 units and has vertex $(0, 3)$.

Since x can be any real number, the domain is $(-\infty, \infty)$. The value of y is always greater than or equal to 3, so the range is $[3, \infty)$.

(b)

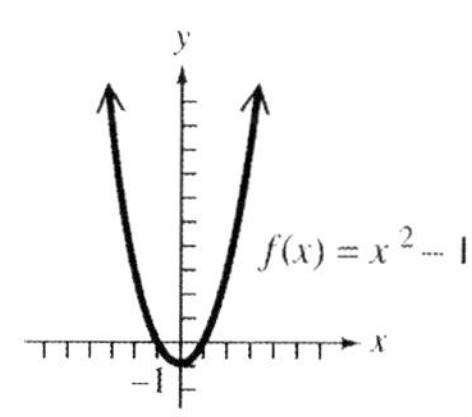

The graph of $f(x) = x^2 - 1$ has the same shape as the graph of $f(x) = x^2$. Here $k = -1$, so the graph is shifted 1 unit down and has vertex $(0, -1)$.

Since x can be any real number, the domain is $(-\infty, \infty)$. The value of y is always greater than or equal to -1, so the range is $[-1, \infty)$.

2. **(a)**

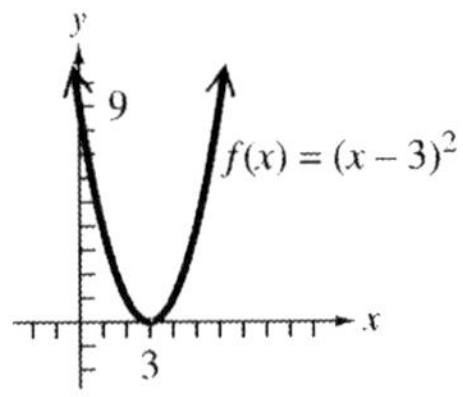

The parabola $f(x) = (x - 3)^2$ has the same shape as $f(x) = x^2$ but is shifted 3 units to the right, since 3 would cause $x - 3$ to equal 0. The vertex is $(3, 0)$ and the axis is $x = 3$. The domain is $(-\infty, \infty)$ and the range is $[0, \infty)$.

(b)

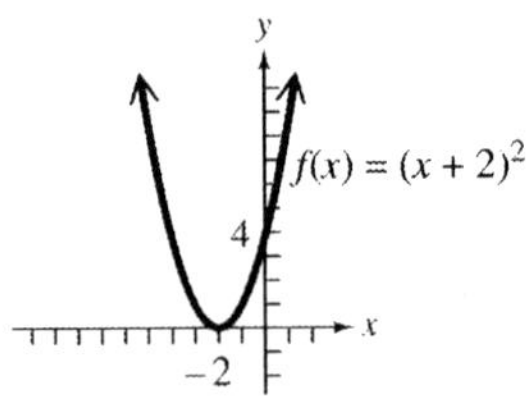

The parabola $f(x) = (x + 2)^2$ has the same shape as $f(x) = x^2$ but is shifted 2 units to the left, since -2 would cause $x + 2$ to equal 0. The vertex is $(-2, 0)$ and the axis is $x = -2$. The domain is $(-\infty, \infty)$ and the range is $[0, \infty)$.

3. **(a)**

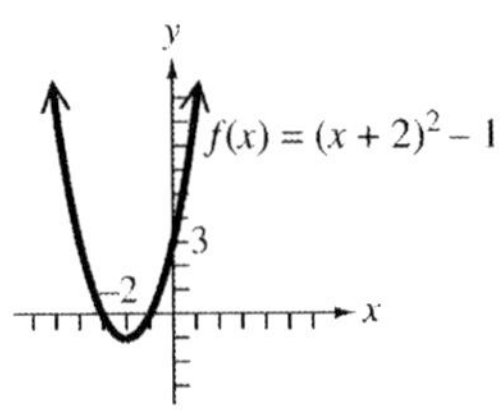

$f(x) = (x + 2)^2 - 1$ has the same shape as $f(x) = x^2$ but is shifted 2 units to the left (since $x + 2 = 0$ if $x = -2$) and 1 unit down (because of the -1). The vertex is $(-2, -1)$ and the axis is $x = -2$. The domain is $(-\infty, \infty)$ and the range is $[-1, \infty)$.

(b)

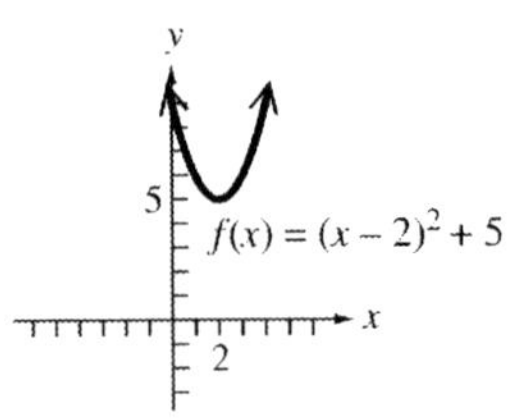

$f(x) = (x - 2)^2 + 5$ has the same shape as $f(x) = x^2$ but is shifted 2 units to the right (since $x - 2 = 0$ if $x = 2$) and 5 units up (because of the $+5$). The vertex is $(2, 5)$ and the axis is $x = 2$. The domain is $(-\infty, \infty)$ and the range is $[5, \infty)$.

4. An equation of the form $f(x) = a(x - h)^2 + k$ has a parabola for its graph. If $a > 0$, the parabola opens up; if $a < 0$, the parabola opens down.

(a) $f(x) = -\dfrac{2}{3}x^2$

Here, $a = -\frac{2}{3} < 0$, so the parabola opens *down.*

(b) $f(x) = \dfrac{3}{4}x^2 + 1$

Here, $a = \frac{3}{4} > 0$, so the parabola opens *up.*

(c) $f(x) = -2x^2 - 3$

Here, $a = -2 < 0$, so the parabola opens *down.*

(d) $f(x) = 3x^2 + 2$

Here, $a = 3 > 0$, so the parabola opens *up.*

5. For $f(x) = a(x - h)^2 + k$; if $0 < |a| < 1$, then the graph is wider than the graph of $f(x) = x^2$; if $|a| > 1$, then the graph is narrower.

(a) $f(x) = -\dfrac{2}{3}x^2$

Here, $|a| = \left|-\frac{2}{3}\right| = \frac{2}{3} < 1$, so the graph is *wider.*

(b) $f(x) = \dfrac{3}{4}x^2 + 1$

Here, $|a| = \left|\frac{3}{4}\right| = \frac{3}{4} < 1$, so the parabola is *wider.*

(c) $f(x) = -2x^2 - 3$

Here, $|a| = |-2| = 2 > 1$, so the parabola is *narrower.*

(d) $f(x) = 3x^2 + 2$

Here, $|a| = |3| = 3 > 1$, so the parabola is *narrower.*

6. $f(x) = \dfrac{1}{2}(x - 2)^2 + 1$

This equation has a graph like $f(x) = x^2$ but is shifted 2 units to the right and 1 unit up. It has vertex $(2, 1)$. Since $a = \frac{1}{2} > 0$, the parabola opens up. Since $|a| = \left|\frac{1}{2}\right| = \frac{1}{2} < 1$, the parabola is wider than the graph of $f(x) = x^2$. If $x = 0$, then $f(x) = 3$, so $(0, 3)$ is on the graph. By symmetry about the axis $x = 2$, the point $(4, 3)$ is also on the graph.

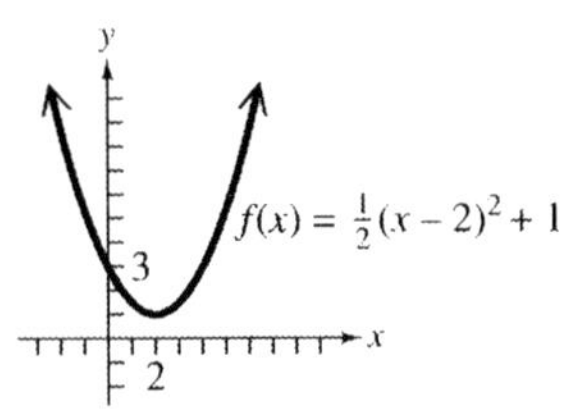

7. **(a)** The points appear to lie on a line, so a *linear* function would be a more appropriate model. The line would rise, so it would have a *positive* slope.

(b) The points appear to lie on a parabola, so a *quadratic* function would be a more appropriate model. The parabola would open up, so a would be *positive*.

8. Use the points with

$$y = ax^2 + bx + c.$$

$$\begin{aligned}
&\text{Use } (1, 29.1): && a(1)^2 + b(1) + c = 29.1 \\
&\text{Use } (6, 35): && a(6)^2 + b(6) + c = 35 \\
&\text{Use } (26, 152.6): && a(26)^2 + b(26) + c = 152.6
\end{aligned}$$

Simplifying the system gives us:

$$\begin{array}{rrrll}
a & +b & +c & = 29.1 & (1) \\
36a & +6b & +c & = 35 & (2) \\
676a & +26b & +c & = 152.6 & (3)
\end{array}$$

To eliminate c, multiply equation (1) by -1 and add the result to equation (2).

$$\begin{array}{rrrrl}
-a & -\ b & -\ c & = -29.1 & -1 \times (1) \\
36a & +\ 6b & +\ c & = 35 & (2) \\
\hline
35a & +\ 5b & & = 5.9 & (4)
\end{array}$$

To eliminate c again, multiply equation (2) by -1 and add the result to equation (3).

$$\begin{array}{rrrrl}
-36a & -\ 6b & -\ c & = -35 & -1 \times (2) \\
676a & +\ 26b & +\ c & = 152.6 & (3) \\
\hline
640a & +\ 20b & & = 117.6 & (5)
\end{array}$$

To eliminate b, multiply equation (4) by -4 and add the result to equation (5).

$$\begin{array}{rrrl}
-140a & -\ 20b & = -23.6 & -4 \times (4) \\
640a & +\ 20b & = 117.6 & (5) \\
\hline
500a & & = 94 &
\end{array}$$

$$a = \frac{94}{500} = .188$$

To find b, substitute .188 for a in (4).

$$\begin{aligned}
35a + 5b &= 5.9 \qquad (4) \\
35(.188) + 5b &= 5.9 \\
6.58 + 5b &= 5.9 \\
5b &= -.68 \\
b &= -.136
\end{aligned}$$

To find c, use (1).

$$\begin{aligned}
c &= 29.1 - a - b \\
&= 29.1 - .188 - (-.136) = 29.048
\end{aligned}$$

So using these three points gives us the quadratic model

$$y = .188x^2 - .136x + 29.048.$$

10.5 Section Exercises

1. A parabola with equation $f(x) = a(x - h)^2 + k$ has vertex $V(h, k)$. We'll identify the vertex for each quadratic function.

(a) $f(x) = (x + 2)^2 - 1$

$V(-2, -1)$, choice **B**

(b) $f(x) = (x + 2)^2 + 1$

$V(-2, 1)$, choice **C**

(c) $f(x) = (x - 2)^2 - 1$

$V(2, -1)$, choice **A**

(d) $f(x) = (x - 2)^2 + 1$

$V(2, 1)$, choice **D**

2. See Exercise 1.

(a) $f(x) = -x^2 + 2$

$V(0, 2)$, choice **D**

(b) $f(x) = -x^2 - 2$

$V(0, -2)$, choice **C**

(c) $f(x) = -(x + 2)^2$

$V(-2, 0)$, choice **B**

(d) $f(x) = -(x - 2)^2$

$V(2, 0)$, choice **A**

For Exercises 3–10, we write $f(x)$ in the form $f(x) = a(x - h)^2 + k$ and then list the vertex (h, k).

3. $f(x) = -3x^2 = -3(x - 0)^2 + 0$
The vertex (h, k) is $(0, 0)$.

4. $f(x) = \frac{1}{2}x^2 = \frac{1}{2}(x - 0)^2 + 0$
The vertex (h, k) is $(0, 0)$.

5. $f(x) = x^2 + 4 = 1(x - 0)^2 + 4$
The vertex (h, k) is $(0, 4)$.

6. $f(x) = x^2 - 4 = 1(x - 0)^2 - 4$
The vertex (h, k) is $(0, -4)$.

7. $f(x) = (x - 1)^2 = 1(x - 1)^2 + 0$
The vertex (h, k) is $(1, 0)$.

8. $f(x) = (x + 3)^2 = 1[x - (-3)]^2 + 0$
The vertex (h, k) is $(-3, 0)$.

9. $f(x) = (x + 3)^2 - 4 = 1[x - (-3)]^2 - 4$
The vertex (h, k) is $(-3, -4)$.

10. $f(x) = (x - 5)^2 - 8 = 1(x - 5)^2 - 8$
The vertex (h, k) is $(5, -8)$.

11. In Exercise 9, the parabola is shifted 3 units to the left and 4 units down. The parabola in Exercise 10 is shifted 5 units to the right and 8 units down.

12. If $a > 0$, the parabola opens up like $y = x^2$. If $a < 0$, the parabola opens down. If $|a| > 1$, the graph will be narrower than the graph of $y = x^2$, whereas if $0 < |a| < 1$, the graph will be wider than the graph of $y = x^2$. If $|a| = 1$, the graph has the same shape as $y = x^2$.

13. $f(x) = -\frac{2}{5}x^2$

Since $a = -\frac{2}{5} < 0$, the graph opens down. Since $|a| = \left|-\frac{2}{5}\right| = \frac{2}{5} < 1$, the graph is wider than the graph of $f(x) = x^2$.

14. $f(x) = -2x^2$

Since $a = -2 < 0$, the graph opens down. Since $|a| = |-2| = 2 > 1$, the graph is narrower than the graph of $f(x) = x^2$.

15. $f(x) = 3x^2 + 1$

Since $a = 3 > 0$, the graph opens up. Since $|a| = |3| = 3 > 1$, the graph is narrower than the graph of $f(x) = x^2$.

16. $f(x) = \frac{2}{3}x^2 - 4$

Since $a = \frac{2}{3} > 0$, the graph opens up. Since $|a| = \left|\frac{2}{3}\right| = \frac{2}{3} < 1$, the graph is wider than the graph of $f(x) = x^2$.

17. Consider $f(x) = a(x - h)^2 + k$.

(a) If $h > 0$ and $k > 0$ in $f(x) = a(x - h)^2 + k$, the shift is to the right and upward, so the vertex is in quadrant I.

(b) If $h > 0$ and $k < 0$, the shift is to the right and downward, so the vertex is in quadrant IV.

(c) If $h < 0$ and $k > 0$, the shift is to the left and upward, so the vertex is in quadrant II.

(d) If $h < 0$ and $k < 0$, the shift is to the left and downward, so the vertex is in quadrant III.

18. **(a)** $f(x) = (x - 4)^2 - 2 = 1(x - 4)^2 - 2$ has vertex $(4, -2)$. Because $a = 1 > 0$, the graph opens up. The correct answer is **D**.

(b) $f(x) = (x - 2)^2 - 4 = 1(x - 2)^2 - 4$ has vertex $(2, -4)$. Because $a = 1 > 0$, the graph opens up. The correct answer is **B**.

(c) $f(x) = -(x - 4)^2 - 2 = -1(x - 4)^2 - 2$ has vertex $(4, -2)$. Because $a = -1 < 0$, the graph opens down. The correct answer is **C**.

(d) $f(x) = -(x - 2)^2 - 4 = -1(x - 2)^2 - 4$ has vertex $(2, -4)$. Because $a = -1 < 0$, the graph opens down. The correct answer is **A**.

19. $f(x) = -2x^2$ written in the form
$f(x) = a(x - h)^2 + k$ is
$f(x) = -2(x - 0)^2 + 0$.

Here, $h = 0$ and $k = 0$, so the vertex (h, k) is $(0, 0)$. Since $a = -2 < 0$, the graph opens down. Since $|a| = |-2| = 2 > 1$, the graph is narrower than the graph of $f(x) = x^2$. By evaluating the function with $x = 2$ and $x = -2$, we see that the points $(2, -8)$ and $(-2, -8)$ are on the graph.

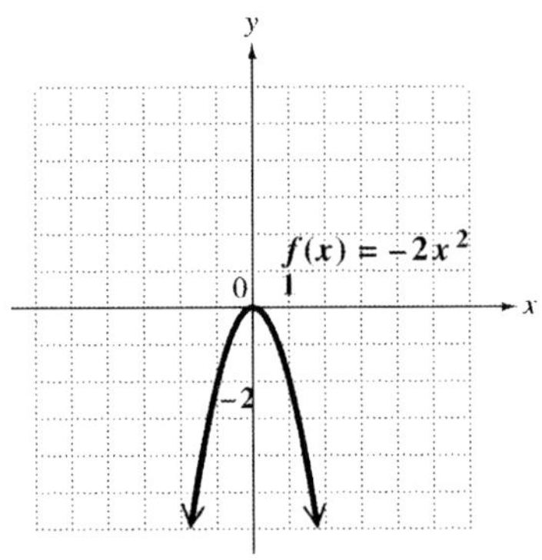

20. The graph of $f(x) = \frac{1}{3}x^2$ is wider than the graph of $f(x) = x^2$ and opens up because of the coefficient $\frac{1}{3}$. There is no shift, so the vertex is at $(0, 0)$. The points $(3, 3)$ and $(-3, 3)$ are on the graph.

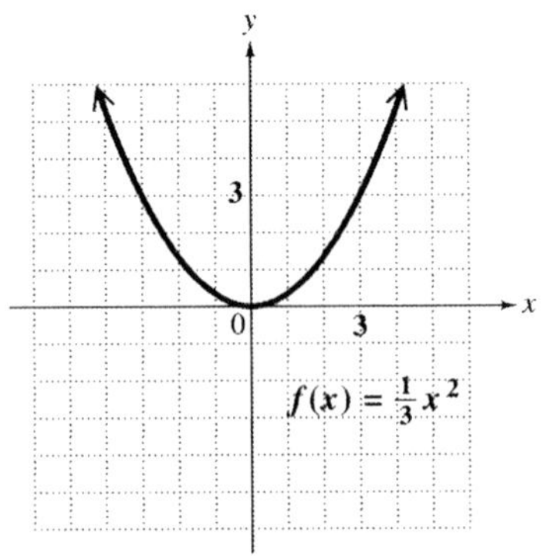

21. $f(x) = x^2 - 1$ written in the form
$f(x) = a(x - h)^2 + k$ is
$f(x) = 1(x - 0)^2 + (-1)$.

Here, $h = 0$ and $k = -1$, so the vertex is $(0, -1)$. The graph opens up and has the same shape as $f(x) = x^2$ because $a = 1$. Two other points on the graph are $(-2, 3)$ and $(2, 3)$.

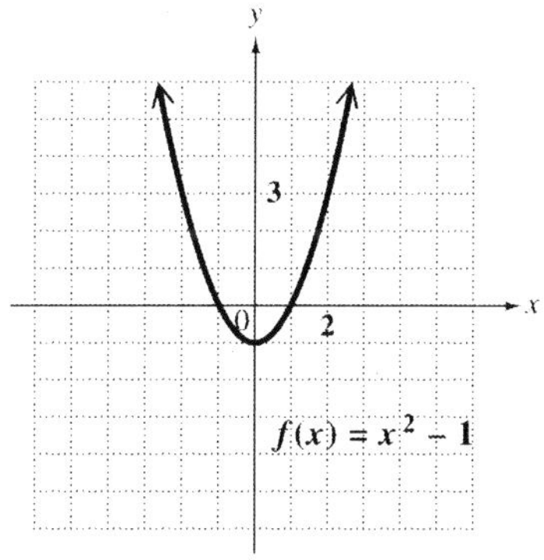

22. $f(x) = x^2 + 3$

The graph has the same shape as $f(x) = x^2$. Because $k = 3$, the graph is shifted up 3 units, so the vertex is at $(0, 3)$. The points $(1, 4)$ and $(2, 7)$ are on the graph.
The axis is $x = 0$, so by symmetry $(-1, 4)$ and $(-2, 7)$ are also points on the graph.

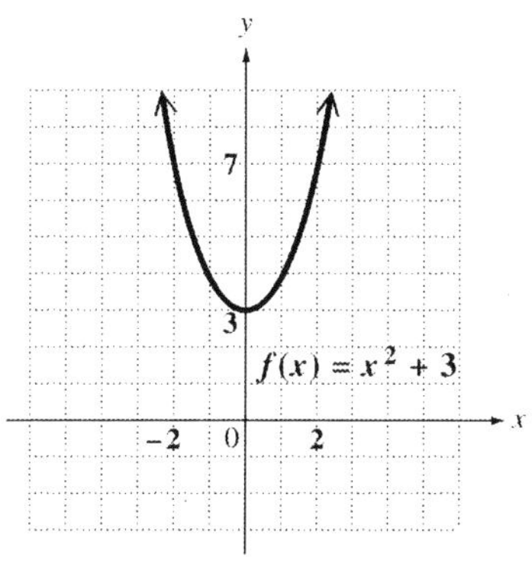

23. $f(x) = -x^2 + 2$ written in the form
$f(x) = a(x - h)^2 + k$ is
$f(x) = -1(x - 0)^2 + 2$.

Here, $h = 0$ and $k = 2$, so the vertex (h, k) is $(0, 2)$. Since $a = -1 < 0$, the graph opens down. Since $|a| = |-1| = 1$, the graph has the same shape as $f(x) = x^2$. The points $(2, -2)$ and $(-2, -2)$ are on the graph.

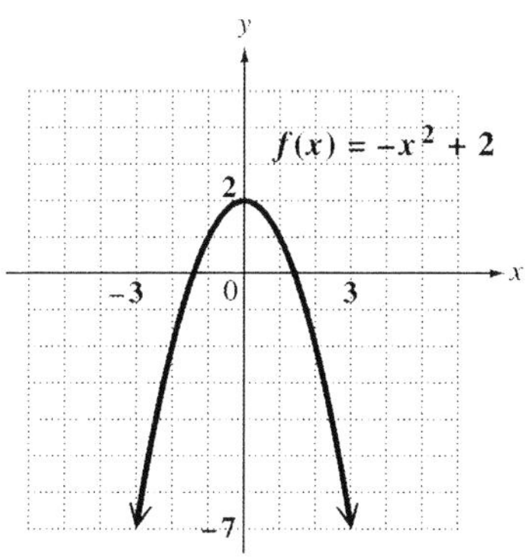

24. $f(x) = 2x^2 - 2$

Because $a = 2$, the graph is narrower than the graph of $f(x) = x^2$. Because $k = -2$, the graph is shifted 2 units downward. The vertex is at $(0, -2)$. The points $(2, 6)$ and $(-2, 6)$ are on the graph.

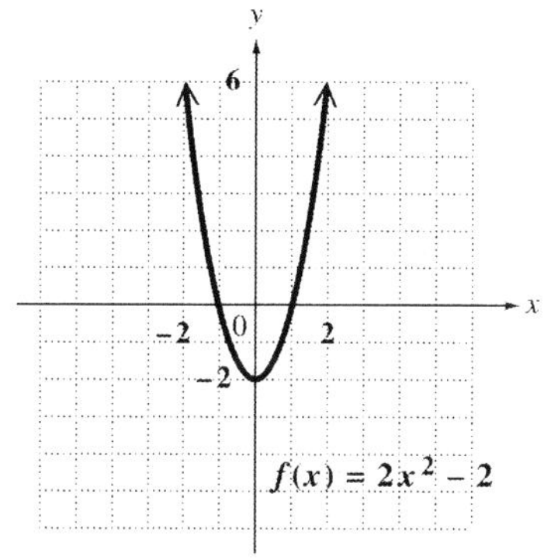

25. $f(x) = \frac{1}{2}(x - 4)^2$ written in the form
$f(x) = a(x - h)^2 + k$ is
$f(x) = \frac{1}{2}(x - 4)^2 + 0$.

Here, $h = 4$ and $k = 0$, so the vertex (h, k) is $(4, 0)$ and the axis is $x = 4$. The graph opens up since a is positive and is wider than the graph of $f(x) = x^2$ because $|a| = \frac{1}{2} < 1$. Two other points on the graph are $(2, 2)$ and $(6, 2)$. We can substitute any value for x, so the domain is $(-\infty, \infty)$. The range is $[0, \infty)$ since the smallest y-value is 0.

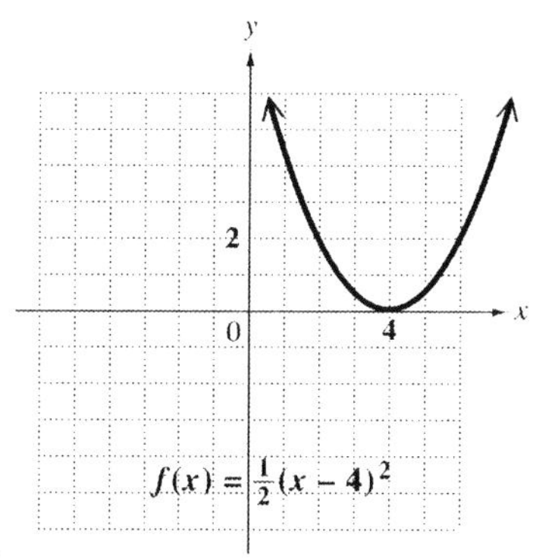

26. $f(x) = -2(x + 1)^2$

Because $a = -2$, the graph opens down and is narrower than the graph of $f(x) = x^2$. Because $h = -1$, the graph is shifted 1 unit to the left. The vertex is at $(-1, 0)$ and the axis is $x = -1$. The points $(-3, -8)$ and $(1, -8)$ are on the graph. The domain is $(-\infty, \infty)$. The range is $(-\infty, 0]$ since the largest y-value is 0.

continued

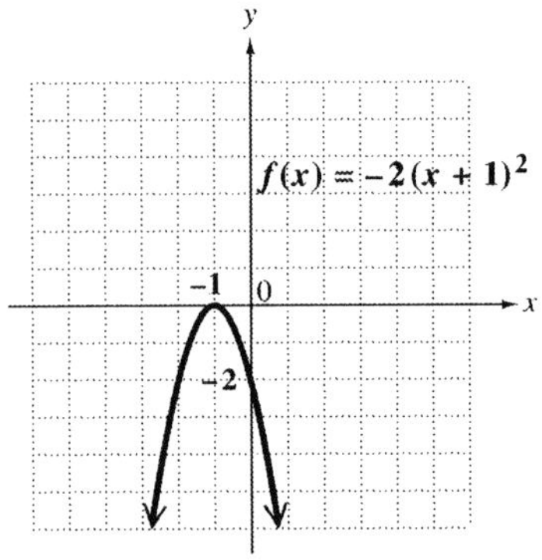

27. $f(x) = (x + 2)^2 - 1$ written in the form
$f(x) = a(x - h)^2 + k$ is
$f(x) = 1[x - (-2)]^2 + (-1)$.

Since $h = -2$ and $k = -1$, the vertex (h, k) is $(-2, -1)$ and the axis is $x = -2$. Here, $a = 1$, so the graph opens up and has the same shape as $f(x) = x^2$. The points $(-1, 0)$ and $(-3, 0)$ are on the graph. The domain is $(-\infty, \infty)$. The range is $[-1, \infty)$ since the smallest y-value is -1.

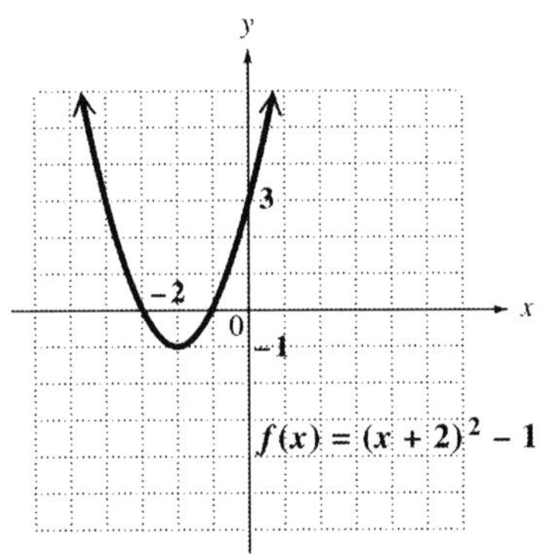

28. $f(x) = (x - 1)^2 + 2$

The graph has the same shape as $f(x) = x^2$ and opens up because $a = 1$. The graph is shifted 1 unit to the right and up 2 units because $h = 1$ and $k = 2$. The vertex is at $(1, 2)$ and the axis is $x = 1$. The points $(-1, 6)$ and $(3, 6)$ are on the graph. The domain is $(-\infty, \infty)$. The range is $[2, \infty)$ since the smallest y-value is 2.

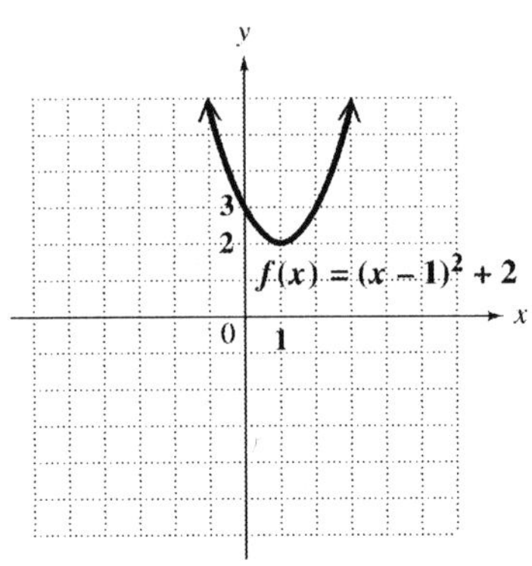

29. $f(x) = -2(x + 3)^2 + 4$ written in the form
$f(x) = a(x - h)^2 + k$ is
$f(x) = -2[x - (-3)]^2 + 4$.

Since $h = -3$ and $k = 4$, the vertex (h, k) is $(-3, 4)$ and the axis is $x = -3$. Here, $a = -2 < 0$, so the graph opens down. Also, $|a| = 2 > 1$, so the graph is narrower than the graph of $f(x) = x^2$. The points $(-5, -4)$ and $(-1, -4)$ are on the graph.
We can substitute any value for x, so the domain is $(-\infty, \infty)$. The value of y is less than or equal to 4, so the range is $(-\infty, 4]$.

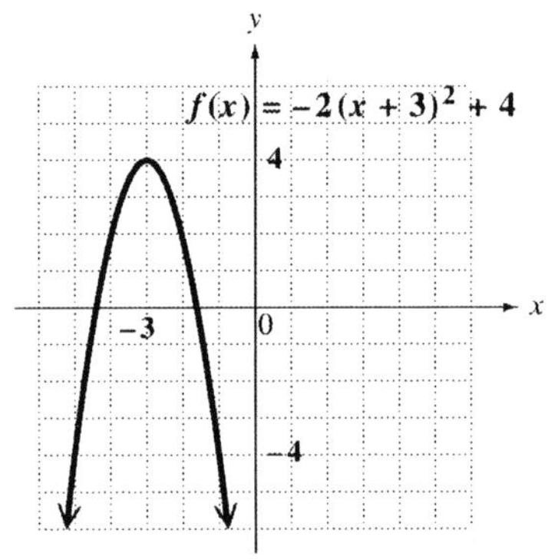

30. $f(x) = 2(x - 2)^2 - 3$ written in the form
$f(x) = a(x - h)^2 + k$ is
$f(x) = 2(x - 2)^2 + (-3)$.

Here, $h = 2$ and $k = -3$, so the vertex (h, k) is $(2, -3)$ and the axis is $x = 2$. The graph opens up since a is positive and the graph is narrower than the graph of $f(x) = x^2$ because $|a| = |2| > 1$. Two other points on the graph are $(0, 5)$ and $(4, 5)$.
We can substitute any value for x, so the domain is $(-\infty, \infty)$. The value of y is greater than or equal to -3, so the range is $[-3, \infty)$.

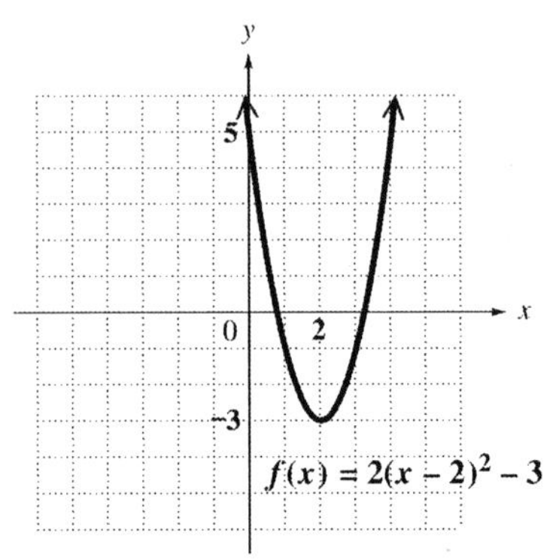

31. $f(x) = -\frac{2}{3}(x+2)^2 + 1$

Because $a = -\frac{2}{3}$, the graph opens down and is wider than the graph of $f(x) = x^2$. Because $h = -2$ and $k = 1$, the graph is shifted 2 units to the left and 1 unit upward. The vertex is at $(-2, 1)$ and the axis is $x = -2$. Two other points on the graph are $\left(-4, -\frac{5}{3}\right)$ and $\left(0, -\frac{5}{3}\right)$.
We can substitute any value for x, so the domain is $(-\infty, \infty)$. The value of y is less than or equal to 1, so the range is $(-\infty, 1]$.

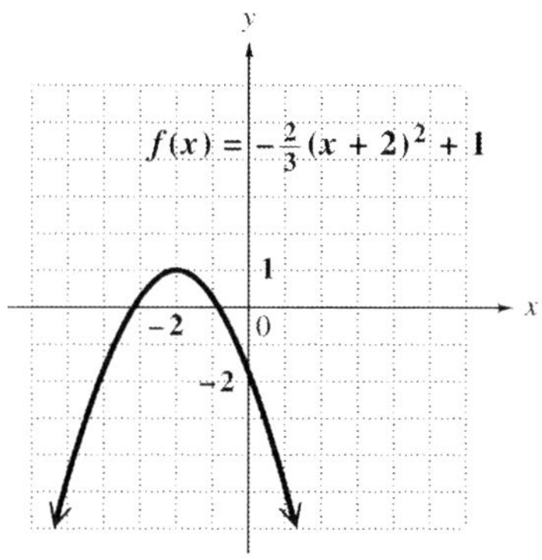

32. $f(x) = -\frac{1}{2}(x+1)^2 + 2$ written in the form
$f(x) = a(x-h)^2 + k$ is
$f(x) = -\frac{1}{2}[x - (-1)]^2 + 2.$

Since $h = -1$ and $k = 2$, the vertex (h, k) is $(-1, 2)$ and the axis is $x = -1$. Here, $a = -.5 < 0$, so the graph opens down. Also, $|a| = |-.5| = .5 < 1$, so the graph is wider than the graph of $f(x) = x^2$. The points $(1, 0)$ and $(-3, 0)$ are on the graph.
We can substitute any value for x, so the domain is $(-\infty, \infty)$. The value of y is less than or equal to 2, so the range is $(-\infty, 2]$.

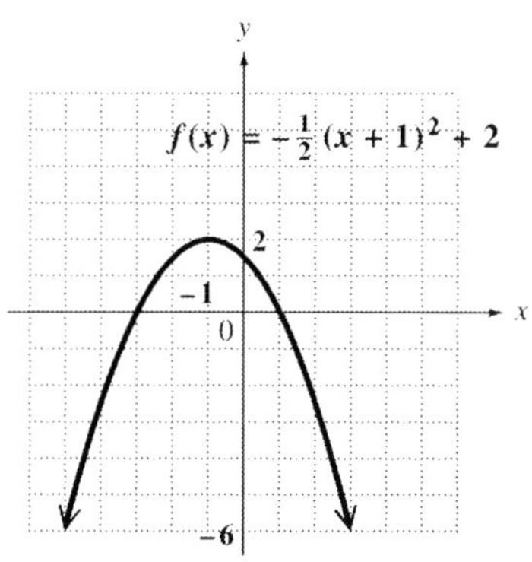

33. The graph of $F(x) = x^2 + 6$ would be shifted 6 units upward from the graph of $f(x) = x^2$.

34. To graph $G(x) = x + 6$, plot the intercepts $(-6, 0)$ and $(0, 6)$, and draw the line through them.

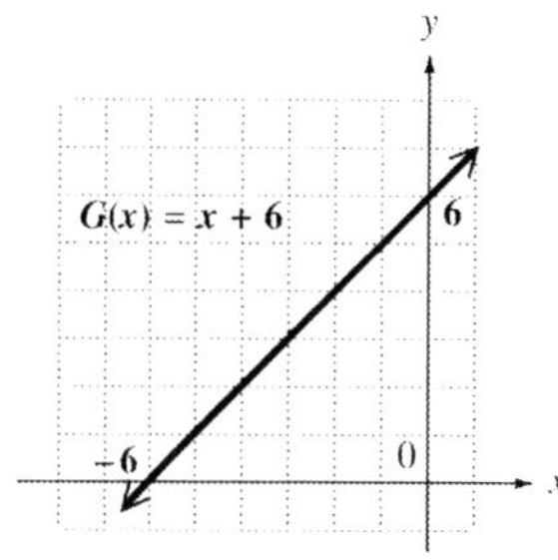

35. When considering the graph of $G(x) = x + 6$, the y-intercept is 6. The graph of $g(x) = x$ has y-intercept 0. Therefore, the graph of $G(x) = x + 6$ is shifted 6 units upward compared to the graph of $g(x) = x$.

36. The graph of $F(x) = (x-6)^2$ is shifted 6 units to the right compared to the graph of $f(x) = x^2$.

37. To graph $G(x) = x - 6$, plot the intercepts $(6, 0)$ and $(0, -6)$, and draw the line through them.

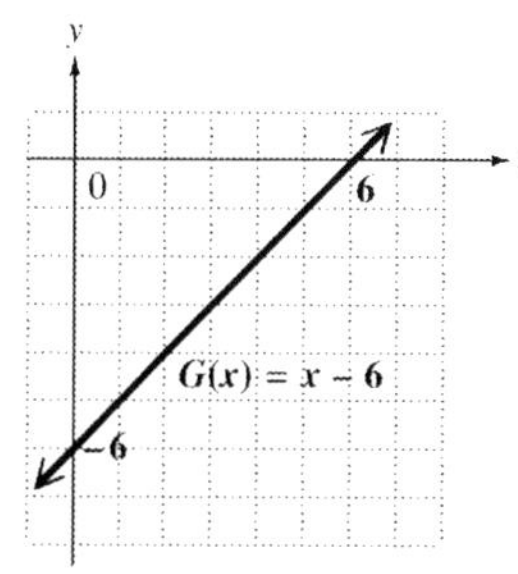

38. When considering the graph of $G(x) = x - 6$, its x-intercept is 6 as compared to the graph of $g(x) = x$ with x-intercept 0. The graph of $G(x) = x - 6$ is shifted 6 units to the right compared to the graph of $g(x) = x$.

39. Since the arrangement of the data points is approximately parabolic, a quadratic function would be the more appropriate model for the data set. The coefficient of x^2 should be positive, since the roughly parabolic shape of the graphed data set opens upward.

40. Since the points lie approximately along a straight line, a linear function would be the more appropriate model for the data set. As the x-values of the points increase, the y-values decrease, so the slope should be negative.

41. Since the arrangement of the data points is approximately parabolic, a quadratic function would be the more appropriate model for the data set. The coefficient of x^2 should be negative, since the roughly parabolic shape of the graphed data set opens downward.

42. Since the arrangement of the data points is approximately parabolic, a quadratic function would be the more appropriate model for the data set. The coefficient of x^2 should be negative, since the roughly parabolic shape of the graphed data set opens downward.

43. Since the points lie approximately along a straight line, a linear function would be the more appropriate model for the data set. As the x-values of the points increase, the y-values increase, so the slope should be positive.

44. Since the arrangement of the data points is approximately parabolic, a quadratic function would be the more appropriate model for the data set. The coefficient of x^2 should be positive, since the roughly parabolic shape of the graphed data set opens upward.

45. (a)

COMPANY BANKRUPTCY FILINGS

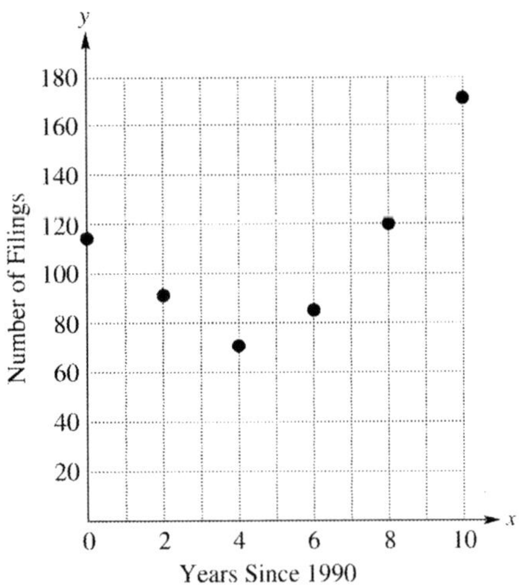

(b) Since the arrangement of the data points is approximately parabolic, a quadratic function would be the more appropriate model for the data set. The coefficient of x^2 should be positive, since the roughly parabolic shape of the graphed data set opens upward.

(c) Use $ax^2 + bx + c = y$ with $(0, 115)$, $(4, 70)$, and $(8, 120)$.

$$\begin{aligned} 0a + 0b + c &= 115 \quad (1) \\ 16a + 4b + c &= 70 \quad (2) \\ 64a + 8b + c &= 120 \quad (3) \end{aligned}$$

From (1), $c = 115$, so the system becomes

$$\begin{aligned} 16a + 4b &= -45 \quad (4) \\ 64a + 8b &= 5 \quad (5) \end{aligned}$$

Now eliminate b.

$$\begin{aligned} -32a - 8b &= 90 \qquad -2 \times (4) \\ 64a + 8b &= 5 \quad (5) \\ \hline 32a &= 95 \\ a &= \tfrac{95}{32} = 2.96875 \approx 2.969 \end{aligned}$$

From (5) with $a = \frac{95}{32}$,

$$\begin{aligned} 64\left(\tfrac{95}{32}\right) + 8b &= 5 \\ 190 + 8b &= 5 \\ 8b &= -185 \\ b &= -\tfrac{185}{8} = -23.125 \end{aligned}$$

The quadratic function is approximately

$$y = f(x) = 2.969x^2 - 23.125x + 115.$$

(d) $x = 2002 - 1990 = 12$ and $f(12) \approx 265$.

(e) No. About 16 companies filed for bankruptcy each month, so at this rate, filings for 2002 would be about 192. The approximation from the model seems high.

46. (a)

PERCENT OF HIGH SCHOOL STUDENTS WHO SMOKE

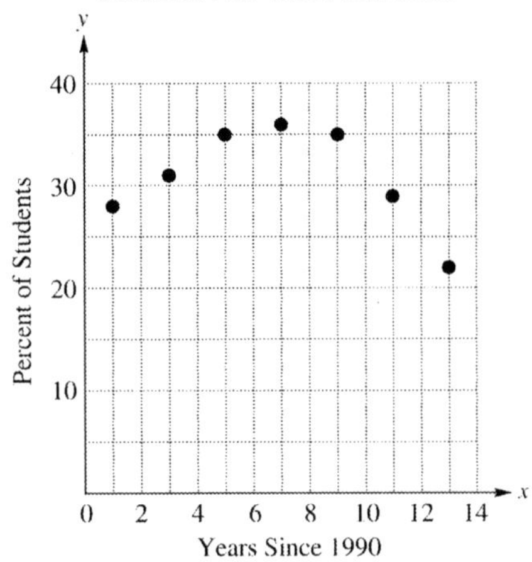

(b) Since the arrangement of the data points is approximately parabolic, a quadratic function would be the more appropriate model for the data set.

(c) Since the parabola would be opening downward, the value of a would be *negative*.

(d) Use $ax^2 + bx + c = y$ with $(1, 28)$, $(7, 36)$, and $(11, 29)$.

$$\begin{aligned} a + b + c &= 28 \quad (1) \\ 49a + 7b + c &= 36 \quad (2) \\ 121a + 11b + c &= 29 \quad (3) \end{aligned}$$

Eliminate c twice.

$$\begin{array}{rcl} -a - b - c &=& -28 \quad -1\times(1) \\ 49a + 7b + c &=& 36 \quad (2) \\ \hline 48a + 6b &=& 8 \\ 24a + 3b &=& 4 \quad (4) \end{array}$$

$$\begin{array}{rcl} -a - b - c &=& -28 \quad -1\times(1) \\ 121a + 11b + c &=& 29 \quad (3) \\ \hline 120a + 10b &=& 1 \quad (5) \end{array}$$

Now eliminate a from (4) and (5).

$$\begin{array}{rcl} -120a - 15b &=& -20 \quad -5\times(4) \\ 120a + 10b &=& 1 \quad (5) \\ \hline -5b &=& -19 \\ b &=& 3.8 \end{array}$$

Use (4) to find a.

$$\begin{aligned} 24a + 3b &= 4 \quad (4) \\ 24a + 3(3.8) &= 4 \\ 24a + 11.4 &= 4 \\ 24a &= -7.4 \\ a &= -\tfrac{7.4}{24} = -\tfrac{37}{120} = -.308\overline{3} \end{aligned}$$

Use (1) to find c.

$$\begin{aligned} a + b + c &= 28 \quad (1) \\ -\tfrac{37}{120} + 3.8 + c &= 28 \\ c &= \tfrac{2941}{120} = 24.508\overline{3} \end{aligned}$$

So the model using exact values is

$$y = -.308\overline{3}x^2 + 3.8x + 24.508\overline{3}. \quad (*)$$

Answers may vary due to roundoff.

(e) Using $(*)$ with $x = 5$ gives us $y = 35.8\% \approx 36\%$. Using $(*)$ with $x = 13$ gives us $y = 21.8\% \approx 22\%$. Since the values are very close to the ones given in the table, the model approximates the data very well.

47. **(a)** $y = f(x) = .2455x^2 - 1.856x + 30.7105$
$x = 2002 - 1970 = 32$ and
$f(32) = 222.7105 \approx 222.7$ (per 100,000)

(b) The approximation using the model is high.

48. No; Since the model did not approximate the 2002 rate very closely, it will probably not be valid for later years. Notice also that the actual rate for 2002 did not follow the same pattern and increase as in former years.

10.6 More About Parabolas; Applications

10.6 Margin Exercises

1. **(a)** $f(x) = x^2 - 6x + 7$

To find the vertex, first complete the square on $x^2 - 6x$.

$$\left[\tfrac{1}{2}(-6)\right]^2 = (-3)^2 = 9$$

Now add and subtract 9 on the right.

$$\begin{aligned} f(x) &= \left(x^2 - 6x + 9 - 9\right) + 7 \\ &= \left(x^2 - 6x + 9\right) - 9 + 7 \\ f(x) &= (x-3)^2 - 2 \end{aligned}$$

The vertex of this parabola is $(3, -2)$.

(b) $f(x) = x^2 + 4x - 9$

To find the vertex, first complete the square on $x^2 + 4x$.

$$\left[\tfrac{1}{2}(4)\right]^2 = 2^2 = 4$$

Now add and subtract 4 on the right.

$$\begin{aligned} f(x) &= \left(x^2 + 4x + 4 - 4\right) - 9 \\ &= \left(x^2 + 4x + 4\right) - 4 - 9 \\ f(x) &= (x+2)^2 - 13 \end{aligned}$$

The vertex of this parabola is $(-2, -13)$.

2. **(a)** $f(x) = 2x^2 - 4x + 1$

Factor out 2 from the first two terms.

$$f(x) = 2(x^2 - 2x) + 1$$

Complete the square on $x^2 - 2x$.

$$\left[\tfrac{1}{2}(-2)\right]^2 = (-1)^2 = 1$$

Add and subtract 1.

$$\begin{aligned} f(x) &= 2\left(x^2 - 2x + 1 - 1\right) + 1 \\ &= 2\left(x^2 - 2x + 1\right) + 2(-1) + 1 \\ &= 2\left(x^2 - 2x + 1\right) - 2 + 1 \\ f(x) &= 2(x-1)^2 - 1 \end{aligned}$$

The vertex of this parabola is $(1, -1)$.

(b) $f(x) = -\dfrac{1}{2}x^2 + 2x - 3$

Factor out $-\frac{1}{2}$ from the first two terms.

$$f(x) = -\tfrac{1}{2}(x^2 - 4x) - 3$$

Complete the square on $x^2 - 4x$.

$$\left[\tfrac{1}{2}(-4)\right]^2 = (-2)^2 = 4$$

continued

Add and subtract 4.

$$\begin{aligned} f(x) &= -\tfrac{1}{2}(x^2 - 4x + 4 - 4) - 3 \\ &= -\tfrac{1}{2}(x^2 - 4x + 4) - \tfrac{1}{2}(-4) - 3 \\ &= -\tfrac{1}{2}(x^2 - 4x + 4) + 2 - 3 \\ f(x) &= -\tfrac{1}{2}(x - 2)^2 - 1 \end{aligned}$$

The vertex of this parabola is $(2, -1)$.

3. **(a)** For $f(x) = -2x^2 + 3x - 1$, $a = -2$, $b = 3$, and $c = -1$.

The vertex is $\left(\frac{-b}{2a}, f\left(\frac{-b}{2a}\right)\right)$, so the x-coordinate is

$$\frac{-b}{2a} = \frac{-3}{2(-2)} = \frac{-3}{-4} = \frac{3}{4},$$

and the y-coordinate is

$$\begin{aligned} f\left(\tfrac{3}{4}\right) &= -2\left(\tfrac{3}{4}\right)^2 + 3\left(\tfrac{3}{4}\right) - 1 \\ &= -2\left(\tfrac{9}{16}\right) + \tfrac{9}{4} - 1 \\ &= -\tfrac{9}{8} + \tfrac{18}{8} - \tfrac{8}{8} \\ &= \tfrac{1}{8}. \end{aligned}$$

The vertex is $\left(\frac{3}{4}, \frac{1}{8}\right)$.

(b) For $f(x) = 4x^2 - x + 5$, $a = 4$, $b = -1$, and $c = 5$.

The vertex is $\left(\frac{-b}{2a}, f\left(\frac{-b}{2a}\right)\right)$, so the x-coordinate is

$$\frac{-b}{2a} = \frac{-(-1)}{2(4)} = \frac{1}{8},$$

and the y-coordinate is

$$\begin{aligned} f\left(\tfrac{1}{8}\right) &= 4\left(\tfrac{1}{8}\right)^2 - \tfrac{1}{8} + 5 \\ &= 4\left(\tfrac{1}{64}\right) - \tfrac{1}{8} + 5 \\ &= \tfrac{1}{16} - \tfrac{2}{16} + \tfrac{80}{16} \\ &= \tfrac{79}{16}. \end{aligned}$$

The vertex is $\left(\frac{1}{8}, \frac{79}{16}\right)$.

4. $f(x) = x^2 - 6x + 5$

Step 1
The graph opens up because $a = 1 > 0$.

Step 2
To find the vertex, complete the square on $x^2 - 6x$.

$$\left[\tfrac{1}{2}(-6)\right]^2 = (-3)^2 = 9$$

Add and subtract 9.

$$\begin{aligned} f(x) &= x^2 - 6x + 9 - 9 + 5 \\ f(x) &= (x - 3)^2 - 4 \end{aligned}$$

The vertex is at $(3, -4)$.

Step 3
Find any x-intercepts. Let $f(x) = 0$.

$$\begin{aligned} f(x) &= x^2 - 6x + 5 \\ 0 &= x^2 - 6x + 5 \\ 0 &= (x - 5)(x - 1) \end{aligned}$$

$$\begin{aligned} x - 5 &= 0 \quad \text{or} \quad x - 1 = 0 \\ x &= 5 \quad \text{or} \quad x = 1 \end{aligned}$$

The x-intercepts are $(5, 0)$ and $(1, 0)$.

Find the y-intercept. Let $x = 0$.

$$f(0) = 0^2 - 6(0) + 5 = 5$$

The y-intercept is $(0, 5)$.

Step 4
The axis is $x = 3$. By symmetry, another point on the graph is $(6, 5)$. Since x can be any real number, the domain is $(-\infty, \infty)$. Since the lowest point of the parabola is $(3, -4)$, the range is $[-4, \infty)$.

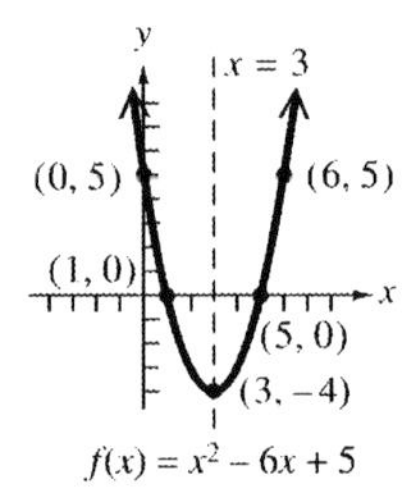

5. **(a)** $f(x) = 4x^2 - 20x + 25$

Here $a = 4$, $b = -20$, and $c = 25$. The discriminant is

$$\begin{aligned} b^2 - 4ac &= (-20)^2 - 4(4)(25) \\ &= 400 - 400 \\ &= 0. \end{aligned}$$

Since the discriminant is 0, the graph has only one x-intercept.

(b) $f(x) = 2x^2 + 3x + 5$

Here $a = 2$, $b = 3$, and $c = 5$. The discriminant is

$$\begin{aligned} b^2 - 4ac &= 3^2 - 4(2)(5) \\ &= 9 - 40 \\ &= -31. \end{aligned}$$

Since the discriminant is negative, the graph has no x-intercepts.

(c) $f(x) = -3x^2 - x + 2$

Here $a = -3$, $b = -1$, and $c = 2$. The discriminant is

$$\begin{aligned} b^2 - 4ac &= (-1)^2 - 4(-3)(2) \\ &= 1 + 24 \\ &= 25. \end{aligned}$$

Since the discriminant is positive, the graph has two x-intercepts.

6. In Example 6, replace 120 with 100, so

$$A(x) = -2x^2 + 100x.$$

Here $a = -2$, $b = 100$, and $c = 0$, so the x-coordinate of the vertex is

$$h = \frac{-b}{2a} = \frac{-100}{2(-2)} = 25,$$

and the y-coordinate of the vertex, $f(h)$, is

$$\begin{aligned} f(25) &= -2(25)^2 + 100(25) \\ &= -2(625) + 2500 \\ &= -1250 + 2500 \\ &= 1250. \end{aligned}$$

The graph is a parabola that opens down with vertex at $(25, 1250)$. The vertex of the graph shows that the maximum area will be 1250 ft^2. This area will occur if the width, x, is 25 ft and the length is $100 - 2x$, or 50 ft.

7. For $s(t) = -16t^2 + 208t$, $a = -16$, $b = 208$, and $c = 0$.

The vertex is $\left(\frac{-b}{2a}, f\left(\frac{-b}{2a}\right)\right)$, so the x-coordinate of the vertex is

$$\frac{-b}{2a} = \frac{-208}{2(-16)} = \frac{13}{2} = 6.5,$$

and the y-coordinate of the vertex is

$$\begin{aligned} f\left(\tfrac{13}{2}\right) &= -16\left(\tfrac{13}{2}\right)^2 + 208\left(\tfrac{13}{2}\right) \\ &= -16\left(\tfrac{169}{4}\right) + 1352 \\ &= -676 + 1352 \\ &= 676. \end{aligned}$$

Therefore, the toy rocket reaches a maximum height of 676 ft in 6.5 sec.

8. $x = (y + 1)^2 - 4$

Since the roles of x and y are reversed, this graph has its vertex at $(-4, -1)$. It opens to the right, the positive x-direction, and has the same shape as $y = x^2$. The axis is $y = -1$, the domain is $[-4, \infty)$, and the range is $(-\infty, \infty)$. The points $(0, 1)$ and $(0, -3)$ are on the graph.

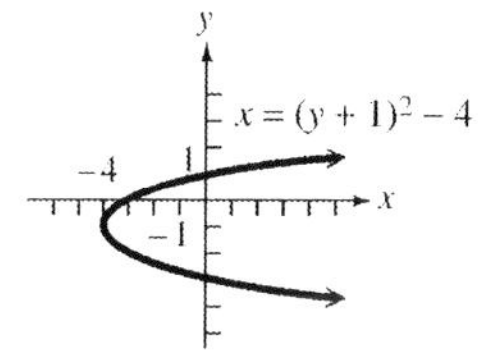

9. **(a)** $x = 2y^2 - 6y + 5$
$x = 2(y^2 - 3y) + 5$

Complete the square on $y^2 - 3y$.

$$\left[\tfrac{1}{2}(-3)\right]^2 = \left(-\tfrac{3}{2}\right)^2 = \tfrac{9}{4}$$

Add and subtract $\frac{9}{4}$.

$$\begin{aligned} x &= 2\left(y^2 - 3y + \tfrac{9}{4} - \tfrac{9}{4}\right) + 5 \\ &= 2\left(y^2 - 3y + \tfrac{9}{4}\right) - 2\left(\tfrac{9}{4}\right) + 5 \\ &= 2\left(y^2 - 3y + \tfrac{9}{4}\right) - \tfrac{9}{2} + 5 \\ x &= 2\left(y - \tfrac{3}{2}\right)^2 + \tfrac{1}{2} \end{aligned}$$

The vertex is $\left(\frac{1}{2}, \frac{3}{2}\right)$. Since $a = 2 > 0$, the graph opens to the right. Since y can be any real number, the range is $(-\infty, \infty)$. The domain is $\left[\frac{1}{2}, \infty\right)$.

(b) $x = -y^2 + 2y + 5$
$x = -(y^2 - 2y) + 5$

Complete the square on $y^2 - 2y$.

$$\left[\tfrac{1}{2}(-2)\right]^2 = (-1)^2 = 1$$

Add and subtract 1.

$$\begin{aligned} x &= -\left(y^2 - 2y + 1 - 1\right) + 5 \\ &= -\left(y^2 - 2y + 1\right) - (-1) + 5 \\ x &= -(y - 1)^2 + 6 \end{aligned}$$

The vertex is $(6, 1)$. Since $a = -1 < 0$, the graph opens to the left. The domain is $(-\infty, 6]$, and the range is $(-\infty, \infty)$.

10. From Margin Exercise 9(b), where we found the vertex, domain, and range, we have

$$x = -(y - 1)^2 + 6,$$

so the axis is $y = 1$.

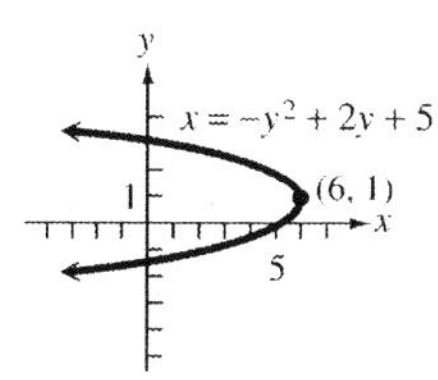

11. **(a)** **A.** $y = -x^2 + 20x + 80$ is of the form $y = ax^2 + bx + c$, so it has a *vertical* parabola as its graph.

B. $x = 2y^2 + 6y + 5$ is of the form $x = ay^2 + by + c$, so it has a *horizontal* parabola as its graph.

C.
$$\begin{aligned} x + 1 &= (y + 2)^2 \\ x + 1 &= y^2 + 4y + 4 \\ x &= y^2 + 4y + 3 \end{aligned}$$

As in **B**, this is a *horizontal* parabola.

D.
$$\begin{aligned} f(x) &= (x - 4)^2 \\ y &= x^2 - 8x + 16 \end{aligned}$$

As in **A**, this is a *vertical* parabola.

(b) Only vertical parabolas are functions, so equations **A** and **D** represent functions.

10.6 Section Exercises

1. If there is an x^2-term in the equation, the axis is vertical. If there is a y^2-term, the axis is horizontal.

2. The graph of a quadratic function cannot be a parabola with a horizontal axis because a function would have to pass the vertical line test and a vertical line would intersect the parabola in two points.

3. Use the discriminant, $b^2 - 4ac$, of the corresponding quadratic equation. If it is positive, there are two x-intercepts. If it is zero, there is one x-intercept (at the vertex), and if it is negative, there is no x-intercept.

4. If the vertex is at $(1, -3)$ and the graph opens down, the vertex is the highest point of the graph. Because the y-coordinate of the vertex is negative, the parabola lies below the x-axis so the graph has no x-intercepts.

5. As in Example 2, we'll complete the square to find the vertex.

$$\begin{aligned} y &= 2x^2 + 4x + 5 \\ &= 2(x^2 + 2x) + 5 \\ &= 2(x^2 + 2x + 1 - 1) + 5 \\ &= 2(x^2 + 2x + 1) + 2(-1) + 5 \\ &= 2(x + 1)^2 - 2 + 5 \\ y &= 2(x + 1)^2 + 3 \end{aligned}$$

The vertex is $(-1, 3)$.
Because $a = 2 > 1$, the graph opens up and is narrower than the graph of $y = x^2$.

For $y = 2x^2 + 4x + 5$, $a = 2$, $b = 4$, and $c = 5$. The discriminant is

$$\begin{aligned} b^2 - 4ac &= 4^2 - 4(2)(5) \\ &= 16 - 40 = -24. \end{aligned}$$

The discriminant is negative, so the parabola has no x-intercepts.

6. As in Example 2, we'll complete the square to find the vertex.

$$\begin{aligned} y &= 3x^2 - 6x + 4 \\ &= 3(x^2 - 2x) + 4 \\ &= 3(x^2 - 2x + 1 - 1) + 4 \\ &= 3(x^2 - 2x + 1) + 3(-1) + 4 \\ &= 3(x - 1)^2 - 3 + 4 \\ y &= 3(x - 1)^2 + 1 \end{aligned}$$

The vertex is $(1, 1)$.
Because $a = 3 > 1$, the graph opens up and is narrower than the graph of $y = x^2$.

For $y = 3x^2 - 6x + 4$, $a = 3$, $b = -6$, and $c = 4$. The discriminant is

$$\begin{aligned} b^2 - 4ac &= (-6)^2 - 4(3)(4) \\ &= 36 - 48 = -12. \end{aligned}$$

The discriminant is negative, so the parabola has no x-intercepts.

7. $y = f(x) = -x^2 + 5x + 3$

Use the vertex formula with $a = -1$ and $b = 5$.

The x-coordinate of the vertex is

$$\frac{-b}{2a} = \frac{-5}{2(-1)} = \frac{5}{2}.$$

The y-coordinate of the vertex is

$$\begin{aligned} f\left(\frac{-b}{2a}\right) &= f\left(\frac{5}{2}\right) \\ &= -\left(\frac{5}{2}\right)^2 + 5\left(\frac{5}{2}\right) + 3 \\ &= -\frac{25}{4} + \frac{25}{2} + 3 \\ &= \frac{-25 + 50 + 12}{4} = \frac{37}{4}. \end{aligned}$$

The vertex is

$$\left(\frac{-b}{2a}, f\left(\frac{-b}{2a}\right)\right) = \left(\frac{5}{2}, \frac{37}{4}\right).$$

Because $a = -1$, the parabola opens down and has the same shape as the graph of $y = x^2$.

$$\begin{aligned} b^2 - 4ac &= 5^2 - 4(-1)(3) \\ &= 25 + 12 = 37 \end{aligned}$$

The discriminant is positive, so the parabola has two x-intercepts.

8. Complete the square on the y-terms to find the vertex.

$$\begin{aligned} x &= -y^2 + 7y - 2 \\ &= -1(y^2 - 7y) - 2 \\ &= -1\left(y^2 - 7y + \tfrac{49}{4} - \tfrac{49}{4}\right) - 2 \\ &= -1\left(y^2 - 7y + \tfrac{49}{4}\right) + (-1)\left(-\tfrac{49}{4}\right) - 2 \\ &= -1\left(y - \tfrac{7}{2}\right)^2 + \tfrac{49}{4} - 2 \\ x &= -1\left(y - \tfrac{7}{2}\right)^2 + \tfrac{41}{4} \end{aligned}$$

The vertex is $\left(\frac{41}{4}, \frac{7}{2}\right)$. The graph opens to the left since the equation is of the form

$$x = a(y - k)^2 + h$$

and $a = -1 < 0$. Since $|a| = |-1| = 1$, the graph is the same shape as $y = x^2$.

9. Complete the square on the y-terms to find the vertex.

$$\begin{aligned} x &= \tfrac{1}{3}y^2 + 6y + 24 \\ &= \tfrac{1}{3}(y^2 + 18y) + 24 \\ &= \tfrac{1}{3}(y^2 + 18y + 81 - 81) + 24 \\ &= \tfrac{1}{3}(y^2 + 18y + 81) + \tfrac{1}{3}(-81) + 24 \\ &= \tfrac{1}{3}(y + 9)^2 - 27 + 24 \\ x &= \tfrac{1}{3}(y + 9)^2 - 3 \end{aligned}$$

The vertex is $(-3, -9)$.
The graph is a horizontal parabola. The graph opens to the right since $a = .\overline{3} > 0$ and is wider than the graph of $y = x^2$ since $|a| = |.\overline{3}| < 1$.

10. Complete the square on the y-terms to find the vertex.

$$\begin{aligned} x &= \tfrac{1}{2}y^2 + 10y - 5 \\ &= \tfrac{1}{2}(y^2 + 20y) - 5 \\ &= \tfrac{1}{2}(y^2 + 20y + 100 - 100) - 5 \\ &= \tfrac{1}{2}(y^2 + 20y + 100) + \tfrac{1}{2}(-100) - 5 \\ &= \tfrac{1}{2}(y + 10)^2 - 50 - 5 \\ x &= \tfrac{1}{2}(y + 10)^2 - 55 \end{aligned}$$

The equation is of the form

$$x = a(y - k)^2 + h,$$

so this is a horizontal parabola with vertex at $(h, k) = (-55, -10)$. The graph opens to the right since $a = .5 > 0$ and is wider than the graph of $y = x^2$ since $|a| = |.5| < 1$.

11. $y = f(x) = x^2 + 4x + 3$

Step 1
Because $a = 1$, the graph opens up and has the same shape as the graph of $f(x) = x^2$.

Step 2
Here $a = 1$, $b = 4$, and $c = 3$. The vertex formula gives the x-coordinate as

$$\frac{-b}{2a} = \frac{-4}{2(1)} = -2,$$

and the y-coordinate as

$$f(-2) = (-2)^2 + 4(-2) + 3 = -1.$$

The vertex is at $(-2, -1)$. Since the graph opens up, the axis goes through the x-coordinate of the vertex—its equation is $x = -2$.

Step 3
Evaluate $f(0)$ to find the y-intercept.

$$f(0) = 0^2 + 4(0) + 3 = 3$$

The y-intercept is $(0, 3)$.

Solve $f(x) = 0$ to find any x-intercepts.

$$\begin{aligned} 0 &= x^2 + 4x + 3 \\ 0 &= (x + 3)(x + 1) \end{aligned}$$

$$\begin{aligned} x + 3 &= 0 \quad \text{or} \quad x + 1 = 0 \\ x &= -3 \quad \text{or} \quad x = -1 \end{aligned}$$

The x-intercepts are $(-3, 0)$ and $(-1, 0)$.

Step 4
By symmetry, another point on the graph is $(-4, 3)$. Since x can be any real number, the domain is $(-\infty, \infty)$. Since $(-2, -1)$ is the lowest point of the parabola, the range is $[-1, \infty)$.

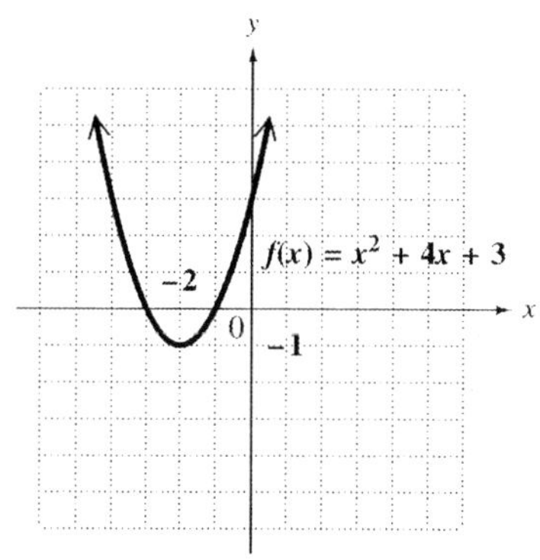

12. $y = f(x) = x^2 + 2x - 2$

Step 1
Because $a = 1$, the graph opens upward and has the same shape as the graph of $f(x) = x^2$.

Step 2
Use the vertex formula to find the vertex.

$$x = \frac{-b}{2a} = \frac{-2}{2(1)} = -1$$

$$f(-1) = (-1)^2 + 2(-1) - 2 = -3$$

The vertex is at $(-1, -3)$. Since the graph opens up, the axis goes through the x-coordinate of the vertex—its equation is $x = -1$.

Step 3
Evaluate $f(0)$ to find the y-intercept.

$$f(0) = 0^2 + 2(0) - 2 = -2$$

The y-intercept is $(0, -2)$.

Solve $f(x) = 0$ to find any x-intercepts.

$$0 = x^2 + 2x - 2$$

Use the quadratic formula with $a = 1$, $b = 2$, and $c = -2$.

$$x = \frac{-b \pm \sqrt{b^2 - 4ac}}{2a}$$

$$x = \frac{-2 \pm \sqrt{2^2 - 4(1)(-2)}}{2(1)}$$

$$= \frac{-2 \pm \sqrt{12}}{2} = \frac{-2 \pm 2\sqrt{3}}{2}$$

$$= -1 \pm \sqrt{3}$$

$x = -1 + \sqrt{3} \approx .7$ or $x = -1 - \sqrt{3} \approx -2.7$.

The x-intercepts are approximately $(-2.7, 0)$ and $(.7, 0)$.

Step 4
Since x can be any real number, the domain is $(-\infty, \infty)$. Since the parabola opens upward with lowest point $(-1, -3)$, the range is $[-3, \infty)$.

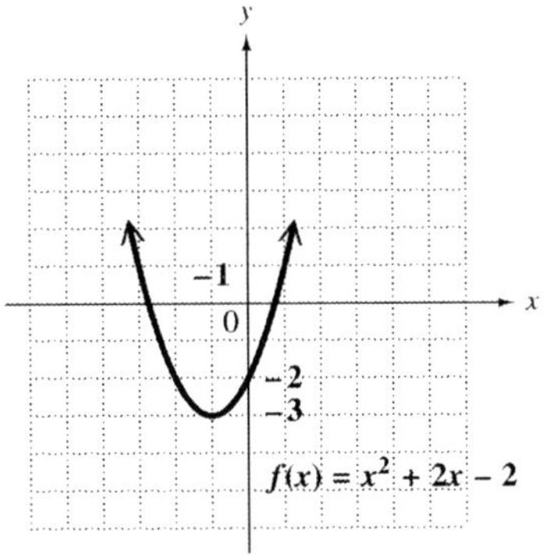

13. $y = f(x) = -2x^2 + 4x - 5$

Step 1
Since $a = -2$, the graph opens down and is narrower than the graph of $y = x^2$.

Step 2
Use the vertex formula to find the vertex.

$$x = \frac{-b}{2a} = \frac{-4}{2(-2)} = 1$$

$$f(1) = -2(1)^2 + 4(1) - 5 = -3$$

The vertex is at $(1, -3)$. Since the graph opens down, the axis goes through the x-coordinate of the vertex—its equation is $x = 1$.

Step 3
If $x = 0$, $y = -5$, so the y-intercept is $(0, -5)$. To find the x-intercepts, let $y = 0$.

$$0 = -2x^2 + 4x - 5$$

Here $a = -2$, $b = 4$, and $c = -5$.

$$x = \frac{-4 \pm \sqrt{16 - 40}}{2(-2)}$$

The discriminant is negative, so there are no x-intercepts.

Step 4
By symmetry, $(2, -5)$ is also on the graph.

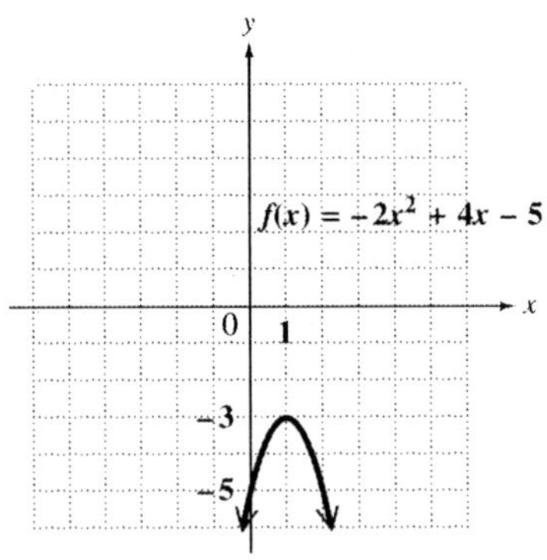

From the graph, we see that the domain is $(-\infty, \infty)$ and the range is $(-\infty, -3]$.

14. $y = f(x) = -3x^2 + 12x - 8$

Step 1
Because $a = -3$, the graph opens down and is narrower than the graph of $f(x) = x^2$.

Step 2
Use the vertex formula to find the vertex.

$$x = \frac{-b}{2a} = \frac{-12}{2(-3)} = 2$$

$$f(2) = -3(2)^2 + 12(2) - 8 = 4$$

The vertex is at $(2, 4)$. Since the graph opens down, the axis goes through the x-coordinate of the vertex—its equation is $x = 2$.

Step 3
If $x = 0$, $y = -8$, so the y-intercept is $(0, -8)$.
To find the x-intercepts, let $y = 0$.

$$0 = -3x^2 + 12x - 8$$

Here $a = -3$, $b = 12$, and $c = -8$.

$$x = \frac{-12 \pm \sqrt{12^2 - 4(-3)(-8)}}{2(-3)}$$
$$= \frac{-12 \pm \sqrt{48}}{-6} = \frac{-12 \pm 4\sqrt{3}}{-6}$$
$$= \frac{-6 \pm 2\sqrt{3}}{-3} = 2 \mp \tfrac{2}{3}\sqrt{3}$$

$x \approx .8$ or $x \approx 3.2$

The x-intercepts are approximately $(.8, 0)$ and $(3.2, 0)$.

Step 4
By symmetry, $(4, -8)$ is also on the graph.

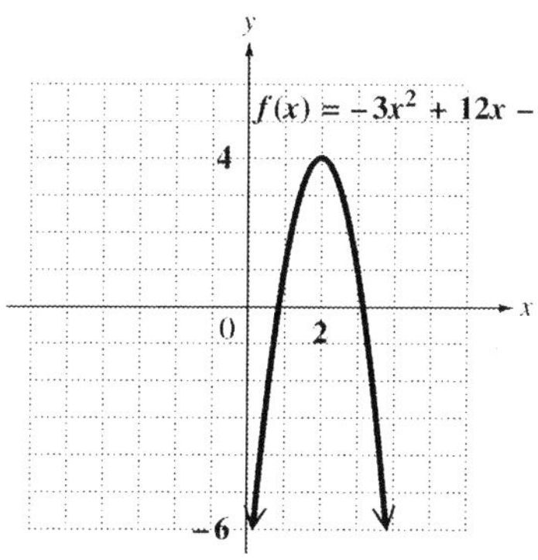

From the graph, we see that the domain is $(-\infty, \infty)$ and the range is $(-\infty, 4]$.

15. $x = -\frac{1}{5}y^2 + 2y - 4$

The roles of x and y are reversed, so this is a horizontal parabola.

Step 1
Since $a = -\frac{1}{5} < 0$, the graph opens to the left and is wider than the graph of $y - x^2$.

Step 2
The y-coordinate of the vertex is

$$\frac{-b}{2a} = \frac{-2}{2\left(-\frac{1}{5}\right)} = \frac{-2}{-\frac{2}{5}} = 5.$$

The x-coordinate of the vertex is

$$-\tfrac{1}{5}(5)^2 + 2(5) - 4 = -5 + 10 - 4 = 1.$$

Thus, the vertex is $(1, 5)$. Since the graph opens left, the axis goes through the y-coordinate of the vertex—its equation is $y = 5$.

Step 3
To find the x-intercept, let $y = 0$.
If $y = 0$, $x = -4$, so the x-intercept is $(-4, 0)$.
To find the y-intercepts, let $x = 0$.

$$0 = -\tfrac{1}{5}y^2 + 2y - 4$$
$$0 = y^2 - 10y + 20 \quad \textit{Multiply by } -5.$$
$$y = \frac{10 \pm \sqrt{100 - 80}}{2} = \frac{10 \pm \sqrt{20}}{2}$$
$$= \frac{10 \pm 2\sqrt{5}}{2} = 5 \pm \sqrt{5}$$

The y-intercepts are approximately $(0, 7.2)$ and $(0, 2.8)$.

Step 4
For an additional point on the graph, let $y = 7$ (two units above the axis) to get $x = \frac{1}{5}$. So the point $\left(\frac{1}{5}, 7\right)$ is on the graph. By symmetry, the point $\left(\frac{1}{5}, 3\right)$ (two units below the axis) is on the graph.

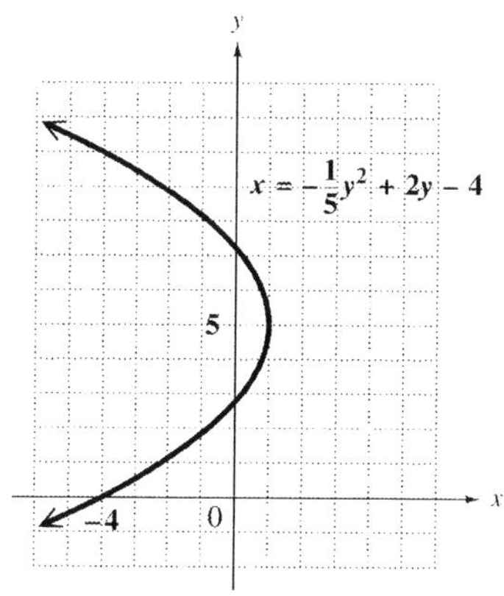

From the graph, we see that the domain is $(-\infty, 1]$ and the range is $(-\infty, \infty)$.

16. $x = -\frac{1}{2}y^2 - 4y - 6$

The roles of x and y are reversed, so this is a horizontal parabola.

Step 1
Since $a = -.5 < 0$, the graph opens to the left and is wider than the graph of $y = x^2$.

Step 2
The y-coordinate of the vertex is

$$\frac{-b}{2a} = \frac{-(-4)}{2\left(-\frac{1}{2}\right)} = \frac{4}{-1} = -4.$$

The x-coordinate of the vertex is

$$-\tfrac{1}{2}(-4)^2 - 4(-4) - 6 = -8 + 16 - 6 = 2.$$

Thus, the vertex is $(2, -4)$. Since the graph opens left, the axis goes through the y-coordinate of the vertex—its equation is $y = -4$.

continued

Step 3
If $y = 0$, $x = -6$, so the x-intercept is $(-6, 0)$.
To find the y-intercepts, let $x = 0$.

$$0 = -\tfrac{1}{2}y^2 - 4y - 6$$
$$0 = y^2 + 8y + 12 \quad \textit{Multiply by } -2.$$
$$0 = (y + 6)(y + 2)$$
$$y = -6 \text{ or } y = -2$$

The y-intercepts are $(0, -6)$ and $(0, -2)$.

Step 4
Additional points on the graph are $(1.5, -3)$ and $(1.5, -5)$.

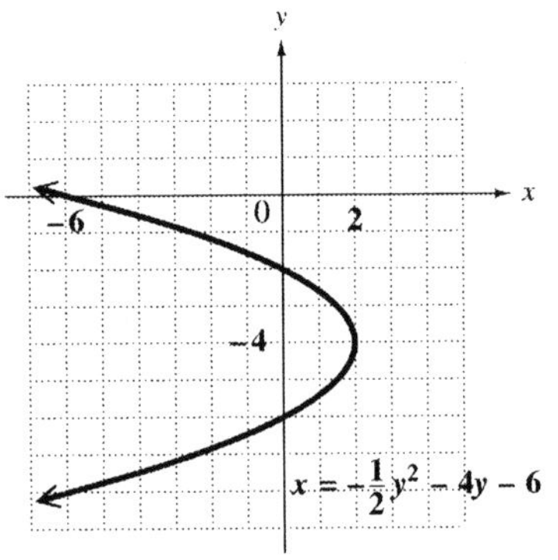

From the graph, we see that the domain is $(-\infty, 2]$ and the range is $(-\infty, \infty)$.

17. $x = 3y^2 + 12y + 5$

The roles of x and y are reversed, so this is a horizontal parabola.

Step 1
Since $a = 3 > 0$, the graph opens to the right and is narrower than the graph of $y = x^2$.

Step 2
Use the formula to find the y-value of the vertex.

$$\frac{-b}{2a} = \frac{-12}{2(3)} = -2$$

If $y = -2$, $x = -7$, so the vertex is $(-7, -2)$.
Since the graph opens right, the axis goes through the y-coordinate of the vertex—its equation is $y = -2$.

Step 3
If $y = 0$, $x = 5$, so the x-intercept is $(5, 0)$.
To find the y-intercepts, let $x = 0$.

$$0 = 3y^2 + 12y + 5$$

Here $a = 3$, $b = 12$, and $c = 5$.

$$y = \frac{-12 \pm \sqrt{144 - 60}}{6} = \frac{-12 \pm \sqrt{84}}{6}$$
$$= \frac{-12 \pm 2\sqrt{21}}{6} = -2 \pm \tfrac{1}{3}\sqrt{21}$$

The y-intercepts are approximately $(0, -.5)$ and $(0, -3.5)$.

Step 4
By symmetry, $(5, -4)$ is also on the graph.

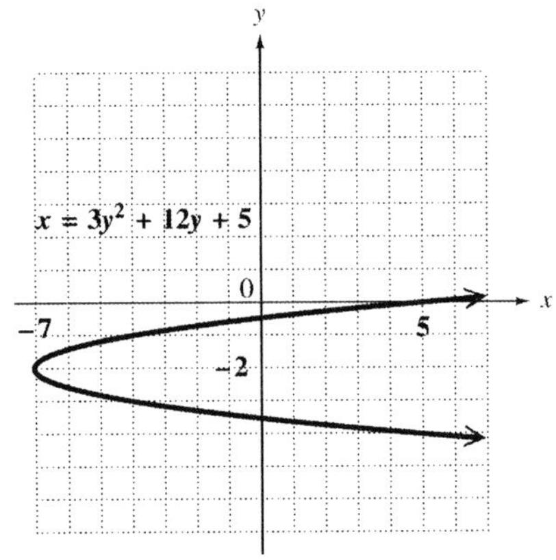

From the graph, we see that the domain is $[-7, \infty)$ and the range is $(-\infty, \infty)$.

18. $x = 4y^2 + 16y + 11$

The roles of x and y are reversed, so this is a horizontal parabola.

Step 1
Since $a = 4 > 0$, the graph opens to the right and is narrower than the graph of $y = x^2$.

Step 2
Use the formula to find the y-value of the vertex.

$$\frac{-b}{2a} = \frac{-16}{2(4)} = -2$$

If $y = -2$, $x = -5$, so the vertex is $(-5, -2)$.
Since the graph opens right, the axis goes through the y-coordinate of the vertex—its equation is $y = -2$.

Step 3
If $y = 0$, $x = 11$, so the x-intercept is $(11, 0)$.
To find the y-intercepts, let $x = 0$.

$$0 = 4y^2 + 16y + 11$$

Here $a = 4$, $b = 16$, and $c = 11$.

$$y = \frac{-16 \pm \sqrt{256 - 176}}{8} = \frac{-16 \pm \sqrt{80}}{8}$$
$$= \frac{-16 \pm 4\sqrt{5}}{8} = -2 \pm \tfrac{1}{2}\sqrt{5}$$

The y-intercepts are approximately $(0, -.9)$ and $(0, -3.1)$.

Step 4
By symmetry, $(11, -4)$ is also on the graph.

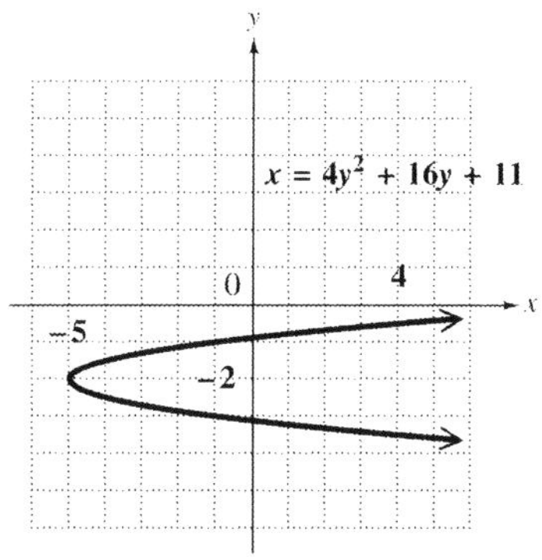

From the graph, we see that the domain is $[-5, \infty)$ and the range is $(-\infty, \infty)$.

19. The graph of $y = 2x^2 + 4x - 3$ is a vertical parabola opening up, so choice **F** is correct. **(F)**

20. The graph of $y = -x^2 + 3x + 5$ is a vertical parabola opening down, so choices **A** and **C** are possibilities. The graph in **A** is narrower than the graph in **C**, so it must correspond to $a = -1$ while the graph in **C** must correspond to $a = -\frac{1}{2}$. **(A)**

21. The graph of $y = -\frac{1}{2}x^2 - x + 1$ is a vertical parabola opening down, so choices **A** and **C** are possibilities. The graph in **C** is wider than the graph in **A**, so it must correspond to $a = -\frac{1}{2}$ while the graph in **A** must correspond to $a = -1$. **(C)**

22. The graph of $x = y^2 + 6y + 3$ is a horizontal parabola opening to the right, so choices **B** and **E** are possibilities. One of the graphs has y-intercepts and the other one doesn't, so check the discriminant.

$$b^2 - 4ac = 6^2 - 4(1)(3) = 24$$

A positive discriminant says we have 2 y-intercepts, so choice **B** is correct. **(B)**

23. The graph of $x = -y^2 - 2y + 4$ is a horizontal parabola opening to the left, so choice **D** is correct. **(D)**

24. The graph of $x = 3y^2 + 6y + 5$ is a horizontal parabola opening to the right. (See Exercise 22.) The discriminant is

$$b^2 - 4ac = 6^2 - 4(3)(5) = -24.$$

A negative discriminant says that there are no y-intercepts. Choice **E** is correct. **(E)**

25. Let $x =$ one number, $60 - x =$ the other number, and $P =$ the product.

$$\begin{aligned} P &= x(60 - x) \\ &= 60x - x^2 \quad \text{or} \quad -x^2 + 60x \end{aligned}$$

This parabola opens down so the maximum occurs at the vertex.

Here $a = -1$, $b = 60$, and $c = 0$. The x-coordinate of the vertex is

$$\frac{-b}{2a} = \frac{-60}{2(-1)} = 30.$$

$x = 30$ when the product is a maximum.

Since $x = 30$, $60 - x = 30$, and the two numbers are 30 and 30.

26. See the solution for Exercise 25. In general, if the sum is S, then the numbers are $\frac{1}{2}S$ and $\frac{1}{2}S$. In this case, the numbers are 5 and 5.

27. Let $x =$ the width of the lot.
Then $640 - 2x =$ the length of the lot.

Area A is length times width.

$$\begin{aligned} A &= x(640 - 2x) \\ A(x) &= 640x - 2x^2 = -2x^2 + 640x \end{aligned}$$

Use the vertex formula.

$$x = \frac{-b}{2a} = \frac{-640}{2(-2)} = 160$$

$$\begin{aligned} A(160) &= -2(160)^2 + 640(160) \\ &= -51{,}200 + 102{,}400 \\ &= 51{,}200 \end{aligned}$$

The graph is a parabola that opens down, so the maximum occurs at the vertex $(160, 51{,}200)$. The maximum area is $51{,}200$ ft^2 if the width x is 160 ft and the length is $640 - 2x = 640 - 2(160) = 320$ ft.

28. Let $x =$ the width of the rectangle, and
$y =$ the length of the rectangle.

The sum of the sides is 100 m, so

$$\begin{aligned} x + x + y + y &= 100. \\ 2x + 2y &= 100 \\ 2y &= 100 - 2x \\ y &= 50 - x \qquad \textit{Divide by 2.} \end{aligned}$$

Since Area = Width • Length,
$A = xy$.
Substitute $50 - x$ for y.

$$\begin{aligned} A &= x(50 - x) \\ &= 50x - x^2 \quad \text{or} \quad -x^2 + 50x \end{aligned}$$

The width x of the rectangle with maximum area will occur at the vertex of $A = -x^2 + 50x$. Use the vertex formula with $a = -1$ and $b = 50$ to get

$$x = \frac{-b}{2a} = \frac{-50}{2(-1)} = 25.$$

continued

A width of 25 m will produce the maximum area. Note that the length, $50 - x$, is also 25 m, so the rectangle is a square.

29. $h(t) = 32t - 16t^2$ or $h(t) = -16t^2 + 32t$

Here, $a = -16 < 0$, so the parabola opens down. The time it takes to reach the maximum height and the maximum height are given by the vertex of the parabola. Use the vertex formula to find that

$$t = \frac{-b}{2a} = \frac{-32}{2(-16)} = \frac{-32}{-32} = 1,$$

and $h(t) = -16(1)^2 + 32(1) = -16 + 32 = 16$.

The vertex is $(1, 16)$, so the maximum height is 16 feet which occurs when the time is 1 second. The object hits the ground when $h = 0$.

$$\begin{aligned} 0 &= -16t^2 + 32t \\ 0 &= -16t(t - 2) \end{aligned}$$

$$\begin{aligned} -16t = 0 \quad &\text{or} \quad t - 2 = 0 \\ t = 0 \quad &\text{or} \quad t = 2 \end{aligned}$$

It takes 2 seconds for the object to hit the ground.

30. $s(t) = -16t^2 + 400t$
The parabola opens down so the maximum height occurs at the vertex.

$$t = \frac{-b}{2a} = \frac{-400}{2(-16)} = \frac{400}{32} = \frac{25}{2} = 12.5$$

$$\begin{aligned} s(12.5) &= -16(12.5)^2 + 400(12.5) \\ &= -2500 + 5000 = 2500 \end{aligned}$$

The maximum height is 2500 feet, and it occurs at 12.5 seconds.

31. The number of people on the plane is $100 - x$ since x is the number of unsold seats. The price per seat is $200 + 4x$.

(a) The total revenue received for the flight is found by multiplying the number of seats by the price per seat. Thus, the revenue is

$$\begin{aligned} R(x) &= (100 - x)(200 + 4x) \\ &= 20{,}000 + 400x - 200x - 4x^2 \\ \text{or} \quad R(x) &= 20{,}000 + 200x - 4x^2. \end{aligned}$$

(b) Find the vertex.
Here $a = -4$, $b = 200$, and $c = 20{,}000$.

$$x = \frac{-b}{2a} = \frac{-200}{2(-4)} = 25$$

The number of unsold seats that will produce the maximum revenue is 25.

(c)
$$\begin{aligned} R(25) &= 20{,}000 + 200(25) - 4(25)^2 \\ &= 20{,}000 + 5000 - 2500 \\ &= 22{,}500 \end{aligned}$$

The maximum revenue is \$22,500.

32. The number of people on the bus is $42 - x$ since x is the number of unsold seats. The price per seat is $48 + 2x$.

(a) Thus, the revenue is

$$\begin{aligned} R(x) &= (42 - x)(48 + 2x) \\ &= 2016 + 84x - 48x - 2x^2 \\ &= 2016 + 36x - 2x^2 \\ \text{or} \quad R(x) &= -2x^2 + 36x + 2016. \end{aligned}$$

(b) The parabola opens downward, so the maximum occurs at the vertex. Here, $a = -2$, $b = 36$, and $c = 2016$.

$$x = \frac{-b}{2a} = \frac{-36}{2(-2)} = 9$$

9 unsold seats produce the maximum revenue.

(c)
$$\begin{aligned} R(9) &= -2(9)^2 + 36(9) + 2016 \\ &= 2178 \end{aligned}$$

The maximum revenue is \$2178.

33. **(a)** Since the parabola opens up, the y-value of the vertex of this graph is a minimum.

(b) $f(x) = .228x^2 - 2.57x + 8.97$

To find the vertex, use $\left(\frac{-b}{2a}, f\left(\frac{-b}{2a}\right)\right)$, where $a = .228$, $b = -2.57$, and $c = 8.97$.

$$x = \frac{-b}{2a} = -\frac{-2.57}{2(.228)} \approx 5.6$$

Rounding down to the nearest year, the minimum percent increase occurred in $1990 + 5 = 1995$.

$$\begin{aligned} f(5.6) &= .228(5.6)^2 - 2.57(5.6) + 8.97 \\ &\approx 1.73 \end{aligned}$$

To the nearest tenth, the percent increase in 1995 was 1.7%.

34. $f(x) = -.386x^2 + 1.28x + 11.3$

(a) Since the graph opens down, the vertex is a maximum.

(b) The x-value of the vertex is given by

$$x = \frac{-b}{2a} = \frac{-1.28}{2(-.386)} \approx 1.658 \approx 1.7$$

The year was $1998 + 1 = 1999$.

$f(1.7) \approx 12.4$ million, which is the maximum number of tickets sold.

35. $f(x) = -20.57x^2 + 758.9x - 3140$

(a) The coefficient of x^2 is negative because the parabola opens down.

(b) Use the vertex formula.

$$x = \frac{-b}{2a} = \frac{-758.9}{2(-20.57)} \approx 18.45$$
$$f(18.45) \approx 3860$$

The vertex is approximately $(18.45, 3860)$.

(c) 18 corresponds to 2018, so in 2018 social security assets will reach their maximum value of \$3860 billion.

36. **(a)** The graph is not the graph of a function since it clearly does not pass the vertical line test.

(b) The leftmost dot represents the vertex. It corresponds to an investment mixture of 80% U.S. stocks and 20% foreign stocks. The relative risk at this point appears to be about 15.8%. The return on investment at this point appears to be about 14.2%.

(c) The point on the graph that represents the riskiest investment mixture corresponds to 100% foreign stocks. The average annual total return is about 16%.

10.7 Quadratic and Rational Inequalities

10.7 Margin Exercises

1. **(a)** $x^2 + 6x + 8 > 0$

The x-intercepts are $(-2, 0)$ and $(-4, 0)$.

Notice from the graph that x-values less than -4 or greater than -2 result in y-values *greater than* 0. Therefore, the solution set is $(-\infty, -4) \cup (-2, \infty)$.

(b) $x^2 + 6x + 8 < 0$

Notice from the graph that x-values between -4 and -2 result in y-values *less than* 0. Therefore, the solution set is $(-4, -2)$.

2. $f(x) = x^2 + 3x - 4 = (x + 4)(x - 1)$

$$x = \frac{-b}{2a} = \frac{-3}{2} \text{ and } f\left(\frac{-3}{2}\right) = -\frac{25}{4}.$$

The vertex is $\left(-\frac{3}{2}, -\frac{25}{4}\right)$.

The y-intercept is $(0, -4)$ and the x-intercepts are $(-4, 0)$ and $(1, 0)$. The parabola opens up.

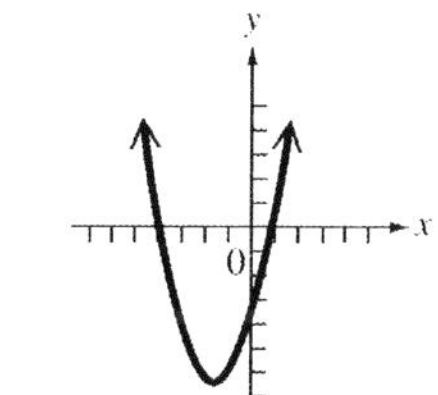

$f(x) = x^2 + 3x - 4$

(a) $x^2 + 3x - 4 \geq 0$

Notice from the graph that x-values less than or equal to -4 or greater than or equal to 1 result in y-values *greater than or equal to* 0. Therefore, the solution set is $(-\infty, -4] \cup [1, \infty)$.

(b) $x^2 + 3x - 4 \leq 0$

Notice from the graph that x-values between -4 and 1, including -4 and 1, result in y-values *less than or equal to* 0. Therefore, the solution set is $[-4, -1]$.

3. $x^2 - x - 12 > 0$

Let $x = 5$ in the inequality.

$$5^2 - 5 - 12 > 0 \quad ?$$
$$25 - 5 - 12 > 0 \quad ?$$
$$8 > 0 \quad \textit{True}$$

Yes, 5 satisfies $x^2 - x - 12 > 0$.

4. **(a)** $x^2 + x - 6 > 0$

Use factoring to solve the quadratic equation

$$x^2 + x - 6 = 0.$$
$$(x + 3)(x - 2) = 0$$
$$x + 3 = 0 \quad \text{or} \quad x - 2 = 0$$
$$x = -3 \quad \text{or} \quad x = 2$$

Locate the numbers -3 and 2 that divide the number line into three intervals A, B, and C.

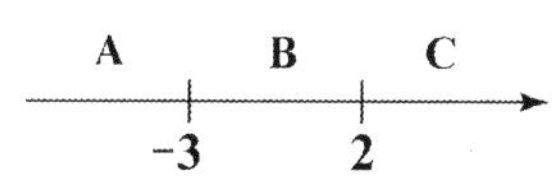

Choose a number from each interval to substitute in the inequality

$$x^2 + x - 6 > 0.$$

Interval A: Let $x = -4$.

$$(-4)^2 + (-4) - 6 > 0 \quad ?$$
$$16 - 4 - 6 > 0 \quad ?$$
$$6 > 0 \quad \textit{True}$$

Interval B: Let $x = 0$.

$$0^2 + 0 - 6 > 0 \quad ?$$
$$-6 > 0 \quad \textit{False}$$

continued

Interval C: Let $x = 3$.

$$3^2 + 3 - 6 > 0 \quad ?$$
$$9 + 3 - 6 > 0 \quad ?$$
$$6 > 0 \quad \textit{True}$$

The numbers in Intervals A and C are solutions. The numbers -3 and 2 are not included because of $>$.

Solution set: $(-\infty, -3) \cup (2, \infty)$

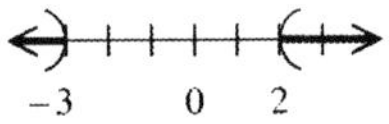

(b) $3m^2 - 13m - 10 \le 0$
Solve the quadratic equation
$3m^2 - 13m - 10 = 0$.
$(3m + 2)(m - 5) = 0$

$$3m + 2 = 0 \quad \text{or} \quad m - 5 = 0$$
$$m = -\tfrac{2}{3} \quad \text{or} \quad m = 5$$

Locate these numbers on a number line.

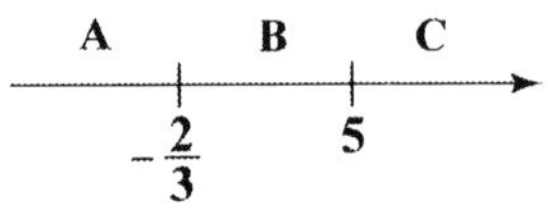

Choose a number from each interval to substitute in the inequality

$$3m^2 - 13m - 10 \le 0.$$

Interval A: Let $m = -1$.

$$3(-1)^2 - 13(-1) - 10 \le 0 \quad ?$$
$$3 + 13 - 10 \le 0 \quad ?$$
$$6 \le 0 \quad \textit{False}$$

Interval B: Let $m = 0$.

$$3(0)^2 - 13(0) - 10 \le 0 \quad ?$$
$$-10 \le 0 \quad \textit{True}$$

Interval C: Let $m = 6$.

$$3(6)^2 - 13(6) - 10 \le 0 \quad ?$$
$$108 - 78 - 10 \le 0 \quad ?$$
$$20 \le 0 \quad \textit{False}$$

The numbers in interval B, including $-\frac{2}{3}$ and 5 because of $\le$, are solutions.

Solution set: $\left[-\frac{2}{3}, 5\right]$

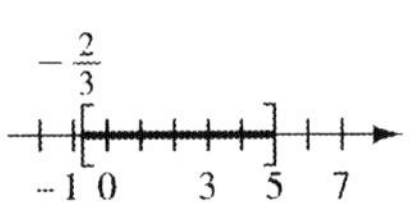

5. **(a)** $(3k - 2)^2 > -2$

The square of any real number is always greater than or equal to 0, so any real number satisfies this inequality.

Solution set: $(-\infty, \infty)$

(b) $(5z + 3)^2 < -3$

The square of a real number is never negative.

Solution set: $\emptyset$

6. **(a)** $(x - 3)(x + 2)(x + 1) > 0$
Solve the equation
$(x - 3)(x + 2)(x + 1) = 0$.

$$x - 3 = 0 \quad \text{or} \quad x + 2 = 0 \quad \text{or} \quad x + 1 = 0$$
$$x = 3 \quad \text{or} \quad x = -2 \quad \text{or} \quad x = -1$$

Locate these numbers and the intervals A, B, C, and D on a number line.

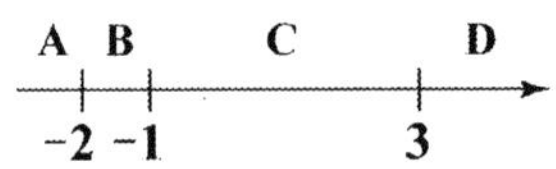

Test a number from each interval in the inequality

$$(x - 3)(x + 2)(x + 1) > 0.$$

Interval A: Let $x = -3$.

$$(-3 - 3)(-3 + 2)(-3 + 1) > 0 \quad ?$$
$$-6(-1)(-2) > 0 \quad ?$$
$$-12 > 0 \quad \textit{False}$$

Interval B: Let $x = -\frac{3}{2}$.

$$\left(-\tfrac{3}{2} - 3\right)\left(-\tfrac{3}{2} + 2\right)\left(-\tfrac{3}{2} + 1\right) > 0 \quad ?$$
$$-\tfrac{9}{2}\left(\tfrac{1}{2}\right)\left(-\tfrac{1}{2}\right) > 0 \quad ?$$
$$\tfrac{9}{8} > 0 \quad \textit{True}$$

Interval C: Let $x = 0$.

$$(0 - 3)(0 + 2)(0 + 1) > 0 \quad ?$$
$$-3(2)(1) > 0 \quad ?$$
$$-6 > 0 \quad \textit{False}$$

Interval D: Let $x = 4$.

$$(4 - 3)(4 + 2)(4 + 1) > 0 \quad ?$$
$$1(6)(5) > 0 \quad ?$$
$$30 > 0 \quad \textit{True}$$

The numbers in Intervals B and D, not including $-2, -1$, or 3 because of $>$, are solutions.

Solution set: $(-2, -1) \cup (3, \infty)$

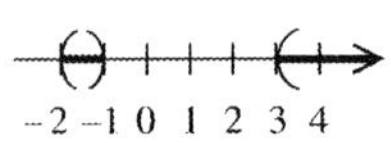

(b) $(k-5)(k+1)(k-3) \le 0$
Solve the equation
$(k-5)(k+1)(k-3) = 0.$

$$\begin{array}{lllll} k-5=0 & \text{or} & k+1=0 & \text{or} & k-3=0 \\ k=5 & \text{or} & k=-1 & \text{or} & k=3 \end{array}$$

Locate these numbers on a number line.

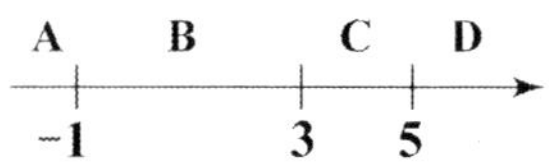

Test a number from each interval in the inequality

$$(k-5)(k+1)(k-3) \le 0.$$

Interval A: Let $k=-2$.

$$(-2-5)(-2+1)(-2-3) \le 0 \quad ?$$
$$-7(-1)(-5) \le 0 \quad ?$$
$$-35 \le 0 \quad \textit{True}$$

Interval B: Let $k=0$.

$$(0-5)(0+1)(0-3) \le 0 \quad ?$$
$$-5(1)(-3) \le 0 \quad ?$$
$$15 \le 0 \quad \textit{False}$$

Interval C: Let $k=4$.

$$(4-5)(4+1)(4-3) \le 0 \quad ?$$
$$-1(5)(1) \le 0 \quad ?$$
$$-5 \le 0 \quad \textit{True}$$

Interval D: Let $k=6$.

$$(6-5)(6+1)(6-3) \le 0 \quad ?$$
$$1(7)(3) \le 0 \quad ?$$
$$21 \le 0 \quad \textit{False}$$

The numbers in Intervals A and C, including -1, 3, and 5 because of $\le$, are solutions.

Solution set: $(-\infty,-1] \cup [3,5]$

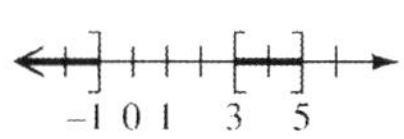

7. **(a)** $$\frac{2}{x-4} < 3$$

Write the inequality so that 0 is on one side.

$$\frac{2}{x-4} - 3 < 0$$
$$\frac{2}{x-4} - \frac{3(x-4)}{x-4} < 0$$
$$\frac{2-3x+12}{x-4} < 0$$
$$\frac{-3x+14}{x-4} < 0$$

The number $\frac{14}{3}$ makes the numerator 0, and 4 makes the denominator 0. These two numbers determine three intervals.

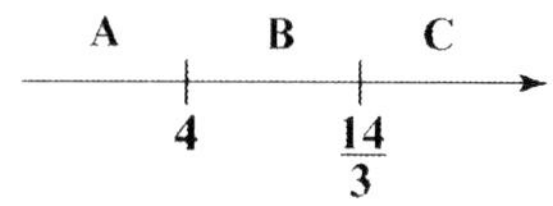

Test a number from each interval in the inequality

$$\frac{2}{x-4} < 3.$$

Interval A: Let $x=0$.

$$\frac{2}{0-4} < 3 \quad ?$$
$$-\tfrac{1}{2} < 3 \quad \textit{True}$$

Interval B: Let $x=\frac{13}{3}$.

$$\frac{2}{\frac{13}{3}-4} < 3 \quad ?$$
$$6 < 3 \quad \textit{False}$$

Interval C: Let $x=5$.

$$\frac{2}{5-4} < 3 \quad ?$$
$$2 < 3 \quad \textit{True}$$

The solution set includes numbers in Intervals A and C, excluding endpoints.
Solution set: $(-\infty,4) \cup \left(\frac{14}{3},\infty\right)$

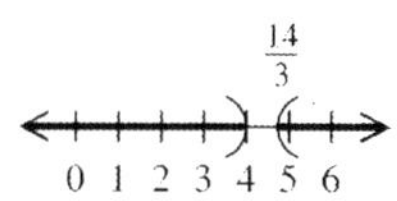

(b) $$\frac{5}{z+1} > 4$$

Write the inequality so that 0 is on one side.

$$\frac{5}{z+1} - 4 > 0$$
$$\frac{5}{z+1} - \frac{4(z+1)}{z+1} > 0$$
$$\frac{5-4z-4}{z+1} > 0$$
$$\frac{-4z+1}{z+1} > 0$$

The number $\frac{1}{4}$ makes the numerator 0, and -1 makes the denominator 0. These two numbers determine three intervals.

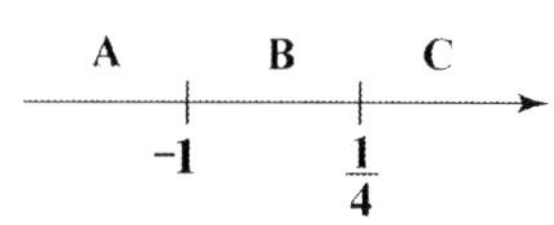

Test a number from each interval in the inequality

$$\frac{5}{z+1} > 4.$$

continued

Interval A: Let $z = -2$.

$$\frac{5}{-2+1} > 4 \quad ?$$

$$-5 > 4 \quad \textit{False}$$

Interval B: Let $z = 0$.

$$\frac{5}{0+1} > 4 \quad ?$$

$$5 > 4 \quad \textit{True}$$

Interval C: Let $z = 1$.

$$\frac{5}{1+1} > 4 \quad ?$$

$$\frac{5}{2} > 4 \quad \textit{False}$$

The numbers in Interval B, not including -1 or $\frac{1}{4}$, are solutions.

Solution set: $\left(-1, \frac{1}{4}\right)$

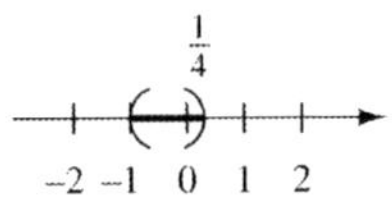

8.
$$\frac{k+2}{k-1} \leq 5$$

Write the inequality so that 0 is on one side.

$$\frac{k+2}{k-1} - 5 \leq 0$$

$$\frac{k+2}{k-1} - \frac{5(k-1)}{k-1} \leq 0$$

$$\frac{k+2-5k+5}{k-1} \leq 0$$

$$\frac{-4k+7}{k-1} \leq 0$$

The number $\frac{7}{4}$ makes the numerator 0, and 1 makes the denominator 0. These two numbers determine three intervals.

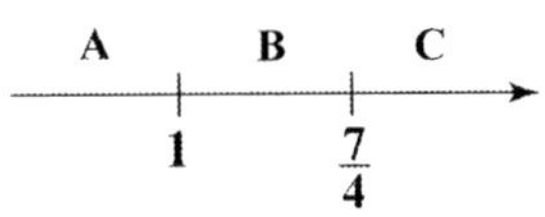

Test a number from each interval in the inequality

$$\frac{k+2}{k-1} \leq 5.$$

Interval A: Let $k = 0$.

$$\frac{0+2}{0-1} \leq 5 \quad ?$$

$$-2 \leq 5 \quad \textit{True}$$

Interval B: Let $k = \frac{3}{2}$.

$$\frac{\frac{3}{2}+2}{\frac{3}{2}-1} \leq 5 \quad ?$$

$$7 \leq 5 \quad \textit{False}$$

Interval C: Let $k = 2$.

$$\frac{2+2}{2-1} \leq 5 \quad ?$$

$$4 \leq 5 \quad \textit{True}$$

The numbers in Intervals A and C are solutions. 1 is not in the solution set (since it makes the denominator 0), but $\frac{7}{4}$ is.

Solution set: $(-\infty, 1) \cup \left[\frac{7}{4}, \infty\right)$

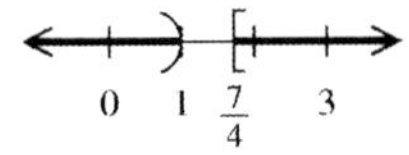

10.7 Section Exercises

1. **(a)** The x-intercepts determine the solutions of the equation $x^2 - 4x + 3 = 0$. From the graph, the solution set is $\{1, 3\}$.

(b) The x-values of the points on the graph that are *above* the x-axis form the solution set of the inequality $x^2 - 4x + 3 > 0$. From the graph, the solution set is $(-\infty, 1) \cup (3, \infty)$.

(c) The x-values of the points on the graph that are *below* the x-axis form the solution set of the inequality $x^2 - 4x + 3 < 0$. From the graph, the solution set is $(1, 3)$.

2. **(a)** The x-intercepts determine the solutions of the equation $3x^2 + 10x - 8 = 0$. From the graph, the solution set is $\left\{-4, \frac{2}{3}\right\}$.

(b) The x-values of the points on the graph that are *above* the x-axis form the solution set of the inequality $3x^2 + 10x - 8 > 0$. From the graph, the solution set for $3x^2 + 10x - 8 \geq 0$ is $(-\infty, -4] \cup \left[\frac{2}{3}, \infty\right)$.

(c) The x-values of the points on the graph that are *below* the x-axis form the solution set of the inequality $3x^2 + 10x - 8 < 0$. From the graph, the solution set is $\left(-4, \frac{2}{3}\right)$.

3. **(a)** The x-intercepts determine the solutions of the equation $-2x^2 - x + 15 = 0$. From the graph, the solution set is $\left\{-3, \frac{5}{2}\right\}$.

(b) The x-values of the points on the graph that are *above* the x-axis form the solution set of the inequality $-2x^2 - x + 15 > 0$. From the graph, the solution set for $-2x^2 - x + 15 \geq 0$ is $\left[-3, \frac{5}{2}\right]$.

(c) The x-values of the points on the graph that are *below* the x-axis form the solution set of the inequality $-2x^2 - x + 15 < 0$. From the graph, the solution set for $-2x^2 - x + 15 \leq 0$ is $(-\infty, -3] \cup \left[\frac{5}{2}, \infty\right)$.

4. **(a)** The x-intercepts determine the solutions of the equation $-x^2 + 3x + 10 = 0$. From the graph, the solution set is $\{-2, 5\}$.

(b) The x-values of the points on the graph that are *above* the x-axis form the solution set of the inequality $-x^2 + 3x + 10 > 0$. From the graph, the solution set for $-x^2 + 3x + 10 \geq 0$ is $[-2, 5]$.

(c) The x-values of the points on the graph that are *below* the x-axis form the solution set of the inequality $-x^2 + 3x + 10 < 0$. From the graph, the solution set for $-x^2 + 3x + 10 \leq 0$ is $(-\infty, -2] \cup [5, \infty)$.

5. Include the endpoints if the symbol is $\leq$ or $\geq$. Exclude the endpoints if the symbol is $<$ or $>$.

6. The solution set of the inequality $x^2 + x - 12 \geq 0$ includes all the numbers *not* in the solution set of the inequality $x^2 + x - 12 < 0$, which is given as $(-4, 3)$. So the solution set is

$$(-\infty, -4] \cup [3, \infty).$$

7. $(x + 1)(x - 5) > 0$
Solve the equation
$(x + 1)(x - 5) = 0$.

$x + 1 = 0$ or $x - 5 = 0$
$x = -1$ or $x = 5$

The numbers -1 and 5 divide a number line into three intervals: A, B, and C.

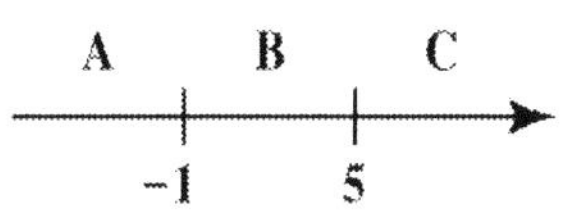

Test a number from each interval in the original inequality.

Interval A: Let $x = -2$.
$(x + 1)(x - 5) > 0$
$(-2 + 1)(-2 - 5) > 0$?
$-1(-7) > 0$?
$7 > 0$ *True*

Interval B: Let $x = 0$.
$(0 + 1)(0 - 5) > 0$?
$-5 > 0$ *False*

Interval C: Let $x = 6$.
$(6 + 1)(6 - 5) > 0$?
$7 > 0$ *True*

The solution set includes the numbers in Intervals A and C, excluding -1 and 5 because of $>$.
Solution set: $(-\infty, -1) \cup (5, \infty)$

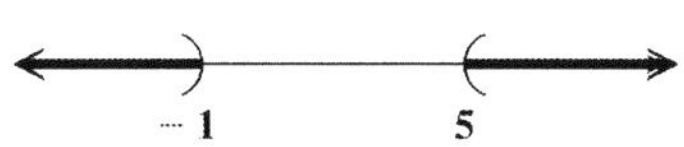

8. $(m + 6)(m - 2) > 0$
Solve the equation
$(m + 6)(m - 2) = 0$.

$m + 6 = 0$ or $m - 2 = 0$
$m = -6$ or $m = 2$

The numbers -6 and 2 divide a number line into three intervals: A, B, and C.

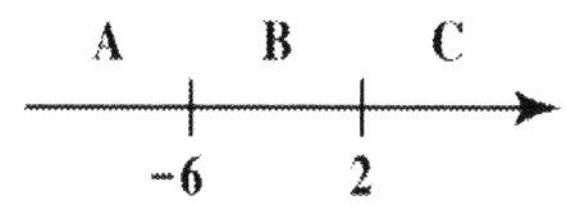

Test a number from each interval in the original inequality.

Interval A: Let $m = -7$.
$(m + 6)(m - 2) > 0$
$(-7 + 6)(-7 - 2) > 0$?
$-1(-9) > 0$?
$9 > 0$ *True*

Interval B: Let $m = 0$.
$6(-2) > 0$?
$-12 > 0$ *False*

Interval C: Let $m = 3$.
$9(1) > 0$?
$9 > 0$ *True*

The solution set includes the numbers in Intervals A and C, excluding -6 and 2 because of $>$.
Solution set: $(-\infty, -6) \cup (2, \infty)$

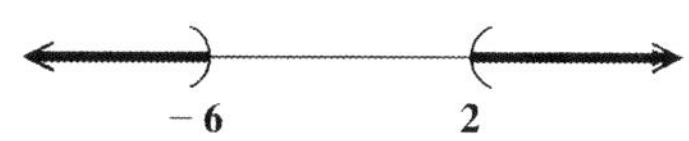

9. $(r + 4)(r - 6) < 0$
Solve the equation
$(r + 4)(r - 6) = 0$.

$r + 4 = 0$ or $r - 6 = 0$
$r = -4$ or $r = 6$

These numbers divide a number line into three intervals: A, B, and C.

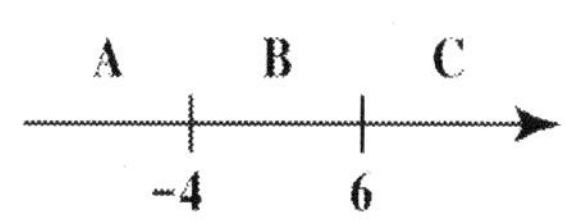

Test a number from each interval in the original inequality.

continued

Interval A: Let $r = -5$.

$(r+4)(r-6) < 0$

$(-5+4)(-5-6) < 0$?

$-1(-11) < 0$?

$11 < 0$ *False*

Interval B: Let $r = 0$.

$4(-6) < 0$?

$-24 < 0$ *True*

Interval C: Let $r = 7$.

$(7+4)(7-6) < 0$?

$11(1) < 0$?

$11 < 0$ *False*

The solution set includes Interval B, excluding -4 and 6 because of $<$.
Solution set: $(-4, 6)$

10. $(x+4)(x-8) < 0$

$(x+4)(x-8) = 0$

$x+4=0$ or $x-8=0$

$x=-4$ or $x=8$

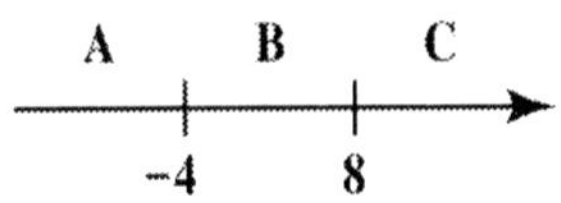

Test a number from each interval in the original inequality.

Interval A: Let $x = -5$.

$(x+4)(x-8) < 0$

$-1(-13) < 0$?

$13 < 0$ *False*

Interval B: Let $x = 0$.

$4(-8) < 0$?

$-32 < 0$ *True*

Interval C: Let $x = 9$.

$13(1) < 0$?

$13 < 0$ *False*

The solution set includes Interval B, excluding -4 and 8 because of $<$.
Solution set: $(-4, 8)$

11. $x^2 - 4x + 3 \geq 0$

Solve the equation

$x^2 - 4x + 3 = 0$.

$(x-1)(x-3) = 0$

$x-1=0$ or $x-3=0$

$x=1$ or $x=3$

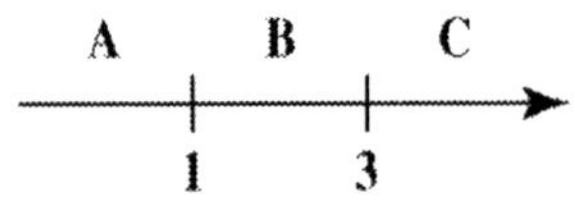

Test a number from each interval in the original inequality.

Interval A: Let $x = 0$.

$3 \geq 0$ *True*

Interval B: Let $x = 2$.

$2^2 - 4(2) + 3 \geq 0$?

$-1 \geq 0$ *False*

Interval C: Let $x = 4$.

$4^2 - 4(4) + 3 \geq 0$?

$3 \geq 0$ *True*

The solution set includes the numbers in Intervals A and C, including 1 and 3 because of $\geq$.
Solution set: $(-\infty, 1] \cup [3, \infty)$

12. $m^2 - 3m - 10 \geq 0$

Solve the equation

$m^2 - 3m - 10 = 0$

$(m+2)(m-5) = 0$

$m+2=0$ or $m-5=0$

$m=-2$ or $m=5$

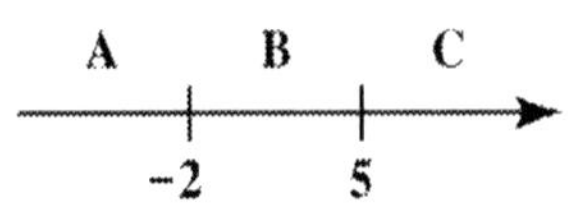

Test a number from each interval in the original inequality.

Interval A: Let $m = -3$.

$(-3)^2 - 3(-3) - 10 \geq 0$?

$8 \geq 0$ *True*

Interval B: Let $m = 0$.

$-10 \geq 0$ *False*

Interval C: Let $m = 6$.

$6^2 - 3(6) - 10 \geq 0$?

$8 \geq 0$ *True*

The solution set includes the numbers in Intervals A and C, including -2 and 5 because of $\geq$.
Solution set: $(-\infty, -2] \cup [5, \infty)$

13. $10t^2 + 9t \geq 9$

$10t^2 + 9t - 9 \geq 0$

Solve the equation

$10t^2 + 9t - 9 = 0.$

$(2t + 3)(5t - 3) = 0$

$2t + 3 = 0$ or $5t - 3 = 0$

$t = -\frac{3}{2}$ or $t = \frac{3}{5}$

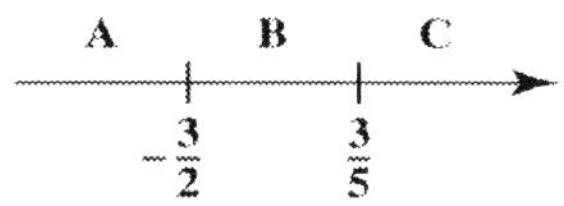

Test a number from each interval in the original inequality.

Interval A: Let $t = -2$.

$10(-2)^2 + 9(-2) \geq 9$?

$40 - 18 \geq 9$?

$22 \geq 9$ *True*

Interval B: Let $t = 0$.

$0 \geq 9$ *False*

Interval C: Let $t = 1$.

$10(1)^2 + 9(1) \geq 9$?

$10 + 9 \geq 9$?

$19 \geq 9$ *True*

The solution set includes the numbers in Intervals A and C, including $-\frac{3}{2}$ and $\frac{3}{5}$ because of $\geq$.
Solution set: $\left(-\infty, -\frac{3}{2}\right] \cup \left[\frac{3}{5}, \infty\right)$

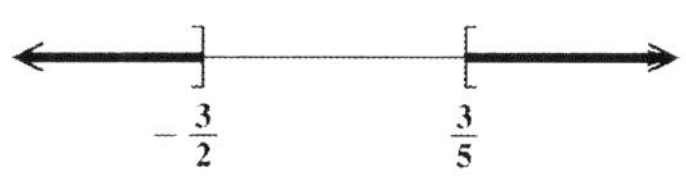

14. $3r^2 + 10r \geq 8$

$3r^2 + 10r = 8$

$3r^2 + 10r - 8 = 0$

$(r + 4)(3r - 2) = 0$

$r + 4 = 0$ or $3r - 2 = 0$

$r = -4$ or $r = \frac{2}{3}$

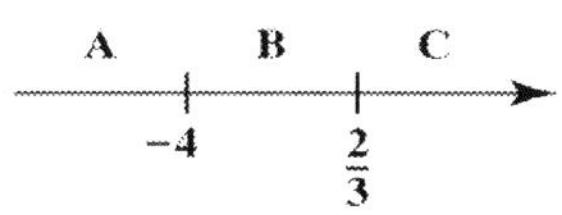

Test a number from each interval in the inequality

$3r^2 + 10r \geq 8.$

Interval A: Let $r = -5$.

$75 - 50 \geq 8$?

$25 \geq 8$ *True*

Interval B: Let $r = 0$.

$0 \geq 8$ *False*

Interval C: Let $r = 1$.

$3 + 10 \geq 8$?

$13 \geq 8$ *True*

The solution set includes the numbers in Intervals A and C, including -4 and $\frac{2}{3}$ because of $\geq$.
Solution set: $(-\infty, -4] \cup \left[\frac{2}{3}, \infty\right)$

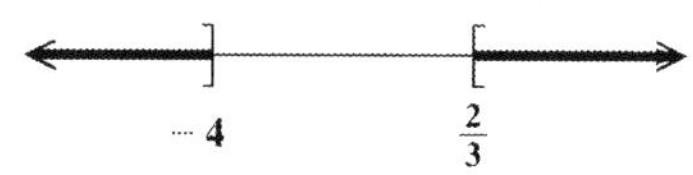

15. $9p^2 + 3p < 2$

Solve the equation

$9p^2 + 3p = 2.$

$9p^2 + 3p - 2 = 0$

$(3p - 1)(3p + 2) = 0$

$3p - 1 = 0$ or $3p + 2 = 0$

$p = \frac{1}{3}$ or $p = -\frac{2}{3}$

These numbers divide a number line into three intervals: A, B, and C.

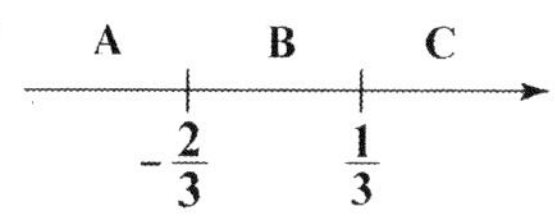

Test a number from each interval in the original inequality.

Interval A: Let $p = -1$.

$9p^2 + 3p < 2$

$9(-1)^2 + 3(-1) < 2$?

$9 - 3 < 2$?

$6 < 2$ *False*

Interval B: Let $p = 0$.

$0 < 2$ *True*

Interval C: Let $p = 1$.

$9(1)^2 + 3(1) < 2$?

$9 + 3 < 2$?

$12 < 2$ *False*

The solution set includes Interval B, excluding $-\frac{2}{3}$ and $\frac{1}{3}$ because of $<$.

Solution set: $\left(-\frac{2}{3}, \frac{1}{3}\right)$

16. $2x^2 + x < 15$

Solve the equation

$$2x^2 + x = 15.$$
$$2x^2 + x - 15 = 0$$
$$(2x - 5)(x + 3) = 0$$

$2x - 5 = 0$ or $x + 3 = 0$

$x = \frac{5}{2}$ or $x = -3$

These numbers divide a number line into three intervals: A, B, and C.

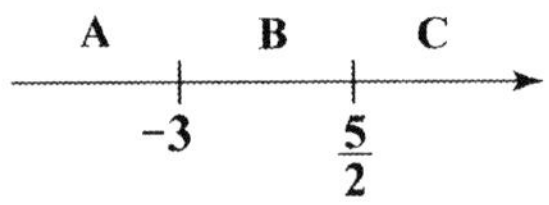

Test a number from each interval in the original inequality.

Interval A: Let $x = -4$.

$2x^2 + x < 15$

$2(-4)^2 + (-4) < 15$?

$28 < 15$ *False*

Interval B: Let $x = 0$.

$0 < 15$ *True*

Interval C: Let $x = 4$.

$2(4)^2 + 4 < 15$?

$36 < 15$ *False*

The solution set includes Interval B, excluding -3 and $\frac{5}{2}$ because of $<$.

Solution set: $\left(-3, \frac{5}{2}\right)$

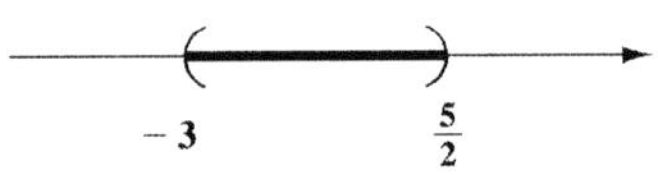

17. $6x^2 + x \geq 1$

$6x^2 + x - 1 \geq 0$

Solve the equation

$$6x^2 + x - 1 = 0.$$
$$(2x + 1)(3x - 1) = 0$$

$2x + 1 = 0$ or $3x - 1 = 0$

$x = -\frac{1}{2}$ or $x = \frac{1}{3}$

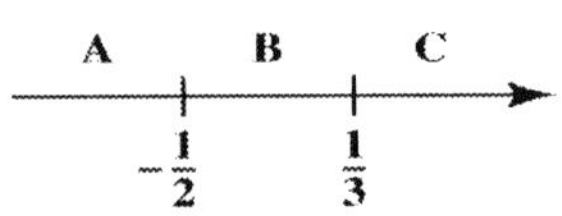

Test a number from each interval in the inequality.

$$6x^2 + x \geq 1.$$

Interval A: Let $x = -1$.

$6(-1)^2 + (-1) \geq 1$?

$5 \geq 1$ *True*

Interval B: Let $x = 0$.

$0 \geq 1$ *False*

Interval C: Let $x = 1$.

$6(1)^2 + 1 \geq 1$?

$7 \geq 1$ *True*

The solution set includes the numbers in Intervals A and C, including $-\frac{1}{2}$ and $\frac{1}{3}$ because of $\geq$.

Solution set: $\left(-\infty, -\frac{1}{2}\right] \cup \left[\frac{1}{3}, \infty\right)$

18. $4m^2 + 7m \geq -3$

$$4m^2 + 7m = -3$$
$$4m^2 + 7m + 3 = 0$$
$$(m + 1)(4m + 3) = 0$$

$m + 1 = 0$ or $4m + 3 = 0$

$m = -1$ or $m = -\frac{3}{4}$

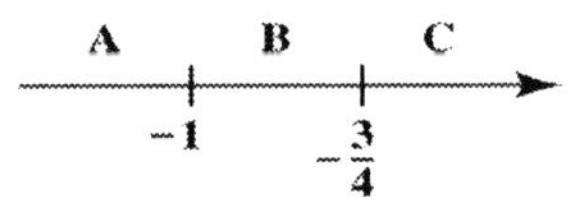

Test a number from each interval in the inequality

$$4m^2 + 7m \geq -3.$$

Interval A: Let $m = -2$.

$4(-2)^2 + 7(-2) \geq -3$?

$2 \geq -3$ *True*

Interval B: Let $m = -\frac{7}{8}$.

$4\left(-\frac{7}{8}\right)^2 + 7\left(-\frac{7}{8}\right) \geq -3$?

$\frac{49}{16} - \frac{49}{8} \geq -3$?

$-\frac{49}{16} \geq -3$ *False*

Interval C: Let $m = 0$.

$0 \geq -3$ *True*

The solution set includes the numbers in Intervals A and C, including -1 and $-\frac{3}{4}$ because of $\geq$.

Solution set: $(-\infty, -1] \cup \left[-\frac{3}{4}, \infty\right)$

19. $x^2 - 6x + 6 \geq 0$
Solve the equation

$$x^2 - 6x + 6 = 0.$$

Since $x^2 - 6x + 6$ does not factor, let $a = 1$, $b = -6$, and $c = 6$ in the quadratic formula.

$$\begin{aligned} x &= \frac{-(-6) \pm \sqrt{(-6)^2 - 4(1)(6)}}{2(1)} \\ &= \frac{6 \pm \sqrt{12}}{2} = \frac{6 \pm 2\sqrt{3}}{2} \\ &= \frac{2\left(3 \pm \sqrt{3}\right)}{2} = 3 \pm \sqrt{3} \end{aligned}$$

$x = 3 + \sqrt{3} \approx 4.7$ or
$x = 3 - \sqrt{3} \approx 1.3$

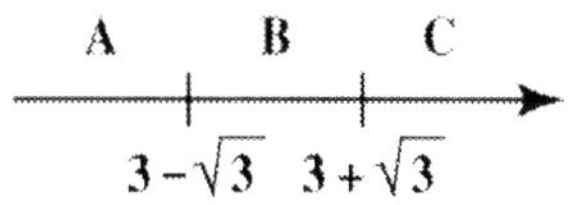

Test a number from each interval in the inequality

$$x^2 - 6x + 6 \geq 0.$$

Interval A: Let $x = 0$.
$6 \geq 0$ *True*
Interval B: Let $x = 3$.
$3^2 - 6(3) + 6 \geq 0$?
$-3 \geq 0$ *False*
Interval C: Let $x = 5$.
$5^2 - 6(5) + 6 \geq 0$?
$1 \geq 0$ *True*

The solution set includes the numbers in Intervals A and C, including $3 - \sqrt{3}$ and $3 + \sqrt{3}$ because of $\geq$.

Solution set: $\left(-\infty, 3 - \sqrt{3}\right] \cup \left[3 + \sqrt{3}, \infty\right)$

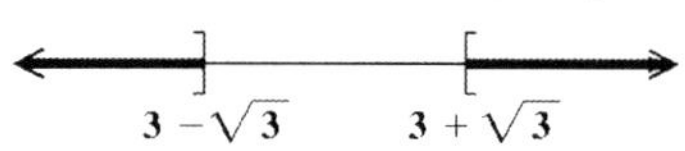

20. $3k^2 - 6k + 2 \leq 0$
Solve the equation $3k^2 - 6k + 2 = 0$ by the quadratic formula. The solutions are
$\frac{3 + \sqrt{3}}{3} \approx 1.6$ and $\frac{3 - \sqrt{3}}{3} \approx .4$.

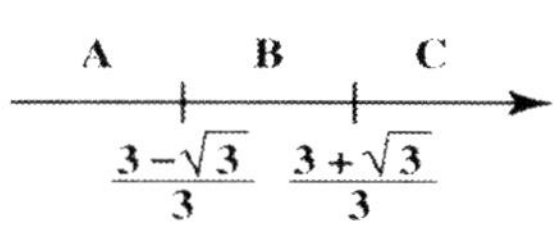

Test a number from each interval in the inequality

$$3k^2 - 6k + 2 \leq 0.$$

Interval A: Let $k = 0$.
$2 \leq 0$ *False*
Interval B: Let $k = 1$.
$3(1)^2 - 6(1) + 2 \leq 0$?
$-1 \leq 0$ *True*
Interval C: Let $k = 2$.
$3(2)^2 - 6(2) + 2 \leq 0$?
$2 \leq 0$ *False*

The solution set includes Interval B, including the endpoints.

Solution set: $\left[\frac{3 - \sqrt{3}}{3}, \frac{3 + \sqrt{3}}{3}\right]$

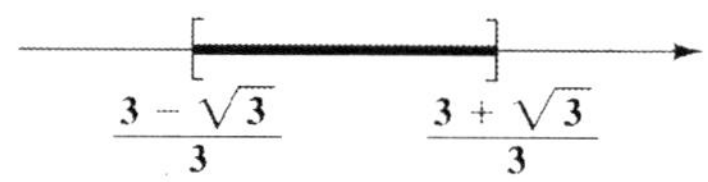

21. $(4 - 3x)^2 \geq -2$

Since $(4 - 3x)^2$ is either 0 or positive, $(4 - 3x)^2$ will always be greater than -2. Therefore, the solution set is $(-\infty, \infty)$.

22. $(6p + 7)^2 \geq -1$

Since $(6p + 7)^2$ is either 0 or positive, $(6p + 7)^2$ will always be greater than -1. Therefore, the solution set is $(-\infty, \infty)$.

23. $(3x + 5)^2 \leq -4$

Since $(3x + 5)^2$ is never negative, $(3x + 5)^2$ will never be less than or equal to a negative number. Therefore, the solution set is $\emptyset$.

24. $(8t + 5)^2 \leq -5$

Since $(8t + 5)^2$ is never negative, $(8t + 5)^2$ will never be less than or equal to a negative number. Therefore, the solution set is $\emptyset$.

25. $(p - 1)(p - 2)(p - 4) < 0$
The numbers 1, 2, and 4 are solutions of the cubic equation

$$(p - 1)(p - 2)(p - 4) = 0.$$

These numbers divide a number line into four intervals.

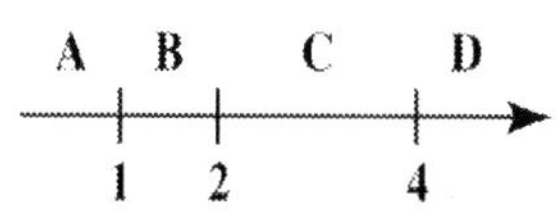

Test a number from each interval in the inequality

$$(p - 1)(p - 2)(p - 4) < 0.$$

continued

Interval A:	Let $p = 0$.	
	$-1(-2)(-4) < 0$?	
	$-8 < 0$	*True*
Interval B:	Let $p = 1.5$.	
	$(1.5-1)(1.5-2)(1.5-4) < 0$?	
	$.5(-.5)(-2.5) < 0$?	
	$.625 < 0$	*False*
Interval C:	Let $p = 3$.	
	$(3-1)(3-2)(3-4) < 0$?	
	$2(1)(-1) < 0$?	
	$-2 < 0$	*True*
Interval D:	Let $p = 5$.	
	$(5-1)(5-2)(5-4) < 0$?	
	$4(3)(1) < 0$?	
	$12 < 0$	*False*

The numbers in Intervals A and C, not including 1, 2, or 4, are solutions.

Solution set: $(-\infty, 1) \cup (2, 4)$

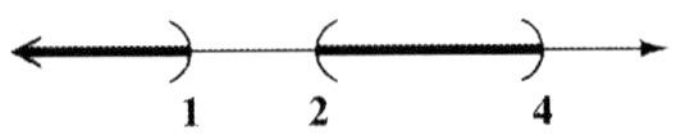

26. $(2r+1)(3r-2)(4r+7) < 0$
The numbers $-\frac{1}{2}$, $\frac{2}{3}$, and $-\frac{7}{4}$ are solutions of the cubic equation

$$(2r+1)(3r-2)(4r+7) = 0.$$

These numbers divide a number line into 4 intervals.

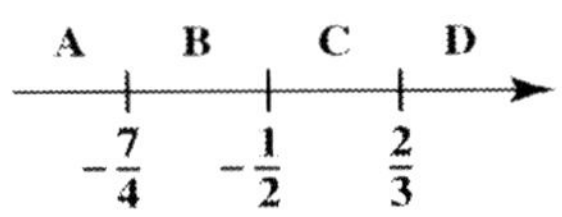

Test a number from each interval in the original inequality.

Interval A:	Let $r = -2$.	
	$-3(-8)(-1) < 0$?	
	$-24 < 0$	*True*
Interval B:	Let $r = -1$.	
	$-1(-5)(3) < 0$?	
	$15 < 0$	*False*
Interval C:	Let $r = 0$.	
	$1(-2)(7) < 0$?	
	$-14 < 0$	*True*
Interval D:	Let $r = 1$.	
	$3(1)(11) < 0$?	
	$33 < 0$	*False*

The solution set includes numbers in Intervals A and C, excluding endpoints.

Solution set: $\left(-\infty, -\frac{7}{4}\right) \cup \left(-\frac{1}{2}, \frac{2}{3}\right)$

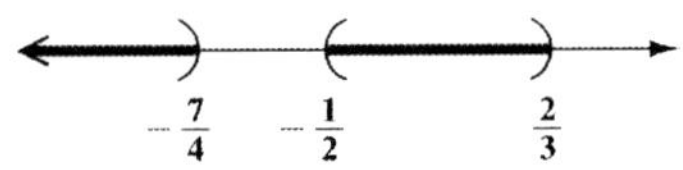

27. $(x-4)(2x+3)(3x-1) \geq 0$
The numbers 4, $-\frac{3}{2}$, and $\frac{1}{3}$ are solutions of the cubic equation

$$(x-4)(2x+3)(3x-1) = 0.$$

These numbers divide a number line into 4 intervals.

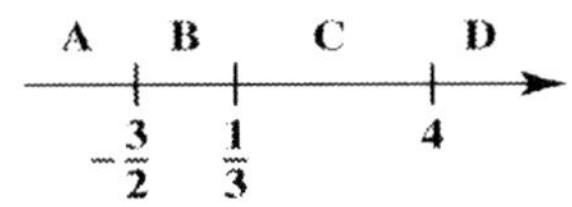

Test a number from each interval in the original inequality.

Interval A:	Let $x = -2$.	
	$-6(-1)(-7) \geq 0$?	
	$-42 \geq 0$	*False*
Interval B:	Let $x = 0$.	
	$-4(3)(-1) \geq 0$?	
	$12 \geq 0$	*True*
Interval C:	Let $x = 1$.	
	$-3(5)(2) \geq 0$?	
	$-30 \geq 0$	*False*
Interval D:	Let $x = 5$.	
	$1(13)(14) \geq 0$?	
	$182 \geq 0$	*True*

The solution set includes numbers in Intervals B and D, including the endpoints.

Solution set: $\left[-\frac{3}{2}, \frac{1}{3}\right] \cup [4, \infty)$

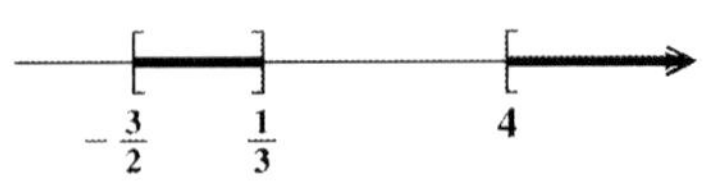

28. $(z+2)(4z-3)(2z+7) \geq 0$
The numbers -2, $\frac{3}{4}$, and $-\frac{7}{2}$ are solutions of the cubic equation

$$(z+2)(4z-3)(2z+7) = 0.$$

These numbers divide a number line into 4 intervals.

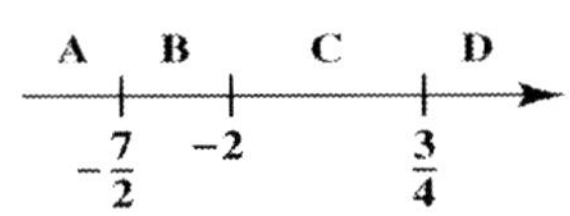

Test a number from each interval in the original inequality.

Interval A: Let $z = -4$.

$-2(-19)(-1) \geq 0$?

$-38 \geq 0$ *False*

Interval B: Let $z = -3$.

$-1(-15)(1) \geq 0$?

$15 \geq 0$ *True*

Interval C: Let $z = 0$.

$2(-3)(7) \geq 0$?

$-42 \geq 0$ *False*

Interval D: Let $z = 1$.

$3(1)(9) \geq 0$?

$27 \geq 0$ *True*

The solution set includes numbers in Intervals B and D, including the endpoints.

Solution set: $\left[-\frac{7}{2}, -2\right] \cup \left[\frac{3}{4}, \infty\right)$

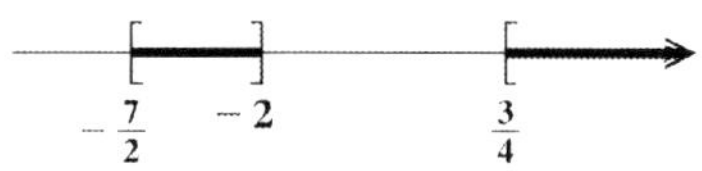

29. $\dfrac{x-1}{x-4} > 0$

The number 1 makes the numerator 0, and 4 makes the denominator 0. These two numbers determine three intervals.

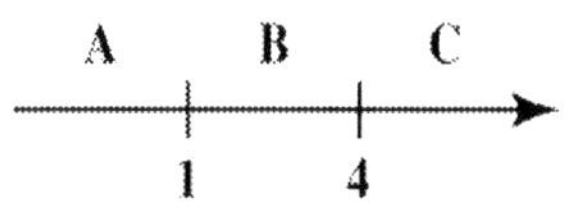

Test a number from each interval in the inequality

$$\frac{x-1}{x-4} > 0.$$

Interval A: Let $x = 0$.

$\dfrac{0-1}{0-4} > 0$?

$\dfrac{1}{4} > 0$ *True*

Interval B: Let $x = 2$.

$\dfrac{2-1}{2-4} > 0$?

$\dfrac{1}{-2} > 0$ *False*

Interval C: Let $x = 5$.

$\dfrac{5-1}{5-4} > 0$?

$4 > 0$ *True*

The solution set includes numbers in Intervals A and C, excluding endpoints.

Solution set: $(-\infty, 1) \cup (4, \infty)$

30. $\dfrac{x+1}{x-5} > 0$

The number -1 makes the numerator 0, and 5 makes the denominator 0. These two numbers determine three intervals.

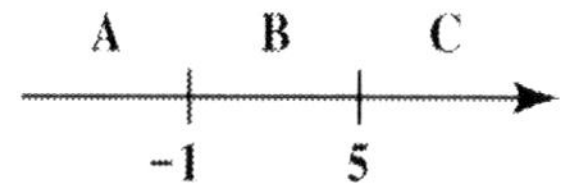

Test a number from each interval in the inequality

$$\frac{x+1}{x-5} > 0.$$

Interval A: Let $x = -2$.

$\dfrac{-2+1}{-2-5} > 0$?

$\dfrac{1}{7} > 0$ *True*

Interval B: Let $x = 0$.

$\dfrac{0+1}{0-5} > 0$?

$-\dfrac{1}{5} > 0$ *False*

Interval C: Let $x = 6$.

$\dfrac{6+1}{6-5} > 0$?

$7 > 0$ *True*

The solution set includes numbers in Intervals A and C, excluding endpoints.

Solution set: $(-\infty, -1) \cup (5, \infty)$

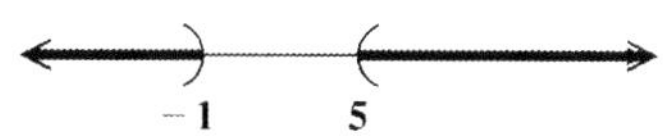

31. $\dfrac{2n+3}{n-5} \leq 0$

The number $-\frac{3}{2}$ makes the numerator 0, and 5 makes the denominator 0. These two numbers determine three intervals.

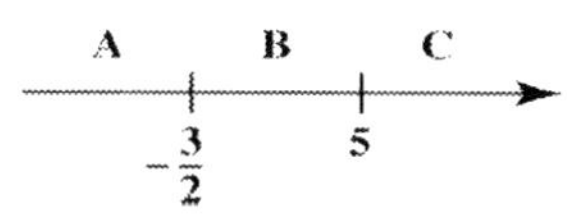

Test a number from each interval in the inequality

$$\frac{2n+3}{n-5} \leq 0.$$

continued

Interval A: Let $n = -2$.

$$\frac{2(-2)+3}{(-2)-5} \le 0 \quad ?$$

$$\frac{1}{7} \le 0 \qquad \textit{False}$$

Interval B: Let $n = 0$.

$$\frac{2(0)+3}{0-5} \le 0 \quad ?$$

$$-\frac{3}{5} \le 0 \qquad \textit{True}$$

Interval C: Let $n = 6$.

$$\frac{2(6)+3}{6-5} \le 0 \quad ?$$

$$15 \le 0 \qquad \textit{False}$$

The solution set includes the points in Interval B. The endpoint 5 is not included since it makes the left side undefined. The endpoint $-\frac{3}{2}$ is included because it makes the left side equal to 0.

Solution set: $\left[-\frac{3}{2}, 5\right)$

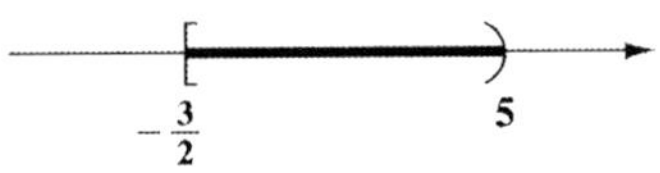

32. $\dfrac{3t+7}{t-3} \le 0$

The number $-\frac{7}{3}$ makes the numerator 0, and 3 makes the denominator 0. These two numbers determine three intervals.

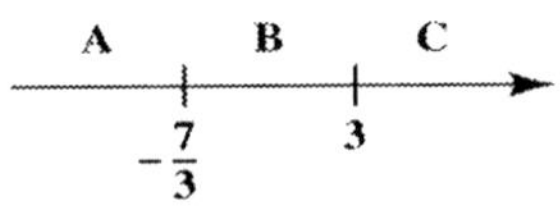

Test a number from each interval in the inequality

$$\frac{3t+7}{t-3} \le 0.$$

Interval A: Let $t = -3$.

$$\frac{3(-3)+7}{(-3)-3} \le 0 \quad ?$$

$$\frac{1}{3} \le 0 \qquad \textit{False}$$

Interval B: Let $t = 0$.

$$\frac{3(0)+7}{0-3} \le 0 \quad ?$$

$$-\frac{7}{3} \le 0 \qquad \textit{True}$$

Interval C: Let $t = 4$.

$$\frac{3(4)+7}{4-3} \le 0 \quad ?$$

$$19 \le 0 \qquad \textit{False}$$

The solution set includes numbers in Interval B, including $-\frac{7}{3}$ but excluding 3, which makes the fraction undefined.

Solution set: $\left[-\frac{7}{3}, 3\right)$

33. $$\frac{8}{x-2} \ge 2$$

Write the inequality so that 0 is on one side.

$$\frac{8}{x-2} - 2 \ge 0$$

$$\frac{8}{x-2} - \frac{2(x-2)}{x-2} \ge 0$$

$$\frac{8-2x+4}{x-2} \ge 0$$

$$\frac{-2x+12}{x-2} \ge 0$$

The number 6 makes the numerator 0, and 2 makes the denominator 0. These two numbers determine three intervals.

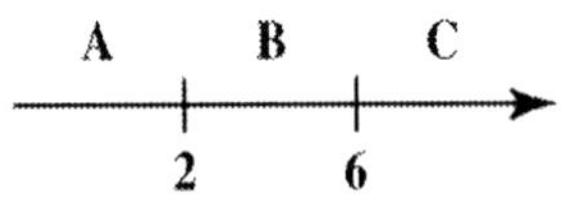

Test a number from each interval in the inequality

$$\frac{8}{x-2} \ge 2.$$

Interval A: Let $x = 0$.

$$\frac{8}{0-2} \ge 2 \quad ?$$

$$-4 \ge 2 \qquad \textit{False}$$

Interval B: Let $x = 3$.

$$\frac{8}{3-2} \ge 2 \quad ?$$

$$8 \ge 2 \qquad \textit{True}$$

Interval C: Let $x = 7$.

$$\frac{8}{7-2} \ge 2 \quad ?$$

$$\frac{8}{5} \ge 2 \qquad \textit{False}$$

The solution set includes numbers in Interval B, including 6 but excluding 2, which makes the fraction undefined.

Solution set: $(2, 6]$

34. $$\frac{20}{x-1} \geq 1$$

Write the inequality so that 0 is on one side.

$$\frac{20}{x-1} - 1 \geq 0$$

$$\frac{20}{x-1} - \frac{1(x-1)}{x-1} \geq 0$$

$$\frac{20 - x + 1}{x-1} \geq 0$$

$$\frac{-x + 21}{x-1} \geq 0$$

The number 21 makes the numerator 0, and 1 makes the denominator 0. These two numbers determine three intervals.

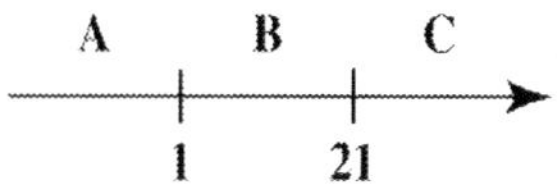

Test a number from each interval in the inequality

$$\frac{20}{x-1} \geq 1.$$

Interval A: Let $x = 0$.

$$\frac{20}{0-1} \geq 1 \quad ?$$

$$-20 \geq 1 \qquad \textit{False}$$

Interval B: Let $x = 2$.

$$\frac{20}{2-1} \geq 1 \quad ?$$

$$20 \geq 1 \qquad \textit{True}$$

Interval C: Let $x = 22$.

$$\frac{20}{22-1} \geq 1 \quad ?$$

$$\frac{20}{21} \geq 1 \qquad \textit{False}$$

The solution set includes numbers in Interval B, including 21, but excluding 1, which makes the fraction undefined.

Solution set: $(1, 21]$

35. $$\frac{3}{2t-1} < 2$$

Write the inequality so that 0 is on one side.

$$\frac{3}{2t-1} - 2 < 0$$

$$\frac{3}{2t-1} - \frac{2(2t-1)}{2t-1} < 0$$

$$\frac{3 - 4t + 2}{2t-1} < 0$$

$$\frac{-4t + 5}{2t-1} < 0$$

The number $\frac{5}{4}$ makes the numerator 0, and $\frac{1}{2}$ makes the denominator 0. These two numbers determine three intervals.

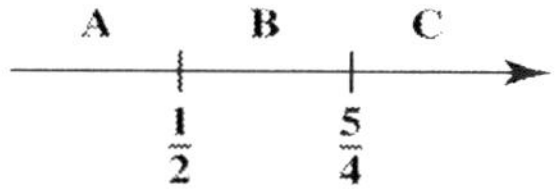

Test a number from each interval in the inequality

$$\frac{3}{2t-1} < 2.$$

Interval A: Let $t = 0$.

$$\frac{3}{2(0)-1} < 2 \quad ?$$

$$-3 < 2 \qquad \textit{True}$$

Interval B: Let $t = 1$.

$$\frac{3}{2(1)-1} < 2 \quad ?$$

$$3 < 2 \qquad \textit{False}$$

Interval C: Let $t = 2$.

$$\frac{3}{2(2)-1} < 2 \quad ?$$

$$1 < 2 \qquad \textit{True}$$

The solution set includes numbers in Intervals A and C, excluding endpoints.

Solution set: $\left(-\infty, \frac{1}{2}\right) \cup \left(\frac{5}{4}, \infty\right)$

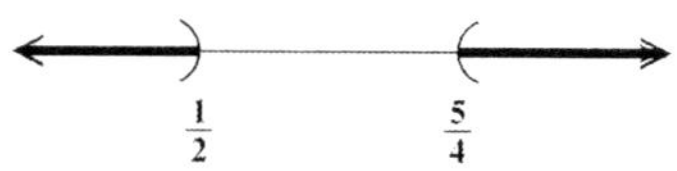

36. $$\frac{6}{m-1} < 1$$

Write the inequality so that 0 is on one side.

$$\frac{6}{m-1} - 1 < 0$$

$$\frac{6}{m-1} - \frac{1(m-1)}{m-1} < 0$$

$$\frac{6 - m + 1}{m-1} < 0$$

$$\frac{-m + 7}{m-1} < 0$$

The number 7 makes the numerator 0, and 1 makes the denominator 0. These two numbers determine three intervals.

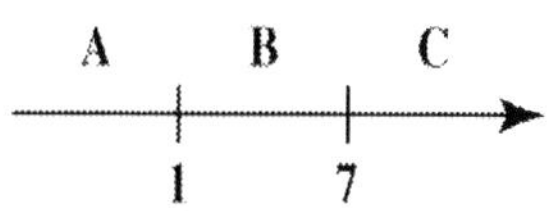

Test a number from each interval in the inequality

$$\frac{6}{m-1} < 1.$$

continued

Interval A: Let $m = 0$.

$$\frac{6}{0-1} < 1 \quad ?$$
$$-6 < 1 \qquad \textit{True}$$

Interval B: Let $m = 2$.

$$\frac{6}{2-1} < 1 \quad ?$$
$$6 < 1 \qquad \textit{False}$$

Interval C: Let $m = 8$.

$$\frac{6}{8-1} < 1 \quad ?$$
$$\frac{6}{7} < 1 \qquad \textit{True}$$

The solution set includes numbers in Intervals A and C, excluding the endpoints.

Solution set: $(-\infty, 1) \cup (7, \infty)$

37.
$$\frac{w}{w+2} \geq 2$$

Write the inequality so that 0 is on one side.

$$\frac{w}{w+2} - 2 \geq 0$$
$$\frac{w}{w+2} - \frac{2(w+2)}{w+2} \geq 0$$
$$\frac{w-2w-4}{w+2} \geq 0$$
$$\frac{-w-4}{w+2} \geq 0$$

The number -4 makes the numerator 0, and -2 makes the denominator 0. These two numbers determine three intervals.

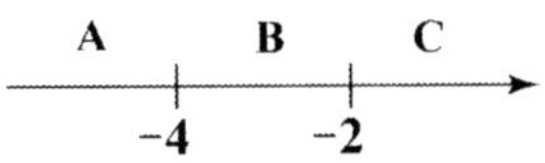

Test a number from each interval in the inequality

$$\frac{w}{w+2} \geq 2.$$

Interval A: Let $w = -5$.

$$\frac{-5}{-3} \geq 2 \quad ?$$
$$\frac{5}{3} \geq 2 \qquad \textit{False}$$

Interval B: Let $w = -3$.

$$\frac{-3}{-1} \geq 2 \quad ?$$
$$3 \geq 2 \qquad \textit{True}$$

Interval C: Let $w = 0$.

$$\frac{0}{2} \geq 2 \quad ?$$
$$0 \geq 2 \qquad \textit{False}$$

The solution set includes numbers in Interval B, including -4 but excluding -2, which makes the fraction undefined.

Solution set: $[-4, -2)$

38.
$$\frac{m}{m+5} \geq 2$$

Write the inequality so that 0 is on one side.

$$\frac{m}{m+5} - 2 \geq 0$$
$$\frac{m}{m+5} - \frac{2(m+5)}{m+5} \geq 0$$
$$\frac{m-2m-10}{m+5} \geq 0$$
$$\frac{-m-10}{m+5} \geq 0$$

The number -10 makes the numerator 0, and -5 makes the denominator 0. These two numbers determine three intervals.

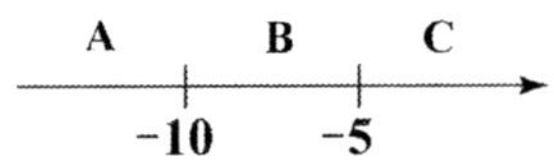

Test a number from each interval in the inequality

$$\frac{m}{m+5} \geq 2.$$

Interval A: Let $m = -15$.

$$\frac{-15}{-10} \geq 2 \quad ?$$
$$1.5 \geq 2 \qquad \textit{False}$$

Interval B: Let $m = -7$.

$$\frac{-7}{-2} \geq 2 \quad ?$$
$$3.5 \geq 2 \qquad \textit{True}$$

Interval C: Let $m = 0$.

$$\frac{0}{5} \geq 2 \quad ?$$
$$0 \geq 2 \qquad \textit{False}$$

The solution set includes numbers in Interval B, including -10 but excluding -5, which makes the fraction undefined.

Solution set: $[-10, -5)$

39.

$$\frac{4k}{2k-1} < k$$

Write the inequality so that 0 is on one side.

$$\frac{4k}{2k-1} - k < 0$$

$$\frac{4k}{2k-1} - \frac{k(2k-1)}{2k-1} < 0$$

$$\frac{4k - 2k^2 + k}{2k-1} < 0$$

$$\frac{-2k^2 + 5k}{2k-1} < 0$$

$$\frac{k(-2k+5)}{2k-1} < 0$$

The numbers 0 and $\frac{5}{2}$ make the numerator 0, and $\frac{1}{2}$ makes the denominator 0. These three numbers determine four intervals.

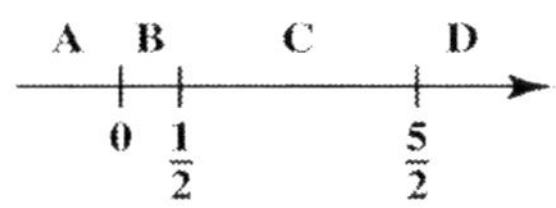

Test a number from each interval in the inequality

$$\frac{4k}{2k-1} < k.$$

Interval A: Let $k = -1$.

$$\frac{4(-1)}{2(-1)-1} < -1 \quad ?$$

$$\frac{4}{3} < -1 \quad \textit{False}$$

Interval B: Let $k = \frac{1}{4}$.

$$\frac{4(\frac{1}{4})}{2(\frac{1}{4})-1} < \frac{1}{4} \quad ?$$

$$-2 < \frac{1}{4} \quad \textit{True}$$

Interval C: Let $k = 1$.

$$\frac{4(1)}{2(1)-1} < 1 \quad ?$$

$$4 < 1 \quad \textit{False}$$

Interval D: Let $k = 3$.

$$\frac{4(3)}{2(3)-1} < 3 \quad ?$$

$$\frac{12}{5} < 3 \quad \textit{True}$$

The solution set includes numbers in Intervals B and D. None of the endpoints are included.

Solution set: $\left(0, \frac{1}{2}\right) \cup \left(\frac{5}{2}, \infty\right)$

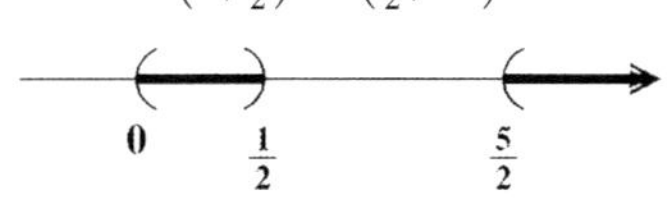

40.

$$\frac{r}{r+2} < 2r$$

Write the inequality so that 0 is on one side.

$$\frac{r}{r+2} - 2r < 0$$

$$\frac{r}{r+2} - \frac{2r(r+2)}{r+2} < 0$$

$$\frac{r - 2r^2 - 4r}{r+2} < 0$$

$$\frac{-2r^2 - 3r}{r+2} < 0$$

$$\frac{r(-2r-3)}{r+2} < 0$$

The numbers $-\frac{3}{2}$ and 0 make the numerator 0, and -2 makes the denominator 0. These three numbers determine four intervals.

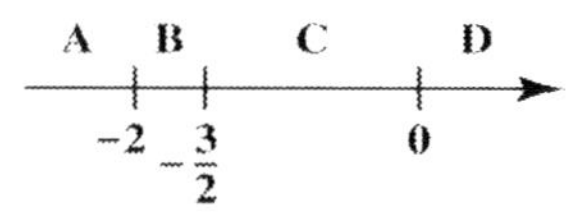

Test a number from each interval in the inequality

$$\frac{r}{r+2} < 2r.$$

Interval A: Let $r = -3$.

$$\frac{-3}{-3+2} < 2(-3) \quad ?$$

$$3 < -6 \quad \textit{False}$$

Interval B: Let $r = -\frac{7}{4}$.

$$\frac{-\frac{7}{4}}{-\frac{7}{4}+2} < 2\left(-\frac{7}{4}\right) \quad ?$$

$$-7 < -\frac{7}{2} \quad \textit{True}$$

Interval C: Let $r = -1$.

$$\frac{-1}{-1+2} < 2(-1) \quad ?$$

$$-1 < -2 \quad \textit{False}$$

Interval D: Let $r = 1$.

$$\frac{1}{1+2} < 2(1) \quad ?$$

$$\frac{1}{3} < 2 \quad \textit{True}$$

The solution set includes numbers in Intervals B and D. None of the endpoints are included.

Solution set: $\left(-2, -\frac{3}{2}\right) \cup (0, \infty)$

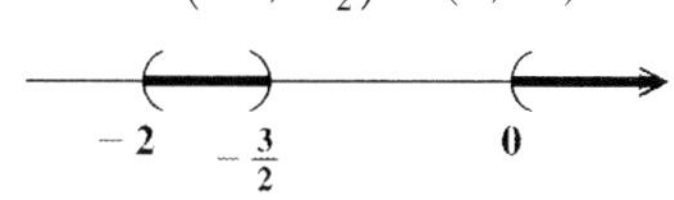

41. $$\frac{x-8}{x-4} \leq 3$$

Write the inequality so that 0 is on one side.

$$\frac{x-8}{x-4} - 3 \leq 0$$
$$\frac{x-8}{x-4} - \frac{3(x-4)}{x-4} \leq 0$$
$$\frac{x-8-3x+12}{x-4} \leq 0$$
$$\frac{-2x+4}{x-4} \leq 0$$

The number 2 makes the numerator 0, and 4 makes the denominator 0. These two numbers determine three intervals.

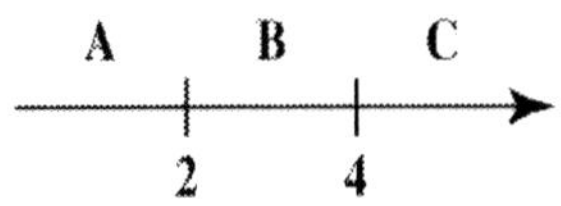

Test a number from each interval in the inequality

$$\frac{x-8}{x-4} \leq 3.$$

Interval A: Let $x = 0$.
$$\frac{-8}{-4} \overset{?}{\leq} 3$$
$$2 \leq 3 \qquad \textit{True}$$

Interval B: Let $x = 3$.
$$\frac{-5}{-1} \overset{?}{\leq} 3$$
$$5 \leq 3 \qquad \textit{False}$$

Interval C: Let $x = 5$.
$$\frac{-3}{1} \leq 3 \qquad \textit{True}$$

The solution set includes numbers in Intervals A and C, including 2 but excluding 4, which makes the fraction undefined.

Solution set: $(-\infty, 2] \cup (4, \infty)$

42. $$\frac{2t-3}{t+1} \geq 4$$

Write the inequality so that 0 is on one side.

$$\frac{2t-3}{t+1} - 4 \geq 0$$
$$\frac{2t-3}{t+1} - \frac{4(t+1)}{t+1} \geq 0$$
$$\frac{2t-3-4t-4}{t+1} \geq 0$$
$$\frac{-2t-7}{t+1} \geq 0$$

The number $-\frac{7}{2}$ makes the numerator 0, and -1 makes the denominator 0. These two numbers determine three intervals.

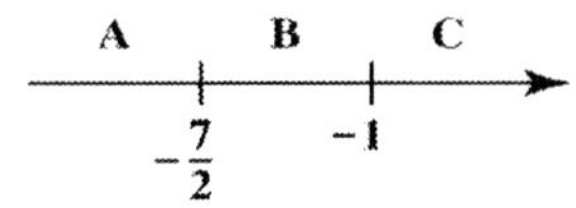

Test a number from each interval in the inequality

$$\frac{2t-3}{t+1} \geq 4.$$

Interval A: Let $t = -4$.
$$\frac{-11}{-3} \overset{?}{\geq} 4$$
$$3\frac{2}{3} \geq 4 \qquad \textit{False}$$

Interval B: Let $t = -2$.
$$\frac{-7}{-1} \geq 4 \qquad \textit{True}$$

Interval C: Let $t = 0$.
$$\frac{-3}{1} \geq 4 \qquad \textit{False}$$

The solution set includes numbers in Interval B, including $-\frac{7}{2}$ but excluding -1, which makes the fraction undefined.

Solution set: $\left[-\frac{7}{2}, -1\right)$

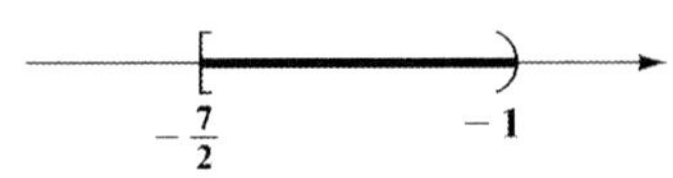

43.
$$s(t) = 624$$
$$-16t^2 + 256t = 624$$
$$t^2 - 16t = -39 \quad \textit{Divide by } -16.$$
$$t^2 - 16t + 39 = 0$$
$$(t-3)(t-13) = 0$$

$$t - 3 = 0 \quad \text{or} \quad t - 13 = 0$$
$$t = 3 \quad \text{or} \quad t = 13$$

The rock will be 624 feet above the ground after 3 seconds and after 13 seconds.

44. See Exercise 43 for the solution to the equation $s(t) = 624$.

For this problem, our common sense tells us that $s(t) > 624$ between 3 seconds and 13 seconds.

45.
$$s(t) = 0$$
$$-16t^2 + 256t = 0$$
$$t^2 - 16t = 0 \quad \textit{Divide by } -16.$$
$$t(t-16) = 0$$

$$t = 0 \quad \text{or} \quad t - 16 = 0$$
$$t = 16$$

The rock is at ground level at 0 seconds (the time when it is initially projected) and at 16 seconds (the time when it hits the ground).

46. Based on the answers for Exercises 43 and 45, we deduce that $s(t) < 624$ between 0 and 3 seconds and between 13 and 16 seconds.

Chapter 10 Review Exercises

1. $t^2 = 121$

$t = 11 \quad \text{or} \quad t = -11$

Solution set: $\{-11, 11\}$

2. $p^2 = 3$

$p = \sqrt{3} \quad \text{or} \quad p = -\sqrt{3}$

Solution set: $\{-\sqrt{3}, \sqrt{3}\}$

3. $(2x+5)^2 = 100$

$2x + 5 = 10 \quad \text{or} \quad 2x + 5 = -10$

$2x = 5 \quad \text{or} \quad 2x = -15$

$x = \frac{5}{2} \quad \text{or} \quad x = -\frac{15}{2}$

Solution set: $\{-\frac{15}{2}, \frac{5}{2}\}$

4. $(3k-2)^2 = -25$

$3k - 2 = \sqrt{-25} \quad \text{or} \quad 3k - 2 = -\sqrt{-25}$

$3k - 2 = 5i \quad \text{or} \quad 3k - 2 = -5i$

$3k = 2 + 5i \quad \text{or} \quad 3k = 2 - 5i$

$k = \dfrac{2+5i}{3} \quad \text{or} \quad k = \dfrac{2-5i}{3}$

Solution set: $\left\{\dfrac{2+5i}{3}, \dfrac{2-5i}{3}\right\}$

5. $x^2 + 4x = 15$

Complete the square.

$$\left[\tfrac{1}{2}(4)\right]^2 = 2^2 = 4$$

Add 4 to each side.

$x^2 + 4x + 4 = 15 + 4$

$(x+2)^2 = 19$

$x + 2 = \sqrt{19} \quad \text{or} \quad x + 2 = -\sqrt{19}$

$x = -2 + \sqrt{19} \quad \text{or} \quad x = -2 - \sqrt{19}$

Solution set: $\left\{-2 + \sqrt{19}, -2 - \sqrt{19}\right\}$

6. $2m^2 - 3m = -1$

$m^2 - \frac{3}{2}m = -\frac{1}{2}$ *Divide by 2.*

Complete the square.

$$\left[\tfrac{1}{2}\left(-\tfrac{3}{2}\right)\right]^2 = \left(-\tfrac{3}{4}\right)^2 = \tfrac{9}{16}$$

Add $\frac{9}{16}$ to each side.

$m^2 - \frac{3}{2}m + \frac{9}{16} = -\frac{1}{2} + \frac{9}{16}$

$\left(m - \frac{3}{4}\right)^2 = -\frac{8}{16} + \frac{9}{16}$

$\left(m - \frac{3}{4}\right)^2 = \frac{1}{16}$

$m - \frac{3}{4} = \sqrt{\frac{1}{16}} \quad \text{or} \quad m - \frac{3}{4} = -\sqrt{\frac{1}{16}}$

$m - \frac{3}{4} = \frac{1}{4} \quad \text{or} \quad m - \frac{3}{4} = -\frac{1}{4}$

$m = \frac{3}{4} + \frac{1}{4} \quad \text{or} \quad m = \frac{3}{4} - \frac{1}{4}$

$m = 1 \quad \text{or} \quad m = \frac{1}{2}$

Solution set: $\{\frac{1}{2}, 1\}$

7. By the square root property, the first step should be

$$x = \sqrt{12} \quad \text{or} \quad x = -\sqrt{12}.$$

8. $16t^2 = d$

$16t^2 = 150$ *Let $d = 150$.*

$t^2 = \dfrac{150}{16}$

$t = \sqrt{\dfrac{150}{16}}$ *$t \geq 0$*

$t \approx 3.1$ seconds

It would take about 3.1 seconds for the wallet to fall 150 feet.

9. $2x^2 + x - 21 = 0$

Here $a = 2$, $b = 1$, and $c = -21$.

$$x = \frac{-b \pm \sqrt{b^2 - 4ac}}{2a}$$

$$x = \frac{-1 \pm \sqrt{1^2 - 4(2)(-21)}}{2(2)}$$

$$= \frac{-1 \pm \sqrt{1 + 168}}{4}$$

$$= \frac{-1 \pm \sqrt{169}}{4} = \frac{-1 \pm 13}{4}$$

$$x = \frac{-1+13}{4} = \frac{12}{4} = 3 \text{ or}$$

$$x = \frac{-1-13}{4} = -\frac{14}{4} = -\frac{7}{2}$$

Solution set: $\{-\frac{7}{2}, 3\}$

10. $k^2 + 5k = 7$

$k^2 + 5k - 7 = 0$

Here $a = 1$, $b = 5$, and $c = -7$.

$$k = \frac{-b \pm \sqrt{b^2 - 4ac}}{2a}$$

$$k = \frac{-5 \pm \sqrt{5^2 - 4(1)(-7)}}{2(1)}$$

continued

$$= \frac{-5 \pm \sqrt{25+28}}{2}$$
$$= \frac{-5 \pm \sqrt{53}}{2}$$

Solution set: $\left\{ \frac{-5+\sqrt{53}}{2}, \frac{-5-\sqrt{53}}{2} \right\}$

11. $(t+3)(t-4) = -2$
$$t^2 - t - 12 = -2$$
$$t^2 - t - 10 = 0$$

Here $a = 1$, $b = -1$, and $c = -10$.

$$t = \frac{-b \pm \sqrt{b^2 - 4ac}}{2a}$$
$$t = \frac{-(-1) \pm \sqrt{(-1)^2 - 4(1)(-10)}}{2(1)}$$
$$= \frac{1 \pm \sqrt{1+40}}{2} = \frac{1 \pm \sqrt{41}}{2}$$

Solution set: $\left\{ \frac{1+\sqrt{41}}{2}, \frac{1-\sqrt{41}}{2} \right\}$

12. $2x^2 + 3x + 4 = 0$

Here $a = 2$, $b = 3$, and $c = 4$.

$$x = \frac{-b \pm \sqrt{b^2 - 4ac}}{2a}$$
$$x = \frac{-3 \pm \sqrt{3^2 - 4(2)(4)}}{2(2)}$$
$$= \frac{-3 \pm \sqrt{9-32}}{4}$$
$$= \frac{-3 \pm \sqrt{-23}}{4}$$
$$= \frac{-3 \pm i\sqrt{23}}{4}$$

Solution set: $\left\{ \frac{-3+i\sqrt{23}}{4}, \frac{-3-i\sqrt{23}}{4} \right\}$

13. $3p^2 = 2(2p-1)$
$$3p^2 = 4p - 2$$
$$3p^2 - 4p + 2 = 0$$

Here $a = 3$, $b = -4$, and $c = 2$.

$$p = \frac{-b \pm \sqrt{b^2 - 4ac}}{2a}$$
$$p = \frac{-(-4) \pm \sqrt{(-4)^2 - 4(3)(2)}}{2(3)}$$
$$= \frac{4 \pm \sqrt{16-24}}{6} = \frac{4 \pm \sqrt{-8}}{6}$$
$$= \frac{4 \pm 2i\sqrt{2}}{6} = \frac{2\left(2 \pm i\sqrt{2}\right)}{6}$$
$$= \frac{2 \pm i\sqrt{2}}{3}$$

Solution set: $\left\{ \frac{2+i\sqrt{2}}{3}, \frac{2-i\sqrt{2}}{3} \right\}$

14. $m(2m-7) = 3m^2 + 3$
$$2m^2 - 7m = 3m^2 + 3$$
$$0 = m^2 + 7m + 3$$

Here $a = 1$, $b = 7$, and $c = 3$.

$$m = \frac{-b \pm \sqrt{b^2 - 4ac}}{2a}$$
$$m = \frac{-7 \pm \sqrt{7^2 - 4(1)(3)}}{2(1)}$$
$$= \frac{-7 \pm \sqrt{49-12}}{2}$$
$$= \frac{-7 \pm \sqrt{37}}{2}$$

Solution set: $\left\{ \frac{-7+\sqrt{37}}{2}, \frac{-7-\sqrt{37}}{2} \right\}$

15. $x^2 + 5x + 2 = 0$

Here $a = 1$, $b = 5$, and $c = 2$.

$$b^2 - 4ac = 5^2 - 4(1)(2)$$
$$= 25 - 8 = 17$$

Since the discriminant is positive, but not a perfect square, there are two distinct irrational number solutions. The answer is **C**.

16. $4t^2 = 3 - 4t$
$$4t^2 + 4t - 3 = 0$$

Here $a = 4$, $b = 4$, and $c = -3$.

$$b^2 - 4ac = 4^2 - 4(4)(-3)$$
$$= 16 + 48$$
$$= 64 \text{ or } 8^2$$

Since the discriminant is positive, and a perfect square, there are two distinct rational number solutions. The answer is **A**.

17. $4x^2 = 6x - 8$

$4x^2 - 6x + 8 = 0$

Here $a = 4$, $b = -6$, and $c = 8$.

$$\begin{aligned} b^2 - 4ac &= (-6)^2 - 4(4)(8) \\ &= 36 - 128 = -92 \end{aligned}$$

Since the discriminant is negative, there are two distinct nonreal complex number solutions. The answer
is **D**.

18. $9z^2 + 30z + 25 = 0$

Here $a = 9$, $b = 30$, and $c = 25$.

$$\begin{aligned} b^2 - 4ac &= 30^2 - 4(9)(25) \\ &= 900 - 900 = 0 \end{aligned}$$

Since the discriminant is zero, there is exactly one rational number solution. The answer is **B**.

19. $\frac{15}{x} = 2x - 1$

$x\left(\frac{15}{x}\right) = x(2x - 1)$ *Multiply by the LCD, x.*

$15 = 2x^2 - x$

$0 = 2x^2 - x - 15$

$0 = (2x + 5)(x - 3)$

$2x + 5 = 0$ or $x - 3 = 0$

$x = -\frac{5}{2}$ or $x = 3$

Check $x = -\frac{5}{2}$: $-6 = -5 - 1$ *True*

Check $x = 3$: $5 = 6 - 1$ *True*

Solution set: $\{-\frac{5}{2}, 3\}$

20. $\frac{1}{n} + \frac{2}{n+1} = 2$

$n(n+1)\left(\frac{1}{n} + \frac{2}{n+1}\right) = n(n+1) \bullet 2$ *Multiply by the LCD, n(n+1).*

$(n + 1) + 2n = 2n^2 + 2n$

$0 = 2n^2 - n - 1$

$0 = (2n + 1)(n - 1)$

$2n + 1 = 0$ or $n - 1 = 0$

$n = -\frac{1}{2}$ or $n = 1$

Check $n = -\frac{1}{2}$: $-2 + 4 = 2$ *True*

Check $n = 1$: $1 + 1 = 2$ *True*

Solution set: $\{-\frac{1}{2}, 1\}$

21. $-2r = \sqrt{\frac{48 - 20r}{2}}$

Square both sides.

$$(-2r)^2 = \left(\sqrt{\frac{48 - 20r}{2}}\right)^2$$

$$4r^2 = \frac{48 - 20r}{2}$$

$$4r^2 = 24 - 10r$$

$4r^2 + 10r - 24 = 0$

$2r^2 + 5r - 12 = 0$

$(r + 4)(2r - 3) = 0$

$r + 4 = 0$ or $2r - 3 = 0$

$r = -4$ or $r = \frac{3}{2}$

Check $r = -4$: $8 = \sqrt{64}$ *True*

Check $r = \frac{3}{2}$: $-3 = \sqrt{9}$ *False*

Solution set: $\{-4\}$

22. $8(3x + 5)^2 + 2(3x + 5) - 1 = 0$

Let $u = 3x + 5$. The equation becomes

$$8u^2 + 2u - 1 = 0.$$

$$(2u + 1)(4u - 1) = 0$$

$2u + 1 = 0$ or $4u - 1 = 0$

$u = -\frac{1}{2}$ or $u = \frac{1}{4}$

To find x, substitute $3x + 5$ for u.

$3x + 5 = -\frac{1}{2}$ or $3x + 5 = \frac{1}{4}$

$3x = -\frac{11}{2}$ or $3x = -\frac{19}{4}$

$x = -\frac{11}{6}$ or $x = -\frac{19}{12}$

Check $x = -\frac{11}{6}$: $2 - 1 - 1 = 0$ *True*

Check $x = -\frac{19}{12}$: $.5 + .5 - 1 = 0$ *True*

Solution set: $\{-\frac{11}{6}, -\frac{19}{12}\}$

23. $2x^{2/3} - x^{1/3} - 28 = 0$

Let $u = x^{1/3}$, so $u^2 = \left(x^{1/3}\right)^2 = x^{2/3}$.

The equation becomes

$2u^2 - u - 28 = 0.$

$(2u + 7)(u - 4) = 0$

$2u + 7 = 0$ or $u - 4 = 0$

$u = -\frac{7}{2}$ or $u = 4$

To find x, substitute $x^{1/3}$ for u.

$x^{1/3} = -\frac{7}{2}$ or $x^{1/3} = 4$

$\left(x^{1/3}\right)^3 = \left(-\frac{7}{2}\right)^3$ or $\left(x^{1/3}\right)^3 = 4^3$

$x = -\frac{343}{8}$ or $x = 64$

Check $x = -\frac{343}{8}$: $24.5 + 3.5 - 28 = 0$ *True*

Check $x = 64$: $32 - 4 - 28 = 0$ *True*

Solution set: $\{-\frac{343}{8}, 64\}$

24. $p^4 - 5p^2 + 4 = 0$

Let $x = p^2$, so $x^2 = p^4$.

$$x^2 - 5x + 4 = 0$$
$$(x-1)(x-4) = 0$$

$x - 1 = 0$ or $x - 4 = 0$
$x = 1$ or $x = 4$

To find p, substitute p^2 for x.

$p^2 = 1$ or $p^2 = 4$
$p = \pm\sqrt{1}$ or $p = \pm\sqrt{4}$
$p = \pm 1$ or $p = \pm 2$

Check $p = \pm 1$: $1 - 5 + 4 = 0$ *True*
Check $p = \pm 2$: $16 - 20 + 4 = 0$ *True*
Solution set: $\{-2, -1, 1, 2\}$

25. Let $x =$ Phong's speed.
Make a table. Use $d = rt$, or $t = \frac{d}{r}$.

	Distance	Rate	Time
Upstream	20	$x-3$	$\frac{20}{x-3}$
Downstream	20	$x+3$	$\frac{20}{x+3}$

Time upstream plus time downstream equals 7 hr.

$$\frac{20}{x-3} + \frac{20}{x+3} = 7$$

Multiply each side by the LCD, $(x+3)(x-3)$.

$$(x+3)(x-3)\left(\frac{20}{x-3} + \frac{20}{x+3}\right) = (x+3)(x-3)(7)$$
$$20(x+3) + 20(x-3) = 7(x^2 - 9)$$
$$20x + 60 + 20x - 60 = 7x^2 - 63$$
$$0 = 7x^2 - 40x - 63$$
$$0 = (x-7)(7x+9)$$

$x - 7 = 0$ or $7x + 9 = 0$
$x = 7$ or $x = -\frac{9}{7}$

Reject $-\frac{9}{7}$ since speed can't be negative. Phong's speed was 7 mph.

26. Let $x =$ Maureen's speed on the trip to pick up Laurie.
Make a chart. Use $d = rt$, or $t = \frac{d}{r}$.

	Distance	Rate	Time
To Laurie	8	x	$\frac{8}{x}$
To the Mall	11	$x+15$	$\frac{11}{x+15}$

Time to pick up Laurie + time to mall = 24 min (or .4 hr).

$$\frac{8}{x} + \frac{11}{x+15} = .4$$

Multiply each side by the LCD, $x(x+15)$.

$$x(x+15)\left(\frac{8}{x} + \frac{11}{x+15}\right) = x(x+15)(.4)$$
$$8(x+15) + 11x = .4x(x+15)$$
$$8x + 120 + 11x = .4x^2 + 6x$$
$$0 = .4x^2 - 13x - 120$$

Multiply by 5 to clear the decimal.

$$0 = 2x^2 - 65x - 600$$
$$0 = (x-40)(2x+15)$$

$x - 40 = 0$ or $2x + 15 = 0$
$x = 40$ or $x = -\frac{15}{2}$

Speed cannot be negative, so $-\frac{15}{2}$ is not a solution. Maureen's speed on the trip to pick up Laurie was 40 mph.

27. Let $x =$ the amount of time for the old machine alone and
$x - 1 =$ the amount of time for the new machine alone.

Make a chart.

Machine	Rate	Time Together	Fractional Part of the Job Done
Old	$\frac{1}{x}$	2	$\frac{2}{x}$
New	$\frac{1}{x-1}$	2	$\frac{2}{x-1}$

Part done by old machine + Part done by new machine = 1 whole job.

$$\frac{2}{x} + \frac{2}{x-1} = 1$$

Multiply by the LCD, $x(x-1)$.

$$x(x-1)\left(\frac{2}{x} + \frac{2}{x-1}\right) = x(x-1) \cdot 1$$
$$2(x-1) + 2x = x^2 - x$$
$$2x - 2 + 2x = x^2 - x$$
$$0 = x^2 - 5x + 2$$

Use the quadratic formula.

$$x = \frac{-b \pm \sqrt{b^2 - 4ac}}{2a}$$
$$x = \frac{-(-5) \pm \sqrt{(-5)^2 - 4(1)(2)}}{2(1)}$$

$$= \frac{5 \pm \sqrt{25-8}}{2} = \frac{5 \pm \sqrt{17}}{2}$$

$$x = \frac{5+\sqrt{17}}{2} \approx 4.6 \text{ or}$$

$$x = \frac{5-\sqrt{17}}{2} \approx .4$$

Reject .4 as the time for the old machine, because that would yield a negative time for the new machine. Thus, the old machine takes about 4.6 hours.

28. Let $x =$ the time for Carter alone and $x - 1 =$ the time for Greg alone.

Worker	Rate	Time Together	Fractional Part of the Job Done
Carter	$\frac{1}{x}$	1.5	$\frac{1.5}{x}$
Greg	$\frac{1}{x-1}$	1.5	$\frac{1.5}{x-1}$

Part by Carter + part by Greg = 1 whole job.

$$\frac{1.5}{x} + \frac{1.5}{x-1} = 1$$

Multiply by the LCD, $x(x-1)$.

$$x(x-1)\left(\frac{1.5}{x} + \frac{1.5}{x-1}\right) = x(x-1) \cdot 1$$

$$1.5(x-1) + 1.5x = x^2 - x$$

$$1.5x - 1.5 + 1.5x = x^2 - x$$

$$0 = x^2 - 4x + 1.5$$

Use the quadratic formula.

$$x = \frac{-b \pm \sqrt{b^2 - 4ac}}{2a}$$

$$x = \frac{-(-4) \pm \sqrt{(-4)^2 - 4(1)(1.5)}}{2(1)}$$

$$= \frac{4 \pm \sqrt{16-6}}{2} = \frac{4 \pm \sqrt{10}}{2}$$

$$= \frac{4 \pm \sqrt{10}}{2}$$

$$x = \frac{4+\sqrt{10}}{2} \approx 3.6 \text{ or}$$

$$x = \frac{4-\sqrt{10}}{2} \approx .4$$

Reject .4 as Carter's time, because that would make Greg's time negative. Thus, Carter's time alone is about 3.6 hours and Greg's time alone is about $x - 1 = 2.6$ hours.

29. Solve $k = \frac{rF}{wv^2}$ for v.

Multiply both sides by v^2, then divide by k.

$$v^2 = \frac{rF}{kw}$$

$$v = \pm\sqrt{\frac{rF}{kw}} = \frac{\pm\sqrt{rF}}{\sqrt{kw}}$$

$$= \frac{\pm\sqrt{rF}}{\sqrt{kw}} \cdot \frac{\sqrt{kw}}{\sqrt{kw}}$$

$$v = \frac{\pm\sqrt{rFkw}}{kw}$$

30. Solve $p = \sqrt{\frac{yz}{6}}$ for y.

Square both sides.

$$p^2 = \left(\sqrt{\frac{yz}{6}}\right)^2$$

$$p^2 = \frac{yz}{6}$$

$$\frac{6p^2}{z} = y \quad \text{or} \quad y = \frac{6p^2}{z}$$

31. Solve $mt^2 = 3mt + 6$ for t.

$$mt^2 - 3mt - 6 = 0$$

Use the quadratic formula with $a = m$, $b = -3m$, and $c = -6$.

$$t = \frac{-b \pm \sqrt{b^2 - 4ac}}{2a}$$

$$t = \frac{3m \pm \sqrt{(-3m)^2 - 4(m)(-6)}}{2m}$$

$$= \frac{3m \pm \sqrt{9m^2 + 24m}}{2m}$$

32. Let $x =$ the length of the longer leg;
$\frac{3}{4}x =$ the length of the shorter leg;
$2x - 9 =$ the length of the hypotenuse.

Use the Pythagorean formula.

$$c^2 = a^2 + b^2$$

$$(2x-9)^2 = x^2 + \left(\tfrac{3}{4}x\right)^2$$

$$4x^2 - 36x + 81 = x^2 + \tfrac{9}{16}x^2$$

$$16(4x^2 - 36x + 81) = 16\left(x^2 + \tfrac{9}{16}x^2\right)$$

$$64x^2 - 576x + 1296 = 16x^2 + 9x^2$$

$$39x^2 - 576x + 1296 = 0$$

$$13x^2 - 192x + 432 = 0 \qquad \textit{Divide by 3.}$$

$$(13x - 36)(x - 12) = 0$$

$$13x - 36 = 0 \quad \text{or} \quad x - 12 = 0$$

$$x = \tfrac{36}{13} \quad \text{or} \quad x = 12$$

continued

Reject $x = \frac{36}{13}$ since $2\left(\frac{36}{13}\right) - 9$ is negative.
Since $x = 12$, $\frac{3}{4}x = \frac{3}{4}(12) = 9$, and
$2x - 9 = 2(12) - 9 = 15$.

The lengths of the three sides are 9 feet, 12 feet, and 15 feet.

33. Let $x =$ the amount removed from one dimension.

The area of the square is 256 cm^2, so the length of one side is $\sqrt{256}$ or 16 cm. The dimensions of the new rectangle are $16 + x$ and $16 - x$ cm. The area of the new rectangle is 16 cm^2 less than the area of the square.

$$\begin{aligned}(16 + x)(16 - x) &= 256 - 16\\ 256 - x^2 &= 240\\ -x^2 &= -16\\ x^2 &= 16\\ x &= \pm\sqrt{16} = \pm 4\end{aligned}$$

Length cannot be negative, so reject -4. If $x = 4$, then $16 + x = 20$, and $16 - x = 12$. The dimensions are 20 cm by 12 cm.

34. Let $x =$ the width of the border.

$$\begin{aligned}\text{Area of mat} &= \text{length} \cdot \text{width}\\ 352 &= (2x + 20)(2x + 14)\\ 352 &= 4x^2 + 68x + 280\\ 0 &= 4x^2 + 68x - 72\\ 0 &= x^2 + 17x - 18\\ 0 &= (x + 18)(x - 1)\end{aligned}$$

$x + 18 = 0$ or $x - 1 = 0$
$x = -18$ or $x = 1$

Reject the negative answer for length. The border is 1 inch wide.

35. $f(t) = 100t^2 - 300t$ is the distance of the light from the starting point at t minutes. When the light returns to the starting point, the value of $f(t)$ will be 0.

$$\begin{aligned}0 &= 100t^2 - 300t\\ 0 &= t^2 - 3t && \textit{Divide by 100.}\\ 0 &= t(t - 3)\end{aligned}$$

$t = 0$ or $t - 3 = 0$
$t = 3$

Since $t = 0$ represents the starting time, the light will return to the starting point in 3 minutes.

36.
$$\begin{aligned}f(t) &= -16t^2 + 45t + 400\\ 200 &= -16t^2 + 45t + 400 && \textit{Let } f(t) = 200.\\ 0 &= -16t^2 + 45t + 200\\ 0 &= 16t^2 - 45t - 200\end{aligned}$$

Here $a = 16$, $b = -45$, and $c = -200$.

$$\begin{aligned}t &= \frac{-b \pm \sqrt{b^2 - 4ac}}{2a}\\ t &= \frac{-(-45) \pm \sqrt{(-45)^2 - 4(16)(-200)}}{2(16)}\\ &= \frac{45 \pm \sqrt{2025 + 12{,}800}}{32}\\ &= \frac{45 \pm \sqrt{14{,}825}}{32}\\ t &= \frac{45 + \sqrt{14{,}825}}{32} \approx 5.2 \quad \text{or}\\ t &= \frac{45 - \sqrt{14{,}825}}{32} \approx -2.4\end{aligned}$$

Reject the negative solution since time cannot be negative. The ball will reach a height of 200 ft above the ground after about 5.2 seconds.

37.
$$\begin{aligned}s(t) &= -16t^2 + 75t + 407\\ 450 &= -16t^2 + 75t + 407 && \textit{Let } s(t) = 450.\\ 0 &= -16t^2 + 75t - 43\end{aligned}$$

Here $a = -16$, $b = 75$, and $c = -43$.

$$\begin{aligned}t &= \frac{-b \pm \sqrt{b^2 - 4ac}}{2a}\\ t &= \frac{-75 \pm \sqrt{75^2 - 4(-16)(-43)}}{2(-16)}\\ &= \frac{-75 \pm \sqrt{5625 - 2752}}{-32}\\ &= \frac{-75 \pm \sqrt{2873}}{-32}\\ t &= \frac{-75 + \sqrt{2873}}{-32} \approx .7 \quad \text{or}\\ t &= \frac{-75 - \sqrt{2873}}{-32} \approx 4.0\end{aligned}$$

The ball will be 450 ft above the ground after .7 second (going up) and again after 4.0 seconds (coming down).

38. Let $p =$ the price.

$$\begin{aligned}\text{Demand} &= \text{Supply}\\ \frac{25}{p} &= 70p + 15\\ p\left(\frac{25}{p}\right) &= p(70p + 15) && \textit{Multiply by p.}\\ 25 &= 70p^2 + 15p\\ 0 &= 70p^2 + 15p - 25\\ 0 &= 14p^2 + 3p - 5 && \textit{Divide by 5.}\\ 0 &= (2p - 1)(7p + 5)\end{aligned}$$

$$2p - 1 = 0 \quad \text{or} \quad 7p + 5 = 0$$
$$p = \tfrac{1}{2} \quad \text{or} \quad p = -\tfrac{5}{7}$$

Reject the negative answer for price. Supply and demand are equal when p is $\frac{1}{2}$ dollar or \$.50.

39. Let $A = 10{,}920.25$, $P = 10{,}000$, and solve for r.

$$A = P(1+r)^2$$
$$10{,}920.25 = 10{,}000(1+r)^2$$
$$1.092025 = (1+r)^2$$

$$1 + r = \sqrt{1.092025} \quad \text{or} \quad 1 + r = -\sqrt{1.092025}$$
$$1 + r = 1.045 \quad \text{or} \quad 1 + r = -1.045$$
$$r = .045 \quad \text{or} \quad r = -2.045$$

Reject the negative rate. The required interest rate is $.045 = 4.5\%$.

40. **(a)** Use $f(x) = 3.29x^2 - 10.4x + 21.6$ with $x = 11$.

$$f(11) = 3.29(11)^2 - 10.4(11) + 21.6$$
$$f(11) \approx 305$$

The value of 305 is close to the number shown on the graph.

(b) $200 = 3.29x^2 - 10.4x + 21.6$ *Let f(x) = 200.*

$0 = 3.29x^2 - 10.4x - 178.4$

Here $a = 3.29$, $b = -10.4$, and $c = -178.4$.

$$x = \frac{-b \pm \sqrt{b^2 - 4ac}}{2a}$$
$$x = \frac{-(-10.4) \pm \sqrt{(-10.4)^2 - 4(3.29)(-178.4)}}{2(3.29)}$$
$$= \frac{10.4 \pm \sqrt{2455.904}}{6.58}$$
$$x = \frac{10.4 + \sqrt{2455.904}}{6.58} \approx 9.1 \quad \text{or}$$
$$x = \frac{10.4 - \sqrt{2455.904}}{6.58} \approx -6.0$$

Since time cannot be negative, the correct answer is $x \approx 9$, which represents 1999. Based on the graph, the number of e-mail boxes reached 200 million in 2000.

41. $f(x) = -(x-1)^2$
Write in $y = a(x-h)^2 + k$ form as $y = -1(x-1)^2 + 0$. The vertex (h, k) is $(1, 0)$.

42. $y = (x-3)^2 + 7$
The equation is in the form $y = a(x-h)^2 + k$, so the vertex (h, k) is $(3, 7)$.

43. $y = f(x) = -3x^2 + 4x - 2$

Use the vertex formula with $a = -3$ and $b = 4$.

The x-coordinate of the vertex is

$$\frac{-b}{2a} = \frac{-4}{2(-3)} = \frac{2}{3}.$$

The y-coordinate of the vertex is

$$f\left(\frac{-b}{2a}\right) = f\left(\frac{2}{3}\right)$$
$$= -3\left(\tfrac{2}{3}\right)^2 + 4\left(\tfrac{2}{3}\right) - 2$$
$$= -\tfrac{4}{3} + \tfrac{8}{3} - 2$$
$$= -\tfrac{2}{3}.$$

The vertex is $\left(\frac{2}{3}, -\frac{2}{3}\right)$.

44. $x = (y-3)^2 - 4$
Since the roles of x and y are reversed, this is a horizontal parabola.
$x = (y-3)^2 + (-4)$
The equation is in the form

$$x = a(y-k)^2 + h,$$

so the vertex (h, k) is $(-4, -3)$.

45. $y = 2(x-2)^2 - 3$

The graph opens up since $a = 2 > 0$.

The vertex is $(2, -3)$ and the axis is $x = 2$. If $x = 0$, we get the y-intercept $(0, 5)$, and by symmetry, the point $(4, 5)$ is also on the graph. The domain is $(-\infty, \infty)$. The smallest y-value is -3, so the range is $[-3, \infty)$.

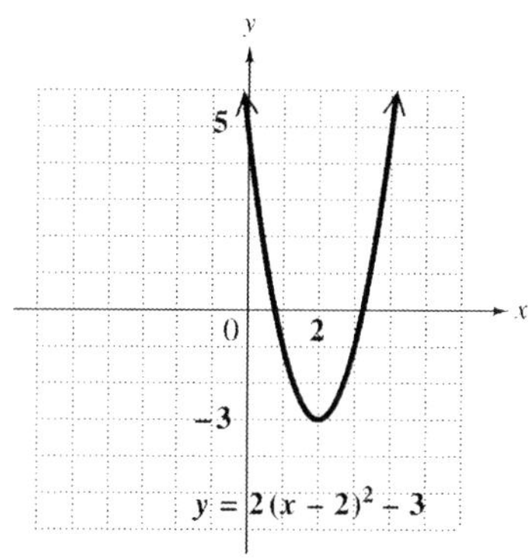

46. $f(x) = -2x^2 + 8x - 5$
Complete the square to find the vertex.

$$f(x) = -2(x^2 - 4x) - 5$$
$$= -2(x^2 - 4x + 4 - 4) - 5$$
$$= -2(x^2 - 4x + 4) - 2(-4) - 5$$
$$= -2(x-2)^2 + 8 - 5$$
$$= -2(x-2)^2 + 3$$

The equation is in the form $y = a(x-h)^2 + k$, so the vertex (h, k) is $(2, 3)$ and the axis is $x = 2$. Here, $a = -2 < 0$, so the parabola opens down.

continued

Also, $|a| = |-2| = 2 > 1$, so the graph is narrower than the graph of $y = x^2$. The points $(0, -5)$, $(1, 1)$, and $(3, 1)$ are on the graph.

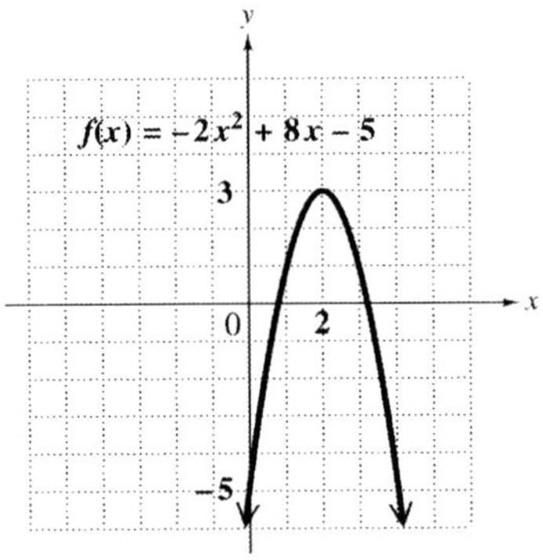

The domain is $(-\infty, \infty)$. The largest y-value is 3, so the range is $(-\infty, 3]$.

47. $x = 2(y + 3)^2 - 4$

Since the roles of x and y are reversed, this is a horizontal parabola.

$x = 2[y - (-3)]^2 + (-4)$

The equation is in the form

$$x = a(y - k)^2 + h,$$

so the vertex (h, k) is $(-4, -3)$ and the axis is $y = -3$. Here, $a = 2 > 0$, so the parabola opens to the right and is narrower than the graph of $y = x^2$.

Two other points on the graph are $(4, -1)$ and $(4, -5)$.

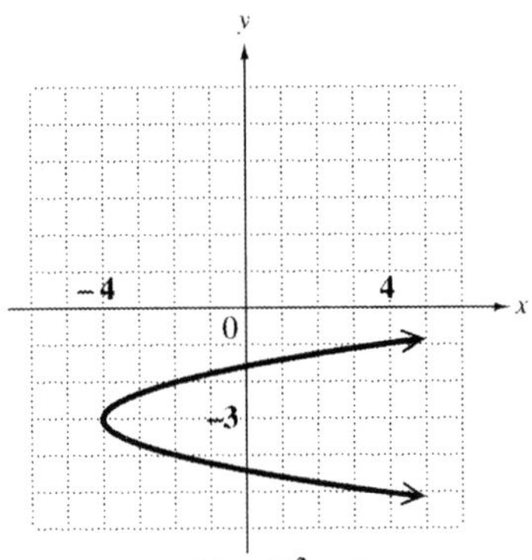

The smallest x-value is -4, so the domain is $[-4, \infty)$. The range is $(-\infty, \infty)$.

48. $x = -\dfrac{1}{2}y^2 + 6y - 14$

Since the roles of x and y are reversed, this is a horizontal parabola. Complete the square to find the vertex.

$$\begin{aligned} x &= -\tfrac{1}{2}y^2 + 6y - 14 \\ &= -\tfrac{1}{2}(y^2 - 12y) - 14 \\ &= -\tfrac{1}{2}(y^2 - 12y + 36 - 36) - 14 \\ &= -\tfrac{1}{2}(y^2 - 12y + 36) - \tfrac{1}{2}(-36) - 14 \\ &= -\tfrac{1}{2}(y - 6)^2 + 18 - 14 \\ x &= -\tfrac{1}{2}(y - 6)^2 + 4 \end{aligned}$$

The equation is in the form $x = a(y - k)^2 + h$, so the vertex (h, k) is $(4, 6)$ and the axis is $y = 6$. Here, $a = -\frac{1}{2} < 0$, so the parabola opens to the left.

Also, $|a| = \left|-\frac{1}{2}\right| = \frac{1}{2} < 1$, so the graph is wider than the graph of $y = x^2$. The points $(-14, 0)$, $(2, 4)$, and $(2, 8)$ are on the graph.

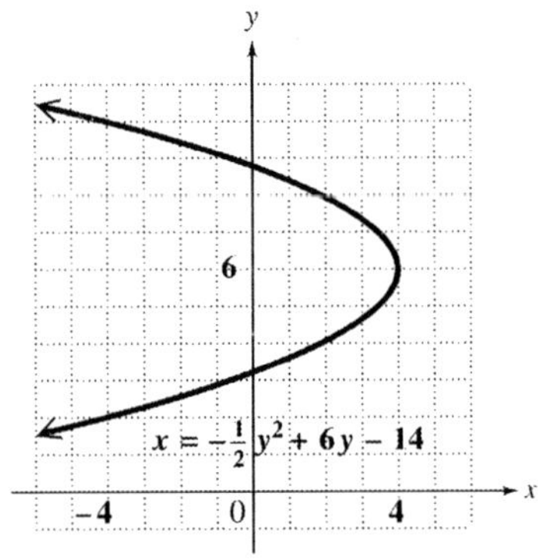

The largest x-value is 4, so the domain is $(-\infty, 4]$. The range is $(-\infty, \infty)$.

49. **(a)** Use $ax^2 + bx + c = y$ with $(0, 10.54)$, $(3, 18.49)$, and $(6, 27.69)$.

$$\begin{aligned} c &= 10.54 \quad (1) \\ 9a + 3b + c &= 18.49 \quad (2) \\ 36a + 6b + c &= 27.69 \quad (3) \end{aligned}$$

(b) Rewrite equations (2) and (3) with $c = 10.54$.

$$\begin{aligned} 9a + 3b + 10.54 &= 18.49 \\ 36a + 6b + 10.54 &= 27.69 \end{aligned}$$

$$\begin{aligned} 9a + 3b &= 7.95 \quad (4) \\ 36a + 6b &= 17.15 \quad (5) \end{aligned}$$

Now eliminate b.

$$\begin{array}{rcrcl} -18a & - & 6b & = & -15.90 \qquad -2 \times (4) \\ 36a & + & 6b & = & 17.15 \qquad (5) \\ \hline 18a & & & = & 1.25 \\ & & a & = & \frac{1.25}{18} = \frac{5}{72} \end{array}$$

Use (4) to find b.

$$\begin{aligned} 9a + 3b &= 7.95 \qquad (4) \\ 9\left(\tfrac{5}{72}\right) + 3b &= 7.95 \\ \tfrac{5}{8} + 3b &= 7.95 \\ 3b &= 7.325 \\ b &= \tfrac{7.325}{3} = \tfrac{293}{120} \end{aligned}$$

Rounding a and b to four decimal places gives us the quadratic function

$$f(x) = .0694x^2 + 2.4417x + 10.54.$$

(c) Use $f(x) = .0694x^2 + 2.442x + 10.54$ with $x = 5$ for the year 2000.

$$\begin{aligned} f(5) &= .0694(5)^2 + 2.442(5) + 10.54 \\ &= 24.485 \approx \$24.49 \end{aligned}$$

The result using the model is close to the table value of \$24.65, but slightly low.

50. $s(t) = -16t^2 + 160t$
The equation represents a parabola. Since $a = -16 < 0$, the parabola opens down. The time and maximum height occur at the vertex (h, k) of the parabola, given by

$$\left(\frac{-b}{2a}, s\left(\frac{-b}{2a}\right)\right).$$

Using the standard form of the equation, $a = -16$ and $b = 160$, so

$$h = \frac{-b}{2a} = \frac{-160}{2(-16)} = 5,$$

and $k = s(h) = -16(5)^2 + 160(5)$
$= -400 + 800 = 400.$

The vertex is $(5, 400)$. The time at which the maximum height is reached is 5 seconds. The maximum height is 400 feet.

51. Let L = the length of the rectangle and W = the width.

The perimeter of the rectangle is 200 m, so

$$\begin{aligned} 2L + 2W &= 200 \\ 2W &= 200 - 2L \\ W &= 100 - L. \end{aligned}$$

Since the area is length times width, substitute $100 - L$ for W.

$$\begin{aligned} A &= LW \\ &= L(100 - L) \\ &= 100L - L^2 \quad \text{or} \quad -L^2 + 100L \end{aligned}$$

Use the vertex formula.

$$L = \frac{-b}{2a} = \frac{-100}{2(-1)} = 50$$

So $L = 50$ meters and
$W = 100 - L = 100 - 50 = 50$ meters.

52. $(x - 4)(2x + 3) > 0$
Solve the equation
$(x - 4)(2x + 3) = 0.$
$x - 4 = 0$ or $2x + 3 = 0$
$x = 4$ or $x = -\frac{3}{2}$
The numbers $-\frac{3}{2}$ and 4 divide a number line into three intervals.

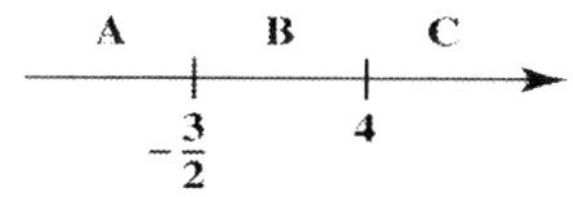

Test a number from each interval in the inequality

$$(x - 4)(2x + 3) > 0.$$

Interval A: Let $x = -2$.
$-6(-1) > 0$?
$6 > 0$ *True*

Interval B: Let $x = 0$.
$-4(3) > 0$?
$-12 > 0$ *False*

Interval C: Let $x = 5$.
$1(13) > 0$?
$13 > 0$ *True*

The solution set includes numbers in Intervals A and C, excluding endpoints.

Solution set: $\left(-\infty, -\frac{3}{2}\right) \cup (4, \infty)$

53. $x^2 + x \le 12$
Solve the equation
$x^2 + x = 12.$
$x^2 + x - 12 = 0$
$(x + 4)(x - 3) = 0$
$x + 4 = 0$ or $x - 3 = 0$
$x = -4$ or $x = 3$
The numbers -4 and 3 divide a number line into three intervals.

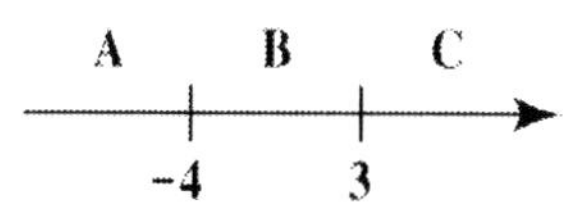

Test a number from each interval in the inequality

$$x^2 + x \le 12.$$

continued

Interval A: Let $x = -5$.

$25 - 5 \leq 12$?

$20 \leq 12$ *False*

Interval B: Let $x = 0$.

$0 \leq 12$ *True*

Interval C: Let $x = 4$.

$16 + 4 \leq 12$?

$20 \leq 12$ *False*

The numbers in Interval B, including -4 and 3, are solutions.

Solution set: $[-4, 3]$

54. $(x+2)(x-3)(x+5) \leq 0$
The numbers -2, 3, and -5 are solutions of the cubic equation

$$(x+2)(x-3)(x+5) = 0.$$

These numbers divide a number line into four intervals.

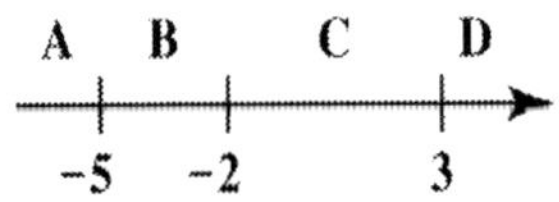

Test a number from each interval in the inequality

$$(x+2)(x-3)(x+5) \leq 0.$$

Interval A: Let $x = -6$.

$(-4)(-9)(-1) \leq 0$ *True*

Interval B: Let $x = -3$.

$(-1)(-6)(2) \leq 0$ *False*

Interval C: Let $x = 0$.

$(2)(-3)(5) \leq 0$ *True*

Interval D: Let $x = 4$.

$(6)(1)(9) \leq 0$ *False*

The numbers in Intervals A and C, including -5, -2, and 3, are solutions.

Solution set: $(-\infty, -5] \cup [-2, 3]$

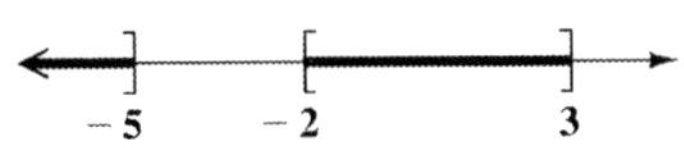

55. $(4m+3)^2 \leq -4$
The square of a real number is never negative. So, the solution set of this inequality is $\emptyset$.

56. $$\frac{6}{2z-1} < 2$$

Write the inequality so that 0 is on one side.

$$\frac{6}{2z-1} - 2 < 0$$

$$\frac{6}{2z-1} - \frac{2(2z-1)}{2z-1} < 0$$

$$\frac{6-4z+2}{2z-1} < 0$$

$$\frac{-4z+8}{2z-1} < 0$$

The number 2 makes the numerator 0, and $\frac{1}{2}$ makes the denominator 0. These two numbers determine three intervals.

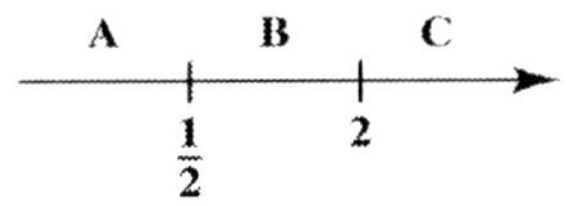

Test a number from each interval in the inequality

$$\frac{6}{2z-1} < 2.$$

Interval A: Let $z = 0$.

$-6 < 2$ *True*

Interval B: Let $z = 1$.

$6 < 2$ *False*

Interval C: Let $z = 3$.

$\frac{6}{5} < 2$ *True*

The solution set includes numbers in Intervals A and C, excluding endpoints.

Solution set: $(-\infty, \frac{1}{2}) \cup (2, \infty)$

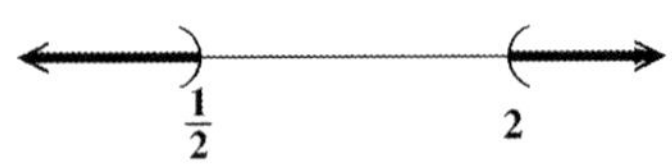

57. $$\frac{3t+4}{t-2} \leq 1$$

Write the inequality so that 0 is on one side.

$$\frac{3t+4}{t-2} - 1 \leq 0$$

$$\frac{3t+4}{t-2} - \frac{1(t-2)}{t-2} \leq 0$$

$$\frac{3t+4-t+2}{t-2} \leq 0$$

$$\frac{2t+6}{t-2} \leq 0$$

The number -3 makes the numerator 0, and 2 makes the denominator 0. These two numbers determine three intervals.

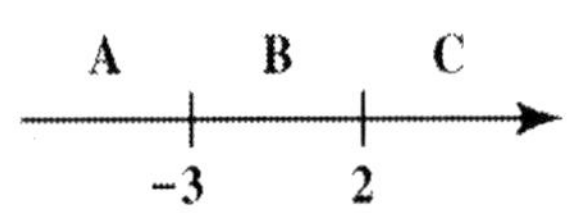

Test a number from each interval in the inequality

$$\frac{3t+4}{t-2} \le 1.$$

Interval A: Let $t = -4$.

$\frac{-8}{-6} \le 1$?

$\frac{4}{3} \le 1$ *False*

Interval B: Let $t = 0$.

$\frac{4}{-2} \le 1$?

$-2 \le 1$ *True*

Interval C: Let $t = 3$.

$\frac{13}{1} \le 1$?

$13 \le 1$ *False*

The numbers in Interval B, including -3 but not 2, are solutions.

Solution set: $[-3, 2)$

58. **[10.4]** Solve $V = r^2 + R^2h$ for R.

$$V - r^2 = R^2h$$
$$R^2h = V - r^2$$
$$R^2 = \frac{V - r^2}{h}$$
$$R = \pm\sqrt{\frac{V - r^2}{h}} = \frac{\pm\sqrt{V - r^2}}{\sqrt{h}}$$
$$= \frac{\pm\sqrt{V - r^2}}{\sqrt{h}} \cdot \frac{\sqrt{h}}{\sqrt{h}}$$
$$= \frac{\pm\sqrt{Vh - r^2h}}{h}$$

59. **[10.2]** $3t^2 - 6t = -4$

$3t^2 - 6t + 4 = 0$

Use the quadratic formula.

$$t = \frac{-b \pm \sqrt{b^2 - 4ac}}{2a}$$
$$t = \frac{-(-6) \pm \sqrt{(-6)^2 - 4(3)(4)}}{2(3)}$$
$$= \frac{6 \pm \sqrt{-12}}{6} = \frac{6 \pm 2i\sqrt{3}}{6}$$
$$= \frac{2\left(3 \pm i\sqrt{3}\right)}{6} = \frac{3 \pm i\sqrt{3}}{3}$$

Solution set: $\left\{\frac{3 + i\sqrt{3}}{3}, \frac{3 - i\sqrt{3}}{3}\right\}$

60. **[10.3]** $\left(x^2 - 2x\right)^2 = 11\left(x^2 - 2x\right) - 24$

Let $u = x^2 - 2x$. The equation becomes

$$u^2 = 11u - 24.$$
$$u^2 - 11u + 24 = 0$$
$$(u - 8)(u - 3) = 0$$

$u - 8 = 0$ or $u - 3 = 0$

$u = 8$ or $u = 3$

To find x, substitute $x^2 - 2x$ for u.

$x^2 - 2x = 8$ or $x^2 - 2x = 3$

$x^2 - 2x - 8 = 0$ $\quad$ $x^2 - 2x - 3 = 0$

$(x - 4)(x + 2) = 0$ $\quad$ $(x - 3)(x + 1) = 0$

$x - 4 = 0$ or $x + 2 = 0$ $\quad$ $x - 3 = 0$ or $x + 1 = 0$

$x = 4$ or $x = -2$ or $x = 3$ or $x = -1$

The potential solutions all check.

Solution set: $\{-2, -1, 3, 4\}$

61. **[10.7]** $(r - 1)(2r + 3)(r + 6) < 0$

Solve the equation

$(r - 1)(2r + 3)(r + 6) = 0.$

$r - 1 = 0$ or $2r + 3 = 0$ or $r + 6 = 0$

$r = 1$ or $r = -\frac{3}{2}$ or $r = -6$

The numbers -6, $-\frac{3}{2}$, and 1 divide a number line into four intervals.

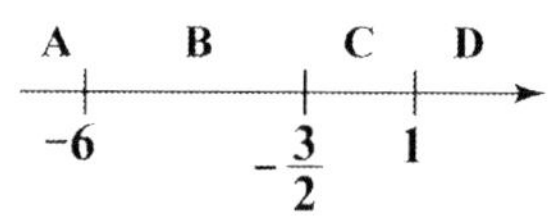

Test a number from each interval in the inequality

$$(r - 1)(2r + 3)(r + 6) < 0.$$

Interval A: Let $r = -7$.

$-8(-11)(-1) < 0$?

$-88 < 0$ *True*

Interval B: Let $r = -2$.

$-3(-1)(4) < 0$?

$12 < 0$ *False*

Interval C: Let $r = 0$.

$-1(3)(6) < 0$?

$-18 < 0$ *True*

Interval D: Let $r = 2$.

$1(7)(8) < 0$?

$56 < 0$ *False*

The numbers in Intervals A and C, not including -6, $-\frac{3}{2}$, or 1, are solutions.

Solution set: $(-\infty, -6) \cup \left(-\frac{3}{2}, 1\right)$

62. **[10.1]** $(3k+11)^2 = 7$

$$3k+11 = \sqrt{7} \quad \text{or} \quad 3k+11 = -\sqrt{7}$$
$$3k = -11+\sqrt{7} \quad \text{or} \quad 3k = -11-\sqrt{7}$$
$$k = \frac{-11+\sqrt{7}}{3} \quad \text{or} \quad k = \frac{-11-\sqrt{7}}{3}$$

Solution set: $\left\{\frac{-11+\sqrt{7}}{3}, \frac{-11-\sqrt{7}}{3}\right\}$

63. **[10.4]** Solve $S = \frac{Id^2}{k}$ for d.

Multiply both sides by k, then divide by I.

$$\frac{Sk}{I} = d^2$$
$$d = \pm\sqrt{\frac{Sk}{I}} = \frac{\pm\sqrt{Sk}}{\sqrt{I}}$$
$$= \frac{\pm\sqrt{Sk}}{\sqrt{I}} \cdot \frac{\sqrt{I}}{\sqrt{I}}$$
$$d = \frac{\pm\sqrt{SkI}}{I}$$

64. **[10.3]** $2x - \sqrt{x} = 6$

$$2x - \sqrt{x} - 6 = 0$$

Let $u = \sqrt{x}$, so $u^2 = x$.

$$2u^2 - u - 6 = 0$$
$$(2u+3)(u-2) = 0$$

$$2u+3 = 0 \quad \text{or} \quad u-2 = 0$$
$$u = -\tfrac{3}{2} \quad \text{or} \quad u = 2$$
$$\sqrt{x} = -\frac{3}{2} \quad \text{or} \quad \sqrt{x} = 2$$

Since $\sqrt{x} \geq 0$, we must have $\sqrt{x} = 2$, or $x = 4$.

Check $x = 4$: $8 - 2 = 6$ *True*

Solution set: $\{4\}$

65. **[10.3]** $6 + \frac{15}{s^2} = -\frac{19}{s}$

Multiply by the LCD, s^2.

$$s^2\left(6 + \frac{15}{s^2}\right) = s^2\left(-\frac{19}{s}\right)$$
$$6s^2 + 15 = -19s$$
$$6s^2 + 19s + 15 = 0$$
$$(3s+5)(2s+3) = 0$$

$$3s+5 = 0 \quad \text{or} \quad 2s+3 = 0$$
$$s = -\tfrac{5}{3} \quad \text{or} \quad s = -\tfrac{3}{2}$$

Check $s = -\frac{5}{3}$: $6 + \frac{27}{5} = \frac{57}{5}$ *True*

Check $s = -\frac{3}{2}$: $6 + \frac{20}{3} = \frac{38}{3}$ *True*

Solution set: $\left\{-\frac{5}{3}, -\frac{3}{2}\right\}$

66. **[10.7]** $\frac{-2}{x+5} \leq -5$

Write the inequality so that 0 is on one side.

$$\frac{-2}{x+5} + 5 \leq 0$$
$$\frac{-2}{x+5} + \frac{5(x+5)}{x+5} \leq 0$$
$$\frac{-2+5x+25}{x+5} \leq 0$$
$$\frac{5x+23}{x+5} \leq 0$$

The number $-\frac{23}{5}$ makes the numerator 0, and -5 makes the denominator 0. These two numbers determine three intervals.

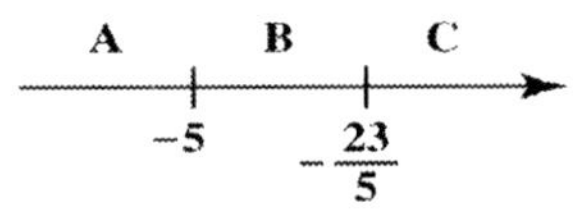

Test a number from each interval in the inequality

$$\frac{-2}{x+5} \leq -5.$$

Interval A: Let $x = -6$.

$$\frac{-2}{-1} \leq -5 \quad ?$$
$$2 \leq -5 \quad \textit{False}$$

Interval B: Let $x = -\frac{24}{5}$.

$$-\frac{2}{\frac{1}{5}} \leq -5 \quad ?$$
$$-10 \leq -5 \quad \textit{True}$$

Interval C: Let $x = 0$.

$$\frac{-2}{5} \leq -5 \quad \textit{False}$$

The numbers in Interval B, including $-\frac{23}{5}$ but not -5, are solutions.

Solution set: $\left(-5, -\frac{23}{5}\right]$

67. **[10.5]** $y = f(x) = 4x^2 + 4x - 2$

Complete the square to find the vertex.

$$y = 4(x^2 + x) - 2$$
$$= 4\left(x^2 + x + \tfrac{1}{4} - \tfrac{1}{4}\right) - 2$$
$$= 4\left(x^2 + x + \tfrac{1}{4}\right) + 4\left(-\tfrac{1}{4}\right) - 2$$
$$= 4\left(x + \tfrac{1}{2}\right)^2 - 1 - 2$$
$$= 4\left(x + \tfrac{1}{2}\right)^2 - 3$$
$$y = 4\left(x + \tfrac{1}{2}\right)^2 - 3$$

The equation is now in the form $y = a(x-h)^2 + k$, so the vertex (h, k) is $\left(-\frac{1}{2}, -3\right)$ and the axis is $x = -\frac{1}{2}$. Since $a = 4 > 0$, the parabola opens up. Also, $|a| = |4| = 4 > 1$, so the graph is narrower than the graph of $y = x^2$. The points $(-2, 6)$, $(0, -2)$, and $(1, 6)$ are on the graph.

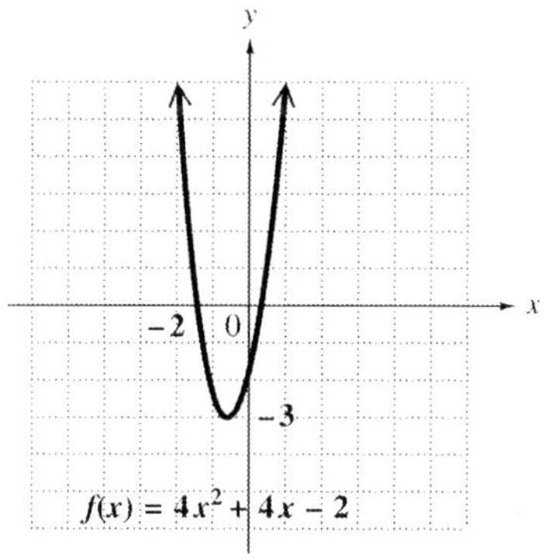

The domain is $(-\infty, \infty)$.
The smallest y-value is -3, so the range is $[-3, \infty)$.

68. **[10.6] (a)** Use $f(x) = .014x^2 - .396x + 21.2$ with $x = 30$.

$$f(30) = .014(30)^2 - .396(30) + 21.2 = 21.92$$

The natural gas use in 2000 was approximately 21.92 trillion ft^3.

(b) The result using the model, 21.92 trillion ft^3, is a little low compared to the actual use, 23.33 trillion ft^3.

(c) $25 = .014x^2 - .396x + 21.2$ *Let f(x) = 25.*
$0 = .014x^2 - .396x - 3.8$

Here $a = .014$, $b = -.396$, and $c = -3.8$.

$$x = \frac{-b \pm \sqrt{b^2 - 4ac}}{2a}$$

$$x = \frac{-(-.396) \pm \sqrt{(-.396)^2 - 4(.014)(-3.8)}}{2(.014)}$$

$$x = \frac{.396 + \sqrt{.369616}}{.028} \approx 35 \text{ (rounding down)} \quad \text{or}$$

$$x = \frac{.396 - \sqrt{.369616}}{.028} \approx -8$$

Since time cannot be negative, the solution is 35 and the year is $1970 + 35 = 2005$.

Chapter 10 Test

1. $t^2 = 54$

$t = \sqrt{54}$ or $t = -\sqrt{54}$
$t = 3\sqrt{6}$ or $t = -3\sqrt{6}$

Solution set: $\{3\sqrt{6}, -3\sqrt{6}\}$

2. $(7x + 3)^2 = 25$

$7x + 3 = 5$ or $7x + 3 = -5$
$7x = 2$ or $7x = -8$
$x = \frac{2}{7}$ or $x = -\frac{8}{7}$

Solution set: $\{-\frac{8}{7}, \frac{2}{7}\}$

3. $x^2 + 2x = 1$

Complete the square on $x^2 + 2x$.

$$\left[\tfrac{1}{2}(2)\right]^2 = 1^2 = 1$$

Add 1 to each side.

$$x^2 + 2x + 1 = 1 + 1$$
$$(x + 1)^2 = 2$$

$x + 1 = \sqrt{2}$ or $x + 1 = -\sqrt{2}$
$x = -1 + \sqrt{2}$ or $x = -1 - \sqrt{2}$

Solution set: $\{-1 + \sqrt{2}, -1 - \sqrt{2}\}$

4. $2x^2 - 3x - 1 = 0$

Here $a = 2$, $b = -3$, and $c = -1$.

$$x = \frac{-b \pm \sqrt{b^2 - 4ac}}{2a}$$

$$x = \frac{-(-3) \pm \sqrt{(-3)^2 - 4(2)(-1)}}{2(2)} = \frac{3 \pm \sqrt{17}}{4}$$

Solution set: $\left\{\frac{3 + \sqrt{17}}{4}, \frac{3 - \sqrt{17}}{4}\right\}$

5. $3t^2 - 4t = -5$
$3t^2 - 4t + 5 = 0$

Here $a = 3$, $b = -4$, and $c = 5$.

$$t = \frac{-b \pm \sqrt{b^2 - 4ac}}{2a}$$

$$t = \frac{-(-4) \pm \sqrt{(-4)^2 - 4(3)(5)}}{2(3)}$$

$$= \frac{4 \pm \sqrt{-44}}{6} = \frac{4 \pm 2i\sqrt{11}}{6}$$

$$= \frac{2\left(2 \pm i\sqrt{11}\right)}{6} = \frac{2 \pm i\sqrt{11}}{3}$$

Solution set: $\left\{\frac{2 + i\sqrt{11}}{3}, \frac{2 - i\sqrt{11}}{3}\right\}$

6. $$3x = \sqrt{\frac{9x+2}{2}}$$

Square both sides.

$$9x^2 = \frac{9x+2}{2}$$
$$18x^2 = 9x + 2$$
$$18x^2 - 9x - 2 = 0$$

As directed, use the quadratic formula with $a = 18$, $b = -9$, and $c = -2$.

$$x = \frac{-b \pm \sqrt{b^2 - 4ac}}{2a}$$
$$x = \frac{-(-9) \pm \sqrt{(-9)^2 - 4(18)(-2)}}{2(18)}$$
$$= \frac{9 \pm \sqrt{225}}{36} = \frac{9 \pm 15}{36}$$
$$x = \frac{9+15}{36} = \frac{24}{36} = \frac{2}{3} \text{ or}$$
$$x = \frac{9-15}{36} = \frac{-6}{36} = -\frac{1}{6}$$

Check $x = \frac{2}{3}$: $2 = \sqrt{4}$ *True*

Check $x = -\frac{1}{6}$: $-\frac{1}{2} = \sqrt{\frac{1}{4}}$ *False*

Solution set: $\left\{\frac{2}{3}\right\}$

7. If k is a negative number, then $4k$ is also negative, so the equation $x^2 = 4k$ will have two nonreal complex solutions. The answer is **A**.

8. $2x^2 - 8x - 3 = 0$

$$b^2 - 4ac = (-8)^2 - 4(2)(-3)$$
$$= 64 + 24 = 88$$

The discriminant, 88, is positive but not a perfect square, so there will be two distinct irrational number solutions.

9. $$3 - \frac{16}{x} - \frac{12}{x^2} = 0$$

Multiply by the LCD, x^2.

$$x^2\left(3 - \frac{16}{x} - \frac{12}{x^2}\right) = x^2 \bullet 0$$
$$3x^2 - 16x - 12 = 0$$
$$(3x+2)(x-6) = 0$$

$3x + 2 = 0$ or $x - 6 = 0$

$x = -\frac{2}{3}$ or $x = 6$

Check $x = -\frac{2}{3}$: $3 + 24 - 27 = 0$ *True*

Check $x = 6$: $3 - \frac{8}{3} - \frac{1}{3} = 0$ *True*

Solution set: $\left\{-\frac{2}{3}, 6\right\}$

10. $4x^2 + 7x - 3 = 0$

Use the quadratic formula with $a = 4$, $b = 7$, and $c = -3$.

$$x = \frac{-b \pm \sqrt{b^2 - 4ac}}{2a}$$
$$x = \frac{-7 \pm \sqrt{7^2 - 4(4)(-3)}}{2(4)}$$
$$= \frac{-7 \pm \sqrt{97}}{8}$$

Solution set: $\left\{\frac{-7+\sqrt{97}}{8}, \frac{-7-\sqrt{97}}{8}\right\}$

11. $$9x^4 + 4 = 37x^2$$
$$9x^4 - 37x^2 + 4 = 0$$

Let $u = x^2$, so $u^2 = \left(x^2\right)^2 = x^4$.
The equation becomes

$$9u^2 - 37u + 4 = 0.$$
$$(9u - 1)(u - 4) = 0$$

$9u - 1 = 0$ or $u - 4 = 0$

$u = \frac{1}{9}$ or $u = 4$

To find x, substitute x^2 for u.

$x^2 = \frac{1}{9}$ or $x^2 = 4$

$x = \pm\sqrt{\frac{1}{9}}$ or $x = \pm\sqrt{4}$

$x = \pm\frac{1}{3}$ or $x = \pm 2$

Check $x = \pm\frac{1}{3}$: $\frac{1}{9} + 4 = \frac{37}{9}$ *True*

Check $x = \pm 2$: $144 + 4 = 37(4)$ *True*

Solution set: $\left\{-2, -\frac{1}{3}, \frac{1}{3}, 2\right\}$

12. $12 = (2n+1)^2 + (2n+1)$

Let $u = 2n + 1$. The equation becomes

$$12 = u^2 + u.$$
$$0 = u^2 + u - 12$$
$$0 = (u+4)(u-3)$$

$u + 4 = 0$ or $u - 3 = 0$

$u = -4$ or $u = 3$

To find n, substitute $2n + 1$ for u.

$2n + 1 = -4$ or $2n + 1 = 3$

$2n = -5$ or $2n = 2$

$n = -\frac{5}{2}$ or $n = 1$

Check $n = -\frac{5}{2}$: $12 = 16 - 4$ *True*

Check $n = 1$: $12 = 9 + 3$ *True*

Solution set: $\left\{-\frac{5}{2}, 1\right\}$

13. Solve $S = 4\pi r^2$ for r.

$$\frac{S}{4\pi} = r^2$$

$$r = \pm\sqrt{\frac{S}{4\pi}} = \frac{\pm\sqrt{S}}{2\sqrt{\pi}}$$

$$= \frac{\pm\sqrt{S}}{2\sqrt{\pi}} \cdot \frac{\sqrt{\pi}}{\sqrt{\pi}}$$

$$r = \frac{\pm\sqrt{\pi S}}{2\pi}$$

14. Let x = Andrew's time alone.
Then $x - 2$ = Kent's time alone.

Make a table.

	Rate	Time Together	Fractional Part of the Job Done
Andrew	$\frac{1}{x}$	5	$\frac{5}{x}$
Kent	$\frac{1}{x-2}$	5	$\frac{5}{x-2}$

Part done by Andrew plus part done by Kent equals 1 whole job.

$$\frac{5}{x} + \frac{5}{x-2} = 1$$

Multiply both sides by the LCD, $x(x-2)$.

$$x(x-2)\left(\frac{5}{x} + \frac{5}{x-2}\right) = x(x-2) \cdot 1$$

$$5x - 10 + 5x = x^2 - 2x$$

$$0 = x^2 - 12x + 10$$

Use the quadratic formula with $a = 1$, $b = -12$, and $c = 10$.

$$x = \frac{-b \pm \sqrt{b^2 - 4ac}}{2a}$$

$$x = \frac{-(-12) \pm \sqrt{(-12)^2 - 4(1)(10)}}{2(1)}$$

$$= \frac{12 \pm \sqrt{104}}{2} = \frac{12 \pm 2\sqrt{26}}{2}$$

$$= \frac{2\left(6 \pm \sqrt{26}\right)}{2} = 6 \pm \sqrt{26}$$

$x = 6 + \sqrt{26} \approx 11.1$ or

$x = 6 - \sqrt{26} \approx .9$

Reject .9 for Andrew's time, because that would yield a negative time for Kent. Thus, Andrew's time is about 11.1 hours and Kent's time is $x - 2 \approx 9.1$ hours.

15. Let x = Abby's rate.

Make a table. Use $d = rt$, or $t = \frac{d}{r}$.

	Distance	Rate	Time
Upstream	10	$x - 3$	$\frac{10}{x-3}$
Downstream	10	$x + 3$	$\frac{10}{x+3}$

Time upstream plus Time downstream equals 3.5 hr.

$$\frac{10}{x-3} + \frac{10}{x+3} = \frac{7}{2}$$

Multiply both sides by the LCD, $2(x+3)(x-3)$.

$$2(x+3)(x-3)\left(\frac{10}{x-3} + \frac{10}{x+3}\right) = 2(x+3)(x-3)\left(\tfrac{7}{2}\right)$$

$$20(x+3) + 20(x-3) = 7\left(x^2 - 9\right)$$

$$20x + 60 + 20x - 60 = 7x^2 - 63$$

$$0 = 7x^2 - 40x - 63$$

$$0 = (x-7)(7x+9)$$

$x - 7 = 0$ or $7x + 9 = 0$

$x = 7$ or $x = -\frac{9}{7}$

Reject $-\frac{9}{7}$ since the rate can't be negative. Abby's rate was 7 mph.

16. Let x = the width of the walk.
The area of the walk is equal to the area of the outer figure minus the area of the pool.

$$152 = (10 + 2x)(24 + 2x) - (24)(10)$$

$$152 = 240 + 68x + 4x^2 - 240$$

$$0 = 4x^2 + 68x - 152$$

$$0 = x^2 + 17x - 38$$

$$0 = (x + 19)(x - 2)$$

$x + 19 = 0$ or $x - 2 = 0$

$x = -19$ or $x = 2$

Reject -19 since width can't be negative. The walk is 2 feet wide.

17. Let $x =$ the height of the tower. Then
$2x + 2 =$ the distance from the point to the top.

The distance from the base to the point is 30 m. These three segments form a right triangle, so the Pythagorean formula applies.

$$\begin{aligned} a^2 + b^2 &= c^2 \\ x^2 + 30^2 &= (2x+2)^2 \\ x^2 + 900 &= 4x^2 + 8x + 4 \\ 0 &= 3x^2 + 8x - 896 \\ 0 &= (x-16)(3x+56) \end{aligned}$$

$$x - 16 = 0 \quad \text{or} \quad 3x + 56 = 0$$
$$x = 16 \quad \text{or} \quad x = -\tfrac{56}{3}$$

Reject $-\frac{56}{3}$ since height can't be negative. The tower is 16 m high.

18. **(a)** Use $f(x) = .156x^2 - 2.05x + 10.2$ with $x = 11$.

$$\begin{aligned} f(11) &= .156(11)^2 - 2.05(11) + 10.2 \\ &\approx 6.5 \text{ (to the nearest tenth)} \end{aligned}$$

In 2001, there was a 6.5% increase in tuition.

(b) Find the vertex.

$$x = \frac{-b}{2a} = \frac{-(-2.05)}{2(.156)} \approx 6 \text{ (rounding down)}$$

$$\begin{aligned} f(6) &= .156(6)^2 - 2.05(6) + 10.2 \\ &\approx 3.5 \text{ (to the nearest tenth)} \end{aligned}$$

The minimum tuition increase was 3.5%, which occurred in $1990 + 6 = 1996$.

19. $f(x) = a(x-h)^2 + k$
Since $a < 0$, the parabola opens down. Since $h > 0$ and $k < 0$, the x-coordinate of the vertex is positive and the y-coordinate of the vertex is negative. Therefore, the vertex is in quadrant IV. The correct graph is **A**.

20. $f(x) = \frac{1}{2}x^2 - 2$
$f(x) = \frac{1}{2}(x-0)^2 - 2$
The graph is a parabola in $f(x) = a(x-h)^2 + k$ form with vertex (h, k) at $(0, -2)$. The axis is $x = 0$. Since $a = \frac{1}{2} > 0$, the parabola opens up. Also, $|a| = \left|\frac{1}{2}\right| = \frac{1}{2} < 1$, so the graph of the parabola is wider than the graph of $f(x) = x^2$. The points $(2, 0)$ and $(-2, 0)$ are on the graph.

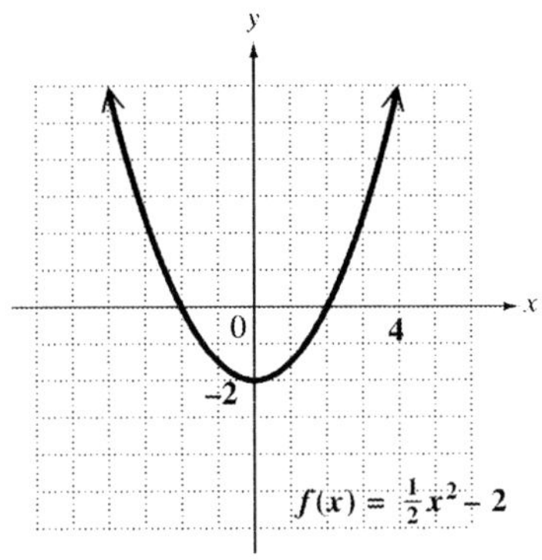

From the graph, we see that the x-values can be any real number, so the domain is $(-\infty, \infty)$. The y-values are greater than or equal to -2, so the range is $[-2, \infty)$.

21. $f(x) = -x^2 + 4x - 1$

The x-coordinate of the vertex is

$$x = \frac{-b}{2a} = \frac{-4}{2(-1)} = 2.$$

The y-coordinate of the vertex is

$$f(2) = -4 + 8 - 1 = 3.$$

The graph is a parabola with vertex (h, k) at $(2, 3)$ and axis $x = 2$. Since $a = -1 < 0$, the parabola opens down.
Also, $|a| = |-1| = 1$, so the graph has the same shape as the graph of $f(x) = x^2$. The points $(0, -1)$ and $(4, -1)$ are on the graph.

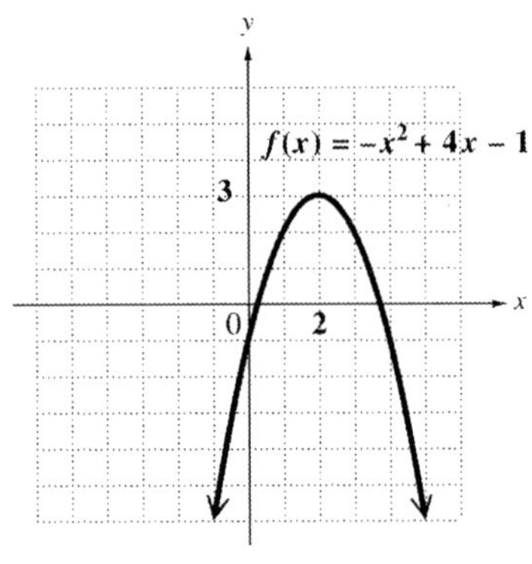

From the graph, we see that the x-values can be any real number, so the domain is $(-\infty, \infty)$.

The y-values are less than or equal to 3, so the range is $(-\infty, 3]$.

22. $x = 2y^2 + 8y + 3$

The graph of this equation is a horizontal parabola. Complete the square to find the vertex.

$$\begin{aligned} x &= 2(y^2 + 4y) + 3 \\ &= 2(y^2 + 4y + 4 - 4) + 3 \\ &= 2(y^2 + 4y + 4) + 2(-4) + 3 \\ &= 2(y + 2)^2 - 5 \end{aligned}$$

The equation is in the form $x = a(y - k)^2 + h$, so the vertex (h, k) is $(-5, -2)$ and the axis is $y = -2$. If $y = 0$, $x = 3$, so the x-intercept is $(3, 0)$. By symmetry, $(3, -4)$ is also on the graph.

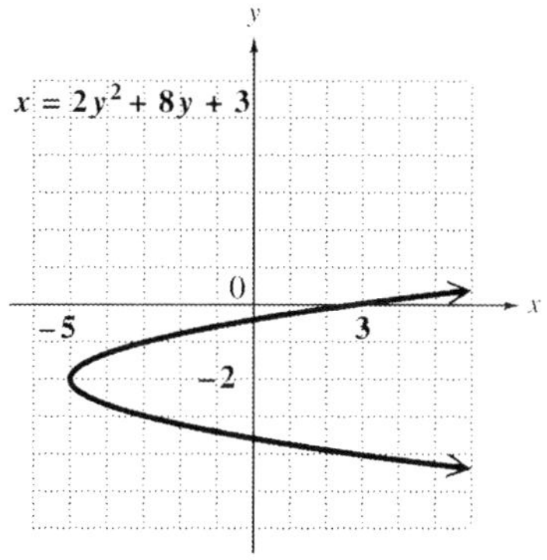

Since x is always greater than or equal to -5, the domain is $[-5, \infty)$; y can be any real number, so the range is $(-\infty, \infty)$.

23. Let x represent the length of the two equal sides, and let $280 - 2x$ represent the length of the remaining side (the side parallel to the highway). Then substitute x for L and $280 - 2x$ for W in the formula for the area of a rectangle, $A = LW$.

$$\begin{aligned} A &= x(280 - 2x) \\ &= 280x - 2x^2 \text{ or } -2x^2 + 280x \end{aligned}$$

The maximum area will occur at the vertex.

$$x = \frac{-b}{2a} = \frac{-280}{2(-2)} = 70$$

When $x = 70$, $280 - 2x = 140$, and $A = 9800$. Thus, the dimensions of the lot with maximum area are 140 feet by 70 feet and the maximum area is 9800 square feet.

24.

$$\begin{aligned} 2x^2 + 7x &> 15 \\ 2x^2 + 7x - 15 &> 0 \end{aligned}$$

Solve the equation

$$\begin{aligned} 2x^2 + 7x - 15 &= 0. \\ (2x - 3)(x + 5) &= 0 \end{aligned}$$

$$2x - 3 = 0 \quad \text{or} \quad x + 5 = 0$$
$$x = \tfrac{3}{2} \quad \text{or} \quad x = -5$$

The numbers -5 and $\frac{3}{2}$ divide a number line into three intervals.

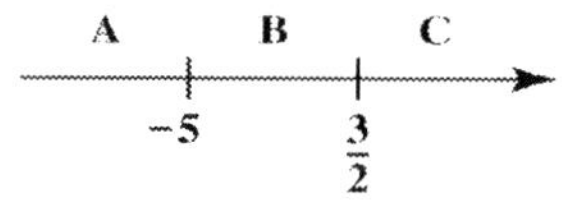

Test a number from each interval in the inequality

$$2x^2 + 7x > 15.$$

Interval A: Let $x = -6$.
$72 - 42 > 15$?
$30 > 15$ *True*

Interval B: Let $x = 0$.
$0 > 15$ *False*

Interval C: Let $x = 2$.
$8 + 14 > 15$?
$22 > 15$ *True*

The numbers in Intervals A and C, not including -5 and $\frac{3}{2}$, are solutions.

Solution set: $(-\infty, -5) \cup \left(\frac{3}{2}, \infty\right)$

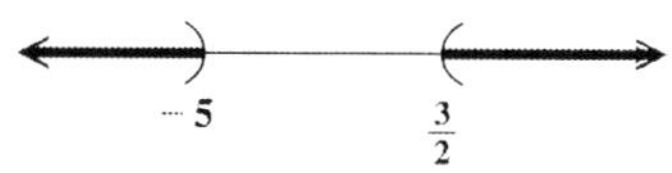

25.

$$\frac{5}{t - 4} \le 1$$

Write the inequality so that 0 is on one side.

$$\begin{aligned} \frac{5}{t - 4} - 1 &\le 0 \\ \frac{5}{t - 4} - \frac{1(t - 4)}{t - 4} &\le 0 \\ \frac{5 - t + 4}{t - 4} &\le 0 \\ \frac{-t + 9}{t - 4} &\le 0 \end{aligned}$$

The number 9 makes the numerator 0, and 4 makes the denominator 0. These two numbers determine three intervals.

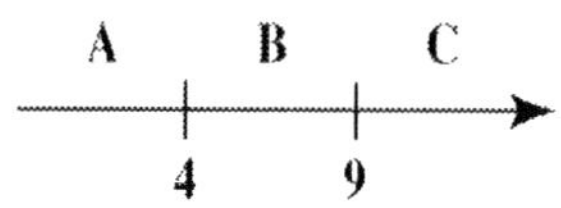

continued

Test a number from each interval in the inequality

$$\frac{5}{t-4} \le 1.$$

Interval A: Let $t = 0$.

$\frac{5}{-4} \le 1$ *True*

Interval B: Let $t = 7$.

$\frac{5}{3} \le 1$ *False*

Interval C: Let $t = 10$.

$\frac{5}{6} \le 1$ *True*

The numbers in Intervals A and C, including 9 but not 4, are solutions.

Solution set: $(-\infty, 4) \cup [9, \infty)$

Cumulative Review Exercises (Chapters 1–10)

1. $S = \left\{-\frac{7}{3}, -2, -\sqrt{3}, 0, .7, \sqrt{12}, \sqrt{-8}, 7, \frac{32}{3}\right\}$

(a) The elements of S that are integers are -2, 0, and 7.

(b) The elements of S that are rational numbers are $-\frac{7}{3}$, -2, 0, .7, 7, and $\frac{32}{3}$.

(c) All the elements of S except $\sqrt{-8}$ are real numbers.

(d) All the elements of S are complex numbers.

2.
$$\begin{aligned}|-3| + 8 - |-9| - (-7+3) &= 3 + 8 - 9 - (-4)\\ &= 3 + 8 - 9 + 4\\ &= 6\end{aligned}$$

3.
$$\begin{aligned}&2(-3)^2 + (-8)(-5) + (-17)\\ &= 2(9) + (-8)(-5) + (-17)\\ &= 18 + 40 - 17 = 41\end{aligned}$$

4. Since 31% of the people in the United States curse the ATM, the number of people in the United States who curse the ATM in a sample of 3000 is

$$.31(3000) = 930.$$

5. $.24(3000) = 720$

6. $.33(3000) = 990$

7. German cursers: $.53(3000) = 1590$
United States thankers: $.22(3000) = 660$

The difference is

$$1590 - 660 = 930.$$

8.
$$\begin{aligned}-2x + 4 &= 5(x-4) + 17\\ -2x + 4 &= 5x - 20 + 17\\ -2x + 4 &= 5x - 3\\ 7 &= 7x\\ 1 &= x\end{aligned}$$

Solution set: $\{1\}$

9.
$$\begin{aligned}-2x + 4 &\le -x + 3\\ -x &\le -1\end{aligned}$$

Multiply by -1, and reverse the direction of the inequality.

$$x \ge 1$$

Solution set: $[1, \infty)$

10.
$$\begin{aligned}|3x - 7| &\le 1\\ -1 \le 3x - 7 &\le 1\\ 6 \le 3x &\le 8\\ 2 \le x &\le \frac{8}{3}\end{aligned}$$

Solution set: $\left[2, \frac{8}{3}\right]$

11. $2x - 4y = 7$

Solve the equation for y to get the slope-intercept form, $y = mx + b$.

$$\begin{aligned}2x - 4y &= 7\\ -4y &= -2x + 7\\ y &= \tfrac{1}{2}x - \tfrac{7}{4}\end{aligned}$$

The slope m is $\frac{1}{2}$ and the y-intercept, since $b = -\frac{7}{4}$, is $\left(0, -\frac{7}{4}\right)$.

12. Through $(2, -1)$; perpendicular to $-3x + y = 5$

Find the slope of the given line.

$$\begin{aligned}-3x + y &= 5\\ y &= 3x + 5\end{aligned}$$

The slope is 3. The negative reciprocal of 3 is $-\frac{1}{3}$, so use the point-slope form with $m = -\frac{1}{3}$ and $(x_1, y_1) = (2, -1)$ to find the equation of the desired line.

$$\begin{aligned}y - y_1 &= m(x - x_1)\\ y - (-1) &= -\tfrac{1}{3}(x-2)\\ y + 1 &= -\tfrac{1}{3}(x-2)\\ 3y + 3 &= -1(x-2) \quad \textit{Multiply by 3.}\\ 3y + 3 &= -x + 2\\ x + 3y &= -1\end{aligned}$$

13. $4x - 5y = 15$
Draw the line through its intercepts, $\left(\frac{15}{4}, 0\right)$ and $(0, -3)$. The graph passes the vertical line test, so the relation is a function. As with any line that is not horizontal or vertical, the domain and range are both $(-\infty, \infty)$.

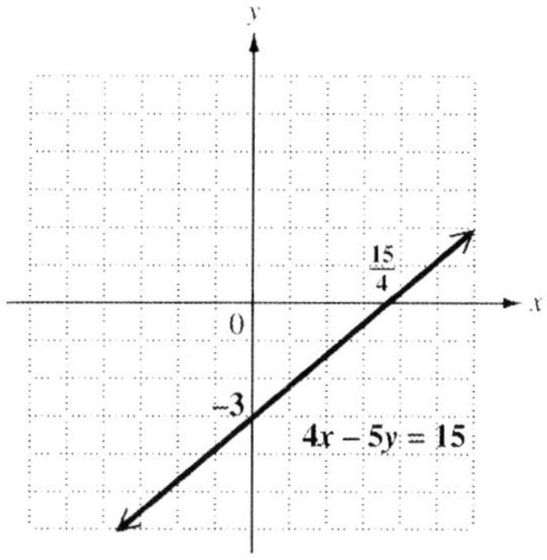

14. $4x - 5y < 15$
Draw a dashed line through the points $\left(\frac{15}{4}, 0\right)$ and $(0, -3)$. Check the origin:

$$4(0) - 5(0) < 15 \quad ?$$
$$0 < 15 \qquad \textit{True}$$

Shade the region that contains the origin.

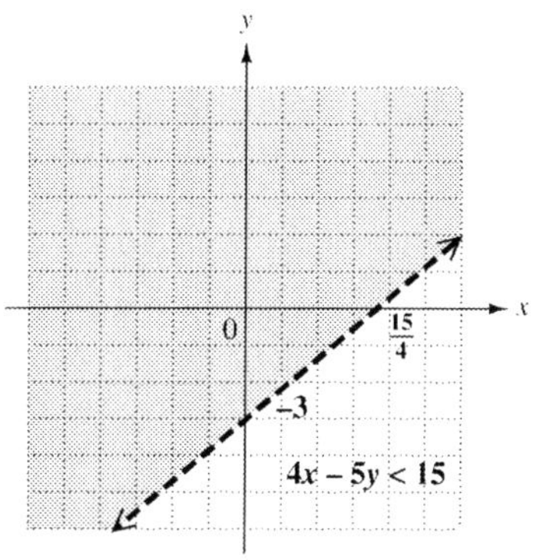

The relation is not a function since for any value of x, there is more than one value of y.

15. $f(x) = -2(x-1)^2 + 3$
The equation is in $f(x) = a(x-h)^2 + k$ form, so the graph is a parabola with vertex (h, k) at $(1, 3)$. Since $a = -2 < 0$, the parabola opens down. Also $|a| = |-2| = 2 > 1$, so the graph is narrower than the graph of $f(x) = x^2$. The points $(0, 1)$ and $(2, 1)$ are on the graph.

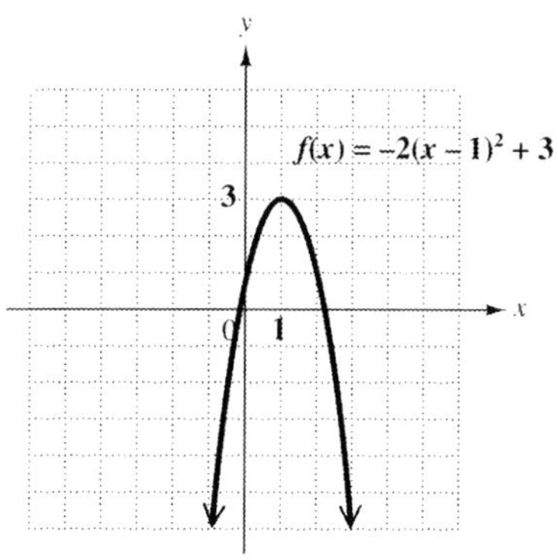

The relation is a function since it passes the vertical line test. The domain is $(-\infty, \infty)$. The largest value of y is 3, so the range is $(-\infty, 3]$.

16. **(a)** We are given the y-intercept, $(0, 600)$, so the slope-intercept form of the line is

$$y = mx + 600.$$

Substitute 1340 for y and 6 for x to solve for the slope m.

$$\begin{aligned} 1340 &= m(6) + 600 \\ 740 &= 6m \\ m &= \tfrac{740}{6} = 123.\overline{3} \end{aligned}$$

Rounding to the nearest whole number gives us the linear model

$$y = 123x + 600.$$

(b) For 2002, use $x = 7$.

$$\begin{aligned} y &= 123(7) + 600 \\ &= 861 + 600 \\ &= 1461 \end{aligned}$$

From the model, the approximate sales were \$1461 million, which is a little high compared to the table value of \$1400 million.

17. The relation $x = 5$ does not define a function because its graph is a vertical line, which is not the graph of a function by the vertical line test.

18. $2x - 4y = 10 \quad (1)$
$9x + 3y = 3 \quad (2)$
Simplify the equations.

$$\begin{aligned} x - 2y &= 5 \quad (3) \qquad \tfrac{1}{2} \times (1) \\ 3x + y &= 1 \quad (4) \qquad \tfrac{1}{3} \times (2) \end{aligned}$$

To eliminate y, multiply (4) by 2 and add the result to (3).

$$\begin{aligned} x - 2y &= 5 \quad (3) \\ 6x + 2y &= 2 \qquad 2 \times (4) \\ \hline 7x \qquad &= 7 \\ x &= 1 \end{aligned}$$

Substitute $x = 1$ into (4).

$$\begin{aligned} 3x + y &= 1 \qquad (4) \\ 3(1) + y &= 1 \\ y &= -2 \end{aligned}$$

Solution set: $\{(1, -2)\}$

19.
$$\begin{aligned} x + y + 2z &= 3 \quad (1) \\ -x + y + z &= -5 \quad (2) \\ 2x + 3y - z &= -8 \quad (3) \end{aligned}$$

Eliminate z by adding (2) and (3).

$$\begin{aligned} -x + y + z &= -5 \quad (2) \\ 2x + 3y - z &= -8 \quad (3) \\ \hline x + 4y \quad &= -13 \quad (4) \end{aligned}$$

To get another equation without z, multiply equation (3) by 2 and add the result to equation (1).

$$\begin{aligned} x + y + 2z &= 3 \quad (1) \\ 4x + 6y - 2z &= -16 \quad 2 \times (3) \\ \hline 5x + 7y \quad &= -13 \quad (5) \end{aligned}$$

To eliminate x, multiply (4) by -5 and add the result to (5).

$$\begin{aligned} -5x - 20y &= 65 \quad -5 \times (4) \\ 5x + 7y &= -13 \quad (5) \\ \hline -13y &= 52 \\ y &= -4 \end{aligned}$$

Use (4) to find x.

$$\begin{aligned} x + 4y &= -13 \quad (4) \\ x + 4(-4) &= -13 \\ x - 16 &= -13 \\ x &= 3 \end{aligned}$$

Use (2) to find z.

$$\begin{aligned} -x + y + z &= -5 \quad (2) \\ -3 - 4 + z &= -5 \\ -7 + z &= -5 \\ z &= 2 \end{aligned}$$

Solution set: $\{(3, -4, 2)\}$

20. (a) Let $x =$ the amount of sales (in billions) for AOL and $y =$ the amount of sales for Time Warner.

The combined sales were \$34.2 billion, so

$$x + y = 34.2. \quad (1)$$

Sales for AOL were \$.3 billion less than 4 times the sales of Time Warner, so

$$x = 4y - .3. \quad (2)$$

(b) Substitute $4y - .3$ for x in (1), then solve for y.

$$\begin{aligned} (4y - .3) + y &= 34.2 \\ 5y &= 34.5 \\ y &= 6.9 \end{aligned}$$

From (2), $x = 4(6.9) - .3 = 27.3$. Thus, AOL had \$27.3 billion in sales and Time Warner had \$6.9 billion in sales.

21.
$$\begin{aligned} \left(\frac{x^{-3}y^2}{x^5y^{-2}}\right)^{-1} &= \left(x^{-3-5}y^{2-(-2)}\right)^{-1} \\ &= \left(x^{-8}y^4\right)^{-1} \\ &= x^8y^{-4} \\ &= \frac{x^8}{y^4} \end{aligned}$$

22.
$$\begin{aligned} \frac{(4x^{-2})^2(2y^3)}{8x^{-3}y^5} &= \frac{16x^{-4}(2y^3)}{8x^{-3}y^5} \\ &= \frac{4x^{-4}y^3}{x^{-3}y^5} \\ &= 4x^{-4-(-3)}y^{3-5} \\ &= 4x^{-1}y^{-2} \\ &= \frac{4}{xy^2} \end{aligned}$$

23.
$$\begin{aligned} \left(\tfrac{2}{3}t + 9\right)^2 &= \left(\tfrac{2}{3}t\right)^2 + 2\left(\tfrac{2}{3}t\right)(9) + 9^2 \\ &= \tfrac{4}{9}t^2 + 12t + 81 \end{aligned}$$

24.
$$\begin{aligned} &(3t^3 + 5t^2 - 8t + 7) - (6t^3 + 4t - 8) \\ &= 3t^3 + 5t^2 - 8t + 7 - 6t^3 - 4t + 8 \\ &= -3t^3 + 5t^2 - 12t + 15 \end{aligned}$$

25. Divide $4x^3 + 2x^2 - x + 26$ by $x + 2$.

$$\begin{array}{r|l} & \quad\; 4x^2 - 6x + 11 \\ x + 2 & 4x^3 + 2x^2 - x + 26 \\ & \underline{4x^3 + 8x^2} \\ & \quad\; -6x^2 - x \\ & \quad\; \underline{-6x^2 - 12x} \\ & \qquad\qquad 11x + 26 \\ & \qquad\qquad \underline{11x + 22} \\ & \qquad\qquad\qquad 4 \end{array}$$

The answer is

$$4x^2 - 6x + 11 + \frac{4}{x+2}.$$

26. 4.6×10^9

Move the decimal point to the right 9 places.

$$4.6 \times 10^9 = 4{,}600{,}000{,}000$$

27.
$$\begin{aligned} 16x - x^3 &= x(16 - x^2) \\ &= x(4 + x)(4 - x) \end{aligned}$$

28. $24m^2 + 2m - 15$

The two integers whose product is $24(-15) = -360$ and whose sum is 2 are 20 and -18.

$$\begin{aligned} &24m^2 + 2m - 15 \\ &= 24m^2 + 20m - 18m - 15 \\ &= 4m(6m + 5) - 3(6m + 5) \\ &= (6m + 5)(4m - 3) \end{aligned}$$

29. $9x^2 - 30xy + 25y^2$

Use the perfect square formula,

$$a^2 - 2ab + b^2 = (a - b)^2,$$

with $a = 3x$ and $b = 5y$.

$$\begin{aligned} &9x^2 - 30xy + 25y^2 \\ &= \left[(3x)^2 - 2(3x)(5y) + (5y)^2\right] \\ &= (3x - 5y)^2 \end{aligned}$$

30. $\dfrac{5t+2}{-6} \div \dfrac{15t+6}{5}$

Multiply by the reciprocal.

$$\begin{aligned} &= \frac{5t+2}{-6} \cdot \frac{5}{15t+6} \\ &= \frac{5t+2}{-6} \cdot \frac{5}{3(5t+2)} \\ &= \frac{5}{-18} \text{ or } -\frac{5}{18} \end{aligned}$$

31. $\dfrac{3}{2-k} - \dfrac{5}{k} + \dfrac{6}{k^2-2k}$

$$\begin{aligned} &= \frac{3}{2-k} - \frac{5}{k} + \frac{6}{k(k-2)} \\ &= \frac{-3}{k-2} - \frac{5}{k} + \frac{6}{k(k-2)} \end{aligned}$$

The LCD is $k(k-2)$.

$$\begin{aligned} &= \frac{-3k}{(k-2)k} - \frac{5(k-2)}{k(k-2)} + \frac{6}{k(k-2)} \\ &= \frac{-3k - 5(k-2) + 6}{k(k-2)} \\ &= \frac{-3k - 5k + 10 + 6}{k(k-2)} \\ &= \frac{-8k+16}{k(k-2)} \\ &= \frac{-8(k-2)}{k(k-2)} = -\frac{8}{k} \end{aligned}$$

32. $\dfrac{\frac{r}{s} - \frac{s}{r}}{\frac{r}{s} + 1}$

Multiply the numerator and denominator by the LCD of all the fractions, rs.

$$\begin{aligned} &= \frac{\left(\frac{r}{s} - \frac{s}{r}\right)rs}{\left(\frac{r}{s} + 1\right)rs} = \frac{r^2 - s^2}{r^2 + rs} \\ &= \frac{(r-s)(r+s)}{r(r+s)} = \frac{r-s}{r} \end{aligned}$$

33.
$$\begin{aligned} \sqrt[3]{\frac{27}{16}} &= \frac{\sqrt[3]{27}}{\sqrt[3]{16}} = \frac{\sqrt[3]{3^3}}{\sqrt[3]{8 \cdot 2}} = \frac{3}{2\sqrt[3]{2}} \\ &= \frac{3 \cdot \sqrt[3]{4}}{2\sqrt[3]{2} \cdot \sqrt[3]{4}} = \frac{3\sqrt[3]{4}}{2\sqrt[3]{8}} \\ &= \frac{3\sqrt[3]{4}}{2 \cdot 2} = \frac{3\sqrt[3]{4}}{4} \end{aligned}$$

34.
$$\begin{aligned} \frac{2}{\sqrt{7} - \sqrt{5}} &= \frac{2\left(\sqrt{7} + \sqrt{5}\right)}{\left(\sqrt{7} - \sqrt{5}\right)\left(\sqrt{7} + \sqrt{5}\right)} \\ &= \frac{2\left(\sqrt{7} + \sqrt{5}\right)}{7 - 5} \\ &= \frac{2\left(\sqrt{7} + \sqrt{5}\right)}{2} = \sqrt{7} + \sqrt{5} \end{aligned}$$

35.
$$2x = \sqrt{\frac{5x+2}{3}}$$

Square both sides.

$$\begin{aligned} (2x)^2 &= \left(\sqrt{\frac{5x+2}{3}}\right)^2 \\ 4x^2 &= \frac{5x+2}{3} \\ 12x^2 &= 5x + 2 \\ 12x^2 - 5x - 2 &= 0 \\ (3x-2)(4x+1) &= 0 \end{aligned}$$

$$\begin{aligned} 3x - 2 = 0 \quad &\text{or} \quad 4x + 1 = 0 \\ x = \tfrac{2}{3} \quad &\text{or} \quad x = -\tfrac{1}{4} \end{aligned}$$

Check $x = -\frac{1}{4}$: $\quad -\frac{1}{2} = \frac{1}{2} \quad$ *False*

Check $x = \frac{2}{3}$: $\quad \frac{4}{3} = \sqrt{\frac{16}{9}} \quad$ *True*

Solution set: $\left\{\frac{2}{3}\right\}$

36. $2x^2 - 4x - 3 = 0$

Use $a = 2$, $b = -4$, and $c = -3$ in the quadratic formula.

$$\begin{aligned} x &= \frac{-b \pm \sqrt{b^2 - 4ac}}{2a} \\ &= \frac{-(-4) \pm \sqrt{(-4)^2 - 4(2)(-3)}}{2(2)} \\ &= \frac{4 \pm \sqrt{16 + 24}}{4} \\ &= \frac{4 \pm \sqrt{40}}{4} = \frac{4 \pm 2\sqrt{10}}{4} \\ &= \frac{2\left(2 \pm \sqrt{10}\right)}{4} = \frac{2 \pm \sqrt{10}}{2} \end{aligned}$$

Solution set: $\left\{\dfrac{2 + \sqrt{10}}{2}, \dfrac{2 - \sqrt{10}}{2}\right\}$

37. $z^2 - 2z = 15$

To complete the square on $z^2 - 2z$, take half the coefficient of z and square it.

$$\left[\tfrac{1}{2}(-2)\right]^2 = (-1)^2 = 1$$

Add 1 to each side.

$$z^2 - 2z + 1 = 15 + 1$$
$$(z-1)^2 = 16$$

$$z - 1 = 4 \quad \text{or} \quad z - 1 = -4$$
$$z = 5 \quad \text{or} \quad z = -3$$

Solution set: $\{-3, 5\}$

38. $\dfrac{3}{x-3} - \dfrac{2}{x-2} = \dfrac{3}{x^2 - 5x + 6}$

$$\frac{3}{x-3} - \frac{2}{x-2} = \frac{3}{(x-3)(x-2)}$$

Multiply each side by the LCD, $(x-3)(x-2)$. $(x \neq 2, 3)$

$$(x-3)(x-2)\left(\frac{3}{x-3} - \frac{2}{x-2}\right) = (x-3)(x-2)\left(\frac{3}{(x-3)(x-2)}\right)$$

$$3(x-2) - 2(x-3) = 3$$
$$3x - 6 - 2x + 6 = 3$$
$$x = 3$$

But $x \neq 3$ since it makes a denominator 0.

Solution set: $\emptyset$

39. $p^4 - 10p^2 + 9 = 0$

Let $u = p^2$, so $u^2 = p^4$.

$$u^2 - 10u + 9 = 0$$
$$(u-1)(u-9) = 0$$

$$u - 1 = 0 \quad \text{or} \quad u - 9 = 0$$
$$u = 1 \qquad\qquad u = 9$$
$$p^2 = 1 \quad \text{or} \quad p^2 = 9$$
$$p = \pm 1 \qquad\qquad p = \pm 3$$

Solution set: $\{-3, -1, 1, 3\}$

40. Let $x =$ the distance traveled by the southbound car and
$2x - 38 =$ the distance traveled by the eastbound car.

Since the cars are traveling at right angles with one another, the Pythagorean formula can be applied.

$$a^2 + b^2 = c^2$$
$$x^2 + (2x - 38)^2 = 95^2$$
$$x^2 + 4x^2 - 152x + 1444 = 9025$$
$$5x^2 - 152x - 7581 = 0$$

$$x = \frac{-(-152) \pm \sqrt{(-152)^2 - 4(5)(-7581)}}{2(5)}$$
$$= \frac{152 \pm \sqrt{174{,}724}}{10} = \frac{152 \pm 418}{10}$$

Thus, $x = \dfrac{152 + 418}{10} = 57$ (the other value is negative). The southbound car traveled 57 miles, and the eastbound car traveled $2x - 38 = 2(57) - 38 = 76$ miles.

CHAPTER 11 EXPONENTIAL AND LOGARITHMIC FUNCTIONS

11.1 Inverse Functions

11.1 Margin Exercises

1. **(a)** $\{(1,2),(2,4),(3,3),(4,5)\}$

Each x-value in the relation corresponds to only one y-value, so the relation is indeed a function. Furthermore, each y-value corresponds to only one x-value, so this is a one-to-one function.

The inverse function is found by interchanging the x- and y-values in each ordered pair.

The inverse is

$$\{(2,1),\ (4,2),\ (3,3),\ (4,5)\}.$$

(b) $\{(0,3),(-1,2),(1,3)\}$

Since the y-value 3 corresponds to x-values 0 and 1, the function is not one-to-one. Therefore, it does not have an inverse.

(c) Since each time corresponds to only one distance, the function is one-to-one. The inverse of the function defined in the table is found by interchanging the columns.

Time	Distance
4:22	1.5 K
9:18	3 K
16:00	5 K
33:40	10 K

2. **(a)** Since a horizontal line will intersect the graph in no more than one point, the function is one-to-one.

(b) Since a horizontal line could intersect the graph in two points, the function is not one-to-one.

3. To find the equation of the inverse of a function,

(1) determine if the function is one-to-one;
(2) replace $f(x)$ with y;
(3) interchange x and y;
(4) solve for y;
(5) replace y with $f^{-1}(x)$.

(a) $f(x) = 3x - 4$

The graph of f is a line. By the horizontal line test, it is a one-to-one function. To find the inverse, first replace $f(x)$ with y.

Step 1 Interchange x and y.

$$y = 3x - 4$$
$$x = 3y - 4$$

Step 2 Solve for y.

$$x + 4 = 3y$$
$$\frac{x+4}{3} = y$$

Step 3 Replace y with $f^{-1}(x)$.

$$f^{-1}(x) = \frac{x+4}{3}$$

(b) $f(x) = x^3 + 1$

Refer to Section 6.3 to see from its graph that a cubing function like this is a one-to-one function. To find the inverse, first replace $f(x)$ with y.

Step 1 Interchange x and y.

$$y = x^3 + 1$$
$$x = y^3 + 1$$

Step 2 Solve for y.

$$x - 1 = y^3$$

Take the cube root on each side.

$$\sqrt[3]{x-1} = y$$

Step 3 Replace y with $f^{-1}(x)$.

$$f^{-1}(x) = \sqrt[3]{x-1}$$

(c) $f(x) = (x-3)^2$

This equation has a vertical parabola as its graph, so some horizontal lines will intersect the graph at two points. For example, both $x = 1$ and $x = 5$ correspond to $y = 4$. Thus, the function is not a one-to-one function and does not have an inverse.

4. **(a)** The given graph goes through $(-5,1)$ and $(1,-1)$. So the inverse will go through $(1,-5)$ and $(-1,1)$. Use these points and symmetry about $y = x$ to complete the graph of the inverse.

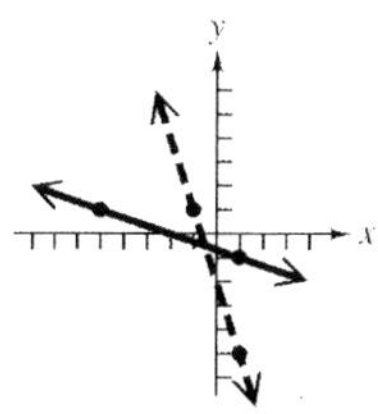

(b) Points on the given graph are $(-3,0)$, $(-2,2)$, and $(0,3)$. So, the inverse will contain $(0,-3)$, $(2,-2)$, and $(3,0)$. Use these points and symmetry about $y = x$ to complete the graph of the inverse.

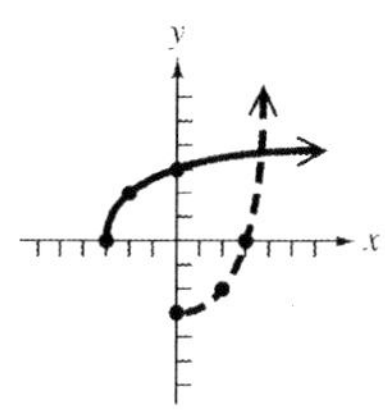

(c) The given graph goes through $\left(-1, \frac{1}{2}\right)$, $(0, 1)$, $(1, 2)$, and $(2, 4)$. So, the inverse will go through $\left(\frac{1}{2}, -1\right)$, $(1, 0)$, $(2, 1)$, and $(4, 2)$. Use these points and symmetry about $y = x$ to complete the graph of the inverse.

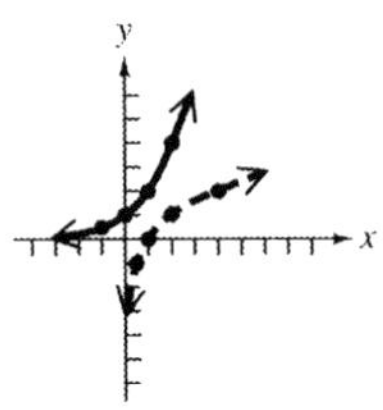

11.1 Section Exercises

1. The function is one-to-one because no two states have the same number of uncontrolled waste sites.

2. This function is one-to-one because each year corresponds to only one amount and each amount corresponds to only one year.

3. Consider the ordered pair (x, y) where x is a student in a math class and y is that student's mother. If two siblings were attending that class, the function would not be one-to-one since there would exist two ordered pairs that have different x-values but the same y-values.

4. The function in the table that pairs a city with a distance is a one-to-one function because for each city there is one distance and each distance has only one city to which it is paired.

   If the distance from Indianapolis to Denver had 1 mile added to it, it would be $1058 + 1 = 1059$ mi, the same as the distance from Los Angeles to Denver. In this case, one distance would have two cities to which it is paired, and the function would not be one-to-one.

5. If a function is made up of ordered pairs in such a way that the same y-value appears in a correspondence with two different x-values, then the function is not one-to-one. Choice **B**

6. **A.** $y = f(x) = x$ is one-to-one because each x-value corresponds to only one y-value and each y-value corresponds to just one x-value.

   B. $y = f(x) = x^2$ is not one-to-one since any positive y-value corresponds to two x-values. For example, $f(3) = 9$ and $f(-3) = 9$.

   C. $y = f(x) = |x|$ is not one-to-one since any positive y-value corresponds to two x-values. For example, $f(3) = 3$ and $f(-3) = 3$.

   D. $y = f(x) = -x^2 + 2x - 1$ is not one-to-one since its graph is a parabola and doesn't pass the horizontal line test. The x-coordinate of the vertex is

$$x = \frac{-b}{2a} = \frac{-2}{2(-1)} = 1,$$

so the axis of the parabola is $x = 1$. Pick a number to the right of 1, say 2, and to the left of 1 by the same amount—in this case 0, to find an example. $f(0) = -1$ and $f(2) = -1$.

7. All of the graphs pass the vertical line test, so they all represent functions. The graph in choice **A** is the only one that passes the horizontal line test, so it is the one-to-one function.

8. f is a one-to-one function with the point (p, q) on its graph. To form the inverse, the x- and y-coordinates are interchanged. Then (q, p) must be on the graph of the inverse of f. The correct answer is **D**.

9. $\{(3, 6), (2, 10), (5, 12)\}$
   This function is one-to-one since each y-value corresponds to only one x-value. To find the inverse, interchange x and y in each ordered pair. The inverse is

$$\{(6, 3), (10, 2), (12, 5)\}.$$

10. $\left\{(-1, 3), (0, 5), (5, 0), \left(7, -\frac{1}{2}\right)\right\}$
    This function is one-to-one since each y-value corresponds to only one x-value. To find the inverse, interchange x and y in each ordered pair. The inverse is

$$\left\{(3, -1), (5, 0), (0, 5), \left(-\tfrac{1}{2}, 7\right)\right\}$$

11. $\{(-1, 3), (2, 7), (4, 3), (5, 8)\}$ is not a one-to-one function. The ordered pairs $(-1, 3)$ and $(4, 3)$ have the same y-values for two different x-values.

12. $\{(-8, 6), (-4, 3), (0, 6), (5, 10)\}$ is not a one-to-one function. The ordered pairs $(-8, 6)$ and $(0, 6)$ have the same y-values for two different x-values.

13. The graph of $f(x) = 2x + 4$ is a nonvertical, nonhorizontal line. By the horizontal line test, $f(x)$ is a one-to-one function. To find the inverse, first replace $f(x)$ with y.

$$y = 2x + 4$$

Interchange x and y.

$$x = 2y + 4$$

Solve for y.

$$2y = x - 4$$

$$y = \frac{x - 4}{2}$$

Replace y with $f^{-1}(x)$.

$$f^{-1}(x) = \frac{x - 4}{2}$$

14. The graph of $f(x) = 3x + 1$ is a nonvertical, nonhorizontal line. By the horizontal line test, $f(x)$ is a one-to-one function. To find the inverse, first replace $f(x)$ with y.

$$y = 3x + 1$$

Interchange x and y.

$$x = 3y + 1$$

Solve for y.

$$3y = x - 1$$
$$y = \frac{x-1}{3}$$

Replace y with $f^{-1}(x)$.

$$f^{-1}(x) = \frac{x-1}{3}$$

15. Write $g(x) = \sqrt{x-3}$ as $y = \sqrt{x-3}$.
Since $x \geq 3$, $y \geq 0$. The graph of g is half of a horizontal parabola that opens to the right. The graph passes the horizontal line test, so g is one-to-one. To find the inverse, interchange x and y to get

$$x = \sqrt{y-3}.$$

Note that now $y \geq 3$, so $x \geq 0$.
Solve for y by squaring both sides.

$$x^2 = y - 3$$
$$x^2 + 3 = y$$

Replace y with $g^{-1}(x)$.

$$g^{-1}(x) = x^2 + 3, x \geq 0$$

16. Write $g(x) = \sqrt{x+2}$ as $y = \sqrt{x+2}$.
Since $x \geq -2$, $y \geq 0$. The graph of g is half of a horizontal parabola that opens to the right. The graph passes the horizontal line test, so g is one-to-one. To find the inverse, interchange x and y to get

$$x = \sqrt{y+2}.$$

Note that now $y \geq -2$, so $x \geq 0$.
Solve for y by squaring both sides.

$$x^2 = y + 2$$
$$x^2 - 2 = y$$

Replace y with $g^{-1}(x)$.

$$g^{-1}(x) = x^2 - 2, x \geq 0$$

17. $f(x) = 3x^2 + 2$ is not a one-to-one function because two x-values, such as 1 and -1, both have the same y-value, in this case 5. The graph of this function is a vertical parabola which does not pass the horizontal line test.

18. $f(x) = -4x^2 - 1$ is not a one-to-one function because two x-values, such as 1 and -1, both have the same y-value, in this case -5. The graph of this function is a vertical parabola which does not pass the horizontal line test.

19. The graph of $f(x) = x^3 - 4$ is the graph of $g(x) = x^3$ shifted down 4 units. (Recall that $g(x) = x^3$ is the elongated S-shaped curve.) The graph of f passes the horizontal line test, so f is one-to-one.

Replace $f(x)$ with y.

$$y = x^3 - 4$$

Interchange x and y.

$$x = y^3 - 4$$

Solve for y.

$$x + 4 = y^3$$

Take the cube root of each side.

$$\sqrt[3]{x+4} = y$$

Replace y with $f^{-1}(x)$.

$$f^{-1}(x) = \sqrt[3]{x+4}$$

20. The graph of $f(x) = x^3 - 3$ is the graph of $g(x) = x^3$ shifted down 3 units. (Recall that $g(x) = x^3$ is the elongated S-shaped curve.) The graph of f passes the horizontal line test, so f is one-to-one.

Replace $f(x)$ with y.

$$y = x^3 - 3$$

Interchange x and y.

$$x = y^3 - 3$$

Solve for y.

$$x + 3 = y^3$$

Take the cube root of each side.

$$\sqrt[3]{x+3} = y$$

Replace y with $f^{-1}(x)$.

$$f^{-1}(x) = \sqrt[3]{x+3}$$

In Exercises 21–24, $f(x) = 2^x$ is a one-to-one function.

21. **(a)** To find $f(3)$, substitute 3 for x.

$f(x) = 2^x$, so $f(3) = 2^3 = 8$.

(b) Since f is one-to-one and $f(3) = 8$, it follows that $f^{-1}(8) = 3$.

22. **(a)** $f(x) = 2^x$, so $f(4) = 2^4 = 16$.

(b) $f^{-1}(16) = 4$

23. **(a)** To find $f(0)$, substitute 0 for x.

$f(x) = 2^x$, so $f(0) = 2^0 = 1$.

(b) Since f is one-to-one and $f(0) = 1$, it follows that $f^{-1}(1) = 0$.

24. **(a)** $f(x) = 2^x$, so $f(-2) = 2^{-2} = \frac{1}{2^2} = \frac{1}{4}$.

(b) $f^{-1}(\frac{1}{4}) = -2$

25. **(a)** The function is one-to-one since any horizontal line intersects the graph at most once.

(b) In the graph, the two points marked on the line are $(-1, 5)$ and $(2, -1)$. Interchange x and y in each ordered pair to get $(5, -1)$ and $(-1, 2)$. Plot these points, then draw a dashed line through them to obtain the graph of the inverse function.

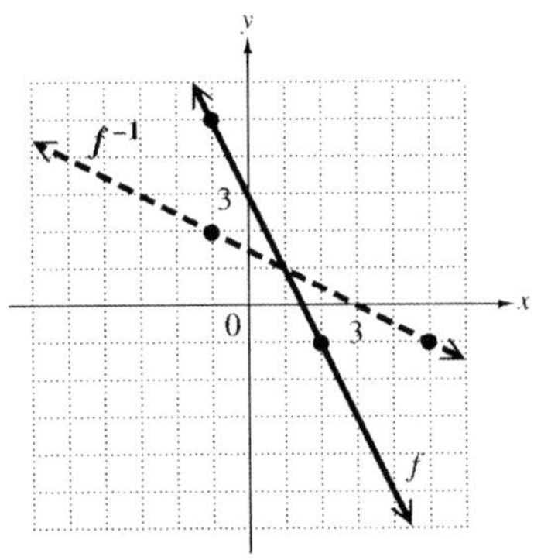

26. **(a)** The function is one-to-one since any horizontal line intersects the graph at most once.

(b) The points $(-3, 0)$, $(-1, 2)$, and $(5, 3)$ are points of f, so $(0, -3)$, $(2, -1)$, and $(3, 5)$ are points of f^{-1}. Draw these points and a smooth dashed curve connecting them.

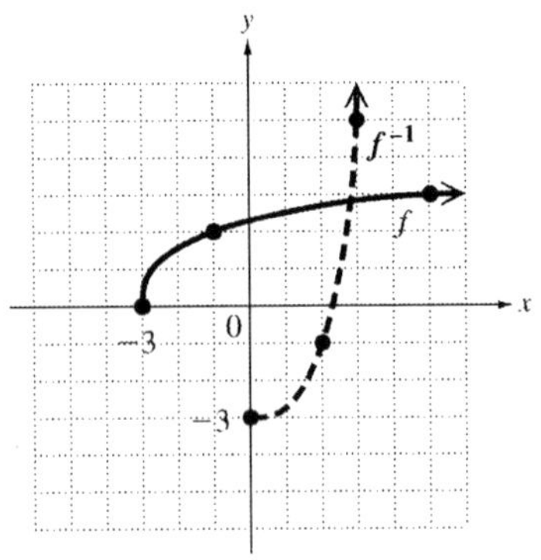

27. **(a)** The function is not one-to-one since there are horizontal lines that intersect the graph more than once. For example, the line $y = 1$ intersects the graph twice.

28. **(a)** The function is not one-to-one since there are horizontal lines that intersect the graph more than once. For example, the line $y = 1$ intersects the graph twice.

29. **(a)** The function is one-to-one since any horizontal line intersects the graph at most once.

(b) In the graph, the four points marked on the curve are $(-4, 2)$, $(-1, 1)$, $(1, -1)$, and $(4, -2)$. Interchange x and y in each ordered pair to get $(2, -4)$, $(1, -1)$, $(-1, 1)$, and $(-2, 4)$. Plot these points, then draw a dashed curve (symmetric to the original graph about the line $y = x$) through them to obtain the graph of the inverse.

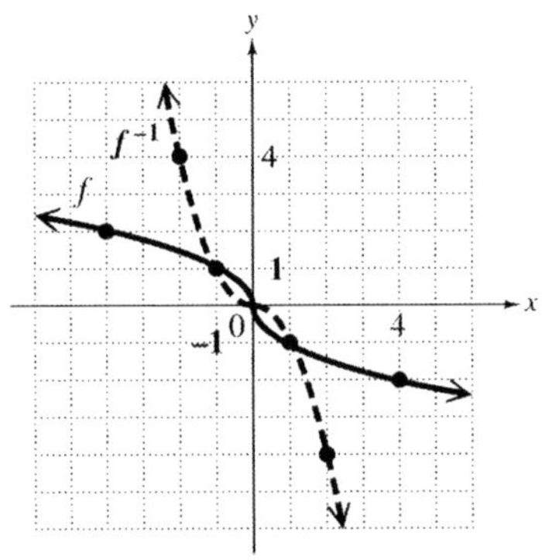

30. **(a)** The function is one-to-one since any horizontal line intersects the graph at most once.

(b) The points $(-2, -5)$, $(-1, -1)$, $(0, 0)$, $(1, 1)$, and $(2, 5)$ are points of f, so $(-5, -2)$, $(-1, -1)$, $(0, 0)$, $(1, 1)$, and $(5, 2)$ are points of f^{-1}. Draw these points and a smooth dashed curve connecting them. Use the fact that the graph of f^{-1} is symmetric to the graph of f with respect to the line $y = x$.

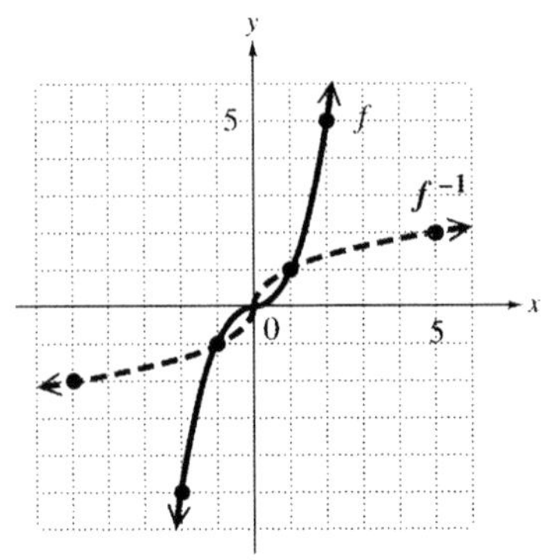

31. $f(x) = 2x - 1$ or $y = 2x - 1$
The graph is a line through $(-2, -5)$, $(0, -1)$, and $(3, 5)$. Plot these points and draw the solid line through them. Then the inverse will be a line through $(-5, -2)$, $(-1, 0)$, and $(5, 3)$. Plot these points and draw the dashed line through them.

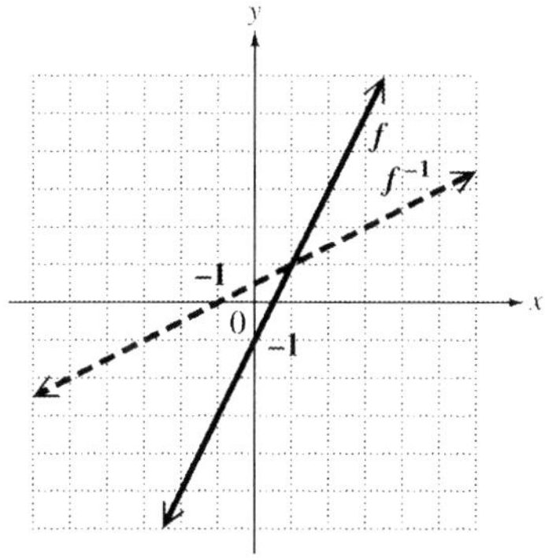

32. $f(x) = 2x + 3$ or $y = 2x + 3$
This is a line with x-intercept $\left(-\frac{3}{2}, 0\right)$ and y-intercept $(0, 3)$. Draw a solid line through the intercepts for f. f^{-1} is the line through $\left(0, -\frac{3}{2}\right)$ and $(3, 0)$. Draw it as a dashed line.

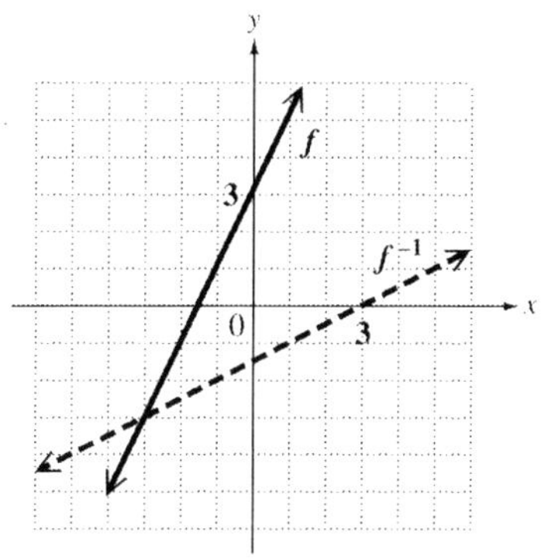

33. $g(x) = -4x$ or $y = -4x$
The graph is a line through $(0, 0)$ and $(1, -4)$. For the inverse, interchange x and y in each ordered pair to get the points $(0, 0)$ and $(-4, 1)$. Draw a dashed line through these points to obtain the graph of the inverse function.

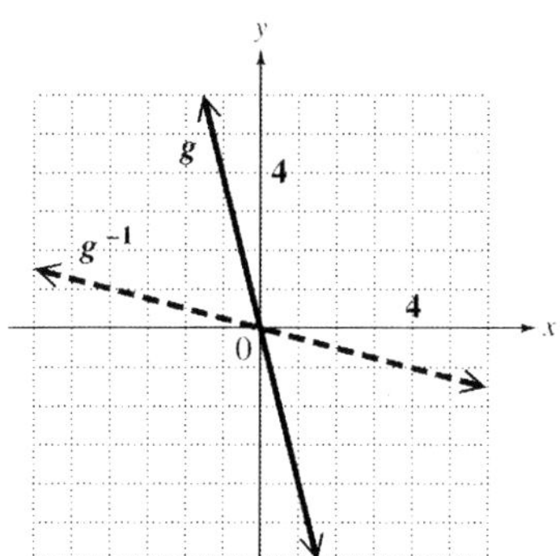

34. $g(x) = -2x$ or $y = -2x$
Draw a solid line through $(0, 0)$ and $(2, -4)$ for g. Draw a dashed line through $(0, 0)$ and $(-4, 2)$ for g^{-1}.

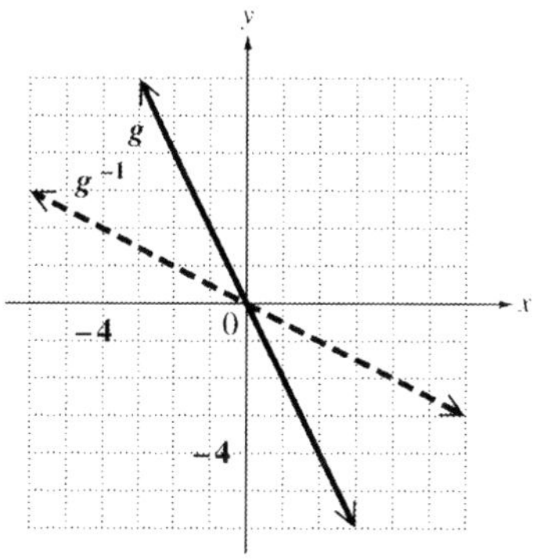

35. $f(x) = \sqrt{x}$, $x \geq 0$
Make a table of values such that the x-values are perfect squares.

x	$f(x)$
0	0
1	1
4	2

Plot these points and connect them with a solid smooth curve.
Since $f(x)$ is one-to-one, make a table of values for $f^{-1}(x)$ by interchanging x and y.

x	$f^{-1}(x)$
0	0
1	1
2	4

Plot these points and connect them with a dashed smooth curve. Use the fact that the graph of f^{-1} is symmetric to the graph of f with respect to the line $y = x$.

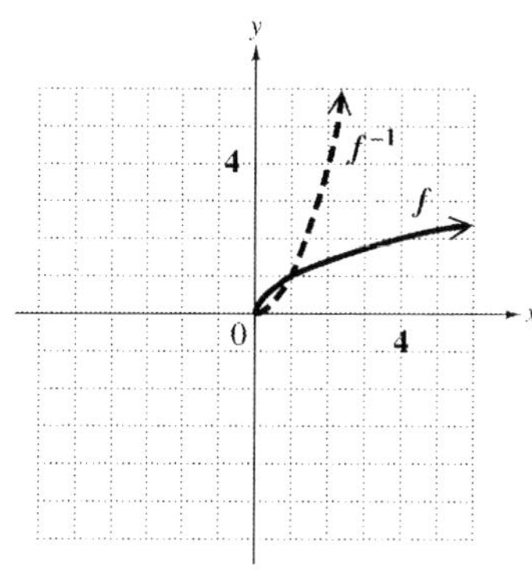

36. $f(x) = -\sqrt{x}, x \geq 0$

Make a table of values such that the x-values are perfect squares.

x	$f(x)$
0	0
1	−1
4	−2

Plot these points and connect them with a solid smooth curve.
For $f^{-1}(x)$, draw a dashed curve through the points in the table. Note that $f^{-1}(x) \geq 0$.

x	$f^{-1}(x)$
0	0
−1	1
−2	4

Plot these points and connect them with a dashed smooth curve. Use the fact that the graph of f^{-1} is symmetric to the graph of f with respect to the line $y = x$.

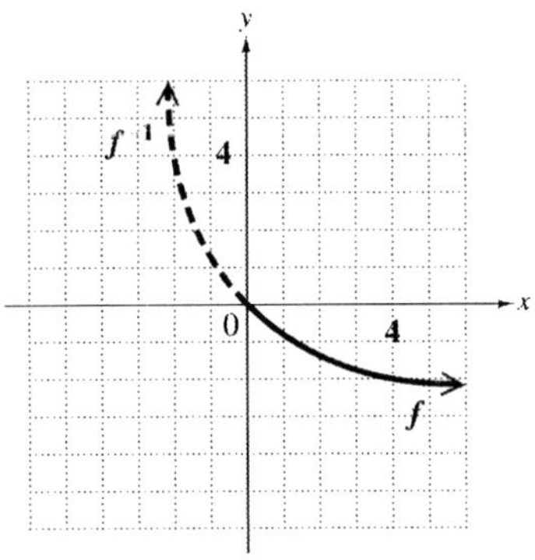

37. $f(x) = x^3 - 2$
Complete the table of values.

x	$f(x)$
−1	−3
0	−2
1	−1
2	6

Plot these points and connect them with a solid smooth curve.
Make a table of values for f^{-1}.

x	$f^{-1}(x)$
−3	−1
−2	0
−1	1
6	2

Plot these points and connect them with a dashed smooth curve. Use the fact that the graph of f^{-1} is symmetric to the graph of f with respect to the line $y = x$.

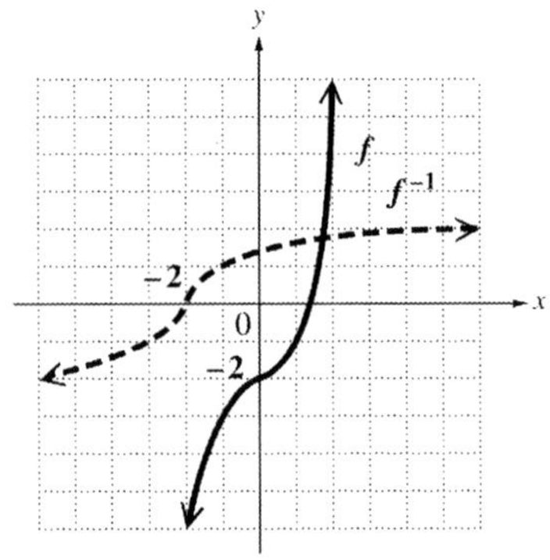

38. $f(x) = x^3 + 3$
For $f(x)$, draw a solid curve through the points in the table.

x	$f(x)$
−2	−5
−1	2
0	3
1	4

For $f^{-1}(x)$, draw a dashed curve through the points in the table. Use the fact that the graph of f^{-1} is symmetric to the graph of f with respect to the line $y = x$.

x	$f^{-1}(x)$
−5	−2
2	−1
3	0
4	1

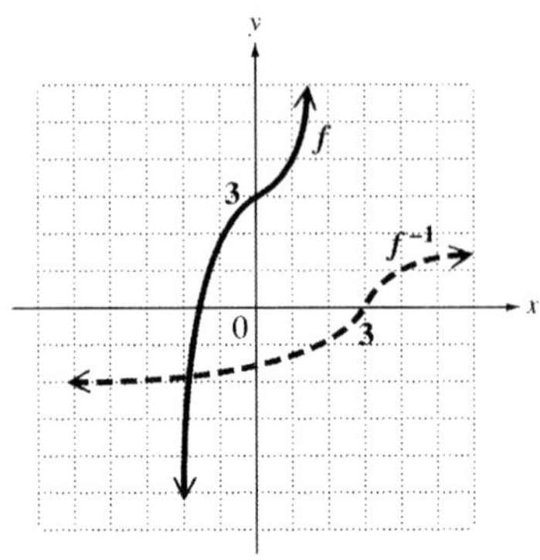

11.2 Exponential Functions

11.2 Margin Exercises

In Exercises 1 and 2, choose values of x and find the corresponding values of y. Then plot the points, and draw a smooth curve through them.

1. **(a)** $f(x) = 10^x$

x	-2	-1	0	1	2
$f(x) = 10^x$	.01	.1	1	10	100

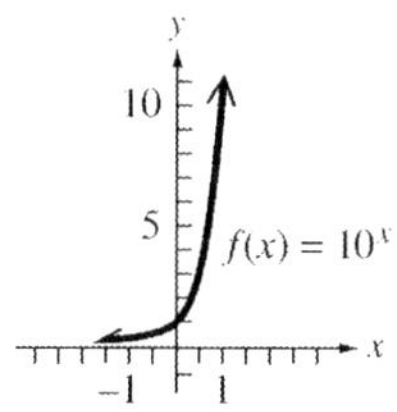

(b) $g(x) = \left(\frac{1}{4}\right)^x$

x	-2	-1	0	1	2
$g(x) = \left(\frac{1}{4}\right)^x$	16	4	1	$\frac{1}{4}$	$\frac{1}{16}$

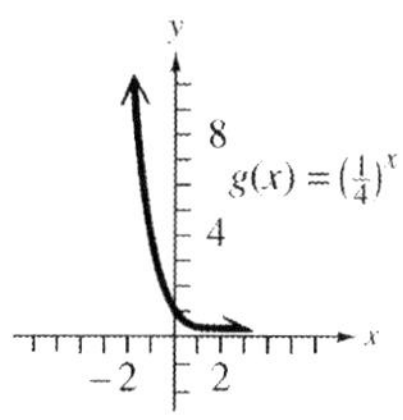

2. $y = 2^{4x-3}$

Make a table of values. It will help to find values for $4x - 3$ before you find y.

x	$4x - 3$	$y = 2^{4x-3}$
0	-3	$\frac{1}{8}$
$\frac{3}{4}$	0	1
1	1	2
-1	-7	$\frac{1}{128}$

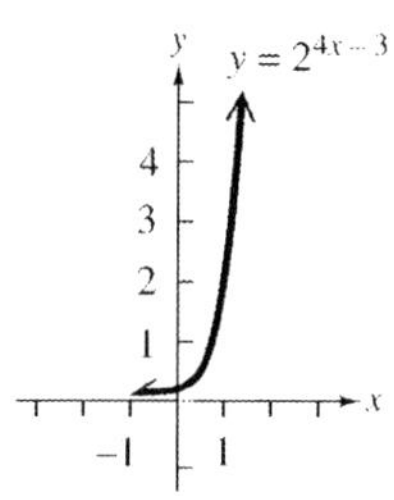

3. **(a)** $25^x = 125$

Step 1 Write each side with the base 5.

$(5^2)^x = 5^3$

Step 2 Simplify exponents.

$5^{2x} = 5^3$

Step 3 Set the exponents equal.

$2x = 3$

Step 4 Solve.

$x = \frac{3}{2}$

Check $x = \frac{3}{2}$ in the original equation.

$$25^{3/2} = 125 \quad ?$$
$$(25^{1/2})^3 = 125 \quad ?$$
$$5^3 = 125 \quad ?$$
$$125 = 125 \quad \textit{True}$$

Solution set: $\left\{\frac{3}{2}\right\}$

(b) $4^x = 32$

Step 1 Write each side with the base 2.

$(2^2)^x = 2^5$

Step 2 Simplify exponents.

$2^{2x} = 2^5$

Step 3 Set the exponents equal.

$2x = 5$

Step 4 Solve.

$x = \frac{5}{2}$

Check $x = \frac{5}{2}$: $4^{5/2} = 2^5 = 32$

Solution set: $\left\{\frac{5}{2}\right\}$

(c) $81^p = 27$

Step 1 Write each side with the base 3.

$(3^4)^p = 3^3$

Step 2 Simplify exponents.

$3^{4p} = 3^3$

Step 3 Set the exponents equal.

$4p = 3$

Step 4 Solve.

$p = \frac{3}{4}$

Check $p = \frac{3}{4}$: $81^{3/4} = 3^3 = 27$

Solution set: $\left\{\frac{3}{4}\right\}$

4. (a)
$$\begin{aligned} 25^{x-2} &= 125^x \\ (5^2)^{x-2} &= (5^3)^x && \textit{Same base} \\ 5^{2(x-2)} &= 5^{3x} && \textit{Simplify exponents.} \\ 2(x-2) &= 3x && \textit{Set exponents equal.} \\ 2x-4 &= 3x \\ -4 &= x \end{aligned}$$

Check $x=-4$:

$$\begin{aligned} 25^{-4-2} &= 125^{-4} && ? \\ 25^{-6} &= 125^{-4} && ? \\ 4.096\times 10^{-9} &= 4.096\times 10^{-9} && \textit{True} \end{aligned}$$

Solution set: $\{-4\}$

(b)
$$\begin{aligned} 4^x &= \frac{1}{32} \\ (2^2)^x &= \frac{1}{2^5} \\ 2^{2x} &= 2^{-5} && \textit{Same base} \\ 2x &= -5 && \textit{Set exponents equal.} \\ x &= -\tfrac{5}{2} \end{aligned}$$

Check $x=-\frac{5}{2}$: $4^{-5/2} = \dfrac{1}{4^{5/2}} = \dfrac{1}{2^5} = \dfrac{1}{32}$

Solution set: $\{-\frac{5}{2}\}$

(c)
$$\begin{aligned} \left(\tfrac{3}{4}\right)^x &= \tfrac{16}{9} \\ \left(\tfrac{3}{4}\right)^x &= \left(\tfrac{4}{3}\right)^2 \\ \left(\tfrac{3}{4}\right)^x &= \left(\tfrac{3}{4}\right)^{-2} \\ x &= -2 \end{aligned}$$

Check $x=-2$: $\left(\frac{3}{4}\right)^{-2} = \left(\frac{4}{3}\right)^2 = \frac{16}{9}$

Solution set: $\{-2\}$

5. (a) $x = 1925 - 1750 = 175$

$$\begin{aligned} f(x) &= 278(1.00084)^x \\ f(175) &= 278(1.00084)^{175} \\ &\approx 322 \text{ parts per million} \end{aligned}$$

(b)
$$\begin{aligned} f(x) &= 1038(1.000134)^{-x} \\ f(8000) &= 1038(1.000134)^{-8000} \\ &\approx 355 \end{aligned}$$

The pressure is approximately 355 millibars.

11.2 Section Exercises

1. Since the graph of $F(x) = a^x$ always contains the point $(0, 1)$, the correct response is **C**.

2. **A.** The y-intercept of the graph of $f(x) = 10^x$ is $(0,1)$, not $(0,10)$. *False*

B. For $a > 1$, the graph of $f(x) = a^x$ *rises* from left to right, not *falls*. *False*

C. If $f(x) = 5^x$, then $f\left(\frac{1}{2}\right) = 5^{1/2} = \sqrt{5}$. Thus the point $\left(\frac{1}{2}, \sqrt{5}\right)$ lies on the graph of $f(x) = 5^x$ and the statement is *true*.

D. The graph of $y = 10^x$ rises at a faster rate than the graph of $y = 4^x$, not the other way around. *False*

The correct response is **C**.

3. Since the graph of $F(x) = a^x$ always approaches the x-axis, the correct response is **A**.

4. The graphs of **A** and **B** fall from left to right, so exclude them. The graph for **D**, $y = 1000^x$, would rise too fast to fit on this graph—for example, if $x = 2$, $y = 1000^2$ or 1 million! That leaves choice **C** as the only candidate. Substituting $x = 0$ and $x = 10$ gives us $y = 1000$ and $y = 8000$, which seem reasonable. The correct response is **C**.

5. $f(x) = 3^x$
Make a table of values.

$$f(-2) = 3^{-2} = \frac{1}{3^2} = \frac{1}{9},$$
$$f(-1) = 3^{-1} = \frac{1}{3^1} = \frac{1}{3}, \text{ and so on.}$$

x	-2	-1	0	1	2
$f(x)$	$\frac{1}{9}$	$\frac{1}{3}$	1	3	9

Plot the points from the table and draw a smooth curve through them.

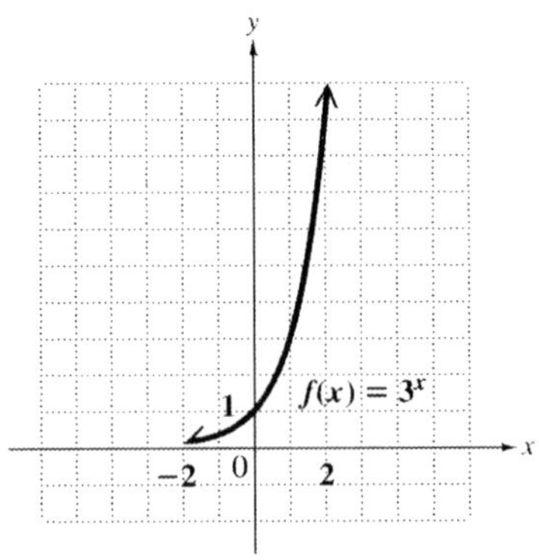

6. $f(x) = 5^x$
Make a table of values.

x	-2	-1	0	1	2
$f(x)$	$\frac{1}{25}$	$\frac{1}{5}$	1	5	25

Plot the points from the table and draw a smooth curve through them.

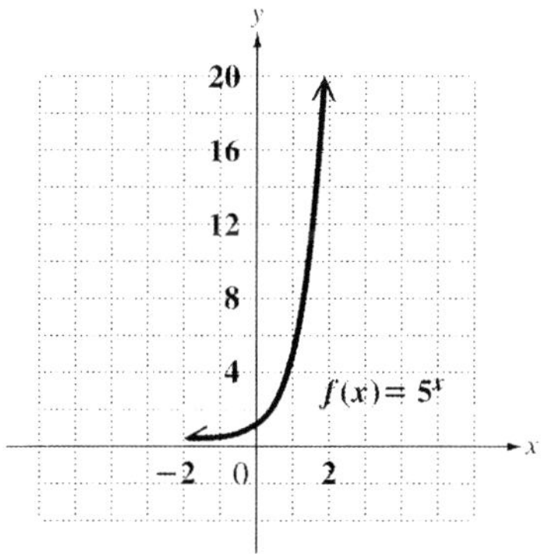

7. $g(x) = \left(\frac{1}{3}\right)^x$

Make a table of values.

$g(-2) = \left(\frac{1}{3}\right)^{-2} = \left(\frac{3}{1}\right)^2 = 9,$

$g(-1) = \left(\frac{1}{3}\right)^{-1} = \left(\frac{3}{1}\right)^1 = 3$, and so on.

x	-2	-1	0	1	2
$g(x)$	9	3	1	$\frac{1}{3}$	$\frac{1}{9}$

Plot the points from the table and draw a smooth curve through them.

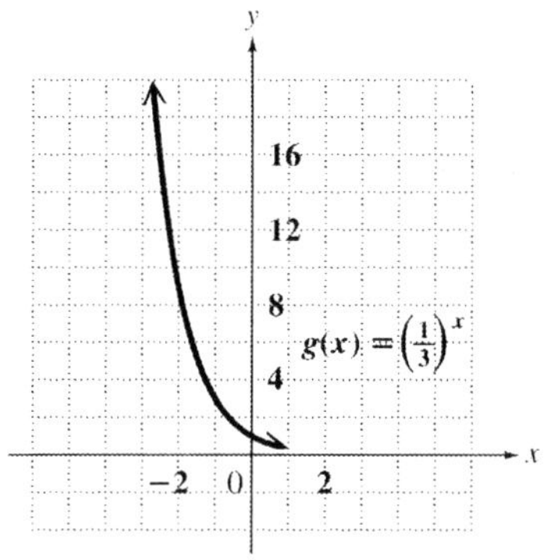

8. $g(x) = \left(\frac{1}{5}\right)^x$

Make a table of values.

x	-2	-1	0	1	2
$g(x)$	25	5	1	$\frac{1}{5}$	$\frac{1}{25}$

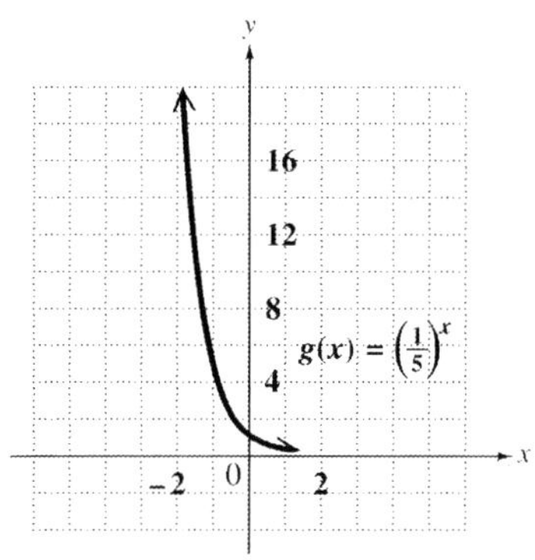

9. $y = 2^{2x-2}$

Make a table of values. It will help to find values for $2x - 2$ before you find y.

x	-2	-1	0	1	2	3
$2x-2$	-6	-4	-2	0	2	4
y	$\frac{1}{64}$	$\frac{1}{16}$	$\frac{1}{4}$	1	4	16

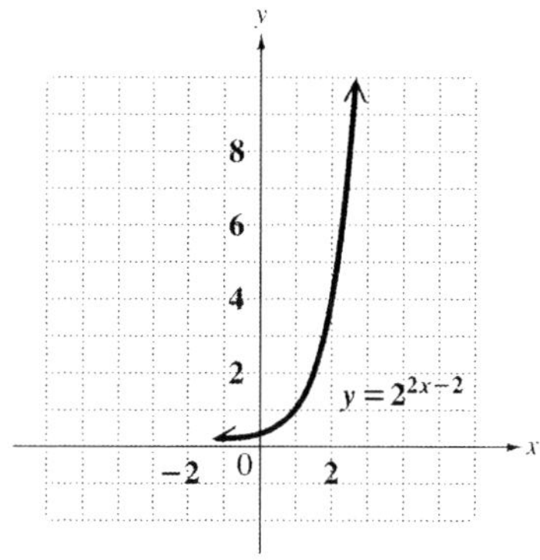

10. $y = 2^{2x+1}$

Make a table of values. It will help to find values for $2x + 1$ before you find y.

x	-2	-1	0	1
$2x+1$	-3	-1	1	3
y	$\frac{1}{8}$	$\frac{1}{2}$	2	8

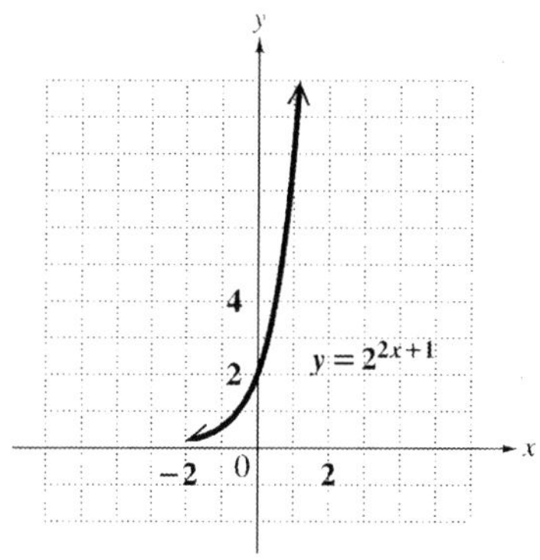

11. $6^x = 36$

Write each side as a power of 6.

$6^x = 6^2$

For $a > 0$ and $a \neq 1$, if $a^x = a^y$, then $x = y$. Set the exponents equal to each other.

$x = 2$

Check $x = 2$: $6^2 = 36$ *True*

Solution set: $\{2\}$

12. $8^x = 64$

Write each side as a power of 8.

$8^x = 8^2$

For $a > 0$ and $a \neq 1$, if $a^x = a^y$, then $x = y$. Set the exponents equal to each other.

$x = 2$

Check $x = 2$: $8^2 = 64$ *True*

Solution set: $\{2\}$

13. $100^x = 1000$

Write each side as a power of 10.

$(10^2)^x = 10^3$

$10^{2x} = 10^3$

For $a > 0$ and $a \neq 1$, if $a^x = a^y$, then $x = y$. Set the exponents equal to each other.

$2x = 3$

$x = \frac{3}{2}$

Check $x = \frac{3}{2}$: $100^{3/2} = 1000$ *True*

Solution set: $\{\frac{3}{2}\}$

14. $8^x = 4$

Write each side as a power of 2.

$(2^3)^x = 2^2$

$2^{3x} = 2^2$

For $a > 0$ and $a \neq 1$, if $a^x = a^y$, then $x = y$. Set the exponents equal to each other.

$3x = 2$

$x = \frac{2}{3}$

Check $x = \frac{2}{3}$: $8^{2/3} = 4$ *True*

Solution set: $\{\frac{2}{3}\}$

15. $16^{2x+1} = 64^{x+3}$

Write each side as a power of 4.

$(4^2)^{2x+1} = (4^3)^{x+3}$

$4^{4x+2} = 4^{3x+9}$

Set the exponents equal.

$4x + 2 = 3x + 9$

$x = 7$

Check $x = 7$: $16^{15} = 64^{10}$ *True*

Solution set: $\{7\}$

16. $9^{2x-8} = 27^{x-4}$

$(3^2)^{2x-8} = (3^3)^{x-4}$

$3^{4x-16} = 3^{3x-12}$

$4x - 16 = 3x - 12$

$x = 4$

Check $x = 4$: $9^0 = 27^0$ *True*

Solution set: $\{4\}$

17. $5^x = \frac{1}{125}$

$5^x = \left(\frac{1}{5}\right)^3$

Write each side as a power of 5.

$5^x = 5^{-3}$

Set the exponents equal.

$x = -3$

Check $x = -3$: $5^{-3} = \frac{1}{125}$ *True*

Solution set: $\{-3\}$

18. $3^x = \frac{1}{81}$

$3^x = \frac{1}{3^4}$

$3^x = 3^{-4}$

$x = -4$

Check $x = -4$: $3^{-4} = \frac{1}{81}$ *True*

Solution set: $\{-4\}$

19. $5^x = .2$

$5^x = \frac{2}{10} = \frac{1}{5}$

Write each side as a power of 5.

$5^x = 5^{-1}$

Set the exponents equal.

$x = -1$

Check $x = -1$: $5^{-1} = .2$ *True*

Solution set: $\{-1\}$

20. $10^x = .1$

$10^x = \frac{1}{10}$

$10^x = 10^{-1}$

$x = -1$

Check $x = -1$: $10^{-1} = .1$ *True*

Solution set: $\{-1\}$

21. $\left(\frac{3}{2}\right)^x = \frac{8}{27}$

$\left(\frac{3}{2}\right)^x = \left(\frac{2}{3}\right)^3$

Write each side as a power of $\frac{3}{2}$.

$\left(\frac{3}{2}\right)^x = \left(\frac{3}{2}\right)^{-3}$

Set the exponents equal.

$x = -3$

Check $x = -3$: $\left(\frac{3}{2}\right)^{-3} = \frac{8}{27}$ *True*

Solution set: $\{-3\}$

22. $\left(\frac{4}{3}\right)^x = \frac{27}{64}$

$\left(\frac{4}{3}\right)^x = \left(\frac{3}{4}\right)^3$

$\left(\frac{4}{3}\right)^x = \left(\frac{4}{3}\right)^{-3}$

$x = -3$

Check $x = -3$: $\left(\frac{4}{3}\right)^{-3} = \frac{27}{64}$ *True*

Solution set: $\{-3\}$

23. **(a)** For an exponential function defined by $f(x) = a^x$, if $a > 1$, the graph rises from left to right. (See Example 1, $f(x) = 2^x$, in your text.) If $0 < a < 1$, the graph falls from left to right. (See Example 2, $g(x) = \left(\frac{1}{2}\right)^x = 2^{-x}$, in your text.)

(b) An exponential function defined by $f(x) = a^x$ is one-to-one and has an inverse, since each value of $f(x)$ corresponds to one and only one value of x.

24. **(a)** The increase for the exponential-type curve in the year 2000 is about .5°C.

(b) The increase for the linear graph in the year 2000 is about .35°C.

25. **(a)** The increase for the exponential-type curve in the year 2010 is about 1.0°C.

(b) The increase for the linear graph in the year 2010 is about .4°C.

26. **(a)** The increase for the exponential-type curve in the year 2020 is about 1.6°C.

(b) The increase for the linear graph in the year 2020 is about .5°C.

27. **(a)** The increase for the exponential-type curve in the year 2040 is about 3.0°C.

(b) The increase for the linear graph in the year 2040 is about .7°C.

28. $V(t) = 5000(2)^{-.15t}$

(a) The original value is found when $t = 0$.

$$\begin{aligned} V(0) &= 5000(2)^{-.15(0)} \\ &= 5000(2)^0 \\ &= 5000(1) = 5000 \end{aligned}$$

The original value is \$5000.

(b) The value after 5 years is found when $t = 5$.

$$\begin{aligned} V(5) &= 5000(2)^{-.15(5)} \\ &= 5000(2)^{-.75} \approx 2973.02 \end{aligned}$$

The value after 5 years is about \$2973.

(c) The value after 10 years is found when $t = 10$.

$$\begin{aligned} V(10) &= 5000(2)^{-.15(10)} \\ &= 5000(2)^{-1.5} \approx 1767.77 \end{aligned}$$

The value after 10 years is about \$1768.

(d) Use the results of parts (a) – (c) to make a table of values.

t	0	5	10
$V(t)$	5000	2973	1768

Plot the points from the table and draw a smooth curve through them.

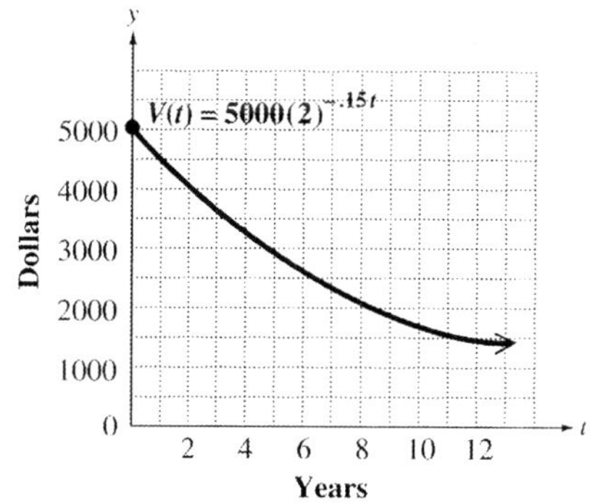

11.3 Logarithmic Functions

11.3 Margin Exercises

1.

Exponential Form	Logarithmic Form
$2^5 = 32$	$\log_2 32 = 5$
$100^{1/2} = 10$	$\log_{100} 10 = \frac{1}{2}$
$8^{2/3} = 4$	$\log_8 4 = \frac{2}{3}$
$6^{-4} = \frac{1}{1296}$	$\log_6 \frac{1}{1296} = -4$

2. **(a)** $\log_3 27 = x$

Change to exponential form.

$3^x = 27$

Write with the same base.

$3^x = 3^3$

Set the exponents equal.

$x = 3$

Check $x = 3$:

$\log_3 27 = 3$ is *true* since $3^3 = 27$.

Solution set: $\{3\}$

(b) $\log_5 p = 2$

Change to exponential form.

$5^2 = p$

$25 = p$

Check $p = 25$:

$\log_5 25 = 2$ is *true* since $5^2 = 25$.

Solution set: $\{25\}$

(c) $\log_m \frac{1}{16} = -4$

Change to exponential form.

$$m^{-4} = \frac{1}{16}$$
$$\frac{1}{m^4} = \frac{1}{2^4}$$
$$m^4 = 2^4$$

Take the principal fourth root.

$$m = 2$$

Check $m = 2$:

$\log_2 \frac{1}{16} = -4$ is *true* since $2^{-4} = \frac{1}{16}$.

Solution set: $\{2\}$

(d) $\log_x 12 = 3$

Change to exponential form.

$$x^3 = 12$$
$$x = \sqrt[3]{12}$$

Check $x = \sqrt[3]{12}$:

$\log_{\sqrt[3]{12}} 12 = 3$ is *true* since $\left(\sqrt[3]{12}\right)^3 = 12$.

Solution set: $\left\{\sqrt[3]{12}\right\}$

3. Use the properties of logarithms,

$$\log_b b = 1 \quad \text{and} \quad \log_b 1 = 0,$$

for $b > 0$, $b \neq 1$.

(a) $\log_{2/5} \frac{2}{5} = 1$

(b) $\log_\pi \pi = 1$

(c) $\log_{.4} 1 = 0$

(d) $\log_6 1 = 0$

4. **(a)** $y = \log_3 x$ is equivalent to $x = 3^y$. Choose values for y and find x.

x	$\frac{1}{9}$	$\frac{1}{3}$	1	3	9
y	-2	-1	0	1	2

Plot the points, and draw a smooth curve through them.

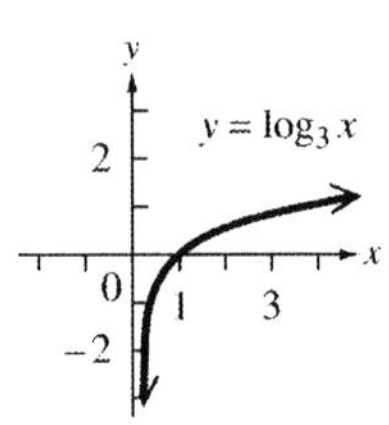

(b) $y = \log_{1/10} x$ is equivalent to $x = \left(\frac{1}{10}\right)^y$.

Choose values for y and find x.

x	10	1	$\frac{1}{10}$	$\frac{1}{100}$
y	-1	0	1	2

Plot the points, and draw a smooth curve through them.

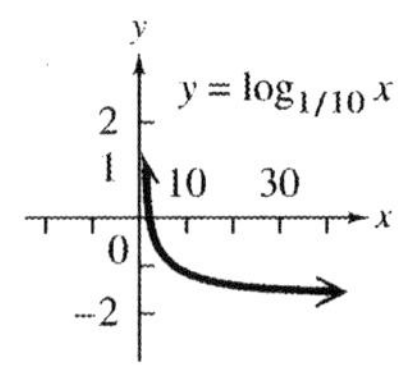

5. $P(t) = 80 \log_{10} (t + 10)$

(a)
$$\begin{aligned} P(0) &= 80 \log_{10} (0 + 10) && \textit{Let } t = 0. \\ &= 80 \log_{10} 10 \\ &= 80(1) && \log_{10} 10 = 1 \\ &= 80 \end{aligned}$$

The number of mites at the beginning of the study is 80.

(b)
$$\begin{aligned} P(90) &= 80 \log_{10} (90 + 10) && \textit{Let } t = 90. \\ &= 80 \log_{10} 100 \\ &= 80(2) && \log_{10} 100 = 2 \\ &= 160 \end{aligned}$$

The population after 90 days is 160 mites.

(c)
$$\begin{aligned} P(990) &= 80 \log_{10} (990 + 10) && \textit{Let } t = 990. \\ &= 80 \log_{10} 1000 \\ &= 80(3) && \log_{10} 1000 = 3 \\ &= 240 \end{aligned}$$

The population after 990 days is 240 mites.

11.3 Section Exercises

1. **(a)** $\log_4 16$ is equal to 2, because 2 is the exponent to which 4 must be raised to obtain 16. **(C)**

(b) $\log_3 81$ is equal to 4, because 4 is the exponent to which 3 must be raised to obtain 81. **(F)**

(c) $\log_3 \left(\frac{1}{3}\right)$ is equal to -1, because -1 is the exponent to which 3 must be raised to obtain $\frac{1}{3}$. **(B)**

(d) $\log_{10} .01$ is equal to -2, because -2 is the exponent to which 10 must be raised to obtain .01. **(A)**

(e) $\log_5 \sqrt{5}$ is equal to $\frac{1}{2}$, because $\frac{1}{2}$ is the exponent to which 5 must be raised to obtain $\sqrt{5}$. **(E)**

(f) $\log_{13} 1$ is equal to 0, because 0 is the exponent to which 13 must be raised to obtain 1. **(D)**

2. **(a)** $\log_{1/3} 3 = -1$ is equivalent to $\left(\frac{1}{3}\right)^{-1} = 3$. **(B)**

(b) $\log_5 1 = 0$ is equivalent to $5^0 = 1$. **(E)**

(c) $\log_2 \sqrt{2} = \frac{1}{2}$ is equivalent to $2^{1/2} = \sqrt{2}$. **(D)**

(d) $\log_{10} 1000 = 3$ is equivalent to $10^3 = 1000$. **(F)**

(e) $\log_8 \sqrt[3]{8} = \frac{1}{3}$ is equivalent to $8^{1/3} = \sqrt[3]{8}$. **(A)**

(f) $\log_4 4 = 1$ is equivalent to $4^1 = 4$. **(C)**

3. The base is 4, the exponent (logarithm) is 5, and the number is 1024, so $4^5 = 1024$ becomes $\log_4 1024 = 5$ in logarithmic form.

4. $3^6 = 729$ becomes $\log_3 729 = 6$ in logarithmic form.

5. $\frac{1}{2}$ is the base and -3 is the exponent, so $\left(\frac{1}{2}\right)^{-3} = 8$ becomes $\log_{1/2} 8 = -3$ in logarithmic form.

6. $\left(\frac{1}{6}\right)^{-3} = 216$ becomes $\log_{1/6} 216 = -3$ in logarithmic form.

7. The base is 10, the exponent (logarithm) is -3, and the number is .001, so $10^{-3} = .001$ becomes $\log_{10} .001 = -3$ in logarithmic form.

8. $36^{1/2} = 6$ becomes $\log_{36} 6 = \frac{1}{2}$ in logarithmic form.

9. $\sqrt[4]{625} = 625^{1/4} = 5$

The base is 625, the exponent (logarithm) is $\frac{1}{4}$, and the number is 5, so $\sqrt[4]{625} = 5$ becomes $\log_{625} 5 = \frac{1}{4}$ in logarithmic form.

10. $\sqrt[3]{343} = 7$ or $343^{1/3} = 7$ becomes $\log_{343} 7 = \frac{1}{3}$ in logarithmic form.

11. In $\log_4 64 = 3$, 4 is the base and 3 is the logarithm (exponent), so $\log_4 64 = 3$ becomes $4^3 = 64$ in exponential form.

12. $\log_2 512 = 9$ becomes $2^9 = 512$ in exponential form.

13. In $\log_{10} \frac{1}{10{,}000} = -4$, the base is 10, the logarithm (exponent) is -4, and the number is $\frac{1}{10{,}000}$, so $\log_{10} \frac{1}{10{,}000} = -4$ becomes $10^{-4} = \frac{1}{10{,}000}$ in exponential form.

14. $\log_{100} 100 = 1$ becomes $100^1 = 100$ in exponential form.

15. In $\log_6 1 = 0$, 6 is the base and 0 is the logarithm (exponent), so $\log_6 1 = 0$ becomes $6^0 = 1$ in exponential form.

16. $\log_\pi 1 = 0$ becomes $\pi^0 = 1$ in exponential form.

17. In $\log_9 3 = \frac{1}{2}$, the base is 9, the logarithm (exponent) is $\frac{1}{2}$, and the number is 3, so $\log_9 3 = \frac{1}{2}$ becomes $9^{1/2} = 3$ in exponential form.

18. $\log_{64} 2 = \frac{1}{6}$ becomes $64^{1/6} = 2$ in exponential form.

19. To evaluate $\log_9 3$, one has to ask "9 raised to what power gives you a result of 3?" We know that the square root (a radical) of 9 is 3. Therefore, the teacher's hint was to see what root of 9 equals 3. The answer is the reciprocal of the root index.

20. Because $a^0 = 1$, $\log_a 1 = 0$ for any value of a that is allowed as the base of a logarithm. Because $2^0 = 1$, the value of $\log_2 1$ is 0.

21.
$$x = \log_{27} 3$$
Write in exponential form.
$$27^x = 3$$
Write each side as a power of 3.
$$(3^3)^x = 3$$
$$3^{3x} = 3^1$$
Set the exponents equal.
$$3x = 1$$
$$x = \frac{1}{3}$$
Solution set: $\left\{\frac{1}{3}\right\}$

22.
$$x = \log_{125} 5$$
Write in exponential form.
$$125^x = 5$$
Write each side as a power of 5.
$$(5^3)^x = 5^1$$
$$5^{3x} = 5^1$$
$$3x = 1$$
$$x = \frac{1}{3}$$
Solution set: $\left\{\frac{1}{3}\right\}$

23. $\log_x 9 = \frac{1}{2}$

Change to exponential form.
$$x^{1/2} = 9$$
$$(x^{1/2})^2 = 9^2 \qquad \textit{Square}$$
$$x^1 = 81$$
$$x = 81$$

$x = 81$ is an acceptable base since it is a positive number (not equal to 1).

Check $x = 81$: $81^{1/2} = 9$ *True*

Solution set: $\{81\}$

24. $\log_x 5 = \frac{1}{2}$ becomes

$$x^{1/2} = 5.$$
$$\left(x^{1/2}\right)^2 = 5^2 \quad \textit{Square}$$
$$x = 25$$

$x = 25$ is an acceptable base since it is a positive number (not equal to 1).

Check $x = 25$: $25^{1/2} = 5$ *True*

Solution set: $\{25\}$

25. $\log_x 125 = -3$

Write in exponential form.

$$x^{-3} = 125$$
$$\frac{1}{x^3} = 125$$
$$1 = 125\left(x^3\right)$$
$$\frac{1}{125} = x^3$$

Take the cube root of each side.

$$\sqrt[3]{\frac{1}{125}} = \sqrt[3]{x^3}$$
$$x = \sqrt[3]{\frac{1}{5^3}} = \frac{1}{5}$$

$x = \frac{1}{5}$ is an acceptable base since it is a positive number (not equal to 1).

Solution set: $\left\{\frac{1}{5}\right\}$

26. $\log_x 64 = -6$ becomes

$$x^{-6} = 64.$$
$$\frac{1}{x^6} = 64$$
$$64\left(x^6\right) = 1$$
$$x^6 = \frac{1}{64}$$

Take the sixth root of each side.

$$x = \pm\sqrt[6]{\frac{1}{64}} = \pm\frac{1}{2}$$

Since x is the base of a logarithm, it must be positive and not equal to 1, so reject $x = -\frac{1}{2}$.

Solution set: $\left\{\frac{1}{2}\right\}$

27. $\log_{12} x = 0$

Write in exponential form.

$$12^0 = x$$
$$1 = x$$

The argument (the input of the logarithm) must be a positive number, so $x = 1$ is acceptable.

Solution set: $\{1\}$

28. $\log_4 x = 0$ becomes

$$4^0 = x.$$
$$1 = x$$

The argument (the input of the logarithm) must be a positive number, so $x = 1$ is acceptable.

Solution set: $\{1\}$

29. $\log_x x = 1$

Write in exponential form.

$$x^1 = x$$

This equation is true for all the numbers x that are allowed as the base of a logarithm; that is, all positive numbers x, $x \neq 1$.

Solution set: $\{x \mid x > 0,\ x \neq 1\}$

30. $\log_x 1 = 0$ becomes

$$x^0 = 1.$$

This equation is true for all the numbers x that are allowed as the base of a logarithm; that is, all positive numbers x, $x \neq 1$.

Solution set: $\{x \mid x > 0,\ x \neq 1\}$

31. $\log_x \frac{1}{25} = -2$

Write in exponential form.

$$x^{-2} = \frac{1}{25}$$
$$\frac{1}{x^2} = \frac{1}{25}$$
$$x^2 = 25 \quad \textit{Denominators must be equal.}$$
$$x = \pm 5$$

Reject $x = -5$ since the base of a logarithm must be positive and not equal to 1.

Solution set: $\{5\}$

32. $\log_x \frac{1}{10} = -1$

Write in exponential form.

$$x^{-1} = \frac{1}{10}$$
$$\frac{1}{x} = \frac{1}{10}$$
$$x = 10 \quad \textit{Denominators must be equal.}$$

$x = 10$ is an acceptable base since it is a positive number (not equal to 1).

Solution set: $\{10\}$

33. $\log_8 32 = x$

$8^x = 32$ *Exponential form*

Write each side as a power of 2.

$(2^3)^x = 2^5$

$2^{3x} = 2^5$

$3x = 5$ *Equate exponents*

$x = \frac{5}{3}$

Solution set: $\{\frac{5}{3}\}$

34. $\log_{81} 27 = x$

$81^x = 27$ *Exponential form*

Write each side as a power of 3.

$(3^4)^x = 3^3$

$3^{4x} = 3^3$

$4x = 3$ *Equate exponents*

$x = \frac{3}{4}$

Solution set: $\{\frac{3}{4}\}$

35. $\log_\pi \pi^4 = x$

$\pi^x = \pi^4$ *Exponential form*

$x = 4$ *Equate exponents*

Solution set: $\{4\}$

36. $\log_{\sqrt{2}} \sqrt{2}^9 = x$

$\sqrt{2}^x = \sqrt{2}^9$ *Exponential form*

$x = 9$ *Equate exponents*

Solution set: $\{9\}$

37. $\log_6 \sqrt{216} = x$

$\log_6 216^{1/2} = x$ *Equivalent form*

$6^x = 216^{1/2}$ *Exponential form*

$6^x = (6^3)^{1/2}$ *Same base*

$6^x = 6^{3/2}$

$x = \frac{3}{2}$ *Equate exponents*

Solution set: $\{\frac{3}{2}\}$

38. $\log_4 \sqrt{64} = x$

$\log_4 64^{1/2} = x$ *Equivalent form*

$4^x = 64^{1/2}$ *Exponential form*

$4^x = (4^3)^{1/2}$ *Same base*

$4^x = 4^{3/2}$

$x = \frac{3}{2}$ *Equate exponents*

Solution set: $\{\frac{3}{2}\}$

39. $y = \log_3 x$

Change to exponential form.

$3^y = x$

Refer to Section 11.2, Exercise 5, for the graph of $f(x) = 3^x$. Since $y = \log_3 x$ (or $3^y = x$) is the inverse of $f(x) = y = 3^x$, its graph is symmetric about the line $y = x$ to the graph of $f(x) = 3^x$. The graph can be plotted by reversing the ordered pairs in the table of values belonging to $f(x) = 3^x$.

x	$\frac{1}{9}$	$\frac{1}{3}$	1	3	9
y	-2	-1	0	1	2

Plot the points, and draw a smooth curve through them.

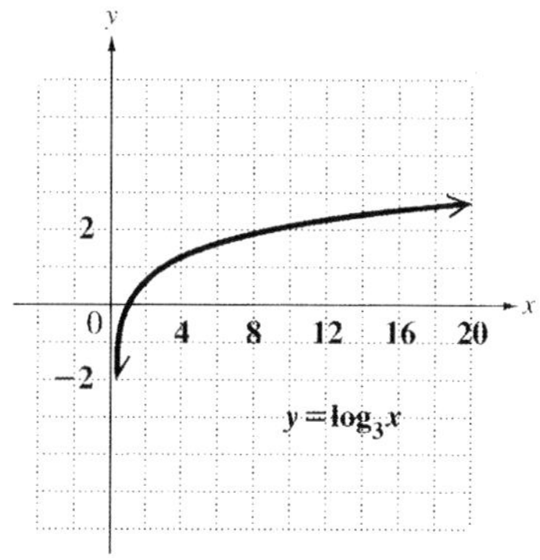

40. $y = \log_5 x$

From Section 11.2, Exercise 6, use the table for $f(x) = 5^x$ to construct the table for $y = f^{-1}(x)$ by reversing x- and y-values.

x	$\frac{1}{5}$	1	5	25
y	-1	0	1	2

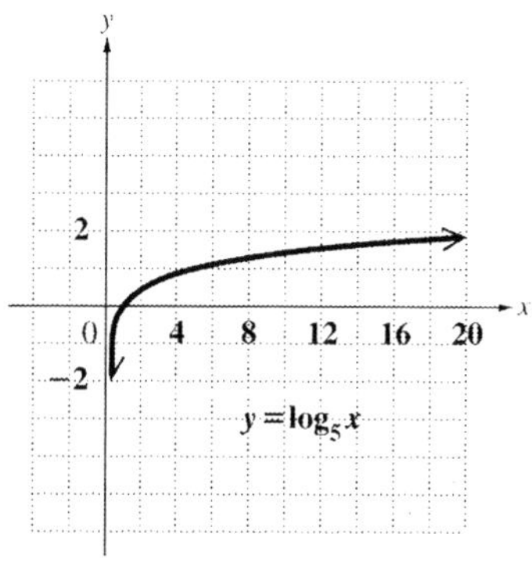

41. $y = \log_{1/3} x$
Change to exponential form.

$$\left(\tfrac{1}{3}\right)^y = x$$

Refer to Section 11.2, Exercise 7, for the graph of $g(x) = \left(\frac{1}{3}\right)^x$. Since $y = \log_{1/3} x$ (or $\left(\frac{1}{3}\right)^y = x$) is the inverse of $y = \left(\frac{1}{3}\right)^x$, its graph is symmetric about the line $y = x$ to the graph of $y = \left(\frac{1}{3}\right)^x$. The graph can be plotted by reversing the ordered pairs in the table of values belonging to $g(x) = \left(\frac{1}{3}\right)^x$.

x	9	3	1	$\frac{1}{3}$	$\frac{1}{9}$
y	−2	−1	0	1	2

Plot the points, and draw a smooth curve through them.

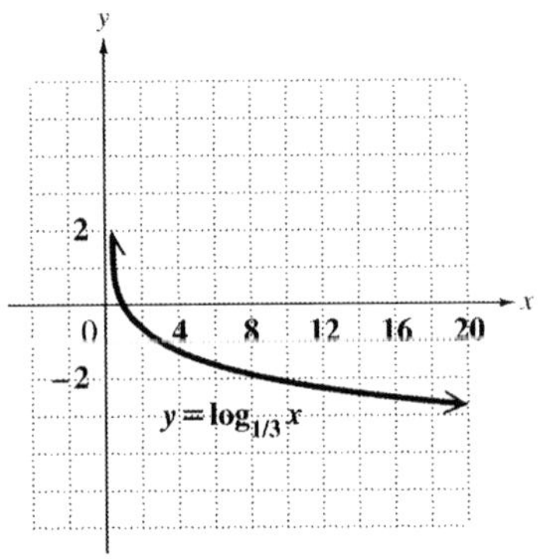

42. $y = \log_{1/5} x$
From Section 11.2, Exercise 8, use the table for $g(x) = \left(\frac{1}{5}\right)^x$ to construct the table for $y = g^{-1}(x)$ by reversing x- and y-values.

x	25	5	1	$\frac{1}{5}$
y	−2	−1	0	1

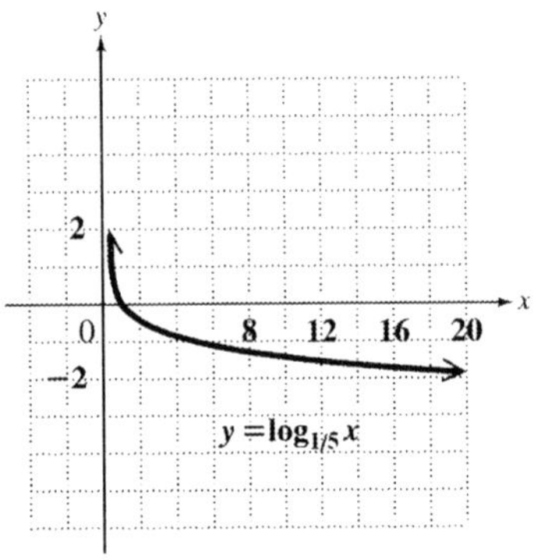

43. The following characteristics reinforce the concept that $F(x) = a^x$ and $G(x) = \log_a x$ are inverse functions.
(1) The graph of F will always contain $(0, 1)$; the graph of G will always contain $(1, 0)$.

(2) When $a > 1$, the graph of F will rise from left to right from the second quadrant to the first. When $a > 1$, the graph of G will rise from left to right from the fourth quadrant to the first, thus making G a mirror image of F with respect to the line $y = x$. Also, when $0 < a < 1$, the graph of F will fall from left to right from the second quadrant to the first, while the graph of G will fall from left to right from the first quadrant to the fourth, also making G a mirror image of F with respect to the line $y = x$.

(3) The graph of F approaches the x-axis but never touches it. The graph of G approaches the y-axis but never touches it.

(4) The domain of F is $(-\infty, \infty)$, and the range is $(0, \infty)$. The domain of G is $(0, \infty)$, and the range is $(-\infty, \infty)$.

44. The range of $F(x) = a^x$ is the domain of $G(x) = \log_a x$, that is, $\underline{(0, \infty)}$.
The domain of $F(x) = a^x$ is the range of $G(x) = \log_a x$, that is, $\underline{(-\infty, \infty)}$.

45. The values of t are on the horizontal axis, and the values of $f(t)$ are on the vertical axis. Read the value of $f(t)$ from the graph for the given value of t. At $t = 0$, $f(0) = 8$.

46. From the graph, when $t = 10$, $f(t) = 16$.

47. To find $f(60)$, find 60 on the t-axis, then go up to the graph and across to the $f(t)$ axis to read the value of $f(60)$. At $t = 60$, $f(60) = 24$.

48. From the graph, when $t = 0$, $f(t) = 8$, when $t = 10$, $f(t) = 16$, and , when $t = 60$, $f(t) = 24$.

$$f(t) = 8 \log_5 (2t + 5)$$
$$f(0) = 8 \log_5 (2 \cdot 0 + 5)$$
$$= 8 \log_5 5 = 8 \cdot 1 = 8$$

$$f(10) = 8 \log_5 (2 \cdot 10 + 5)$$
$$= 8 \log_5 25$$
$$(\log_5 25 = 2 \text{ since } 5^2 = 25)$$
$$= 8 \cdot 2 = 16$$

$$\begin{aligned} f(60) &= 8\log_5(2\cdot 60+5) \\ &= 8\log_5 125 \end{aligned}$$
$(\log_5 125 = 3 \text{ since } 5^3 = 125)$
$$= 8\cdot 3 = 24$$

49. Since every real number power of 1 equals 1, if $y = \log_1 x$, then $x = 1^y$ and so $x = 1$ for every y. This contradicts the definition of a function.

50. $x = \log_a 1$ is equivalent to $a^x = 1$. The only value of x that makes $a^x = 1$ is 0. (Recall that $a \neq 1$.)

51. The graph of $f(x) = 3^x$ (see Exercise 5 in Section 11.2) rises at a faster rate from left to right than the graph of $g(x) = \log_3 x$ (see Exercise 39 in this section).

52. Let $f(x) = \left(1+\dfrac{1}{x}\right)^x$ and make a table of values.

x	1	10	100	1000	10,000
$f(x)$	2	2.594	2.705	2.717	2.718

As x gets larger and larger, the expression gets closer and closer to a number that is approximately 2.718.

53. $f(x) = 3800 + 585\log_2 x$

(a) $x = 1982 - 1980 = 2$

$$\begin{aligned} f(2) &= 3800 + 585\log_2 2 \\ &= 3800 + 585(1) \\ &= 4385 \end{aligned}$$

The model gives an approximate withdrawal of 4385 billion ft^3 of natural gas from crude oil wells in the United States for 1982.

(b) $x = 1988 - 1980 = 8$

$$\begin{aligned} f(8) &= 3800 + 585\log_2 8 \\ &= 3800 + 585(3) \\ &= 5555 \end{aligned}$$

The model gives an approximate withdrawal of 5555 billion ft^3 of natural gas from crude oil wells in the United States for 1988.

(c) $x = 1996 - 1980 = 16$

$$\begin{aligned} f(16) &= 3800 + 585\log_2 16 \\ &= 3800 + 585(4) \\ &= 6140 \end{aligned}$$

The model gives an approximate withdrawal of 6140 billion ft^3 of natural gas from crude oil wells in the United States for 1996.

54. $f(x) = 51.47 + 6.044\log_2 x$

(a) 1980 corresponds to $x = 1$.

$$\begin{aligned} f(1) &= 51.47 + 6.044\log_2 1 \\ &= 51.47 + 6.044(0) \\ &= 51.47 \end{aligned}$$

The model gives an approximate consumption of 51.47 trillion cubic feet of dry natural gas in 1980.

(b) 1987 corresponds to $x = 8$.

$$\begin{aligned} f(8) &= 51.47 + 6.044\log_2 8 \\ &= 51.47 + 6.044(3) \\ &= 51.47 + 18.132 = 69.602 \end{aligned}$$

The model gives an approximate consumption of 69.602 trillion cubic feet of dry natural gas in 1987.

(c) 1995 corresponds to $x = 16$.

$$\begin{aligned} f(16) &= 51.47 + 6.044\log_2 16 \\ &= 51.47 + 6.044(4) \\ &= 51.47 + 24.176 = 75.646 \end{aligned}$$

The model gives an approximate consumption of 75.646 trillion cubic feet of dry natural gas in 1995.

55. $R = \log_{10}\dfrac{x}{x_0}$

Change to exponential form.

$$10^R = \frac{x}{x_0}, \text{ so } x = x_0\,10^R.$$

Let $R = 6.7$ for the Northridge earthquake, x_1.

$$x_1 = x_0 10^{6.7}$$

Let $R = 7.3$ for the Landers earthquake, x_2.

$$x_2 = x_0 10^{7.3}$$

The ratio of x_2 to x_1 is

$$\frac{x_2}{x_1} = \frac{x_0 10^{7.3}}{x_0 10^{6.7}} = 10^{.6} \approx 3.98.$$

The Landers earthquake was about 4 times more powerful than the Northridge earthquake.

56. As in the previous exercise, the ratio is

$$\frac{x_0 10^{7.3}}{x_0 10^{4.8}} = 10^{2.5} \approx 316.$$

The Landers earthquake was about 300 times more powerful than the smallest rated earthquake.

11.4 Properties of Logarithms

11.4 Margin Exercises

1. Use the Product rule for Logarithms.

$$\log_b xy = \log_b x + \log_b y$$

(a) $\log_6 (5 \cdot 8) = \log_6 5 + \log_6 8$

(b) $\log_4 3 + \log_4 7 = \log_4 (3 \cdot 7)$
$= \log_4 21$

(c) $\log_8 8k = \log_8 8 + \log_8 k$
$= 1 + \log_8 k \quad (k > 0)$

(d) $\log_5 m^2 = \log_5 (m \cdot m)$
$= \log_5 m + \log_5 m$
$= 2\log_5 m \quad (m > 0)$

2. Use the Quotient rule for Logarithms.

$$\log_b \frac{x}{y} = \log_b x - \log_b y$$

(a) $\log_7 \frac{9}{4} = \log_7 9 - \log_7 4$

(b) $\log_3 p - \log_3 q = \log_3 \frac{p}{q} \quad (p > 0,\ q > 0)$

(c) $\log_4 \frac{3}{16} = \log_4 3 - \log_4 16$
$= \log_4 3 - 2$

3. Use the Power rule for Logarithms.

$$\log_b x^r = r \log_b x$$

(a) $\log_3 5^2 = 2 \log_3 5$

(b) $\log_a x^4 = 4 \log_a x \quad (x > 0)$

(c) $\log_b \sqrt{8} = \log_b 8^{1/2}$
$= \frac{1}{2} \log_b 8 \quad (b > 0)$

(d) $\log_2 \sqrt[3]{2} = \log_2 2^{1/3}$
$= \frac{1}{3} \log_2 2$
$= \frac{1}{3}(1) = \frac{1}{3}$

4. Use the Special Properties,

$$b^{\log_b x} = x \quad \text{or} \quad \log_b b^x = x.$$

(a) By the second property,

$$\log_{10} 10^3 = 3.$$

(b) Using the second property,

$$\log_2 8 = \log_2 2^3 = 3.$$

(c) By the first property,

$$5^{\log_5 3} = 3.$$

5. Use the Properties of Logarithms.

(a) $\log_6 36m^5$
$= \log_6 36 + \log_6 m^5$ *Product rule*
$= 2 + 5 \log_6 m$ *$\log_6 6^2 = 2$; Power rule*

(b) $\log_2 \sqrt{9z}$
$= \log_2 (9z)^{1/2}$
$= \frac{1}{2} \log_2 (9z)$ *Power rule*
$= \frac{1}{2}(\log_2 9 + \log_2 z)$ *Product rule*
$= \frac{1}{2} \log_2 3^2 + \frac{1}{2} \log_2 z$
$= \frac{1}{2}(2 \log_2 3) + \frac{1}{2} \log_2 z$ *Power rule*
$= \log_2 3 + \frac{1}{2} \log_2 z$

(c) $\log_q \frac{8r^2}{m-1} \quad (m > 1, q \neq 1)$
$= \log_q (8r^2) - \log_q (m-1)$
Quotient rule
$= \log_q 8 + \log_q r^2 - \log_q (m-1)$
Product rule
$= \log_q 8 + 2 \log_q r - \log_q (m-1)$
Power rule

(d) $2 \log_a x + 3 \log_a y \quad (a \neq 1)$
$= \log_a x^2 + \log_a y^3$ *Power rule*
$= \log_a x^2 y^3$ *Product rule*

(e) $\log_4 (3x + y)$ cannot be written as a sum of logarithms, since

$$\log_b (M + N) \neq \log_b M + \log_b N.$$

There is no property of logarithms to rewrite the logarithm of a sum.

11.4 Section Exercises

1. $\log_b x + \log_b y = \log_b (x + y)$ is *false*;
$\log_b x + \log_b y = \log_b xy$

2. $\log_b \frac{x}{y} = \log_b x - \log_b y$ is *true.*

3. $\log_b b^x = x$ is *true.*

4. $\log_b x^r = \log_b rx$ is *false*; $\log_b x^r = r \log_b x$

5. $\log_7 \frac{4}{5} = \log_7 4 - \log_7 5$ *Quotient rule*

6. $\log_8 \frac{9}{11} = \log_8 9 - \log_8 11$ *Quotient rule*

7. $\log_2 8^{1/4} = \frac{1}{4} \log_2 8$ *Power rule*

We can simplify as follows:

$$\frac{1}{4} \log_2 8 = \frac{1}{4} \log_2 2^3$$
$$= \frac{1}{4}(3) = \frac{3}{4}$$

8. $\log_3 9^{3/4} = \frac{3}{4}\log_3 9$ *Power rule*

We can simplify as follows:

$$\frac{3}{4}\log_3 9 = \frac{3}{4}\log_3 3^2 = \frac{3}{4}(2) = \frac{3}{2}$$

9. $\log_4 \frac{3\sqrt{x}}{y}$

$= \log_4 (3\sqrt{x}) - \log_4 y$ *Quotient rule*

$= \log_4 3 + \log_4 x^{1/2} - \log_4 y$ *Product rule*

$= \log_4 3 + \frac{1}{2}\log_4 x - \log_4 y$ *Power rule*

10. $\log_5 \frac{6\sqrt{z}}{w}$

$= \log_5 (6\sqrt{z}) - \log_5 w$ *Quotient rule*

$= \log_5 6 + \log_5 z^{1/2} - \log_5 w$ *Product rule*

$= \log_5 6 + \frac{1}{2}\log_5 z - \log_5 w$ *Power rule*

11. $\log_3 \frac{\sqrt[3]{4}}{x^2 y} = \log_3 \frac{4^{1/3}}{x^2 y}$

Use the quotient rule for logarithms.

$$= \log_3 4^{1/3} - \log_3 (x^2 y)$$

Use the product rule for logarithms.

$$= \log_3 4^{1/3} - (\log_3 x^2 + \log_3 y)$$

$$= \log_3 4^{1/3} - \log_3 x^2 - \log_3 y$$

Use the power rule for logarithms.

$$= \frac{1}{3}\log_3 4 - 2\log_3 x - \log_3 y$$

12. $\log_7 \frac{\sqrt[3]{13}}{pq^2}$

$= \log_7 \sqrt[3]{13} - (\log_7 p + \log_7 q^2)$

$= \log_7 13^{1/3} - \log_7 p - \log_7 q^2$

$= \frac{1}{3}\log_7 13 - \log_7 p - 2\log_7 q$

13. $\log_3 \sqrt{\frac{xy}{5}} = \log_3 \left(\frac{xy}{5}\right)^{1/2}$

$= \frac{1}{2}\log_3 \left(\frac{xy}{5}\right)$ *Power rule*

$= \frac{1}{2}[\log_3 (xy) - \log_3 5]$ *Quotient rule*

$= \frac{1}{2}(\log_3 x + \log_3 y - \log_3 5)$ *Product rule*

$= \frac{1}{2}\log_3 x + \frac{1}{2}\log_3 y - \frac{1}{2}\log_3 5$

14. $\log_6 \sqrt{\frac{pq}{7}} = \log_6 \left(\frac{pq}{7}\right)^{1/2}$

$= \frac{1}{2}\log_6 \left(\frac{pq}{7}\right)$ *Power rule*

$= \frac{1}{2}[\log_6 (pq) - \log_6 7]$ *Quotient rule*

$= \frac{1}{2}(\log_6 p + \log_6 q - \log_6 7)$ *Product rule*

$= \frac{1}{2}\log_6 p + \frac{1}{2}\log_6 q - \frac{1}{2}\log_6 7$

15. $\log_2 \frac{\sqrt[3]{x} \cdot \sqrt[5]{y}}{r^2}$

$= \log_2 \frac{x^{1/3}y^{1/5}}{r^2}$

$= \log_2 (x^{1/3}y^{1/5}) - \log_2 r^2$ *Quotient rule*

$= \log_2 x^{1/3} + \log_2 y^{1/5} - \log_2 r^2$ *Product rule*

$= \frac{1}{3}\log_2 x + \frac{1}{5}\log_2 y - 2\log_2 r$ *Power rule*

16. $\log_4 \frac{\sqrt[4]{z} \cdot \sqrt[5]{w}}{s^2}$

$= \log_4 z^{1/4} + \log_4 w^{1/5} - \log_4 s^2$

$= \frac{1}{4}\log_4 z + \frac{1}{5}\log_4 w - 2\log_4 s$

17. The distributive property tells us that the *product* $a(x+y)$ equals the sum $ax + ay$. In the notation $\log_a (x+y)$, the parentheses do not indicate multiplication. They indicate that $x + y$ is the result of raising a to some power.

18. Logarithms of products and quotients behave exactly as the exponents in products and quotients of exponentials. To find the logarithm of a *product*, we *add* the logarithms of the factors. This is analogous to adding the exponents when we multiply with like bases.
Similarly, to find the logarithm of a *quotient*, we *subtract* the logarithm of the denominator from the logarithm of the numerator. This is analogous to subtracting the exponents when we divide with like bases.

19. By the product rule for logarithms,

$$\log_b x + \log_b y = \log_b xy.$$

20. $\log_b 2 + \log_b z = \log_b 2z$

21. $3\log_a m - \log_a n$

$= \log_a m^3 - \log_a n$ *Power rule*

$= \log_a \frac{m^3}{n}$ *Quotient rule*

22. $5\log_b x - \log_b y$

$= \log_b x^5 - \log_b y$ *Power rule*

$= \log_b \frac{x^5}{y}$ *Quotient rule*

23. $(\log_a r - \log_a s) + 3\log_a t$

Use the quotient and power rules for logarithms.

$$= \log_a \frac{r}{s} + \log_a t^3$$

$$= \log_a \frac{rt^3}{s} \quad \textit{Product rule}$$

24. $(\log_a p - \log_a q) + 2\log_a r$

$$= \log_a \frac{p}{q} + \log_a r^2$$

$$= \log_a \frac{pr^2}{q}$$

25. $3\log_a 5 - 4\log_a 3$

$$= \log_a 5^3 - \log_a 3^4 \quad \textit{Power rule}$$

$$= \log_a \frac{5^3}{3^4} \quad \textit{Quotient rule}$$

$$= \log_a \frac{125}{81}$$

26. $3\log_a 5 + \frac{1}{2}\log_a 9 = \log_a 5^3 + \log_a 9^{1/2}$

$$= \log_a (5^3 \cdot 9^{1/2})$$

$$= \log_a (125 \cdot 3) = \log_a 375$$

27. $\log_{10}(x+3) + \log_{10}(x-3)$

$$= \log_{10}(x+3)(x-3) \quad \textit{Product rule}$$

$$= \log_{10}(x^2 - 9)$$

28. $\log_{10}(y+4) + \log_{10}(y-4)$

$$= \log_{10}(y+4)(y-4)$$

$$= \log_{10}(y^2 - 16)$$

29. By the power rule for logarithms,

$$3\log_p x + \frac{1}{2}\log_p y - \frac{3}{2}\log_p z - 3\log_p a$$

$$= \log_p x^3 + \log_p y^{1/2} - \log_p z^{3/2} - \log_p a^3$$

Group the terms into sums.

$$= (\log_p x^3 + \log_p y^{1/2}) - (\log_p z^{3/2} + \log_p a^3)$$

$$= \log_p x^3 y^{1/2} - \log_p z^{3/2} a^3 \quad \textit{Product rule}$$

$$= \log_p \frac{x^3 y^{1/2}}{z^{3/2} a^3} \quad \textit{Quotient rule}$$

30. $\frac{1}{3}\log_b x + \frac{2}{3}\log_b y - \frac{3}{4}\log_b s - \frac{2}{3}\log_b t$

$$= \log_b x^{1/3} + \log_b y^{2/3} - \log_b s^{3/4} - \log_b t^{2/3}$$

$$= \log_b \frac{x^{1/3} y^{2/3}}{s^{3/4} t^{2/3}}$$

31. The rule $\log_b x^r = r(\log_b x)$ applies only when x is positive because $\log_b x$ is not a real number when x is zero or negative.

32. In the proof when $\log_2(-4)(-4)$ is equated with $\log_2(-4) + \log_2(-4)$, you are imitating the product rule for logarithms. The product rule, however, states that the factors must be positive, and here $-4 < 0$. Thus, this is a false "proof."

33. $\log_3 81 = \log_3 3^4 = 4$

34. $\log_3 81$ is the exponent to which 3 must be raised to obtain 81.

35. Using the result from Exercise 33,

$$3^{\log_3 81} = 3^4 = 81.$$

36. $\log_2 19$ is the exponent to which 2 must be raised to obtain 19.

37. Keeping in mind the result from Exercise 35,

$$2^{\log_2 19} = 19.$$

38. To find $k^{\log_k m}$, first assume $\log_k m = y$. This means, changing to an exponential equation, $k^y = m$. Therefore,

$$k^{\log_k m} = k^y = m.$$

11.5 Common and Natural Logarithms

11.5 Margin Exercises

1. Evaluate each logarithm using a calculator.

(a) $\log 41{,}600 \approx 4.6191$

(b) $\log 43.5 \approx 1.6385$

(c) $\log .442 \approx -.3546$

2.

$$\begin{aligned} \text{pH} &= -\log[\text{H}_3\text{O}^+] \\ &= -\log(1.2 \times 10^{-3}) \\ &= -(\log 1.2 + \log 10^{-3}) \\ &= -(\log 1.2 - 3\log 10) \\ &\approx -[.0792 - 3(1)] \\ &= -.0792 + 3 \\ &= 2.9208 \approx 2.9 \end{aligned}$$

Since the pH is less than 3.0, the wetland is a bog.

3. (a)

$$\begin{aligned} \text{pH} &= -\log[\text{H}_3\text{O}^+] \\ 4.6 &= -\log[\text{H}_3\text{O}^+] \quad \textit{Let pH} = 4.6. \\ -4.6 &= \log[\text{H}_3\text{O}^+] \\ -4.6 &= \log_{10}[\text{H}_3\text{O}^+] \\ 10^{-4.6} &= [\text{H}_3\text{O}^+] \quad \textit{Exponential form} \\ [\text{H}_3\text{O}^+] &\approx 2.5 \times 10^{-5} \end{aligned}$$

(b)

$$\begin{aligned} \text{pH} &= -\log[\text{H}_3\text{O}^+] \\ 7.5 &= -\log[\text{H}_3\text{O}^+] && \textit{Let pH = 7.5.} \\ -7.5 &= \log[\text{H}_3\text{O}^+] \\ -7.5 &= \log_{10}[\text{H}_3\text{O}^+] \\ 10^{-7.5} &= [\text{H}_3\text{O}^+] && \textit{Exponential form} \\ [\text{H}_3\text{O}^+] &\approx 3.2 \times 10^{-8} \end{aligned}$$

4. Evaluate each logarithm using a calculator.

(a) $\ln .01 \approx -4.6052$

(b) $\ln 27 \approx 3.2958$

(c) $\ln 529 \approx 6.2710$

5.

$$\begin{aligned} f(x) &= 51{,}600 - 7457 \ln x \\ f(700) &= 51{,}600 - 7457 \ln 700 \\ &\approx 2748.6 \approx 2700 \end{aligned}$$

Atmospheric pressure is 700 millibars at approximately 2700 m.

11.5 Section Exercises

1. Since $\log x = \log_{10} x$, the base is 10. The correct response is **C**.

2. The base that is understood in the expression $\ln x$ is the base e. The correct response is **A**.

3. $10^0 = 1$ and $10^1 = 10$, so $\log 1 = 0$ and $\log 10 = 1$. Thus, the value of $\log 5.6$ must lie between 0 and 1. The correct response is **C**.

4. $e^1 \approx 2.718$ so $\ln 2.718 \approx 1$ and $e^2 \approx 7.389$ so $\ln 7.389 \approx 2$.

$$\begin{aligned} 2.718 &< 5.6 < 7.389 \\ \ln 2.718 &< \ln 5.6 < \ln 7.389 \end{aligned}$$

Because we found that $\ln 2.718 \approx 1$ and $\ln 7.389 \approx 2$, we can say that $\ln 5.6$ is between 1 and 2. The correct response is **C**.

5. $\log 10^{19.2} = \log_{10} 10^{19.2} = 19.2$ by the special property, $\log_b b^x = x$.

6. $\ln e^{\sqrt{2}} = \log_e e^{\sqrt{2}} = \sqrt{2}$ by the special property, $\log_b b^x = x$.

7. To four decimal places,

$$\log 43 \approx 1.6335.$$

8. To four decimal places,

$$\log 98 \approx 1.9912.$$

9. $\log 328.4 \approx 2.5164$

10. $\log 457.2 \approx 2.6601$

11. $\log .0326 \approx -1.4868$

12. $\log .1741 \approx -.7592$

13. $\log(4.76 \times 10^9) \approx 9.6776$
On a TI-83, enter

[LOG] 4.76 [2nd] [EE] 9).

14. $\log(2.13 \times 10^4) \approx 4.3284$

15. $\ln 7.84 \approx 2.0592$

16. $\ln 8.32 \approx 2.1187$

17. $\ln .0556 \approx -2.8896$

18. $\ln .0217 \approx -3.8304$

19. $\ln 388.1 \approx 5.9613$

20. $\ln 942.6 \approx 6.8486$

21. $\ln(8.59 \times e^2) \approx 4.1506$
On a TI-83, enter

[LN] 8.59 [X] [2nd] [e^x] 2)).

22. $\ln(7.46 \times e^3) \approx 5.0096$

23. $\ln 10 \approx 2.3026$

24. $\log e \approx .4343$

25. Let m = the number of letters in your first name and n = the number of letters in your last name.

Answers will vary, but suppose the name is Paul Bunyan, with $m = 4$ and $n = 6$.

(a) $\log_m n = \log_4 6$ is the exponent to which 4 must be raised in order to obtain 6.

(b) Use the change-of-base rule.

$$\begin{aligned} \log_4 6 &= \frac{\log 6}{\log 4} \\ &\approx 1.29248125 \end{aligned}$$

(c) Here, $m = 4$. Use the power key (y^x, x^y, ^) on your calculator.

$$4^{1.29248125} \approx 6$$

The result is 6, the value of n.

26. **(a)** $\log 356.8 \approx 2.552424846$

(b) $\log 35.68 \approx 1.552424846$

(c) $\log 3.568 \approx 0.552424846$

(d) The whole number part of the answers (2, 1, or 0) varies, whereas the decimal part (.552424846) remains the same, indicating that the whole number part corresponds to the placement of the decimal point and the decimal part corresponds to the digits 3, 5, 6, and 8.

27. When you try to find $\log(-1)$ on a calculator, an error message is displayed. This is because the domain of the logarithmic function is $(0, \infty)$; -1 is not in the domain.

28. $\text{pH} = -\log[\text{H}_3\text{O}^+]$
$= -\log(2.5 \times 10^{-5}) \approx 4.6$
Since the pH is between 3.0 and 6.0, the wetland is classified as a *poor fen*.

29. $\text{pH} = -\log[\text{H}_3\text{O}^+]$
$= -\log(2.5 \times 10^{-2}) \approx 1.6$
Since the pH is less than 3.0, the wetland is classified as a *bog*.

30. Ammonia has a hydronium ion concentration of 2.5×10^{-12}.

$$\text{pH} = -\log[\text{H}_3\text{O}^+]$$
$$\text{pH} = \log(2.5 \times 10^{-12}) \approx 11.6$$

31. Tuna has a hydronium ion concentration of 1.3×10^{-6}.

$$\text{pH} = -\log[\text{H}_3\text{O}^+]$$
$$\text{pH} = -\log(1.3 \times 10^{-6}) \approx 5.9$$

32. Human blood plasma has a pH of 7.4.

$$\text{pH} = -\log[\text{H}_3\text{O}^+]$$
$$7.4 = -\log[\text{H}_3\text{O}^+]$$
$$\log_{10}[\text{H}_3\text{O}^+] = -7.4$$
$$[\text{H}_3\text{O}^+] = 10^{-7.4} \approx 4.0 \times 10^{-8}$$

33. Human gastric contents have a pH of 2.0.

$$\text{pH} = -\log[\text{H}_3\text{O}^+]$$
$$2.0 = -\log[\text{H}_3\text{O}^+]$$
$$\log_{10}[\text{H}_3\text{O}^+] = -2.0$$
$$[\text{H}_3\text{O}^+] = 10^{-2} \approx 1.0 \times 10^{-2}$$

34. Spinach has a pH value of 5.4.

$$\text{pH} = -\log[\text{H}_3\text{O}^+]$$
$$5.4 = -\log[\text{H}_3\text{O}^+]$$
$$\log_{10}[\text{H}_3\text{O}^+] = -5.4$$
$$[\text{H}_3\text{O}^+] = 10^{-5.4} \approx 4.0 \times 10^{-6}$$

35. Bananas have a pH of 4.6.

$$\text{pH} = -\log[\text{H}_3\text{O}^+]$$
$$4.6 = -\log[\text{H}_3\text{O}^+]$$
$$\log_{10}[\text{H}_3\text{O}^+] = -4.6$$
$$[\text{H}_3\text{O}^+] = 10^{-4.6} \approx 2.5 \times 10^{-5}$$

36. $N(r) = -5000 \ln r$

(a) 85% (or .85)

$N(.85) = -5000 \ln .85 \approx 813 \approx 800$

The number of years since the split, if the percent of words common to both languages is 85%, is about 800 years.

(b) 35% (or .35)

$N(.35) = -5000 \ln .35 \approx 5249 \approx 5200$

The number of years since the split, if the percent of words common to both languages is 35%, is about 5200 years.

(c) 10% (or .10)

$N(.10) = -5000 \ln .10 \approx 11{,}513 \approx 11{,}500$

The number of years since the split, if the percent of words common to both languages is 10%, is about 11,500 years.

37. $t = t(r) = \dfrac{\ln 2}{\ln(1 + r)}$

(a) 2% (or .02); $t(.02) = \dfrac{\ln 2}{\ln(1 + .02)} \approx 35.0$

The doubling time for 2% is about 35.0 years.

(b) 5% (or .05); $t(.05) = \dfrac{\ln 2}{\ln(1 + .05)} \approx 14.2$

The doubling time for 5% is about 14.2 years.

(c) 8% (or .08); $t(.08) = \dfrac{\ln 2}{\ln(1 + .08)} \approx 9.0$

The doubling time for 8% is about 9.0 years.

38. $T = \dfrac{1}{k} \ln \dfrac{C_2}{C_1}$

$T = \dfrac{1}{\frac{1}{3}} \ln \dfrac{5}{2} = 3 \ln 2.5 \approx 2$ (rounded down)

It should be administered every 2 hours.

39. $D = 10 \log \left(\frac{I}{I_0} \right)$

(a) $D = 10 \log \left(\frac{5.012 \times 10^{10} I_0}{I_0} \right)$
$= 10 \log \left(5.012 \times 10^{10} \right) \approx 107$

The average decibel level for *Spider-Man 2* is about 107 dB.

(b) $D = 10 \log \left(\frac{10^{10} I_0}{I_0} \right)$
$= 10 \log 10^{10} = 10 \cdot 10 = 100$

The average decibel level for *Finding Nemo* is 100 dB.

(c) $D = 10 \log \left(\frac{6{,}310{,}000{,}000\, I_0}{I_0} \right)$
$= 10 \log (6{,}310{,}000{,}000) \approx 98$

The average decibel level for *National Treasure* is about 98 dB.

40. $f(x) = -1317 + 304 \ln x$

(a) $x = 1998 - 1900 = 98$

$f(98) = -1317 + 304 \ln (98) \approx 77$

The prediction for 1998 is 77%.

(b) Let $y = f(x) = 50$.

$$\begin{aligned} 50 &= -1317 + 304 \ln x \\ 1367 &= 304 \ln x \\ \ln x &= \frac{1367}{304} \\ x &= e^{1367/304} \\ x &\approx 89.72 \end{aligned}$$

The outpatient surgeries reached 50% in 1989.

41. $p(h) = 86.3 \ln h - 680$

(a) $p(5000) = 86.3 \ln 5000 - 680 \approx 55$

The percent of moisture at 5000 feet that falls as snow rather than rain is 55%.

(b) $p(7500) = 86.3 \ln 7500 - 680 \approx 90$

The percent of moisture at 7500 feet that falls as snow rather than rain is 90%.

42. $T = -.642 - 189 \ln (1 - p)$

(a) $T = -.642 - 189 \ln (1 - .25)$
$= -.642 - 189 \ln (.75) \approx 53.73$

About \$54 per ton will reduce emissions by 25%.

(b) If $p = 0$, then $\ln (1 - p) = \ln 1 = 0$, so T would be negative. If $p = 1$, then $\ln (1 - p) = \ln 0$, but the domain of $\ln x$ is $(0, \infty)$.

43. $t = -2.57 \ln \left(\frac{87 - L}{63} \right)$

(a) $t = -2.57 \ln \left(\frac{87 - 80}{63} \right) \approx 5.6$

The age of a female blue whale that measures 80 feet is 5.6 years.

(b) t must be positive since it represents years. Because t is the product of a negative number, -2.57, and a natural logarithm, the natural logarithm must be negative to make it positive. For the natural logarithm to be negative (and still defined), its argument (the input of the logarithm) must be between 0 and 1, that is,

$$\begin{aligned} 0 &< \frac{87 - L}{63} < 1 \\ 0(63) &< 87 - L < 1(63) \\ -87 &< -L < -24 \\ 87 &> L > 24 \end{aligned}$$

Thus, $t > 0$ and $\frac{87 - L}{63}$ is positive and in the domain of the function only if $24 < L < 87$.

11.6 Exponential and Logarithmic Equations; Further Applications

11.6 Margin Exercises

1. **(a)**

$$\begin{aligned} 2^x &= 9 \\ \log 2^x &= \log 9 && \textit{Property 3} \\ x \log 2 &= \log 9 && \textit{Power rule} \\ x &= \frac{\log 9}{\log 2} \approx 3.170 \end{aligned}$$

Check $x = 3.170$: $2^{3.170} \approx 9$

Solution set: $\{3.170\}$

(b)

$$\begin{aligned} 10^x &= 4 \\ \log 10^x &= \log 4 && \textit{Property 3} \\ x \log 10 &= \log 4 && \textit{Power rule} \\ x &\approx .602 && \textit{log 10 = 1} \end{aligned}$$

Check $x = .602$: $10^{.602} \approx 4$

Solution set: $\{.602\}$

2. $e^{-.01t} = .38$

Take base e logarithms on both sides.

$$\begin{aligned} \ln e^{-.01t} &= \ln .38 \\ -.01t \ln e &= \ln .38 && \textit{Power rule} \\ -.01t &= \ln .38 && \textit{ln e = 1} \\ t &= \frac{\ln .38}{-.01} \approx 96.8 \end{aligned}$$

Check $t = 96.8$: $e^{-.01(96.8)} \approx .38$

Solution set: $\{96.8\}$

3. $\log_3 (x+1)^5 = 3$

$$\begin{aligned} (x+1)^5 &= 3^3 && \textit{Exponential form} \\ (x+1)^5 &= 27 \\ x+1 &= \sqrt[5]{27} && \textit{Fifth root} \\ x &= -1 + \sqrt[5]{27} && \textit{Subtract 1.} \end{aligned}$$

Check $x = -1 + \sqrt[5]{27}$:

$$\begin{aligned} \log_3 \left(-1 + \sqrt[5]{27} + 1\right)^5 &= \log_3 \left(\sqrt[5]{27}\right)^5 \\ &= \log_3 27 = \log_3 3^3 = 3 \end{aligned}$$

Solution set: $\{-1 + \sqrt[5]{27}\}$

4. $\log_8 (2x+5) + \log_8 3 = \log_8 33$

$$\begin{aligned} \log_8 [3(2x+5)] &= \log_8 33 && \textit{Product rule} \\ 3(2x+5) &= 33 && \textit{Property 4} \\ 2x+5 &= 11 && \textit{Divide by 3.} \\ 2x &= 6 \\ x &= 3 \end{aligned}$$

Check $x = 3$: $\log_8 11 + \log_8 3 = \log_8 33$ *True*

Solution set: $\{3\}$

5. $\log_3 2x - \log_3 (3x+15) = -2$

$$\log_3 \frac{2x}{3x+15} = -2$$

Quotient rule

$$3^{-2} = \frac{2x}{3x+15}$$

Exponential form

$$\frac{1}{9} = \frac{2x}{3x+15}$$

$$3x + 15 = 18x$$

Multiply by 9(3x + 15).

$$\begin{aligned} 15 &= 15x \\ 1 &= x \end{aligned}$$

Check $x = 1$:

$$\begin{aligned} \log_3 2 - \log_3 18 &= \log_3 \tfrac{2}{18} \\ &= \log_3 \tfrac{1}{9} = \log_3 3^{-2} = -2 \end{aligned}$$

Solution set: $\{1\}$

6. Use $A = P\left(1 + \frac{r}{n}\right)^{nt}$ with $P = 2000$, $r = 5\% = .05$, $n = 1$ (compounded annually), and $t = 10$.

$$\begin{aligned} A &= 2000\left(1 + \frac{.05}{1}\right)^{1(10)} \\ &= 2000(1.05)^{10} \approx 3257.79 \end{aligned}$$

The value is about \$3257.79.

7.

$$\begin{aligned} A &= P\left(1 + \frac{r}{n}\right)^{nt} \\ 750 &= 500\left(1 + \frac{.04}{2}\right)^{2 \cdot t} \end{aligned}$$

Semiannually (n = 2)

$$\begin{aligned} 750 &= 500(1.02)^{2t} \\ 1.5 &= (1.02)^{2t} && \textit{Divide by 500.} \\ \log 1.5 &= \log (1.02)^{2t} && \textit{Property 3} \\ \log 1.5 &= 2t \log 1.02 && \textit{Power rule} \\ 2t &= \frac{\log 1.5}{\log 1.02} && \textit{Divide by log 1.02.} \\ t &= \frac{\log 1.5}{2 \log 1.02} && \textit{Divide by 2.} \\ t &\approx 10.24 \end{aligned}$$

It will take about 10.24 yr for \$500 in an account paying 4% interest compounded semiannually to increase to \$750.

8. **(a)**

$$\begin{aligned} A &= Pe^{rt} \\ A &= 2500e^{(.04)3} \\ A &\approx 2818.74 \end{aligned}$$

It will grow to \$2818.74.

(b)

$$\begin{aligned} A &= Pe^{rt} \\ 2(2500) &= 2500e^{.04t} \\ 2 &= e^{.04t} && \textit{Divide by 2500.} \\ \ln 2 &= .04t && \textit{Property 3 with ln} \\ t &= \frac{\ln 2}{.04} && \textit{Divide by .04.} \\ t &\approx 17.33 \end{aligned}$$

It would take about 17.33 yr for the initial investment to double.

9. $y = y_0 e^{-.0239t}$

(a) $y = 12e^{-.0239(35)}$ *Let t = 35.*

$y \approx 5.20$

After 35 yr, there will be about 5.20 g.

(b) $\frac{1}{2}(12) = 12e^{-.0239t}$

$$\frac{1}{2} = e^{-.0239t}$$
$$\ln\frac{1}{2} = -.0239t$$
$$t = \frac{\ln\frac{1}{2}}{-.0239} \approx 29$$

The half-life is about 29 years.

10. **(a)** $\log_3 17 = \dfrac{\log 17}{\log 3} \approx 2.5789$

(b) $\log_3 17 = \dfrac{\ln 17}{\ln 3} \approx 2.5789$

11. $x = 1990 - 1980 = 10$

$$\begin{aligned} f(x) &= 38.83 + 4.208 \log_2 x \\ f(10) &= 38.83 + 4.208 \log_2 10 \\ &= 38.83 + 4.208\left(\frac{\log 10}{\log 2}\right) \\ &\approx 52.8\% \end{aligned}$$

This is very close to the actual value of 53%.

11.6 Section Exercises

1.
$$5^x = 125$$
$$\log 5^x = \log 125$$

2. $x \log 5 = \log 125$

3.
$$\frac{x \log 5}{\log 5} = \frac{\log 125}{\log 5}$$
$$x = \frac{\log 125}{\log 5}$$

4. $\dfrac{\log 125}{\log 5} = 3$ (from calculator)

Solution set: $\{3\}$

5. $7^x = 5$

Take the logarithm of each side.

$\log 7^x = \log 5$

Use the power rule for logarithms.

$x \log 7 = \log 5$

$$x = \frac{\log 5}{\log 7} \approx .827$$

Solution set: $\{.827\}$

6.
$$\begin{aligned} 4^x &= 3 \\ \log 4^x &= \log 3 \\ x \log 4 &= \log 3 \\ x &= \frac{\log 3}{\log 4} \approx .792 \end{aligned}$$

Solution set: $\{.792\}$

7.
$$\begin{aligned} 9^{-x+2} &= 13 \\ \log 9^{-x+2} &= \log 13 \\ (-x+2)\log 9 &= \log 13 \; (*) \\ -x\log 9 + 2\log 9 &= \log 13 \\ -x \log 9 &= \log 13 - 2\log 9 \\ x\log 9 &= 2\log 9 - \log 13 \\ x &= \frac{2\log 9 - \log 13}{\log 9} \\ &\approx .833 \end{aligned}$$

$(*)$ Alternative solution steps:

$$\begin{aligned} (-x+2)\log 9 &= \log 13 \\ -x+2 &= \frac{\log 13}{\log 9} \\ 2 - \frac{\log 13}{\log 9} &= x \end{aligned}$$

Solution set: $\{.833\}$

8.
$$\begin{aligned} 6^{-t+1} &= 22 \\ \log 6^{-t+1} &= \log 22 \\ (-t+1)\log 6 &= \log 22 \\ -t+1 &= \frac{\log 22}{\log 6} \\ t &= 1 - \frac{\log 22}{\log 6} \\ &\approx -.725 \end{aligned}$$

Solution set: $\{-.725\}$

9.
$$\begin{aligned} 3^{2x} &= 14 \\ \log 3^{2x} &= \log 14 \\ 2x\log 3 &= \log 14 \\ x &= \frac{\log 14}{2\log 3} \approx 1.201 \end{aligned}$$

Solution set: $\{1.201\}$

10.
$$\begin{aligned} 5^{.3x} &= 11 \\ \log 5^{.3x} &= \log 11 \\ .3x\log 5 &= \log 11 \\ x &= \frac{\log 11}{.3\log 5} \approx 4.966 \end{aligned}$$

Solution set: $\{4.966\}$

11.
$$\begin{aligned} 2^{y+3} &= 5^y \\ \log 2^{y+3} &= \log 5^y \\ (y+3)\log 2 &= y\log 5 \\ y\log 2 + 3\log 2 &= y\log 5 \end{aligned}$$
Get y-terms on one side.
$$\begin{aligned} y\log 2 - y\log 5 &= -3\log 2 \\ y(\log 2 - \log 5) &= -3\log 2 \quad \textit{Factor out } y. \\ y &= \frac{-3\log 2}{\log 2 - \log 5} \\ &\approx 2.269 \end{aligned}$$
Solution set: $\{2.269\}$

12.
$$\begin{aligned} 6^{m+3} &= 4^m \\ \log 6^{m+3} &= \log 4^m \\ (m+3)\log 6 &= m\log 4 \\ m\log 6 + 3\log 6 &= m\log 4 \\ m\log 6 - m\log 4 &= -3\log 6 \\ m(\log 6 - \log 4) &= -3\log 6 \\ m &= \frac{-3\log 6}{\log 6 - \log 4} \\ &\approx -13.257 \end{aligned}$$
Solution set: $\{-13.257\}$

13.
$$\begin{aligned} 2^{x+3} &= 3^{x-4} \\ \log 2^{x+3} &= \log 3^{x-4} \\ (x+3)\log 2 &= (x-4)\log 3 \\ x\log 2 + 3\log 2 &= x\log 3 - 4\log 3 \end{aligned}$$
Distributive property

Get x-terms on one side.
$$x\log 2 - x\log 3 = -3\log 2 - 4\log 3$$
Factor out x.
$$\begin{aligned} x(\log 2 - \log 3) &= -3\log 2 - 4\log 3 \\ x &= \frac{-3\log 2 - 4\log 3}{\log 2 - \log 3} \\ &\approx 15.967 \end{aligned}$$
Solution set: $\{15.967\}$

14.
$$e^{.006x} = 30$$
Take base e logarithms on both sides.
$$\begin{aligned} \ln e^{.006x} &= \ln 30 \\ .006x(\ln e) &= \ln 30 \\ .006x &= \ln 30 \qquad \textit{ln e = 1} \\ x &= \frac{\ln 30}{.006} \approx 566.866 \end{aligned}$$
Solution set: $\{566.866\}$

15.
$$\begin{aligned} e^{.012x} &= 23 \\ \ln e^{.012x} &= \ln 23 \\ .012x(\ln e) &= \ln 23 \\ .012x &= \ln 23 \qquad \textit{ln e = 1} \\ x &= \frac{\ln 23}{.012} \approx 261.291 \end{aligned}$$
Solution set: $\{261.291\}$

16.
$$\begin{aligned} e^{-.103x} &= 7 \\ \ln e^{-.103x} &= \ln 7 \\ -.103x(\ln e) &= \ln 7 \\ -.103x &= \ln 7 \qquad \textit{ln e = 1} \\ x &= \frac{\ln 7}{-.103} \approx -18.892 \end{aligned}$$
Solution set: $\{-18.892\}$

17.
$$\begin{aligned} e^{-.205x} &= 9 \\ \ln e^{-.205x} &= \ln 9 \\ -.205x(\ln e) &= \ln 9 \\ -.205x &= \ln 9 \qquad \textit{ln e = 1} \\ x &= \frac{\ln 9}{-.205} \approx -10.718 \end{aligned}$$
Solution set: $\{-10.718\}$

18.
$$\begin{aligned} \ln e^x &= 4 \\ x &= 4 \end{aligned}$$
Solution set: $\{4\}$

19.
$$\begin{aligned} \ln e^{3x} &= 9 \\ 3x &= 9 \\ x &= 3 \end{aligned}$$
Solution set: $\{3\}$

20.
$$\begin{aligned} \ln e^{.04x} &= \sqrt{3} \\ .04x &= \sqrt{3} \\ x &= \frac{\sqrt{3}}{.04} \approx 43.301 \end{aligned}$$
Solution set: $\{43.301\}$

21.
$$\begin{aligned} \ln e^{.45x} &= \sqrt{7} \\ .45x &= \sqrt{7} \\ x &= \frac{\sqrt{7}}{.45} \approx 5.879 \end{aligned}$$
Solution set: $\{5.879\}$

22.
$$\begin{aligned} \ln e^{2x} &= \pi \\ 2x &= \pi \\ x &= \frac{\pi}{2} \approx 1.571 \end{aligned}$$
Solution set: $\{1.571\}$

23. Let's try Exercise 14.

$$\begin{aligned} e^{.006x} &= 30 \\ \log e^{.006x} &= \log 30 \\ .006x(\log e) &= \log 30 \\ x &= \frac{\log 30}{.006 \log e} \approx 566.866 \end{aligned}$$

The natural logarithm is easier because $\ln e = 1$, whereas $\log e$ needs to be calculated.

24. To solve $10^{.0025x} = 75$, it would be better to use common logarithms because $\log 10 = 1$, and the solution would involve only one approximate value, $\log 75$. Using natural logarithms would give a solution that involves two approximate values.

25. $\log_3(6x+5) = 2$

$$\begin{aligned} 6x+5 &= 3^2 && \textit{Exponential form} \\ 6x+5 &= 9 \\ 6x &= 4 \\ x &= \tfrac{4}{6} = \tfrac{2}{3} \end{aligned}$$

$x = \frac{2}{3}$ is acceptable since $6x + 5 = 9 > 0$; that is, it yields a logarithm of a positive number in the original equation.

Solution set: $\left\{\frac{2}{3}\right\}$

26. $\log_5(12x-8) = 3$

$$\begin{aligned} 12x-8 &= 5^3 && \textit{Exponential form} \\ 12x &= 125+8 \\ 12x &= 133 \\ x &= \tfrac{133}{12} \end{aligned}$$

$x = \frac{133}{12}$ is acceptable since $12x - 8 = 125 > 0$; that is, it yields a logarithm of a positive number in the original equation.

Solution set: $\left\{\frac{133}{12}\right\}$

27. $\log_2(2x-1) = 5$

$$\begin{aligned} 2x-1 &= 2^5 && \textit{Exponential form} \\ 2x-1 &= 32 \\ 2x &= 33 \\ x &= \tfrac{33}{2} \end{aligned}$$

$x = \frac{33}{2}$ is acceptable since $2x - 1 = 32 > 0$.

Solution set: $\left\{\frac{33}{2}\right\}$

28. $\log_6(4x+2) = 2$

$$\begin{aligned} 4x+2 &= 6^2 && \textit{Exponential form} \\ 4x+2 &= 36 \\ 4x &= 34 \\ x &= \tfrac{34}{4} = \tfrac{17}{2} \end{aligned}$$

$x = \frac{17}{2}$ is acceptable since $4x + 2 = 36 > 0$.

Solution set: $\left\{\frac{17}{2}\right\}$

29. $\log_7(x+1)^3 = 2$

$$\begin{aligned} (x+1)^3 &= 7^2 && \textit{Exponential form} \\ x+1 &= \sqrt[3]{49} && \textit{Cube root} \\ x &= -1 + \sqrt[3]{49} \end{aligned}$$

$x = -1 + \sqrt[3]{49}$ is acceptable since $(x+1)^3 = 49 > 0$.

Solution set: $\left\{-1 + \sqrt[3]{49}\right\}$

30. $\log_4(x-3)^3 = 4$

$$\begin{aligned} (x-3)^3 &= 4^4 && \textit{Exponential form} \\ x-3 &= \sqrt[3]{4^4} && \textit{Cube root} \\ x-3 &= 4\sqrt[3]{4} \\ x &= 3 + 4\sqrt[3]{4} \end{aligned}$$

$x = 3 + 4\sqrt[3]{4}$ is acceptable since $(x-3)^3 = 256 > 0$.

Solution set: $\left\{3 + 4\sqrt[3]{4}\right\}$

31. 2 cannot be a solution because $\log(2-3) = \log(-1)$, and -1 is not in the domain of $\log x$.

32. No, because -4 leads to a positive value of $3 - x$: $3 - (-4) = 7$. The number -4 is in the domain of $\log(3-x)$.

33. $\log(6x+1) = \log 3$

$$\begin{aligned} 6x+1 &= 3 && \textit{Property 4} \\ 6x &= 2 \\ x &= \tfrac{2}{6} = \tfrac{1}{3} \end{aligned}$$

$x = \frac{1}{3}$ is acceptable since $6x + 1 = 3 > 0$.

Solution set: $\left\{\frac{1}{3}\right\}$

34. $\log(7-x) = \log 12$

$$\begin{aligned} 7-x &= 12 && \textit{Property 4} \\ -5 &= x \end{aligned}$$

$x = -5$ is acceptable since $7 - x = 12 > 0$.

Solution set: $\{-5\}$

35.
$$\begin{aligned}\log_5(3t+2) - \log_5 t &= \log_5 4\\ \log_5 \frac{3t+2}{t} &= \log_5 4\\ \frac{3t+2}{t} &= 4\\ 3t+2 &= 4t\\ 2 &= t\end{aligned}$$

$t=2$ is acceptable since $3t+2=8>0$ and $t=2>0$.

Solution set: $\{2\}$

36.
$$\begin{aligned}\log_2(x+5) - \log_2(x-1) &= \log_2 3\\ \log_2 \frac{x+5}{x-1} &= \log_2 3\\ \frac{x+5}{x-1} &= 3\\ x+5 &= 3(x-1)\\ x+5 &= 3x-3\\ 8 &= 2x\\ 4 &= x\end{aligned}$$

$x=4$ is acceptable since $x+5=9>0$ and $x-1=3>0$.

Solution set: $\{4\}$

37.
$$\begin{aligned}\log 4x - \log(x-3) &= \log 2\\ \log \frac{4x}{x-3} &= \log 2\\ \frac{4x}{x-3} &= 2\\ 4x &= 2(x-3)\\ 4x &= 2x-6\\ 2x &= -6\\ x &= -3\end{aligned}$$

Reject $x=-3$, because $4x=-12$, which yields an equation in which the logarithm of a negative number must be found.

Solution set: $\emptyset$

38.
$$\begin{aligned}\log(-x) + \log 3 &= \log(2x-15)\\ \log(-3x) &= \log(2x-15)\\ -3x &= 2x-15\\ 15 &= 5x\\ 3 &= x\end{aligned}$$

Reject $x=3$, because $-x=-3$, which yields an equation in which the logarithm of a negative number must be found.

Solution set: $\emptyset$

39.
$$\begin{aligned}\log_2 x + \log_2(x-7) &= 3\\ \log_2[x(x-7)] &= 3\\ x(x-7) &= 2^3 \quad \textit{Exponential form}\\ x^2-7x &= 8\\ x^2-7x-8 &= 0\\ (x-8)(x+1) &= 0\end{aligned}$$

$$x-8=0 \quad \text{or} \quad x+1=0$$
$$x=8 \quad \text{or} \quad x=-1$$

Reject $x=-1$, because it yields an equation in which the logarithm of a negative number must be found.

Solution set: $\{8\}$

40.
$$\begin{aligned}\log(2x-1) + \log 10x &= \log 10\\ \log[(2x-1)(10x)] &= \log 10\\ (2x-1)(10x) &= 10\\ 20x^2-10x &= 10\\ 20x^2-10x-10 &= 0\\ 2x^2-x-1 &= 0\\ (2x+1)(x-1) &= 0\end{aligned}$$

$$2x+1=0 \quad \text{or} \quad x-1=0$$
$$x=-\tfrac{1}{2} \quad \text{or} \quad x=1$$

Reject $x=-\frac{1}{2}$, because $2x-1=-2$, which yields an equation in which the logarithm of a negative number must be found.

Solution set: $\{1\}$

41.
$$\begin{aligned}\log 5x - \log(2x-1) &= \log 4\\ \log \frac{5x}{2x-1} &= \log 4\\ \frac{5x}{2x-1} &= 4\\ 5x &= 8x-4\\ 4 &= 3x\\ \tfrac{4}{3} &= x\end{aligned}$$

$x=\frac{4}{3}$ is acceptable since $5x=\frac{20}{3}>0$ and $2x-1=\frac{5}{3}>0$.

Solution set: $\{\frac{4}{3}\}$

42.
$$\begin{aligned}\log_3 x + \log_3(2x+5) &= 1\\ \log_3[x(2x+5)] &= 1\\ x(2x+5) &= 3^1 \quad \textit{Exponential form}\\ 2x^2+5x &= 3\\ 2x^2+5x-3 &= 0\\ (2x-1)(x+3) &= 0\end{aligned}$$

$$2x-1=0 \quad \text{or} \quad x+3=0$$
$$x=\tfrac{1}{2} \quad \text{or} \quad x=-3$$

Reject $x=-3$, because it yields an equation in which the logarithm of a negative number must be found.

Solution set: $\{\frac{1}{2}\}$

43.
$$\begin{aligned}
\log_2 x + \log_2 (x-6) &= 4 \\
\log_2 [x(x-6)] &= 4 \\
x(x-6) &= 2^4 \quad \textit{Exponential form} \\
x^2 - 6x &= 16 \\
x^2 - 6x - 16 &= 0 \\
(x-8)(x+2) &= 0
\end{aligned}$$

$$x - 8 = 0 \quad \text{or} \quad x + 2 = 0$$
$$x = 8 \quad \text{or} \quad x = -2$$

Reject $x = -2$, because it yields an equation in which the logarithm of a negative number must be found.

Solution set: $\{8\}$

44.
$$\begin{aligned}
\log_2 x + \log_2 (x+4) &= 5 \\
\log_2 [x(x+4)] &= 5 \\
x(x+4) &= 2^5 \quad \textit{Exponential form} \\
x^2 + 4x &= 32 \\
x^2 + 4x - 32 &= 0 \\
(x-4)(x+8) &= 0
\end{aligned}$$

$$x - 4 = 0 \quad \text{or} \quad x + 8 = 0$$
$$x = 4 \quad \text{or} \quad x = -8$$

Reject $x = -8$, because it yields an equation in which the logarithm of a negative number must be found.

Solution set: $\{4\}$

45. **(a)** Use the formula $A = P\left(1 + \frac{r}{n}\right)^{nt}$ with $P = 2000$, $r = .04$, $n = 4$, and $t = 6$.

$$\begin{aligned}
A &= 2000\left(1 + \frac{.04}{4}\right)^{4 \cdot 6} \\
&= 2000(1.01)^{24} \approx 2539.47
\end{aligned}$$

The account will contain \$2539.47.

(b)
$$\begin{aligned}
3000 &= 2000\left(1 + \frac{.04}{4}\right)^{4t} \\
\tfrac{3000}{2000} &= (1.01)^{4t} \\
\log\left(\tfrac{3}{2}\right) &= \log (1.01)^{4t} \\
\log\left(\tfrac{3}{2}\right) &= 4t \log (1.01) \\
t &= \frac{\log\left(\frac{3}{2}\right)}{4 \log (1.01)} \approx 10.2
\end{aligned}$$

It will take about 10.2 years for the account to grow to \$3000.

46. **(a)** Use the formula $A = P\left(1 + \frac{r}{n}\right)^{nt}$ with $P = 3000$, $r = .035$, $n = 4$, and $t = 7$.

$$\begin{aligned}
A &= 3000\left(1 + \frac{.035}{4}\right)^{4 \cdot 7} \\
&= 3000(1.00875)^{28} \approx 3828.78
\end{aligned}$$

The account will contain \$3828.78.

(b)
$$\begin{aligned}
5000 &= 3000\left(1 + \frac{.035}{4}\right)^{4t} \\
\tfrac{5000}{3000} &= (1.00875)^{4t} \\
\log\left(\tfrac{5}{3}\right) &= \log (1.00875)^{4t} \\
\log\left(\tfrac{5}{3}\right) &= 4t \log (1.00875) \\
t &= \frac{\log\left(\frac{5}{3}\right)}{4 \log (1.00875)} \approx 14.7
\end{aligned}$$

It will take about 14.7 years for the account to grow to \$5000.

47. Use the formula $A = Pe^{rt}$ with $P = 4000$, $r = .035$, and $t = 6$.

$$\begin{aligned}
A &= 4000e^{(.035)(6)} \\
&= 4000e^{.21} \approx 4934.71
\end{aligned}$$

There will be \$4934.71 in the account.

48. Use $A = Pe^{rt}$ with $P = 3000$, $r = .03$, and $t = 7$ to find the amount with continuously compounded interest.

$$A = 3000e^{.03(7)} = 3000e^{.21} \approx 3701.03$$

With continuously compounded interest, the account will contain \$3701.03. The account in Exercise 46 will amount to \$3828.78 – \$3701.03 = \$127.75 more, so those conditions produce a larger value.

49. Here $r = 4.5\%$ (.045), and if the principal doubles, then $A = 2P$.

$$\begin{aligned}
A &= Pe^{rt} \\
2P &= Pe^{.045t} \\
2 &= e^{.045t} \quad \textit{Divide by P.} \\
\ln 2 &= \ln e^{.045t} \quad \textit{Property 4 with ln} \\
\ln 2 &= .045t \\
t &= \frac{\ln 2}{.045} \approx 15.4
\end{aligned}$$

The money will double in about 15.4 yr.

50.
$$\begin{aligned} A &= Pe^{rt} \\ 2(4000) &= 4000e^{.0325t} \\ 2 &= e^{.0325t} \\ \ln 2 &= .0325t \\ t &= \frac{\ln 2}{.0325} \approx 21.3 \end{aligned}$$

It would take about 21.3 yr for $4000 to double when invested at 3.5% compounded continuously.

51.
$$\begin{aligned} A(t) &= 400e^{-.032t} \\ A(25) &= 400e^{-.032(25)} \quad \textit{Let } t = 25. \\ &= 400e^{-.8} \approx 179.73 \end{aligned}$$

About 180 g of lead will be left in the sample after 25 yr.

52.
$$\begin{aligned} A(t) &= 400e^{-.032t} \\ \tfrac{1}{2}(400) &= 400e^{-.032t} \\ \tfrac{1}{2} &= e^{-.032t} \\ \ln \tfrac{1}{2} &= -.032t \\ t &= \frac{\ln \frac{1}{2}}{-.032} \approx 21.66 \end{aligned}$$

It would take about 21.7 yr for the sample to decay to half of its original amount.

53. $\log_6 13 = \dfrac{\log 13}{\log 6} \approx 1.4315$

54. $\log_7 19 = \dfrac{\log 19}{\log 7} \approx 1.5131$

55. $\log_{\sqrt{2}} \pi = \dfrac{\log \pi}{\log \sqrt{2}} \approx 3.3030$

56. $\log_\pi \sqrt{2} = \dfrac{\ln \sqrt{2}}{\ln \pi} \approx .3028$

57. $\log_{21} .7496 = \dfrac{\log .7496}{\log 21} \approx -.0947$

58. $\log_{19} .8325 = \dfrac{\log .8325}{\log 19} \approx -.0623$

59. $\log_{1/2} 5 = \dfrac{\ln 5}{\ln \frac{1}{2}} \approx -2.3219$

60. $\log_{1/3} 7 = \dfrac{\ln 7}{\ln \frac{1}{3}} \approx -1.7712$

61. $\log_{.3} 12 = \dfrac{\ln 12}{\ln .3} \approx -2.0639$

62. There are 90 of one species and 10 of another, so

$$p_1 = \tfrac{90}{100} = .9 \text{ and } p_2 = \tfrac{10}{100} = .1.$$

Thus, the index of diversity is

$$\begin{aligned} &-(p_1 \ln p_1 + p_2 \ln p_2) \\ &= -(.9 \ln .9 + .1 \ln .1) \\ &\approx .325. \end{aligned}$$

63. There are 60 of one species and 40 of another, so

$$p_1 = \tfrac{60}{100} = .6 \text{ and } p_2 = \tfrac{40}{100} = .4.$$

Thus, the index of diversity is

$$\begin{aligned} &-(p_1 \ln p_1 + p_2 \ln p_2) \\ &= -(.6 \ln .6 + .4 \ln .4) \\ &\approx .673. \end{aligned}$$

Chapter 11 Review Exercises

1. Since a horizontal line intersects the graph in two points, the function is not one-to-one.

2. Since every horizontal line intersects the graph in no more than one point, the function is one-to-one.

3. This function is not one-to-one because two sodas in the list have 41 mg of caffeine.

4. The function $f(x) = -3x + 7$ is a linear function. By the horizontal line test, it is a one-to-one function. To find the inverse, first replace $f(x)$ with y.

$$y = -3x + 7$$

Interchange x and y.

$$x = -3y + 7$$

Solve for y.

$$x - 7 = -3y$$
$$\frac{x-7}{-3} = y \quad \text{or} \quad \frac{7-x}{3} = y$$

Replace y with $f^{-1}(x)$.

$$f^{-1}(x) = \frac{x-7}{-3} \quad \text{or} \quad f^{-1}(x) = \frac{7-x}{3}$$

5. $f(x) = \sqrt[3]{6x-4}$

The graph of f is similar to the graph in Example 3(b) of Section 9.1. The graph passes the horizontal line test, so the function is one-to-one. To find the inverse, first replace $f(x)$ with y.

$$y = \sqrt[3]{6x-4}$$

Interchange x and y.

$$x = \sqrt[3]{6y-4}$$

Solve for y. Cube both sides first.

$$\begin{aligned} x^3 &= 6y - 4 \\ x^3 + 4 &= 6y \\ \frac{x^3+4}{6} &= y \end{aligned}$$

Replace y with $f^{-1}(x)$.

$$\frac{x^3+4}{6} = f^{-1}(x)$$

6. $f(x) = -x^2 + 3$
This is an equation of a vertical parabola which opens down. Since a horizontal line will intersect the graph in two points, the function is not one-to-one.

7. The graph is a linear function through $(0, 1)$ and $(3, 0)$. The graph of $f^{-1}(x)$ will include the points $(1, 0)$ and $(0, 3)$, found by interchanging x and y. Plot these points, and draw a straight line through them.

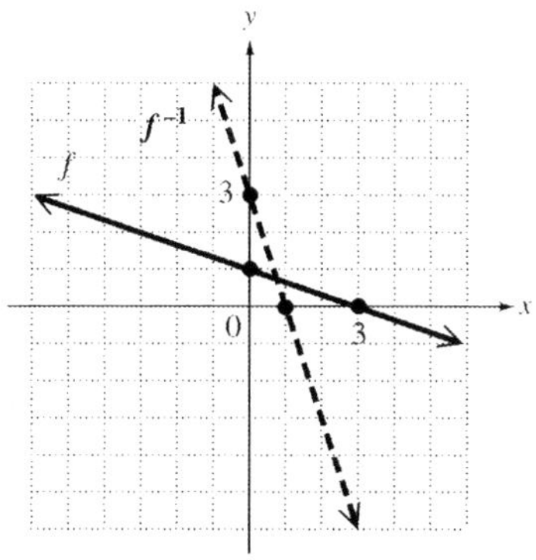

8. The graph is a curve through $(1, 2)$, $(0, 1)$, and $\left(-1, \frac{1}{2}\right)$. Interchange x and y to get $(2, 1)$, $(1, 0)$, and $\left(\frac{1}{2}, -1\right)$, which are on the graph of $f^{-1}(x)$. Plot these points, and draw a smooth curve through them. Use the fact that the graph of f^{-1} is symmetric to the graph of f with respect to the line $y = x$.

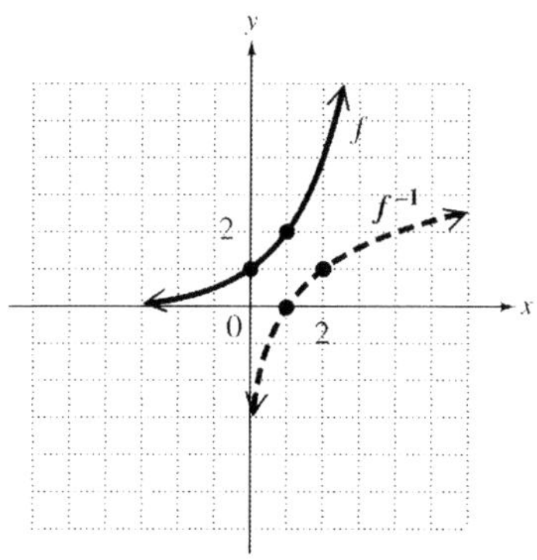

9. $f(x) - 3^x$
Make a table of values.

x	-2	-1	0	1	2
$f(x)$	$\frac{1}{9}$	$\frac{1}{3}$	1	3	9

Plot the points from the table and draw a smooth curve through them.

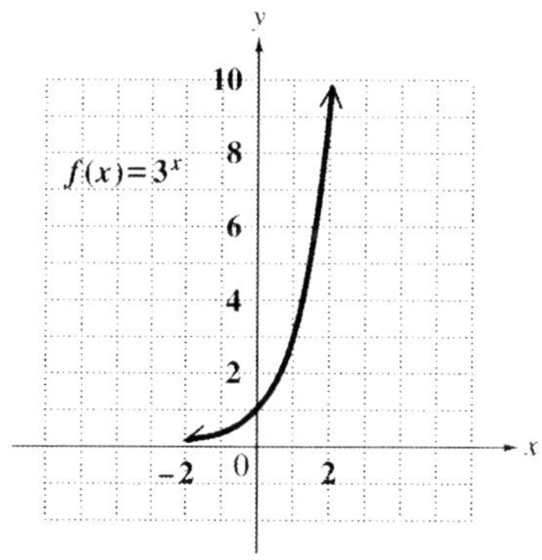

10. $f(x) = \left(\frac{1}{3}\right)^x$
Make a table of values.

x	-2	-1	0	1	2
$f(x)$	9	3	1	$\frac{1}{3}$	$\frac{1}{9}$

Plot the points from the table and draw a smooth curve through them.

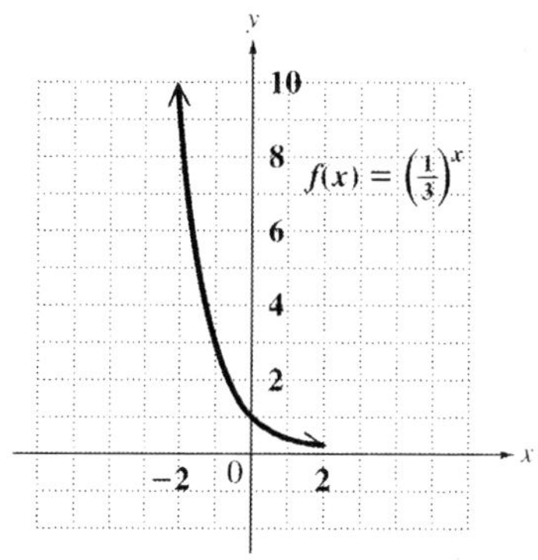

11. $y = 3^{x+1}$
Make a table of values.

x	-3	-2	-1	0	1
y	$\frac{1}{9}$	$\frac{1}{3}$	1	3	9

Plot the points from the table and draw a smooth curve through them.

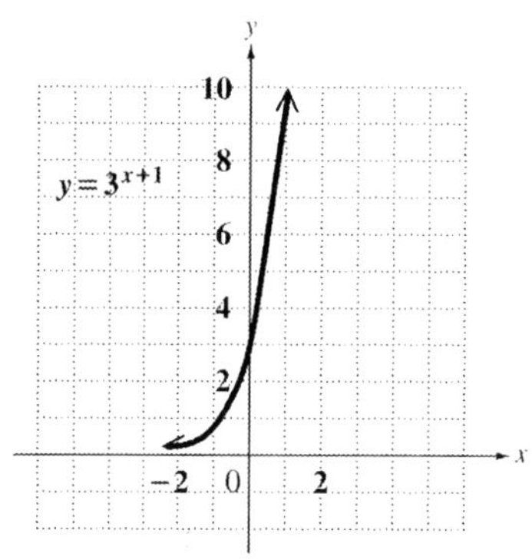

12. $4^{3x} = 8^{x+4}$

Write each side as a power of 2.

$$(2^2)^{3x} = (2^3)^{(x+4)}$$
$$2^{6x} = 2^{(3x+12)}$$
$$6x = 3x + 12 \quad \textit{Equate exponents}$$
$$3x = 12$$
$$x = 4$$

Solution set: $\{4\}$

13. $\left(\frac{1}{27}\right)^{x-1} = 9^{2x}$

$$\left[\left(\tfrac{1}{3}\right)^3\right]^{x-1} = (3^2)^{2x}$$

Write each side as a power of 3.

$$(3^{-3})^{x-1} = (3^2)^{2x}$$
$$3^{(-3x+3)} = 3^{4x}$$
$$-3x + 3 = 4x \quad \textit{Equate exponents}$$
$$3 = 7x$$
$$\tfrac{3}{7} = x$$

Solution set: $\left\{\frac{3}{7}\right\}$

14. $5^x = 1$

Write each side with the same base.

$5^x = 5^0$

Set the exponents equal.

$x = 0$

Solution set: $\{0\}$

15. $f(x) = 26.7(2)^{.0332x}$

(a) 2010: $x = 2010 - 1995 = 15$

$$f(15) = 26.7(2)^{.0332(15)} \approx 37.7$$

The U.S. Hispanic population estimate for 2010 is 37.7 million.

(b) 2015: $x = 2015 - 1995 = 20$

$$f(20) = 26.7(2)^{.0332(20)} \approx 42.3$$

The U.S. Hispanic population estimate for 2015 is 42.3 million.

16. **(a)** The base is 5, the logarithm (exponent) is 4, and the number is 625, so $\log_5 625 = 4$ becomes $5^4 = 625$ in exponential form.

(b) The base is 5, the exponent (logarithm) is -2, and the number is .04, so $5^{-2} = .04$ becomes $\log_5 .04 = -2$ in logarithmic form.

17. **(a)** $\log_b a$ is the exponent to which b must be raised to obtain a.

(b) From part (a),

$$b^{\log_b a} = a.$$

18. $g(x) = \log_3 x$

Replace $g(x)$ with y, and write in exponential form.

$$y = \log_3 x$$
$$3^y = x$$

Make a table of values. Since $x = 3^y$ is the inverse of $f(x) = y = 3^x$ in Exercise 9, simply reverse the ordered pairs in the table of values belonging to $f(x) = 3^x$.

x	$\frac{1}{9}$	$\frac{1}{3}$	1	3	9
y	-2	-1	0	1	2

Plot the points from the table and draw a smooth curve through them.

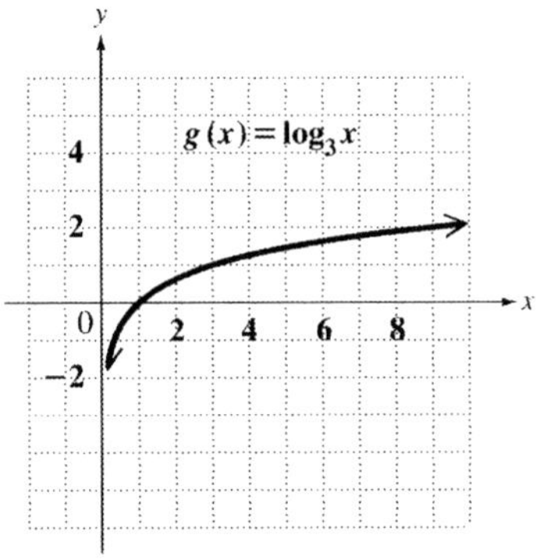

19. $g(x) = \log_{1/3} x$

Replace $g(x)$ with y, and write in exponential form.

$$y = \log_{1/3} x$$
$$\left(\tfrac{1}{3}\right)^y = x$$

Make a table of values. Since $x = \left(\frac{1}{3}\right)^y$ is the inverse of $f(x) = y = \left(\frac{1}{3}\right)^x$ in Exercise 10, simply reverse the ordered pairs in the table of values belonging to $f(x) = \left(\frac{1}{3}\right)^x$.

x	9	3	1	$\frac{1}{3}$	$\frac{1}{9}$
y	-2	-1	0	1	2

Plot the points from the table and draw a smooth curve through them.

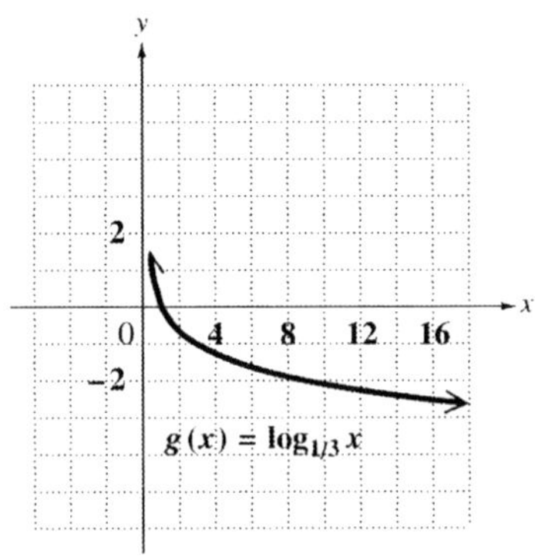

20. $\log_8 64 = x$

$8^x = 64$ *Exponential form*

Write each side as a power of 8.

$8^x = 8^2$

$x = 2$ *Equate exponents*

Solution set: $\{2\}$

21. $\log_7\left(\frac{1}{49}\right) = x$

$$7^x = \tfrac{1}{49}$$
$$7^x = \frac{1}{7^2}$$
$$7^x = 7^{-2}$$
$$x = -2$$

Solution set: $\{-2\}$

22. $\log_4 x = \frac{3}{2}$

$x = 4^{3/2}$ *Exponential form*

$x = \left(\sqrt{4}\right)^3 = 2^3 = 8$

The argument (the input of the logarithm) must be a positive number, so $x = 8$ is acceptable.

Solution set: $\{8\}$

23. $\log_b b^2 = 2$

$b^2 = b^2$ *Exponential form*

This is an identity. Thus, b can be any real number, $b > 0$ and $b \neq 1$.

Solution set: $\{b \mid b > 0,\ b \neq 1\}$

24. $\log_4 3x^2$

$= \log_4 3 + \log_4 x^2$ *Product rule*

$= \log_4 3 + 2\log_4 x$ *Power rule*

25. $\log_2 \dfrac{p^2 r}{\sqrt{z}}$

$= \log_2 \dfrac{p^2 r}{z^{1/2}}$

$= \log_2 (p^2 r) - \log_2 z^{1/2}$ *Quotient rule*

$= \log_2 p^2 + \log_2 r - \log_2 z^{1/2}$ *Product rule*

$= 2\log_2 p + \log_2 r - \frac{1}{2}\log_2 z$ *Power rule*

26. $\log_b 3 + \log_b x - 2\log_b y$

Use the product and power rules for logarithms.

$= \log_b (3 \cdot x) - \log_b y^2$

$= \log_b \dfrac{3x}{y^2}$ *Quotient rule*

27. $\log_3 (x+7) - \log_3 (4x+6)$

$= \log_3 \left(\dfrac{x+7}{4x+6}\right)$ *Quotient rule*

28. $\log 28.9 \approx 1.4609$

29. $\log .257 \approx -.5901$

30. $\ln 28.9 \approx 3.3638$

31. $\ln .257 \approx -1.3587$

32. Milk has a hydronium ion concentration of 4.0×10^{-7}.

$$\text{pH} = -\log\left[H_3O^+\right]$$
$$\text{pH} = -\log\left(4.0 \times 10^{-7}\right) \approx 6.4$$

33. Crackers have a hydronium ion concentration of 3.8×10^{-9}.

$$\text{pH} = -\log\left[H_3O^+\right]$$
$$\text{pH} = -\log\left(3.8 \times 10^{-9}\right) \approx 8.4$$

34. Orange juice has a pH of 4.6.

$$\text{pH} = -\log\left[H_3O^+\right]$$
$$4.6 = -\log\left[H_3O^+\right]$$
$$\log_{10}\left[H_3O^+\right] = -4.6$$
$$\left[H_3O^+\right] = 10^{-4.6} \approx 2.5 \times 10^{-5}$$

35. $t = t(r) = \dfrac{\ln 2}{\ln(1+r)}$

(a) $4\% = .04;\ t(.04) = \dfrac{\ln 2}{\ln(1+.04)} \approx 18$

At 4%, it would take about 18 years.

(b) $6\% = .06;\ t(.06) = \dfrac{\ln 2}{\ln(1+.06)} \approx 12$

At 6%, it would take about 12 years.

(c) $10\% = .10;\ t(.10) = \dfrac{\ln 2}{\ln(1+.10)} \approx 7$

At 10%, it would take about 7 years.

(d) $12\% = .12;\ t(.12) = \dfrac{\ln 2}{\ln(1+.12)} \approx 6$

At 12%, it would take about 6 years.

(e) Each comparison shows approximately the same number. For example, in part (a) the doubling time is 18 yr (rounded) and $\frac{72}{4} = 18$. Thus, the formula $t = \dfrac{72}{100r}$ (called the *rule of 72*) is an excellent approximation of the doubling time formula. (See the Focus on Real Data Application following Section 11.6 exposition.)

36. $g(x) = -650 + 143\ln x$

(a) $g(92) = -650 + 143\ln 92 \approx -3.4$

$g(99) = -650 + 143\ln 99 \approx 7.1$

(b) $g(92)$ agrees closely with the y-value from the graph of about -3; $g(99)$ also is reasonably close to the y-value from the graph of about 7.5.

37.
$$\begin{aligned} 3^x &= 9.42 \\ \log 3^x &= \log 9.42 \\ x \log 3 &= \log 9.42 \\ x &= \frac{\log 9.42}{\log 3} \approx 2.042 \end{aligned}$$

Solution set: $\{2.042\}$

38.
$$\begin{aligned} 2^{x-1} &= 15 \\ \log 2^{x-1} &= \log 15 \\ (x-1)\log 2 &= \log 15 \\ x - 1 &= \frac{\log 15}{\log 2} \\ x &= \frac{\log 15}{\log 2} + 1 \approx 4.907 \end{aligned}$$

Solution set: $\{4.907\}$

39. $e^{.06x} = 3$

Take base e logarithms on both sides.

$$\begin{aligned} \ln e^{.06x} &= \ln 3 \\ .06x \ln e &= \ln 3 \\ .06x &= \ln 3 && \textit{ln e = 1} \\ x &= \frac{\ln 3}{.06} \approx 18.310 \end{aligned}$$

Solution set: $\{18.310\}$

40.
$$\begin{aligned} \log_3 (9x+8) &= 2 \\ 9x + 8 &= 3^2 && \textit{Exponential form} \\ 9x + 8 &= 9 \\ 9x &= 1 \\ x &= \tfrac{1}{9} \end{aligned}$$

$x = \frac{1}{9}$ is acceptable since $9x + 8 = 9 > 0$.

Solution set: $\{\frac{1}{9}\}$

41. $\log_5 (x+6)^3 = 2$

Change to exponential form.

$$\begin{aligned} (x+6)^3 &= 5^2 \\ (x+6)^3 &= 25 \end{aligned}$$

Take the cube root of each side.

$$\begin{aligned} x + 6 &= \sqrt[3]{25} \\ x &= \sqrt[3]{25} - 6 \end{aligned}$$

$x = -6 + \sqrt[3]{25}$ is acceptable since $(x+6)^3 = 25 > 0$.

Solution set: $\left\{-6 + \sqrt[3]{25}\right\}$

42.
$$\begin{aligned} \log_3 (p+2) - \log_3 p &= \log_3 2 \\ \log_3 \frac{p+2}{p} &= \log_3 2 && \textit{Quotient rule} \\ \frac{p+2}{p} &= 2 && \textit{Property 4} \\ p + 2 &= 2p \\ 2 &= p \end{aligned}$$

$p = 2$ is acceptable since $p + 2 = 4 > 0$ and $p = 2 > 0$.

Solution set: $\{2\}$

43.
$$\begin{aligned} \log (2x+3) - \log x &= 1 \\ \log_{10} (2x+3) - \log_{10} x &= 1 \\ \log_{10} \frac{2x+3}{x} &= 1 && \textit{Quotient rule} \\ 10^1 &= \frac{2x+3}{x} && \textit{Exponential form} \\ 10x &= 2x + 3 \\ 8x &= 3 \\ x &= \tfrac{3}{8} \end{aligned}$$

$x = \frac{3}{8}$ is acceptable since $2x + 3 = \frac{15}{4} > 0$ and $x = \frac{3}{8} > 0$.

Solution set: $\{\frac{3}{8}\}$

44.
$$\begin{aligned} \log_4 x + \log_4 (8-x) &= 2 \\ \log_4 [x(8-x)] &= 2 && \textit{Product rule} \\ x(8-x) &= 4^2 && \textit{Exponential form} \\ 8x - x^2 &= 16 \\ x^2 - 8x + 16 &= 0 \\ (x-4)(x-4) &= 0 \\ x - 4 &= 0 \\ x &= 4 \end{aligned}$$

$x = 4$ is acceptable since $x = 4 > 0$ and $8 - x = 4 > 0$.

Solution set: $\{4\}$

45.
$$\begin{aligned} \log_2 x + \log_2 (x+15) &= 4 \\ \log_2 [x(x+15)] &= 4 && \textit{Product rule} \\ x(x+15) &= 2^4 && \textit{Exponential form} \\ x^2 + 15x &= 16 \\ x^2 + 15x - 16 &= 0 \\ (x+16)(x-1) &= 0 \end{aligned}$$

$$\begin{aligned} x + 16 &= 0 \quad \text{or} \quad x - 1 = 0 \\ x &= -16 \quad \text{or} \quad x = 1 \end{aligned}$$

Reject $x = -16$, because it yields an equation in which the logarithm of a negative number must be found.

Solution set: $\{1\}$

46. Use $A = P\left(1 + \frac{r}{n}\right)^{nt}$ with $P = 6500$, $r = 3\% = .03$, $n = 365$ (compounded daily), and $t = 3$.

$$A = 6500\left(1 + \frac{.03}{365}\right)^{365(3)} \approx 7112.11$$

In 3 yr, the value would be about \$7112.11.

47. Use $A = P\left(1 + \frac{r}{n}\right)^{nt}$.

Plan A:
Let $P = 1000$, $r = .04$, $n = 4$, and $t = 3$.

$$A = 1000\left(1 + \frac{.04}{4}\right)^{4 \cdot 3} \approx 1126.83$$

Plan B:
Let $P = 1000$, $r = .039$, $n = 12$, and $t = 3$.

$$A = 1000\left(1 + \frac{.039}{12}\right)^{12 \cdot 3} \approx 1123.91$$

Plan A is the better plan by \$2.92.

48. $S = C(1 - r)^n$

Let $C = 30{,}000$, $r = .15$, and $n = 12$.

$$\begin{aligned} S &= 30{,}000(1 - .15)^{12} \\ &= 30{,}000(.85)^{12} \approx 4267 \end{aligned}$$

The scrap value is about \$4267.

49. $S = C(1 - r)^n$

Let $S = \frac{1}{2}C$ and $n = 6$.

$$\begin{aligned} S &= C(1 - r)^n \\ \tfrac{1}{2}C &= C(1 - r)^6 \\ .5 &= (1 - r)^6 \qquad (*) \\ \ln .5 &= \ln (1 - r)^6 \\ \ln .5 &= 6 \ln (1 - r) \\ \ln (1 - r) &= \frac{\ln .5}{6} \\ \ln (1 - r) &\approx -.1155 \\ 1 - r &= e^{-.1155} \\ 1 - r &\approx .89 \\ r &= .11 \end{aligned}$$

The rate is approximately 11%.

$(*)$ Alternative solution steps without logarithms:

$$\begin{aligned} .5 &= (1 - r)^6 \\ \sqrt[6]{.5} &= 1 - r \\ r &= 1 - \sqrt[6]{.5} \approx .11 \end{aligned}$$

Note that $1 - r$ must be positive, so $\pm\sqrt[6]{.5}$ is not needed.

In Exercises 50–52, use the change-of-base rule,

$$\log_a x = \frac{\log_b x}{\log_b a}.$$

50. $\log_{16} 13 = \frac{\log 13}{\log 16} \approx .9251$

51. $\log_4 12 = \frac{\log 12}{\log 4} \approx 1.7925$

52. $\log_{\sqrt{6}} \sqrt{13} = \frac{\ln \sqrt{13}}{\ln \sqrt{6}} \approx 1.4315$

53. **[11.6]** $\log_3 (x + 9) = 4$

$$\begin{aligned} x + 9 &= 3^4 && \textit{Exponential form} \\ x + 9 &= 81 \\ x &= 72 \end{aligned}$$

$x = 72$ is acceptable since $x + 9 = 81 > 0$.

Solution set: $\{72\}$

54. **[11.3]** $\log_2 32 = x$

$$\begin{aligned} 2^x &= 32 && \textit{Exponential form} \\ 2^x &= 2^5 && \textit{Same base} \\ x &= 5 && \textit{Set the exponents equal.} \end{aligned}$$

Solution set: $\{5\}$

55. **[11.3]** $\log_x \frac{1}{81} = 2$

$$\begin{aligned} x^2 &= \tfrac{1}{81} && \textit{Exponential form} \\ x &= \pm\sqrt{\tfrac{1}{81}} && \textit{Square root property} \\ x &= \tfrac{1}{9} && \textit{Reject } -\tfrac{1}{9}. \end{aligned}$$

$x = \frac{1}{9}$ is an acceptable base since it is a positive number (not equal to 1).

Solution set: $\left\{\frac{1}{9}\right\}$

56. **[11.2]** $27^x = 81$

Write each side as a power of 3.

$$\begin{aligned} (3^3)^x &= 3^4 \\ 3^{3x} &= 3^4 \\ 3x &= 4 && \textit{Equate exponents} \\ x &= \tfrac{4}{3} \end{aligned}$$

Solution set: $\left\{\frac{4}{3}\right\}$

57. **[11.2]** $2^{2x-3} = 8$

Write each side as a power of 2.

$$\begin{aligned} 2^{2x-3} &= 2^3 \\ 2x - 3 &= 3 && \textit{Equate exponents} \\ 2x &= 6 \\ x &= 3 \end{aligned}$$

Solution set: $\{3\}$

58. **[11.6]**

$$\log_3(x+1) - \log_3 x = 2$$
$$\log_3 \frac{x+1}{x} = 2 \quad \textit{Quotient rule}$$
$$\frac{x+1}{x} = 3^2 \quad \textit{Exponential form}$$
$$9x = x + 1$$
$$8x = 1$$
$$x = \tfrac{1}{8}$$

$x = \frac{1}{8}$ is acceptable since $x + 1 = \frac{9}{8} > 0$ and $x = \frac{1}{8} > 0$.

Solution set: $\{\frac{1}{8}\}$

59. **[11.6]** $\log(3x-1) = \log 10$
$$3x - 1 = 10$$
$$3x = 11$$
$$x = \tfrac{11}{3}$$

$x = \frac{11}{3}$ is acceptable since $3x - 1 = 10 > 0$.

Solution set: $\{\frac{11}{3}\}$

60. **[11.6]** $S = C(1-r)^n$

Let $S = 10{,}000$, $C = 30{,}000$, and $r = 15\% = .15$.
$$10{,}000 = 30{,}000(1-.15)^n$$
$$\tfrac{1}{3} = (.85)^n$$
$$\log \tfrac{1}{3} = \log(.85)^n$$
$$\log \tfrac{1}{3} = n \log(.85)$$
$$n = \frac{\log \frac{1}{3}}{\log(.85)} \approx 6.8$$

The useful life is 6.8 years.

Chapter 11 Test

1. **(a)** $f(x) = x^2 + 9$

This function is not one-to-one. The graph of $f(x)$ is a vertical parabola. A horizontal line will intersect the graph more than once.

(b) This function is one-to-one. A horizontal line will not intersect the graph in more than one point.

2. $f(x) = \sqrt[3]{x+7}$

Replace $f(x)$ with y.
$$y = \sqrt[3]{x+7}$$
Interchange x and y.
$$x = \sqrt[3]{y+7}$$
Solve for y.
$$x^3 = y + 7$$
$$x^3 - 7 = y$$
Replace y with $f^{-1}(x)$.
$$f^{-1}(x) = x^3 - 7$$

3. By the horizontal line test, $f(x)$ is a one-to-one function and has an inverse. Choose some points on the graph of $f(x)$, such as $(4, 0)$, $(3, -1)$, and $(0, -2)$. To graph the inverse, interchange the x- and y-values to get $(0, 4)$, $(-1, 3)$, and $(-2, 0)$. Plot these points and draw a smooth curve through them.

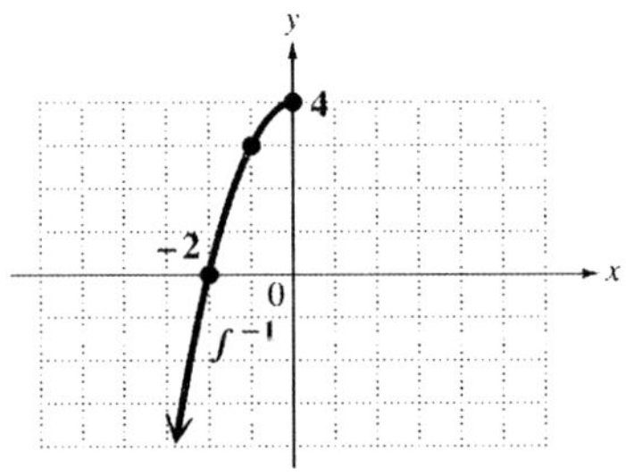

4. $y = f(x) = 6^x$

Make a table of values.

x	-2	-1	0	1
$f(x)$	$\frac{1}{36}$	$\frac{1}{6}$	1	6

Plot these points and draw a smooth exponential curve through them.

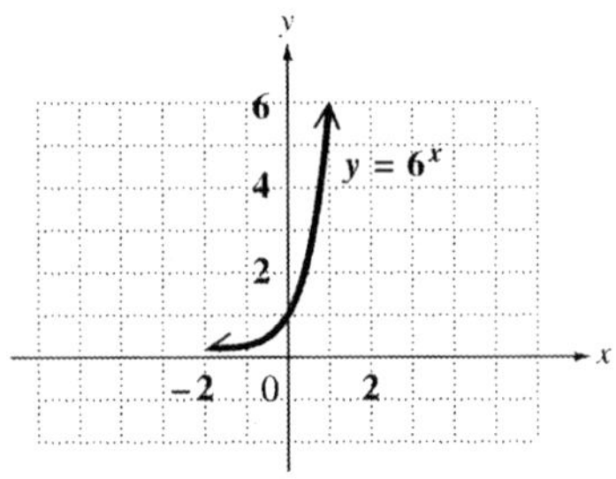

5. $y = g(x) = \log_6 x$

Make a table of values.

Powers of 6	6^{-2}	6^{-1}	6^0	6^1
x	$\frac{1}{36}$	$\frac{1}{6}$	1	6
$g(x)$	-2	-1	0	1

Plot these points and draw a smooth logarithmic curve through them.

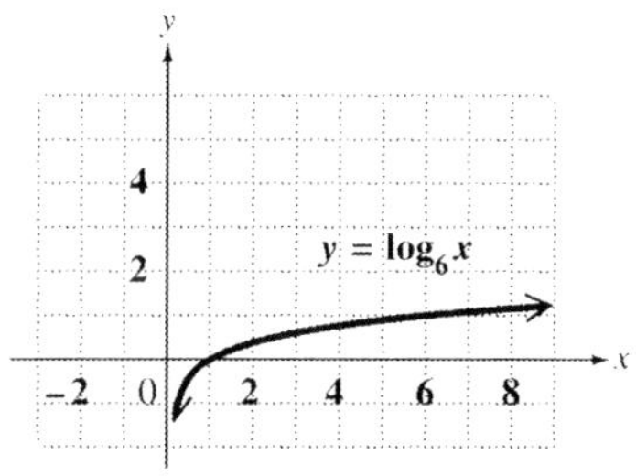

6. $y = 6^x$ and $y = \log_6 x$ are inverse functions. To use the graph from Exercise 4 to obtain the graph of the function in Exercise 5, interchange the x- and y-coordinates of the ordered pairs $\left(-2, \frac{1}{36}\right)$, $\left(-1, \frac{1}{6}\right)$, $(0, 1)$, and $(1, 6)$ to get $\left(\frac{1}{36}, -2\right)$, $\left(\frac{1}{6}, -1\right)$, $(1, 0)$, and $(6, 1)$. Plot these points and draw a smooth logarithmic curve through them.

7. $5^x = \dfrac{1}{625}$

$5^x = \left(\frac{1}{5}\right)^4$

Write each side as a power of 5.

$5^x = 5^{-4}$

$x = -4$ *Equate exponents*

Solution set: $\{-4\}$

8. $2^{3x-7} = 8^{2x+2}$

Write each side as a power of 2.

$2^{3x-7} = 2^{3(2x+2)}$

$3x - 7 = 3(2x + 2)$ *Equate exponents*

$3x - 7 = 6x + 6$

$-13 = 3x$

$-\frac{13}{3} = x$

Solution set: $\left\{-\frac{13}{3}\right\}$

9. $f(x) = 1013e^{-.0001341x}$

(a) $f(2000) = 1013e^{-.0001341(2000)}$
≈ 775

The atmospheric pressure at 2000 m is about 775 millibars.

(b) $f(10{,}000) = 1013e^{-.0001341(10{,}000)}$
≈ 265

The atmospheric pressure at 10,000 m is about 265 millibars.

10. The base is 4, the exponent (logarithm) is -2, and the number is .0625, so $4^{-2} = .0625$ becomes $\log_4 .0625 = -2$ in logarithmic form.

11. The base is 7, the logarithm (exponent) is 2, and the number is 49, so $\log_7 49 = 2$ becomes $7^2 = 49$ in exponential form.

12. $\log_{1/2} x = -5$

$x = \left(\frac{1}{2}\right)^{-5}$ *Exponential form*

$x = \left(\frac{2}{1}\right)^5 = 32$

$x = 32$ is acceptable since it is positive.

Solution set: $\{32\}$

13. $x = \log_9 3$

$9^x = 3$ *Exponential form*

Write each side as a power of 3.

$(3^2)^x = 3$

$3^{2x} = 3^1$

$2x = 1$ *Equate exponents*

$x = \frac{1}{2}$

Solution set: $\left\{\frac{1}{2}\right\}$

14. $\log_x 16 = 4$

$x^4 = 16$ *Exponential form*

$x^2 = \pm 4$ *Square root property*

Reject -4 since $x^2 \geq 0$.

$x^2 = 4$

$x = \pm 2$ *Square root property*

Reject -2 since the base cannot be negative.

Solution set: $\{2\}$

15. $\log_3 x^2 y$

$= \log_3 x^2 + \log_3 y$ *Product rule*

$= 2\log_3 x + \log_3 y$ *Power rule*

16. $\frac{1}{4}\log_b r + 2\log_b s - \frac{2}{3}\log_b t$

Use the power rule for logarithms.

$= \log_b r^{1/4} + \log_b s^2 - \log_b t^{2/3}$

Use the product and quotient rules for logarithms.

$= \log_b \dfrac{r^{1/4}s^2}{t^{2/3}}$

17. **(a)** $\log 21.3 \approx 1.3284$

(b) $\ln .43 \approx -.8440$

(c) $\log_6 45 = \dfrac{\log 45}{\log 6} \approx 2.1245$

18. $3^x = 78$

$\ln 3^x = \ln 78$

$x \ln 3 = \ln 78$ *Power rule*

$x = \dfrac{\ln 78}{\ln 3} \approx 3.9656$

Solution set: $\{3.9656\}$

19. $\log_8(x+5) + \log_8(x-2) = \log_8 8$

$\log_8(x+5) + \log_8(x-2) = 1$

Use the product rule for logarithms.

$$\log_8[(x+5)(x-2)] = 1$$
$$(x+5)(x-2) = 8^1 \quad \textit{Exp. form}$$
$$x^2+3x-10 = 8$$
$$x^2+3x-18 = 0$$
$$(x+6)(x-3) = 0$$

$$x+6=0 \quad \text{or} \quad x-3=0$$
$$x=-6 \quad \text{or} \quad x=3$$

Reject $x=-6$, because $x+5=-1$, which yields an equation in which the logarithm of a negative number must be found.

Solution set: $\{3\}$

20. **(a)** $A = P\left(1+\frac{r}{n}\right)^{nt}$

$$A = 10{,}000\left(1+\frac{.045}{4}\right)^{4\cdot 5} \approx 12{,}507.51$$

\$10,000 invested at 4.5% annual interest, compounded quarterly, will increase to \$12,507.51 in 5 years.

(b) $2(10{,}000) = 10{,}000\left(1+\frac{.045}{4}\right)^{4\cdot t}$

$$2 = (1.01125)^{4t}$$
$$\log 2 = \log(1.01125)^{4t}$$
$$\log 2 = 4t\log(1.01125)$$
$$t = \frac{\log 2}{4\log(1.01125)}$$
$$t \approx 15.5$$

It will take about 15.5 years for the initial principal to double.

Cumulative Review Exercises (Chapters 1–11)

For Exercises 1–3,

$S = \left\{-\frac{9}{4}, -2, -\sqrt{2}, 0, .6, \sqrt{11}, \sqrt{-8}, 6, \frac{30}{3}\right\}$.

1. The integers are -2, 0, 6, and $\frac{30}{3}$ (or 10).

2. The rational numbers are $-\frac{9}{4}$, -2, 0, $.6$, 6, and $\frac{30}{3}$ (or 10). Each can be expressed as a quotient of two integers.

3. The irrational numbers are $-\sqrt{2}$ and $\sqrt{11}$.

4. $|-8|+6-|-2|-(-6+2)$
$= 8+6-2-(-4)$
$= 14-2+4 = 16$

5. $2(-5)+(-8)(4)-(-3)$
$= -10-32+3 = -39$

6. $7-(3+4x)+2x = -5(x-1)-3$

$$7-3-4x+2x = -5x+5-3$$
$$4-2x = -5x+2$$
$$3x = -2$$
$$x = -\tfrac{2}{3}$$

Solution set: $\left\{-\frac{2}{3}\right\}$

7. $2x+2 \le 5x-1$

$-3x \le -3$

Divide by -3; reverse the inequality.

$x \ge 1$

Solution set: $[1, \infty)$

8. $|2x-5| = 9$

$$2x-5=9 \quad \text{or} \quad 2x-5=-9$$
$$2x=14 \quad \text{or} \quad 2x=-4$$
$$x=7 \quad \text{or} \quad x=-2$$

Solution set: $\{-2, 7\}$

9. $|3x|-4 = 12$

$|3x| = 16$

$$3x=16 \quad \text{or} \quad 3x=-16$$
$$x=\tfrac{16}{3} \quad \text{or} \quad x=-\tfrac{16}{3}$$

Solution set: $\left\{\pm\frac{16}{3}\right\}$

10. $|3x-8| \le 1$

$$-1 \le 3x-8 \le 1$$
$$7 \le 3x \le 9$$
$$\tfrac{7}{3} \le x \le 3$$

Solution set: $\left[\frac{7}{3}, 3\right]$

11. $|4x+2| > 10$

$$4x+2>10 \quad \text{or} \quad 4x+2<-10$$
$$4x>8 \quad \text{or} \quad 4x<-12$$
$$x>2 \quad \text{or} \quad x<-3$$

Solution set: $(-\infty, -3) \cup (2, \infty)$

12. $y = -2.5x+5$

Since the equation is in slope-intercept form, the slope is -2.5 and the y-intercept is $(0, 5)$. Plot $(0, 5)$. To find another point on the line, move down 2.5 units and to the right 1 unit (or, equivalently, move down 5 units and to the right 2 units). Draw a line through the two points.

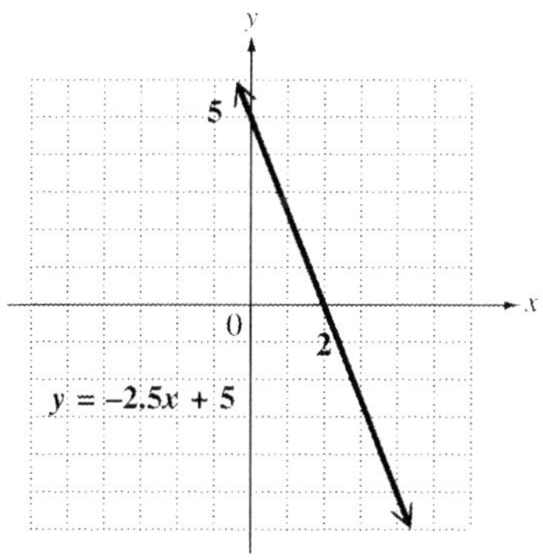

13. $-4x + y \leq 5$

Graph the line $-4x + y = 5$, which has intercepts $(0, 5)$ and $\left(-\frac{5}{4}, 0\right)$, as a solid line because the inequality involves $\leq$. Test $(0, 0)$, which yields $0 \leq 5$, a true statement. Shade the region that includes $(0, 0)$.

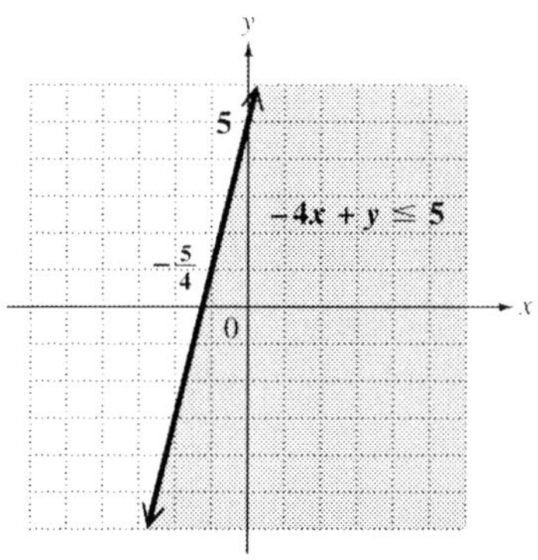

14. **(a)** Yes, this graph is the graph of a function because it passes the vertical line test.

(b) $(x_1, y_1) = (1990, 44{,}623)$ and $(x_2, y_2) = (2000, 60{,}816)$.

$$m = \frac{y_2 - y_1}{x_2 - x_1} = \frac{60{,}816 - 44{,}623}{2000 - 1990} = \frac{16{,}193}{10} = 1619.3$$

The slope of the line in the graph is 1619.3 and can be interpreted as follows: The number of travelers increased by an average of 1619.3 thousand per year during the period 1990-2000.

15. Through $(5, -1)$; parallel to $3x - 4y = 12$
Find the slope of the line with equation

$$\begin{aligned} 3x - 4y &= 12 \\ -4y &= -3x + 12 \\ y &= \tfrac{3}{4}x - 3. \end{aligned}$$

The slope is $\frac{3}{4}$, so a line parallel to it also has slope $\frac{3}{4}$.

Let $m = \frac{3}{4}$ and $(x_1, y_1) = (5, -1)$ in the point-slope form.

$$\begin{aligned} y - y_1 &= m(x - x_1) \\ y - (-1) &= \tfrac{3}{4}(x - 5) \\ y + 1 &= \tfrac{3}{4}x - \tfrac{15}{4} \\ y &= \tfrac{3}{4}x - \tfrac{19}{4} \end{aligned}$$

16. $5x - 3y = 14 \quad (1)$
$2x + 5y = 18 \quad (2)$

Multiply equation (1) by 5 and equation (2) by 3. Then add the results.

$$\begin{array}{rcrcll} 25x & - & 15y & = & 70 & 5 \times (1) \\ 6x & + & 15y & = & 54 & 3 \times (2) \\ \hline 31x & & & = & 124 & \\ x & & & = & 4 & \end{array}$$

Substitute 4 for x in equation (1) to find y.

$$\begin{aligned} 5x - 3y &= 14 \quad (1) \\ 5(4) - 3y &= 14 \\ 20 - 3y &= 14 \\ -3y &= -6 \\ y &= 2 \end{aligned}$$

Solution set: $\{(4, 2)\}$

17. $x + 2y + 3z = 11 \quad (1)$
$3x - y + z = 8 \quad (2)$
$2x + 2y - 3z = -12 \quad (3)$

To eliminate z, add equations (1) and (3).

$$\begin{array}{rcrcrcrl} x & + & 2y & + & 3z & = & 11 & (1) \\ 2x & + & 2y & - & 3z & = & -12 & (3) \\ \hline 3x & + & 4y & & & = & -1 & (4) \end{array}$$

To eliminate z again, multiply equation (2) by 3 and add the result to equation (3).

$$\begin{array}{rcrcrcrl} 9x & - & 3y & + & 3z & = & 24 & 3 \times (2) \\ 2x & + & 2y & - & 3z & = & -12 & (3) \\ \hline 11x & - & y & & & = & 12 & (5) \end{array}$$

Multiply equation (5) by 4 and add the result to equation (4).

$$\begin{array}{rcrcrl} 44x & - & 4y & = & 48 & 4 \times (5) \\ 3x & + & 4y & = & -1 & (4) \\ \hline 47x & & & = & 47 & \\ x & & & = & 1 & \end{array}$$

Substitute 1 for x in equation (5) to find y.

continued

$$\begin{aligned} 11x - y &= 12 \quad (5) \\ 11(1) - y &= 12 \\ 11 - y &= 12 \\ -y &= 1 \\ y &= -1 \end{aligned}$$

Substitute 1 for x and -1 for y in equation (2) to find z.

$$\begin{aligned} 3x - y + z &= 8 \quad (2) \\ 3(1) - (-1) + z &= 8 \\ 3 + 1 + z &= 8 \\ 4 + z &= 8 \\ z &= 4 \end{aligned}$$

Solution set: $\{(1, -1, 4)\}$

18. Let x = the number of pounds of \$1.00/lb candy and y = the number of pounds of \$1.96/lb candy.

Price per Pound	Number of Pounds	Value
\$1.00	x	$1x$
\$1.96	y	$1.96y$
\$1.60	16	$1.6(16) = 25.6$

We obtain a system of equations from the last two columns in the table.

$$\begin{aligned} x + y &= 16 \quad (1) \\ x + 1.96y &= 25.6 \quad (2) \end{aligned}$$

Solve equation (1) for y: $y = 16 - x$ (3)

Substitute $16 - x$ for y in equation (2).

$$\begin{aligned} x + 1.96(16 - x) &= 25.6 \\ x + 31.36 - 1.96x &= 25.6 \\ -.96x &= -5.76 \\ x &= 6 \end{aligned}$$

From (3) with $x = 6$, $y = 16 - 6 = 10$.

6 pounds of \$1.00-per-lb candy should be mixed with 10 lb of \$1.96-per-lb candy.

19. $$\begin{aligned} (2p + 3)(3p - 1) &= 6p^2 - 2p + 9p - 3 \\ &= 6p^2 + 7p - 3 \end{aligned}$$

20. $$\begin{aligned} (4k - 3)^2 &= (4k)^2 - 2(4k)(3) + 3^2 \\ &= 16k^2 - 24k + 9 \end{aligned}$$

21. $$\begin{aligned} &(3m^3 + 2m^2 - 5m) - (8m^3 + 2m - 4) \\ &= 3m^3 + 2m^2 - 5m - 8m^3 - 2m + 4 \\ &= 3m^3 - 8m^3 + 2m^2 - 5m - 2m + 4 \\ &= -5m^3 + 2m^2 - 7m + 4 \end{aligned}$$

22.

$$\begin{array}{r|rrrrr} & 2t^3 & +\,5t^2 & -\,3t & +\,4 & \\ \hline 3t + 1 & 6t^4 + 17t^3 & -\,4t^2 & +\,9t & +\,4 & \\ & 6t^4 + 2t^3 & & & & \\ \hline & 15t^3 & -\,4t^2 & & & \\ & 15t^3 & +\,5t^2 & & & \\ \hline & & -9t^2 & +\,9t & & \\ & & -9t^2 & -\,3t & & \\ \hline & & & 12t & +\,4 & \\ & & & 12t & +\,4 & \\ \hline & & & & 0 & \end{array}$$

The quotient is

$$2t^3 + 5t^2 - 3t + 4.$$

23. $8x + x^3 = x(8 + x^2)$

24. $24y^2 - 7y - 6 = (8y + 3)(3y - 2)$

25. $$\begin{aligned} 5z^3 - 19z^2 - 4z &= z(5z^2 - 19z - 4) \\ &= z(5z + 1)(z - 4) \end{aligned}$$

26. $16a^2 - 25b^4$

Use the difference of squares formula,

$$x^2 - y^2 = (x + y)(x - y),$$

where $x = 4a$ and $y = 5b^2$.

$$16a^2 - 25b^4 = (4a + 5b^2)(4a - 5b^2)$$

27. $8c^3 + d^3$

Use the sum of cubes formula,

$$x^3 + y^3 = (x + y)(x^2 - xy + y^2),$$

where $x = 2c$ and $y = d$.

$$8c^3 + d^3 = (2c + d)(4c^2 - 2cd + d^2)$$

28. $$\begin{aligned} &16r^2 + 56rq + 49q^2 \\ &= (4r)^2 + 2(4r)(7q) + (7q)^2 \end{aligned}$$

Use the perfect square formula,

$$x^2 + 2xy + y^2 = (x + y)^2,$$

where $x = 4r$ and $y = 7q$.

$$16r^2 + 56rq + 49q^2 = (4r + 7q)^2$$

29. $$\begin{aligned} \frac{(5p^3)^4(-3p^7)}{2p^2(4p^4)} &= \frac{(5^4p^{12})(-3p^7)}{8p^6} \\ &= \frac{(625)(-3)p^{19}}{8p^6} \\ &= -\frac{1875p^{13}}{8} \end{aligned}$$

30. $\dfrac{x^2-9}{x^2+7x+12} \div \dfrac{x-3}{x+5}$

Multiply by the reciprocal.

$$= \frac{x^2-9}{x^2+7x+12} \cdot \frac{x+5}{x-3}$$

$$= \frac{(x+3)(x-3)}{(x+3)(x+4)} \cdot \frac{(x+5)}{(x-3)} \quad \textit{Factor.}$$

$$= \frac{x+5}{x+4}$$

31. $\dfrac{2}{k+3} - \dfrac{5}{k-2}$

The LCD is $(k+3)(k-2)$.

$$= \frac{2(k-2)}{(k+3)(k-2)} - \frac{5(k+3)}{(k-2)(k+3)}$$

$$= \frac{2k-4-5k-15}{(k+3)(k-2)}$$

$$= \frac{-3k-19}{(k+3)(k-2)}$$

32. $\dfrac{3}{p^2-4p} - \dfrac{4}{p^2+2p}$

$$= \frac{3}{p(p-4)} - \frac{4}{p(p+2)}$$

The LCD is $p(p-4)(p+2)$.

$$= \frac{3(p+2)}{p(p-4)(p+2)} - \frac{4(p-4)}{p(p+2)(p-4)}$$

$$= \frac{3p+6-4p+16}{p(p-4)(p+2)}$$

$$= \frac{22-p}{p(p-4)(p+2)}$$

33. $\dfrac{1}{x} - \dfrac{3}{2x} = \dfrac{1}{x+1}$

Multiply by the LCD, $2x(x+1)$. $(x \neq 0, -1)$

$$2x(x+1)\left(\frac{1}{x} - \frac{3}{2x}\right) = 2x(x+1)\left(\frac{1}{x+1}\right)$$

$$2(x+1) - 3(x+1) = 2x$$

$$2x+2-3x-3 = 2x$$

$$-x-1 = 2x$$

$$-1 = 3x$$

$$-\tfrac{1}{3} = x$$

Solution set: $\left\{-\frac{1}{3}\right\}$

34. $\sqrt{288} = \sqrt{144 \cdot 2} = \sqrt{144}\sqrt{2} = 12\sqrt{2}$

35. $\dfrac{-8^{4/3}}{8^2} = -8^{4/3-2} = -8^{4/3-6/3}$

$$= -8^{-2/3} = -\frac{1}{8^{2/3}}$$

$$= -\frac{1}{\left(\sqrt[3]{8}\right)^2} = -\frac{1}{2^2} = -\frac{1}{4}$$

36. $2\sqrt{32} - 5\sqrt{98} = 2\sqrt{16 \cdot 2} - 5\sqrt{49 \cdot 2}$

$$= 2 \cdot 4\sqrt{2} - 5 \cdot 7\sqrt{2}$$

$$= 8\sqrt{2} - 35\sqrt{2}$$

$$= -27\sqrt{2}$$

37. $\sqrt{2x+1} - \sqrt{x} = 1$

$$\sqrt{2x+1} = 1 + \sqrt{x}$$

$$\left(\sqrt{2x+1}\right)^2 = \left(1+\sqrt{x}\right)^2$$

$$2x+1 = 1 + 2\sqrt{x} + x$$

$$x = 2\sqrt{x}$$

$$(x)^2 = \left(2\sqrt{x}\right)^2$$

$$x^2 = 4x$$

$$x^2 - 4x = 0$$

$$x(x-4) = 0$$

$$x = 0 \quad \text{or} \quad x = 4$$

Check $x = 0$: $\sqrt{1} - \sqrt{0} = 1$ *True*

Check $x = 4$: $\sqrt{9} - \sqrt{4} = 1$ *True*

Solution set: $\{0, 4\}$

38. $(5+4i)(5-4i) = 5^2 - (4i)^2$

$$= 25 - 16i^2$$

$$= 25 - 16(-1)$$

$$= 25 + 16 = 41$$

39. $3x^2 = x + 1$

$$3x^2 - x - 1 = 0$$

Use $a = 3$, $b = -1$, and $c = -1$ in the quadratic formula.

$$x = \frac{-b \pm \sqrt{b^2-4ac}}{2a}$$

$$x = \frac{-(-1) \pm \sqrt{(-1)^2 - 4(3)(-1)}}{2(3)}$$

$$= \frac{1 \pm \sqrt{1+12}}{6} = \frac{1 \pm \sqrt{13}}{6}$$

Solution set: $\left\{\dfrac{1+\sqrt{13}}{6}, \dfrac{1-\sqrt{13}}{6}\right\}$

40. $x^2 + 2x - 8 > 0$

Solve the equation

$$x^2 + 2x - 8 = 0.$$

$$(x+4)(x-2) = 0$$

$$x + 4 = 0 \quad \text{or} \quad x - 2 = 0$$

$$x = -4 \quad \text{or} \quad x = 2$$

The numbers -4 and 2 divide a number line into three intervals.

continued

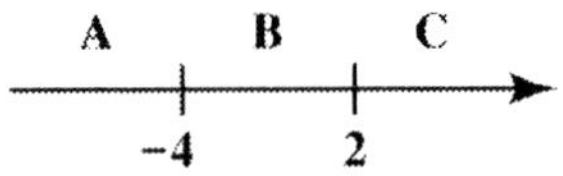

Test a number from each interval in the inequality

$$x^2 + 2x - 8 > 0.$$

Interval A: Let $x = -5$.

$$25 - 10 - 8 > 0 \quad ?$$
$$7 > 0 \quad \textit{True}$$

Interval B: Let $x = 0$.

$$-8 > 0 \quad \textit{False}$$

Interval C: Let $x = 3$.

$$9 + 6 - 8 > 0 \quad ?$$
$$7 > 0 \quad \textit{True}$$

The numbers in Intervals A and C, not including -4 or 2 because of $>$, are solutions.

Solution set: $(-\infty, -4) \cup (2, \infty)$

41. $$x^4 - 5x^2 + 4 = 0$$

Let $u = x^2$, so $u^2 = (x^2)^2 = x^4$.

$$u^2 - 5u + 4 = 0$$
$$(u-1)(u-4) = 0$$

$u - 1 = 0$ or $u - 4 = 0$
$u = 1$ or $u = 4$

To find x, substitute x^2 for u.

$x^2 = 1$ or $x^2 = 4$
$x = \pm 1$ or $x = \pm 2$

Solution set: $\{\pm 1, \pm 2\}$

42. $y = f(x) = \frac{1}{3}(x-1)^2 + 2$ is in $f(x) = a(x-h)^2 + k$ form. The graph is a vertical parabola with vertex (h, k) at $(1, 2)$. Since $a = \frac{1}{3} > 0$, the graph opens up.
Also, $|a| = \left|\frac{1}{3}\right| = \frac{1}{3} < 1$, so the graph is wider than the graph of $f(x) = x^2$. The points $\left(0, 2\frac{1}{3}\right)$, $(-2, 5)$, and $(4, 5)$ are also on the graph.

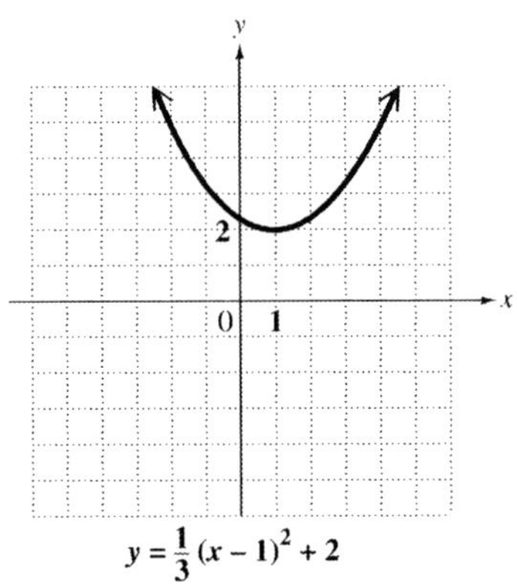

43. $f(x) = 2^x$
Make a table of values.

x	-2	-1	0	1	2
$f(x)$	$\frac{1}{4}$	$\frac{1}{2}$	1	2	4

Plot the ordered pairs from the table, and draw a smooth exponential curve through the points.

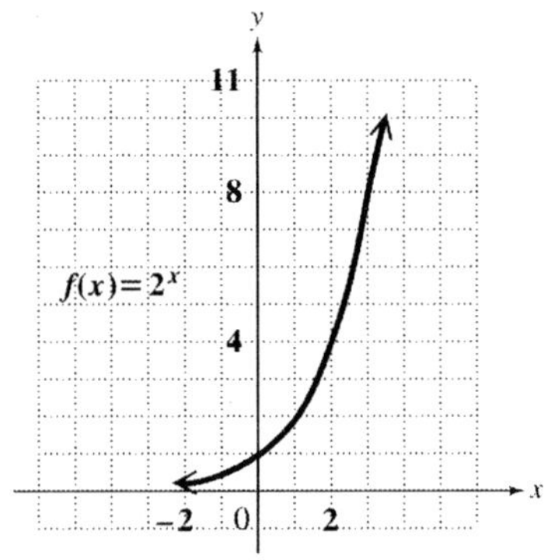

44. $f(x) = \log_3 x$
Make a table of values.

Powers of 3	3^{-2}	3^{-1}	3^0	3^1	3^2
x	$\frac{1}{9}$	$\frac{1}{3}$	1	3	9
y	-2	-1	0	1	2

Plot the ordered pairs and draw a smooth logarithmic curve through the points.

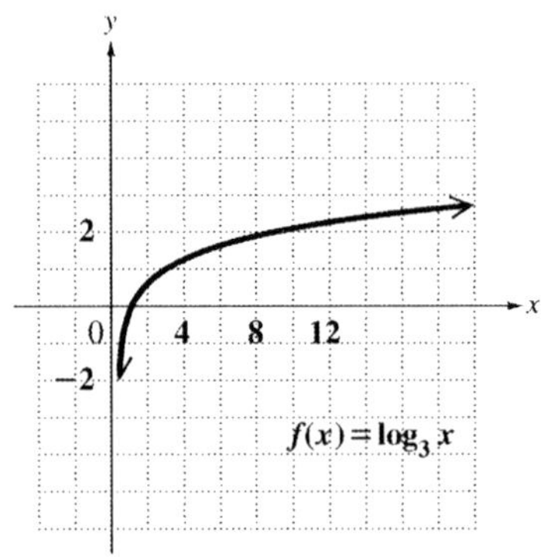

45. $$5^{x+3} = \left(\tfrac{1}{25}\right)^{3x+2}$$
$$5^{x+3} = \left[\left(\tfrac{1}{5}\right)^2\right]^{3x+2}$$

Write each side to the power of 5.

$$5^{x+3} = \left(5^{-2}\right)^{(3x+2)}$$
$$5^{x+3} = 5^{-2(3x+2)}$$
$$x + 3 = -2(3x+2)$$
$$x + 3 = -6x - 4$$
$$7x = -7$$
$$x = -1$$

Solution set: $\{-1\}$

46. $\log_5 x + \log_5 (x+4) = 1$

$\log_5 [x(x+4)] = 1$ *Product rule*

$x(x+4) = 5^1$ *Exponential form*

$x^2 + 4x = 5$

$x^2 + 4x - 5 = 0$

$(x+5)(x-1) = 0$

$x + 5 = 0$ or $x - 1 = 0$

$x = -5$ or $x = 1$

Check both potential solutions. The number 1 makes the equation true. The number -5 must be rejected because it yields an equation in which the logarithm of a negative number must be found.

Solution set: $\{1\}$

47. $\log_5 125 = 3$ *Logarithmic form*

$5^3 = 125$ *Exponential form*

48. $\log \dfrac{x^3\sqrt{y}}{z}$

$= \log \dfrac{x^3 y^{1/2}}{z}$

$= \log\left(x^3 y^{1/2}\right) - \log z$ *Quotient rule*

$= \log x^3 + \log y^{1/2} - \log z$ *Product rule*

$= 3\log x + \frac{1}{2}\log y - \log z$ *Power rule*

49. $A = Pe^{rt}$

$3P = Pe^{r\cdot 10}$ *triple, A = 3P*

$3 = e^{10r}$ *Divide by P.*

$\ln 3 = \ln e^{10r}$ *Take ln of both sides.*

$\ln 3 = 10r$

$r = \dfrac{\ln 3}{10} \approx .110$

A growth rate of about 11.0% will triple the value of P in 10 yr.

50. Let $a = 1{,}073{,}741{,}824$.

Solve $2^x = a$ for x.

$\log 2^x = \log a$

$x \log 2 = \log a$ *Power rule*

$x = \dfrac{\log a}{\log 2} = 30$

Solution set: $\{30\}$

Thus, 30 is the exponent such that

$$2^{30} = 1{,}073{,}741{,}824.$$

CHAPTER 12 NONLINEAR FUNCTIONS, CONIC SECTIONS, AND NONLINEAR SYSTEMS

12.1 Additional Graphs of Functions; Composition

12.1 Margin Exercises

1. $f(x) = \sqrt{x+4}$

   The graph of this function is found by shifting the graph of the square root function, $f(x) = \sqrt{x}$, 4 units to the left.

   Since x must be at least -4 for f to be defined, the domain of this function is $[-4, \infty)$. Since y is nonnegative, the range of this function is $[0, \infty)$.

x	y
-4	0
-3	1
0	2

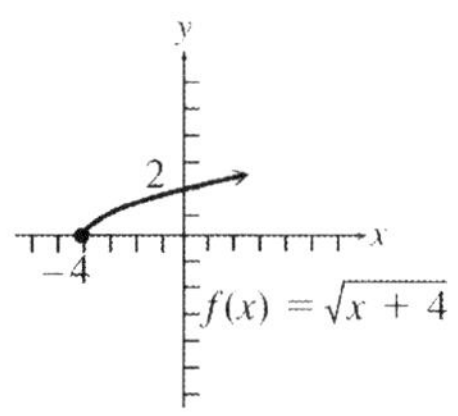

2. $f(x) = \dfrac{1}{x} - 2$

   This is the graph of the reciprocal function, $f(x) = \frac{1}{x}$, shifted 2 units down. Since $x \neq 0$ (or a denominator of 0 results), the line $x = 0$ is a vertical asymptote and the domain of this function is $(-\infty, 0) \cup (0, \infty)$.

   Since $\frac{1}{x} \neq 0$, the line $y = -2$ is a horizontal asymptote, and the range of this function is $(-\infty, -2) \cup (-2, \infty)$.

x	$\frac{1}{x}$	y
-2	$-\frac{1}{2}$	$-\frac{5}{2}$
-1	-1	-3
$-\frac{1}{2}$	-2	-4
$\frac{1}{2}$	2	0
1	1	-1
2	$\frac{1}{2}$	$-\frac{3}{2}$

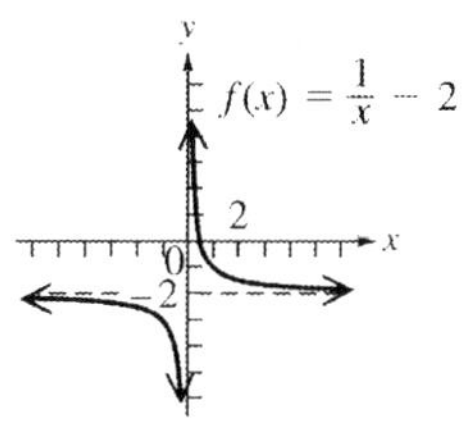

3. $f(x) = |x+2| + 1$

   This is the graph of the absolute value function, $f(x) = |x|$, shifted 2 units to the left (since $x + 2 = 0$ if $x = -2$) and 1 unit upward (because of the $+1$).

   The domain of this function is $(-\infty, \infty)$ and its range is $[1, \infty)$.

x	y
-4	3
-3	2
-2	1
-1	2
0	3

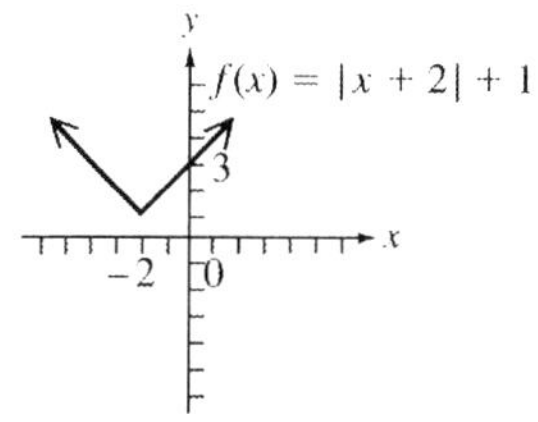

4. In each case, the answer is the largest integer that is less than or equal to the number in the greatest integer symbol.

   (a) $[\![18]\!] = 18 \quad (18 \leq 18)$

   (b) $[\![8.7]\!] = 8 \quad (8 \leq 8.7)$

   (c) $[\![-5]\!] = -5 \quad (-5 \leq -5)$

   (d) $[\![-6.9]\!] = -7 \quad (-7 \leq -6.9)$

   (e) $[\![\frac{1}{2}]\!] = 0 \quad (0 \leq \frac{1}{2})$

5. $f(x) = [\![x+1]\!]$

   The graph of f is the same as the graph of $y = [\![x]\!]$ shifted one unit to the left. Shifting the graph of $y = [\![x]\!]$ (right, left, up, down) does not change the domain and range. The domain of f is $(-\infty, \infty)$ and the range of f is the set of integers, that is, $\{\ldots, -2, -1, 0, 1, 2, \ldots\}$.

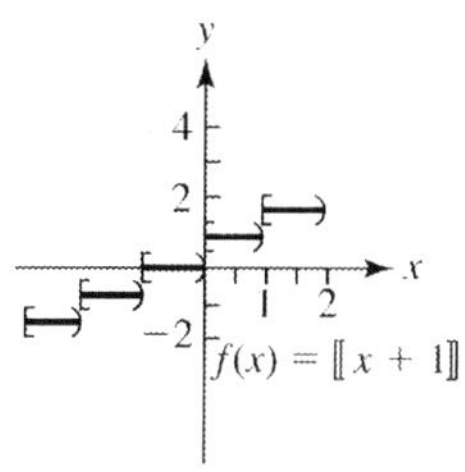

6. This function is similar to the greatest integer function, but in this case, we use the integer that is *greater than* or equal to the number. For example, for a $\frac{1}{2}$ oz letter, you would be charged for 1 oz.

Interval	Ounces Charged for	Cost
$(0, 1]$	1	\$ .80
$(1, 2]$	2	\$1.60
$(2, 3]$	3	\$2.40
$(3, 4]$	4	\$3.20

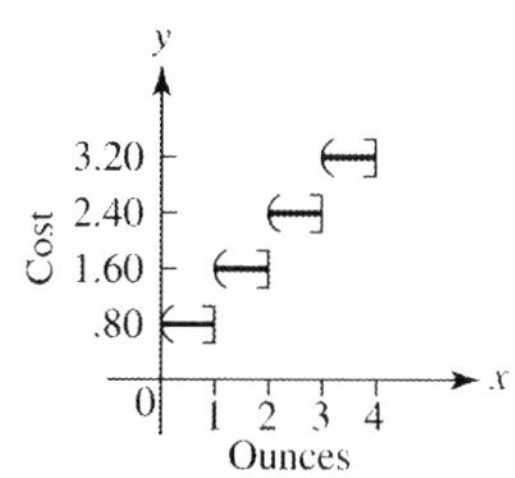

7. $f(x) = 3x + 6$, $g(x) = x^3$

(a) $(f \circ g)(2) = f[g(2)]$
$= f(2^3) \quad g(x) = x^3$
$= f(8)$
$= 3(8) + 6 \quad f(x) = 3x + 6$
$= 30$

(b) $(g \circ f)(2) = g[f(2)]$
$= g[3(2) + 6] \quad f(x) = 3x + 6$
$= g(12)$
$= 12^3 \quad g(x) = x^3$
$= 1728$

(c) $(f \circ g)(x) = f[g(x)]$
$= 3(g(x)) + 6 \quad f(x) = 3x + 6$
$= 3(x^3) + 6 \quad g(x) = x^3$
$= 3x^3 + 6$

(d) $(g \circ f)(x) = g[f(x)]$
$= [f(x)]^3 \quad g(x) = x^3$
$= (3x + 6)^3 \quad f(x) = 3x + 6$

12.1 Section Exercises

1. For the reciprocal function defined by $f(x) = 1/x$, $\underline{0}$ is the only real number not in the domain since division by 0 is undefined.

2. The range of the square root function, given by $f(x) = \sqrt{x}$, is $\underline{[0, \infty)}$ since the resulting values of the square root are positive or zero.

3. The lowest point on the graph of $f(x) = |x|$ has coordinates $(\underline{0}, \underline{0})$.

4. The range of $f(x) = [\![x]\!]$, the greatest integer function, is

$$\underline{\{\ldots, -2, -1, 0, 1, 2, \ldots\}}.$$

5. $f(x) = |x - 2| + 2$

The graph of this function has its "vertex" at $(2, 2)$, so the correct graph is **B**.

6. $f(x) = |x + 2| + 2$

The graph of this function has its "vertex" at $(-2, 2)$, so the correct graph is **C**.

7. $f(x) = |x - 2| - 2$

The graph of this function has its "vertex" at $(2, -2)$, so the correct graph is **A**.

8. $f(x) = |x + 2| - 2$

The graph of this function has its "vertex" at $(-2, -2)$, so the correct graph is **D**.

9. $f(x) = |x + 1|$

Since x can be any real number, the domain is $(-\infty, \infty)$.

The value of y is always greater than or equal to 0, so the range is $[0, \infty)$.

The graph of $y = |x + 1|$ looks like the graph of the absolute value function $y = |x|$, but the graph is translated 1 unit to the left. The x-value of its "vertex" is obtained by setting $x + 1 = 0$ and solving for x:

$$x + 1 = 0$$
$$x = -1.$$

Since the corresponding y-value is 0, the "vertex" is $(-1, 0)$. The axis of symmetry is the vertical line $x = -1$. Some additional points are $(-3, 2)$, $(-2, 1)$, $(0, 1)$, and $(1, 2)$.

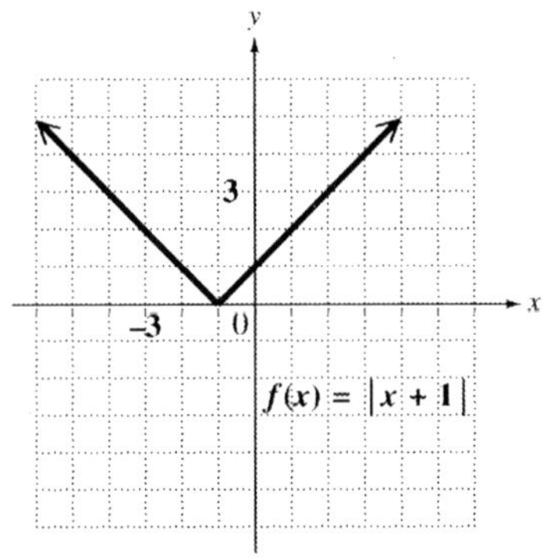

10. $f(x) = |x - 1|$

Since x can be any real number, the domain is $(-\infty, \infty)$.

The value of y is always greater than or equal to 0, so the range is $[0, \infty)$.

The graph of $y = |x - 1|$ looks like the graph of the absolute value function $y = |x|$, but the graph is translated 1 unit to the right. The x-value of its "vertex" is obtained by setting $x - 1 = 0$ and solving for x:

$$x - 1 = 0$$
$$x = 1.$$

Since the corresponding y-value is 0, the "vertex" is $(1, 0)$. The axis of symmetry is the vertical line $x = 1$. Some additional points are $(-1, 2)$, $(0, 1)$, $(2, 1)$, and $(3, 2)$.

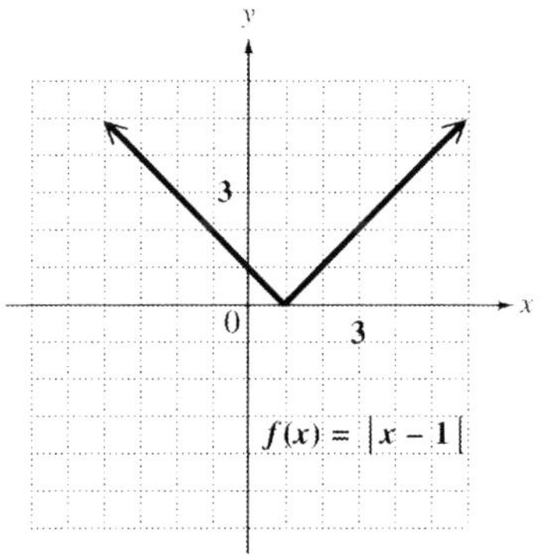

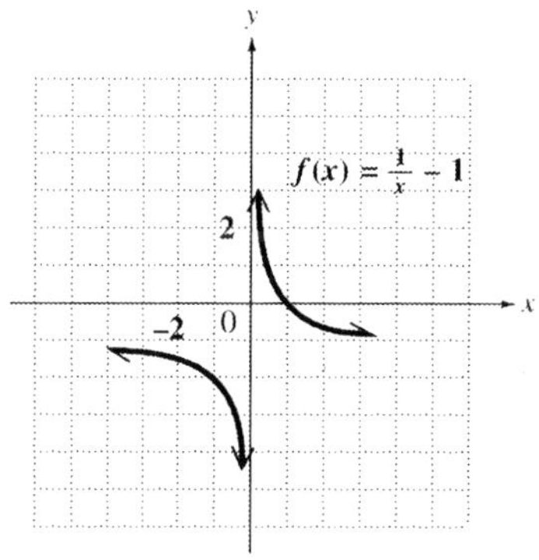

11. $f(x) = \dfrac{1}{x} + 1$

The graph of this function is similar to the graph of $g(x) = \dfrac{1}{x}$, except that each point is translated 1 unit upward. Just as with $g(x) = \dfrac{1}{x}$, $x = 0$ is the vertical asymptote, but this graph has $y = 1$ as its horizontal asymptote.

The domain is all real numbers except 0, that is, $(-\infty, 0) \cup (0, \infty)$.

The range is all real numbers except 1, that is, $(-\infty, 1) \cup (1, \infty)$.

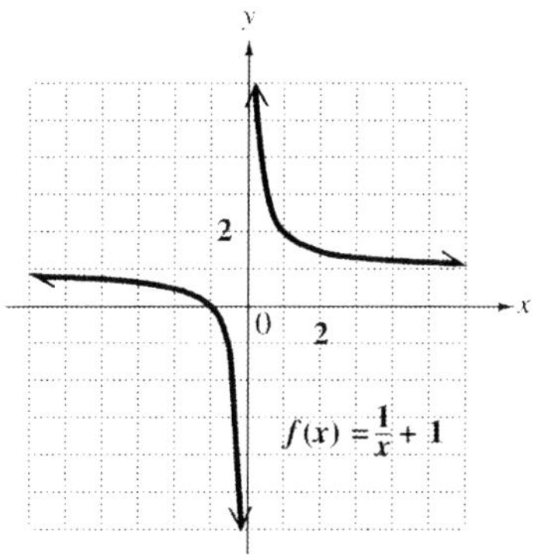

12. $f(x) = \dfrac{1}{x} - 1$

The graph of this function is similar to the graph of $g(x) = \dfrac{1}{x}$, except that each point is translated 1 unit downward. Just as with $g(x) = \dfrac{1}{x}$, $x = 0$ is the vertical asymptote, but this graph has $y = -1$ as its horizontal asymptote.

The domain is all real numbers except 0, that is, $(-\infty, 0) \cup (0, \infty)$.

The range is all real numbers except -1, that is, $(-\infty, -1) \cup (-1, \infty)$.

13. $f(x) = \sqrt{x - 2}$

The graph is found by shifting the graph of $g(x) = \sqrt{x}$ two units to the right. The following table of ordered pairs gives some specific points the graph passes through.

x	2	3	6
y	0	1	2

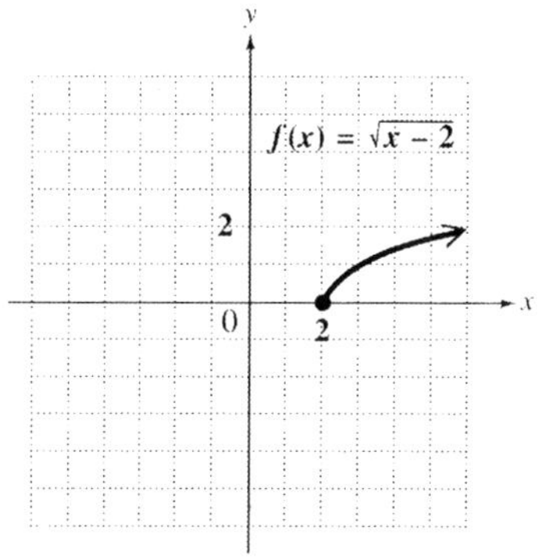

The domain of the function is $[2, \infty)$ and its range is $[0, \infty)$.

14. $f(x) = \sqrt{x + 5}$

This is the graph of the square root function $g(x) = \sqrt{x}$ shifted 5 units to the left.

x	-5	-4	-1
y	0	1	2

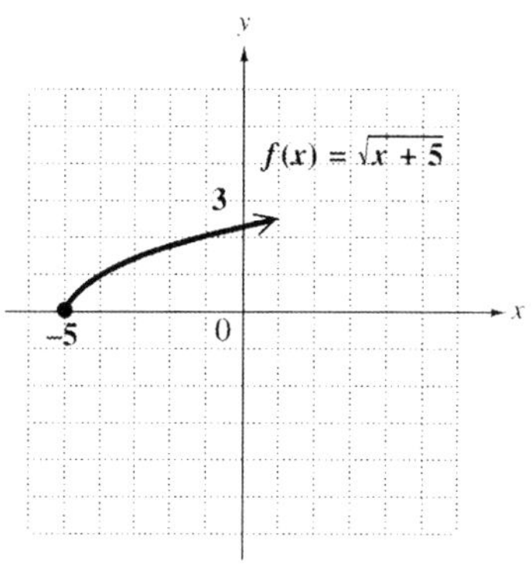

The domain is $[-5, \infty)$ and the range is $[0, \infty)$.

15. $f(x) = \dfrac{1}{x-2}$

This is the graph of the reciprocal function, $g(x) = \dfrac{1}{x}$, shifted 2 units to the right. Since $x \neq 2$ (or a denominator of 0 results), the line $x = 2$ is a vertical asymptote. Since $\dfrac{1}{x-2} \neq 0$, the line $y = 0$ is a horizontal asymptote.

x	$\frac{5}{2}$	3	4	5
y	2	1	$\frac{1}{2}$	$\frac{1}{3}$

x	-1	0	1	$\frac{3}{2}$
y	$-\frac{1}{3}$	$-\frac{1}{2}$	-1	-2

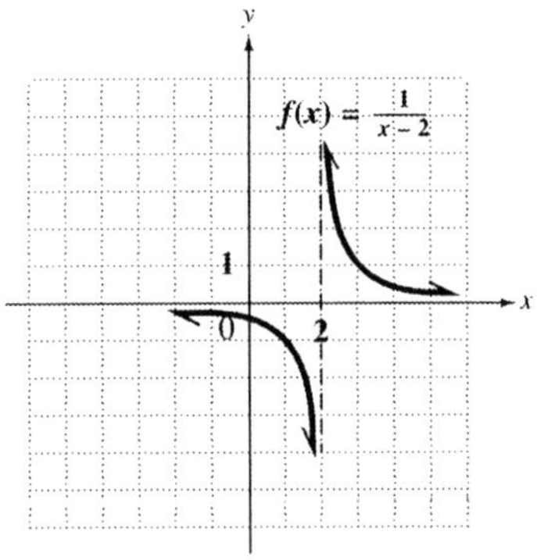

The domain of the function is $(-\infty, 2) \cup (2, \infty)$ and its range is $(-\infty, 0) \cup (0, \infty)$.

16. $f(x) = \dfrac{1}{x+2}$

This is the graph of the reciprocal function, $g(x) = \dfrac{1}{x}$, shifted 2 units to the left. Since $x \neq -2$ (or a denominator of 0 results), the line $x = -2$ is a vertical asymptote. Since $\dfrac{1}{x+2} \neq 0$, the line $y = 0$ is a horizontal asymptote.

x	-5	-4	-3	$-\frac{5}{2}$
y	$-\frac{1}{3}$	$-\frac{1}{2}$	-1	-2

x	1	0	-1	$-\frac{3}{2}$
y	$\frac{1}{3}$	$\frac{1}{2}$	1	2

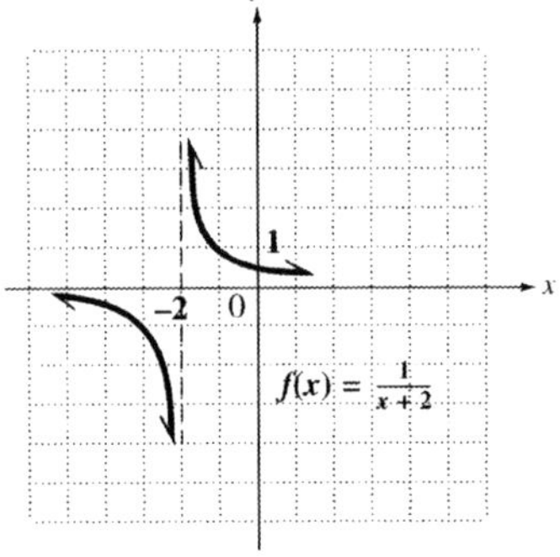

The domain of the function is $(-\infty, -2) \cup (-2, \infty)$ and the range is $(-\infty, 0) \cup (0, \infty)$.

17. $f(x) = \sqrt{x+3} - 3$

The graph is found by shifting the graph of $g(x) = \sqrt{x}$ three units to the left and three units down. The following table of ordered pairs gives some specific points the graph passes through.

x	-3	-2	1
y	-3	-2	-1

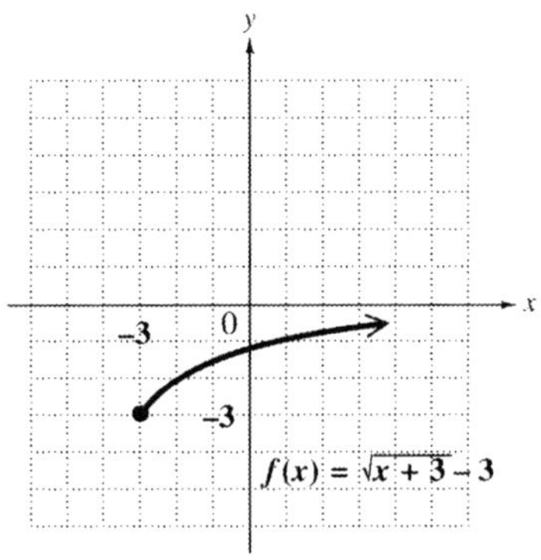

The domain of the function is $[-3, \infty)$ and its range is $[-3, \infty)$.

18. Shift the graph of $g(x) = \dfrac{1}{x}$ three units to the right and two units up to get the graph of $f(x) = \dfrac{1}{x-3} + 2$.

19. $f(x) = [\![x - 3]\!]$

This is the graph of the greatest integer function, $g(x) = [\![x]\!]$, shifted 3 units to the right.

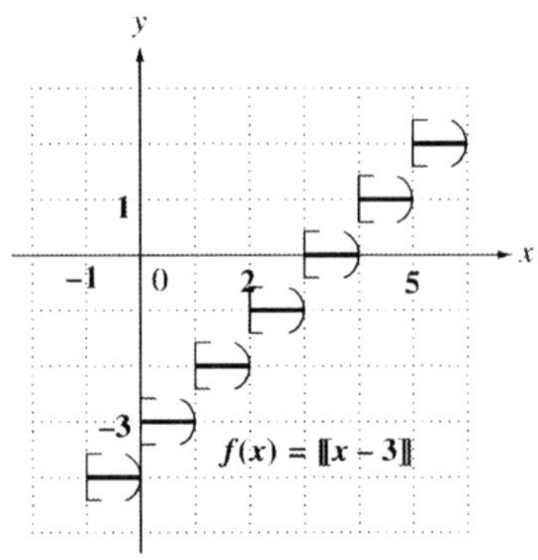

20. $g(x) = [\![x+2]\!]$

This is the graph of the greatest integer function, $f(x) = [\![x]\!]$, shifted 2 units to the left.

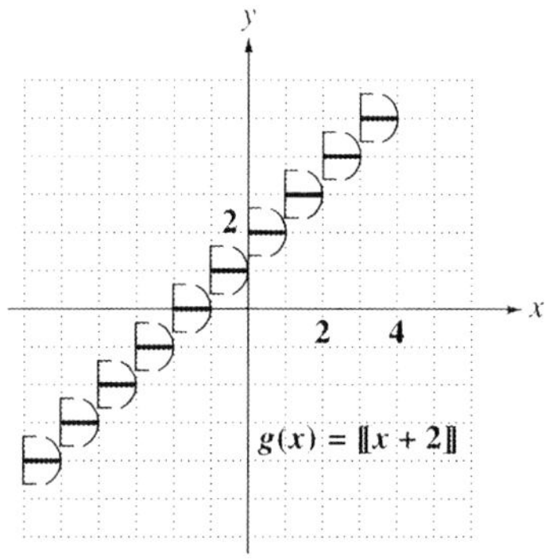

21. For any portion of the first ounce, the cost will be one 37¢ stamp. If the weight exceeds one ounce (up to two ounces), an additional 23¢ stamp is required. The following table summarizes the weight of a letter, x, and the number of stamps required, $p(x)$, on the interval $(0, 5]$.

x	$(0,1]$	$(1,2]$	$(2,3]$	$(3,4]$	$(4,5]$
$p(x)$	1	2	3	4	5

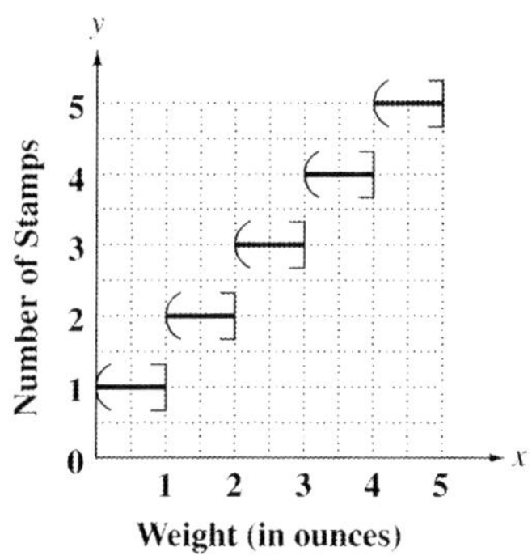

22. The following table summarizes the number of hours, x, and the cost of parking, $f(x)$, on the interval $(0, 2]$.

x	$(0,.5]$	$(.5,1]$	$(1,1.5]$	$(1.5,2]$
$f(x)$	\$3	\$5	\$7	\$9

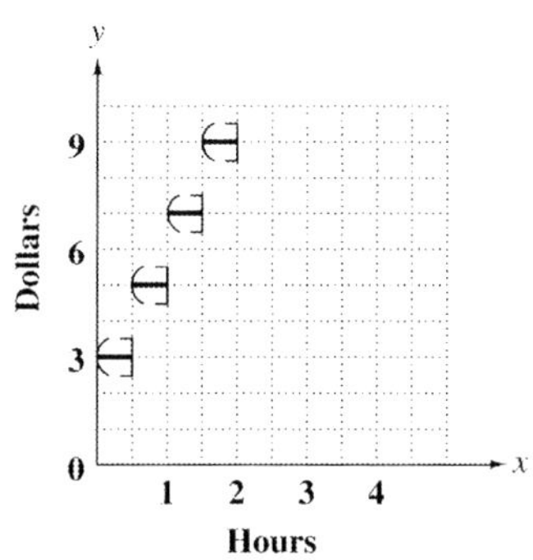

For Exercises 23–34,

$$f(x) = x^2 + 4,\ g(x) = 2x + 3, \text{ and } h(x) = x + 5.$$

23. $$\begin{aligned}(h \circ g)(4) &= h[g(4)] \\ &= h(2 \cdot 4 + 3) \\ &= h(11) \\ &= 11 + 5 \\ &= 16\end{aligned}$$

24. $$\begin{aligned}(f \circ g)(4) &= f[g(4)] \\ &= f(2 \cdot 4 + 3) \\ &= f(11) \\ &= 11^2 + 4 \\ &= 125\end{aligned}$$

25. $$\begin{aligned}(g \circ f)(6) &= g[f(6)] \\ &= g(6^2 + 4) \\ &= g(40) \\ &= 2 \cdot 40 + 3 \\ &= 83\end{aligned}$$

26. $$\begin{aligned}(h \circ f)(6) &= h[f(6)] \\ &= h(6^2 + 4) \\ &= h(40) \\ &= 40 + 5 \\ &= 45\end{aligned}$$

27. $$\begin{aligned}(f \circ h)(-2) &= f[h(-2)] \\ &= f(-2 + 5) \\ &= f(3) \\ &= 3^2 + 4 \\ &= 13\end{aligned}$$

28. $$\begin{aligned}(h \circ g)(-2) &= h[g(-2)] \\ &= h[2(-2) + 3] \\ &= h(-1) \\ &= -1 + 5 \\ &= 4\end{aligned}$$

29. $$\begin{aligned}(f \circ g)(x) &= f[g(x)] \\ &= f(2x + 3) \\ &= (2x + 3)^2 + 4 \\ &= 4x^2 + 12x + 9 + 4 \\ &= 4x^2 + 12x + 13\end{aligned}$$

30. $$\begin{aligned}(g \circ h)(x) &= g[h(x)] \\ &= g(x + 5) \\ &= 2(x + 5) + 3 \\ &= 2x + 10 + 3 \\ &= 2x + 13\end{aligned}$$

31. $$\begin{aligned}(f \circ h)(x) &= f[h(x)] \\ &= f(x+5) \\ &= (x+5)^2 + 4 \\ &= x^2 + 10x + 25 + 4 \\ &= x^2 + 10x + 29\end{aligned}$$

32. $$\begin{aligned}(g \circ f)(x) &= g[f(x)] \\ &= g(x^2 + 4) \\ &= 2(x^2 + 4) + 3 \\ &= 2x^2 + 8 + 3 \\ &= 2x^2 + 11\end{aligned}$$

33. $$\begin{aligned}(h \circ g)(x) &= h[g(x)] \\ &= h(2x+3) \\ &= 2x + 3 + 5 \\ &= 2x + 8\end{aligned}$$

34. $$\begin{aligned}(h \circ f)(x) &= h[f(x)] \\ &= h(x^2 + 4) \\ &= x^2 + 4 + 5 \\ &= x^2 + 9\end{aligned}$$

35. $f(x) = 12x$, $g(x) = 5280x$

$$\begin{aligned}(f \circ g)(x) &= f[g(x)] \\ &= f(5280x) \\ &= 12(5280x) \\ &= 63{,}360x\end{aligned}$$

$(f \circ g)(x)$ computes the number of inches in x mi.

36. **(a)** $$\begin{aligned}x &= 4s \\ \frac{x}{4} &= \frac{4s}{4} \\ \frac{x}{4} &= s\end{aligned}$$

(b) $$\begin{aligned}y &= s^2 \\ y &= \left(\frac{x}{4}\right)^2 \quad \textit{from part (a)} \\ y &= \frac{x^2}{16}\end{aligned}$$

(c) $$\begin{aligned}y &= \frac{6^2}{16} \quad \textit{Let } x = 6. \\ y &= \frac{36}{16} = 2.25\end{aligned}$$

The area is 2.25 square units.

37. $r(t) = 2t$, $A(r) = \pi r^2$

$$\begin{aligned}(A \circ r)(t) &= A[r(t)] \\ &= A(2t) \\ &= \pi(2t)^2 \\ &= 4\pi t^2\end{aligned}$$

This is the area of the circular layer as a function of time.

38. $r(t) = 4t$, $A(r) = \pi r^2$

$$\begin{aligned}(A \circ r)(t) &= A[r(t)] \\ &= A(4t) \\ &= \pi(4t)^2 \\ &= 16\pi t^2\end{aligned}$$

This is the area of the circular oil slick as a function of time.

12.2 The Circle and the Ellipse

12.2 Margin Exercises

1. If the point (x, y) is on the circle, the distance from (x, y) to the center $(0, 0)$ is 4. By the distance formula,

$$\begin{aligned}\sqrt{(x_2 - x_1)^2 + (y_2 - y_1)^2} &= d \quad \textit{Distance formula} \\ \sqrt{(x-0)^2 + (y-0)^2} &= 4 \\ \sqrt{x^2 + y^2} &= 4 \\ x^2 + y^2 &= 16. \quad \textit{Square both sides.}\end{aligned}$$

An equation of this circle is $x^2 + y^2 = 16$.

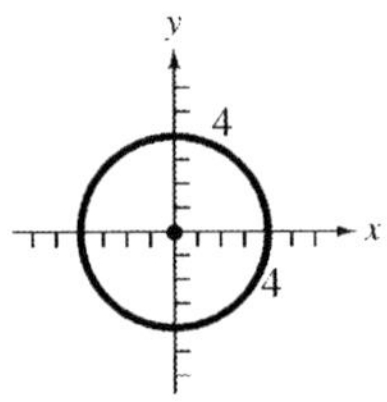

2. **(a)** Center at $(3, -2)$; radius 4

$$\begin{aligned}\sqrt{(x_2 - x_1)^2 + (y_2 - y_1)^2} &= d \\ \sqrt{(x-3)^2 + [y-(-2)]^2} &= 4 \\ \sqrt{(x-3)^2 + (y+2)^2} &= 4 \\ (x-3)^2 + (y+2)^2 &= 16 \quad \textit{Square both sides.}\end{aligned}$$

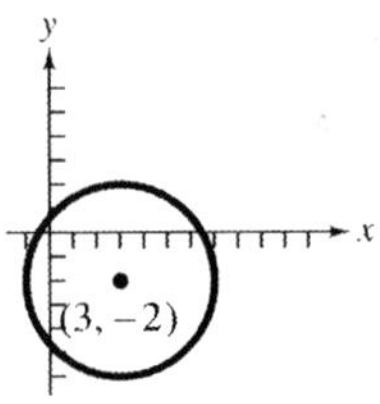

(b) $(x-5)^2+(y+2)^2=9$

The equation of the circle is an equation of the form

$$(x-h)^2+(y-k)^2=r^2$$
$$(x-5)^2+[y-(-2)]^2=3^2.$$

The center (h,k) is $(5,-2)$. The radius r is 3.

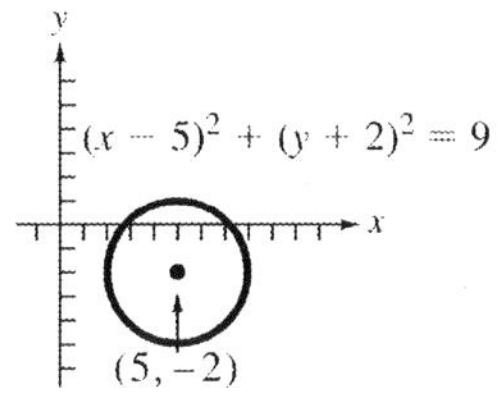

3. To find the center and radius, add the appropriate constants to complete the squares on x and y.

$$x^2+y^2-6x+8y-11=0$$
$$\left(x^2-6x\right)+\left(y^2+8y\right)=11$$
$$\left(x^2-6x+9\right)+\left(y^2+8y+16\right)=11+9+16$$
$$(x-3)^2+(y+4)^2=36$$
$$(x-3)^2+[y-(-4)]^2=6^2$$

The circle has center at $(3,-4)$ and radius 6.

4. **(a)** $\dfrac{x^2}{4}+\dfrac{y^2}{25}=1$

This ellipse has an equation of the form

$$\frac{x^2}{a^2}+\frac{y^2}{b^2}=1.$$

Here, $a^2=4$, so $a=2$ and the x-intercepts are $(2,0)$ and $(-2,0)$. Similarly, $b^2=25$, so $b=5$ and the y-intercepts are $(0,5)$ and $(0,-5)$. Plot the intercepts, and draw the ellipse through them.

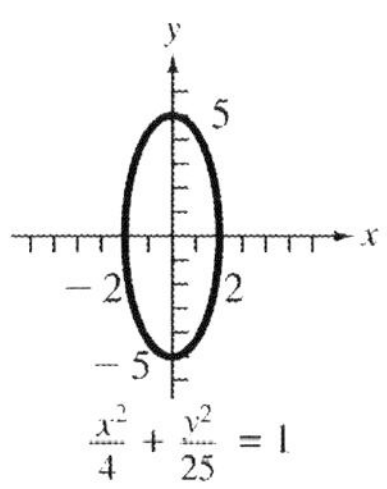

(b) $\dfrac{x^2}{64}+\dfrac{y^2}{49}=1$ or $\dfrac{x^2}{8^2}+\dfrac{y^2}{7^2}=1$

The x-intercepts for this ellipse are $(8,0)$ and $(-8,0)$. The y-intercepts are $(0,7)$ and $(0,-7)$. Plot the intercepts, and draw the ellipse through them.

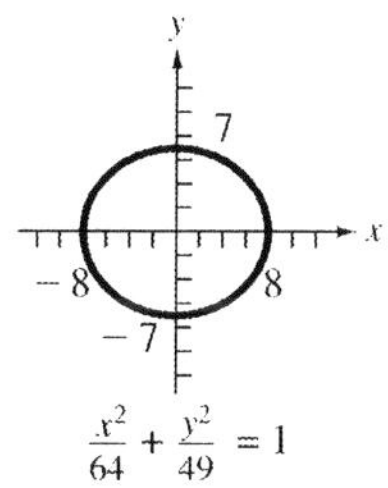

5.
$$\frac{(x+4)^2}{16}+\frac{(y-1)^2}{36}=1$$
$$\frac{[x-(-4)]^2}{4^2}+\frac{(y-1)^2}{6^2}=1$$

The center is $(-4,1)$.

The ellipse passes through the four points:

$$\begin{aligned}(-4-4,1)&=(-8,1)\\(-4+4,1)&=(0,1)\\(-4,1+6)&=(-4,7)\\(-4,1-6)&=(-4,-5)\end{aligned}$$

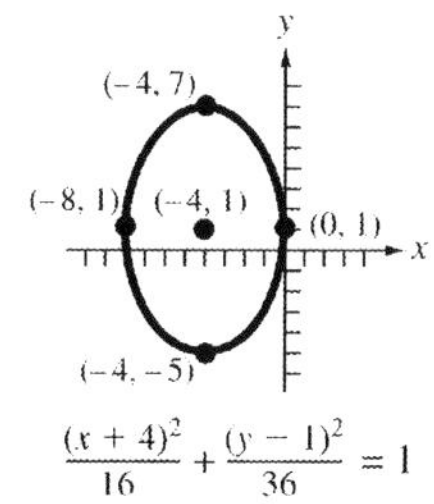

12.2 Section Exercises

1. **(a)** $x^2+y^2=25$ can be written in the center-radius form as

$$(x-0)^2+(y-0)^2=5^2.$$

The center is the point $(0,0)$.

(b) The radius is 5.

(c) The x-intercepts are $(5,0)$ and $(-5,0)$. The y-intercepts are $(0,5)$ and $(0,-5)$.

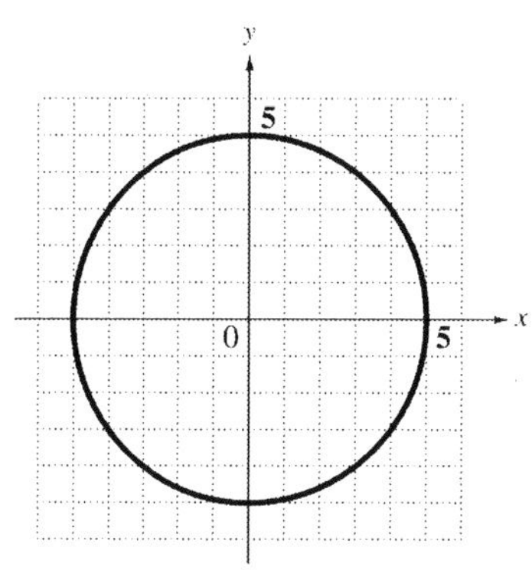

2. There will always be domain values that yield more than one range value. A circle fails the vertical line test.

3. $(x-3)^2+(y-2)^2=25$ is an equation of a circle with center $(3,2)$ and radius 5, choice **B**.

4. $(x-3)^2+(y+2)^2=25$ is an equation of a circle with center $(3,-2)$ and radius 5, choice **C**.

5. $(x+3)^2+(y-2)^2=25$ is an equation of a circle with center $(-3,2)$ and radius 5, choice **D**.

6. $(x+3)^2+(y+2)^2=25$ is an equation of a circle with center $(-3,-2)$ and radius 5, choice **A**.

7. Center: $(-4,3)$; radius: 2

Substitute $h=-4$, $k=3$, and $r=2$ in the center-radius form of the equation of a circle.

$$\begin{aligned}(x-h)^2+(y-k)^2&=r^2\\ [x-(-4)]^2+(y-3)^2&=2^2\\ (x+4)^2+(y-3)^2&=4\end{aligned}$$

8. Center: $(5,-2)$; radius: 4

Substitute $h=5$, $k=-2$, and $r=4$ in the center-radius form of the equation of a circle.

$$\begin{aligned}(x-h)^2+(y-k)^2&=r^2\\ (x-5)^2+[y-(-2)]^2&=4^2\\ (x-5)^2+(y+2)^2&=16\end{aligned}$$

9. Center: $(-8,-5)$; radius: $\sqrt{5}$

Substitute $h=-8$, $k=-5$, and $r=\sqrt{5}$ in the center-radius form of the equation of a circle.

$$\begin{aligned}(x-h)^2+(y-k)^2&=r^2\\ [x-(-8)]^2+[y-(-5)]^2&=\left(\sqrt{5}\right)^2\\ (x+8)^2+(y+5)^2&=5\end{aligned}$$

10. Center: $(-12,13)$; radius: $\sqrt{7}$

Substitute $h=-12$, $k=13$, and $r=\sqrt{7}$ in the center-radius form of the equation of a circle.

$$\begin{aligned}(x-h)^2+(y-k)^2&=r^2\\ [x-(-12)]^2+(y-13)^2&=\left(\sqrt{7}\right)^2\\ (x+12)^2+(y-13)^2&=7\end{aligned}$$

11. $$x^2+y^2+4x+6y+9=0$$

Rewrite the equation keeping only the variable terms on the left and grouping the x-terms and y-terms.

$$x^2+4x+y^2+6y=-9$$

Complete the squares on x and y and add the same constants to the right.

$$\begin{aligned}(x^2+4x+\underline{4})+(y^2+6y+\underline{9})&=-9+\underline{4}+\underline{9}\\ (x+2)^2+(y+3)^2&=4\end{aligned}$$

From the form $(x-h)^2+(y-k)^2=r^2$, we have $h=-2$, $k=-3$, and $r=2$. The center is $(-2,-3)$, and the radius r is 2.

12. $$x^2+y^2-8x-12y+3=0$$

To find the center and radius, complete the square on x and the square on y.

$$\begin{aligned}(x^2-8x\quad\;)+(y^2-12y\quad\;)&=-3\\ (x^2-8x+\underline{16})+(y^2-12y+\underline{36})&\\ &=-3+\underline{16}+\underline{36}\\ (x-4)^2+(y-6)^2&=49\end{aligned}$$

The center is $(4,6)$, and the radius is $\sqrt{49}=7$.

13. $x^2+y^2+10x-14y-7=0$

$$\begin{aligned}(x^2+10x\quad\;)+(y^2-14y\quad\;)&=7\\ (x^2+10x+\underline{25})+(y^2-14y+\underline{49})&\\ &=7+\underline{25}+\underline{49}\\ (x+5)^2+(y-7)^2&=81\end{aligned}$$

The center is $(-5,7)$, and the radius is $\sqrt{81}=9$.

14. $$\begin{aligned}x^2+y^2-2x+4y-4&=0\\ (x^2-2x\quad\;)+(y^2+4y\quad\;)&=4\\ (x^2-2x+\underline{1})+(y^2+4y+\underline{4})&=4+\underline{1}+\underline{4}\\ (x-1)^2+(y+2)^2&=9\end{aligned}$$

The center is $(1,-2)$, and the radius is $\sqrt{9}=3$.

15. $$\begin{aligned}3x^2+3y^2-12x-24y+12&=0\\ 3(x^2-4x\quad\;)+3(y^2-8y\quad\;)&=-12\end{aligned}$$

Divide by 3.

$$\begin{aligned}(x^2-4x\quad\;)+(y^2-8y\quad\;)&=-4\\ (x^2-4x+\underline{4})+(y^2-8y+\underline{16})&\\ &=-4+\underline{4}+\underline{16}\\ (x-2)^2+(y-4)^2&=16\end{aligned}$$

The center is $(2,4)$, and the radius is $\sqrt{16}=4$.

16. $$2x^2 + 2y^2 + 20x + 16y + 10 = 0$$
$$x^2 + y^2 + 10x + 8y + 5 = 0 \quad \textit{Divide by 2.}$$

Complete the square on x and the square on y.

$$(x^2 + 10x \qquad) + (y^2 + 8y \qquad) = -5$$
$$(x^2 + 10x + \underline{25}) + (y^2 + 8y + \underline{16}) = -5 + \underline{25} + \underline{16}$$
$$(x + 5)^2 + (y + 4)^2 = 36$$

The center is $(-5, -4)$, and the radius is $\sqrt{36} = 6$.

17. This method works because the pencil is always the same distance from the fastened end. The fastened end works as the center, and the length of the string from the fastened end to the pencil is the radius.

18. $$x^2 + y^2 = 9$$
$$(x - 0)^2 + (y - 0)^2 = 3^2$$

Here, $h = 0$, $k = 0$, and $r = 3$, so the graph is a circle with center $(0, 0)$ and radius 3.

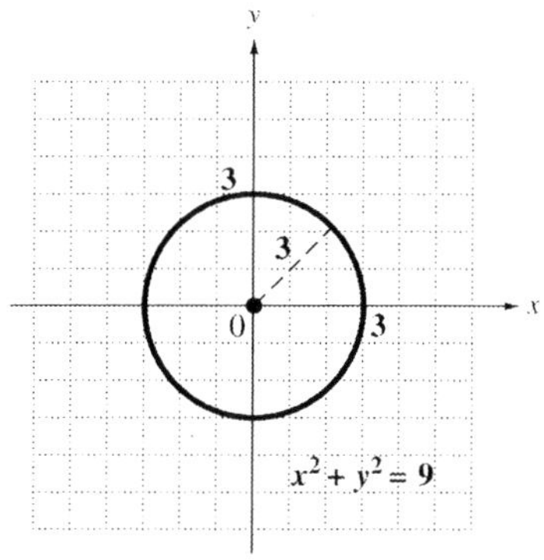

19. $$x^2 + y^2 = 4$$
$$(x - 0)^2 + (y - 0)^2 = 2^2$$

The center of the circle is $(0, 0)$ and $r = 2$.

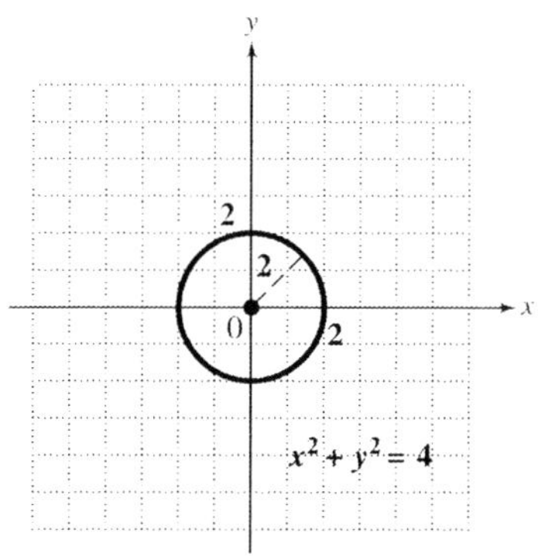

20. $$2y^2 = 10 - 2x^2$$
$$2x^2 + 2y^2 = 10$$
$$x^2 + y^2 = 5 \quad \textit{Divide by 2.}$$
$$(x - 0)^2 + (y - 0)^2 = \left(\sqrt{5}\right)^2$$

Here, $h = 0$, $k = 0$, and $r = \sqrt{5} \approx 2.2$, so the graph is a circle with center $(0, 0)$ and radius $\sqrt{5}$.

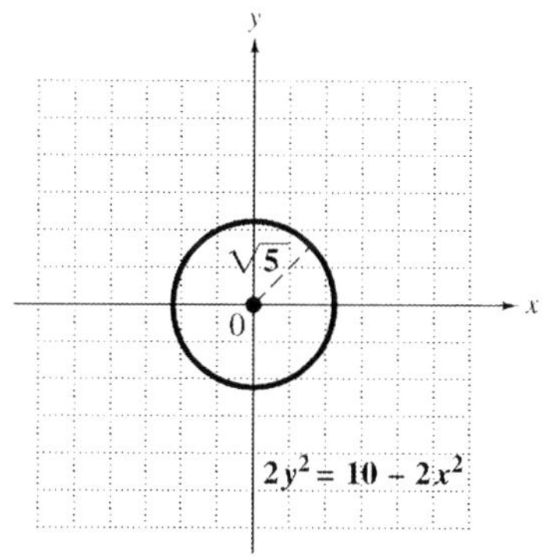

21. $$3x^2 = 48 - 3y^2$$
$$3x^2 + 3y^2 = 48$$
$$x^2 + y^2 = 16 = 4^2 \quad \textit{Divide by 3.}$$

This is an equation of a circle with center $(0, 0)$ and radius 4.

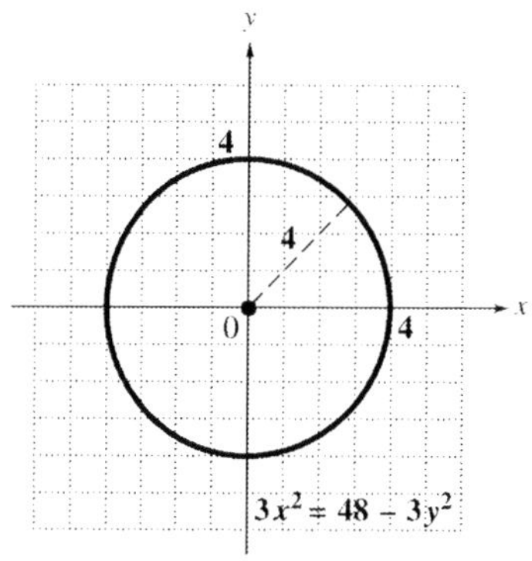

22. $(x + 3)^2 + (y - 2)^2 = 9$

Here, $h = -3$, $k = 2$, and $r = \sqrt{9} = 3$. The graph is a circle with center $(-3, 2)$ and radius 3.

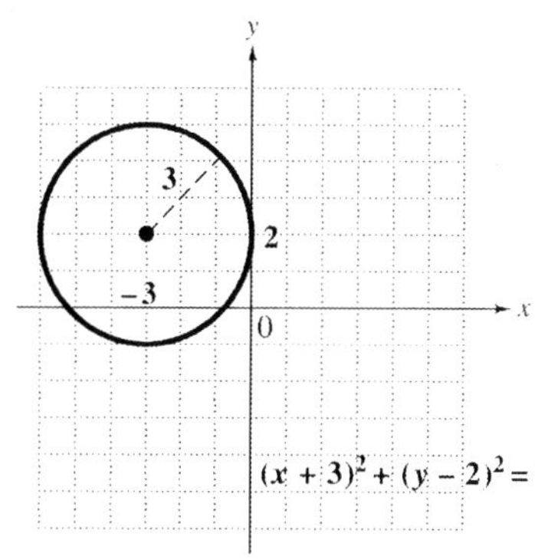

23. $(x-1)^2 + (y+3)^2 = 16$

Here, $h = 1$, $k = -3$, and $r = \sqrt{16} = 4$. The graph is a circle with center $(1, -3)$ and radius 4.

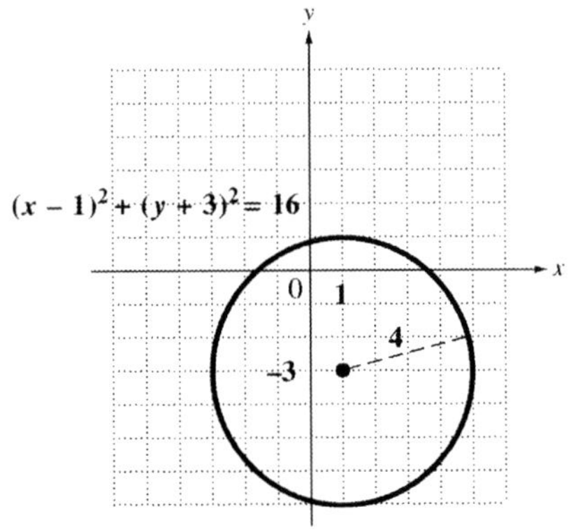

24.

$$x^2 + y^2 - 4x - 6y + 9 = 0$$

$$(x^2 - 4x \quad) + (y^2 - 6y \quad) = -9$$

Complete the square for each variable.

$$(x^2 - 4x + \underline{4}) + (y^2 - 6y + \underline{9}) = -9 + \underline{4} + \underline{9}$$

$$(x-2)^2 + (y-3)^2 = 4$$

Here, $h = 2$, $k = 3$, and $r = \sqrt{4} = 2$. The graph is a circle with center $(2, 3)$ and radius 2.

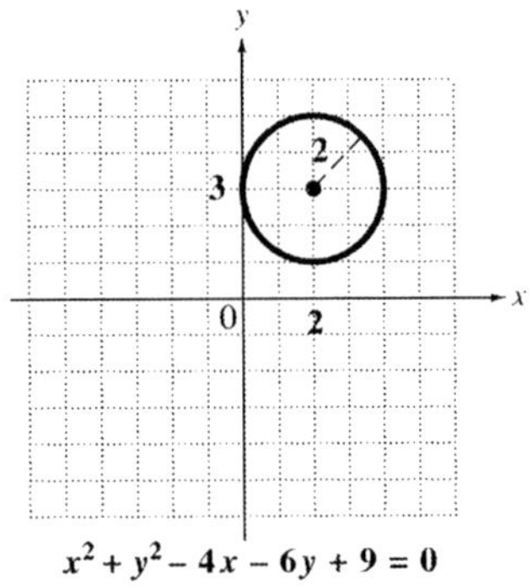

25.

$$x^2 + y^2 + 8x + 2y - 8 = 0$$

$$(x^2 + 8x \quad) + (y^2 + 2y \quad) = 8$$

$$(x^2 + 8x + \underline{16}) + (y^2 + 2y + \underline{1}) = 8 + \underline{16} + \underline{1}$$

$$(x+4)^2 + (y+1)^2 = 25$$

Here, $h = -4$, $k = -1$, and $r = \sqrt{25} = 5$. The graph is a circle with center $(-4, -1)$ and radius 5.

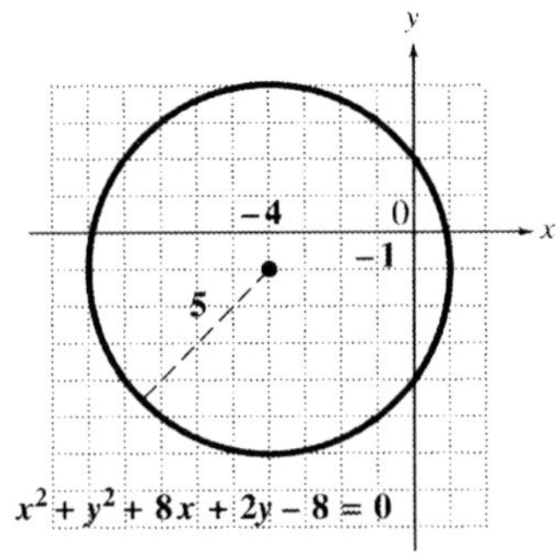

26. As the pencil traces the graph along the taut string, notice that the distances from the pencil to the two fastened ends continue to change, but the sum of the two distances is always the same, namely, the length of the string between the two fastened ends. This is exactly the definition of an ellipse—the set of all points in a plane the sum of whose distances from two fixed points (the fastened ends) is constant.

27. The equation $\frac{x^2}{9} + \frac{y^2}{25} = 1$ is of the form $\frac{x^2}{a^2} + \frac{y^2}{b^2} = 1$. The graph is an ellipse with $a^2 = 9$ and $b^2 = 25$, so $a = 3$ and $b = 5$. The x-intercepts are $(3, 0)$ and $(-3, 0)$. The y-intercepts are $(0, 5)$ and $(0, -5)$. Plot the intercepts, and draw the ellipse through them.

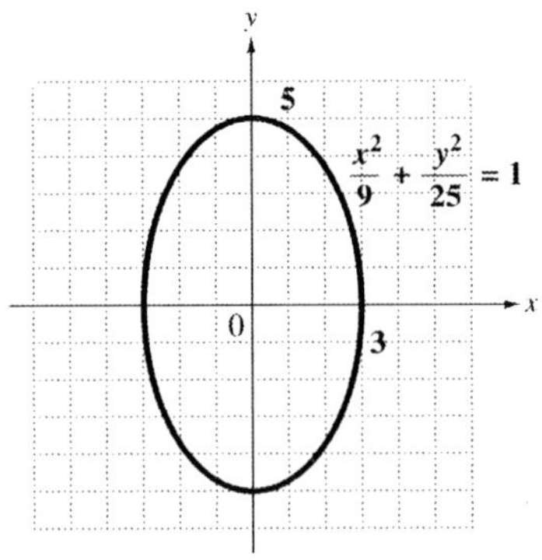

28. $\frac{x^2}{9} + \frac{y^2}{16} = 1$ is in the form

$$\frac{x^2}{a^2} + \frac{y^2}{b^2} = 1,$$

so $a = 3$ and $b = 4$. Its x-intercepts are $(3, 0)$ and $(-3, 0)$, and its y-intercepts are $(0, 4)$ and $(0, -4)$.

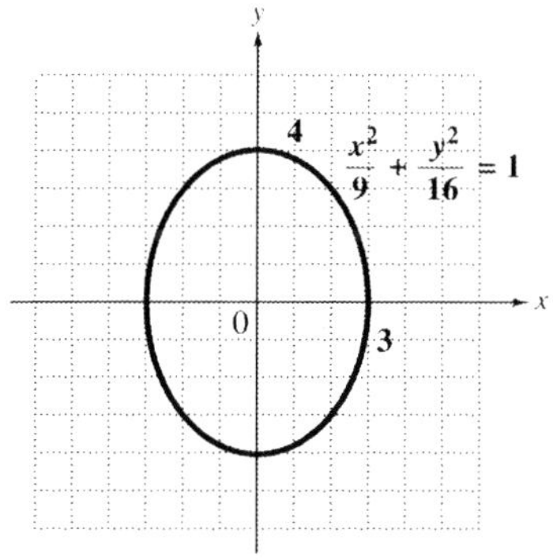

29. $\dfrac{x^2}{36} + \dfrac{y^2}{16} = 1$ is in the form $\dfrac{x^2}{a^2} + \dfrac{y^2}{b^2} = 1$.

The graph is an ellipse with $a^2 = 36$ and $b^2 = 16$, so $a = 6$ and $b = 4$. The x-intercepts are $(6, 0)$ and $(-6, 0)$. The y-intercepts are $(0, 4)$ and $(0, -4)$. Plot the intercepts, and draw the ellipse through them.

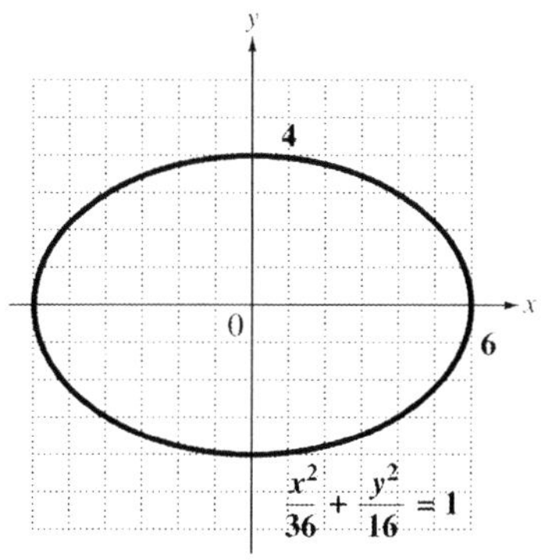

30. $\dfrac{x^2}{9} + \dfrac{y^2}{4} = 1$ is in the form $\dfrac{x^2}{a^2} + \dfrac{y^2}{b^2} = 1$,

so $a = 3$ and $b = 2$. The x-intercepts are $(3, 0)$ and $(-3, 0)$. The y-intercepts are $(0, 2)$ and $(0, -2)$.

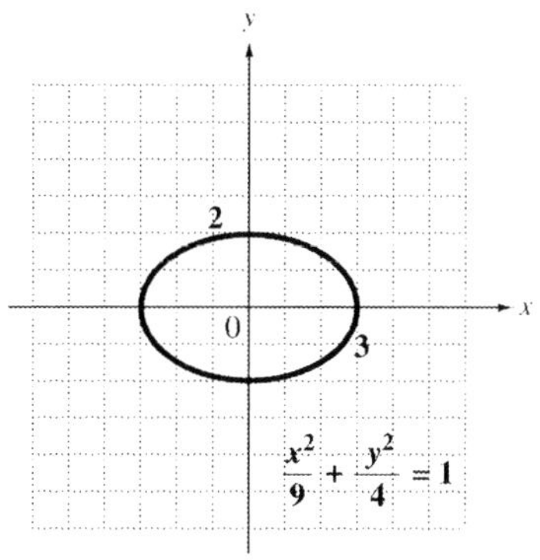

31. $\dfrac{x^2}{49} + \dfrac{y^2}{25} = 1$ is in the form $\dfrac{x^2}{a^2} + \dfrac{y^2}{b^2} = 1$.

The graph is an ellipse with $a^2 = 49$ and $b^2 = 25$, so $a = 7$ and $b = 5$. The x-intercepts are $(7, 0)$ and $(-7, 0)$. The y-intercepts are $(0, 5)$ and $(0, -5)$. Plot the intercepts, and draw the ellipse through them.

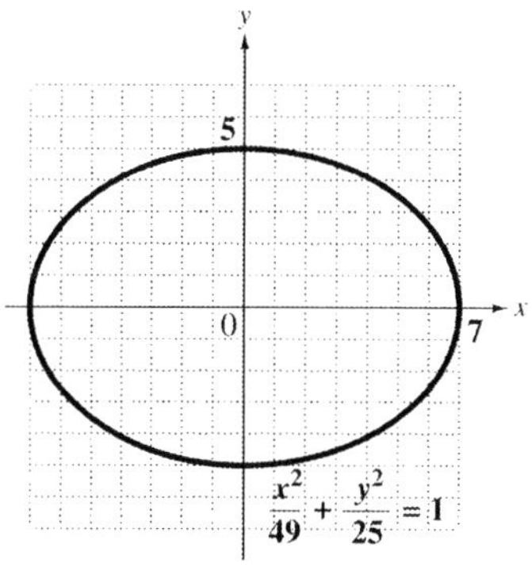

32. $\dfrac{x^2}{16} + \dfrac{y^2}{9} = 1$ is in the form $\dfrac{x^2}{a^2} + \dfrac{y^2}{b^2} = 1$,

so $a = 4$ and $b = 3$. The x-intercepts are $(4, 0)$ and $(-4, 0)$. The y-intercepts are $(0, 3)$ and $(0, -3)$.

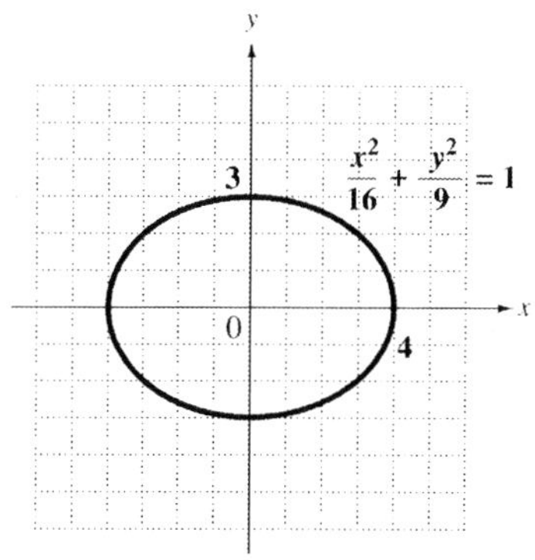

33. $\dfrac{(x-2)^2}{16} + \dfrac{(y-1)^2}{9} = 1$

The center of the ellipse is at $(2, 1)$. Since $a^2 = 16$, $a = 4$. Since $b^2 = 9$, $b = 3$. Add ± 4 to 2, and add ± 3 to 1 to find the points $(6, 1)$, $(-2, 1)$, $(2, 4)$, and $(2, -2)$. Plot the points, and draw the ellipse through them.

continued

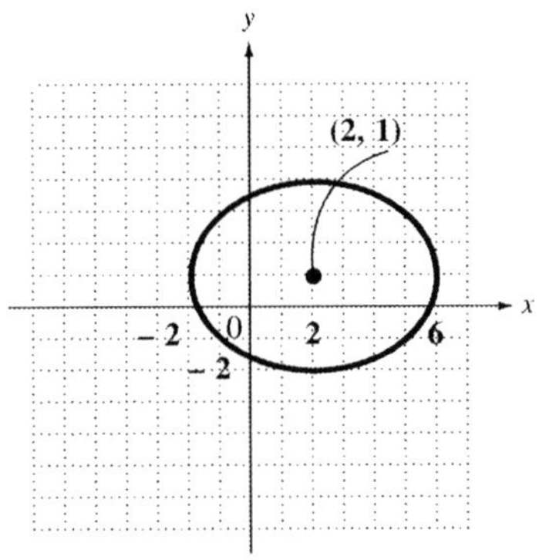

$\frac{(x-2)^2}{16} + \frac{(y-1)^2}{9} = 1$

34. $\frac{(x-4)^2}{9} + \frac{(y+2)^2}{4} = 1$

This equation is of the form

$$\frac{(x-h)^2}{a^2} + \frac{(y-k)^2}{b^2} = 1,$$

so the center of the ellipse is at $(4, -2)$. Since $a^2 = 9$, $a = 3$. Since $b^2 = 4$, $b = 2$. The graph includes the points $(4 \pm 3, -2)$ and $(4, -2 \pm 2)$ or $(7, -2)$, $(1, -2)$, $(4, -4)$, and $(4, 0)$.

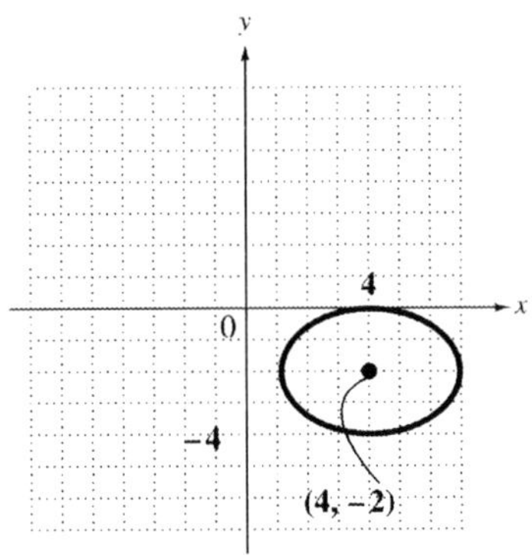

$\frac{(x-4)^2}{9} + \frac{(y+2)^2}{4} = 1$

35. **(a)** $100x^2 + 324y^2 = 32{,}400$
Divide by 32,400.

$$\frac{x^2}{324} + \frac{y^2}{100} = 1$$
$$\frac{x^2}{18^2} + \frac{y^2}{10^2} = 1$$

The height in the center is the y-coordinate of the positive y-intercept. The height is 10 meters.

(b) The width of the ellipse is the distance between the x-intercepts, $(-18, 0)$ and $(18, 0)$. The width across the bottom of the arch is $18 + 18 = 36$ meters.

36. We can represent the entire ellipse with x-intercepts $(10, 0)$ and $(-10, 0)$ and y-intercepts $(0, 15)$ and $(0, -15)$ by the equation

$$\frac{x^2}{10^2} + \frac{y^2}{15^2} = 1.$$

A 12-foot-wide truck can be represented by having its leftmost point at $x = -6$ and its rightmost point at $x = 6$. Let $x = 6$ and solve for y.

$$\frac{6^2}{10^2} + \frac{y^2}{15^2} = 1$$
$$\frac{y^2}{15^2} = 1 - \frac{36}{100}$$
$$y^2 = 15^2\left(\tfrac{64}{100}\right)$$
$$y = \pm 15\left(\tfrac{8}{10}\right) = \pm 12$$

The maximum possible height of the truck is 12 feet.

37. $\frac{x^2}{141.7^2} + \frac{y^2}{141.1^2} = 1$

(a) $c^2 = a^2 - b^2$, so

$$c = \sqrt{a^2 - b^2} = \sqrt{141.7^2 - 141.1^2}$$
$$= \sqrt{169.68} \approx 13.0$$

From the figure, the *apogee* is $a + c = 141.7 + 13.0 = 154.7$ million miles.

(b) The *perigee* is $a - c = 141.7 - 13.0 = 128.7$ million miles.

38. $\frac{x^2}{5013} + \frac{y^2}{4970} = 1$

(a) $c^2 = a^2 - b^2$, so

$$c = \sqrt{a^2 - b^2} = \sqrt{5013 - 4970}$$
$$= \sqrt{43} \approx 6.6$$

The greatest distance between Venus and the sun is $a + c = \sqrt{5013} + \sqrt{43} \approx 77.4$ million miles.

(b) The smallest distance is $a - c = \sqrt{5013} - \sqrt{43} \approx 64.2$ million miles.

12.3 The Hyperbola and Other Functions Defined by Radicals

12.3 Margin Exercises

1. $\frac{x^2}{4} - \frac{y^2}{25} = 1$ or $\frac{x^2}{2^2} - \frac{y^2}{5^2} = 1$

The graph is a hyperbola with $a = 2$ and $b = 5$. The x-intercepts are $(2, 0)$ and $(-2, 0)$. There are no y-intercepts. The points $(2, 5)$, $(2, -5)$, $(-2, -5)$, and $(-2, 5)$ are the vertices (corners) of the fundamental rectangle that determines the asymptotes. Graph a branch of the hyperbola through each intercept and approaching the asymptotes.

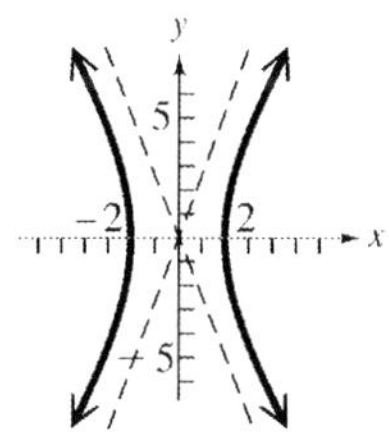

$\frac{x^2}{4} - \frac{y^2}{25} = 1$

2. $\frac{y^2}{81} - \frac{x^2}{64} = 1$ or $\frac{y^2}{9^2} - \frac{x^2}{8^2} = 1$

The graph is a hyperbola with $a = 8$ and $b = 9$. The y-intercepts are $(0, 9)$ and $(0, -9)$. There are no x-intercepts. The corners of the fundamental rectangle are $(8, 9)$, $(8, -9)$, $(-8, -9)$, and $(-8, 9)$. Extend the diagonals of the rectangle through these points to get the asymptotes. Graph a branch of the hyperbola through each intercept and approaching the asymptotes.

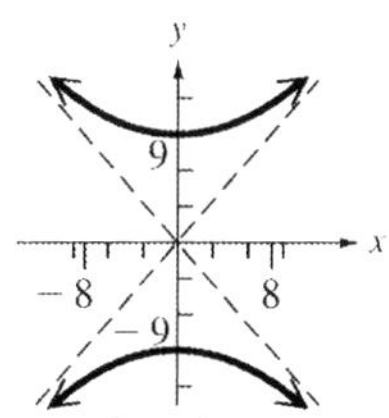

$\frac{y^2}{81} - \frac{x^2}{64} = 1$

3. **(a)**

$$3x^2 = 27 - 4y^2$$
$$3x^2 + 4y^2 = 27$$
$$\frac{x^2}{9} + \frac{4y^2}{27} = 1$$

Since the x^2- and y^2-terms have different positive coefficients, the graph of the equation is an ellipse.

(b)

$$6x^2 = 100 + 2y^2$$
$$6x^2 - 2y^2 = 100$$
$$\frac{x^2}{\frac{50}{3}} - \frac{y^2}{50} = 1$$

Because of the minus sign and since both variables are squared, the graph of the equation is a hyperbola.

(c) $3x^2 = 27 - 4y$

$$4y = -3x^2 + 27$$

This is an equation of a vertical parabola since only one variable, x, is squared.

(d)

$$3x^2 = 27 - 3y^2$$
$$3x^2 + 3y^2 = 27$$
$$x^2 + y^2 = 9$$

This is an equation of the form $(x - h)^2 + (y - k)^2 = r^2$, where $(h, k) = (0, 0)$ and $r = 3$, so the graph of the equation is a circle.

4.

$$f(x) = \sqrt{36 - x^2}$$
$$y = \sqrt{36 - x^2} \qquad f(x) = y$$
$$y^2 = 36 - x^2 \qquad \textit{Square both sides.}$$
$$x^2 + y^2 = 36 \qquad \textit{Add } x^2.$$

This equation is the graph of a circle with center $(0, 0)$ and radius 6. Since $f(x)$ represents a principal square root in the original equation, $f(x)$ is nonnegative and its graph is the upper half of the circle.

The domain of the function is $[-6, 6]$ and its range is $[0, 6]$.

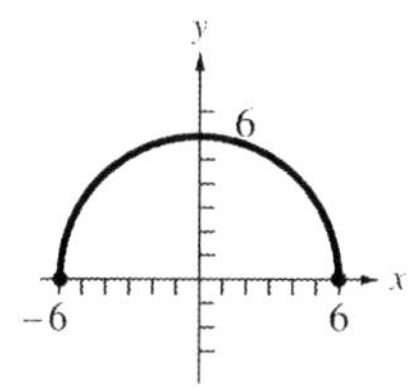

$f(x) = \sqrt{36 - x^2}$

5.

$$\frac{y}{3} = -\sqrt{1 - \frac{x^2}{4}}$$
$$\frac{y^2}{9} = 1 - \frac{x^2}{4} \qquad \textit{Square both sides.}$$
$$\frac{x^2}{4} + \frac{y^2}{9} = 1 \qquad \textit{Add } \frac{x^2}{4}.$$

This is the equation of an ellipse with intercepts $(2, 0)$, $(-2, 0)$, $(0, 3)$, and $(0, -3)$. Since $\frac{y}{3}$ equals a negative square root in the original equation, y must be nonpositive, restricting the graph to the lower half of the ellipse.

The domain of the function is $[-2, 2]$ and its range is $[-3, 0]$.

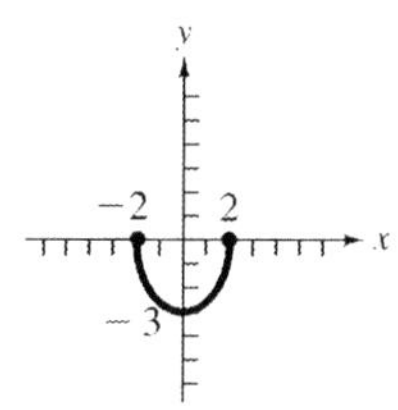

$\frac{y}{3} = -\sqrt{1 - \frac{x^2}{4}}$

12.3 Section Exercises

1. $\frac{x^2}{25} + \frac{y^2}{9} = 1$

This is the standard form for the equation of an ellipse with x-intercepts $(5, 0)$ and $(-5, 0)$ and y-intercepts $(0, 3)$ and $(0, -3)$. This is graph **C**.

2. $\frac{x^2}{9} + \frac{y^2}{25} = 1$

This is the standard form for the equation of an ellipse with x-intercepts $(3, 0)$ and $(-3, 0)$ and y-intercepts $(0, 5)$ and $(0, -5)$. This is graph **B**.

3. $\frac{x^2}{9} - \frac{y^2}{25} = 1$

This is the standard form for the equation of a hyperbola that opens left and right. Its x-intercepts are $(3, 0)$ and $(-3, 0)$. This is graph **D**.

4. $\frac{x^2}{25} - \frac{y^2}{9} = 1$

This is the standard form for the equation of a hyperbola that opens left and right. Its x-intercepts are $(5, 0)$ and $(-5, 0)$. This is graph **A**.

5. If the equation of a hyperbola is in standard form (that is, equal to one), the hyperbola would open to the left and right if the x^2-term was positive. It would open up and down if the y^2-term was positive.

6. The points (a, b), $(a, -b)$, $(-a, -b)$, and $(-a, b)$ are used as vertices of a rectangle. The extended diagonals of the rectangle are the asymptotes of the hyperbola.

7. The equation $\frac{x^2}{16} - \frac{y^2}{9} = 1$ is in the form $\frac{x^2}{a^2} - \frac{y^2}{b^2} = 1$. The graph is a hyperbola with $a = 4$ and $b = 3$. The x-intercepts are $(4, 0)$ and $(-4, 0)$. There are no y-intercepts. The vertices of the fundamental rectangle are $(4, 3)$, $(4, -3)$, $(-4, -3)$, and $(-4, 3)$. Extend the diagonals of the rectangle through these points to get the asymptotes. Graph a branch of the hyperbola through each intercept and approaching the asymptotes.

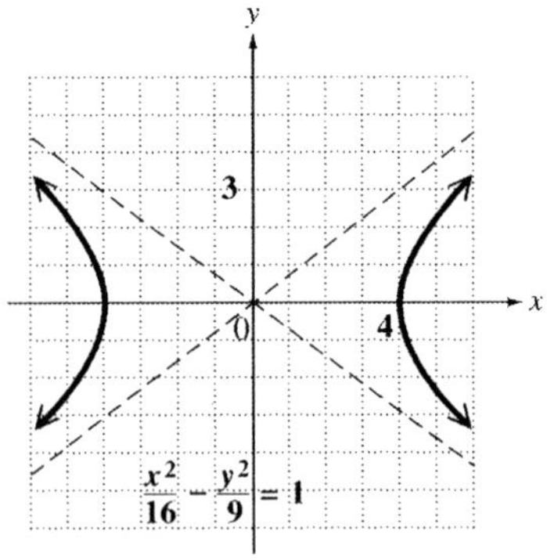

8. $\frac{y^2}{4} - \frac{x^2}{25} = 1$ is in the form $\frac{y^2}{b^2} - \frac{x^2}{a^2} = 1$, where $b = 2$ and $a = 5$. The branches open upward and downward, so the hyperbola has y-intercepts $(0, 2)$ and $(0, -2)$. The fundamental rectangle has vertices $(5, 2)$, $(5, -2)$, $(-5, -2)$, and $(-5, 2)$. Sketch the extended diagonals, which are the asymptotes of this hyperbola. Graph a branch of the hyperbola through each intercept and approaching the asymptotes.

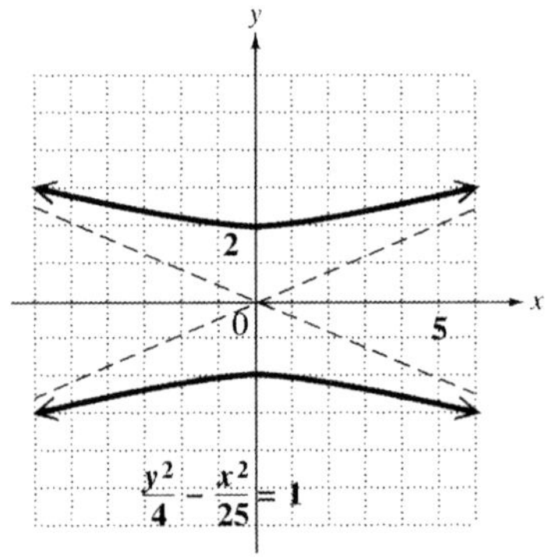

9. $\frac{y^2}{9} - \frac{x^2}{9} = 1$ is a hyperbola with $a = 3$ and $b = 3$. The y-intercepts are $(0, 3)$ and $(0, -3)$. There are no x-intercepts. One asymptote passes through $(3, 3)$ and $(-3, -3)$. The other asymptote passes through $(-3, 3)$ and $(3, -3)$. Draw the asymptotes and sketch the hyperbola through the intercepts and approaching the asymptotes.

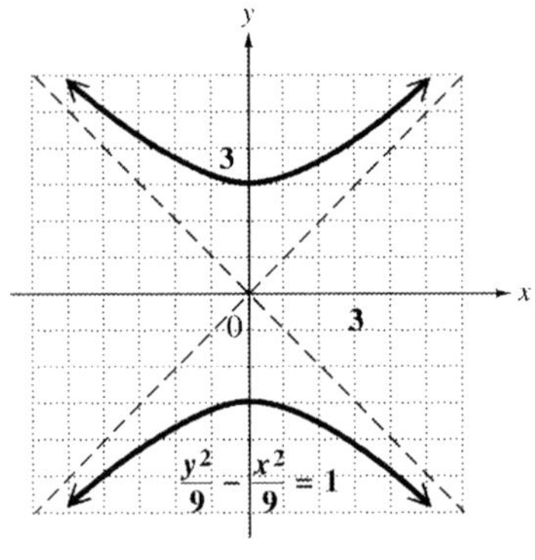

10. $\frac{x^2}{49} - \frac{y^2}{16} = 1 \quad (a = 7, b = 4)$

The branches open to the right and left, so the hyperbola has x-intercepts $(7, 0)$ and $(-7, 0)$. The asymptotes go through $(7, 4)$ and $(-7, -4)$ and through $(-7, 4)$ and $(7, -4)$.

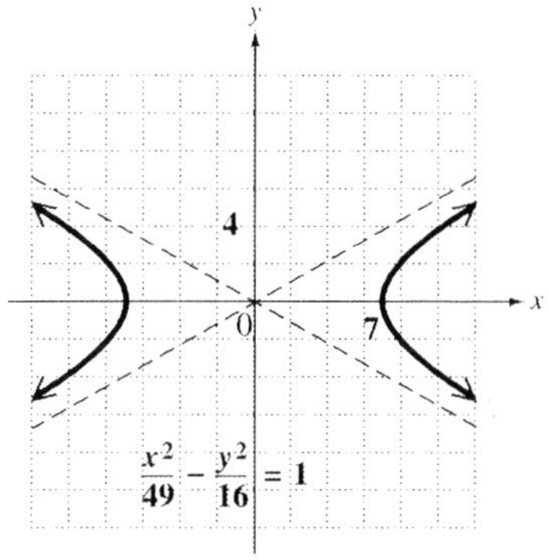

11. $\frac{x^2}{25} - \frac{y^2}{36} = 1$ is a hyperbola with $a = 5$ and $b = 6$. The x-intercepts are $(5, 0)$ and $(-5, 0)$. There are no y-intercepts. To sketch the graph, draw the diagonals of the fundamental rectangle with vertices $(5, 6)$, $(5, -6)$, $(-5, -6)$ and $(-5, 6)$. Graph a branch of the hyperbola through each intercept and approaching the asymptotes.

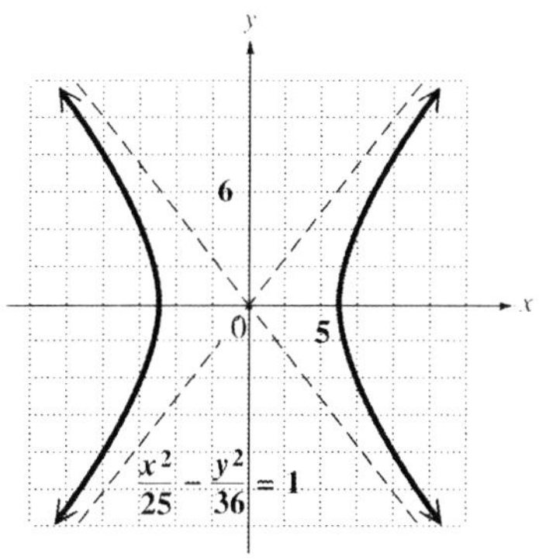

12. $\frac{y^2}{9} - \frac{x^2}{4} = 1 \quad (b = 3, a = 2)$

The branches open upward and downward, so the hyperbola has y-intercepts $(0, 3)$ and $(0, -3)$. The asymptotes go through $(2, 3)$ and $(-2, -3)$ and through $(-2, 3)$ and $(2, -3)$.

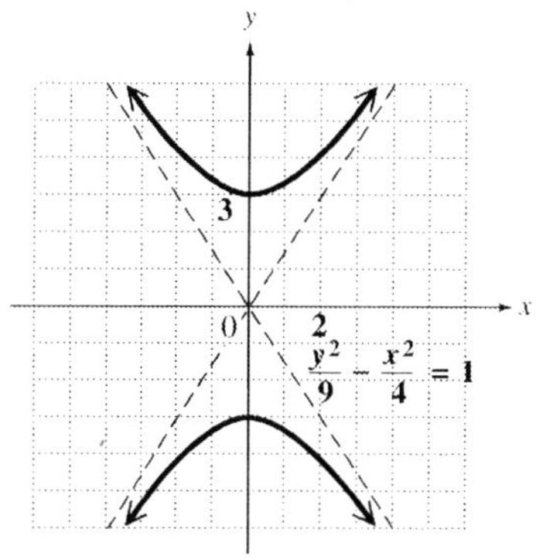

13. $x^2 - y^2 = 16$

$$\frac{x^2}{16} - \frac{y^2}{16} = 1 \qquad \textit{Divide by 16.}$$

This equation is in the form $\frac{x^2}{a^2} - \frac{y^2}{b^2} = 1$ with $a = 4$ and $b = 4$. The graph is a hyperbola with x-intercepts $(4, 0)$ and $(-4, 0)$ and no y-intercepts. One asymptote passes through $(4, 4)$ and $(-4, -4)$. The other asymptote passes through $(-4, 4)$ and $(4, -4)$. Sketch the graph through the intercepts and approaching the asymptotes.

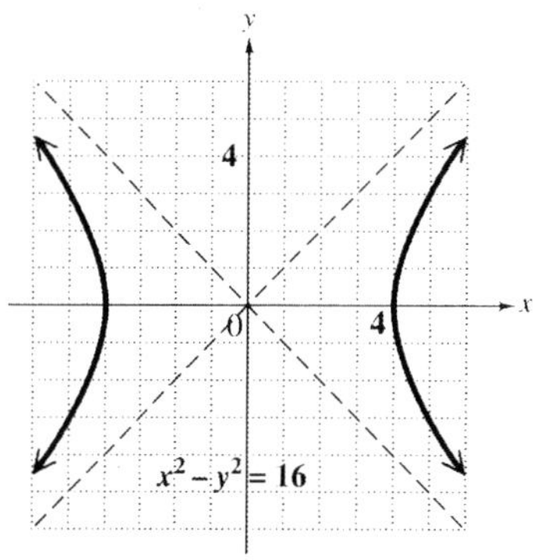

14. $x^2 + y^2 = 16$ is in the form $(x - h)^2 + (y - k)^2 = r^2$, with $h = 0$, $k = 0$, $r = 4$, so the graph is a circle with center $(0, 0)$ and radius 4.

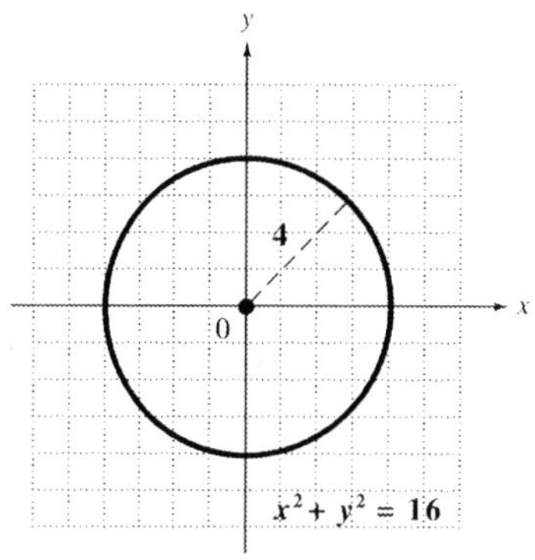

15. $4x^2 + y^2 = 16$

$$\frac{x^2}{4} + \frac{y^2}{16} = 1 \qquad \textit{Divide by 16.}$$

This equation is in the form $\frac{x^2}{a^2} + \frac{y^2}{b^2} = 1$ with $a = 2$ and $b = 4$. The graph is an ellipse. The x-intercepts $(2, 0)$ and $(-2, 0)$. The y-intercepts are $(0, 4)$ and $(0, -4)$. Plot the intercepts and draw the ellipse through them.

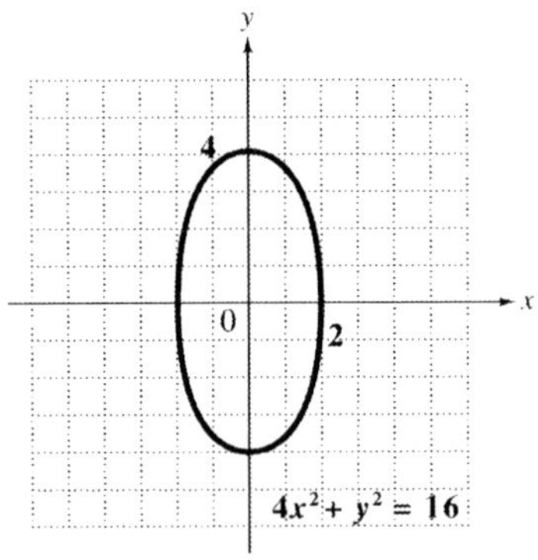

16. $x^2 - 2y = 0$

$$x^2 = 2y$$
$$\tfrac{1}{2}x^2 = y$$

The equation is in the form

$$f(x) = a(x-h)^2 + k,$$

with $a = \frac{1}{2}$, $h = 0$, and $k = 0$. The graph is a parabola that opens up since $a = \frac{1}{2} > 0$. The graph is wider than the graph of $f(x) = x^2$ because $|a| = \frac{1}{2} < 1$. The vertex is at $(0, 0)$.

x	1	2	3
y	$\frac{1}{2}$	2	4.5

Use symmetry about $x = 0$ to draw the graph.

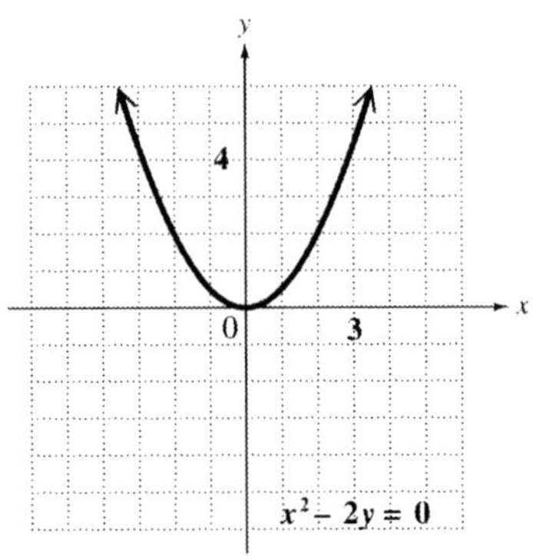

17.

$$y^2 = 36 - x^2$$
$$x^2 + y^2 = 36$$
$$(x-0)^2 + (y-0)^2 = 36$$

The graph is a circle with center at $(0, 0)$ and radius $\sqrt{36} = 6$.

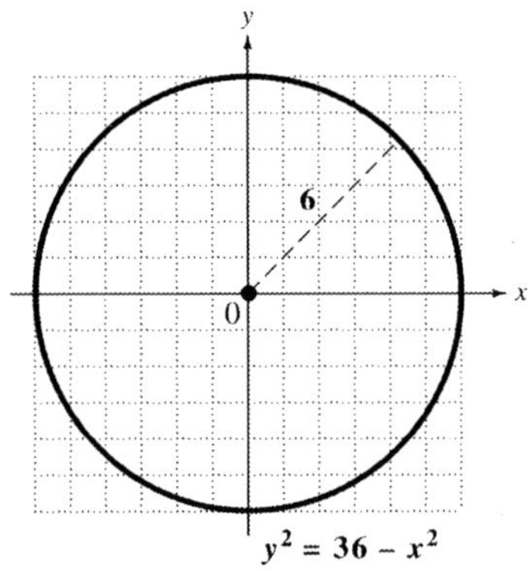

18. $9x^2 + 25y^2 = 225$

$$\frac{x^2}{25} + \frac{y^2}{9} = 1 \qquad \textit{Divide by 225.}$$

The graph is an ellipse with x-intercepts $(5, 0)$ and $(-5, 0)$ and y-intercepts $(0, 3)$ and $(0, -3)$.

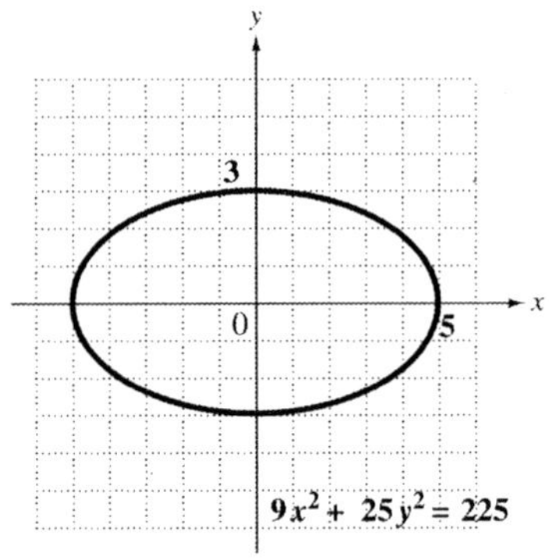

19.

$$9x^2 = 144 + 16y^2$$
$$9x^2 - 16y^2 = 144$$
$$\frac{x^2}{16} - \frac{y^2}{9} = 1 \qquad \textit{Divide by 144.}$$

The equation is a hyperbola in the form $\frac{x^2}{a^2} - \frac{y^2}{b^2} = 1$ with $a = 4$ and $b = 3$. The x-intercepts are $(4, 0)$ and $(-4, 0)$. There are no y-intercepts. To sketch the graph, draw the diagonals of the fundamental rectangle with vertices $(4, 3)$, $(4, -3)$, $(-4, -3)$ and $(-4, 3)$. These are the asymptotes. Graph a branch of the hyperbola through each intercept approaching the asymptotes.

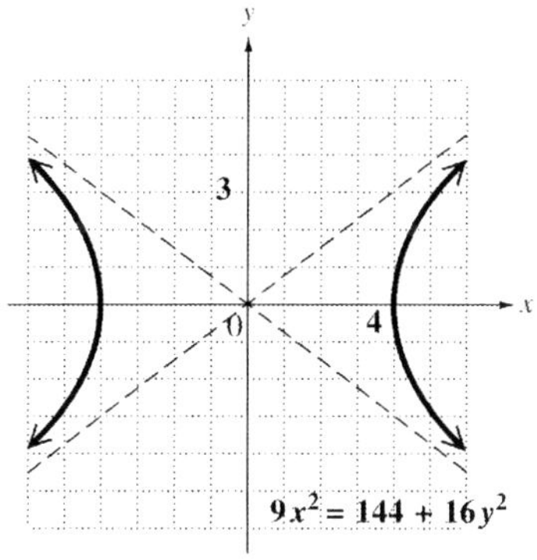

20.

$$\begin{aligned} y^2 &= 4 + x^2 \\ y^2 - x^2 &= 4 \\ \frac{y^2}{4} - \frac{x^2}{4} &= 1 \qquad \textit{Divide by 4.} \end{aligned}$$

The graph is a hyperbola with y-intercepts $(0, 2)$ and $(0, -2)$. One asymptote passes through $(2, 2)$ and $(-2, -2)$. The other asymptote passes through $(-2, 2)$ and $(2, -2)$.

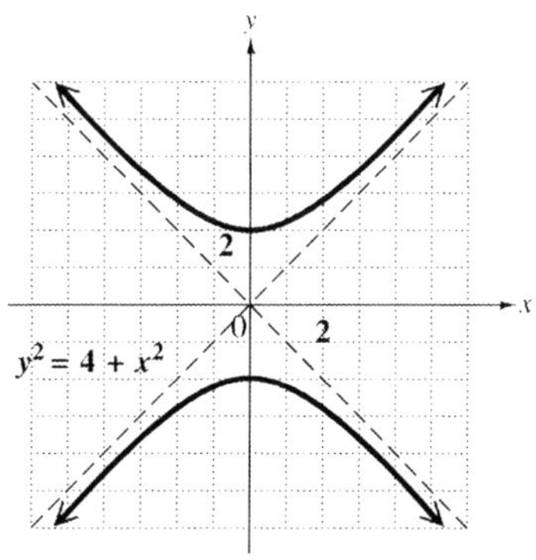

21. $x^2 + 9y^2 = 9$

$$\frac{x^2}{9} + \frac{y^2}{1} = 1 \qquad \textit{Divide by 9.}$$

The graph is an ellipse with x-intercepts $(3, 0)$ and $(-3, 0)$ and y-intercepts $(0, 1)$ and $(0, -1)$.

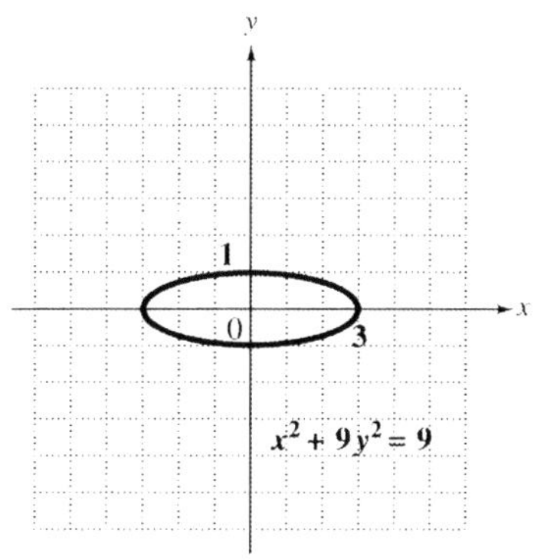

22. $f(x) = \sqrt{16 - x^2}$

Replace $f(x)$ with y and square both sides to get the equation

$$y^2 = 16 - x^2 \quad \text{or} \quad x^2 + y^2 = 16.$$

This is the graph of a circle with center $(0, 0)$ and radius 4. Since $f(x)$, or y, represents a principal square root in the original equation, $f(x)$ must be nonnegative. This restricts the graph to the upper half of the circle.

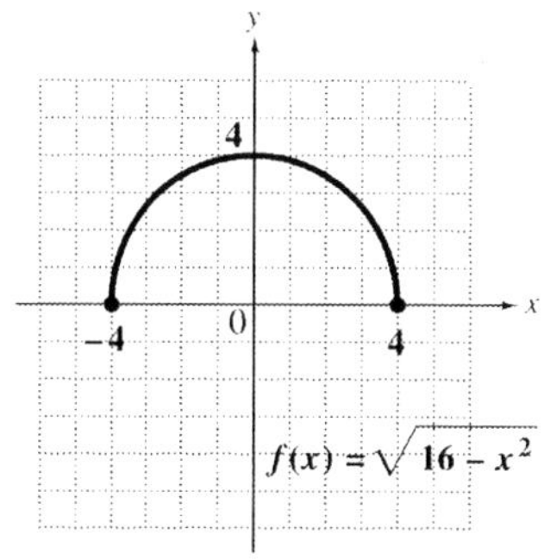

The domain is $[-4, 4]$, and the range is $[0, 4]$.

23. $f(x) = \sqrt{9 - x^2}$

Replace $f(x)$ with y and square both sides to get the equation

$$y^2 = 9 - x^2 \quad \text{or} \quad x^2 + y^2 = 9.$$

This is the graph of a circle with center $(0, 0)$ and radius 3. Since $f(x)$, or y, represents a principal square root in the original equation, $f(x)$ must be nonnegative. This restricts the graph to the upper half of the circle.

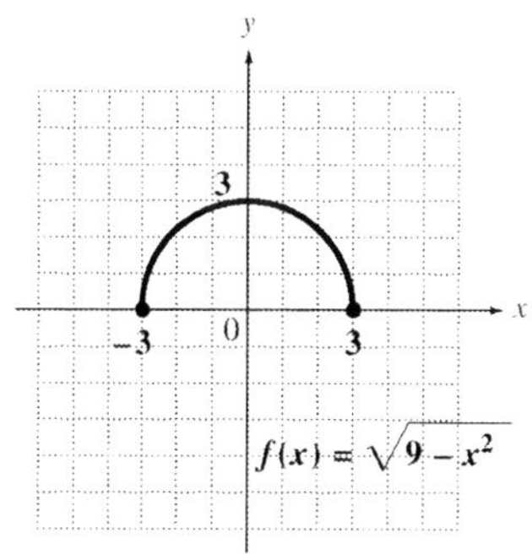

The domain is $[-3, 3]$, and the range is $[0, 3]$.

24. $$f(x) = -\sqrt{36 - x^2}$$

Replace $f(x)$ with y, and square both sides of the equation.

$$\begin{aligned} y &= -\sqrt{36 - x^2} \\ y^2 &= 36 - x^2 \\ x^2 + y^2 &= 36 \end{aligned}$$

This is a circle centered at the origin with radius $\sqrt{36} = 6$. Since $f(x)$, or y, represents a nonpositive square root in the original equation, $f(x)$ must be nonpositive. This restricts the graph to the bottom half of the circle.

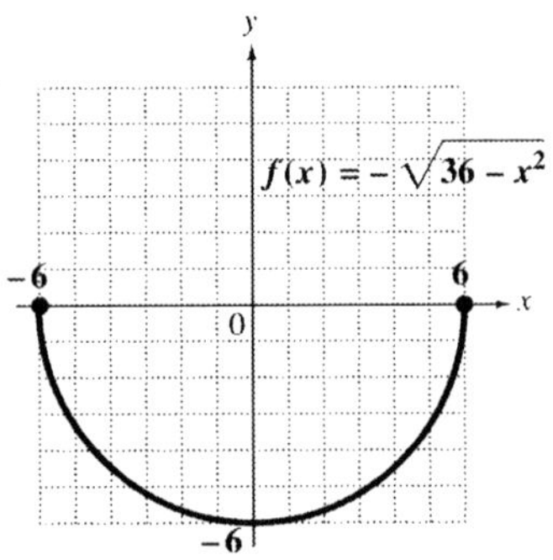

The domain is $[-6, 6]$, and the range is $[-6, 0]$.

25. $$f(x) = -\sqrt{25 - x^2}$$

Replace $f(x)$ with y, and square both sides of the equation.

$$\begin{aligned} y &= -\sqrt{25 - x^2} \\ y^2 &= 25 - x^2 \\ x^2 + y^2 &= 25 \end{aligned}$$

This is a circle centered at the origin with radius $\sqrt{25} = 5$. Since $f(x)$, or y, represents a nonpositive square root in the original equation, $f(x)$ must be nonpositive. This restricts the graph to the bottom half of the circle.

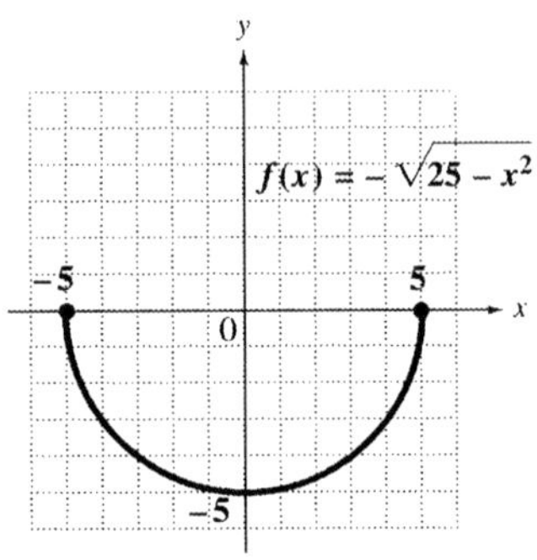

The domain is $[-5, 5]$, and the range is $[-5, 0]$.

26. $$\frac{y}{3} = \sqrt{1 + \frac{x^2}{9}}$$

Square both sides.

$$\begin{aligned} \frac{y^2}{9} &= 1 + \frac{x^2}{9} \\ \frac{y^2}{9} - \frac{x^2}{9} &= 1 \end{aligned}$$

This is a hyperbola opening up and down with y-intercepts $(0, 3)$ and $(0, -3)$. The four points $(3, 3)$, $(3, -3)$, $(-3, 3)$, and $(-3, -3)$ are the vertices of the rectangle that determine the asymptotes. Since $f(x)$, or y, represents a square root in the original equation, $f(x)$ must be nonnegative. This restricts the graph to the upper branch of the hyperbola.

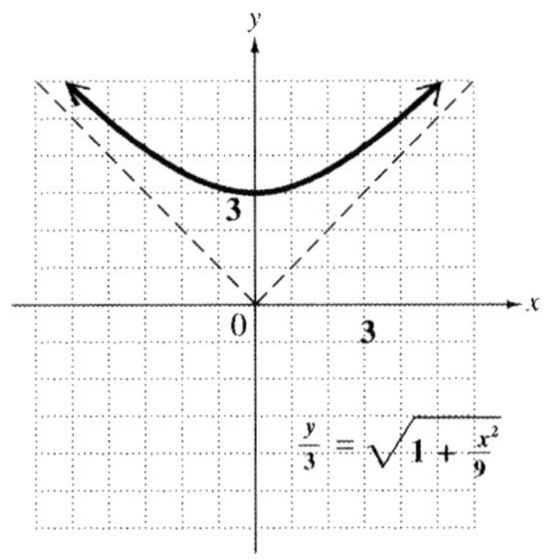

The domain is $(-\infty, \infty)$, and the range is $[3, \infty)$.

27. $$y = \sqrt{\frac{x + 4}{2}}$$

Square both sides.

$$\begin{aligned} y^2 &= \frac{x + 4}{2} \\ 2y^2 &= x + 4 \\ 2y^2 - 4 &= x \\ 2(y - 0)^2 - 4 &= x \end{aligned}$$

This is a parabola that opens to the right with vertex $(-4, 0)$. However, y is nonnegative in the original equation, so only the top half of the parabola is included in the graph.

x	-2	0	4
y	1	$\sqrt{2}$	2

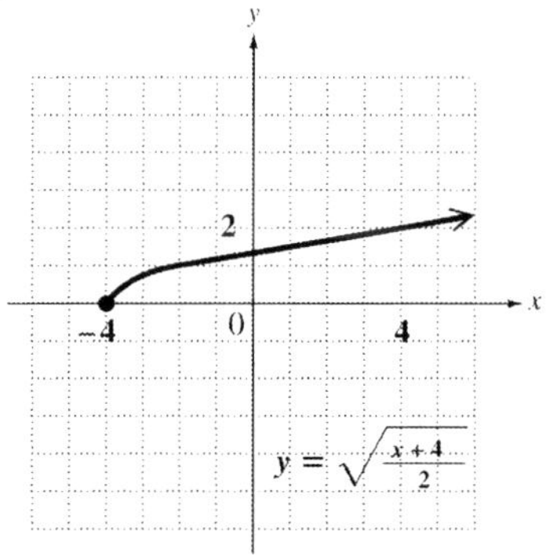

The domain is $[-4, \infty)$, and the range is $[0, \infty)$.

28.

$$y = -2\sqrt{\frac{9 - x^2}{9}}$$

$$y^2 = 4\left(\frac{9 - x^2}{9}\right) \quad \textit{Square both sides.}$$

$$\frac{y^2}{4} = \frac{9 - x^2}{9} \quad \textit{Divide by 4.}$$

$$\frac{y^2}{4} = 1 - \frac{x^2}{9}$$

$$\frac{x^2}{9} + \frac{y^2}{4} = 1$$

This is the equation of an ellipse with intercepts $(3, 0)$, $(-3, 0)$, $(0, 2)$, and $(0, -2)$. Since y equals a negative square root in the original equation, y must be nonpositive, restricting the graph to the lower half of the ellipse.

The domain of the function is $[-3, 3]$ and its range is $[-2, 0]$.

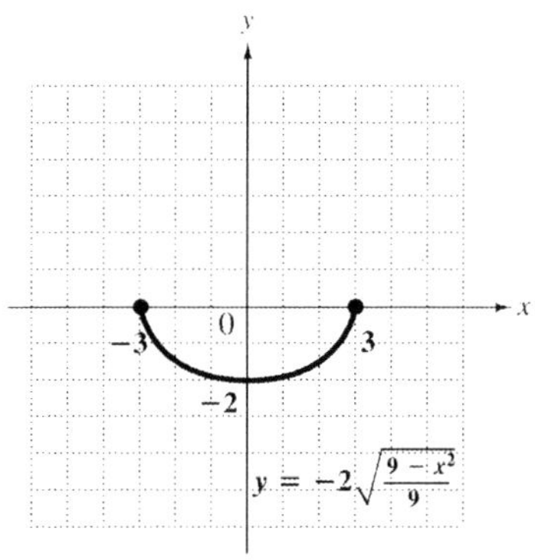

29. **(a)** $400x^2 - 625y^2 = 250{,}000$

Divide by 250,000.

$$\frac{x^2}{625} - \frac{y^2}{400} = 1$$

$$\frac{x^2}{25^2} - \frac{y^2}{20^2} = 1$$

The x-intercepts are $(25, 0)$ and $(-25, 0)$. The distance between the buildings is the distance between the x-intercepts. The buildings are $25 + 25 = 50$ meters apart at their closest point.

(b) At $x = 50$, $y = \frac{d}{2}$, so $d = 2y$.

$$400(50)^2 - 625y^2 = 250{,}000$$

$$1{,}000{,}000 - 625y^2 = 250{,}000$$

$$-625y^2 = -750{,}000$$

$$y^2 = 1200$$

$$y = \sqrt{1200}$$

The distance d is $2\sqrt{1200} \approx 69.3$ meters.

30. $\frac{x^2}{g^2} - \frac{y^2}{g^2} = 1 \quad (*)$

Think of the crossbar as the x-axis and the goal posts at $(g, 0)$ and $(-g, 0)$ as the x-intercepts. The asymptotes go through the center of the goal posts, $(0, 0)$, and on the line from $(-g, -g)$ to (g, g) and on the line from $(g, -g)$ to $(-g, g)$. In the hyperbola equation $(*)$, $a = b = g$, the asymptotes are

$$y = \pm\frac{b}{a}x = \pm\frac{g}{g}x = \pm x.$$

These form a $45°$ angle with the line through the goal posts. Most people can estimate a $45°$ angle fairly easily.

31. $V > \frac{2.82 \times 10^7}{\sqrt{D}}$

$$V > \frac{2.82 \times 10^7}{\sqrt{4.25 \times 10^7}} \approx 4325.68$$

The trajectory will be hyperbolic for V greater than 4325.68 meters per second.

32. **(a)** According to the graph, the percent for 1999 is about 60%.

(b) $y = .607\sqrt{383.9 + x^2}$

$y = .607\sqrt{383.9 + 99^2}$ *Let $x = 99$.*

$y \approx 61.26$

According to the equation, the percent is about 61%.

12.4 Nonlinear Systems of Equations

12.4 Margin Exercises

1. **(a)** $x^2 + y^2 = 10 \quad (1)$
$x = y + 2 \quad (2)$

Substitute $y + 2$ for x in equation (1).

$$\begin{aligned} x^2 + y^2 &= 10 \quad (1) \\ (y+2)^2 + y^2 &= 10 \\ y^2 + 4y + 4 + y^2 &= 10 \\ 2y^2 + 4y - 6 &= 0 \\ y^2 + 2y - 3 &= 0 \\ (y+3)(y-1) &= 0 \end{aligned}$$

$$y + 3 = 0 \quad \text{or} \quad y - 1 = 0$$
$$y = -3 \quad \text{or} \quad y = 1$$

From (2) with $y = -3$, $x = -3 + 2 = -1$.
From (2) with $y = 1$, $x = 1 + 2 = 3$.

Check $(-1, -3)$: $1 + 9 = 10$, $-1 = -3 + 2$
Check $(3, 1)$: $9 + 1 = 10$, $3 = 1 + 2$

Solution set: $\{(3, 1), (-1, -3)\}$

(b) $x^2 - 2y^2 = 8 \quad (1)$
$y + x = 6 \quad (2)$

Solve equation (2) for y.

$$y = 6 - x \quad (3)$$

Substitute $6 - x$ for y in equation (1).

$$\begin{aligned} x^2 - 2y^2 &= 8 \quad (1) \\ x^2 - 2(6-x)^2 &= 8 \\ x^2 - 2(x^2 - 12x + 36) &= 8 \\ x^2 - 2x^2 + 24x - 72 &= 8 \\ -x^2 + 24x - 80 &= 0 \\ x^2 - 24x + 80 &= 0 \\ (x-4)(x-20) &= 0 \end{aligned}$$

$$x - 4 = 0 \quad \text{or} \quad x - 20 = 0$$
$$x = 4 \quad \text{or} \quad x = 20$$

From (3) with $x = 4$, $y = 6 - 4 = 2$.
From (3) with $x = 20$, $y = 6 - 20 = -14$.

Check $(4, 2)$: $16 - 8 = 8$, $2 + 4 = 6$
Check $(20, -14)$: $400 - 392 = 8$, $-14 + 20 = 6$

Solution set: $\{(4, 2), (20, -14)\}$

2. **(a)** $xy = 8 \quad (1)$
$x + y = 6 \quad (2)$

Solve equation (2) for x.

$$x = 6 - y \quad (3)$$

Substitute $6 - y$ for x in equation (1).

$$\begin{aligned} xy &= 8 \quad (1) \\ (6-y)y &= 8 \\ 6y - y^2 &= 8 \\ -y^2 + 6y - 8 &= 0 \\ y^2 - 6y + 8 &= 0 \\ (y-2)(y-4) &= 0 \end{aligned}$$

$$y - 2 = 0 \quad \text{or} \quad y - 4 = 0$$
$$y = 2 \quad \text{or} \quad y = 4$$

From (3) with $y = 2$, $x = 6 - 2 = 4$.
From (3) with $y = 4$, $x = 6 - 4 = 2$.

Check $(4, 2)$: $4(2) = 8$, $4 + 2 = 6$
Check $(2, 4)$: $2(4) = 8$, $2 + 4 = 6$

Solution set: $\{(4, 2), (2, 4)\}$

(b) $xy + 10 = 0 \quad (1)$
$4x + 9y = -2 \quad (2)$

Solve equation (1) for y.

$$y = -\frac{10}{x} \quad (3)$$

Substitute $-\frac{10}{x}$ for y in equation (2).

$$\begin{aligned} 4x + 9y &= -2 \quad (2) \\ 4x + 9\left(-\frac{10}{x}\right) &= -2 \\ 4x - \frac{90}{x} &= -2 \\ 4x^2 - 90 &= -2x \quad \textit{Multiply by x.} \\ 4x^2 + 2x - 90 &= 0 \\ 2x^2 + x - 45 &= 0 \\ (2x-9)(x+5) &= 0 \end{aligned}$$

$$2x - 9 = 0 \quad \text{or} \quad x + 5 = 0$$
$$x = \tfrac{9}{2} \quad \text{or} \quad x = -5$$

From (3) with $x = \dfrac{9}{2}$, $y = -\dfrac{10}{\frac{9}{2}} = -\dfrac{20}{9}$.

From (3) with $x = -5$, $y = -\frac{10}{-5} = 2$.

Check $\left(\frac{9}{2}, -\frac{20}{9}\right)$: $-10 + 10 = 0$, $18 - 20 = -2$
Check $(-5, 2)$: $-10 + 10 = 0$, $-20 + 18 = -2$

Solution set: $\left\{(-5, 2), \left(\frac{9}{2}, -\frac{20}{9}\right)\right\}$

3. **(a)**

$$\begin{array}{rcrcl} x^2 & + & y^2 & = & 41 \quad (1) \\ x^2 & - & y^2 & = & 9 \quad (2) \\ \hline 2x^2 & & & = & 50 \quad \textit{Add} \\ & & x^2 & = & 25 \end{array}$$

$x = 5$ or $x = -5$

Substitute each value of x in equation (1) to find y. First substitute 5 for x.

$$\begin{aligned} x^2 + y^2 &= 41 \quad (1) \\ 5^2 + y^2 &= 41 \\ 25 + y^2 &= 41 \\ y^2 &= 16 \end{aligned}$$

$y = 4$ or $y = -4$

Now substitute -5 for x in equation (1).

$$\begin{aligned} x^2 + y^2 &= 41 \quad (1) \\ (-5)^2 + y^2 &= 41 \\ 25 + y^2 &= 41 \\ y^2 &= 16 \end{aligned}$$

$y = 4$ or $y = -4$

Check: If $x = \pm 5$, $x^2 = 25$, and if $y = \pm 4$, $y^2 = 16$. Thus, in any case, we get $25 + 16 = 41$ in (1) and $25 - 16 = 9$ in (2). All four ordered pairs check.

Solution set: $\{(5, 4), (5, -4), (-5, 4), (-5, -4)\}$

(b) $x^2 + 3y^2 = 40 \quad (1)$
$4x^2 - y^2 = 4 \quad (2)$

Multiply equation (2) by 3 and add the result to equation (1).

$$\begin{array}{rcrcl} x^2 & + & 3y^2 & = & 40 \quad (1) \\ 12x^2 & - & 3y^2 & = & 12 \quad 3 \times (2) \\ \hline 13x^2 & & & = & 52 \\ & & x^2 & = & 4 \end{array}$$

$x = 2$ or $x = -2$

Substitute 2 for x in equation (1) to find y.

$$\begin{aligned} x^2 + 3y^2 &= 40 \quad (1) \\ 2^2 + 3y^2 &= 40 \\ 4 + 3y^2 &= 40 \\ 3y^2 &= 36 \\ y^2 &= 12 \end{aligned}$$

$y = 2\sqrt{3}$ or $y = -2\sqrt{3}$

Since x is squared in equation (1), the values of y, $2\sqrt{3}$ or $-2\sqrt{3}$, are the same for $x = 2$ and $x = -2$. So we get four solutions as in part (a).

Solution set:

$\{(2, 2\sqrt{3}), (2, -2\sqrt{3}), (-2, 2\sqrt{3}), (-2, -2\sqrt{3})\}$

4. **(a)** $x^2 + xy + y^2 = 3 \quad (1)$
$x^2 + y^2 = 5 \quad (2)$

Multiply equation (2) by -1 and add the result to equation (1).

$$\begin{array}{rcrcrcl} x^2 & + & xy & + & y^2 & = & 3 \quad (1) \\ -x^2 & & & - & y^2 & = & -5 \quad -1 \times (2) \\ \hline & & xy & & & = & -2 \quad (3) \end{array}$$

Solve equation (3) for y to get

$$y = -\frac{2}{x}. \quad (4)$$

Substitute $-\frac{2}{x}$ for y in equation (2).

$$\begin{aligned} x^2 + y^2 &= 5 \quad (2) \\ x^2 + \left(-\frac{2}{x}\right)^2 &= 5 \\ x^2 + \frac{4}{x^2} &= 5 \\ x^4 + 4 &= 5x^2 \quad \textit{Multiply by } x^2. \\ x^4 - 5x^2 + 4 &= 0 \\ (x^2 - 4)(x^2 - 1) &= 0 \end{aligned}$$

$x^2 - 4 = 0$ or $x^2 - 1 = 0$
$x^2 = 4$ or $x^2 = 1$
$x = \pm 2$ or $x = \pm 1$

Substitute these values for x in equation (4) to find the values of y.

If $x = 1$, then $y = -2$. If $x = -1$, then $y = 2$. If $x = 2$, then $y = -1$. If $x = -2$, then $y = 1$.

Solution set: $\{(1, -2), (-1, 2), (2, -1), (-2, 1)\}$

(b) $x^2 + 7xy - 2y^2 = -8 \quad (1)$
$-2x^2 + 4y^2 = 16 \quad (2)$

Multiply equation (1) by 2 and add the result to equation (2).

$$\begin{array}{rcrcrcl} 2x^2 & + & 14xy & - & 4y^2 & = & -16 \quad 2 \times (1) \\ -2x^2 & & & + & 4y^2 & = & 16 \quad (2) \\ \hline & & 14xy & & & = & 0 \\ & & & & xy & = & 0 \end{array}$$

If $xy = 0$, then either $x = 0$ or $y = 0$.

If $x = 0$, then substitute 0 for x in equation (1).

$$\begin{aligned} x^2 + 7xy - 2y^2 &= -8 \quad (1) \\ 0 + 0 - 2y^2 &= -8 \\ y^2 &= 4 \end{aligned}$$

$y = 2$ or $y = -2$

continued

If $y = 0$, then substitute 0 for y in equation (1).

$$x^2 + 7xy - 2y^2 = -8 \qquad (1)$$
$$x^2 + 0 + 0 = -8$$

$$x = \sqrt{-8} \quad \text{or} \quad x = -\sqrt{-8}$$
$$x = 2i\sqrt{2} \quad \text{or} \quad x = -2i\sqrt{2}$$

Solution set:

$$\left\{(0,2), (0,-2), (2i\sqrt{2}, 0), \left(-2i\sqrt{2}, 0\right)\right\}$$

12.4 Section Exercises

1. Substitute $x - 1$ for y in the first equation. Then solve for x. Find the corresponding y-values by substituting back into $y = x - 1$. In the first equation, both variables are squared and in the second, both variables are to the first power, so the elimination method is not appropriate.

2. Add the two equations to eliminate y^2. Solve for the values of x. Then go back to either original equation to solve for the values of y.

3. The line intersects the ellipse in exactly one point, so there is *one* point in the solution set of the system.

4. The line intersects the circle twice, so there are *two* points in the solution set of the system.

5. The line does not intersect the hyperbola, so there are no points in the solution set of the system.

6. The ellipse and the hyperbola intersect four times, so there are *four* points in the solution set of the system.

7. A line and a circle; no points

   Draw any circle, and then draw a line that does not cross the circle.

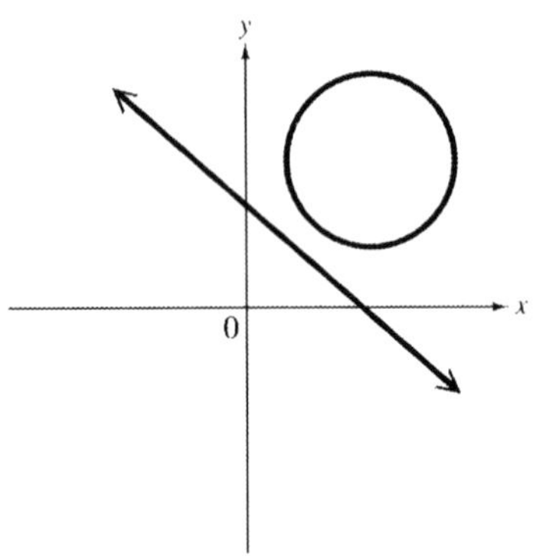

8. A line and a circle; one point

   The line must be tangent to the circle; i.e., it must touch the circle at only one point.

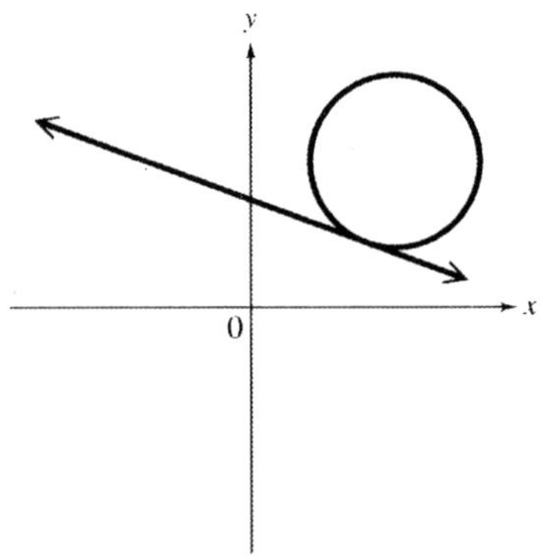

9. A line and an ellipse; two points

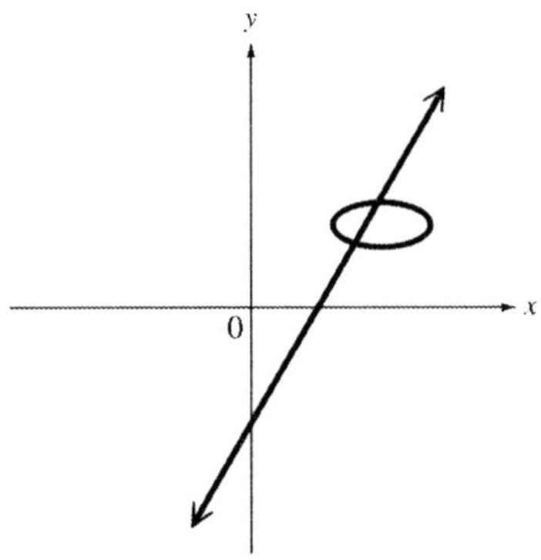

10. A line and a hyperbola; no points

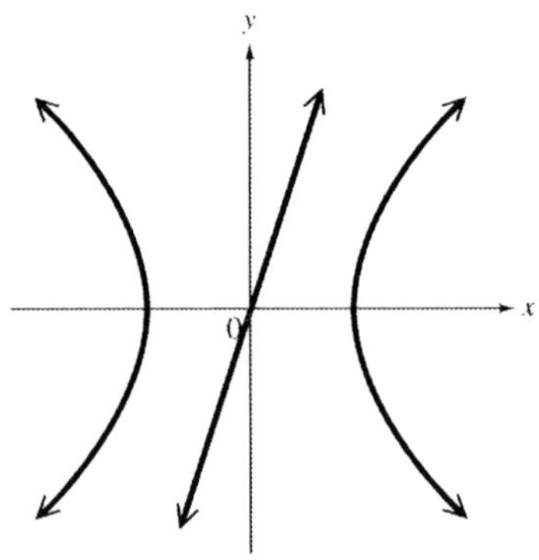

11. A circle and an ellipse; four points

    Draw any ellipse, and then draw a circle with the same center whose radius is just large enough so that there are four points of intersection. (If the radius of the circle is too large or too small, there may be fewer points of intersection.)

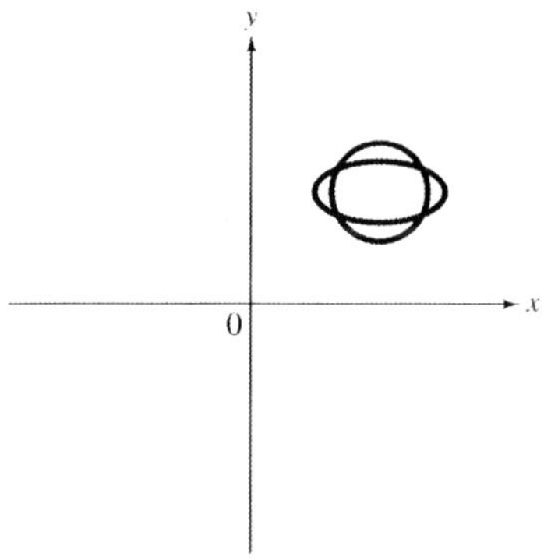

12. A parabola and an ellipse; one point

The parabola must be tangent to the ellipse.

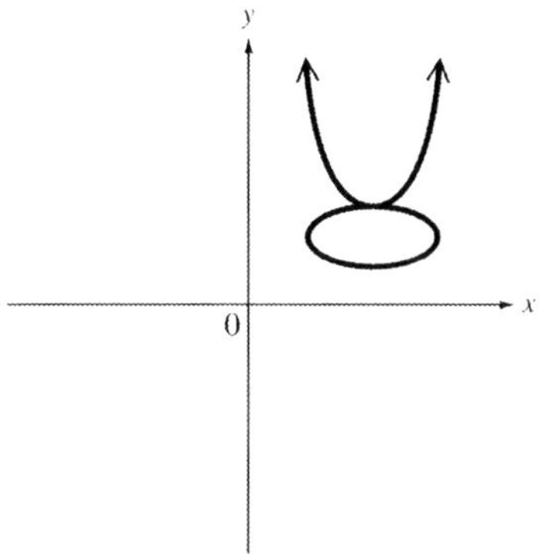

13. $y = 4x^2 - x \quad (1)$
$y = x \quad (2)$

Substitute x for y in equation (1).

$$\begin{aligned} y &= 4x^2 - x \quad (1) \\ x &= 4x^2 - x \\ 0 &= 4x^2 - 2x \\ 0 &= 2x(2x - 1) \end{aligned}$$

$$2x = 0 \quad \text{or} \quad 2x - 1 = 0$$
$$x = 0 \quad \text{or} \quad x = \tfrac{1}{2}$$

Use equation (2) to find y for each x-value.

If $x = 0$, then $y = 0$.
If $x = \frac{1}{2}$, then $y = \frac{1}{2}$.

Solution set: $\{(0, 0), (\frac{1}{2}, \frac{1}{2})\}$

14. $y = x^2 + 6x \quad (1)$
$3y = 12x \quad (2)$

Replace y with $x^2 + 6x$ in (2).

$$\begin{aligned} 3(x^2 + 6x) &= 12x \\ 3x^2 + 18x &= 12x \\ 3x^2 + 6x &= 0 \\ 3x(x + 2) &= 0 \end{aligned}$$

$$3x = 0 \quad \text{or} \quad x + 2 = 0$$
$$x = 0 \quad \text{or} \quad x = -2$$

Let $x = 0$ in (1).

$$y = 0^2 + 6(0) = 0$$

Let $x = -2$ in (1).

$$\begin{aligned} y &= (-2)^2 + 6(-2) \\ &= 4 - 12 = -8 \end{aligned}$$

Solution set: $\{(0, 0), (-2, -8)\}$

15. $y = x^2 + 6x + 9 \quad (1)$
$x + y = 3 \quad (2)$

Substitute $x^2 + 6x + 9$ for y in equation (2).

$$\begin{aligned} x + y &= 3 \quad (2) \\ x + (x^2 + 6x + 9) &= 3 \\ x^2 + 7x + 9 &= 3 \\ x^2 + 7x + 6 &= 0 \\ (x + 6)(x + 1) &= 0 \end{aligned}$$

$$x + 6 = 0 \quad \text{or} \quad x + 1 = 0$$
$$x = -6 \quad \text{or} \quad x = -1$$

Substitute these values for x in equation (2) and solve for y.

If $x = -6$, then

$$\begin{aligned} x + y &= 3 \quad (2) \\ -6 + y &= 3 \\ y &= 9. \end{aligned}$$

If $x = -1$, then

$$\begin{aligned} x + y &= 3 \quad (2) \\ -1 + y &= 3 \\ y &= 4. \end{aligned}$$

Solution set: $\{(-6, 9), (-1, 4)\}$

16. $y = x^2 + 8x + 16 \quad (1)$
$x - y = -4 \quad (2)$

Replace y with $x^2 + 8x + 16$ in (2).

$$\begin{aligned} x - (x^2 + 8x + 16) &= -4 \\ x - x^2 - 8x - 16 &= -4 \\ -x^2 - 7x - 12 &= 0 \\ x^2 + 7x + 12 &= 0 \\ (x + 3)(x + 4) &= 0 \end{aligned}$$

$$x + 3 = 0 \quad \text{or} \quad x + 4 = 0$$
$$x = -3 \quad \text{or} \quad x = -4$$

continued

Let $x = -3$ in (2).

$$\begin{aligned} -3 - y &= -4 \\ -y &= -1 \\ y &= 1 \end{aligned}$$

Let $x = -4$ in (2).

$$\begin{aligned} -4 - y &= -4 \\ -y &= 0 \\ y &= 0 \end{aligned}$$

Solution set: $\{(-3, 1), (-4, 0)\}$

17. $x^2 + y^2 = 2 \quad (1)$
$2x + y = 1 \quad (2)$

Solve equation (2) for y.

$$y = 1 - 2x \qquad (3)$$

Substitute $1 - 2x$ for y in equation (1).

$$\begin{aligned} x^2 + y^2 &= 2 \qquad (1) \\ x^2 + (1 - 2x)^2 &= 2 \\ x^2 + 1 - 4x + 4x^2 &= 2 \\ 5x^2 - 4x - 1 &= 0 \\ (5x + 1)(x - 1) &= 0 \end{aligned}$$

$$5x + 1 = 0 \quad \text{or} \quad x - 1 = 0$$
$$x = -\tfrac{1}{5} \quad \text{or} \quad x = 1$$

Use equation (3) to find y for each x-value.

If $x = -\frac{1}{5}$, then

$$y = 1 - 2\left(-\tfrac{1}{5}\right) = 1 + \tfrac{2}{5} = \tfrac{7}{5}.$$

If $x = 1$, then

$$y = 1 - 2(1) = -1.$$

Solution set: $\left\{\left(-\frac{1}{5}, \frac{7}{5}\right), (1, -1)\right\}$

18. $2x^2 + 4y^2 = 4 \quad (1)$
$x = 4y \quad (2)$

Replace x with $4y$ in (1).

$$\begin{aligned} 2(4y)^2 + 4y^2 &= 4 \\ 32y^2 + 4y^2 &= 4 \\ 36y^2 &= 4 \\ y^2 &= \frac{1}{9} \\ y &= \pm\tfrac{1}{3} \end{aligned}$$

Let $y = \frac{1}{3}$in (2).

$$x = 4\left(\tfrac{1}{3}\right) = \tfrac{4}{3}$$

Let $y = -\frac{1}{3}$ in (2).

$$x = 4\left(-\tfrac{1}{3}\right) = -\tfrac{4}{3}$$

Solution set: $\left\{\left(\frac{4}{3}, \frac{1}{3}\right), \left(-\frac{4}{3}, -\frac{1}{3}\right)\right\}$

19. $xy = 4 \quad (1)$
$3x + 2y = -10 \quad (2)$

Solve equation (1) for y to get

$$y = \frac{4}{x}. \quad (3)$$

Substitute $\frac{4}{x}$ for y in equation (2) to find x.

$$\begin{aligned} 3x + 2y &= -10 \qquad (2) \\ 3x + 2\left(\frac{4}{x}\right) &= -10 \end{aligned}$$

Multiply by the LCD, x.

$$\begin{aligned} 3x^2 + 8 &= -10x \\ 3x^2 + 10x + 8 &= 0 \\ (3x + 4)(x + 2) &= 0 \end{aligned}$$

$$3x + 4 = 0 \quad \text{or} \quad x + 2 = 0$$
$$x = -\tfrac{4}{3} \quad \text{or} \quad x = -2$$

Substitute these values for x in equation (3).
If $x = -\frac{4}{3}$, then $y = \frac{4}{-\frac{4}{3}} = -3$.
If $x = -2$, then $y = \frac{4}{-2} = -2$.

Solution set: $\left\{(-2, -2), \left(-\frac{4}{3}, -3\right)\right\}$

20. $xy = -5 \quad (1)$
$2x + y = 3$
or $y = 3 - 2x \quad (2)$

Replace y with $3 - 2x$ in (1).

$$\begin{aligned} x(3 - 2x) &= -5 \\ 3x - 2x^2 &= -5 \\ 2x^2 - 3x - 5 &= 0 \\ (x + 1)(2x - 5) &= 0 \end{aligned}$$

$$x + 1 = 0 \quad \text{or} \quad 2x - 5 = 0$$
$$x = -1 \quad \text{or} \quad x = \tfrac{5}{2}$$

Let $x = -1$ in (2).

$$y = 3 - 2(-1) = 5$$

Let $x = \frac{5}{2}$ in (2).

$$y = 3 - 2\left(\tfrac{5}{2}\right) = -2$$

Solution set: $\left\{(-1, 5), \left(\frac{5}{2}, -2\right)\right\}$

21. $xy = -3 \quad (1)$
$x + y = -2 \quad (2)$

Solve equation (2) for y.

$$y = -x - 2 \qquad (3)$$

Substitute $-x - 2$ for y in equation (1).

$$\begin{aligned} xy &= -3 \qquad (1) \\ x(-x-2) &= -3 \\ -x^2 - 2x &= -3 \\ -x^2 - 2x + 3 &= 0 \\ x^2 + 2x - 3 &= 0 \qquad \textit{Multiply by } -1. \\ (x+3)(x-1) &= 0 \end{aligned}$$

$$x + 3 = 0 \quad \text{or} \quad x - 1 = 0$$
$$x = -3 \quad \text{or} \quad x = 1$$

Use equation (3) to find y for each x-value.

If $x = -3$, then
$$y = -(-3) - 2 = 1.$$

If $x = 1$, then
$$y = -(1) - 2 = -3.$$

Solution set: $\{(-3, 1), (1, -3)\}$

22. $xy = 12 \quad (1)$
$x + y = 8$
or $y = 8 - x \quad (2)$

Replace y by $8 - x$ in (1).

$$\begin{aligned} x(8-x) &= 12 \\ 8x - x^2 &= 12 \\ x^2 - 8x + 12 &= 0 \\ (x-6)(x-2) &= 0 \end{aligned}$$

$$x - 6 = 0 \quad \text{or} \quad x - 2 = 0$$
$$x = 6 \quad \text{or} \quad x = 2$$

Let $x = 6$ in (2).

$$y = 8 - 6 = 2$$

Let $x = 2$ in (2).

$$y = 8 - 2 = 6$$

Solution set: $\{(6, 2), (2, 6)\}$

23. $y = 3x^2 + 6x \quad (1)$
$y = x^2 - x - 6 \quad (2)$

Substitute $x^2 - x - 6$ for y in equation (1) to find x.

$$\begin{aligned} y &= 3x^2 + 6x \qquad (1) \\ x^2 - x - 6 &= 3x^2 + 6x \\ 0 &= 2x^2 + 7x + 6 \\ 0 &= (2x+3)(x+2) \end{aligned}$$

$$2x + 3 = 0 \quad \text{or} \quad x + 2 = 0$$
$$x = -\tfrac{3}{2} \quad \text{or} \quad x = -2$$

Substitute $-\frac{3}{2}$ for x in equation (1) to find y.

$$\begin{aligned} y &= 3x^2 + 6x \qquad (1) \\ y &= 3\left(-\tfrac{3}{2}\right)^2 + 6\left(-\tfrac{3}{2}\right) \\ &= 3\left(\tfrac{9}{4}\right) + 6\left(-\tfrac{6}{4}\right) \\ &= \tfrac{27}{4} - \tfrac{36}{4} = -\tfrac{9}{4} \end{aligned}$$

Substitute -2 for x in equation (1) to find y.

$$\begin{aligned} y &= 3x^2 + 6x \qquad (1) \\ y &= 3(-2)^2 + 6(-2) \\ &= 12 - 12 = 0 \end{aligned}$$

Solution set: $\left\{\left(-\frac{3}{2}, -\frac{9}{4}\right), (-2, 0)\right\}$

24. $y = 2x^2 + 1 \quad (1)$
$y = 5x^2 + 2x - 7 \quad (2)$

Replace y by $2x^2 + 1$ in (2).

$$\begin{aligned} 2x^2 + 1 &= 5x^2 + 2x - 7 \\ 0 &= 3x^2 + 2x - 8 \\ 0 &= (x+2)(3x-4) \end{aligned}$$

$$x + 2 = 0 \quad \text{or} \quad 3x - 4 = 0$$
$$x = -2 \quad \text{or} \quad x = \tfrac{4}{3}$$

Let $x = -2$ in (1).

$$y = 2(-2)^2 + 1 = 9$$

Let $x = \frac{4}{3}$ in (1).

$$y = 2\left(\tfrac{4}{3}\right)^2 + 1 = \tfrac{32}{9} + 1 = \tfrac{41}{9}$$

Solution set: $\left\{(-2, 9), \left(\frac{4}{3}, \frac{41}{9}\right)\right\}$

25. $2x^2 - y^2 = 6 \quad (1)$
$y = x^2 - 3 \quad (2)$

Solve equation (2) for x^2.

$$x^2 = y + 3 \quad (3)$$

Substitute $y + 3$ for x^2 in equation (1).

$$\begin{aligned} 2x^2 - y^2 &= 6 \qquad (1) \\ 2(y+3) - y^2 &= 6 \\ 2y + 6 - y^2 &= 6 \\ 0 &= y^2 - 2y \\ 0 &= y(y-2) \end{aligned}$$

$$y = 0 \quad \text{or} \quad y = 2$$

continued

From (3) with $y = 0$, $x^2 = 3$, so $x = \pm\sqrt{3}$.

From (3) with $y = 2$, $x^2 = 5$, so $x = \pm\sqrt{5}$.

Solution set:

$$\left\{\left(-\sqrt{3}, 0\right), \left(\sqrt{3}, 0\right), \left(-\sqrt{5}, 2\right), \left(\sqrt{5}, 2\right)\right\}$$

26. $x^2 + y^2 = 4$ (1)

$y = x^2 - 2$ (2)

Replace y by $x^2 - 2$ in (1).

$$\begin{aligned} x^2 + \left(x^2 - 2\right)^2 &= 4 \\ x^2 + x^4 - 4x^2 + 4 &= 4 \\ x^4 - 3x^2 &= 0 \\ x^2\left(x^2 - 3\right) &= 0 \end{aligned}$$

$$\begin{aligned} x^2 = 0 \quad &\text{or} \quad x^2 - 3 = 0 \\ x = 0 \quad &\text{or} \quad x^2 = 3 \\ & \quad\quad x = \pm\sqrt{3} \end{aligned}$$

Let $x = 0$ in (2).

$$y = (0)^2 - 2 = -2$$

If $x = \sqrt{3}$ or $x = -\sqrt{3}$ in (2), then

$$y = \left(\pm\sqrt{3}\right)^2 - 2 = 3 - 2 = 1.$$

Solution set: $\left\{(0, -2), \left(\sqrt{3}, 1\right), \left(-\sqrt{3}, 1\right)\right\}$

27. $3x^2 + 2y^2 = 12$ (1)

$x^2 + 2y^2 = 4$ (2)

Multiply equation (2) by -1 and add the result to equation (1).

$$\begin{aligned} 3x^2 + 2y^2 &= 12 \quad (1) \\ -x^2 - 2y^2 &= -4 \quad -1 \times (2) \\ \hline 2x^2 &= 8 \\ x^2 &= 4 \\ x &= \pm 2 \end{aligned}$$

Substitute ± 2 for x in equation (2) to find y.

$$\begin{aligned} x^2 + 2y^2 &= 4 \quad (2) \\ (\pm 2)^2 + 2y^2 &= 4 \\ 4 + 2y^2 &= 4 \\ 2y^2 &= 0 \\ y^2 &= 0 \\ y &= 0 \end{aligned}$$

Solution set: $\{(-2, 0), (2, 0)\}$

28. $2x^2 + y^2 = 28$ (1)

$4x^2 - 5y^2 = 28$ (2)

Multiply (1) by 5 and add the result to equation (2).

$$\begin{aligned} 10x^2 + 5y^2 &= 140 \quad 5 \times (1) \\ 4x^2 - 5y^2 &= 28 \quad (2) \\ \hline 14x^2 &= 168 \\ x^2 &= 12 \\ x &= \pm\sqrt{12} = \pm 2\sqrt{3} \end{aligned}$$

Let $x = \pm 2\sqrt{3}$ in (1).

$$\begin{aligned} 2\left(\pm 2\sqrt{3}\right)^2 + y^2 &= 28 \\ 2(12) + y^2 &= 28 \\ y^2 &= 4 \\ y &= \pm 2 \end{aligned}$$

Solution set: $\left\{\left(2\sqrt{3}, 2\right), \left(2\sqrt{3}, -2\right), \left(-2\sqrt{3}, 2\right), \left(-2\sqrt{3}, -2\right)\right\}$

29. $xy = 6$ (1)

$3x^2 - y^2 = 12$ (2)

Solve equation (1) for y to get

$$y = \frac{6}{x}. \quad (3)$$

Substitute $\frac{6}{x}$ for y in equation (2).

$$\begin{aligned} 3x^2 - \left(\frac{6}{x}\right)^2 &= 12 \\ 3x^2 - \frac{36}{x^2} &= 12 \\ 3x^4 - 36 &= 12x^2 \\ 3x^4 - 12x^2 - 36 &= 0 \\ x^4 - 4x^2 - 12 &= 0 \\ \left(x^2 - 6\right)\left(x^2 + 2\right) &= 0 \end{aligned}$$

$$\begin{aligned} x^2 - 6 = 0 \quad &\text{or} \quad x^2 + 2 = 0 \\ x^2 = 6 \quad &\text{or} \quad x^2 = -2 \\ x = \pm\sqrt{6} \quad &\text{or} \quad x = \pm i\sqrt{2} \end{aligned}$$

Use equation (3) to find y.

If $x = i\sqrt{2}$,

$$y = \frac{6}{i\sqrt{2}} = \frac{6 \cdot i\sqrt{2}}{i\sqrt{2} \cdot i\sqrt{2}} = \frac{6i\sqrt{2}}{2i^2} = -3i\sqrt{2}.$$

If $x = -i\sqrt{2}$, $y = 3i\sqrt{2}$.

If $x = \sqrt{6}$, $y = \dfrac{6}{\sqrt{6}} = \dfrac{6 \cdot \sqrt{6}}{\sqrt{6} \cdot \sqrt{6}} = \dfrac{6\sqrt{6}}{6} = \sqrt{6}$.

If $x = -\sqrt{6}$, $y = -\sqrt{6}$.

Solution set: $\{(i\sqrt{2}, -3i\sqrt{2}), (-i\sqrt{2}, 3i\sqrt{2}), (-\sqrt{6}, -\sqrt{6}), (\sqrt{6}, \sqrt{6})\}$

30.
$$\begin{aligned} xy &= 5 \quad (1) \\ 2y^2 - x^2 &= 5 \quad (2) \end{aligned}$$

Solve equation (1) for y to get

$$y = \frac{5}{x}. \quad (3)$$

Substitute $\frac{5}{x}$ for y in equation (2).

$$\begin{aligned} 2\left(\frac{5}{x}\right)^2 - x^2 &= 5 \\ \frac{50}{x^2} - x^2 &= 5 \\ 50 - x^4 &= 5x^2 \\ x^4 + 5x^2 - 50 &= 0 \\ (x^2 + 10)(x^2 - 5) &= 0 \end{aligned}$$

$$\begin{aligned} x^2 + 10 &= 0 \quad \text{or} \quad x^2 - 5 = 0 \\ x^2 &= -10 \quad \text{or} \quad x^2 = 5 \\ x &= \pm i\sqrt{10} \quad \text{or} \quad x = \pm\sqrt{5} \end{aligned}$$

Use equation (3) to find y.

If $x = \sqrt{5}$, $y = \dfrac{5}{\sqrt{5}} = \dfrac{5\sqrt{5}}{\sqrt{5} \cdot \sqrt{5}} = \dfrac{5\sqrt{5}}{5} = \sqrt{5}$.

If $x = -\sqrt{5}$, $y = -\sqrt{5}$.

If $x = i\sqrt{10}$,

$$y = \frac{5}{i\sqrt{10}} = \frac{5 \cdot i\sqrt{10}}{i\sqrt{10} \cdot i\sqrt{10}} = \frac{5i\sqrt{10}}{-10} = -\frac{i\sqrt{10}}{2}.$$

If $x = -i\sqrt{10}$, $y = \dfrac{i\sqrt{10}}{2}$.

Solution set: $\left\{(\sqrt{5}, \sqrt{5}), (-\sqrt{5}, -\sqrt{5}) \left(i\sqrt{10}, -\frac{i\sqrt{10}}{2}\right), \left(-i\sqrt{10}, \frac{i\sqrt{10}}{2}\right)\right\}$

31.
$$\begin{aligned} 2x^2 + 2y^2 &= 8 \quad (1) \\ 3x^2 + 4y^2 &= 24 \quad (2) \end{aligned}$$

Multiply equation (1) by -2 and add the result to equation (2).

$$\begin{aligned} -4x^2 - 4y^2 &= -16 \qquad -2 \times (1) \\ 3x^2 + 4y^2 &= 24 \quad (2) \\ \hline -x^2 &= 8 \\ x^2 &= -8 \\ x &= \pm\sqrt{-8} = \pm 2i\sqrt{2} \end{aligned}$$

Substitute $\pm 2i\sqrt{2}$ for x in equation (1).

$$\begin{aligned} 2x^2 &= 8 - 2y^2 \qquad (1) \\ 2\left(\pm 2i\sqrt{2}\right)^2 &= 8 - 2y^2 \\ 2(-8) &= 8 - 2y^2 \\ -16 &= 8 - 2y^2 \\ 2y^2 &= 24 \\ y^2 &= 12 \\ y &= \pm\sqrt{12} = \pm 2\sqrt{3} \end{aligned}$$

Since $2i\sqrt{2}$ can be paired with either $2\sqrt{3}$ or $-2\sqrt{3}$ and $-2i\sqrt{2}$ can be paired with either $2\sqrt{3}$ or $-2\sqrt{3}$, there are four possible solutions.

Solution set:
$\{(-2i\sqrt{2}, -2\sqrt{3}), (-2i\sqrt{2}, 2\sqrt{3}), (2i\sqrt{2}, -2\sqrt{3}), (2i\sqrt{2}, 2\sqrt{3})\}$

32.
$$\begin{aligned} 5x^2 + 5y^2 &= 20 \quad (1) \\ x^2 + 2y^2 &= 2 \quad (2) \end{aligned}$$

Multiply equation (2) by -5 and add to equation (1).

$$\begin{aligned} 5x^2 + 5y^2 &= 20 \quad (1) \\ -5x^2 - 10y^2 &= -10 \qquad -5 \times (2) \\ \hline -5y^2 &= 10 \\ y^2 &= -2 \\ y &= \pm\sqrt{-2} = \pm i\sqrt{2} \end{aligned}$$

Let $y = \pm i\sqrt{2}$ in equation (2).

$$\begin{aligned} x^2 + 2\left(\pm i\sqrt{2}\right)^2 &= 2 \\ x^2 + 2(-2) &= 2 \\ x^2 - 4 &= 2 \\ x^2 &= 6 \\ x &= \pm\sqrt{6} \end{aligned}$$

Solution set:

$$\left\{\left(\sqrt{6}, i\sqrt{2}\right), \left(-\sqrt{6}, i\sqrt{2}\right), \left(\sqrt{6}, -i\sqrt{2}\right), \left(-\sqrt{6}, -i\sqrt{2}\right)\right\}$$

33. $x^2 + xy + y^2 = 15 \quad (1)$
$x^2 + y^2 = 10 \quad (2)$

Multiply equation (2) by -1 and add the result to equation (1).

$$\begin{array}{rcrcrcrl} x^2 & + & xy & + & y^2 & = & 15 & (1) \\ -x^2 & & & - & y^2 & = & -10 & \quad -1 \times (2) \\ \hline & & xy & & & = & 5 & \\ & & & & y & = & \frac{5}{x} & (3) \end{array}$$

Substitute $\frac{5}{x}$ for y in equation (2).

$$x^2 + y^2 = 10 \qquad (2)$$
$$x^2 + \left(\frac{5}{x}\right)^2 = 10$$
$$x^2 + \frac{25}{x^2} = 10$$
$$x^4 + 25 = 10x^2 \quad \textit{Multiply by } x^2.$$
$$x^4 - 10x^2 + 25 = 0$$

Let $z = x^2$, so $z^2 = x^4$.

$$z^2 - 10z + 25 = 0$$
$$(z-5)^2 = 0$$
$$z - 5 = 0$$
$$z = 5$$

Since $z = x^2$, $x^2 = 5$, so $x = \pm\sqrt{5}$.

Using equation (3) we get the following.

If $x = -\sqrt{5}$, then

$$y = \frac{5}{-\sqrt{5}} = \frac{5 \cdot \sqrt{5}}{-\sqrt{5} \cdot \sqrt{5}} = \frac{5\sqrt{5}}{-5} = -\sqrt{5}.$$

Similarly, if $x = \sqrt{5}$, then $y = \sqrt{5}$.

Solution set: $\left\{\left(-\sqrt{5}, -\sqrt{5}\right), \left(\sqrt{5}, \sqrt{5}\right)\right\}$

34. $2x^2 + 3xy + 2y^2 = 21 \quad (1)$
$x^2 + y^2 = 6 \quad (2)$

Multiply equation (2) by -2 and add the result to equation (1).

$$\begin{array}{rcrcrcrl} 2x^2 & + & 3xy & + & 2y^2 & = & 21 & (1) \\ -2x^2 & + & & - & 2y^2 & = & -12 & \quad -2 \times (2) \\ \hline & & 3xy & & & = & 9 & \\ & & & & x & = & \frac{9}{3y} = \frac{3}{y} & (3) \end{array}$$

Substitute $\frac{3}{y}$ for x in equation (2).

$$\left(\frac{3}{y}\right)^2 + y^2 = 6$$
$$\frac{9}{y^2} + y^2 = 6$$
$$9 + y^4 = 6y^2 \quad \textit{Multiply by } y^2.$$
$$y^4 - 6y^2 + 9 = 0 \quad \textit{Rearrange terms.}$$
$$\left(y^2 - 3\right)^2 = 0$$
$$y^2 - 3 = 0$$
$$y = \pm\sqrt{3}$$

Use (3), $x = \frac{3}{y}$.

If $y = \sqrt{3}$, then $x = \frac{3}{\sqrt{3}} = \sqrt{3}$.

If $y = -\sqrt{3}$, then $x = \frac{3}{-\sqrt{3}} = -\sqrt{3}$.

Solution set: $\left\{\left(\sqrt{3}, \sqrt{3}\right), \left(-\sqrt{3}, -\sqrt{3}\right)\right\}$

35. $3x^2 + 2xy - 3y^2 = 5 \quad (1)$
$-x^2 - 3xy + y^2 = 3 \quad (2)$

Multiply equation (2) by 3 and add the result to equation (1).

$$\begin{array}{rcrcrcrl} 3x^2 & + & 2xy & - & 3y^2 & = & 5 & (1) \\ -3x^2 & - & 9xy & + & 3y^2 & = & 9 & \quad 3 \times (2) \\ \hline & & -7xy & & & = & 14 & \\ & & & & x & = & \frac{14}{-7y} = -\frac{2}{y} & (3) \end{array}$$

Substitute $-\frac{2}{y}$ for x in equation (2).

$$-x^2 - 3xy + y^2 = 3 \qquad (2)$$
$$-\left(-\frac{2}{y}\right)^2 - 3\left(-\frac{2}{y}\right)y + y^2 = 3$$
$$-\left(\frac{4}{y^2}\right) + 6 + y^2 = 3$$
$$y^2 + 3 - \frac{4}{y^2} = 0$$
$$y^4 + 3y^2 - 4 = 0 \quad \textit{Multiply by } y^2.$$
$$(y^2 + 4)(y^2 - 1) = 0$$

$y^2 + 4 = 0$ or $y^2 - 1 = 0$
$y^2 = -4$ $\qquad y^2 = 1$
$y = \pm 2i$ or $y = \pm 1$

Substitute these values for y in equation (3) to find the values of x.

If $y = 2i$, then

$$x = -\frac{2}{2i} = -\frac{1}{i} = -\frac{1}{i} \cdot \frac{i}{i} = \frac{-i}{-1} = i.$$

If $y = -2i$, then

$$x = -\frac{2}{-2i} = \frac{1}{i} = \frac{1}{i} \bullet \frac{i}{i} = \frac{i}{-1} = -i.$$

If $y = 1$, then $x = -\frac{2}{1} = -2$.

If $y = -1$, then $x = -\frac{2}{-1} = 2$.

Solution set: $\{(i, 2i), (-i, -2i), (2, -1), (-2, 1)\}$

36.
$$\begin{array}{rcll} -2x^2 + 7xy - 3y^2 &=& 4 & (1) \\ 2x^2 - 3xy + 3y^2 &=& 4 & (2) \\ \hline 4xy &=& 8 & \textit{Add} \\ x &=& \frac{8}{4y} = \frac{2}{y} & (3) \end{array}$$

Substitute $\frac{2}{y}$ for x in equation (2).

$$\begin{aligned} 2\left(\frac{2}{y}\right)^2 - 3\left(\frac{2}{y}\right)y + 3y^2 &= 4 \\ \left(\frac{8}{y^2}\right) - 6 + 3y^2 &= 4 \\ 3y^2 - 10 + \frac{8}{y^2} &= 0 \\ 3y^4 - 10y^2 + 8 &= 0 \quad \textit{Multiply by } y^2. \\ (3y^2 - 4)(y^2 - 2) &= 0 \end{aligned}$$

$$\begin{array}{lcl} 3y^2 - 4 = 0 & \text{or} & y^2 - 2 = 0 \\ 3y^2 = 4 & \text{or} & y^2 = 2 \\ y^2 = \frac{4}{3} & & \\ y = \pm\sqrt{\frac{4}{3}} & & \\ \quad = \pm\frac{2\sqrt{3}}{3} & \text{or} & y = \pm\sqrt{2} \end{array}$$

Use (3), $x = \frac{2}{y}$.

If $y = \sqrt{2}$, then $x = \sqrt{2}$.

If $y = -\sqrt{2}$, then $x = -\sqrt{2}$.

If $y = \frac{2\sqrt{3}}{3}$, then

$$x = \frac{2}{\frac{2\sqrt{3}}{3}} = 2 \bullet \frac{3}{2\sqrt{3}} = \frac{3}{\sqrt{3}} = \sqrt{3}.$$

If $y = -\frac{2\sqrt{3}}{3}$, then $x = -\sqrt{3}$.

Solution set:

$$\left\{\left(\sqrt{2}, \sqrt{2}\right), \left(-\sqrt{2}, -\sqrt{2}\right), \left(\sqrt{3}, \frac{2\sqrt{3}}{3}\right), \left(-\sqrt{3}, -\frac{2\sqrt{3}}{3}\right)\right\}$$

37. Let $W =$ the width, and
$L =$ the length.

The formula for the area of a rectangle is $LW = A$, so

$$LW = 84. \quad (1)$$

The perimeter of a rectangle is given by $2L + 2W = P$, so

$$2L + 2W = 38. \quad (2)$$

Solve equation (2) for L to get

$$L = 19 - W. \quad (3)$$

Substitute $19 - W$ for L in equation (1).

$$\begin{aligned} LW &= 84 \quad (1) \\ (19 - W)W &= 84 \\ 19W - W^2 &= 84 \\ -W^2 + 19W - 84 &= 0 \\ W^2 - 19W + 84 &= 0 \qquad \textit{Multiply by } -1. \\ (W - 7)(W - 12) &= 0 \end{aligned}$$

$$\begin{array}{lcl} W - 7 = 0 & \text{or} & W - 12 = 0 \\ W = 7 & \text{or} & W = 12 \end{array}$$

Using equation (3), with $W = 7$,

$$L = 19 - 7 = 12.$$

If $W = 12$, then $L = 7$, which are the same two numbers. Length must be greater than width, so the length is 12 feet and the width is 7 feet.

38. Let L be the length and W be the width. The perimeter is 50 meters and the area is 100 square meters.

$$\begin{aligned} 2L + 2W &= 50 \quad (1) \\ LW &= 100 \quad (2) \end{aligned}$$

Solve (1) for L to get

$$L = 25 - W.$$

Substitute $25 - W$ for L in (2).

$$\begin{aligned} (25 - W)W &= 100 \\ 25W - W^2 &= 100 \\ W^2 - 25W + 100 &= 0 \\ (W - 20)(W - 5) &= 0 \end{aligned}$$

$$\begin{array}{lcl} W - 20 = 0 & \text{or} & W - 5 = 0 \\ W = 20 & \text{or} & W = 5 \end{array}$$

If $W = 20$, then $L = 25 - W = 25 - 20 = 5$. If $W = 5$, then $L = 25 - W = 25 - 5 = 20$. The length is 20 meters and the width is 5 meters.

39. $px = 16 \quad (1)$
$p = 10x + 12 \quad (2)$

Substitute $10x + 12$ for p in equation (1).

$$\begin{aligned} px &= 16 \quad (1) \\ (10x+12)x &= 16 \\ 10x^2 + 12x &= 16 \\ 10x^2 + 12x - 16 &= 0 \\ 5x^2 + 6x - 8 &= 0 \\ (5x-4)(x+2) &= 0 \end{aligned}$$

$$5x - 4 = 0 \quad \text{or} \quad x + 2 = 0$$
$$x = \tfrac{4}{5} \quad \text{or} \quad x = -2$$

Since x cannot be negative, eliminate -2 as a value of x. Substitute $\frac{4}{5}$ for x in equation (2) to find p.

$$\begin{aligned} p &= 10x + 12 \quad (2) \\ p &= 10\left(\tfrac{4}{5}\right) + 12 \\ &= 8 + 12 = 20 \end{aligned}$$

Since the demand x is in thousands,

$$\text{demand} = \tfrac{4}{5}(1000) = 800.$$

The equilibrium price is \$20. The supply/demand at that price is 800 calculators.

40. The cost y to make x thousand calculators is

$$y = 4x^2 + 36x + 20.$$

The revenue y from the sale of x thousand calculators is

$$\begin{aligned} 36x^2 - 3y &= 0. \\ 36x^2 &= 3y \\ 12x^2 &= y \end{aligned}$$

The *break-even point* is where cost equals revenue,

$$\begin{aligned} 4x^2 + 36x + 20 &= 12x^2 \\ 8x^2 - 36x - 20 &= 0 \\ 2x^2 - 9x - 5 &= 0 \\ (2x+1)(x-5) &= 0 \end{aligned}$$

$$2x + 1 = 0 \quad \text{or} \quad x - 5 = 0$$
$$x = -\tfrac{1}{2} \quad \text{or} \quad x = 5$$

Discard the negative solution. The number of calculators at the break-even point is 5 thousand, or 5000. Let $x = 5$ in either the cost or revenue equation to find the cost and the revenue at the break-even point. We use the simplified revenue equation.

$$\begin{aligned} y &= 12x^2 \\ y &= 12(5)^2 = 300 \end{aligned}$$

The cost and the revenue are each \$300.

41. Men: $y = .138x^2 + .064x + 451 \quad (1)$
Women: $y = 12.1x + 334 \quad (2)$

Substitute $12.1x + 334$ for y in (1).

$$\begin{aligned} 12.1x + 334 &= .138x^2 + .064x + 451 \\ 0 &= .138x^2 - 12.036x + 117 \end{aligned}$$

Use the quadratic formula.

$$\begin{aligned} x &= \frac{-(-12.036) \pm \sqrt{(-12.036)^2 - 4(.138)(117)}}{2(.138)} \\ &= \frac{12.036 \pm \sqrt{80.28196}}{.276} \end{aligned}$$

These values are approximately 76 and 11. $x = 11$ corresponds to 1981 and $x = 76$ is out of the range for these models.

Substituting $x = 11$ into (2) gives us approximately 467 thousand bachelor's degrees (or 470 thousand to the nearest ten thousand) awarded to both men and women in 1981.

42. **(a)**
$$\begin{pmatrix}\text{Number}\\ \text{of}\\ \text{months}\end{pmatrix} \cdot \begin{pmatrix}\text{Cost}\\ \text{per}\\ \text{month}\end{pmatrix} = \begin{pmatrix}\text{Total}\\ \text{cost}\end{pmatrix}$$
$$x \cdot y = 1000$$
$$y = \frac{1000}{x} \quad (1)$$

(b) "The average monthly on-line cost is about \$20" translates as "$y = 20$."

(c) Substitute 20 for y in (1).

$$\begin{aligned} 20 &= \frac{1000}{x} \\ 20x &= 1000 \\ x &= 50 \end{aligned}$$

The solution set of the system is $\{(50, 20)\}$. The solution tells us that in 50 months, the monthly cost of the computer will be \$20, the same as the on-line cost.

12.5 Second-Degree Inequalities and Systems of Inequalities

12.5 Margin Exercises

1. $y \geq (x+1)^2 - 5$

The boundary, $y = (x+1)^2 - 5$ is a parabola opening upward with its vertex at $(-1,-5)$. It is drawn as a solid curve because of the $\geq$ symbol. Use the point $(0,0)$ as a test point.

$$\begin{aligned} 0 &\geq (0+1)^2 - 5 \quad ? \\ 0 &\geq 1 - 5 \quad ? \\ 0 &\geq -4 \quad \textit{True} \end{aligned}$$

Shade the region that contains $(0,0)$. This is the region inside the parabola.

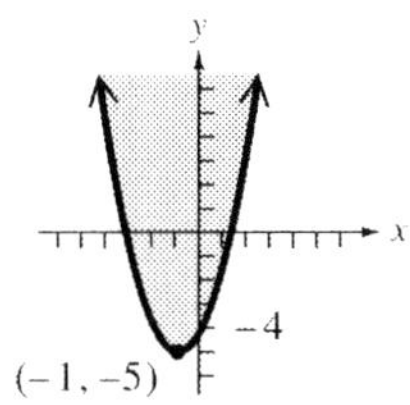

2. $x^2 + 4y^2 > 36$

$\frac{x^2}{36} + \frac{y^2}{9} > 1$ *Divide by 36.*

The boundary is the ellipse

$$\frac{x^2}{36} + \frac{y^2}{9} = 1,$$

which has intercepts $(6,0)$, $(-6,0)$, $(0,3)$, and $(0,-3)$. It is drawn as a dashed curve because of the $>$ symbol. Test $(0,0)$.

$$\begin{aligned} 0^2 + 4(0)^2 &> 36 \quad ? \\ 0 &> 36 \quad \textit{False} \end{aligned}$$

Shade the region that does not contain $(0,0)$. This is the region outside the ellipse.

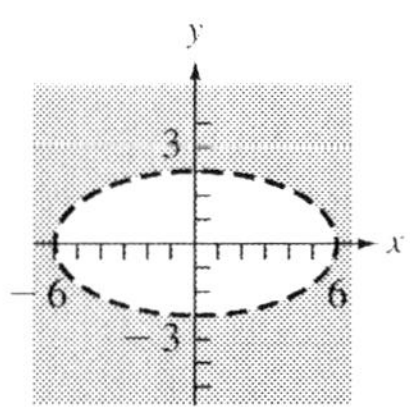

3. $x^2 + y^2 \leq 25$
$x + y \leq 3$

The boundary $x^2 + y^2 = 25$ is a circle with center $(0,0)$ and radius 5. Draw as a solid curve. Test $(0,0)$.

$$\begin{aligned} 0^2 + 0^2 &\leq 25 \quad ? \\ 0 &\leq 25 \quad \textit{True} \end{aligned}$$

Shade the region inside the circle.

The boundary $x + y = 3$ is a line with x-intercept $(3,0)$ and y-intercept $(0,3)$. Draw as a solid line. Test $(0,0)$.

$$\begin{aligned} 0 + 0 &\leq 3 \quad ? \\ 0 &\leq 3 \quad \textit{True} \end{aligned}$$

Shade the region on the side of the line containing $(0,0)$.

The graph of the solution set of the system includes all points common to the two graphs, that is, the intersection of the two shaded regions.

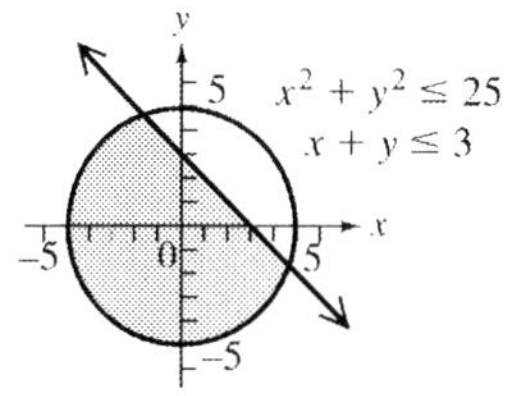

4. $3x - 4y \geq 12$
$x + 3y \geq 6$
$y \leq 2$

The boundary $3x - 4y = 12$ is a solid line with intercepts $(4,0)$ and $(0,-3)$. Test $(0,0)$.

$$\begin{aligned} 3(0) - 4(0) &\geq 12 \quad ? \\ 0 &\geq 12 \quad \textit{False} \end{aligned}$$

Shade the side of the line that does not contain $(0,0)$.

The boundary $x + 3y = 6$ is a solid line with intercepts $(6,0)$ and $(0,2)$. Test $(0,0)$.

$$\begin{aligned} 0 + 3(0) &\geq 6 \quad ? \\ 0 &\geq 6 \quad \textit{False} \end{aligned}$$

Shade the side of the line that does not contain $(0,0)$.

The boundary $y = 2$ is a solid horizontal line through $(0,2)$. Shade the region below the line $y \leq 2$.

The intersection of the three shaded regions is the graph of the solution set of the system.

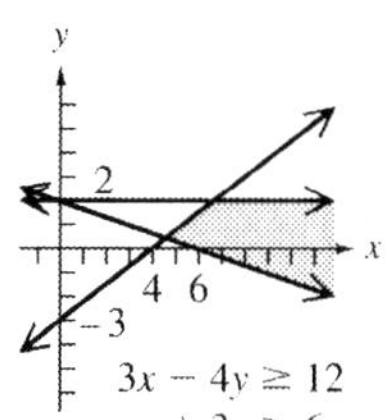

5.

$$y \geq x^2 + 1$$
$$\frac{x^2}{9} + \frac{y^2}{4} \geq 1$$
$$y \leq 5$$

The boundary $y = x^2 + 1$ is a parabola opening upward with vertex at $(0, 1)$. It is drawn as a solid curve because of the $\geq$ symbol. Test $(0, 0)$.

$$0 \geq 0^2 + 1 \quad ?$$
$$0 \geq 1 \qquad \textit{False}$$

Shade the region that does not contain $(0, 0)$. This is the region inside the parabola.

The boundary $\frac{x^2}{9} + \frac{y^2}{4} = 1$ is an ellipse with intercepts $(3, 0)$, $(-3, 0)$, $(0, 2)$, and $(0, -2)$. It is drawn as a solid curve because of the $\geq$ symbol. Test $(0, 0)$.

$$\frac{0^2}{9} + \frac{0^2}{4} \geq 1 \quad ?$$
$$0 \geq 1 \qquad \textit{False}$$

Shade the region that does not contain $(0, 0)$. This is the region outside the ellipse.

The boundary $y = 5$ is a solid horizontal line through $(0, 5)$. Shade the region below the line since $y \leq 5$.

The graph of the solution set of the system includes all points common to the three graphs, that is, the intersection of the three shaded regions.

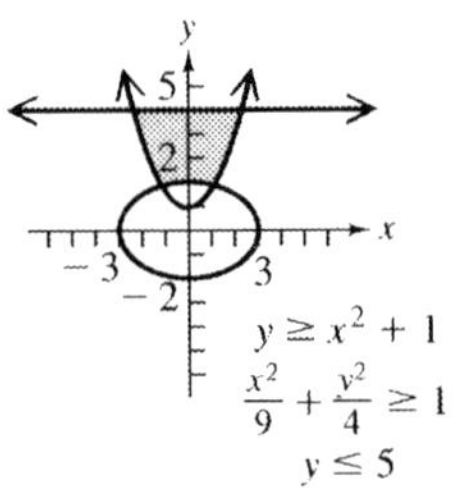

12.5 Section Exercises

1.

$$x^2 + y^2 < 25$$
$$y > -2$$

The boundary, $x^2 + y^2 = 25$, is a circle (dashed because of the $<$ sign) with center $(0, 0)$ and radius 5. When $(0, 0)$ is tested, a true statement, $0 < 25$, results, so the inside of the circle is shaded. The graph of $y = -2$ is a horizontal line through $(0, -2)$ with shading above the dashed line, since $y > -2$. The correct answer is **C**.

2. Investigate the graph of each of the inequalities before answering the question about the graph of the system, which is the intersection of the graphs of the inequalities. Consider the graph of $y > x^2 + 1$. Since the test point $(0, 0)$ gives

$$0 > 0^2 + 1 \quad ?$$
$$0 > 1, \qquad \textit{False}$$

it does not satisfy the inequality. The boundary is the parabola

$$y = x^2 + 1,$$

which has its vertex at $(0, 1)$ and opens upward. The test point $(0, 0)$ lies below the vertex so the points of the graph of the inequality lie *above* the parabola.

Testing the point $(0, 0)$ in

$$\frac{x^2}{9} + \frac{y^2}{4} > 1$$

gives

$$\frac{0^2}{9} + \frac{0^2}{4} > 1 \quad ?$$
$$0 > 1. \qquad \textit{False}$$

The boundary is the ellipse

$$\frac{x^2}{9} + \frac{y^2}{4} = 1,$$

which is centered at $(0, 0)$, so the points of the graph of

$$\frac{x^2}{9} + \frac{y^2}{4} > 1$$

lie *outside* the ellipse.

Testing $(0, 0)$ in

$$y < 5$$

gives

$$0 < 5, \quad \textit{True}$$

so the points of the graph lie on the side of the graph of $y = 5$ that contains $(0, 0)$, that is, *below* the line.

Therefore, the graph of the system

$$y > x^2 + 1$$
$$\frac{x^2}{9} + \frac{y^2}{4} > 1$$
$$y < 5$$

consists of all points <u>above</u> the parabola $y = x^2 + 1$, <u>outside</u> the ellipse $\frac{x^2}{9} + \frac{y^2}{4} = 1$, and <u>below</u> the line $y = 5$.

3. To graph the solution set of a nonlinear inequality, first graph the corresponding equality. This graph should be a dashed curve for $<$ or $>$ inequalities or a solid curve for $\leq$ or $\geq$ inequalities. Next, decide which region to shade by substituting any point not on the boundary (usually $(0, 0)$ is the easiest) into the inequality. If the statement is true, then shade that area. If the statement is false, then shade the other area.

4. Graph each inequality separately, and determine the intersection of the shaded regions. This is the solution set of the system.

5. $y > x^2 - 1$

The boundary, $y = x^2 - 1$, is a dashed parabola opening up with vertex $(0, -1)$. Its shape is the same as the graph of $f(x) = x^2$. Use $(0, 0)$ as a test point.

$$\begin{aligned} 0 &> 0^2 - 1 \quad ? \\ 0 &> -1 \qquad \textit{True} \end{aligned}$$

The point $(0, 0)$ satisfies the inequality so the graph includes it. Shade the region that includes $(0, 0)$.

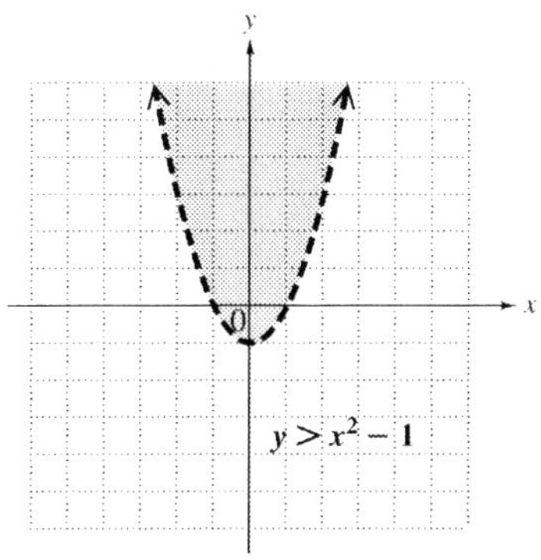

6.
$$\begin{aligned} y^2 &> 4 + x^2 \\ y^2 - x^2 &> 4 \\ \frac{y^2}{4} - \frac{x^2}{4} &> 1 \end{aligned}$$

The boundary, $\frac{y^2}{4} - \frac{x^2}{4} = 1$, is a hyperbola with y-intercepts $(0, 2)$ and $(0, -2)$ and asymptotes formed by the diagonals of the fundamental rectangle with vertices at $(2, 2)$, $(2, -2)$, $(-2, -2)$, and $(-2, 2)$. The hyperbola has dashed branches because of $>$. Test $(0, 0)$.

$$\begin{aligned} 0^2 &> 4 + 0^2 \quad ? \\ 0 &> 4 \qquad \textit{False} \end{aligned}$$

Shade the sides of the hyperbola that do not contain $(0, 0)$. These are the regions inside the branches of the hyperbola.

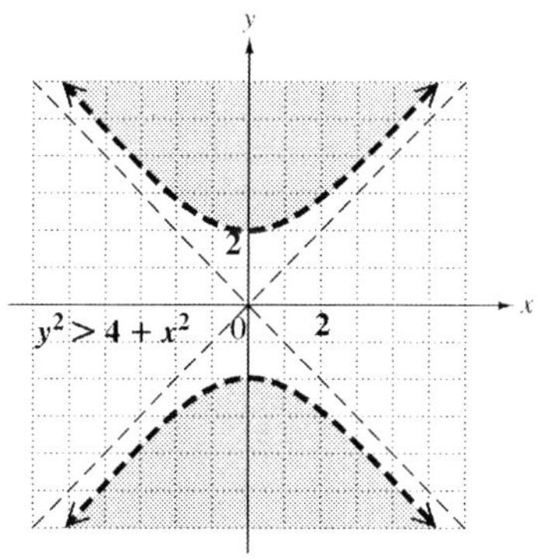

7. $y^2 \leq 4 - 2x^2$

The boundary is

$$\begin{aligned} y^2 &= 4 - 2x^2 \\ 2x^2 + y^2 &= 4 \\ \frac{x^2}{2} + \frac{y^2}{4} &= 1. \end{aligned}$$

The boundary is a solid ellipse with x-intercepts $(\sqrt{2}, 0)$ and $(-\sqrt{2}, 0)$ and y-intercepts $(0, 2)$ and $(0, -2)$. Test $(0, 0)$.

$$\begin{aligned} 0^2 &\leq 4 - 2(0)^2 \quad ? \\ 0 &\leq 4 \qquad \textit{True} \end{aligned}$$

Shade the region that includes $(0, 0)$.

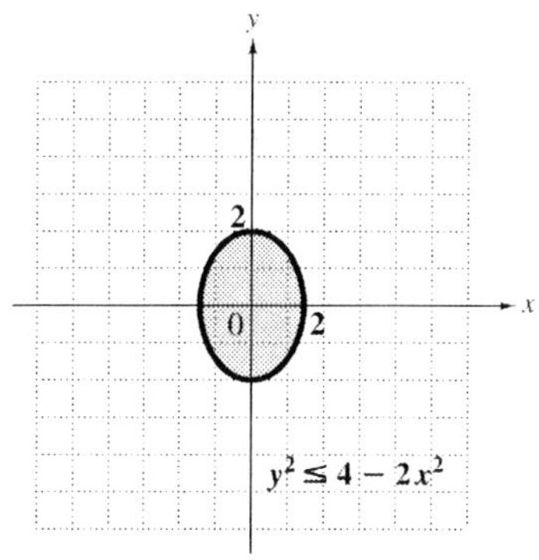

8.
$$\begin{aligned} y + 2 &\geq x^2 \\ y &\geq x^2 - 2 \\ y &\geq (x - 0)^2 - 2 \end{aligned}$$

Graph the solid vertical parabola $y = x^2 - 2$ with vertex $(0, -2)$. Two other points on the parabola are $(2, 2)$ and $(-2, 2)$. Test a point not on the parabola, say $(0, 0)$, in $y \geq x^2 - 2$ to get $0 \geq -2$, a true statement. Shade that portion of the graph that contains the point $(0, 0)$. This is the region inside the parabola.

continued

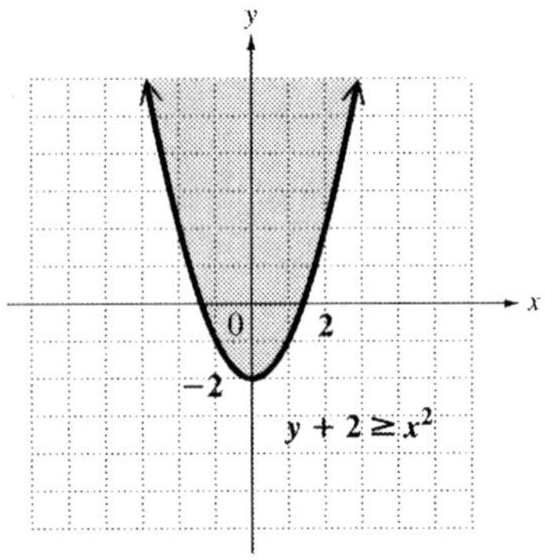

9. $x^2 \le 16 - y^2$

The boundary is

$$x^2 = 16 - y^2$$
$$x^2 + y^2 = 16.$$

The boundary is a solid circle with center $(0, 0)$ and radius 4. Test $(0, 0)$.

$$0^2 \le 16 - 0^2 \quad ?$$
$$0 \le 16 \qquad \textit{True}$$

Shade the region that includes $(0, 0)$.

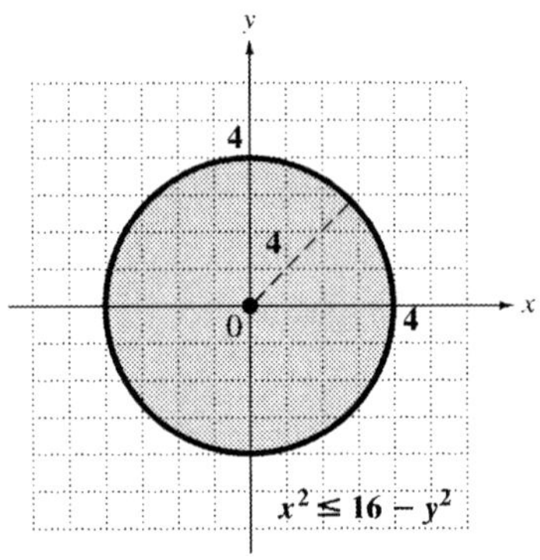

10.
$$2y^2 \ge 8 - x^2$$
$$x^2 + 2y^2 \ge 8$$
$$\frac{x^2}{8} + \frac{y^2}{4} \ge 1$$

The boundary, $\frac{x^2}{8} + \frac{y^2}{4} = 1$, is the ellipse with intercepts $\left(2\sqrt{2}, 0\right)$, $\left(-2\sqrt{2}, 0\right)$, $(0, 2)$, and $(0, -2)$, drawn as a solid curve because of $\ge$. Test $(0, 0)$.

$$2(0)^2 \ge 8 - 0^2 \quad ?$$
$$0 \ge 8 \qquad \textit{False}$$

Shade the region of the ellipse that does not contain $(0, 0)$. This is the region outside the ellipse.

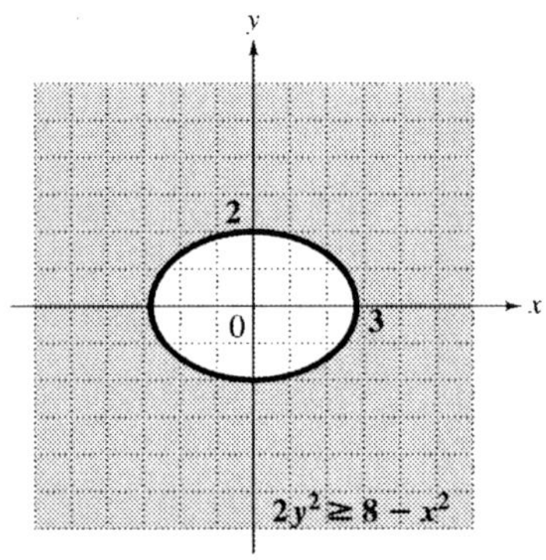

11. $x^2 \le 16 + 4y^2$

The boundary is

$$x^2 = 16 + 4y^2$$
$$x^2 - 4y^2 = 16$$
$$\frac{x^2}{16} - \frac{y^2}{4} = 1.$$

The boundary is a solid hyperbola with x-intercepts $(4, 0)$ and $(-4, 0)$. The asymptotes go through $(4, 2)$ and $(-4, -2)$ and through $(4, -2)$ and $(-4, 2)$. Test $(0, 0)$.

$$0^2 \le 16 + 4(0)^2 \quad ?$$
$$0 \le 16 \qquad \textit{True}$$

Shade the region between the branches of the hyperbola, which includes $(0, 0)$.

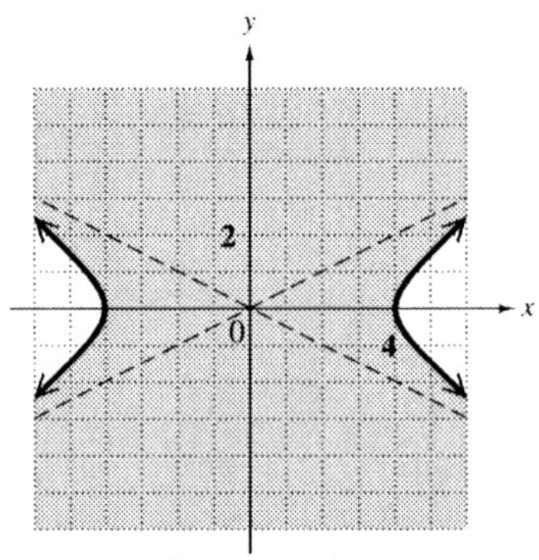

12. $y \le x^2 + 4x + 2$

Graph the solid vertical parabola $y = x^2 + 4x + 2$. Use the vertex formula $x = \frac{-b}{2a}$ to obtain the vertex $(-2, -2)$. Two other points on the parabola are $(0, 2)$ and $(-4, 2)$. Test a point not on the parabola, say $(0, 0)$, in $y \le x^2 + 4x + 2$ to get $0 \le 2$, a true statement. Shade outside the parabola, since this region contains $(0, 0)$.

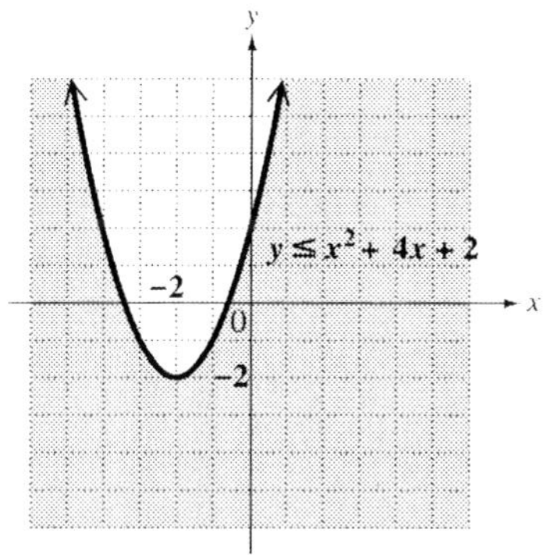

13. $9x^2 < 16y^2 - 144$

The boundary is

$$\begin{aligned} 9x^2 &= 16y^2 - 144 \\ 144 &= 16y^2 - 9x^2 \\ 1 &= \frac{y^2}{9} - \frac{x^2}{16}. \end{aligned}$$

The boundary is a dashed hyperbola with y-intercepts $(0, 3)$ and $(0, -3)$. The asymptotes go through $(4, 3)$ and $(-4, -3)$ and through $(4, -3)$ and $(-4, 3)$. Test $(0, 0)$.

$$\begin{aligned} 9(0)^2 &< 16(0)^2 - 144 \quad ? \\ 0 &< -144 \qquad \textit{False} \end{aligned}$$

Shade the region that does *not* include $(0, 0)$.

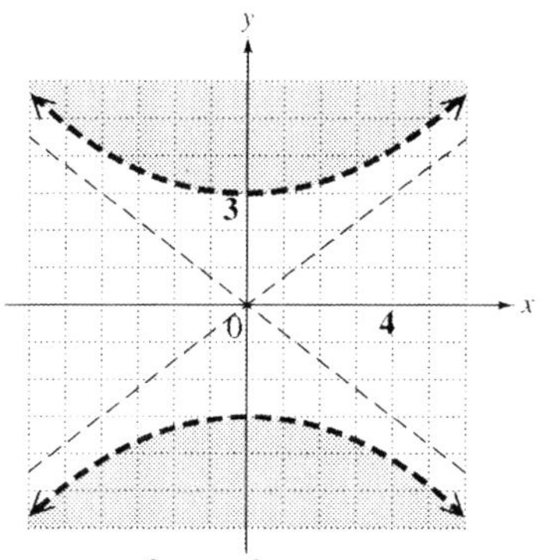

$9x^2 < 16y^2 - 144$

14. $$\begin{aligned} 9x^2 &> 16y^2 + 144 \\ 9x^2 - 16y^2 &> 144 \\ \frac{x^2}{16} - \frac{y^2}{9} &> 1 \end{aligned}$$

The boundary, $\frac{x^2}{16} - \frac{y^2}{9} = 1$, is a hyperbola with x-intercepts $(4, 0)$ and $(-4, 0)$ and asymptotes formed by the diagonals of the fundamental rectangle with vertices at $(4, 3)$, $(4, -3)$, $(-4, -3)$, and $(-4, 3)$. The hyperbola has dashed branches because of $>$. Test $(0, 0)$.

$$\begin{aligned} 9(0)^2 &> 16(0)^2 + 144 \quad ? \\ 0 &> 144 \qquad \textit{False} \end{aligned}$$

Shade the sides of the hyperbola that do not contain $(0, 0)$. These are the regions inside the branches of the hyperbola.

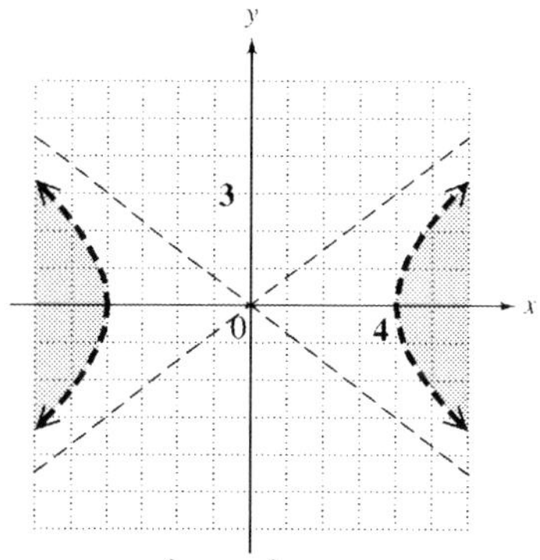

$9x^2 > 16y^2 + 144$

15. $4y^2 \le 36 - 9x^2$

The boundary is

$$\begin{aligned} 4y^2 &= 36 - 9x^2 \\ 9x^2 + 4y^2 &= 36 \\ \frac{x^2}{4} + \frac{y^2}{9} &= 1. \end{aligned}$$

The boundary is a solid ellipse with x-intercepts $(2, 0)$ and $(-2, 0)$ and y-intercepts $(0, 3)$ and $(0, -3)$. Test $(0, 0)$.

$$\begin{aligned} 4(0)^2 &\le 36 - 9(0)^2 \quad ? \\ 0 &\le 36 \qquad \textit{True} \end{aligned}$$

Shade the region that includes $(0, 0)$.

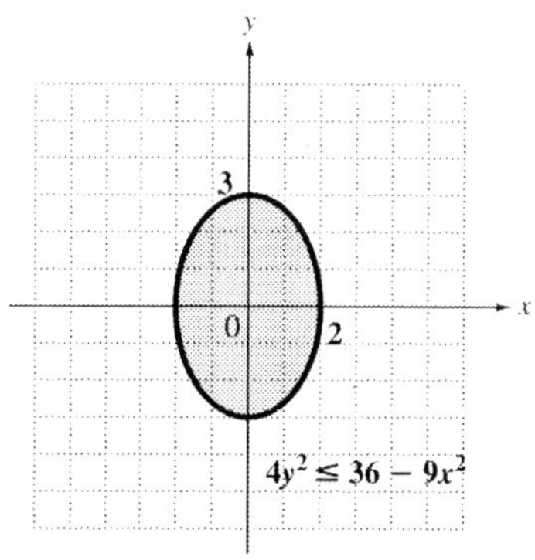

16. $x^2 - 4 \geq -4y^2$

$$x^2 + 4y^2 \geq 4$$

$$\frac{x^2}{4} + \frac{y^2}{1} \geq 1$$

Graph the solid ellipse $\frac{x^2}{4} + \frac{y^2}{1} = 1$ through the x-intercepts $(2, 0)$ and $(-2, 0)$ and y-intercepts $(0, 1)$ and $(0, -1)$. Test a point not on the ellipse, say $(0, 0)$, in $x^2 - 4 \geq -4y^2$ to get $-4 \geq 0$, a false statement. Shade outside the ellipse, since this region does *not* include $(0, 0)$.

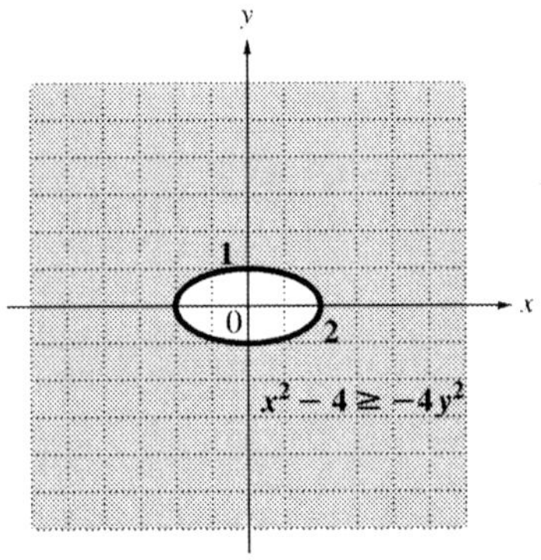

17. $x \geq y^2 - 8y + 14$

The boundary is

$$x = y^2 - 8y + 14$$
$$x = y^2 - 8y + 16 - 16 + 14$$
$$x = (y - 4)^2 - 2.$$

The boundary is a solid horizontal parabola with vertex $(-2, 4)$ opening to the right. Its shape is the same as the graph of $x = y^2$. Test $(0, 0)$.

$0 \geq 0^2 - 8(0) + 14$?

$0 \geq 14$ *False*

Shade the region that does *not* include $(0, 0)$.

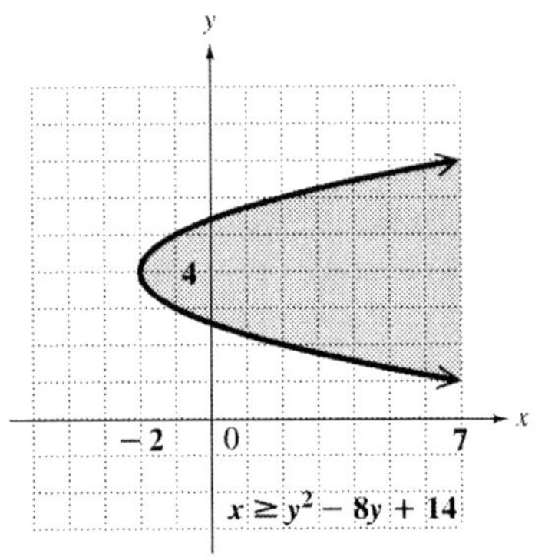

18. $x \leq -y^2 + 6y - 7$

Complete the square to find the vertex.

$$\begin{aligned} x &= -y^2 + 6y - 7 \\ &= -(y^2 - 6y) - 7 \\ &= -(y^2 - 6y + 9 - 9) - 7 \\ &= -(y^2 - 6y + 9) - (-9) - 7 \\ x &= -(y - 3)^2 + 2 \end{aligned}$$

The boundary, $x = -(y - 3)^2 + 2$, is a solid horizontal parabola with vertex at $(2, 3)$ that opens to the left. Test $(0, 0)$.

$0 \leq -0^2 + 6(0) - 7$?

$0 \leq -7$ *False*

Shade the region of the parabola that does not contain $(0, 0)$. This is the region inside the parabola.

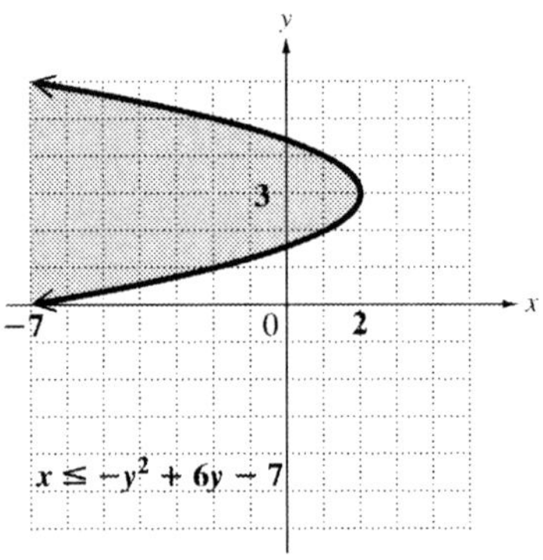

19. $25x^2 \leq 9y^2 + 225$

$$25x^2 - 9y^2 \leq 225$$

$$\frac{x^2}{9} - \frac{y^2}{25} \leq 1$$

Graph the solid hyperbola $\frac{x^2}{9} - \frac{y^2}{25} = 1$ through the x-intercepts $(3, 0)$ and $(-3, 0)$. One asymptote passes through $(3, 5)$ and $(-3, -5)$. The other asymptote passes through $(-3, 5)$ and $(3, -5)$.

Test a point not on the hyperbola, say $(0, 0)$, in $25x^2 \leq 9y^2 + 225$ to get $0 \leq 225$, a true statement.

Shade the part of the graph that contains $(0, 0)$. This is the region between the two branches of the hyperbola.

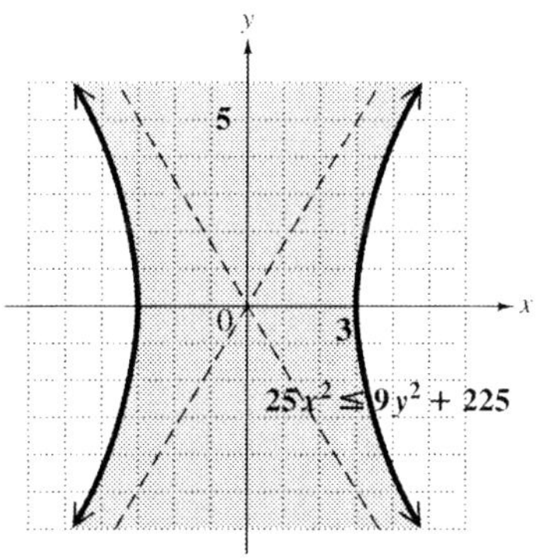

20. $2x + 5y < 10$
$x - 2y < 4$

Graph $2x + 5y = 10$ as a dashed line through $(5, 0)$ and $(0, 2)$. Test $(0, 0)$.

$$2x + 5y < 10$$
$$2(0) + 5(0) < 10 \quad ?$$
$$0 < 10 \qquad \textit{True}$$

Shade the region containing $(0, 0)$.
Graph $x - 2y = 4$ as a dashed line through $(4, 0)$ and $(0, -2)$. Test $(0, 0)$.

$$x - 2y < 4$$
$$0 - 2(0) < 4 \quad ?$$
$$0 < 4 \qquad \textit{True}$$

Shade the region containing $(0, 0)$.
The graph of the system is the intersection of the two shaded regions.

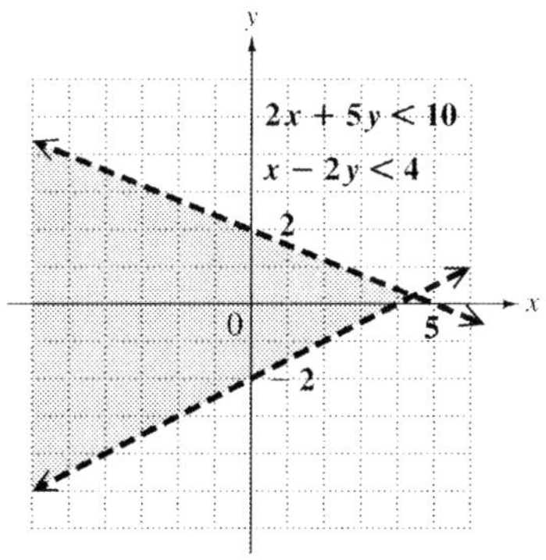

21. $3x - y > -6$ (1)
$4x + 3y > 12$ (2)

The graph of (1) has dashed boundary line $3x - y = -6$ through $(-2, 0)$ and $(0, 6)$. Since $3(0) - 0 > -6$ is true, the graph contains the region on the side of the line that includes the test point $(0, 0)$.
The graph of (2) has dashed boundary line $4x + 3y = 12$ through $(3, 0)$ and $(0, 4)$. Since $4(0) + 3(0) > 12$ is false, the graph contains the region that does not include $(0, 0)$. The graph of the system is the intersection of these graphs.

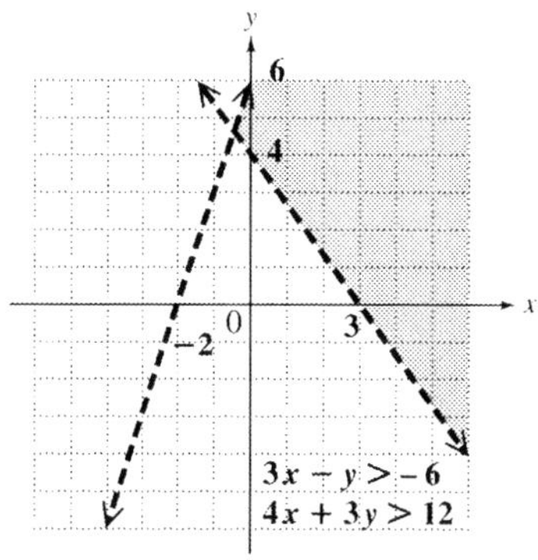

22. $5x - 3y \le 15$
$4x + y \ge 4$

The boundary, $5x - 3y = 15$, is a solid line with intercepts $(3, 0)$ and $(0, -5)$. Test $(0, 0)$.

$$5(0) - 3(0) \le 15 \quad ?$$
$$0 \le 15 \qquad \textit{True}$$

Shade the side of the line that contains $(0, 0)$.
The boundary, $4x + y = 4$, is a solid line with intercepts $(1, 0)$ and $(0, 4)$. Test $(0, 0)$.

$$4(0) + 0 \ge 4 \quad ?$$
$$0 \ge 4 \qquad \textit{False}$$

Shade the side of the line that does not contain $(0, 0)$.
The graph of the system is the intersection of the two shaded regions.

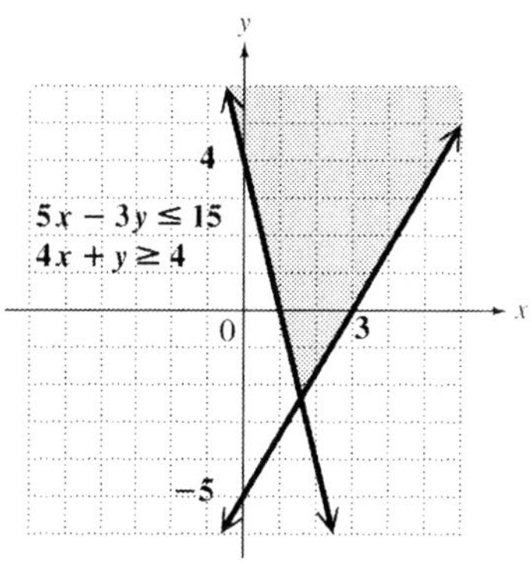

23. $4x - 3y \le 0 \quad (1)$
$x + y \le 5 \quad (2)$

The graph of (1) has solid boundary line $4x - 3y = 0$ through $(0,0)$ and $(3,4)$. Since $4(-1) - 3(0) \le 0$ is true, the graph contains the region that includes $(-1,0)$.
The graph of (2) has solid boundary line $x + y = 5$ through $(5,0)$ and $(0,5)$. Since $0 + 0 \le 5$ is true, the graph contains the region that includes $(0,0)$.
The graph of the system is the intersection of these graphs.

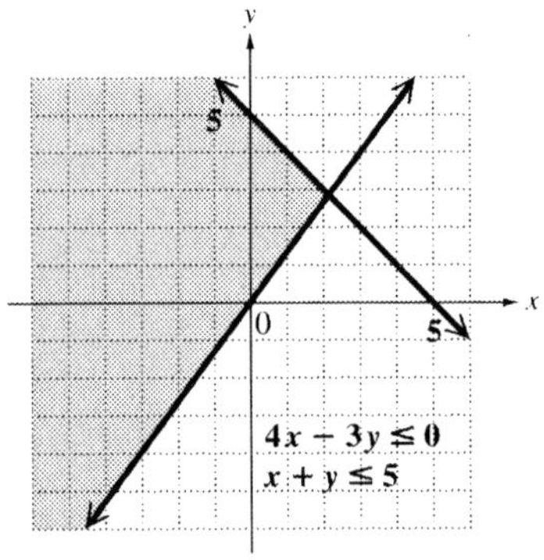

24. $x \le 5$
$y \le 4$

Graph $x = 5$ as a solid vertical line through $(5,0)$. Since $x \le 5$, shade the left side of the line.
Graph $y = 4$ as a solid horizontal line through $(0,4)$. Since $y \le 4$, shade below the line.
The graph of the system is the intersection of the two shaded regions.

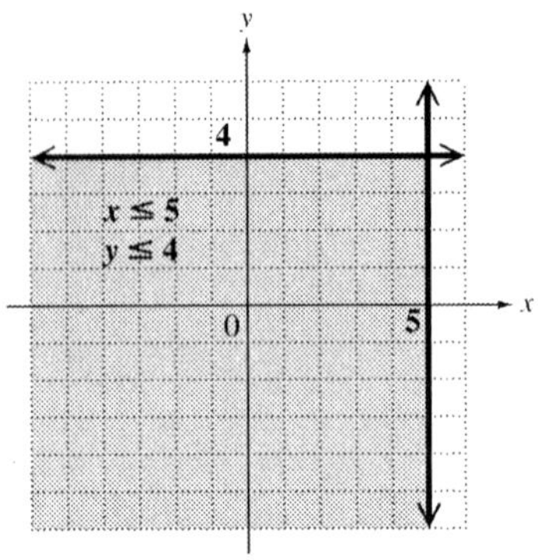

25. $x \ge -2 \quad (1)$
$y \le 4 \quad (2)$

The graph of (1) has a solid vertical boundary line, $x = -2$, and includes points to the right of the line.
The graph of (2) has a solid horizontal boundary line, $y = 4$, and includes points below the line.
The graph of the system is the intersection of these graphs.

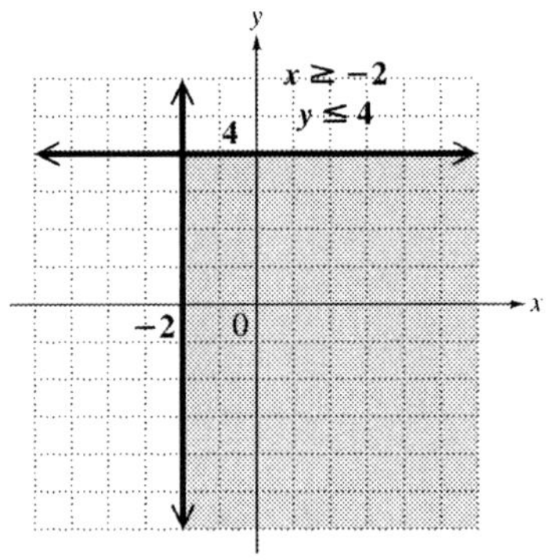

26. $y > x^2 - 4$
$y < -x^2 + 3$

The boundary, $y = x^2 - 4$, is a dashed parabola with vertex $(0,-4)$ that opens up. Test $(0,0)$.

$$0 > 0^2 - 4 \quad ?$$
$$0 > -4 \qquad \textit{True}$$

Shade the side of the parabola that contains $(0,0)$. This is the region inside the parabola.
The boundary, $y = -x^2 + 3$, is a dashed parabola with vertex $(0,3)$ that opens down. Test $(0,0)$.

$$0 < -0^2 + 3 \quad ?$$
$$0 < 3 \qquad \textit{True}$$

Shade the side of the parabola that contains $(0,0)$. This is the region inside the parabola.
The graph of the system is the intersection of the two shaded regions.

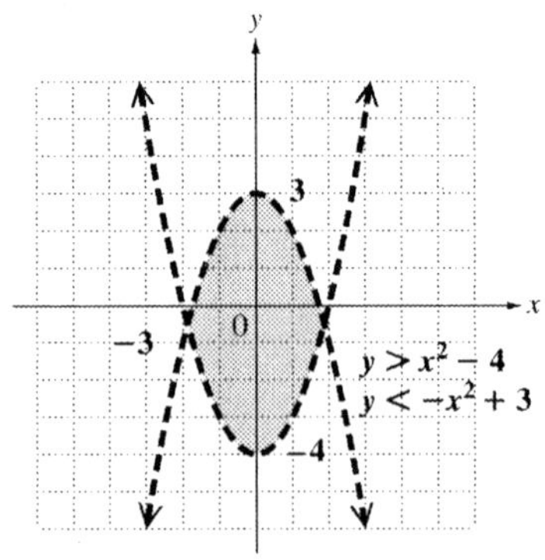

27. $x^2 - y^2 \ge 9 \quad (1)$
$\dfrac{x^2}{16} + \dfrac{y^2}{9} \le 1 \quad (2)$

The graph of (1) has the solid hyperbola $\dfrac{x^2}{9} - \dfrac{y^2}{9} = 1$ with x-intercepts $(3,0)$ and $(-3,0)$ as its boundary. The asymptotes go through $(3,3)$ and $(-3,-3)$ and through $(3,-3)$ and $(-3,3)$. Since $0^2 - 0^2 \ge 9$ is false, the graph includes points to the right of the right branch of the hyperbola and to the left of the left branch of the hyperbola.

The graph of (2) has the solid ellipse $\frac{x^2}{16} + \frac{y^2}{9} = 1$ with x-intercepts $(4, 0)$ and $(-4, 0)$ and y-intercepts $(0, 3)$ and $(0, -3)$ as its boundary. Since $\frac{0^2}{16} + \frac{0^2}{9} \leq 1$ is true, the graph includes points inside the ellipse.
The graph of the system is the intersection of these graphs.

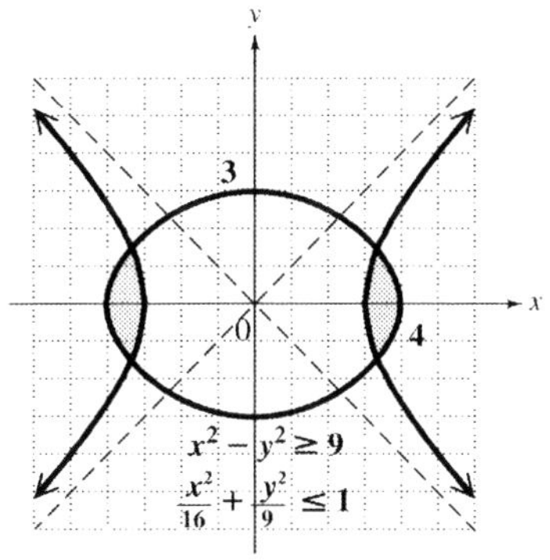

28. $y^2 - x^2 \geq 4$
$-5 \leq y \leq 5$

Rewrite $y^2 - x^2 \geq 4$ as $\frac{y^2}{4} - \frac{x^2}{4} \geq 1$. Graph the solid hyperbola through the y-intercepts $(0, 2)$ and $(0, -2)$. The asymptotes go through $(2, 2)$ and $(-2, -2)$ and through $(-2, 2)$ and $(2, -2)$. Test $(0, 0)$ in $y^2 - x^2 \geq 4$ to get $0 \geq 4$, a false statement. Shade above and below the two branches of the hyperbola.
Graph solid horizontal lines through $(0, 5)$ and $(0, -5)$. Since $-5 \leq y \leq 5$, shade the region between the two horizontal lines.
The graph of the system is the intersection of the shaded regions.

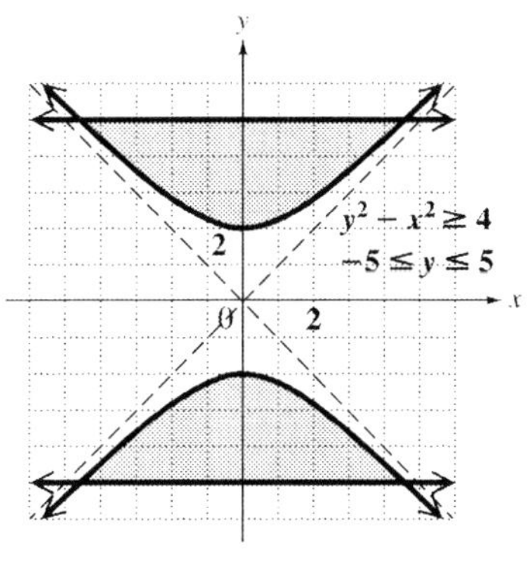

29.
$$\begin{aligned} x &\geq 0 \quad (1) \\ y &\geq 0 \quad (2) \\ x^2 + y^2 &\geq 4 \quad (3) \\ x + y &\leq 5 \quad (4) \end{aligned}$$

The intersection of the graphs of (1) and (2) is quadrant I where both x and y are positive. The positive x- and y-axes are included.
The boundary for (3) is the solid circle $x^2 + y^2 = 4$ with center $(0, 0)$ and radius 2.
Test $(0, 0)$.

$$\begin{aligned} 0^2 + 0^2 &\geq 4 \quad ? \\ 0 &\geq 4 \quad \textit{False} \end{aligned}$$

Shade outside the circle.
The boundary for (4) is the solid line $x + y = 5$ with x- and y-intercepts, $(5, 0)$ and $(0, 5)$.
Test $(0, 0)$.

$$\begin{aligned} 0 + 0 &\leq 5 \quad ? \\ 0 &\leq 5 \quad \textit{True} \end{aligned}$$

Shade the side of the line that contains $(0, 0)$.
The graph of the system is the intersection of the graphs of (3) and (4) in quadrant I.

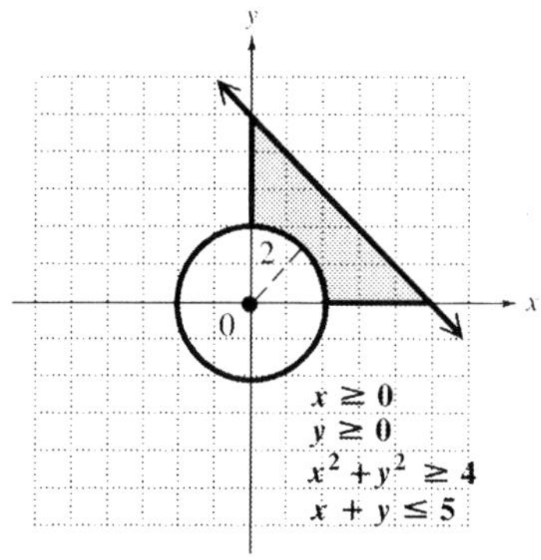

30. $y \leq -x^2$
$y \geq x - 3$
$y \leq -1$
$x < 1$

The boundary, $y = -x^2$, is a solid parabola with vertex at $(0, 0)$ that opens down. Test $(0, -1)$.

$$\begin{aligned} -1 &\leq -(0)^2 \quad ? \\ -1 &\leq 0 \quad \textit{True} \end{aligned}$$

Shade the side of the parabola that contains $(0, -1)$. This is the region inside the parabola.
The boundary, $y = x - 3$, is a solid line with intercepts $(3, 0)$ and $(0, -3)$. Test $(0, 0)$.

$$\begin{aligned} 0 &\geq 0 - 3 \quad ? \\ 0 &\geq -3 \quad \textit{True} \end{aligned}$$

Shade the side of the line that contains $(0, 0)$.
For $y \leq -1$, shade below the solid horizontal line $y = -1$.
For $x < 1$, shade to the left of the dashed vertical line $x = 1$.
The intersection of the four shaded regions is the graph of the system.

continued

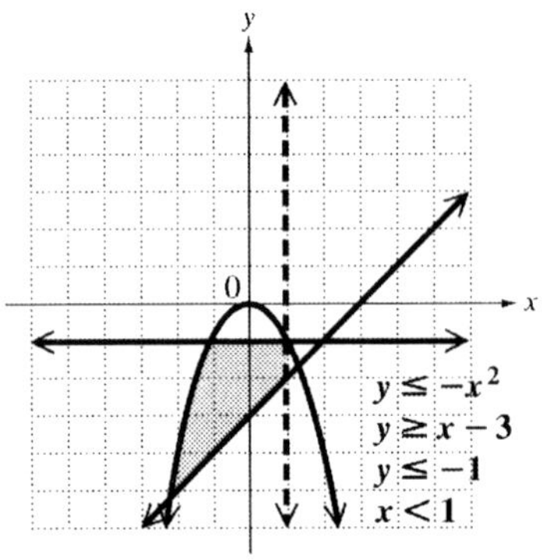

31.

$$\begin{aligned} y &< x^2 \\ y &> -2 \\ x + y &< 3 \\ 3x - 2y &> -6 \end{aligned} \quad \textbf{OR} \quad \begin{aligned} y &< x^2 \\ y &> -2 \\ y &< -x + 3 \\ y &< \tfrac{3}{2}x + 3 \end{aligned}$$

The graph of the first inequality is the region below the dashed parabola with vertex $(0, 0)$ opening up. The graph of the second inequality is the region above the dashed horizontal line $y = -2$. The graph of the third inequality is the region below the dashed line with intercepts $(3, 0)$ and $(0, 3)$. The graph of the fourth inequality is the region below the dashed line with intercepts $(-2, 0)$ and $(0, 3)$.
The graph of the system is the intersection of the four graphs.

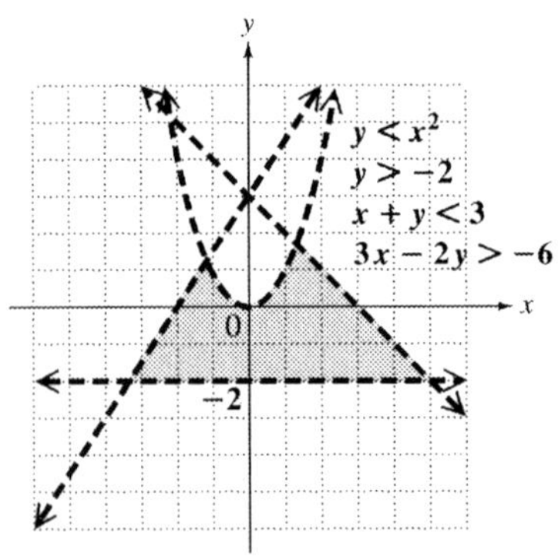

Chapter 12 Review Exercises

1. $f(x) = |x + 4|$

This is the graph of the absolute value function $g(x) = |x|$ shifted 4 units to the left (since $x + 4 = 0$ if $x = -4$).

x	y
-6	2
-5	1
-4	0
-3	1
-2	2

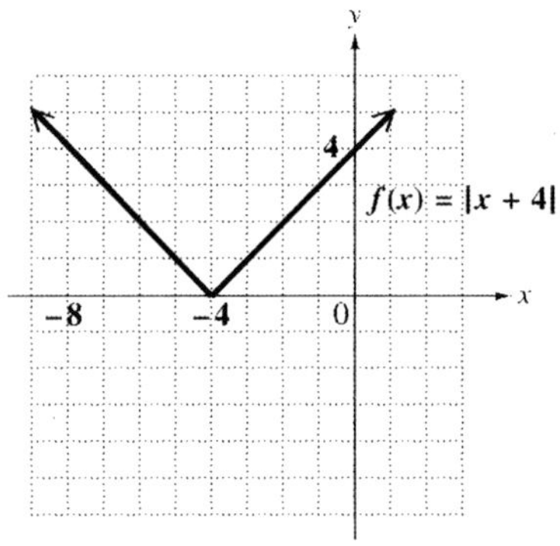

2. $f(x) = \dfrac{1}{x - 4}$

This is the graph of the reciprocal function, $g(x) = \dfrac{1}{x}$, shifted 4 units to the right. Since $x \neq 4$ (or a denominator of 0 results), the line $x = 4$ is a vertical asymptote. Since $\dfrac{1}{x - 4} \neq 0$, the line $y = 0$ is a horizontal asymptote.

x	y
2	$-\frac{1}{2}$
3	-1
$3\frac{1}{2}$	-2
$4\frac{1}{2}$	2
5	1
6	$\frac{1}{2}$

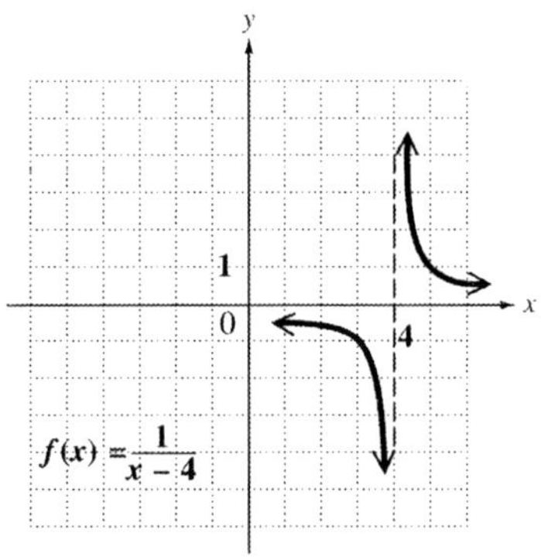

3. $f(x) = \sqrt{x} + 3$

This is the graph of the square root function, $g(x) = \sqrt{x}$, shifted 3 units upward.

x	y
0	3
1	4
4	5

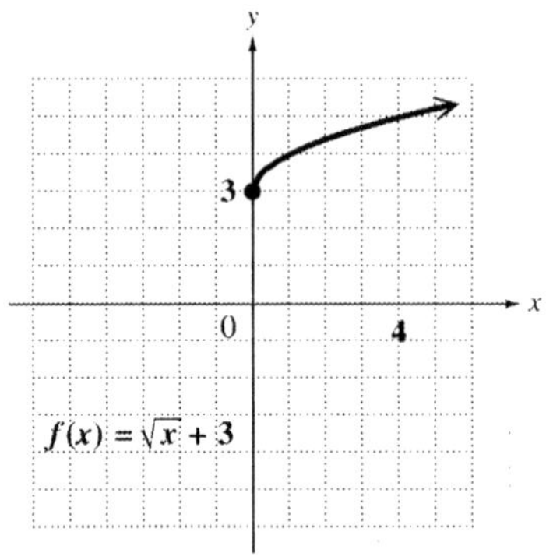

4. $[\![12]\!] = 12 \quad (12 \le 12)$

5. $[\![2\frac{1}{4}]\!] = 2 \quad (2 \le 2\frac{1}{4})$

6. $[\![-4.75]\!] = -5 \quad (-5 \le -4.75)$

In Exercises 7–9, $f(x) = 3x^2 + 2x - 1$ and $g(x) = 5x + 7$.

7. **(a)** $$\begin{aligned}(g \circ f)(3) &= g[f(3)] \\ &= g(3 \cdot 3^2 + 2 \cdot 3 - 1) \\ &= g(32) \\ &= 5 \cdot 32 + 7 \\ &= 167\end{aligned}$$

(b) $$\begin{aligned}(f \circ g)(3) &= f[g(3)] \\ &= f(5 \cdot 3 + 7) \\ &= f(22) \\ &= 3 \cdot 22^2 + 2 \cdot 22 - 1 \\ &= 1495\end{aligned}$$

8. **(a)** $$\begin{aligned}(f \circ g)(-2) &= f[g(-2)] \\ &= f[5(-2) + 7] \\ &= f(-3) \\ &= 3(-3)^2 + 2(-3) - 1 \\ &= 20\end{aligned}$$

(b) $$\begin{aligned}(g \circ f)(-2) &= g[f(-2)] \\ &= g[3(-2)^2 + 2(-2) - 1] \\ &= g(7) \\ &= 5 \cdot 7 + 7 \\ &= 42\end{aligned}$$

9. **(a)** $$\begin{aligned}&(f \circ g)(x) \\ &= f[g(x)] \\ &= f(5x + 7) \\ &= 3(5x + 7)^2 + 2(5x + 7) - 1 \\ &= 3(25x^2 + 70x + 49) + 10x + 14 - 1 \\ &= 75x^2 + 210x + 147 + 10x + 13 \\ &= 75x^2 + 220x + 160\end{aligned}$$

(b) $$\begin{aligned}(g \circ f)(x) &= g[f(x)] \\ &= g(3x^2 + 2x - 1) \\ &= 5(3x^2 + 2x - 1) + 7 \\ &= 15x^2 + 10x - 5 + 7 \\ &= 15x^2 + 10x + 2\end{aligned}$$

10. No, composition of functions is not a commutative operation. For example, the results of Exercise 9 show that $(f \circ g)(x) \neq (g \circ f)(x)$ in this case.

11. Center $(-2, 4)$, $r = 3$
Here $h = -2$, $k = 4$, and $r = 3$, so an equation of the circle is

$$\begin{aligned}(x - h)^2 + (y - k)^2 &= r^2 \\ [x - (-2)]^2 + (y - 4)^2 &= 3^2 \\ (x + 2)^2 + (y - 4)^2 &= 9.\end{aligned}$$

12. Center $(-1, -3)$, $r = 5$
Here $h = -1$, $k = -3$, and $r = 5$, so an equation of the circle is

$$\begin{aligned}(x - h)^2 + (y - k)^2 &= r^2 \\ [x - (-1)]^2 + [y - (-3)]^2 &= 5^2 \\ (x + 1)^2 + (y + 3)^2 &= 25.\end{aligned}$$

13. Center $(4, 2)$, $r = 6$
Here $h = 4$, $k = 2$, and $r = 6$, so an equation of the circle is

$$\begin{aligned}(x - h)^2 + (y - k)^2 &= r^2 \\ (x - 4)^2 + (y - 2)^2 &= 6^2 \\ (x - 4)^2 + (y - 2)^2 &= 36.\end{aligned}$$

14. $x^2 + y^2 + 6x - 4y - 3 = 0$
Write the equation in center-radius form,
$$(x - h)^2 + (y - k)^2 = r^2,$$
by completing the squares on x and y.

$$\begin{aligned}(x^2 + 6x \quad\;) + (y^2 - 4y \quad\;) &= 3 \\ (x^2 + 6x + \underline{9}) + (y^2 - 4y + \underline{4}) & \\ &= 3 + \underline{9} + \underline{4} \\ (x + 3)^2 + (y - 2)^2 &= 16 \\ [x - (-3)]^2 + (y - 2)^2 &= 16\end{aligned}$$

The circle has center (h, k) at $(-3, 2)$ and radius $\sqrt{16} = 4$.

15. $x^2 + y^2 - 8x - 2y + 13 = 0$
Write the equation in center-radius form,
$$(x - h)^2 + (y - k)^2 = r^2,$$
by completing the squares on x and y.

$$\begin{aligned}(x^2 - 8x \quad\;) + (y^2 - 2y \quad\;) &= -13 \\ (x^2 - 8x + \underline{16}) + (y^2 - 2y + \underline{1}) & \\ &= -13 + \underline{16} + \underline{1} \\ (x - 4)^2 + (y - 1)^2 &= 4\end{aligned}$$

The circle has center (h, k) at $(4, 1)$ and radius $\sqrt{4} = 2$.

16. $$\begin{aligned}2x^2 + 2y^2 + 4x + 20y &= -34 \\ x^2 + y^2 + 2x + 10y &= -17\end{aligned}$$
Write the equation in center-radius form,
$$(x - h)^2 + (y - k)^2 = r^2,$$
by completing the squares on x and y.

$$\begin{aligned}(x^2 + 2x \quad\;) + (y^2 + 10y \quad\;) &= -17 \\ (x^2 + 2x + \underline{1}) + (y^2 + 10y + \underline{25}) & \\ &= -17 + \underline{1} + \underline{25} \\ (x + 1)^2 + (y + 5)^2 &= 9 \\ [x - (-1)]^2 + [y - (-5)]^2 &= 9\end{aligned}$$

The circle has center (h, k) at $(-1, -5)$ and radius $\sqrt{9} = 3$.

17. $4x^2 + 4y^2 - 24x + 16y = 48$
$x^2 + y^2 - 6x + 4y = 12$

Write the equation in center-radius form,
$$(x-h)^2 + (y-k)^2 = r^2,$$
by completing the squares on x and y.

$$(x^2 - 6x \quad\ ) + (y^2 + 4y \quad\ ) = 12$$
$$(x^2 - 6x + \underline{9}) + (y^2 + 4y + \underline{4}) = 12 + \underline{9} + \underline{4}$$
$$(x-3)^2 + (y+2)^2 = 25$$
$$(x-3)^2 + [y-(-2)]^2 = 25$$

The circle has center (h, k) at $(3, -2)$ and radius $\sqrt{25} = 5$.

18. $x^2 + y^2 = 16$
$(x-0)^2 + (y-0)^2 = 4^2$

This is a circle with center $(0, 0)$ and radius 4.

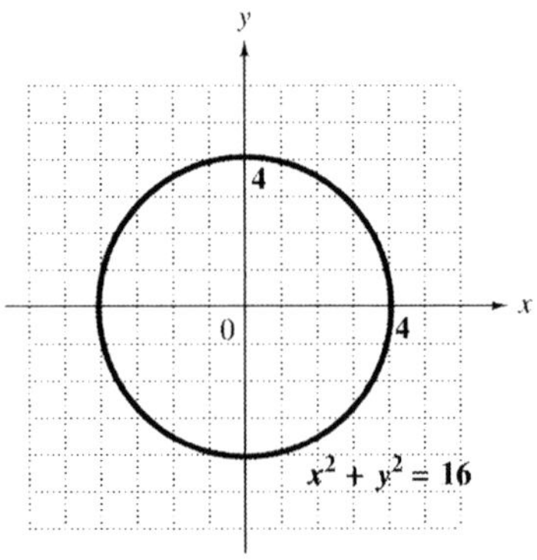

19. $\dfrac{x^2}{16} + \dfrac{y^2}{9} = 1$ is in $\dfrac{x^2}{a^2} + \dfrac{y^2}{b^2} = 1$ form with $a = 4$ and $b = 3$. The graph is an ellipse with x-intercepts $(4, 0)$ and $(-4, 0)$ and y-intercepts $(0, 3)$ and $(0, -3)$. Plot the intercepts, and draw the ellipse through them.

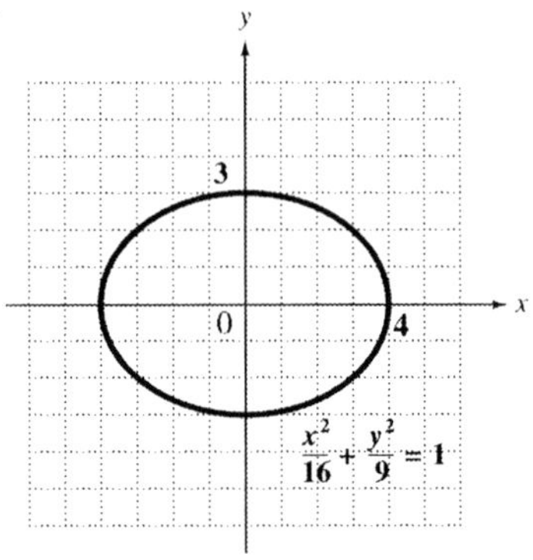

20. $\dfrac{x^2}{49} + \dfrac{y^2}{25} = 1$ is in $\dfrac{x^2}{a^2} + \dfrac{y^2}{b^2} = 1$ form with $a = 7$ and $b = 5$. The graph is an ellipse with x-intercepts $(7, 0)$ and $(-7, 0)$ and y-intercepts $(0, 5)$ and $(0, -5)$. Plot the intercepts, and draw the ellipse through them.

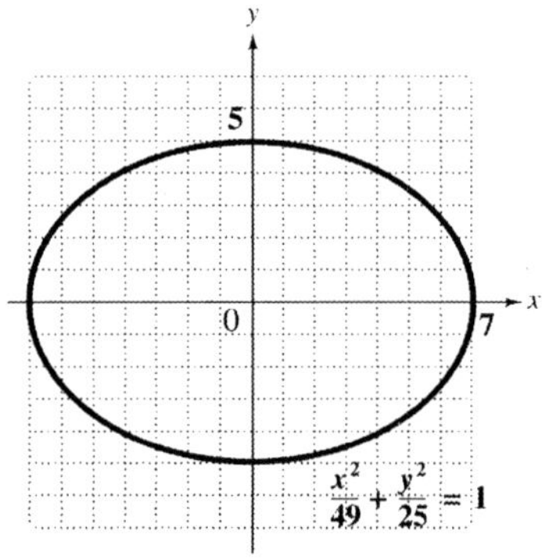

21. The total distance on the horizontal axis is $160 + 16{,}000 = 16{,}160$ km. This represents $2a$, so $a = \frac{1}{2}(16{,}160) = 8080$. The distance from Earth to the center of the ellipse is
$$8080 - 160 = 7920,$$
which is the value of c. From Exercise 37 in Section 12.2, we know that $c^2 = a^2 - b^2$, so
$$b^2 = a^2 - c^2.$$
$$b^2 = 8080^2 - 7920^2$$
$$= 2{,}560{,}000$$

Thus, $b = \sqrt{2{,}560{,}000} = 1600$ and the equation is
$$\frac{x^2}{8080^2} + \frac{y^2}{1600^2} = 1$$
$$\text{or} \quad \frac{x^2}{65{,}286{,}400} + \frac{y^2}{2{,}560{,}000} = 1.$$

22. **(a)** The distance between the foci is $2c$, where c can be found using the relationship
$$c = \sqrt{a^2 - b^2}.$$
$$= \sqrt{310^2 - (513/2)^2}$$
$$\approx 174.1 \text{ feet}$$

So the distance is about 348.2 feet.

(b) The approximate circumference of the Roman Colosseum is
$$C \approx 2\pi\sqrt{\frac{a^2 + b^2}{2}}.$$
$$= 2\pi\sqrt{\frac{310^2 + (513/2)^2}{2}}$$
$$\approx 1787.6 \text{ feet}$$

23. $\frac{x^2}{16} - \frac{y^2}{25} = 1$ is in $\frac{x^2}{a^2} - \frac{y^2}{b^2} = 1$ form with $a = 4$ and $b = 5$. The graph is a hyperbola with x-intercepts $(4, 0)$ and $(-4, 0)$ and asymptotes that are the extended diagonals of the rectangle with vertices $(4, 5)$, $(4, -5)$, $(-4, -5)$, and $(-4, 5)$. Graph a branch of the hyperbola through each intercept approaching the asymptotes.

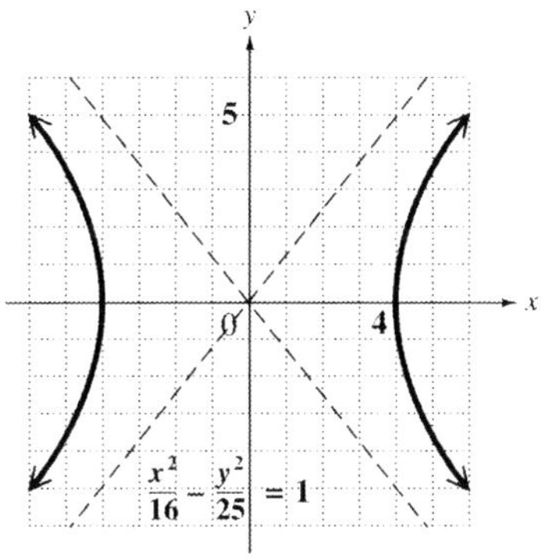

24. $\frac{y^2}{25} - \frac{x^2}{4} = 1$ is in $\frac{y^2}{b^2} - \frac{x^2}{a^2} = 1$ form with $a = 2$ and $b = 5$. The graph is a hyperbola with y-intercepts $(0, 5)$ and $(0, -5)$ and asymptotes that are the extended diagonals of the rectangle with vertices $(2, 5)$, $(2, -5)$, $(-2, -5)$, and $(-2, 5)$. Graph a branch of the hyperbola through each intercept approaching the asymptotes.

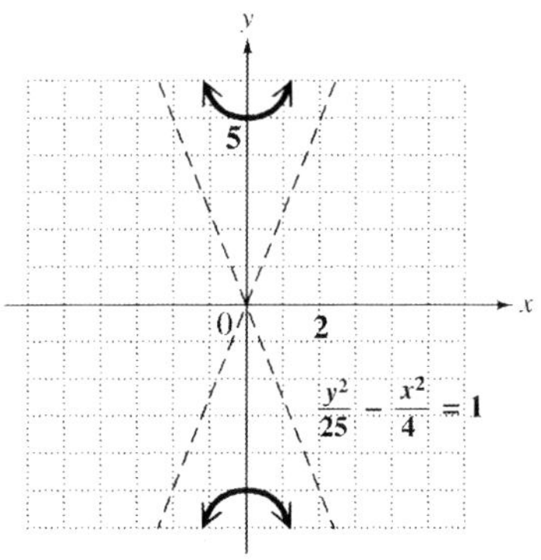

25.
$$f(x) = -\sqrt{16 - x^2}$$
Replace $f(x)$ with y.
$$y = -\sqrt{16 - x^2}$$
Square both sides.
$$y^2 = 16 - x^2$$
$$x^2 + y^2 = 16$$

This equation is the graph of a circle with center $(0, 0)$ and radius 4. Since $f(x)$ represents a nonpositive square root, $f(x)$ is nonpositive and its graph is the lower half of the circle.

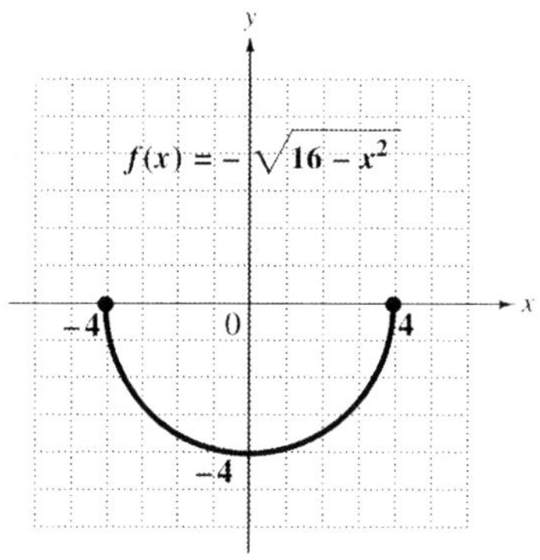

26.
$$x^2 + y^2 = 64$$
$$(x - 0)^2 + (y - 0)^2 = 8^2$$

The last equation is in $(x - h)^2 + (y - k)^2 = r^2$ form. The graph is a *circle*.

27. $y = 2x^2 - 3$
$$y = 2(x - 0)^2 - 3$$

The last equation is in $y = a(x - h)^2 + k$ form. The graph is a *parabola*.

28.
$$y^2 = 2x^2 - 8$$
$$2x^2 - y^2 = 8$$
$$\frac{x^2}{4} - \frac{y^2}{8} = 1$$

The last equation is in $\frac{x^2}{a^2} - \frac{y^2}{b^2} = 1$ form, so the graph is a *hyperbola*.

29.
$$y^2 = 8 - 2x^2$$
$$2x^2 + y^2 = 8$$
$$\frac{x^2}{4} + \frac{y^2}{8} = 1$$

The last equation is in $\frac{x^2}{a^2} + \frac{y^2}{b^2} = 1$ form, so the graph is an *ellipse*.

30. $x = y^2 + 4$
$$x = (y - 0)^2 + 4$$

The last equation is in $x = a(y - k)^2 + h$ form, so the graph is a *parabola*.

31. $x^2 - y^2 = 64$
$$\frac{x^2}{64} - \frac{y^2}{64} = 1$$

The last equation is in $\frac{x^2}{a^2} - \frac{y^2}{b^2} = 1$ form, so the graph is a *hyperbola*.

32. A hyperbola is defined as the set of all points in a plane such that the absolute value of the *difference* of the distances from two fixed points (called *foci*) is constant. The hyperbola shown in the text will have an equation of the form

$$\frac{x^2}{a^2} - \frac{y^2}{b^2} = 1. \quad (1)$$

The constant difference for this hyperbola is

$$|d_1 - d_2| = |80 - 30| = 50.$$

Let Q be the point on the right branch of the hyperbola that is on $\overline{MS}$. Let $x = \overline{MQ}$ and $y = \overline{QS}$. Since $\overline{MS} = 100$ and $\overline{MQ} - \overline{QS}$ must be 50, we have the following system.

$$\begin{aligned} x + y &= 100 \quad (2) \\ x - y &= 50 \\ \hline 2x &= 150 \quad \textit{Add} \\ x &= 75 \end{aligned}$$

From (2), we see that $y = 25$. The distance from the center of the hyperbola, C, to S, is 50, so $a = \overline{CQ} = 50 - 25 = 25$.
To find b^2, we'll find a point on the hyperbola, substitute for a, x, and y, and then solve for b^2. Let P have coordinates (d, e). If we draw a perpendicular line from P to $\overline{MS}$, we see two right triangles, PQM and PQS. Using the Pythagorean formula, we get the following system of equations.

$$\begin{aligned} (50 + d)^2 + e^2 &= 80^2 \\ (50 - d)^2 + e^2 &= 30^2 \quad (3) \\ \hline (2500 + 100d + d^2) - (2500 - 100d + d^2) & \\ &= 6400 - 900 \quad \textit{Subtract} \\ 200d &= 5500 \\ d &= 27.5 \end{aligned}$$

From (3) with $d = 27.5$, we get $e^2 = 30^2 - 22.5^2$, so $e = \sqrt{393.75}$. Now substitute 25 for a, 27.5 for x, and $\sqrt{393.75}$ for y in (1) to solve for b^2.

$$\begin{aligned} \frac{27.5^2}{25^2} - \frac{393.75}{b^2} &= 1 \\ \frac{756.25}{625} - \frac{625}{625} &= \frac{393.75}{b^2} \\ \frac{131.25}{625} &= \frac{393.75}{b^2} \\ \frac{625}{131.25} &= \frac{b^2}{393.75} \\ b^2 &= \frac{625}{131.25}(393.75) \\ b^2 &= 1875 \end{aligned}$$

Thus, equation (1) becomes

$$\frac{x^2}{625} - \frac{y^2}{1875} = 1.$$

33.
$$\begin{aligned} 2y &= 3x - x^2 \quad (1) \\ x + 2y &= -12 \quad (2) \end{aligned}$$

Substitute $3x - x^2$ for $2y$ in equation (2).

$$\begin{aligned} x + 2y &= -12 \quad (2) \\ x + (3x - x^2) &= -12 \\ -x^2 + 4x + 12 &= 0 \\ x^2 - 4x - 12 &= 0 \\ (x - 6)(x + 2) &= 0 \end{aligned}$$

$$x - 6 = 0 \quad \text{or} \quad x + 2 = 0$$
$$x = 6 \quad \text{or} \quad x = -2$$

Substitute these values for x in equation (2) to find y.

If $x = 6$, then

$$\begin{aligned} x + 2y &= -12 \quad (2) \\ 6 + 2y &= -12 \\ 2y &= -18 \\ y &= -9. \end{aligned}$$

If $x = -2$ then

$$\begin{aligned} x + 2y &= -12 \quad (2) \\ -2 + 2y &= -12 \\ 2y &= -10 \\ y &= -5 \end{aligned}$$

Solution set: $\{(6, -9), (-2, -5)\}$

34.
$$\begin{aligned} y + 1 &= x^2 + 2x \quad (1) \\ y + 2x &= 4 \quad (2) \end{aligned}$$

Solve equation (2) for y.

$$y = 4 - 2x \quad (3)$$

Substitute $4 - 2x$ for y in (1).

$$\begin{aligned} (4 - 2x) + 1 &= x^2 + 2x \\ 0 &= x^2 + 4x - 5 \\ 0 &= (x + 5)(x - 1) \end{aligned}$$

$$x + 5 = 0 \quad \text{or} \quad x - 1 = 0$$
$$x = -5 \quad \text{or} \quad x = 1$$

Substitute these values for x in equation (3) to find y.

If $x = -5$, then $y = 4 - 2(-5) = 14$.
If $x = 1$, then $y = 4 - 2(1) = 2$.

Solution set: $\{(1, 2), (-5, 14)\}$

35. $x^2 + 3y^2 = 28 \quad (1)$
$y - x = -2 \quad (2)$

Solve equation (2) for y.

$$y = x - 2 \quad (3)$$

Substitute $x - 2$ for y in equation (1).

$$\begin{aligned} x^2 + 3y^2 &= 28 \quad (1) \\ x^2 + 3(x-2)^2 &= 28 \\ x^2 + 3(x^2 - 4x + 4) - 28 &= 0 \\ x^2 + 3x^2 - 12x + 12 - 28 &= 0 \\ 4x^2 - 12x - 16 &= 0 \\ 4(x^2 - 3x - 4) &= 0 \\ 4(x-4)(x+1) &= 0 \end{aligned}$$

$$x - 4 = 0 \quad \text{or} \quad x + 1 = 0$$
$$x = 4 \quad \text{or} \quad x = -1$$

Substitute these values for x in equation (3) to find y.
If $x = 4$, then $y = 4 - 2 = 2$.
If $x = -1$, then $y = -1 - 2 = -3$.

Solution set: $\{(4,2), (-1,-3)\}$

36. $xy = 8 \quad (1)$
$x - 2y = 6 \quad (2)$

Solve equation (2) for x.

$$x = 2y + 6 \quad (3)$$

Substitute $2y + 6$ for x in equation (1) to find y.

$$\begin{aligned} xy &= 8 \quad (1) \\ (2y+6)y &= 8 \\ 2y^2 + 6y - 8 &= 0 \\ 2(y^2 + 3y - 4) &= 0 \\ 2(y+4)(y-1) &= 0 \end{aligned}$$

$$y + 4 = 0 \quad \text{or} \quad y - 1 = 0$$
$$y = -4 \quad \text{or} \quad y = 1$$

Substitute these values for y in equation (3) to find x.
If $y = -4$, then $x = 2(-4) + 6 = -2$.
If $y = 1$, then $x = 2(1) + 6 = 8$.

Solution set: $\{(-2,-4), (8,1)\}$

37. $x^2 + y^2 = 6 \quad (1)$
$x^2 - 2y^2 = -6 \quad (2)$

Multiply equation (2) by -1 and add the result to equation (1).

$$\begin{aligned} x^2 + y^2 &= 6 \quad (1) \\ -x^2 + 2y^2 &= 6 \quad -1 \times (2) \\ \hline 3y^2 &= 12 \\ y^2 &= 4 \end{aligned}$$

$$y = 2 \quad \text{or} \quad y = -2$$

Substitute these values for y in equation (1) to find x.
If $y = \pm 2$, then

$$\begin{aligned} x^2 + y^2 &= 6 \quad (1) \\ x^2 + (\pm 2)^2 &= 6 \\ x^2 + 4 &= 6 \\ x^2 &= 2. \end{aligned}$$

$$x = \sqrt{2} \quad \text{or} \quad x = -\sqrt{2}$$

Since each value of x can be paired with each value of y, there are four points and the solution set is

$$\{(\sqrt{2},2), (-\sqrt{2},2), (\sqrt{2},-2), (-\sqrt{2},-2)\}.$$

38. $3x^2 - 2y^2 = 12 \quad (1)$
$x^2 + 4y^2 = 18 \quad (2)$

Multiply equation (1) by 2 and add the result to equation (2).

$$\begin{aligned} 6x^2 - 4y^2 &= 24 \quad 2 \times (1) \\ x^2 + 4y^2 &= 18 \quad (2) \\ \hline 7x^2 &= 42 \\ x^2 &= 6 \end{aligned}$$

$$x = \sqrt{6} \quad \text{or} \quad x = -\sqrt{6}$$

Substitute these values for x in equation (2) to find y.
If $x = \pm\sqrt{6}$, then

$$\begin{aligned} x^2 + 4y^2 &= 18. \quad (2) \\ (\pm\sqrt{6})^2 + 4y^2 &= 18 \\ 6 + 4y^2 &= 18 \\ 4y^2 &= 12 \\ y^2 &= 3 \end{aligned}$$

$$y = \sqrt{3} \quad \text{or} \quad y = -\sqrt{3}$$

The solution set is

$$\{(\sqrt{6},\sqrt{3}), (\sqrt{6},-\sqrt{3}), (-\sqrt{6},\sqrt{3}), (-\sqrt{6},-\sqrt{3})\}.$$

39. A circle and a line can intersect in zero, one, or two points, so zero, one, or two solutions are possible.

40. A parabola and a hyperbola can intersect in zero, one, two, three, or four points, so zero, one, two, three, or four solutions are possible.

41.
$$9x^2 \geq 16y^2 + 144$$
$$9x^2 - 16y^2 \geq 144$$
$$\frac{x^2}{16} - \frac{y^2}{9} \geq 1$$

The boundary, $\frac{x^2}{16} - \frac{y^2}{9} = 1$, is a solid hyperbola with x-intercepts $(4,0)$ and $(-4,0)$. The asymptotes are the extended diagonals of the rectangle with vertices $(4,3)$, $(4,-3)$, $(-4,-3)$, and $(-4,3)$. Test $(0,0)$.

$$9(0)^2 \geq 16(0)^2 + 144 \quad ?$$
$$0 \geq 144 \qquad \textit{False}$$

Shade the sides of the hyperbola that do not contain $(0,0)$. These are the regions inside the branches of the hyperbola.

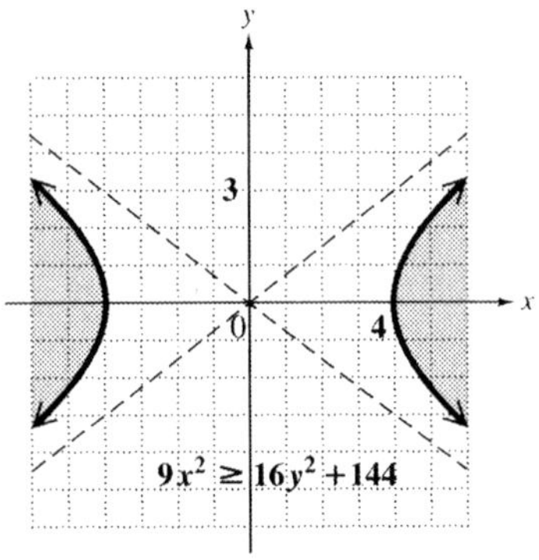

42. $4x^2 + y^2 \geq 16$
$$\frac{x^2}{4} + \frac{y^2}{16} \geq 1$$

The boundary, $\frac{x^2}{4} + \frac{y^2}{16} = 1$, is a solid ellipse with intercepts $(2,0)$, $(-2,0)$, $(0,4)$, and $(0,-4)$. Test $(0,0)$.

$$4(0)^2 + 0^2 \geq 16 \quad ?$$
$$0 \geq 16 \qquad \textit{False}$$

Shade the side of ellipse that does not contain $(0,0)$. This is the region outside the ellipse.

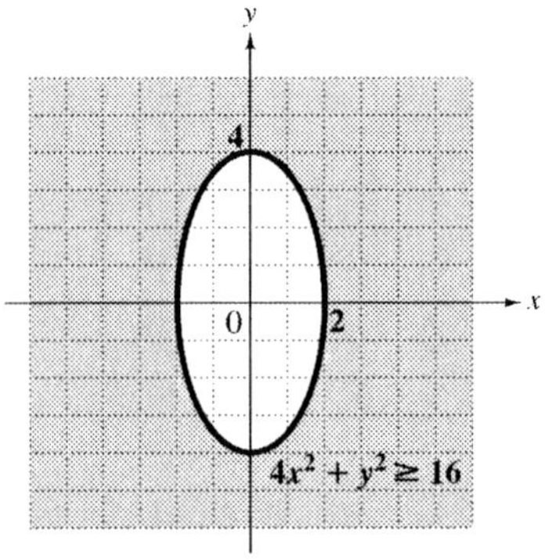

43. $y < -(x+2)^2 + 1$

The boundary, $y = -(x+2)^2 + 1$, is a dashed vertical parabola with vertex $(-2,1)$. Since $a = -1 < 0$, the parabola opens down. Also, $|a| = |-1| = 1$, so the graph has the same shape as the graph of $y = x^2$. Test $(0,0)$.

$$0 < -(0+2)^2 + 1 \quad ?$$
$$0 < -(4) + 1 \quad ?$$
$$0 < -3 \qquad \textit{False}$$

Shade the side of the parabola that does not contain $(0,0)$. This is the region inside the parabola.

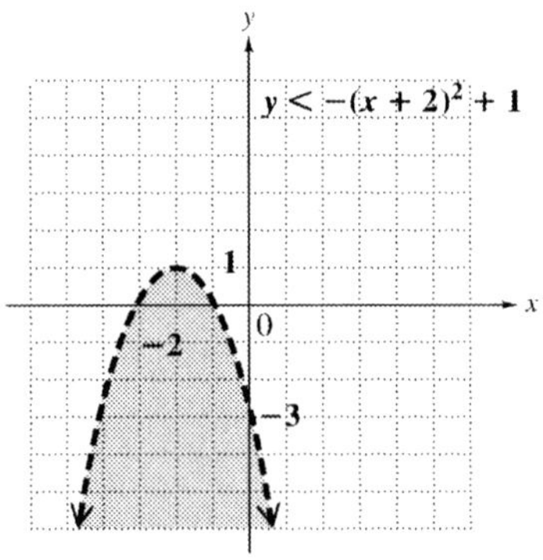

44.
$$2x + 5y \leq 10$$
$$3x - y \leq 6$$

The boundary, $2x + 5y = 10$ is a solid line with intercepts $(5,0)$ and $(0,2)$. Test $(0,0)$.

$$2(0) + 5(0) \leq 10 \quad ?$$
$$0 \leq 10 \qquad \textit{True}$$

Shade the side of the line that contains $(0,0)$. The boundary, $3x - y = 6$, is a solid line with intercepts $(2,0)$ and $(0,-6)$. Test $(0,0)$.

$$3(0) - 0 \leq 6 \quad ?$$
$$0 \leq 6 \qquad \textit{True}$$

Shade the side of the line that contains $(0,0)$. The graph of the system is the intersection of the two shaded regions.

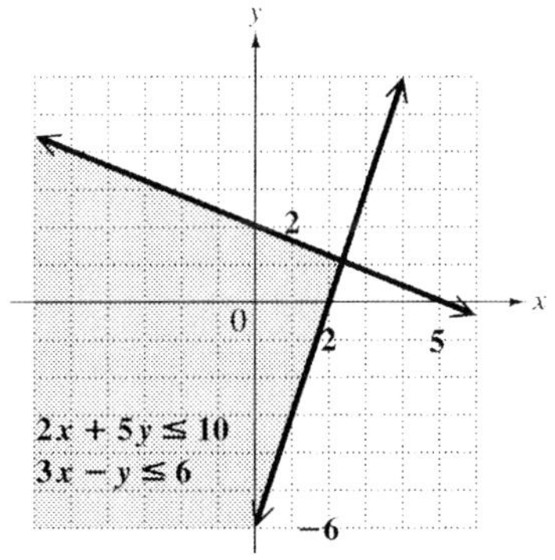

45. $|x| \leq 2$
$|y| > 1$
$4x^2 + 9y^2 \leq 36$

The equation of the boundary, $|x| = 2$, can be written as

$$x = -2 \text{ or } x = 2.$$

The graph is these two solid vertical lines. Since $|0| \leq 2$ is true, the region between the lines, containing $(0, 0)$, is shaded.
The boundary, $|y| = 1$, consists of the two dashed horizontal lines $y = 1$ and $y = -1$. Since $|0| > 1$ is false, the regions above and below the lines, not containing $(0, 0)$, are shaded.
The boundary given by

$$4x^2 + 9y^2 = 36$$
$$\text{or } \frac{x^2}{9} + \frac{y^2}{4} = 1$$

is graphed as a solid ellipse with intercepts $(3, 0)$, $(-3, 0)$, $(0, 2)$, and $(0, -2)$. Test $(0, 0)$.

$$4(0)^2 + 9(0)^2 \leq 36 \quad ?$$
$$0 \leq 36 \quad \textit{True}$$

The region inside the ellipse, containing $(0, 0)$, is shaded.
The graph of the system consists of the regions that include the common points of the three shaded regions.

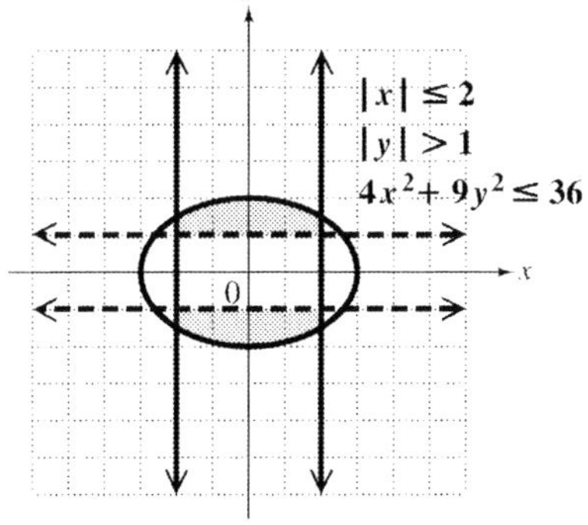

46. $9x^2 \leq 4y^2 + 36$
$x^2 + y^2 \leq 16$

The equation of the first boundary is

$$9x^2 = 4y^2 + 36$$
$$9x^2 - 4y^2 = 36$$
$$\frac{x^2}{4} - \frac{y^2}{9} = 1.$$

The graph is a solid hyperbola with x-intercepts $(2, 0)$ and $(-2, 0)$. The asymptotes are the extended diagonals of the rectangle with vertices $(2, 3)$, $(2, -3)$, $(-2, -3)$, and $(-2, 3)$. Test $(0, 0)$.

$$9(0)^2 \leq 4(0)^2 + 36 \quad ?$$
$$0 \leq 36 \quad \textit{True}$$

Shade the region between the branches of the hyperbola that contains $(0, 0)$.
The equation of the second boundary is $x^2 + y^2 = 16$. This is a solid circle with center $(0, 0)$ and radius 4. Test $(0, 0)$.

$$0^2 + 0^2 \leq 16 \quad ?$$
$$0 \leq 16 \quad \textit{True}$$

Shade the region inside the circle.
The graph of the system is the intersection of the shaded regions which is between the two branches of the hyperbola and inside the circle.

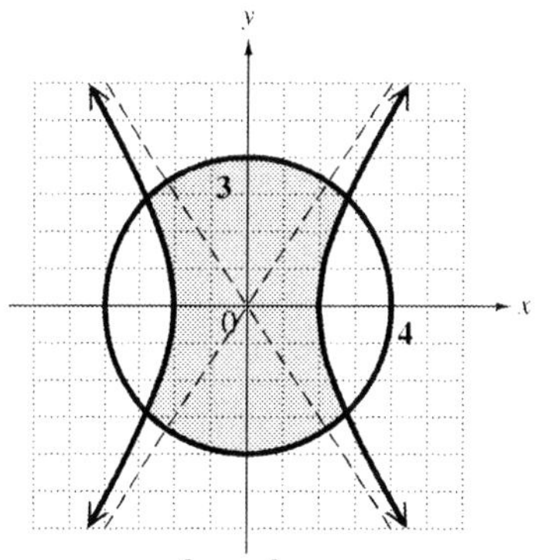

47. **[12.2]** $x^2 + y^2 = 25$ is in the form $(x - h)^2 + (y - k)^2 = r^2$. The graph is a circle with center at $(0, 0)$ and radius 5.

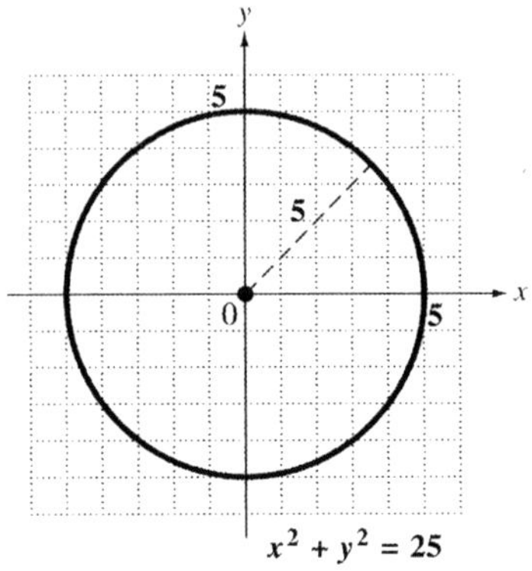

48. **[12.2]** $x^2 + 9y^2 = 9$ in $\frac{x^2}{a^2} + \frac{y^2}{b^2} = 1$ form is $\frac{x^2}{9} + \frac{y^2}{1} = 1$ with $a = 3$ and $b = 1$. The graph is an ellipse with x-intercepts $(3, 0)$ and $(-3, 0)$ and y-intercepts $(0, 1)$ and $(0, -1)$. Plot the intercepts, and draw the ellipse through them.

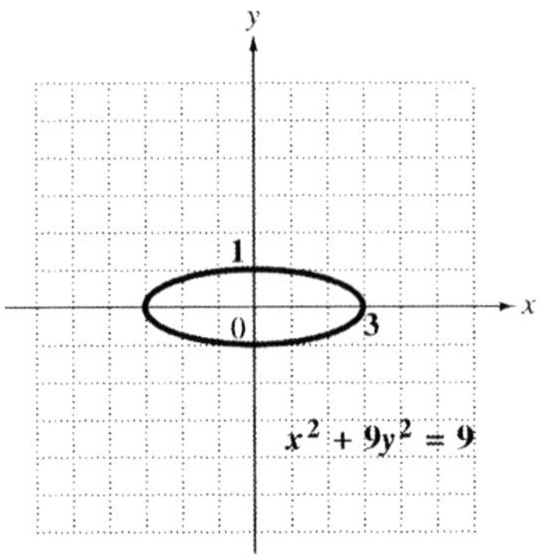

49. **[12.3]** $x^2 - 9y^2 = 9$

$$\frac{x^2}{9} - \frac{y^2}{1} = 1$$

The equation is in the $\frac{x^2}{a^2} - \frac{y^2}{b^2} = 1$ form with $a = 3$ and $b = 1$. The graph is a hyperbola with x-intercepts $(3, 0)$ and $(-3, 0)$ and asymptotes that are the extended diagonals of the rectangle with vertices $(3, 1)$, $(3, -1)$, and $(-3, -1)$, and $(-3, 1)$. Graph a branch of the hyperbola through each intercept and approaching the asymptotes.

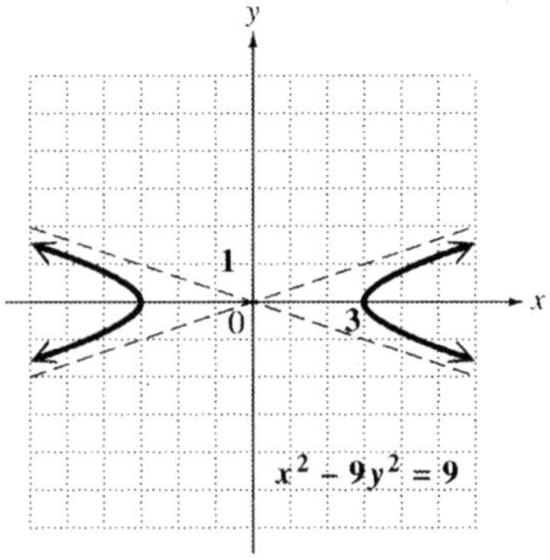

50. **[12.3]** $f(x) = \sqrt{4 - x}$

Replace $f(x)$ with y.

$$y = \sqrt{4 - x}$$

Square both sides.

$$y^2 = 4 - x$$
$$x = -y^2 + 4$$
$$x = -1(y - 0)^2 + 4$$

This equation is the graph of a horizontal parabola with vertex $(4, 0)$. Since $a = -1 < 0$, the graph opens to the left. Also, $|a| = |-1| = 1$, so the graph has the same shape as the graph of $y = x^2$. The points $(0, 2)$ and $(3, 1)$ are on the graph. Since $f(x)$ represents a square root, $f(x)$ is nonnegative and its graph is the upper half of the parabola.

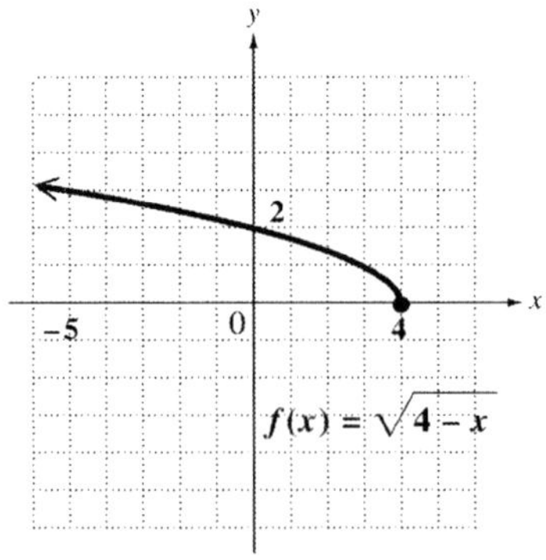

51. **[12.1]** $f(x) = [\![x]\!] - 1$

This is the graph of the greatest integer function, $g(x) = [\![x]\!]$, shifted down 1 unit.

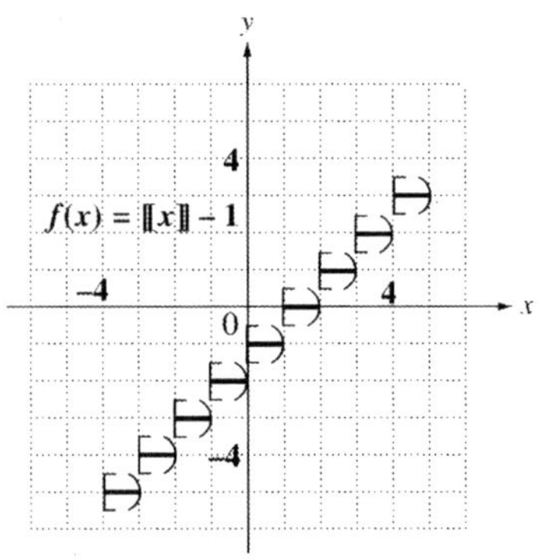

52. **[12.5]** $4y > 3x - 12$
$x^2 < 16 - y^2$

The boundary $4y = 3x - 12$ is a dashed line with intercepts $(0, -3)$ and $(4, 0)$. Test $(0, 0)$.

$$4(0) > 3(0) - 12 \quad ?$$
$$0 > -12 \quad \textit{True}$$

Shade the side of the line that contains $(0, 0)$.
The boundary $x^2 = 16 - y^2$, or $x^2 + y^2 = 16$, is a dashed circle with center at $(0, 0)$ and radius 4. Test $(0, 0)$.

$$0^2 < 16 - 0^2 \quad ?$$
$$0 < 16 \quad \textit{True}$$

Shade the region inside the circle.
The graph of the system is the intersection of the two shaded regions.

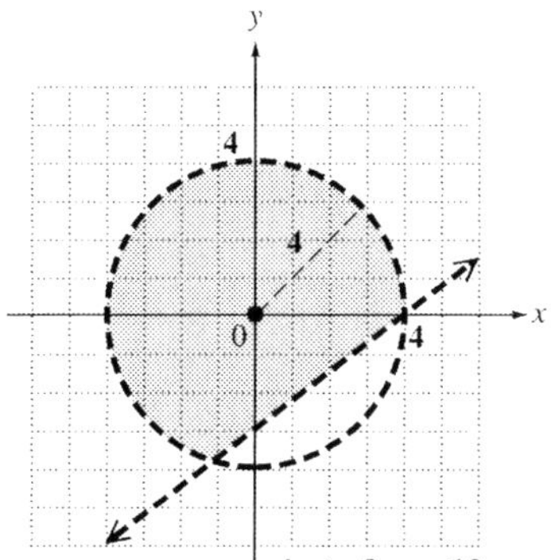

53. **[12.2]** There are cases where one x-value will yield two y-values. In a function, every x yields one and only one y.

Chapter 12 Test

1. $f(x) = \sqrt{x - 2}$ is the graph of $g(x) = \sqrt{x}$ shifted 2 units right. The graph is **C**.

2. $f(x) = \sqrt{x + 2}$ is the graph of $g(x) = \sqrt{x}$ shifted 2 units left. The graph is **A**.

3. $f(x) = \sqrt{x} + 2$ is the graph of $g(x) = \sqrt{x}$ shifted 2 units up. The graph is **D**.

4. $f(x) = \sqrt{x} - 2$ is the graph of $g(x) = \sqrt{x}$ shifted 2 units down. The graph is **B**.

5. $f(x) = |x - 3| + 4$

This is the graph of the absolute value function, $g(x) = |x|$, shifted 3 units to the right (since $x - 3 = 0$ if $x = 3$) and 4 units upward (because of the $+4$). Its vertex is at $(3, 4)$.

x	0	1	2	3	4	5	6
y	7	6	5	4	5	6	7

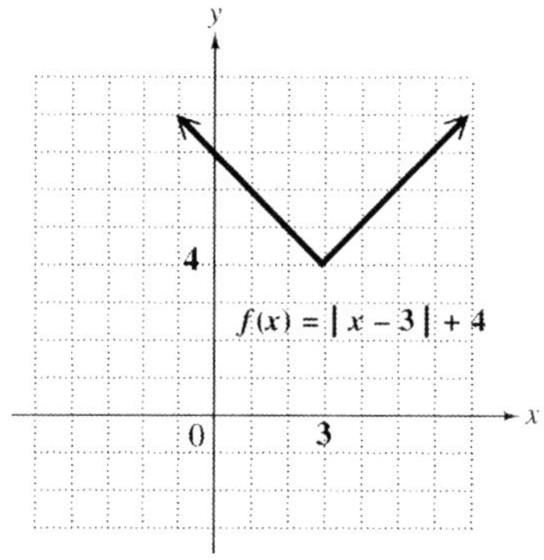

6. $f(x) = 3x + 5$, $g(x) = x^2 + 2$

(a) $(f \circ g)(-2) = f[g(-2)]$
$= f[(-2)^2 + 2]$
$= f(6)$
$= 3 \cdot 6 + 5$
$= 23$

(b) $(f \circ g)(x) = f[g(x)]$
$= f(x^2 + 2)$
$= 3(x^2 + 2) + 5$
$= 3x^2 + 6 + 5$
$= 3x^2 + 11$

(c) $(g \circ f)(x) = g[f(x)]$
$= g(3x + 5)$
$= (3x + 5)^2 + 2$
$= 9x^2 + 30x + 25 + 2$
$= 9x^2 + 30x + 27$

7. $(x - 2)^2 + (y + 3)^2 = 16$
$(x - 2)^2 + [y - (-3)]^2 = 4^2$

The graph is a circle with center $(2, -3)$ and radius 4.

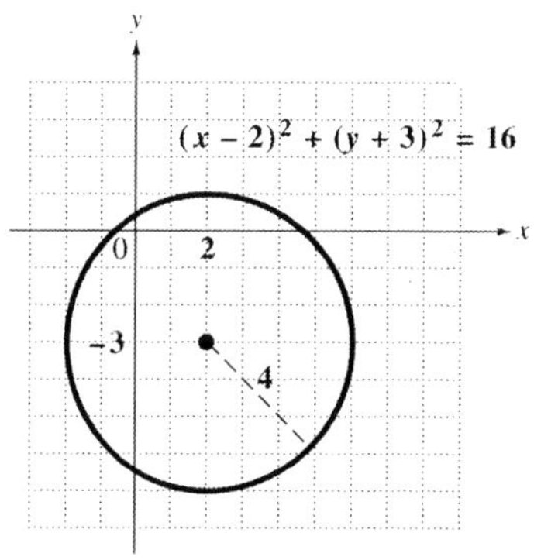

8. $x^2 + y^2 + 8x - 2y = 8$

To find the center and radius, complete the squares on x and y.

$$(x^2 + 8x \qquad) + (y^2 - 2y \qquad) = 8$$
$$(x^2 + 8x + \underline{16}) + (y^2 - 2y + \underline{1}) = 8 + \underline{16} + \underline{1}$$
$$(x + 4)^2 + (y - 1)^2 = 25$$

The graph is a circle with center $(-4, 1)$ and radius $\sqrt{25} = 5$.

9. $$f(x) = \sqrt{9 - x^2}$$

Replace $f(x)$ with y.

$$y = \sqrt{9 - x^2}$$

Square both sides.

$$y^2 = 9 - x^2$$
$$x^2 + y^2 = 9$$

The graph of $x^2 + y^2 = 9$ is a circle of radius $\sqrt{9} = 3$ centered at the origin. Since $f(x)$ is nonnegative, only the top half of the circle is graphed.

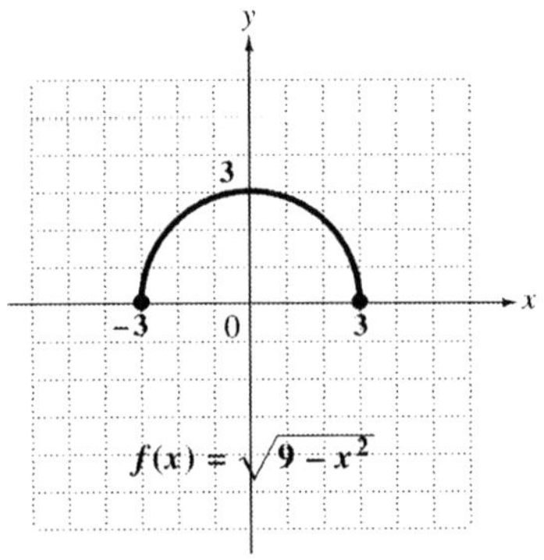

10. $4x^2 + 9y^2 = 36$

$$\frac{x^2}{9} + \frac{y^2}{4} = 1$$

The equation is in $\frac{x^2}{a^2} + \frac{y^2}{b^2} = 1$ form with $a = 3$ and $b = 2$. The graph is an ellipse with intercepts $(3, 0)$, $(-3, 0)$, $(0, 2)$, and $(0, -2)$. Plot these intercepts, and draw the ellipse through them.

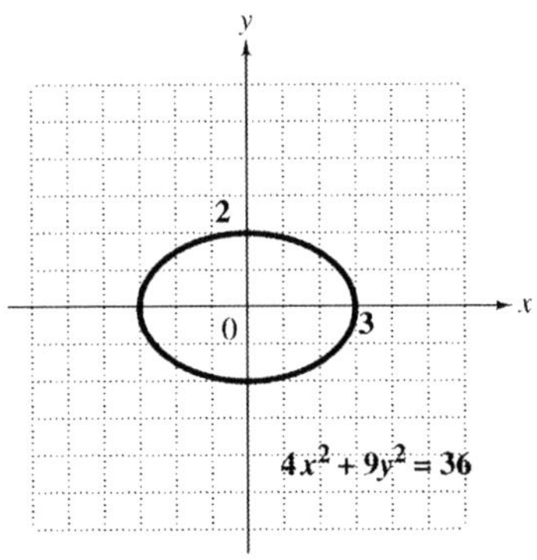

11. $16y^2 - 4x^2 = 64$

$$\frac{y^2}{4} - \frac{x^2}{16} = 1$$

The equation is in $\frac{y^2}{b^2} - \frac{x^2}{a^2} = 1$ form with $a = 4$ and $b = 2$. The graph is a hyperbola with y-intercepts $(0, 2)$ and $(0, -2)$ and asymptotes that are the extended diagonals of the rectangle with vertices $(4, 2)$, $(4, -2)$, $(-4, -2)$, and $(-4, 2)$.

Draw a branch of the hyperbola through each intercept and approaching the asymptotes.

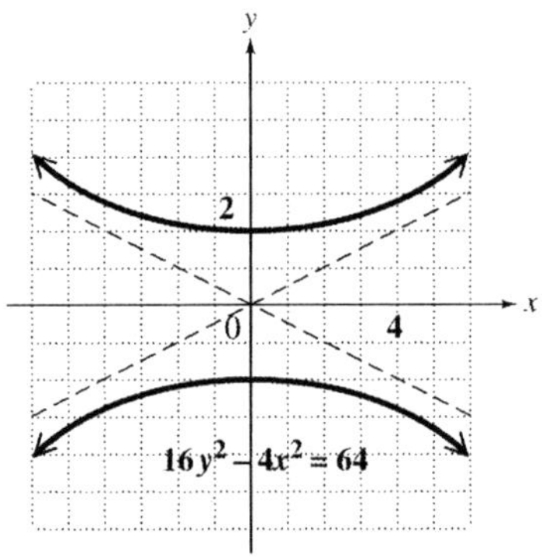

12. $$\frac{y}{2} = -\sqrt{1 - \frac{x^2}{9}}$$

Square both sides.

$$\frac{y^2}{4} = 1 - \frac{x^2}{9}$$
$$\frac{x^2}{9} + \frac{y^2}{4} = 1$$

This is an ellipse with x-intercepts $(3, 0)$ and $(-3, 0)$ and y-intercepts $(0, 2)$ and $(0, -2)$. Since y represents a negative square root in the original equation, y must be nonpositive. This restricts the graph to the lower half of the ellipse.

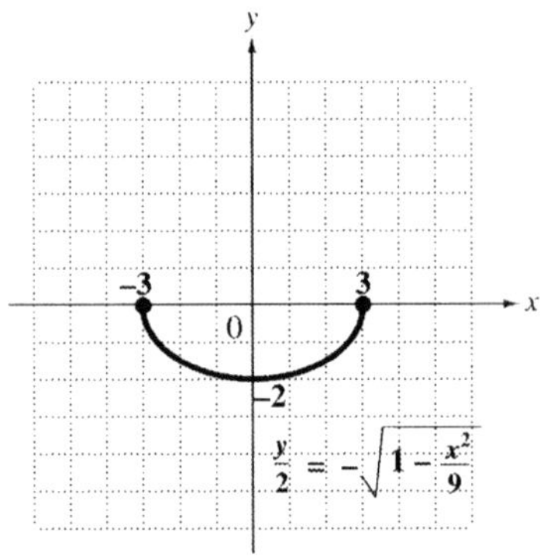

13. $6x^2 + 4y^2 = 12$

We have the *sum* of squares with different coefficients equal to a positive number, so this is an equation of an *ellipse*.

14. $$16x^2 = 144 + 9y^2$$
$$16x^2 - 9y^2 = 144$$

We have the *difference* of squares equal to a positive number, so this is an equation of a *hyperbola.*

15. $$4y^2 + 4x = 9$$
$$4x = -4y^2 + 9$$

We have an x-term and a y^2-term, so this is an equation of a horizontal *parabola.*

16. $$\begin{aligned} 2x - y &= 9 \quad (1) \\ xy &= 5 \quad (2) \end{aligned}$$

Solve equation (1) for y.

$$y = 2x - 9 \quad (3)$$

Substitute $2x - 9$ for y in equation (2).

$$\begin{aligned} xy &= 5 \quad (2) \\ x(2x - 9) &= 5 \\ 2x^2 - 9x &= 5 \\ 2x^2 - 9x - 5 &= 0 \\ (2x + 1)(x - 5) &= 0 \end{aligned}$$

$$2x + 1 = 0 \quad \text{or} \quad x - 5 = 0$$
$$x = -\tfrac{1}{2} \quad \text{or} \quad x = 5$$

Substitute these values for x in equation (3) to find y.

If $x = -\frac{1}{2}$, then $y = 2\left(-\frac{1}{2}\right) - 9 = -10$.

If $x = 5$, then $y = 2(5) - 9 = 1$.

Solution set: $\left\{\left(-\frac{1}{2}, -10\right), (5, 1)\right\}$

17. $$\begin{aligned} x - 4 &= 3y \quad (1) \\ x^2 + y^2 &= 8 \quad (2) \end{aligned}$$

Solve equation (1) for x.

$$x = 3y + 4 \quad (3)$$

Substitute $3y + 4$ for x in equation (2).

$$\begin{aligned} x^2 + y^2 &= 8 \quad (2) \\ (3y + 4)^2 + y^2 &= 8 \\ 9y^2 + 24y + 16 + y^2 &= 8 \\ 10y^2 + 24y + 8 &= 0 \\ 2(5y^2 + 12y + 4) &= 0 \\ 2(5y + 2)(y + 2) &= 0 \end{aligned}$$

$$5y + 2 = 0 \quad \text{or} \quad y + 2 = 0$$
$$y = -\tfrac{2}{5} \quad \text{or} \quad y = -2$$

Substitute these values for y in equation (3) to find x.

If $y = -\frac{2}{5}$, then

$$x = 3\left(-\tfrac{2}{5}\right) + 4 = -\tfrac{6}{5} + 4 = \tfrac{14}{5}.$$

If $y = -2$, then $x = 3(-2) + 4 = -2$.

Solution set: $\left\{(-2, -2), \left(\frac{14}{5}, -\frac{2}{5}\right)\right\}$

18. $$\begin{aligned} x^2 + y^2 &= 25 \quad (1) \\ x^2 - 2y^2 &= 16 \quad (2) \end{aligned}$$

Multiply equation (1) by 2 and add the result to equation (2).

$$\begin{array}{rcrcrl} 2x^2 & + & 2y^2 & = & 50 & \quad 2 \times (1) \\ x^2 & - & 2y^2 & = & 16 & \quad (2) \\ \hline 3x^2 & & & = & 66 & \\ & & x^2 & = & 22 & \end{array}$$

$$x = \sqrt{22} \quad \text{or} \quad x = -\sqrt{22}$$

Substitute 22 for x^2 in equation (1).

$$\begin{aligned} x^2 + y^2 &= 25 \quad (1) \\ 22 + y^2 &= 25 \\ y^2 &= 3 \end{aligned}$$

$$y = \sqrt{3} \quad \text{or} \quad y = -\sqrt{3}$$

Since $\sqrt{22}$ can be paired with either $\sqrt{3}$ or $-\sqrt{3}$ and $-\sqrt{22}$ can be paired with either $\sqrt{3}$ or $-\sqrt{3}$, there are four possible solutions.

Solution set: $\left\{\left(\sqrt{22}, \sqrt{3}\right), \left(\sqrt{22}, -\sqrt{3}\right), \left(-\sqrt{22}, \sqrt{3}\right), \left(-\sqrt{22}, -\sqrt{3}\right)\right\}$

19. $y < x^2 - 2$

The boundary, $y = x^2 - 2$, is a dashed parabola in $y = a(x - h)^2 + k$ form with vertex (h, k) at $(0, -2)$. Since $a = 1 > 0$, the parabola opens up. It also has the same shape as $y = x^2$. The points $(2, 2)$ and $(-2, 2)$ are on the graph. Test $(0, 0)$.

$$0 \le (0)^2 - 2 \quad ?$$
$$0 \le -2 \quad \textit{False}$$

Shade the side of the parabola that does not contain $(0, 0)$. This is the region outside the parabola.

continued

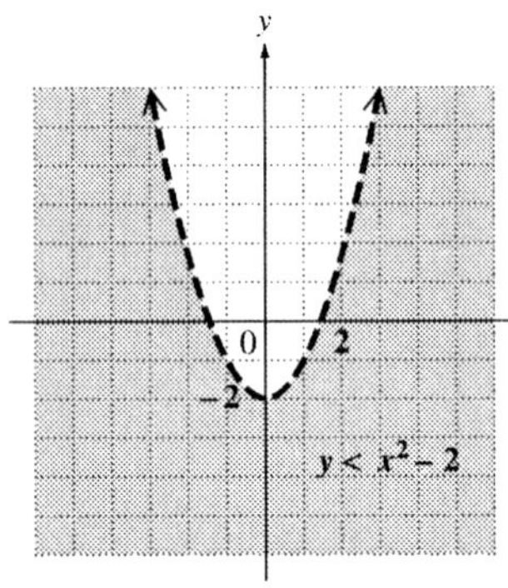

20. $x^2 + 25y^2 \le 25$
$x^2 + y^2 \le 9$

The first boundary, $\frac{x^2}{25} + \frac{y^2}{1} = 1$, is a solid ellipse with intercepts $(5, 0)$, $(-5, 0)$, $(0, 1)$, and $(0, -1)$. Test $(0, 0)$.

$$0^2 + 25 \cdot 0^2 \le 25 \quad ?$$
$$0 \le 25 \quad \textit{True}$$

Shade the region inside the ellipse.
The second boundary, $x^2 + y^2 = 9$, is a solid circle with center $(0, 0)$ and radius 3. Test $(0, 0)$.

$$0^2 + 0^2 \le 9 \quad ?$$
$$0 \le 9 \quad \textit{True}$$

Shade the region inside the circle. The solution of the system is the intersection of the two shaded regions.

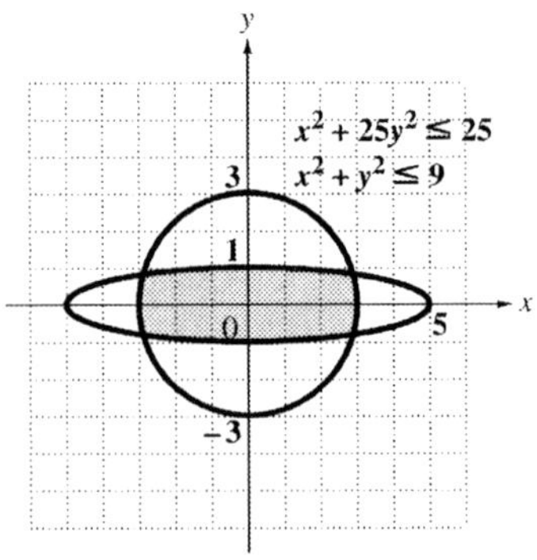

Cumulative Review Exercises (Chapters 1–12)

1. $-10 + |-5| - |3| + 4$
$= -10 + 5 - 3 + 4$
$= -5 - 3 + 4$
$= -8 + 4 = -4$

2. $4 - (2x + 3) + x = 5x - 3$
$4 - 2x - 3 + x = 5x - 3$
$-x + 1 = 5x - 3$
$-6x = -4$
$x = \frac{2}{3}$

Solution set: $\{\frac{2}{3}\}$

3. $-4k + 7 \ge 6k + 1$
$-10k \ge -6$

Divide by -10; reverse the direction of the inequality.

$$k \le \frac{-6}{-10}$$
$$k \le \frac{3}{5}$$

Solution set: $(-\infty, \frac{3}{5}]$

4. $|5m| - 6 = 14$
$|5m| = 20$

$5m = 20$ or $5m = -20$
$m = 4$ or $m = -4$

Solution set: $\{-4, 4\}$

5. $|2p - 5| > 15$

$2p - 5 > 15$ or $2p - 5 < -15$
$2p > 20$ $\qquad 2p < -10$
$p > 10$ or $p < -5$

Solution set: $(-\infty, -5) \cup (10, \infty)$

6. Let $(x_1, y_1) = (2, 5)$ and $(x_2, y_2) = (-4, 1)$.

$$m = \frac{y_2 - y_1}{x_2 - x_1} = \frac{1 - 5}{-4 - 2} = \frac{-4}{-6} = \frac{2}{3}$$

7. Through $(-3, -2)$; perpendicular to $2x - 3y = 7$

Write $2x - 3y = 7$ in slope-intercept form.

$$-3y = -2x + 7$$
$$y = \tfrac{2}{3}x - \tfrac{7}{3}$$

The slope is $\frac{2}{3}$. Perpendicular lines have slopes that are negative reciprocals of each other, so a line perpendicular to the given line will have slope $-\frac{3}{2}$. Let $m = -\frac{3}{2}$ and $(x_1, y_1) = (-3, -2)$ in the point-slope form.

$$y - y_1 = m(x - x_1)$$
$$y - (-2) = -\tfrac{3}{2}[x - (-3)]$$
$$y + 2 = -\tfrac{3}{2}(x + 3)$$

Multiply by 2 to clear the fraction.

$$2y + 4 = -3(x + 3)$$
$$2y + 4 = -3x - 9$$
$$3x + 2y = -13$$

8.
$$\begin{aligned} 3x - y &= 12 \quad (1) \\ 2x + 3y &= -3 \quad (2) \end{aligned}$$

Multiply equation (1) by 3 and add the result to equation (2).

$$\begin{aligned} 9x - 3y &= 36 \quad 3 \times (1) \\ 2x + 3y &= -3 \quad (2) \\ \hline 11x &= 33 \\ x &= 3 \end{aligned}$$

Substitute 3 for x in equation (1) to find y.

$$\begin{aligned} 3x - y &= 12 \quad (1) \\ 3(3) - y &= 12 \\ 9 - y &= 12 \\ -y &= 3 \\ y &= -3 \end{aligned}$$

Solution set: $\{(3, -3)\}$

9.
$$\begin{aligned} x + y - 2z &= 9 \quad (1) \\ 2x + y + z &= 7 \quad (2) \\ 3x - y - z &= 13 \quad (3) \end{aligned}$$

Add equation (2) and equation (3).

$$\begin{aligned} 2x + y + z &= 7 \quad (2) \\ 3x - y - z &= 13 \quad (3) \\ \hline 5x &= 20 \\ x &= 4 \end{aligned}$$

Multiply equation (1) by -1 and add the result to equation (2).

$$\begin{aligned} -x - y + 2z &= -9 \quad -1 \times (1) \\ 2x + y + z &= 7 \quad (2) \\ \hline x + 3z &= -2 \quad (4) \end{aligned}$$

Substitute 4 for x in equation (4) to find z.

$$\begin{aligned} x + 3z &= -2 \quad (4) \\ 4 + 3z &= -2 \\ 3z &= -6 \\ z &= -2 \end{aligned}$$

Substitute 4 for x and -2 for z in equation (2) to find y.

$$\begin{aligned} 2x + y + z &= 7 \quad (2) \\ 2(4) + y - 2 &= 7 \\ y + 6 &= 7 \\ y &= 1 \end{aligned}$$

Solution set: $\{(4, 1, -2)\}$

10.
$$\begin{aligned} xy &= -5 \quad (1) \\ 2x + y &= 3 \quad (2) \end{aligned}$$

Solve equation (2) for y.

$$y = -2x + 3 \quad (3)$$

Substitute $-2x + 3$ for y in equation (1).

$$\begin{aligned} xy &= -5 \quad (1) \\ x(-2x + 3) &= -5 \\ -2x^2 + 3x &= -5 \\ -2x^2 + 3x + 5 &= 0 \\ 2x^2 - 3x - 5 &= 0 \\ (2x - 5)(x + 1) &= 0 \end{aligned}$$

$$\begin{aligned} 2x - 5 = 0 \quad &\text{or} \quad x + 1 = 0 \\ x = \tfrac{5}{2} \quad &\text{or} \quad x = -1 \end{aligned}$$

Substitute these values for x in equation (3) to find y.
If $x = \frac{5}{2}$, then $y = -2\left(\frac{5}{2}\right) + 3 = -2$.

If $x = -1$, then $y = -2(-1) + 3 = 5$.

Solution set: $\left\{(-1, 5), \left(\frac{5}{2}, -2\right)\right\}$

11. Let $x =$ Al's average speed. Then $2x =$ Bev's average speed.

Make a table. Use $t = \frac{d}{r}$.

	Distance	Rate	Time
Al	20	x	$\frac{20}{x}$
Bev	20	$2x$	$\frac{20}{2x} = \frac{10}{x}$

Al's time was $\frac{1}{2}$ hr longer than Bev's, so

$$\begin{aligned} \frac{20}{x} &= \frac{1}{2} + \frac{10}{x} \\ 2x\left(\frac{20}{x}\right) &= 2x\left(\frac{1}{2}\right) + 2x\left(\frac{10}{x}\right) \quad \textit{Multiply by 2x.} \\ 40 &= x + 20 \\ 20 &= x. \end{aligned}$$

So Al's average speed was 20 mph and Bev's average speed was $2 \cdot 20 = 40$ mph.

12.
$$\begin{aligned} f(x) &= .07x + 135 \\ f(2000) &= .07(2000) + 135 = 275 \end{aligned}$$

The weekly fee is \$275 if \$2000 is taken in for the week.

13.
$$\begin{aligned} (5y - 3)^2 &= (5y)^2 - 2(5y)3 + 3^2 \\ &= 25y^2 - 30y + 9 \end{aligned}$$

14. $(2r+7)(6r-1)$
$= 12r^2 - 2r + 42r - 7$
$= 12r^2 + 40r - 7$

15. $\dfrac{8x^4 - 4x^3 + 2x^2 + 13x + 8}{2x+1}$

$$\begin{array}{r|l} & 4x^3 - 4x^2 + 3x + 5 \\ 2x+1 & 8x^4 - 4x^3 + 2x^2 + 13x + 8 \\ & \underline{8x^4 + 4x^3} \\ & \quad -8x^3 + 2x^2 \\ & \quad \underline{-8x^3 - 4x^2} \\ & \qquad 6x^2 + 13x \\ & \qquad \underline{6x^2 + 3x} \\ & \qquad\quad 10x + 8 \\ & \qquad\quad \underline{10x + 5} \\ & \qquad\qquad 3 \end{array}$$

The answer is

$$4x^3 - 4x^2 + 3x + 5 + \frac{3}{2x+1}.$$

16. $12x^2 - 7x - 10 = (4x-5)(3x+2)$

17. $2y^4 + 5y^2 - 3$

Let $p = y^2$, so $p^2 = y^4$.

$2y^4 + 5y^2 - 3 = 2p^2 + 5p - 3$
$= (2p-1)(p+3)$

Now substitute y^2 for p.

$= (2y^2 - 1)(y^2 + 3)$

18. $z^4 - 1 = (z^2+1)(z^2-1)$
$= (z^2+1)(z+1)(z-1)$

19. $a^3 - 27b^3 = a^3 - (3b)^3$
$= (a-3b)(a^2 + 3ab + 9b^2)$

20. $\dfrac{5x-15}{24} \cdot \dfrac{64}{3x-9} = \dfrac{5(x-3)}{3 \cdot 8} \cdot \dfrac{8 \cdot 8}{3(x-3)}$
$= \dfrac{5 \cdot 8}{3 \cdot 3} = \dfrac{40}{9}$

21. $\dfrac{y^2-4}{y^2-y-6} \div \dfrac{y^2-2y}{y-1}$

Multiply by the reciprocal.

$= \dfrac{y^2-4}{y^2-y-6} \cdot \dfrac{y-1}{y^2-2y}$

Factor and simplify.

$= \dfrac{(y+2)(y-2)}{(y-3)(y+2)} \cdot \dfrac{(y-1)}{y(y-2)}$
$= \dfrac{y-1}{y(y-3)}$

22. $\dfrac{5}{c+5} - \dfrac{2}{c+3}$

The LCD is $(c+5)(c+3)$.

$= \dfrac{5(c+3)}{(c+5)(c+3)} - \dfrac{2(c+5)}{(c+3)(c+5)}$
$= \dfrac{5c + 15 - 2c - 10}{(c+5)(c+3)}$
$= \dfrac{3c+5}{(c+5)(c+3)}$

23. $\dfrac{p}{p^2+p} + \dfrac{1}{p^2+p} = \dfrac{p+1}{p^2+p}$
$= \dfrac{p+1}{p(p+1)} = \dfrac{1}{p}$

24. Let $x =$ the time to do the job working together.

Make a chart.

Worker	Rate	Time Together	Part of Job Done
Kareem	$\frac{1}{3}$	x	$\frac{x}{3}$
Jamal	$\frac{1}{2}$	x	$\frac{x}{2}$

Part done by Kareem	plus	part done by Jamal	equals	1 whole job.
$\frac{x}{3}$	$+$	$\frac{x}{2}$	$=$	1

Multiply by the LCD, 6.

$6\left(\frac{x}{3} + \frac{x}{2}\right) = 6 \cdot 1$
$2x + 3x = 6$
$5x = 6$
$x = \frac{6}{5}$ or $1\frac{1}{5}$

It takes $\frac{6}{5}$ or $1\frac{1}{5}$ hours to do the job together.

25. $\left(\frac{4}{3}\right)^{-1} = \left(\frac{3}{4}\right)^{1} = \frac{3}{4}$

26. $\dfrac{(2a)^{-2}a^4}{a^{-3}} = \dfrac{2^{-2}a^{-2}a^4}{a^{-3}} = \dfrac{2^{-2}a^2}{a^{-3}}$
$= \dfrac{a^2a^3}{2^2} = \dfrac{a^5}{4}$

27. $4\sqrt[3]{16} - 2\sqrt[3]{54} = 4\sqrt[3]{8 \cdot 2} - 2\sqrt[3]{27 \cdot 2}$
$= 4 \cdot 2\sqrt[3]{2} - 2 \cdot 3\sqrt[3]{2}$
$= 8\sqrt[3]{2} - 6\sqrt[3]{2} = 2\sqrt[3]{2}$

28. $\dfrac{3\sqrt{5x}}{\sqrt{2x}} = \dfrac{3\sqrt{5x} \cdot \sqrt{2x}}{\sqrt{2x} \cdot \sqrt{2x}} = \dfrac{3\sqrt{10x^2}}{2x}$
$= \dfrac{3x\sqrt{10}}{2x} = \dfrac{3\sqrt{10}}{2}$

29. $\dfrac{5+3i}{2-i}$

Multiply the numerator and denominator by the conjugate of the denominator.

$$=\frac{(5+3i)(2+i)}{(2-i)(2+i)}$$
$$=\frac{10+5i+6i+3i^2}{4-i^2}$$
$$=\frac{10+11i+3(-1)}{4-(-1)}$$
$$=\frac{7+11i}{5}=\frac{7}{5}+\frac{11}{5}i$$

30. $2\sqrt{k}=\sqrt{5k+3}$

Square both sides.

$$4k=5k+3$$
$$-k=3$$
$$k=-3$$

Since k must be nonnegative so that $\sqrt{k}$ is a real number, -3 cannot be a solution. The solution set is $\emptyset$.

31.
$$10q^2+13q=3$$
$$10q^2+13q-3=0$$
$$(5q-1)(2q+3)=0$$

$5q-1=0$ or $2q+3=0$

$q=\frac{1}{5}$ or $q=-\frac{3}{2}$

Solution set: $\left\{\frac{1}{5},-\frac{3}{2}\right\}$

32. $(4x-1)^2=8$

$$4x-1=\pm\sqrt{8}$$
$$4x=1\pm2\sqrt{2}$$
$$x=\frac{1\pm2\sqrt{2}}{4}$$

Solution set: $\left\{\dfrac{1+2\sqrt{2}}{4},\dfrac{1-2\sqrt{2}}{4}\right\}$

33. $3k^2-3k-2=0$

Use the quadratic formula with $a=3$, $b=-3$, and $c=-2$.

$$k=\frac{-b\pm\sqrt{b^2-4ac}}{2a}$$
$$k=\frac{-(-3)\pm\sqrt{(-3)^2-4(3)(-2)}}{2(3)}$$
$$=\frac{3\pm\sqrt{9+24}}{6}=\frac{3\pm\sqrt{33}}{6}$$

Solution set: $\left\{\dfrac{3+\sqrt{33}}{6},\dfrac{3-\sqrt{33}}{6}\right\}$

34. $2\left(x^2-3\right)^2-5\left(x^2-3\right)=12$

Let $u=\left(x^2-3\right)$.

$$2u^2-5u=12$$
$$2u^2-5u-12=0$$
$$(2u+3)(u-4)=0$$

$2u+3=0$ or $u-4=0$

$u=-\frac{3}{2}$ or $u=4$

Substitute x^2-3 for u to find x.

If $u=-\frac{3}{2}$, then

$$x^2-3=-\tfrac{3}{2}$$
$$x^2=\tfrac{3}{2}$$
$$x=\pm\sqrt{\frac{3}{2}}=\pm\frac{\sqrt{3}}{\sqrt{2}}\cdot\frac{\sqrt{2}}{\sqrt{2}}=\pm\frac{\sqrt{6}}{2}.$$

If $u=4$, then

$$x^2-3=4$$
$$x^2=7$$
$$x=\pm\sqrt{7}.$$

Solution set: $\left\{-\dfrac{\sqrt{6}}{2},\dfrac{\sqrt{6}}{2},-\sqrt{7},\sqrt{7}\right\}$

35. Solve $F=\dfrac{kwv^2}{r}$ for v.

$$Fr=kwv^2$$
$$v^2=\frac{Fr}{kw}$$

Take the square root of each side.

$$v=\pm\sqrt{\frac{Fr}{kw}}=\frac{\pm\sqrt{Fr}}{\sqrt{kw}}\cdot\frac{\sqrt{kw}}{\sqrt{kw}}=\frac{\pm\sqrt{Frkw}}{kw}$$

36. $f(x)=y=x^3+4$

Interchange x and y and solve for y.

$$x=y^3+4$$
$$y^3=x-4$$
$$y=\sqrt[3]{x-4}$$
$$f^{-1}(x)=\sqrt[3]{x-4}$$

37. $a^{\log_a x}=x$, so $3^{\log_3 4}=4$.

38. $a^{\log_a x}=x$, so $e^{\ln x}=e^{\log_e 7}=7$.

39. $2\log(3x+7)-\log 4=\log(3x+7)^2-\log 4$

$$=\log\frac{(3x+7)^2}{4}$$

40. $\log(x+2) + \log(x-1) = 1$

$$\log_{10}[(x+2)(x-1)] = 1$$
$$(x+2)(x-1) = 10^1$$
$$x^2 + x - 2 = 10$$
$$x^2 + x - 12 = 0$$
$$(x+4)(x-3) = 0$$

$x + 4 = 0$ or $x - 3 = 0$

$x = -4$ or $x = 3$

The original equation is undefined when $x = -4$.

Solution set: $\{3\}$

41. $P = 10{,}000$, $r = .05$, $t = 4$

(a) $A = P\left(1 + \frac{r}{n}\right)^{nt}$

$= 10{,}000\left(1 + \frac{.05}{4}\right)^{4 \cdot 4}$ *Let $n = 4$.*

$= 10{,}000(1.0125)^{16}$

$\approx 12{,}198.90$

There will be \$12,198.90 in the account.

(b) $A = Pe^{rt}$

$= 10{,}000e^{(.05)(4)}$

$\approx 12{,}214.03$

There will be \$12,214.03 in the account.

42. $y = 1.38(1.65)^x$

$y = 1.38(1.65)^5$ *Let $x = 5$.*

≈ 16.9 billion dollars

The sales are estimated to be \$16.9 billion in 2000.

43. $2003 - 1995 = 8$

$y = 1.38(1.65)^8$ *Let $x = 8$.*

≈ 75.8 billion dollars

The sales are estimated to be \$75.8 billion in 2003.

44. $f(x) = x^2 + 2x - 4$, $g(x) = 3x + 2$

(a) $(g \circ f)(1) = g[f(1)]$

$= g(1^2 + 2 \cdot 1 - 4)$

$= g(-1)$

$= 3(-1) + 2$

$= -1$

(b) $(f \circ g)(x) = f[g(x)]$

$= f(3x + 2)$

$= (3x + 2)^2 + 2(3x + 2) - 4$

$= 9x^2 + 12x + 4 + 6x + 4 - 4$

$= 9x^2 + 18x + 4$

45. $f(x) = -3x + 5$

The equation is in slope-intercept form, so the y-intercept is $(0, 5)$ and $m = -3$ or $\frac{-3}{1}$.

Plot $(0, 5)$. From $(0, 5)$, move down 3 units and right 1 unit. Draw the line through these two points.

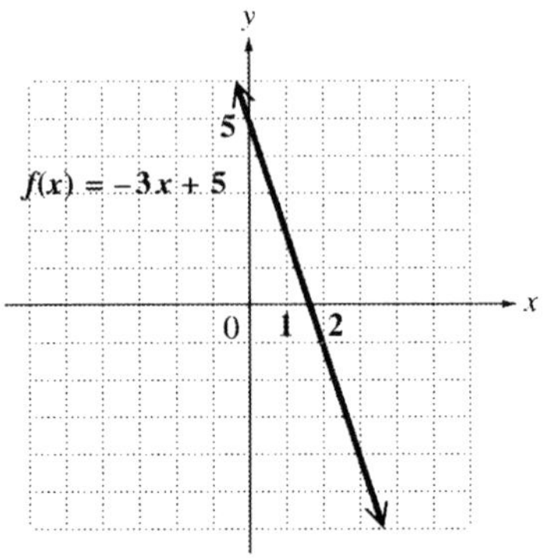

46. $f(x) = -2(x-1)^2 + 3$

The graph is a parabola that has been shifted 1 unit to the right and 3 units upward from $(0, 0)$, so its vertex is at $(1, 3)$. Since $a = -2 < 0$, the parabola opens down. Also $|a| = |-2| = 2 > 1$, so the graph is narrower than the graph of $f(x) = x^2$. The points $(0, 1)$ and $(2, 1)$ are on the graph.

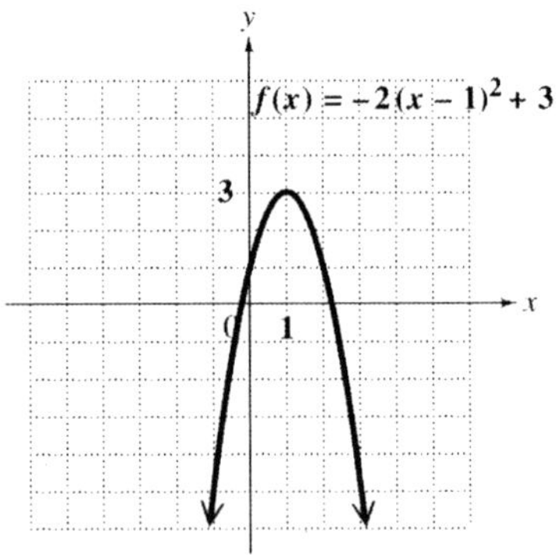

47. $\frac{x^2}{25} + \frac{y^2}{16} \le 1$

The boundary, $\frac{x^2}{25} + \frac{y^2}{16} = 1$, is a solid ellipse in $\frac{x^2}{a^2} + \frac{y^2}{b^2} = 1$ form with intercepts $(5, 0)$, $(-5, 0)$, $(0, 4)$, and $(0, -4)$. Test $(0, 0)$.

$$\frac{0^2}{25} + \frac{0^2}{16} \le 1 \quad ?$$
$$0 \le 1 \qquad \textit{True}$$

Shade the region inside the ellipse.

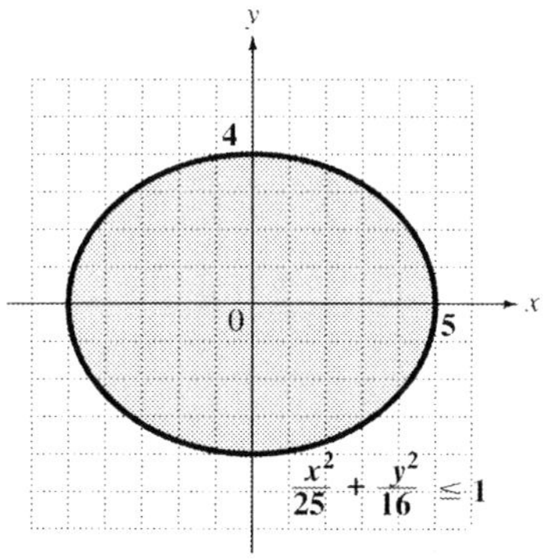

48. $f(x) = \sqrt{x - 2}$

This is the graph of $g(x) = \sqrt{x}$ shifted 2 units right.

x	2	3	6
$f(x)$	0	1	2

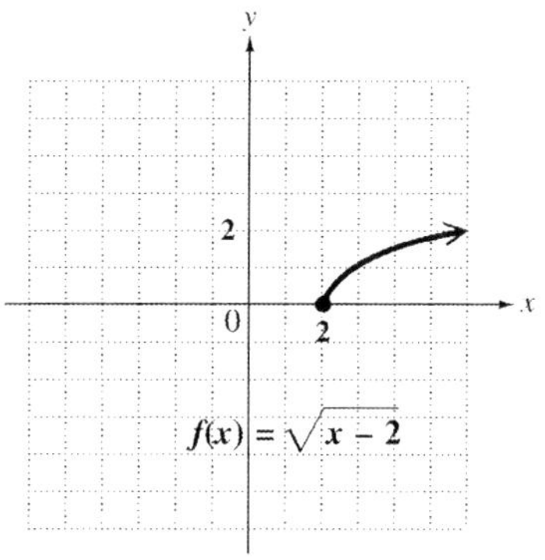

49. $\frac{x^2}{4} - \frac{y^2}{16} = 1$ is in $\frac{x^2}{a^2} - \frac{y^2}{b^2} = 1$ form.

The graph is a hyperbola with x-intercepts $(2, 0)$ and $(-2, 0)$ and asymptotes that are the extended diagonals of the rectangle with vertices $(2, 4)$, $(2, -4)$, $(-2, -4)$, and $(-2, 4)$. Draw a branch of the hyperbola through each intercept approaching the asymptotes.

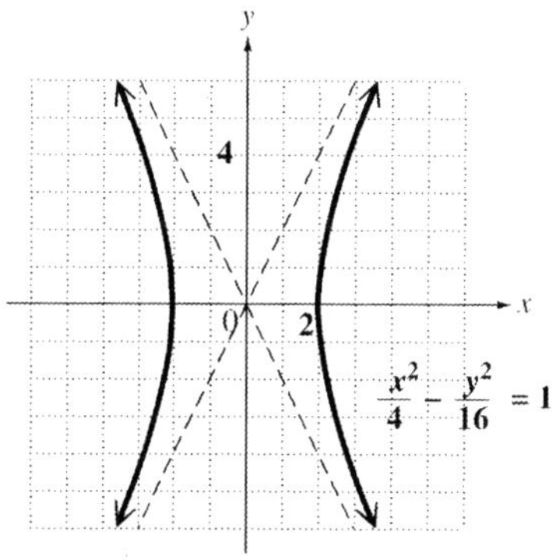

50. $f(x) = 3^x$

The graph of f is an increasing exponential.

x	-1	0	1	2
$f(x)$	$\frac{1}{3}$	1	3	9

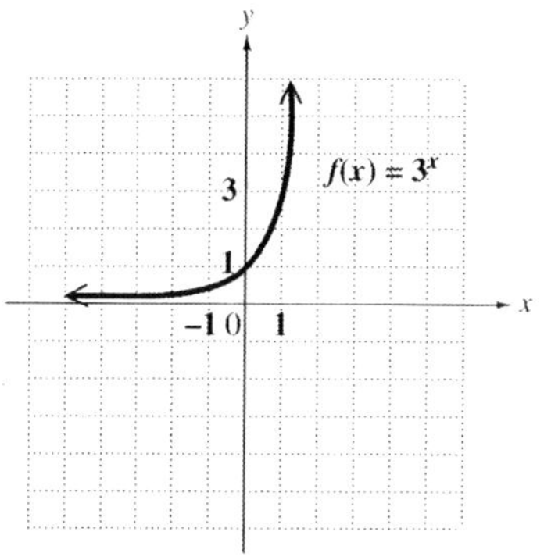

APPENDIX A STRATEGIES FOR PROBLEM SOLVING

Appendix A Margin Exercises

1. Step 1 compares to Polya's first step, Steps 2 and 3 compare to his second step, Step 4 compares to his third step, and Step 6 compares to his fourth step.

2.

Chosen Terms	First chosen term	Fourth chosen term	Product	Second chosen term	Third chosen term	Product
4th–7th	3	13	39	5	8	40
6th–9th	8	34	272	13	21	273
8th–11th	21	89	1869	34	55	1870

The product of the two middle terms is always 1 more than the product of the first and last term.

3. **(a)** Suppose she started with x dollars. She bought a book for \$10, so now she has $x - 10$ dollars. She spent half her remaining money on a train ticket, so she has $\frac{1}{2}(x-10) = \frac{1}{2}x - 5$ dollars remaining. She bought lunch for \$4, so she has $\left(\frac{1}{2}x - 5\right) - 4 = \frac{1}{2}x - 9$ dollars remaining. She spent half her remaining money at a bazaar, so she has $\frac{1}{2}\left(\frac{1}{2}x - 9\right) = \frac{1}{4}x - \frac{9}{2}$ dollars remaining. This must equal the \$20 she left the bazaar with, so solve the equation

$$\frac{1}{4}x - \frac{9}{2} = 20.$$

$$\frac{1}{4}x = \frac{49}{2} \quad \textit{Add } \tfrac{9}{2}.$$

$$x = 98 \quad \textit{Multiply by 4.}$$

She started with \$98.

(b) The given subtraction problem:

$$\begin{array}{r} 7\,a\,2 \\ -\,4\,8\,b \\ \hline c\,7\,3 \end{array}$$

Change to an addition problem:

$$\begin{array}{r} 4\,8\,b \\ +\,c\,7\,3 \\ \hline 7\,a\,2 \end{array}$$

Now $b + 3 = 2$, so $b = 9$ and we carry 1 to the tens column. Thus, $1 + 8 + 7 = 16$, so $a = 6$ and we carry 1 to the hundreds column. Thus, $1 + 4 + c = 7$, so $c = 2$. Hence, $a + b + c = 6 + 9 + 2 = 17$, which is choice **D**.

4. **(a)** By inspection, $44^2 = 1936$, $45^2 = 2025$, and $46^2 = 2116$. Only 2025 makes sense for this problem, and $2025 - 76 = 1949$, which must be the year he was born.

(b) Use trial and error for this problem. Here is one possible solution.

	3	5	
7	1	8	2
	4	6	

5. **(a)** Let x represent the positive number. "If I square it" gives x^2. "Double the result" gives $2x^2$. "Take half of that result" gives $\frac{1}{2}(2x^2) = x^2$ (the last two steps canceled each other). "Then add 12" gives $x^2 + 12$. "I get 21" implies that $x^2 + 12 = 21$. Solve this equation.

$$x^2 + 12 = 21$$

$$x^2 = 9 \quad \textit{Subtract 12.}$$

$$x = 3 \quad \textit{Positive square root}$$

The number is 3.

(b) Condition (2) leads us to a year in the 1960s. If x is the ones digit, then condition (1) gives $1 + 9 + 6 + x = 23$, so $x = 7$, and the year is 1967.

6. **(a)** Examine the units digit in powers of 7.

$7^1 =$	**7**	$7^5 =$	16,80**7**
$7^2 =$	4**9**	$7^6 =$	117,64**9**
$7^3 =$	34**3**	$7^7 =$	823,54**3**
$7^4 =$	240**1**	$7^8 =$	5,764,80**1**

We see that the units digit cycles through 7, 9, 3, and 1. Since $491 = 4 \cdot 122 + 3$ (122 full cycles), we know that 7^{491} and 7^3 have the same units digit. Thus, the units digit in 7^{491} is 3.

(b) Since $\frac{1}{11} = .0909\ldots$, we see that every odd-numbered digit in the decimal representation of $\frac{1}{11}$ is 0. Since 103 is odd, the 103rd digit in the decimal representation of $\frac{1}{11}$ is 0.

7. **(a)** Since there are 3 rows, and there could only be a maximum of 2 crosses per row, we know that there are a maximum of 6 crosses in the 9 squares. In the first row, put a cross in the second and third squares. In the second row, put a cross in the first square (since the first column already has one blank square), and then put a cross in the third square (if we put a cross in the second square, we wouldn't be able to put a cross in the first square of the third row—why?). In the third row, put a cross in the first and second squares. This gives us the maximum number of crosses, 6.

	X	X
X		X
X	X	

(b) The sum of the numbers 1 to 12 on the face of a clock is 78, and 78 divided by 3 is 26. Start with the number 12 and investigate what numbers can be combined with 12 to add up to 26. Moving left, $12 + 11 = 23$. We can't include 10 with 12 and 11 because the sum is greater than 26, so we'll move right and include 1 and 2 for a sum of 26. Draw a line that separates 11, 12, 1, and 2 from the rest of the numbers.

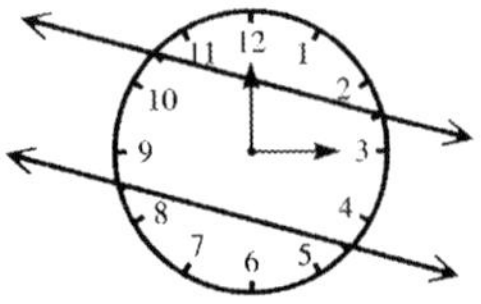

Now we can group the 10 and 9 together (but not with the 8—why?) along with the 3 and the 4 for a sum of 26. Draw another line and check that the remaining numbers add up to 26.

8. **(a)** Neither is correct, since $3^3 = 27$. However, "three cubed *is* twenty-seven" would be correct.

(b) If you take 7 bowling pins from 10 bowling pins, you have <u>7 bowling pins</u>.

(c) If you boil all the eggs at the same time, it still take $7\frac{1}{2}$ minutes.

Appendix A Section Exercises

1. You could choose a sock from the box labeled *red and green socks*. Since it is mislabeled, it contains only red socks or only green socks, determined by the sock you choose. If the sock is green, relabel this box *green socks*. Since the other two boxes were mislabeled, switch the remaining label to the other box and place the label that says *red and green socks* on the unlabeled box. No other choice guarantees a correct relabeling, since you can remove only one sock.

2. Each die has $1 + 2 + 3 + 4 + 5 + 6 = 21$ dots. Three dice have $3(21) = 63$ dots. The number of dots shown are $1 + 2 + 3 + 4 + 6 + 5 + 1 = 22$. Thus, the number of dots not visible in this view is $63 - 22 = 41$.

3. Let b = Mr. Green's birth year and let c = the current year. So the year of Mr. Green's tenth birthday is $(b + 10)$, the year of his fiftieth birthday is $(b + 50)$, and his current age is $(c - b)$. From the statement, write an equation.

$$\begin{aligned} b + c - (b + 10) - (b + 50) + (c - b) &= 80 \\ b + c - b - 10 - b - 50 + c - b &= 80 \\ 2c - 2b &= 140 \\ c - b &= 70 \end{aligned}$$

So his current age, $c - b$, is 70.

4. Since you are the bus driver, the answer is *your* age.

5. There are 4 choices to place your next O.

(1) Row 2, column 2: If you place an O here, then your friend will place an X in row 3, column 3, which forces you to get three Os in a row, and you lose.

(2) Row 2, column 3: If you place an O here, then your friend will place an X in row 3, column 1, which forces you to get three Os in a row, and you lose.

(3) Row 3, column 1: If you place an O here, then your friend can't use row 2, column 2 (Why?), so they must use row 2, column 3 or row 3, column 3.

(A) If they place an X in row 2, column 3, then you would use row 3, column 3 and they are forced to place an X in row 2, column 2, which gives them 3 Xs in a row (2nd column).

(B) If they place an X in row 3, column 3, then you would use row 2, column 3 and they are forced to place an X in row 2, column 2, which gives them 3 Xs in a row (2nd column).

Thus, you win in all cases by placing an O in the *bottom-left square*.

(4) Row 3, column 3: If you place an O here, then your friend would place an X in row 3, column 1, you would place an O in row 2, column 2, your friend would place an X in row 2, column 3, and there would not be a winner.

6. One possible solution is shown.

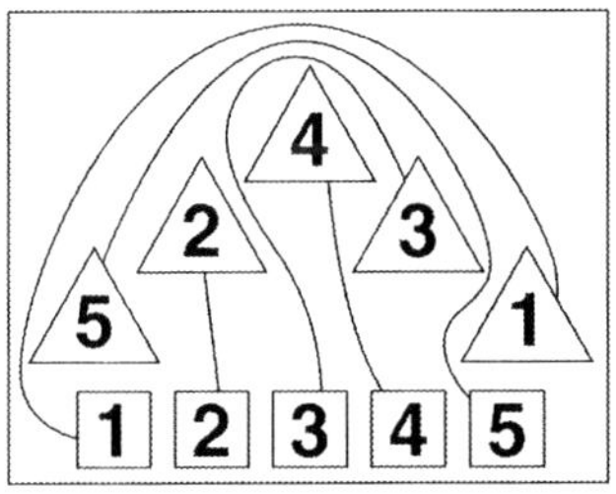

7. One strategy is to organize a table such as the one which follows. Let x = Chris's current age.

	Current Age	Past Age	Elapsed No. of Years
Pat	24	x	$24 - x$
Chris	x	$x - (24 - x)$	

Since Chris's past age can be represented as

$$x - (24 - x) = -24 + 2x = 2x - 24,$$

and Pat's current age, 24, is twice that of Chris's past age, we have

$$\begin{aligned} 24 &= 2(2x - 24) \\ 24 &= 4x - 48 \\ 72 &= 4x \\ 18 &= x. \end{aligned}$$

Thus, Chris's current age is 18 years.

8. Use trial and error. One possible solution is as follows.

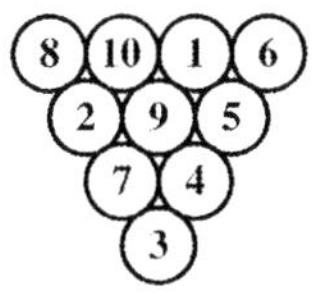

9. Use a calculator to find the square root of each number. Only 329,476 has a square root, 524, without a decimal remainder. Thus,

$$524^2 = 329{,}476.$$

10. This exercise can be solved algebraically. If we let
D = the total distance of the trip, and
x = the distance traveled while asleep,

then

$$\begin{aligned} x + \frac{1}{2}x &= \frac{1}{2}D \\ 2x + x &= D \\ 3x &= D \\ x &= \frac{1}{3}D. \end{aligned}$$

Thus, the distance traveled while asleep is $\frac{1}{3}$ of the total distance traveled.

11. Fill the big bucket. Pour into the small bucket. This leaves 4 gallons in the larger bucket. Empty the small bucket. Pour from the big bucket to fill up the small bucket. This leaves 1 gallon in the big bucket. Empty the small bucket. Pour 1 gallon from the big bucket to the small bucket. Fill up the big bucket. Pour into the small bucket. This leaves 5 gallons in the big bucket. Pour out the small bucket. This leaves exactly 5 gallons in the big bucket to take home. The above sequence is indicated by the following table.

Big	7	4	4	1	1	0	7	5	5
Small	0	3	0	3	0	1	1	3	0

12. This problem can be solved in many ways. Let's solve it in a way that shows the pattern. In order to see the pattern, we will not simplify the expressions that we generate for each day. Let n be the number of acorns on Monday morning.

Day	Number of Acorns Each Morning
M	n
T	$\frac{n}{2} + \frac{32}{2}$
W	$\frac{n}{2^2} + \frac{32}{2^2} + \frac{32}{2}$
T	$\frac{n}{2^3} + \frac{32}{2^3} + \frac{32}{2^2} + \frac{32}{2}$
F	$\frac{n}{2^4} + \frac{32}{2^4} + \frac{32}{2^3} + \frac{32}{2^2} + \frac{32}{2}$

Simplify this last expression, setting it equal to the remaining acorns on Saturday morning, and solving yields:

$$\begin{aligned} \frac{n}{16} + 2 + 4 + 8 + 16 &= 38 \\ \frac{n}{16} &= 8 \\ n &= 128 \end{aligned}$$

13. Similar to Example 5 in the text, we might examine the units place and tens place for repetitive powers of 7 in order to explore possible patterns.

$$\begin{array}{llll} 7^1 = & 07 & 7^5 = & 16{,}807 \\ 7^2 = & 49 & 7^6 = & 117{,}649 \\ 7^3 = & 343 & 7^7 = & 823{,}543 \\ 7^4 = & 2401 & 7^8 = & 5{,}764{,}801 \end{array}$$

Since the final two digits cycle over four values, we might consider dividing the successive exponents by 4 and examining their remainders. (Note: We are using inductive reasoning when we assume that this pattern will continue and will apply when the exponent is 1997.) Dividing the exponent 1997 by 4, we get a remainder of 1. This is the same remainder we get when dividing the exponent 1 (on 7^1) and 5 (on 7^5). Thus, we expect that the last two digits for 7^{1997} would be 07 as well.

14. Similar to Example 5 in the text, we might examine the units place for repetitive powers of 3 in order to explore possible patterns.

$$\begin{array}{llllll} 3^1 = & 3 & 3^5 = & 243 & 3^9 = & 19683 \\ 3^2 = & 9 & 3^6 = & 729 & 3^{10} = & 59049 \\ 3^3 = & 27 & 3^7 = & 2187 & 3^{11} = & 177147 \\ 3^4 = & 81 & 3^8 = & 6561 & 3^{12} = & 531441 \end{array}$$

If we divide the exponent 324 by 4 (since the pattern of the units digit cycles after every 4th power), we get a remainder of 0. Noting that in the line of 3^4, where each exponent, when divide by 4, yields a remainder of 0, we reason inductively that 3^{324} has the same units digit, 1.

15. At the end of 1st day, the frog has a net progression of 1 foot; day 2: 2 feet; day 3: 3 feet; ...; day 16: 16 feet (it crawls up 4 feet from 15 to 19 feet and then falls back 3 feet to 16 feet); on the 17th day it crawls up 4 feet from the 16-foot level, which takes it to the top.

16. Working backward, we see that if the lily pad doubles its size each day so that it completely covers the pond on the twentieth day, the pond was half covered on the previous (or nineteenth) day.

17. Set 4 opposite 12, then 8 is halfway around from 4 to 12. Opposite 8 must be 16 to allow for three equally spaced numbers (children) between each of these values. Note that children 1, 2, and 3 stand between 16 and 4. So there are 16 children in the circle.

18. Use trial and error to find the smallest perfect number. Try making a chart such as the following one.

Number	Divisors Other Than Itself	Sum
1	None	
2	1	1
3	1	1
4	1, 2	3
5	1	1
6	1, 2, 3	6

Six is the smallest perfect number.

19. Solve by drawing a sketch. The following figure satisfies the description. Only three birds are needed.

20. **B**, since either Bill or Bob must be the tallest person.

21. Add the given diagonal elements together to get 15. Each row, column, and other diagonal must also add to 15. This yields the following perfect square.

6	1	8
7	5	3
2	9	4

22. Notice that the first column has three given numbers. Thus,

$$34 - (6 + 11 + 16) = 1$$

is the first number in the second row. Then, in the second row, we must have

$$34 - (1 + 15 + 14) = 4$$

in the third column. The diagonal from upper left to lower right also has three given numbers. Therefore,

$$34 - (6 + 15 + 10) = 3$$

is in the fourth row, fourth column. Continue filling in the missing numbers until the magic square is completed.

6	12	7	9
1	15	4	14
11	5	10	8
16	2	13	3

23. 25 pitches (The visiting team's pitcher retires 24 consecutive batters through the first eight innings, using only one pitch per batter. His team does not score either. Going into the bottom of the ninth inning tied 0–0, the first batter for the home team hits his first pitch for a home run. The pitcher threw 25 pitches and loses the game by a score of 1–0.)

24. Solve this problem by trial and error. Write the names of the natural numbers in the English language. *One* is the least natural number, but its letters are not in alphabetical order. You will find that *forty* is the least natural number that has its letters in alphabetical order.

25. For three weighings, first balance four against four. Of the lighter four, balance two against the other two. Finally, of the lighter two, balance them one against the other.

To find the bad coin in two weighings, divide the eight coins into groups of 3, 3, 2. Weigh the groups of three against each other on the scale. If the groups weigh the same, the fake is in the two left out and can be found in one additional weighing. If the two groups of three do not weigh the same, pick the lighter group. Choose any two of the coins and weigh them. If one of these is lighter, it is the fake; if they weigh the same, then the third coin is the fake.

26. The person takes the goat across and returns alone. On the second trip, the person takes the wolf across and returns with the goat. On the third trip, the goat is left on the first side while the person takes the cabbage across. Then the person returns alone and brings the goat across.

27. Draw a sketch, visualize, or cut a piece of paper to build the cube. The cube may be folded with Z on the front.

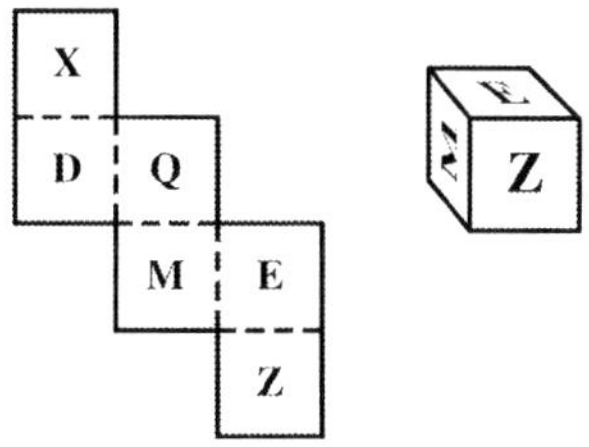

Then, E is on top and M is on the left face. This places Q opposite the face marked Z. (D is on the bottom and X is on the right face.)

28.

29. This may be worked algebraically or in reverse as Example 2 in the text. Multiplying 2 by 10 gives 20. Subtract 8 to give 12 and then square to get 144. Add 52 to get 196. This represents a number times itself. The number is 14, from the fact that $14 \times 14 = 196$ (or the square root of $196 = 14$). The quotient must be 21 since $21 - \frac{1}{3} \times 21 = 14$. Multiplying 21 by 7, we get 147, which represents 3 times the original number plus $\frac{3}{4}$ of that same product. The original number must be 28 since $3 \times 28 = 84$ and $\frac{3}{4}$ of 84 is 63, and $84 + 63 = 147$.

30. A strategy would be to assume the first possible age for the teenage girl, 13, and check by adding 2. This gives 15, not a perfect square. Trying the next possible age, 14, adding 2 gives the perfect square 16. Check by subtracting 10 from 14 to get 4 and note that $4^2 = 16$. Thus, her age is 14 years, and her brother's age is $14 - 5 = 9$ years.

31. A solution, found by trial and error, is shown here.

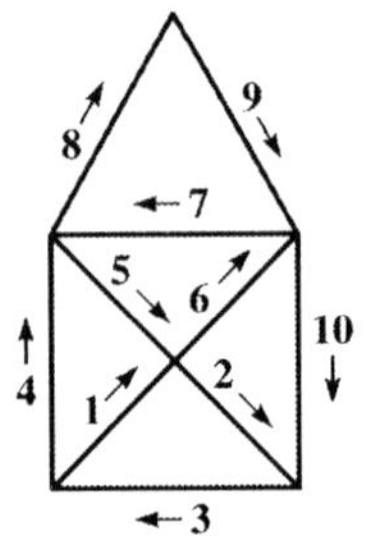

32. A solution, found by trial and error, is shown here.

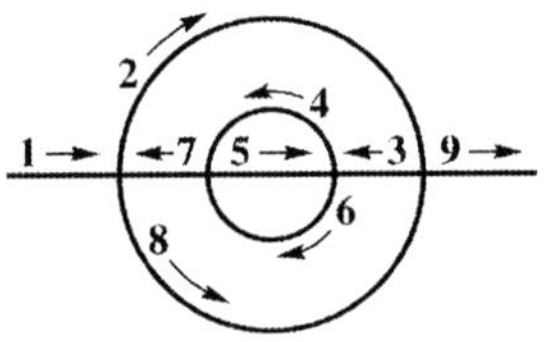

33. Jessica is married to James or Dan. Since Jessica is married to the oldest person in the group, she is not married to James, who is younger than Cathy. So Jessica is married to Dan, and Cathy is married to James. Since Jessica is married to the oldest person, we know that Dan is 36. Since James is older than Jessica but younger than Cathy, we conclude that Cathy is 31, James is 30, and Jessica is 29.

34. They must be February and March in a non-leap year.

35. Common sense tell you that the CEO is a woman.

36. This is a problem with a "catch." The obvious answer is that only one month, February, has 28 days. However, the problem does not specify exactly 28 days, so any month with at least 28 days qualifies. All 12 months have 28 days.

37. This is a problem with a "catch." Someone reading this problem might go ahead and calculate the volume of a cube 6 feet on each side, to get the answer 216 cubic feet. However, common sense tells us that since holes are by definition empty, there is no dirt in a hole.

38. Becky's mother named her third child Becky, since Becky is the only child left after Penny and Nichole have been named.

39. One solution is
$1 + 2 + 3 + 4 + 5 + 6 + 7 + 8 \times 9 = 100$.

40. The answer is $4 + 2 + 7 = 14$, since the correct problem is as follows.

$$\begin{array}{r} \underline{\mathbf{4}}35 \\ 8\underline{\mathbf{2}}6 \\ +14\underline{\mathbf{7}} \\ \hline \underline{\mathbf{1}}408 \end{array}$$

41. The first digit in the answer cannot be 0, 2, 3, or 5, since these digits have already been used. It cannot be more than 3, since one of the factors is a number in the 30's, making it impossible to get a product over 45,000. Thus, the first digit of the answer must be 1. To find the first digit in the 3-digit factor, use estimation. Dividing a number between 15,000 and 16,000 by a number between 30 and 40 could give a result with a first digit of 3, 4, or 5. Since 3 and 5 have already been used, this first digit must be 4. Thus, the 3-digit factor is 402. We now have the following.

$$\begin{array}{r} \underline{\mathbf{4}}\ 0\ 2 \\ \times \quad 3 \\ \hline \underline{\mathbf{1}}\,5, \end{array}$$

To find the units digit of the 2-digit factor, use trial and error with the digits that have not yet been used: 6, 7, 8, and 9.

$36 \times 402 = 14{,}472$ (Too small and reuses 2 and 4)
$37 \times 402 = 14{,}874$ (Too small and reuses 4)
$38 \times 402 = 15{,}276$ (Reuses 2)
$39 \times 402 = 15{,}678$ (Correct)

The correct problem is as follows.

$$\begin{array}{r} \underline{\mathbf{4}}\ 0\ 2 \\ \times \quad 3\ \underline{\mathbf{9}} \\ \hline \underline{\mathbf{1}}\,5,\underline{\mathbf{6}}\ \underline{\mathbf{7}}\ \underline{\mathbf{8}} \end{array}$$

Notice that a combination of strategies was used to solve this problem.

42. Making the following observations

$$\begin{aligned} 9 \times 1 &= 9 \\ 9 \times 2 &= 18\ (1 + 8 = 9) \\ 9 \times 3 &= 27\ (2 + 7 = 9) \\ 9 \times 4 &= 36\ (3 + 6 = 9) \\ 9 \times 5 &= 45\ (4 + 5 = 9) \\ &\vdots \end{aligned}$$

suggests that the sum of the digits in the product will always be 9.

43. To count the triangles, it helps to draw sketches of the figure several times. There are 5 triangles formed by two sides of the pentagon and a diagonal. There are 4 triangles formed with each side of the pentagon as a base, so there are $4 \times 5 = 20$ triangles formed in this way. Each point of the star forms a small triangle, so there are 5 of these. Finally, there are 5 triangles formed with a diagonal as a base. In each, the other two sides are inside the pentagon. (None of these triangles has a side common to the pentagon.) Thus, the total number of triangles in the figure is $5 + 20 + 5 + 5 = 35$.

44. Counting directly we get 9 (1-unit squares) plus 4 (4-unit squares) plus 1 (9-unit square). This equals 14 squares in total.

45. None. To see this result, set two books beside each other and note the location of page 1 of the 1st book and the location of the last page of the 2nd book.

46. If each digit cost \$.75 (as in buying house address numbers), then the sum of the 5 numerals would be $5 \times \$.75 = \$3.75$.

47. By Eve's statement, Adam must have \$2 more than Eve. But according to Adam, a loss of \$1 from Eve to Adam gives Adam twice the amount that Eve has. By trial and error, the counting numbers 5 and 7 are the first to satisfy both conditions. Thus, Eve has \$5 and Adam has \$7.

48. To find the minimum number of socks to pull out, guess and check. Thre are two colors of socks. If you pull out 2 socks, you could have 2 of one color or 1 of each color. You must pull out more than 2 socks. If you pull out 3 socks, you might have 3 of one color or 1 of one color and 2 of the other. In either case, you have a matching pair, so 3 is the minimum number of socks to pull out.

APPENDIX B REVIEW OF FRACTIONS

Appendix B Margin Exercises

1. **(a)** Since 12 can be divided by 2, it has more than two different factors, so it is a *composite* number. Note: There are always two factors of any number–the number and 1.

(b) 13 has exactly two different factors, 1 and 13, so it is a *prime* number.

(c) Since 27 can be divided by 3, it has more than two different factors, so it is *composite.*

(d) 59 has exactly two different factors, 1 and 59, so it is *prime.*

(e) 1806 can be divided by 2, so it is *composite.*

2. **(a)** To write 70 in prime factored form, first divide by the smallest prime, 2, to get

$$70 = 2 \cdot 35.$$

Since 35 can be factored as $5 \cdot 7$, we have

$$70 = 2 \cdot 5 \cdot 7.$$

(b)

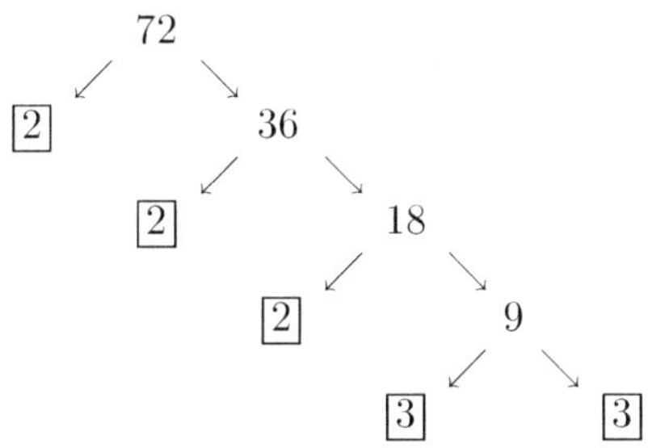

$72 = 2 \cdot 2 \cdot 2 \cdot 3 \cdot 3$

(c)

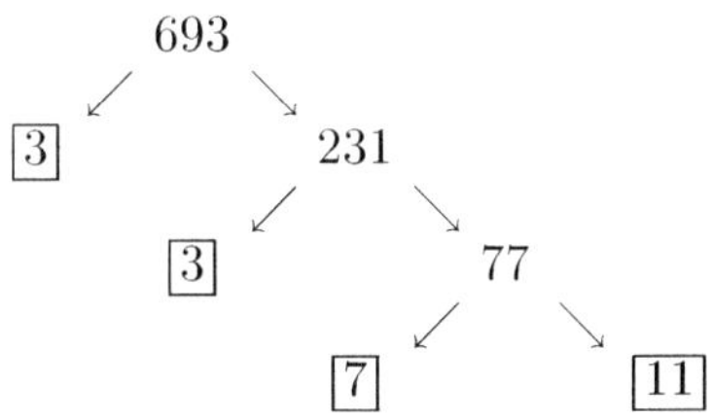

$693 = 3 \cdot 3 \cdot 7 \cdot 11$

(d) Since 97 is a prime number, its prime factored form is just 97.

3. **(a)** $\frac{8}{14} = \frac{4 \cdot 2}{7 \cdot 2} = \frac{4}{7} \cdot \frac{2}{2} = \frac{4}{7} \cdot 1 = \frac{4}{7}$

(b) $\frac{35}{42} = \frac{5 \cdot 7}{6 \cdot 7} = \frac{5}{6} \cdot \frac{7}{7} = \frac{5}{6} \cdot 1 = \frac{5}{6}$

(c) $\frac{120}{72} = \frac{5 \cdot 24}{3 \cdot 24} = \frac{5}{3} \cdot \frac{24}{24} = \frac{5}{3} \cdot 1 = \frac{5}{3}$

4. **(a)** $\frac{5}{8} \cdot \frac{2}{10} = \frac{5 \cdot 2}{8 \cdot 10}$ *Multiply numerators. Multiply denominators.*

$= \frac{5 \cdot 2}{2 \cdot 4 \cdot 2 \cdot 5}$ *Factor.*

$= \frac{1}{2 \cdot 4} = \frac{1}{8}$ *Write in lowest terms.*

(b) $\frac{1}{10} \cdot \frac{12}{5} = \frac{1 \cdot 12}{10 \cdot 5}$ *Multiply numerators. Multiply denominators.*

$= \frac{1 \cdot 2 \cdot 6}{2 \cdot 5 \cdot 5}$ *Factor.*

$= \frac{6}{5 \cdot 5} = \frac{6}{25}$ *Write in lowest terms.*

(c) $\frac{7}{9} \cdot \frac{12}{14} = \frac{7 \cdot 12}{9 \cdot 14}$ *Multiply numerators. Multiply denominators.*

$= \frac{7 \cdot 2 \cdot 2 \cdot 3}{3 \cdot 3 \cdot 2 \cdot 7}$ *Factor.*

$= \frac{2}{3}$ *Write in lowest terms.*

(d) $3\frac{1}{3} \cdot 1\frac{3}{4} = \frac{10}{3} \cdot \frac{7}{4}$ *Write as improper fractions.*

$= \frac{10 \cdot 7}{3 \cdot 4}$ *Multiply numerators. Multiply denominators.*

$= \frac{2 \cdot 5 \cdot 7}{3 \cdot 2 \cdot 2}$ *Factor.*

$= \frac{35}{6}$ or $5\frac{5}{6}$ *Write as a mixed number.*

5. **(a)** $\frac{3}{10} \div \frac{2}{7} = \frac{3}{10} \cdot \frac{7}{2}$ *Multiply by the reciprocal of the second fraction.*

$= \frac{21}{20}$ or $1\frac{1}{20}$

(b) $\frac{3}{4} \div \frac{7}{16} = \frac{3}{4} \cdot \frac{16}{7}$ *Multiply by the reciprocal of the second fraction.*

$= \frac{3 \cdot 4 \cdot 4}{7 \cdot 4}$

$= \frac{12}{7}$ or $1\frac{5}{7}$

(c) $\frac{4}{3} \div 6 = \frac{4}{3} \div \frac{6}{1}$

$= \frac{4}{3} \cdot \frac{1}{6}$ *Multiply by the reciprocal of the second fraction.*

$= \frac{2 \cdot 2}{3 \cdot 2 \cdot 3}$

$= \frac{2}{3 \cdot 3} = \frac{2}{9}$

(d) $3\frac{1}{4} \div 1\frac{2}{5} = \frac{13}{4} \div \frac{7}{5}$ *Change both mixed numbers to improper fractions.*

$= \frac{13}{4} \cdot \frac{5}{7}$ *Multiply by the reciprocal of the second fraction.*

$= \frac{65}{28}$ or $2\frac{9}{28}$

6. **(a)** $\frac{3}{5} + \frac{4}{5} = \frac{3+4}{5}$ *Add numerators; denominator does not change.*

$= \frac{7}{5}$ or $1\frac{2}{5}$

(b) $\frac{5}{14} + \frac{3}{14} = \frac{5+3}{14}$ *Add numerators; denominator does not change.*

$= \frac{8}{14}$

$= \frac{2 \cdot 4}{2 \cdot 7}$ *Factor.*

$= \frac{4}{7}$

7. **(a)** $\frac{7}{30} + \frac{2}{45}$

Since $30 = 2 \cdot 3 \cdot 5$ and $45 = 3 \cdot 3 \cdot 5$, the least common denominator must have one factor of 2 (from 30), two factors of 3 (from 45), and one factor of 5 (from either 30 or 45), so it is $2 \cdot 3 \cdot 3 \cdot 5 = 90$.

Write each fraction with a denominator of 90.

$\frac{7}{30} = \frac{7 \cdot 3}{30 \cdot 3} = \frac{21}{90}$ and $\frac{2}{45} = \frac{2 \cdot 2}{45 \cdot 2} = \frac{4}{90}$

Now add.

$$\frac{7}{30} + \frac{2}{45} = \frac{21}{90} + \frac{4}{90} = \frac{21+4}{90} = \frac{25}{90}$$

$\frac{25}{90}$ can be simplified.

$$\frac{25}{90} = \frac{5 \cdot 5}{5 \cdot 18} = \frac{5}{18}$$

(b) $\frac{17}{10} + \frac{8}{27}$

Since $10 = 2 \cdot 5$ and $27 = 3 \cdot 3 \cdot 3$, the least common denominator is $2 \cdot 5 \cdot 3 \cdot 3 \cdot 3 = 270$.

Write each fraction with a denominator of 270.

$\frac{17}{10} = \frac{17 \cdot 27}{10 \cdot 27} = \frac{459}{270}$ and $\frac{8}{27} = \frac{8 \cdot 10}{27 \cdot 10} = \frac{80}{270}$

Now add.

$$\frac{17}{10} + \frac{8}{27} = \frac{459}{270} + \frac{80}{270} = \frac{539}{270} \text{ or } 1\frac{269}{270}$$

(c) $2\frac{1}{8} + 1\frac{2}{3} = \frac{17}{8} + \frac{5}{3}$ *Change both mixed numbers to improper fractions.*

The least common denominator is 24, so write each fraction with a denominator of 24.

$\frac{17}{8} = \frac{17 \cdot 3}{8 \cdot 3} = \frac{51}{24}$ and $\frac{5}{3} = \frac{5 \cdot 8}{3 \cdot 8} = \frac{40}{24}$

Now add.

$$\frac{17}{8} + \frac{5}{3} = \frac{51}{24} + \frac{40}{24} = \frac{51+40}{24}$$

$$= \frac{91}{24} \text{ or } 3\frac{19}{24}$$

(d) $132\frac{4}{5} + 28\frac{3}{4}$

We will use a vertical method. LCD = 20

$132\frac{4}{5} = 132\frac{16}{20}$ *Add the whole numbers and the fractions separately.*

$+\,28\frac{3}{4} = 28\frac{15}{20}$

$160\frac{31}{20} = 160 + \left(1 + \frac{11}{20}\right) = 161\frac{11}{20}$

8. **(a)** $\frac{9}{11} - \frac{3}{11} = \frac{9-3}{11}$ *Subtract numerators; denominator does not change.*

$= \frac{6}{11}$

(b) $\frac{13}{15} - \frac{5}{6}$

Since $15 = 3 \cdot 5$ and $6 = 2 \cdot 3$, the least common denominator is $3 \cdot 5 \cdot 2 = 30$. Write each fraction with a denominator of 30.

$\frac{13}{15} = \frac{13 \cdot 2}{15 \cdot 2} = \frac{26}{30}$ and $\frac{5}{6} = \frac{5 \cdot 5}{6 \cdot 5} = \frac{25}{30}$

Now subtract.

$$\frac{13}{15} - \frac{5}{6} = \frac{26}{30} - \frac{25}{30} = \frac{1}{30}$$

(c) $2\frac{3}{8} - 1\frac{1}{2} = \frac{19}{8} - \frac{3}{2}$ *Change each mixed number into an improper fraction.*

The least common denominator is 8. Write each fraction with a denominator of 8. $\frac{19}{8}$ remains unchanged, and

$$\frac{3}{2} = \frac{3 \cdot 4}{2 \cdot 4} = \frac{12}{8}.$$

Now subtract.

$$\frac{19}{8} - \frac{3}{2} = \frac{19}{8} - \frac{12}{8} = \frac{19-12}{8} = \frac{7}{8}$$

(d) $50\frac{1}{4} - 32\frac{2}{3} = \frac{201}{4} - \frac{98}{3}$ *Write as improper fractions.*

The least common denominator is 12.

$$\frac{201}{4} = \frac{201 \cdot 3}{4 \cdot 3} = \frac{603}{12} \text{ and } \frac{98}{3} = \frac{98 \cdot 4}{3 \cdot 4} = \frac{392}{12}$$

Now subtract.

$$\frac{603}{12} - \frac{392}{12} = \frac{211}{12} \text{ or } 17\frac{7}{12}$$

9. To find out how much fabric Wei Jen should buy, add the amounts of fabric needed for each article of clothing.

Write each mixed number as an improper fraction.

$$1\frac{1}{4} + 1\frac{2}{3} + 2\frac{1}{2} = \frac{5}{4} + \frac{5}{3} + \frac{5}{2}$$

The LCD is 12. Write each fraction with a denominator of 12, then add and simplify the answer.

$$\frac{5}{4} + \frac{5}{3} + \frac{5}{2} = \frac{15}{12} + \frac{20}{12} + \frac{30}{12} = \frac{65}{12} \text{ or } 5\frac{5}{12}$$

Wei Jen should purchase $5\frac{5}{12}$ yd of fabric.

10. To find out how many gallons of paint Tram should buy, divide the total area to be painted by the area that one gallon of paint covers.

$$\frac{4200}{500} = \frac{42}{5} \text{ or } 8\frac{2}{5}$$

$8\frac{2}{5}$ gal are needed, so he must buy 9 gal.

Appendix B Section Exercises

1. True; the number above the fraction bar is called the numerator and the number below the fraction bar is called the denominator.

2. True; 5 divides the 41 eight times with a remainder of one, so $\frac{41}{5} = 8\frac{1}{5}$.

3. False; the fraction $\frac{17}{51}$ can be reduced to $\frac{1}{3}$ since $\frac{17}{51} = \frac{17 \cdot 1}{17 \cdot 3} = \frac{1}{3}$.

4. False; the reciprocal of $\frac{8}{2} = 4$ is $\frac{2}{8} = \frac{1}{4}$.

5. False; *product* refers to multiplication, so the product of 8 and 2 is 16. The *sum* of 8 and 2 is 10.

6. False; *difference* refers to subtraction, so the difference between 12 and 2 is 10. The *quotient* of 12 and 2 is 6.

7. Since 19 has only itself and 1 as factors, it is a prime number.

8. $$\begin{aligned} 99 &= 3 \cdot 33 \\ &= 3 \cdot 3 \cdot 11 \end{aligned}$$

Since 99 has factors other than itself and 1, it is a composite number.

9. The number 52 is composite since it has factors other than 1 and itself (2, for example).

10. Since 61 has only itself and 1 as factors, it is a prime number.

11. 2468 can be divided by 2. It has more than two different factors, so it is composite.

12. Since 3125 can be divided by 5, it is composite.

13. As stated in the text, the number 1 is neither prime nor composite, by agreement.

14. Since 83 has only itself and 1 as factors, it is a prime number.

15. $$\begin{aligned} 30 &= 2 \cdot 15 \\ &= 2 \cdot 3 \cdot 5 \end{aligned}$$

16. $$\begin{aligned} 40 &= 2 \cdot 20 \\ &= 2 \cdot 2 \cdot 10 \\ &= 2 \cdot 2 \cdot 2 \cdot 5 \end{aligned}$$

17. $$\begin{aligned} 252 &= 2 \cdot 126 \\ &= 2 \cdot 2 \cdot 63 \\ &= 2 \cdot 2 \cdot 3 \cdot 21 \\ &= 2 \cdot 2 \cdot 3 \cdot 3 \cdot 7 \end{aligned}$$

18. $$\begin{aligned} 168 &= 2 \cdot 84 \\ &= 2 \cdot 2 \cdot 42 \\ &= 2 \cdot 2 \cdot 2 \cdot 21 \\ &= 2 \cdot 2 \cdot 2 \cdot 3 \cdot 7 \end{aligned}$$

19. $$\begin{aligned} 124 &= 2 \cdot 62 \\ &= 2 \cdot 2 \cdot 31 \end{aligned}$$

20. $$\begin{aligned} 165 &= 3 \cdot 55 \\ &= 3 \cdot 5 \cdot 11 \end{aligned}$$

21. Since 29 has only itself and 1 as factors, it is a prime number. Its prime factored form is just 29.

22. Since 31 is a prime number, its prime factored form is just 31.

23. $$\frac{8}{16} = \frac{1 \cdot 8}{2 \cdot 8} = \frac{1}{2}$$

24. $$\frac{4}{12} = \frac{1 \cdot 4}{3 \cdot 4} = \frac{1}{3}$$

25. $\frac{15}{18} = \frac{3 \cdot 5}{3 \cdot 6} = \frac{5}{6}$

26. $\frac{16}{20} = \frac{4 \cdot 4}{5 \cdot 4} = \frac{4}{5}$

27. $\frac{15}{75} = \frac{1 \cdot 15}{5 \cdot 15} = \frac{1}{5}$

28. $\frac{24}{64} = \frac{3 \cdot 8}{8 \cdot 8} = \frac{3}{8}$

29. $\frac{144}{120} = \frac{6 \cdot 24}{5 \cdot 24} = \frac{6}{5}$

30. $\frac{132}{77} = \frac{12 \cdot 11}{7 \cdot 11} = \frac{12}{7}$

31. A common denominator for $\frac{p}{q}$ and $\frac{r}{s}$ must be a multiple of both denominators, q and s. Such a number is $q \cdot s$. Therefore, **A** is correct.

32. $\frac{16}{24} = \frac{8 \cdot 2}{8 \cdot 3} = \frac{2}{3}$

Therefore, **C** is correct.

33. $\frac{4}{5} \cdot \frac{6}{7} = \frac{4 \cdot 6}{5 \cdot 7} = \frac{24}{35}$

34. $\frac{5}{9} \cdot \frac{10}{7} = \frac{5 \cdot 10}{9 \cdot 7} = \frac{50}{63}$

35. $\frac{1}{10} \cdot \frac{12}{5} = \frac{1 \cdot 12}{10 \cdot 5} = \frac{1 \cdot 2 \cdot 6}{2 \cdot 5 \cdot 5} = \frac{6}{25}$

36.
$$\begin{aligned} \frac{6}{11} \cdot \frac{2}{3} &= \frac{6 \cdot 2}{11 \cdot 3} \\ &= \frac{2 \cdot 3 \cdot 2}{11 \cdot 3} \\ &= \frac{2 \cdot 2}{11} \\ &= \frac{4}{11} \end{aligned}$$

37.
$$\begin{aligned} \frac{15}{4} \cdot \frac{8}{25} &= \frac{15 \cdot 8}{4 \cdot 25} \\ &= \frac{3 \cdot 5 \cdot 4 \cdot 2}{4 \cdot 5 \cdot 5} \\ &= \frac{3 \cdot 2}{5} \\ &= \frac{6}{5} \quad \text{or} \quad 1\frac{1}{5} \end{aligned}$$

38.
$$\begin{aligned} \frac{4}{7} \cdot \frac{21}{8} &= \frac{4 \cdot 21}{7 \cdot 8} \\ &= \frac{4 \cdot 3 \cdot 7}{7 \cdot 4 \cdot 2} \\ &= \frac{3}{2} \quad \text{or} \quad 1\frac{1}{2} \end{aligned}$$

39. $2\frac{2}{3} \cdot 5\frac{4}{5}$

Change both mixed numbers to improper fractions.

$$2\frac{2}{3} = 2 + \frac{2}{3} = \frac{6}{3} + \frac{2}{3} = \frac{8}{3}$$

$$5\frac{4}{5} = 5 + \frac{4}{5} = \frac{25}{5} + \frac{4}{5} = \frac{29}{5}$$

$$\begin{aligned} 2\frac{2}{3} \cdot 5\frac{4}{5} &= \frac{8}{3} \cdot \frac{29}{5} \\ &= \frac{8 \cdot 29}{3 \cdot 5} \\ &= \frac{232}{15} \quad \text{or} \quad 15\frac{7}{15} \end{aligned}$$

40.
$$\begin{aligned} 3\frac{3}{5} \cdot 7\frac{1}{6} &= \frac{18}{5} \cdot \frac{43}{6} \\ &= \frac{18 \cdot 43}{5 \cdot 6} \\ &= \frac{3 \cdot 6 \cdot 43}{5 \cdot 6} \\ &= \frac{3 \cdot 43}{5} \\ &= \frac{129}{5} \quad \text{or} \quad 25\frac{4}{5} \end{aligned}$$

41. $\frac{5}{4} \div \frac{3}{8} = \frac{5}{4} \cdot \frac{8}{3}$ *Multiply by the reciprocal of the second fraction.*

$$\begin{aligned} &= \frac{5 \cdot 8}{4 \cdot 3} \\ &= \frac{5 \cdot 4 \cdot 2}{4 \cdot 3} \\ &= \frac{5 \cdot 2}{3} \\ &= \frac{10}{3} \quad \text{or} \quad 3\frac{1}{3} \end{aligned}$$

42. $\frac{7}{6} \div \frac{9}{10} = \frac{7}{6} \cdot \frac{10}{9}$ *Multiply by the reciprocal of the second fraction.*

$$\begin{aligned} &= \frac{7 \cdot 10}{6 \cdot 9} \\ &= \frac{7 \cdot 2 \cdot 5}{2 \cdot 3 \cdot 9} \\ &= \frac{35}{27} \quad \text{or} \quad 1\frac{8}{27} \end{aligned}$$

43. $\frac{32}{5} \div \frac{8}{15} = \frac{32}{5} \cdot \frac{15}{8}$ *Multiply by the reciprocal of the second fraction.*

$$= \frac{32 \cdot 15}{5 \cdot 8} = \frac{8 \cdot 4 \cdot 3 \cdot 5}{1 \cdot 5 \cdot 8} = \frac{4 \cdot 3}{1} = 12$$

44. $\frac{24}{7} \div \frac{6}{21} = \frac{24}{7} \cdot \frac{21}{6}$ *Multiply by the reciprocal of the second fraction.*

$$= \frac{24 \cdot 21}{7 \cdot 6} = \frac{4 \cdot 6 \cdot 3 \cdot 7}{1 \cdot 7 \cdot 6} = \frac{4 \cdot 3}{1} = 12$$

45. $\frac{3}{4} \div 12 = \frac{3}{4} \cdot \frac{1}{12}$ *Multiply by the reciprocal of 12.*

$$= \frac{3 \cdot 1}{4 \cdot 12} = \frac{3 \cdot 1}{4 \cdot 3 \cdot 4} = \frac{1}{4 \cdot 4} = \frac{1}{16}$$

46. $\frac{2}{5} \div 30 = \frac{2}{5} \cdot \frac{1}{30}$ *Multiply by the reciprocal of 30.*

$$= \frac{2 \cdot 1}{5 \cdot 30} = \frac{2 \cdot 1}{5 \cdot 2 \cdot 15} = \frac{1}{5 \cdot 15} = \frac{1}{75}$$

47. $2\frac{5}{8} \div 1\frac{15}{32}$

Change both mixed numbers to improper fractions.

$$2\frac{5}{8} = 2 + \frac{5}{8} = \frac{16}{8} + \frac{5}{8} = \frac{21}{8}$$

$$1\frac{15}{32} = 1 + \frac{15}{32} = \frac{32}{32} + \frac{15}{32} = \frac{47}{32}$$

$$2\frac{5}{8} \div 1\frac{15}{32} = \frac{21}{8} \div \frac{47}{32} = \frac{21}{8} \cdot \frac{32}{47} = \frac{21 \cdot 32}{8 \cdot 47} = \frac{21 \cdot 8 \cdot 4}{8 \cdot 47} = \frac{21 \cdot 4}{47} = \frac{84}{47} \text{ or } 1\frac{37}{47}$$

48. $2\frac{3}{10} \div 7\frac{4}{5} = \frac{23}{10} \div \frac{39}{5}$

$$= \frac{23}{10} \cdot \frac{5}{39} = \frac{23 \cdot 5}{2 \cdot 5 \cdot 39} = \frac{23}{78}$$

49. Multiply the first fraction (the dividend) by the reciprocal of the second fraction (the divisor) to divide two fractions.

50. To add two fractions with different denominators, find a common denominator, change each fraction to an equivalent fraction with that denominator, add numerators, and place the sum *over* the common denominator.

51. $\frac{7}{12} + \frac{1}{12} = \frac{7 + 1}{12} = \frac{8}{12} = \frac{2 \cdot 4}{3 \cdot 4} = \frac{2}{3}$

52. $\frac{3}{16} + \frac{5}{16} = \frac{3 + 5}{16} = \frac{8}{16} = \frac{1}{2}$

53. $\frac{5}{9} + \frac{1}{3}$

Since $9 = 3 \cdot 3$, and 3 is prime, the LCD (least common denominator) is $3 \cdot 3 = 9$.

$$\frac{1}{3} = \frac{1}{3} \cdot \frac{3}{3} = \frac{3}{9}$$

Now add the two fractions with the same denominator.

$$\frac{5}{9} + \frac{1}{3} = \frac{5}{9} + \frac{3}{9} = \frac{8}{9}$$

54. $\frac{4}{15} + \frac{1}{5}$

To add $\frac{4}{15}$ and $\frac{1}{5}$, first find the LCD. Since $15 = 3 \cdot 5$ and 5 is prime, the LCD is 15.

$$\frac{4}{15} + \frac{1}{5} = \frac{4}{15} + \frac{1 \cdot 3}{5 \cdot 3} = \frac{4}{15} + \frac{3}{15} = \frac{4+3}{15} = \frac{7}{15}$$

55. $3\frac{1}{8} + \frac{1}{4}$

$$3\frac{1}{8} = 3 + \frac{1}{8} = \frac{24}{8} + \frac{1}{8} = \frac{25}{8}$$

$$3\frac{1}{8} + \frac{1}{4} = \frac{25}{8} + \frac{1}{4}$$

Since $8 = 2 \cdot 2 \cdot 2$ and $4 = 2 \cdot 2$, the LCD is $2 \cdot 2 \cdot 2$ or 8.

$$3\frac{1}{8} + \frac{1}{4} = \frac{25}{8} + \frac{1 \cdot 2}{4 \cdot 2} = \frac{25}{8} + \frac{2}{8} = \frac{27}{8} \text{ or } 3\frac{3}{8}$$

56. $5\frac{3}{4} + \frac{2}{3}$

To add $5\frac{3}{4}$ and $\frac{2}{3}$, first change $5\frac{3}{4}$ to an improper fraction; then find the LCD, which is 12.

$$5\frac{3}{4} + \frac{1}{3} = \frac{23}{4} + \frac{2}{3} = \frac{23 \cdot 3}{4 \cdot 3} + \frac{2 \cdot 4}{3 \cdot 4} = \frac{69}{12} + \frac{8}{12} = \frac{77}{12} \text{ or } 6\frac{5}{12}$$

57. $\frac{7}{12} - \frac{1}{9}$

Since $12 = 2 \cdot 2 \cdot 3$ and $9 = 3 \cdot 3$, the LCD is

$2 \cdot 2 \cdot 3 \cdot 3 = 36$.

$$\frac{7}{12} = \frac{7}{12} \cdot \frac{3}{3} = \frac{21}{36} \text{ and } \frac{1}{9} \cdot \frac{4}{4} = \frac{4}{36}$$

Now subtract fractions with the same denominator.

$$\frac{7}{12} - \frac{1}{9} = \frac{21}{36} - \frac{4}{36} = \frac{17}{36}$$

58. $\frac{11}{16} - \frac{1}{12} = \frac{11 \cdot 3}{16 \cdot 3} - \frac{1 \cdot 4}{12 \cdot 4}$ *The LCD of 12 and 16 is 48.*

$$= \frac{33}{48} - \frac{4}{48} = \frac{29}{48}$$

59. $6\frac{1}{4} - 5\frac{1}{3}$

$$6\frac{1}{4} = 6 + \frac{1}{4} = \frac{24}{4} + \frac{1}{4} = \frac{25}{4}$$

$$5\frac{1}{3} = 5 + \frac{1}{3} = \frac{15}{3} + \frac{1}{3} = \frac{16}{3}$$

Since $4 = 2 \cdot 2$, and 3 is prime, the LCD is $2 \cdot 2 \cdot 3 = 12$.

$$6\frac{1}{4} - 5\frac{1}{3} = \frac{25}{4} - \frac{16}{3} = \frac{25 \cdot 3}{4 \cdot 3} - \frac{16 \cdot 4}{3 \cdot 4} = \frac{75}{12} - \frac{64}{12} = \frac{11}{12}$$

60. $8\frac{4}{5} - 7\frac{4}{9}$

Change both numbers to improper fractions; then add, using 45 as the common denominator.

$$8\frac{4}{5} - 7\frac{4}{9} = \frac{44}{5} - \frac{67}{9} = \frac{44 \cdot 9}{5 \cdot 9} - \frac{67 \cdot 5}{9 \cdot 5} = \frac{396}{45} - \frac{335}{45} = \frac{61}{45} \text{ or } 1\frac{16}{45}$$

61. $\frac{5}{3} + \frac{1}{6} - \frac{1}{2}$

Since 2 and 3 are prime, and $6 = 2 \cdot 3$, the LCD is $2 \cdot 3 = 6$. Write $\frac{5}{3} \cdot \frac{2}{2} = \frac{10}{6}$ and $\frac{1}{2} = \frac{1}{2} \cdot \frac{3}{3} = \frac{3}{6}$.

Now add and subtract, then write the answer in lowest terms.

$$\frac{5}{3} + \frac{1}{6} - \frac{1}{2} = \frac{10}{6} + \frac{1}{6} - \frac{3}{6} = \frac{8}{6} = \frac{4 \cdot 2}{3 \cdot 2} = \frac{4}{3} \text{ or } 1\frac{1}{3}$$

62. $$\frac{7}{15}+\frac{1}{6}-\frac{1}{10}=\frac{7\cdot 2}{15\cdot 2}+\frac{1\cdot 5}{6\cdot 5}-\frac{1\cdot 3}{10\cdot 3}$$
$$=\frac{14}{30}+\frac{5}{30}-\frac{3}{30}$$
$$=\frac{16}{30}=\frac{8}{15}$$

63. Multiply the number of cups of water per serving by the number of servings.

$$\frac{3}{4}\cdot 8=\frac{3}{4}\cdot\frac{8}{1}$$
$$=\frac{3\cdot 8}{4\cdot 1}$$
$$=\frac{3\cdot 2\cdot 4}{4\cdot 1}$$
$$=\frac{3\cdot 2}{1}=6\text{ cups}$$

For 8 microwave servings, 6 cups of water will be needed.

64. Four stove top servings require $\frac{1}{4}$ tsp, or $\frac{2}{8}$ tsp, of salt. Six stove top servings require $\frac{1}{2}$ tsp, or $\frac{4}{8}$ tsp, of salt. Five is halfway between 4 and 6, and $\frac{3}{8}$ is halfway between $\frac{2}{8}$ and $\frac{4}{8}$. Therefore, 5 stove top servings would require $\frac{3}{8}$ tsp of salt.

65. The perimeter is the sum of the measures of the 5 sides.

$$196+98\frac{3}{4}+146\frac{1}{2}+100\frac{7}{8}+76\frac{5}{8}$$
$$=196+98\frac{6}{8}+146\frac{4}{8}+100\frac{7}{8}+76\frac{5}{8}$$
$$=196+98+146+100+76+\frac{6+4+7+5}{8}$$
$$=616+\frac{22}{8}\quad\left(\frac{22}{8}=2\frac{6}{8}=2\frac{3}{4}\right)$$
$$=618\frac{3}{4}\text{ feet}$$

The perimeter is $618\frac{3}{4}$ feet.

66. To find the perimeter of a triangle, add the lengths of the three sides.

$$5\frac{1}{4}+7\frac{1}{2}+10\frac{1}{8}=5\frac{2}{8}+7\frac{4}{8}+10\frac{1}{8}$$
$$=22\frac{7}{8}$$

The perimeter of the triangle is $22\frac{7}{8}$ feet.

67. The difference between the two measures is found by subtracting, using 16 as the LCD.

$$\frac{3}{4}-\frac{3}{16}=\frac{3\cdot 4}{4\cdot 4}-\frac{3}{16}$$
$$=\frac{12}{16}-\frac{3}{16}$$
$$=\frac{12-3}{16}=\frac{9}{16}$$

The difference is $\frac{9}{16}$ inch.

68. The difference between the two measures is found by subtracting, using 16 as a common denominator.

$$\frac{9}{16}-\frac{3}{8}=\frac{9}{16}-\frac{3\cdot 2}{8\cdot 2}$$
$$=\frac{9}{16}-\frac{6}{16}$$
$$=\frac{9-6}{16}=\frac{3}{16}$$

The difference is $\frac{3}{16}$ inch.

69. Subtract $\frac{3}{8}$ from $\frac{11}{16}$ using 16 as the LCD.

$$\frac{11}{16}-\frac{3}{8}=\frac{11}{16}-\frac{3\cdot 2}{8\cdot 2}$$
$$=\frac{11}{16}-\frac{6}{16}$$
$$=\frac{5}{16}$$

Thus, $\frac{3}{8}$ inch is $\frac{5}{16}$ inch smaller than $\frac{11}{16}$ inch.

70. Divide the total amount of tomato sauce by the number of servings.

$$2\frac{1}{3}\div 7=\frac{7}{3}\div\frac{7}{1}=\frac{7}{3}\cdot\frac{1}{7}=\frac{7\cdot 1}{3\cdot 7}=\frac{1}{3}$$

For 1 serving of barbeque sauce, $\frac{1}{3}$ cup of tomato sauce is needed.

71. Multiply the amount of fabric it takes to make one costume by the number of costumes.

$$2\frac{3}{8}\cdot 7=\frac{19}{8}\cdot\frac{7}{1}$$
$$=\frac{19\cdot 7}{8\cdot 1}$$
$$=\frac{133}{8}\text{ or }16\frac{5}{8}\text{ yd}$$

For 7 costumes, $16\frac{5}{8}$ yards of fabric would be needed.

72. To find the number of cakes the caterer can make, divide $15\frac{1}{2}$ by $1\frac{3}{4}$.

$$\begin{aligned} 15\frac{1}{2} \div 1\frac{3}{4} &= \frac{31}{2} \div \frac{7}{4} \\ &= \frac{31}{2} \cdot \frac{4}{7} \\ &= \frac{31 \cdot 2 \cdot 2}{2 \cdot 7} \\ &= \frac{62}{7} = 8\frac{6}{7} \end{aligned}$$

There is not quite enough sugar for 9 cakes. The caterer can make 8 cakes with some sugar left over.

73. The sum of the fractions representing immigrants from Latin America, Asia, or Europe is

$$\begin{aligned} \frac{13}{25} + \frac{3}{10} + \frac{13}{100} &= \frac{13 \cdot 4}{25 \cdot 4} + \frac{3 \cdot 10}{10 \cdot 10} + \frac{13}{100} \\ &= \frac{52 + 30 + 13}{100} \\ &= \frac{95}{100} = \frac{19}{20}. \end{aligned}$$

So the fraction representing immigrants from other regions is

$$\begin{aligned} 1 - \frac{19}{20} &= \frac{20}{20} - \frac{19}{20} \\ &= \frac{1}{20}. \end{aligned}$$

74. The sum of the fractions representing immigrants from Latin America or Asia is

$$\begin{aligned} \frac{13}{25} + \frac{3}{10} &= \frac{13 \cdot 4}{25 \cdot 4} + \frac{3 \cdot 10}{10 \cdot 10} \\ &= \frac{52 + 30}{100} \\ &= \frac{82}{100} = \frac{41}{50}. \end{aligned}$$

75. Multiply the fraction representing immigrants from Europe $\left(\frac{13}{100}\right)$ by the total number of immigrants (more than 8 million).

$$\frac{13}{100} \cdot 8 = \frac{13}{100} \cdot \frac{8}{1} = \frac{104}{100} = \frac{26}{25} = 1\frac{1}{25}$$

There were more than $1\frac{1}{25}$ million immigrants from Europe.

76. From Exercise 73, the fraction that represents "Other" is $\frac{1}{20}$. The sum of the fractions representing immigrants from Asia, Europe, or "Other" is

$$\begin{aligned} \frac{3}{10} + \frac{13}{100} + \frac{1}{20} &= \frac{3 \cdot 10}{10 \cdot 10} + \frac{13}{100} + \frac{1 \cdot 5}{20 \cdot 5} \\ &= \frac{30 + 13 + 5}{100} \\ &= \frac{48}{100} = \frac{12}{25}. \end{aligned}$$

The *fraction* representing the difference between Latin American immigrants and those from all other regions combined is

$$\frac{13}{25} - \frac{12}{25} = \frac{1}{25}.$$

Thus, the *number* representing the difference between Latin American immigrants and those from all other regions combined is

$$\frac{1}{25} \cdot 8 \text{ million} = \frac{8}{25} \text{ million.}$$

APPENDIX C DETERMINANTS AND CRAMER'S RULE

Appendix C Margin Exercises

1. The value of a 2×2 determinant is given by

$$\begin{vmatrix} a & b \\ c & d \end{vmatrix} = ad - bc.$$

(a) $\begin{vmatrix} -4 & 6 \\ 2 & 3 \end{vmatrix} = -4(3) - 6(2)$

$= -12 - 12 = -24$

(b) $\begin{vmatrix} 3 & -1 \\ 0 & 2 \end{vmatrix} = 3(2) - (-1)0$

$= 6 - 0 = 6$

(c) $\begin{vmatrix} -2 & 5 \\ 1 & 5 \end{vmatrix} = -2(5) - 5(1)$

$= -10 - 5 = -15$

2. **(a)** $\begin{vmatrix} 0 & -1 & 0 \\ 2 & 4 & 2 \\ 3 & 1 & 5 \end{vmatrix}$

Expand by minors about column 1. Note that the sign of the number in the second row, first column is changed.

$$\begin{aligned} &= 0\begin{vmatrix} 4 & 2 \\ 1 & 5 \end{vmatrix} - 2\begin{vmatrix} -1 & 0 \\ 1 & 5 \end{vmatrix} + 3\begin{vmatrix} -1 & 0 \\ 4 & 2 \end{vmatrix} \\ &= 0[4(5) - 2(1)] - 2[-1(5) - 0(1)] \\ &\quad + 3[-1(2) - 0(4)] \\ &= 0 - 2(-5) + 3(-2) \\ &= 10 - 6 = 4 \end{aligned}$$

(b) $\begin{vmatrix} 2 & 1 & 4 \\ -3 & 0 & 2 \\ -2 & 1 & 5 \end{vmatrix}$

Expand by minors about column 1. Note that the sign of the number in the second row, first column is changed.

$$\begin{aligned} &= 2\begin{vmatrix} 0 & 2 \\ 1 & 5 \end{vmatrix} - (-3)\begin{vmatrix} 1 & 4 \\ 1 & 5 \end{vmatrix} + (-2)\begin{vmatrix} 1 & 4 \\ 0 & 2 \end{vmatrix} \\ &= 2[0(5) - 2(1)] + 3[1(5) - 4(1)] \\ &\quad - 2[1(2) - 4(0)] \\ &= 2(-2) + 3(1) - 2(2) \\ &= -4 + 3 - 4 = -5 \end{aligned}$$

3. **(a)** $\begin{vmatrix} 2 & 1 & 3 \\ -1 & 0 & 4 \\ 2 & 4 & 3 \end{vmatrix}$

Expand by minors about column 2. Note that the signs of the numbers in the first and third rows are changed.

$$\begin{aligned} &= -1\begin{vmatrix} -1 & 4 \\ 2 & 3 \end{vmatrix} + 0\begin{vmatrix} 2 & 3 \\ 2 & 3 \end{vmatrix} - 4\begin{vmatrix} 2 & 3 \\ -1 & 4 \end{vmatrix} \\ &= -1[-1(3) - 4(2)] + 0[2(3) - 3(2)] \\ &\quad - 4[2(4) - 3(-1)] \\ &= -1(-11) + 0 - 4(11) \\ &= 11 - 44 = -33 \end{aligned}$$

(b) $\begin{vmatrix} 5 & -1 & 2 \\ 0 & 4 & 3 \\ -1 & 2 & 0 \end{vmatrix}$

Expand by minors about column 2. Note that the signs of the numbers in the first and third rows are changed.

$$\begin{aligned} &= -(-1)\begin{vmatrix} 0 & 3 \\ -1 & 0 \end{vmatrix} + 4\begin{vmatrix} 5 & 2 \\ -1 & 0 \end{vmatrix} - 2\begin{vmatrix} 5 & 2 \\ 0 & 3 \end{vmatrix} \\ &= 1[0(0) - 3(-1)] + 4[5(0) - 2(-1)] \\ &\quad - 2[5(3) - 2(0)] \\ &= 1(3) + 4(2) - 2(15) \\ &= 3 + 8 - 30 = -19 \end{aligned}$$

4. **(a)** $\begin{aligned} x + y &= 5 \\ x - y &= 1 \end{aligned}$

$$D = \begin{vmatrix} 1 & 1 \\ 1 & -1 \end{vmatrix} = 1(-1) - 1(1) = -2$$

$$D_x = \begin{vmatrix} 5 & 1 \\ 1 & -1 \end{vmatrix} = 5(-1) - 1(1) = -6$$

$$D_y = \begin{vmatrix} 1 & 5 \\ 1 & 1 \end{vmatrix} = 1(1) - 5(1) = -4$$

$$x = \frac{D_x}{D} = \frac{-6}{-2} = 3; \; y = \frac{D_y}{D} = \frac{-4}{-2} = 2$$

Solution set: $\{(3, 2)\}$

(b) $\begin{aligned} 2x - 3y &= -26 \\ 3x + 4y &= 12 \end{aligned}$

$$D = \begin{vmatrix} 2 & -3 \\ 3 & 4 \end{vmatrix} = 2(4) - (-3)(3) = 17$$

$$D_x = \begin{vmatrix} -26 & -3 \\ 12 & 4 \end{vmatrix} = -26(4) - (-3)(12) = -68$$

$$D_y = \begin{vmatrix} 2 & -26 \\ 3 & 12 \end{vmatrix} = 2(12) - (-26)(3) = 102$$

$$x = \frac{D_x}{D} = \frac{-68}{17} = -4; \; y = \frac{D_y}{D} = \frac{102}{17} = 6$$

Solution set: $\{(-4, 6)\}$

(c) $\begin{aligned} 4x - 5y &= -8 \\ 3x + 7y &= -6 \end{aligned}$

$$D = \begin{vmatrix} 4 & -5 \\ 3 & 7 \end{vmatrix} = 4(7) - (-5)(3) = 43$$

$$D_x = \begin{vmatrix} -8 & -5 \\ -6 & 7 \end{vmatrix} = -8(7) - (-5)(-6) = -86$$

$$D_y = \begin{vmatrix} 4 & -8 \\ 3 & -6 \end{vmatrix} = 4(-6) - (-8)(3) = 0$$

$$x = \frac{D_x}{D} = \frac{-86}{43} = -2;\ y = \frac{D_y}{D} = \frac{0}{43} = 0$$

Solution set: $\{(-2, 0)\}$

5. $D_z = \begin{vmatrix} 1 & 1 & -2 \\ 2 & -1 & -5 \\ 1 & -2 & 4 \end{vmatrix}$

Expand by minors about column 1.

$$= 1\begin{vmatrix} -1 & -5 \\ -2 & 4 \end{vmatrix} - 2\begin{vmatrix} 1 & -2 \\ -2 & 4 \end{vmatrix} + 1\begin{vmatrix} 1 & -2 \\ -1 & -5 \end{vmatrix}$$
$$= 1(-14) - 2(0) + 1(-7)$$
$$= -14 - 7 = -21$$

6. (a) $\begin{aligned} x + y + z &= 2 \\ 2x \qquad - z &= -3 \\ y + 2z &= 4 \end{aligned}$

$$D = \begin{vmatrix} 1 & 1 & 1 \\ 2 & 0 & -1 \\ 0 & 1 & 2 \end{vmatrix}$$

Expand by minors about column 2.

$$= -1\begin{vmatrix} 2 & -1 \\ 0 & 2 \end{vmatrix} + 0\begin{vmatrix} 1 & 1 \\ 0 & 2 \end{vmatrix} - 1\begin{vmatrix} 1 & 1 \\ 2 & -1 \end{vmatrix}$$
$$= -1(4) + 0 - 1(-3)$$
$$= -4 + 3 = -1$$

$$D_x = \begin{vmatrix} 2 & 1 & 1 \\ -3 & 0 & -1 \\ 4 & 1 & 2 \end{vmatrix}$$

Expand by minors about column 2.

$$= -1\begin{vmatrix} -3 & -1 \\ 4 & 2 \end{vmatrix} + 0\begin{vmatrix} 2 & 1 \\ 4 & 2 \end{vmatrix} - 1\begin{vmatrix} 2 & 1 \\ -3 & -1 \end{vmatrix}$$
$$= -1(-2) + 0 - 1(1)$$
$$= 2 - 1 = 1$$

$$D_y = \begin{vmatrix} 1 & 2 & 1 \\ 2 & -3 & -1 \\ 0 & 4 & 2 \end{vmatrix}$$

Expand by minors about column 1.

$$= 1\begin{vmatrix} -3 & -1 \\ 4 & 2 \end{vmatrix} - 2\begin{vmatrix} 2 & 1 \\ 4 & 2 \end{vmatrix} + 0\begin{vmatrix} 2 & 1 \\ -3 & -1 \end{vmatrix}$$
$$= 1(-2) - 2(0) + 0 = -2$$

$$D_z = \begin{vmatrix} 1 & 1 & 2 \\ 2 & 0 & -3 \\ 0 & 1 & 4 \end{vmatrix}$$

Expand by minors about column 2.

$$= -1\begin{vmatrix} 2 & -3 \\ 0 & 4 \end{vmatrix} + 0\begin{vmatrix} 1 & 2 \\ 0 & 4 \end{vmatrix} - 1\begin{vmatrix} 1 & 2 \\ 2 & -3 \end{vmatrix}$$
$$= -1(8) + 0 - 1(-7)$$
$$= -8 + 7 = -1$$

$$x = \frac{D_x}{D} = \frac{1}{-1} = -1;\ y = \frac{D_y}{D} = \frac{-2}{-1} = 2$$
$$z = \frac{D_z}{D} = \frac{-1}{-1} = 1$$

Solution set: $\{(-1, 2, 1)\}$

(b) $\begin{aligned} 3x - 2y + 4z &= 5 \\ 4x + y + z &= 14 \\ x - y - z &= 1 \end{aligned}$

$$D = \begin{vmatrix} 3 & -2 & 4 \\ 4 & 1 & 1 \\ 1 & -1 & -1 \end{vmatrix}$$

Expand by minors about row 3.

$$= 1\begin{vmatrix} -2 & 4 \\ 1 & 1 \end{vmatrix} - (-1)\begin{vmatrix} 3 & 4 \\ 4 & 1 \end{vmatrix} + (-1)\begin{vmatrix} 3 & -2 \\ 4 & 1 \end{vmatrix}$$
$$= -6 + (-13) - 11 = -30$$

$$D_x = \begin{vmatrix} 5 & -2 & 4 \\ 14 & 1 & 1 \\ 1 & -1 & -1 \end{vmatrix}$$

Expand by minors about row 3.

$$= 1\begin{vmatrix} -2 & 4 \\ 1 & 1 \end{vmatrix} - (-1)\begin{vmatrix} 5 & 4 \\ 14 & 1 \end{vmatrix} + (-1)\begin{vmatrix} 5 & -2 \\ 14 & 1 \end{vmatrix}$$
$$= -6 + (-51) - 33 = -90$$

$$D_y = \begin{vmatrix} 3 & 5 & 4 \\ 4 & 14 & 1 \\ 1 & 1 & -1 \end{vmatrix}$$

Expand by minors about row 3.

$$= 1\begin{vmatrix} 5 & 4 \\ 14 & 1 \end{vmatrix} - 1\begin{vmatrix} 3 & 4 \\ 4 & 1 \end{vmatrix} + (-1)\begin{vmatrix} 3 & 5 \\ 4 & 14 \end{vmatrix}$$
$$= -51 - (-13) - 22 = -60$$

$$D_z = \begin{vmatrix} 3 & -2 & 5 \\ 4 & 1 & 14 \\ 1 & -1 & 1 \end{vmatrix}$$

Expand by minors about row 3.

$$= 1\begin{vmatrix} -2 & 5 \\ 1 & 14 \end{vmatrix} - (-1)\begin{vmatrix} 3 & 5 \\ 4 & 14 \end{vmatrix} + 1\begin{vmatrix} 3 & -2 \\ 4 & 1 \end{vmatrix}$$
$$= -33 + 22 + 11 = 0$$

$$x = \frac{D_x}{D} = \frac{-90}{-30} = 3;\ y = \frac{D_y}{D} = \frac{-60}{-30} = 2$$
$$z = \frac{D_z}{D} = \frac{0}{-30} = 0$$

Solution set: $\{(3, 2, 0)\}$

7.
$$\begin{aligned} x - y + z &= 6 \\ 3x + 2y + z &= 4 \\ 2x - 2y + 2z &= 14 \end{aligned}$$

$$D = \begin{vmatrix} 1 & -1 & 1 \\ 3 & 2 & 1 \\ 2 & -2 & 2 \end{vmatrix}$$

Expand by minors about row 3.

$$\begin{aligned} &= 2\begin{vmatrix} -1 & 1 \\ 2 & 1 \end{vmatrix} - (-2)\begin{vmatrix} 1 & 1 \\ 3 & 1 \end{vmatrix} + 2\begin{vmatrix} 1 & -1 \\ 3 & 2 \end{vmatrix} \\ &= 2(-3) + 2(-2) + 2(5) \\ &= -6 - 4 + 10 = 0 \end{aligned}$$

Because $D = 0$, Cramer's rule does not apply.

Appendix C Section Exercises

1. *True*, a matrix is an array of numbers, while a determinant is a single number.

2. *True*, a square matrix has the same number of rows as columns.

3. *False*, the determinant $\begin{vmatrix} a & b \\ c & d \end{vmatrix}$ is equal to $ad - bc$.

4. The value $\begin{vmatrix} 0 & 0 \\ x & y \end{vmatrix}$ is zero for any replacements of x and y.

$$\begin{vmatrix} 0 & 0 \\ x & y \end{vmatrix} = 0(y) - 0(x) = 0$$

No matter what replacements are used for x and y, the value of the determinant is zero since both x and y are being multiplied by zero. The statement is *true*.

5. $$\begin{vmatrix} -2 & 5 \\ -1 & 4 \end{vmatrix} = -2(4) - 5(-1) = -8 + 5 = -3$$

6. $$\begin{vmatrix} 3 & -6 \\ 2 & -2 \end{vmatrix} = 3(-2) - (-6)2 = -6 + 12 = 6$$

7. $$\begin{vmatrix} 1 & -2 \\ 7 & 0 \end{vmatrix} = 1(0) - (-2)7 = 0 + 14 = 14$$

8. $$\begin{vmatrix} -5 & -1 \\ 1 & 0 \end{vmatrix} = -5(0) - (-1)1 = 0 + 1 = 1$$

9. $$\begin{vmatrix} 0 & 4 \\ 0 & 4 \end{vmatrix} = 0(4) - 4(0) = 0 - 0 = 0$$

10. $$\begin{vmatrix} 8 & -3 \\ 0 & 0 \end{vmatrix} = 8(0) - (-3)0 = 0 - 0 = 0$$

11. $$\begin{vmatrix} -1 & 2 & 4 \\ -3 & -2 & -3 \\ 2 & -1 & 5 \end{vmatrix}$$ Expand by minors about the first column.

$$\begin{aligned} &= -1\begin{vmatrix} -2 & -3 \\ -1 & 5 \end{vmatrix} - (-3)\begin{vmatrix} 2 & 4 \\ -1 & 5 \end{vmatrix} + 2\begin{vmatrix} 2 & 4 \\ -2 & -3 \end{vmatrix} \\ &= -1[-2(5) - (-3)(-1)] + 3[2(5) - 4(-1)] + 2[2(-3) - 4(-2)] \\ &= -1(-13) + 3(14) + 2(2) \\ &= 13 + 42 + 4 = 59 \end{aligned}$$

12. $$\begin{vmatrix} 2 & -3 & -5 \\ 1 & 2 & 2 \\ 5 & 3 & -1 \end{vmatrix}$$

$$\begin{aligned} &= 2\begin{vmatrix} 2 & 2 \\ 3 & -1 \end{vmatrix} - 1\begin{vmatrix} -3 & -5 \\ 3 & -1 \end{vmatrix} + 5\begin{vmatrix} -3 & -5 \\ 2 & 2 \end{vmatrix} \\ &= 2[2(-1) - 2(3)] - 1[-3(-1) - (-5)3] + 5[-3(2) - (-5)2] \\ &= 2(-8) - 1(18) + 5(4) \\ &= -16 - 18 + 20 = -14 \end{aligned}$$

13. $$\begin{vmatrix} 1 & 0 & -2 \\ 0 & 2 & 3 \\ 1 & 0 & 5 \end{vmatrix}$$ There are two 0s in column 2. We'll expand about that column since there is only 1 minor to evaluate.

$$\begin{aligned} &= -0\begin{vmatrix} 0 & 3 \\ 1 & 5 \end{vmatrix} + 2\begin{vmatrix} 1 & -2 \\ 1 & 5 \end{vmatrix} - 0\begin{vmatrix} 1 & -2 \\ 0 & 3 \end{vmatrix} \\ &= 0 + 2[1(5) - (-2)(1)] - 0 \\ &= 2[5 + 2] = 2(7) = 14 \end{aligned}$$

14. $$\begin{vmatrix} 2 & -1 & 0 \\ 0 & -1 & 1 \\ 1 & 2 & 0 \end{vmatrix}$$ Expand about column 3 (two 0s).

$$\begin{aligned} &= 0\begin{vmatrix} 0 & -1 \\ 1 & 2 \end{vmatrix} - 1\begin{vmatrix} 2 & -1 \\ 1 & 2 \end{vmatrix} + 0\begin{vmatrix} 2 & -1 \\ 0 & -1 \end{vmatrix} \\ &= 0 - 1[2(2) - (-1)1] + 0 \\ &= -1(4 + 1) = -1(5) = -5 \end{aligned}$$

15. $$\begin{vmatrix} 1 & 0 & 0 \\ 0 & 1 & 0 \\ 0 & 0 & 1 \end{vmatrix}$$ Expand about column 1.

$$= 1\begin{vmatrix} 1 & 0 \\ 0 & 1 \end{vmatrix} - 0 + 0 = 1[(1)(1) - (0)(0)] = 1$$

16. $$\begin{vmatrix} 0 & 0 & 1 \\ 0 & 1 & 0 \\ 1 & 0 & 0 \end{vmatrix}$$ Expand about column 1.

$$= 0 - 0 + 1\begin{vmatrix} 0 & 1 \\ 1 & 0 \end{vmatrix} = 0 - 0 + 1(-1) = -1$$

17. $\begin{vmatrix} 4 & 4 & 2 \\ 1 & -1 & -2 \\ 1 & 0 & 2 \end{vmatrix}$ Expand about column 2.

$$= -4\begin{vmatrix} 1 & -2 \\ 1 & 2 \end{vmatrix} + (-1)\begin{vmatrix} 4 & 2 \\ 1 & 2 \end{vmatrix} - 0\begin{vmatrix} 4 & 2 \\ 1 & -2 \end{vmatrix}$$
$$= -4[1(2) - (-2)1] - 1[4(2) - 2(1)] - 0$$
$$= -4(4) - 1(6)$$
$$= -16 - 6 = -22$$

18. $\begin{vmatrix} 3 & -1 & 2 \\ 1 & 5 & -2 \\ 0 & 2 & 0 \end{vmatrix}$ Expand about row 3.

$$= 0 - 2\begin{vmatrix} 3 & 2 \\ 1 & -2 \end{vmatrix} + 0$$
$$= -2[3(-2) - 2(1)]$$
$$= -2[-6 - 2] = -2[-8] = 16$$

19. $\begin{vmatrix} 2 & 0 & 1 \\ -1 & 0 & 2 \\ 5 & 0 & 4 \end{vmatrix} = 0$ Expand about column 2.

20. $\begin{vmatrix} 2 & -4 & 0 \\ 3 & -5 & 0 \\ 6 & -7 & 0 \end{vmatrix} = 0$ Expand about column 3.

21. $\begin{vmatrix} -6 & 3 & 5 \\ -3 & 2 & 2 \\ 0 & 0 & 0 \end{vmatrix} = 0$ Expand about row 3.

22. $\begin{vmatrix} 0 & 0 & 0 \\ 4 & 0 & -2 \\ 2 & -1 & 3 \end{vmatrix} = 0$ Expand about row 1.

23. $\begin{vmatrix} 3 & 5 & -2 \\ 1 & -4 & 1 \\ 3 & 1 & -2 \end{vmatrix}$ Expand about column 1.

$$= 3\begin{vmatrix} -4 & 1 \\ 1 & -2 \end{vmatrix} - 1\begin{vmatrix} 5 & -2 \\ 1 & -2 \end{vmatrix} + 3\begin{vmatrix} 5 & -2 \\ -4 & 1 \end{vmatrix}$$
$$= 3[-4(-2) - 1(1)] - 1[5(-2) - (-2)1]$$
$$+ 3[5(1) - (-2)(-4)]$$
$$= 3(7) - 1(-8) + 3(-3)$$
$$= 21 + 8 - 9 = 20$$

24. $\begin{vmatrix} 1 & 3 & 2 \\ 3 & -1 & -2 \\ 1 & 10 & 20 \end{vmatrix}$ Expand about column 1.

$$= 1\begin{vmatrix} -1 & -2 \\ 10 & 20 \end{vmatrix} - 3\begin{vmatrix} 3 & 2 \\ 10 & 20 \end{vmatrix} + 1\begin{vmatrix} 3 & 2 \\ -1 & -2 \end{vmatrix}$$
$$= 1[-1(20) - (-2)(10)] - 3[3(20) - 2(10)]$$
$$+ 1[3(-2) - 2(-1)]$$
$$= 1(0) - 3(40) + 1(-4)$$
$$= 0 - 120 - 4 = -124$$

25. $x = \dfrac{D_x}{D} = \dfrac{112}{28} = 4$ and $y = \dfrac{D_y}{D} = \dfrac{168}{28} = 6.$

Solution set: $\{(4, 6)\}$

26. The solution set is $\{(1, 0, -1\}$, so we have $x = 1$, $y = 0$, and $z = -1$. Also $D = -43$.

$$x = \frac{D_x}{D}, \quad \text{so} \quad 1 = \frac{D_x}{-43} \quad \text{gives} \quad D_x = -43.$$
$$y = \frac{D_y}{D}, \quad \text{so} \quad 0 = \frac{D_y}{-43} \quad \text{gives} \quad D_y = 0.$$
$$z = \frac{D_z}{D}, \quad \text{so} \quad -1 = \frac{D_z}{-43} \quad \text{gives} \quad D_z = 43.$$

27.
$$\begin{aligned} 3x + 5y &= -5 \\ -2x + 3y &= 16 \end{aligned}$$

$$D = \begin{vmatrix} 3 & 5 \\ -2 & 3 \end{vmatrix} = 3(3) - 5(-2) = 19$$
$$D_x = \begin{vmatrix} -5 & 5 \\ 16 & 3 \end{vmatrix} = -5(3) - 5(16) = -95$$
$$D_y = \begin{vmatrix} 3 & -5 \\ -2 & 16 \end{vmatrix} = 3(16) - (-5)(-2) = 38$$

$$x = \frac{D_x}{D} = \frac{-95}{19} = -5; \; y = \frac{D_y}{D} = \frac{38}{19} = 2$$

Solution set: $\{(-5, 2)\}$

28.
$$\begin{aligned} 5x + 2y &= -3 \\ 4x - 3y &= -30 \end{aligned}$$

$$D = \begin{vmatrix} 5 & 2 \\ 4 & -3 \end{vmatrix} = -15 - 8 = -23$$
$$D_x = \begin{vmatrix} -3 & 2 \\ -30 & -3 \end{vmatrix} = 9 + 60 = 69$$
$$D_y = \begin{vmatrix} 5 & -3 \\ 4 & -30 \end{vmatrix} = -150 + 12 = -138$$

$$x = \frac{D_x}{D} = \frac{69}{-23} = -3; \; y = \frac{D_y}{D} = \frac{-138}{-23} = 6$$

Solution set: $\{(-3, 6)\}$

29.
$$\begin{aligned} 3x + 2y &= 3 \\ 2x - 4y &= 2 \end{aligned}$$

$$D = \begin{vmatrix} 3 & 2 \\ 2 & -4 \end{vmatrix} = -12 - 4 = -16;$$
$$D_x = D = -16$$
$$D_y = \begin{vmatrix} 3 & 3 \\ 2 & 2 \end{vmatrix} = 6 - 6 = 0$$

$$x = \frac{D_x}{D} = \frac{-16}{-16} = 1; \; y = \frac{D_y}{D} = \frac{0}{-16} = 0$$

Solution set: $\{(1, 0)\}$

30. $7x - 2y = 6$
$4x - 5y = 15$

$$D = \begin{vmatrix} 7 & -2 \\ 4 & -5 \end{vmatrix} = -35 + 8 = -27$$

$$D_x = \begin{vmatrix} 6 & -2 \\ 15 & -5 \end{vmatrix} = -30 + 30 = 0$$

$$D_y = \begin{vmatrix} 7 & 6 \\ 4 & 15 \end{vmatrix} = 105 - 24 = 81$$

$$x = \frac{D_x}{D} = \frac{0}{-27} = 0;\ y = \frac{D_y}{D} = \frac{81}{-27} = -3$$

Solution set: $\{(0, -3)\}$

31. $8x + 3y = 1$
$6x - 5y = 2$

$$D = \begin{vmatrix} 8 & 3 \\ 6 & -5 \end{vmatrix} = 8(-5) - 3(6) = -58$$

$$D_x = \begin{vmatrix} 1 & 3 \\ 2 & -5 \end{vmatrix} = 1(-5) - 3(2) = -11$$

$$D_y = \begin{vmatrix} 8 & 1 \\ 6 & 2 \end{vmatrix} = 8(2) - 1(6) = 10$$

$$x = \frac{D_x}{D} = \frac{-11}{-58} = \frac{11}{58};\ y = \frac{D_y}{D} = \frac{10}{-58} = -\frac{5}{29}$$

Solution set: $\left\{\left(\frac{11}{58}, -\frac{5}{29}\right)\right\}$

32. $3x - y = 9$
$2x + 5y = 8$

$$D = \begin{vmatrix} 3 & -1 \\ 2 & 5 \end{vmatrix} = 15 + 2 = 17$$

$$D_x = \begin{vmatrix} 9 & -1 \\ 8 & 5 \end{vmatrix} = 45 + 8 = 53$$

$$D_y = \begin{vmatrix} 3 & 9 \\ 2 & 8 \end{vmatrix} = 24 - 18 = 6$$

$$x = \frac{D_x}{D} = \frac{53}{17};\ y = \frac{D_y}{D} = \frac{6}{17}$$

Solution set: $\left\{\left(\frac{53}{17}, \frac{6}{17}\right)\right\}$

33. $2x + 3y + 2z = 15$
$x - y + 2z = 5$
$x + 2y - 6z = -26$

$$D = \begin{vmatrix} 2 & 3 & 2 \\ 1 & -1 & 2 \\ 1 & 2 & -6 \end{vmatrix} \text{ Expand about row 1.}$$

$$= 2\begin{vmatrix} -1 & 2 \\ 2 & -6 \end{vmatrix} - 3\begin{vmatrix} 1 & 2 \\ 1 & -6 \end{vmatrix} + 2\begin{vmatrix} 1 & -1 \\ 1 & 2 \end{vmatrix}$$

$$= 2(2) - 3(-8) + 2(3)$$
$$= 4 + 24 + 6 = 34$$

$$D_x = \begin{vmatrix} 15 & 3 & 2 \\ 5 & -1 & 2 \\ -26 & 2 & -6 \end{vmatrix} \text{ Expand about row 2.}$$

$$= -5\begin{vmatrix} 3 & 2 \\ 2 & -6 \end{vmatrix} + (-1)\begin{vmatrix} 15 & 2 \\ -26 & -6 \end{vmatrix}$$
$$- 2\begin{vmatrix} 15 & 3 \\ -26 & 2 \end{vmatrix}$$

$$= -5(-22) - 1(-38) - 2(108)$$
$$= 110 + 38 - 216 = -68$$

$$D_y = \begin{vmatrix} 2 & 15 & 2 \\ 1 & 5 & 2 \\ 1 & -26 & -6 \end{vmatrix} \text{ Expand about column 1.}$$

$$= 2\begin{vmatrix} 5 & 2 \\ -26 & -6 \end{vmatrix} - 1\begin{vmatrix} 15 & 2 \\ -26 & -6 \end{vmatrix}$$
$$+ 1\begin{vmatrix} 15 & 2 \\ 5 & 2 \end{vmatrix}$$

$$= 2(22) - 1(-38) + 1(20)$$
$$= 44 + 38 + 20 = 102$$

$$D_z = \begin{vmatrix} 2 & 3 & 15 \\ 1 & -1 & 5 \\ 1 & 2 & -26 \end{vmatrix} \text{ Expand about column 1.}$$

$$= 2\begin{vmatrix} -1 & 5 \\ 2 & -26 \end{vmatrix} - 1\begin{vmatrix} 3 & 15 \\ 2 & -26 \end{vmatrix}$$
$$+ 1\begin{vmatrix} 3 & 15 \\ -1 & 5 \end{vmatrix}$$

$$= 2(16) - 1(-108) + 1(30)$$
$$= 32 + 108 + 30 = 170$$

$$x = \frac{D_x}{D} = \frac{-68}{34} = -2;\ y = \frac{D_y}{D} = \frac{102}{34} = 3$$

$$z = \frac{D_z}{D} = \frac{170}{34} = 5$$

Solution set: $\{(-2, 3, 5)\}$

34.
$$\begin{aligned} x - y + 6z &= 19 \\ 3x + 3y - z &= 1 \\ x + 9y + 2z &= -19 \end{aligned}$$

$$D = \begin{vmatrix} 1 & -1 & 6 \\ 3 & 3 & -1 \\ 1 & 9 & 2 \end{vmatrix} \text{ Expand about column 1.}$$
$$= 1\begin{vmatrix} 3 & -1 \\ 9 & 2 \end{vmatrix} - 3\begin{vmatrix} -1 & 6 \\ 9 & 2 \end{vmatrix} + 1\begin{vmatrix} -1 & 6 \\ 3 & -1 \end{vmatrix}$$
$$= 1(6+9) - 3(-2-54) + 1(1-18)$$
$$= 15 + 168 - 17 = 166$$

$$D_x = \begin{vmatrix} 19 & -1 & 6 \\ 1 & 3 & -1 \\ -19 & 9 & 2 \end{vmatrix} \text{ Expand about column 3.}$$
$$= 6\begin{vmatrix} 1 & 3 \\ -19 & 9 \end{vmatrix} - (-1)\begin{vmatrix} 19 & -1 \\ -19 & 9 \end{vmatrix} + 2\begin{vmatrix} 19 & -1 \\ 1 & 3 \end{vmatrix}$$
$$= 6(9+57) + 1(171-19) + 2(57+1)$$
$$= 396 + 152 + 116 = 664$$

$$D_y = \begin{vmatrix} 1 & 19 & 6 \\ 3 & 1 & -1 \\ 1 & -19 & 2 \end{vmatrix} \text{ Expand about column 1.}$$
$$= 1\begin{vmatrix} 1 & -1 \\ -19 & 2 \end{vmatrix} - 3\begin{vmatrix} 19 & 6 \\ -19 & 2 \end{vmatrix} + 1\begin{vmatrix} 19 & 6 \\ 1 & -1 \end{vmatrix}$$
$$= 1(2-19) - 3(38+114) + 1(-19-6)$$
$$= -17 - 456 - 25 = -498$$

$$D_z = \begin{vmatrix} 1 & -1 & 19 \\ 3 & 3 & 1 \\ 1 & 9 & -19 \end{vmatrix} \text{ Expand about column 1.}$$
$$= 1\begin{vmatrix} 3 & 1 \\ 9 & -19 \end{vmatrix} - 3\begin{vmatrix} -1 & 19 \\ 9 & -19 \end{vmatrix} + 1\begin{vmatrix} -1 & 19 \\ 3 & 1 \end{vmatrix}$$
$$= 1(-57-9) - 3(19-171) + 1(-1-57)$$
$$= -66 + 456 - 58 = 332$$

$$x = \frac{D_x}{D} = \frac{664}{166} = 4;\ y = \frac{D_y}{D} = \frac{-498}{166} = -3$$
$$z = \frac{D_z}{D} = \frac{332}{166} = 2$$

Solution set: $\{(4, -3, 2)\}$

35.
$$\begin{aligned} 2x + 2y + z &= 10 \\ 4x - y + z &= 20 \\ -x + y - 2z &= -5 \end{aligned}$$

$$D = \begin{vmatrix} 2 & 2 & 1 \\ 4 & -1 & 1 \\ -1 & 1 & -2 \end{vmatrix} \text{ Expand about row 1.}$$
$$= 2\begin{vmatrix} -1 & 1 \\ 1 & -2 \end{vmatrix} - 2\begin{vmatrix} 4 & 1 \\ -1 & -2 \end{vmatrix} + 1\begin{vmatrix} 4 & -1 \\ -1 & 1 \end{vmatrix}$$
$$= 2(1) - 2(-7) + 1(3)$$
$$= 2 + 14 + 3 = 19$$

$$D_x = \begin{vmatrix} 10 & 2 & 1 \\ 20 & -1 & 1 \\ -5 & 1 & -2 \end{vmatrix} \text{ Expand about row 1.}$$
$$= 10\begin{vmatrix} -1 & 1 \\ 1 & -2 \end{vmatrix} - 2\begin{vmatrix} 20 & 1 \\ -5 & -2 \end{vmatrix} + 1\begin{vmatrix} 20 & -1 \\ -5 & 1 \end{vmatrix}$$
$$= 10(1) - 2(-35) + 1(15)$$
$$= 10 + 70 + 15 = 95$$

$$D_y = \begin{vmatrix} 2 & 10 & 1 \\ 4 & 20 & 1 \\ -1 & -5 & -2 \end{vmatrix} \text{ Expand about row 1.}$$
$$= 2\begin{vmatrix} 20 & 1 \\ -5 & -2 \end{vmatrix} - 10\begin{vmatrix} 4 & 1 \\ -1 & -2 \end{vmatrix} + 1\begin{vmatrix} 4 & 20 \\ -1 & -5 \end{vmatrix}$$
$$= 2(-35) - 10(-7) + 1(0)$$
$$= -70 + 70 = 0$$

$$D_z = \begin{vmatrix} 2 & 2 & 10 \\ 4 & -1 & 20 \\ -1 & 1 & -5 \end{vmatrix} \text{ Expand about row 1.}$$
$$= 2\begin{vmatrix} -1 & 20 \\ 1 & -5 \end{vmatrix} - 2\begin{vmatrix} 4 & 20 \\ -1 & -5 \end{vmatrix} + 10\begin{vmatrix} 4 & -1 \\ -1 & 1 \end{vmatrix}$$
$$= 2(-15) - 2(0) + 10(3)$$
$$= -30 + 30 = 0$$

$$x = \frac{D_x}{D} = \frac{95}{19} = 5;\ y = \frac{D_y}{D} = \frac{0}{19} = 0$$
$$z = \frac{D_z}{D} = \frac{0}{19} = 0$$

Solution set: $\{(5, 0, 0)\}$

36. $\begin{aligned} x + 3y - 4z &= -12 \\ 3x + y - z &= -5 \\ 5x - y + z &= -3 \end{aligned}$

$$D = \begin{vmatrix} 1 & 3 & -4 \\ 3 & 1 & -1 \\ 5 & -1 & 1 \end{vmatrix} \text{ Expand about column 3.}$$

$$= -4\begin{vmatrix} 3 & 1 \\ 5 & -1 \end{vmatrix} - (-1)\begin{vmatrix} 1 & 3 \\ 5 & -1 \end{vmatrix} + 1\begin{vmatrix} 1 & 3 \\ 3 & 1 \end{vmatrix}$$

$$= -4(-8) + 1(-16) + 1(-8)$$
$$= 32 - 16 - 8 = 8$$

$$D_x = \begin{vmatrix} -12 & 3 & -4 \\ -5 & 1 & -1 \\ -3 & -1 & 1 \end{vmatrix} \text{ Expand about column 3.}$$

$$= -4\begin{vmatrix} -5 & 1 \\ -3 & -1 \end{vmatrix} - (-1)\begin{vmatrix} -12 & 3 \\ -3 & -1 \end{vmatrix} + 1\begin{vmatrix} -12 & 3 \\ -5 & 1 \end{vmatrix}$$

$$= -4(8) + 1(21) + 1(3)$$
$$= -32 + 21 + 3 = -8$$

$$D_y = \begin{vmatrix} 1 & -12 & -4 \\ 3 & -5 & -1 \\ 5 & -3 & 1 \end{vmatrix} \text{ Expand about column 3.}$$

$$= -4\begin{vmatrix} 3 & -5 \\ 5 & -3 \end{vmatrix} - (-1)\begin{vmatrix} 1 & -12 \\ 5 & -3 \end{vmatrix} + 1\begin{vmatrix} 1 & -12 \\ 3 & -5 \end{vmatrix}$$

$$= -4(16) + 1(57) + 1(31)$$
$$= -64 + 57 + 31 = 24$$

$$D_z = \begin{vmatrix} 1 & 3 & -12 \\ 3 & 1 & -5 \\ 5 & -1 & -3 \end{vmatrix} \text{ Expand about column 2.}$$

$$= -3\begin{vmatrix} 3 & -5 \\ 5 & -3 \end{vmatrix} + 1\begin{vmatrix} 1 & -12 \\ 5 & -3 \end{vmatrix} - (-1)\begin{vmatrix} 1 & -12 \\ 3 & -5 \end{vmatrix}$$

$$= -3(16) + 1(57) + 1(31)$$
$$= -48 + 57 + 31 = 40$$

$$x = \frac{D_x}{D} = \frac{-8}{8} = -1;\ y = \frac{D_y}{D} = \frac{24}{8} = 3$$

$$z = \frac{D_z}{D} = \frac{40}{8} = 5$$

Solution set: $\{(-1, 3, 5)\}$

37. $\begin{aligned} 2x - 3y + 4z &= 8 \\ 6x - 9y + 12z &= 24 \\ -4x + 6y - 8z &= -16 \end{aligned}$

$$D = \begin{vmatrix} 2 & -3 & 4 \\ 6 & -9 & 12 \\ -4 & 6 & -8 \end{vmatrix} \text{ Expand about row 1.}$$

$$= 2\begin{vmatrix} -9 & 12 \\ 6 & -8 \end{vmatrix} + 3\begin{vmatrix} 6 & 12 \\ -4 & -8 \end{vmatrix} + 4\begin{vmatrix} 6 & -9 \\ -4 & 6 \end{vmatrix}$$

$$= 2(0) + 3(0) + 4(0) = 0$$

Because $D = 0$, Cramer's rule does not apply.

38. $\begin{aligned} 7x + y - z &= 4 \\ 2x - 3y + z &= 2 \\ -6x + 9y - 3z &= -6 \end{aligned}$

$$D = \begin{vmatrix} 7 & 1 & -1 \\ 2 & -3 & 1 \\ -6 & 9 & -3 \end{vmatrix} \text{ Expand about column 3.}$$

$$= -1\begin{vmatrix} 2 & -3 \\ -6 & 9 \end{vmatrix} - 1\begin{vmatrix} 7 & 1 \\ -6 & 9 \end{vmatrix} - 3\begin{vmatrix} 7 & 1 \\ 2 & -3 \end{vmatrix}$$

$$= -1(18 - 18) - 1(63 + 6) - 3(-21 - 2)$$
$$= 0 - 69 + 69 = 0$$

Because $D = 0$, Cramer's rule does not apply.

39. $\begin{aligned} 3x \qquad + 5z &= 0 \\ 2x + 3y \qquad &= 1 \\ -y + 2z &= -11 \end{aligned}$

$$D = \begin{vmatrix} 3 & 0 & 5 \\ 2 & 3 & 0 \\ 0 & -1 & 2 \end{vmatrix} \text{ Expand about row 1.}$$

$$= 3\begin{vmatrix} 3 & 0 \\ -1 & 2 \end{vmatrix} - 0 + 5\begin{vmatrix} 2 & 3 \\ 0 & -1 \end{vmatrix}$$

$$= 3(6) + 5(-2)$$
$$= 18 - 10 = 8$$

$$D_x = \begin{vmatrix} 0 & 0 & 5 \\ 1 & 3 & 0 \\ -11 & -1 & 2 \end{vmatrix} \text{ Expand about row 1.}$$

$$= 0 - 0 + 5\begin{vmatrix} 1 & 3 \\ -11 & -1 \end{vmatrix}$$

$$= 5(32) = 160$$

$$D_y = \begin{vmatrix} 3 & 0 & 5 \\ 2 & 1 & 0 \\ 0 & -11 & 2 \end{vmatrix} \text{ Expand about row 1.}$$

$$= 3\begin{vmatrix} 1 & 0 \\ -11 & 2 \end{vmatrix} - 0 + 5\begin{vmatrix} 2 & 1 \\ 0 & -11 \end{vmatrix}$$

$$= 3(2) + 5(-22)$$
$$= 6 - 110 = -104$$

continued

$$D_z = \begin{vmatrix} 3 & 0 & 0 \\ 2 & 3 & 1 \\ 0 & -1 & -11 \end{vmatrix} \text{ Expand about row 1.}$$

$$= 3\begin{vmatrix} 3 & 1 \\ -1 & -11 \end{vmatrix} - 0 + 0$$

$$= 3(-32) = -96$$

$$x = \frac{D_x}{D} = \frac{160}{8} = 20;\ y = \frac{D_y}{D} = \frac{-104}{8} = -13$$

$$z = \frac{D_z}{D} = \frac{-96}{8} = -12$$

Solution set: $\{(20, -13, -12)\}$

40. $$\begin{aligned} -x + 2y &= 4 \\ 3x + y &= -5 \\ 2x + z &= -1 \end{aligned}$$

$$D = \begin{vmatrix} -1 & 2 & 0 \\ 3 & 1 & 0 \\ 2 & 0 & 1 \end{vmatrix} \text{ Expand about column 3.}$$

$$= 0 - 0 + 1\begin{vmatrix} -1 & 2 \\ 3 & 1 \end{vmatrix}$$

$$= 1(-1-6) = -7$$

$$D_x = \begin{vmatrix} 4 & 2 & 0 \\ -5 & 1 & 0 \\ -1 & 0 & 1 \end{vmatrix} \text{ Expand about column 3.}$$

$$= 0 - 0 + 1\begin{vmatrix} 4 & 2 \\ -5 & 1 \end{vmatrix}$$

$$= 1(4+10) = 14$$

$$D_y = \begin{vmatrix} -1 & 4 & 0 \\ 3 & -5 & 0 \\ 2 & -1 & 1 \end{vmatrix} \text{ Expand about column 3.}$$

$$= 0 - 0 + 1\begin{vmatrix} -1 & 4 \\ 3 & -5 \end{vmatrix}$$

$$= 1(5-12) = -7$$

$$D_z = \begin{vmatrix} -1 & 2 & 4 \\ 3 & 1 & -5 \\ 2 & 0 & -1 \end{vmatrix} \text{ Expand about row 3.}$$

$$= 2\begin{vmatrix} 2 & 4 \\ 1 & -5 \end{vmatrix} - 0 - 1\begin{vmatrix} -1 & 2 \\ 3 & 1 \end{vmatrix}$$

$$= 2(-10-4) - 1(-1-6)$$

$$= -28 + 7 = -21$$

$$x = \frac{D_x}{D} = \frac{14}{-7} = -2;\ y = \frac{D_y}{D} = \frac{-7}{-7} = 1$$

$$z = \frac{D_z}{D} = \frac{-21}{-7} = 3$$

Solution set: $\{(-2, 1, 3)\}$

41. $$\begin{aligned} x - 3y &= 13 \\ 2y + z &= 5 \\ -x + z &= -7 \end{aligned}$$

$$D = \begin{vmatrix} 1 & -3 & 0 \\ 0 & 2 & 1 \\ -1 & 0 & 1 \end{vmatrix} \text{ Expand about column 1.}$$

$$= 1\begin{vmatrix} 2 & 1 \\ 0 & 1 \end{vmatrix} - 0 + (-1)\begin{vmatrix} -3 & 0 \\ 2 & 1 \end{vmatrix}$$

$$= 1(2) - 1(-3) = 2 + 3 = 5$$

$$D_x = \begin{vmatrix} 13 & -3 & 0 \\ 5 & 2 & 1 \\ -7 & 0 & 1 \end{vmatrix} \text{ Expand about column 3.}$$

$$= 0 - 1\begin{vmatrix} 13 & -3 \\ -7 & 0 \end{vmatrix} + 1\begin{vmatrix} 13 & -3 \\ 5 & 2 \end{vmatrix}$$

$$= -1(-21) + 1(41)$$

$$= 21 + 41 = 62$$

$$D_y = \begin{vmatrix} 1 & 13 & 0 \\ 0 & 5 & 1 \\ -1 & -7 & 1 \end{vmatrix} \text{ Expand about column 1.}$$

$$= 1\begin{vmatrix} 5 & 1 \\ -7 & 1 \end{vmatrix} - 0 + (-1)\begin{vmatrix} 13 & 0 \\ 5 & 1 \end{vmatrix}$$

$$= 1(12) - 1(13) = 12 - 13 = -1$$

$$D_z = \begin{vmatrix} 1 & -3 & 13 \\ 0 & 2 & 5 \\ -1 & 0 & -7 \end{vmatrix} \text{ Expand about column 1.}$$

$$= 1\begin{vmatrix} 2 & 5 \\ 0 & -7 \end{vmatrix} - 0 - 1\begin{vmatrix} -3 & 13 \\ 2 & 5 \end{vmatrix}$$

$$= 1(-14) - 1(-41) = -14 + 41 = 27$$

$$x = \frac{D_x}{D} = \frac{62}{5};\ y = \frac{D_y}{D} = -\frac{1}{5}$$

$$z = \frac{D_z}{D} = \frac{27}{5}$$

Solution set: $\left\{\left(\frac{62}{5}, -\frac{1}{5}, \frac{27}{5}\right)\right\}$

APPENDIX D SYNTHETIC DIVISION

Appendix D Margin Exercises

1. **(a)** $\dfrac{3z^2+10z-8}{z+4}$

$$\begin{array}{r|rrr} -4 & 3 & 10 & -8 \\ & & -12 & 8 \\ \hline & 3 & -2 & 0 \end{array} \leftarrow \textit{Remainder}$$

Read the quotient from the bottom row.

Answer: $3z-2$

(b) $(2x^2+3x-5)\div(x+1)$

$$\begin{array}{r|rrr} -1 & 2 & 3 & -5 \\ & & -2 & -1 \\ \hline & 2 & 1 & -6 \end{array} \leftarrow \textit{Remainder}$$

Answer: $2x+1+\dfrac{-6}{x+1}$

2. **(a)** $\dfrac{3a^3-2a+21}{a+2}$

Insert a 0 for the missing a^2-term.

$$\begin{array}{r|rrrr} -2 & 3 & 0 & -2 & 21 \\ & & -6 & 12 & -20 \\ \hline & 3 & -6 & 10 & 1 \end{array} \leftarrow \textit{Remainder}$$

Answer: $3a^2-6a+10+\dfrac{1}{a+2}$

(b) $\left(-4x^4+3x^3+18x+2\right)\div(x-2)$

$$\begin{array}{r|rrrrr} 2 & -4 & 3 & 0 & 18 & 2 \\ & & -8 & -10 & -20 & -4 \\ \hline & -4 & -5 & -10 & -2 & -2 \end{array} \leftarrow \textit{Remainder}$$

Answer: $-4x^3-5x^2-10x-2+\dfrac{-2}{x-2}$

3. Let $P(x)=x^3-5x^2+7x-3$.

To find $P(a)$, divide the polynomial by $x-a$ using synthetic division. $P(a)$ will be the remainder.

(a) Find $P(1)$.

Divide $P(x)$ by $x-1$.

$$\begin{array}{r|rrrr} 1 & 1 & -5 & 7 & -3 \\ & & 1 & -4 & 3 \\ \hline & 1 & -4 & 3 & 0 \end{array} \leftarrow \textit{Remainder}$$

The remainder is 0, so by the remainder theorem, $P(1)=0$.

(b) Find $P(-2)$.

Divide $P(x)$ by $x-(-2)=x+2$.

$$\begin{array}{r|rrrr} -2 & 1 & -5 & 7 & -3 \\ & & -2 & 14 & -42 \\ \hline & 1 & -7 & 21 & -45 \end{array} \leftarrow \textit{Remainder}$$

By the remainder theorem, $P(-2)=-45$.

4. To check whether 2 is a solution to $P(x)=0$, divide the polynomial by $x-2$. If the remainder is 0, then 2 is a solution. Otherwise, it is not.

(a) $3x^3-11x^2+17x-14=0$

$$\begin{array}{r|rrrr} 2 & 3 & -11 & 17 & -14 \\ & & 6 & -10 & 14 \\ \hline & 3 & -5 & 7 & 0 \end{array} \leftarrow \textit{Remainder}$$

Since the remainder is 0, 2 is a solution of the given equation.

(b) $4x^5-7x^4-11x^2+2x+6=0$

Insert a 0 for the missing x^3-term.

$$\begin{array}{r|rrrrrr} 2 & 4 & -7 & 0 & -11 & 2 & 6 \\ & & 8 & 2 & 4 & -14 & -24 \\ \hline & 4 & 1 & 2 & -7 & -12 & -18 \end{array} \leftarrow \textit{Remainder}$$

Since the remainder is not 0, 2 is not a solution of the given equation.

Appendix D Section Exercises

1. $\dfrac{x^2+3x-6}{x-2}$

The correct set-up is **C**. $\begin{array}{r|rrr} 2 & 1 & 3 & -6 \end{array}$

2. $\dfrac{x^3-3x^2+2}{x-1}$ Note the missing x-term.

The correct set-up is **C**. $\begin{array}{r|rrrr} 1 & 1 & -3 & 0 & 2 \end{array}$

3. $\dfrac{x^2-6x+5}{x-1}$

$$\begin{array}{r|rrr} 1 & 1 & -6 & 5 \\ & & 1 & -5 \\ \hline & 1 & -5 & 0 \end{array} \leftarrow \textit{Coefficients of numerator}$$

$\downarrow \quad \downarrow$

$x \quad -5$

Write the answer from the bottom row.

Answer: $x-5$

4. $\dfrac{x^2 - 4x - 21}{x + 3}$

$$\begin{array}{r|rrr} -3 & 1 & -4 & -21 \\ & & -3 & 21 \\ \hline & 1 & -7 & 0 \end{array}$$

Answer: $x - 7$

5. $\dfrac{4m^2 + 19m - 5}{m + 5}$

$m + 5 = m - (-5)$, so use -5.

$$\begin{array}{r|rrr} -5 & 4 & 19 & -5 \\ & & -20 & 5 \\ \hline & 4 & -1 & 0 \end{array}$$

Answer: $4m - 1$

6. $\dfrac{3k^2 - 5k - 12}{k - 3}$

$$\begin{array}{r|rrr} 3 & 3 & -5 & -12 \\ & & 9 & 12 \\ \hline & 3 & 4 & 0 \end{array}$$

Answer: $3k + 4$

7. $\dfrac{2a^2 + 8a + 13}{a + 2}$

$a + 2 = a - (-2)$, so use -2.

$$\begin{array}{r|rrr} -2 & 2 & 8 & 13 \\ & & -4 & -8 \\ \hline & 2 & 4 & 5 \end{array} \leftarrow \textit{Remainder}$$

Answer: $2a + 4 + \dfrac{5}{a + 2}$

8. $\dfrac{4y^2 - 5y - 20}{y - 4}$

$$\begin{array}{r|rrr} 4 & 4 & -5 & -20 \\ & & 16 & 44 \\ \hline & 4 & 11 & 24 \end{array} \leftarrow \textit{Remainder}$$

Answer: $4y + 11 + \dfrac{24}{y - 4}$

9. $(p^2 - 3p + 5) \div (p + 1)$

$$\begin{array}{r|rrr} -1 & 1 & -3 & 5 \\ & & -1 & 4 \\ \hline & 1 & -4 & 9 \end{array}$$

Answer: $p - 4 + \dfrac{9}{p + 1}$

10. $(z^2 + 4z - 6) \div (z - 5)$

$$\begin{array}{r|rrr} 5 & 1 & 4 & -6 \\ & & 5 & 45 \\ \hline & 1 & 9 & 39 \end{array} \leftarrow \textit{Remainder}$$

Answer: $z + 9 + \dfrac{39}{z - 5}$

11. $\dfrac{4a^3 - 3a^2 + 2a - 3}{a - 1}$

$$\begin{array}{r|rrrr} 1 & 4 & -3 & 2 & -3 \\ & & 4 & 1 & 3 \\ \hline & 4 & 1 & 3 & 0 \end{array}$$

Answer: $4a^2 + a + 3$

12. $\dfrac{5p^3 - 6p^2 + 3p + 14}{p + 1}$

$$\begin{array}{r|rrrr} -1 & 5 & -6 & 3 & 14 \\ & & -5 & 11 & -14 \\ \hline & 5 & -11 & 14 & 0 \end{array}$$

Answer: $5p^2 - 11p + 14$

13. $(x^5 - 2x^3 + 3x^2 - 4x - 2) \div (x - 2)$

Insert 0 for the missing x^4-term.

$$\begin{array}{r|rrrrrr} 2 & 1 & 0 & -2 & 3 & -4 & -2 \\ & & 2 & 4 & 4 & 14 & 20 \\ \hline & 1 & 2 & 2 & 7 & 10 & 18 \end{array} \leftarrow \textit{Remainder}$$

Answer: $x^4 + 2x^3 + 2x^2 + 7x + 10 + \dfrac{18}{x - 2}$

14. $(2y^5 - 5y^4 - 3y^2 - 6y - 23) \div (y - 3)$

Insert 0 for the missing y^3-term.

$$\begin{array}{r|rrrrrr} 3 & 2 & -5 & 0 & -3 & -6 & -23 \\ & & 6 & 3 & 9 & 18 & 36 \\ \hline & 2 & 1 & 3 & 6 & 12 & 13 \end{array} \leftarrow \textit{Remainder}$$

Answer: $2y^4 + y^3 + 3y^2 + 6y + 12 + \dfrac{13}{y - 3}$

15. $(-4r^6 - 3r^5 - 3r^4 + 5r^3 - 6r^2 + 3r) \div (r - 1)$

$$\begin{array}{r|rrrrrrr} 1 & -4 & -3 & -3 & 5 & -6 & 3 & 0 \\ & & -4 & -7 & -10 & -5 & -11 & -8 \\ \hline & -4 & -7 & -10 & -5 & -11 & -8 & -8 \end{array} \leftarrow \textit{Remainder}$$

Answer:

$-4r^5 - 7r^4 - 10r^3 - 5r^2 - 11r - 8 + \dfrac{-8}{r - 1}$

16. $\left(-3t^5 + 2t^4 - 5t^3 + 6t^2 - 3t - 2\right) \div (t - 2)$

$$\begin{array}{r|rrrrrr} 2 & -3 & 2 & -5 & 6 & -3 & -2 \\ & & -6 & -8 & -26 & -40 & -86 \\ \hline & -3 & -4 & -13 & -20 & -43 & -88 \end{array} \leftarrow \textit{Remainder}$$

Answer:

$$-3t^4 - 4t^3 - 13t^2 - 20t - 43 + \frac{-88}{t-2}$$

17. $\left(-3y^5 + 2y^4 - 5y^3 - 6y^2 - 1\right) \div (y + 2)$

Insert 0 for the missing y-term.

$$\begin{array}{r|rrrrrr} -2 & -3 & 2 & -5 & -6 & 0 & -1 \\ & & 6 & -16 & 42 & -72 & 144 \\ \hline & -3 & 8 & -21 & 36 & -72 & 143 \end{array} \leftarrow \textit{Remainder}$$

Answer:

$$-3y^4 + 8y^3 - 21y^2 + 36y - 72 + \frac{143}{y+2}$$

18. $\left(m^6 + 2m^4 - 5m + 11\right) \div (m - 2)$

Insert zeros for the missing terms.

$$\begin{array}{r|rrrrrrr} 2 & 1 & 0 & 2 & 0 & 0 & -5 & 11 \\ & & 2 & 4 & 12 & 24 & 48 & 86 \\ \hline & 1 & 2 & 6 & 12 & 24 & 43 & 97 \end{array} \leftarrow \textit{Remainder}$$

Answer:

$$m^5 + 2m^4 + 6m^3 + 12m^2 + 24m + 43 + \frac{97}{m-2}$$

19. $\dfrac{y^3 + 1}{y - 1} = \dfrac{y^3 + 0y^2 + 0y + 1}{y - 1}$

$$\begin{array}{r|rrrr} 1 & 1 & 0 & 0 & 1 \\ & & 1 & 1 & 1 \\ \hline & 1 & 1 & 1 & 2 \end{array} \leftarrow \textit{Remainder}$$

Answer: $y^2 + y + 1 + \dfrac{2}{y-1}$

20. $\dfrac{z^4 + 81}{z - 3}$

Insert zeros for the missing terms.

$$\begin{array}{r|rrrrr} 3 & 1 & 0 & 0 & 0 & 81 \\ & & 3 & 9 & 27 & 81 \\ \hline & 1 & 3 & 9 & 27 & 162 \end{array} \leftarrow \textit{Remainder}$$

Answer: $z^3 + 3z^2 + 9z + 27 + \dfrac{162}{z-3}$

21. $P(x) = 2x^3 - 4x^2 + 5x - 3;\ k = 2$

To find $P(2)$, divide the polynomial by $x - 2$. $P(2)$ will be the remainder.

$$\begin{array}{r|rrrr} 2 & 2 & -4 & 5 & -3 \\ & & 4 & 0 & 10 \\ \hline & 2 & 0 & 5 & 7 \end{array} \leftarrow \textit{Remainder}$$

By the remainder theorem, $P(2) = 7$.

22. $P(x) = x^3 + 3x^2 - x + 5;\ k = -1$

Divide by $x + 1$. The remainder is equal to $P(-1)$.

$$\begin{array}{r|rrrr} -1 & 1 & 3 & -1 & 5 \\ & & -1 & -2 & 3 \\ \hline & 1 & 2 & -3 & 8 \end{array} \leftarrow \textit{Remainder}$$

By the remainder theorem, $P(-1) = 8$.

23. $P(r) = -r^3 - 5r^2 - 4r - 2;\ k = -4$

Divide by $r + 4$. The remainder is equal to $P(-4)$.

$$\begin{array}{r|rrrr} -4 & -1 & -5 & -4 & -2 \\ & & 4 & 4 & 0 \\ \hline & -1 & -1 & 0 & -2 \end{array} \leftarrow \textit{Remainder}$$

By the remainder theorem, $P(-4) = -2$.

24. $P(z) = -z^3 + 5z^2 - 3z + 4;\ k = 3$

Divide by $z - 3$. The remainder is equal to $P(3)$.

$$\begin{array}{r|rrrr} 3 & -1 & 5 & -3 & 4 \\ & & -3 & 6 & 9 \\ \hline & -1 & 2 & 3 & 13 \end{array} \leftarrow \textit{Remainder}$$

By the remainder theorem, $P(3) = 13$.

25. $P(x) = 2x^3 - 4x^2 + 5x - 33;\ k = 3$

Divide by $x - 3$. The remainder is equal to $P(3)$.

$$\begin{array}{r|rrrr} 3 & 2 & -4 & 5 & -33 \\ & & 6 & 6 & 33 \\ \hline & 2 & 2 & 11 & 0 \end{array} \leftarrow \textit{Remainder}$$

By the remainder theorem, $P(3) = 0$.

26. $P(x) = x^3 - 3x^2 + 4x - 4;\ k = 2$

Divide by $x - 2$. The remainder is equal to $P(2)$.

$$\begin{array}{r|rrrr} 2 & 1 & -3 & 4 & -4 \\ & & 2 & -2 & 4 \\ \hline & 1 & -1 & 2 & 0 \end{array} \leftarrow \textit{Remainder}$$

By the remainder theorem, $P(2) = 0$.

27. Is $x = -2$ a solution of

$$x^3 - 2x^2 - 3x + 10 = 0?$$

To decide whether -2 is a solution to the given equation, divide the polynomial by $x + 2$.

$$\begin{array}{r|rrrr} -2 & 1 & -2 & -3 & 10 \\ & & -2 & 8 & -10 \\ \hline & 1 & -4 & 5 & 0 \end{array} \leftarrow \textit{Remainder}$$

Since the remainder is 0, -2 is a solution of the equation.

28. Is $x = -2$ a solution of

$$x^3 - 3x^2 - x + 10 = 0?$$

To decide whether -2 is a solution to the given equation, divide the polynomial by $x + 2$.

$$\begin{array}{r|rrrr} -2 & 1 & -3 & -1 & 10 \\ & & -2 & 10 & -18 \\ \hline & 1 & -5 & 9 & -8 \end{array} \leftarrow \textit{Remainder}$$

Since the remainder is not 0, -2 is not a solution of the equation.

29. Is $m = -2$ a solution of

$$m^4 + 2m^3 - 3m^2 + 8m - 8 = 0?$$

To decide whether -2 is a solution to the given equation, divide the polynomial by $m + 2$.

$$\begin{array}{r|rrrrr} -2 & 1 & 2 & -3 & 8 & -8 \\ & & -2 & 0 & 6 & -28 \\ \hline & 1 & 0 & -3 & 14 & -36 \end{array} \leftarrow \textit{Remainder}$$

Since the remainder is not 0, -2 is not a solution of the equation.

30. Is $r = -2$ a solution of

$$r^4 - r^3 - 6r^2 + 5r + 10 = 0?$$

$$\begin{array}{r|rrrrr} -2 & 1 & -1 & -6 & 5 & 10 \\ & & -2 & 6 & 0 & -10 \\ \hline & 1 & -3 & 0 & 5 & 0 \end{array} \leftarrow \textit{Remainder}$$

Since the remainder is 0, -2 is a solution of the equation.

31. Is $x = -2$ a solution of

$$3x^3 + 2x^2 - 2x + 11 = 0?$$

$$\begin{array}{r|rrrr} -2 & 3 & 2 & -2 & 11 \\ & & -6 & 8 & -12 \\ \hline & 3 & -4 & 6 & -1 \end{array} \leftarrow \textit{Remainder}$$

Since the remainder is not 0, -2 is not a solution of the equation.

32. Is $z = -2$ a solution of

$$3z^3 + 10z^2 + 3z - 9 = 0?$$

$$\begin{array}{r|rrrr} -2 & 3 & 10 & 3 & -9 \\ & & -6 & -8 & 10 \\ \hline & 3 & 4 & -5 & 1 \end{array} \leftarrow \textit{Remainder}$$

Since the remainder is not 0, -2 is not a solution of the equation.

33. Since the variables are not present, a missing term will not be noticed in synthetic division, so the quotient will be wrong if placeholders are not inserted.

34. By the remainder theorem, a zero remainder means that $P(k) = 0$; that is, k is a number that makes $P(x) = 0$.